Related titles on radar:

Advances in Bistatic Radar Willis and Griffiths
Airborne Early Warning System Concepts, 3rd Edition Long
Bistatic Radar, 2nd Edition Willis
Design of Multi-Frequency CW Radars Jankiraman
Digital Techniques for Wideband Receivers, 2nd Edition Tsui
Electronic Warfare Pocket Guide Adamy
Foliage Penetration Radar: Detection and characterisation of objects under trees Davis
Fundamentals of Ground Radar for ATC Engineers and Technicians Bouwman
Fundamentals of Systems Engineering and Defense Systems Applications Jeffrey
Introduction to Electronic Warfare Modeling and Simulation Adamy
Introduction to Electronic Defense Systems Neri
Introduction to Sensors for Ranging and Imaging Brooker
Microwave Passive Direction Finding Lipsky
Microwave Receivers with Electronic Warfare Applications Tsui
Phased-Array Radar Design: Application of radar fundamentals Jeffrey
Pocket Radar Guide: Key facts, equations, and data Curry
Principles of Modern Radar, Volume 1: Basic principles Richards, Scheer and Holm
Principles of Modern Radar, Volume 2: Advanced techniques Melvin and Scheer
Principles of Modern Radar, Volume 3: Applications Scheer and Melvin
Principles of Waveform Diversity and Design Wicks *et al.*
Principles of Space-Time Adaptive Processing, 3rd edition Klemm
Pulse Doppler Radar Alabaster
Radar Cross Section Measurements Knott
Radar Cross Section, 2nd Edition Knott *et al.*
Radar Design Principles: Signal processing and the environment, 2nd Edition Nathanson *et al.*
Radar Detection DiFranco and Ruby
Radar Essentials: A concise handbook for radar design and performance Curry
Radar Foundations for Imaging and Advanced Concepts Sullivan
Radar Principles for the Non-Specialist, 3rd Edition Toomay and Hannan
Test and Evaluation of Aircraft Avionics and Weapons Systems McShea
Understanding Radar Systems Kingsley and Quegan
Understanding Synthetic Aperture Radar Images Oliver and Quegan
Radar and Electronic Warfare Principles for the Non-specialist, 4th Edition Hannen
Inverse Synthetic Aperture Radar Imaging: Principles, algorithms and applications Chen and Martorella
Stimson's Introduction to Airborne Radar, 3rd Edition Griffiths, Baker and Adamy
Test and Evaluation of Avionics and Weapon Systems, 2nd Edition McShea
Angle-of-Arrival Estimation Using Radar Interferometry: Methods and applications Holder

Novel Radar Techniques and Applications

Volume 1: Real Aperture Array Radar, Imaging Radar, and Passive and Multistatic Radar

Edited by

Richard Klemm

Volume 1 Editors

Part I: Real aperture array radar
Ulrich Nickel
Fraunhofer FKIE, Germany

Part II: Imaging radar
Christoph Gierull
DRDC, Canada

Part III: Passive and multistatic radar
Pierfrancesco Lombardo
University of Rome, Italy

SciTech Publishing
an imprint of the IET

SciTech Publishing

Published by The Institution of Engineering and Technology, London, United Kingdom

The Institution of Engineering and Technology is registered as a Charity in England & Wales (no. 211014) and Scotland (no. SC038698).

First published 2017

The Institution of Engineering and Technology
Michael Faraday House
Six Hills Way, Stevenage
Herts SG1 2AY, United Kingdom

www.theiet.org

British Library Cataloguing in Publication Data
A catalogue record for this product is available from the British Library

ISBN 978-1-61353-225-6 (Hardback Volume 1)
ISBN 978-1-61353-227-0 (PDF Volume 1)
ISBN 978-1-61353-226-3 (Hardback Volume 2)
ISBN 978-1-61353-228-7 (PDF Volume 2)
ISBN 978-1-61353-229-4 (Hardback Volume 1 & 2)

Typeset in India by MPS Limited
Printed in the UK by CPI Group (UK) Ltd, Croydon

Contents

Preface to Volumes 1 and 2

The title 'Novel Radar Techniques and Applications (NRTA)' suggests that the content of these two volumes is twofold. On the one hand, the team of editors together with the authors aimed at presenting a variety of radar techniques that have not yet found their way into operational use. Techniques such as MIMO, compressive sensing, cognitive radar operation, radar management, radar networks and waveform diversity are presented in these volumes as well as tracking, data fusion, passive radar operation and new imaging techniques.

Moreover, a large number of applications demonstrate the usefulness, the potential and the limits of performance of the presented techniques. In order to reach this multifold goal, a large number of authors have been invited from universities, research laboratories and industry, so as to cover as many different aspects arising between theory, practice and operational use as possible.

These volumes are subdivided into five parts, each of them including a number of chapters. Each part is dedicated to a specific area: Vol. 1: Part I. Array Radar (interference and clutter cancellation, target parameter estimation, etc.); Part II. Imaging Radar; Part III. Passive and Multistatic Radar; Vol. 2: Part IV. Waveform Diversity; Part V. Data Fusion and Tracking. Each part has been taken care of by a co-editor, typically a renowned expert in the respective area. Each part starts with an introduction written by the associated co-editor. In their introductions, the co-editors give overviews of the current state of the art in the respective area and point out the relevance of the subsequent chapters.

These volumes would never have been completed without the enthusiastic effort of a large number of persons. First of all I have to thank the five co-editors Ulrich Nickel, Germany; Christoph Gierull, Canada; Pierfrancesco Lombardo, Italy; Hugh Griffiths, UK and Wolfgang Koch, Germany for their outstanding co-operation on this project. Based on the expertise in their respective fields, they assisted me in composing the contents of these volumes and identifying potential authors and reviewers who were selected according to their expertise in the specific fields. In this way an important feature of these volumes is its multiplicity of facets of modern radar technology and associated applications. It makes these volumes a deep source of information and inspiration for teachers, students, researchers and system designers, in summary all people involved in the development of the radar of tomorrow.

I would like to thank the authors for their excellent work over a long period of time and the reviewers whose critical comments contributed to the quality of the book. Finally the excellent cooperation with Jennifer Grace, Nikki Tarplett and Paul Deards of IET Publishers as well as Vijay Ramalingam of MPS Ltd. is gratefully acknowledged.

Richard Klemm, Editor NRTA
08.09.2017, Bonn, Germany

Volume 1 Editor Biographies

Richard Klemm received his Dipl.-Ing. and Dr.-Ing. degrees in communications from the Technical University of Berlin in 1968 and 1974, respectively. Since 1968, he has been with FGAN FHR (now Fraunhofer FHR), a research institute working in the areas defence and security. From 1977 to 1980, he was with SACLANT ASW Centre, La Spezia, Italy. Richard Klemm's main fields of activity have been research into adaptive clutter and jammer suppression for radar, array processing for active and passive sonar, with emphasis on matched field processing for shallow water applications and detection of moving targets by moving sensor platforms. Richard Klemm has published numerous articles on various aspects of radar and sonar signal processing, and a book (3rd edition) on space-time adaptive processing. He is editor of a book on 'Applications of STAP', including contributions by 45 international authors. He is a permanent reviewer of renowned journals and has provided seminars and consultancy to various organizations in different countries (European Commission, several European countries, USA, Canada, Russia, China, Turkey). Richard Klemm has been a member of the NATO AGARD AVP and RTO-SET panels and chaired various AGARD and RTO symposia. He initialized and chaired the European Conference on Synthetic Aperture Radar EUSAR in 1996 which takes place every 2 years since then. He received several awards in recognition of his scientific achievements, among them the NATO RTO von Karman medal. Richard Klemm gave seminars to different Chinese institutions (Tsinghua University Beijing; Institute of Electronics CAS, Beijing; Xi'an University, Xi'an; UESTC Chengdu; Jiaotong University, Shanghai; NRIET Nanjing). He is honorary professor of UESTC (University of Electronics Science and Technology of China) in Chengdu. His book 'Principles of Space-Time Adaptive Processing 3rd edition' was translated into Chinese by members of NRIET, Nanjing, and has been published by Higher Education Press, Beijing. In his spare time, Richard Klemm is a passionate classical pianist. Under the motto Science and Music, he likes to give piano recitals at technical conferences. He is married, has three children and seven grandchildren.

Ulrich Nickel received the Diploma in Mathematics from the University of Cologne (Germany) in 1975 and the Dr. rer. nat. degree from the Technical University of Aachen (Germany) in 1983. Since 1975, he was at the Array-Based Radar Imaging Department of the Research Institute for High Frequency Physics and Radar Techniques (FHR) of FGAN (German Defence Research Establishment) in Wachtberg, Germany. Since 2007 he is with the Sensor Data and Information Fusion Department (SDF) of the Fraunhofer Institute for Communication, Information Processing and Ergonomics (FKIE) in Wachtberg. His interests include all aspects of array signal processing, especially radar detection and parameter estimation with adaptive beamforming and superresolution methods. He was visiting scientist at the Defence R&D Canada, Ottawa, in 1987 and at the University of Connecticut, CT, USA, in 2009. He received the VDE-ITG paper prize 1989 and a conference paper prize at CIE RADAR 2006 (Shanghai). He is contributing author of the books 'Radar Array Processing' (eds. S. Haykin *et al.*, Springer-Verlag, 1993) and 'Applications of Space-Time Adaptive Processing' (ed. R. Klemm, IEE Publishers, 2004). From 1996 to 2010, he was member of the Editorial Board of the *International Journal of Electronics and Communications* (AEÜ) and from 2010 to 2013 Associate Editor of the *IEEE Transactions on Aerospace and Electronic Systems*.

Christoph H. Gierull received the Dr.-Ing. degree from the Ruhr-University Bochum, Germany, in 1995. Since 2011, he has been appointed as Adjunct Professor at Laval University, Quebec and since 2017 also at Simon Fraser University, British Columbia. From 1991 to 1994, he was a scientist with the research establishment FGAN, Germany. In 1994, he joined the German Aerospace Center DLR, where he headed the SAR Simulation Group. As part of a DLR team, he assured X-band interferometric SAR performance during the Space-Shuttle Radar Topography Mission (SRTM) at NASA's Mission Control in Houston. Since 2000, he has been a Senior Defence Scientist with DRDC Ottawa assuming the duties of Group Leader, Space-Based Radar. From 2006 to 2009, he was on assignment to Fraunhofer FHR, Germany. He has been a Technical Advisor to the Canadian and the European Space Agency on several SAR missions including RADARSAT-2, RCM and Sentinel-1. Dr. Gierull is author and coauthor of numerous journal publications, scientific reports and a chapter in Application of Space-Time Adaptive Processing (IEE, 2004). He served as Associate Editor for EURASIP's Signal Processing in 2004 and for IEEE Trans. Geoscience and Remote Sensing 2010–2013. He initiated and co-organized the first Special Issue on Multi-channel Spaced-Based SAR in IEEE J-STARS, Nov. 2015. He has been granted a CAN/US patent on vessel detection in SAR imagery. Dr. Gierull received the Best Annual Paper Award of the Association of German Electrical Engineers in 1998 and jointly the Best Paper Awards at the Int. Radar Conf. 2004, the EUSAR 2006 as well as EUSAR 2016. He is recipient of DRDC's S&T Performance Excellence Award 2013 and winner of the IEEE Geoscience and Remote Sensing Society 2016 J-STARS Paper Award. He is a Fellow of the IET.

Pierfrancesco Lombardo graduated in1991 at the University of Rome 'La Sapienza', Italy. After serving at the Official Test Center of the Italian Air Force, he was Associate at Birmingham University (UK) and at Defense Research Agency in Malvern. In 1995, he received his Ph.D. and was research associate at Syracuse University (NY, USA). In 1996 he joined the University of Rome 'La Sapienza', where he is presently Full Professor. Dr. Lombardo is involved in, and coordinates, research projects funded by European and National Research Agencies and national industries. He leads the 'Radar, Remote Sensing and Navigation' (RRSN) group at the University of Rome 'La Sapienza'. He chairs the Cosmo SkyMed consulting group for the Italian Space Agency. His main interests are radar adaptive signal processing, radar clutter modelling, radar coherent detection, passive radar and multistatic radar, SAR processing and radio-localization systems. Dr. Lombardo's research has been reported in over 250 publications in international technical journals and conferences. He is co-recipient of the Barry Carlton award (best paper) of IEEE Trans. on AES for year 2001 and of the best paper award for the IEEE Trans. on Geoscience and Remote Sensing for year 2003. He served in the technical committee of many international conferences on radar systems and signal processing. He was Technical Committee Chairman of the IEEE/ISPRS Workshop on Remote Sensing and Data Fusion over Urban Areas URBAN'2001, Rome, URBAN'2003, Berlin, and URBAN'2005, Tempe (US). He was also Technical Chairman of the IEEE Radar Conference 2008. Dr. Lombardo is associate Editor for Radar Systems for the IEEE Transactions on Aerospace and Electronic Systems (AES) since June 2001 and Technical Editor for radar System since January 2016. He is member of the IEEE AES Radar System Panel and the Editorial board of *IET Proceedings on Radar Sonar and Navigation*.

List of Authors

Ahmad, Fauzia	Temple University	USA
Allen, Christopher	University of Kansas	USA
Amin, Moeness	Villanova University	USA
Antoniou, Michail	University of Birmingham	UK
Aubry, Augusto	Università degli studi di Napoli Fed. II	Italy
Baker, Chris	Aveillant Ltd.	UK
Balleri, Alessio	Cranfield University	UK
Blunt, Shannon	University of Kansas	USA
Brötje, Martina	Fraunhofer FKIE	Germany
Bucciarelli, Marta	Sympas S.r.l.	Italy
Bürger, Wolfram	Fraunhofer FHR	Germany
Cerutti-Maori, Delphine	Fraunhofer FHR	Germany
Charlish, Alexander	Fraunhofer FKIE	Germany
Cherniakov, Mikhail	University of Birmingham	UK
Colone, Fabiola	DIET University of Rome 'La Sapienza'	Italy
Cristallini, Diego	Fraunhofer FHR	Germany
Damini, Anthony	DRDC Ottawa	Canada
Daniel, Liam	University of Birmingham	UK
Dawidowicz, Bartek	Warsaw University of Technology	Poland
Fargetton, Hervé	DGA Defense	France
Farina, Alfonso	Selex-ES ret. and ELETTRONICA S.p.A	Italy
Feldmann, Michael	Fraunhofer FKIE	Germany
Fitzgerald, Dennis	Information Directorate AFRL	USA
Fornaro, Gianfranco	IREA-CNR	Italy
Fränken, Dietrich	Hensoldt Sensors GmbH	Germany
Gashinova, Marina	University of Birmingham	UK
Georgiev, Krasin	Cranfield University	UK
Gierull, Christoph	DRDC Ottawa	Canada
Govaers, Felix	Fraunhofer FKIE	Germany
Griffiths, Hugh	University College London	UK
Gromek, Damian	Warsaw University of Technology	Poland
Hack, Daniel E.	Matrix Research Inc.	USA
Hernandez, Marcel	Hernandez Technical Solutions Ltd.	UK
Himed, Braham	AFRL	USA
Hoffmann, Fabienne	Airbus Defence & Space	Germany
Hoffmann, Folker	Fraunhofer FKIE	Germany
Holderied, Marc	University of Bristol	UK
Jakabosky, John	University of Kansas	USA
Jovanoska, Snezhana	Fraunhofer FKIE	Germany
Katsilieris, Fotios	Fraunhofer FKIE	Germany
Klemm, Richard	Fraunhofer FHR ret.	Germany

(Continues)

(*Continued*)

Koch, Wolfgang	Fraunhofer FKIE	Germany
Kohlleppel, Robert	Fraunhofer FHR	Germany
Kulpa, Krzysztof	Warsaw University of Technology	Poland
Kuschel, Heiner	Fraunhofer FHR	Germany
Linderman, Richard	Office of the Secretary of Defense	USA
Lombardo, Pierfrancesco	DIET University of Rome 'La Sapienza'	Italy
Maio, Antonio de	Università degli studi di Napoli Fed. II	Italy
Maslikowski, Lukas	Warsaw University of Technology	Poland
Myakinkov, Alexander	Nishny Nowgorod Technical State University	Russia
Monti-Guarnieri, Andrea	DEI POLIMI	Italy
Nadjiasngar, Roaldje	Fraunhofer FKIE	Germany
Nickel, Ulrich	Fraunhofer FHR/FKIE	Germany
Opitz, Felix	Airbus Space & Defence	Germany
Palamà, Riccardo	University College London	UK
Pastina, Debora	DIET University of Rome 'La Sapienza'	Italy
Patton, Lee K.	Matrix Research Inc.	USA
Pauciullo, Antonio	IREA – CNR	Italy
Piezzo, Marco	Università degli studi di Napoli Fed. II	Italy
Ristic, Branko	RMIT University	Australia
Rocca, Fabio	DEI POLIMI	Italy
Rosenberg, Luke	Defence Science and Technology Group	Australia
Schroeder, Alexander	Airbus Defence & Space	Germany
Sen, Satyabrata	Oak Ridge National Laboratory	USA
Sikaneta, Ishuwa	DRDC Ottawa	Canada
Sletten, Mark	NRL	USA
Stove, Andy	Stove Specialities, Sussex	UK
Tebaldini, Stefano	DEI POLIMI	Italy
Thomä, Reiner	Technical University Ilmenau	Germany
Toporkov, Jakov	NRL	USA
Walterscheid, Ingo	Fraunhofer FHR	Germany

List of reviewers

Reviewer	Affiliation	Country
Baumgartner, Stefan	German Aerospace Center DLR	Germany
Bucciarelli, Marta	Sympas S.r.l.	Italy
Charlish, Alexander	Fraunhofer FKIE	Germany
Cristallini, Diego	Fraunhofer FHR	Germany
Damini, Anthony	DRDC Ottawa	Canada
Demissie, Bruno	Fraunhofer FKIE	Germany
Erricolo, Danilo	University of Illinois at Chicago	USA
Fabrizio, Joe	Defence Science and Technology Group	Australia
Farina, Alfonso	Selex-ES ret. and ELETTRONICA S.p.A	Italy
Fornaro, Gianfranco	IREA – CNR	Italy
Geudtner, Dirk	European Space Agency ESA	Netherlands
Gierull, Christoph	DRDC Ottawa	Canada
Gini, Fulvio	University of Pisa	Italy
Govaers, Felix	Fraunhofer FKIE	Germany
Griffiths, Hugh	University College London	UK
Klemm, Richard	Fraunhofer FHR ret.	Germany
Koch, Wolfgang	Fraunhofer FKIE	Germany
Kohlleppel, Robert	Fraunhofer FHR	Germany
Lesturgie, Marc	ONERA	France
Livingstone, Chuck	DRDC Ottawa	Canada
Lombardo, Pierfrancesco	University of Rome La Sapienza	Italy
Mertens, Michael	Eletronica, Meckenheim	Germany
Monti-Guarnieri, Andrea	DEI POLIMI	Italy
Mihaylova, Mila	University of Sheffield	UK
Nickel, Ulrich	Fraunhofer FHR/FKIE	Germany
Pastina, Debora	DIET University of Rome 'La Sapienza'	Italy
Pauciullo, Antonio	IREA – CNR	Italy
Prati, Claudio Maria	Politecnico di Milano	Italy
Rosenberg, Luke	Defence Science and Technology Group	Australia
Savy, Laurent	ONERA DEMR	France
Schlangen,Isabel	Heriot-Watt University	UK
Sikaneta, Ishuwa	DRDC Ottawa	Canada
Sletten, Mark	NRL	USA
Smith, Graeme	The Ohio State University	USA
Soldovieri, Francesco	IREA	Italy

(Continues)

(*Continued*)

Reviewer	Affiliation	Country
Ulmke, Martin	Fraunhofer FKIE	Germany
Wiedmann, Michael	Airbus Defence & Space	Germany
Walterscheid, Ingo	Fraunhofer FHR	Germany
Wirth, Wulf-Dieter	Fraunhofer FKIE	Germany
Woodbridge, Karl	University College London	UK

Part I

Real aperture array radar

Introduction to real aperture array radar

Ulrich Nickel[1]

The evolution of radar proceeded in a number of steps, of which each had a fundamental effect on the performance. In the beginning, the radar was built using all-analogue technology with an analogue display. Information extraction was performed at that time by adjusting the dynamic range of the cathode-ray display. A big improvement was obtained when the output data after analogue filtering and signal processing were sampled and digitized. This allowed some post processing of these data. A number of detection techniques with different algorithms and adaptive thresholds could be implemented. The algorithms could be changed by software if necessary. Digital displays provided a clear extraction of the relevant information. With the increasing availability of fast analogue-to-digital converters and computers, the point of digitization could then be shifted more and more towards the antenna such that programmable filters and signal processing algorithms could be applied. This allowed to apply all the flexibility of digital signal processing in the time domain.

In parallel, a major step was achieved by sampling and digitizing data in the spatial domain as well. This started the era of antenna-array processing. Even if certain reduced forms of spatial samples are digitally available, this concept may offer new options: digital beamforming with different weightings and pattern shaping, adaptive beamforming, mismatched spatial filtering, multi-target filtering and super-resolution. These are the concepts that are considered in Part I of this book 'Real Aperture Array Radar'. The efficiency of these real-aperture-array-processing methods depends strongly on the kind of spatial sampling. We may have fully filled arrays, sparse arrays, arrays with partial digital beamforming using sub-arrays and linear, planar or volume arrays. Therefore, the array configuration has to be considered in combination with the spatial processing algorithms. Furthermore, the processing methods in time and space have an implication on the resulting parameter estimates as well as on the target detection performance. Therefore, the array configuration and the aspects of detection and estimation have to be considered jointly. With digital signal processing, different methods resulting from different optimality criteria can be applied for each radar application, and it is possible to switch rapidly between different applications or radar modes. This is the concept of multi-function radar, which is a key application of real aperture array radar.

[1]Fraunhofer FKIE, Germany

The third jump in radar technology was by overriding the fixed spatial sampling scheme given by the mechanical setup, and this was achieved by spatial sampling sequentially in time in a defined sequence. This opens a great flexibility and leads to the big field of synthetic aperture radar (SAR) and inverse SAR. Applying these principles also on transmit signals opens the field of multiple-input–multiple-output radar. The theory of compressive sensing provides the theoretical background of proper and efficient sampling and processing of such 2D data fields. All these novel radar techniques will be considered in Parts II and III of this book. Actually, most techniques for real aperture array radar will also be applied in a modified form for the novel techniques, too. In this sense, the first part of this book provides the basis for the subsequent parts.

The task of modern radar is not to simply detect a target. The radar should provide relevant information of all features of interest of a target, where it depends on the specific application what the relevant information is. This requires the development of a mathematical model of the target and then to estimate the feature parameters of this model. The target dynamic behaviour is of particular interest, but other features like the radar cross-section, micro-Doppler fluctuation may be of interest too, e.g. for classification. Estimation of the dynamic target parameters is the topic of target tracking which is considered in Part II in Volume 2 of this book. Modern Bayesian tracking algorithms offer the possibility to exploit all kinds of information about the target and the environment. For optimum performance, all this information should be exploited. Such information can be gained by learning with the radar itself, or from other additional sensors or radars. Thus, we are led to netted radar systems and multiple sensor systems for which one has to apply intelligent sensor-data-fusion techniques, which is a topic of Part II in Volume 2 of this book.

A radar system with digital processing is able to switch between different modes of operation. A system that exploits adaptively all information about the target and the environment and which controls the radar operational modes adaptively according to this information has been termed 'cognitive radar' [1]. This feature is in fact the objective of a 'novel-radar' system to be considered in this book. It comprises the tasks of information extraction from different sensors and from big data bases, information processing and sensor-fusion techniques and adaptive control of all involved sensors. There is no cookbook on cognitive radar, but this present book can provide various aspects of such a system. The critical feature of a cognitive radar is that various constraints have to be fulfilled resulting from the hardware, the platform or the environment. In reality, the constraining requirements are often contradicting. The art-of-radar-system engineering consists of handling these constraints by matching the different hardware and software components of the system to fulfil the requirements in a good compromise. The antenna, signal processing, data processing, operational modes, resources and time management must be carefully designed in this sense.

What has all this to do with real aperture array radar? In fact, the real aperture radar already includes many aspects of novel, cognitive radar in a nutshell. It is the objective of the first part of this book to give an introduction to these problems for the more simple case of a real aperture.

Chapter 1 is quite comprehensive, and it serves to provide the relevant knowledge about array radar. It does not contain too much detail but covers the basic knowledge and interrelations between the processing blocks. The objective is to give a unified view of the related problems rather than introducing the latest novel algorithms. The issues of interest are the generic algorithms, the relations to the antenna design, the influence of hardware components, an analysis of the parameter estimation and detection problem, the relations to target tracking, the special features of applications and system operational modes and system control. Much emphasis is laid on the aspects of adaptive processing. For this case, particular relationships to the antenna configuration and the hardware features are evaluated and the detection and tracking procedures are reviewed. From a system design viewpoint, a key requirement of any adaptive processing is a well-defined behaviour and system predictability.

Chapter 2 considers combined space-time adaptive processing (STAP) for airborne radar application. Usually, adaptation is based on a second independent-training data set and on the assumption that the interference scenario is stationary over the processing time. In this section, an alternative interference cancellation technique is presented with minimal training data taken from the same range cell. Such 'single-snapshot' or 'direct-data-domain' techniques can overcome the problems of stationarity and limited sample support of standard adaptive processing, but they are also prone to deficiencies like target self-nulling. Direct data domain STAP is thus a good example of finding an improved solution satisfying different contradicting constraints. This is again a case where the simple real-aperture-array radar can provide examples of novel approaches that may be applicable to the more complicated techniques in Parts II and III in Volume 1 and Parts I and II in Volume 2.

Chapter 3 provides the state-of-the-art of multi-function radar and of the management of the radar operational modes. Multiple-mode operation is today the most attractive feature of real aperture array radar. The management of all possible modes, full exploitation of the hardware resources, efficient use of the data and communication resources is a problem of high complexity. At this time, we are just at the beginning of understanding and exploiting all the possibilities in this area. Consequently, this topic will be revisited and extended in some sections of Parts I and II of Volume 2.

With the mixture of relevant topics in the five parts of this book, I am confident that this book will give a valuable contribution to the radar community. If this book will eventually become a success, this will be to a great deal due to my colleague Richard Klemm who had the idea of compiling this kind of book and who managed the editorial process together with the excellent support of the IET publishing department.

Reference

[1] S. Haykin: Cognitive Radar, *IEEE Signal Processing Magazine*, January 2006, pp. 30–40.

Chapter 1

Target parameter estimation and array features

*Ulrich Nickel**

Abstract

The primary objective of this chapter is to introduce the basic array signal processing methods. If such methods are to be implemented in a radar one has to take into account the particular interrelationships of the different radar processing blocks and one has to adapt the related processing. Therefore, another objective of this chapter is to analyse these problems. The antenna array has to be designed to fulfil all requirements of the radar which is typically a multitasking system. In fact, the requirements can often only be fulfilled in a compromise. These problems are pointed out. The principal array processing methods of deterministic pattern shaping and adaptive beamforming are presented as complementary techniques to be considered together. It depends on the application whether more emphasis is put on low sidelobes or on adaptive nulling. Angular super-resolution methods are presented as a special mode for critical tasks. It is shown that combining detection and monopulse estimation with adapted beamforming results in mutual influence that has to be accounted for. Most important are the implications on tracking algorithms. From these a number of rules for the radar management are derived. Results and conclusions presented in this chapter are not definite design rules but should be considered as exemplary to provide awareness where problems may arise.

1.1 Introduction

Publications on array signal processing have appeared for decades. Books, conferences and special sessions are devoted to this topic. However, there is a discrepancy between the large number of published papers and the few methods implemented in existing radar systems. One of the reasons may be that all

*Fraunhofer Institute for Communication, Information Processing and Ergonomics (FKIE)[1], Germany

[1]The main part of this work was performed while the author was with the Fraunhofer Institute for High Frequency Physics and Radar Techniques (FHR), Germany.

processing elements of a radar system are mutually dependent. It turned out that it is not a trivial task to implement advanced algorithms and to obtain also a definite improvement. In fact, simply implementing an isolated algorithm is not very useful. The performance and the properties of any algorithm often depend critically on the preceding and subsequent processing steps.

First of all, the antenna design has an impact. There is an incredible variety of ways to realize a number of spatial digital channels which constitute the basis for array signal processing. Regularity or irregularity of the array elements or, more general, of the spatial channel characteristics are a critical feature and have a decisive impact on the performance of the implemented algorithm. The antenna design must be flexible enough to support all other desired applications of signal and data processing.

The second key element of radar processing is the detection step, where the large amount of sampled complex array data is reduced to a low data rate of the relevant data of interest. In classical radar engineering, the detector is considered as a single component in the processing chain. However, in modern digital processing the data reduction by the detector may be a multi-step procedure with increasing confidence about the presence of a target. This offers a variety of methods to control the amount of data and the complexity of the signal processing and the tracking algorithms. In the different stages of detection, there may also be a feedback to the previous processing stages requesting an improved processing. With digitally stored data, this is always an option. This means that additional system adaptivity can be implemented. Moreover, all the algorithms have to be compatible with the radar operational modes. These modes have to be adapted for advanced and adaptive signal processing. In fact, these are all building blocks for a radar system that uses in an intelligent way all sources of information. This is also known under the name ‘cognitive radar’ [5,6]. Predictability of the system performance with new algorithms is a key issue for the radar designer. This is a critical feature of all adaptive algorithms and of cognitive radar.

In a nutshell, in order to achieve a real improvement with advanced array processing one has to take into account the particular inter-relationships of the radar processing blocks and one has to adapt the related processing. Standard handbooks on radar [7,8] mention this problem only by the way. The objective of this chapter is to provide the background on radar array signal processing in a unified notation including the problems of implementation and of contradicting requirements. These may be relevant issues in the subsequent parts of this book. Some of the results presented in this chapter have been published in [1]. This chapter differs in a more generalized presentation with emphasis on providing an overview. Further issues of implementation and system aspects of radar array processing are discussed in the books [1,6,9,10].

We adopt a signal processing viewpoint to analyse the associated problems of implementation. In Section 1.2, the notation of array data is introduced and the basic concepts and results of arrays are presented. Then, we take a look back on the preceding stage of processing which is spatial sampling and antenna technology. The antenna array has to be designed to fulfil all requirements of the radar system. Modern radar is typically a multi-tasking system. So, the design of the antenna has to fulfil these purposes in a compromise. Requirements on the array antenna that arise from array processing methods are pointed out in Sections 1.3 and 1.4. In Sections 1.5–1.7, we present the array processing methods of interest. We do not present any new or sophisticated algorithms, but state the established algorithms as examples. If the reader wants to know details, he should consult the references. For the given examples, we point out the relationships between array processing and the preceding and subsequent radar processing blocks. In Section 1.5, we briefly review the approaches for deterministic pattern shaping which is the standard approach of antenna-based interference mitigation. It has the advantage of requiring little knowledge about the interference scenario, but requires very precise knowledge about the array transfer function ('the array manifold'). Adaptive beamforming (ABF) is presented in Section 1.6. This approach requires little knowledge about the array manifold but needs a good estimate of the interference scenario, which is gained from some training data. So, deterministic pattern shaping and ABF are two complementary techniques that have to be considered together. It depends on the particular application whether the highest emphasis is put on deterministic low sidelobes or on adaptive nulling. Section 1.7 is devoted to super-resolution. Super-resolution for best resolution of multiple targets is sometimes also subsumed under ABF as it resolves everything, interference and targets. However, the implementation for active radar is normally different. Super-resolution is typically a special mode for critical tasks in the list of available radar-processing techniques. In Section 1.9, we examine how some established radar-processing techniques like detection and monopulse estimation have to be modified if adapted beamforming is used. In fact, ABF has a significant influence on these techniques. Most important in this section is the consideration of the implications on tracking algorithms. The tracking stage is the final and most effective step of information extraction with a radar system. In this section, we point out how with modified direction estimation with adaptive beams the tracking algorithm and the track management can be adapted for ABF. This results in a number of rules for the radar management.

Of course, multi-function radar is a complex and dynamically evolving technology and with any new algorithm new problems will arise. In this sense, the material presented here will not be complete and there are many issues that are not covered in this chapter. The results and conclusions presented in this chapter have to be considered as exemplary to provide awareness where problems may arise.

1.2 Basic concepts and results of array antennas

The array principle, which consists of synthesizing a plane wave by a number of elementary spherical waves in transmit or receive mode, is well known. We assume that the elements of the array are positioned in the three-dimensional space with (x, y, z)-co-ordinates, and we denote the plane wave directions by a unit direction vector $\mathbf{u}$ in the (x, y, z)-co-ordinate system with $\|\mathbf{u}\| = 1$. Vectors are noted as column vectors and the components of the direction vector are denoted by $\mathbf{u} = (u, v, w)^T$. For a suitable definition of azimuth and elevation angles, this direction vector can also be parameterized by the corresponding angles, but the unit vector notation is a parameterization-free description and therefore more convenient.

1.2.1 Plane wave at single frequency

Suppose, we have an array with N elements at the positions $\mathbf{r}_i = (x_i, y_i, z_i)^T$, $i = 1 \ldots N$. A plane wave from direction $\mathbf{u}_0 = (u_0, v_0, w_0)^T$ at frequency f sampled at time t at these positions can then be written as:

$$s_i(t) = Ae^{j2\pi ft}\, e^{j2\pi f\, \mathbf{r}_i^T \mathbf{u}_0/c} \tag{1.1}$$

where c is the velocity of light and A is a complex amplitude with an initial phase φ, $A = |A|e^{j\varphi}$. This is a notation viewed from a receiving antenna for an incoming wave. The signal that is actually measured at the antenna elements $z_i(t)$ is of course corrupted by some noise, which is inevitable with any real hardware:

$$\begin{aligned} z_i(t) &= s_i(t) + n_i, \quad i = 1 \ldots N, \quad \text{or in vector notation} \\ \mathbf{z}(t) &= Ae^{j2\pi ft} \cdot \left(e^{j2\pi f \mathbf{r}_i^T \mathbf{u}_0/c}\right)_{i=1\ldots N} + \mathbf{n} \end{aligned} \tag{1.2}$$

This is the standard signal model for array antennas.

For a planar or linear array with elements at positions $\mathbf{r}_i = (x_i, y_i)^T$, or $\mathbf{r}_i = x_i$, $i = 1 \ldots N$, respectively, a signal with a frequency deviation $f = \alpha f_0$ is indistinguishable from a signal at displaced angles $(\alpha u_0, \alpha v_0)$ at frequency f_0, because $f \cdot (x_i u_0 + y_i v_0) = f_0 \cdot (x_i \alpha u_0 + y_i \alpha v_0)$. Another interpretation is that the array positions seem to be stretched if a signal at higher frequency is observed with receivers at the original frequency. This may introduce bias and grating effects which we will discuss below. Clearly, such errors are zero for the boresight direction $\mathbf{u}_0 = (0, 0)^T$ and the errors will increase with the angle of incidence $\mathbf{u}_0$.

The frequency dependent angle shift does not appear for arrays with a true three-dimensional element distribution (volume array or crow's nest antenna, [1 Section 4.6.1]), because the third component of $\mathbf{u}$, w, is not independent of u, v, because $\|\mathbf{u}\| = 1$. Therefore, there is no way of aliasing $\alpha\mathbf{u}_0$ with αf_0. It is a fundamental property of 3D volume arrays that squint and bias effects due to frequency do not appear. The 3D volume array has an inherent bandpass-filtering property.

For the detection of such an elementary plane wave, we have to sum up all array elements phase coherently in a beamforming operation $S_{\text{sum}} = \sum_{i=1}^{N} a_i^* s_i$ with $a_i = e^{j2\pi\alpha f\, \mathbf{r}_i^T \mathbf{u}_0/c}$ to maximize the signal power. At this point, we ignore any noise contribution. The asterisk means complex conjugate. Of course, the true signal frequency is in general unknown and usually the centre frequency or different frequencies of a frequency filter bank are used. Beamforming with the phases at the centre frequency produces for a planar array the afore mentioned angle shift by the factor α or a relative angle error $(\|\mathbf{u}_\alpha - \mathbf{u}_0\|)/\|\mathbf{u}_0\| = \alpha - 1$.

The relation between bandwidth and the angle error for planar arrays can be characterized by the bandwidth factor, [7 p. 13.39], which is defined as:

$$K_B = \frac{\text{bandwidth } (\%)}{\text{boresight beamwidth (deg)}} \tag{1.3}$$

The bandwidth factor is used to calculate the admissible bandwidth from the maximum scan angle and the tolerable angle error in beamwidth at this scan angle, if narrowband beamforming at centre frequency is performed. The magnitude of an acceptable K_B typically ranges from 1 to 2. For example, if one accepts a maximum squint angle of one-fourth of the beamwidth at a maximum scan angle of 60°, then one obtains $K_B = 1$, [7 p. 13.39]. This is the famous rule of thumb that the relative bandwidth in per cent should be equal to the beamwidth in degrees. For example, according to this rule one can use 2% bandwidth with an antenna with 2° beamwidth.

1.2.2 Band-limited signals

If a broadband signal is received, we have a super-position of all frequencies in the band. Such a signal can be described by a homogeneous plane wave field, see [11 p. 5], which is a special stochastic process in time and space. It can be written in the following notation (Cramér representation):

$$s(t, \mathbf{r}, \mathbf{u}) = \int_{-f_0-B/2}^{-f_0+B/2} e^{j2\pi f(t+\mathbf{r}^T\mathbf{u}/c)} dZ_s(f) + \int_{f_0-B/2}^{f_0+B/2} e^{j2\pi f(t+\mathbf{r}^T\mathbf{u}/c)} dZ_s(f) \tag{1.4}$$

where f_0 is the centre frequency, and B is the receiving bandwidth. $Z_s(f)$ is a complex stochastic process with independent increments, i.e. with the following properties: $E\{Z_s(f)\} = 0$ and $E\{dZ_s(f)dZ_s^*(g)\} = P_s(f)\delta(f-g)dfdg$ and $E\{dZ_s(f)dZ_s(g)\} = 0$. $P_s(f)$ denotes the power spectral density of the signal, the asterisk denotes complex conjugation. As the process s is assumed to be real, we have $dZ_s(-f) = dZ_s(f)^*$, and the signal can also be written as:

$$s(t, \mathbf{r}, \mathbf{u}) = 2\mathrm{Re}\left\{ \int_{f_0-B/2}^{f_0+B/2} e^{j2\pi f(t+\mathbf{r}^T\mathbf{u}/c)} dZ_s(f) \right\} \tag{1.5}$$

Such a stochastic integral cannot be solved as in classical analysis. This is only a representation of the stochastic process, but the moments of the process can be conveniently calculated with the rules for the stochastic differentials given above.

The complex baseband outputs (the *I*- and *Q*-components) are generated by the complex demodulation procedure which can ideally be written as:

$$\begin{aligned} I(t, \mathbf{r}, \mathbf{u}) &= \mathrm{LP}\{s(t, \mathbf{r}, \mathbf{u})\cos(2\pi f_0 t)\} \quad \text{and} \\ Q(t, \mathbf{r}, \mathbf{u}) &= \mathrm{LP}\{-s(t, \mathbf{r}, \mathbf{u})\sin(2\pi f_0 t)\} \end{aligned} \tag{1.6}$$

where LP{.} denotes the ideal low-pass filtering operation. As described in [12] the complex baseband signal $s(t, \mathbf{r}, \mathbf{u}) = I(t, \mathbf{r}, \mathbf{u}) + jQ(t, \mathbf{r}, \mathbf{u})$ can then be written as:

$$s(t, \mathbf{r}, \mathbf{u}) = e^{j2\pi f_0 \mathbf{r}^T\mathbf{u}/c} \int_{-B/2}^{B/2} e^{j2\pi\xi(t+\mathbf{r}^T\mathbf{u}/c)} dZ_s(\xi + f_0) \tag{1.7}$$

Such a stochastic signal is not typical for the desired radar echo. In fact, the transmit signal is a special deterministically modulated signal (e.g. linear/non-linear frequency modulated), but such a signal is included in this model by means of a suitable deterministic measure dZ_s. The model (1.7) is very useful in describing clutter and interference signals.

1.2.3 Narrowband and broadband beamforming

Beamforming is a linear filtering operation which extracts maximum signal energy for all possible signal spectra. This is achieved by coherently summing up the delayed signals, see [11 p. 9]. If the wave field is sampled by an array at positions $\mathbf{r}_i = (x_i, y_i, z_i)$, the sum beam output power is:

$$|S_{\text{sum}}|^2 = E\left\{\left|\sum_{i=1}^{N} s(t-\tau_i, \mathbf{r}_i, \mathbf{u})\right|^2\right\} \tag{1.8}$$

with $\tau_i = \mathbf{r}_i^T \mathbf{u}/c$. This is the classical delay and sum beamformer.

For narrowband processing, one assumes the delays to be small, $B\tau_i \ll f_0\tau_i$ or $B/f_0 \ll 1$, such one can write this summation with a steering vector $\mathbf{w} = (w_i)_{i=1\ldots N}$ containing only phase shifts $w_i = e^{j2\pi f_0 \mathbf{r}_i^T \mathbf{u}/c}$:

$$\begin{aligned} |S_{\text{sum}}|^2 &= E\left\{\left|\sum_{i=1}^{N} w_i^* s(t, \mathbf{r}_i, \mathbf{u})\right|^2\right\} \\ &= \mathbf{w}^H E\left\{(s(t, \mathbf{r}_i, \mathbf{u}) s^*(t, \mathbf{r}_k, \mathbf{u}))_{i,k=1\ldots N}\right\}\mathbf{w} \\ &= \mathbf{w}^H \mathbf{Q}_s \mathbf{w} \end{aligned} \tag{1.9}$$

The entries of the covariance matrix $\mathbf{Q}_s$ are given by:

$$(Q_s)_{i,k} = e^{j2\pi f_0 \mathbf{r}_i^T \mathbf{u}/c} e^{-j2\pi f_0 \mathbf{r}_k^T \mathbf{u}/c} \int_{-B/2}^{B/2} P_s(\xi + f_o) e^{j2\pi\xi(\mathbf{r}_i - \mathbf{r}_k)^T \mathbf{u}/c} d\xi \tag{1.10}$$

or in matrix notation:

$$\mathbf{Q}_s = \left(\mathbf{a}\mathbf{a}^H\right) \odot \mathbf{C}_s \tag{1.11}$$

with $\mathbf{a} = \left(e^{j2\pi f_0 \mathbf{r}_i^T \mathbf{u}/c}\right)_{i=1\ldots N}$, $\mathbf{C}_s = \left(\int_{-B/2}^{B/2} P_s(\xi + f_o) e^{j2\pi\xi(\mathbf{r}_i - \mathbf{r}_k)^T \mathbf{u}/c} d\xi\right)_{i,k=1\ldots N}$, and $\odot$ denoting the element-wise (Schur–Hadamard) matrix product. P_s is the signal spectral power density at the output of the antenna element. For example, for rectangular spectral power density we have:

$$\mathbf{C}_s = P_s(\text{sinc}(\pi B(\tau_i - \tau_k)))_{i,k=1\ldots N} \tag{1.12}$$

In the narrowband case, one has $B/f_0 \ll 1$ and $\mathbf{C}_s \approx P_s \cdot (1)_{i,k=1\ldots N}$, such that with (1.11):

$$\mathbf{Q}_s = P_s \mathbf{a}\mathbf{a}^H \tag{1.13}$$

The element receiver noise is assumed to be independent in each channel and independent of the direction, i.e. the delays $\tau = \mathbf{r}^T \mathbf{u}/c$ in (1.7) are zero such that $n(t, \mathbf{r}, \mathbf{u}) = \int_{-B/2}^{B/2} e^{j2\pi\xi t} dZ_n\,(\xi + f_0)$. The noise-alone covariance matrix is then:

$$\mathbf{Q}_n = \text{diag}\left\{\left(\int_{-B/2}^{B/2} P_{n,i}(\xi + f_o) d\xi\right)_{i=1\ldots N}\right\} \tag{1.14}$$

with noise power spectral densities $P_{n,i}(f)$ for $i = 1 \ldots N$.

In general, the received signal may be composed of a sum of M plane waves plus receiver noise such that one obtains the array outputs:

$$z(t, \mathbf{r}_i) = \sum_{k=1}^{M} s_k(t, \mathbf{r}_i, \mathbf{u}_k) + n_i(t), \quad i = 1 \ldots N, \quad \text{or in a short notation}$$
$$\mathbf{z}(t) = \sum_{k=1}^{M} \mathbf{s}_k(t) + \mathbf{n}(t) \tag{1.15}$$

with $n_i = \int_{-B/2}^{B/2} e^{j2\pi\xi t} dZ_{n;i}(\xi + f_o)$ and with obvious definition of the vectors $\mathbf{s}_k$. According to (1.11), (1.14) one obtains for uncorrelated plane waves the array output covariance matrix $\mathbf{R} = \sum_{k=1}^{M} \mathbf{Q}_{s;k} + \mathbf{Q}_n$. The narrowband version of this covariance matrix is:

$$\mathbf{R} = \sum_{k=1}^{M} P_k \mathbf{a}_k \mathbf{a}_k^H + \mathbf{Q}_n = \mathbf{A}\mathbf{B}\mathbf{A}^H + \mathbf{Q}_n \tag{1.16}$$

where P_k denotes the power, $\mathbf{a}_k = \mathbf{a}(\mathbf{u}_k)$, and $\mathbf{u}_k$ indicates the direction of the kth plane wave. For uncorrelated narrowband plane waves, we have $\mathbf{B} = \text{diag}_{k=1}^{M}\{P_k\}$; for correlated waves $\mathbf{B}$ is a full matrix, but possibly of rank lower than M if there are fully correlated sources. Broadband beamforming and narrowband beamforming as in (1.9) is performed with these matrices.

The delay operation for broadband beamforming may be too costly if it is applied at each element of a large array. Therefore, a hybrid solution is often preferred, where sub-arrays are summed up with phase shifting at the elements (the delays are small within the sub-arrays) and where the sub-array outputs are summed up with delays (*time delayed sub-arrays*). For the application of digital multichannel (*array processing*) techniques, A/D conversion of the sub-array outputs is needed (*digital sub-arrays*). These digital sub-arrays may be larger than the time delay sub-arrays, depending on receiver cost, bandwidth and error considerations. However, the preferred solution is to have a sufficiently large number of time delay sub-arrays which coincide with the digital sub-arrays.

In the sequel, we will assume narrow-band beamforming. The case of broadband beamforming and advanced broadband array signal processing methods with stepwise increasing complexity has been described in [2 Chapter 16].

1.2.4 Difference beamforming and monopulse estimation

Difference beamforming followed by monopulse estimation is mentioned in nearly all radar textbooks as a standard beamforming procedure, but in most cases, it is not mentioned why this is a good procedure and how it is related to optimum processing.

In fact, difference beams and the monopulse procedure are only needed for accurate angle estimation. In the radar search mode, the beam-pointing direction is taken as a rough direction estimate if a detection of a target has occurred. In a second step, this detection is confirmed and a more accurate angle estimate is performed. This is a statistical parameter estimation problem and not a matter of beamforming. The problem will be considered in more detail in Sections 1.7 and 1.9. We mention here briefly the underlying principle and the rationale of difference beamforming, because difference beamforming is a feature that influences heavily the array design.

Radar parameter estimation can be formulated as a maximum likelihood (ML) estimation procedure, see [1 Section 11.1, 13 Section 3.2.1] or [14]. For a set of measured array data vectors (often called array snapshots) $\mathbf{z}_1, \ldots \mathbf{z}_K$, $k = 1 \ldots K$, $\mathbf{z}_k \in \mathbb{C}^N$, modelled as in Section 1.2.1, one can define an appropriate statistical model for the noise and the signal amplitude fluctuation. The probability density of the snapshots depending on all the desired (unknown) parameters $\boldsymbol{\theta}$ can then be written as $p(\mathbf{z}_1, \ldots, \mathbf{z}_K | \boldsymbol{\theta})$. In general, no globally optimum procedure exists for estimating the parameters $\boldsymbol{\theta}$. However, the ML procedure is a realisable way that can produce at least an asymptotically optimum estimator. Asymptotic refers here to the number of array elements or the number of time samples K going to infinity. For assumptions fulfilled in a reasonable practical system, the ML estimator is asymptotically unbiased and has asymptotically the smallest variance, i.e. it attains the Cramér–Rao bound. Angle estimation for a single target by the ML procedure with a given set of snapshots requires forming the sum beam for all possible directions and taking the value with maximum power as the angle estimate, as explained in [13]. This is nothing else than maximizing what we call the antenna scan pattern. The procedure of sequential lobing mentioned in standard radar textbooks [7 p. 9.16], [8] is an abbreviated way of maximizing the antenna scan pattern.

An alternative way of finding the maximum is to extrapolate the shape of the scan pattern in the vicinity of the rough direction estimate obtained from detection. One can approximate the beam shape by a parabola and determine the parabola parameters from a set of neighbouring sum beams or from the sum beam in the search direction plus its derivatives. The difference beams are just estimates of the derivatives of the sum beam. Note that for planar or volume arrays we have derivatives for azimuth and elevation, i.e. we have multiple difference beams. The extrapolation is typically done by Taylor expansion. This Taylor expansion finally provides a direct formula for the unknown directions depending only on the sum and difference beam outputs. This is the basis of the famous monopulse formula for the estimated angle $\hat{u} = u_0 - \mu \operatorname{Re}\{D/S\}$. D, S denote the difference and sum beam outputs, respectively, and μ is a proportionality constant which depends on the second derivative and which is determined approximately or experimentally. The estimate for the elevation $\hat{v}$ is obtained analogous by an elevation difference beam. The relationship with the Taylor expansion is of importance when the monopulse procedure has to be generalized to ABF, where the beam shape is dependent on the unknown interference scenario. This will be considered in Section 1.9.

The optimum difference beam (in the ML sense) is taken as the derivative of the sum beam $S_{\text{sum}}(u, v) = \mathbf{a}(u, v)^H \mathbf{z}$, which gives:

$$S_{diff,az} = \frac{\partial S_{\text{sum}}}{\partial u} = \gamma \left(\sum_{k=1}^{N} x_k a_k^*(u, v) z_k \right) = \gamma \mathbf{d}_x^H \mathbf{z} = \gamma \mathbf{a}^H \operatorname{diag}\{x_k\} \mathbf{z}$$

for $d_k = x_k a_k$, $k = 1 \ldots N$. The constant factor $\gamma = -j2\pi f_0/c$ is irrelevant and can be omitted. So, the optimum difference beam weighting is the sum beam weighting with an additional amplitude tapering according to the element positions. For a uniform linear array (ULA), this means to form the beam by summing the differences of the opposing element pairs, hence the name difference beam. If we take as a crude approximation the amplitude weighting sign(x_k), then we just subtract the outputs of the left from the right half of the array which for a planar circular array is also known as four-quadrant monopulse. In any case, the beam pattern is zero in the array steering direction.

1.3 Design factors for arrays

The basis of all array processing is the array antenna. The antenna has to be designed and manufactured to fulfil all requirements of size, weight, manufacturability and of signal-processing performance. A once selected construction cannot be altered without considerable cost. Digital array processing requires digital array outputs. The number and quality of the compact receivers providing these outputs (e.g. the linearity and the number of ADC bits) determines the performance and the cost of the whole system. It may be desirable to design a fully digital array with AD-converters at each antenna element. For a moderate number of array elements, this may be feasible. However, system weight and cost will often lead to a design with reduced number of digital receivers. On the other hand, radar needs an antenna with a high gain, because of the $1/R^4$ decay of the received power. Also a high directional discrimination is desired. This leads to arrays with many elements. A compromise between these contradicting features has to be found.

The features of array antennas like gain, beamwidth and sidelobes have been discussed in detail in [1]. In many cases, one is not interested in the real antenna pattern but on the effect of the spatial (array) processing only. This effect is described by the array factor which is defined as the antenna pattern $f(u, v)$ with ideal omni-directional array elements with gain equal to one, i.e.

$$f(u,v) = \left|\mathbf{w}^H \mathbf{a}(u,v)\right|^2 \tag{1.17}$$

If all array elements have the same element pattern $g(u,v)$, then the array antenna pattern can be derived from (1.17) as $|g(u,v)|^2 f(u,v)$. From [1], we cite the following basic properties of the array factor:

1. **Half-power beamwidth:** For a uniform linear array (ULA), the diameter of the beam measured at half peak power of the array factor is:

   $$\text{BW} = 0.887 \frac{\lambda}{D \cos u_0} [\text{rad}] \text{ or } \text{BW} \approx 50 \frac{\lambda}{D \cos u_0} [^\circ]$$

   where D is the antenna aperture diameter, and λ is the centre frequency wavelength. For a circular planar array with elements on a $\lambda/2$ grid, one has:

   $$\text{BW} \approx 1.02 \frac{\lambda}{D \cos u_0} [\text{rad}]$$

   Amplitude tapering of the elements will result in a larger beamwidth.
2. **Acceptable bandwidth:** From (1.3), we have the rule of thumb for beam steering by phase shifts that the relative bandwidth B/f in per cent must be

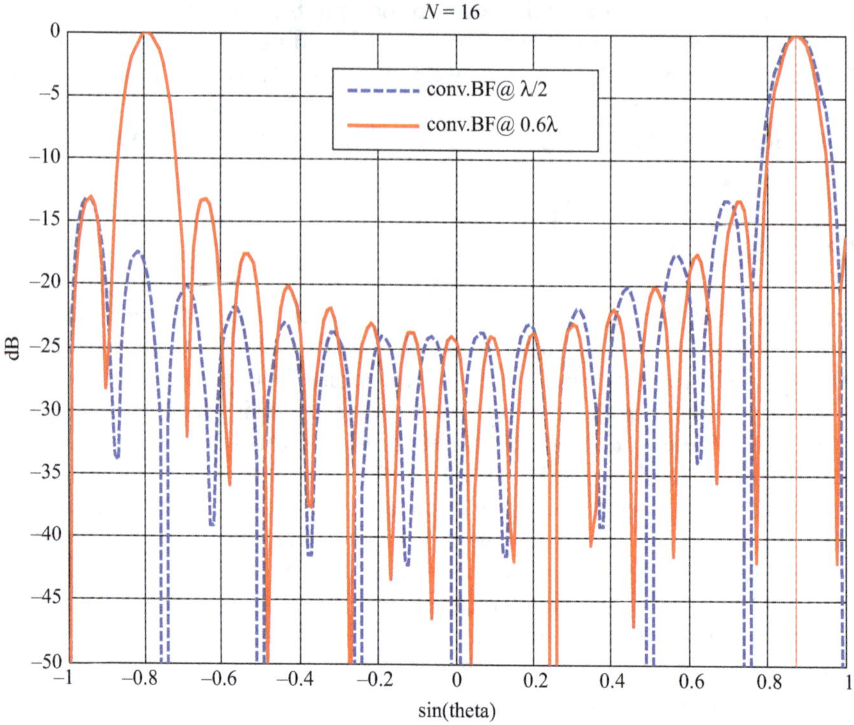

Figure 1.1 Array factor for 16 element ULA at $\lambda/2$ spacing and 0.6λ spacing and look direction 60° ($u_0 = 0.875$)

equal to the beamwidth at boresight, if an angle error below one-fourth BW at scan angle of 60° is required. For example, if one wants 10% relative bandwidth one can only realize a phased array with beamwidth BW $\geq$ 10°.

3. **Grating lobes and element spacing:** For a ULA with element spacing greater than $\lambda/2$, a secondary mainlobe (grating lobe) will enter the visible region of $[-1, 1]$ of the direction cosine because of the periodicity of the phase. This is illustrated in Figure 1.1 showing that a grating lobe at $u = -0.8$ appears for an element spacing of 0.6λ with look direction 60°. Note that also the beamwidth is slightly reduced due to the increased aperture. Ambiguities due to grating lobes can be mitigated by limiting the admissible scan angle of the array and by an appropriately modified element pattern.
4. **Gain and sidelobe level:** The array factor in boresight direction is from (1.13) given by $G_{\text{array}} = |\mathbf{w}^H \mathbf{1}|^2 = \left|\sum_{i=1}^N w_i\right|^2$. The maximum array factor (peak value) is obtained if $\mathbf{w} \parallel \mathbf{1}$, i.e. if all $w_i = \gamma$ resulting in $G_{\text{array}} = \gamma^2 N^2$. The sidelobe level of the array factor is defined as:

$$\bar{SL} = \frac{1}{|\Omega|} \iint_{\Omega} |f(u,v)|^2 du\, dv = \frac{1}{|\Omega|} \iint_{\Omega} \mathbf{w}^H \mathbf{a}\mathbf{a}^H \mathbf{w}\, du\, dv = \mathbf{w}^H \mathbf{C} \mathbf{w} \tag{1.18}$$

with $\mathbf{C} = 1/|\Omega| \iint_{\Omega} \mathbf{a}\mathbf{a}^H \, du \, dv$ and Ω denoting the set of direction cosines for the visible region $u^2 + v^2 \leq 1$. This matrix can be well approximated by $\mathbf{C} \approx \mathbf{I}$ (in fact $\mathbf{C} \rightarrow \mathbf{I}$ for $\Omega \rightarrow \mathbb{R}^2$). Therefore, we have for any weighting $\bar{SL} \approx \mathbf{w}^H \mathbf{w}$. The sidelobe level of the weighting maximizing the array factor is $\bar{SL} = \gamma^2 N$ or $\bar{SL} = 1/N$ if we normalize the peak to one. The highest sidelobe is for a ULA at $\lambda/2$ spacing for all N equal to -13 dB (this is a property of the sinc-function) and for a regular circular planar array with $\lambda/2$ grid at -17.5 dB.

The statements about linear arrays can be immediately generalized to planar arrays and the associated antenna pattern in a certain angular cut: if we consider a cut through the two-dimensional azimuth-elevation plane along the line $(u(t), v(t)) = (t \cos \alpha, t \sin \alpha)$, then this pattern cut is the same as the pattern of the array with a rotated co-ordinate system $\begin{pmatrix} \tilde{x} \\ \tilde{y} \end{pmatrix} = \begin{pmatrix} \cos \alpha & \sin \alpha \\ -\sin \alpha & \cos \alpha \end{pmatrix} \begin{pmatrix} x \\ y \end{pmatrix}$ at elevation $\tilde{v} = 0$ in the new co-ordinate system. The antenna pattern cut is the same function as the pattern of the linear array with the co-ordinates $(\tilde{x}_i)_{i=1 \dots N}$, which is also called the equivalent linear array (ELA). Ambiguities and pattern properties of planar and volume arrays have been analysed in depth with tools of differential geometry in the book [15]. It is shown there that the patterns and associated features like ambiguities, beamwidth etc. of arbitrary arrays can be analysed by all possible cuts and the associated ELAs. In particular, this says that a regular structure appearing in any projection of the array element positions on a line will result in grating effects.

In reality, real grating effects are not so much important, but quasi-grating effects arising from antenna positions which are close to regular. Figure 1.2 shows such a case where we have modified the element positions with a slight density tapering (the elements at the border of the aperture are a bit closer and in the centre a bit wider separated). The grating lobe now has become an unacceptably high sidelobe.

The concept of ELA is a convenient tool to analyse grating effects of planar arrays. As a typical example, we consider a hexagonal regular array which displays periodicities in three directions. Figure 1.3 shows this fact for a small array with seven elements. To show the periodicities, the function is evaluated beyond the visible region over the square $(u, v) = [-2, 2] \times [-2, 2]$. The visible region $u^2 + v^2 < 1$ is shown by the circle, the steering direction of the array factor is shown by the small circle at $(u, v) = (0.7, 0)$. The periodicity can be broken up by configuring the array with some irregularity, e.g. by saving one element of the outer ring as in Figure 1.3(c) and (d).

Antenna element design and the need for fixing elements mechanically lead to element patterns which are never omni-directional. Moreover, the elements have to be designed with patterns that allow a unique identification of the direction. A planar array can observe unambiguously only angles in a hemisphere. Antenna elements mounted on a conducting plane or with significant back attenuation are therefore a suitable solution for such an antenna. To achieve full spherical

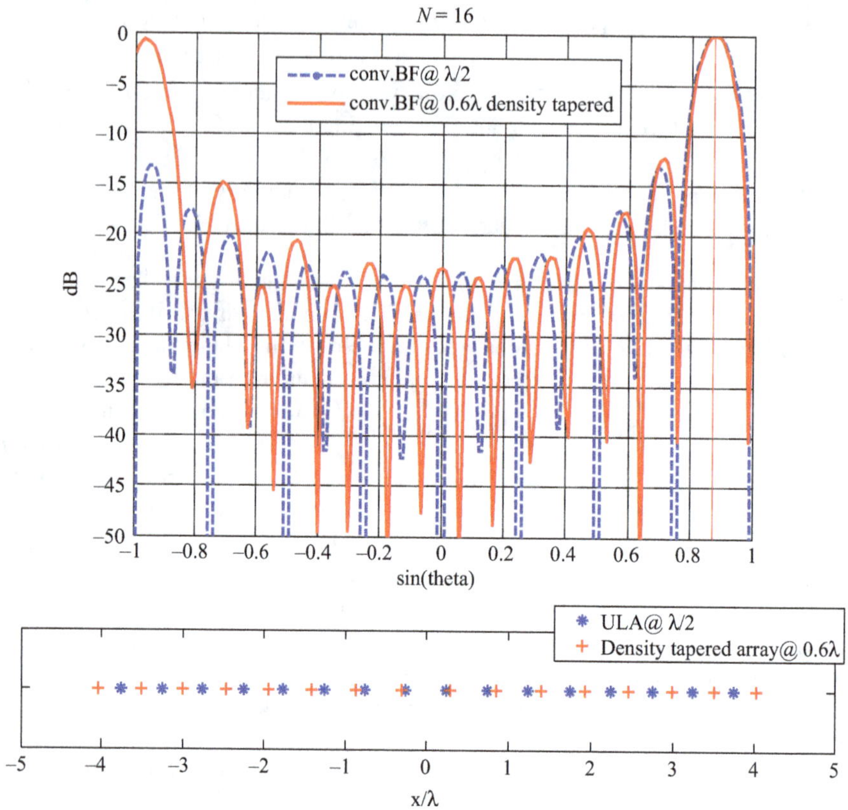

Figure 1.2 Array of Figure 1.1 with 0.6λ spacing but slightly density tapered with positions shown in the lower sub-plot

coverage, several arrays can be combined (multi-facetted array) or elements may be placed on a curved surface or in a full volume. Antenna elements may be composed of different linear arrays (3D configurations of one-dimensional arrays), on a plane (a ring or a two-dimensional planar array), on a curved surface (conformal array), or within a volume (3D-array, also called crow's nest antenna, [1 Section 4.6.1]). A one-dimensional array can only measure one independent angle, 2D and 3D arrays can measure jointly azimuth and elevation, which may be restricted to a limited angular sector. The array element design should be compatible with these restrictions.

The basic solutions to solve the contradicting requirements between the large number of elements with digital receivers and cost, weight and field of view are as follows:

- Thinned arrays. The angular discrimination of an array with a number of elements can be improved by increasing the separation between the elements. Thus, the aperture of the antenna is increased.

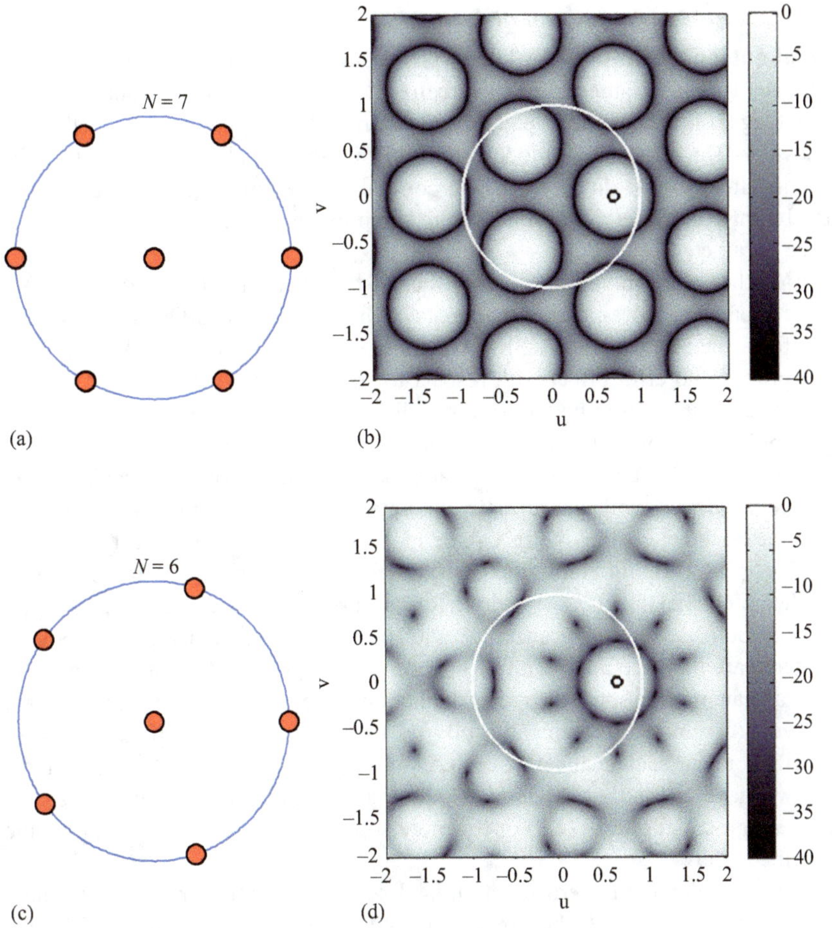

Figure 1.3 Planar arrays showing periodicities along symmetry axes. Steering direction of array factor is (u, v) = (0.7, 0). (a) Hexagon array with 2λ diameter, (b) array factor of hexagon array expanded on (u, v) = [−2,2] × [−2,2], (c) pentagon array of 2λ diameter and (d) array factor of pentagon array expanded on (u, v) = [−2,2] × [−2,2]

- Sub-arrays. To save digital receivers, some element outputs are summed up analogue which are then AD-converted and processed digitally. The size and the shape of the sub-arrays are an important design criterion. The notion of an array with sub-arrays is very general and includes the case of directional array elements which may be even independently steerable.

1.3.1 Influence of element patterns

Equal antenna elements. For linear, planar and volume arrays, antenna elements with nearly equal patterns can be realized. These have the advantage that the knowledge of the element pattern is for many array-processing methods not necessary. The array signal model vector in free space and at the element output is pointing in the same direction. Equal complex values of the element patterns can also be interpreted as a modified target complex amplitude, which is often a nuisance parameter.

Another more important point is that, if the element patterns are really absolutely equal, any cross-polar components of the signal are in all channels equal and thus fulfil the array model (1.5) in the same way as the co-polar components, i.e. they produce no error effect with respect to spatial (array) processing. This means that patterns nulls from beamforming will be present in the co- as well as in the cross-polar channel. Thus, cross-polar jamming may be ineffective.

In reality, the antenna element patterns are never absolutely equal or exactly known. However, with appropriate theoretical models together with calibration procedures, the element patterns can be modelled as known or at least being equal.

Unequal antenna elements. Unequal element patterns typically arise by tilting the antenna elements as it is done for conformal arrays. For a planar array, this tilt may be used to realize an array with polarization diversity. Single polarized elements are then mounted with orthogonal alignment at different positions. Such an array can provide some degree of dual polarization reception with single channel receivers (contrary to more costly fully polarimetric arrays with receivers for both polarizations for each channel). Common to arrays with unequal element patterns is that we have to know the element patterns for applying any kind of spatially coherent processing as in (1.9) or for more sophisticated array-processing methods described later.

The full element pattern function is also called the array manifold. If unequal array elements are used, the array manifold must be exactly known for correct beamforming (coherent processing). In particular, the cross-polar (or short x-pol) component has a different influence for each element. If the x-pol component is not sufficiently attenuated, it can be a significant source of error.

1.3.2 Thinned arrays

To save the cost of receiving modules, sparse arrays are considered, i.e. arrays with fewer elements than with the fully populated $\lambda/2$ grid. If such a 'thinned array' spans the same aperture as the fully filled ($\lambda/2$ grid) array, it has the same beamwidth. Hence, the angular accuracy and resolution is the same. Due to the gaps, ambiguities or at least high sidelobes may arise. In early publications like [7 p. 13.23], it was advocated to simply take out elements of the fully filled array. It was soon recognized that this kind of thinning does not give the best results because this does not generate 'sufficiently random' positions. Note that the array gain of a thinned array with N elements is always N and the average sidelobe level is $1/N$. Today, the theory of compressed sensing tells on a solid statistical basis that

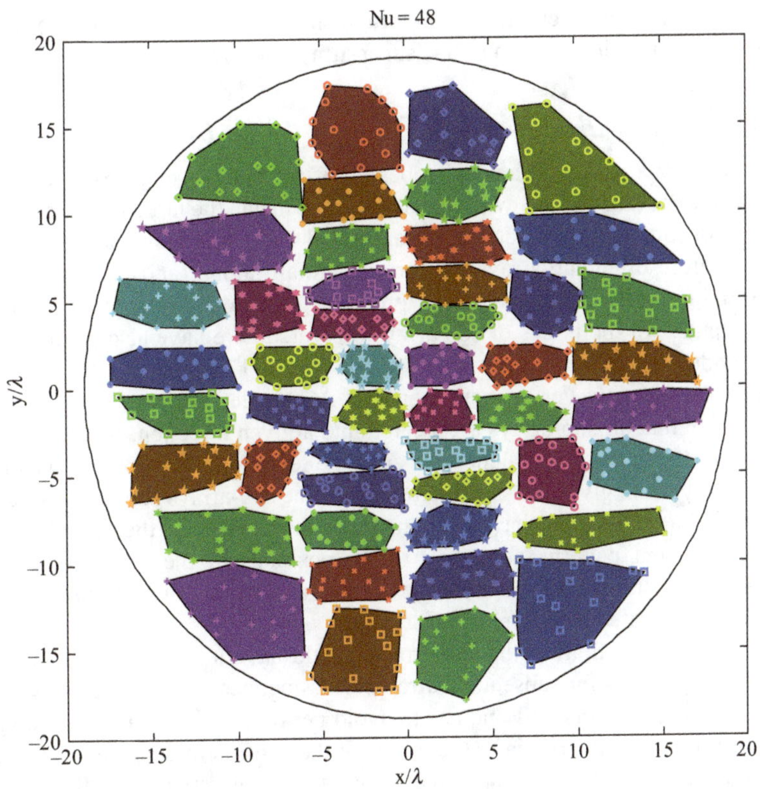

Figure 1.4 Experimental thinned array ELRA with 768 elements configured into 48 sub-arrays

a selection of sufficiently random positions can produce a unique reconstruction of a not too large number of impinging wave fields with high probability [16].

A good example of a thinned array is the ELRA antenna of Fraunhofer FHR in Figure 1.4, [1 Chapter 17]. The element positions are selected random with a parabolic density tapering constrained to a $\lambda/16$ grid. The thinning factor is 18% which means that the fully populated array would have 4,185 elements instead of the 768 elements of this antenna. This is a significant cost saving while the beamwidth of the fully populated aperture of 2° is preserved. The mean sidelobe level is at −29 dB and this corresponds exactly to $1/N$.

For an array with sufficiently many elements it is usually easy to generate a thinned array with no real secondary main lobes (true grating lobes). The practical problem is to avoid high quasi-grating lobes. Such quasi-grating lobes may not only alias a target direction if we detect an echo through the high sidelobe, but it will also create blind directions if we apply ABF. In Section 1.6, it will be explained that ABF creates pattern notches inversely to the sidelobe pattern. Quasi-grating effects are even more critical if we apply ABF at sub-array level. As will be explained in the next section, the sub-array centres (at which ABF is applied) can be considered as elements of a sparse super-array which may create grating effects.

1.3.3 Arrays with sub-arrays

If a high antenna gain with low sidelobes is desired one has to go back to the fully filled array. For large arrays with thousands of elements the large number of digital channel constitutes a significant cost factor and a challenge for the resulting data rate. Therefore often sub-arrays are formed and all digital (and perhaps adaptive) beamforming and sophisticated array processing methods are applied to the sub-array outputs. All the issues of sub-arrays presented here have also been reviewed and extended to further applications by Hang Hu in [17].

Sub-arraying is a very general concept. At the elements we may have phase shifters such that all sub-arrays are steered into a given direction. In addition some attenuation (tapering) to influence the sidelobe level may be applied. The sum of the sub-arrays then gives the sum beam output. The sub-arrays can be considered as elements of a super-array having elements at the centres of the sub-arrays with different element patterns according to the positions of the involved elements. Typically, the sub-arrays are all steered into the same selected direction. The principle is indicated in Figure 1.5.

For a fully filled array the sub-arrays should have unequal size and shape to avoid grating effects for subsequent array processing, because the sub-array centres constitute a sparse array and one has to avoid periodic super-array configurations. The question is: 'what criteria are relevant for the configuration of the sub-arrays'? This is a serious problem because the sub-array configuration is the basis for the mechanical and in most cases also the electrical design of the full array. A decision for a specific concept cannot be easily altered.

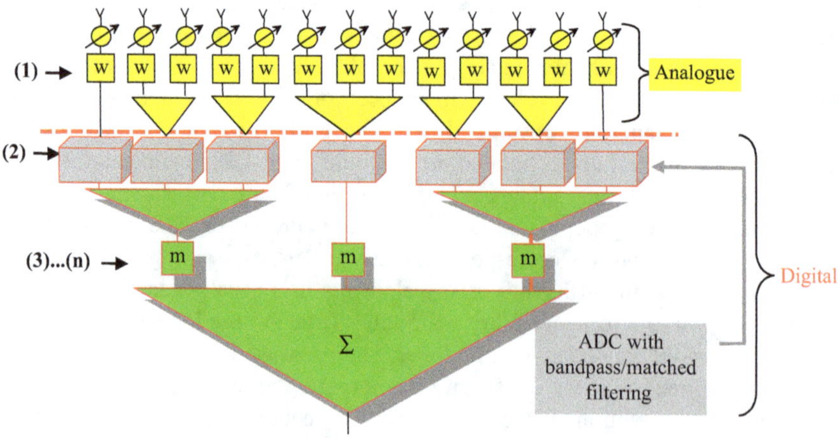

Figure 1.5 Principle of forming sub-arrays

For a given sub-array configuration there is still a small degree of flexibility, because, once we are digital, we can combine new sub-arrays at the digital level. This is nothing else than a certain matrix transformation of the sub-array outputs. This will be explained in the last paragraph of this section. One can also distribute the desired element tapering over the analogue level (1) and possibly many digital levels (2) ... (*n*) as indicated in Figure 1.5.

Criteria to configure sub-arrays have been studied in [18,19]. Below the essential ideas are described. In the studies, it turned out that a critical design factor is the requirement to apply a taper function for low sidelobes at the elements, which must be valid for both, sum and difference beamforming, and to apply then at the sub-array outputs the necessary final weights for the sum and the azimuth and elevation difference beams. In addition, these weights may be chosen adaptively to achieve further interference suppression.

Beamforming using sub-arrays can be mathematically described by a simple matrix operation. Let the complex array element outputs be denoted by $\mathbf{z}$. The sub-array forming operation is described by a matrix $\mathbf{T}$ by which the element outputs are summed up as $\tilde{\mathbf{z}} = \mathbf{T}^H\mathbf{z}$. For L sub-arrays and N antenna elements $\mathbf{T}$ is of size $N \times L$. Vectors and matrices at the sub-array outputs are denoted by the tilde. Suppose we steer the array into a look direction $(u_0, v_0) =: \mathbf{u}_0$ by applying phase shifts $a_i(\mathbf{u}_0) = e^{j2\pi f_0(x_i u_0 + y_i v_0)/c}$ and apply additionally amplitude weighting g_i at the elements (a real vector of length N) for a sum beam with low sidelobes. Both weightings can be combined in a complex weighting $g_i a_i(\mathbf{u}_0)$ which can be included in the elements of the matrix $\mathbf{T}$. The beams are formed digitally with the sub-array outputs by applying a final weighting $\tilde{m}_i$ $(i = 1, \ldots L)$ as:

$$S_{\text{sum}} = \tilde{\mathbf{m}}^H\tilde{\mathbf{z}} \tag{1.19}$$

In the simplest case, when the complete weighting for the sum beam is applied at the elements, i.e. when the complete weighting is incorporated in the matrix $\mathbf{T}$, $\tilde{\mathbf{m}}$ consists of only ones. However, the weighting can also be partly applied digitally in $\tilde{\mathbf{m}}$, i.e. we split up $\mathbf{T}$ into $\mathbf{T}_0\mathbf{D}$ and form the sub-arrays with $\mathbf{T}_0$ and use a beamforming vector $\tilde{\mathbf{m}}_0 = \mathbf{D}\tilde{\mathbf{m}}$. This allows to lower the dynamic range of the AD-converters or to generate a sub-array transformation matrix $\mathbf{T}$ with certain properties as we will show later in Section 1.6. The antenna pattern of a beam formed with sub-arrays is written as:

$$f(\mathbf{u}) = \tilde{\mathbf{m}}^H\mathbf{T}^H\mathbf{a}(\mathbf{u}) = \tilde{\mathbf{m}}^H\tilde{\mathbf{a}}(\mathbf{u}), \tag{1.20}$$

where $\mathbf{a}(\mathbf{u}) = (a_i(\mathbf{u}))_{i=1\ldots N}$; and $\tilde{\mathbf{a}}(\mathbf{u}) = \mathbf{T}^H\mathbf{a}(\mathbf{u})$ denotes the plane wave response at the sub-array outputs or, in other words, it describes the array manifold of the super-array. With digital beamforming all kinds of beams (sum, azimuth and elevation difference, guard channel etc.) can be formed from these sub-array outputs.

For characterizing the super-array, let us denote the positions of the super-array elements by $(\rho_{x,l}, \rho_{y,l})$, $l = 1 \ldots L$, and suppose we have a (possibly partial) weighting w_i, $i = 1 \ldots N$, at the elements. For a steering direction $\mathbf{u}_0$, the sub-array patterns are defined by:

$$F_l(\mathbf{u}) := \sum_{v \in U_l} w_v \exp(j2\pi f_0(x_v(u - u_0) + y_v(v - v_0))/c)$$

where U_l denotes the set of indices of the elements belonging to the lth sub-array. The positions of the super-array elements are now calculated by the property that the sub-array patterns $F_l(\mathbf{u})$ should be equal to $f_l(u, v)\exp(2\pi f_0(\rho_l(u - u_0) + \rho_l(v - v_0))/c)$ (super-array notation) for suitable super-array element patterns $f_l(\mathbf{u})$. Using the condition that for both notations, the sub-array patterns should have a maximum in direction $\mathbf{u}_0$, one obtains the conditions:

$$F_l(\mathbf{u}_0) = f_l(\mathbf{u}_0) = \sum_{v \in U_l} w_v \quad \text{and} \quad \frac{\partial F_l}{\partial u}(\mathbf{u}_0) = \sum_{v \in U_l} w_v j2\pi f_0 x_v/c \quad \text{and}$$

$$\frac{\partial F_l}{\partial v}(\mathbf{u}_0) = \sum_{v \in U_l} w_v j2\pi f_0 y_v/c$$

Using $(\partial F_l/\partial \alpha)(\mathbf{u}_0) = f_l(\mathbf{u}_0) j2\pi f_0 \rho_{\varsigma,l}/c$ (for $\alpha = u$ or v and $\zeta = x$ or y) the conditions yield the super-array positions:

$$\rho_{\varsigma,l} = \frac{\sum_{k \in U_l} w_k \varsigma_k}{\sum_{k \in U_l} w_k}, \quad \text{for coordinates} \quad \varsigma = x, y \tag{1.21}$$

We can define the array factor of the super-array by

$$f_{super}(\mathbf{u}) = \sum_{l=1}^{L} \tilde{m}_i e^{j2\pi f_0(\rho_{x,l} u + \rho_{y,l} v)/c} \tag{1.22}$$

The extreme cases of sub-arrays are:

- Each sub-array contains only one element. Then we have the fully digital array with digital channels at each array element and the super-array is the same as the array itself.
- Each sub-array uses all elements. Then the super-array positions are all the same in the centre of the original array and the elements outputs are split up L times.

The first question for the sub-array design is: should the sub-arrays be overlapping or disjoint? A first kind of overlap can be achieved by using some elements simultaneously in different sub-arrays. This requires a splitter at those array elements (or m-fold splitter, if the element is used in m sub-arrays). This is a cost factor and an issue of calibration if high performance array processing methods should be applied. More important is that the receiver noise, which is generated by the front end amplifiers, will be correlated at the sub-array outputs. This is because $\mathrm{E}\{\tilde{\mathbf{n}}\tilde{\mathbf{n}}^H\} = \mathbf{T}^H \mathrm{E}\{\mathbf{n}\mathbf{n}^H\}\mathbf{T} = \mathbf{T}^H\mathbf{T}$, if $\mathrm{E}\{\mathbf{n}\mathbf{n}^H\} = \mathbf{I}$, and $\mathbf{T}^H\mathbf{T}$ is not diagonal for overlapping sub-arrays. This causes problems in ABF as will be explained in Section 1.6. The reason for using overlapping sub-arrays is mainly to generate patterns with low sidelobes for the sum as well as for both difference beams. However, for a sufficiently large number of sub-arrays acceptable low sidelobe patterns can also be achieved with disjoint sub-arrays by the procedure described in [18].

A second kind of overlap can be realized with single element use. This is achieved if the convex hulls of the sub-arrays element positions are overlapping. This means e.g. for a fully filled array that the sub-arrays constitute a kind of sparse array. The sub-array aperture is then larger than a corresponding sub-array with fully filled elements. This results in a narrow sub-array beamwidth. However, for most applications, we want multiple usages of the digital sub-array outputs and the sub-arrays are all steered into the same direction. In this case a maximally broad sub-array beamwidth is desired. Therefore, such a sub-array configuration is not desirable.

In the sequel we will consider only non-overlapping sub-arrays with a unique mapping of each element to only one sub-array.

1.3.3.1 Sub-array design procedure for low sidelobe sum and difference beams

The design procedure of [18] is based on the observation that the difference beam is more sensitive to grating lobes than the sum pattern. The difference beam is therefore taken as the design criterion. Suppose we apply a weighting $\mathbf{w}$ for sum beamforming at the elements. This may be a partial weighting only which needs a corresponding sub-array weighting $\mathbf{m}$ for a sum beam with the correct weighting and hence with the desired sidelobes.

We can now apply a difference weighting only at the sub-array outputs given this partial element weighting. This is not a serious restriction, because the typical low sidelobe difference weightings like Bayliss consist of a bell shaped amplitude taper similar to the sum beam taper times the optimal difference weighting according to the element positions, see e.g. [13 p. 54] or [14]. We now want to find a sub-array configuration that allows to realize two desired element level difference weightings $\mathbf{d}_x, \mathbf{d}_y \in \mathbb{C}^N$ as good as possible at sub-array level. These desired element level taperings $\mathbf{d}$ are functions of the element positions $d_i = f(x_i, y_i)$. We may quantize the range of this function into q steps. This produces in the (x, y)-plane q sets, see Figure 1.6 for the case of 6 steps.

Note that, as we have already a fixed taper $\mathbf{w}$ applied at the elements, we have to apply this set generating procedure for the taper $\mathbf{d}_0 = (d_i/w_i)_{i=1,N}$. We do this for both, the azimuth and elevation difference beam weightings. The inter-sections of the resulting sets represent sub-arrays that represent the best sets of elements for reconstructing the original weighting at sub-array level. However, to avoid some undesirable side effects the following refinement steps are necessary:

- The inter-sections of the two contour sets may create sub-arrays with only a single or very few elements. These sub-arrays can be ignored by merging them with a bigger neighbour sub-array.

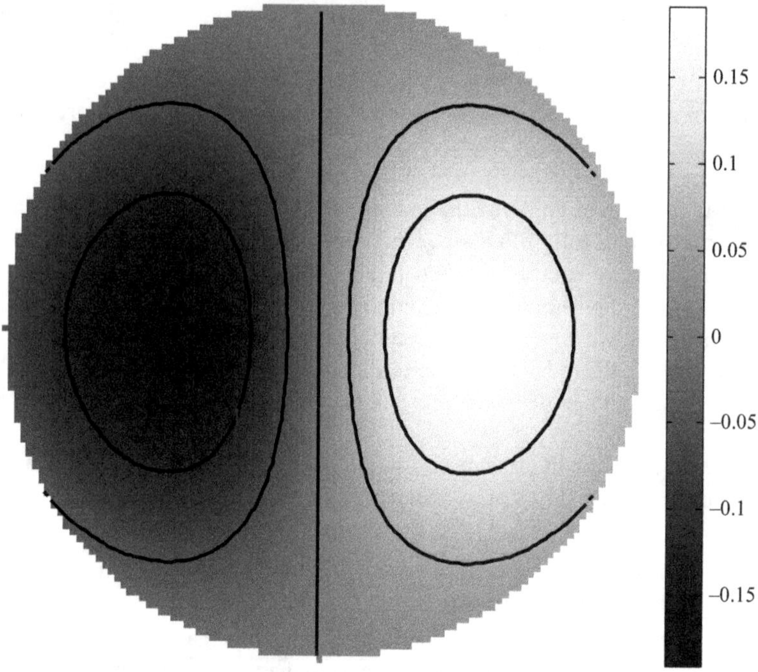

Figure 1.6 Plot of azimuth difference beam taper for planar circular antenna with low sidelobes. Contour lines for 6 levels induce 6 corresponding sets in the (x, y)-plane

- In the centre of the array fairly quadratic sub-arrays of nearly equal size will appear. These give rise to quasi-grating lobes and also, more important, quasi-grating notches with ABF. In a manual fine tuning step these regularities should be broken up.

Figure 1.7 shows an example of a design of a planar array with 902 elements on a triangular grid with 32 sub-arrays. The shape of the sub-arrays was optimized by the strategy of [20] such that the difference beams have low sidelobes approximating a desired −35 dB Bayliss weighting applied at the elements. At the elements a fixed partial −35 dB Taylor sum beam weighting is applied such that $\mathbf{T}^H\mathbf{T} = \mathbf{I}$. One can see that the regularity in the centre is broken up here by turning one ring of the inner sub-arrays by a small angle.

The positions of the super-array according to (1.21) are indicated by the larger symbols of the corresponding sub-array element symbols. The resulting super-array has sufficiently good symmetry $\sum_{l=1}^{L} \rho_{x,l} \approx \sum_{l=1}^{L} \rho_{y,l} \approx 0$ and acceptable balance $\sum_{l=1}^{L} \rho_{x,l}\rho_{y,l} \approx 0$, which is required for good monopulse performance, [14]. The resulting super-array pattern and the 32 sub-array patterns are seen in Figure 1.8 (azimuth cuts only). One can see that due to the optimized design the super-array

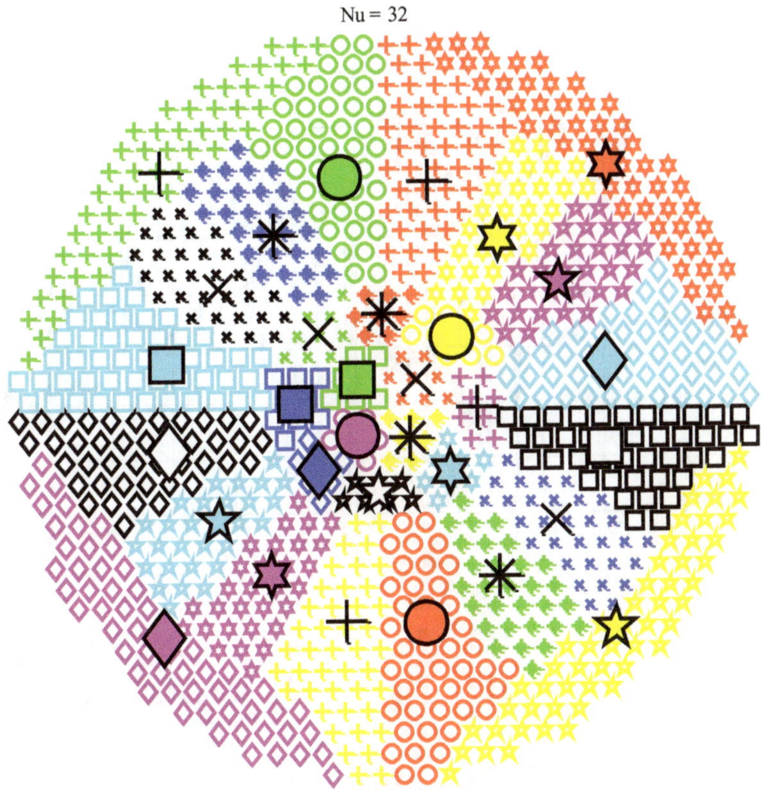

Figure 1.7 *2D generic array with 902 elements grouped into 32 sub-arrays. Super-array centres from −35 dB Taylor weighting are shown by a large symbol of the corresponding sub-array*

pattern has no grating lobes. The sub-array pattern cuts have of course quite different beamwidths in the azimuth direction due to the different extent of the sub-arrays. We will use this configuration as a generic array in the sequel for presenting various examples.

With the sub-array configuration given the remaining question is: what is the best difference weighting to apply at the sub-arrays for approximating the desired weighting $\mathbf{d}_0$? One can make a least squares approximation and seek a sub-array weighting $\tilde{\mathbf{d}}$ with:

$$\tilde{\mathbf{d}} = \arg\min_{\mathbf{r}} \left\{ \|\mathbf{T}\mathbf{r} - \mathbf{d}_0\|^2 \right\} \tag{1.23}$$

The solution of this problem is known to be $\tilde{\mathbf{d}} = (\mathbf{T}^H\mathbf{T})^{-1}\mathbf{T}^H\mathbf{d}_0$. One can refine this solution by additional constraints, e.g. by requiring $\mathbf{a}_0^H\mathbf{T}\mathbf{r} = 0$, which makes the difference beam exactly zero in direction $\mathbf{u}_0$, or by using, instead of the Euclidean

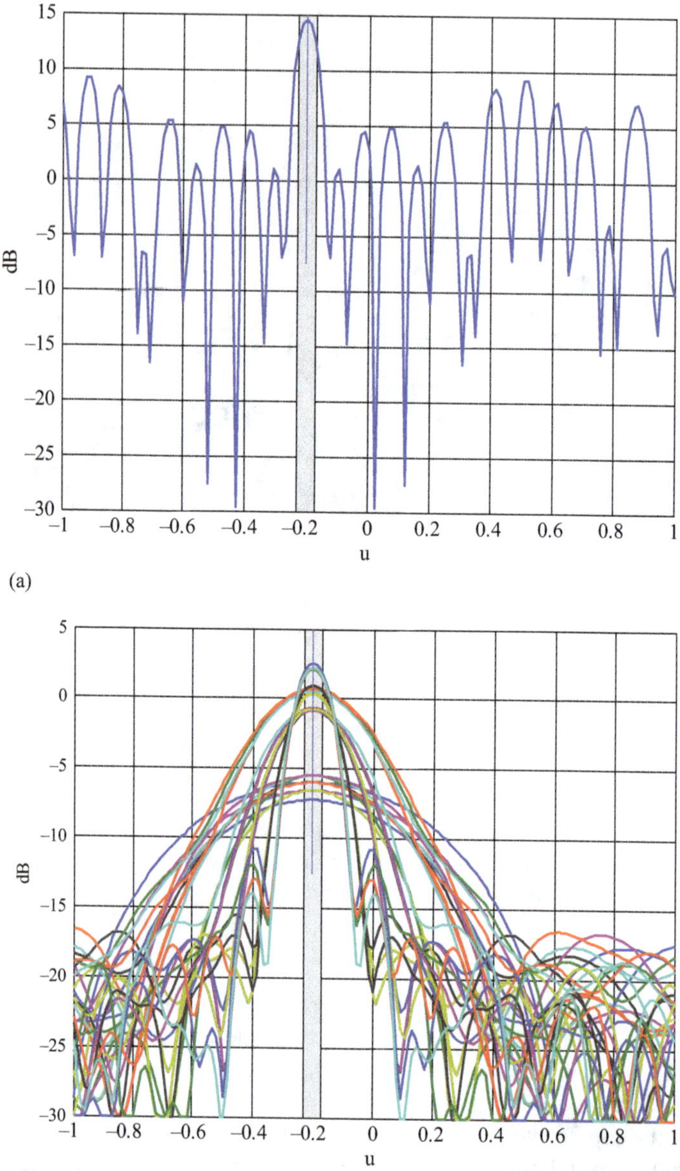

Figure 1.8 Azimuth cuts of super-array and sub-array patterns of generic array of Figure 1.7. Phase shifter setting is for direction (u, v) = (−0.2, 0), the 3 dB sum beamwidth is shown by the shaded area. (a) Super-array pattern and (b) 32 sub-array patterns

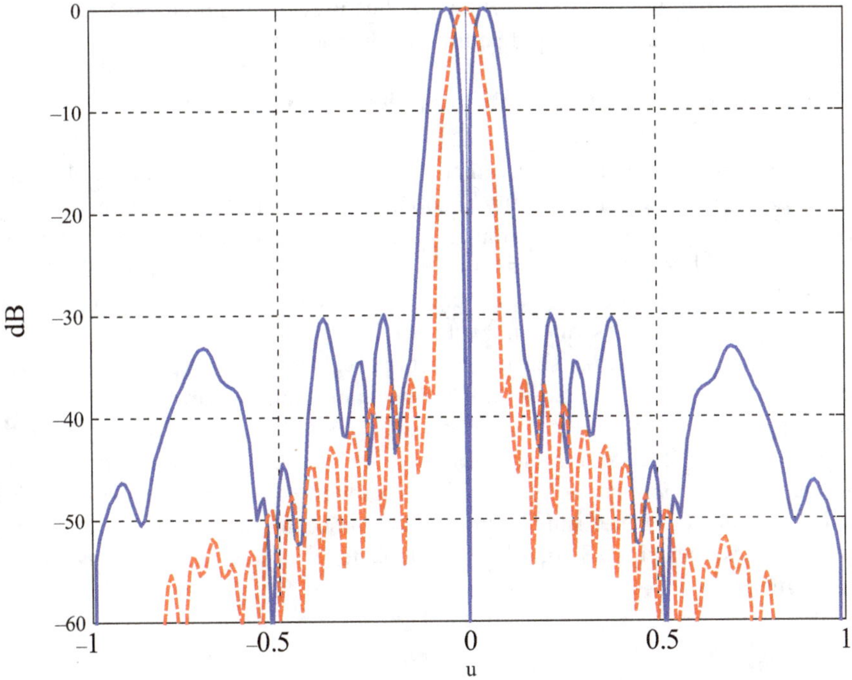

Figure 1.9 Sum and difference pattern formed with sub-arrays resulting from configuration of Figure 1.7

square norm, the norm of the weighted patterns, i.e. $\|\mathbf{d}\|^2 = \int_\Omega |\mathbf{a}(\mathbf{u})^H \mathbf{d}|^2 p(\mathbf{u})\, d\mathbf{u}$, where p is a directional weighting function. The solution of this refined approach is:

$$\tilde{\mathbf{d}} = \left(\mathbf{T}^H\mathbf{C}\mathbf{T}\right)^{-1}\mathbf{T}^H\left(\mathbf{I} - \frac{\mathbf{a}_0\mathbf{a}_0^H\mathbf{T}\left(\mathbf{T}^H\mathbf{C}\mathbf{T}\right)^{-1}\mathbf{T}^H}{\mathbf{a}_0^H\mathbf{T}\left(\mathbf{T}^H\mathbf{C}\mathbf{T}\right)^{-1}\mathbf{T}^H\mathbf{a}_0}\right)\mathbf{C}\mathbf{d}_0 \tag{1.24}$$

with the matrix $\mathbf{C} = \int_\Omega \mathbf{a}(\mathbf{u})\mathbf{a}(\mathbf{u})^H p(\mathbf{u})\, d\mathbf{u}$ and Ω the directions of view. For the full visible region and with $p \equiv 1$ one obtains $\mathbf{C} \approx \mathbf{I}$.

Figure 1.9 shows the resulting azimuth patterns for this array. Clearly, the sum beam is unaffected by the weighting and has exactly the −35 dB Taylor shape. The difference pattern does not fully achieve the −35 dB Bayliss sidelobe level. However, a maximum sidelobe level of −30 dB is achieved. This is the price to be paid for the coarse discretization of only sub-array difference weighting.

An important feature of digital beamforming with sub-arrays is that the weighting for beamforming can be distributed between the element level (the weighting incorporated in the matrix **T**) and the digital sub-array level (the weighting $\tilde{\mathbf{m}}$). This yields some freedom in designing the dynamic range of the amplifiers at the elements and of the level of the AD-converter input. This freedom also allows to normalize the power of the sub-array outputs such that

$\mathbf{T}^H\mathbf{T} = \mathbf{I}$. As will be shown in Section 1.6, this is also a requirement for adaptive interference suppression to avoid pattern distortions.

1.3.3.2 Beam scanning at sub-array level and sub-array modification

For a given pointing direction according to the phase shifts applied at the elements we can also scan the beam digitally at sub-array level into another direction, [19]. This is of interest for forming simultaneous multiple receiving beams which are useful for rapid search or for tracking extended targets, see Section 1.9. Beams scanned digitally at sub-array level are also of interest to apply the multi-step monopulse procedure described in Section 1.9.1.

In general the sub-array shapes are not optimal for beam scanning at sub-array level. However, there is a possibility to generate digitally new sub-arrays by a suitable transformation of the digital outputs. This offers some flexibility to mitigate deleterious effects of the sub-array shapes. By an $L \times L$ transformation matrix $\mathbf{M}$ one can create new sub-arrays resulting in a modified sub-array matrix $\tilde{\mathbf{T}} = \mathbf{TM}$ which may possess better suited patterns. The procedure to determine this matrix has been described in [18,20]. We may look for a desired sub-array pattern of the form:

$$g_l(\mathbf{u}) = h_l(\mathbf{u}) \exp\left[j2\pi f_0(\rho_{x,l}u + \rho_{y,l}v)/c\right] \tag{1.25}$$

with prescribed functions $h_l(\mathbf{u})$. One can realize only sub-array patterns by choosing a vector $\mathbf{r}_l$ with:

$$f_l(\mathbf{u}) = \mathbf{r}_l^H \mathbf{T}^H \mathbf{a}(\mathbf{u}) \tag{1.26}$$

So, we have to minimize:

$$I = \int_V |f_l(\mathbf{u}) - g_l(\mathbf{u})|^2 p(\mathbf{u})\, d\mathbf{u} \tag{1.27}$$

with respect to the $\mathbf{r}_l$ where we allow a directional weighting function $p(\mathbf{u})$ to put different emphasis on this criterion for different angles. This is a classical least squares problem and it has been shown in [18] that the solution is given by:

$$\begin{gathered}\mathbf{M} = (\mathbf{r}_1, \ldots, \mathbf{r}_L) = \left(\mathbf{T}^H\mathbf{T}\right)^{-1}\mathbf{T}^H\mathbf{C}^{-1}(\mathbf{v}_1, \ldots, \mathbf{v}_L) = \left(\mathbf{T}^H\mathbf{T}\right)^{-1}\mathbf{T}^H\mathbf{C}^{-1}\mathbf{V} \text{ with} \\ \mathbf{C} = \int_V \mathbf{a}(\mathbf{u})\mathbf{a}(\mathbf{u})^H p(\mathbf{u})\, d\mathbf{u} \in \mathbb{C}^{N\times N} \quad \text{and} \\ \mathbf{v}_l = \left(\int_V h_l(\mathbf{u}) e^{j2\pi f_0((x_i-\rho_{x,l})u+(y_i-\rho_{y,l})v)/c} p(\mathbf{u})\, d\mathbf{u}\right)_{i=1\ldots N}, \quad l = 1\ldots L.\end{gathered} \tag{1.28}$$

As a result we obtain the total sub-array transformation $\tilde{\mathbf{T}} = \mathbf{TM} = \mathbf{P}_\mathbf{T}\mathbf{C}^{-1}\mathbf{V}$, where $\mathbf{P}_\mathbf{T} = \mathbf{T}(\mathbf{T}^H\mathbf{T})\mathbf{T}^H$ is a projection on the space spanned by the columns of $\mathbf{T}$ and $\mathbf{V} = (\mathbf{v}_1, \ldots, \mathbf{v}_L) \in \mathbb{C}^{N\times L}$.

1.3.4 Space-time arrays

Coherent processing of a time series $z_1, \ldots, z_K$ is mathematically a form of beamforming. For a time series of array snapshots $\mathbf{z}_1, \ldots, \mathbf{z}_K$ we have therefore a double beamforming procedure of the space-time data matrix $\mathbf{Z} = (\mathbf{z}_1, \ldots \mathbf{z}_K)$ of the form:

$$S = \mathbf{w}_s^H \mathbf{Z} \mathbf{w}_t \tag{1.29}$$

and $\mathbf{w}_s$, $\mathbf{w}_t$ denote the weight vectors for spatial and temporal beamforming, respectively. Using the rule of Kronecker products (1.29) can be written as a single beamforming operation:

$$S = (\mathbf{w}_t \otimes \mathbf{w}_s)^H \text{vec}\{\mathbf{Z}\} \tag{1.30}$$

where vec$\{\mathbf{Z}\}$ is a vector obtained by stacking all columns of the matrix $\mathbf{Z}$ on top. So, mathematically it does not matter whether the data come from spatial or temporal sampling. Coherent processing is in both cases a beamforming-type operation with the correspondingly modified beamforming vector. Relation (1.30) is often exploited when spatial and temporal parameters are dependent (e.g. direction and Doppler frequency as it is the case in airborne radar, see [2,3]). One can then solve the parameter estimation and interference suppression task jointly in a higher dimensional space.

In particular, partial preprocessing like sub-arraying can be generalized to space-time processing. If we denote spatial sub-arraying by a matrix $\mathbf{T}_s$ and final spatial beamforming by $\mathbf{m}_s$ and the corresponding quantities for time processing by $\mathbf{T}_t$, $\mathbf{m}_t$, then space-time beamforming gives:

$$S = \mathbf{m}_s^H \mathbf{T}_s^H \mathbf{Z} \mathbf{T}_t \mathbf{m}_t \tag{1.31}$$

With the rules of Kronecker products this can be written as $S = ((\mathbf{T}_t \mathbf{m}_t) \otimes (\mathbf{m}_s^H \mathbf{T}_s^H))\text{vec}\{\mathbf{Z}\}$ or:

$$S = (\mathbf{m}_t \otimes \mathbf{m}_s)^H (\mathbf{T}_t \otimes \mathbf{T}_s)^H \text{vec}\{\mathbf{Z}\} \tag{1.32}$$

with a compound space-time sub-array beamforming vector $\mathbf{m}_t \otimes \mathbf{m}_s$ and space-time sub-array transformation $\mathbf{T}_t \otimes \mathbf{T}_s$.

Because of the high dimensionality and the need to reduce the numerical effort, sub-arraying is of particular interest for space-time processing. Currently proposed space-time adaptive processing (STAP) schemes for airborne clutter suppression and detection are mainly based on different sub-array configurations, see [3 Section 5, 21 Section 10.6]. Figure 1.10 shows an example of an efficient space-time sub-arraying scheme used for STAP clutter cancellation for airborne radar. One can see that this is a cascaded main and auxiliary channel configuration similar to the one shown in Figure 1.14.

Sub-arraying in both, time and spatial domain is an important tool to reduce the numerical complexity for STAP. Furthermore, the statements about grating effects, sub-arrays and super-arrays of this section can be applied analogously to space-time processing using this formalism of transforming 2D-beamforming with a data matrix into a usual beamforming operation. The ABF and super-resolution

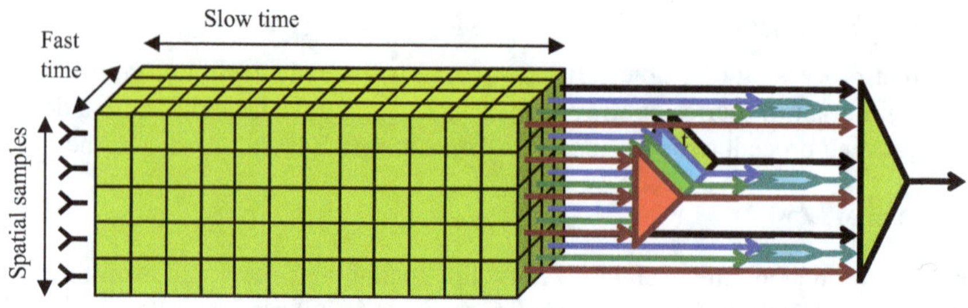

Figure 1.10 Symmetric auxiliary sensor/echo processor of Klemm [3], as an example for forming space-time sub-arrays (from [4] by courtesy of W. Bürger)

methods presented in Sections 1.6 and 1.7.2 can be similarly transformed into corresponding sub-arrayed space-time methods.

1.4 Array accuracy requirements

The Cramér signal representation introduced in Section 1.2.2 is also well suited to analyse some error effects. This has been done in [12] and in [2 Section 16.2.3]. We recall the main results here. These results are not meant as a method of predicting errors appearing in a real system, where we have a super-position of a multitude of effects. Rather such models can demonstrate the typical consequences of some kinds of error (especially the eigenvalue leakage effect) and they are useful for rapid system simulations. For demonstration purposes we confine the presentation to analogue de-modulation (de-modulation before digitization) which is rather used for broadband signals. Digital de-modulation with a high sampling rate and the resulting typical errors are described in [1 Section 6.3].

1.4.1 IQ-de-modulation errors

An amplification error R (or H) in the I- (or Q-) channel, respectively, an orthogonality error φ (or ψ) and offset error F (G), and delay error δ appearing by IQ-de-modulation can be described by extending (1.6) to the form:

$$\begin{aligned} I(t,\mathbf{r},\mathbf{u}) &= R\,\mathrm{LP}\{s(t-\delta,\mathbf{r},\mathbf{u})\cos(2\pi f_0 t-\varphi)\}+F \quad \text{and} \\ Q(t,\mathbf{r},\mathbf{u}) &= H\,\mathrm{LP}\{-s(t-\delta,\mathbf{r},\mathbf{u})\sin(2\pi f_0 t-\psi)\}+G \end{aligned} \tag{1.33}$$

This leads to a complex base-band signal description, [12],

$$\begin{aligned} s(t,\mathbf{r},\mathbf{u}) =& \frac{1}{2}e^{j2\pi f_0(\mathbf{r}^T\mathbf{u}/c+\delta)}p\int_{-B/2}^{B/2} e^{j2\pi\xi(t+\mathbf{r}^T\mathbf{u}/c+\delta)}dZ_s(\xi+f_o) \\ &+\frac{1}{2}e^{-j2\pi f_0(\mathbf{r}^T\mathbf{u}/c+\delta)}q\int_{-B/2}^{B/2} e^{-j2\pi\xi(t+\mathbf{r}^T\mathbf{u}/c+\delta)}dZ_s^*(\xi+f_o)+g \end{aligned} \tag{1.34}$$

With $p = Re^{j\varphi} + He^{j\psi}$, $q = Re^{-j\varphi} - He^{-j\psi}$ and $g = F + jG$.

For an array with different errors $R_i, H_i, F_i, G_i, \varphi_i, \psi_i, \delta_i$ in each channel, the covariance-correlation matrix $\mathbf{Q}_s(\delta_0) = \mathrm{E}\{s(t+\delta_i, \mathbf{r}_i, \mathbf{u})s(t+\delta_k+\delta_0, \mathbf{r}_k, \mathbf{u})\}_{i,k=1\ldots N}$ can then be written as described in [2 p. 555] by:

$$\mathbf{Q}_s(\delta_0) = \frac{1}{4}[(\mathbf{bb}^H) \odot (\mathbf{pp}^H) + (\mathbf{bb}^H)^* \odot (\mathbf{qq}^H)] \odot \mathbf{C}_s(\delta_0) \tag{1.35}$$

where $\odot$ denotes the Hadamard (element-wise) matrix product. The quantities $\mathbf{b}$, $\mathbf{p}$, $\mathbf{q}$, $\mathbf{C}_s$ depend on the specific signal power spectrum. As an illustrative example we present here the results for the special case of a rectangular signal power density with centre frequency f_s and bandwidth B_s with $[f_s - B_s/2, f_s + B_s/2] \subset [f_0 - B/2, f_0 + B/2]$. Then one obtains $b_i = e^{j2\pi f_s \mathbf{r}_i^T \mathbf{u}/c}$, $p_i = e^{j2\pi f_s \delta_i}(R_i e^{j\varphi_i} + H_i e^{j\psi_i})$, $q_i = e^{-j2\pi f_s \delta_i}(R_i e^{-j\varphi_i} - H_i e^{-j\psi_i})$ for $i = 1 \ldots N$, and:

$$\mathbf{C}_s(\delta_0) = P_s\left(\mathrm{sinc}\left(\pi B_s \frac{(\mathbf{r}_i - \mathbf{r}_k)^T \mathbf{u}}{c} + \delta_i - \delta_k + \delta_0\right) \cdot \mathrm{e}^{j2\pi\Delta f \delta_0}\right)_{i,k=1\ldots N} \tag{1.36}$$

where $\Delta f = f_s - f_0$. For the receiver noise covariance matrix with independent white noise in each channel we obtain:

$$\mathbf{Q}_n(\delta_0) = \sigma^2 \mathrm{sinc}(\pi B_s \delta_0) e^{j2\pi\Delta f \delta_0} \operatorname*{diag}_{i=1}^{N} \{(R_i^2 + H_i^2)/2\}$$

A complete array output covariance matrix for M targets then results in:

$$\mathbf{R}(\delta_0) = \sum_{i=1}^{M} \mathbf{Q}_{s,i}(\delta_0) + \mathbf{Q}_n(\delta_0) + \mathbf{gg}^H \tag{1.37}$$

With these formulas one can build all kinds of sub-band and space-time covariance matrices. Special cases:

- No errors ($R = H = 1, F = G = 0, \varphi = \psi = 0$), zero bandwidth ($B_s = 0$), $\delta_0 = 0$: Then $\mathbf{Q}_s = \mathbf{bb}^H P_s$ and $\mathbf{b}$ contains the frequency shifted steering vector (e.g. Doppler).
- No errors ($R = H = 1, F = G = 0, \varphi = \psi = 0$) and bandwidth B_s only, $\delta_0 = 0$: Then $\mathbf{Q}_s = \mathbf{bb}^H \odot \mathbf{C}_s$.
- With errors present:
  Then $\mathbf{Q}_s$ of (1.35) can be written with the rules of the Hadamard product as:

$$\mathbf{Q}_s = \frac{1}{4}\left[(\mathbf{b} \odot \mathbf{p})(\mathbf{b} \odot \mathbf{p})^H + (\mathbf{b}^* \odot \mathbf{q})(\mathbf{b}^* \odot \mathbf{q})^H\right] \odot \mathbf{C}_s \tag{1.38}$$

Comparing this with the error-free narrow-band covariance matrix shows that a single target with steering vector $\mathbf{b}$ will result in a tilted steering vector $\mathbf{a}_{\mathrm{error}}(\mathbf{u}) = \mathbf{b} \odot \mathbf{p}$ plus an 'image' target $\mathbf{a}_{\mathrm{image}}(-\mathbf{u}) = \mathbf{b}^* \odot \mathbf{q}$.

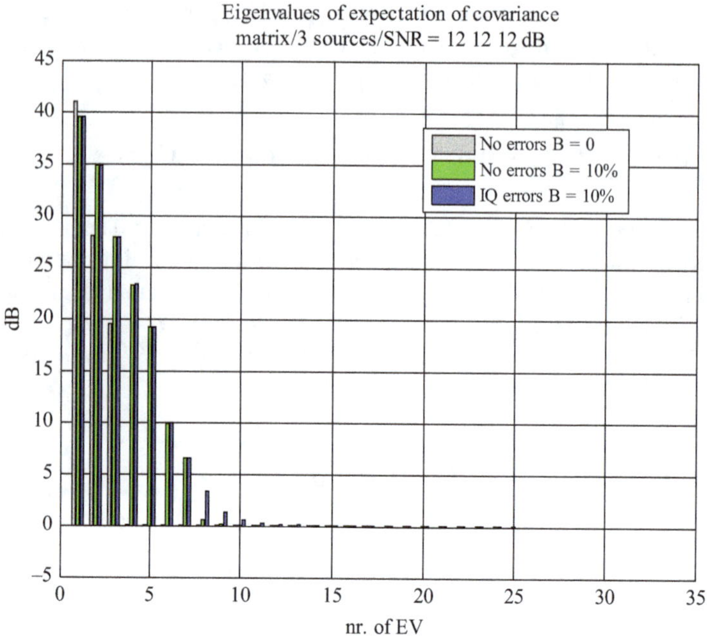

Figure 1.11 Eigenvalues of covariance matrix at sub-array level for generic array with 32 sub-arrays, narrowband beamforming by phase shifting at elements. Eigenvalues are calculated from the theoretical covariance model of (1.38). Three cases: no errors and zero bandwidth, no errors and 10% relative bandwidth, IQ errors with 1 dB std amplitude fluctuation and 10% bandwidth

Figure 1.11 shows the eigenvalues of a covariance matrix for narrowband beamforming (phase shifting only) with and without errors and also the case of zero bandwidth for reference. The scenario consists of three targets in azimuth directions $-53°$, $-24°$, $0°$ ($u = -0.8, -0.4, 0.0$) with equal element signal-to-noise ratio (SNR) of 12 dB. The generic antenna of Figure 1.7 was used. In the case of no errors and zero bandwidth one can see the three dominant eigenvalues and the noise eigenvalues at constant level of 0 dB. Additional IQ-errors were modelled here by an additional complex-normal distributed component with an amplitude standard deviation (std) of 1 dB. With bandwidth and additional I- and Q-errors the typical dominant eigenvalue leakage effect can be seen. The source power, which in the narrowband error-free case is concentrated in the three dominant eigenvalues, here leaks into the smaller eigenvalues. Also the noise eigenvalues become unequal. The bandwidth and random channel errors lead to an approximate doubling of the dominant eigenvalues. The amount of leakage depends on the signal power, the bandwidth and the amount of error. It is important to note that eigenvalue leakage is typically a strong signal effect. The leakage eigenvalues are at a certain level below

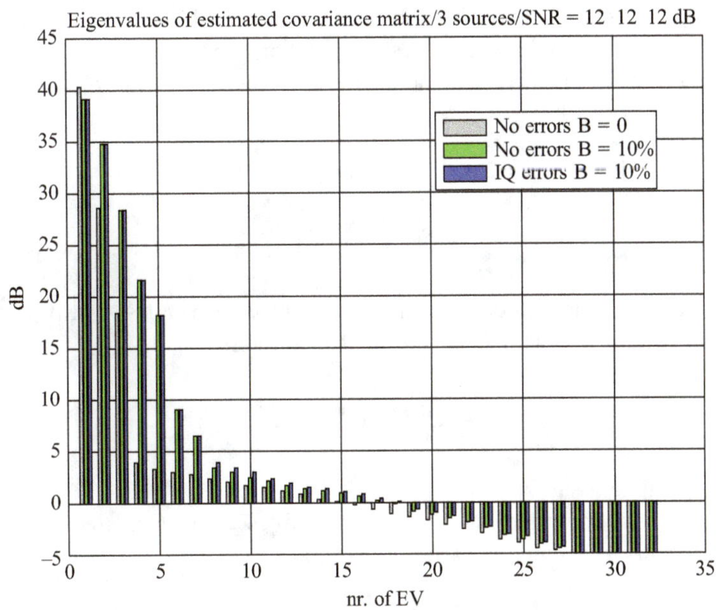

Figure 1.12 Eigenvalues of estimated covariance matrices at sub-array level from 64 snapshots, generic array with 32 sub-arrays, narrowband beamforming by phase shifting at elements. Three cases: no errors and zero bandwidth, no errors and 10% relative bandwidth, IQ errors with 1 dB std amplitude fluctuation and 10% bandwidth

the dominant eigenvalues (defined by the vectors $\mathbf{b} \odot \mathbf{p}$, and $\mathbf{b}^* \odot \mathbf{q}$). For weak sources they will merge into the noise eigenvalues. This shows that the common belief that with high signal-to-noise ratio performance converges to the ideal error-free case is obviously not true for realistic types of error.

Eigenvalue leakage is also particularly strong for small number of samples for estimating the covariance matrix. Figure 1.12 shows the eigenvalues for estimated covariance matrices corresponding to the cases of Figure 1.11. The covariance matrices are estimated from $K = 64$ snapshots. This is the value that would give a 3-dB loss according to a rule of thumb termed 'Brennan's rule' which says that $K \geq 2N$ is required for a 3 dB loss, [22,23].

The eigenvalue leakage will in particular complicate the determination of the number of dominant eigenvalues, which is important for the application of subspace methods for jammer suppression or super-resolution methods in Sections 1.6 and 1.7.2.

1.4.2 Bandpass filter errors

For general bandpass filter functions the integrals cannot be solved analytically. However, a special technically important case is that of a sinusoidal ripple over the bandwidth and this case can be solved. In order to show what effects can be observed with this error model the results are shortly presented.

Let the baseband filter functions have the form $H_i(f) = a_i(f)e^{j\phi_i(f)}$, $i = 1 \ldots N$, with:

$$a_i(f) = \begin{cases} 1 + \varepsilon_i \cos 2\pi f\, h_i, & \text{for } |f| < B/2 \\ 0, & \text{else} \end{cases} \qquad (1.39)$$
$$\phi_i(f) = 2\pi f \delta_i$$

$\varepsilon_i < 1$ is the ripple amplitude, h_i counts the number of ripples, and δ_i is a linear phase slope or a delay. This results in the signal model:

$$s_i(t) = e^{j2\pi f_0 \tau_i} \int_{-B/2}^{B/2} e^{j2\pi\xi(t+\tau_i)} a_i(\xi) e^{j\phi_i(\xi)} dZ_s(\xi + f_0) \qquad (1.40)$$

Again we assume a sub-bandwidth $[f_s - B_s/2, f_s + B_s/2] \subset [f_0 - B/2, f_0 + B/2]$, which may be incorporated into the definition of dZ_s. As calculated in [2 p. 556] this model leads to a signal covariance matrix:

$$\mathbf{Q}_s(\delta_0) = (\mathbf{b}\mathbf{b}^H) \odot (\mathbf{p}_\delta \mathbf{p}_\delta^H) \odot \mathbf{C}_s(\delta_0) = \left[(\mathbf{b} \odot \mathbf{p}_\delta)(\mathbf{b} \odot \mathbf{p}_\delta)^H\right] \odot \mathbf{C}_s(\delta_0) \qquad (1.41)$$

with $\mathbf{b} = \left(e^{j2\pi f_s \tau_i}\right)_{i=1\ldots N}$, $\tau_i = \mathbf{r}_i^T \mathbf{u}/c$, $\mathbf{p}_\delta = \left(e^{j2\pi f_s \delta_i}\right)_{i=1\ldots N}$, and $C_{s;ik}(\delta_0) = P_s e^{j2\pi\Delta f \delta_0}[\text{sinc}(\pi B_s d_{ik}) + \varepsilon_i H_{ik} + \varepsilon_k J_{ik} + \varepsilon_i \varepsilon_k K_{ik}]$, $\Delta f = f_s - f_0$, $d_{ik} = \tau_i - \tau_k + \delta_i - \delta_k + \delta_0$ and

$$H_{ik} = \frac{1}{2}\left[e^{j2\pi\Delta f h_i} \text{sinc}\, \pi B_s(d_{ik} + h_i) + e^{-j2\pi\Delta f h_i} \text{sinc}\, \pi B_s(d_{ik} - h_i)\right]$$

$$J_{ik} = \frac{1}{2}\left[e^{j2\pi\Delta f h_k} \text{sinc}\, \pi B_s(d_{ik} + h_k) + e^{-j2\pi\Delta f h_k} \text{sinc}\, \pi B_s(d_{ik} - h_k)\right]$$

$$H_{ik} = \frac{1}{4}[e^{j2\pi\Delta f(h_i - h_k)} \text{sinc}\, \pi B_s(d_{ik} + h_i - h_k) + e^{-j2\pi\Delta f(h_i - h_k)}$$
$$\times \text{sinc}\, \pi B_s(d_{ik} - (h_i - h_k)) + e^{j2\pi\Delta f(h_i + h_k)} \text{sinc}\, \pi B_s(d_{ik} + h_i + h_k)$$
$$+ e^{-j2\pi\Delta f(h_i + h_k)} \text{sinc}\, \pi B_s(d_{ik} - (h_i + h_k))]$$

From this formula, the receiver noise covariance can be obtained as a diagonal matrix $\mathbf{Q}_n(\delta_0)$ by replacing in $\mathbf{Q}_s(\delta_0)$ the quantities $P_s = \sigma^2$ and $d_{ik} = \delta_0$ (see [2 p. 556] for the explicit formula).

1.4.3 AD-converter limitation

A third kind of error, which is however difficult to model, arises due to limiting effects. All digital signal processing algorithms assume a linear receiver. Limiting effects are the most frequent non-linear effects appearing in real systems and the AD converters are most sensitive to limiting because radar has typically very high dynamic range: the target echo may be below receiver noise while the clutter may be more than 60 dB above receiver noise. Suppose we have an AD-converter with b bits then the classical formula for the maximum possible SNR with respect to quantization noise is $\mathrm{SNR}_{\max} = 6.02b + 1.76$ dB. The receiver noise should be slightly higher than the quantization noise level to allow an integration effect for a signal below receiver noise. This means that for a 14 bit ADC we have a maximum SNR of 86 dB which is a value that can be easily attained in reality. Normally an automatic gain control (AGC), which is an SNR adaptive attenuation, is inserted into the digital receivers to avoid limiting. If we have strong clutter the AGC may make small targets to be sub-merged in noise, or in other words, can reduce the radar range. So, the system is not completely distorted in the case of limiting, but it will suffer a soft degradation with respect to the maximum range.

If the non-linearity is known in its functional form, it is possible to compensate the non-linear effects. This has been shown in [24] for non-linearities up to third order. This technique may be applied for analogue limiting. For hard clipping as produced by AD converter limiting there is no correction possible and only an AGC can help avoiding this.

In any case, whether produced in the analogue or digital domain, any uncompensated limiting effects will give rise to dispersive effects leading to additional eigenvalues in the covariance matrix similar to the effects described in Sections 1.4.1 and 1.4.2.

1.5 Antenna pattern shaping

Conventional beamforming means nothing else than coherent integration of the spatially sampled data, i.e. the array elements outputs are phase compensated and coherently summed up. This results in a pronounced main beam when the phase differences match with the direction of the plane wave, and results in sidelobes otherwise. The beam shape and the sidelobes can be further influenced by additional amplitude weighting. Let us consider the complex beamforming weights $w_i = g_i e^{j2\pi f \mathbf{r}_i^T \mathbf{u}/c}$, $i = 1 \ldots N$. Examples of common weightings for sidelobe reduction are:

- Some bell-shaped amplitude weighting over the aperture, like $g_i = \cos^\nu(\pi x_i/A) + \alpha$ (for suitable constants ν, α), or $g_i = e^{-\nu x_i^2}$. This is the simplest way of pattern shaping and the foundation is quite heuristic.
- Taylor weighting, [7 p. 13.29]: This weighting imposes minimal constraints in the sense that it tries to leave the conventional (uniformly weighted) pattern undistorted except for a reduction of the first n sidelobes to a prescribed level.
- The Dolph–Chebyshev weighting [7 p. 13.29]: This creates a pattern with all sidelobes equal to a prescribed level.

There are some reasons to require an equal sidelobe level (uniform discrimination of all sidelobe directions) as well as for a decay of the sidelobe level (maximum suppression of interference outside the field of interest). Hence, it is not clear what the optimum design is and only an analysis at system level comparing false alarms could answer this. Actually, deterministic pattern shaping is applied if nothing about the expected interference is known. The price to be paid for all these sidelobe reduction is a broadening of the main beam.

Figure 1.13(a) shows examples of such patterns for a ULA with 40 elements. The taper functions for low sidelobes were selected such that the 3 dB beamwidth of all patterns are equal. The conventional pattern is plotted for reference showing how tapering increases the beamwidth. Which tapering may be preferred depends on the emphasis on close-in and far-off low sidelobes. Another point of interest is the dynamic range of the weights and the SNR loss, because at the array elements only attenuations can be applied. Obviously the weighting of a few boundary array elements determine the shape of the sidelobes as seen from Figure 1.13(b). For planar arrays the efficiency of the tapering is slightly different. Robustness and loss properties of low sidelobe taperings have been analysed in detail in [25].

Optimum deterministic pattern shaping. The rationale for low sidelobes is that we want to minimize some unknown interference power coming over the sidelobes. This can be achieved by solving the following optimization problem [25 Chapter 5.5, 26].

$$\begin{aligned}&\min_{\mathbf{w}} \int_{\Omega} \left|\mathbf{w}^H \mathbf{a}(\mathbf{u})\right|^2 p(\mathbf{u}) d\mathbf{u} \quad \text{subject to} \quad \mathbf{w}^H \mathbf{a}_0 = \mathbf{1}, \text{ or equivalently} \\ &\min_{\mathbf{w}} \mathbf{wCw} \quad s.t. \quad \mathbf{w}^H \mathbf{a}_0 = \mathbf{1}, \text{ with } \mathbf{C} = \int_{\Omega} \mathbf{a}(\mathbf{u})\, \mathbf{a}(\mathbf{u})^H p(\mathbf{u}) d\mathbf{u}\end{aligned} \tag{1.42}$$

Ω denotes the angular sector where we want to influence the pattern, e.g. the whole visible region $u^2 + v^2 \leq 1$, and $p(\mathbf{u})$ is a weighting function which allows to put different emphasis on the criterion in different angular regions. The solution of this constrained least squares optimization is:

$$\mathbf{w} = \frac{\mathbf{C}^{-1}\mathbf{a}_0}{\mathbf{a}_0^H \mathbf{C}^{-1}\mathbf{a}_0} \tag{1.43}$$

For the choice of the function p we remark that, for a global reduction of the sidelobes when $\Omega = \{\mathbf{u} \in \mathbb{R}^2 \,|\, u^2 + v^2 \leq 1\}$, one should exclude the main beam from the minimization by setting $p = 0$ on this set of directions (in fact, a slightly larger region is recommended, e.g. the null-to-null width) to allow a certain mainbeam broadening. This makes the solution more efficient. This technique has been applied in [27] to generate a pattern with reduced sidelobes only at negative elevations as a mean to reduce ground clutter for an airborne nose radar.

One may also form discrete nulls in directions $\mathbf{u}_1, \ldots, \mathbf{u}_M$ by setting $p(\mathbf{u}) = \sum_{k=1}^{M} \delta(\mathbf{u} - \mathbf{u}_k)$. The solution of (1.42) then can be shown to be:

$$\mathbf{w} = \frac{\mathbf{P}\mathbf{a}_0}{\mathbf{a}_0^H \mathbf{P}\mathbf{a}_0} \text{ with } \mathbf{P} = \mathbf{I} - \mathbf{A}\left(\mathbf{A}^H\mathbf{A}\right)^{-1}\mathbf{A}^H \quad \text{and} \quad \mathbf{A} = (\mathbf{a}(\mathbf{u}_1), \ldots \mathbf{a}(\mathbf{u}_M)) \tag{1.44}$$

This is just the weight for deterministic nulling because the projection $\mathbf{P}$ creates true nulls in the array pattern. To avoid insufficient suppression due to channel inaccuracies one may also create small extended nulls using a set of neighbouring

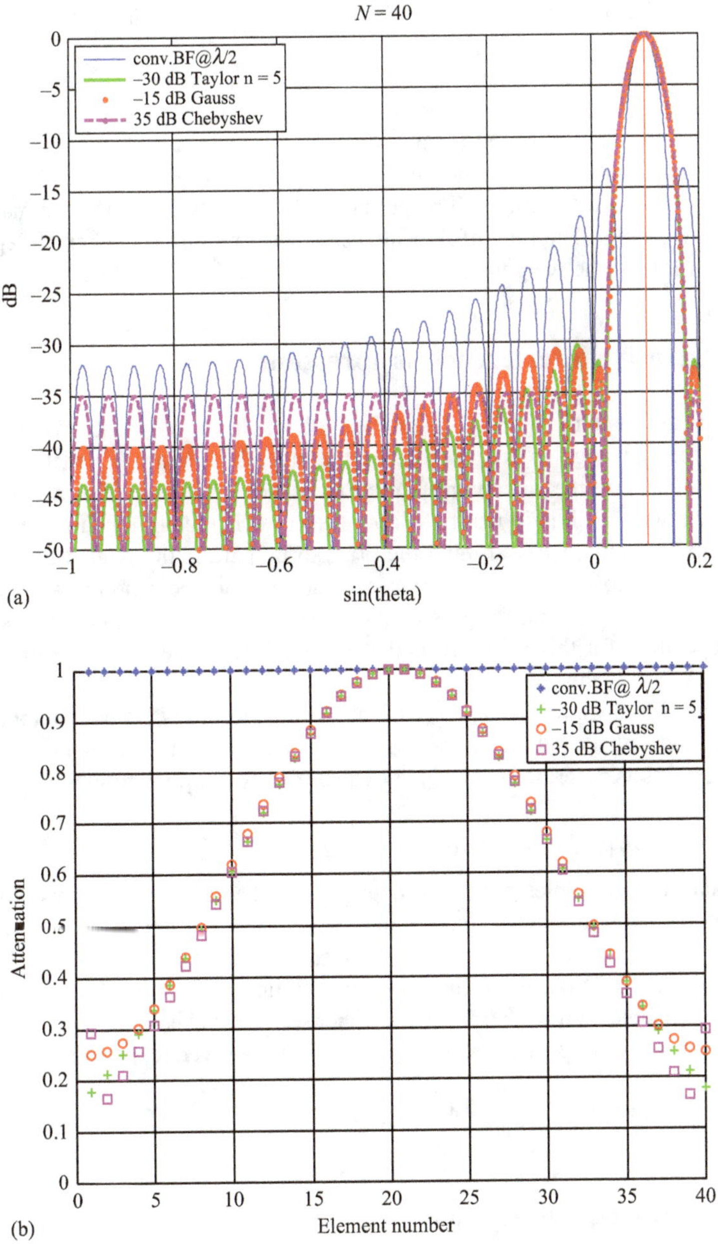

Figure 1.13 Low sidelobes by Gauss, Taylor and Chebyshev amplitude tapering (from [4]). (a) Low sidelobe patterns of equal beamwidth and (b) amplitude taper values over element number corresponding to patterns in (a)

directions. The form of these weightings shows the close relationship to ABF in (1.49) and (1.55).

The form of the projection also shows the relation between the number of elements and the number of pattern nulls. If the directions $\mathbf{u}_1, \ldots, \mathbf{u}_M$ are all different and if the array has no ambiguities, then $\mathrm{rank}\{\mathbf{A}\} = M$. Therefore, if M is equal to the number of elements then **A** is square and regular, then **P** is the null matrix and no beamforming will be produced. With N elements one can maximally produce N-1 nulls. The rank of **A** is also called the required number of degrees of freedom (dof) of the weighting **w**.

1.6 Adaptive interference suppression

Deterministic pattern shaping is applied if we have rough knowledge about the interference angular distribution. In the sidelobe region this method can be inefficient because the antenna response to a plane wave (the vector $\mathbf{a}(\mathbf{u})$) must be exactly known which is in reality seldom the case. Typically much more suppression is applied than necessary and the price to be paid is the related beam broadening and SNR loss. Adaptive interference suppression needs no knowledge of the directional behaviour and suppresses the interference only as much as necessary. The proposition for this approach is that we are able to measure or learn in some way the ABF weights.

In the sequel we formulate the ABF algorithms for array element outputs. With the formalism of the sub-array transformation matrix **T** and the concept of super-array elements as described in (1.19) this can be directly applied to sub-array outputs.

1.6.1 Adaptive beamforming principles

To explain the principles and basic approaches let us first suppose we know the interference situation in the sense that we know the interference covariance matrix **Q**. What is then an optimum beamforming vector **w**?

Maximizing SNIR. From the Likelihood Ratio test criterion we know that the probability of detection is maximized if we choose a weight vector $\mathbf{w}_{\mathrm{opt}}$ that maximizes the signal-to-noise-plus-interference ratio (SNIR) for a given (expected) signal $\mathbf{a}_0$,

$$\max_{\mathbf{w}} \frac{|\mathbf{w}^H \mathbf{a}_0|^2}{\mathrm{E}\left\{|\mathbf{w}^H \mathbf{n}|^2\right\}} = \max_{\mathbf{w}} \frac{|\mathbf{w}^H \mathbf{a}_0|^2}{\mathbf{w}^H \mathbf{Q}\ \mathbf{w}} \tag{1.45}$$

The solution of this optimization is:

$$\mathbf{w}_{\mathrm{opt}} = \mu\, \mathbf{Q}^{-1} \mathbf{a}_0 \quad \text{with} \quad \mathbf{Q} = \mathrm{E}\{\mathbf{n}\mathbf{n}^H\} \tag{1.46}$$

μ is a free normalization constant and **n** denotes interference and receiver noise. This weighting has a very intuitive interpretation. If we decompose $\mathbf{Q}^{-1} = \mathbf{L}\mathbf{L}^H$ and apply this weight to the data, we have $\mathbf{w}_{\mathrm{opt}}^H \mathbf{z} = \mathbf{a}_0^H \mathbf{Q}^{-1} \mathbf{z} = \mathbf{a}_0^H \mathbf{L}^H \mathbf{L} \mathbf{z} = (\mathbf{L}\mathbf{a}_0)^H (\mathbf{L}\mathbf{z})$. This reveals that ABF does nothing else but a pre-whiten and match operation: If $\mathbf{z}$ contains only interference, i.e. if $\mathrm{E}\{\mathbf{z}\mathbf{z}^H\} = \mathbf{Q}$, then $\mathrm{E}\{(\mathbf{L}\mathbf{z})(\mathbf{L}\mathbf{z})^H\} = \mathbf{I}$, and the multiplication by $\mathbf{L}$ is a pre-whitening operation. The operation of $\mathbf{L}$ on the (matched) signal vector $\mathbf{a}_0$ accounts for the distortion from the pre-whitening operation and restores just the matching.

Sub-arrays. The formulation of weight vectors at the array elements can be easily extended to sub-arrays with digital outputs. As mentioned in Section 1.3.3, a sub-arrayed array can be viewed as a super-array with directive elements positioned at the centres of the sub-arrays. This means that we have only to replace the quantities $\mathbf{a}$, $\mathbf{n}$ by $\tilde{\mathbf{a}} = \mathbf{T}^H\mathbf{a}$ and $\tilde{\mathbf{n}} = \mathbf{T}^H\mathbf{n}$. However, there is a difference with respect to receiver noise. If the noise at the elements is white with covariance matrix $\sigma^2\mathbf{I}$ then it will be at the sub-array outputs $\tilde{\mathbf{Q}} = \sigma^2\mathbf{T}^H\mathbf{T}$ which is for overlapping sub-arrays not diagonal and which has for non-overlapping unequal sub-arrays unequal diagonal elements. Adaptive processing will turn this into white noise. In particular, if we apply at the elements some weighting for low sidelobes, which are contained in the matrix $\mathbf{T}$, then ABF will reverse this operation by the pre-whiten and match principle. Then after ABF at sub-array level the residual noise will be white and the desired low sidelobe pattern is reversed at sub-array level. This effect can be avoided by simply normalizing the matrix $\mathbf{T}$ such that $\mathbf{T}^H\mathbf{T} = \mathbf{I}$. For non-overlapping sub-arrays this can be achieved by normalizing the element weights as mentioned in Section 1.3.3. We call configurations with interference suppression by adaptively weighting the sub-array outputs a *direct sub-array weighting* (DSW) configuration.

Sidelobe canceller configurations. A very particular sub-array configuration for ABF is the sidelobe canceller (SLC). The basic idea is that any (even a reflector) antenna can be turned into an adaptive antenna by adding some auxiliary antennas. This is a cheap and simple method to provide any antenna with adaptivity. The additional antennas (called auxiliary antennas) are used to estimate the interference power and this is subtracted from the main antenna. Figure 1.14 shows the principle. The SLC has been described [7 p. 24.11] and has been studied in detail in [25 Chapter 4].

Actually, this is a way of post-beamforming adaptation. The adaptive weights are estimated by minimizing $\mathrm{E}\{|S_{\mathrm{sum}} - \hat{\mathbf{w}}_{\mathrm{aux}}^H\mathbf{z}_{\mathrm{aux}}|^2\}$. The solution of this problem is:

$$\hat{\mathbf{w}}_{\mathrm{aux}} = \mathbf{Q}_{\mathrm{aux}}^{-1}\mathbf{r}, \quad \text{where } \mathbf{Q}_{\mathrm{aux}} = \mathrm{E}\{\mathbf{z}_{\mathrm{aux}}\mathbf{z}_{\mathrm{aux}}^H\} \quad \text{and} \quad \mathbf{r} = \mathrm{E}\{\mathbf{z}_{\mathrm{aux}}S_{\mathrm{sum}}^*\}$$

One can show that this is the same as the SNIR-optimum solution of (1.46) if the desired signal has the form $\mathbf{a}_0 = (1, 0, \ldots 0)^T$, i.e. if we assume that the signal to be detected is only present in the main channel and not in the auxiliary channels:

$$\begin{aligned}\mathbf{Q}^{-1} &= \mathrm{E}\left\{\begin{pmatrix} S_{\mathrm{sum}} \\ \mathbf{z}_{\mathrm{aux}} \end{pmatrix}\begin{pmatrix} S_{\mathrm{sum}} \\ \mathbf{z}_{\mathrm{aux}} \end{pmatrix}^H\right\}^{-1} = \begin{pmatrix} \mathrm{E}\left\{|S_{\mathrm{sum}}|^2\right\} & \mathbf{r}^H \\ \mathbf{r} & \mathbf{Q}_{\mathrm{aux}} \end{pmatrix}^{-1} \\ &= \begin{pmatrix} (\mathbf{Q}^{-1})_{11} & -(\mathbf{Q}_{\mathrm{aux}}^{-1}\mathbf{r})^H(\mathbf{Q}^{-1})_{11} \\ -\mathbf{Q}_{\mathrm{aux}}^{-1}\mathbf{r}(\mathbf{Q}^{-1})_{11} & (\mathbf{Q}^{-1})_{22} \end{pmatrix}\end{aligned} \tag{1.47}$$

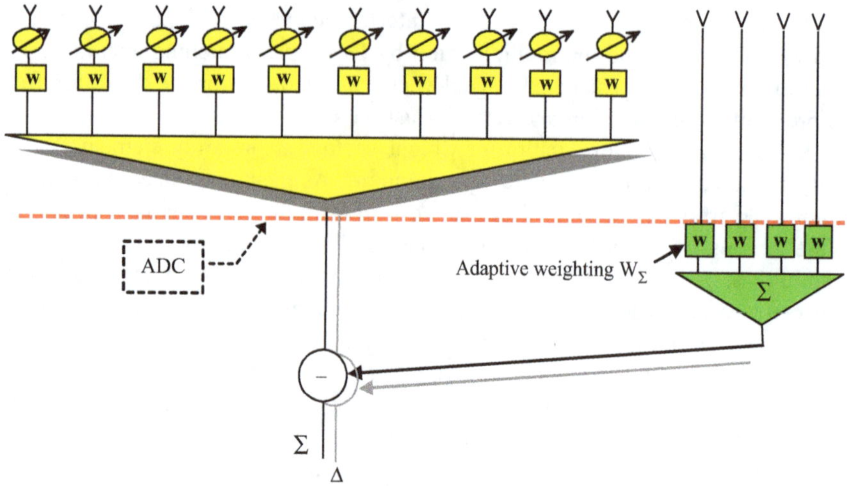

Figure 1.14 Principle of sidelobe canceller (SLC)

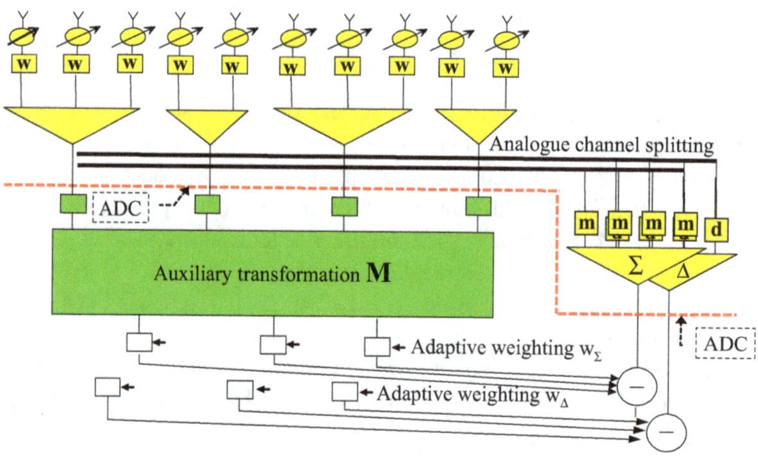

Figure 1.15 Generalized sidelobe canceller (GSLC)

Therefore $\hat{\mathbf{w}} = \mathbf{Q}^{-1}\mathbf{a}_0 = (\mathbf{Q}^{-1})_{11}\begin{pmatrix} 1 \\ -\mathbf{Q}_{\text{aux}}^{-1}\mathbf{r} \end{pmatrix} = (\mathbf{Q}^{-1})_{11}\begin{pmatrix} 1 \\ -\mathbf{w}_{\text{aux}} \end{pmatrix}$ is the solution (1.46) for a special choice of μ.

If, as indicated in Figure 1.14, the main antenna is also an array, it is not necessary to take extra-auxiliary elements. These elements can be taken out of the whole array. Then one arrives at a configuration as in Figure 1.15 called the generalized SLC (GSLC). Like the SLC the GSLC is a jammer suppression method after beamforming. The key feature is that the main channels (sum and difference beams) and the auxiliary channels are all generated from the sub-arrays. Note that the sub-arrays in Figure 1.15 could be also single elements.

The important feature of the GSLC system is the option to form the sum and difference beams by analogue beamforming networks. This reduces the danger of ADC limiting for strong sidelobe jammers. However, the danger of ADC limiting for mainbeam jammers is increased by this concept.

The auxiliary channels for the GSLC need not to be single elements or sub-arrays, but may be generated from the whole array by an auxiliary transformation matrix $\mathbf{M}$ as indicated in Figure 1.15. Suppose we have L sub-arrays and want to have P auxiliary channels. The $L \times P$ auxiliary transformation matrix $\mathbf{M}$ can be any matrix, from a simple selection matrix for single sub-arrays to a matrix that forms full beams with all sub-array outputs. The auxiliary channel outputs are $\mathbf{z}_{\text{aux}} = \mathbf{M}^H \tilde{\mathbf{z}} = \mathbf{M}^H \mathbf{T}^H \mathbf{z}$ and the main channel output is $S_{\text{sum}} = \tilde{\mathbf{m}}^H \mathbf{T}^H \mathbf{z}$. The optimum GSLC weight is then:

$$\mathbf{w}_{\text{aux}} = \mathbf{Q}_{\text{aux}}^{-1}\mathbf{r}, \quad \text{where } \mathbf{Q}_{\text{aux}} = \mathbf{M}^H \tilde{\mathbf{Q}} \mathbf{M} \text{ and } \mathbf{r} = \mathbf{M}^H \tilde{\mathbf{Q}} \tilde{\mathbf{m}}$$

Several observations can be made for the GSLC:

(i) The columns of $\mathbf{M}$ and $\tilde{\mathbf{m}}$ must be linear independent, i.e. the auxiliaries must provide additional information. In particular the number of columns of $\mathbf{M}$ must be smaller than the number of sub-arrays L.
[Proof: If $\tilde{\mathbf{m}} = \mathbf{M}\mathbf{c}$ for some coefficients $\mathbf{c}$, then $\mathbf{w}_{\text{aux}} = \mathbf{c}$ and therefore $S_{\text{sum}} - \mathbf{w}_{\text{aux}}^H \mathbf{z}_{\text{aux}} = 0$].

(ii) If the auxiliary channels block the vector used for beamforming, i.e. if $\mathbf{M}^H \tilde{\mathbf{m}} = 0$ (reference blocking condition), and if the auxiliaries preserve all dof, i.e. if $\mathbf{M}$ is of rank L-1, then one can calculate that the GSLC weight is the same as the DSW weight for the transformed sub-array configuration. More precisely, for the beam space sub-arrays $\tilde{\mathbf{z}}_{\text{BS}} = \begin{pmatrix} \tilde{\mathbf{m}}^H \tilde{\mathbf{z}} \\ \mathbf{M}^H \tilde{\mathbf{z}} \end{pmatrix}$ one has $\tilde{\mathbf{Q}}_{\text{BS}} = \begin{pmatrix} \tilde{\mathbf{m}}^H \tilde{\mathbf{Q}} \tilde{\mathbf{m}} & \tilde{\mathbf{m}}^H \tilde{\mathbf{Q}} \mathbf{M} \\ \mathbf{M}^H \tilde{\mathbf{Q}} \tilde{\mathbf{m}} & \mathbf{M}^H \tilde{\mathbf{Q}} \mathbf{M} \end{pmatrix}$ and $\tilde{\mathbf{s}}_{\text{BS}} = \begin{pmatrix} \tilde{\mathbf{m}}^H \tilde{\mathbf{m}} \\ \mathbf{M}^H \tilde{\mathbf{m}} \end{pmatrix} = \tilde{\mathbf{m}}^H \tilde{\mathbf{m}} \begin{pmatrix} 1 \\ \mathbf{0} \end{pmatrix}$ and therefore

$$\mathbf{w}_{\text{DSW,BS}} = \tilde{\mathbf{Q}}_{\text{BS}}^{-1} \tilde{\mathbf{s}}_{\text{BS}} = \mu \begin{pmatrix} 1 \\ -(\mathbf{M}^H \tilde{\mathbf{Q}} \mathbf{M})^{-1} \mathbf{M}^H \tilde{\mathbf{Q}} \tilde{\mathbf{m}} \end{pmatrix} = \mu \begin{pmatrix} 1 \\ -\mathbf{w}_{\text{aux}} \end{pmatrix}$$

This means that under these conditions the GSLC is exactly the same as the SNIR-optimum weight applied to a special (modified) sub-array configuration.

(iii) For the special case that the auxiliaries are adapted to the interference in the sense that the columns of the matrix $\mathbf{M}$ span the interference sub-space, then the GSLC is the same as the DSW interference suppression by a projection.

[Proof: If the columns of the matrix $\mathbf{M}$ span the interference sub-space then the covariance matrix can be written as $\tilde{\mathbf{Q}} = \mathbf{I} + \mathbf{MBM}^H$ with a certain interference cross-correlation matrix $\mathbf{B}$ and one obtains:

$$\begin{aligned}\mathbf{w}_{aux} &= (\mathbf{M}^H\mathbf{M} - \mathbf{M}^H\mathbf{MBM}^H\mathbf{M})^{-1}(\mathbf{M}^H - \mathbf{M}^H\mathbf{MBM}^H)\tilde{\mathbf{m}} \\ &= (\mathbf{M}^H\mathbf{M})^{-1}(\mathbf{I} - \mathbf{M}^H\mathbf{MB})^{-1}(\mathbf{I} - \mathbf{M}^H\mathbf{MB})\mathbf{M}^H\tilde{\mathbf{m}} \\ &= (\mathbf{M}^H\mathbf{M})^{-1}\mathbf{M}^H\tilde{\mathbf{m}}\end{aligned}$$

from which follows that $S_{sum} - \mathbf{w}_{aux}^H\mathbf{z}_{aux} = \tilde{\mathbf{m}}^H(\mathbf{I} - \mathbf{M}(\mathbf{M}^H\mathbf{M})^{-1}\mathbf{M}^H)\tilde{\mathbf{z}}$, which is the projection operation.]

As mentioned before, $\tilde{\mathbf{m}}$ needs not to be equal to the desired signal, i.e. the reference blocking condition is not necessarily a signal blocking condition. The term signal blocking is often used in the literature.

The shape of the sub-arrays of the equivalent beamspace representation should be irregular to avoid grating effects, i.e. possible periodic repetitions of the jammer nulls. The main difference between GSLC and DSW lies in the dynamic range of the adaptive channels because of the different point of AD-conversion in the processing chain. This leads to differences in the sensitivity and the suppression of strong jammers. Limiting of the AD-converters must be avoided, because any nonlinearity degrades the adaptive suppression. For the GSLC strong sidelobe jammers are attenuated before adaptation by the sidelobe level and limiting effects will occur only for very strong jammers, whereas with DSW the jammer will in general be located within the sub-array mainlobe. Conversely, for mainbeam jammers the GSLC will soon come to limiting in the mainbeam, while for DSW with the much lower sub-array gain adaptive suppression may still be possible. The number and size of the sub-arrays thus determine the performance against mainbeam jammers. However, both systems do not completely fail if ADC limiting occurs. In this case, the gain of the sub-arrays or the main beam would be reduced by an AGC device. This produces some SNR degradation resulting in a range reduction, but the losses are different for DSW and GSLC for sidelobe and mainlobe jammers.

The GSLC is not suited to reduce the necessary dof for nulling the interference. The required number of dof depends on the number of jammers and the system errors, as has been shown in [12] and in Section 1.4. Any reduction of the dof below the necessary number will result in some performance loss. As a rule of thumb, it has been found in [12] that a number of dof of two to three times the number of jammers is necessary to compensate for channel errors. This is primarily a requirement for strong jammers, especially mainbeam jammers, because only then leakage eigenvalues due to the channels errors emerge from noise. Another effect of this feature is that an attempt to reduce cost by reducing the number of adaptive channels may fail because of the higher accuracy requirements for each channel.

With respect to channel errors there is another difference between both concepts, DSW and GSLC. The analogue beamforming networks of the GSLC tend to be more broadband. For DSW all bandpass filtering and AD-conversion errors have an impact on beamforming, in particular with respect to the null depth and the low sidelobe level. DSW with channel errors can perform significantly worse than the GSLC.

This error sensitivity is basically a problem of digital beamforming and is not specific to ABF. It is known that suitable calibration procedures are the key solution to this problem. Simple phase and amplitude calibration is not the problem. Channel inequalities have to be reduced over the receiving (signal) bandwidth to a sufficiently low level, if high jammer suppression is desired. Passband equalization techniques may be required.

ABF as a constrained optimization. Sometimes interference suppression is realized by minimizing only the jamming power subject to additional constraints, e.g. $\mathbf{w}^H\mathbf{c}_i = k_i$, for suitable vectors $\mathbf{c}_i$ and numbers k_i, $i = 1\ldots r$. Although this is an intuitively reasonable criterion, it does not necessarily give the maximum SNIR. For certain constraints however, both solutions are equivalent. The constrained optimization problem can be written in general terms as:

$$\min_{\mathbf{w}} \mathbf{w}^H\mathbf{Q}\mathbf{w} \quad \text{s.t.} \quad \mathbf{w}^H\mathbf{C} = \mathbf{k} \quad \left(\text{or } \mathbf{w}^H\mathbf{c}_i = k_i, i = 1\ldots r\right) \tag{1.48}$$

This minimization problem has the solution:

$$\mathbf{w} = \sum_{i=1}^{r} \lambda_i \mathbf{Q}^{-1}\mathbf{c}_i = \mathbf{Q}^{-1}\mathbf{C}\left(\mathbf{C}^H\mathbf{Q}^{-1}\mathbf{C}\right)^{-1}\mathbf{k} \tag{1.49}$$

Examples of special cases:

- Single unit gain directional constraint: $\mathbf{w}^H\mathbf{a}_0 = 1 \Rightarrow \mathbf{w} = (\mathbf{a}_0^H\mathbf{Q}^{-1}\mathbf{a}_0)^{-1}\mathbf{Q}^{-1}\mathbf{a}_0$. This is obviously equivalent to the SNIR-optimum solution (1.46) with the specific normalization $\mu = (\mathbf{a}_0^H\mathbf{Q}^{-1}\mathbf{a}_0)^{-1}$.
- Gain and derivative constraint: $\mathbf{w}^H\mathbf{a}_0 = 1$, $\mathbf{w}^H\mathbf{a}'_0 = 0 \Rightarrow \mathbf{w} = \mu\,\mathbf{Q}^{-1}\mathbf{a}_0 + \kappa\mathbf{Q}^{-1}\mathbf{a}'_0$ with suitable values of the Lagrange parameters μ, λ. A derivative constraint is added to make the weight less sensitive against directional mismatch of the steering direction.
- Gain and norm constraint: $\mathbf{w}^H\mathbf{a}_0 = 1$, $\mathbf{w}^H\mathbf{w} = c \Rightarrow \mathbf{w} = \mu(\mathbf{Q} + \delta\mathbf{I})^{-1}\mathbf{a}_0$. The norm constraint is added to make the weight numerically stable. In fact, this is equivalent to the famous diagonal loading technique which we will consider later.
- Norm constraint only: $\mathbf{w}^H\mathbf{w} = 1 \Rightarrow \mathbf{w} = \min EV(\mathbf{Q})$. Without a directional constraint the weight vector produces a nearly omni-directional pattern which posseses only nulls in the interference directions. This is also called the power inversion weight, because the pattern displays the inverted interference power.

As we mentioned before, fulfilling the constraints may imply a loss in SNIR. Therefore several techniques have been proposed to mitigate the loss. The first idea is to allow a compromise between power minimization and constraints by introducing coupling factors b_i and solve a soft constraint optimization:

$$\begin{aligned} &\min_{\mathbf{w}} \mathbf{w}^H\mathbf{Q}\mathbf{w} + \sum_{i=1}^{r} b_i \left|\mathbf{w}^H\mathbf{c}_i - k_i\right|^2 \text{ or equivalently} \\ &\min_{\mathbf{w}} \mathbf{w}^H\mathbf{Q}\mathbf{w} + (\mathbf{w}^H\mathbf{C} - \mathbf{k})^H\mathbf{B}(\mathbf{w}^H\mathbf{C} - \mathbf{k}) \end{aligned} \tag{1.50}$$

with $\mathbf{B} = \text{diag}\{b_1, \ldots b_r\}$. The solution of this soft-constraint optimization is:

$$\mathbf{w} = (\mathbf{Q} + \mathbf{CBC}^H)^{-1}\mathbf{CBk} \tag{1.51}$$

One may extend the constrained optimization by adding inequality constraints. This leads to additional solutions with improved robustness properties but which often cannot be solved analytically. A number of methods of this kind have been proposed, see [28–31]. As we are only presenting the principles here we do not go into further details.

Performance criteria and displays. The performance of ABF is often displayed by the adapted antenna pattern. A typical adapted antenna pattern with 3 jammers of 20 dB SNR is shown in Figure 1.16(a) for the generic array of Figure 1.7. This pattern does not show how the actual jamming power is compensated by the null depth.

Plots of the SNIR are better suited for displaying this effect. The SNIR is typically plotted for varying target direction while the interference scenario is held fixed, as seen in Figure 1.16(b). The SNIR is normalized to the SNR in the clear (i.e. in absence of any jamming) and without ABF. In other words, this pattern shows the insertion loss arising from the jamming scenario with applied ABF. One can see that in this ideal case an insertion loss only occurs for targets close to the interference direction, i.e. when the jammer lies on the skirt of the main beam. The 3 dB width of the sum beam is indicated by the shaded area. For sidelobe jammers the insertion loss is virtually zero.

The effect of target and steering direction mismatch is not accounted for in the SNIR plot. This effect is displayed by the scan pattern, i.e. the pattern arising if the adapted beam scans over a fixed target and interference scenario. Such a plot is rarely shown because of the many parameters to be varied. In this context, we note that for the case that the training data contains only the interference plus noise the main beam of the adapted pattern is fairly broad similar to the unadapted sum beam and is therefore fairly insensitive to pointing mismatch. Obtaining an interference alone covariance matrix is a matter of proper selection of the training data as discussed in the following section.

Figure 1.16 shows the case of an untapered (uniformly weighted) planar antenna. The first sidelobes of the unadapted antenna pattern are at the typical level of -17 dB and they are nearly unaffected by the adaptation process. If we have an antenna with low sidelobes, the peak sidelobe level is affected and increases in this case by 15 dB as see in Figure 1.17. Due to the tapering we have a loss in SNIR of 1.2 dB compared to the reference antenna (untapered without ABF and jamming). For comparison we have also plotted the insertion loss with a non-adaptive antenna which reproduces virtually the inverted antenna pattern. This shows the significant SNIR loss in spite of the low sidelobes.

1.6.2 Estimation of adaptive weights

In reality the interference covariance matrix is not known and must be estimated from some training data $\mathbf{Z} = (\mathbf{z}_1, \cdots \mathbf{z}_K)$. To avoid signal cancellation the training

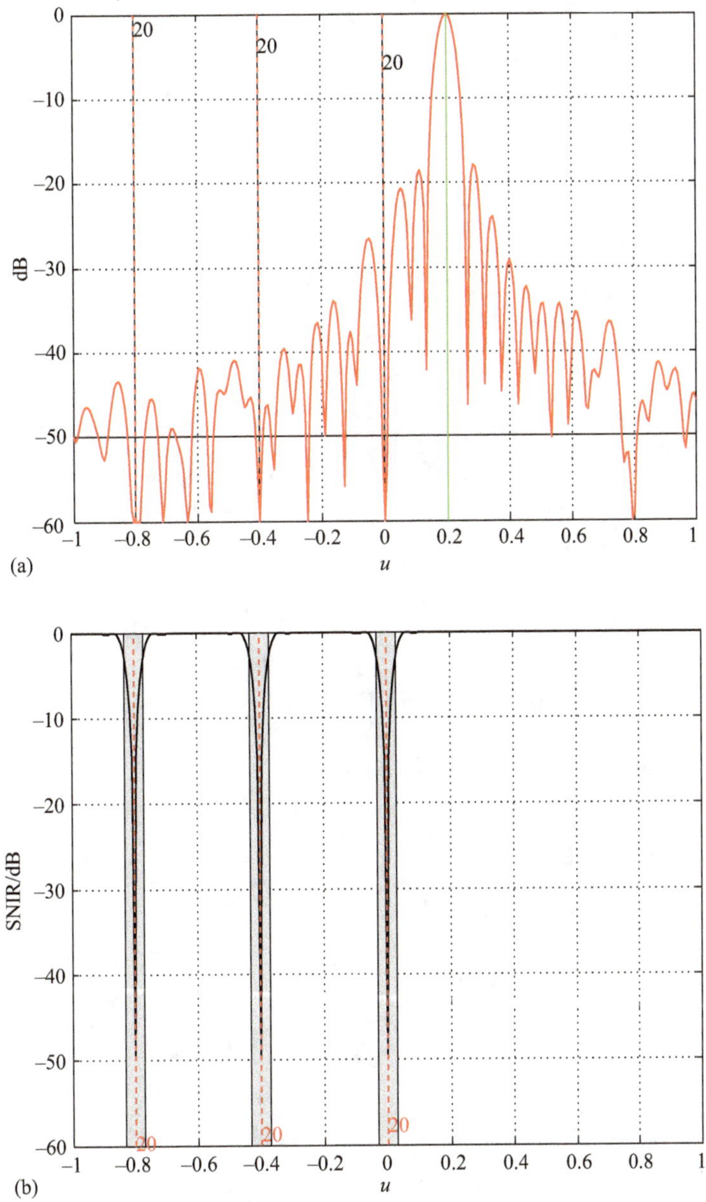

Figure 1.16 Untapered beamforming: Antenna and normalized SNIR patterns for a three jammer configuration and generic array (from [4]). (a) Adapted antenna pattern and (b) SNIR

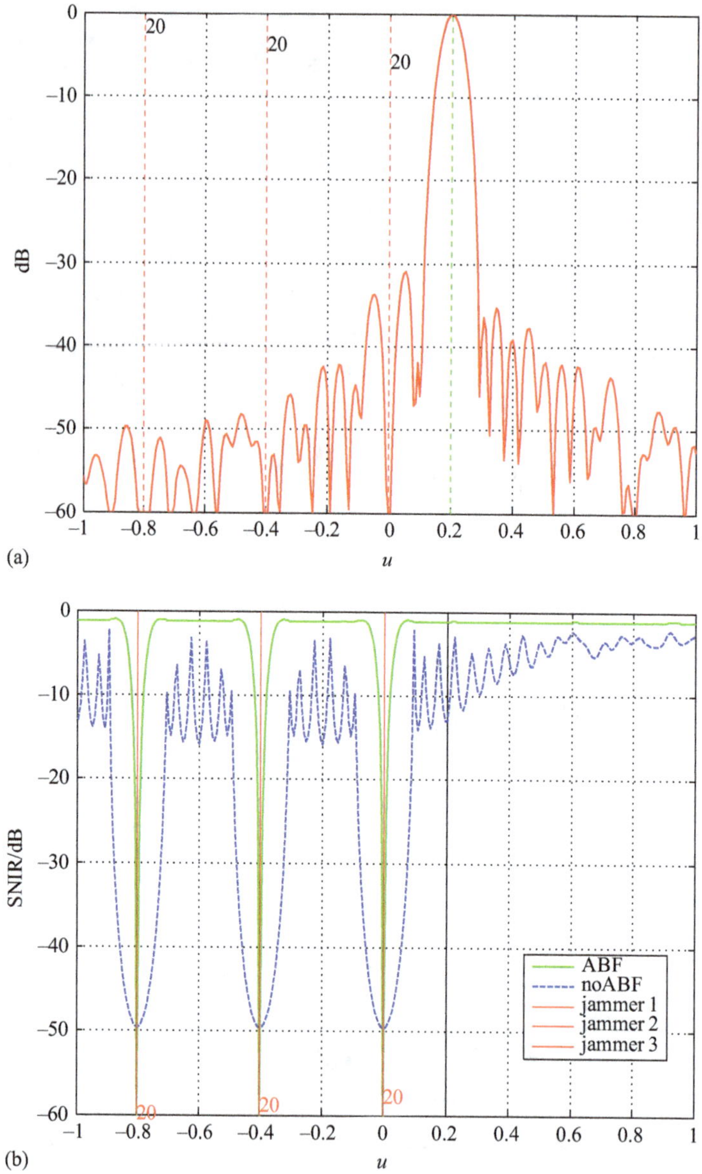

Figure 1.17 Tapered beamforming: Antenna and normalized SNIR patterns for a three jammer configuration for generic array with low sidelobes, −35 dB Taylor weighting, (from [4]). (a) Adapted antenna pattern and (b) SNIR

data should only contain the interference alone. Techniques how to achieve this will be considered in Section 1.6.4.

The ML estimate of the covariance matrix is:

$$\hat{\mathbf{Q}}_{\mathrm{SMI}} = \frac{1}{K}\sum_{k=1}^{K} \mathbf{z}_k \mathbf{z}_k^H \tag{1.52}$$

This is called the Sample Matrix Inversion algorithm (SMI). The SMI method is only asymptotically a good estimate. For small sample size it is known to be not very stable. To guarantee matrix inversion we need at least $K = N$ samples. According to Brennan's Rule, [22,23], one needs $K = 2N$ samples to obtain an average SNIR loss below 3 dB. For smaller sample size the performance can be considerably worse. However, by simply adding a multiple of the identity matrix to the SMI estimate a close to optimum performance can be achieved. This is called the loaded sample matrix inversion (LSMI) method:

$$\hat{\mathbf{Q}}_{\mathrm{LSMI}} = \frac{1}{K}\sum_{k=1}^{K} \mathbf{z}_k \mathbf{z}_k^H + \delta \cdot \mathbf{I} \tag{1.53}$$

The drastic difference between SMI and LSMI is shown in Figure 1.18 for the generic array (Figure 1.7) for three jammers of 20 dB input JNR with 32 sub-arrays and only 32 data snapshots (the minimum number). It can be shown that for a 'reasonable' choice of the loading factor (a rule of thumb is $\delta = 2\sigma^2 \ldots 4\sigma^2$ for an

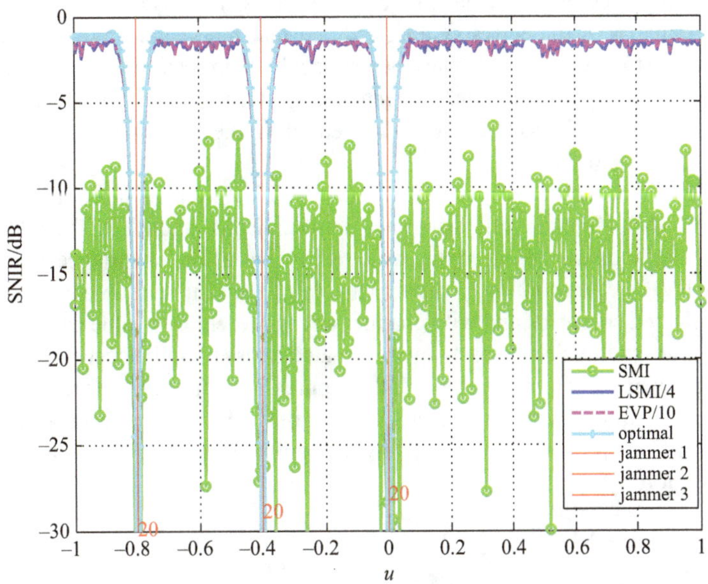

Figure 1.18 SNIR for SMI, LSMI ($\delta = 4\sigma^2$) and eigenvector projection with dimJSS = 3 (from [4])

untapered antenna) we need only $2M$ snapshots to obtain a 3 dB SNIR loss, if M denotes the number of jammers (or better the number of dominant eigenvalues) present, [23,32]. So, with diagonal loading the sample size can be considerably lower than the dimension of the matrix. The beneficial effect of the loading factor is based on the fact that the dynamic range of the small eigenvalues is compressed. The small eigenvalues possess the largest statistical fluctuation, but have the greatest influence on the weight fluctuation due to the matrix inversion.

One may go even further and ignore the small eigenvalue estimates completely, i.e. one tries to find an estimate of the inverse covariance matrix based only on the dominant eigenvectors and eigenvalues. For high JNR we can replace the inverse covariance matrix by a projection matrix. Suppose we have M jammers with amplitudes $b_1(t_k), \ldots, b_M(t_k)$ in directions $\mathbf{u}_1, \ldots, \mathbf{u}_M$ with received data $\mathbf{z}_k = \mathbf{A}\mathbf{b}_k + \mathbf{n}_k$, then:

$$\mathrm{E}\{\mathbf{z}\mathbf{z}^H\} = \mathbf{Q} = \mathbf{A}\mathbf{B}\mathbf{A}^H + \mathbf{I} \tag{1.54}$$

For convenience we have normalized the noise power to 1. With this normalization the diagonal elements of $\mathbf{B} = \mathrm{E}\{\mathbf{b}\mathbf{b}^H\}$ represent the signal–noise ratio. Using the matrix inversion lemma we have that:

$$\mathbf{Q}^{-1} = \mathbf{I} - \mathbf{A}(\mathbf{B}^{-1} + \mathbf{A}^H\mathbf{A})^{-1}\mathbf{A}^H \xrightarrow[\mathbf{B}\to\infty]{} \mathbf{I} - \mathbf{A}(\mathbf{A}^H\mathbf{A})^{-1}\mathbf{A}^H = \mathbf{P}_\mathbf{A}^\perp \tag{1.55}$$

$\mathbf{P}_\mathbf{A}^\perp$ is a projection on the space orthogonal to the columns of $\mathbf{A}$. For strong jammers the space spanned by the columns of $\mathbf{A}$ will be the same as the space spanned by the dominant eigenvectors, if we have no channel errors. The matrix $\mathbf{X} = \mathbf{A}(\mathbf{A}^H\mathbf{A})^{-1/2}$ is just the matrix of the corresponding orthonormalized vectors. We may therefore replace the estimated inverse covariance matrix by a projection $\mathbf{P}_X^\perp = \mathbf{I} - \mathbf{X}\mathbf{X}^H$ and approximate $\mathbf{X}$ by the dominant eigenvectors because the eigenvectors $\mathbf{X}$ are orthonormalized. This is called the EVP method. In the asymptotic case one has up to a complex factor $\mathbf{X} = \mathbf{A}(\mathbf{A}^H\mathbf{A})^{-1/2}$.

Figure 1.18 shows the performance of the EVP method in comparison with SMI and LSMI. Note the little difference between LSMI and EVP. The results with the three methods are based on the same realization of the covariance estimate.

For EVP we have to know the dimension of the jammer sub-space (dimJSS). In complicated scenarios and with channel errors present this value can be difficult to determine. If dimJSS is grossly overestimated a loss in SNIR occurs. If dimJSS is underestimated the jammers are not fully suppressed. One is therefore interested in sub-space methods with low sensitivity against the choice of the sub-space dimension. This property is achieved by a 'weighted projection', i.e. by replacing the projection by:

$$\mathbf{P}_{LMI} = \mathbf{I} - \mathbf{X}\mathbf{D}\mathbf{X}^H \tag{1.56}$$

where $\mathbf{D}$ is a diagonal weighting matrix and $\mathbf{X}$ is a set of orthonormal vectors spanning the interference sub-space. $\mathbf{P}_{LMI}$ is of course no projection. Methods of this type of are called lean matrix inversion (LMI) methods. Comparing (1.56) with

(1.55) one can see that this is just a simplified (lean) estimate of the inverse covariance matrix. A number of methods have been proposed that can be interpreted as an LMI method with different weighting matrices **D**. The LMI matrix can also be economically calculated by an eigenvector-free QR-decomposition method, [32,33].

One of the most efficient methods for pattern stabilization while maintaining a low desired sidelobe level is the constrained adaptive pattern synthesis (CAPS) algorithm, [34], which is also a sub-space method. Let **m** be the vector for beamforming with low sidelobes in a certain direction. In full generality the CAPS weight can be written as:

$$\mathbf{w}_{\mathrm{CAPS}} = \frac{1}{\mathbf{m}^H \hat{\mathbf{Q}}_{\mathrm{SMI}}^{-1} \mathbf{m}} \hat{\mathbf{Q}}_{\mathrm{SMI}}^{-1} \mathbf{m} - \mathbf{X}_\perp \left(\mathbf{X}_\perp^H \mathbf{C} \mathbf{X}_\perp\right)^{-1} \mathbf{X}_\perp^H \mathbf{C} \left(\frac{1}{\mathbf{m}^H \hat{\mathbf{Q}}_{\mathrm{SMI}}^{-1} \mathbf{m}} \hat{\mathbf{Q}}_{\mathrm{SMI}}^{-1} \mathbf{m} - \mathbf{m} \right) \tag{1.57}$$

where the columns of the matrix $\mathbf{X}_\perp$ span the space orthogonal to $[\mathbf{X}, \mathbf{m}]$ and $\mathbf{X}$ is again a unitary $L \times M$ matrix with columns spanning the interference sub-space which is assumed to be of dimension M. $\mathbf{C}$ is a directional weighting matrix, $\mathbf{C} = \int_\Omega \mathbf{a}(\mathbf{u})\mathbf{a}(\mathbf{u})^H p(\mathbf{u})\, d\mathbf{u}$, and Ω denotes the set of directions of interest and $p(\mathbf{u})$ is a directional weighting function. If we use no directional weighting, $\mathbf{C} \approx \mathbf{I}$, the CAPS weight vector simplifies to:

$$\mathbf{w}_{\mathrm{CAPS}} = \mathbf{m} + \mathbf{P}_{[\mathbf{X},\mathbf{m}]} \left(\frac{1}{\mathbf{m}^H \hat{\mathbf{Q}}_{\mathrm{SMI}}^{-1} \mathbf{m}} \hat{\mathbf{Q}}_{\mathrm{SMI}}^{-1} \mathbf{m} - \mathbf{m} \right) \tag{1.58}$$

where $\mathbf{P}_{[\mathbf{X},\mathbf{m}]}$ denotes the projection onto the space spanned by the columns of $\mathbf{X}$ and $\mathbf{m}$. This method is particularly effective for very low sidelobes and mainbeam jammers.

1.6.3 Determination of the dimension of jammer sub-space (dimJSS)

Sub-space methods require an estimate of the dimension of the interference sub-space. Usually this is derived from the sample eigenvalues. For complicated scenarios and small sample size a clear decision of what constitutes a dominant eigenvalue may be difficult. There are two principle approaches to determine the number of dominant eigenvalues, information theoretic criteria and noise power tests.

The information theoretic criteria are often based on the sphericity test criterion; see e.g. [35],

$$T(m) = \frac{\frac{1}{N-m} \sum_{i=m+1}^{N} \lambda_i}{\left(\prod_{i=m+1}^{N} \lambda_i \right)^{1/(N-m)}} \tag{1.59}$$

where λ_i denote the eigenvalues of the estimated covariance matrix ordered in decreasing magnitude. This ratio of the arithmetic to geometric mean of the eigenvalues is a measure of the equality of the eigenvalues. If all eigenvalues are equal this ratio is equal to one. The information theoretic criteria select dimSS by minimizing this ratio with a penalty function added. Many criteria of this kind have been proposed, e.g. the Akaike Information Criterion (AIC) and Minimum Description Length (MDL) choose dimJSS$=\hat{M}$ as the minimum of the following functions:

$$\begin{aligned} \mathrm{AIC}(m) &= K(N-m)\log[T(m)] + m(2N-m) \\ \mathrm{MDL}(m) &= K(N-m)\log[T(m)] + \left(\frac{m}{2}\right)(2N-m)\log K \end{aligned} \tag{1.60}$$

With the penalty function the criteria try to model the decay of the small eigenvalues (see e.g. Figure 1.12) for finite sample size.

Another group of criteria assume that the noise power σ^2 is known and just check the estimated noise power against this value, [35]. One of these approaches is the white noise test (WNT) which uses the statistic

$$L(m) = \frac{2K}{\sigma^2} \sum_{i=m+1}^{N} \lambda_i \tag{1.61}$$

If the data consist only of M signals plus white Gaussian receiver noise, the sum can be considered as the trace of a certain Wishart matrix. The decision for the correct white noise part can be found sequentially, starting with all eigenvalues considered as noise eigenvalues and by reducing the noise part in each step. This results in a sequence of tests which stops if the test statistic $L(m)$ is for the first time below a given threshold:

$$\begin{aligned} &\text{for } i = 1 \ldots N \text{ do} \\ &\qquad \text{if } L(m) \le \chi^2_{2K(N-m);\alpha} : \hat{M} = m; \text{STOP} \\ &\text{end} \end{aligned} \tag{1.62}$$

The symbol $\chi^2_{r;\alpha}$ denotes the α-percentage point of the χ^2-distribution with r dof. This threshold is possible due to the white noise assumption. The probability to overestimate dimJSS is then asymptotically bounded by α. More modern versions of this test have also been derived, e.g. [36].

For small sample size AIC and MDL are known for grossly overestimating the number of sources. In addition, bandwidth and array channels errors lead to a leakage of the dominant eigenvalues into the small eigenvalues, [37]. Improved eigenvalue estimates for small sample size can mitigate this effect. The simplest way could be to use the asymptotic approximation using the well-known linkage factors [38]:

$$\hat{\hat{\lambda}}_i = \hat{\lambda}_i - \frac{1}{K}\hat{\lambda}_i \sum_{\substack{j=1 \\ j\neq i}}^{N} \frac{\hat{\lambda}_i}{\hat{\lambda}_i - \hat{\lambda}_j} \tag{1.63}$$

More refined methods are also possible, e.g. minimum risk estimators with shrinking-expansion (SE) property, see [35,39]. However, as explained in [35] simple diagonal loading can effectively improve AIC and MDL for small sample size and can make these criteria robust against errors. For WNT such loading is

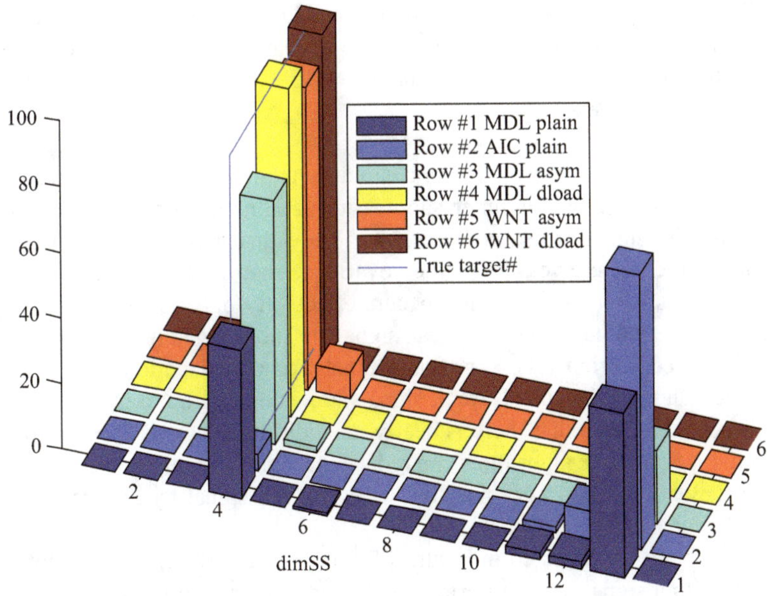

Figure 1.19 Comparison of tests for linear array with N = 14 elements, K = 14 snapshots, with and without asymptotic correction or diagonal loading of $1\sigma^2$ (from [4])

already contained in the setting of the assumed noise level σ^2. Figure 1.19 shows as an example a comparison of MDL and AIC without any corrections, MDL and WNT with asymptotic correction (1.63), and MDL and WNT with diagonal loading of $\mu = 1\sigma^2$. The threshold for WNT was set for a probability to overestimate the target number of $\alpha = 10\%$. The scenario consists of four sources at $u = -0.7$, -0.55, -0.31, -0.24 with SNR of 18, 6, 20, 20.4 dB and a uniform linear antenna with 14 elements and 10% relative bandwidth leading to some eigenvalue leakage. The empirical probabilities were determined from 100 Monte Carlo trials. Note that the asymptotic correction seems to work better for WNT than for MDL. With diagonal loading all decisions with both, MDL and WNT, were correct (equal to 4).

A more thorough study of the small sample size dimJSS estimation problem considering the 'effective number of identifiable signals' has been performed in [40] and a new modified information theoretic criterion has been derived.

1.6.4 Other aspects of implementation

The key problem of implementation is in which time window and how many secondary data should be sampled for learning the adaptive weights. For ideal interference suppression, the secondary data used for adaptation should

(i) not contain the desired signal to avoid signal cancellation,
(ii) not contain clutter because this would consume unnecessary dof as clutter can be suppressed effectively in the spectral domain after beamforming.

(iii) The whole adaptation time window should be short for rapid adaptation against non-stationary scenarios while a degradation of the coherent integration by weight/pattern fluctuation in the coherent processing interval (CPI) should be avoided.

Signal inclusion: The inclusion of the desired signal in the training data results in a suppression of the signal. Jammer suppression is normally done before matched filtering for the signal (pulse compression). The signal power is then fairly low, often below the noise level. So, if the covariance matrix is estimated from finite (few) snapshots, signal cancellation occurs effectively only for the strong signals because the eigenvalues due to the weak signal are sub-merged in noise and leakage eigenvalues (see Section 1.4). For strong signals some degradation in SNR may be acceptable. On the other hand, weak signals are estimated quite inaccurate in the covariance matrix such that there is nearly no additional degradation due to signal cancellation. Sub-space algorithms can reduce the signal cancellation effect further, because they allow ignoring a possible weak desired signal by the threshold that defines the interference sub-space.

The problem is illustrated in Figure 1.20 for the case of DSW. This plot shows the SNIR with a signal present during adaptation, normalized to the optimum case (no jammer, no adaptation, no tapering) for various signal directions (the values of the x-axis) and for element SNR values of -10, 0, and 10 dB (powers are taken before pulse compression). The dashed lines show the SNIR using the exact (asymptotic) covariance matrix; the solid lines show the case for an estimated covariance matrix from 64 snapshots. We used the generic antenna of Figure 1.7 with 902 elements and 32 sub-arrays. A 40 dB Taylor weighting is applied at the elements, leading to a loss in gain of 1.7 dB. The receiving bandwidth is set to 1% (resulting in some leakage eigenvalues); 4 jammers are present with powers of 18, 3, 20, 20 dB. One can see that the maximum loss appears for the strong 10 dB target. For the target in the look direction this loss is about 6 dB for the estimated covariance matrix while we have 11 dB loss for the true covariance. This 5 dB loss may be tolerable for such a strong target, which corresponds to a SNR of 40 dB at the sum beam output. For the weak target cases the loss is less than 1 dB in the look direction and about 3 dB for target at the skirt of the main beam at 0.5 BW. Thus signal cancellation does not seem to be a serious problem. In the tracking mode the range bin of the target is approximately known and there should be no problem to find target-free secondary data windows.

Inclusion of clutter: The inclusion of clutter in the training data consumes dof needed for jammer suppression. Training data with clutter can be avoided by sampling at ranges without clutter, e.g. for a ground based radar before transmitting or at large ranges, or, for an airborne radar before the ground clutter returns arrive. If there are no range bins without clutter (e.g. for a high PRF radar with range ambiguity), simple pre-clutter filters may be necessary for all sub-array channels. This is of course more expensive. The optimum solution would be to suppress clutter and jammer together by STAP. Much research has

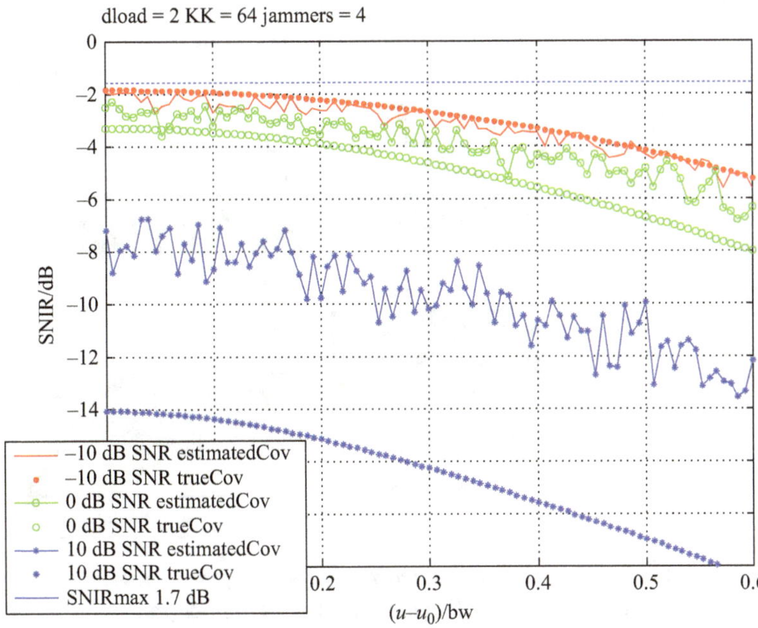

Figure 1.20 *SNIR with DSW and signal present. Input SNR −10, 0, 10 dB, dashed lines: true covariance, solid lines: covariance estimated from 64 snapshots, generic array with 902 elements and 32 sub-arrays, −40 dB Taylor weighting. © DGON 1998. Reprinted with permission, from [41]*

been done in this field. It is however a much more expensive solution than the concepts discussed here.

Non-stationary scenarios: The usual procedure for adaptive processing is to sample the training data in clutter-free intervals and then freeze the weights to suppress the interference in the subsequently incoming data. This is only reasonable for stationary situations. For rapidly changing scenarios, e.g. moving jammers or a moving platform, the jammer may move out of the adaptively formed pattern null. In particular for airborne radar a platform roll movement can lead to high changes in angle. One solution is to apply additional ('derivative') constraints which produce broader pattern nulls [29,42].

Another solution is to apply the weights to the data in the same window where the training data have been sampled. This means that these data have to be stored. The training data may be selected uniformly out of the time window and have to be pre-clutter filtered. This way, moving jammers will also produce broader nulls. The null depth may be insufficient with this sampling scheme due to the different jamming power at the time of weight learning and weight application, but this can be mitigated by using projection algorithms which tend to be independent of the

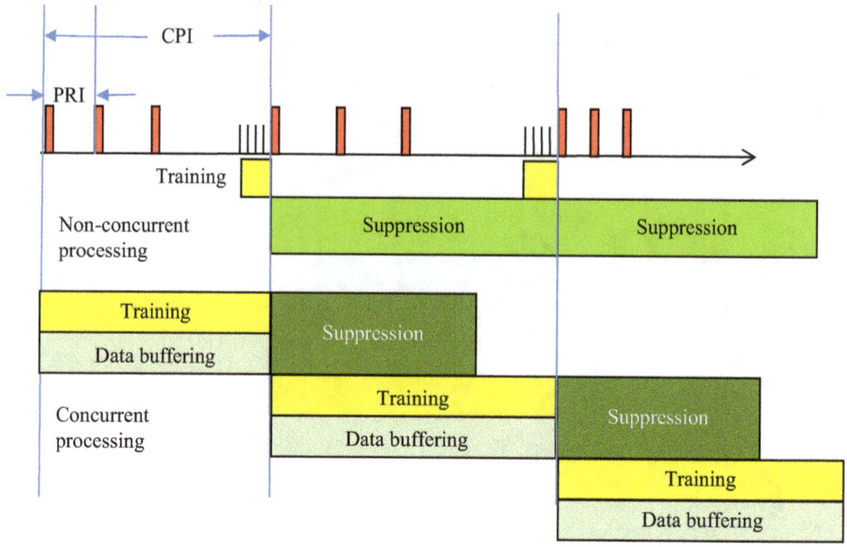

Figure 1.21 Concurrent and non-concurrent processing schemes: Time windows for training data collection and weight application (interference suppression)

actual jamming power. These approaches are only successful for not too rapidly changing situations. For very rapidly changing scenarios the adaptation intervals have to be made as small as possible. Figure 1.21 illustrates the two concepts for training data collection and weight application ('suppression'). In these figures we have assumed that one weight vector is used for each CPI.

A better but more expensive solution is to extend the procedure for time dependent adaptive weights as proposed in [43]. This is achieved by expanding $\mathbf{w}(t) = \mathbf{w}_{t=0} + t\dot{\mathbf{w}}_{t=0} + \frac{1}{2}t^2\ddot{\mathbf{w}}_{t=0} + \cdots$. This leads to estimators of the form $\hat{\mathbf{w}}_k = \mathbf{w}_0 + k\mathbf{w}_\Delta + \frac{1}{2}k^2\mathbf{w}_{\Delta\Delta} + \cdots$. In most cases a linear approximation is sufficient and then quantities $\mathbf{w}_0$ and $\mathbf{w}_\Delta$ have to be estimated based on a covariance matrix

$$\mathbf{R} = \begin{pmatrix} \mathbf{R}^{(0)} & \mathbf{R}^{(\Delta)} \\ \mathbf{R}^{(\Delta)H} & \mathbf{R}^{(\Delta\Delta)} \end{pmatrix}$$

So the additional expense lies in estimating a covariance matrix of the double dimension than for estimating $\mathbf{w}_0$ alone.

Preserving coherency: If adaptation is done as indicated Figure 1.21, then we have one adaptive weight vector for the whole CPI and there is no problem with coherent processing (Doppler filtering). If however adaptation has to be faster than the CPI, because the jammer is changing so fast (e.g. against blinking jammer, hot clutter), then the antenna pattern will be modulated by the changing adaptive weights and this may lead to a degradation of the coherent integration. For sidelobe jammers the mainlobe perturbation is small and the loss in coherency may be negligible. For mainlobe jamming coherency has to be ensured by additional constraints. Attempts to keep the mainlobe shape fixed over the CPI lead to a severe degradation of the SNIR. A solution to this dilemma is the so-called stochastic constraints approach which allows some variation of the main beam, but constrains the statistical properties of the clutter to remain constant, [44,45]. In this approach the clutter statistics are described by an AR model. Restricting the AR coefficients to be constant requires not too many constraints.

1.7 Parameter estimation and super-resolution

The objective of radar processing is not to maximize the SNR, but to detect targets and to determine their parameters. If we consider the detection problem as the first step of radar processing, then it can be shown that the common likelihood ratio test for a single target contains the beam output SNR as a sufficient statistic. That means, if we maximize the output SNR this will also give the maximum probability of detection. The radar detection problem as a likelihood ratio test has been discussed in detail in [13 Chapter 2].

The detection procedure is the first step of data reduction and it is applied on a course grid of the parameter values of interest in direction, range and possibly Doppler. Only for the few bins with detected targets the target parameters are then subsequently estimated in a refined procedure.

1.7.1 Maximum likelihood estimation and monopulse

Standard radar parameter estimation can be traced back to ML estimation of a single target. Single target ML estimation leads to the matched filter in different domains [13 Section 3.2.1]. The properties of the matched filter can be judged by the correlation of the signal model used in the likelihood function and the received data. This leads to the beam pattern for angle estimation and to the ambiguity function for range and Doppler estimation. If the ambiguity function has a narrow beam and sufficiently low sidelobes, then the model of a single target is a good approximation as other targets are attenuated by the sidelobes.

Let us illustrate this fact for the simple case of direction estimation based on a single snapshot $\mathbf{z} \in \mathbb{C}^N$ with an N element array. We assume $\mathbf{z} \sim \mathcal{N}_{\mathbb{C}^N}(\mathbf{a}(\mathbf{u})b, \mathbf{I})$ with some complex amplitude b which may be random or deterministic. We assume the receiver noise is normalized to a standard deviation one. The ML estimate of the target direction then is given by

$$\begin{aligned}\hat{\mathbf{u}} &= \arg\max_{\mathbf{u},b} p(\mathbf{z}|\mathbf{u},b)\\ &= \arg\max_{\mathbf{u},b} \frac{1}{\pi^N} e^{-(\mathbf{z}-\mathbf{a}(\mathbf{u})b)^H(\mathbf{z}-\mathbf{a}(\mathbf{u})b)}\end{aligned} \tag{1.64}$$

Maximization of this expression is equivalent to minimizing the exponent and this is just a least-squares problem. For all values of $\mathbf{u}$ the minimum over b is attained for $\hat{b} = (\mathbf{a}(\mathbf{u})^H\mathbf{a}(\mathbf{u}))^{-1}\mathbf{a}(\mathbf{u})^H\mathbf{z}$ which is the conventional sum beam output normalized by the length of the beamforming vector. Inserting this expression for $\hat{b}$ into the exponent of (1.64) we find that the ML estimate will minimize $\|\mathbf{z} - \mathbf{a}(\mathbf{u})\mathbf{a}(\mathbf{u})^H\mathbf{z}/\mathbf{a}(\mathbf{u})^H\mathbf{a}(\mathbf{u})\|^2$. Multiplying this out and omitting constant terms this is equivalent to maximizing $|\mathbf{a}(\mathbf{u})^H\mathbf{z}|^2/\mathbf{a}(\mathbf{u})^H\mathbf{a}(\mathbf{u})$. This means that the ML estimate is given by the maximum of the magnitude pattern by scanning the antenna beam over the target (the scan pattern). If there is only one target present, $\mathbf{z} = \mathbf{a}(\mathbf{u}_0)$, we see that this is just the same as an ambiguity function. The shape of this function determines the accuracy and resolution.

The resolution limit for classical beamforming, i.e. the single target ML estimator, is the 3 dB beamwidth. This is the separation of two closely spaced point targets at which the ambiguity function (beam pattern) begins to produce two local maxima. If we have targets at closer spacing or patterns with high sidelobes, this simple criterion is inadequate and multiple target models have to be used for parameter estimation. A variety of such multiple target estimation methods have been introduced and we denote these as *super-resolution methods*. These methods will be considered in the next section.

Maximization of the scan pattern for a single snapshot $\mathbf{z}$ may be impossible for analogue beamforming and can be time consuming for digital beamforming depending on the number of directions. A rapid procedure for this task is desired. One could approximate the measured sum beam output by a parabola by approximating $|\mathbf{a}(u)^H\mathbf{z}|^2 \approx a + bu + cu^2$ (for simplicity we note this here only for one angle). The beam parameters a, b and c could be measured from a set of beam directions or from derivatives of the beam and one can then determine the location of the maximum. If we maximize instead the function $F(u) = \ln(|\mathbf{a}(u)^H\mathbf{z}|^2)$, then one can obtain an approximation in the form of a linear equation. Assuming that the target direction $\hat{u}$ is close to the antenna look direction u_0 such that $F(\hat{u}) \approx F(u_0)$, one can approximate the derivative of F in a first order Taylor series at $\hat{u}$ by:

$$F'(u_0) \approx F'(\hat{u}) + F''(\hat{u})(u_0 - \hat{u})$$

As the derivative vanishes at the maximum, we obtain the linear approximation

$$\hat{u} \approx u_0 - (F'')^{-1}(\hat{u})F'(u_0) \tag{1.65}$$

Writing this explicitly shows that this leads to the well-known monopulse formula for angle estimation:

$$F' = (\mathbf{a}'^H\mathbf{z}\mathbf{z}^H\mathbf{a} + \mathbf{a}^H\mathbf{z}\mathbf{z}^H\mathbf{a}')/\mathbf{a}^H\mathbf{z}\mathbf{z}^H\mathbf{a} = 2\mathrm{Re}\left\{\mathbf{a}'^H\mathbf{z}/\mathbf{a}^H\mathbf{z}\right\} \tag{1.66}$$

In fact, this is the monopulse ratio because the weighting with $\partial a_i/\partial u = j(2\pi f/c)x_i\, a_i$ produces a difference pattern. This also tells us how to form the optimum difference pattern with phased arrays: with an amplitude weighting x_i. A simplified weighting with $\mathrm{sign}\{x_i\}$ would give the classical difference beam and is a rough approximation. The first derivative is taken for the antenna look direction with the measured data inserted. The second derivative is taken at the unknown target direction and this is approximated by a constant value. For vanishing receiver noise one obtains:

$$F''(\hat{u}) \approx \frac{\mathbf{a}''^H\mathbf{a} + \mathbf{a}^H\mathbf{a}''}{\mathbf{a}^H\mathbf{a}}(\hat{u}) \tag{1.67}$$

Thus F'' is a fixed slope correction quantity and is only determined by the antenna configuration. For a centred array we have $\mathbf{a}'^H\mathbf{a} = -j2\pi f/c \ \sum_{i=1}^N x_i = 0$ and $\mathbf{a}''^H\mathbf{a} = -(2\pi f/c)^2 \ \sum_{i=1}^N x_i^2$. As the Taylor expansion is the best linear approximation this formula also tells us the optimum slope correction factor. This procedure can be extended to planar and volume arrays, see [14]. The optimal slope correction formula also gives an indication in which way good angle estimates can be obtained: the array should be dynamically balanced by (a) $\sum_{i=1}^N x_i = 0$, $\sum_{i=1}^N y_i = 0$, (b) $\sum_{i=1}^N x_i y_i = 0$, and (c) $\sum_{i=1}^N x_i^2 = \sum_{i=1}^N y_i^2$. Condition (a) ensures unbiasedness and condition (b) gives independence of the azimuth and elevation estimates.

Let us emphasize that the monopulse technique and all other kinds of inter- or extra-polation of the scan pattern are numerical methods to approximate the single target ML-estimate, i.e. are in general not adequate for multiple targets.

1.7.2 Super-resolution

An antenna array provides spatial samples of the impinging wavefronts and one may define a multi-target model for this case. Such refined target models are the basis for enhanced resolution beyond the classical resolution limit (the 3 dB width of the beam or of the ambiguity function). Historically these methods have often been introduced to improve the limited resolution of the matched filter. We call these super-resolution methods and distinguish it from high resolution, which may be achieved also by the classical method of using a very large antenna or time aperture. Super-resolution methods have been discussed since decades, and textbooks on this topic are available, e.g. [11].

We confine our development to the angle parameter estimation problem (spatial domain), but corresponding versions can be applied in the time domain as well. The spatial resolution is described by the classical Rayleigh limit which is the 3 dB beamwidth and this is determined by the antenna aperture. The antenna is a once defined hardware design resulting from several electrical, mechanical and operational constraints. Therefore there is a high interest to overcome this limitation by signal processing techniques. In the spatial domain the super-resolution is faced with the challenges of irregular sampling and sub-array processing.

The intention of this chapter is to point out the principles, inter-relations and aspects of implementation of these methods. Therefore we mention here from the many proposed methods only some most popular methods which are also applicable to sub-arrays and irregular arrays. To give an overview we like to classify the methods into two categories: spectral methods, which generate a spiky estimate of the angular spectral density, and parametric methods, which deliver only a set of 'optimal' parameter estimates which explain in a sense best the data for the inserted model. This classification is not exclusive; there are methods in between (e.g. the ESPRIT method mentioned below).

The most popular spectral methods with an angular power density estimate $S(\mathbf{u})$ are (see [13 Sections 3.2.4 and 3.2.5])

Capon's method (Capon 1969):

$$S_C(\mathbf{u}) = \left(\mathbf{a}(\mathbf{u})^H \hat{\mathbf{R}}_{\mathrm{ML}}^{-1} \mathbf{a}(\mathbf{u})\right)^{-1} \quad \text{with } \hat{\mathbf{R}}_{\mathrm{ML}} = \frac{1}{K}\sum_{k=1}^{K} \mathbf{z}_k \mathbf{z}_k^H \tag{1.68}$$

MUSIC method (Multiple Signal Classification, Bienvenu/Kopp 1986, Schmidt 1987):

$$S_{\mathrm{MUSIC}}(\mathbf{u}) = \left(\mathbf{a}(\mathbf{u})^H \mathbf{P}^{\perp} \mathbf{a}(\mathbf{u})\right)^{-1} \tag{1.69}$$

with $\mathbf{P}^{\perp} = \mathbf{I} - \mathbf{X}\mathbf{X}^H$, and $\mathbf{X}$ spanning the dominant sub-space, which is usually determined by the eigenvectors corresponding to the dominant eigenvalues of the estimated array data covariance matrix $\hat{\mathbf{R}}_{\mathrm{ML}}$. As we have shown in (1.55) this projection is an estimate of the inverse covariance matrix for high SNR. Therefore the MUSIC method incorporates an artificial SNR enhancement compared with Capon's method. If the projection would be built with true steering vectors of the targets the power density would have poles in these directions. Parameter estimation, i.e. the target directions, is then done by finding the M highest maxima of these spectra (M 1-dimensional maximizations for a linear array or M 2-dimensional maximizations for a planar array). For special arrays this numerical maximization can be replaced by an analytic solution of the maxima: for a ULA by solving for the roots of a complex polynomial (*Root-MUSIC*, [46], Pisarenko method, [47]) and for an array consisting of two sub-arrays separated by a fixed shift vector by solving a certain eigen-decomposition problem (*ESPRIT*, [48]). Of course, an estimate of the signal sub-space can also be obtained without eigen-decomposition, e.g. by rank revealing QR decomposition or by a suitable transformation of the data, see [49] for the Hung-Turner projection (HTP), Yeh–Brandwood projection (YBT) and Matrix Transform projection (MTP).

The Capon density function can be generalized to the form $S_C(\mathbf{u}) = (\mathbf{a}(\mathbf{u})^H \hat{\mathbf{R}}_{ML}^{-r} \mathbf{a}(\mathbf{u}))^{-r}$ which results in an enhancement of the peaks as with MUSIC, if we normalize the noise eigenvalues to one. If the true covariance matrix $\mathbf{R}$ can be written as $\mathbf{R} = \mathbf{U}_s \operatorname{diag}\{\lambda_i\} \mathbf{U}_s^H + \mathbf{U}_n \mathbf{U}_n^H$, then it can be shown that $\mathbf{R}^{-r} = \mathbf{I} - \mathbf{U}_s \operatorname{diag}\left\{1 - \frac{1}{\lambda_i^r}\right\} \mathbf{U}_s^H$. This estimate has been introduced by Pisarenko [50].

An LMI-version as in (1.56) instead of MUSIC would also be possible, but this is not very common. A weighting of the sub-space components is counter-productive to the desired high peaks of the density. The spiky density is a desired feature for super-resolution. Adding some small eigenvalues to a given MUSIC projection is a good way to check the angular relevance of these eigenvalues and can be used to confirm the selected sub-space dimension, see also Section 1.7.4.

The most popular parametric methods are:

Deterministic ML (detML) method

For ML estimation with a multiple target model with directions $\mathbf{u}_1, \ldots \mathbf{u}_M$ with deterministic complex amplitudes $\mathbf{b}$ under Gaussian assumptions we assume that the array output data $\mathbf{z}_k$, $k = 1 \ldots K$, are assumed to be distributed as $\mathbf{z}_k \sim \mathbf{N}_{\mathbb{C}^N}(\mathbf{A}(\boldsymbol{\theta})\mathbf{b}_k, \mathbf{I})$. Similar to the maximization problem in (1.64) this leads to minimizing the objective function:

$$F_{\text{det}}(\boldsymbol{\theta}) = \frac{1}{K}\sum_{k=1}^{K} \|\mathbf{z}_k - \mathbf{A}(\mathbf{A}^H\mathbf{A})^{-1}\mathbf{A}^H\mathbf{z}_k\|^2 = \text{tr}(\mathbf{P}_{\mathbf{A}}^{\perp}\hat{\mathbf{R}}_{ML}) \tag{1.70}$$

with $\mathbf{P}_{\mathbf{A}}^{\perp} = \mathbf{I} - \mathbf{A}(\mathbf{A}^H\mathbf{A})^{-1}\mathbf{A}^H$ and $\mathbf{A} = (\mathbf{a}(\mathbf{u}_1), \cdots, \mathbf{a}(\mathbf{u}_M))$, [13 p. 74]. Actually, the measured data enter here only via the estimated covariance matrix. One could replace this estimated covariance matrix by the direction information carrying part, i.e. by the dominant (signal) sub-space $\mathbf{X}\mathbf{X}^H$. One may additionally apply a positive weighting for the sub-space components with a matrix $\mathbf{W} = \text{diag}\{w_i\}$. This results in the weighted sub-space fitting (WSSF) method $F_{\text{WSSF}}(\boldsymbol{\theta}) = tr(\mathbf{P}_{\mathbf{A}}^{\perp}\mathbf{X}\mathbf{W}\mathbf{X}^H)$, [51].

The detML method has some intuitive interpretations which give rise to some efficient numerical procedures:

1. $F_{\text{det}}(\boldsymbol{\theta}) = \frac{1}{K}\sum_{k=1}^{K}\mathbf{z}_k^H\mathbf{P}_{\mathbf{A}}^{\perp}\mathbf{z}_k = \frac{1}{K}\sum_{k=1}^{K}\|\mathbf{P}_{\mathbf{A}}^{\perp}\mathbf{z}_k\|^2 = \frac{1}{K}\sum_{k=1}^{K}\|\mathbf{z}_k - \mathbf{A}\underbrace{(\mathbf{A}^H\mathbf{A})^{-1}\mathbf{A}^H\mathbf{z}_k}_{=\hat{\mathbf{b}}}\|^2$
   means that the mean squared residual error after signal extraction is minimized, which is a useful criterion for target number estimation.
2. Writing $F_{\text{det}}(\boldsymbol{\theta}) = C - \sum_{k=1}^{K}\mathbf{z}_k^H\mathbf{A}(\mathbf{A}^H\mathbf{A})^{-1}\mathbf{A}^H\mathbf{z}_k$ shows that this can be interpreted as maximizing a set of decoupled sum beams $(\mathbf{a}^H(\mathbf{u}_1)\mathbf{z}_k, \ldots, \mathbf{a}^H(\mathbf{u}_M)\mathbf{z}_k)$ with a decoupling matrix $(\mathbf{A}^H\mathbf{A})^{-1}$. Maximization of these decoupled beams could numerically be realized by a gradient iteration which results in decoupled difference beams similar to the monopulse technique. A refined iteration of this kind is used in Figure 1.24.
3. One can write the detML criterion as $F_{\text{det}}(\boldsymbol{\theta}) = C - \mathbf{a}_{\text{null}}^H\hat{\mathbf{R}}\,\mathbf{a}_{\text{null}}/\mathbf{a}_{\text{null}}^H\mathbf{a}_{\text{null}}$ with $\mathbf{a}_{\text{null}} = \mathbf{P}_{\breve{\mathbf{A}}}^{\perp}\mathbf{a}(\mathbf{u})$, where we have partitioned the matrix of steering vectors into $\mathbf{A} = (\mathbf{a}(\mathbf{u}), \breve{\mathbf{A}})$. This property is valid due to the projection decomposition lemma which says that for any partitioning $\mathbf{A} = (\mathbf{F}, \mathbf{G})$ we can write $\mathbf{P}_{\mathbf{A}}^{\perp} = \mathbf{P}_{\mathbf{G}}^{\perp} - \mathbf{P}_{\mathbf{G}}^{\perp}\mathbf{F}(\mathbf{F}^H\mathbf{P}_{\mathbf{G}}^{\perp}\mathbf{F})^{-1}\mathbf{F}^H\mathbf{P}_{\mathbf{G}}^{\perp}$. If we keep the directions in $\breve{\mathbf{A}}$ fixed, this relation says that one may maximize the scan pattern over $\mathbf{u}$ while the source directions in $\breve{\mathbf{A}}$ are deterministically nulled (see (1.44)). One can now perform the multi-dimensional maximization by alternating one-dimensional maximizations of this kind while keeping the remaining directions fixed. This is the basis of the alternating projection (AP) method [52] which is equivalent to the IMP (Incremental Multi-Parameter) method, [11 p. 105], which was developed independently. AP and IMP are iterative numerical minimizations along alternating slices parallel to the co-ordinate axes. For these methods inspection of the residual scan patterns $\mathbf{a}_{\text{null}}^H\hat{\mathbf{R}}\,\mathbf{a}_{\text{null}}$ is also useful to check the relevance of some additional target directions, see also Section 1.7.4.

There are other methods similar the AP/IMP method to solve the detML estimation by a sequence of 1-dimensional maximizations: the Expectation Maximization (EM) method and the Space Alternating Generalized EM (SAGE) algorithm ([1, p. 316], [53]). These methods are based on a decomposed signal model $\mathbf{z}_m = \mathbf{a}(\mathbf{u}_m)b_m + \mathbf{n}/\sqrt{M}, m = 1 \ldots M$. The associated data and likelihood functions are sequentially estimated and maximized.

Stochastic ML (stoML) method

For ML estimation with a multiple target model with directions $\mathbf{u}_1, \ldots \mathbf{u}_M$ with complex normal distributed amplitudes $\mathbf{b}$ under Gaussian assumptions we assume that the array output data are distributed i.i.d. as $\mathbf{z} \sim \mathcal{N}_{\mathbb{C}^N}(\mathbf{0}, \mathbf{A}(\boldsymbol{\theta})\mathbf{B}\mathbf{A}(\boldsymbol{\theta})^H + \mathbf{I})$. One can show that this leads to minimizing the objective function:

$$F_{sto}(\boldsymbol{\theta}) = \log \det(\mathbf{R}(\boldsymbol{\theta})) + \mathrm{tr}\left(\mathbf{R}(\boldsymbol{\theta})^{-1}\hat{\mathbf{R}}_{ML}\right) \tag{1.71}$$

where $\mathbf{R}(\theta)$ denotes the completely parameterized covariance matrix, w.r.t. angles, powers and correlations. A formulation depending only on the unknown directions can be given by, [54],

$$F_{sto}(\boldsymbol{\theta}) = \det\{\mathbf{A}\hat{\mathbf{B}}(\boldsymbol{\theta})\mathbf{A}^H + \hat{\sigma}^2(\boldsymbol{\theta})\mathbf{I}\} \quad \text{with} \tag{1.72}$$

$$\hat{\mathbf{B}}(\boldsymbol{\theta}) = (\mathbf{A}^H\mathbf{A})^{-1}\mathbf{A}^H(\hat{\mathbf{R}}_{\mathrm{ML}} - \hat{\sigma}^2(\boldsymbol{\theta})\mathbf{I})\mathbf{A}(\mathbf{A}^H\mathbf{A})^{-1} \quad \text{and} \quad \hat{\sigma}^2(\boldsymbol{\theta}) = \frac{1}{N-M}\mathrm{tr}\{\mathbf{P}_{\mathbf{A}}^{\perp}\hat{\mathbf{R}}_{\mathrm{ML}}\} \tag{1.73}$$

and for $\mathbf{A} = \mathbf{A}(\theta)$. The behaviour of the function (1.72) is quite difficult to check.

A first general comment concerning all super-resolution methods is that these methods in different ways try to interpret the fine structure of the signal. To achieve any super-resolution effect with these methods we have therefore to require a higher SNR than for the detection of a single target.

To characterize the performance of these methods we show three examples. A typical feature of the MUSIC method is illustrated in Figure 1.22 obtained with the experimental system DESAS of Fraunhofer FHR, [1 p. 309], or [13 p. 81]. Sub-plot (a) shows the excellent resolution in simulations for an azimuth cut for 2 targets separated at 0.6 BW while for real data in sub-plot (b) with the same scenario the pattern looks almost the same as with Capon's method. This degradation is due to the channel errors in the used experimental system leading to a mismatch between the real and the assumed signal model. Note that the used calibration method has also a decisive effect. In this case we focused the array on each of the two sources using each emitted signal alone and averaged the correction quantities. Clearly, one can improve the results by more clever calibration procedures.

A result with the deterministic ML method with real data is shown in Figure 1.24. The scenario consisted of a vertical 32 element ULA observing an approaching aircraft at constant low height as indicated in Figure 1.23. The problem of this scenario over sea is that the electromagnetic energy will be reflected from the sea surface which may create a highly correlated image target. This result

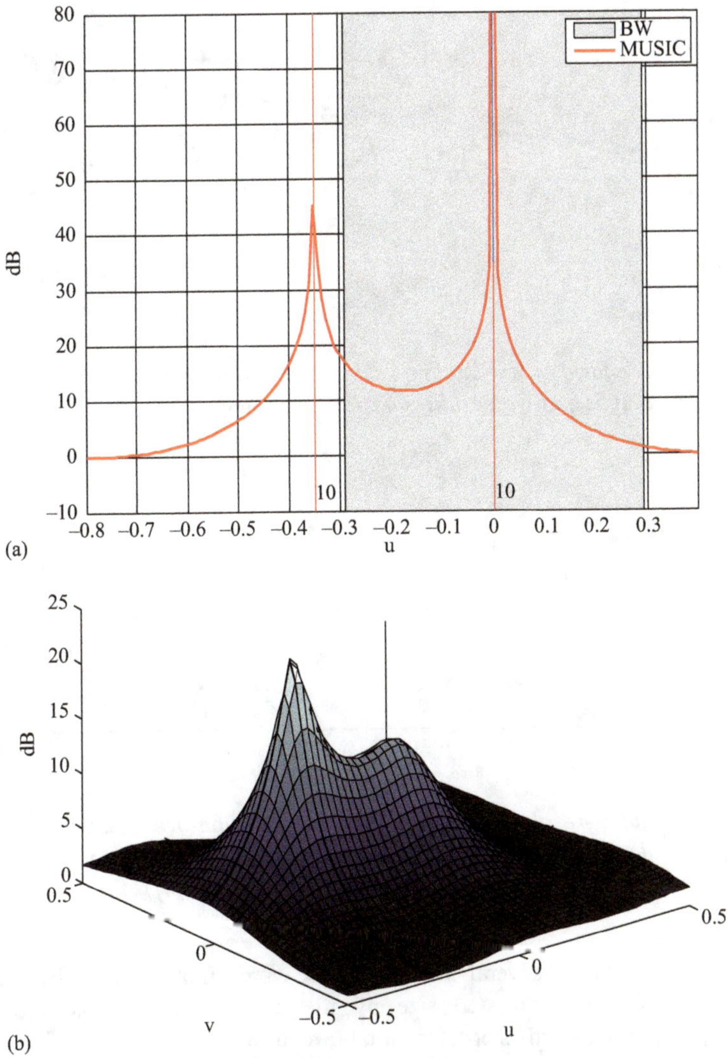

Figure 1.22 MUSIC spectra of two targets at 0.6BW separation with a planar array of 8 elements (7 elements irregularly distributed on a ring and with one in the centre). (a) Simulated data and (b) measured data

has also been presented in [1 p. 307] under the name 'Parametric Target Model Fitting' (PTMF).

The numerical minimization of the decoupled sum beam function (interpretation 2) was performed by a stochastic approximation method, i.e. by using in each iteration step a new measurement $\mathbf{z}_k$ [55]. However, the standard Robbins–Monro iteration converged unacceptably slow and this has also been reported by other

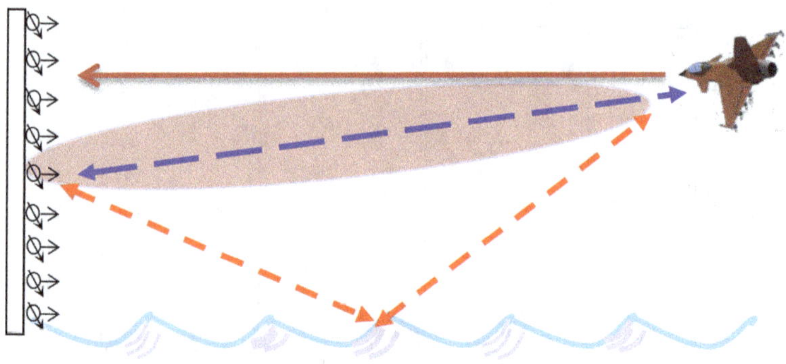

Figure 1.23 Scenario of low-angle tracking over sea with target flying at constant height and with multi-path (dashed lines)

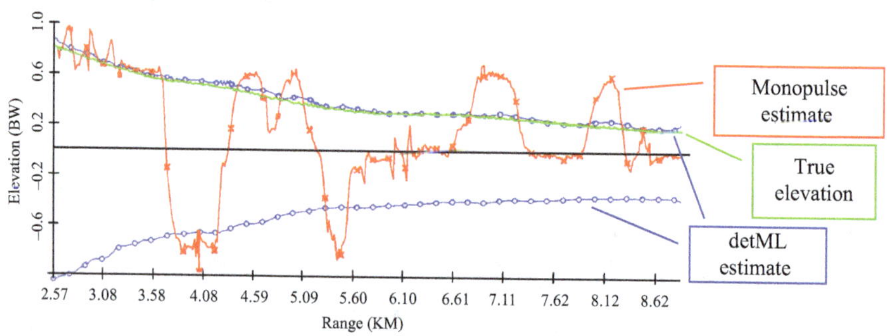

Figure 1.24 Super-resolution of multi-path propagation over sea with deterministic ML method, real data from vertical linear array with 32 elements, scenario of Figure 1.23, (from [4])

researchers. Therefore several modifications were implemented. Besides the reduction of the iteration step size by the classical Robbins–Monro factor $a_k = \mu/(\beta + k)$ in the kth step (for suitable constants μ, β), a modified update vector was introduced because we observed strong fluctuations in the length of the gradient. Instead of the gradient of the function (1.70) we used an objective function which is oriented along the derivation of the monopulse estimator in [14], namely the logarithm of the decoupled sum beam function $\ln(S_k(\boldsymbol{\theta})) = \ln(\mathbf{z}_k^H \mathbf{A}(\mathbf{A}^H\mathbf{A})^{-1}\mathbf{A}^H\mathbf{z}_k)$. This leads to an iteration with an update vector consisting of the gradient divided by the decoupled sum beam, i.e. $\text{grad}\{S_k(\boldsymbol{\theta})\}/S_k(\boldsymbol{\theta})$, which is similar to the monopulse ratio. Also a clipping of the length of this gradient was introduced. The combination of these modifications resulted in an accelerated convergence. We also mention here that any modification of the stochastic approximation by using in addition an estimate of the Hessian (the matrix of second derivatives) of the objective function did not result in an improvement because the fluctuations of this matrix were so strong that they rather distorted the convergence.

This real data result shows that the detML-method is able to resolve highly correlated targets which arise due to the reflections on the sea surface for low angle tracking. The behaviour of the classical monopulse estimates in Figure 1.24 reflects the variation of the phase differences between the direct and reflected path from 0° to 180° (glint effect). For a phase difference of 0° the monopulse points into the centre, for 180° it points outside the 2-target configuration. These phase differences can be well tracked by the associated complex amplitude estimates $\hat{\mathbf{b}}_k = (\mathbf{A}^H\mathbf{A})^{-1}\mathbf{A}^H\mathbf{z}_k$ and this is an advantage of the detML-method.

The general problems of super-resolution methods are described in [13,56]. One of the key problems is the numerical effort of finding the M maxima (one M-dimensional optimization or M 1-dimensional optimizations for a linear antenna). For the deterministic ML method the stochastic approximation algorithm and the IMP or alternating projection method have been successfully proven with real data. The IMP method is an iteration of maximizations of an adaptively formed beam pattern. Therefore the generalized monopulse method of (1.75) can be used for this purpose, see Section 1.9.1 and [14].

Another problem is the exact knowledge of the signal model for all possible directions, i.e. the vector function $\mathbf{a}(\mathbf{u})$. The co-domain of this function is sometimes called the array manifold. The correctness of this model is mainly a problem of the receiving system accuracy and of proper calibration. While the transmission of a plane wave in the main beam direction can be locally quite accurately modelled (using calibration) this can be difficult in the sidelobe region. In particular this can lead to problems if super-resolution methods are applied to sub-arrayed arrays.

More refined parametric methods with higher asymptotic resolution property have been suggested (e.g. COMET, Covariance Matching Estimation Technique [57]). However, application of such methods to real data often revealed no improvement (similar to the case with MUSIC in Figure 1.22). The reason is that these methods are much more sensitive to the signal model than the accuracy of the system provides. We have observed that a very sensitive matching criterion with a very sharp ideal minimum of the objective function may lead to a measured data objective function where the minimum has completely disappeared.

1.7.3 Super-resolution applied to sub-arrays

For an array with digital sub-arrays super-resolution has to be performed only with the sub-array outputs. The array manifold taken at the sub-array outputs has then to be considered, see Section 1.3.3. This manifold is essentially described by the sub-array patterns which can be well modelled in the main beam region. However, in the sidelobe region the random sub-array sidelobes produce in general no reasonable results. If e.g. we consider the sidelobes of the generic array as in Figure 1.8(b) it is quite reasonable that beams formed into the sub-array sidelobe region will lead to patterns with terrible sidelobes and an insignificant main beam.

In that case it is advantageous to use a simplified array manifold model which is only locally valid in the mainbeam of the sub-arrays. The simplest and most effective model is to use the sub-array centres (the super-array) and consider these as omni-directional elements but with the associated sub-array gains. We called this the Direct Uniform Manifold model (DUM). This simplified model has been successfully applied to MUSIC and to the deterministic ML method. In [58] the application of this approach to MUSIC was called 'Spotlight MUSIC', because it provides good super-resolution results within the mainbeam (the spot) and no results outside the sub-array (mean) beamwidth. Using the DUM model requires little calibration effort and gives improved performance in the sense that artificial sub-array sidelobe effects are suppressed, [58].

A practical implementation of this method in a phased array radar could be achieved by scanning over a grid of directions, e.g. in search mode, and calculating super-resolution patterns for each look direction. To get a picture of the whole field of view one has then to fuse these local super-resolution results.

1.7.4 Super-resolution combined with adaptive interference suppression

If one is able to measure the interference by an interference alone covariance matrix $\hat{\mathbf{Q}}$, it is straightforward to extend the super-resolution methods to include ABF. According to the pre-whiten and match principle this is achieved by replacing the data vector by the pre-whitened data vector $\tilde{\mathbf{z}}_{pre-w} = \hat{\mathbf{Q}}^{-1/2}\tilde{\mathbf{z}}$ and by replacing the multi-target model by a pre-whitened model $\tilde{\mathbf{a}}_{pre-w} = \hat{\mathbf{Q}}^{-1/2}\tilde{\mathbf{a}}$ or $\tilde{\mathbf{A}}_{pre-w} = \hat{\mathbf{Q}}^{-1/2}\tilde{\mathbf{A}}$. This will result in new super-resolution methods whose properties have to be studied and which depend on the individual method.

For the detML method this approach leads to some interesting extensions. Because detML can be interpreted as a generalized beamforming procedure, much of the results given for ABF can be applied. In fact, this pre-whiten and match approach is equivalent to the deterministic ML estimation procedure for a non-white Gaussian data model $\mathbf{z}_k \sim \mathcal{N}_{\mathbb{C}^N}(\mathbf{A}(\boldsymbol{\theta})\mathbf{b}_k, \hat{\mathbf{Q}})$. One can show that this leads to an objective function as in (1.70) but with an oblique projection $\mathbf{P}_{\mathbf{A}}^{\perp} = \hat{\mathbf{Q}}^{-1} - \hat{\mathbf{Q}}^{-1}\mathbf{A}(\mathbf{A}^H\hat{\mathbf{Q}}^{-1}\mathbf{A})^{-1}\mathbf{A}^H\hat{\mathbf{Q}}^{-1}$. The corresponding notations of this method as 'parallel decoupled beamforming' or 'alternating projections' can be derived analogously based on the notions of $\tilde{\mathbf{z}}_{pre-w}$ and $\tilde{\mathbf{a}}_{pre-w}$ or $\tilde{\mathbf{A}}_{pre-w}$.

1.7.5 Adaptive target number determination

Super-resolution is a combined target number and target parameter estimation problem. As a starting point all the methods of Section 1.6.3 can be used for target number estimation. If we use the detML method we can exploit that the objective function can be interpreted as the residual error between the model and the data. The test statistic for the WNT criterion (1.61) is just an estimate of this quantity. The detML residual can therefore be used for this test instead of the sum of the eigenvalues. Actually, in this form the WNT criterion turns into a true Likelihood Ratio test, see [35].

The performance of such nested multi-hypotheses detection procedures can be described by the two error probabilities, the probability of over and under-estimating the target number. It depends on the application how serious these errors are and if some control of these errors is desired. The information theoretic criteria are known to tend to overestimate the target number. The WNT criterion provides an approximate level to overestimate the number. It is useful to consider an additional criterion to confirm or reduce the target number in a following step. This can be done by checking the power allocated to each target with the estimated directions $\mathbf{A}(\boldsymbol{\theta})$ by a refined ML power estimate as used in (1.73). This estimate can even reveal correlations between the targets. This has been successfully demonstrated with the low angle tracking data of Figure 1.24 and in other experiments. In case that some target power is too low, the target number can be reduced and the angle estimates can be updated with the corrected target number. Thus we arrive at an iterative procedure of direction estimation, target number estimation, target confirmation or target number reduction. This way, all target modelling can be matched to the data with a desired accuracy.

The deterministic ML method (1.70) with the stochastic approximation method together with the white noise test (1.62) is particularly suited for this kind of iterative model fitting. It has been implemented in an experimental system with an 8-element planar array at Fraunhofer FHR and was first reported in [13, p. 81]. A snapshot of the resulting display movie is shown in Figure 1.25. The estimated

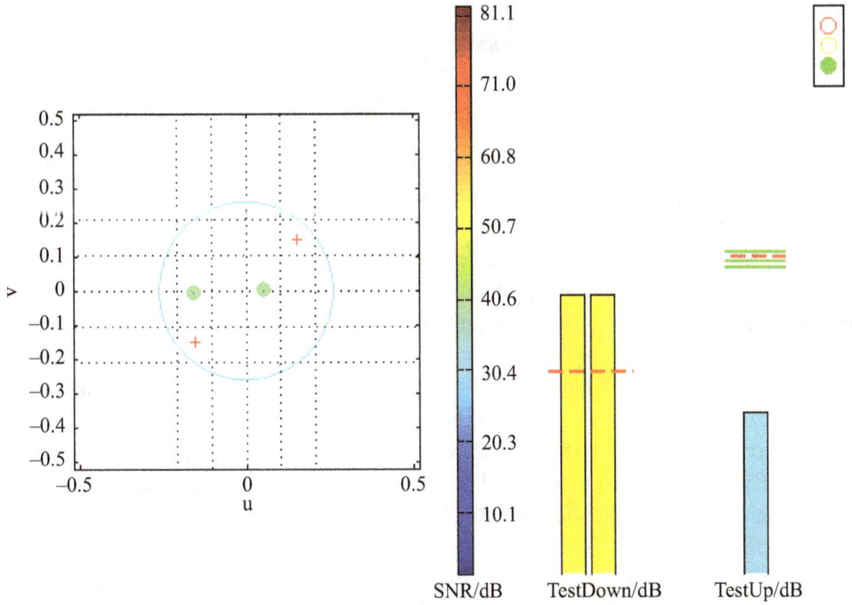

Figure 1.25 Combined target number and direction estimation for 2 targets with 7-element planar array, (from [4])

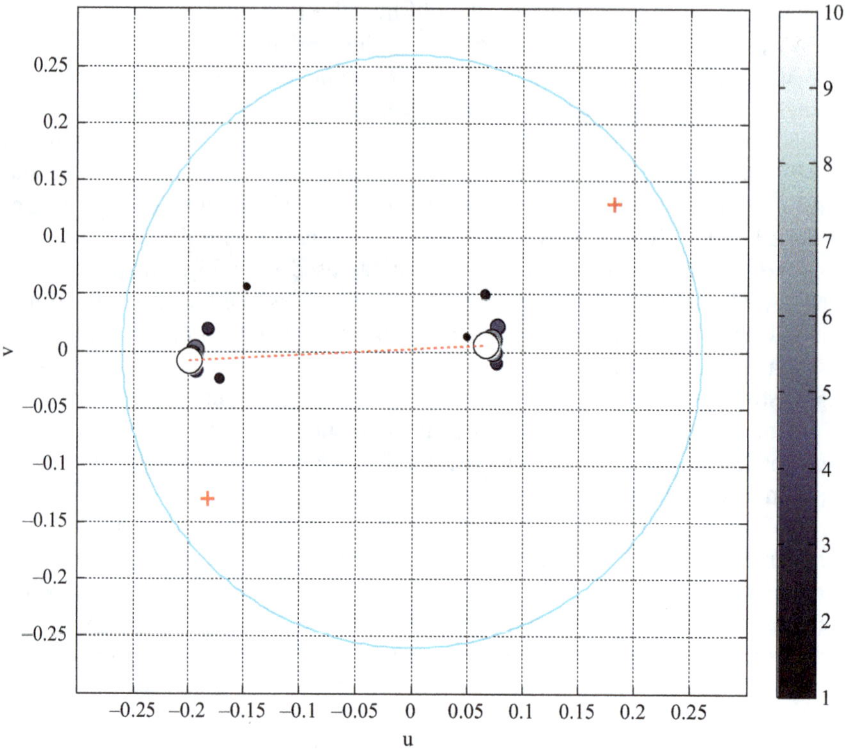

Figure 1.26 Convergence of stochastic approximation for two targets as in Figure 1.25

directions in the u,v-plane are shown by small dishes having a colour according to the estimated target SNR corresponding to the colour bar. The circle indicates the 3 dB contour of the sum beam. The current estimates by the two dots represent in fact the two targets directions which are at about 0.5 beamwidth separation. Direction estimation was done by the refined stochastic approximation algorithm as in Figure 1.24. The test statistic for increasing the target number (the WNT statistic (1.62)) is shown by the right most bar. The thresholds for increasing the number according to the χ^2-distribution in (1.62) are indicated by lines. The dashed line is the actually active threshold; in the plot it is shown for the 2 target case. The target number can be reduced if the estimated power of some target falls below a threshold which is shown by two bars in the middle. The whole estimation and testing procedure can also be performed adaptively with changing target situations. Actually, Figure 1.25 displays a snapshot from a movie with two blinking targets at fixed positions, alternating between the states ‘only target 1 on’, ‘both targets on’, ‘only target 2 on’, ‘both targets on’ etc. Clearly, these multiple tests work only if the estimation procedure has converged and this is indicated by the traffic light in the right up corner. We used a fixed empirically determined iteration number to switch the test procedure on (=green traffic light). Our experience was that in these scenarios ten iterations delivered an accuracy that was sufficient for target number testing. The starting values for the iteration are shown by the crosses opposite on a diagonal through the main beam. The selection of the starting values is motivated by the requirement that the directions should be widely separated within the admissible search region. Figure 1.26 indicates the convergence for this case of two targets. One can see that indeed after five iterations the positions are approximately found (the grey shade indicates the iteration number and the marker size increases with iteration number).

It is clear that all thresholds and iteration numbers have to be selected carefully and the choice of all these values depends on the specific application. Otherwise situations may arise where this adaptive procedure switches between two target models e.g. between 2 and 3 targets.

The problem of resolution of two closely spaced targets becomes a particular problem in the so called threshold region, which denotes configurations where the SNR or the separation of the targets lead to an angular variance departing significantly from the Cramér–Rao bound (CRB). The design of the tests must be compatible with this threshold region to give consistent joint estimation-detection resolution results. These problems have been further studied in [59,60]. One way to achieve consistency and improving resolution proposed in [59] is to detect and remove outliers in the data, which are basically responsible for the threshold effect. A general discussion about the achievable resolution and the best realistic representation of a target cluster can be found in [60].

1.8 Extension to space-time arrays

As mentioned in Section 1.3.4, there is mathematically no difference between spatial and temporal samples as long as the distributional assumptions are the same. The adaptive methods and super-resolution methods presented in the previous sections can therefore be applied analogously in the time or space-time domain. For example, the monopulse procedure can be applied in the time domain for accurate Doppler frequency estimation. This has been described in [1 p. 183].

In particular we can form sub-arrays in the spatial or time domain to reduce the dof. Using the notation as in Section 1.3.4, the sub-array data matrix $\tilde{\mathbf{Z}}$ is formed from the element data matrix $\mathbf{Z}$ as $\tilde{\mathbf{Z}} = \mathbf{T}_s^H \mathbf{Z} \mathbf{T}_t$, which can be written with a joint space-time sub-array forming matrix as $\text{vec}\{\tilde{\mathbf{Z}}\} = (\mathbf{T}_t^H \otimes \mathbf{T}_s^H)\text{vec}\{\mathbf{Z}\} = \mathbf{T}_{t-s}^H \text{vec}\{\mathbf{Z}\}$. The space-time covariance matrix formed from the sub-arrays then is

$$\begin{aligned}\tilde{\mathbf{R}}_{t-s} &= \mathrm{E}\left\{\text{vec}\{\tilde{\mathbf{Z}}\}\text{vec}\{\tilde{\mathbf{Z}}\}^H\right\} = \mathbf{T}_{t-s}^H \mathrm{E}\left\{\text{vec}\{\mathbf{Z}\}\text{vec}\{\mathbf{Z}\}^H\right\}\mathbf{T}_{t-s} \\ &= \mathbf{T}_{t-s}^H \mathbf{R}_{t-s} \mathbf{T}_{t-s}\end{aligned} \tag{1.74}$$

The space-time sub-array forming matrix $\mathbf{T}_{t\text{-}s}$ determines completely the performance of ABF in terms of grating lobes, grating notches or quasi-grating notches. The concept of the super-array of Section 1.3.3 and of the pattern properties with the ELA concept can be applied. Space-time ABF is typically applied for ground clutter suppression in airborne radar (STAP, see [3]) and for broadband interference suppression with arrays, [2].

For narrowband adaptive space-time beamforming (STAP) as used for airborne radar the theory as developed in Section 1.6 can be applied. The key point for this application is to design the most efficient space-time sub-array configurations and weight estimation algorithms. The challenge is to find a solution satisfying the joint constraints of the given antenna hardware on the airborne platform, the field of view resulting in a special type of clutter and the fast and robust estimation algorithm. A typical sub-array configuration of this kind is given in Figure 1.10.

For broadband applications we can apply combinations of narrowband beamforming (with phase shifts) and broadband beamforming (with time delays). A good approximation for moderate bandwidth is to use within each sub-array narrowband beamforming, but broadband beamforming for the super-array, i.e. for combing the sub-arrays. The influence of these configurations on the eigenvalue spectrum, in particular on the amount of leakage eigenvalues, and the efficiency of the various sub-space methods has been discussed in detail in [2 Chapter 16].

1.9 Embedding of array processing into full radar data processing

If the objective is to improve a radar system one has to recognize that e.g. maximizing the SNR or determining just a direction is not the task of a radar. The improvement with new processing has to be judged evaluating the resulting system features. As a starting point this means that when implementing refined methods we have also to consider the effect on the subsequent processing. Known properties of the advanced methods should be exploited in subsequent processing stages to get optimum performance. This requires some mutual communication between the different kinds of processing, in particular between the signal processing and data processing stage. Also, the system management (beam steering, waveform control etc.) should be adapted according to super-resolution or ABF results. We have studied these problems for the case of adaptive interference suppression for the case of three special tasks: detection, angle estimation and tracking. With the modifications introduced below significant improvements for the system could be demonstrated.

1.9.1 Adaptive monopulse

For large arrays angle estimation is usually implemented by monopulse estimation which is an established technique for rapid and precise angle estimation. It is based on two beams formed in parallel, a sum beam and a difference beam. The difference beam is zero at the position of the maximum of the sum beam. The ratio of both beams gives an error value that indicates the offset of a target direction from the sum beam pointing direction. In fact, it can be shown that the monopulse estimator is an approximation of the Maximum-Likelihood angle estimator, see Section 1.7 and [14], which is considered as the optimum estimator. The monopulse estimator has been generalized in [14] to arrays with arbitrary sub-arrays and arbitrary sum and difference beams.

With irregular sub-arrays many different forms of difference and sum beams can be formed resulting in different monopulse functions. In particular, adaptive beams will distort the shape of the sum beam due to the interference that is to be suppressed. The difference beam, which must adaptively suppress the interference as well, will suffer another and different distortion. The ratio of both beams which is used in conventional monopulse as described in Section 1.2.4 will then no more indicate the target direction. Several techniques have been proposed to counter this problem, e.g. [61–63], to mention a few. The generalized monopulse procedure described in [14] provides a general approach by providing correction values to compensate these distortions.

The generalized monopulse formula for estimating angles $(\hat{u},\hat{v})^T$ with a planar array and a sum and two difference beams formed into direction $(u_0,v_0)^T$ is given by:

$$\begin{pmatrix}\hat{u}\\ \hat{v}\end{pmatrix}=\begin{pmatrix}u_0\\ v_0\end{pmatrix}-\begin{pmatrix}c_{xx} & c_{xy}\\ c_{yx} & c_{yy}\end{pmatrix}\begin{pmatrix}R_x-\mu_x\\ R_y-\mu_y\end{pmatrix} \tag{1.75}$$

$\mathbf{C}=\begin{pmatrix}c_{xx} & c_{xy}\\ c_{yx} & c_{yy}\end{pmatrix}$ is a slope correction matrix and $\boldsymbol{\mu}=\begin{pmatrix}\mu_x\\ \mu_y\end{pmatrix}$ is a bias correction.

$R=\mathrm{Re}\{D/S\}$ is the monopulse ratio formed with the measured difference beam $D_x=\mathbf{d}_x^H\mathbf{z}$ for azimuth beamforming weight $\mathbf{d}_x$, or $D_y=\mathbf{d}_y^H\mathbf{z}$ for elevation beamforming weight $\mathbf{d}_y$, divided by the sum beam output $S=\mathbf{w}^H\mathbf{z}$. For standard beamforming one uses $\mathbf{w}=\mathbf{a}$, $\mathbf{d}_x=\mathrm{diag}\{x_i\}\mathbf{a}$ and $\mathbf{d}_y=\mathrm{diag}\{y_i\}\mathbf{a}$. However, the formula can be applied as well for any kind of beams, in particular adapted beams. The point is that we can determine the correction values for all kinds of beams.

The monopulse ratio is a function of the unknown target directions (u, v). Let the vector of monopulse ratios be denoted by $\mathbf{R}(u, v)=(R_x(u, v), R_y(u, v))^T$. The correction quantities are determined such that the expectation of the error is zero and such that a linear function with slope one is approximated. More precisely, for the function of the unknown target direction:

$$\mathbf{M}(u,v)=\mathbf{C}\cdot(\mathrm{E}\{\mathbf{R}(u,v)\}-\boldsymbol{\mu}) \tag{1.76}$$

we require:

$$\mathbf{M}(u_0,v_0)=\mathbf{0}\quad\text{and}\quad\frac{\partial\mathbf{M}}{\partial u}(u_0,v_0)=\begin{pmatrix}1\\ 0\end{pmatrix},\quad\frac{\partial\mathbf{M}}{\partial v}(u_0,v_0)=\begin{pmatrix}0\\ 1\end{pmatrix}\quad\text{or}$$
$$\mathbf{C}\begin{pmatrix}\dfrac{\partial\mathbf{R}}{\partial u} & \dfrac{\partial\mathbf{R}}{\partial v}\end{pmatrix}(u_0,v_0)=\mathbf{I} \tag{1.77}$$

These conditions can only approximately be fulfilled for sufficiently high SNR. One obtains for the bias correction μ for a pointing direction $\mathbf{a}_0 = \mathbf{a}(u_0,v_0)$, [14]

$$\mu_\alpha = \mathrm{Re}\left\{\frac{\mathbf{d}_\alpha^H \mathbf{a}_0}{\mathbf{w}^H \mathbf{a}_0}\right\}, \quad \text{for } \alpha = x, y \tag{1.78}$$

For the elements of the inverse slope correction matrix $\mathbf{C}^{-1} = \left(c^{\alpha,h}\right)_{\substack{\alpha=x,y \\ h=u,v}}$ one obtains:

$$c^{\alpha,h} = \frac{\mathrm{Re}\left\{\mathbf{d}_\alpha^H \mathbf{a}_{h,0} \mathbf{a}_0^H \mathbf{w} + \mathbf{d}_\alpha^H \mathbf{a}_0 \mathbf{a}_{h,0}^H \mathbf{w}\right\}}{\left|\mathbf{w}^H \mathbf{a}_0\right|^2} - \mu_\alpha 2\mathrm{Re}\left\{\frac{\mathbf{w}^H \mathbf{a}_{h,0}}{\mathbf{w}^H \mathbf{a}_0}\right\} \tag{1.79}$$

with $\alpha = x$ or y and $\mathbf{a}_{h,0}$ denoting the derivative $(\partial \mathbf{a}/\partial h)\big|_{(u_0,v_0)}$, where $h = u$ or v. In the non-adaptive case the slope corrections are fixed quantities determined by the antenna configuration. For example, for omni-directional antenna elements and phase steering at the elements we have $\mathbf{a}_0 = G_e\,(1, \ldots, 1)^T$, where G_e is the antenna element gain, and $\mathbf{a}_{u,0}^T = G_e(j2\pi f/c)(x_1, \ldots, x_N)$. For sub-arrayed antennas we have to calculate the correction values for the sub-array manifolds as in (1.20). The important point is that for the adapted case the correction values can be calculated from known quantities: the adaptive weight vectors and array transmission vectors. Also note that this formula is independent of any scaling of the difference and sum weights. Constant factors in the difference and sum weight will be cancelled by the corresponding slope correction.

This formula can also be extended to finding the maximum of the non-coherently averaged scan pattern $\sum_{k=1}^K |S_k|^2$. This leads to an averaged monopulse ratio also called Mosca-monopulse which is formed as $\mathbf{R}_K = \mathrm{Re}\left\{\sum_{k=1}^K \mathbf{D}_k S_k^* / \sum_{k=1}^K |S_k|^2\right\}$.

The statistical performance of this generalized monopulse estimator has been described in [64] for the case of Rayleigh fluctuating targets by calculation of the mean and covariance. This fluctuation model is also known as the Swerling-I model for slow fluctuations from scan to scan, or as the Swerling-II model for rapidly fluctuating targets from pulse to pulse. Moreover, in [65] the corresponding mean and covariance have been given for the case of targets with deterministic amplitude variation, sometimes also called the Swerling-0 case. These results have been extended in [66] to a mixture of Swerling-0, I, II models and to extended targets. To indicate the general performance the results for the Rayleigh targets (Swerling-I case), which is the simplest solution, is given here. We are looking for the statistical description of:

$$\mathrm{E}\{\mathbf{u}\} = \mathbf{u}_0 - \mathbf{C}(\mathrm{E}\{\mathbf{R}\} - \boldsymbol{\mu}) \quad \text{and} \quad \mathrm{cov}\{\mathbf{u}\} = \mathbf{C}\,\mathrm{cov}\{\mathbf{R}\}\mathbf{C}^T \tag{1.80}$$

For the expectations of the monopulse ratio $\mathbf{R}$ we consider only those measurements which have an integrated sum beam output above a certain threshold η, i.e. for which $P_s = \sum_{k=1}^{K} |S_k|^2 > \eta$. Then one obtains, [14],

$$E\{\mathbf{R}|P_s > \eta\} = \mathrm{Re}\left\{\frac{\mathbf{G}_{\mathrm{DS}}}{G_S}\right\}$$
$$\mathrm{cov}\{\mathbf{R}|P_s > \eta\} = \begin{cases} 0.5\mathbf{V} \cdot \mathrm{E}_1\left(\dfrac{\eta}{G_S}\right) e^{\eta/G_S}, & \text{for } K = 1 \\ 0.5\mathbf{V} \cdot \dfrac{1}{K-1} \dfrac{\mathrm{e}_{K-2}(\eta/G_S)}{\mathrm{e}_{K-1}(\eta/G_S)}, & \text{for } K > 1 \end{cases} \tag{1.81}$$

The matrix $\mathbf{V}$ is given by:

$$\mathbf{V} = \mathrm{Re}\{\mathbf{G}_D - \mathbf{G}_{\mathrm{DS}} G_S^{-1} \mathbf{G}_{\mathrm{DS}}^H\}/G_S \tag{1.82}$$

E_1 denotes the exponential integral $\mathrm{E}_1(x) = \int_x^\infty e^{-t}/t \, dt$ which can be well approximated for high SNR by $\log(1 + 1/x)$, and $\mathrm{e}_n\ (x)$ is defined as $e_n(x) = \sum_{k=0}^{n} x^k/k!$. The matrices $\mathbf{G}_D$, $\mathbf{G}_{\mathrm{DS}}$, G_S denote the blocks of the joint covariance matrix of the sum and difference beams. More precisely, if we assume the vector of beam outputs $(D_x, D_y, S)^T = \begin{pmatrix} \mathbf{D} \\ S \end{pmatrix}$ to be complex Gaussian distributed with mean $\mathbf{m}_{\mathrm{DS}}$ and covariance matrix $\mathbf{G}$, then we partition $\mathbf{G} = \begin{pmatrix} \mathbf{G}_D & \mathbf{G}_{\mathrm{DS}} \\ \mathbf{G}_{\mathrm{DS}}^H & G_S \end{pmatrix}$.

Clearly, the detection threshold η controls the magnitude of the variance. If we consider the classical monopulse with $K = 1$ and $\eta = 0$ we obtain the well-known result that the variance is infinite.

To indicate the performance of this approach we consider the generic array of Figure 1.7 with 902 elements and 32 sub-arrays and a scenario with a jammer slightly less than a beamwidth away from the antenna look direction. Figure 1.27 shows the resulting biases and variances for different target directions within the beamwidth. A jammer is located at the asterisk symbol direction with JNR = 27 dB. The hypothetical target has a sum beam output SNR of 22.8 dB. The 3 dB contour of the unadapted sum beam is shown by a dashed circle. The 3 dB contour of the adapted beam will be of course different. No averaging was performed (true monopulse) and the detection threshold was set to the classical level of 13 dB SNR. The biases are shown by arrows for different possible target positions with the standard deviation ellipses at the tip corresponding to (1.81). One can see that in the beam pointing direction (0,0) the bias is zero by construction and the variance is small. The errors increase for target directions on the skirt of the main beam and close to the jammer.

To get an impression of the relevance of this accuracy we show the corresponding results with the Cramér–Rao bound (CRB) for comparison. We use the CRB formulation of [67]. As we are using sum and difference beams, we have to take the CRB based on the distribution of the three input quantities (D_x, D_y, S). The result is seen in Figure 1.28. This would be the asymptotic accuracy of an ideal unbiased estimator using the three beam outputs. One can see that the uncertainty

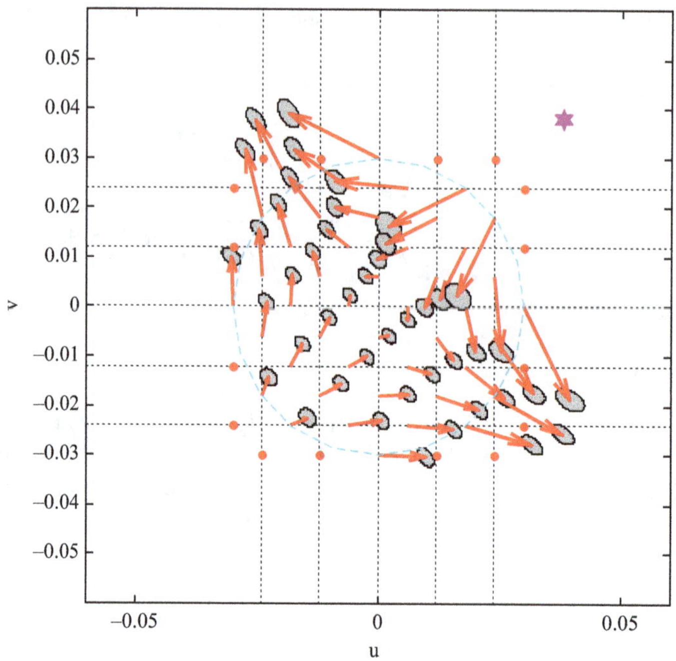

Figure 1.27 Bias and standard deviation ellipses of generalized monopulse with adapted sum and difference beams for different target positions, (from [4])

ellipses are somewhat larger than with our generalized monopulse. However, the CRB is a rather meaningless measure in this case because the asymptotic case which the CRB describes is not attained for finite SNR and finite element number. In realistic scenarios the estimates are biased and the real overall error is a combination of bias and variance. So, angle estimation with ABF is a good example that the CRB is not always a good indicator of the performance of an estimator. In fact, if we consider the CRB for biased estimates then the results will be more similar to our generalized monopulse estimator performance of Figure 1.27, see the plots in [14]. The variances are a bit larger because the CRB does not account for the detection threshold. The important point is that for this result with modified CRB we need to know the bias which we have taken from our result (1.81).

The large bias may not be satisfying. However, one may repeat the monopulse estimate with a look direction steered at sub-array level into the new estimated direction. This is an all-offline processing with the given sub-array data. No new transmit pulse is needed. We have called this the *multi-step monopulse procedure* [14]. Figure 1.29 shows that the multi-step monopulse procedure with only one additional iteration can reduce the bias considerably. The variances appearing in Figure 1.27 are virtually not changed with the multi-step monopulse procedure and they are omitted here for better visibility.

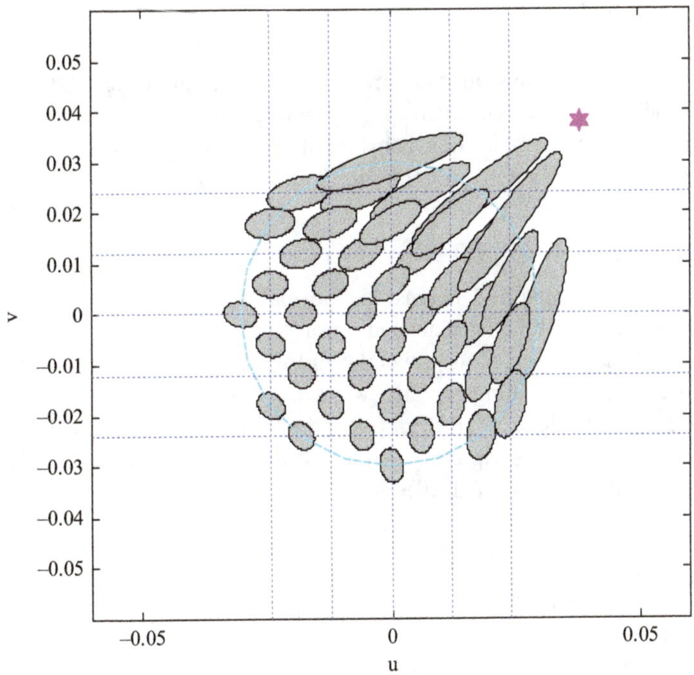

Figure 1.28 Cramér–Rao bound corresponding to scenario of Figure 1.27

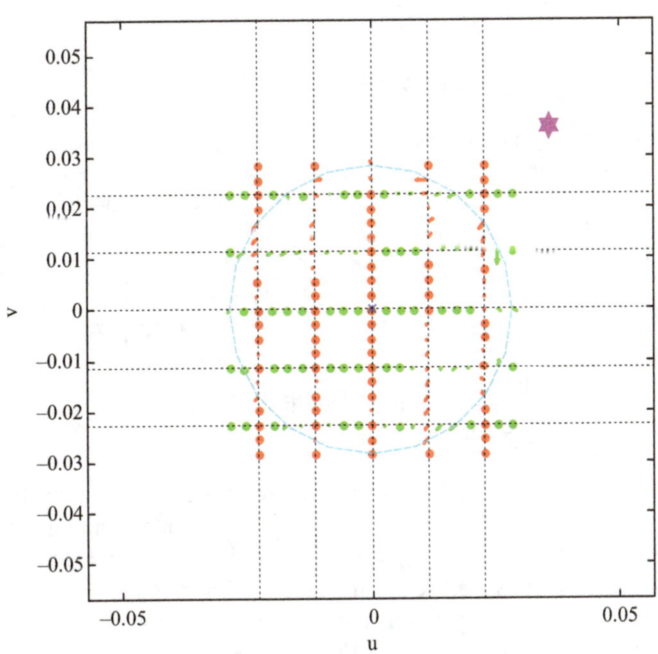

Figure 1.29 Bias for 2-step monopulse for different target positions and jammer scenario of Figure 1.27, (from [4])

An application of this adaptive monopulse technique to space-time adaptive processing (STAP) has been given in [68], where the case of ground target detection from an airborne radar was considered. Another application to tracking in jammed scenarios will be given below in Section 1.9.3.

1.9.2 Adaptive detection

For detection with adaptive beams the normal test procedure is not adequate because we have a test statistic depending on two different kinds of random data: the training data for the adaptive weight and the data under test. The pioneering first test statistics accounting for this two data set structure were the Generalized Likelihood test (GLRT) of Kelly, [69], the Adaptive Matched Filter (AMF) detector, [70], and the Adaptive Cosine Estimator (ACE) detector, [71]. In our sub-array output formulation (1.20) these tests have the form:

$$T_{\mathrm{GLRT}}(\tilde{\mathbf{z}}) = \frac{\left|\tilde{\mathbf{a}}_0^H \hat{\mathbf{Q}}_{\mathrm{SMI}}^{-1} \tilde{\mathbf{z}}\right|^2}{\tilde{\mathbf{a}}_0^H \hat{\mathbf{Q}}_{\mathrm{SMI}}^{-1} \tilde{\mathbf{a}}_0 \left(1 + \frac{1}{K} \tilde{\mathbf{z}}^H \hat{\mathbf{Q}}_{\mathrm{SMI}}^{-1} \tilde{\mathbf{z}}\right)} \tag{1.83}$$

$$T_{\mathrm{AMF}}(\tilde{\mathbf{z}}) = \frac{\left|\tilde{\mathbf{a}}_0^H \hat{\mathbf{Q}}_{\mathrm{SMI}}^{-1} \tilde{\mathbf{z}}\right|^2}{\tilde{\mathbf{a}}_0^H \hat{\mathbf{Q}}_{\mathrm{SMI}}^{-1} \tilde{\mathbf{a}}_0} \tag{1.84}$$

$$T_{\mathrm{ACE}}(\tilde{\mathbf{z}}) = \frac{\left|\tilde{\mathbf{a}}_0^H \hat{\mathbf{Q}}_{\mathrm{SMI}}^{-1} \tilde{\mathbf{z}}\right|^2}{\tilde{\mathbf{a}}_0^H \hat{\mathbf{Q}}_{\mathrm{SMI}}^{-1} \tilde{\mathbf{a}}_0 \cdot \tilde{\mathbf{z}}^H \hat{\mathbf{Q}}_{\mathrm{SMI}}^{-1} \tilde{\mathbf{z}}} \tag{1.85}$$

This formulation admits that the quantities $\tilde{\mathbf{z}}, \tilde{\mathbf{a}}_0, \hat{\mathbf{Q}}$ are all generated at the sub-array outputs. $\hat{\mathbf{Q}}_{\mathrm{SMI}}$ is the SMI estimate of the covariance matrix (1.52) applied at the subarray outputs. $\tilde{\mathbf{a}}_0$ denotes the plane wave model for a direction $\mathbf{u}_0$. Basic relations between these tests are:

(i) $$T_{\mathrm{GLRT}} = \frac{T_{\mathrm{AMF}}}{\left(1 + \frac{1}{K} \tilde{\mathbf{z}}^H \hat{\mathbf{Q}}_{\mathrm{SMI}}^{-1} \tilde{\mathbf{z}}\right)} \quad \text{and} \quad T_{\mathrm{ACE}} = \frac{T_{\mathrm{AMF}}}{\tilde{\mathbf{z}}^H \hat{\mathbf{Q}}_{\mathrm{SMI}}^{-1} \tilde{\mathbf{z}}} \tag{1.86}$$

(ii) $$T_{\mathrm{AMF}} = \frac{\left|\tilde{\mathbf{w}}^H \tilde{\mathbf{z}}\right|^2}{\tilde{\mathbf{w}}^H \hat{\mathbf{Q}}_{\mathrm{SMI}} \tilde{\mathbf{w}}} \quad \text{if we set } \tilde{\mathbf{w}} = \hat{\mathbf{Q}}_{\mathrm{SMI}}^{-1} \tilde{\mathbf{a}}_0 \tag{1.87}$$

Property (ii) shows that the AMF detector represents an estimate of the signal-to-noise ratio and this provides a meaningful physical interpretation. A complete statistical description of these tests has been given in a very compact form in [72,73]. These results are valid as well for planar arrays with irregular sub-arrays and also with a mismatched weighting vector (not pointing in the true target direction).

Actually, all these detectors use the adaptive weight of the SMI algorithm which has unsatisfactory performance as mentioned in Section 1.6.2. The unsatisfactory finite sample performance is just the motivation for introducing advanced weight estimators like LSMI, LMI or CAPS. Clutter, insufficient adaptive suppression and surprise interference are the motivation for requiring low sidelobes. Recently several more complicated adaptive detectors have been introduced with the aim of achieving additional robustness properties, [74–78]. In some cases well know adaptive techniques like diagonal loading have been re-invented by this laborious detour. In particular it has already been shown in [75] that diagonal loading provides significantly better detection performance.

In this context it would be desirable to generalize the tests of (1.83)–(1.85) to arbitrary weight vectors with the aim of inserting the already well-known and robust weights derived in Section 1.6.1. This has been done in [79]. The first observation is that the formulation of the AMF test (1.87) can be used for any weight vector $\tilde{\mathbf{w}}$. Secondly, one can observe that ACE and GLRT have the form of a sidelobe blanking device because the detector $T_{\mathrm{ACE}} > \eta$ is equivalent to a detector $T_{\mathrm{AMF}} > \eta \cdot \tilde{\mathbf{z}}^H \hat{\mathbf{Q}}_{\mathrm{SMI}}^{-1} \tilde{\mathbf{z}}$ and we have a similar relation for T_{GLRT}. These two observations allow the generalization to arbitrary weights. In the following we show that the forms $\tilde{\mathbf{z}}^H \hat{\mathbf{Q}}_{\mathrm{SMI}}^{-1} \tilde{\mathbf{z}}$ and $1 + \frac{1}{K} \tilde{\mathbf{z}}^H \hat{\mathbf{Q}}_{\mathrm{SMI}}^{-1} \tilde{\mathbf{z}}$ can be considered as generalized guard channels.

Adaptive guard channel and adaptive sidelobe blanking. A guard channel is implemented in radar systems to eliminate impulsive interference, may be hostile or from other neighbouring radars, using the sidelobe blanking (SLB) device. The guard channel receives data from a separate omni-directional antenna element which is amplified such that its power level is above the sidelobe level of the highly directional radar antenna, but below the power of the radar main beam. If the received signal power in the guard channel is above the power of the main channel, this must be a signal coming via the sidelobes. Such a signal will be blanked. If the guard channel power is below the main channel power it must result from a signal in the main beam and this is considered as a valid detection. We call the property of the guard channel pattern to be above the sidelobes and below the main beam the *SLB condition*. The principle of sidelobe blanking has been described by Farina in [7 p. 24.11] and has been studied in detail in [25]. Also, in [7 p. 24.17] a concept is described how for an adaptive antenna in a SLC configuration the SLB device can be realized.

The first observation is that with phased arrays it is not necessary to provide a separate and external omni-directional guard channel. Such a channel can be generated from the antenna itself; all the required information is in the antenna. In fact, one can use the non-coherent sum of the sub-arrays as a guard channel G. This is the same as the average omni-directional power. Some additional shaping of the guard pattern may be achieved by using a weighting for the non-coherent sum:

$$G = \sum_{i=1}^{L} \tilde{g}_i |\tilde{z}_i|^2 \tag{1.88}$$

The directivity pattern of such guard channel is given by $S_G(\mathbf{u}) = \sum_{i=1}^{L} \tilde{g}_i |\tilde{a}_i(\mathbf{u})|^2$. If all sub-arrays are equal, a uniform weighting $\tilde{\mathbf{g}} = (1, .., 1)^T$ may be suitable.

For unequal irregular sub-arrays as for the generic array (Figure 1.7) the different contributions of the sub-arrays may be compensated. Recall that the notion of sub-arrays is fairly general and includes the case of fully digital arrays, arrays with auxiliary antennas and multi-beam antennas.

More generally, we may use a combination of coherent and non-coherent sums of the sub-arrays. For example, one can also take differences of pairs of sub-arrays to generate a guard channel with a kind of difference beam pattern. Mathematically this can be written with a diagonal matrix $\mathbf{D}$ containing the weights for non-coherent integration and by a beamforming matrix $\mathbf{K}$ for the coherent integration. This results in a generalized guard channel

$$G = \tilde{\mathbf{z}}^H \mathbf{K}\mathbf{D}\mathbf{K}^H \tilde{\mathbf{z}} \text{ with the directivity pattern } S_G(\mathbf{u}) = \tilde{\mathbf{a}}^H(\mathbf{u})\mathbf{K}\mathbf{D}\mathbf{K}^H \tilde{\mathbf{a}}(\mathbf{u}) \quad (1.89)$$

These two kinds of weightings allow a precise matching of the guard channel to fulfil the SLB condition. Examples of such generalized guard channels are shown in Figure 1.30 for the generic array with −35 dB Taylor element tapering. Obviously the SLB condition is always fulfilled. The nice feature of these guard channels is that

(i) they automatically scan together with the antenna look direction, and
(ii) they can be easily made adaptive.

If we use adaptive beams, we must also use an adaptive guard channel. A CW jammer would make the SLB blank all range cells, i.e. would just switch off the radar. The simultaneous presence of continuous interference with impulsive interference requires an adaptive SLB device. To generate an adaptive guard channel one can replace in (1.89) the data vector of the cell under test (CUT) by the pre-whitened data vector $\tilde{\mathbf{z}}_{\text{pre}-w} = \hat{\mathbf{Q}}^{-1/2}\tilde{\mathbf{z}}$. Then the test statistic can be written as

$$T = \frac{T_{\text{AMF}}(\tilde{\mathbf{z}})}{G_{\text{adapt}}(\tilde{\mathbf{z}})} \text{ with } G_{\text{adapt}}(\tilde{\mathbf{z}}) = \tilde{\mathbf{z}}^H \hat{\mathbf{Q}}^{-1/2}\mathbf{K}\mathbf{D}\mathbf{K}^H \hat{\mathbf{Q}}^{-1/2}\tilde{\mathbf{z}} \text{ for ACE}$$
$$\text{and } G_{\text{adapt}}(\tilde{\mathbf{z}}) = 1 + \frac{1}{K}\tilde{\mathbf{z}}^H \hat{\mathbf{Q}}^{-1/2}\mathbf{K}\mathbf{D}\mathbf{K}^H \hat{\mathbf{Q}}^{-1/2}\tilde{\mathbf{z}} \text{ for GLRT} \quad (1.90)$$

Hence T_{ACE} can be modified to an AMF detector with a generalized adaptive guard channel and similarly T_{GLRT} with a generalized adaptive guard channel on a pedestal. Figure 1.31 shows examples of some generalized adapted guard channels generated with the generic array of Figure 1.7 with −35 dB Taylor weighting and a jammer at the first sidelobes of the main beam. The unadapted patterns are shown by dashed lines.

Typically the SLB device is used as a second detection step only in range or angle cells where a detection has occurred, i.e. where $T_{\text{AMF}} > \eta$. For the adaptive case this two-step detection procedure with the AMF, ACE and GLR tests has been introduced in [72], called there the 2D adaptive sidelobe blanking (ASB) detector. The generalization of the 2D ASB to arbitrary weights has been introduced in [79]. The generalized AMF is testing the presence of a potential target and

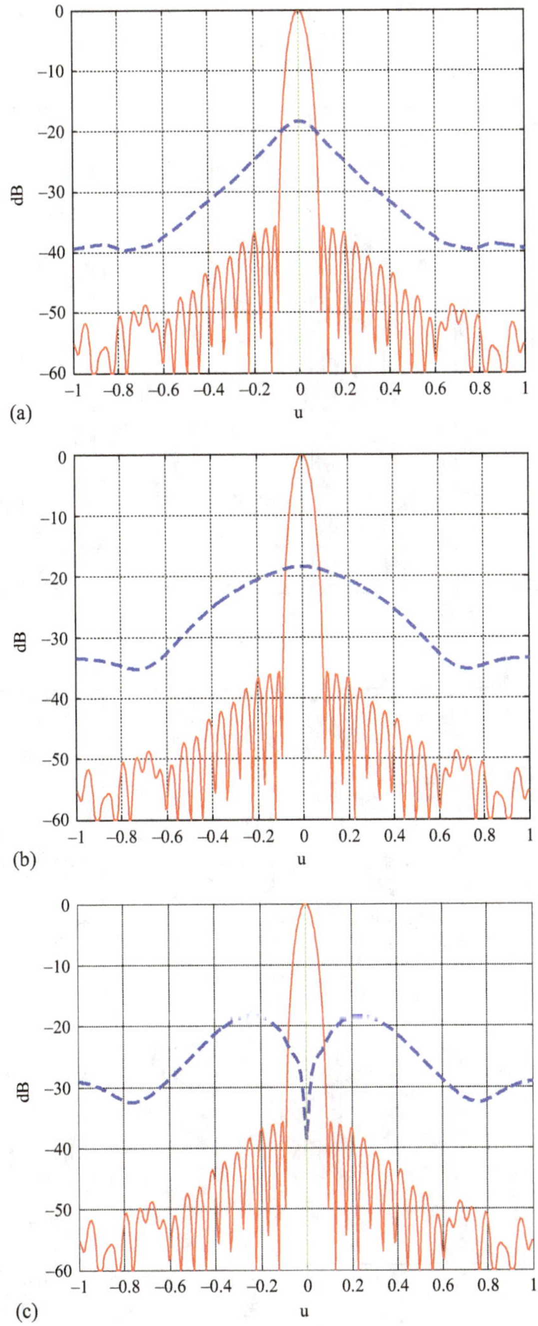

Figure 1.30 Generalized guard channels patterns and sum beam patterns for generic array (from [4]). (a) Uniform sub-array weighting, (b) non-coherent weighting for equal sub-array power and (c) power equalized plus sub-array difference beam weighting

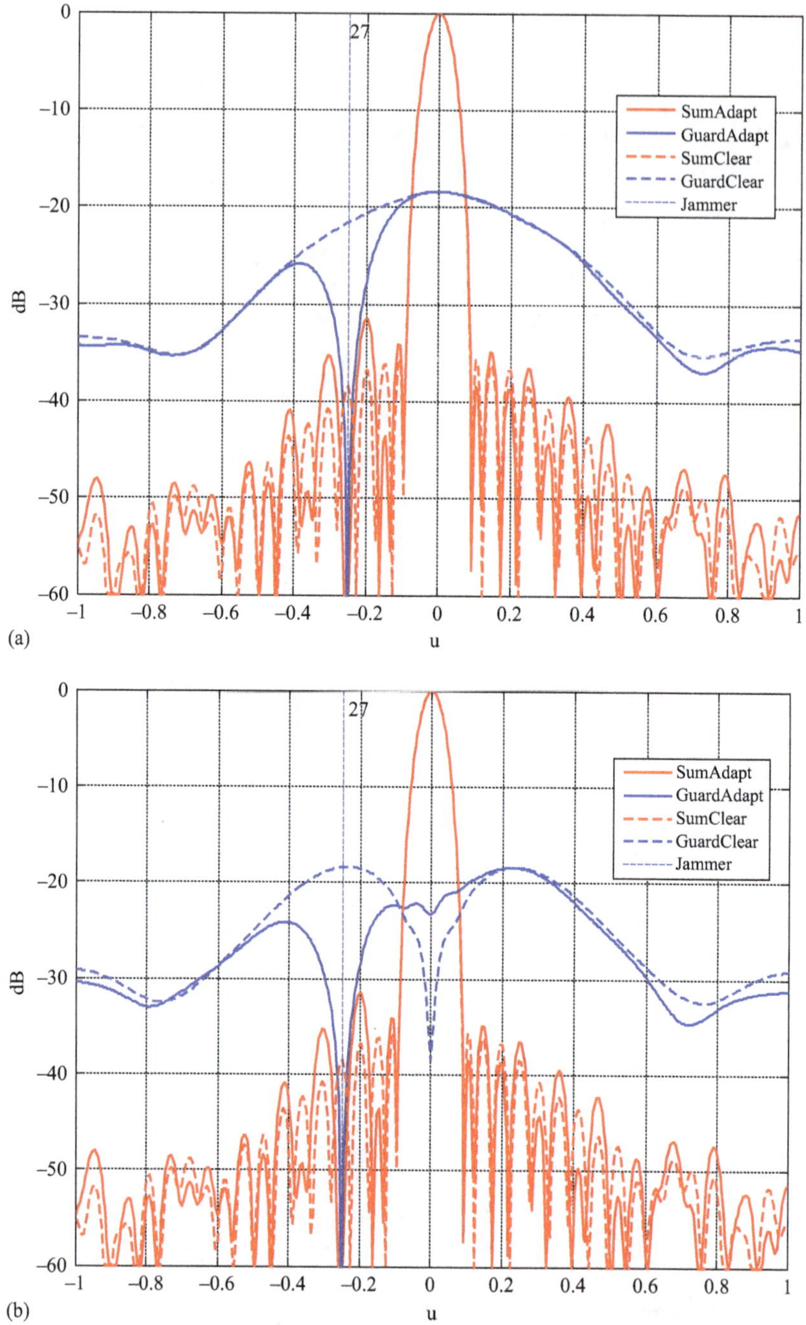

Figure 1.31 Adapted guard patterns for a jammer with JNR of 34 dB at u = −0.27 (−15.7°). The unadapted patterns are shown by dashed lines. (a) Weighting for equal sub-array power and (b) difference type guard with weighting for equal sub-array power

the generalized ACE or GRL tests are confirming this target or blank it. By using two different thresholds for the two tests one can control the properties of the overall performance. An analytical statistical analysis of the features of the 2D ASB test with the original AMF, ACE and GLR test of (1.83)–(1.85) has been given in [72,73].

It turns out that for adapted beams and adapted guard channel it is a challenge to fulfil the SLB condition and to define the suitable thresholds for the modified tests. For an arbitrary weight vector it is nearly impossible to determine this analytically. In [79] the detection margin has been introduced as an empirical tool for judging a good balance between the AMF and ASB thresholds for given jammer scenarios. The detection margin is defined as the difference between the expectation of the AMF statistic and the guard channel, where the expectation is taken only over the interference entering the adaptation for a known interference scenario. This conditional expectation can be calculated in case studies for different scenarios. More precisely, let $\tilde{\mathbf{n}}$ denote the interference that is present in the training data with a covariance matrix $\tilde{\mathbf{Q}}$. Let $\tilde{\mathbf{s}}$ be a test signal, e.g. a surprise jammer or a target to be detected. Then we calculate:

$$\bar{T}_{\text{AMF}} = \frac{\mathrm{E}_{\tilde{\mathbf{n}}}\{\tilde{\mathbf{w}}^H(\tilde{\mathbf{s}}\tilde{\mathbf{s}}^H + \tilde{\mathbf{n}}\tilde{\mathbf{n}}^H)\tilde{\mathbf{w}}\}}{\tilde{\mathbf{w}}^H\hat{\mathbf{Q}}\tilde{\mathbf{w}}} = \frac{\tilde{\mathbf{w}}^H\tilde{\mathbf{s}}\tilde{\mathbf{s}}^H\tilde{\mathbf{w}}}{\tilde{\mathbf{w}}^H\hat{\mathbf{Q}}\tilde{\mathbf{w}}} + \frac{\tilde{\mathbf{w}}^H\tilde{\mathbf{Q}}\tilde{\mathbf{w}}}{\tilde{\mathbf{w}}^H\hat{\mathbf{Q}}\tilde{\mathbf{w}}} \approx \frac{\tilde{\mathbf{w}}^H\tilde{\mathbf{s}}\tilde{\mathbf{s}}^H\tilde{\mathbf{w}}}{\tilde{\mathbf{w}}^H\hat{\mathbf{Q}}\tilde{\mathbf{w}}} + 1 \tag{1.91}$$

Using the approximation $\tilde{\mathbf{Q}} \approx \hat{\mathbf{Q}}$. Similarly, we take this expectation for the guard channel which results as shown in detail in [79]:

$$\begin{aligned} \bar{G}_{\text{ACE}} &= \mathrm{E}_{\tilde{\mathbf{n}}}\left\{\mathrm{tr}\left\{\mathbf{KDK}^H\hat{\mathbf{Q}}^{-1/2}(\tilde{\mathbf{s}}\tilde{\mathbf{s}}^H + \tilde{\mathbf{n}}\tilde{\mathbf{n}}^H)\hat{\mathbf{Q}}^{-1/2}\right\}\right\} \\ &\approx \tilde{\mathbf{s}}^H\hat{\mathbf{Q}}^{-1/2}\mathbf{KDK}^H\hat{\mathbf{Q}}^{-1/2}\tilde{\mathbf{s}} + \mathrm{tr}\{\mathbf{KDK}^H\} \\ \bar{G}_{\text{GLR}} &\approx 1 + \frac{1}{K}\tilde{\mathbf{s}}^H\hat{\mathbf{Q}}^{-1/2}\mathbf{KDK}^H\hat{\mathbf{Q}}^{-1/2}\tilde{\mathbf{s}} + \frac{1}{K}\mathrm{tr}\{\mathbf{KDK}^H\} \end{aligned} \tag{1.92}$$

The detection margin is now $\bar{T}_{\text{AMF}}/\eta_{\text{ACE}} - \bar{G}_{\text{ACE}}$ or $\bar{T}_{\text{AMF}}/\eta_{\text{GLR}} - G_{\text{GLR}}$ and this indicates the fulfilment of the SLB condition on the average. If $\tilde{\mathbf{s}}$ is in the main beam it should be detected, if it is in the sidelobe region it should be blanked. In addition, we can calculate the variance of $\bar{T}_{\text{AMF}}$, $\bar{G}_{\text{ACE}}$ and $\bar{G}_{\text{GLR}}$. Inspecting the mean levels together with standard deviations allows characterizing the probability of a random threshold excess if the estimated weights are applied and this characterizes the SLB condition.

The critical feature that has to be checked is the performance against jammers close to the main beam. An example of the detection margin is shown in Figure 1.32 (same antenna and weighting as in Figure 1.30) for the AMF and ACE guard channel. The scenario for which the antenna (the generic array) and the guard channel are adapted consists of a CW jammer at azimuth $u = -0.4$ with JNR at the elements of 40 dB and an intermittent interference with 50% duty cycle at azimuth $u = -0.8$ with JNR of 35 dB. We have used the LSMI algorithm for adaptation with diagonal load of $3\sigma^2$. All patterns are normalized to the ACE threshold η_{ACE}.

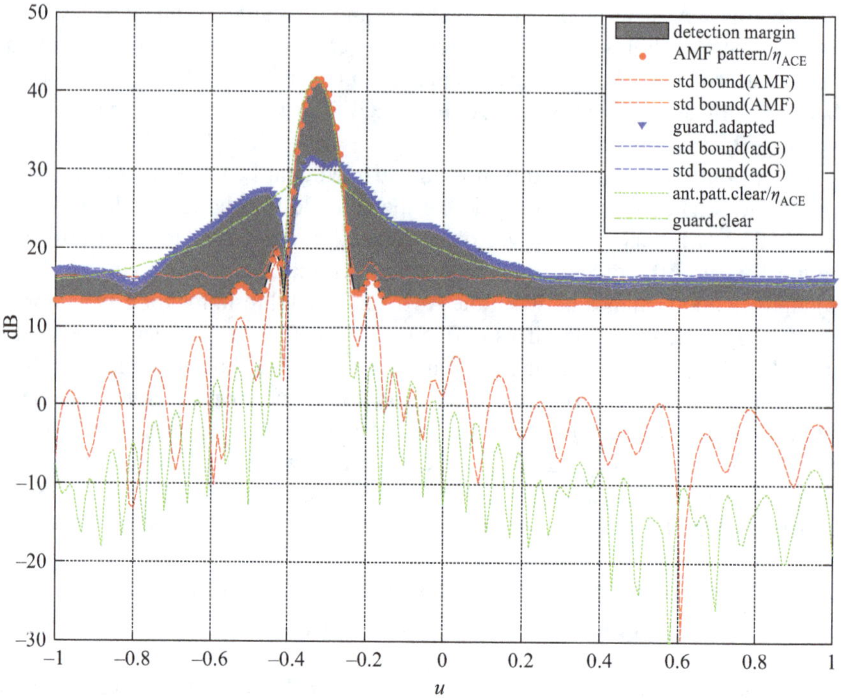

Figure 1.32 Detection margin (grey shading) between AMF with LSMI (64 snapshots and $3\sigma^2$ diagonal loading) and non-weighted adapted guard channel of ACE-type with confidence bounds

The parameter of the x-axis is the direction of the test signal. For reference the unadapted patterns for sum beam and guard channel are also plotted by dotted lines. One can see that the mean of T_{AMF} (marked with dots only) follows closely the unadapted beam shape while the sidelobe region is filled up with the residual noise after adaptation. The mean of the adapted guard channel (marked with triangles only) shows the typical notch for the CW jammer and deviates slightly from the unadapted pattern. The standard deviation (std) of the guard channel is so small that the $\pm$std confidence bounds are hardly to be seen. The main fluctuation occurs for T_{AMF} i.e. the adapted antenna pattern. The grey shaded region shows the difference between guard and antenna pattern that enable sidelobe blanking; this is the detection margin. For comparison the performance of the corresponding quantities with the GLR guard channel is plotted in Figure 1.33. The shape of the patterns is very similar, but the detection threshold in the sidelobe region is larger. The GLR guard channel has much less fluctuations and is less distorted by the adaptation process. Thus the GLR ASB may be preferred. More details can be found in [79].

This adaptive detection procedure with the aid of the detection margin has been studied in an application in [80], where the detection of ground targets with an airborne radar with STAP was considered.

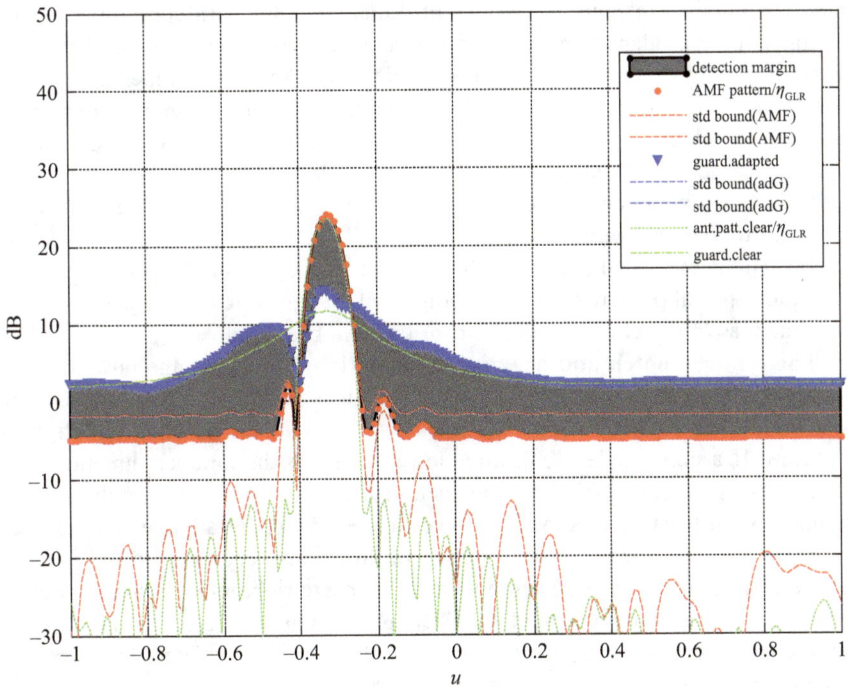

Figure 1.33 *Detection margin (grey shading) between AMF with LSMI (64 snapshots and $3\sigma^2$ diagonal loading) and non-weighted adapted guard pattern of GLR-Type with confidence bounds*

1.9.3 Adaptive tracking

A key feature of detection and angle estimation with ABF is that the overall performance is dramatically influenced by the proximity of the main beam to an interference source. For sidelobe interference, detection and angle estimation is often not seriously affected and target tracking will suffer little degradation. On the other hand, target tracking in the proximity of a jammer is of high operational relevance. The jammer direction is normally not known from the adaptation process. However, if the direction of a jammer is known, this can be valuable information which can be exploited in various areas. The information on the jammer direction can be made available by a jammer mapping mode, which determines the direction of the interferences by a background procedure using already available data such that no additional radar dwell time is required. We named tracking techniques which exploit all these information about adaptive arrays as 'adaptive tracking'.

Jammer mapping. Jammers are typically strong emitters and thus easy to detect even in the presence of targets (radar echoes). In particular, the Spotlight MUSIC method (see Section 1.7.3 and [58]) working with sub-array outputs is suited for jammer mapping with a multi-function radar with sub-arrays. If the training data for adaptation are sampled before pulse compression then they represent mainly the jammers because at this stage the target echoes are merely below noise. One can estimate for each antenna look direction a jammer covariance matrix from which a Spotlight MUSIC pattern can be generated and jammer directions can be extracted. This pattern is only meaningful within the beamwidth around this look direction. However, in the search mode of a multifunction radar all directions of interest will be scanned. Then all the resulting Spotlight MUSIC patterns can be evaluated and the results can be fused. For this fusion process one can exploit special properties of the Spotlight MUSIC pattern, e.g. some knowledge about the bias of directions on the skirt of the sum beam.

The jammer notch and negative information. Let us assume now for simplicity that the jammer directions are known. This is highly important information for a multi-function radar where the tracker determines the pointing direction of the beam. If a commanded look direction is close to the jammer direction it is better to use a squinted look direction such that the jammer is more on the skirt of the main beam because this induces lower losses and better angle estimates with ABF. One cannot expect target echoes from directions very close to the jammer and therefore it does not make sense to steer the beam right into the jammer notch. The width of the null generated by ABF processing is related to the antenna aperture (for direct sub-array weighting DSW) and can be calculated. Furthermore, in the case of a missing measurement when tracking a target through the jammer notch, the lack of a successful detection supports the conclusion that this is a direct result of jammer nulling by ABF. The knowledge that one cannot expect a detection in the jammer null is called negative information. This concept is explained in more detail in [81]. However, if we cannot expect a detection the tracking algorithm must be supported with measurements. In this situation one can use the direction of the jammer as a pseudo-measurement to update and maintain the track file because we know that the target must be hidden in the jammer notch. The width of the jammer notch defines the uncertainty of this pseudo measurement. Of course, the track uncertainty increases also with time if only pseudo measurements are delivered, but after some time the target may leave the jammer notch and the track identity may be preserved. This concept will also be explained in Part II of Vol. 2.

The jammer direction can provide even more information for the tracker. Let us assume that the adaptive monopulse procedure of Section 1.9.1 is used for estimating the target angle. The adaptive monopulse estimator exhibits quite specific error covariances as seen in Figure 1.27. One can use the theoretically calculated variances for the adaptive monopulse estimate as prior information in the tracking filter. In particular the shape and orientation of the eccentric uncertainty ellipses are highly relevant for the tracker. The large bias appearing in Figure 1.27 is not known by the tracker, but this can be reduced by applying the multi-step monopulse procedure.

These techniques have been implemented in a tracking algorithm and have been refined by a number of other stabilization measures as reported in [81,82].

From these studies the following set of special measures for ABF tracking has been derived and proven to be particularly useful:

1. **Look direction stabilization**: The monopulse estimate may deliver measurements outside of the 3 dB contour of the sum beam. We know that monopulse estimates for targets on the border of the 3 dB beamwidth are heavily biased, especially for look directions close to the jammer, despite the use of the multi-step monopulse procedure. Estimates outside the sum beam contour are therefore corrected by projecting them onto the boundary circle of the beam.
2. **Detection threshold**: In the update step of the tracking algorithm only those measurements are considered whose sum beam power is above a certain detection threshold which is given by a reasonable monopulse estimate as in (1.81). It is well known that the variance of the monopulse estimate decreases monotonically with this threshold increasing. Typically an SNR level of 13 dB gives useful estimates. This avoids random outliers.
3. **Adjustment of the antenna look direction**: Generally any beam steering into the jammer notch should be avoided due to the expected lack of good measurements. In case that the proposed look direction lies in the jammer notch, we select an adjusted direction on the skirt of the jammer notch.
4. **Variable measurement covariance for the tracking algorithm**: The tracking algorithm uses a model for the expected accuracy of the parameter estimates. Normally this is a fixed estimate of the corresponding standard deviation. However, for adaptive beams the accuracies can vary considerably depending on the proximity of the target to the jammer. A variable covariance matrix of the adaptive monopulse estimation according to (1.81) is considered, but only for a mainlobe jammer situation. For jammers in the sidelobes there is little effect on the angle estimates and one can use the fixed covariance matrix of the non-jammed case.
5. **QuadSearch and pseudo-measurements**: If the predicted target direction lies inside the jammer notch and if, despite all adjustments of the antenna look direction, the target is not detected, a specific search pattern is initiated named QuadSearch. It uses a set of look directions on the skirt of the jammer notch to obtain detections and acceptable monopulse estimates. If this procedure does not lead to detection, we know that the target is hidden in the jammer notch and we cannot see it. We use then the direction of the jammer as a pseudo-bearing measurement to maintain the track file. The pseudo-measurement noise is determined by the width of the jammer notch.
6. **LocSearch**: In case of a permanent lack of detections (e.g. for three consecutive scans) while the track position lies outside the jammer notch, a specific search pattern is initiated (named LocSearch) that is similar to the QuadSearch. The new look directions lie on the circle of a certain radius around the predicted target direction. This measure aims at enlarging the expected range of the target position.
7. **Modelling of target dynamics**: The selection of a suitable dynamics model plays a major role for the quality of tracking results. In this context, the

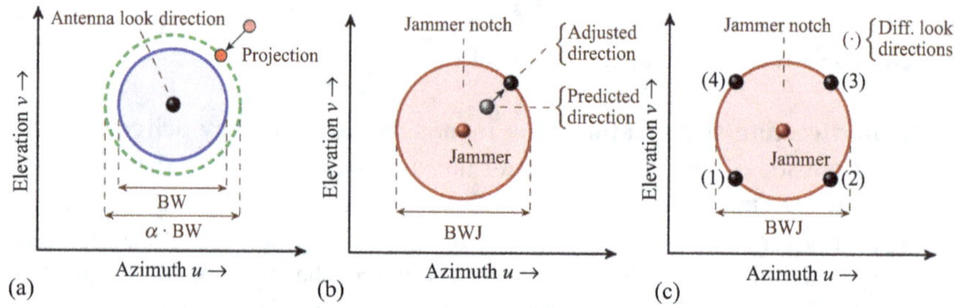

Figure 1.34 Illustration of different stabilization measures to improve track stability and track continuity (from [4] courtesy of M. Feldmann) (a) Look direction stabilization, (b) adjustment in jammer notch and (c) QuadSearch

so-called interacting multiple model (IMM) is a well-known method to reliably track even those objects whose dynamic behaviour remains constant only during certain periods.

8. **Gating**: In the vicinity of the jammer, the predicted target direction (as an approximation of the true direction) is used to compute the variable angle measurement covariance. Strictly speaking, this is only valid exactly in the particular look direction. Moreover, the tracking algorithm regards all incoming sensor data as unbiased measurements. To avoid track instabilities, an acceptance region is defined for each measurement depending on the predicted target state and the assumed measurement accuracy. Sensor reports lying outside this gate are considered as invalid.

The rules 'Look direction stabilization' (nr. 1), 'Adjustment of look direction in jammer notch' (nr. 3) and 'QuadSearch' (nr. 5) are illustrated by the graphic of Figure 1.34.

The effect of these stabilization measures on the tracking performance was evaluated in different scenarios. One example of an air-to-air target tracking scenario from [82] is shown in Figure 1.35. In this scenario, the sensor is a forward looking airborne radar flying at constant speed of 265 m/s using the generic array of Figure 1.7 (sum beamwidth BW= 3.4°, field of view 120°, scan interval 1 s). The sensor platform approaches a target heading towards the sensor at velocity 300 m/s, which thereupon veers away after a short time. A stand-off jammer (SOJ) is present at larger distance flying a race track at 235 m/s. All these platforms are flying at the same constant altitude. The signal-to-noise ratio was set to 26 dB and the jammer-to-noise ratio to 27 dB at a reference range of 70 km. The whole simulation lasted for 240 s. During this simulation, the target is hidden twice in the jammer notch of the SOJ – first for 3 s and then again for 4 s.

Figure 1.36 shows a period of the azimuth measurements and tracking estimates over a time window where the target first passes through the jammer notch. The true target azimuth is shown by a dashed line. The time period where the target

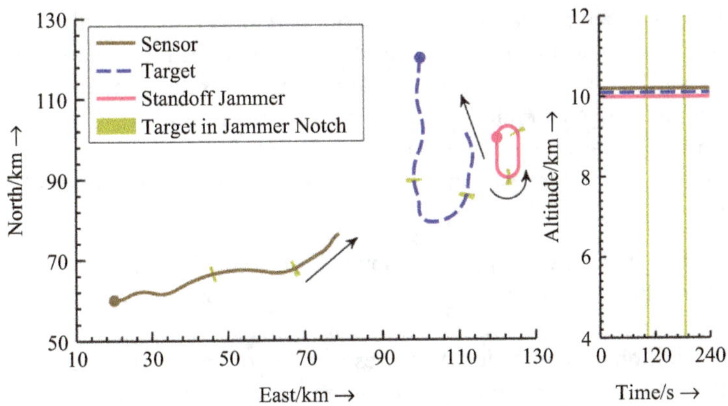

Figure 1.35 Target tracking scenario with standoff jammer: Geographic plot of platform trajectories (from [4] courtesy of M. Feldmann)

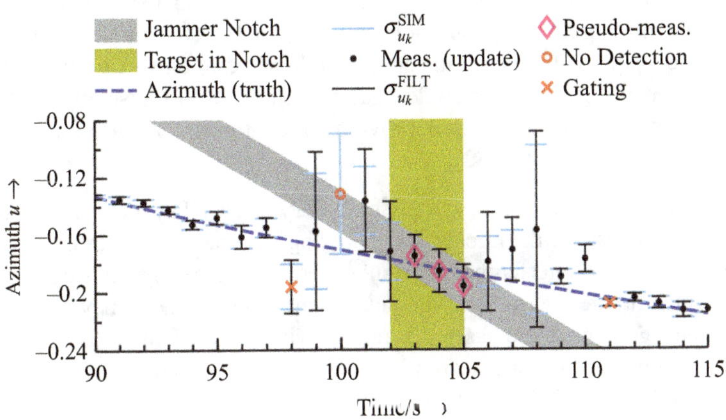

Figure 1.36 Exemplary azimuth measurements and models for a specific time window for tracking scenario of Figure 1.35 (from [4] courtesy of M. Feldmann)

enters the jammer notch is shown by grey shadings. For the tracking performance it is now critical that the estimated error covariance matrices which are updated by the tracker represent in an acceptable way the true covariance matrices which we create in the simulation. These values are indicated by the different error bars representing the std values: $\sigma_{u_k}^{\text{SIM}}$ denotes the true azimuth standard deviation (std) from the simulation; $\sigma_{u_k}^{\text{FILT}}$ corresponds to the std which is used in the tracking algorithm. According to rule nr. 4 the tracking program computes the adaptive angle measurement covariance only in the vicinity of the jammer with a diameter of this zone of 8.5°. Outside of this region, the tracking algorithm uses a constant std of 0.004 for both components of the angle measurement (values of direction cosine *u,v*). One can see that the approximated std often over-estimates the true std, but represents well its magnitude. The constant std for the other parameters are 75 m and 7.5 m/s for range and range-rate measurements. From Figure 1.36 the benefits of using pseudo-bearing measurements become apparent.

In summary, these investigations showed some interesting features of adaptive tracking:

1. If we use ABF and adaptive monopulse only, track loss will almost surely occur in the vicinity of the jammer. Of course, for jammers in the sidelobe region tracks will be extracted and maintained.
2. With additional stabilization measures that do not require the knowledge of the jammer direction (projection of the monopulse estimate, detection threshold, LocSearch, gating) still track instabilities occurred culminating finally in track loss.
3. An advanced tracking version which uses in addition to (1) and (2) pseudo-measurements mitigates this problem to some degree.
4. Finally, the additional consideration of (1), (2) and (3) together with the variable measurement covariance providing a better estimate of the highly variable shape of the angle uncertainty ellipse resulted in significantly fewer measurements that were excluded due to gating. In this case all the stabilization measures could not only improve track continuity, but also track accuracy and thus track stability, [82].

The lesson learned is that it is absolutely necessary to use all available information of the adaptive process for the tracker for achieving the goal of detection and tracking in the vicinity of the interference. The modified beam steering rules developed for adaptive tracking are measures of the radar management. There may be other rules of radar management in the case of jamming with known jammer directions. This should be a topic of further research in the context of cognitive radar. Chapter 1.3 and Part I and II of Vol. 2, review this topic again.

1.10 Conclusions and final remarks

This chapter should provide a survey on the general methods of array signal processing and on the links to antenna design, hardware constraints and target detection, parameter estimation and tracking. Specifically we have discussed the following features:

- Array signal processing is based on the availability of spatial samples of the incoming wavefront. Such samples can be conveniently obtained from arrays with digital sub-arrays. We have described how all the array processing methods can be extended to sub-array outputs, which problems arise for sub-array design, what the accuracy requirements are and how different sub-array features can be obtained by post-processing. In particular the concept of the super-array associated with every sub-array configuration provides a tool to analyse the feature of the sub-array configuration. Sub-arrays are most important for large arrays, not only because of the requirement of cost, weight and processing time, but also because sub-array solutions for large arrays can be more flexible and efficient.

- Interference suppression by deterministic and adaptive pattern shaping are key techniques of array processing. Both approaches can be reasonably combined, where ABF should be applied against interference that can be learned in a training phase, while deterministic pattern shaping should counter all remaining surprise interference. Applying ABF after deterministic sidelobe reduction allows reducing the requirements on the low sidelobe level. So, a compromise between both techniques must be achieved. Special and effective techniques are available to make ABF preserve the low sidelobe level (e.g. CAPS method).
- General principles and relationships between ABF algorithms and super-resolution methods have been discussed, like dependency on the sample number, dependency on signal fluctuations, robustness, the benefits of sub-space methods, problems of determining the signal/interference sub-space, resolution limit.
- Array signal processing methods like ABF and super-resolution methods can be applied to sub-arrays generated from a large fully filled array. This means applying these methods to the sparse super-array formed by the sub-array centres. We have pointed out problems and solutions for this special array problem. Once a sub-array forming matrix **T** has been defined, the properties of the super-array and the sub-arrays can be analysed and from this performance features can be derived.
- A special technique of surprise interference suppression is side lobe blanking aiming at blanking impulsive interference. We have shown a general approach to combine side lobe blanking with adaptive interference suppression for the case of joint impulsive and continuous interference, where each of these techniques alone fails.
- ABF can be combined with super-resolution in a canonical way by applying the pre-whiten and match principle to the data and the signal model vector.
- All array signal processing methods can be extended to space-time processing (arrays) by defining a corresponding space-time plane wave model and a space-time sub-array transformation matrix. Again, this space-time sub-array transformation matrix defines the super-array pattern and the sub-array patterns and from these some features can be derived.
- Super-resolution is a joint detection-estimation problem. Many methods have been proposed for both aspects of the problem, but one must also consider the joint problem. One has to determine a multi-target model which contains the number, directions and powers of the targets and these parameters are strongly coupled. A practical iterative joint estimation and detection procedure has been presented. For a robust solution of the joint detection-estimation problem a sequence of criteria has been proposed which step-by-step refines the estimates.

- The problems of implementation in real system have been discussed, in particular the effects of limited knowledge of the array manifold, effect of channel errors, eigenvalue leakage, unequal noise power in array channels and dynamic range of AD-converters.
- The problem of embedding signal processing into the whole radar system has been exemplarily studied for ABF. For achieving best performance an adaptation of the processing subsequent to ABF is necessary. We have shown how improved performance can be achieved by modifying some subsequent processing blocks: direction estimation by ABF-monopulse, the detector by adaptive detection with ASLB, and the tracking algorithms by several adaptive tracking techniques and track management with jammer mapping.

With a single advanced array signal processing method alone a significant improvement of a radar system will often not be obtained. All functionalities have to be mutually tuned and balanced. This is an important task for future research. Of course, this tuning of functionalities depends strongly on the application of interest. Therefore general rules cannot be given and the applications presented in the following chapters will require additional special considerations. The approaches presented here constitute only first ad hoc steps providing examples of effects that have to be considered. Note that in most cases tuning the functionalities is often a software problem. So, there is the possibility to upgrade existing systems softly and step-wise. In many cases estimates of the limitations of the signal processing methods (of the antenna beamwidth, width of the jammer notch, admissible region for monopulse, convergence of super-resolution methods, dependency of bias values etc.) is already important information to be used for radar data processing and radar management.

In the future, radar systems will increasingly collect information about the environment in a background procedure, similar to the jammer mapping mode presented here. Using a clutter map together with a geographic map for supporting waveform selection, the detector and the tracking procedure is already an established technique. Multi-function radar may also operate in an electronic intelligence mode to collect information about its emitter environment. Information from other 'sensors' (e.g. weather data, GPS data, TV/IR data) together with data from big data bases may be fed into the radar system to improve signal or data processing and thus the overall performance. A meaningful model of the additional sensor data is the basis of successful processing of this information in the future.

This chapter is far from being exhaustive and a number of radar issues have not been considered, like clutter, passive radar, multistatic radar, or polarization. Still it is hoped that a set of useful tools has been provided.

Acknowledgements

This chapter summarizes some decades of experience and research in array radar collected in two research departments at the Fraunhofer institutes in Wachtberg, the department of Array Radar Imaging (ARB, formerly called 'Electronics Dept') of

Fraunhofer FHR (Fraunhofer Institute of High Frequency Physics and Radar) and the department of Sensor Data and Information Fusion (SDF) of Fraunhofer FKIE (Fraunhofer Institute of Communication, Information Processing and Ergonomics). In FHR all viewpoints of radar signal processing with the associated hardware problems are studied while FKIE represents the associated tracking problems and system aspects. Together both perspectives offer a quite general view on radar technology and should be considered jointly in the future.

Without the continuing and stimulating spirit of the research groups in Wachtberg, the knowledge for writing this chapter would not be available. I am indebted to my colleagues in all these years, first of all to my former department head Wulf-Dieter Wirth, to the director of FHR Joachim Ender and to my current department head Wolfgang Koch. My special thanks also go to the excellent colleague scientists with whom I had the pleasure to work for many years: Richard Klemm, Christoph Gierull, Wolfram Bürger and Helmut Wilden. As a consequence, this chapter includes work that has been published in many preceding papers as may be seen in the references. The most noteworthy works on radar research from Fraunhofer institutes in Wachtberg are concentrated in the excellent radar book of W.D. Wirth [1] and the books of R. Klemm on space-time adaptive processing [2,3]. The review paper [4] which appeared in 2013 is a kind of precursor of this chapter, and it gave the motivation to write this extended view. For the preparation of this chapter, I had to rely on the generous support of my team colleagues. Of these, I want to thank especially Reda Zemmari and Michael Feldmann, as well as Martina Brötje and Alexander Charlish.

List of symbols and functions

$a/\mathbf{a}/\mathbf{A}$	Scalar/vector/matrix variable
$\lvert a \rvert$	Magnitude of a (complex) variable
$\lVert \mathbf{a} \rVert$	Norm of a vector
$\mathbf{a}^T/\mathbf{a}^*/\mathbf{a}^H$	Transpose/complex-conjugate/complex-conjugate transpose (Hermitean operation) of a vector
$\mathrm{diag}\{\mathbf{a}\}$	Diagonal matrix composed of the components of vector **a**
$\underset{k=1}{\overset{M}{\mathrm{diag}}}\{a_k\}$	$M \times M$ diagonal matrix with diagonal elements a_k
$\mathrm{vec}\{\mathbf{A}\}$	Vector generated from matrix **A** by stacking all columns on top
$(a_i)_{i=1\ldots N}$	Column vector with components a_i
$(A_{ik})_{i,k=1\ldots N}$	Matrix with components A_{ik}
$\otimes$	Kronecker matrix product
$\odot$	Schur–Hadamard (element wise) matrix product
f, λ	Frequency of a signal and corresponding wavelength
B	Bandwidth
c	Velocity of light

N	Number of array elements
K	Number of time samples
L	Number of sub-arrays
$\mathbf{u} = (u, v, w)$	Direction vector of length one, for a planar array $\mathbf{u} = (u, v)$ is also called direction cosines (for azimuth and elevation)
$\mathbf{r} = (x, y, z)$	Co-ordinates of the array element positions
$\mathbf{a}(u, v)$	Plane wave array model vector for a plane wave from direction $\mathbf{u} = (u, v)$
$\mathbf{s}$	Signal vector
$\mathbf{n}$	Noise vector
$\mathbf{z}$	Array output data vector (data snapshot)
$\mathbf{1}, \mathbf{0}$	Vector containing only ones or zeros
$\mathbf{T}$	The sub-array forming matrix
$\hat{\mathbf{z}}$	Estimated quantity
$\tilde{\mathbf{z}}$	Vector at sub-array output level
$\mathrm{E}\{\ldots\}$	Expectation operation
$\mathrm{cov}\{\ldots\}$	Covariance operation
$\mathrm{LP}\{\ldots\}$	Low pass filtering operation
$\mathrm{Re}\{\ldots\}$	Real part of a complex quantity
sinc	The function $\mathrm{sinc}\, x = (\sin x)/x$

List of acronyms

ABF	adaptive beamforming
ACE	adaptive cosine estimator
ADC	analogue-to-digital converter
AIC	Akaike information criterion
AMF	adaptive matched filter
AP	alternating projections
AR	auto-regressive
ASB	adaptive sidelobe blanking
BW	3 dB antenna beamwidth
CAPS	constrained adaptive pattern synthesis
CPI	coherent processing interval
CRB	Cramér–Rao bound
DimSS	dimension of jammer or interference sub-space
Dof	degrees of freedom
DSW	direct sub-array weighting
ELA	equivalent linear array

ELRA	electronically steerable radar (name of the experimental phased array radar of Fraunhofer FHR)
FHR	Fraunhofer Institute for High Frequency Physics and Radar Technology
FKIE	Fraunhofer Institute for Communications, Information Processing and Ergonomics
EM	expectation maximization method
EVP	eigenvector projection
GLRT	generalized likelihood ratio test
GSLC	generalized sidelobe canceller
IMP	incremental multi-parameter
LMI	lean matrix inversion
LSMI	loaded sample matrix inversion
MDL	minimum description length
ML	maximum likelihood
MUSIC	multiple signal classification
PRF	pulse repetition frequency
SAGE	space alternating generalized EM algorithm
SLB	sidelobe blanking
SLC	sidelobe canceller
Std	standard deviation
SNIR	signal-to-noise plus interference ratio
STAP	space-time adaptive processing
ULA	uniform linear array
WNT	white noise test

References

[1] W.-D. Wirth *Radar Techniques Using Array Antennas.* 2nd edition, The Institution of Engineering and Technology (IET), London, 2013, ISBN 9781849196987.

[2] R. Klemm (ed.) *Applications of Space-Time Adaptive Beamforming.* The Institution of Electrical Engineers (IEE), London, 2004, ISBN 0852969244.

[3] R. Klemm *Principles of Space-Time Adaptive Processing.* 3rd edition, The Institution of Engineering and Technology (IET), London, 2006, ISBN 0863415660.

[4] U. Nickel 'Array processing for radar – achievements and challenges'. *Int. J. Antennas Propag.* 2013, http://dx.doi.org/10.1155/2013/261230.

[5] S. Haykin 'Cognitive radar'. *IEEE Signal Process. Mag.* January 2006, pp. 30–40.

[6] J.R. Guerci *Cognitive Radar: The Knowledge-Aided Fully Adaptive Approach.* Artech House, Boston, 2010, ISBN 978-1-59693-364-4.

[7] M.I. Skolnik (ed.) *Radar Handbook*. 3rd edition, McGraw Hill Publishing Company, New York, 2008, ISBN 9780071485470.

[8] M.A. Richards, J.A. Scheer, W.A. Holden (eds.) *Principles of Modern Radar: Basic Principles*. The Institution of Engineering and Technology (IET), London, 2010, ISBN 978-1-891121-52-4.

[9] M.A. Richards, J.A. Scheer, W.A. Holden (eds.) *Principles of Modern Radar: Advanced Techniques*. The Institution of Engineering and Technology (IET), London, 2012, ISBN 978-1-891121-53-1.

[10] P. Lacomme, J.P. Hardange, J.C. Marchais, E. Normant *Air and Spaceborne Radar Systems – An Introduction*. William Andrew Publishing, Norwich NY, 2001, ISBN 9781891121135, co-published by The Institution of Electrical Engineers (IEE), London.

[11] S. Haykin (ed.) *Advances in Spectrum Analysis and Array Processing*, Vol. II. Prentice Hall, Englewood Cliffs, USA, 1991, ISBN 013008574X.

[12] U. Nickel 'On the influence of channel errors on array signal processing methods'. *AEU Int. J. Electron. Commun.* 47 (4), 1993, pp. 209–219.

[13] S. Haykin, J. Litva, T.J. Shepherd (eds.) *Radar Array Processing*. Springer Verlag, New York, 1993, ISBN 0387552243, pp. 47–98.

[14] U. Nickel 'Overview of generalized monopulse estimation'. *IEEE AES Mag.* 21 (6), 2006, pp. 27–55, Part 2: Tutorials.

[15] A. Manikas *Differential Geometry in Array Processing*. Imperial College Press, London, 2004, ISBN 1860944221.

[16] J.H.G. Ender 'On compressive sensing applied to radar'. *Signal Process.* 90 (5), 2010, pp. 1402–1414.

[17] Hang Hu 'Aspects of the subarrayed array processing for the phased array radar'. *Int. J. Antennas Propag*, Hindawi Pub, Corp., Vol. 2015, Article ID 797352.

[18] U. Nickel 'Digital formation of subarrays'. *Int. Radar Symp. IRS2005*, Berlin, Germany, German Institute of Navigation (DGON) 2005, pp. 331–335.

[19] U. Nickel 'Properties of digital beamforming with subarrays'. *CIE Int. Conf. RADAR 2006*, Shanghai, China, 6–19 October 2006, IEEE Cat. Nr. 06TH8857.

[20] U. Nickel 'Subarray configurations for digital beamforming with low sidelobes and adaptive interferences suppression'. *Proc. IEEE Int. Radar Conference*, 1995, IEEE Cat. No. 95CH-3571-0, pp. 714–719.

[21] W.L. Melvin, J.A. Scheer *Principles of Modern Radar: Advanced Techniques*. The Institution of Engineering and Technology (IET), London, 2013, ISBN 9781891121531.

[22] I.S. Reed, J.D. Mallett, L.E. Brennan 'Rapid convergence rate in adaptive arrays'. *IEEE Trans. AES* 16, 1980, pp. 853–863.

[23] C.H. Gierull 'Statistical analysis of the eigenvector projection method for adaptive spatial filtering of interference'. *IEE Proc. Radar Sonar Navig.* 144 (2), 1997, pp. 57–63.

[24] M. Grimm, R.K. Sharma, M. Hein, R.S. Thomä, R. Zemmari 'Improved BER performance in GSM by mitigating non-linear distortions in the receiver', *Proc. European Microwave Week*, 2013.

[25] A. Farina *Antenna-Based Signal Processing Techniques for Radar Systems*. Artech House Inc., Norwood, USA, 1992, ISBN 0890063966.

[26] I. Gröger *Nebenkeulenformung der Richtcharakteristik von Gruppenantennen*. PhD dissertation D83, Technical University of Berlin, Department of Electrical Engineering, 1979.

[27] W. Bürger 'Sidelobe forming for ground clutter and jammer suppression for airborne active array radar'. *IEEE Int. Symp. on Array Systems & Technology*, Boston, 2003.

[28] J. Liu, A.B. Gershman, Z.-Q. Luo, K.M. Wong 'Adaptive beamforming with sidelobe control: a second order cone programming approach'. *IEEE SP Lett.* 10 (11), 2003, pp. 331–334.

[29] A.B. Gershman, U. Nickel, J.F. Böhme 'Adaptive beamforming algorithms with robustness against jammer motion'. *IEEE Trans. SP-45*, No. 7, July 1997, pp. 1878–1885.

[30] A.B. Gershman 'Uplink robust adaptive beamforming'. Chapter 9 in T. Kaiser, A. Bourdoux, H. Boche, J.R. Fonollosa, J.B. Andersen, W. Utschick (eds) *Smart Antennas – State of the Art*. EURASIP Book Series on Signal Processing and Communications, Vol. 3, Hindawi Publishing Company, New York, 2005, ISBN 9775945097.

[31] Y. Hua, A.B. Gershman, Q. Cheng *High Resolution and Robust Signal Processing*. Marcel Dekker, New York, 2004, ISBN 0-8247-4752-6.

[32] C.H. Gierull 'A fast subspace estimation method for adaptive beamforming based on covariance matrix transformation'. *AEU Int. J. Electron. Commun.* 51 (4), 1997, pp. 196–205.

[33] C.H. Gierull 'Fast and effective method for low-rank interference suppression in presence of channel errors'. *IEE Electron. Lett.* 34 (6), 1998, pp. 518–520.

[34] G.M. Herbert 'A new projection based algorithm for low sidelobe pattern synthesis in adaptive arrays'. *IEE Int. Conf. Radar 97*, October 1997, IEE Publ. Nr. 449, pp. 396–400.

[35] U. Nickel 'Determination of the dimension of the signal subspace for small sample size'. *IASTED Int. Conf. on Signal Processing and Comm.* IASTED/Acta Press 1998, pp. 119–122.

[36] S. Kritchman, B. Nadler 'Non-parametric detection of the number of signals: hypothesis testing and random matrix theory'. *IEEE Trans. SP* 57 (10), 2009, pp. 3930–3940.

[37] U. Nickel 'On the influence of channel errors on array signal processing methods'. *AEU Int. J. Electron. Commun.* 47 (4), 1993, pp. 209–219.

[38] R.J. Muirhead *Aspects of Multivariate Analysis Theory*. John Wiley, New York, 1982, ISBN 0471094420, p. 405.

[39] D.K. Dey, C. Srinivasan 'Estimation of a covariance matrix under Stein's loss'. *Ann. Stat.* 13 (4), 1985, pp. 1581–1591.

[40] R.R. Nadakuditi, A. Edelman 'Sample eigenvalue based detection of high-dimensional signals in white noise using relatively few samples'. *IEEE Trans. SP* 56 (7), 2008, pp. 2625–2638.

[41] U. Nickel 'Adaptive beamforming for phased array radars'. *Ortung und Navigation*, 2/1998 pp. 99–108 (Reprint of Conf. Contribution at IRS'98, Munich, 1998), DGON, Germany.

[42] S.A. Vorobyov, A.B. Gershman, Z.-Q. Luo, N. Ma 'Adaptive beamforming with joint robustness against mismatched signal steering vector and interference nonstationarity'. *IEEE SP Lett.* 11 (2), 2004, pp. 108–111.

[43] S.D. Hayward 'Adaptive beamforming for rapidly moving arrays'. *CIE Int. Conf. Radar ICR96*, Beijing, IEEE Cat. Nr. 96TH8117.

[44] Y.I. Abramovich, N. Spencer, S.J. Anderson 'Stochastic constraint method in non-stationary hot clutter cancellation – Part 1: Fundamentals and supervised training applications'. *IEEE Trans. AES* 34 (4), 1998, pp. 1271–1292.

[45] Y.I. Abramovich, N. Spencer, S.J. Anderson 'Stochastic constraint method in non-stationary hot clutter cancellation – Part 2: Unsupervised training applications'. *IEEE Trans. AES* 36 (1), 2000, pp. 132–150.

[46] B.D. Rao, K.V.S Hari 'Performance analysis of root MUSIC'. *IEEE Trans. ASSP* 37 (12), 1989, pp. 1939–1949.

[47] V.F. Pisarenko 'The retrieval of harmonics from a covariance function'. *Geophys. J. R. Astron. Soc.* 33, 1973, pp. 347–366.

[48] R. Roy, T. Kailath 'ESPRIT – estimation of signal parameters via rotational invariance techniques'. *IEEE Trans. ASSP* 37 (7), 1989, pp. 984–995.

[49] U. Nickel 'On the application of subspace methods for small sample size'. *AEU Int. J. Electron. Commun.* 51 (6), 1997, pp. 279–289.

[50] V.F. Pisarenko 'On the estimation of spectra by means of non-linear functions of the covariance matrix'. *Geophys. J. R. Astron. Soc.* 28, 1972, pp. 511–531.

[51] M. Viberg, B. Otttersten, T. Kailath 'Detection and estimation in sensor arrays using weighted subspace fitting'. *IEEE Trans. SP* 39 (11), 1991, pp. 2436–2449.

[52] I. Ziskind, M. Wax 'Maximum likelihood localization of multiple sources by alternating projection'. *IEEE Trans. ASSP* 36 (10), 1988, pp.1553–1560.

[53] P.J. Chung, J.F. Böhme 'Comparative convergence analysis of EM and SAGE algorithms in DOA estimation'. *Proc. IEEE Conf. ICASSP-2001*, pp. 2993–2996.

[54] P. Stoica, A. Nehorai 'Performance study of conditional und unconditional direction-of-arrival estimation'. *IEEE Trans. ASSP* 38 (10), 1990, pp. 1783–1795.

[55] M.T. Wasan *Stochastic Approximation*. Cambridge University Press, Cambridge, UK, 1969, ISBN 521073685.

[56] U. Nickel 'Aspects of implementing super-resolution methods into phased array radar'. *AEU Int. J. Electron. Commun.* 53 (6), 1999, pp. 315–323.

[57] B. Ottersten, P. Stoica, R. Roy 'Covariance matching estimation technique for array signal processing applications'. *Dig. Signal Process.* 8, 1998, pp. 185–210.

[58] U. Nickel 'Spotlight MUSIC: super-resolution with subarrays with low calibration effort'. *IEE Proc. Radar Sonar Navig.* 149 (4), 2002, pp. 166–173.
[59] Y.I. Abramovich, N.K. Spencer, A.Y. Gorokhov 'GLRT-based threshold detection-estimation performance improvement and application to uniform circular antenna arrays'. *IEEE Trans. SP* 55 (1), 2007, pp. 20–31.
[60] Y.I. Abramovich, B.A. Johnson 'Detection-estimation of very close emitters: performance breakdown, ambiguity, and general statistical analysis of maximum-likelihood estimation'. *IEEE Trans. SP* 58 (7), 2010, pp. 3647–3660.
[61] A.S. Paine 'Minimum variance monopulse technique for an adaptive phased array radar'. *IEE Proc. Radar Sonar Navig.* 145 (6), 1998, pp. 374–380.
[62] Y. Seliktar, E.J. Holder, D.B. Williams 'An adaptive monopulse processor for angle estimation in a mainbeam jamming and coherent interference scenario'. *Proc. ICASSP*, Seattle 1998, IEEE Cat. Nr. 0-7803-4428-6/98, pp. 2037–2040.
[63] R.C. Davis, L.E. Brennan, I.S. Reed 'Angle estimation with adaptive arrays in external noise fields'. *IEEE Trans. AES* 12 (2), 1976, pp. 179–186.
[64] U. Nickel 'Performance of corrected adaptive monopulse estimation'. *IEE Proc. Radar Sonar Navig.* 146 (1), 1999, pp. 17–24.
[65] U. Nickel, E. Chaumette, P. Larzabal 'Statistical performance prediction of generalized monopulse estimation'. *IEEE Trans. AES* 47 (1), 2011, pp. 381–404.
[66] U. Nickel, E. Chaumette, P. Larzabal 'Estimation of mixed Swerling 0, and Swerling I–II extended targets from monopulse ratio'. *IEEE Trans. AES* 49 (3), 2013, pp. 2084–2096.
[67] A. Doganzić, A. Nehorai 'Cramér–Rao bounds for estimating range, velocity and direction with an active array'. *IEEE Trans. SP* 49 (6), 2001, pp. 1122–1137.
[68] R. Klemm 'Recognition of convoys with airborne adaptive monopulse radar'. *IET Int. Conf. Radar Systems*. 2007.
[69] E.J. Kelly 'Performance of an adaptive detection algorithm; rejection of unwanted signals'. *IEEE Trans. AES* 25 (2), 1989, pp. 122–133.
[70] F.C. Robey, D.R. Fuhrmann, E.J. Kelly, R. Nitzberg 'A CFAR adaptive matched filter detector'. *IEEE Trans. AES* 28 (1), 1992, pp. 208–216.
[71] S. Kraut, L.L. Scharf, L.T. McWhorter 'Adaptive subspace detectors'. *IEEE Trans. SP* 49 (1), 2001, pp. 1–16.
[72] C.D. Richmond 'Performance of the adaptive sidelobe blanker detection algorithm in homogeneous environments'. *IEEE Trans. SP* 48 (5), 2000, pp. 1235–1247.
[73] C.D. Richmond 'Performance of a class of adaptive detection algorithms in non-homogeneous environments'. *IEEE Trans. SP* 48 (5), 2000, pp. 1248–1262.
[74] T.F. Ayoub, A.M. Haimovich 'Modified GLRT signal detection algorithm'. *IEEE Trans. AES* 36 (3), 2000, pp. 810–818.

[75] Y.I. Abramovich, N.K. Spencer, A.Y. Gorokhov 'Modified GLRT and AMF framework for adaptive detectors'. *IEEE Trans. AES* 43 (3), 2007, pp. 1017–1051.

[76] O. Besson, J.-Y. Tourneret, S. Bidon 'Knowledge-aided bayesian detection in heterogeneous environments'. *IEEE Trans. SP Lett.* 14 (5), 2007, pp. 355–358.

[77] A. De Maio, S. De Nicola, A. Farina, S. Iomelli 'Adaptive detection of a signal with angle uncertainty'. *IET Radar Sonar Navig.* 4 (4), 2010, pp. 537–547.

[78] A. De Maio, E. Conte 'Adaptive detection in gaussian interference with unknown covariance after reduction by invariance'. *IEEE Trans. SP* 58 (6), 2010, pp. 2925–2933.

[79] U. Nickel 'Design of generalised 2D adaptive sidelobe blanking detectors using the detection margin'. *Signal Process.* 90, 2010, pp. 1357–1372.

[80] W. Bürger, U. Nickel 'Space-time adaptive detection with array antennas'. *European Conf. Synthetic Aperture Radar EUSAR 2008*, Germany, VDE-Verlag Berlin, ISBN 978-3-8007-3084-1.

[81] W.R. Blanding, W. Koch, U. Nickel 'Adaptive phased-array tracking in ECM using negative information'. *IEEE Trans. AES* 45 (1), 2009, pp. 152–166.

[82] M. Feldmann, U. Nickel 'Target parameter estimation and tracking with adaptive beamforming'. *Proc. Int. Radar Symposium IRS2011*, Leipzig, 2011.

Chapter 2

Robust direct data domain processing for MTI

Diego Cristallini, Wolfram Bürger* and Richard Klemm**

Abstract

Direct data domain space–time adaptive processing (D^3-STAP) is an interesting approach for multi-channel radar moving target indication. The main difference with respect to conventional stochastic STAP resides in the capability to cancel the interference (both jammers and clutter) using only the information contained in the single range gate under test. Therefore, being implicitly a single snapshot interference cancellation technique, D^3-STAP shows several advantages compared to stochastic STAP in fast varying interference scenarios, where the availability of secondary data for interference statistics estimation is limited. In this chapter, a novel approach for D^3-STAP is described. This amelioration overcomes the main limitation of D^3-STAP in its classical derivation. In fact, target detection performance of D^3-STAP is severely deteriorated in case of inaccurate knowledge of target parameters, namely direction of arrival (DOA) and Doppler frequency. To overcome this problem, a robust D^3-STAP (viz. RD^3-STAP) implementation is shown which takes into account a possible mismatch between nominal and actual target parameters. The approach reformulates the D^3-STAP problem in the context of convex optimization, and it can be applied to the different implementations of D^3-STAP, namely forward, backward and forward–backward methods. In addition to that, an implementation of RD^3-STAP with dimension reducing transformations is shown which limits the number of degrees of freedom. The effectiveness of the proposed approach is shown both in simulated scenarios and by direct application to real data taken from the experimental multi-channel radar system Phased-Array Multi-functional Imaging Radar (PAMIR) developed at Fraunhofer-Institute for high frequency physics and radar techniques; in German: Fraunhofer-Institut für Hochfrequenzphysik und Radartechnik (FHR). Finally, possible applications of RD^3-STAP to multi-channel synthetic aperture radar and to target DOA estimation cases are presented.

*Fraunhofer FHR, Germany

2.1 Introduction

Space–time adaptive processing (STAP) identifies the framework of radar signal processing where multiple parallel digitized receiving channels and the echoes of multiple successive transmitted pulses are jointly exploited to guarantee reliable moving target indication (MTI). STAP is a tool for detecting moving objects with moving radar (for instance, airborne or space based) in presence of stationary clutter background or other strong interfering scenarios. In theory, the optimum STAP processor is given by the well known rule 'pre-whiten and match' (likelihood ratio test) where 'whitening' is performed by multiplying the inverse of the space–time covariance matrix of clutter and noise with the data vector. Matching the target signal in space and time is done in a second step by a beamformer cascaded with a Doppler filter bank. In particular, STAP facilitates the detection of low Doppler targets. For more details the reader is referred to [1,2]. In this context, adaptivity refers to the need for estimating the unknown interference environment through the received data [3]. Range dependency of clutter characteristics requires the STAP filter to be adaptive also over range. In the past literature, two macro classes of approaches have been proposed to make such a range dependent adaptivity feasible: stochastic STAP and direct data domain STAP (D^3-STAP) [4].[1]

In stochastic STAP, secondary data are collected from neighbouring range gates to estimate the interference statistics (space–time clutter+noise covariance matrix) of the range gate under test. By doing so, it is assumed that secondary range cells do not contain the potential target and that they are statistically homogeneous to the range gate under test. The well-known Reed Mallett Brennan (RMB) rule [9] sets the amount of secondary data samples needed to limit adaptivity losses, which is directly related to the dimensions of the corresponding interference covariance matrix. Large covariance matrix dimensions and/or fast range varying interference environments might experience shortage of available secondary data having the same statistics as the cell under test (CUT) as well as problems associated with arithmetic accuracy, computing time and storage. In the context of stochastic STAP, sub-optimal approaches have been derived which drastically reduce the dimensions of the covariance matrix while retaining good interference cancellation capabilities [1,2]. However, sub-optimal stochastic STAP still suffers several drawbacks. In particular, the range adaptivity of the STAP filter can hardly be performed on a single range gate basis, as required in the presence of clutter discretes. Even stabilization techniques like diagonal loading, proposed to limit the shortcomings of stochastic STAP, still require some secondary data samples support. In addition, the need for multiple range gates to estimate the interference characteristics might also cause inter-target nulling effects, meaning that echoes from one target are erroneously added to the

[1]It is worth noticing that also other methods than D^3-STAP might be mentioned as good candidates in case of no available secondary data [5–8].

secondary data corresponding to the other target, thus preventing reliable target detection [10].

D^3-STAP offers a different viable solution for effective interference cancellation in situations with a limited amount of secondary data [4, 11 Chapter 12]. Here, interference cancellation is performed on a single-range gate basis with no need for secondary data. Interference cancellation is performed cancelling out from the received signal all contributions except the expected target one. This is done under the assumption of exact knowledge of target parameters, namely direction of arrival (DOA) and Doppler frequency. To this end, one should notice that assuming a coarse knowledge of target parameters is not completely unrealistic, especially in target tracking and/or confirmation situations; however, their exact knowledge represents a strong limitation for the application of this technique. In fact, a target signal contribution with parameters only slightly deviating from the nominal one will lead to severe performance degradation, since the target signal will be treated as interference (i.e. the so-called target self-nulling effect). To overcome this problem, [12] proposed to bind the D^3-STAP filter with multiple constraints covering the region of uncertainty of target parameters. This solution preserves the STAP filter gain for the target at the expense of a reduced number of degrees of freedom (DOFs) available to suppress the interference.

This chapter addresses the problem of application of D^3-STAP in case of uncertainty in the knowledge of target parameters. In particular, a robust implementation of the D^3-STAP (viz. RD^3-STAP) filter is described to prevent the aforementioned target self-nulling effect [13]. Such an approach has also the nice property of maintaining adequately low sidelobes in the RD^3-STAP pulse response. This is obtained by designing the filter-pulse response to minimize the output interference power while maintaining a pre-determined gain over a region of uncertainty around the nominal target direction[2]. In other words, the RD^3-STAP filter is constrained to have a pre-determined gain over the target for all possible target directions within the region of uncertainty. This is obtained by imposing the constraint for the biggest possible mismatch between nominal and potential target parameter values. By following this approach, the RD^3-STAP filter design is reformulated in the framework of convex optimization [15]. It is worth noticing that a similar approach has already been investigated by [16,17] in the framework of stochastic STAP.

This chapter is organized as follows. In Section 2.2, the D^3-STAP filter derivation is briefly recalled, and the notation is defined. Section 2.3 is devoted to the mathematical derivation of the RD^3-STAP filter. Section 2.4 presents some results of the RD^3-STAP filter, both simulative and with an application to real data. Some applications of the RD^3-STAP filter for synthetic aperture radar (SAR) and for target DOA estimation are presented in Section 2.5. Finally in Section 2.6, we draw some conclusions.

[2]In [14], the D^3-STAP pulse response sidelobe level is controlled by synthesizing multiple constraints in the sidelobe region, thus reducing the available DOFs for interference suppression.

2.2 Notation and signal model

Let us define the complex column vector $\mathbf{x}$ containing the digitized complex voltages measured at a specific time (i.e. for a given range gate) at the N elements of an array and in M successive pulses. The overall signal received from the nth antenna element at the mth pulse can be written as follows:

$$x_{m,n} = \alpha_s \cdot \exp\left[\mathrm{j}2\pi\left(\frac{nd}{\lambda}\sin\theta_s + \frac{mf_s}{f_r}\right)\right] + \sum_{i=1}^{K-1}\alpha_i \cdot \exp\left[\mathrm{j}2\pi\left(\frac{nd}{\lambda}\sin\theta_i + \frac{mf_i}{f_r}\right)\right] + r_{m,n} \tag{2.1}$$

where α_s is the unknown complex amplitude of the target, θ_s is the target DOA (supposed to be known in the ideal case of perfect knowledge of target parameters), d is the inter-element distance between the elements of the array, f_s is the target Doppler frequency (also supposed to be known in the ideal case of perfect knowledge of target parameters), f_r is the pulse repetition frequency (PRF) and λ is the carrier wavelength. $K-1$ is the number of narrowband interferences impinging on the array, α_i, θ_i and f_i are the corresponding unknown complex amplitudes, DOAs and Doppler frequencies ($i = 1, \ldots, K-1$). Finally, $r_{m,n}$ is the complex contribution of thermal noise at the nth antenna element at the mth time instant.

The signal model defined in (2.1) can be conveniently reformulated in vectorial form as follows:

$$\mathbf{x} = \alpha_s \cdot \mathbf{a}_s + \sum_{i=1}^{K-1}\alpha_i \cdot \mathbf{g}_i + \mathbf{r} \tag{2.2}$$

where $\mathbf{a}_s$ is the $NM \times 1$ target space–time steering vector, $\mathbf{g}_i$ is the $NM \times 1$ space–time steering vector of the ith interference and $\mathbf{r}$ contains the thermal noise contributions. As is apparent, both the useful target and the interference contributions are modelled as narrowband signal sources impinging on the antenna array from specific localized positions in the angle/Doppler plane. The thermal noise contribution can be considered as a white normal distributed random process.

Let us now define the two scalar complex variables z_s and z_t as follows:

$$\begin{aligned} z_s &= \exp\left[\mathrm{j}2\pi\frac{d}{\lambda}\sin\theta_s\right] \\ z_t &= \exp\left[\mathrm{j}2\pi\frac{f_s}{f_r}\right] \end{aligned} \tag{2.3}$$

that are uniquely determined if the target parameters θ_s and f_s are assumed to be known. In addition, let us define $K_m \leq (M+1)/2$ and $K_n \leq (N+1)/2$, which are the available DOF in the spatial and temporal domains, respectively.

Then, a space–time steering vector in the nominal target direction $\mathbf{z}$ can be written as (see (2.4)):

$$\mathbf{z} = \begin{bmatrix} 1 & z_s & z_s^2 & \dots & z_s^{K_n-1} & z_t & z_t z_s & \dots & z_t z_s^{K_n-1} & \dots & z_t^{K_m-1} & \dots & z_t^{K_m-1} z_s^{K_n-1} \end{bmatrix} \quad (2.4)$$

To retrieve the expression of the D^3-STAP filter, it is useful to re-arrange the column vector $\mathbf{x}$ in a $M \times N$ matrix form $\mathbf{X}$, where $x_{m,n}$ is the mth row and nth column element of $\mathbf{X}$. Let us also define the matrices:

$$\mathbf{X}_{m,n}^{(K_m K_n)} = \mathbf{X}(m : m + K_m - 1, n : n + K_n - 1) \quad (2.5)$$

where $m = 1, \dots, K_m$ and $n = 1, \dots, K_n$. That is $\mathbf{X}_{m,n}^{(K_m K_n)}$ is a $K_m \times K_n$ truncated version of $\mathbf{X}$, obtained taking K_m rows starting from the mth row and K_n columns starting from the nth column. In addition, let $\mathbf{x}_{m,n}^{(K_m K_n)}$ be a $1 \times K_m K_n$ row vector that contains the elements of $\mathbf{X}_{m,n}^{(K_m K_n)}$ read columnwise.

We observe that three sets of row vectors can be derived from the conventional single canceller principle (i.e. the 2-pulse Doppler filter), which contain interference contributions only (the useful signal contribution is completely removed by the weighted subtraction), namely:

$$\begin{aligned} {}^s\mathbf{x}_{m,n} &= \mathbf{x}_{m,n}^{(K_m K_n)} - z_s^{-1} \mathbf{x}_{m,n+1}^{(K_m K_n)} \\ {}^t\mathbf{x}_{m,n} &= \mathbf{x}_{m,n}^{(K_m K_n)} - z_t^{-1} \mathbf{x}_{m+1,n}^{(K_m K_n)} \\ {}^{s,t}\mathbf{x}_{m,n} &= \mathbf{x}_{m,n}^{(K_m K_n)} - z_t^{-1} z_s^{-1} \mathbf{x}_{m+1,n+1}^{(K_m K_n)} \end{aligned} \quad (2.6)$$

where $m = 1, \dots, K_m$ and $n = 1, \dots, K_n$. The vectors in (2.6) are in total $3K_m K_m$ and they have dimensions $1 \times K_m K_n$ each. They can be arranged in a $3K_m K_m \times K_m K_n$ interference only matrix $\mathbf{F}_2$, see also [4, 11 Chapter 12]. Specifically, $\mathbf{F}_2$ can be built as follows:

$$\mathbf{F}_2 = \begin{bmatrix} {}^s\mathbf{x}_{1,1} \\ \vdots \\ {}^s\mathbf{x}_{K_m,K_n} \\ {}^t\mathbf{x}_{1,1} \\ \vdots \\ {}^t\mathbf{x}_{K_m,K_n} \\ {}^{s,t}\mathbf{x}_{1,1} \\ \vdots \\ {}^{s,t}\mathbf{x}_{K_m,K_n} \end{bmatrix} \quad (2.7)$$

A linear system matrix $\mathbf{F}$ can be obtained simply concatenating the row vector $\mathbf{z}$ with the matrix $\mathbf{F}_2$. That is:

$$\mathbf{F} = \begin{bmatrix} \mathbf{z} \\ \mathbf{F}_2 \end{bmatrix} \tag{2.8}$$

By exploiting $\mathbf{F}$ in (2.8), the $K_m K_n \times 1$ D^3-STAP filter $\mathbf{w}$ is obtained by solving in the least squares sense the following linear system:

$$\mathbf{F} \cdot \begin{bmatrix} w_1 \\ w_2 \\ \vdots \\ w_{K_m K_n} \end{bmatrix} = \begin{bmatrix} 1 \\ 0 \\ \vdots \\ 0 \end{bmatrix} \Leftrightarrow \mathbf{F} \cdot \mathbf{w} = \mathbf{y} \tag{2.9}$$

As is apparent, the linear system in (2.9) tends to null the interference only matrix $\mathbf{F}_2$, while preserving a non-zero look direction constraint in the nominal target direction (first row of $\mathbf{F}$). In addition, one should note that the linear system in (2.9) has to be solved in a least square sense, since $\mathbf{F}$ is generally non-squared.

It has also to be noticed that the D^3-STAP filter deriving from solving (2.9) corresponds to the so-called forward (FW) implementation [11 Chapter 12], where the term *FW* comes from the observation that vector $\mathbf{x}$ is spanned from the first to the last element in the weighted subtractions in (2.6). As mentioned in [11 Chapter 12], an analogous D^3-STAP filter design can be obtained by considering a reversed complex conjugated version of the received data vector $\mathbf{x}$, and then building signal free weighted subtractions like in (2.6), where the vector is now spanned from the last to the first element. The resulting D^3-STAP filter will lead to the so-called backward (BK) implementation. Moreover, FW and BK weighted subtractions can be combined leading to the forward–backward (FB) implementation which gives additional equations to the linear system (2.9).

It should be clear that any uncertainty in the knowledge of the target parameters θ_s and f_s would directly affect the scalar quantities in (2.3) and consequently the matrix $\mathbf{F}$. As a consequence, the matrix $\mathbf{F}_2$ will no more contain only interference contributions but also a useful signal contribution, and also the look direction constraint (first row of $\mathbf{F}$) will be no more exact. The final results of such a mismatch (e.g. of this error) is the so-called target self-nulling effect, see [11 Chapter 12, 13]. In other words, if the target parameters differ from the nominal ones, the filter vector determined through (2.9) will treat the true target as interference, thus nulling it. This is a major drawback of D^3-STAP, since an uncertainty in the target parameters determination is always to be expected in real scenarios. The first idea proposed to overcome this problem is to synthesize multiple constraints covering the region of uncertainty of target parameters [12]. This can be easily obtained by replacing one or more rows in the matrix $\mathbf{F}$ in (2.8) with constraints of the form:

$$\begin{bmatrix} 1 & z_{s,l} & z_{s,l}^2 & \cdots & z_{s,l}^{K_n-1} & \cdots & z_{t,l} & z_{t,l}z_{s,l} & \cdots & z_{t,l}z_{s,l}^{K_n-1} \\ & \cdots & z_{t,l}^{K_m-1} & z_{t,l}^{K_m-1}z_{s,l} & \cdots & z_{t,l}^{K_m-1}z_{s,l}^{K_n-1} \end{bmatrix} \tag{2.10}$$

where:

$$\begin{aligned} z_{s,l} &= \exp\left[\mathrm{j}2\pi\frac{d}{\lambda}\sin\theta_{s,l}\right] \\ z_{t,l} &= \exp\left[\mathrm{j}2\pi\frac{f_{s,l}}{f_r}\right] \end{aligned} \tag{2.11}$$

$\theta_{s,l}$ and $f_{s,l}$ being the DOA and the Doppler frequency values of the lth constraint, and replacing with non-zero entries the corresponding elements of the vector $\mathbf{y}$. As apparent, every additional main beam constraint reduces the number of available DOFs reserved for the interference suppression. In the following section, a robustification of D^3-STAP (i.e. RD^3-STAP) is presented to overcome this problem.

As a final general comment, the D^3-STAP approach has been here presented for the uniform linear array (ULA) case. Under this hypothesis, the derivation of the interference only quantities in ${}^{s}\mathbf{x}_{m,n}$ and in ${}^{s,t}\mathbf{x}_{m,n}$ is simplified. Nevertheless, a generalization to planar arrays with irregular sub-array structures is possible.

2.3 Robust D^3-STAP

If the exact target parameters are unknown, but their values can be confined in a bounded region of uncertainty, the D^3-STAP filter in (2.9) can be constrained to have a pre-determined gain over the target for all possible pairs of target parameters within the region of uncertainty, see [13]. This can be easily fulfilled by imposing the constraint for the worst case situation or, which is equivalent, for the highest possible mismatch between assumed and potential target parameter values, following the approach of robust Capon beamforming presented in [17]. Subject to this constraint, the filter is then designed to minimize the output interference power. In this case, the D^3-STAP filter design will be robust to any target parameters inaccuracy within the region of uncertainty.

In case of uncertainty in target parameter knowledge, the true (T) target steering vector $\mathbf{c}^{(K_mK_n)}$ (which is steered at θ_s^T and f_s^T) can be expressed as the sum of the nominal (N) one $\mathbf{a}^{(K_mK_n)}$ (steered at θ_s^N and f_s^N) plus a norm-bounded error vector $\mathbf{e}^{(K_mK_n)}$. That is:

$$\mathbf{c}^{(K_mK_n)} = \mathbf{a}^{(K_mK_n)} + \mathbf{e}^{(K_mK_n)}, \text{ with } \|\mathbf{e}^{(K_mK_n)}\| \leq \epsilon \tag{2.12}$$

where the superscript '(K_mK_n)' indicates that $K_mK_n \times 1$ complex truncated column vectors are considered and will be omitted in the following to simplify the notation. Clearly, under the target parameters uncertainty assumption, the first equation of (2.9) (i.e. the look direction constraint) can be conveniently reformulated as follows:

$$|\mathbf{w}^H\mathbf{c}| = |\mathbf{w}^H(\mathbf{a}+\mathbf{e})| \geq 1 \tag{2.13}$$

for all vectors $\mathbf{c}$ satisfying (2.12). As already shown in [17], the inequality in (2.13) can be rewritten for small ϵ as follows:

$$|\mathbf{w}^H\mathbf{a}| - \epsilon \cdot ||\mathbf{w}^H|| \geq 1 \tag{2.14}$$

As a consequence, a convenient robust formulation of the D^3-STAP problem (viz. robust D^3-STAP, RD^3-STAP) is:

$$\begin{aligned} &\min_{\mathbf{w}} ||\mathbf{F}_2^H\mathbf{w}|| \\ &\text{subject to}: |\mathbf{w}^H\mathbf{a}| - \epsilon \cdot ||\mathbf{w}|| \geq 1 \end{aligned} \tag{2.15}$$

Several comments are in order. First, it is worth noticing that the RD^3-STAP method, as well as D^3-STAP, designs an adapted filter vector $\mathbf{w}$ of dimensions $K_mK_n \times 1$. Second, RD^3-STAP reformulates the D^3-STAP filter design in terms of convex optimization, [15], which can be easily solved with ready-to-use toolboxes [18,19]. Third, it has to be highlighted that the actual convergence of the RD^3-STAP method directly depends upon the value of ϵ.

Finally, one has to notice that the RD^3-STAP filter design, which has been described here for brevity sole in the FW implementation, can be applied straightforward to the BK and to the FB cases.

Convergence of the RD^3-STAP method

We now investigate the admissible values for ϵ. For the sake of simplicity, the derivation is confined to the spatial-only case (i.e. only the spatial dimension is considered). Extension to the space–time case is then straightforward. Let us express the worst case error vector $\mathbf{e}$ (as in [17]) as:

$$\mathbf{e} = -\frac{\mathbf{w}}{||\mathbf{w}||}\epsilon \cdot e^{j\Phi} \tag{2.16}$$

where $\mathbf{\Phi} = \text{angle}(\mathbf{w}^H\mathbf{a})$, and substituting it in (2.13), it can be easily shown that (2.13) is equivalent to (2.14) if and only if:

$$|\mathbf{w}^H\mathbf{a}| \geq \epsilon||\mathbf{w}|| \Leftrightarrow \epsilon \leq \frac{|\mathbf{w}^H\mathbf{a}|}{||\mathbf{w}||} \tag{2.17}$$

According to [17], (2.17) defines the constraint that ensures (2.13) and (2.14) to be equivalent, so that the convex optimization problem in (2.15) is correctly formulated and will give a stable numerical solution. Nevertheless, if no restriction is made on the possible form of the weight vector $\mathbf{w}$, no specific closed form expression for the upper bound error can be derived. Fortunately, for the particular case of $||\mathbf{w}|| = ||\mathbf{a}|| = \sqrt{K_n}$, (2.17) can be refined thus deriving a closed form expression for the upper bound error ϵ. To understand this point, one should again consider (2.14) and make explicit the inequality with respect to ϵ:

$$|\mathbf{w}^H \mathbf{a}| - \epsilon||\mathbf{w}|| \geq 1 \Leftrightarrow \epsilon \leq \frac{|\mathbf{w}^H \mathbf{a}|}{||\mathbf{w}||} - \frac{1}{||\mathbf{w}||} \tag{2.18}$$

By comparing (2.17) with (2.18), it is clear that (2.18) poses a stricter requirement on ϵ. In particular, recalling that both $\mathbf{w}$ and $\mathbf{a}$ are $K_n \times 1$ vectors and applying the Cauchy–Schwartz inequality to (2.18), we can find an upper bound for ϵ. That is:

$$\epsilon \leq \epsilon_{\text{MAX}} = \sqrt{K_n} - \frac{1}{\sqrt{K_n}} = \frac{K_n - 1}{\sqrt{K_n}} \tag{2.19}$$

Please note that the upper bound in (2.19) is indirectly a function of N. In particular, for the spatial-only case, we will choose the maximum number of DOFs, leading to $K_N = \lfloor (N+1)/2 \rfloor$, where $\lfloor \cdot \rfloor$ is the floor integer operator. To prove that (2.19) is actually an upper bound for ϵ, let us recall that for hypothesis $||\mathbf{w}|| = \sqrt{K_n}$ and let us consider the following expression for ϵ:

$$\epsilon = \epsilon_{\text{MAX}} + p = \frac{K_n - 1}{\sqrt{K_n}} + p \tag{2.20}$$

where p is a real scalar value. Plugging (2.20) into (2.14) we get:

$$|\mathbf{w}^H \mathbf{a}| \geq 1 + \left(\frac{K_n - 1}{\sqrt{K_n}} + p\right) \cdot ||\mathbf{w}|| = K_n + p \cdot \sqrt{K_n} \tag{2.21}$$

Depending on the value of p, (2.21) leads to the following three different cases:

$$\begin{cases} p < 0 \Rightarrow |\mathbf{w}^H \mathbf{a}| < K_n & \text{several solutions are admissible} \\ p = 0 \Rightarrow |\mathbf{w}^H \mathbf{a}| = K_n \Leftrightarrow \mathbf{w} = \mathbf{a} & \text{only one solution is admissible} \\ p > 0 \Rightarrow |\mathbf{w}^H \mathbf{a}| > K_n & \text{no solutions are admissible} \end{cases} \tag{2.22}$$

By observing (2.22), it should be clear that ϵ_{MAX} represents an upper bound for ϵ and that (2.19) is a sufficient condition for ϵ that ensures the existence of an admissible solution for our convex problem. In addition, in the case limit $\epsilon = \epsilon_{\text{MAX}}$, the only admissible solution is $\mathbf{w} = \mathbf{a}$, which clearly does not have any effect in reducing the output interference power. As a consequence, in practical situations, one will always choose $\epsilon < \epsilon_{\text{MAX}}$, in order to maintain some freedom in the selection of the weight vector $\mathbf{w}$ aiming at reducing the interference power.

2.3.1 RD^3-STAP with dimension reducing transformations

It has been shown in [1,2] and other publications that certain dimension reducing transformations (DRTs) can minimize the computational workload of adaptive space–time clutter filtering without significant losses in clutter suppression performance. Typical examples for such transforms are illustrated in Figure 2.1. It shows how a linear equispaced array can be modified by forming equispaced

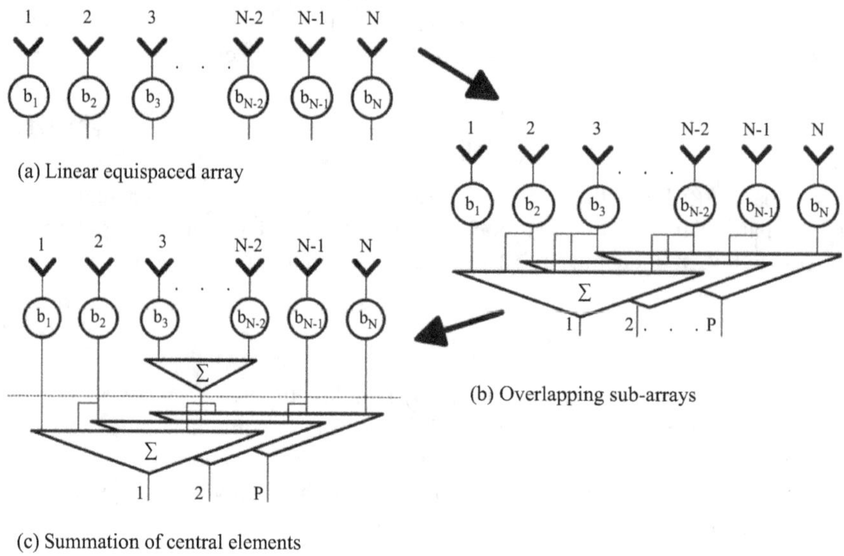

Figure 2.1 Examples of dimension reducing transforms. © 2006 IET. After [1]

sub-arrays overlapping in such a way that the distance of phase centres is the same as in the original array. In this way, one obtains an equispaced array whose channels have a signal gain given by the number of sub-array elements. If $P \ll N$, this transform results in dramatic saving of the subsequent operations (adaptation, filter weight calculation, filtering).

DRTs with overlapping sub-arrays (as those shown in Figure 2.1b) can be also applied in the RD^3-STAP case, as first introduced in [20]. Here the main advantages and limitations of RD^3-STAP with DRT are described.

The number of elements per sub-array is N_{SA}, while P is the number of channels after DRT, and N_B indicates the spacing between adjacent sub-arrays in element unit distance. The signal from the pth sub-array $(p = 1, \ldots, P = ((N - N_{SA})/N_B) + 1)$ becomes:

$$\Omega_p = \sum_{n=1}^{N_{SA}} x_{n+N_B\cdot(p-1)} \tag{2.23}$$

so that an interference matrix $\mathbf{F}_2$ can be built by weighted subtractions:

$$\Omega_p - \zeta^{-1}\Omega_{p+1} \qquad p = 1, \ldots, P \tag{2.24}$$

where ζ is the target signal phase shift between adjacent sub-arrays. From (2.24), the RD^3-STAP filter after DRT can be derived as shown in the preceding paragraphs. Clearly, the application of DRT reduces the dimensions of the interference matrix $\mathbf{F}_2$ in (2.15), and hence also reduces the number of DOFs to $Q = (P + 1)/2$. As a consequence of this DOF reduction, the RD^3-STAP filter pulse response might be unable to synthesize sufficiently deep notches in the near sidelobe region, as already highlighted in [20]. A solution to this resides in the modification of the weighted subtractions (2.24) when DRTs are considered. Specifically, [20] observes that (2.24) identifies a two-pulse canceller, and therefore it proposes to modify the weighted subtractions to those of a double canceller. The idea behind is that a double canceller is expected to have a sharper and faster transition between the filter notch and the pass band region.

A double canceller RD^3-STAP can then be derived starting from the following (interference only) weighted subtractions:

$$-\zeta\Omega_{p-1} + 2\Omega_p - \zeta^{-1}\Omega_{p+1} \tag{2.25}$$

and following the same RD^3-STAP filter design as described above. It is important to notice that in the double canceller case the dimensions of the matrix $\mathbf{F}_2$ (and of the weight vector $\mathbf{w}$) are further reduced, so that now the number of available DOFs is $Q = P/2$. In the following paragraph (see also [20]), a simulative example is shown which demonstrate how DRTs can be effectively applied to double canceller RD^3-STAP filtering.

Single and double RD^3-STAP performance metrics

We here consider an ULA of $N = 15$ half-wavelength spaced elements. N_B is fixed to 1, while we vary N_{SA} from 1 to 13 (only odd integers in this interval). The number of channels P after DRT follows directly from N_{SA}. We assume the useful target to impinge the antenna from broadside with a signal to noise ratio of 13 dB, and one interference to be present with a jammer to noise ratio of 31 dB. Performance is analysed in terms of normalized $\mathrm{SINR}_{\mathrm{out}}$, defined as follows:

$$\mathrm{SINR}_{\mathrm{out}} = \frac{E\left\{|\mathbf{w}^H\mathbf{s}^{(Q)}|^2\right\}}{E\left\{|\mathbf{w}^H(\mathbf{g}^{(Q)} + \mathbf{r}^{(Q)})|^2\right\}} \cdot \frac{1}{\mathbf{s}^H\mathbf{s}} \tag{2.26}$$

where $\mathbf{w}$ is the RD^3-STAP filter vector obtained applying (17) in [13] to the reduced dimension vector $\{\Omega_p\}_{p=1,\ldots,P}$ in (2.23), $E\{\}$ indicates the expected value operator, the superscript (Q) indicates that only the first Q elements of the corresponding vector are considered.

In the first analysis, a MonteCarlo simulation with 10^2 trials has been conducted to evaluate the normalized $\mathrm{SINR}_{\mathrm{out}}$ against the varying interference DOA. In this context, the target DOA has been kept fixed with a mismatch between nominal and true DOA equal to half antenna beamwidth (for $N = 15$, such mismatch guarantees the existence of a numerical solution for the RD^3-STAP problem, as previously discussed). Such a mismatch corresponds a worst case analysis for the more general case of a target DOA uniformly distributed

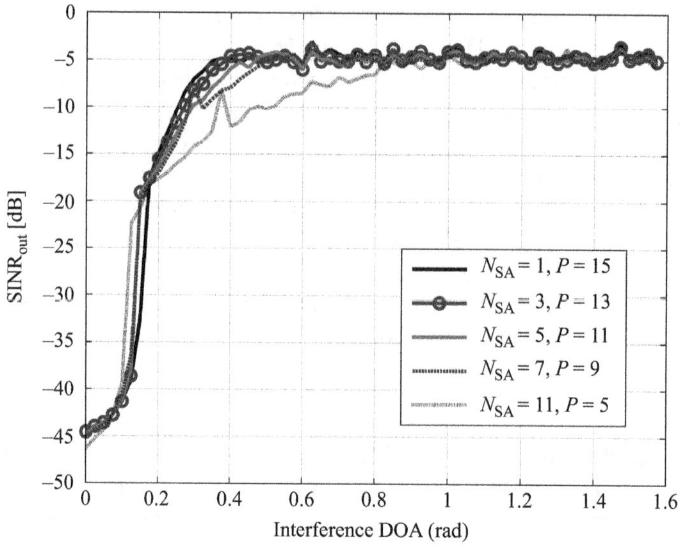

Figure 2.2 Normalized SINR$_{out}$ against varying interference DOA for single RD^3-STAP filter after DRT. © 2014 VDE VERLAG, Berlin, Offenbach. After [20]

between the antenna main beam. Obtained results are shown in Figure 2.2, for all different sub-array configurations. Note that for $N_{SA} = 1$, we have the conventional full DOFs case (i.e. one channel per antenna element), hence this can be considered as a benchmark to evaluate the performance of other DRT RD^3-STAP cases. As is apparent, all DRT configurations exhibit a wide notch, which is the unavoidable consequence of the robustness to potential target DOA mismatches. Apart from that, it is clear that the complete performance recovery is faster for higher number of channels, as one could easily foresee. Nevertheless, even for the case $P = 15$, performance recovery is slow also beyond the DOA region of uncertainty (see e.g. the region between 0.2 and 0.4 rad). This means that the notch in the filter pulse response is too large to place a sufficiently deep null in the near sidelobe region.

Considering the same ULA and the same scenario as before, and varying N_{SA} from 2 to 12 (only even integers, to account for the changed dimensions of $\mathbf{F}_2$), analogous overlapping sub-array configurations can be obtained for the double canceller. Figure 2.3 shows the performance of the double RD^3-STAP after DRT in terms of normalized SINR$_{out}$ against a varying interference DOA. As is apparent, apart from the $P = 4$ case, all other robust double cancellers exhibit much nicer behaviour in the shoulder region (i.e. between 0.2 and 0.4 rad). This means that DRT can be conveniently applied to robust double RD^3-STAP to reduce the number of receiving channels at no performance degradation.

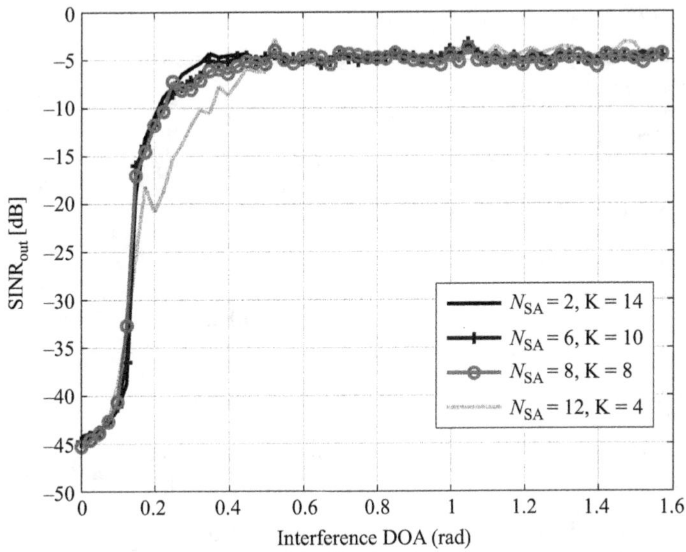

Figure 2.3 *Normalized $SINR_{out}$ against varying interference DOA for double RD^3-STAP filter after DRT. © 2014 VDE VERLAG, Berlin, Offenbach. After [20]*

2.4 Results of RD^3-STAP

In this section, the performance of the RD^3-STAP filter is analysed first in a simulative case study, and then using real data.

2.4.1 Simulative case study

This first analysis compares the performance improvement of RD^3-STAP with respect to the conventional D^3-STAP approach for a specific simulated case study. The main parameters used in the simulation are described in Table 2.1. In particular, only an error in the target Doppler frequency knowledge is considered, whereas the target DOA knowledge is assumed to be error free. A dual situation, with a mismatch in the spatial domain, would also be possible. Clearly, mismatches in the temporal and in the spatial domain originate from different causes. Here, given the higher available DOFs that we have in the temporal domain (see Table 2.1), the effects of the proposed robustification are more evident in the temporal domain, so that a more tutorial analysis can be conducted. As one can see, a challenging interference scenario has been chosen with both a strong jammer and clutter. A target has been placed between the clutter ridge and the jammer DOA.

The resulting input signal power spectrum in shown in Figure 2.4. The target complex amplitude is much weaker than the interference amplitudes, so that the useful signal is hidden in the interference and cannot be detected without adequate adaptive interference suppression. The uncertainty in the knowledge of target

Table 2.1 Case study parameters

Parameter	Value
Wavelength (λ)	0.03 m
Antenna element distance (d)	$\lambda/2$
PRF (f_r)	1,550 Hz
Number of channels (N)	11
K_n	6
Number of pulses (M)	65
K_m	25
Target	$\theta_s^N = -0.1$ rad; $\theta_s^T = -0.1$ rad
	$f_s^N = 100$ Hz; $f_s^T = 120$ Hz
	$\alpha_s = 12.8 \cdot e^{j0.67}$
Jammer	$\theta_1 = -0.6$ rad
	$\alpha_1 = 51 \cdot e^{j1.37}$
Clutter	$\theta_i = [-0.2, 0.2]$ rad
	$f_i = [-220, 220]$ Hz
	$\alpha_i = 42 \cdot e^{j0.78}$
Thermal noise ($r_{m,n}$)	AWGN, zero mean, unit variance

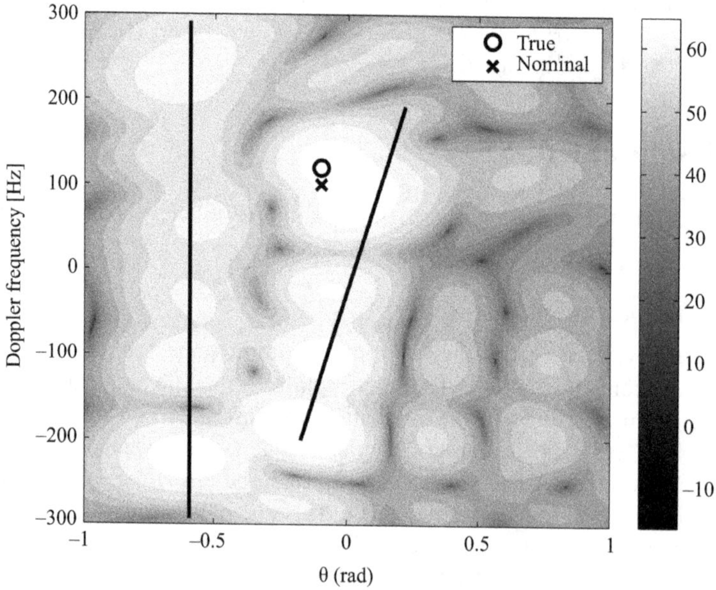

Figure 2.4 Input signal power spectrum (x: nominal target position; o: true target position) in dB. © 2012 IEEE. After [13]

Doppler frequency is shown in Figure 2.4, where the nominal target position is marked with a white 'x', while the real target position is marked with an 'o'. For convenience, in Figure 2.4, also the jammer and clutter contributions have been highlighted with black lines.

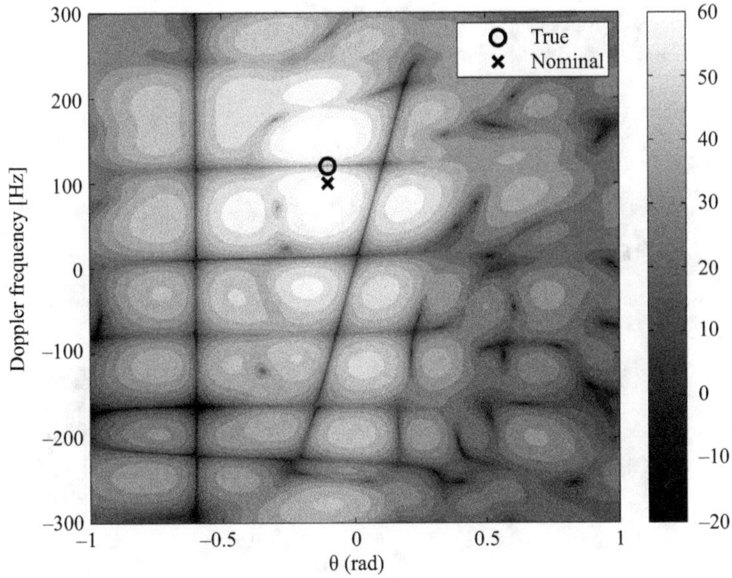

Figure 2.5 D^3*-STAP output signal power spectrum (x: nominal target position; o: true target position) in dB. © 2012 IEEE. After [13]*

The output signal power spectrum obtained by applying the conventional D^3-STAP filter is shown in Figure 2.5. As is apparent, deep notches are placed both at the jammer DOA and along the clutter ridge. However, the target self-nulling effect can be clearly seen. That is the slight mismatch in target Doppler frequency knowledge causes the synthesis of a notch in the real target position (black circle).

A different result is obtained by applying the RD^3-STAP approach. In this case, the adaptive filter pulse response is designed to tolerate a maximum mismatch in the target Doppler frequency knowledge of 25 Hz.

As a consequence, the output signal power spectrum shows no target self-nulling effect, as is clearly visible in Figure 2.6. The real target position (black circle) is now preserved, while strong depressions are synthesized to suppress both jammer and clutter.

2.4.2 *Application of RD*^{3}*-STAP filter to real data*

The RD^3-STAP approach is here applied to a real multi-channel data set acquired with the experimental radar system PAMIR developed at Fraunhofer FHR [21]. Experimental data have been acquired during a flight campaign in 2008. PAMIR main system parameters used in the acquisition are reported in Table 2.2. During the experiment, two signal repeaters have been used to emulate the echoes of two different targets whose characteristics are described in Table 2.2.

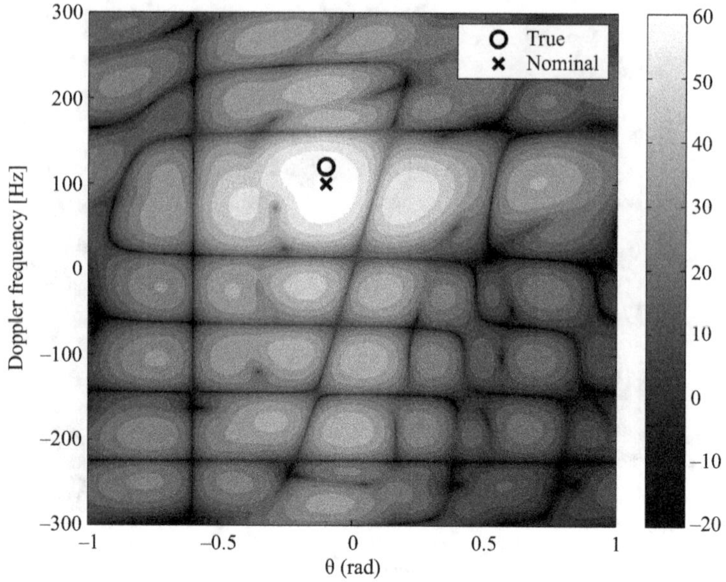

Figure 2.6 RD³-STAP output signal power spectrum (x: nominal target position; o: true target position) in dB. © 2012 IEEE. After [13]

Table 2.2 Main PAMIR parameters

Parameter	Value
Wavelength (λ)	0.03 m
Sub-array distance (d)	0.26 m
PRF (f_r)	1,550 Hz
Signal bandwidth	Lowered to 38 MHz
Number of channels (N)	3
K_n	2
Number of pulses (M)	51
K_m	26
Target 1 (strong)	$\theta_s^T = 0$ rad $f_s^T = 165$ Hz RCS = 100 m^2 Range = 2,664 m
Target 2 (weak)	$\theta_s^T = 0$ rad $f_s^T = 165$ Hz RCS = 10 m^2 Range = 2,694 m

The range-Doppler map of the sum channel is reported in Figure 2.7. The zoom in Figure 2.8 highlights the positions of the two targets just at the edge of the main clutter region.

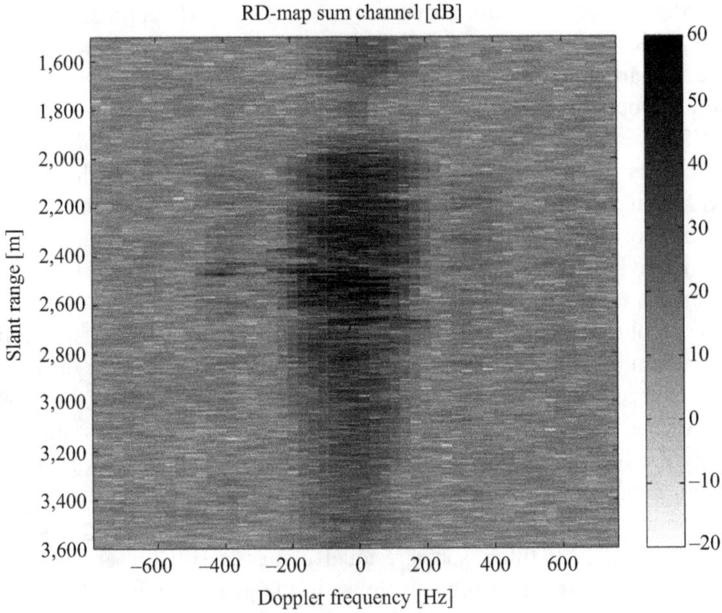

Figure 2.7 Range/Doppler map of the sum channel [dB]. © 2012 IEEE. After [13]

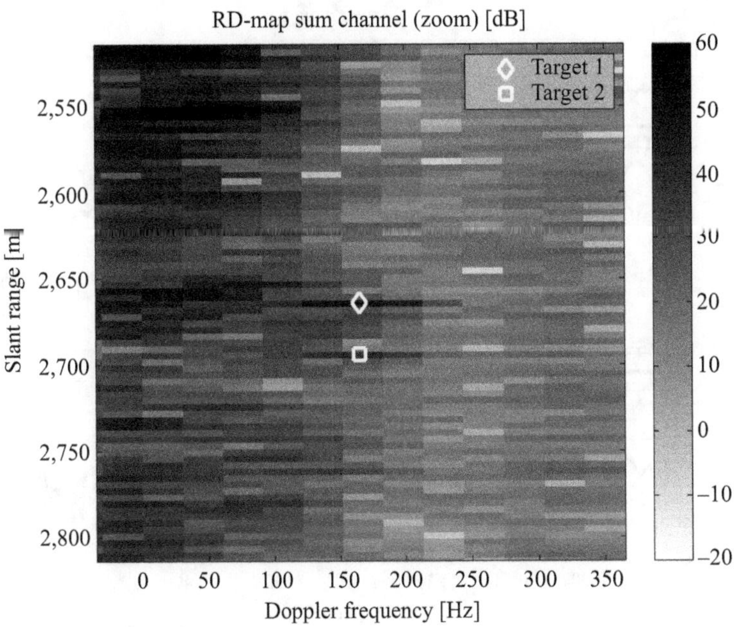

Figure 2.8 Zoom of the range/Doppler map of the sum channel over the two targets [dB]. © 2012 IEEE. After [13]

In order to test the effectiveness of the RD^3-STAP approach in this real situation, some uncertainty in the knowledge of the target parameters has been emulated. In particular, given the limited number of spatial DOFs, only an error in the target Doppler frequency has been assumed of half a Doppler bin ($f_e = f_r/(2M) = 16$ Hz), leading to a nominal Doppler frequency of $f_s^N \approx 150$ Hz for both targets. In the following analysis, the RD^3-STAP approach will be compared with both conventional D^3-STAP and with stochastic STAP. In particular, the adjacent bin post-Doppler STAP technique [2] will be considered, applied with all available receiving channels and with three Doppler bins. The interference covariance matrix is estimated by means of local learning strategy, using 36 range gates (twice the number required by the RMB rule). The strong target is analysed first, by extracting from the PAMIR data cube the corresponding space–time data snapshot.

D^3-STAP and RD^3-STAP filters have then been applied to the input data snapshot of the strong target, leading to the output signal spectra reported in the upper sub-plot of Figure 2.9. For comparison, the input signal power spectrum is reported as well. As one can see, the target is strong and separated from the clutter contribution at 0 Hz so that it is visible even after D^3-STAP filtering. However, by applying the D^3-STAP filter, some residual clutter is still present in the output signal due to the presence of high sidelobes in the pulse response. In particular, a residual clutter contribution at about 40 Hz is still visible after D^3-STAP filtering, which is not present if the RD^3-STAP filter is applied.

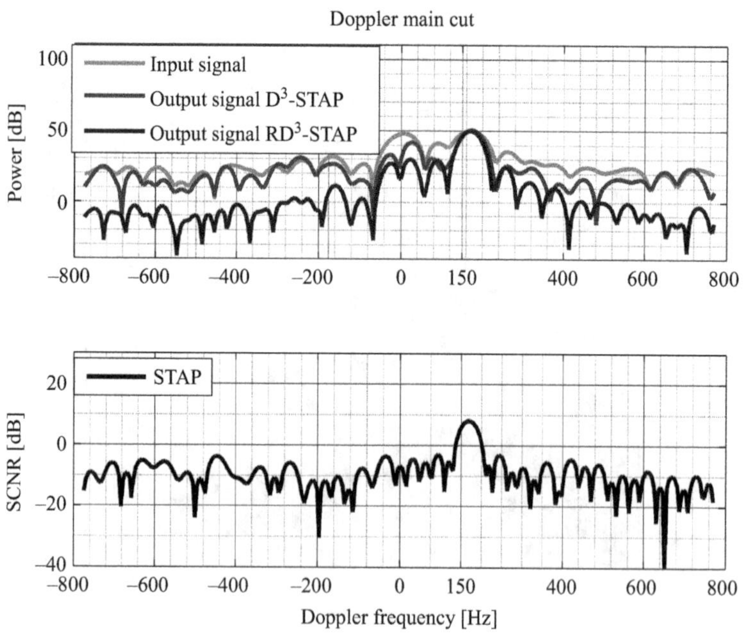

Figure 2.9 Output signal spectrum [dB] – target 1 range gate. © 2012 IEEE. After [13]

Observing Figure 2.9, one can see that the residual clutter contribution in the D^3-STAP output signal spectrum is due to the presence of a high sidelobe in the corresponding pulse response located at the same Doppler frequency. This does not occur if the RD^3-STAP filter is applied, due to the low sidelobes shape of the pulse response. This property directly originates from the solution of the convex minimization problem in (2.15) (see [13] for more details). In the second sub-plot of Figure 2.9, the Doppler cut of the strong target range gate is reported after the application of the adjacent-bin post-Doppler STAP filter. Please note that the output signals obtained after D^3-STAP filtering are not normalized to the interference power, thus leading to a different scaling of the y axis with respect to the stochastic STAP filter output, and hence requiring the use of two different sub-plots for proper comparison. The normalization to the interference power has not been performed in order to show the unit gain of the two D^3-STAP filters at the nominal target Doppler frequency. As mentioned before, the stochastic STAP filter is made adaptive by estimating the interference covariance matrix of the target range gate using secondary data taken from range gates close to the CUT. Due to the vicinity of the two targets, the secondary data corresponding to one target range gate include also the range gate of the other target, thus determining inter-target nulling. This effect is only partly visible when the strong target range gate is considered. In fact, the presence of the weak target in the secondary data has only a limited effect on the strong target detection. In contrast, the inter-target nulling is much more evident when the weak target range gate is considered. In this latter case, the presence of the strong target polarizes the statistics of the secondary data, making them similar to the range gate under test and making the weak target detection more difficult (if not even impossible).

This effect is shown in the range-Doppler map after stochastic STAP filtering reported in Figure 2.10, where the output power has been normalized to the average interference power. As is apparent, the strong target is clearly detectable, with an estimated signal-to-interference-plus-noise ratio (SINR) of about 15 dB. On the other hand, the weak target is completely nulled by the presence of the strong target in the corresponding secondary data. It is clear that an inter-target nulling effect is intrinsically impossible when D^3-STAP or RD^3-STAP filtering is applied. As a consequence, we expect it to be able to detect the weak target. To verify this, the previous analysis has been repeated extracting the weak target range gate from the PAMIR data cube. The corresponding Doppler main cut of the input signal spectrum is reported in the upper plot Figure 2.11. We can easily observe how the target detection is now more challenging with respect to the strong target range gate case.

In particular, the clutter is now mainly characterized by three different contributions located at about -100, 30 and 120 Hz. In particular, the latter clutter contribution is close to the target position. The first sub-plot in Figure 2.11 reports also the output power spectra after the application of D^3-STAP and RD^3-STAP filtering. As one can see, when the D^3-STAP filter is applied, the sidelobe level is unacceptable for reliable target detection, whereas a much better result is obtained when the RD^3-STAP filter is considered.

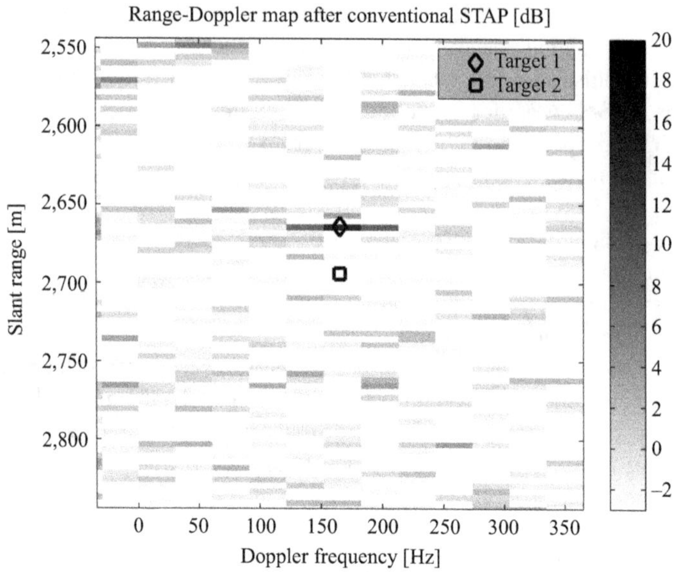

Figure 2.10 Range/Doppler map after stochastic STAP (zoom) [dB]. © 2012 IEEE. After [13]

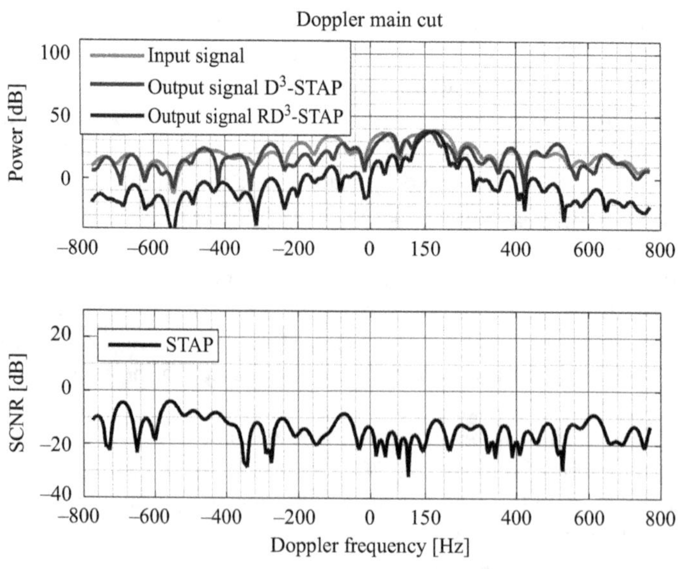

Figure 2.11 Output signal spectrum [dB] target 2 range gate. © 2012 IEEE. After [13]

The second sub-plot of Figure 2.11 reports, for comparison, the output power spectrum after adjacent-bin post-Doppler STAP filtering normalized to interference power level. As is apparent, the weak target is completely cancelled due to the inter-target nulling effect.

2.5 Applications of RD^3-STAP

2.5.1 RD^3-STAP filter in the SAR-GMTI case

In this subsection, RD^3-STAP is extended to the case where GMTI capabilities are required together with SAR. We expect RD^3-STAP to be an effective tool for space–time adaptive interference cancellation in this case, the clutter characteristics here being far from homogeneous from range gate to range gate. In addition, the RD^3-STAP can be integrated with a bank of focusing filters matched to different possible target parameter configurations in order to provide also target imaging capabilities, thus fully exploiting the high resolution characteristics of SAR. It is worth noticing that the idea of joint combination of STAP and imaging has already been investigated in recent works such as [22,23], and references therein. With respect to these contributions, which reside on conventional stochastic STAP, the approach here proposed exploits RD^3-STAP, thus enabling single snapshot clutter suppression.

2.5.1.1 Signal model

The signal model already introduced in (2.1) has to be slightly adapted to account for peculiarities of the SAR case. In particular, a side-looking ULA is considered, moving at a constant speed V in a rectilinear path. For a large number of coherently integrated pulses M, as it is usually the case for SAR systems, the received space–time samples can be conveniently processed in the Doppler domain, that is after temporal fast Fourier transform (FFT). This operation guarantees a decoupling of the different spectral components of the clutter [24], thus making possible a factorization of the overall space–time cancellation problem in independent spatial only cancellation sub-problems [25] without losing clutter cancellation capability. The signal model in (2.1) expressed after Doppler FFT can be written as follows:

$$\begin{aligned}\mathcal{X}_{\mu,n} &= \sigma_\mu(V_{\text{at}}, V_r)\cdot\alpha_{s,\mu}\cdot\exp[\mathrm{j}\,knd\sin\theta_s(\mu)]\\ &\quad+\sigma_\mu(0,0)\cdot\sum_{i=1}^{K-1}\alpha_{i,\mu}\cdot\exp[\mathrm{j}\,knd\sin\theta_i(\mu)]+\mathcal{R}_{\mu,n}\end{aligned} \tag{2.27}$$

where $\mu = 1, \ldots, M$ now indicates the Doppler bin index, $k = 2\pi/\lambda$ indicates the wavenumber, and $\alpha_{s,\mu}$ is the unknown moving target complex amplitude including also the sub-array receiving gain $g(\mu)$, $\alpha_{i,\mu}$ is the complex amplitude of the ith interfering signal present in the μth Doppler bin and $\mathcal{R}_{\mu,n}$ is the thermal noise component assumed to be uncorrelated among channels and Doppler bins and complex normal distributed with zero mean and unit variance (this term is just the Doppler counterpart of $r_{m,n}$ in (2.1)). $\theta_s(\mu)$ and $\theta_i(\mu)$ describe the varying DOA over Doppler of the target and of the ith clutter patch, respectively. $K-1$ is the number of interfering signals.

The peculiarity of the SAR case resides in the terms σ_μ which express, both for the target and for the clutter contribution, the quadratic variation of the phase over Doppler frequency. Specifically, we have:

$$\sigma_\mu(V_{\mathrm{at}}, V_r) = \exp\left[\mathrm{j}\frac{\pi R_0}{2(V - V_{\mathrm{at}})}\left(f_D(\mu) - \frac{2V_r}{\lambda}\right)^2\right] \tag{2.28}$$

where V_{at} represents the target along track velocity component and V_r is the target radial velocity. R_0 is the target slant range at broadside.

While the terms σ_μ account for inter-pulse phase variations, the inter-channel phase differences can be expressed exploiting the variation of the DOA θ over Doppler frequency f_D:

$$f_D = \frac{2V\sin\theta}{\lambda} \tag{2.29}$$

In particular, for the clutter, (2.29) is inverted and $\theta_i(\mu)$ is expressed as follows:

$$\theta_i(\mu) = \sin^{-1}\left(\frac{\lambda\Phi_{D,\mu}(i)}{2V}\right) \tag{2.30}$$

where $\Phi_{D,\mu}$ is a $(K-1)\times 1$ vector of equally spaced values ranging from $f_D(\mu) - 2\sigma_D$ to $f_D(\mu) + 2\sigma_D$, where $f_D(\mu)$ represents the μth Doppler bin and σ_D is the standard deviation of the clutter distribution over the angle/Doppler plane, assumed to be Gaussian. Accordingly, $\alpha_{i,\mu}$ can be expressed as follows:

$$\alpha_{i,\mu} = \alpha_{0,\mu}\cdot\exp\left[-\frac{(\Phi_{D,\mu}(i) - f_D(\mu))^2}{2\sigma_D^2}\right] \tag{2.31}$$

where $\alpha_{0,\mu}$ accounts for all other non-mentioned factors affecting clutter reflectivity including the sub-array receiving gain $g(\mu)$. Concerning the target, (2.29) can be inverted after adequately accounting for target motion parameters, leading to:

$$\theta_s(\mu) = \sin^{-1}\left(\frac{\lambda}{2(V - V_{\mathrm{at}})}\cdot\left(f_D(\mu) - \frac{2V_r}{\lambda}\right)\right) \tag{2.32}$$

If $\theta_s(\mu)$ is known exactly, we can define the scalar complex variable $z_s(\mu) = \exp[jkd\sin\theta_s(\mu)]$. FW cancellation quantities for the μth Doppler bin can then written as follows:

$$\mathcal{X}_{\mu,n} - z_s(\mu)^{-1}\cdot\mathcal{X}_{\mu,n+1}, \qquad n = 1,\ldots,N-1 \tag{2.33}$$

and they can be arranged in an interference-only matrix $\mathbf{F}_2(\mu)$, from which the D^3-STAP filter for the μth Doppler bin an be derived following a derivation analogous to the one already presented in Section 2.2.

In its robust implementation (RD^3-STAP for SAR [26]), the hypothesis of exact knowledge of target parameters ($\theta_s(\mu)$ in our case) is removed as long as it is possible to confine $\theta_s(\mu)$ to a bounded region of uncertainty. Indicating with $\mathbf{c}(\mu)$ the $K_n \times 1$ complex vector steered in the true direction $\theta_s(\mu)$ and with $\mathbf{a}(\mu)$ the $K_n \times 1$ complex vector steered in the nominal direction $\theta_s^{(N)}(\mu)$, the bounded region of uncertainty can be expressed as follows:

$$\mathbf{c}(\mu) = \mathbf{a}(\mu) + \mathbf{e}(\mu), \quad \text{with } \|\mathbf{e}(\mu)\| \leq \epsilon_m \tag{2.34}$$

where $\mathbf{e}(\mu)$ is the norm bounded error vector. As a consequence, the RD^3-STAP problem for the μth Doppler bin is:

$$\begin{aligned} &\min_{\mathbf{w}(\mu)} \|\mathbf{F}_2(\mu)^H \mathbf{w}(\mu)\| \\ &\text{subject to: } |\mathbf{w}(\mu)^H \mathbf{a}(\mu)| - \epsilon \cdot \|\mathbf{w}(\mu)\| \geq 1 \end{aligned} \tag{2.35}$$

Once the RD^3-STAP filter for the μth Doppler bin is obtained, the corresponding spectral clutter free component $y(\mu)$ can be derived as

$$y(\mu) = \mathbf{w}(\mu)^H \cdot \left[\mathcal{X}_{\mu,1} \mathcal{X}_{\mu,2} \ldots \mathcal{X}_{\mu,K_n}\right]^T \tag{2.36}$$

An important comment is in order. As already mentioned in [17], to ensure the minimization problem in (2.35) to be convex, the weight vector $\mathbf{w}(\mu)$ undergoes a phase rotation which locks its phase to the phase of $\mathbf{a}(\mu)$, that is the phase of the nominal target steering vector for the μth Doppler bin including the term $\sigma_\mu(V_{\text{at}}, V_r)$. This phase rotation does not affect the minimization of the cost function in (2.35) and also guarantees control over the phases of the different RD^3-STAP filters. This characteristic is extremely important if we aim at successively focusing the outputs $y(\mu)$. Specifically, note that locking the phase of $\mathbf{w}(\mu)$ to the phase of $\mathbf{a}(\mu)$ automatically ensures that the application of the weight filter to the data in (2.36) not only suppresses the interference, but also compensates the moving target phase history over Doppler. As a consequence, the azimuth compressed response from the moving target can simply be obtained by performing an inverse FFT (IFFT) over $y(\mu)$.

2.5.1.2 Integration with the focusing step

The aforementioned RD^3-STAP filter can be integrated with the SAR focusing technique as sketched in Figure 2.12. Specifically, multi-channel SAR raw data are first processed independently channel by channel in the range domain. Range processing includes range compression, range cell migration correction, and eventually range walk correction and secondary range compression. Subsequently, an azimuth FFT is performed. Range compressed data in the Doppler domain undergo the RD^3-STAP filtering described in the previous paragraph, which suppresses the interference and also corrects for Doppler phase history. Finally, an inverse azimuth FFT is performed leading to a clutter free focused response of the moving target. The entire processing has to be parallelized for all possible target motion parameters V_{at} and V_r, since both the focusing and the clutter cancellation

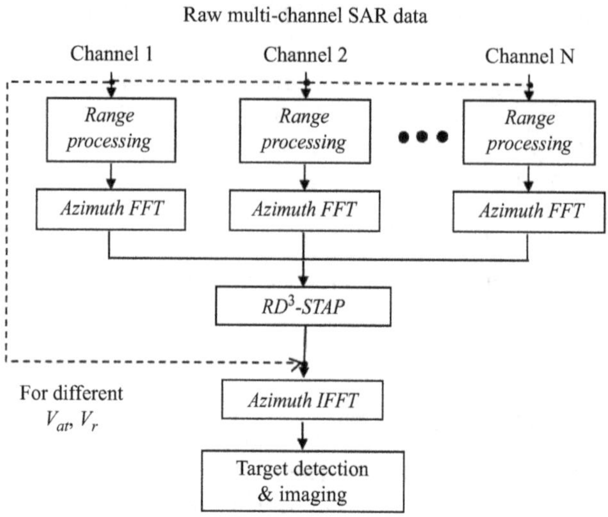

Figure 2.12 Block diagram of the integrated technique. © 2012 IEEE. After [26]

Table 2.3 Case study parameters

Parameter	Value
Carrier frequency	X-band
Sub-array distance (d)	0.26 m
Number of sub-arrays (N)	5
Number of pulses (M)	450
Platform velocity (V)	100 m/s
Random target amplitude ($\alpha_{S,m}$)	$5 \cdot \exp(j\pi/5) \cdot g(m)$
Target along-track velocity (V_{at})	0 m/s
Target radial velocity (V_{rad})	3 m/s
Random clutter amplitude ($\alpha_{0,m}$)	$36 \cdot \exp(j\pi/3) \cdot g(m)$
Clutter dispersion (σ_D)	1
Thermal noise ($r_{m,n}$)	AWGN, zero mean, unit variance

steps highly depend on the target state. The result is then a bank of filters that both suppress the interference and focus the moving target. The expected output signal to interference plus noise ratio is high since not only the interference is suppressed, but also the target energy is maximized by proper azimuth processing [22,23].

2.5.1.3 Case study analysis

The effectiveness of the proposed technique is tested against a simulated dataset, whose main parameters are reported in Table 2.3. Please note that having $N = 5$ receiving channels, only up to $K - 1 = 2$ interferences can be simultaneously suppressed for each Doppler bin. The main aim of this analysis is to verify the effectiveness of the clutter cancellation Doppler bin per Doppler bin, and

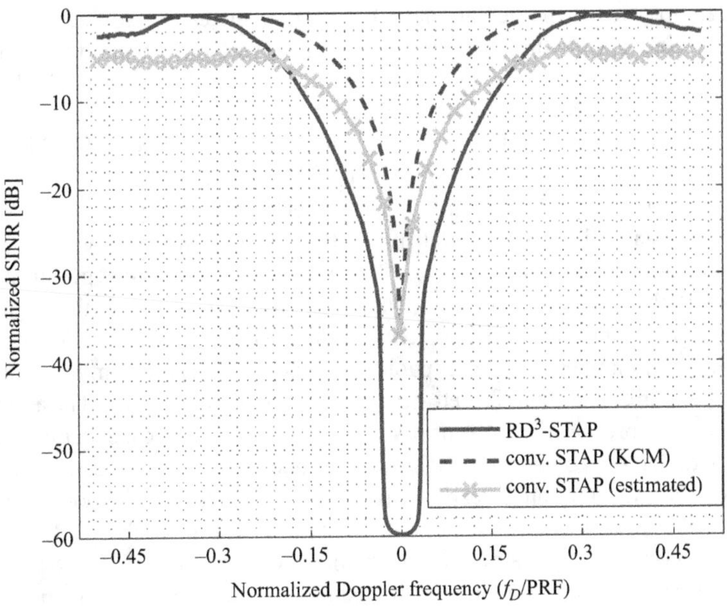

Figure 2.13 Comparison of normalized SINR [dB]. © 2012 IEEE. After [26]

consequently to verify the target imaging capability. With reference to the block diagram in Figure 2.12, for simplicity, the entire processing is matched to the exact target velocity parameters.

Clutter cancellation

To verify the clutter cancellation capability, two different investigations have been conducted.

In the first analysis, we aimed at comparing the clutter cancellation capability of RD^3-STAP in comparison with conventional stochastic STAP, for the case study scenario depicted in Table 2.3. In particular, as conventional stochastic STAP, the single bin post-Doppler technique has been considered because, as the RD^3-STAP technique here presented, it allows a partitioning of the overall STAP problem. Comparison has been conducted in terms of output SINR over normalized target Doppler frequency, and results are reported in Figure 2.13. Both for conventional stochastic STAP and for RD^3-STAP, the SINR is measured as the ratio between the adapted patterns (in power) in the target and in the interference directions. Obviously, for the conventional stochastic STAP, this requires the definition of the corresponding covariance matrix, while for the RD^3-STAP, the simple implementation of (2.1) is required. Concerning RD^3-STAP, the region of uncertainty is set to one-fifth of the antenna beamwidth, which is a reasonable value in target tracking or confirmation cases. Performance of conventional stochastic STAP filter is evaluated under the clairvoyant hypothesis of known covariance matrix, and for the more realistic case of a very limited sample support for interference covariance matrix estimation. In particular, only five homogeneous range gates have been used, and diagonal loading has been applied to robustify matrix estimation. As is apparent, conventional STAP is theoretically able to outperform RD^3-STAP, but only if a fine estimation of the interference is conducted, otherwise adaptivity losses have to be accepted. Please note that the scenario here considered is actually favourable for conventional STAP (over RD^3-STAP), since a sample support (although small) is still available. By further reducing such sample support, additional instabilities are expected up to the situation where covariance matrix estimation is not possible anymore. Concerning the achievable performance of RD^3-STAP, the following considerations are in order. The width of the clutter filter notch is directly introduced by the robustness constraint in (2.15) which preserves target self-nulling in case of target parameter mismatch. Apart from this, consistent degradations are also experienced nearby the central blind zone. These degradations simply represent the limit of a single snapshot clutter cancellation. In other words, they have to be accepted if no covariance matrix can be estimated due to extremely fast varying interference characteristics. Alternatively, a double canceller RD^3-STAP approach such as the one in (2.25) can be exploited.

The second analysis, aimed at evaluating the behaviour of RD^3-STAP in the SAR case. To this end, following the procedure described in the preceding sections, the weight vector $\mathbf{w}(f_D)$ obtained by solving (2.15) has been evaluated for every Doppler bin independently, for a region of uncertainty corresponding to the antenna main beamwidth. After applying a FFT in the spatial domain, the RD^3-STAP filter transfer function is obtained, see Figure 2.14. For convenience the behaviours of

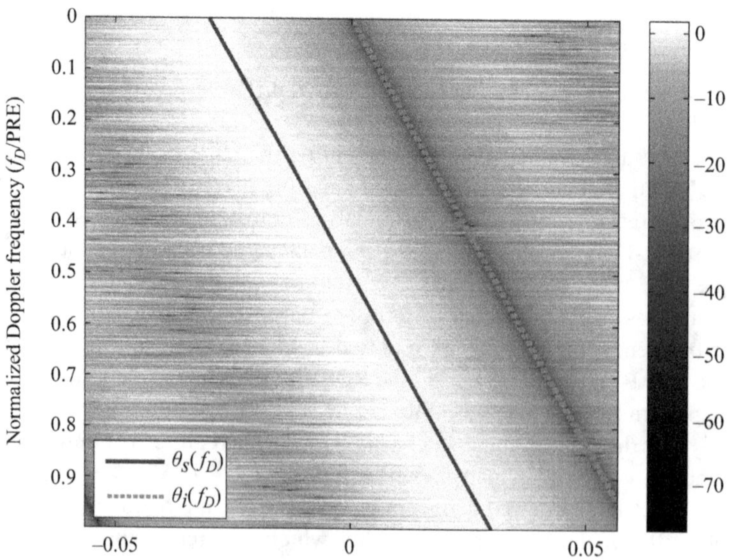

Figure 2.14 RD^3-STAP transfer function [dB]. © 2012 IEEE. After [26]

$\theta_S(f_D)$ (solid line) and of $\theta_i(f_D)$ $(i = 1, 2)$ (two adjacent dashed lines) are also reported. As is apparent, the RD^3-STAP filter maintains over Doppler a constant unitary gain in the target direction, while strong depressions are always ensured to adequately mitigate the interferences. Please note that the two green dashed lines in Figure 2.14 represent the clutter ridge. While all interfering signals along this line are constantly present at the same time, the target signal is moving along the red solid line over slow-time. For this reason and due to the clutter Doppler decoupling, if target parameters are known with some acceptable level of accuracy, the RD^3-STAP filter has to cope, Doppler bin per Doppler bin, only with the few interferences associated with the corresponding target Doppler location.

In Figure 2.15, the output of the RD^3-STAP filter is reported as a function of the Doppler frequency, that is $y(f_D)$. Specifically, the first sub-plot reports the amplitude of $y(f_D)$, while the phase is shown in the second sub-plot. As one can see, the amplitude of $y(f_D)$ nicely replicates the behaviour of the antenna pattern over Doppler or equivalently of $a_S(f_D)$ (superimposed markers), while the linear phase of $y(f_D)$ clearly shows that the phase locking of $\mathbf{w}(f_D)$ to $\mathbf{a}(f_D)$ automatically compensates the moving target phase history. Finally, also note the smooth behaviour of the RD^3-STAP filter transfer function in the surroundings of the target position in Figure 2.14. Such smooth behaviour is the peculiar characteristic of the robust implementation of the D^3-STAP filter. In fact, it allows proper reconstruction of the behaviour of $a_S(f_D)$ even in presence of mismatch between the nominal and the true target DOA (here not considered). Even in normal STAP applications

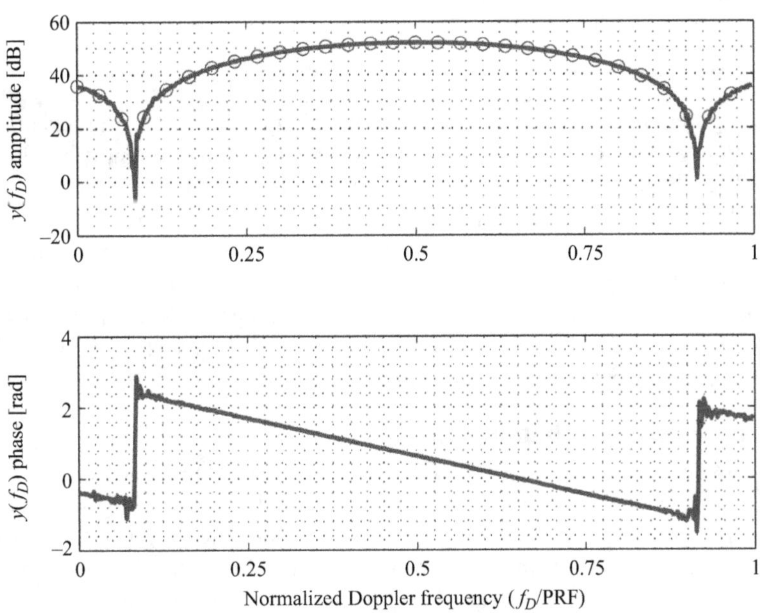

Figure 2.15 Filter output $y(f_D)$ in amplitude and phase. © 2012 IEEE. After [26]

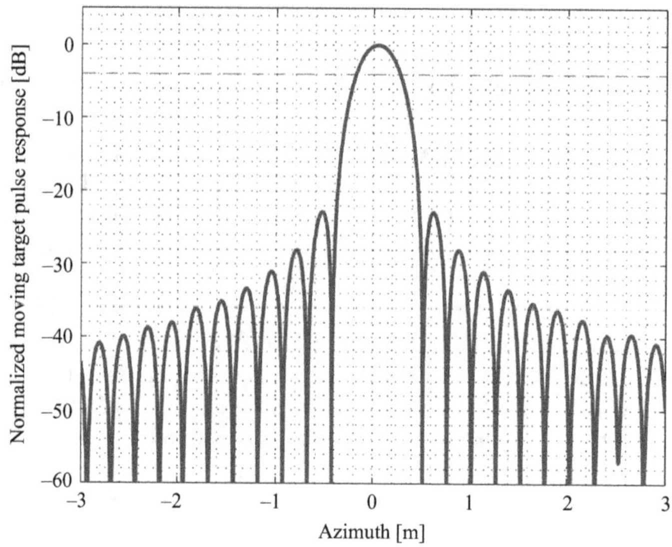

Figure 2.16 Azimuth pulse response of the moving target. © 2012 IEEE. After [26]

[e.g. with short coherent processing intervals (CPIs)], this is an important feature since it allows to prevent target self-nulling effects (see [13]), but in the SAR case, it becomes an extremely important feature since a trusty reconstruction of $\alpha_S(f_D)$ is necessary to avoid distortions in the following imaging step.

Moving target imaging

Following the previous considerations, the imaging of the moving targets after the application of the RD^3-STAP filters in the Doppler domain simply requires an IFFT. The obtained pulse response, for the sample case study under analysis, is reported in Figure 2.16. The remaining noise level is well below the values here shown (compare also markers and solid line in the upper sub-plot of Figure 2.15), so that the behaviour looks noise free but it is not. The obtained azimuth resolution is about 46 cm, which agrees with the target Doppler bandwidth shown in Figure 2.15. The sidelobe level is also increased with respect to the conventional 13 dB of a sinc function due to the azimuth antenna taper.

2.5.1.4 High signal bandwidth

Up to now, we have proposed RD^3-STAP as an MTI solution for SAR, but we have not yet considered the effects of a high signal bandwidth on the formulation of our RD^3-STAP problem. If the bandwidth of the transmitted signal (i.e. chirp) is high, the narrowband assumption does not hold meaning that the wavenumber k in (2.27) is no longer constant. This might require a partition of the signal bandwidth in multiple sub-apertures prior to range compression, so that the narrowband assumption holds for each sub-aperture independently. Subsequently, the RD^3-STAP problem, as described in the preceding paragraph, can be applied to each sub-aperture. This might lead to slight modifications of the diagram sketched in Figure 2.12.

2.5.2 Target DOA estimation with RD^3-STAP

In this section, the RD^3-STAP filter is applied to the space-only case (thus becoming a RD^3 adaptive beam forming, RD^3-ABF, filter) to retrieve an accurate estimation of the actual target DOA based on a closed form formula, see [27]. Such target DOA estimates may serve in tracking scenarios to keep updated the target parameters' knowledge in successive CPIs. It is worth noticing that an approach to refine the target parameters' estimates for D^3-STAP filtering based on the variations of the scalar quantities z_s and z_t in (2.3) can be found in [28].

As already mentioned in the previous sections and restricting the analysis to the space-only case, RD^3-ABF designs a $K_n \times 1$ adapted weight vector starting from the $N \times 1$ received signal vector expressed in (2.1). Thus, cancellation of the interference components can be obtained by vector multiplication between $\mathbf{w}^H$ and a truncated version of the received signal:

$$y_A = \mathbf{w}^H \mathbf{x}_A^{(K_n)} \tag{2.37}$$

where $\mathbf{x}_A^{(K_n)} = [x_1 \cdots x_{K_n}]^\dagger$ is a $K_n \times 1$ vector containing only the first K_n elements of (2.1), and where † indicates the transpose operator. Analogously, one can apply the same-weight vector $\mathbf{w}^H$ to the last K_n elements of (2.1), obtaining a different cancellation of the interference:

$$y_B = \mathbf{w}^H \mathbf{x}_B^{(K_n)} \tag{2.38}$$

being $\mathbf{x}_B^{(K_n)} = [x_{K_n} \cdots x_N]^\dagger$. The situation is depicted in Figure 2.17, for the sample case of $K_n = 3$, $N = 5$. Recalling (2.15), the two quantities y_A and y_B contain only a target signal component with the same amplitude (the same weight vector has been applied to $\mathbf{x}_A$ and $\mathbf{x}_B$), but with a different phase which is directly related to the true target DOA θ_s^T. In other words, if the RD^3-ABF interference cancellation is effective, y_A and y_B contain two different spatial samples of the target signal only that can be exploited to retrieve a refined estimate of θ_s^T using an approach similar to phase monopulse.

To this end, making explicit the target component in (2.37) and (2.38), y_A and y_B can be rewritten as follows:

$$\begin{aligned} y_A &= \alpha_s \cdot \sum_{n=1}^{K_n} w_n^* \exp\left[\mathrm{j}\, knd \sin \theta_s^T\right] + \delta_A \\ y_B &= \alpha_s \cdot \sum_{n=1}^{K_n} w_n^* \exp\left[\mathrm{j}\, k(n + K_n - 1)d \sin \theta_s^T\right] + \delta_B \end{aligned} \tag{2.39}$$

where δ_A and δ_B indicate the residual interference terms after application of the RD^3-ABF filter over $\mathbf{x}_A$ and $\mathbf{x}_B$, respectively. We can now define an equivalent

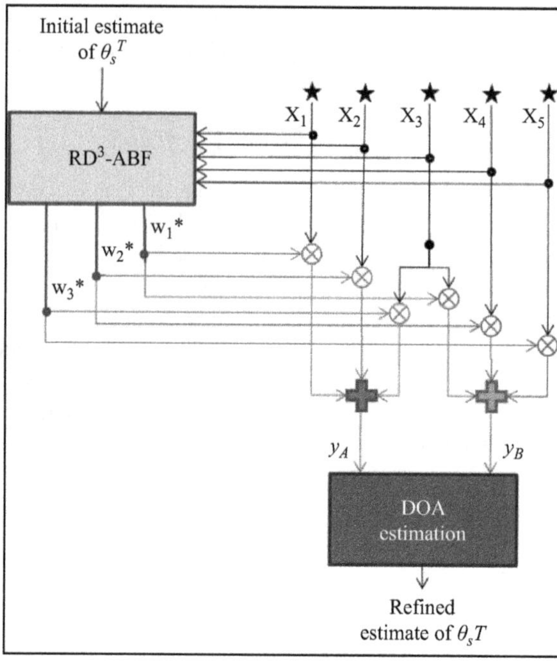

Figure 2.17 DOA estimation block diagram

beam obtained by subtracting y_A and y_B after phase shifting of θ_s: $\Delta(\theta_s) \doteq y_A - \exp[j\bar{\theta}_s] \cdot y_B$, where:

$$\bar{\theta}_s = -k(K_n - 1)d \sin\theta_s \tag{2.40}$$

By simple algebraic manipulations and defining $\eta(\theta_s)$ as:

$$\begin{aligned}\eta(\theta_s) &\doteq \frac{\Delta(\theta_s)}{y_A} \\ &= \left\{1 - \exp\left[jk(K_n - 1)d \cdot (\sin\theta_s^T - \sin\theta_s)\right]\right\} \\ &\quad + \frac{1}{y_A} \cdot \exp[-jk(K_n - 1)d \sin\theta_s] \\ &\quad \cdot \left(\delta_A \cdot \exp\left[jk(K_n - 1)d \sin\theta_s^T\right] - \delta_B\right)\end{aligned} \tag{2.41}$$

a RD³-ABF based DOA estimator can be derived by searching for the θ_s which nulls the absolute value of $\eta(\theta_s)$, that is:

$$|\eta(\theta_s)|_{\theta_s = \hat{\theta}_{s,\mathrm{RD^3-ABF}}} = 0 \tag{2.42}$$

The term in braces in (2.41) is equal to zero for $\theta_s = \theta_s^T$. The remaining contribution to $\eta(\theta_s)$ in (2.41), which is due to the residual interference terms δ_A and δ_B, is responsible for target DOA estimation errors. However, the effect of this error contribution is expected to be limited due to the presence of y_A in the denominator. (2.42) can be solved in closed form by imposing that $Re[\eta(\theta_s)] = Im[\eta(\theta_s)] = 0$. After some tedious algebraic derivations one gets:

$$\sin \bar{\theta}_s = \frac{y_{Ar}^2 + y_{Ai}^2}{\left((y_{Ai}y_{Bi} + y_{Ar}y_{Br})^2/(y_{Ai}y_{Br} - y_{Ar}y_{Bi})\right) - y_{Ar}y_{Bi} + y_{Ai}y_{Br}} \tag{2.43}$$

where $y_A = y_{Ar} + \mathrm{j}y_{Ai}$ and $y_B = y_{Br} + \mathrm{j}y_{Bi}$. By simple inversion of (2.40), the closed expression of the RD^3-ABF based DOA estimate can be retrieved. Note that such a target DOA estimation approach could also be applied to the conventional D^3-ABF method (**w** derived from (2.9)), by solving (2.42) varying θ_s^N. This would however require an increased computational load, since (2.9) should have to be solved for each different θ_s^N in order to overcome the target self-nulling effect.

2.5.2.1 Comparison with MLE

In order to test the effectiveness of the RD^3-ABF based estimator in (2.42), a comparison has been conducted with respect to achievable performances of a maximum likelihood estimator (MLE). Specifically, the MLE has been evaluated exploiting all N available spatial DOFs and under the assumption of known interference plus noise covariance matrix (clairvoyant hypothesis):

$$\hat{\theta}_{s,\mathrm{MLE}} = \underset{\theta_s}{\operatorname{argmax}} \left\{ \left| \frac{\left|\mathbf{s}(\theta_s)^H \mathbf{Q}^{-1}\mathbf{x}\right|^2}{\mathbf{s}(\theta_s)^H \mathbf{Q}^{-1}\mathbf{s}(\theta_s)} \right| \right\} \tag{2.44}$$

where $\mathbf{s}(\theta_s)$ is a $N \times 1$ DOA vector steered at θ_s, $\mathbf{x} = [x_1 \cdots x_N]^\dagger$ and $\mathbf{Q}$ is the $N \times N$ interference plus noise covariance matrix. It is worth noticing that the considered MLE is a target DOA estimation approach belonging to the class of stochastic STAP, since it requires the knowledge of the second order statistics of the interference (i.e. of the disturbance covariance matrix). As a consequence, in practical situations it will suffer all aforementioned drawbacks concerning potential lack of secondary data for covariance matrix estimation. Nevertheless, in the simulated case study here considered, clairvoyant MLE represents the best possible target DOA estimation one can get using stochastic STAP approaches, thus making it a valid benchmark for the RD^3-ABF based estimator. Comparison has been conducted by Monte-Carlo simulation with 1,000 trials for the case study of Table 2.4 varying the input SINR on the single sub-array, measured as:

$$\mathrm{SINR}_{\mathrm{input}} = \frac{|\alpha|^2}{\sum_{i=1}^{K-1} |\alpha_i|^2 + N_{\mathrm{SA}}} \tag{2.45}$$

Table 2.4 Case study parameters

Parameter	Value
Wavelength (λ)	0.03 m
Sub-array distance (d)	0.26 m
Number of sub-arrays (N)	5
Elements per sub-array (N_{SA})	16
Nominal target DOA (θ_s^N)	0 rad
True target DOA (θ_s^T)	$\theta_s^N - \theta_{3\,dB}/2$
Random target amplitude (α)	$5 \cdot e^{j\pi/5}$
Interference DOA (θ_1)	0.03 rad
Random interference amplitude (α_1)	$36 \cdot e^{j\pi/3}$
Thermal noise (r_n)	$\mathcal{N}^{\mathbb{C}}(0, N_{SA})$
DOA region of uncertainty (ϵ)	Within main beam

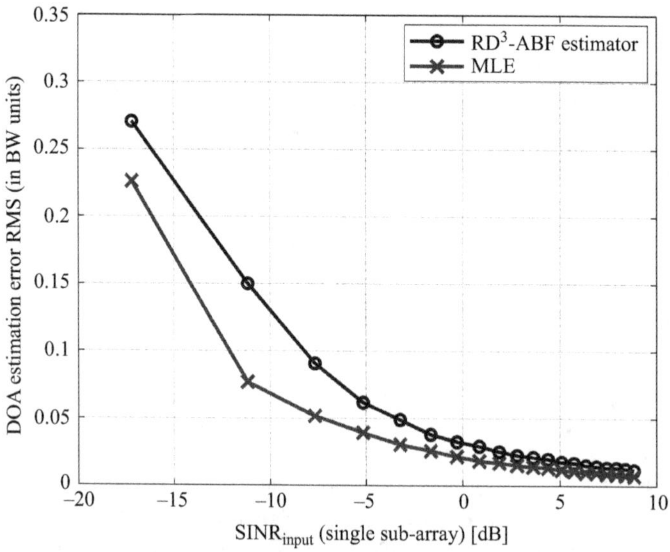

Figure 2.18 DOA estimations with RD³-ABF and MLE. © 2012 VDE VERLAG, Berlin, Offenbach. After [27]

Target DOA estimation errors expressed in terms of error root mean square normalized to the antenna beamwidth are shown in Figure 2.18 for both RD^3-ABF and MLE approaches. As is apparent, target DOA estimation accuracy of one tenth of the antenna beamwidth can be achieved even in presence of strong interference signals impinging on the antenna array. The estimation accuracy of RD^3-ABF based monopulse is only slightly worse than MLE. Such a performance degradation can be thought as the price to be paid for a single snapshot target DOA estimation.

2.6 Conclusions

In this chapter, the robust implementation of the D^3-STAP (viz. RD^3-STAP) filtering has been described. This approach solves the main drawback of D^3-STAP in its classical derivation, which is the need for exact knowledge of target parameters. This limit is overcome by replacing the look direction constraint with a more general constraint over the region of uncertainty of target parameters' knowledge. As a consequence, RD^3-STAP is reformulated as a convex optimization problem, leading to a filter design that minimizes the output interference power. The formulation of RD^3-STAP is provided for the general space–time case and it can be applied to FW, BK and FB implementations of D^3-STAP. The application of DRTs to allow application even with limited number of DOFs is also shown. RD^3-STAP exhibits nice properties for the adapted weight pattern both in terms of robustness to the target self-nulling effect and in terms of low sidelobes. The effectiveness of the proposed approach is tested against simulated and real data. Applications of RD^3-STAP filtering to the SAR case and to target DOA estimation problems are also presented.

2.7 Glossary

The following list includes the acronyms used in this chapter:

BK	backward
CPI	coherent processing interval
CUT	cell under test
DOA	direction of arrival
DOF	degrees of freedom
DRT	dimension reducing transformations
D^3-STAP	direct data domain space–time adaptive processing
FB	forward–backward
FFT	fast Fourier transform
FW	forward
IFFT	inverse fast Fourier transform
KCM	known covariance matrix
MLE	maximum likelihood estimator
MTI	moving target indication
PRF	pulse repetition frequency
RD^3-ABF	robust direct data domain adaptive beam forming
RD^3-STAP	robust direct data domain space–time adaptive processing
RMB	Reed Mallett Brennan
RMS	root mean square

SAR	synthetic aperture radar
SINR	signal-to-interference-plus-noise ratio
STAP	space–time adaptive processing
ULA	uniform linear array

References

[1] Klemm, R.: *Principles of Space-Time Adaptive Processing*, IET, 2006, 3rd Edition, London, UK.

[2] Ward, J.: *Space-Time Adaptive Processing for Airborne Radar*, MIT Lincoln Laboratory, 1994, Lexington, MA, USA.

[3] Melvin, W.L.: 'A STAP overview', *IEEE Aerospace and Electronic Systems Magazine*, 2004, **19**, 1, pp. 19–35.

[4] Sarkar, T.K., Wang, H., Park, S., *et al.*: 'A deterministic least-squares approach to space-time adaptive processing (STAP)', *IEEE Transactions on Antennas and Propagation*, 2001, **49**, 1, pp. 91–103.

[5] Aboutanios, E. and Mulgrew, B.: 'A STAP algorithm for radar target detection in heterogeneous environments', *2005 IEEE/SP 13th Workshop on Statistical Signal Processing*, Jul 2005, pp. 966–971.

[6] Aboutanios, E. and Mulgrew, B.: 'Evaluation of the single and two data set STAP detection algorithms using measured data', *IEEE International Geoscience and Remote Sensing Symposium*, Jul 2007, pp. 494–498.

[7] Degurse, J.-F., Savy, L., Perenon, R. and Marcos, S.: 'An extended formulation of the Maximum Likelihood Estimation algorithm. Application to space-time adaptive processing', *Proceedings International Radar Symposium (IRS)*, Sep 2011, pp. 763–768.

[8] Degurse, J.-F., Savy, L. and Marcos, S.: 'Single dataset methods and deterministic-aided STAP for heterogeneous environments', *International Radar Conference*, Oct 2014, pp. 1–6.

[9] Reed, I.S., Mallett, J.D. and Brennan, L.E.: 'Rapid convergence rate in adaptive arrays', *IEEE Transactions on Aerospace and Electronic Systems*, 1974, **AES-10**, Nov (6), pp. 853–863.

[10] Griesbach, J.D., Krich, S., Pulsone, N. and Rader, C.: 'Power variable training STAP', *Adaptive Sensor Array Processing Workshop*, Mar 2004, MIT Lincoln Laboratory, Lexington, MA, USA.

[11] Klemm, R.: *Applications of Space-Time Adaptive Processing*, IEE, 2004, London, UK.

[12] Park, S., Sarkar, T.K. and Schneible, R.A.: 'Prevention of signal cancellation in an adaptive nulling problem', *IEEE National Radar Conference*, May 1997, pp. 191–195.

[13] Cristallini, D. and Bürger, W.: 'A robust direct data domain approach for STAP', *IEEE Transactions on Signal Processing*, 2012, **60**, 3, pp. 1283–1294.

[14] Gaoxiang, X. and Cai, Z.: 'A sidelobe-constraint direct data domain least square algorithm', *2nd International Congress on Image and Signal Processing (CISP 2009)*, Oct 2009, pp. 1–4.

[15] Boyd, S. and Vandenberghe, L.: *Convex Optimization*, Cambridge University Press, 2004, UK.

[16] Li, J., Stoica, P. and Wang, Z.: 'On robust Capon beamforming and diagonal loading', *IEEE Transactions on Signal Processing*, 2003, **51**, 7, pp. 1702–1715.

[17] Vorobyov, S.A., Gershman, A.B. and Luo, Z.Q.: 'Robust adaptive beamforming using worst-case performance optimization: a solution to the signal mismatch problem', *IEEE Transactions on Signal Processing*, 2003, **51**, 2, pp. 313–324.

[18] Grant, M. and Boyd, S.: 'CVX: Matlab software for disciplined convex programming, version 2.1', *http://cvxr.com/cvx*, Mar 2014.

[19] Grant, M. and Boyd, S.: 'Graph implementations for nonsmooth convex programs', *Recent Advances in Learning and Control*, Springer-Verlag Limited, 2008, Lecture Notes in Control and Information Sciences, pp. 95–110.

[20] Klemm, R. and Cristallini, D.: 'STAP techniques for real-time processing in heterogeneous clutter', *Proceedings of 10th European Conference on Synthetic Aperture Radar, EUSAR 2014*, Jun 2014, pp. 1–4.

[21] Brenner, A.R. and Ender, J.H.G.: 'Demonstration of advanced reconnaissance techniques with the airborne SAR/GMTI sensor PAMIR', *IEE Proceedings Radar, Sonar and Navigation*, 2006, **153**, Apr (2), pp. 152–162.

[22] Cerutti-Maori, D., Sikaneta, I. and Gierull, C.H.: 'Optimum SAR/GMTI processing and its application to the radar satellite RADARSAT-2 for traffic monitoring', *IEEE Transactions on Geoscience and Remote Sensing*, 2012, **50**, 10, pp. 3868–3881.

[23] Cristallini, D., Pastina, D., Colone, F. and Lombardo, P.: 'Efficient detection and imaging of moving targets in SAR images based on chirp scaling', *IEEE Transactions on Geoscience and Remote Sensing*, 2013, **51**, 4, pp. 2403–2416.

[24] Wu, W.: 'Fourier transforms of stationary processes', *Proceedings of the American Mathematical Society*, 2005, **133**, 1, pp. 285–293.

[25] Ender, J.H.G.: 'Space-time processing for multichannel synthetic aperture radar', *Electronics Communication Engineering Journal*, 1999, **11**, Feb (1), pp. 29–38.

[26] Cristallini, D.: 'Exploiting robust direct data domain STAP for GMTI in very high resolution SAR', *IEEE Radar Conference (RADAR)*, May 2012, pp. 0348–0353.

[27] Cristallini, D.: 'Target DOA estimation based on robust deterministic STAP', *Proceedings of 9th European Conference on Synthetic Aperture Radar, EUSAR 2012*, Apr 2012, pp. 271–274.

[28] Choi, W. and Sarkar, T.K.: 'Minimum norm property for the sum of the adaptive weights for a direct data domain least squares algorithm', *IEEE Transactions on Antennas and Propagation*, 2006, **54**, 3, pp. 1045–1050.

Chapter 3
Array radar resource management

Alexander Charlish and Fotios Katsilieris**

Abstract

The operation of an electronically steered array (ESA) antenna is highly flexible in that a range of control parameters can be reconfigured nearly instantaneously. Consequently, an ESA is capable of executing numerous tasks supporting multiple functions, multiplexed in time and angle. However, this flexibility creates a challenging operation and resources management problem, in that a new radar dwell complete with beam direction and transmit waveform must be chosen within the time taken to execute the previous radar dwell. As controlling the operation of an ESA is beyond the capability of a human operator, automated management techniques are required, which are emerging as key performance factors for the next generation of multifunction radar systems. This chapter gives an overview of automated techniques for managing the operation and resources of an ESA.

3.1 Management architecture

Primarily, the radar must decide on a beam position and waveform for the next dwell while the previous dwell is being executed. However, as radar management techniques become more sophisticated, this decision may be a complex process whose computation takes longer than the duration of the previous dwell. Therefore, radar resource management architectures are typically hierarchical, such that the management processes are split over multiple levels, which each have differing management decision epoch time scales.

Many examples of radar resource management architectures can be found in the literature [1–5]; these architectures are comprised of similar components but with slight variations. Based on various approaches, the key components of a radar resource manager can be identified and are illustrated in the simplified

*Fraunhofer FKIE, Germany

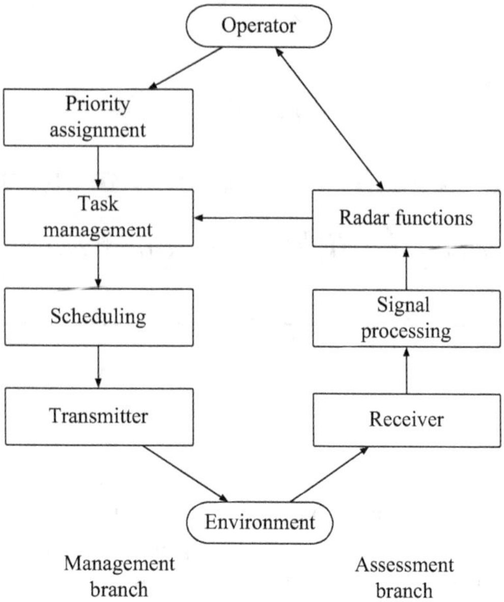

Figure 3.1 Core components for an electronically steered phased array antenna. The management modules are priority assignment, task management and scheduling

architecture in Figure 3.1. The core radar management components are as follows:

Priority assignment: The priority assignment module assigns a priority value to each radar task, which represents the task's entitlement to antenna usage relative to other tasks. Priorities can be defined by the operator based on the operational context and needs, or automatically for certain tasks, such as weapon guidance.

Task management: The task manager is responsible for selecting control parameters for each radar task based on its priority and other task-specific requirements. Examples of control parameters include the next time to execute a radar dwell and the corresponding transmit waveform to use. The task manager issues job requests to the scheduler, based on the selected control parameters for the task and its priority.

Scheduler: The scheduler is responsible for creating a timeline of jobs from the multiple job requests. Time conflicts between the job requests can be resolved using the task priority.

The architecture can be adapted to include an interface to other sensors or an interface to an environment or knowledge database [5]. Each of these core resource management modules is discussed in the following sections. In the following, a radar function is defined as a specific sensing purpose or objective (e.g. tracking in general), a task is a single realization within a function (e.g. tracking a specific target) and a job is a single usage of the antenna (e.g. a track beam at a specific time instance).

3.2 Task management

The task management module is responsible for selecting control parameters for each of the radar tasks and issuing job requests to the scheduler. Traditionally, control parameters for radar tasks have been fixed at design time. However, the flexibility of an ESA antenna in combination with modern computational processing capability enables the control parameters to be optimized in runtime, to give the best system performance based on the current mission requirements, scenario and environment.

This section discusses rules and heuristics for parameter optimization for the core radar functions of search, confirmation and tracking. Optimization methods for radar management, such as quality of service approaches or stochastic control are described in Vol. II, Chapter 5 ‘Cognitive radar management’.

3.2.1 Search management

The search function can be configured by varying the desired search volume, the search lattice, the beam separation, as well as the revisit interval time and transmit waveform (and hence dwell duration) in each beam position. For search, the revisit interval time is the time interval between dwells in the same beam position.

3.2.1.1 Search lattice and beam spacing

To search a specified angular area, it is necessary to decide where to position the individual radar beams. Beams are positioned to cover the angular search area in either a triangular or rectangular lattice with a specified separation between beam centres, that is in units of the 3 dB beamwidth. Spacing beams by the 3 dB beamwidth results in a non-uniform beam spacing in angle, due to the broadening of the radar beamwidth with off-boresight scan angle that results from the loss of effective antenna aperture area. Therefore, it is convenient to position beams equally spaced in $u - v$ space [6], which has the property that the beam pattern does not vary with off-boresight scan angle. The $u - v$ coordinates for a given direction are as follows:

$$u = \sin(\theta) \cdot \cos(\psi) \tag{3.1}$$

$$v = \sin(\theta) \cdot \sin(\psi) \tag{3.2}$$

where θ is the angle between the antenna boresight and the beam centre, and ψ is the orientation of the beam direction, as illustrated in Figure 3.2(a). Beam positions with a specific 3-dB separation can be found by projecting beams that are equally spaced on the $u - v$ plane, onto a hemisphere. This principle is illustrated in Figure 3.2(b).

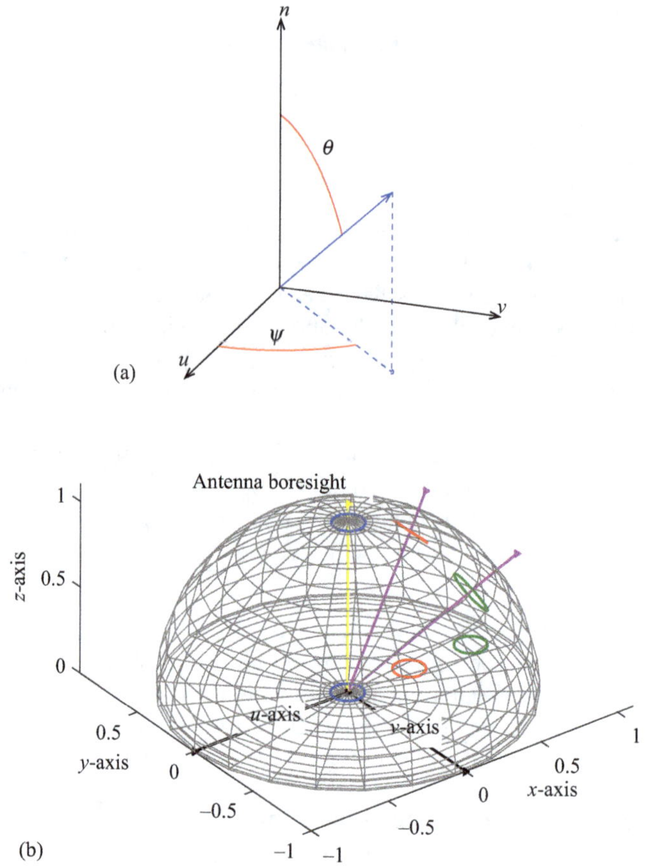

Figure 3.2 Antenna (u − v) coordinate system (a) u − v coordinates relative to antenna plane. The boresight of the antenna is in direction n. (b) Projection of radar beams onto hemisphere

The beams can be positioned in a rectangular or triangular lattice. A triangular search lattice in $u - v$ space is illustrated in Figure 3.3(a) for a static, non-rotating array face used for a ground to air application. For the beam positions off the antenna boresight, a loss in gain is encountered due to the loss of effective aperture in the steering direction. To compensate for the loss of gain the dwell time, which is the time spent in each beam position in one scan, can be increased as a function of the off-boresight scan angle. This is illustrated in Figure 3.3(b) when the antenna boresight is directed at $u = 0$ and $v = 0$.

If the separation between adjacent beams is increased, then less beams are required to search the area and therefore more time can be spent in each beam

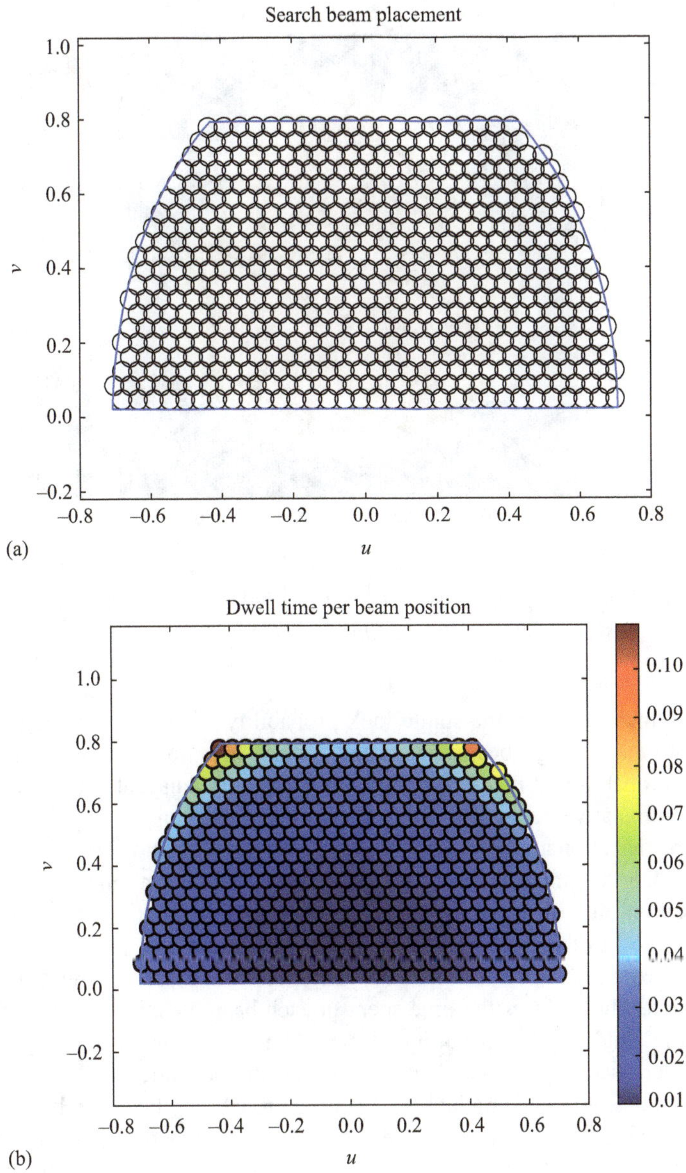

Figure 3.3 Example of beam placement for ground based non-rotating ESA. Loss of effective aperture is compensated by increasing the dwell time per beam position. (a) Beam placement and (b) dwell time per beam in seconds

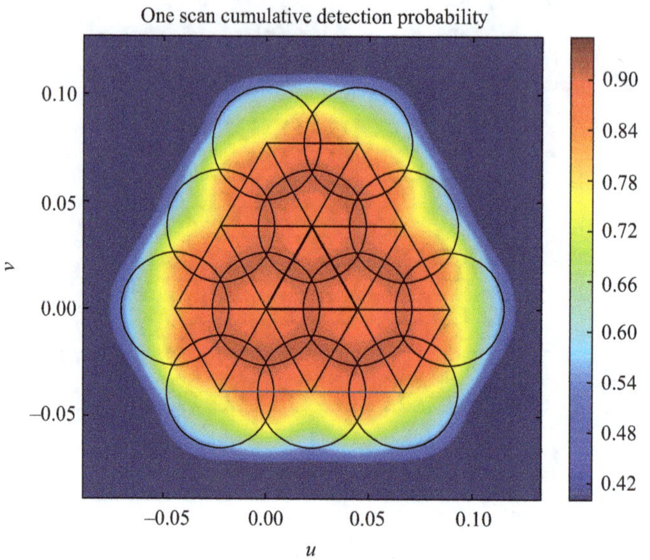

Figure 3.4 One-scan cumulative detection probability for a cluster of beam positions with beam spacing $\theta_s = 0.85$

position, which increases the single look probability of detection through integration. However, a large beam separation results in a low one-scan cumulative detection probability at the points equidistant between adjacent beam centres. The one-scan cumulative detection probability for a specified angle is the probability of at least one detection at that angle in a scan of all beam positions in the lattice. This is higher than the single look probability of detection at that angle from a single beam position, due to multiple detection opportunities resulting from the overlap between beams in the lattice.

If the beam separation is decreased, then more beams are required to search the angular area. This reduces the time spent in each beam position and therefore also reduces the single look probability of detection. However, decreasing the beam spacing increases the overlap between beams in the lattice and therefore also increases the one-scan cumulative detection probability between beams. Eventually, a gain is achieved, whereby the one-scan cumulative detection probability at the equidistant points between beam centres is higher than the one-scan cumulative detection probabilities at each beam centre. This effect is illustrated for a cluster of beam positions in Figure 3.4 with beam spacing $\theta_s = 0.85$ beamwidths.

The trade-off in choosing the beam separation is illustrated in Figure 3.5, where the one-scan cumulative detection range is plotted for varying beam separations and target ranges. It can be seen that for targets at farther range, it is better to use a wide separation between beams with longer dwells. This is because at greater range, longer dwells are required to achieve a single look probability of detection at the beam centre that is high enough to detect the target, whereas the

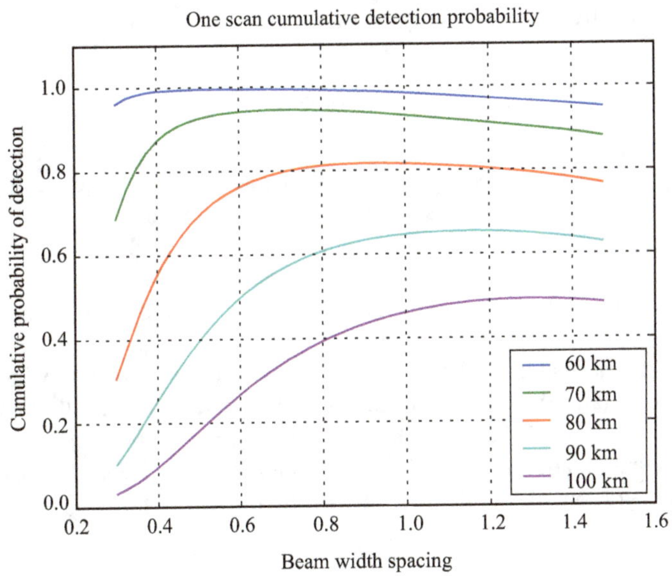

Figure 3.5 One scan cumulative detection probability as a function of beam separation for varying target ranges

one-scan cumulative detection probability drops significantly between the beams. For closer range targets, shorter dwells can be used resulting in closely spaced beams without significant drops in the cumulative detection probability in angle. However, it can be seen that the maxima are all broad and a choice between 0.8 and 1.0 performs reasonably well for all target ranges. The trade-off has been widely studied [7,8,9, Section 14.3], with similar conclusions of a very broad optimum beam separation at approximately 0.8 times the 3 dB beamwidth. The difference between a rectangular or triangular search lattice is not significant [10]; however, a triangular lattice has the benefit that the difference between the maximum and minimum one-scan cumulative detection probability in angle is smaller than for a rectangular lattice.

3.2.1.2 Revisit interval time and dwell duration

Potentially, the revisit interval t_r can be varied for each beam position in the search during runtime. Assuming that the search loading $l_s = \tau_d/t_r$ allowed for a beam position remains constant, then increasing the revisit interval also increases the dwell length τ_d in the beam position.

An objective for the search function is to maximize the multi-scan cumulative detection range, which is the range at which the cumulative probability of at least one detection of a specific target and trajectory over multiple scans exceeds a specified value, such as 0.9 (denoted R_{90}). The objective is then to select the revisit interval and hence dwell length that maximizes the cumulative detection range.

The cumulative detection probability P_c of at least one detection is:

$$P_c(r) = E_{\Delta_s}\left\{1 - \prod_{n\in N}[1 - P_s(r + (n + \Delta_s)t_r v_r)]\right\} \tag{3.3}$$

where $P_s(r)$ is the single look probability of detection for a target at range r, E_{Δ_s} denotes the expectation with respect to Δ_s, v_r is the radial velocity of the target, $N = \{0, \ldots, \lfloor (r_{\text{pu}} - r)/t_r v_r \rfloor\}$ is a set representing the number of looks until the target reaches range r, and r_{pu} is the target pop-up range. The offset $\Delta_s \in [0, 1]$ is a uniformly distributed random variable that represents the possibility of the target arrival at any time during a revisit interval. In the following, the range r and radial velocity v_r are normalized with respect to R_0, which is the range at which a unity signal-to-noise ratio (SNR) is achieved using a representative dwell length τ_d.

The single look probability of detection depends on the target radar cross-section fluctuation and the type of integration applied. For a Swerling 1 target and coherent integration, the single look probability of detection is:

$$P_s(r) = P_{\text{FA}}^{\left(\frac{1}{1+\text{SNR}(r)}\right)} \tag{3.4}$$

where the expected SNR at range r can be calculated based on the radar range equation. If a fixed temporal loading of l_s is available for the beam position, then the revisit interval also determines the dwell time. Assuming coherent integration over the dwell, the SNR is subsequently:

$$\text{SNR}(r) = \left(\frac{\Delta r}{\Delta r_e}\right)\left(\frac{1}{r}\right)^4 \tag{3.5}$$

where $\Delta r_e = \tau_d v_r / l_s$ is a representative closure range (which is the distance travelled by the target in one scan normalized by the temporal loading) that specifies the search problem. Once the cumulative detection probability as a function of range is calculated, it is possible to find the cumulative detection range R_{90}.

This model for the cumulative detection probability results in a trade-off for selecting the revisit interval and hence dwell duration. Reducing the revisit interval time for the beam position increases the number of opportunities to detect the target in a given time frame. These multiple opportunities to observe the target give a 'sampling gain' [7], whereby the 90% cumulative detection range is greater than the 90% single look detection range. Conversely, increasing the revisit interval time for each beam position allows for more time in each beam position, which enables a gain through integration. Therefore, the trade-off when selecting the revisit interval can be thought of as a balance between sampling gain and integration gain.

The trade-off in selecting the revisit interval is illustrated in Figure 3.6(a), where the multi-scan cumulative detection range is plotted against the distance that a target can close in one revisit interval time. In Figure 3.6(a), an inbound target approaching from infinite range is considered. It can be seen that the maximum in

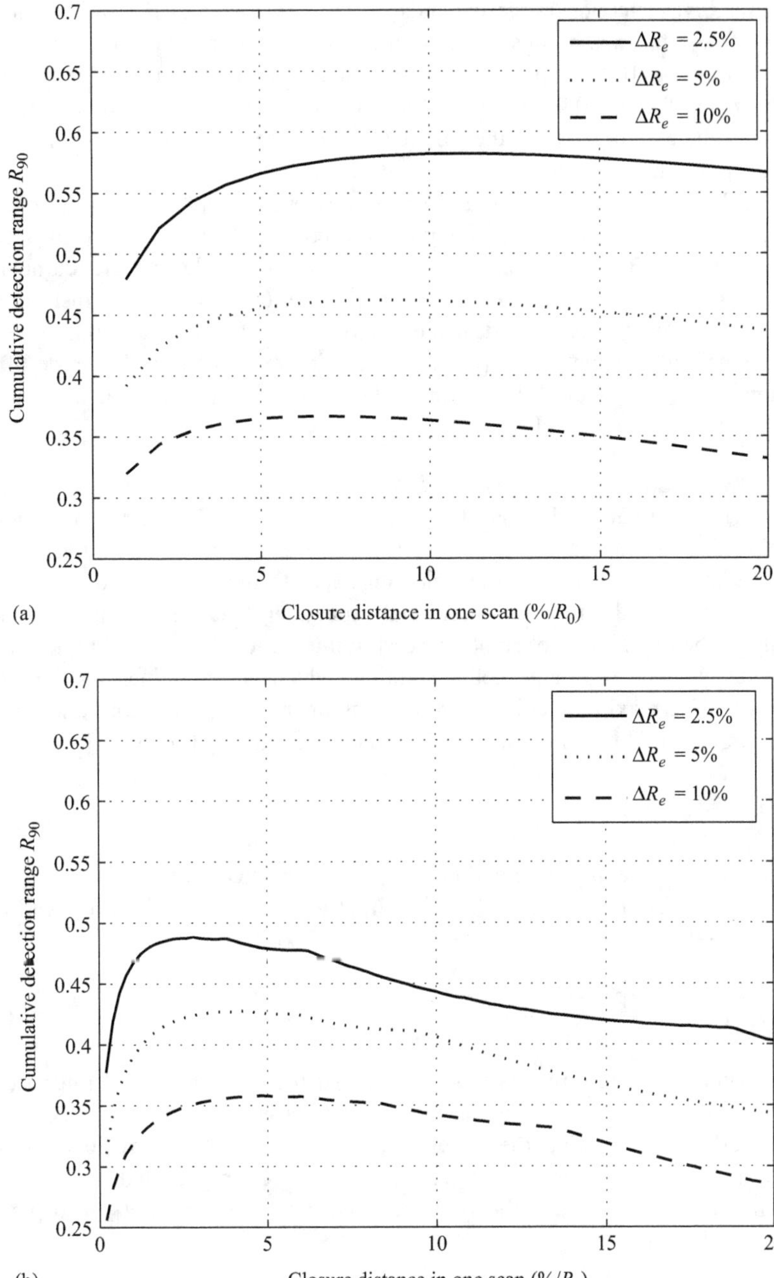

Figure 3.6 Multi-scan cumulative detection range as a function of the target closure range in one search revisit interval. (a) Infinite Range Inbound Target. (b) Pop-up target at 60% of R_0

the cumulative detection range depends on the value of Δr_e. For the considered values of Δr_e, the maximums in the cumulative detection range occur when the target closes 5%–10% of the instrumental range R_0 in one revisit interval period. For example, if $\Delta r_e = 0.05$, the instrumental range is 150 km and the target has a 300 m/s radial velocity then the target should close 7.5% of R_0 in one revisit interval time. This corresponds to a revisit interval of 37.5 s.

The 37.5 s revisit interval found in the previous example is much larger than commonly used. Shorter revisit intervals are commonly used, partly to accommodate targets that pop-up at closer range. In Figure 3.6(b), the cumulative detection range is plotted for a target that pop-ups at 60% of the radar instrumental range R_0. For $\Delta r_e = 0.05$, the ideal target closure range is now around 3% of the instrumental range, which corresponds to a revisit interval of 15 s. The trade-off for selecting the revisit interval has been studied many times [7,11,12], with similar conclusions to those presented here.

Limits on the coherent processing interval

Long revisit intervals and hence long coherent processing intervals may not be desirable due to range and Doppler cell migration. Range cell migration occurs when the target travels through multiple range cells leading to a spread in the signal energy. Consequently, to avoid additional processing that compensates for range migration, the maximum coherent processing interval can be limited by the target radial velocity and the range cell resolution, which is determined by the signal bandwidth. The maximum coherent processing interval T_{CV} for a constant velocity target can be found by combining (2.7) and (2.8) from [13, Chapter 2]:

$$T_{\mathrm{CV}} < \frac{c}{2Bv_r} \tag{3.6}$$

where B is the signal bandwidth, c is the speed of the electromagnetic propagation and v_r is the target radial velocity. If the target is accelerating with acceleration a, then the maximum coherent processing interval T_{ACC} is:

$$T_{\mathrm{ACC}} = \frac{v_r - \sqrt{v_r^2 + 2 \cdot a \cdot c/2B}}{-a} \tag{3.7}$$

The maximum coherent processing interval that prevents range cell migration is illustrated in Figure 3.7(a) for a constant velocity target.

In addition to limiting the coherent processing interval to avoid range cell migration, it should also be limited to avoid Doppler cell migration due to the target acceleration. The maximum allowable coherent integration time T_{DOP} to prevent Doppler cell migration is:

$$T_{\mathrm{DOP}} < \sqrt{\frac{\lambda}{2a}} \tag{3.8}$$

where λ is the wavelength of the transmitted waveform. The maximum coherent processing interval times due to Doppler cell migration are illustrated in Figure 3.7(b).

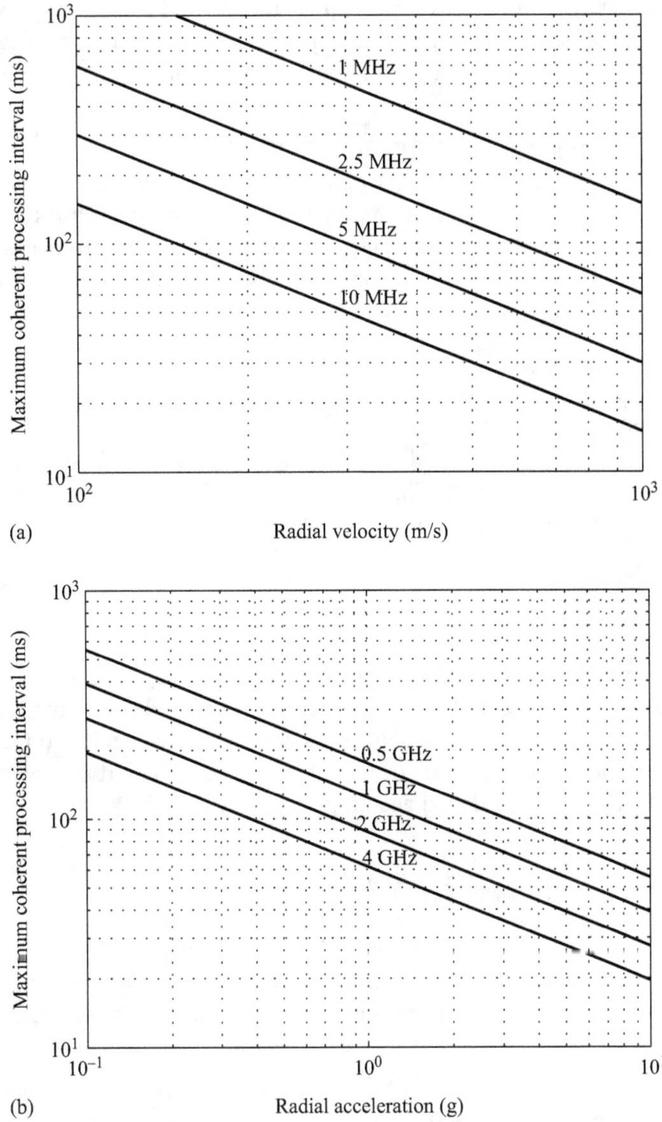

Figure 3.7 Limitations on the coherent processing interval to avoid range and Doppler cell migration. (a) Limits on range cell migration for varying signal bandwidth. (b) Limit on Doppler cell migration for varying operating frequency

Based on these models, it is possible to configure the search based on the range resolution required by the application and then the dwell duration and revisit interval to match the expected target motion. Consequently, it can be possible to specify multiple search modes that are suited to different target types, such as a short-range search for manoeuvrable pop-up targets and a long-range search for less manoeuvrable targets approaching from greater range. The parameters of the search can be adapted online, to match the current requirements of the operator.

3.2.1.3 Pulse repetition frequency selection

The pulse repetition frequency (PRF) can be varied depending on the expected target kinematic parameters. By definition a low PRF mode uses a PRF that is low enough to give an unambiguous range measurement. Therefore, a low PRF for search can be selected based on the maximum range of target or clutter returns of interest R_{max}:

$$\mathrm{PRF} < \frac{c}{2 \cdot R_{max}} \tag{3.9}$$

In contrast, a high PRF is defined as a PRF that is high enough to give an unambiguous velocity measurement. Therefore, a high PRF can be selected for search based on the maximum radial velocity of interest V_{max}:

$$\mathrm{PRF} > \frac{2 \cdot V_{max}}{\lambda} \tag{3.10}$$

In addition to ambiguities in range and Doppler domains, the choice of the PRF also results in blind zones in these domains. In the range domain, the blind zones are caused by the fact that a monostatic radar cannot receive while it transmits. The width and location of blind zones depend both on the PRF and the selected pulse width (PW) and can be modelled according to:

$$r_{\mathrm{blind}} \in \left[n\frac{c}{2\mathrm{PRF}}, n\frac{c}{2\mathrm{PRF}} + c\mathrm{PW} \right], \quad n = 0, 1, 2\ldots \tag{3.11}$$

where r_{blind} is the distance between the radar and a point in a blind zone.

In the Doppler domain, there exist blind zones due to filter notches used for rejecting clutter. The centre of the filter notches depends on the selected PRF, according to

$$d_{\mathrm{blind}} = \pm n\frac{\lambda \cdot \mathrm{PRF}}{2}, n = 0, 1, 2\ldots \tag{3.12}$$

and the extent of the blindness in Doppler depends on the clutter notch width. Based on these principles, high and low PRF sets can be selected considering the required visibility for targets of interest.

Equations (3.9)–(3.12) demonstrate that there exists a trade-off between range and Doppler ambiguities and detectability when selecting a high or low PRF. Due to the large unambiguous range, low PRF provides a good ranging ability but poor Doppler measurement ability due to frequent repetitions of clutter in the Doppler domain. Conversely, due to the large unambiguous velocity, high PRF provides good Doppler measurement but poor range measurement due to frequent repetitions of the clutter in the range domain. In order to balance these complementary strengths and weaknesses, sets of medium PRFs can be selected. A medium PRF set is a sequence of coherent bursts at different PRFs, thus enabling the ambiguities to be resolved.

The need to transmit sets of differing PRFs necessitates methods for selecting sets that give a good performance for targets of interest. Traditionally, PRF sets were calculated at design time, but advances in computing have enabled online adaptive PRF set selection based on geometry and look angle. PRF set selection can be based on optimization criteria such as the average number of clear PRFs, probability of target visibility or the size of the area that is clear on at least a specified number of PRFs.

The selection of PRF sets is dependent on satisfying a number of constraints. First, PRF selection cannot exceed the maximum duty ratio, based on the PW being used. Second, the PRF set should be decodable for all possible combinations of PRFs in the set that can generate a detection. Therefore, the lowest common multiple of possible pulse repetition interval (PRI) combinations should be greater than the longest signal round trip time of interest. Likewise, the lowest common multiple of possible PRF combinations should be greater than the greatest Doppler of interest. Constraints on the PRF set are discussed extensively in [14, Chapter 9].

Generally, PRF set selection methods rely on simple rules; however, sophisticated optimization can be applied, if it is possible to implement under the strict real-time requirements. The following is a summary of common methods that are discussed in detail in [14, Chapter 15]:

- *Maximum visibility:* PRF sets can be selected to minimize target blindness, either using rules [14, Section 15.2] or more sophisticated optimization methods such as neural networks [15].
- *Major–minor method:* Initially, three major PRFs are selected such that the Doppler band of interest is sufficiently covered and the main lobe clutter is staggered. Then two minor PRFs are selected for each major PRF to resolve range-Doppler ambiguities.
- *M out of N:* An M of N set can be created starting from the maximum PRF depending on the PW and duty ratio. Then PRFs are selected below the maximum at multiples of a PRF stagger value that depends on the bandwidth of the mainlobe clutter notch.
- *Exhaustive search:* A fixed maximum and minimum PRI as well as a specified PRI timing resolution results in a discrete number of possible PRIs that can be included in the set. All combinations can be evaluated based on visibility criteria. However, the number of combinations is typically large and therefore exhaustive search is not suitable for online PRF set selection.
- *Evolutionary algorithms:* Evolutionary algorithms can be applied [16] to reduce the search time in comparison to an exhaustive search. Evolutionary algorithms have been shown to generate good PRF sets with reasonable computational cost [17].

3.2.1.4 Non-uniform search parameters

In the previous sections, it was assumed that search control parameters are uniform over the entire surveillance region. However, it is common for a naval or ground-based radar to define search sectors. Each sector can have varied resource allocations or required performances, based on the sector's threat level [4]. Consequently, the revisit interval times and transmit waveforms can be varied per sector. It is possible, but less common, to vary the revisit interval time and transmit waveform in each beam position. This enables the distribution of search energy in angle to be matched to information on the current environment [18,19] or expected threatening trajectories [20]. Matching the search angular energy distribution to the current scenario improves the efficiency of the search resource allocation.

An alternative approach to scheduling search dwells is to apply a Bayesian approach to maintain an 'undetected target density'. This density describes where undetected targets are expected to be located, given the previously executed search dwells and a model of the expected target dynamics. Such an undetected target density can be represented by particles [21] or as a grid [22,23]. The undetected target density can be used as a basis for scheduling search dwells that maximize the probability of detecting previously undetected targets [24].

3.2.2 *Confirmation management*

An ESA can execute a rapid 'look-back' confirmation dwell to determine whether a search detection was due to the presence of a target or a random false alarm. Scheduling confirmation dwells has the benefit of enabling rapid track acquisition, as it is not necessary to wait for the next search dwell in order to confirm the presence of a target.

Executing a rapid confirmation dwell has the additional benefit that the radar cross section (RCS) between the initial search dwell and the confirm dwell is correlated. Consequently, the probability of detection in the confirm dwell given that a detection has occurred in the initial search dwell (in this context called 'alert') is significantly higher than the original search dwell detection probability [25]. As the beam is directed to the angle of the previous alert, the beam positioning losses can be reduced. In addition, the alert detection enables the detection space for the confirmation to be reduced. Therefore, a lower detection threshold and hence higher false alarm probability can be handled in each cell, which increases the confirm probability of detection.

The waveform used in the confirmation dwell can be matched to the alert generated by the original search dwell. For example, the confirm transmit energy and hence dwell length can be varied based on the measured SNR in the original alert. The desired SNR in the confirmation dwell would also be set higher than the SNR in the alert, so that a good quality measurement is obtained to initialize a track. The confirm waveform may also be more complex than the initial search dwell, for example, when an ambiguous search mode is used then the confirm waveform must be capable of resolving the radar ambiguity.

The benefit of using alert-confirm is illustrated in Figure 3.8(a), which plots the target confirmation probability against the normalized target range. In the figure, the following target confirmation processes are compared:

- *2/2 Confirmation:* Two detections from two search dwells are required to confirm the target. Therefore, no alert-confirm is applied.
- *Alert-confirm:* An alert detection is followed by a confirmation dwell that is identical to the original search dwell.
- *Adaptive alert-confirm:* An alert detection is followed by an adaptive confirmation dwell, which adapts the dwell length to give an expected 22 dB SNR on the confirm dwell based on the original alert SNR. The length of the confirmation dwell is limited to a maximum of ten times the original search dwell.

Figure 3.8(a) shows that using alert-confirm increases the target confirmation probability for a given range or increases the range at which targets are confirmed with a specified probability.

The benefit associated with scheduling confirmation dwells comes with the cost of increasing the time taken to perform the search. Based on the choice of the search dwell probability of false alarm and the dwell length of the confirmation, the expected time per search dwell T_s is [9, Section 14.3.1]:

$$T_s = \tau_a + P_{\text{FA}} N_b \tau_c \tag{3.13}$$

where N_b is the number of range-Doppler cells, τ_a is the search dwell time decided upon by the search management, τ_c is the confirmation dwell time and P_{FA} is the probability of false alarm per range-Doppler cell in the search dwell. Here, the number of detected and tracked targets is not taken into account because it is not known *a priori*.

Figure 3.8(b) shows the increase in time taken to perform the search, for the alert-confirm and adaptive alert-confirm strategies. When using the 2/2 conformation strategy, there is no increase in search time because there are no confirmation dwells. Alert-confirm increases the search time by a constant value irrespective of range, which for these example parameters was 1%. As adaptive alert-confirm adapts the confirmation dwell length, the increase in search time depends on the target range. In this example, the confirm dwell length was limited to a maximum increase of 10% of the original search dwell length.

This analysis assumes that the radar is operating in a noise-limited environment. Confirmation is not as effective in environments with time correlated clutter, as the clutter time correlation prevents the immediate discrimination between clutter and a target.

3.2.3 Track management

An ESA antenna can simultaneously track a large number of targets either using measurements from the search function, which is known as track-while-scan/search (TWS), or by scheduling dedicated radar dwells that are optimized to the target, which is known as active tracking. The track manager must decide for each target

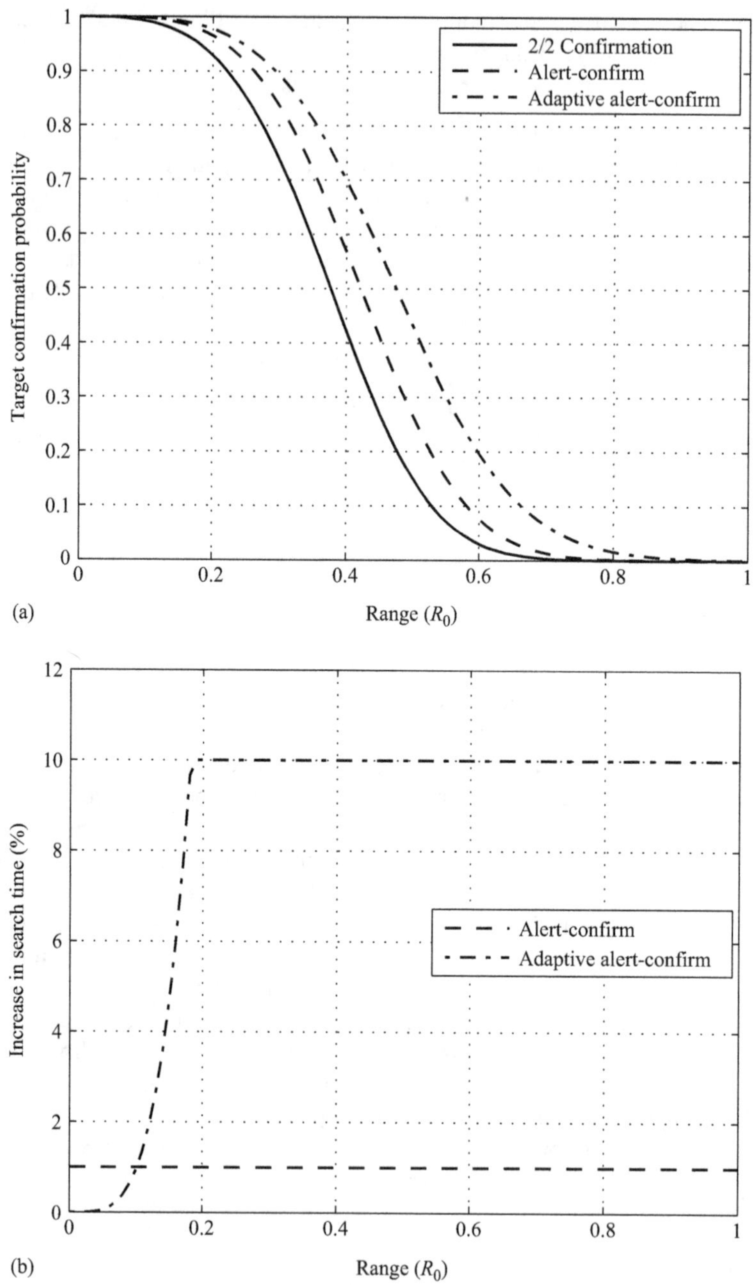

Figure 3.8 Improvement in the track confirmation range due to the confirmation function and corresponding increase in search time. (a) Target confirmation probability. (b) Increase in search time

whether to track using active tracking or TWS. If the target is actively tracked, the track manager must also decide the revisit interval time as well as the transmit waveform to use. For tracking, the revisit interval time is the time between measurements of a target. When an active track update is executed, the radar beam is directed towards the estimated position of the target in angle space. A beam positioning power loss occurs when the true target angle is offset from the estimated target angle.

3.2.3.1 Active tracking

For actively tracked targets, adaptive tracking approaches can be used, which aim to match the revisit interval time and hence the next measurement time based on the target manoeuvres. In this way, when a target follows trajectories without sharp changes in direction, track updates are less frequent than when a target changes direction rapidly. Following such a strategy for adapting the track revisit interval can allow for other tasks to obtain more radar time in benign scenarios while making it harder for a target to induce a track loss by manoeuvring.

Rules/Heuristics

A simple approach to track management is to define a set of rules. For example, a specified number of the highest priority targets can be maintained by active tracking and the rest using TWS. An arbitrary revisit interval time can be chosen for active tracking, or the revisit interval time can be varied based on the target threat or priority. The number of pulses transmitted can be varied depending on the target angle relative to the antenna boresight, to compensate for the loss in effective aperture. Defining such a rule set is simple to implement and solutions can be obtained in real time; however, as the rules are general and not well matched to the specific target, the track updates are not adaptive and therefore wasteful of radar resources.

Van Keuk and Blackman approach

The standard approach [26,27] for adaptive tracking is to select the revisit interval time and hence the next measurement time for when the filter predicted estimation error standard deviation in angle is equal to a specified fraction of the beamwidth, as illustrated in Figure 3.9. This strategy minimizes the resource loading of the track by balancing the trade-off between using short and long revisit interval times. Short revisit intervals exert a high track load as the track dwells are highly frequent. Long revisit intervals also exert a high track load as an increased number of track reacquisitions are required. Track reacquisitions are required as a longer revisit interval results in a degraded angular estimation error, which increases the beam positioning loss and hence lowers the probability of detection.

A variety of studies [9, Section 14.4.1–2, 26,27] indicate that a broad optimum for the minimum track loading occurs when the filter predicted estimation error standard deviation in angle is around 0.15 of the 3 dB beamwidth. Van Keuk and Blackman [26] studied this adaptive tracking approach and presented formulas enabling the revisit interval time to be calculated, based on the relative track accuracy in $u - v$ space, which they called track sharpness.

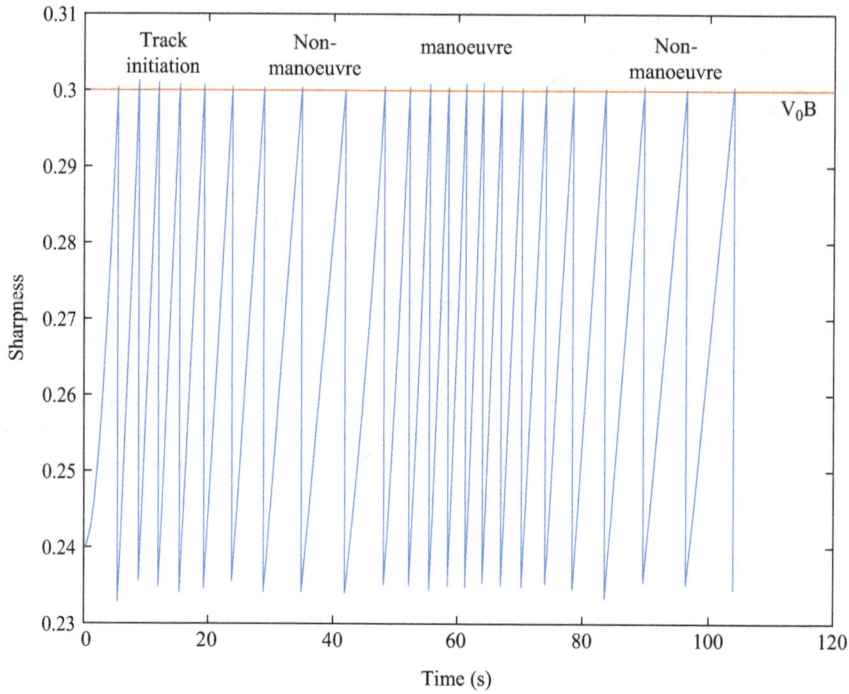

Figure 3.9 Adaptive tracking process. The revisit interval time is selected for when the filter predicted estimation error standard deviation in angle is equal to a specified fraction of the beamwidth. Increased process noise in track initiation and manoeuvre stages results in faster increases in the angular track uncertainty, resulting in more frequent measurements

Van Keuk and Blackman [26] assume a Singer target motion model, whereby the motion is driven by independent Gaussian Markov acceleration processes q_i in each spatial dimension with specified correlation time Θ and standard deviation Σ. Accordingly, the noise autocorrelation is given by:

$$R(\tau) = E\left[q_i(t)q_j(t+\tau)\right] = \delta_{ij}\Sigma^2 \exp\left(\frac{-\tau}{\Theta}\right) \tag{3.14}$$

where $\delta_{ij} = 1$ for $i = j$ and 0 otherwise. Speed and position then follow by integration with respect to time. More on the Singer model, including the resulting discrete time state transition matrix, can be found in standard texts [9,28]. In addition, it is assumed that the multiple targets are well-separated point targets, each occupying a single range-Doppler resolution cell. The power loss is modelled by a Gaussian loss function that is matched to the antenna beamwidth. Coherent integration followed by a square law detector is modelled, and the target amplitude is assumed to fluctuate according to a Swerling 1 model.

Based on these assumptions, the track revisit interval time t_r which achieves a specified track sharpness ν_0 can be calculated according to [26]:

$$t_r = 0.4\left(\frac{r\sigma\sqrt{\Theta}}{\Sigma}\right)^{0.4} \frac{U^{2.4}}{1 + (1/2)U^2} \tag{3.15}$$

where r is the target range. This is an approximation to the revisit interval time when the tracking filter is in steady state, given the specified target parameters. The variance reduction ratio U is the ratio of the track estimation error to the measurement error σ:

$$U = \frac{\theta_B \nu_0}{\sigma} \tag{3.16}$$

where θ_B is the antenna half beamwidth, i.e. half of the antenna 3 dB beamwidth.

The measurement noise standard deviation σ is calculated assuming that unbiased measurements of the angular position of a target are corrupted by additive Gaussian noise that in high SNR has a standard deviation according to (8.29) in [29, Section 8.2.4]:

$$\sigma = \frac{2 \cdot \theta_B}{k_m \sqrt{2 \cdot \text{SNR}}} \tag{3.17}$$

where k_m is the slope of the normalized monopulse error curve, which is taken by Van Keuk and Blackman as $k_m = \sqrt{2}$. SNR is the SNR encountered in the range-Doppler cell.

Due to a non-unity probability of detection, it may be necessary to have numerous looks for the target on a single track update. Van Keuk and Blackman describe a search strategy that minimizes the number of looks required at each update. They present a formula for the expected number of looks, which is a function of track sharpness.

Based on these formulas, the relative track loading for a target can be plotted for varying choices of track sharpness parameter and received SNR, as illustrated in Figure 3.10. It can be seen that the minimum track loading can be found by selecting the probability of false alarm P_{FA}, the track sharpness ν_0 and the desired SNR without beam positioning loss, which is denoted SN_0. Van Keuk and Blackman recommend selecting the probability of false alarm between 10^{-4} and 10^{-5}, the track sharpness to $\nu_0 \approx 0.3\theta_B$ and a coherent dwell length which achieves $\text{SN}_0 \approx 16$ dB.

The Van Keuk and Blackman formulas are a set of rules for selecting active track control parameters with low computational complexity; however, the approach is limited. The formulas only describe the asymptotic performance and therefore it may take many measurements before the actual angular error achieves the angular error predicted by the model. Moreover, the formulas are specific to the use of a Singer target motion model. Finally, they enable the track to be maintained with the minimum resource; however, they cannot be easily adapted for other criteria, such as a Cartesian error requirement or operational requirements.

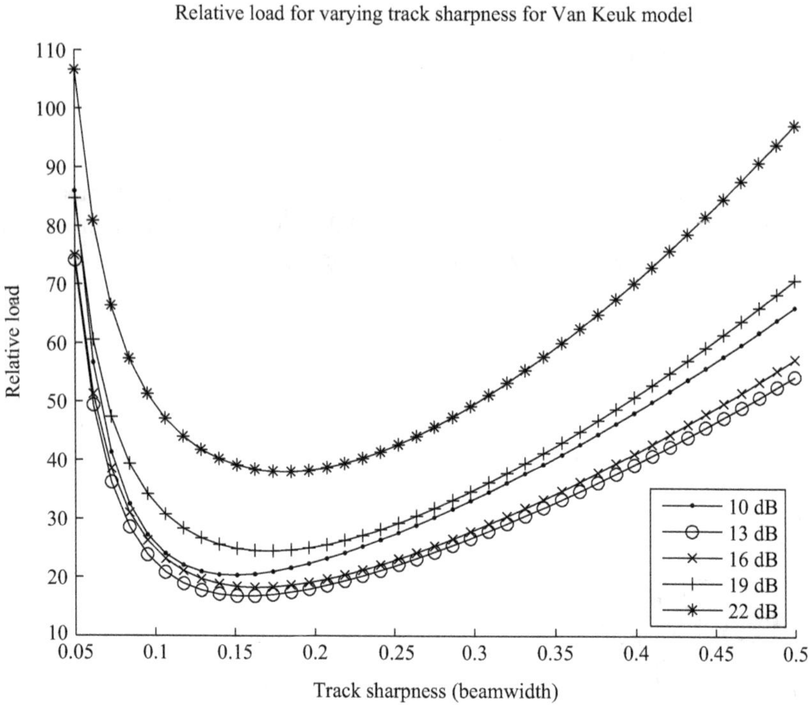

Figure 3.10 Relative tracking loading for varying choice of track sharpness parameter and SNR. Image taken from [30]

Covariance analysis

Instead of using the Van Keuk and Blackman rules of thumb, the current track can be predicted forwards in time to find when the track sharpness threshold is reached, which gives the track revisit interval time [31]. This approach has the benefit that the actual track at the current time is used, and it is therefore not based on the asymptotic performance. In addition, this approach is more flexible, as it can be applied with any target motion model or set of models.

Let $\mathbf{x}_{t_k}$ be the target state vector at time t_k, commonly comprising position and velocity in two or three dimensions, e.g. $\mathbf{x}_{t_k} = [x_{t_k}, y_{t_k}, x'_{t_k}, y'_{t_k}]^T$. Then, the track at time t_k based on all the measurements up to time t_k is comprised of the state estimate $\mathbf{x}_{t_k|t_k} = [\hat{x}_{t_k|t_k}, \hat{y}_{t_k|t_k}, \hat{x}'_{t_k|t_k}, \hat{y}'_{t_k|t_k}]^T$ and the filter posterior error covariance matrix $\mathbf{P}_{t_k|t_k}$, which represents the uncertainty in the state estimate $\mathbf{x}_{t_k|t_k}$, e.g.:

$$\mathbf{P}_{t_k|t_k} = \begin{bmatrix} E\left[\left(x_{t_k} - \hat{x}_{t_k|t_k}\right)\left(x_{t_k} - \hat{x}_{t_k|t_k}\right)\right] & \cdots & E\left[\left(x_{t_k} - \hat{x}_{t_k|t_k}\right)\left(y'_{t_k} - \hat{y}'_{t_k|t_k}\right)\right] \\ \vdots & \ddots & \vdots \\ E\left[\left(y'_{t_k} - \hat{y}'_{t_k|t_k}\right)\left(x_{t_k} - \hat{x}_{t_k|t_k}\right)\right] & \cdots & E\left[\left(y'_{t_k} - \hat{y}'_{t_k|t_k}\right)\left(y'_{t_k} - \hat{y}'_{t_k|t_k}\right)\right] \end{bmatrix} \tag{3.18}$$

The error covariance matrix $\mathbf{P}_{t_k|t_k}$ can be predicted to time $t_k + t_r$, using the standard Kalman filter prediction equation:

$$\mathbf{P}_{t_k+t_r|t_k} = \mathbf{F}(t_r)\mathbf{P}_{t_k|t_k}\mathbf{F}^T(t_r) + \mathbf{Q}(t_r) \tag{3.19}$$

where $\mathbf{F}(t_r)$ and $\mathbf{Q}(t_r)$ are the target state transition and process noise covariance matrices, respectively, for the revisit interval t_r. The state transition matrix describes the assumed change to the target state during the revisit interval t_r and is therefore derived based on an assumed motion model [9,28,32]. Deviations to the assumed model are accounted for by assuming that the state transition also includes a noise sequence that has an associated covariance $\mathbf{Q}(t_r)$.

The predicted track at time $t_k + t_r$ can be converted into spherical coordinates (angles, range and Doppler) to give the predicted error covariance $\mathbf{P}_{t_k+t_r}$. Let $\sigma^b_{t_k+t_r}$ and $\sigma^e_{t_k+t_r}$ be the filter predicted error standard deviation in bearing and elevation, respectively. For a set of possible revisit intervals $t_r \in T_r$, for example $T_r = \{0.5, 1.0, 1.5, 2.0, 2.5, 3.0\}$s, then the following procedure can be applied [31]:

1. Start with t_r as the greatest revisit interval in T_r
2. Predict track covariance $\mathbf{P}_{t_k+t_r|t_k}$
3. Convert covariance into spherical coordinates and extract $\sigma^b_{t_k+t_r}$ and $\sigma^e_{t_k+t_r}$
4. If either $\sigma^b_{t_k+t_r}$ or $\sigma^e_{t_k+t_r}$ exceeds v_0, then decrease t_r, else use the current t_r

By following this procedure, the largest revisit interval is found that does not exceed the specified track sharpness. This procedure can easily be adapted to be based upon alternative track criteria that may be more operationally relevant, such as the Cartesian error. Other popular measures are also based on the covariance of the track estimate, e.g. the trace of the covariance matrix, see the analysis in [33].

An important aspect of this adaptive tracking procedure is the use of interacting multiple model (IMM) filtering [34]. IMM filtering maintains a number of parallel filters each with a different motion model. Based on the observed measurement innovations, the probability of each model being currently active can be evaluated. The estimates and covariances for the all models are then mixed at each time step. The use of multiple models ensures that the filter-predicted covariance is consistent given varied target manoeuvres. When using IMM, the track revisit interval selection procedure is the same; however, step 2 is based on IMM prediction [34].

Benchmark problems

The benchmark problems [35,36] were a set of common scenarios, radar model and performance assessment criteria, that enabled the comparative assessment of adaptive tracking approaches.

As discussed above, adaptive tracking matches the revisit interval time to the estimated manoeuvre state of the target under track. Consequently, the quality of the estimator has a strong influence on the selection of the track control parameters and the subsequent resource allocation. This was clearly demonstrated through the numerous solutions [31,37,38] to the benchmark problems. It was shown that IMM filters are crucial for adaptive tracking, as they ensure that the target dynamic model in the filter is matched to the current target manoeuvre. Consequently, the revisit interval time decreases as the target manoeuvres and increases as the target follows predictable motion. From the benchmark problems, the combination of IMM and multiple hypothesis tracking [9, Section 14.8] emerged as the tracking methodology that resulted in lowest radar loading for each target track [37,38].

3.2.3.2 Dwell length adaptation

In addition to the adaptive selection of the next measurement time, the transmit energy and hence radar dwell time can also be matched to each target. The required radar dwell time can be calculated based on a required received SNR, the target range and the target RCS. In practice, the range and RCS are not known but can be estimated from previous measurements. The range is estimated by the tracking filter, and the RCS can be estimated using an alpha filter, a sliding window [39], or by extending the track state vector [37].

As an example, consider a tracking mode that achieves a 16 dB SNR for a 1 m^2 RCS target at 80 km using 128 pulses. If the RCS is not estimated, then this mode is used for all active track updates, resulting in excessive SNR for targets that are closer in range. For example, if the target is at 40 km then transmitting 128 pulses results in an SNR of 28 dB. If an accurate RCS estimate is available, then it can be known that just 8 pulses are necessary to achieve 16 dB SNR, resulting in a 93.75% shorter dwell.

Zwaga and Boers [40–42] extend the adaptive tracking approach by minimizing the track loading for a required track accuracy without the need to specify a received SNR. This approach results in a lower transmit energy spent on the maintenance of each track.

3.2.3.3 Waveform selection and adaptation

In addition to matching the revisit interval time and the dwell energy to the target, it is also possible to match waveform parameters to the target, such as the intra-pulse modulation and the PRF.

Methods for selecting or adapting the intra-pulse waveform modulation for tracking differ from the search problem, as it is necessary to optimize track performance and not the search performance. Consequently, waveforms with shorter time-bandwidth products or traditionally 'bad' ambiguity properties may be preferred, when they fulfil the specific information need of the tracker. This approach was first taken by Kershaw and Evans [43], where closed form solutions for selecting waveforms that minimize the tracking mean squared error were presented, and later extended to include clutter [44].

Waveform selection for tracking depends on the current knowledge of the target state as well as the measurement accuracy of candidate waveforms. In a tracker, the existing knowledge is described by the posterior probability density function $p(\mathbf{x}_k|Z^k)$, where $\mathbf{x}_k$ is the target state of interest (typically kinematic states), and Z^k is the set of measurements received up to and including time step k. In addition, the measurement accuracy is modelled by a likelihood function $p(\mathbf{z}_k|\mathbf{x}_k)$. Often, as with the Kalman filter and its variants, the posterior and likelihood probability density functions are assumed to be (or approximated as) Gaussian, denoted $p(\mathbf{x}_k|Z^k) \sim \mathcal{N}(\mathbf{x}_k; \hat{\mathbf{x}}_k, \mathbf{P}_{k|k})$ and $p(\mathbf{z}_k|\mathbf{x}_k) \sim \mathcal{N}(\mathbf{z}_k; \mathbf{H}_k\mathbf{x}_k, \mathbf{R}_k(\theta_k))$ respectively, where $\mathbf{H}_k$ is the measurement matrix that relates states to measurements, and $\mathbf{R}_k(\theta_k)$ is the measurement error covariance when waveform θ_k is used.

The accuracy of the delay and Doppler measurement $\mathbf{z}_k = \{\tau, v\}$ can be derived from the narrowband ambiguity function[1]:

$$\phi(\tau, v) = \int_{-\infty}^{\infty} s_{\theta_k}\left(t + \frac{\tau}{2}\right) s_{\theta_k}^{*}\left(t - \frac{\tau}{2}\right) \exp(-j2\pi vt)dt \tag{3.20}$$

where τ is the time delay, v is the Doppler shift, s_{θ_k} is the complex envelope of waveform θ_k and * denotes the complex conjugate.

The Fisher information matrix describes the information that the observations $\mathbf{z}_k$ carry on the state of interest $\mathbf{x}_k$. The Fisher information matrix is related to the ambiguity function [45, Section 10.2]:

$$\mathbf{J} = c \begin{bmatrix} \left.\dfrac{\partial^2 \phi(\tau, v)}{\partial \tau^2}\right|_{\substack{\tau=\tau_0,\\ v=v_0}} & \left.\dfrac{\partial^2 \phi(\tau, v)}{\partial \tau \partial v}\right|_{\substack{\tau=\tau_0,\\ v=v_0}} \\ \left.\dfrac{\partial^2 \phi(\tau, v)}{\partial \tau \partial v}\right|_{\substack{\tau=\tau_0,\\ v=v_0}} & \left.\dfrac{\partial^2 \phi(\tau, v)}{\partial v^2}\right|_{\substack{\tau=\tau_0,\\ v=v_0}} \end{bmatrix} \tag{3.21}$$

where c is a normalization constant, see [45, Section 10.2] or [43], τ_0 is the target time delay and v_0 is the target Doppler shift.

This Fisher information matrix determines the Cramér–Rao Lower Bound (CRLB) on the estimation of τ and v using the waveform with complex envelope s_{θ_k}. It can be argued that for measurements with high SNR, the CRLB can serve as an approximation to the measurement error covariance $\mathbf{R}_k(\theta_k)$.

Given a waveform library Θ comprising multiple waveforms, the best waveform to schedule for the next time step can be based on track performance criteria. As it is desired to minimize the error in the track state estimate, a logical choice is to select a waveform that minimizes the trace or determinant of the expected error covariance $\bar{\mathbf{P}}_{k|k}$ after a measurement has been generated with the candidate waveform at the next time step [46]:

$$\theta_k^* = \arg\min_{\theta_k \in \Theta} \operatorname{tr}\left(\bar{\mathbf{P}}_{k|k}(\theta_k)\right) \tag{3.22}$$

This criterion is used for scheduling waveforms in [47,48], where it is shown scheduling waveforms in this manner can significantly reduce tracking errors.

Waveforms can also be selected based on information theoretic criteria [49], motivated by the assumption that the radar must maximize information production in order to minimize the uncertainty in the surveillance picture. The mutual

[1]Equation (3.20) is denoted as 'time-frequency autocorrelation function' in [45, Sec. 10.1] but as 'ambiguity function' in [43,46].

information between the expected measurement generated using a candidate waveform and the target state can be used for waveform selection. The mutual information $\mathrm{I}(\mathbf{x}_k;\mathbf{z}_k)$, is the reduction in entropy of $\mathbf{x}_k$ due to the expected measurement $\mathbf{z}_k(\theta_k)$ generated using candidate waveform θ_k:

$$\mathrm{I}(\mathbf{x}_k;\mathbf{z}_k(\theta_k)) = H(\mathbf{x}_k) - H(\mathbf{x}_k|\mathbf{z}_k(\theta_k)) \tag{3.23}$$

where $H(\cdot)$ is the entropy of a random variable. Assuming that the target state is represented by a Gaussian, as with a Kalman filter and its variants, the mutual information is:

$$\mathrm{I}(\mathbf{x}_k;\mathbf{z}_k(\theta_k)) = \frac{1}{2}\ln\left(\left|\mathbf{I} + \mathbf{P}_{k|k-1}\mathbf{H}_k^T\mathbf{R}_k^{-1}(\theta_k)\mathbf{H}_k\right|\right) \tag{3.24}$$

where $\mathbf{I}$ is an identity matrix of proper dimensions. Consequently, waveform selection can be performed based on mutual information according to:

$$\theta_k^* = \arg\max_{\theta_k\in\Theta}\left[\mathbf{I}(\mathbf{x}_k;\mathbf{z}_k(\theta_k))\right] \tag{3.25}$$

Alternative information theoretic criteria have also been proposed, such as the alpha-divergence between the prior and posterior estimates [50] and the Kullback–Leibler divergence, which is a single case of the alpha-divergence. In addition, it has been shown that using the conditional entropy, mutual information or Kullback–Leibler divergence for sensor management purposes has the same sensor selection results, for more details see the discussion [51].

Information theoretic criteria are valuable for waveform selection as they act as a surrogate function, in that regardless of higher level objectives, it is always desirable to maximize the information content of the measurement with respect to already acquired information in the tracker. However, care should be taken [52] when using information theoretic criteria for radar management problems. The assumption that the radar wants to maximize information may not always be valid, or at least, the value of different information sources may vary significantly, as also shown in [53].

Ideally, a waveform library should be small but well designed [54]. A small but effective library is motivated due to the possible computation explosion of online waveform design, especially when waveforms are scheduled non-myopically, that is, considering possible waveform sequences extending over a time horizon into the future. It has been shown that including non-linear frequency modulation (FM) waveforms [46,55] or fractional Fourier transformed waveforms [47,54] in the library can enhance tracking performance, especially in clutter.

In addition to selecting or adapting the intra-pulse modulation to optimize tracking performance, the PRF can be selected based on the target. This can be performed using a number of rules. First, a PRF should be selected such that the target is not eclipsed or in a clutter notch, based on the estimated target kinematic parameters. Also, as track information is already available, active track updates can be performed using an ambiguous PRF set, which reduces the total number of PRFs used in the dwell [56–58].

3.3 Priority assignment

The task management module makes requests to use the array antenna for specific intervals of time. A priority is assigned to each task's job requests, to reflect the fact that different tasks have differing importance or differing sensitivity to scheduling delays. When the radar scheduler is under-loaded priority has little influence; however, when the radar scheduler is overloaded, the priority determines which jobs are not executed by the radar. Existing approaches to priority assignment are based on rule sets or alternatively using fuzzy logic.

3.3.1 Rule-based assignment

The simplest approach to priority assignment is to decide priority based on the function to which the task/job belongs. This reflects the fact that some functions, such as track maintenance, are more mission critical than other functions, such as calibration. This approach also reflects the difference in the fluidity of the functions' time constraints. For example, plot confirmation, which requires a rapid revisit for a correlated RCS, has stricter timing constraints than a search job, which can cope with longer delays in job scheduling. Consequently, plot confirmation would be given a higher priority than search. An example of such a function priority list is shown in Table 3.1, which is taken from [59, page 93]. Many examples of such function-based priority assignment can be found in the literature [1,60–64].

Rules or heuristics can also be used to assign different priority to tasks/jobs within the same function. A simple approach for assigning tracking priority is based on the target 'time to go', which is the target range rate over the target range with respect to an asset of interest. Targets with a low 'time to go' are assigned a higher priority, as this implies a more threatening trajectory with respect to an asset of interest. An example of priority assignment for search [61] performed by an airborne fighter aircraft is to give a higher priority to beam positions closer to the antenna boresight. This reflects the fact that targets on the boresight are closing quicker, due to the platform motion. For a ground-based radar, search priority can be based on the possible most threatening trajectories [20].

Table 3.1 Example of function-based priority table, taken from [59, page 93]

Priority	**Function**	
7	Track maintenance	*(Highest priority)*
6	Plot confirmation	
5	Track initiation	
4	Track update	
3	Surveillance	
2	Slow track map/surface picture	
1	Receiver calibration	*(Lowest priority)*

3.3.2 *Fuzzy logic*

Fuzzy logic methods [9, Chapter 12 and Section 15.5.6, 65, 66, 67, Chapter 5] improve upon hard rules for priority assignment by allowing the priority to take continuous values depending on a wide range of task assessment criteria. Fuzzy values can be assigned to variables representing attributes of the surveillance sector or target track. Fuzzy if-then rules can then be applied to determine the priority of the target track or surveillance sector. The attributes considered in [68] are shown in Tables 3.2 and 3.3. In Table 3.3, the higher order levels reflect the criteria that influence each lower level. For example, whether or not a target is deemed hostile depends on the target approach, which itself depends on the heading and range rate of the target. The final priority is based on membership to the first level of attributes.

Table 3.2 Decision tree for search sector priority assignment using fuzzy logic, taken from [68]

First level
New targets rate
Number of threatening targets
Threatening targets rate
Original priority

Table 3.3 Decision tree for track priority assignment using fuzzy logic, taken from [68]

<table>
<tr><th>First level</th><th>Second level</th><th>Third level</th></tr>
<tr><td colspan="3">Track quality</td></tr>
<tr><td rowspan="4">Hostile</td><td rowspan="2">Approach</td><td>Heading</td></tr>
<tr><td>Range rate</td></tr>
<tr><td colspan="2">Velocity</td></tr>
<tr><td colspan="2">Target ID</td></tr>
<tr><td>Weapons system</td><td colspan="2">WS capabilities</td></tr>
<tr><td rowspan="6">Threat</td><td colspan="2">Target ID</td></tr>
<tr><td rowspan="5">Trajectory</td><td>RR/R</td></tr>
<tr><td>Manoeuvre</td></tr>
<tr><td>Height</td></tr>
<tr><td>RR/V</td></tr>
<tr><td>Range</td></tr>
<tr><td rowspan="2">Position</td><td colspan="2">Range</td></tr>
<tr><td colspan="2">Azimuth coherence</td></tr>
</table>

Hard logic approaches and fixed priority approaches are less computationally demanding than a fuzzy approach. However, the fixed approach allows for no variations in different target or surveillance sector types, and a hard logic approach has priority transitions, which tend to jump suddenly between values. The fuzzy logic approach shows smooth transitions allowing greater variations in priority [68]. This is a result of including all possible information into the priority decision-making process. However, this approach is only worthwhile when the priority value has an influence on the radar behaviour. Fuzzy logic approaches can also be applied directly in the resource management process instead of just determining the priority assignment. For example, task control parameters can be selected based on membership to fuzzy sets [9, Chapter 12 and Chapter 15].

3.4 Scheduling

The scheduler is responsible for creating a transmittable timeline of jobs from multiple potentially conflicting job requests. Each job request consists of the job priority, job duration τ_c and the job timing constraints. The job timing constraints are the earliest time t_e, the desired time t_d and the latest time t_l that the job can be scheduled. These parameters are shown in Figure 3.11. The objective of the scheduler is to maximize the radar time utilization, while satisfying the job request constraints.

3.4.1 Queue schedulers

Queue-based schedulers operate by selecting the next best job from a queue or set of queues. The next best job is decided upon an ordered list of the jobs that are eligible to be executed. Such a list can be ordered based on the timing constraints to give the following schedulers:

- Earliest deadline first (EDF)
- Desired time first (DTF)
- Earliest time first (ETF)

Another possibility is to schedule jobs based on highest priority first or shortest job first, but this would not result in desirable scheduling behaviour. However, job priority can be incorporated into a scheduler, by ensuring that lower priority jobs are not able to delay higher priority jobs. Once the next best job is selected it can be

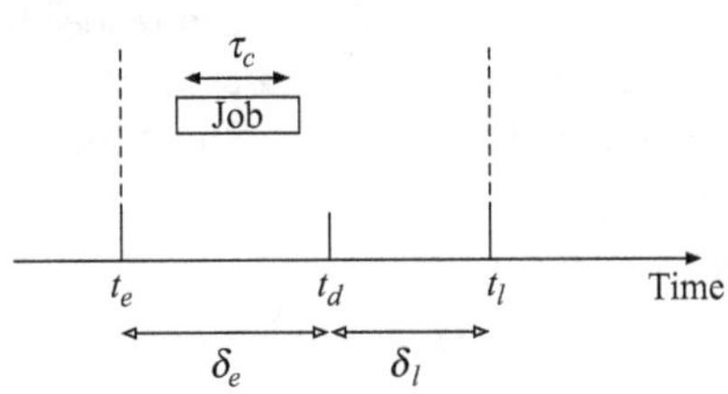

Figure 3.11 Example of parameters defining a job request

directly executed, or added to a scheduling frame of fixed duration. This queue-based scheduling process can be summarized in the following steps:

1. Add new job requests to queue
2. Select job request according to selected criterion
3. If currently selected job delays a higher priority job in the queue, then switch to higher priority job. Repeat until no higher priority jobs are delayed
4. Execute job or add job to scheduling frame
5. Remove jobs that have missed scheduling deadlines
6. Repeat from step 1

As the queue can be maintained with low computational burden, queue-based schedulers are computationally light and easy to implement. As shown in the comparison by Miranda [66], queue-based schedulers are able to obtain close to 100% occupancy for highly fluid jobs. As each allocation frame is not optimized, the delays for the jobs can be greater than the delays encountered with locally optimized schedulers. Reducing scheduling delays is desirable; however, the impact on the task quality of scheduling delays varies for different functions and may in fact be minor.

3.4.2 Frame-based schedulers

Frame-based schedulers generate a timeline by arranging jobs in a time allocation slot of fixed duration. Whilst the previous allocation frame is being executed, the next allocation frame is being calculated. For a given measure of optimality, an exhaustive search could be used, but often heuristics are used to guide the placing of jobs within the allocation frame. Orman *et al.* [62] present a variety of heuristics, based on the allowed delay for tracking and search jobs. Winter *et al.* [69,70] present a method based on linear programming that schedules jobs to minimize the overall cost given cost functions for each job. The cost functions describe the cost associated with scheduling a job early or late. It is found that the computation is too great and therefore heuristics are proposed. Moo [71] presents an approach based on benefit slopes that also describe the penalty associated with scheduling a task early or late.

As frame-based schedulers optimize the placing of the job in the allocation frame, they can generate good quality schedules that minimize the delay of the jobs as well as considering the cost of delaying the jobs. This is evident in the comparison by Miranda [66] where the frame-based schedulers exhibit a smaller delay in comparison to the queue-based schedulers. However, frame-based schedulers incur a higher computational burden. Unfortunately, due to the strict real time requirement of the scheduler, this computational complexity can render frame-based scheduling solutions impractical. In addition, as they construct firm timelines over the allocation frame, they can have difficulties in rapidly scheduling jobs for pop-up tasks.

3.4.3 Comparison of schedulers

In this subsection, different schedulers are compared in terms of occupancy and the number of jobs that missed deadlines. The occupancy is the percentage of the available radar time that is used for jobs, and it is desirable that the occupancy is as close to 100% as possible. The occupancy depends on the fluidity of the jobs which is the time between the desired time and the earliest and latest times (denoted as δ_e and δ_l, respectively, in Figure 3.11) that the job can be scheduled. The number of jobs that missed deadlines depends on the loading of the radar, in overload situations, high priority jobs should be scheduled while low priority jobs are dropped.

To demonstrate the influence of job fluidity, a random set of search and tracking jobs were created for fluidity values in $[1, 2, \ldots, 50]$ ms. It was assumed that there were 200 search beams with revisit intervals sampled from $[0.25, 0.35, \ldots, 3.5]$ s, and dwell length sampled from $[0.1, 0.2, \ldots, 10]$ ms. In addition, there were 50 tracking beams with revisit intervals sampled from $[0.5, 0.6, \ldots, 4]$ s, and dwell length sampled from $[0.1, 0.2, \ldots, 5]$ ms. A priority was randomly assigned to each task: search tasks were assigned priorities in $\{1, 2, 3\}$ and tracking tasks in $\{3, 4, 5\}$ with 1 being the lowest and 5 the highest priority. The total load varied based on the realization in the sampling process of the chosen revisit intervals and dwell lengths and in general the job load was between 80% and 120% of the permissible load.

In addition, for the considered frame-based scheduler [71], the peak benefits, as defined in [71], were set to be equal to the corresponding job priorities. The early and late slopes were equal for each job and randomly sampled from [5,20,50]. The frame length was 100 ms, and if all jobs were finished before the end of a frame, then this frame was shortened accordingly so that the next frame could start.

The considered example mimics a situation where an ESA radar is tracking multiple targets while searching for new ones. Therefore, different dwell lengths, revisit intervals and priorities are sampled randomly from sets suitable to tracking and search functions. In this way, the randomness in targets' locations and in their corresponding priority within different operational contexts is modelled.

The compared schedulers were:

- EDF, without using priorities, marked as EDF;
- EDF, with priorities, marked as EDF-WP;
- DTF, without using priorities, marked as DTF;
- DTF, with priorities, marked as DTF-WP;
- ETF, without using priorities, marked as ETF;
- ETF, with priorities, marked as ETF-WP;
- The frame scheduler described in [71], where jobs with priority ≥ 3 were considered as primary and the rest as secondary jobs. The secondary jobs where scheduled in time gaps using EDF-WP. This algorithm is denoted as Frame scheduler v1;
- A degenerate version of the frame scheduler described in [71], where all jobs were considered as primary jobs. This algorithm is denoted as Frame scheduler v2.

Figure 3.12 shows the occupancy results for each scheduler, averaged over different Monte Carlo runs. It can be seen that as job fluidity increases, more jobs can be scheduled and the occupancy increases irrespective of the use of priorities.

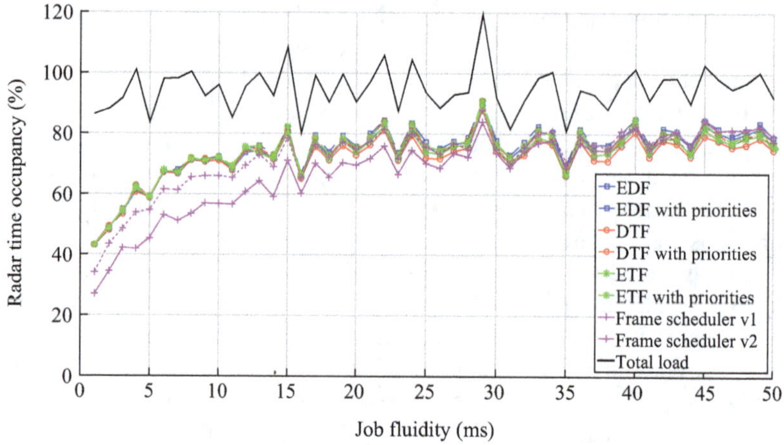

Figure 3.12 Achieved occupancy of each scheduler as function of job fluidity

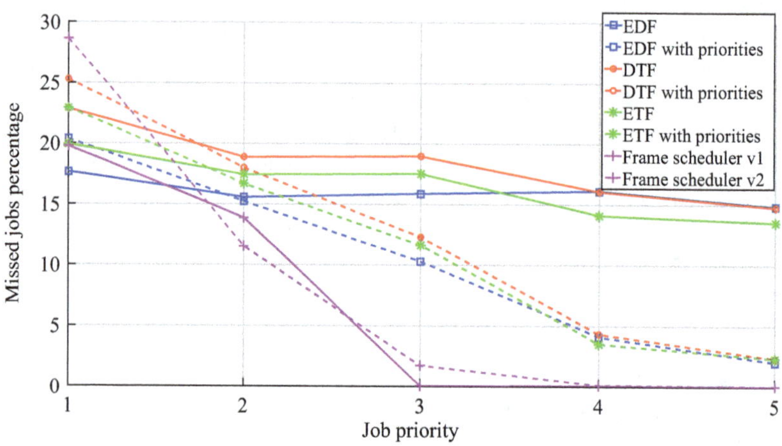

Figure 3.13 Percentage of missed job deadlines as a function of job priority. Dashed lines illustrate schedulers with priorities

On the other hand, the use of priorities results in lower occupancy for low fluidity values. Figure 3.13 shows the benefit of using priorities: tasks with higher priority get dropped (or miss their deadline) much less often than the tasks with lower priority.

For low fluidity values, queue schedulers result in higher radar occupancy but as the job fluidity increases, all schedulers result in similar occupancy. The benefit of using frame schedulers is that jobs with higher priorities miss their deadlines extremely rarely, at the cost of lower priority jobs missing their deadlines much more often when compared to queue schedulers.

3.5 Summary

Traditionally, the configuration of a radar, such as the scan strategy and the waveforms used, was fixed when the radar was designed. ESA, combined with advances in computing and arbitrary waveform generation, enable the radar to reconfigure during operation, based on the encountered situation. This creates the radar resource management problem, whereby it is desired to allocate finite radar resources between multiple sensing tasks as well as optimize the execution of each task. A radar resource management architecture comprises multiple components that are split hierarchically, as each component is required to make decisions on different time scales. Typical components for an ESA resource manager are priority assignment, scheduling and task management.

Task priority assignment is necessary as different tasks can have differing sensitivity to scheduling delays or differing importance with respect to the mission. Typically, task priority is based on a functional ranking with functions such as track maintenance and plot confirmation receiving a high priority and functions such as calibration a low priority. Alternatively, methods such as fuzzy logic can be applied to incorporate more information into the priority assignment leading to smooth variations in priority. However, this is only worthwhile when the resulting priority has a strong influence on the radar behaviour.

Scheduling can be performed using queue or frame-based schedulers, which are both capable of considering job timing constraints and priority. Queue-based schedulers can achieve a high occupancy but incur greater scheduling delays. Frame-based schedulers can achieve low scheduling delays, especially for high priority jobs, but do so at the cost of occupancy. In addition, frame-based schedulers are much more computationally complex than queue-based schedulers.

The task manager must decide how often to execute a radar task and which waveform to use, which also determines the resource allocated to the task. This is performed for the multiple functions that are supported by an ESA:

- *Search function:* The search lattice and beam spacing used for searching a volume can be varied depending on the target range of interest, although the difference in performance is not great. The revisit interval and hence dwell time for the beam positions in a search volume can be optimized to maximize the cumulative detection range based on an expected target pop-up range; however, the coherent processing interval may be limited to avoid range and Doppler cell migration. PRFs for bursts can be selected online to maximize visibility in the search based on the current geometry. Context information, such as terrain information, can be exploited so that the search is matched well to the radar's surroundings.
- *Confirmation function:* Executing rapid 'look-back' confirmation dwells extends the track acquisition range as it is not necessary to wait for an entire scan before confirming the presence of a target. The dwell length of the confirmation dwell can be adapted to the SNR of the original alert, which further extends the track acquisition range. However, both alert-confirm procedures

cost radar time budget and are not so effective in the presence of time correlated clutter.

- *Tracking function:* Revisit intervals and waveforms for targets maintained with active tracking can be selected with rules and heuristics; however, this approach is wasteful of radar resources. Alternatively, the revisit interval can be selected to ensure that the target does not manoeuvre out of the radar beam, and the dwell length can be selected so that a specified SNR is achieved based on an estimate of the target's RCS. Consequently, the revisit interval and dwell time vary depending on target manoeuvres and range. In addition, the intrapulse waveform modulation can be selected to complement the existing information on the target in the tracker.

The influence of many of these radar control parameters have been considered previously in radar system design; however, the benefit with ESA radar is that the radar resource manager can vary these control parameters during operation in response to operational requirements and the encountered scenario. These ESA resource management methods are extended in Vol. II, Chapter 5 'Cognitive radar management', where optimization methods such as quality-of-service optimization and stochastic control are applied.

Glossary

CRLB	Cramér–Rao lower bound
DTF	desired time first
EDF	earliest deadline first
ESA	electronically steered array
ETF	earliest time first
FM	frequency modulation
IMM	interacting multiple model
PRF	pulse repetition frequency
PRI	pulse repetition interval
RCS	radar cross section
SNR	signal-to-noise ratio
TWS	track-while-scan/search

References

[1] W. K. Stafford, 'Real time control of a multifunction electronically scanned adaptive radar (MESAR),' in *IEE Colloquium on Real-Time Management of Adaptive Radar Systems*, London, UK, 1990, pp. 1–7.

[2] S. Musick and R. Malhotra, 'Chasing the elusive sensor manager,' in *IEEE National Aerospace and Electronics Conference*, Dayton, OH, USA, 1994, pp. 606–613.

[3] S. Miranda, C. J. Baker, K. Woodbridge, and H. Griffiths, 'Knowledge-based resource management for multifunction radar: a look at scheduling and task prioritization,' *IEEE Signal Processing Magazine*, vol. 23, no. 1, pp. 66–76, 2006.

[4] F. Barbaresco, J. C. Deltour, G. Desodt, B. Durand, T. Guenais, and C. Labreuche, 'Intelligent M3R radar time resources management: Advanced cognition, agility & autonomy capabilities,' in *SEE International Radar Conference (RADAR 2009)*, Bordeaux, France, 2009, pp. 1–6.

[5] B. Dawber and B. Wallace, 'The advanced radar technology integrated system testbed (ARTIST) and the need for cognitive radar management and control,' in *Defence Applications of Signal Processing (DASP)*, Coolum, Australia, 2011.

[6] W. H. Von Aulock, 'Properties of phased arrays,' *Proceedings of the IRE*, vol. 48, no. 10, pp. 1715–1727, 1960.

[7] E. R. Billam, 'Parameter optimisation in phased array radar,' in *IET International Radar Conference*, Brighton, UK, 1992, pp. 34–37.

[8] G. V. Trunk, J. D. Wilson, and P. K. Hughes, 'Phased array parameter optimization for low-altitude targets,' in *IEEE International Radar Conference*, Alexandria, VA, USA, May 1995, pp. 196–200.

[9] S. Blackman and R. Popoli, *Design and Analysis of Modern Tracking Systems*. Dedham, MA: Artech House, 1999.

[10] P. Hahn and S. Gross, 'Beam shape loss and surveillance optimization for pencil beam arrays,' *IEEE Transactions on Aerospace and Electronic Systems*, vol. AES-5, no. 4, pp. 674–675, Jul. 1969.

[11] J. D. Mallett and L. E. Brennan, 'Cumulative probability of detection for targets approaching a uniformly scanning search radar,' *Proceedings of the IEEE*, vol. 51, no. 4, pp. 596–601, 1963.

[12] I. Rusnak, 'Search radar evaluation by the normalized cumulative probability of detection curves,' *IEEE Transactions on Aerospace and Electronic Systems*, vol. AES-22, no. 4, pp. 461–465, Jul. 1986.

[13] R. Wiley, *ELINT: The Interception and Analysis of Radar Signals*. Norwood, MA: Artech House, 2006.

[14] C. Alabaster, *Pulse Doppler Radar: Principles, Technology, Applications*. Edison, NJ: SciTech Publishing, 2012.

[15] J. W. Yi and Y. J. Byun, 'Real-time PRF selection for medium PRF airborne pulsed-Doppler radars in tracking phase,' in *2007 IEEE International Waveform Diversity and Design Conference*, Pisa, Italy, 2007, pp. 116–121.

[16] P. Davies and E. Hughes, 'Medium PRF set selection using evolutionary algorithms,' *IEEE Transactions on Aerospace and Electronic Systems*, vol. 38, no. 3, pp. 933–939, 2002.

[17] C. Alabaster, E. Hughes, and J. Matthew, 'Medium PRF radar PRF selection using evolutionary algorithms,' *IEEE Transactions on Aerospace and Electronic Systems*, vol. 39, no. 3, pp. 990–1001, Jul. 2003.

[18] F. Hoffmann and A. B. Charlish, 'A resource allocation model for the radar search function,' in *SEE International Radar Conference*, Lille, France, 2014, pp. 1–6.

[19] A. B. Charlish, K. Woodbridge, and H. Griffiths, 'Agent based multifunction radar surveillance control,' in *IEEE Radar Conference (RadarConf)*, Kansas City, MO, USA, 2011, pp. 824–829.

[20] F. Barbaresco, 'Radar resources optimization by adaptive search domains priority assignment based on most threatening trajectories computation,' in *IET Seminar on Intelligent Sensor Management*, Quebec, Que., Canada, May 2007, pp. 1–10.

[21] F. Katsilieris, Y. Boers, and H. Driessen, 'Optimal search: a practical interpretation of information-driven sensor management,' in *5th International Conference on Information Fusion*, Singapore, July 2012, pp. 439–446.

[22] K. White, J. Williams, and P. Hoffensetz, 'Radar sensor management for detection and tracking,' in *11th International Conference on Information Fusion*, Cologne, Germany, 2008, pp. 1–8.

[23] J. Williams, 'Search theory approaches to radar resource allocation,' in *Defense Applications of Signal Processing (DASP)*, Coolum, Australia, July 2011.

[24] F. Katsilieris, A. B. Charlish, and Y. Boers, 'Towards an online, adaptive algorithm for radar surveillance control,' in *Workshop on Sensor Data Fusion: Trends, Solutions, Applications*. Bonn, Germany: IEEE, 2012, pp. 66–71.

[25] R. A. Dana and D. Moraitis, 'Probability of detecting a Swerling I target on two correlated observations,' *IEEE Transactions on Aerospace and Electronic Systems*, vol. 17, no. 5, pp. 727–730, 1981.

[26] G. van Keuk and S. S. Blackman, 'On phased-array radar tracking and parameter control,' *IEEE Transactions on Aerospace and Electronic Systems*, vol. 29, no. 1, pp. 186–194, 1993.

[27] W. H. Gilson, 'Minimum power requirements of tracking,' in *IEEE International Radar Conference*, Arlington, VA, USA, May 1990, pp. 417–421.

[28] X. R. Li and V. P. Jilkov, 'Survey of maneuvering target tracking. Part I. Dynamic models,' *IEEE Transactions on Aerospace and Electronic Systems*, vol. 39, no. 4, pp. 1333–1364, Oct. 2003.

[29] D. K. Barton, *Radar Systems Analysis and Modelling*. Dedham, MA: Artech House, 2004.

[30] A. B. Charlish, 'Autonomous agents for multi-function radar resource management,' Ph.D. thesis, University College London, Nov. 2011.

[31] T. Kirubarajan, Y. Bar-Shalom, W. D. Blair, and G. A. Watson, 'IMMPDAF for radar management and tracking benchmark with ECM,' *IEEE Transactions on Aerospace and Electronic Systems*, vol. 34, no. 4, pp. 1115–1134, Oct. 1998.

[32] Y. Bar-Shalom, X. Rong Li, and T. Kirubarajan, *Estimation with Applications to Tracking and Navigation: Theory Algorithms and Software*. Wiley, 2001.

[33] C. Yang, L. Kaplan, and E. Blasch, 'Performance measures of covariance and information matrices in resource management for target state estimation,' *IEEE Transactions on Aerospace and Electronic Systems*, vol. 48, no. 3, pp. 2594–2613, 2012.

[34] H. A. P. Blom and Y. Bar-Shalom, 'The interacting multiple model algorithm for systems with Markovian switching coefficients,' *IEEE Transactions on Automatic Control*, vol. 33, no. 8, pp. 780–783, 1988.

[35] W. D. Blair, G. A. Watson, S. A. Hoffman, and G. L. Gentry, 'Benchmark problem for beam pointing control of phased array radar against maneuvering targets,' in *American Control Conference*, vol. 4, Baltimore, MD, USA, 1994, pp. 2071–2075.

[36] W. D. Blair, G. A. Watson, T. Kirubarajan, and Y. Bar-Shalom, 'Benchmark for radar allocation and tracking in ECM,' *IEEE Transactions on Aerospace and Electronic Systems*, vol. 34, no. 4, pp. 1097–1114, 1998.

[37] W. Koch, 'Adaptive parameter control for phased-array tracking,' in *SPIE Signal and Data Processing of Small Targets*, Denver, CO, USA, Oct. 1999, pp. 444–455.

[38] S. S. Blackman, R. J. Dempster, M. T. Busch, and R. F. Popoli, 'IMM/MHT solution to radar benchmark tracking problem,' *IEEE Transactions on Aerospace and Electronic Systems*, vol. 35, no. 2, pp. 730–738, 1999.

[39] R. F. Popoli, S. S. Blackman, and M. T. Busch, 'Application of multiple-hypothesis tracking to agile beam radar tracking,' in *SPIE Signal and Data Processing of Small Targets*, Orlando, FL, USA, May 1996, pp. 418–428.

[40] J. H. Zwaga, Y. Boers, and H. Driessen, 'On tracking performance constrained MFR parameter control,' in *6th International Conference of Information Fusion*, Cairns, Australia, July 2003, pp. 712–718.

[41] J. H. Zwaga and H. Driessen, 'Tracking performance constrained MFR parameter control: applying constraints on prediction accuracy,' in *8th International Conference on Information Fusion*, Philadelphia, PA, USA, July 2005, pp. 546–551.

[42] Y. Boers, H. Driessen, and J. Zwaga, 'Adaptive MFR parameter control: fixed against variable probabilities of detection,' *IET Radar, Sonar and Navigation*, vol. 153, no. 1, pp. 2–6, 2006.

[43] D. J. Kershaw and R. J. Evans, 'Optimal waveform selection for tracking systems,' *IEEE Transaction on Information Theory*, vol. 40, no. 5, pp. 1536–1550, 1994.

[44] D. J. Kershaw and R. J. Evans, 'Waveform selective probabilistic data association,' *IEEE Transactions on Aerospace and Electronic Systems*, vol. 33, no. 4, pp. 1180–1188, 1997.

[45] H. Van Trees, *Detection, Estimation, and Modulation Theory, Part III*. Wiley, 2001.

[46] W. Moran, S. Suvorova, and S. Howard, *Application of Sensor Scheduling Concepts to Radar*. Springer, 2008, pp. 221–256.

[47] S. P. Sira, A. Papandreou-Suppappola, and D. Morrell, 'Dynamic configuration of time-varying waveforms for agile sensing and tracking in clutter,'

IEEE Transactions on Signal Processing, vol. 55, no. 7, pp. 3207–3217, 2007.

[48] C. O. Savage and B. Moran, 'Waveform selection for maneuvering targets within an IMM framework,' *IEEE Transactions on Aerospace and Electronic Systems*, vol. 43, no. 3, pp. 1205–1214, Jul. 2007.

[49] M. R. Bell, 'Information theory and radar waveform design,' *IEEE Transactions on Information Theory*, vol. 39, pp. 1578–1597, 1993.

[50] A. O. Hero, D. A. Castanon, D. Cochran, and K. D. Kastella, 'Information theoretic approaches to sensor Management,' in *Foundations and Applications of Sensor Management*. Springer US, 2008, pp. 33–57.

[51] E. H. Aoki, A. Bagchi, P. Mandal, and Y. Boers, 'A theoretical look at information-driven sensor management criteria,' in *14th International Conference on Information Fusion*, Chicago, IL, USA, 2011, pp. 1–8.

[52] E. H. Aoki, A. Bagchi, P. Mandal, and Y. Boers, 'On the "near-universal proxy" argument for theoretical justification of information-driven sensor management,' in *IEEE Statistical Signal Processing Workshop*, Nice, France, 2011, pp. 245–248.

[53] J. M. Aughenbaugh and B. R. L. Cour, 'Metric selection for information theoretic sensor management,' in *11th International Conference on Information Fusion*, Cologne, Germany, June 2008, pp. 1–8.

[54] D. Cochran, S. Suvorova, S. Howard, and B. Moran, 'Waveform libraries,' *IEEE Signal Processing Magazine*, vol. 26, no. 1, pp. 12–21, 2009.

[55] S. P. Sira, Y. Li, A. Papandreou-Suppappola, D. Morrell, D. Cochran, and M. Rangaswamy, 'Waveform-agile sensing for tracking,' *IEEE Signal Processing Magazine*, vol. 26, no. 1, pp. 53–64, 2009.

[56] A. Charlish, 'A case study of target visibility using ambiguous MPRF measurement data,' in *IET International Conference on Radar Systems (Radar 2012)*, Glasgow, UK, Oct. 2012, pp. 1–6.

[57] F. Katsilieris, Y. Boers, and H. Driessen, 'Sensor management for PRF selection in the track-before-detect context,' in *IEEE Radar Conference (RadarConf)*, Atlanta, GA, USA, 2012.

[58] J. Bae and N. Goodman, 'Adaptive PRF selection technique for multiple targets in track-before-detect,' in *5th IEEE International Workshop on Computational Advances in Multi-Sensor Adaptive Processing (CAMSAP)*, Saint Martin, French West Indies, France, 2013, pp. 448–451.

[59] J. M. Butler, 'Tracking and control in multi-function radar,' Ph.D. Thesis, University College London, 1998.

[60] A. G. Huizing and A. A. F. Bloemen, 'An efficient scheduling algorithm for a multifunction radar,' in *IEEE International Symposium on Phased Array Systems and Technology*. Boston, MA, USA: IEEE, 1996, pp. 359–364.

[61] B. Gillespie, E. Hughes, and M. Lewis, 'Scan scheduling of multi-function phased array radars using heuristic techniques,' in *IEEE Radar Conference*, Arlington, VA, USA, May 2005, pp. 513–518.

[62] A. J. Orman, C. N. Potts, A. K. Shahani, and A. R. Moore, 'Scheduling for a multifunction phased array radar system,' *European Journal of Operational Research*, vol. 90, no. 1, pp. 13–25, 1996.

[63] A. Nelander and D. Stromberg, 'Energy management in multiple-beam phased array radars,' in *IET International Radar Conference*, Edinburgh, UK, 1997, pp. 553–557.

[64] A. P. Stoffel, 'Heuristic energy management for active array multifunction radars,' in *IEEE National Telesystems Conference*, San Diego, May 1994, pp. 71–74.

[65] M. T. Vine, 'Fuzzy logic in radar resource management,' in *IEE Multi-function Radar and Sonar Sensor Management Techniques*. Stevenage: IET, 2001.

[66] S. L. C. Miranda, C. J. Baker, K. Woodbridge, and H. D. Griffiths, 'Comparison of scheduling algorithms for multifunction radar,' *IET Radar, Sonar & Navigation*, vol. 1, no. 6, pp. 414–424, 2007.

[67] S. L. C. Miranda, 'Resource management in multifunction array radar,' Ph.D. dissertation, University College London, 2004.

[68] S. L. C. Miranda, C. J. Baker, K. Woodbridge, and H. D. Griffiths, 'Fuzzy logic approach for prioritisation of radar tasks and sectors of surveillance in multifunction radar,' *IET Radar, Sonar & Navigation*, vol. 1, no. 2, pp. 131–141, 2007.

[69] E. Winter and L. Lupinski, 'On scheduling the dwells of a multifunction radar,' in *2006 CIE International Conference on Radar*, China, Oct. 2006, pp. 1–4.

[70] E. Winter and P. Baptiste, 'On scheduling a multifunction radar,' *Aerospace Science and Technology*, vol. 11, no. 4, pp. 289–294, 2007.

[71] P. Moo, 'Scheduling for multifunction radar via two-slope benefit functions,' *IET Radar, Sonar & Navigation*, vol. 5, no. 8, pp. 884–894, Oct. 2011.

Part II

Imaging radar

Introduction to imaging radar

Christoph H. Gierull[1]

Onset

It took almost exactly 100 years from Maxwell's first conception of the theory of electro-magnetic waves and the subsequent experimental proof by Hertz in the nineteenth century until the first patent on Doppler Beam Sharpening (DBS) was issued to Wiley in 1965 [1]. DBS was the original attempt to improve the Doppler resolution of moving pulsed radar by exploitation of a synthetically spanned aperture compared to the physical antenna with its very limited resolution. Although somewhat rudimentary because of the neglected quadratic phase history and thereby still a range-dependent azimuth resolution, it is considered the cornerstone of Synthetic Aperture or imaging Radar (SAR). The fundamental idea here is that through the motion of the radar-carrying platform a long synthetic aperture is spanned, along which the radar raw data are focussed, to achieve a high resolution in the flight direction. Besides the essential platform motion, the other vital attribute of SAR is that the recorded data by no means constitute the desired image but instead must first be processed using suitable focussing algorithms, which among others require a precise knowledge of the platform trajectory.

At the onset of SAR, owing to the absence of digital signal processors, the focussing could only be achieved using systems of optical lenses whereby the SAR raw data were recorded on film. This technology was not flexible and only permitted rather moderate results that typically suffered from low resolution, defocussing and other artefacts.

Twentieth century

After advancements for many years in the second half of the twentieth century, SAR has become an indispensable tool for civilian remote sensing as well as military surveillance and reconnaissance. Since the radar sensor illuminates the scene with its own energy and since its microwaves penetrate through clouds, SAR becomes independent from daylight and weather. Further, the SAR resolution is independent of the stand-off range adding to the palpable advantages in contrast to

[1]Defence R&D Canada – Ottawa Research Centre, Canada

electro-optical systems particularly when mounted on a spacecraft with its global reach and large coverage.

Information about the underlying scene cannot only be derived from the intensity images, but nowadays also by exploiting the information contained in the complex measurement such as polarimetric signatures, the phase history or the variation of the scatterer reflectivity over the frequency. At the end of the twentieth century, there was already a multitude of airborne SAR sensors exploiting the frequency bands from very high frequency all the way up to millimetre waves. Polarimetric and interferometric SAR was already operationally established on airborne platforms, whereby the latter is applied to reconstruct the scene topography (cross-track interferometry) or to indicate scene motion (along-track interferometry). Inverse SAR (ISAR) techniques on ground-based and airborne sensors were devised to generate high-resolution images of moving objects. Following the first ever space-based SAR Sea-SAT (1978), several very successful experimental space-based SAR missions, like SIR-C/X-SAR, ERS-1/2, J-ERS-1 and RADARSAT-1, have been accomplished in the 1990s and resulted in an eruption of research in this field.

Aptly, the turn of the millennium culminated in a milestone of radar-image-based remote sensing. With the spectacular Space Shuttle Radar Topography Mission (SRTM), 80 per cent of the topography of the entire Earth landmass was surveyed in only 11 days achieving an unprecedented three-dimensional resolution; a feat that would have taken an 'army' of land surveyors several decades if done using traditional geodetic techniques. This mission not only meant to overcome tremendous scientific challenges but also pushed the technological limits among others by deploying a 60-m-long extractable boom, the longest structure ever flown in space before construction of the International Space Station. The basics and scientific challenges of single-pass across-track interferometry, the technology employed for SRTM, are introduced in **Chapter 6 by Monti-Guarnieri, Rocca and Tebaldini**.

Twenty-first century

The early twenty-first century saw mostly evolutionary developments, for instance, an expansion of the used frequency bands and a dramatically improved quality of the SAR imagery in terms of highest resolutions in the order of centimetres as well as persistence over the area of interest. As an imposing example of these advancements **Damini, Linderman and Fitzgerald** present the VideoSAR concept in **Chapter 4**. The notion VideoSAR descriptively expresses its goal, namely to continuously monitor a scene of interest over a long period of time producing SAR imagery of quasi-photographic quality in real-time. This kind of performance can only be achieved with enormous technological expenditure employing extreme bandwidth, flexible antenna beam steering along with complex digital signal processing hard- and software.

New algorithms were developed capable of extracting more information about the scattering process, for instance, by exploiting the polarimetric properties together with the across-track interferometric height information (Pol-InSAR). These techniques produce volumetric information, e.g. about forests, an essential step for inferring biophysical parameters such as the biomass. At about the same time, the first SAR systems with multiple parallel receive channels (array antenna) have been established on airborne platforms. A multichannel capability is the prerequisite for a SAR to be sensitive to motion, either to objects travelling on the ground (Ground Moving Target Indication – GMTI) or the surface motion itself such as ocean currents. One promising approach not only to indicate object motion but also to be able to distinguish various motion components within each pixel is based on the Fourier-transform along a stack of SAR images that are co-located in space but slightly staggered in time. This technique is dubbed Velocity SAR and is presented in detail in **Chapter 11 by Rosenberg, Sletten and Toporkov** for a large phased array consisting of 32 elements mounted underneath an aircraft.

During the first decade of the new millennium, the transition of electronically steerable phased-array technology into the most recent generation of commercial/ civilian SAR satellites took place, like ENVISAT/ASAR, ALOS, TerraSAR-X, COSMO-Skymed and RADARSAT-2. For a small subset of these satellites that carry two parallel receive chains (TerraSAR-X and RADARSAT-2), the relatively mature single-channel SAR imaging capability is today rapidly evolving into advanced multi-aperture SAR concepts. The spatial diversity of multiple parallel receive channels can, for instance, be used to discriminate moving objects from the stationary background, to determine the underlying topography of the back-scattering terrain or the large-scale surface motion using single-pass interferometry. The flexible programmability of this newest generation of SAR satellites offers quasi seamless antenna beam steering at a rate equivalent to the pulse repetition frequency and in conjunction with sophisticated array signal processing algorithms opens up entirely novel radar concepts and capabilities from space, which were not previously possible. Although still rather rudimentary, even two parallel apertures offer palpable advantages compared to classical SAR that specialized modes have been designed and implemented on present on-orbit systems after launch. This way swaths widths of the order of 500 km for space-based SAR have been achieved which are today used operationally to aid in the monitoring of maritime approaches. **Chapter 7 by Gierull, Sikaneta and Cerutti-Maori** provides in-depth knowledge of the state-of-the-art of porting mature airborne SAR-GMTI onto a spacecraft, highlighting the technological and scientific challenges but also the tremendous benefits resulting from being able to do simultaneous GMTI and imaging from space.

Another remarkable evolution in radar imaging is the ever increasing number of nations operating SAR satellites. This feat was in the past a luxury only affordable by G7 economic powerhouses in North-America, Europe and Japan. This group is rapidly expanding comprising more and more smaller developed countries such as Spain and Korea as well as developing nations in Asia like China and India, in South America (Argentina, Brazil) and the Middle East. The maturity and affordability of space-based systems has reached now a point that African countries like Ghana, Nigeria and South Africa are also planning to operate their own SAR missions. In the next few years, more than two dozen SAR satellites will be orbiting and remotely sensing the Earth marking the beginning of a golden age for SAR imaging from space [2].

The next decades of the twenty-first century will be seeing the realization of qualitative leaps and revolutionary developments for which the scientific and technological cornerstones are currently emerging. These developments can be rather loosely divided into three different groups, smart technologies, multi-dimensional configurations and novel innovative signal processing approaches. Regarding the first, high-frequency technologies have been hallmarked by vast development leaps. For a few years, imaging radars can now be demonstrated for highest millimetre waves of up to 300 GHz permitting applications of Ultra-Wide Band imaging systems. The exponential advancement and miniaturizing of computing equipment has now reached a point that 'radar-on-chip' becomes viable and permits ever shrinking SAR systems. Cost-effective short-range systems based on this technology can now be mounted on small unmanned platforms such as model planes or drones starting to make experimental exploration of imaging radar affordable to everyman.

As mentioned before, the fast-paced development of digital components started a technological revolution for electronically steered phased array antennas. Instead of the thitherto used complex high frequency equipment with phase-shifters and combiner networks, it is nowadays possible to convert the signals at each element from analog to digital individually and to record them simultaneously. These 'fully digital arrays' will permit the utilization of extremely flexible digital beamforming and sophisticated adaptive radar signal processing opening the door to innovative future radar concepts including many parallel channels and waveform diversity. Fully adaptive digital beamforming in azimuth and elevation can dramatically increase SNR and hence the achievable coverage, overcome stringent ambiguity problems, enhance resilience against electronic interferences stemming from other competing radar sensors or deliberate jamming, or resolve individual scatterers within each resolution cell. One striking example is the High-Resolution Wide-Swath (HRWS) imaging technique which combines digital beamforming on receive in elevation with multiple channels in flight direction to overcome the inherent physical coverage-versus-resolution limitations of conventional single-channel SAR systems. In **Chapter 5 Sikaneta and Cerutti-Maori** present a theoretical approach to HRWS that applies to narrow or wide band aerospace SAR systems including squinted acquisition geometries thereby inherently comprising the required signal processing for moving target signals. Within the operational community HRWS is considered the 'Holy Grail of SAR' as the resulting imagery satisfies the partly contrary requirements of various user groups, i.e. monitoring of vast areas with a simultaneous zoom-in capability for recognition and identification. These advanced space systems will be so effective that a relatively small number of them are sufficient to cover gigantic areas in short revisit times, especially if placed on orbits with different inclinations. Although currently not fully realized, application of fully digital arrays for aerospace imaging radars is only a matter of time. Owing to the palpable benefits there is no doubt that payloads with more than two parallel receive paths will materialize in the next generation of space-based SAR [3].

SAR Interferometry based on two spatial channels was in the twenty-first century extended to multi-dimensional configurations. For instance, one can try to combine the individual data sets of several overpasses from slightly different altitudes in order to build a synthetic aperture across the flight direction. This virtual array in elevation direction may be used to truly image the third dimension (topographic height) in contrast to the classical pixel-based two-dimensional SAR Interferometry, allowing distinguishing a small number of scatterers residing in one and the same range-azimuth cell. This method was named SAR Tomography and has recently demonstrated impressive results particularly from space using various orbits. By merging stacks of across-track interferometric data sets from multiple overpasses, called repeat pass differential interferometry, the slightest height variations of stationary permanent scatterers on the Earth surface over long periods of time can be monitored with accuracy in the order of millimetres. **Fornaro and Pauciullo dedicated Chapter 8** to this important and innovative field of 3D SAR imaging research. It is self-evident that these techniques have great potential for the monitoring of slow-paced geological processes or problems provoked by man-made interference such as land subsidence caused by mining activities.

Comparison of SAR images taken from exactly the same location in space with a considerable time lag, e.g. perfectly revisited orbits over tens of days, leads to clear indications of change within the scattering behaviour of the otherwise stationary scenes. This comparison can be done based on the amplitude image or the complex SAR image; we talk about amplitude change detection or coherent change detection each with its individual pros and cons, both with obvious benefits to the intelligence community. These concepts are visited in the context of VideoSAR in **Chapter 4 by Damini, Linderman and Fitzgerald**.

For bistatic SAR the transmit and receive systems are spatially separated, i.e. mounted on two different carrier platforms, which offers a multitude of advantages compared to classic monostatic SAR sensors. Since the receiving unit does not emit any energy it inherently possesses a low probability of interception for instance against deliberate jamming. The flexible arrangement of the aspect angles to objects offers the possibility to infer complementary information particularly from man-made structures useful for instance for enhanced target recognition or detection of stealthy targets. Specific bistatic formations even provide sufficient resolution in flight direction of the receiver, a feat that is physically impossible for conventional SAR. This property may lead to radar image-based bad weather landing aids for aircrafts or obstruction warning sensors for low-flying helicopters. In the first years of the twenty-first century, the mathematical foundation for bistatic SAR processing algorithms was developed and the technological problems with synchronization of time, phase and beam illumination resolved. **Chapter 9 by Walterscheid, Cristallini and Kohlleppel** deals with the innovative attempt to exploit bistatic SAR configurations to improve GMTI capability of imaging radars. A milestone of bistatic SAR has been the launch of the TanDEM-X satellite in 2010, a clone to TerraSAR-X, intended to generate a Digital Elevation Model

(DEM) of the entire Earth with almost an order improvement in resolution and accuracy compared to SRTM. This can be accomplished by a much longer and variably adjustable baseline between the antenna phase centres ranging from hundreds of metres to kilometres. In order to derive the required metre-level accurate DEMs the distance or baseline vector between the two satellites must be determined and continuously adjusted with sub-millimetre precision.

Recently, bistatic SAR has been extended to multi-static SAR configurations and multi-input/multi-output (MIMO) systems. Independent operation of each transmit element with optimized waveforms permit a flexible creation of space-time diversity giving rise to a variety of novel applications. MIMO radars oriented parallel to the flight trajectory can be used to eliminate range and azimuth ambiguities, e.g. already employed for HRWS imaging, while those mounted perpendicular to the flight path permit new forms of interferometry and tomography. If mounted underneath the wings of an aircraft, MIMO arrays can directly image below and ahead, a property physically unfeasible for conventional side-looking SAR. Although suitable merely for restrictive stand-off ranges these radars can provide high-resolution images and three-dimensional maps completely unhindered by typical detrimental SAR effects such as shadowing and foreshortening, a major advantage for imaging in urban settings consisting of street canyons. Stationary MIMO radars with thousands of TR-modules, capable of real-time imaging with millimetre resolution, have already been realized in commercially available body scanners for airports. The innovative approach to combine multistatic SAR configurations with the MIMO concept to advance ISAR imaging capability is being addressed by **Pastina and Bucciarelli in Chapter 10**.

Outlook

A global, quasi-persistent, three-dimensional, high-resolution radar imaging capability that covers vast areas independent from weather and daylight and that allows to observe slow- and fast dynamic processes on the Earth is becoming increasingly technologically feasible and economically affordable. It offers the potential to radically improve the safety, security and prosperity of mankind, putting today's dream of the radar remote sensing community for the first time into the realm of possibility.

Although the research presented in this book part encompasses the breadth of the present state-of-the art of novel techniques and application of imaging radar science and technology, one exciting recently emerging research branch, Compressive Sensing (CS) for imaging radars, is missing and will most certainly be included in a future edition of this text book. CS based on sparse sampling is finally breaking the dependence on the sampling theorem changing the almost century-old seemingly irrevocable paradigm of digital signal processing, thereby pushing the door wide open to cognitive radar systems. Cognitive imaging radars will function with utmost autonomy, process the observed context in real-time and adapt the transmitted waveform and the radar task depending on the results. By applying knowledge databases, CS will present the image content in an information enriched form, reduced to the essential, thereby extracting value added information already during image formation rather than via subsequent post-proceeding steps.

Acknowledgement

With kind permission of my dear colleague, scientific role model and career-long mentor Prof. Dr.-Ing. Joachim Ender, a pioneer of multi-channel SAR technology and signal processing, this Introduction has adopted many of his foresightful thoughts expressed during speeches at various occasions, e.g. [4,5].

References

[1] C. A. Wiley, Pulsed Doppler Radar Methods and Apparatus, US patent no. 3,196,436, 1965 (originally filed 1954).

[2] A. Moreira, P. Prats-Iraola, M. Younis, G. Krieger, I. Hajnsek, and K. P. Papathanassiou, A Tutorial on Synthetic Aperture Radar, *IEEE Geoscience and Remote Sensing Magazine*, No. 1, March, pp. 6–43, 2013.

[3] C. H. Gierull and P. Vachon, First Special Issue on Multichannel Space-Based SAR, *IEEE Journal of Selected Topics in Applied Earth Observations and Remote Sensing*, Vol. 8, No. 11, pp. 4995–5115, 2015.

[4] J. H. G. Ender, Bildgebenes Radar: Vergangenheit – Gegenwart – Zukunft (Imaging Radar: Past – Present – Future), Presentation at Wachtberg-Forum, Germany, June 2014.

[5] J. H. G. Ender, Bildgebenes Radar im 21. Jahrhundert (Imaging Radar in the 21st Century), Inaugural Lecture at University Siegen, Germany, February 2012.

Chapter 4

VideoSAR imaging for real-time persistent surveillance

Anthony Damini[1], *Richard W. Linderman*[2] *and Dennis Fitzgerald*[3]

Abstract

VideoSAR is a land-imaging mode where the synthetic aperture radar (SAR) is operated in a spotlight configuration for an extended period of time. A sequence of images is continuously formed to a common Cartesian grid, while the radar is either flying towards, by or circling a target area. In general, VideoSAR imaging maintains antenna illumination on a target regardless of changes in squint angle, within theoretical and practical limitations. The video-like nature is a result of the imagery being produced using overlapped synthetic apertures such that the output frame rate can be commensurate with that of a video system. Enhanced exploitation of this product typically requires that the imagery be phase preserving, and that the data acquisition geometries be highly controlled for generation of secondary products such as coherence images (coherence maps) for change detection. One of the most accurate image formation algorithms for the formation of phase preserved imagery is the back-projection algorithm. Though computationally intensive, its attraction lies in its simplicity, the fact that the relative location of each pixel in the output imagery is precisely known and the complex imagery is phase preserved. The back-projection algorithm is presented in terms of its application to VideoSAR imaging. Mathematical decomposition techniques for improving the computational efficiency of the algorithm are reviewed. The application of change detection between pairs of VideoSAR images and 'stacks' of VideoSAR images where the benefits of trading off spatial image filtering and temporal image filtering by averaging coherence in slow-time is also examined. Finally, computer topologies applicable to back projection are outlined from the perspectives of both the signal processing architecture and advances in massively parallel computing.

[1]Defence R&D Canada, Ottawa Research Center, Canada
[2]Office of the Secretary of Defense, USA
[3]Information Directorate, Air Force Research Laboratory, USA

4.1 Introduction

VideoSAR is a land-imaging mode where the synthetic aperture radar (SAR) is operated in a spotlight configuration for an extended period of time. A sequence of images is continuously formed to a common Cartesian grid, while the radar is either flying towards, by or circling a target area. The only constraint on the data acquisition geometry is that the steering of the radar beam be maintained on the area of interest. Circular SAR is a specific instantiation of VideoSAR where the radar platform is circling the target area. In general, VideoSAR imaging maintains antenna illumination on a target regardless of changes in squint angle, within theoretical and practical limitations. Theoretical limitations arise when the antenna pointing approaches alignment with the radar platform velocity vector as the synthetic aperture required for achievement of a constant azimuth resolution increases inversely with the sine of the Doppler-cone angle of the antenna boresight as measured from the radar platform velocity vector. Practical limitations include pointing restrictions due to how the sensor is installed on the platform and maximum sensing range due to limited output power. Figure 4.1(a) illustrates the VideoSAR concept in the Circular SAR mode in which the Doppler-cone angle and range to the centre of the circle are ideally constant, and Figure 4.1(b) in the racetrack mode in which the Doppler-cone angle and range to the centre of the target are changing. In this latter geometry, persistence on the target is maintained as long as the target area remains within the range and angular field of view of the SAR antenna. Note that η_{d-c}, the Doppler-cone angle, and R, the range, both to the pixel of interest, are projected from the three-dimensional (x,y,z) volume to the two-dimensional (x,y) plane.

The theory and approach to spotlight SAR imaging is well documented [1,2]. VideoSAR produces high-resolution spotlight SAR imagery in a video-like format. The video-like nature is a result of the imagery being produced using overlapped synthetic apertures such that the output frame rate can be commensurate with that of a video system, computational resources permitting [3]. Enhanced exploitation of this product typically requires that the imagery be phase preserving, and that the data acquisition geometries be highly controlled for generation of secondary products such as coherence images (coherence maps) for change detection. One of the most accurate image formation algorithms for the formation of phase preserved imagery is the back-projection algorithm. In its native form, this processor is computationally intensive and, for large images, presents a significant challenge even to modern computing systems. Its attraction lies in its simplicity, the fact that the relative location of each pixel in the output imagery is precisely known and the complex imagery is phase preserved.

Within this chapter: the back-projection algorithm is presented in terms of its application to VideoSAR imaging; mathematical decomposition techniques for improving the computational efficiency of the algorithm are reviewed; the application of change detection between pairs of VideoSAR images and 'stacks' of VideoSAR images where the benefits of trading off spatial image filtering and temporal image filtering by averaging coherence in slow-time is examined and

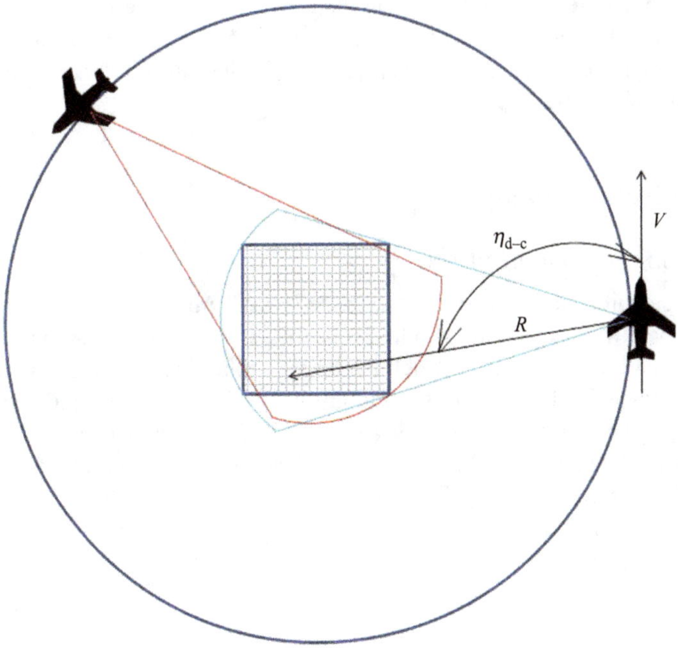

Figure 4.1(a) VideoSAR with radar platform circling the target

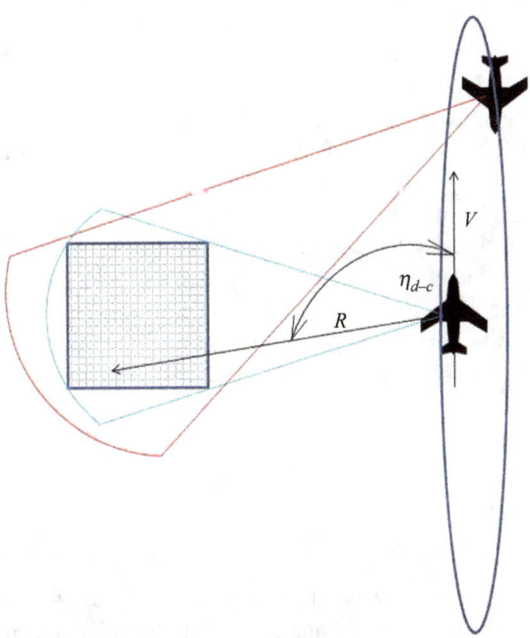

Figure 4.1(b) VideoSAR with changing Doppler-cone angle and range to target

computer topologies for implementing the compute intensive image formation are outlined from the perspectives of both the signal processing architecture inherent to practical implementation of the back-projection algorithm and advances in massively parallel computing.

4.2 VideoSAR imaging

4.2.1 Image formation theory

Target separability in range and cross-range in SAR imagery is directly dependent upon the bandwidth of the transmitted pulse in range and the illumination time or synthetic aperture time during which the target is illuminated in cross-range. In the case of a single SAR image, this paradigm of resolution is well understood [1,2]. However, in the case of multiple images, generated from viewing angles which can easily vary by factors of 10 × the antenna beamwidth or greater, and then non-coherently averaged, this paradigm changes.

For a system of transmitted signal bandwidth β, the ground range resolution is well known to be:

$$\rho_{gr} = \frac{K_{w-r}c}{2\beta\cos(\Upsilon)} \tag{4.1}$$

where c is the speed of light, K_{w-r} is the effective range broadening factor which characterizes the broadening of the impulse response due to the selected signal processing window, and Υ is the angle between the radar line-of-sight and the plane tangential to the earth's surface (similarly, the angle formed from horizontal by the radar line-of-sight). The $\cos(\Upsilon)$ factor converts slant range resolution to ground range resolution. Slant range is the line-of-sight distance between the radar antenna and the illuminated target. Ground range is the projection of slant range to the earth's surface. Ground range resolution becomes coarser at steeper depression angles Υ and thus at near range for wide elevation beamwidths. The compression of the transmitted signal's bandwidth is efficiently performed by applying the signal processing window in the frequency domain and then transforming the product back to the time domain. When a uniform spectral weighting is used, $K_{w-r} = 0.886$, and the impulse response in the time domain is the theoretical sinc function for which the first sidelobes are -13.2 dB down from the mainlobe.

Similarly, for a system integrating in azimuth over a synthetic aperture of angular duration $\Delta\theta$ radians, the azimuth resolution is as follows:

$$\rho_{az} = \frac{K_{w-az}R\lambda}{2\,\mathrm{SAL}\,\sin(\eta_{d-c})} \tag{4.2}$$

where R is the slant range to the point of interest, λ is the wavelength of the carrier frequency, η_{d-c} is the Doppler-cone angle between the radar antenna phase centre's velocity vector and the radar line-of-sight, K_{w-az} is the effective broadening of the mainlobe in azimuth after application of the selected signal processing window during azimuth compression, and SAL is the synthetic aperture length. With approximations, the azimuth resolution can be expressed in terms of $\Delta\theta$ as follows:

$$\rho_{az} \approx \frac{K_{w-az}\lambda}{4\sin(\frac{\Delta\theta}{2})} \approx \frac{K_{w-az}\lambda}{2\Delta\theta} \tag{4.3}$$

The resolution of an ideal point reflector as measured by a SAR, in either range or azimuth, is typically measured at its -3 dB points and is referred to as its impulse response. Ideally, two ideal point scatterers are resolvable when their impulse responses meet at or below the -6 dB points, allowing -3 dB resolvability in the combined response.

For systems which integrate in azimuth over small angles, the two-dimensional impulse response of the SAR image can be approximated by calculating the outer product of the range and azimuth impulse responses. However, as the synthetic aperture angle increases, the two-dimensional impulse response is more accurately found by describing the combined range and azimuth spectrums utilizing the two-dimensional frequency domain and then performing a two-dimensional Fourier Transform on this two-dimensional spectrum. Figure 4.2 comparatively illustrates the frequency support of both a conventional spotlight mode data acquisition and that which is processed in a VideoSAR mode. The frequency support in Figure 4.2(a) is seen to be in the form of an annulus whose angle is that of the synthetic aperture. The inner and outer radial edges of the annulus correspond to the lower and upper spatial frequencies of the transmitted signal pulse. The VideoSAR frequency support in Figure 4.2(b) generally exceeds the bandwidth in the azimuth direction which can be effectively processed in a single SAR image without incurring range migration and azimuth focussing errors. This is due to the fact that this frequency support is intended to support the generation of multiple images. Note that conventional spotlight mode image formation can be performed using synthetic apertures longer than is graphically illustrated in Figure 4.2(a) [4].

Figure 4.3 illustrates the two-dimensional impulse response of a transmitted signal of 1 GHz bandwidth, generated in a circular SAR mode. An image formed from one extremely long synthetic aperture is compared to two images derived from multiple shorter synthetic apertures from which the images have been non-coherently integrated.

For the upper two images, a 160° aperture was specifically chosen to illustrate that the uniformity of the two-dimensional impulse response was not an artefact of integrated observation from both sides of the image under investigation. For the two lower images, the number and angular excursions of the narrower aperture images were specifically chosen to span exactly 180°, thus ensuring uniformity in the impulse responses for analysis purposes. The impulse response for the smaller aperture in the bottom-right is dominated by the 5° limitation on coherent integration which yields an azimuth resolution which is commensurate with the

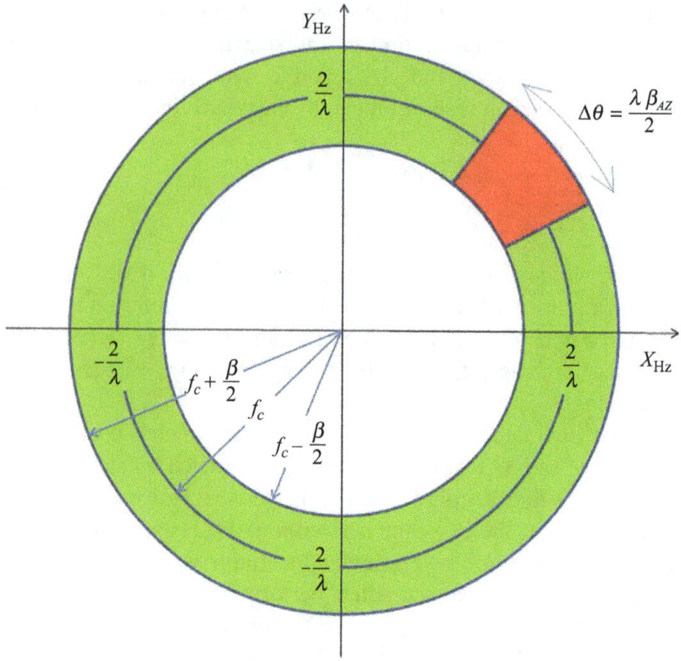

Figure 4.2(a) Typical spotlight mode frequency support

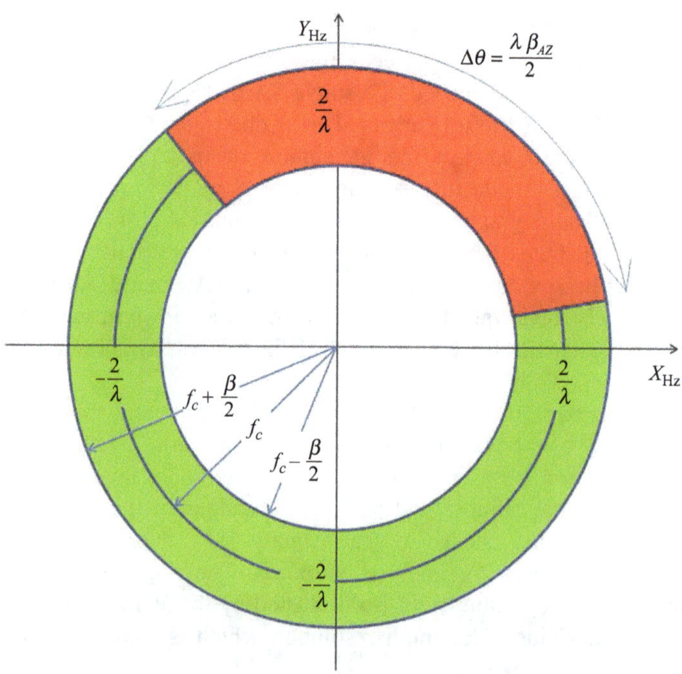

Figure 4.2(b) Typical VideoSAR mode frequency support

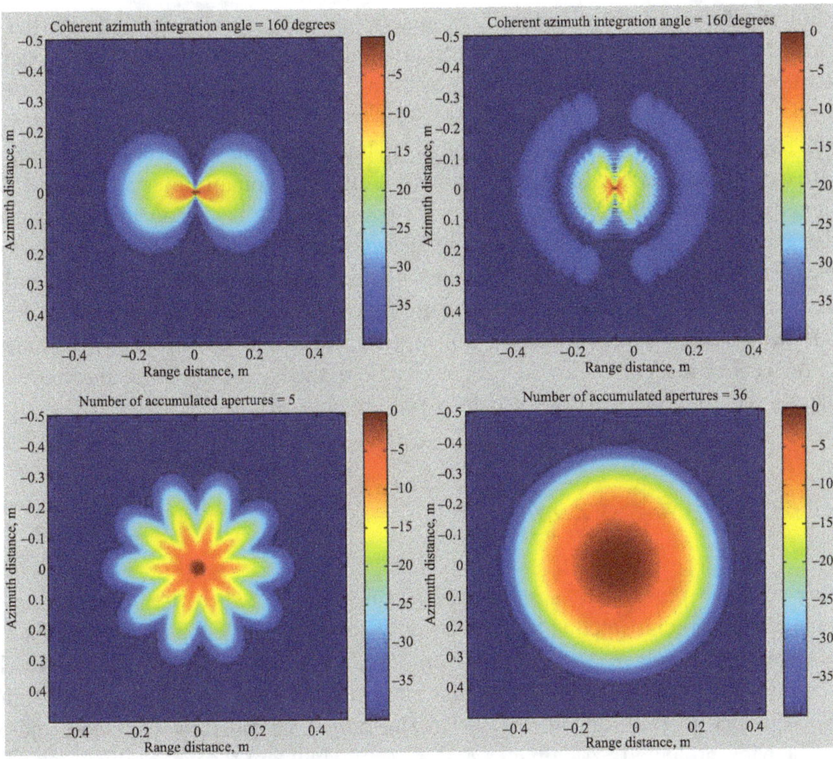

Figure 4.3 Two-dimensional circular SAR impulse responses illustrating coherent vs non-coherent integration; top-left 160° aperture coherently integrated image with Hamming weighting; top-right 160° aperture coherently integrated image with uniform weighting; bottom-left five non-coherently integrated 36° aperture images; bottom-right 36 non-coherently integrated 5° aperture images

range resolution. Increasing the azimuth integration angle results in finer azimuth resolution as shown in the top-left and bottom-left images. For extremely large coherent integration angles, the range and azimuth resolutions converge to the finer azimuth resolution resulting in a narrow mainlobe.

4.2.2 Back-projection for VideoSAR image formation

The back-projection algorithm can be implemented with no approximations, or it can be tailored to trade-off image fidelity with computational requirements. In its native form, the algorithm projects the compressed echoes from each transmitted pulse back onto a three-dimensional model of the earth. It allows full flexibility in the generation of the pixel resolution, pixel spacing and location of the image from within the antenna beam's surface footprint. The output image projection can also be tailored to support options such as generating imagery in a ground stabilized format, in rotating formats such as North up or aligned with the co-ordinate system used for data acquisition (where range is along the radar line-of-sight), or draped across a digital elevation model (DEM). The generation of imagery on a pixel by pixel basis eliminates invariance regions in both range and azimuth, and hence phase discontinuities due to approximations. Constraints on aircraft motion are also relieved as the matched filtering for each pixel is fully based on the associated pulse echo's phase history as the antenna phase centre location is used precisely for the coherent integration. That is, there are no approximations which are inherently necessary for patch-based image formation approaches. Given SAR echo returns $S(k,n)$ where k is the index into the transmitted pulses and n is the fast-time index into the echoes received from the reflectivity field, the back-projection algorithm to generate the VideoSAR image stream is presented as follows:

$$p(A,x,y,z) = \sum_{k=(K-m)(A-1)+1}^{(K-m)(A-1)+K} S(k,R) \cdot e^{-\frac{j4\pi R}{\lambda}} \tag{4.4}$$

where for each pulse k, R is calculated as the range between the antenna phase centre and the pixel located at the grid co-ordinate (x,y,z), where z is altitude. $S(k,R)$ is the return at range R interpolated from the $S(k,n)$ digitized echo returns, which are already range compressed. Any radar platform deviation from the ideal straight line trajectory is fully accounted for in the range interpolation. A is the index into the back-projected SAR images $p(A,x,y,z)$, which are formed to the pixel grid (x,y,z). K is the synthetic aperture length in pulses and m is the overlap between synthetic apertures also measured in radar pulses. It is noted that when $A=1$, $m=0$, that is, for the first aperture there is no overlap.

The (x,y) output pixels are initialized to be consistent in geometric spacing in range and azimuth. Assuming an ellipsoid earth model, or minimally a spherical one, each (x,y) pixel is initialized in a plane tangential to the Earth's surface. The (x,y) pixels are projected vertically down onto the Earth model to give the final (x,y,z) positions in three dimensions. This is where elevation data from a DEM is incorporated.

It is also desirable to output imagery for which the azimuth resolution is constant as a function of range and Doppler-cone angle. This necessitates that the synthetic aperture length, K, be updated for each image as a function of R and η_{d-c}. Further, if the Pulse Repetition Frequency (PRF) is slaved to the antenna phase centre velocity and Doppler-cone angle to the target, the overlap between synthetic apertures defined as m pulses may have significantly different values from aperture to aperture.

In its native form, back-projection has the disadvantage of requiring more compute resources than other phase preserving algorithms such as polar format. Its advantages lie in its straightforward presentation and its simplicity for implementation within parallel computing architectures. Multiple approaches have been investigated to reduce the computational requirements of back-projection.

Fast back-projection is often used to describe algorithms which require less computing capability as they incorporate approximations to the original algorithm. One variant of fast back-projection proposed by Yegulalp [5] includes dividing the synthetic aperture into non-overlapping sub-apertures from each of which an image is formed using conventional back-projection. These images are formed to a pixel grid in a polar co-ordinate system which has fine spacing in range and coarse spacing in azimuth commensurate with the finest resolution allowed by the length of the sub-aperture. Each of these images is then upsampled in two dimensions to the pixel spacing of the desired product. Finally, the final full resolution image is formed by coherently summing the corresponding pixels across the images formed from the sub-apertures. The coherent integration factors can be applied during the final summation or applied to the individual images during the two-dimensional upsampling. The actual reduction in the computational throughput requirement is dependent upon several factors including the size and number of sub-apertures, the accuracy of the kernel selected to perform the two-dimensional upsampling, and the resolution of the final image. Further variants of fast back-projection have been described by Ulander *et al.* [6]. They propose a generalized approach to Yegulalp's in which the original aperture is factored into multiple levels, and the image formation/coherent sub-aperture addition repeated for each level to form images of progressively finer resolution, culminating in the final back-projected image. There is some flexibility in the number of levels into which the original aperture is factorized, and the corresponding number of sub-apertures for each level. An immediate parallel can be drawn to the speed increase achieved using the Fast Fourier Transform vs the Discrete Fourier Transform. It is noted, however, that the computational savings due to the reduction in back-projection operations with the addition of multiple levels of factorization is somewhat offset by the additional two-dimensional interpolations which accompany each level. To compound that, from the perspective of multi-processor implementation, the required inter-processor communication bandwidth increases significantly for such multi-level processing strategies, and may quickly become the limiting factor for system requirements. Further computational savings can be found in that the back-projection of the m pulses which overlap between synthetic apertures need not be repeated as the process moves forward through the data to create the VideoSAR stream. In the case of image formation by coherent sub-aperture addition, for simplicity the overlap between synthetic apertures can be fixed to correspond to that required for an integer number of sub-apertures either in the fast back-projection approach or the first level of factorization in the generalized back-projection approach. An inherent advantage of the back-projection algorithm is the ability to incorporate pulse diversity. With all fast back-projection variants, the computational savings achieved will need to be traded off with any corresponding reduction in the image fidelity.

Antenna elevation pattern gain and range spreading (R^4) loss compensations are usually applied during pulse compression. Implementation of these corrections within the back-projection image formation algorithm would yield theoretically more accurate radiometric calibrated imagery. Similarly, compensation of the output image in azimuth for normalization of image brightness due to the antenna azimuth pattern gain is typically done after back-projection, though can be applied during the image formation.

4.2.3 Non-coherent integration of back-projected images

Non-coherent image integration can take several forms. The parameter space which affects the final image includes the number of images to be combined, the overlap between the individual synthetic apertures contributing to each image and the method of combination of the registered pixels. Examples of the method of combination include averaging the pixel powers or adopting a scheme such as selecting the maximum pixel powers across the VideoSAR image stream. This integration of multiple images yields an image product generated from a much wider radar illumination angle. The non-coherent integration introduces the effect of speckle reduction, allowing more visibility of low signal-to-noise-ratio targets, while the wide observation angle (changing radar line-of-sight) yields a narrower radar shadow and hence more information regarding target reflectivity. Structures such as fences and building perimeters can become more visible. The general case where L images are summed to produce each p' image in a new image stream with an overlap of m' of the original p images contributing to each consecutive new image can be written as

$$p'(A',x,y,z) = \sum_{A=(L-m')\cdot(A'-1)+1}^{(L-m')\cdot(A'-1)+L} p(A,x,y,z) \tag{4.5}$$

$p'(A',x,y,z)$ takes the form of either the original image stream $p(A,x,y,z)$ in the case of $L=1$ and $m'=0$, or a series of images where a running average of $L>1$ images is maintained across the original VideoSAR image stream. In the case of extreme non-coherent image integration where $L=A_{\max}$, then one image is output which is the cumulative sum of all of the input images.

The SAR data in all subsequent figures is from the Defence Research and Development Canada (DRDC) X-band wideband experimental airborne radar (XWEAR) operating in VideoSAR mode [3,7]. The imagery in Figure 4.4 is taken from a VideoSAR image stream. The data was collected using a constant depression angle, of the order of 5°, in a circular SAR mode. A flat earth was assumed during image formation. There are no natural height variations in the terrain, and the man-made variations are five m or less. Foreshortening or layover of the imagery was not an issue given the shallow depression angle. The dynamic range presented within the individual images does change as speckle is reduced with additional non-coherent averaging. Figure 4.4(a) is the first image from the stream with no non-coherent image to image integration. Figure 4.4(b) is the non-coherent average of the first ten images. Figure 4.4(c) is the non-coherent average of images 45 to 54. Figure 4.4(d) is the non-coherent average of all 54 images. The frames input to Figure 4.4(b) and (c) were chosen so that the radar line-of-sight rotated

Figure 4.4(a) Image 1 from VideoSAR image stream

Figure 4.4(b) Images 1 to 10 non-coherently averaged

Figure 4.4(c) Images 45 to 54 non-coherently averaged (radar line-of-sight rotated 90° from Figure 4.4(b))

Figure 4.4(d) Images 1 to 54 non-coherently averaged

90° between image 1 and image 45. Of particular note is the speckle reduction in Figure 4.4(b) and (c) when compared to Figure 4.4(a). Also evident is the image diversity between the two speckle-reduced images in Figure 4.4(b) and (c) in terms of the shadowing and reflectivity of structures in the image. The 90° change in viewing angle between Figure 4.4(b) and (c) has changed where the shadows are cast, and illuminated different sides of the buildings. The non-coherent average of all 54 images in Figure 4.4(d) presents an effective composite view with good shadow elimination and good reflectivity information on the buildings.

Accurate frame-to-frame registration is necessary at higher pixel resolutions for non-coherent integration of the imagery. Translation of the imagery in both the range and azimuth directions is primarily due to sources of error in motion measurement or errors in the processing parameters of radar wavelength and speed of light which can vary with atmospheric conditions during the imaging interval or vary on a pulse basis. An error in the knowledge of the radar carrier centre frequency can be equated to an error in the knowledge of the speed of light in the atmosphere or an error in the knowledge of the radar wavelength. Manifesting these errors in terms of the radar wavelength, a change in wavelength, $\Delta\lambda$, shifts the location of the formed image in both the range (R) and azimuth (A_z) dimensions according to

$$\Delta R = R\frac{\Delta\lambda}{\lambda} \tag{4.6}$$

and

$$\Delta A_z = R\tan(\eta_{d-c})\frac{\Delta\lambda}{\lambda} \tag{4.7}$$

It can be seen that as long as there is an error in the assumed wavelength (either changing or constant $\Delta\lambda$), a changing shift ΔR in the assumed range location will occur with a change in R, and a changing shift ΔA_z in the assumed azimuth location will occur with changes in R or η_{d-c}. The speed of propagation of the radar signal through the atmosphere is affected by the index of refraction, which, even in clear air, changes with pressure, temperature and humidity. The effect that this has on SAR imagery is particularly noticeable in geometries which include long ranges, shallow grazing angles and formation of large area imagery [8,9]. In such cases, significant changes in the index of refraction can be encountered over the synthetic aperture, or between VideoSAR apertures. These unknowns are a challenge to the autofocus mechanism and are difficult to interpret when testing and qualifying SAR systems. Propagation speed increases with decreasing air density and, in general, can be assumed to be a function of altitude such that the speed of signal propagation is a function of temperature, pressure and humidity, whether assumed or measured. Again, this would have more of an effect on long-range, shallow grazing angle, long aperture imaging. Hence, not a significant problem for short-range, steep look-down, short aperture imaging.

Two-dimensional correlation of sequential image pairs can be used to measure the relative shift between image frames. The correlation should be localized to high contrast areas with discrete targets. Any measured shift can then be corrected using two-dimensional interpolation of the most recent frame to match the reference to which it was correlated. The primary source of error in frame-to-frame registration is that of motion measurement. The low-frequency drift in the inertial navigation system can be commensurate with the range resolution of a wideband radar over a long synthetic aperture, hence the importance of correcting shifts between frames.

4.2.4 Image ambiguities

Figure 4.5(a) is a VideoSAR example where 163 images have been non-coherently combined. In this case, the radar platform has completed 190° of rotation about the target area. The image is formed to a grid which is larger in azimuth than the terrain illuminated within the −3 dB antenna azimuth beamwidth. The dark regions in the image fall outside both the digitized pulse echo window and the part of the azimuth beam which has gain sufficient for imaging. The radar PRF was set during data acquisition to adequately support the −3 dB antenna azimuth beamwidth. Hence, azimuth ambiguities appear towards the edges of the imagery as the full image grid is larger than this and does not have adequate sampling support. The ambiguities are particularly noticeable at the top of Figure 4.5(a). As the data acquisition geometry rotates about the target, the corresponding azimuth ambiguities are seen to rotate as well within the non-coherently integrated image stream. Incomplete knowledge of the antenna beam pointing angle over the synthetic aperture can also lead to azimuth ambiguities.

Processing a narrower azimuth signal spectrum is one strategy that can be used to suppress the azimuth ambiguities, albeit fewer pixels will be produced in azimuth in each individual image. However, for a wide swath once data has been acquired over 90° of sensor rotation about a target area, the non-coherently integrated image product can have a much larger spatial coverage than that of any of the contributing images and compensate for the smaller azimuth coverage of each individual image. This can be understood in the context of the acquisition of very wide range swaths at near range where the azimuth beamwidth coverage may not match the range swath width. To illustrate the strategy of processing a narrower azimuth signal spectrum, the data used to form the image in Figure 4.5(a) has been reprocessed in Figure 4.5(b), but each individual image has now been formed from only 60% of the originally sampled azimuth signal spectrum. The ambiguities have been effectively eliminated without compromising the geo-spatial coverage. Doppler centroid estimation can also be used to refine the antenna beam pointing knowledge, hence allowing for an image of optimum dimension to be formed in azimuth before azimuth ambiguities are introduced. In the VideoSAR mode, the computational savings associated with the processing of a narrower spectrum cannot be overlooked as producing fewer pixels per image correlates directly to fewer required computational resources.

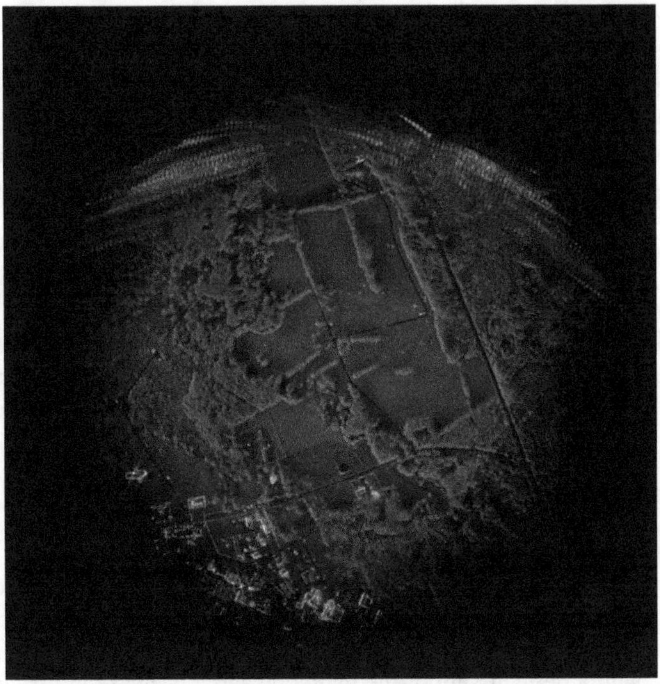

Figure 4.5(a) VideoSAR – full azimuth spectrum processed

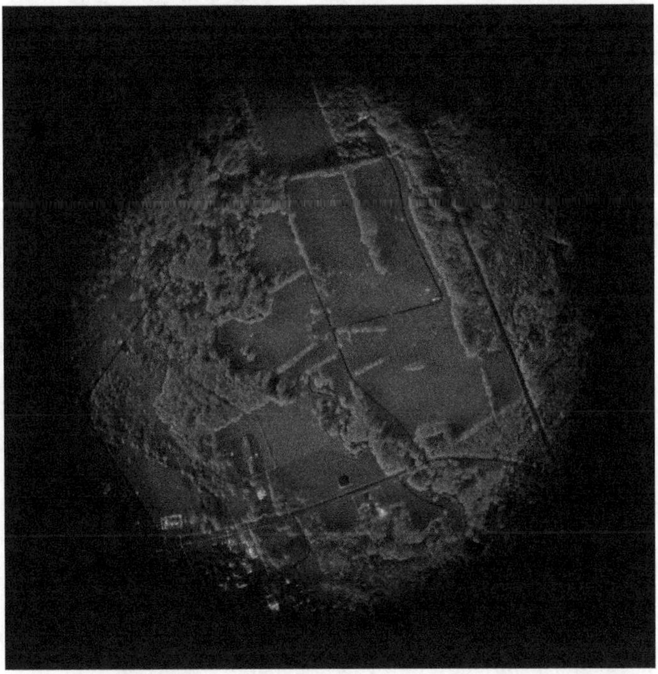

Figure 4.5(b) 60% azimuth spectrum processed

4.2.5 Autofocus considerations

Autofocus considerations are equally applicable whether in the circular SAR or race-track imaging geometry. Sources of error in motion measurement, or inaccurate measurement of the antenna phase centre position, is typically due to drifts in the inertial measurement sensors. Low-frequency errors attributable to long-term drifts result in image shifts and geometrical distortions from the desired grid co-ordinate system. A fixed difference between the measured and true antenna phase centre positions along the radar line-of-sight equates to a positional shift in the imagery of the same difference. A linear drift between the measured and true antenna phase centre positions results in an azimuth shift of the estimated image geometry during image formation. The resulting azimuth shift in the imagery is linearly proportional to the rotational error manifested by the linear drift and can be approximated as

$$\Delta A_z = R\Delta\varepsilon \tag{4.8}$$

where $\Delta\varepsilon$ is the fit to the angular separation between the measured and the true antenna phase centre positions. High-frequency errors in the knowledge of antenna phase centre position result in a degradation of image quality in terms of geometric distortions, with a de-focussing of the mainlobe and a decrease in image contrast. The energy displaced from the mainlobe into the sidelobes is quantified in terms of the integrated-sidelobe-ratio performance metric [10].

Autofocus algorithms for fine resolution SAR imagery need to account for the low-frequency drifts in the antenna phase centre with time and high-frequency errors during data acquisition. High order autofocus which corrects for these errors is well understood [1] and algorithms are designed to produce phase error estimates from the returns of bright targets which are distributed over the reflectivity scene. The process of selecting suitable point targets from which the autofocus phase correction estimate is obtained typically requires the generation of SAR imagery at a resolution which is minimally sufficient for point selection. The trade-off space for generation of this imagery involves range and azimuth pixel spacing, synthetic aperture time, multi-looking in azimuth via shorter apertures for speckle reduction, multi-looking in range via processing of sub-bands within the pulse and the compute time allocated for image generation. High order autofocus corrections are typically in the form of a phase angle corresponding to each transmitted pulse's echo in the synthetic aperture. The phase angle can be applied to the corresponding transmitted pulse echo in one of two ways. It can either be directly applied in the form of a phase error correction or it can be used to compute a corresponding radar line-of-sight error which can then be incorporated into the processing to simultaneously correct the measured range cell migration and phase error. This second method of applying the phase angle for phase error correction is more accurate but only necessary for very fine resolution long aperture SAR where the residual range cell migration, if left uncorrected, would noticeably degrade the imagery. Robust autofocus is important to generate an image product which is suitable for exploitation. Spatially variant autofocus is generally necessary for fine resolution systems which produce imagery over wide swaths in range and cross-range. Significant advances have been made in autofocus based on the phase gradient algorithm [1] towards this problem.

4.3 Change detection

4.3.1 Background

Change detection with SAR is the processing of two geo-registered SAR images, formed at separate times, for the purpose of detecting human activities or changes in the environment. Change detection is conventionally classified as either amplitude change detection (ACD) which operates on the pixel power or coherent change detection (CCD) which operates on the pixel phase. ACD can be thought of as non-CCD. It measures the change in the reflectivity of the scene of interest between two geo-registered SAR images. The detections in ACD can be prone to high false alarms if the images are not filtered first to reduce speckle noise. ACD provides useful information when the change of interest is on the order of the pixel resolution, particularly in the case when the overall coherence between the two images is too low for effective CCD. CCD is particularly effective when the clutter coherence is high as low clutter coherence can be mistaken for change. Under favourable conditions, CCD is sensitive to changes that are on the order of the radar wavelength. These changes may not be readily noticeable from the amplitude imagery.

Table 4.1 outlines how data acquisition for change detection can be accomplished: single-pass data acquisition and dual-pass data acquisition.

For single-pass change detection operating on a VideoSAR image stream, the capability to suppress the reflectivity field which has not changed between image frames is limited because the data may not fully encompass the required specular reflection for both frames. Hence, the scatterers which have not changed may not correlate well and this may mistakenly imply change. From this perspective, the specular change can be used to confirm the presence of targets which are present in both frames, but may not fully appear in both frames. Figure 4.6 illustrates the use

Table 4.1 Change detection topologies

Data acquisition	Single-pass applications	Dual-pass applications
Amplitude change detection	Specular change detection – pixel level	Human activity detection – pixel level Environmental change detection
Coherent change detection	Requires multiple channel SAR for moving target detection	Human activity detection – sub-pixel level Environmental change detection

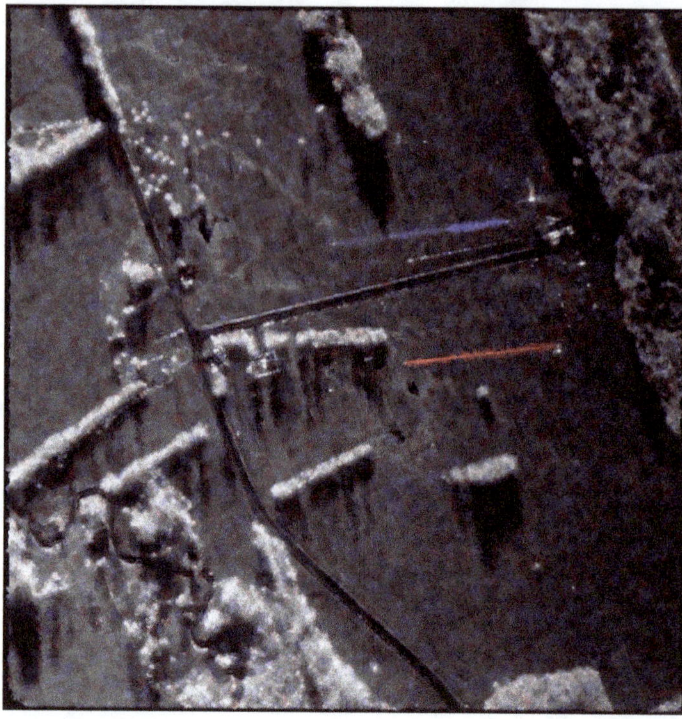

Figure 4.6 Single-pass ACD between two consecutive VideoSAR image frames. Red indicates a wire fence only visible in the first image, and blue indicates a wire fence only visible in the subsequent image

of single-pass ACD for detecting specular change between two consecutive VideoSAR image frames. In this case, the red indicates where a wire fence was only visible in the first image, and the blue indicates where a second wire fence was only visible in the subsequent image. For the moving target indication problem, single-pass change detection is a clutter suppression problem which requires multiple radar antenna channels for simultaneous independent observation of the scene for coherent clutter cancellation of the stationary components [11]. The clutter cancellation is typically performed using either displaced-phase-centre-antenna or space-time-adaptive-processing techniques. Once the clutter is cancelled (or reduced), the moving targets become visible in the remaining signal. In the case of VideoSAR dual-pass change detection, the data collected along the flight path for each pass is segmented into synthetic apertures which correspond between the two passes and ideally observe the target area using the same azimuth and grazing angles. Flight paths which are 180° apart (implying opposing flight directions as in the case of a race-track flight path) are considered identical for this purpose. If the flight paths are identical, this ensures that the azimuth spectra (as described in

Figure 4.2) of the two corresponding images fully overlap in the two-dimensional spatial frequency domain and the image pair will then be fully coherent in the absence of environmental or man-made change. The image pairs must be finely registered to sub-pixel accuracy to account for extremely small errors such as in the antenna phase centre location estimates, jitter in range delay for digitization or due to the atmosphere, because, as previously noted, in the case of CCD the changes being detected can be on the order of the radar wavelength. Details of both ACD and CCD are as follows.

Two-dimensional correlation of corresponding images can be used to measure the relative shift between corresponding image frames from the two passes. The correlation should be localized to high contrast areas with discrete targets that appear in both images. Any measured shift can then be corrected using two-dimensional interpolation of one of the frames to match the other. The primary source of change detection error due to a mismatch in frame registration between passes is that of motion measurement. However, a mismatch in viewing angles can result in a change in reflectivity and thus degradation in coherence which also leads to false change detections.

Matching the azimuth and grazing angles of the synthetic apertures from the two passes is a priority in two-pass change detection. Also important is a close match of the range to the target. Long-range images tend to be noisier which can reduce coherence or allow the introduction of spatially variant phase errors. Hence, the avoidance of noisy long range images, or one of the data collection passes being at a longer range than the other, is important towards avoiding false detections due to non-coherence. Further, small mismatches in azimuth or grazing angles can not only change the relative scene reflectivity and thus coherence, but change the shadows and the layover again yielding non-coherence and false detections. A pixel is the result of the super-position of the reflections from all targets which the wavefront of the signal simultaneously intersects. The layover from terrain with high objects or large changes in topography over short distances will superimpose into different pixels with different grazing and/or squint angles. This leads to the tolerance on the imaging geometry further being a function of the gradient of the natural terrain and the elevations of the man-made structures which are present.

When there is a difference in the imaging geometries between the two data collection passes such that the images do not register consistently across the desired area, additional steps must be taken to register them. Image registration typically involves measuring the displacement of pairs of tie points in the images, and then re-sampling so that the image pixels are registered between the pair. This re-sampling can be spatially variant resulting in one of the images being 'warped' to register with the other. The re-sampling is performed on either pixel detected imagery which is adequately sampled, or imagery within which the pixels have been preserved as complex in value. In the latter case, the accuracy of the re-sampling can be facilitated by ensuring that the frequency spectrum of the complex pixel valued image is centred at 0 Hz in both range and azimuth.

4.3.2 Amplitude change detection

Prior to performing the actual change detection step, each of the images input to the ACD process is low-pass filtered to reduce speckle noise and thus false alarms. The smoothing filters employed are typically square with an odd number of coefficients in the x and y directions. Given the two pass image pair, p_1 and p_2, image smoothing is effected according to the formulae

$$\rho_1(A,x,y) = sf * p_1(A,x,y,z) \tag{4.9}$$

$$\rho_2(A,x,y) = sf * p_2(A,x,y,z) \tag{4.10}$$

where *sf* is the two-dimensional low-pass *s*moothing *f*ilter and * the two-dimensional convolution operator. Increasing the filter dimensions reduces the speckle noise yielding less noise in the change detection process, but this is directly at the expense of spatial resolution [12]. Post-image smoothing, two common approaches to ACD are analysis of either the ratio of the two images or their difference. While differencing subtracts the corresponding pixels' intensity values with the delta being indicative of change, computing the ratio between pairs of pixels allows examination of their relative dynamic values [13]. A ratio = 1 is indicative of no change while ratios significantly greater to or less than 1 are indicative of change. For example, a ratio > 1 can be interpreted as the addition of a target to the area under surveillance while a ratio < 1 as the departure of a target. Similarly, a positive difference can be interpreted as the addition of a target while a negative difference as the departure. The intensities of the image pair under investigation must be calibrated to eliminate any bias introduced by a difference in the mean pixel levels. Image normalization eliminates any bias due to gain or transmitted power differences between the two data collections. Typically, this is performed by computing each image's mean intensity and either normalizing each image or scaling one of the images by first normalizing it and then multiplying it with the mean intensity of the other image. The application of the 'red = fled, blue = new' colour taxonomy effectively supports the interpretation of ACD results with the saturation of the red and blue proportional to the differences between the before and after images, or the difference from ratio = 1 which is neutral or no change.

4.3.3 Coherent change detection

The CCD function takes as its input a pair of single-look-complex (SLC) SAR images generated using the same acquisition geometry at different times, registers them and produces a coherence image which can be used to infer if any change has occurred in the area of interest between the two acquisition times. Change is declared if there is a lack of coherence, and no change is declared for areas where there is coherence. Coherence is a function of multiple factors, the dominant ones being across-track and along-track baseline de-correlation, accuracy of image registration, environmental and man-made change, and radar system noise whether in the radio frequency (RF) chain or due to quantization during sampling. Prior to forming the actual coherence image, the two SLC images are wavenumber filtered [14].

In the ideal case, when the two radar flight paths are identical, the range spectra are fully correlated. When there is a baseline separation between the flight paths, the uncorrelated parts of the two range spectra need to be removed by way of wavenumber filtering prior to formation of the coherence image. Again, this difference in the flight geometries can be due to differences in the Doppler-cone or grazing angles. From the perspective of Fourier space, Figure 4.7 illustrates the frequency support from the before and after passes. The overlapped area comprises the part of each spectrum that can positively contribute to the change detection process. The parts of each image's spectrum which do not overlap will negatively contribute to the change detection process as they will misleadingly reduce coherence. Coherence is therefore improved by filtering the spectra of each data collection to allow just the overlapped spectra to be used for the creation of the final coherence product. One of the consequences of this reduction in spectrum is a corresponding reduction in resolution. As discussed later, this problem can be mitigated by using a stream of VideoSAR coherence images to generate an overall representation of coherence for a much larger aperture. Further, if the frequency domain window weightings are applied before wavenumber filtering, the relative impulse response of the imagery between passes is likely to change depending on the parts of the spectra which were filtered. Hence, it is advisable that the same window is applied to each data set after wavenumber filtering to guarantee identical impulse responses. Note that image registration does not guarantee registration in the Fourier space, hence the importance of wavenumber filtering. Figure 4.7(a) is the direct representation of the two spectrums in the two-dimensional frequency domain. Figure 4.7(b) is the unwrapped frequency domain representation and what will remain after wavenumber filtering is indicated by the hatched area. Note that the unwrapped spectrum is centred at 0 Hz in range and azimuth.

After wavenumber filtering, each of the images is low-pass filtered as with ACD, to reduce speckle noise and hence false detections, but again at the expense of spatial resolution. The images are then registered and a coherence image formed. Given the filtered two pass image pair, $\rho_1(A,x,y)$ and $\rho_2(A,x,y)$, the coherence can be estimated for every (x,y) pixel according to

$$\beta(A,x,y) = \frac{\left| \sum\limits_{m=x-M/2}^{x+M/2} \sum\limits_{n=y-N/2}^{y+N/2} \rho_1(A,m,n)\rho_2^*(A,m,n) \right|}{\sqrt{\left(\sum\limits_{m=x-M/2}^{x+M/2} \sum\limits_{n=y-N/2}^{y+N/2} \rho_1^2(A,m,n) \right) \left(\sum\limits_{m=x-M/2}^{x+M/2} \sum\limits_{n=y-N/2}^{y+N/2} \rho_2^2(A,m,n) \right)}} \tag{4.11}$$

where A is the index into the back-projected VideoSAR images. This estimator is based on $M' \times N'$ samples about the pixel of interest at (x,y), where $M' = M + 1$ and $N' = N + 1$ are odd numbers. The result will increase in accuracy for larger $M' \times N'$ with sufficient image smoothing support. Image smoothing as noted above is typically implemented according to the $M' \times N'$ pixels used to compute the $\beta(A,x,y)$ estimate.

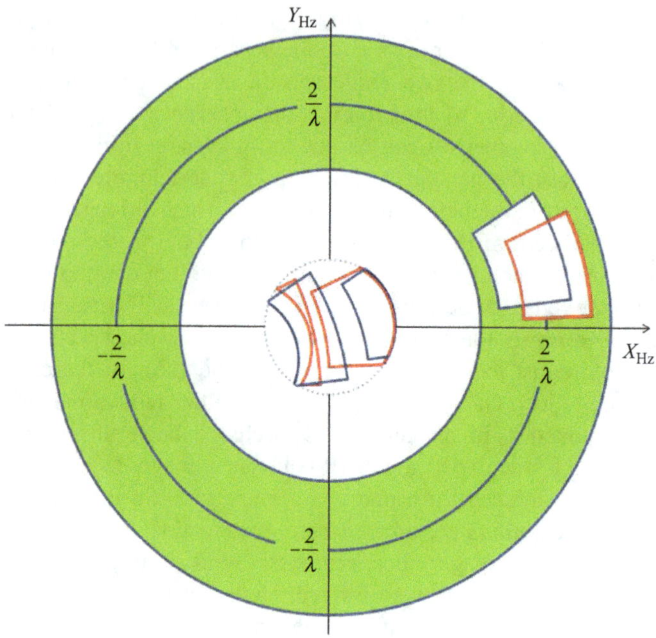

Figure 4.7(a) Two-dimensional frequency spectrums of the two data collections – direct representation

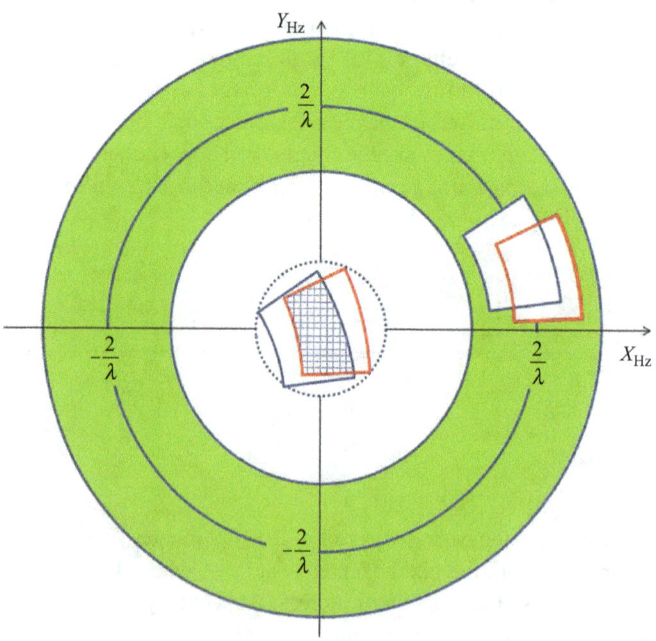

Figure 4.7(b) Unwrapped spectrum – hatched area indicates remaining spectrum post-wavenumber filtering

It usually suffices to set $M' = N'$. $\beta(A,x,y)$ ranges in value from 0 to 1, where 0 indicates no correlation and 1 indicates full correlation between the image pair.

When the coherence image indicates a high coherence (approaching 1) between the first and second data collections, this is interpreted as meaning little to no change. Low coherence (approaching 0) is indicative of change. The desired revisit time between data collections is dependent upon several factors including the desired persistence in observation, the type of change to be detected and the temporal coherence of the existing terrain conditions.

Though CCD was originally designed to look for change between a single pair of SAR images collected from data acquisitions repeated using the same geometry, VideoSAR, as noted above, allows an expansion of CCD to generate a stream of coherence images, $\beta(A,x,y)$, corresponding to the two time-separated image streams of ideally identical data acquisition geometry. The time series of geometrically registered images must be appropriately speckle filtered at the expense of spatial resolution. The coherence images, however, can be further filtered in time for additional speckle reduction to generate a superior coherence product while minimizing impact on spatial resolution [15]. Hence, coherence images can be generated first from corresponding single apertures which are spatially filtered, and then corresponding groups of apertures by temporally filtering the series of coherence images via incoherent averaging to produce a final coherence estimate. The incoherent averaging is performed on the magnitudes of the individual coherence images. For a temporal filter of length J images, assuming a large enough J with small overlap between frames, the speckle will be reduced across the J temporally filtered coherence images reducing false change detection. The coherence image stream does not have to be filtered or corrected for image-to-image phase differences as the temporal filtering is non-coherent. The VideoSAR Temporally Filtered coherence estimate, $\beta_{TF}(A',x,y)$, is expressed as

$$\beta_{TF}(A', x, y) = \sum_{j=-\frac{J}{2}+1}^{\frac{J}{2}} \beta(A' + \text{int}(j), x, y) \tag{4.12}$$

where A' is the index into the temporally filtered coherence image stream and int is the operator which truncates a value to an integer. The direct benefit of this approach is that an equivalent or better coherence estimate can be achieved which maintains the original or slightly degraded image spatial resolution. That is, $M' \times N'$ can be much smaller due to the addition of the temporal filtering of J images. The slightly degraded spatial resolution of the spatially filtered individual coherence images is sufficient to mitigate any error in their temporal registration through the VideoSAR image streams. Maintaining coherence in time across the coherence images is not a direct concern as the final coherence estimate, $\beta_{TF}(A',x,y)$, is derived as the average of the magnitudes of the individual coherence estimates.

To generate a final coherence image that also maintains coherence across the complex estimate would require spatially variant phase adjustments to be made to each individual complex coherence estimate prior to their final coherent integration. The phase adjustments would have to be a function of the pixel positions within each individual coherence image and the spatial centres of their corresponding synthetic apertures, for the entire coherence $\beta(A,x,y)$ image stream. This can equivalently be affected by centring the two-dimensional spectrum of each coherence estimate to zero during formation and then integrating coherently.

Figure 4.8 illustrates the effects on the coherence image of various combinations of spatial and temporal filtering [15]. Figure 4.8(a) is the amplitude image of the target area as collected during the first pass. Visible in the image is a signature of a baseball diamond beside a gravel road. Figures 4.8(b)–(d) are the coherence images resulting from various combinations of spatial and temporal filtering. Bright areas in the coherence image are representative of no change, while dark areas are representative of either man-made change or environmental change such as tree motion, or low reflectivity terrain. The two data collections were approximately 50 min apart, and each spanned a spotlight observation angle in azimuth of slightly greater than 50°. The change that was introduced to the area under observation between passes was a truck driving through the grass and across the baseball diamond while pulling a wooden pallet, and a person walking by and through the baseball diamond while dragging a garden rake. Their disturbances are labelled in Figure 4.8(d). The coherence image in Figure 4.8(b) is based on spatially filtering two images (before and after change), each with a minimal $M' \times N' = 5 \times 5$ boxcar filter. Note that there is only one image in each of the before and after samples, hence the temporal dimension in Figure 4.8(b) is limited to $J = 1$. Figure 4.8(c) uses the same data from Figure 4.8(b); however, a $M' \times N' = 15 \times 15$ spatial boxcar filter is employed, again with no temporal filtering. In Figure 4.8(c), the resulting image noise is reduced at the expense of the spatial resolution in the change detection product. There is a slight improvement in the visibility of the disturbances left by the truck and the person. In Figure 4.8(d), the $M' \times N' = 5 \times 5$ boxcar filter has again been employed for spatial filtering of each of the images, but in this case, a VideoSAR image stream of $J = 24$ images is used in the temporal dimension. The smaller boxcar filter has preserved the spatial resolution while the temporal filtering due to the non-coherent averaging of the 24 change detection images (generated from coincident VideoSAR image pairs) has reduced the noise such that the disturbances left by the truck and the person pulling the rake are clearly visible. The temporal filtering introduces the additional benefit of filling in shadows and hard targets (note the outer fence of the baseball field).

It is immediately inferred that there is a trade-off between spatial and temporal filtering in regards to the quality and usability of the final coherence product. This trade-off can be assessed by examination of an area of the image of uniform coherence which is generally free from artefacts. The metric that is considered is the standard deviation of coherence of the CCD product. Figure 4.8(e) plots the measured standard deviation of the coherence as a function of spatial and temporal filter sizes. Two spatial boxcar filters are examined of size $M' \times N' = 5 \times 5, 7 \times 7$, and five temporal filters are examined of lengths 1, 6, 12, 18, 24. Both increasing the boxcar filter size and increasing the temporal filter length are found to reduce the standard deviation of the coherence. The performance of boxcar filters of size $M' \times N' = 15 \times 15$, 17×17 for a temporal filter length of one (equivalent to no temporal filtering) are included as they illustrate how the same performance can be achieved with much smaller boxcar filters and temporal filtering. Specifically, the $M' \times N' = 5 \times 5$ boxcar filter with a temporal filter length of 24 approximates the performance of the $M' \times N' = 15 \times 15$ boxcar filter with no temporal filtering and the $M' \times N' = 7 \times 7$ boxcar filter with a temporal filter length of 24 approximates the performance of the $M' \times N' = 17 \times 17$ boxcar filter with no temporal filtering. Note again that resolution is not degraded with the use of smaller boxcar filters.

Figure 4.8(a) Image number 1 produced from VideoSAR data collection before change to the reflectivity scene

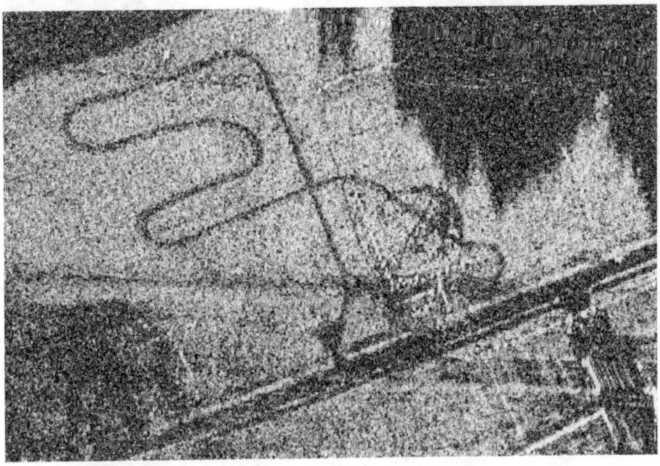

Figure 4.8(b) Coherence image produced using an $M' \times N' = 5 \times 5$ spatial boxcar filter and $J = 1$ temporal filter

The persistence of some shadows presents a problem to the automated detection of change. False detections generally occur in low coherence areas such as those consistent with areas of shadow (and surfaces such as asphalt roads). The detection and masking of low coherence areas reduces false alarms. A simple test statistic for areas of low reflectivity in both of the input images can be derived based on the sum and differences of the images under consideration.

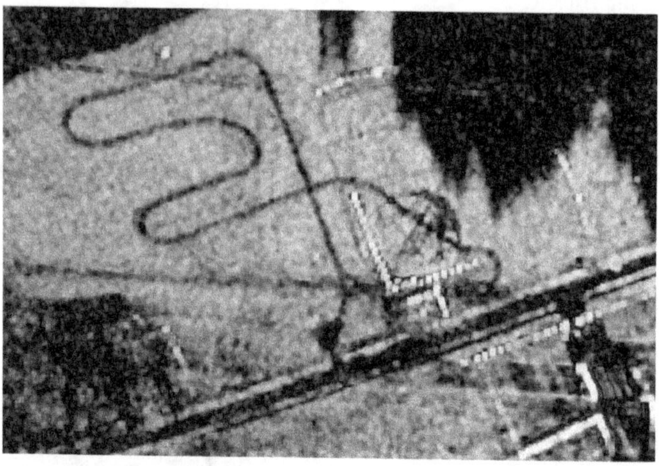

Figure 4.8(c) Coherence image produced using an $M' \times N' = 15 \times 15$ spatial boxcar filter and $J = 1$ temporal filter

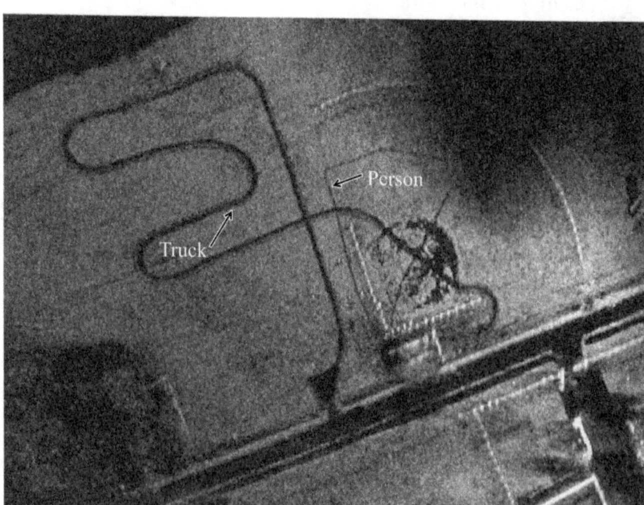

Figure 4.8(d) Coherence image produced using an $M' \times N' = 5 \times 5$ spatial boxcar filter and $J = 24$ temporal filter

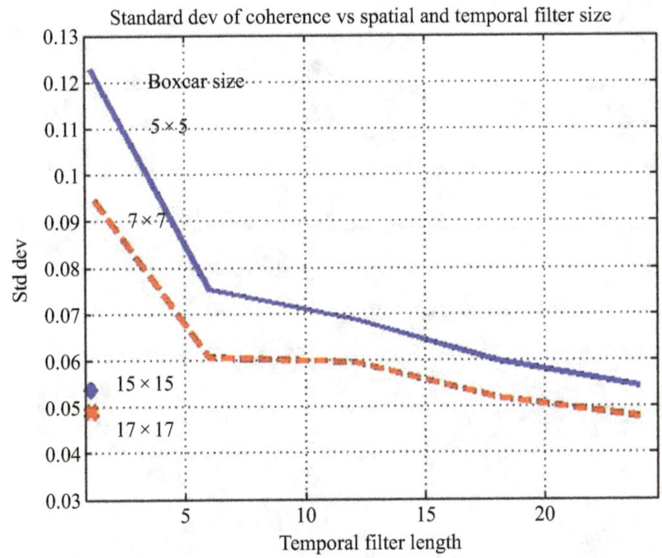

Figure 4.8(e) *Standard deviation of coherence imagery as a function of spatial and temporal filter sizes*

Table 4.2 *When EΔ(x, y) < α mask pixels to reduce false detection*

	$\rho_1(A,x,y)$ **high**	$\rho_1(A,x,y)$ **low**
$\rho_2(A,x,y)$ **high**	$E\Delta(x,y) > \alpha$	$E\Delta(x,y) > \alpha$
$\rho_2(A,x,y)$ **low**	$E\Delta(x,y) > \alpha$	$E\Delta(x,y) < \alpha$ [mask (x,y)]

For example,

$$E\Delta(x,y) = \frac{\sum_{m=x-(M/2)}^{x+(M/2)} \sum_{n=Y-(N/2)}^{y+(N/2)} \left(|\rho_1(A,m,n) + \rho_2(A,m,n)|^2 + |\rho_1(A,m,n) - \rho_2(A,m,n)|^2\right)}{2M'N'} \tag{4.13}$$

where the normalization is done according to the complete number of pixels used to generate the test value. If α is the threshold against which $E\Delta(x,y)$ is compared for determination of an area of low radar cross-section (low coherence), then Table 4.2 reflects when the pixels under test should be masked to reduce false detections for the purposes of change detection.

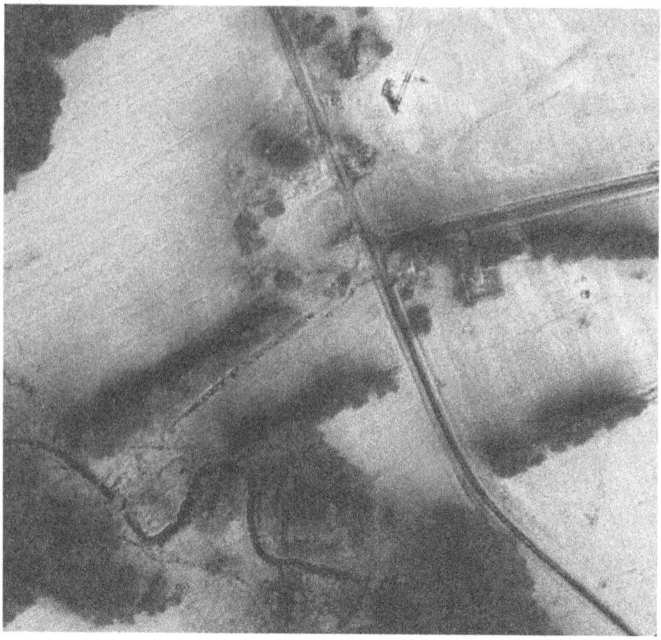

Figure 4.9(a) Result of averaging 19 coherence maps with a boxcar filter of $M' \times N' = 5 \times 5$

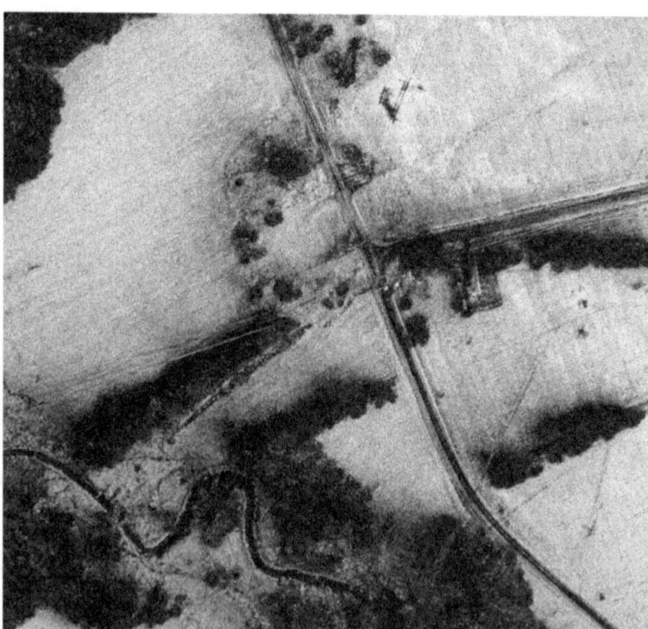

Figure 4.9(b) Result of averaging 19 coherence maps with fully flattened interferometric phase and no boxcar filter

When non-coherently averaging the coherence images, phase flattening across the entire J images may also be applied to ensure a consistent spectrum in lieu of, or to minimize, spatial filtering. Phase flattening is the removal of the phase variation due to the change in the point target's slant range displacement or apparent height between data collects, the latter requiring a DEM to effectively remove. This allows the frequency spectrum to be aligned in range, increasing coherence in unchanged areas and allowing areas of low coherence to stand out. Figure 4.9(a) illustrates CCD achieved via spatial filtering followed by temporal filtering vs Figure 4.9(b) where CCD is achieved via completely phase flattening all of the data followed by temporal filtering. The immediate benefits visible in Figure 4.9(b) are the increase in contrast without any degradation of spatial resolution.

Performance limits for change detection are bound by various sources of error including phase noise, clutter coherence between passes and the sensor resolution compared to the dimensions of the expected change. Temporal filtering is an effective way to preserve spatial resolution in the generation of a coherence map. Increasing the synthetic aperture length is another option, albeit fewer apertures will be available for temporal filtering. Fully flattening the interferometric phase across the multiple aperture data set can also be used to reduce the required spatial filtering. The concept described for CCD of filtering both spatially and temporally can also be applied to the process of ACD. The temporal filtering can again be used to reduce the speckle in the detected image stream prior to ACD.

4.4 Real-time VideoSAR implementation

While the variety of algorithms discussed above have been researched and developed over time by post-processing data collections from SAR systems such as the XWEAR radar, the desired end state is to have the algorithms implemented in a flexible manner to allow for varying operational conditions and yet also produce products in real time to support timely use. This implies a programmable, embedded, high-performance computer capability that is affordable and fits within the size, weight and power constraints of the aircraft. While this has been achieved for traditional SAR-processing algorithms over the years [16], as mentioned above, the straightforward implementation of the back-projection algorithm increases the computational burden akin to the difference between a Discrete Fourier Transform (O(N^2)) and a Fast Fourier Transform (O($N\log N$)). The difference is quite significant for typical applications, moving the problem from the range of 100 billion FLoating point Operations Per Second (100 gigaFLOPS) to 1 trillion FLoating point Operations Per Second (1 teraFLOPS). However, embedded high-performance computing architectures have been doubling in performance per dollar every year for more than a decade, allowing the significant improvements of back-projection in terms of image quality and flexibility to come within reach. In particular, the advent of high end server nodes augmented with general purpose graphic processing units (GPGPUs) has allowed multi-teraFLOPS, embedded high-performance computers to meet cost, size, weight and power constraints with the desired programmability and the ability to leverage considerable commercially available software.

An early real-time implementation of the back-projection VideoSAR algorithm was demonstrated through a joint effort between the US Air Force Research Laboratory (AFRL) Information Directorate and DRDC. This demonstration occurred in November 2010 involving several flights using DRDC's Convair 580 aircraft and the XWEAR, with the real-time processing being performed onboard by an embedded high-performance computer and software provide by AFRL.

The experimental concept was to have the Convair 580 aircraft fly circles around a 10 km target area for persistent VideoSAR imaging at a range of 21 km from the circle's centre. Pulses were transmitted every 18.5 cm along the aircraft's flight path (roughly corresponding to a PRF of 500 Hz given the aircraft velocity). The processing system then converted batches of 2,048 pulses into 1 km in azimuth by 10 km in range swaths of imagery every 4 s.

4.4.1 Challenges mapping core VideoSAR algorithms onto GPGPUs

As discussed in this chapter, VideoSAR presents an interesting and challenging set of algorithms to be implemented under the combined constraints of cost, real-time performance, size, weight, power and time to implement. In addition, there are several variants of the algorithms that might be employed in different circumstances, so a flexible, easily re-programmable implementation is also sought. The advent of GPGPUs provides a workhorse to address these constraints. Stemming from the large commodity marketplace for graphical processing, but with extensions to support general purpose computing (including the signal processing needed for VideoSAR), both cost and programmability are greatly improved. Size, weight and power can also fit within the constraints of an onboard rack of equipment, as these are typically found in home computers. But there remains the challenge of efficiently mapping the computationally intensive portions of the VideoSAR capability to the GPGPU architecture.

GPGPUs feature hundreds of small computational units working in groups on the same sequence of instructions (single instruction, multiple data, or SIMD computing style). Each of the Nvidia C2050 GPGPUs employed here featured 448 processing cores with a total memory size of 3 GB and a peak processing performance of 1 teraFLOPS single precision or 515 gigaFLOPS double precision.

The computationally challenging part of back-projection SAR is converting the information in every sampled radar pulse return into an update of every pixel illuminated by that pulse. Fortunately, in this situation, every pulse and every pixel can be treated independently. This is a tremendous advantage when parallelizing the work to be done across thousands of cores, since they do not have to work together and communicate amongst each other to execute the basic algorithm. The radar returns are pulse compressed by a head node and then multi-cast to the cluster of back-projection nodes which form their assigned portions of the images (in this case broken up into range swaths of 1.25 km).

As new pulses are received, they are segmented in range by the Calisto FPGA card at the head node Xeon processor. They are then concatenated into a single pulse covering the 10 km range to be imaged. At this point, the raw data is archived to a high-speed disc array, as well as pulse compressed prior to distribution to the processing nodes. Each processing node comprises a Xeon processor with two Nvidia C2050 GPGPU boards connected through the PCIe bus. Prior to the implementation of back-projection computations was the critical design and implementation of efficiently distributing data throughout the system. This involved the use of multi-buffering between the nodes using the Xeon processor and then dividing the data and multi-buffering to the two GPGPUs. In addition, the processed imagery from the GPGPUs had to be managed and returned to the head node without causing delays.

With an efficient data distribution implementation, efforts were focussed on the back-projection processing on the GPGPUs which were programmed using the CUDA programming language. Each GPGPU board processed 2,048 pulses to form a 1,024 × 2,048 image. The main loop of the algorithm was written in-lining any procedural routines to reduce the calling overhead and make the code easier to analyse. The aircraft positional data calculations were moved to the main kernel to reduce the amount of required global memory. The computation time at this point was 15 s per image frame. However, with the goal of this experiment to produce a new image every 4 s, the computation for each image would have to be completed in significantly less than 4 s to allow for Input/Output (I/O) and control to occur between frames.

The compute time was reduced to 5.1 s per image frame by changing the order of computation from computing each pixel value across all pulses to computing all the pixel's values one pulse at a time. This allowed a more efficient mapping to the GPGPU. The input pulse data type was modified from double precision complex to single precision complex to reduce required memory and I/O, as well as speed up computations.

Some additional computational variables were evaluated, and it was determined that single-precision floating point would be sufficient. This change allowed the number of required registers to be reduced from 38 to 36 which allowed more threads to be active at a time, reducing the compute time to 3.3 s per image frame.

At this point, the decision to process one pulse at a time was re-evaluated, and it was determined that the available shared memory, which is faster than global memory, was large enough to hold 128 input pulses. The code was modified to process the 2,048 pulses, 128 at a time. In addition, the accumulation matrix containing the output pixel values was changed from double precision complex to single precision complex, reducing memory usage and I/O. The final optimization was based on the observation that there were two division operations occurring that could be replaced with a multiplication operation using the reciprocal of one of the operands. A multiplication operation requires many fewer processor clock cycles to complete than a division operation, reducing the overall compute time.

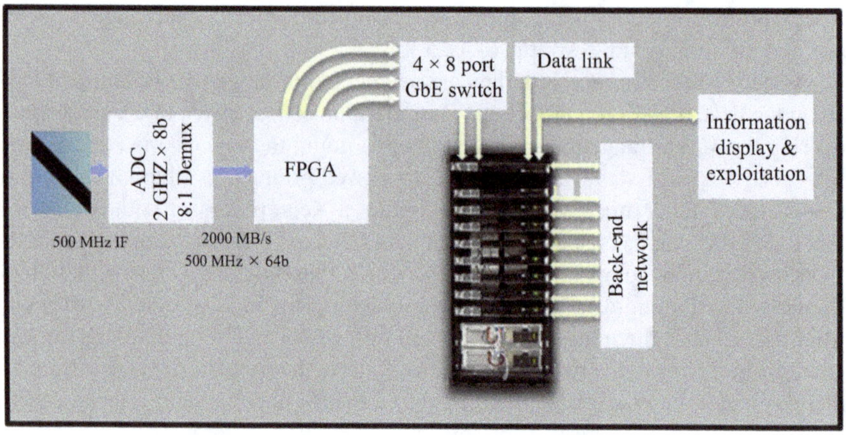

Figure 4.10 Real-time VideoSAR system overview

This combination of changes reduced the compute time to slightly over 2 s per image frame, well within the target of having a 4 s frame rate.

4.4.2 System overview

Figure 4.10 illustrates the system overview. The high-performance computer (HPC) was a Linux cluster consisting of nine nodes each containing a dual quad 2.66 GHz XEON X5550 CPU. One node was the 'head' node, acting as an input/output gateway for the cluster, receiving radar samples and positioning data from the aircraft. This node also performed the pulse compression function and recorded raw radar data, navigation data and radar control data. The remaining eight compute nodes were used to produce the VideoSAR images. Each node processed 1/8 of the 1 × 10 km swath that was formed every 4 s. Each of the eight nodes had two Nvidia C2050 GPGPUs, shown in Figure 4.11. These inexpensive components cost around $2,200 USD each as they are driven by a large commercial market. They performed the back-projection algorithm in a combination of single and double precision floating point operations to produce images. Each of the GPGPUs processed half of the image segment being formed by that node.

4.4.3 High-speed data acquisition

The data acquisition system was built upon the high-speed data capture/distribution processor [16]. The technical challenge was to record and process a wide swath of high resolution radar returns in real time. The data acquisition design served as the bridge between the radar analogue intermediate frequency (IF) signal and the embedded HPC. This necessitated implementing a system accepting a non-standard radar data feed, augmenting it with key timing and control meta-data from the radar control system and then relaying it to the embedded computer via a standard computer interface. The data acquisition sub-system of the radar received an IF signal as input, sampled it with a 2 GHz A/D convertor, captured the 10 km swath

Figure 4.11 NVIDIA C2050 GPGPU card

of interest using discrete timing signals from the radar control unit and output the digital results in a format/structure that was standard to commercially available computers. Commercial-off-the-shelf technology, which was prototyped during a previous project called 'Swathbuckler' [16], was used on this project to digitize and distribute the IF signal to the embedded HPC. Two commercial off-the-shelf boards were integrated into the system, the Quixilica Neptune and Quixilica Callisto cards from Tek Micro Systems. These boards converted the radar data to a universal data packet format acceptable to the embedded HPC and sent segments in range, through eight 1 GB Ethernet connections, via a switch which in turn passed the data on to the head node through a 10 GB Ethernet connection at 78 MB/s.

4.4.4 High-performance computer

The onboard HPC was used to perform two primary tasks. The first was to ingest data from the radar data acquisition system and store it in real-time. The second was to perform near real-time signal processing of that data to produce imagery with a few seconds latency. The experimental objectives went beyond just forming the images in real time. To support algorithms such as change detection between sets of images widely separated in time and to support post-processing in the laboratory to refine future algorithms, a high-performance data acquisition system had to record all salient inputs, outputs and meta-data. This capability also augmented the real-time operations since it allowed users to 'reach back' to the raw data to further examine particular regions of interest as they wish.

The head node was used to receive the IF output from the radar frontend using the 10 GB connection. At this point, the data was saved to a high-speed redundant array of independent discs (RAID) for post-processing after the flight. The head node also concatenated the eight channels of 16 K range samples each into a single array of 128 K complex values, performed pulse compression and associated the data with the ancillary data from the receiver/exciter/processor, which was received through an additional 1 GB Ethernet connection. The resulting 128 K range bins covered the 10 km area to be imaged and were segmented into eight sub-arrays according to the range segment to be processed by each of the processing nodes. To reduce the processing that would be required by the head node to calculate the exact range segment that would be required by each processing node to image its assigned area, the data was segmented with approximately 50% of additional overlap on either side of the sub-swath. Performing the pulse compression operation on the head node and then performing image formation on the processing nodes provided full overlapping of these functions allowing the head node to collect and process a new set of range returns while the processing nodes produced imagery from the previous set in a pipelined fashion. Upon completion of each set of images, the image data was sent back to the head node for storage and display on the operator console. A subset of the imagery could also be sent via RF link to a ground station.

4.4.5 SAR image formation software

Image formation was performed using 16 Nvidia Tesla C2050 GPGPU processor boards installed in the eight Xeon-processing nodes with 2 Tesla boards in each node. Figure 4.12 shows the processor and data communication layout of the system. The radar frontend supplied raw data and its associated ancillary data to a Xeon-based head node. The head node performed pulse compression and associated the raw data with the ancillary data which was then passed to each of the eight processing nodes. Each processing node produced a separate segment of the 1 km × 10 km image, approximately 1 × 1.25 km per node and then sent that

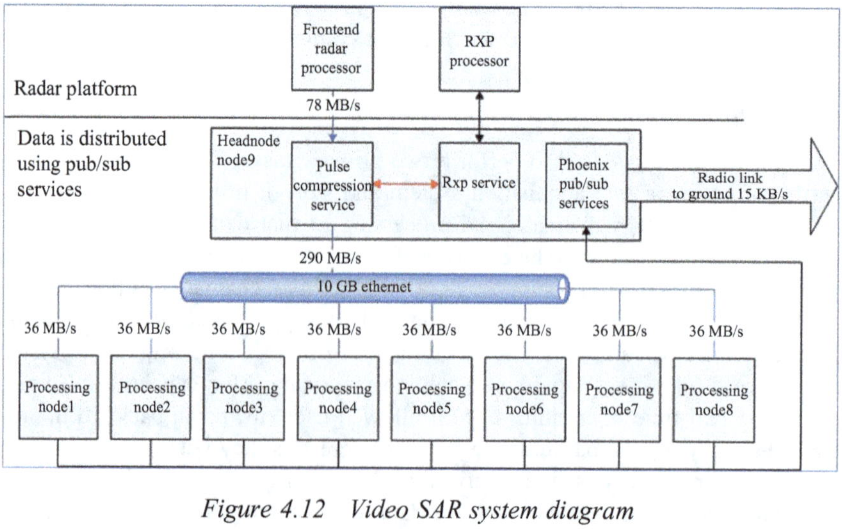

Figure 4.12 Video SAR system diagram

image to the head node for distribution to the visualization stations. A publish/subscribe communication mechanism was employed for the back-end reporting and publishing of results. This information management tool brought flexibility to add new publishers and subscribers during the experiment.

4.4.6 VideoSAR visualization

The VideoSAR experiment produced real-time VideoSAR imagery by producing a large 2,048 × 16,384 pixel image, or 33 megapixels every 4 s. The imagery was produced and stored on the sensor platform while exploiting the real-time capabilities of the system by allowing a subset of the data to be visualized at a ground station within seconds of being produced. The main challenges in accomplishing this task were compressing the images in real time, transmitting over low bandwidth connections and developing a graphical user interface (GUI) that operated on both the airborne platform and remotely at ground stations.

4.4.7 Image compression and scaling

To solve the image compression and scaling challenge, the Intel IPP library was used. By leveraging this library, an image compression utility was created that could scale a 140-megapixel image to 2% of its original size in 45 ms on the Xeon processors. This included converting the original image from complex floating point pixel format and writing a jpeg-compressed image. In the test case, the original image was 1 GB in size and was reduced to 30 kB. Also, when higher resolution images were required, this utility would clip a sub-region from the original image and deliver it in compressed form, allowing the image size to be controlled while displaying more detail. Thus, the compression software supported delivery of both the 'big picture' and high resolution of small areas over affordable air-to-ground communication links.

4.4.8 VideoSAR GUI

In order to visualize the data that was produced onboard the aircraft and sent over the limited bandwidth connection, an information management client was created using the Java 2 Enterprise Edition platform. This consists of an Apache Tomcat 5.5 application server and Java Runtime Environment 6. The viewer ran as an applet inside a web browser and the client placed the geo-referenced imagery thumbnails on a map which allowed for higher resolution imagery to be requested and viewed in the broader context to assist user situational awareness.

The GUI showing some SAR imagery is displayed in Figure 4.13. Each VideoSAR image frame is roughly 1 km wide and 10 km long. These image blocks were continuously displayed as the imagery was produced. To view high-resolution imagery, a request would be made that specified the size of the region and desired resolution. By clicking the map and sending a request, higher resolution imagery would be created, transmitted and then displayed.

Figure 4.14 is a screen shot showing the results of two requests that were made by an operator. The larger square was requested to be 500 m per side with

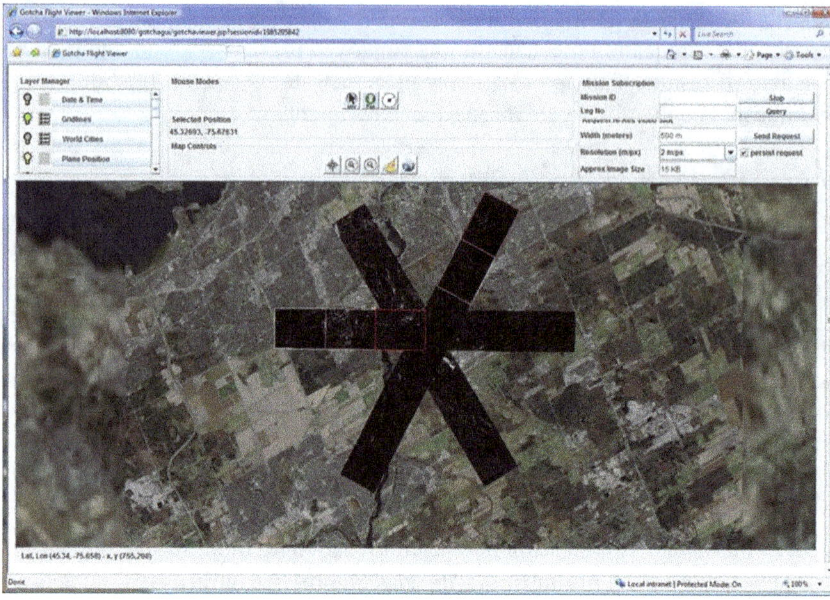

Figure 4.13 VideoSAR GUI

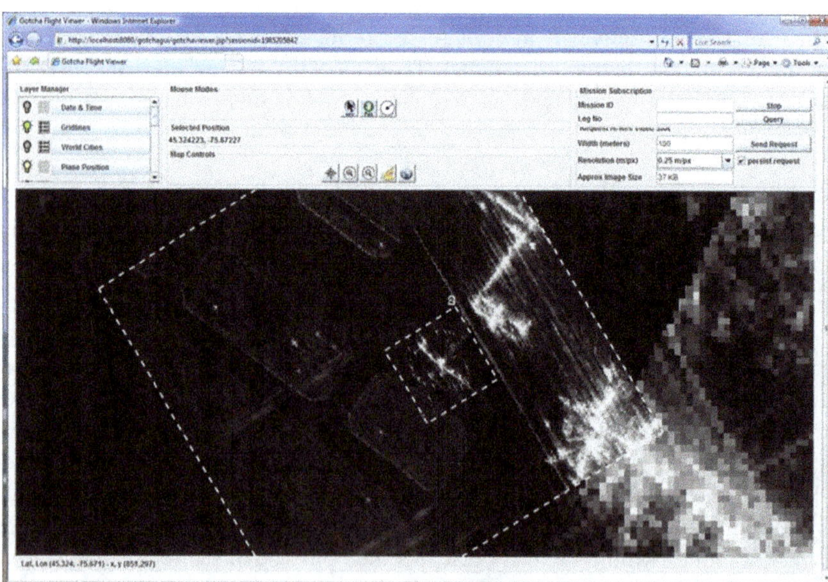

Figure 4.14 VideoSAR GUI processing requests

2 m pixel resolution. The smaller square (overlaid on the larger) was requested to be 100 m per side with 0.25 m pixel resolution. By decreasing the area covered by the image, the image size stayed between 15 and 40 kB, even though the resolution was increased, providing high-resolution VideoSAR. Since the 'persist-request' checkbox had been selected, these images were updated whenever the selected region was imaged in the future – a 'subscription' was posted for that area of interest.

4.5 Summary and outlook

VideoSAR as an imaging mode is also capable of producing products in addition to change detection for exploitation. This includes: slow ground moving target indication where the uninterrupted VideoSAR persistence, when combined with the very precise accuracy of the imagery, allows slow moving targets to be detected in move-stop-move scenarios [17]; and three-dimensional model generation where repeat passes provide multiple elevation angles allowing the development of three-dimensional models of the target [18]. The concept of operation whereby the antenna is in a persistent staring mode is amenable to generating multiple simultaneous surveillance modes by the sheer fact that the persistence yields a richness of information on the target area. The same data stream can be processed for multiple SAR and ground moving target indication functions in parallel. The utilization of antennas with wide azimuth beamwidth, polarimetric channels and multiple phase centres allows the persistent collection of wide swath/wide beam data from which both imaging, slow- and fast-moving target indication and other exploitation products can be derived. There is also a large trade-off space in carrier frequency selection. The examples presented here-in were generated from the X-band airborne radar frequency band. The requirement for foliage penetration leads designers to systems which operate in the VHF and UHF bands. Moving in the other direction on the frequency spectrum, the mid-to-high EHF band affords the possibility of VideoSAR systems which can operate in environments where Electro-Optic system performance is degraded and extremely high image frame rates akin to full motion video are required for monitoring of fast and manoeuvring targets [19]. Real-time VideoSAR imaging of large areas of terrain with update rates which approach that of full motion video will continue to benefit from the advances in computing architectures. Advances in commercial-off-the-shelf computing will not only benefit the core imaging function, but the generation of the exploitable products in time-frames which allow quick turn-around for analysis and re-tasking of the signal processing exploitation function. Finally, the persistent surveillance when coupled with extremely wide beam, wide swath imaging allows target detection and tracking both forward in time and forensically to a higher degree of fidelity than scanning radar systems. Forensically, the persistence and history offer the benefit of allowing the coherent processing interval for slow-moving target detection to be adapted to the clutter and target dynamics, again not always possible with a scanning radar system [20].

List of acronyms and abbreviations

ACD	amplitude change detection
AFRL	Air Force Research Laboratory
CCD	coherent change detection
CPU	central processing unit
DEM	digital elevation model
DRDC	Defence Research and Development Canada
EHF	extremely high frequency
FPGA	field-programmable gate array
GPGPU	general purpose graphical processing unit
GUI	graphical user interface
HPC	high performance computer
HSDC/DP	high speed data capture/distribution processor
IF	intermediate frequency
I/O	input/output
PRF	pulse repetition frequency
R	range
RAID	redundant array of independent discs
RF	radio frequency
RXP	receiver exciter processor
SAL	synthetic aperture length
SAR	synthetic aperture radar
SIMD	single instruction multiple data
UDP	universal data packet
UHF	ultra-high frequency
VideoSAR	video synthetic aperture radar
VHF	very high frequency
XWEAR	X-band wideband experimental airborne radar

References

[1] Jakowatz C.V., Wahl D.E., Eichel P.H., Ghiglia D.C., Thompson P.A. *Spotlight-Mode Synthetic Aperture Radar: A Signal Processing Approach*. Boston: Kluwer Academic Publishers; 1997. pp. 33–50.

[2] Carrara W.G., Goodman R.S., Majewski R.M. *Spotlight Synthetic Aperture Radar Signal Processing Algorithms*. Boston: Artech House; 1995. pp. 31–45.

[3] Damini A., Balaji B., Parry C., Mantle V. 'A VideoSAR Mode for the X-band Wideband Experimental Airborne Radar', *Algorithms for Synthetic*

Aperture Radar Imagery XVII; Orlando, USA, April 2010, SPIE Proceedings Volume 7699; 2010. pp. 76990E-1–76990E-11.

[4] Brenner A.R., Essen H., Stilla U. 'Representation of Stationary Vehicles in Ultra-high Resolution SAR and Turntable ISAR Images', *EUSAR 2012 European Conference on Synthetic Aperture Radar*; Nuremberg, Germany, April 2012, Proceedings of EUSAR 2012; 2012. pp. 147–150.

[5] Yegulalp A. 'Fast Backprojection Algorithm for Synthetic Aperture Radar', *1999 IEEE Radar Conference*; Waltham, Massachusetts, USA, April 1999. pp. 60–65.

[6] Ulander L., Hellsten H., Stenström G. 'Synthetic-Aperture Radar Processing Using Fast Factorized Back-Projection', *IEEE Transaction on Aerospace and Electronic System*. 2003; **39**(3): 760–776.

[7] Damini A., Balaji B., Haslam G., Goulding M. 'A New X-band Experimental Airborne Radar – Phase II: Synthetic Aperture Radar and Ground Moving Target Indication', *Special Issue of IEE Proceedings-Radar, Sonar and Navigation*. 2006; **153**(2): 144–151.

[8] Dickey F.M., Doerry A.W., Romero L.A. 'Degrading Effects of the Lower Atmosphere on Long-range Airborne Synthetic Aperture Radar Imaging', *IET Radar Sonar and Navigation*. 2007; **1**(5): 329–339.

[9] Doerry A. 'Atmospheric Loss Considerations for Synthetic Aperture Radar Design and Operation', *Radar Sensor Technology VIII*; Orlando, USA, August 2004, SPIE Proceedings Volume 5410; 2004. pp. 17–27.

[10] Haslam G.E., Damini A. 'Specifying the Allowable Latencies in the Application of SAR Motion Correction', *EUSAR '96 European Conference on Synthetic Aperture Radar*; Berlin, Germany, March 1996, Proceedings of EUSAR '96; 1996. pp. 403–406.

[11] Gierull C.H., Sikaneta I.C., Cerutti-Maori D. 'Space-based SAR-Ground Moving Target Indication' in Klemm R., Nickel U., Gierull C.H., Lombardo P., Griffiths H.D. and Koch W. (co-eds.). *Chapter II.7 in Novel Radar Techniques and Applications*. London, UK: IET Publishers; 2017.

[12] Touzi R., Lopes A., Bruniquel J., Vachon P. 'Coherence Estimation for SAR Imagery', *IEEE Transactions on Geoscience and Remote Sensing*. 1999, **37**(1): 135–149.

[13] Singh A. 'Digital Change Detection Techniques Using Remotely Sensed Data', *International Journal of Remote Sensing*. 1989; **10**(6): 989–1003.

[14] Gatelli F., Monti Guarnieri A., Parizzi F., Pasquali P., Prati C., Rocca F. 'The Wavenumber Shift in SAR Interferometry', *IEEE Transactions Geoscience and Remote Sensing*. 1994; **32**(4): 855–865.

[15] Damini A., Mantle V., Davidson G. 'A New Approach to Coherent Change Detection in VideoSAR Imagery using Stack Averaged Coherence', *IEEE Radar 2013*; Ottawa, Canada, May 2013, pp. 1–5.

[16] Linderman R.W. 'Swathbuckler: Wide Swath SAR System Architecture', *2006 IEEE Radar Conference*; Verona, New York, USA, April 2006, pp. 465–470.

[17] Page D., Owirka G., Nichols H., Scarborough S., Minardi M., Gorham L. 'Detection and Tracking of Moving Vehicles with Gotcha Radar Systems', *IEEE AES Systems Magazine*. 2014; **29**(1): 50–60.

[18] Moore L., Potter L., Ash J. 'Three-dimensional Position Accuracy in Circular Synthetic Aperture Radar', *IEEE AES Systems Magazine*. 2014; **29**(1): 29–40.

[19] Wallace H.B. 'Development of a Video SAR for FMV through Clouds', *Open Architecture/Open Business Model Net-Centric Systems and Defense Transformation 2015*; Baltimore, USA, June 2015, SPIE Proceedings Volume 9479; 2015. pp. 94790L-1–94790L-2.

[20] Balaji B., Sithiravel R., Damini A., Wang K., Kirubarajan T. 'Radar Resource Management for a Ground Moving Target Indication Radar', *Signal Processing, Sensor/Information Fusion, and Target Recognition XXIV*; Baltimore, USA, June 2015, SPIE Proceedings Volume 9474; 2015. pp. 947402-1–947402-10.

Chapter 5
High-resolution wide-swath SAR

Ishuwa Sikaneta and Delphine Cerutti-Maori***

Abstract

This chapter presents the principle of high-resolution wide-swath synthetic aperture radar (SAR), a means for imaging wide areas at high resolution. The material covers the limitations of achieving wide-swath and high-resolution with a traditional SAR, the basic idea of using a multi-aperture SAR to overcome this limitation and current implementations where multi-aperture (or multiple antenna) systems collect data in an ideal configuration. Overviews of approaches to processing data collected in non-ideal configurations, such as when the data are collected with non-uniform sampling and/or when they are collected with a squinted system, are then introduced.

Armed with an overview, the chapter introduces the theory of multi-aperture SAR processing with the objective of generalizing the concept of high-resolution wide-swath to higher resolution, wider-swath SAR. This enables application of the added degrees of freedom to other modes such as spotlight and high-resolution stripmap. In order to present the theory and the generalizations, and in consideration of possible future systems, the theory is derived in the wavenumber domain for wideband and/or widebeam, space-based systems with special cases for narrow-band systems presented as appropriate. In contrast to much of the current literature, the theory views the antenna patterns as the key provider of the additional degrees of freedom and proposes to utilize other pattern characteristics in addition to the phase-centre separation to improve imaging. For this reason, special care is taken in developing the antenna pattern dependence in the signal model.

The approach for signal reconstruction focuses, mainly, on the minimum mean-square error method as it is quite general and includes, as special cases, the well-known projection approach as well as the space-time adaptive processing (STAP) approach. Further, it inherently, simultaneously improves the geometrical and radiometrical resolution due to favourable weighting by the antenna pattern and a less aggressive ambiguity prescription as compared to other techniques. The approach also naturally incorporates other more generalized system configurations

*Defence R&D Canada, Ottawa Research Centre, Canada
**Fraunhofer FHR, Germany

where, for instance, the antenna patterns have, not only different phase-centres, but also different shapes or different pointing directions. As an added feature, the presented method is robust against matrix inversion problems which can render the projection approach intractable. The special case of a phased-array multi-aperture system is presented.

5.1 Introduction

To set the stage for this chapter, let us consider the following question: is it true, for any imaging system capable of a particular resolution, that there is a limit to the size of area that can be imaged. Two examples, one from nature, one from science suggest that this is true. Just as there is a limit to the height (large field of view) from which an eagle can identify prey on the ground, there is a limit to the field of view in a microscope observing sub-cellular structures.

Of course, a clever imager might, over time, observe a wider area by imaging from different spatial positions and gradually building a mosaic. In the microscope example, the user could move the slide around, while an eagle, gliding as it does, eventually covers a relatively large area. But, as the common saying goes, time is a commodity so one must include this factor into the cost of using the imaging system.

Perhaps a better challenge is to think of system designs where the ratio of area to resolution (which, numerically, is better for smaller values) is improved for acceptable design and time cost. For instance, some might argue for improvements in the design of the human eye since, for the purpose of resolving small objects at large distance, it is inferior to the eagle eye. For the purpose of SAR imaging, this is the challenge undertaken by this chapter: to present system designs and associated processing methods to improve the area to resolution factor. We start by explaining the technical reasons behind the limitation in area to resolution for a SAR. It is assumed that the reader is familiar with the basics of radar imaging and signal processing.

5.1.1 The swath-resolution trade-off

A classical SAR utilizes a single antenna to both transmit and measure radar signals. The pulse repetition interval (PRI) limits the time that the radar can measure reflected echoes from the terrain which thereby limits the swath. Wide-area surveillance SAR modes tend to drive the PRI upwards so that large swaths of data can be collected. The amount of Doppler bandwidth, on the other hand, determines the achievable cross-range resolution of a SAR. High resolution[1] requires large Doppler bandwidths and, per the Nyquist sampling theorem, this requires a high sampling rate or low PRI. These conflicting design specifications on the PRI limit the wide-area surveillance capability of a single-channel SAR. In a sense, a single-channel SAR offers the potential of a single degree of freedom; it is up to the user to either concentrate this degree of freedom in a small area yielding localized

[1]From here onward, resolution means azimuth or cross-range resolution.

high-resolution information, or to dilute it over a wider area yielding poorer resolution. More-or-less, for stripmap SAR, the area to resolution factor remains constant.

Initial publications on the topic [1–6] concentrated on the idea of high-resolution, wide-swath (HRWS). This concept can be generalized to the idea of higher resolution, wider-swath or the idea of improving the swath to resolution ratio. For any given mode, spotlight, stripmap, scansar or otherwise, the challenge is to determine how one can achieve higher resolution for the same swath or a wider swath for the same resolution. For instance, can one conceive of a design to extend a spotlight mode with sub-metre resolution to cover an area spanning several tens of kilometres in azimuth and range?

5.1.2 Approaches to improving the swath to resolution ratio

The literature contains several approaches to improving the swath-to-resolution ratio. Among these are waveform diversity, where orthogonal waveforms are employed from pulse to pulse, and antenna diversity where a multiplicity of SAR antennas is used to measure the transmitted data. The orthogonality of sequentially transmitted waveforms allows a SAR processor to separate a signal composed of pulse echoes from several different ranges allowing measurement of a larger swath without loss in cross-range resolution. For such a system to operate ideally, the waveforms would exhibit uniformly (across all lags) zero cross-correlation functions. As discussed in [7], this is difficult to realize in practice.

Antenna diversity offers a more practical solution but suffers from the drawback of requiring additional hardware and more data recording capability. Multiple antennas have been used for a variety of purposes including ground moving target indication (GMTI) and across-track interferometry. Systems such as RADARSAT-2 and TerraSAR-X are equipped with phased-array radars which are inherently multiple antenna systems. Data recording capability from each antenna is required to exploit the degrees of freedom offered by the multiple antenna systems. In a phased array with several hundred elements, it is sufficient to record the response from several sub-apertures composed of sub-sets of the array.

Figure 5.1 provides a concrete example related to RADARSAT-2. Sixteen vertical columns form the phased-array. With the addition of a signal-routing switch, the system can record two receive patterns simultaneously, albeit with fewer elements, a wider antenna pattern and smaller receive gain than the full aperture single-channel SAR. Further, as we shall see in Section 5.4, by transmitting or receiving with different antenna patterns from pulse to pulse, or time-multiplexing, multiple antenna patterns can be obtained. Recorded data can be grouped into different channels according to the two-way transmit/receive array pattern and can be processed using multi-dimensional techniques such as the HRWS techniques presented in this chapter.

For the purpose of improving the swath to resolution ratio, the multiple antennas, either formed from a phased-array or otherwise, need to be aligned in the along-track direction. An across-track interferometer where antenna elements are aligned in the cross-range direction cannot help to resolve the problem.

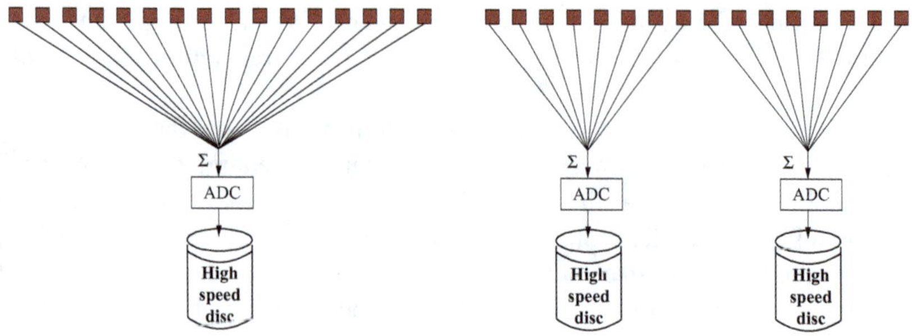

Figure 5.1 Apertures formed by combining the receive array measurements. On the left, a single aperture formed using a single analogue-to-digital converter after combining all antenna element feeds. On the right, two apertures, one formed from all elements on the left half of the array, the second formed from the remaining elements

HRWS systems that use multiple antennas have been investigated in [1–6], and the developed signal-processing algorithms have been applied to measured RADARSAT-2 [4] and TerraSAR-X [6] data yielding results that validate the theory. These algorithms, however, are limited in the resolutions that they can process and apply only in the case of no across-track baseline.[2] Although modern antenna steering practices such as yaw-steering or total zero-Doppler steering minimize the across-track baseline for clutter over the orbit [8], there may be situations, as we shall see, where a large across-track baseline is unavoidable. As will be argued in Section 5.2.3, HRWS-processing methods need to be adapted to process high-resolution data in the case of an across-track baseline; hence, this chapter generalizes the processing approach to accommodate data from a wideband system where an across-track baseline may be present.

An across-track baseline might manifest due to a misalignment between the physical antenna array and the direction of motion of the radar platform. Please note that there will be no across-track baseline when the antennas are aligned in the flight direction with each antenna steered away from broadside (each antenna is squinted). This could occur when using sub-arrays of a phased-array with sub-array weightings such that beams are steered away from broadside. In other words, element and array-squints are different with the only latter leading to an across-track baseline.

As illustrated in Figure 5.2(a) and (b), an across-track baseline results from targets with an across-track velocity.

5.1.3 Outline of the chapter

This chapter presents HRWS signal processing applicable to narrow or wideband systems where an across-track baseline component may be present in the acquisition. As opposed to lower resolution systems, one finds that the signal-processing

[2]A system with the array of antennas exactly aligned in the flight direction.

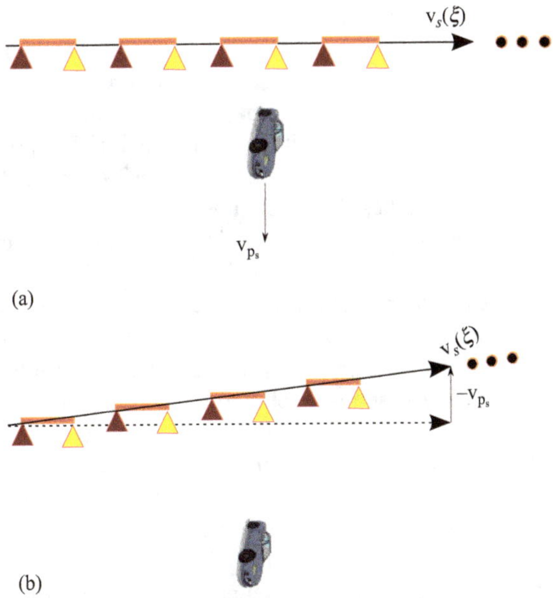

Figure 5.2 (a) Point of view where the clutter is stationary. (b) Point of view where the target is stationary. An across-track baseline is introduced

filters depend on the across-track wavenumber, rather than only on the Doppler frequency (or along-track wavenumber). This somewhat increases the signal processing burden. The other main addition is to the HRWS SAR processing of moving targets. The material presented applies equally to airborne and space-borne SARs.

Section 5.2 presents an introduction to the HRWS problem. This is followed by a derivation of the signal model for a wideband system. In particular, and in contrast to the standard models outlined, for instance, in [9,10], special care is taken to evaluate the antenna pattern gain functions and angles as these play a critical role in the HRWS signal reconstruction algorithms. Section 5.4 implements this model into the HRWS signal processing paradigm. Finally, Section 5.5 presents the phased-array as a candidate for a multi-channel SAR that can be used for HRWS. An example, using measured data from RADARSAT-2, demonstrates the HRWS concept and signal-processing algorithms.

5.2 Under-sampled SAR signals

This section relates the radar antenna patterns to Doppler under-sampling.

Recall that the dynamics of motion between the radar platform and the object or scene being imaged determine the relative rate of change of distance or range.

From this point onward, we shall use the expression velocity vector to mean the relative velocity vector and range vector to mean the vector from an object on the ground to the SAR. As described in the SAR literature (and in Section 5.3.4), the Doppler frequency of the measured signal is proportional to the inner product between the velocity vector and the unit vector that points in the direction of the range vector; thus, if the relative speed of the SAR does not change over time, then the rate of change of range takes extreme values when the velocity vector aligns with the range vector and is zero when the velocity vector is perpendicular to the range vector.

Typically, the SAR antenna pattern measures a finite portion of the possible Doppler frequencies that are available due to the dynamics of motion. One may think of the measured signal as a bandpass filtered product, with the SAR antenna pattern acting as the filter. An omni-directional azimuth pattern would perform no bandpass filtering while a narrow antenna pattern would filter aggressively. A SAR samples the measured signal over both range and azimuth. If one wishes to produce a useful representation of the scene, the azimuth sampling frequency must support the measured bandwidth according to the Nyquist sampling theorem. If the Nyquist sampling theorem is violated, then the imagery will contain azimuth ambiguity distortions. Wider azimuth antenna patterns correspond to a greater bandwidth and require higher azimuth sampling frequencies.

In the situation under consideration, we consider a set of independent measurements of the scene from a set of antenna patterns, each under-sampled with respect to Nyquist. Each of these signals considered in isolation yields a degraded if not useless measurement of the scene due to multiple azimuth ambiguities. Considered together, however, this chapter seeks methods to combine the different measurements into a single adequately sampled signal yielding a useful SAR image.

5.2.1 Note on notation

Two dimensions describe a SAR image: these are the range, r, which represents the across-track or fast-time dimension ($\tau = 2r/c$) and χ which defines the along-track or slow-time dimension ($\xi = \chi/v_e$ where v_e is the effective relative velocity between the SAR platform and the object in the scene, see Section 5.3 for the exact definition). In this chapter, we adopt a two-symbol notation for the measured scalar signal, see Table 5.1; the first symbol corresponds to the domain of the r variable, while the second indicates the domain of the χ variable. For aesthetic reasons, we represent the measured vector and matrix signals using a single symbol; vectors, with lowercase bold, $\mathbf{s}(k_r, k_\chi)$ and matrices, with uppercase bold, $\mathbf{S}(k_r, k_\chi)$. Also note that the symbol $\cdot^\dagger$ is used to denote the complex conjugate of a vector which, for a real-valued vector or matrix, reduces to the regular transpose operation. The magnitude of a vector will be denoted by italicized text; for instance, $|\mathbf{r}(\xi)| = r(\xi)$. Summation without specified limits, $\sum_l$, denotes summation from $-\infty$ to ∞. Similarly, an integration without limits indicates integration over an infinite domain.

Table 5.1 Notation for the scalar signal in various domains. The subscript denotes the nth channel

	Range	Azimuth
$ss_n(\tau, \xi)$	Fast-time (s)	Slow-time (s)
$Ss_n(\omega, \xi)$	Angular-frequency (rad/s)	Slow-time (s)
$SS_n(\omega, f_\xi)$	Angular-frequency (rad/s)	Doppler (Hz)
$\mathbb{S}s_n(k_r, \xi)$	Wavenumber (rad/m)	Slow-time (s)
$\mathbb{S}S_n(k_r, f_\xi)$	Wavenumber (rad/m)	Doppler (Hz)
$\mathbb{S}\mathbb{S}_n(k_r, k_\chi)$	Wavenumber (rad/m)	Wavenumber (rad/m)

5.2.2 *The multi-channel signal as a superposition of vectors*

Without loss of generality, one may consider the measured signals in the 2-D spatial frequency domain. In this domain, if one has measurements from N (possibly different) antennas, the measured signal may be represented as a vector, $\mathbf{z}(k_r, k_\chi)$, with each element corresponding to the measurement from each of the different antennas in the presence of additive white Gaussian noise:

$$
\begin{aligned}
\mathbf{z}(k_r, k_\chi) &= \sum_l \begin{bmatrix} \mathbb{S}\mathbb{S}_1(k_r, k_\chi + lk_{\chi_p}) \\ \mathbb{S}\mathbb{S}_2(k_r, k_\chi + lk_{\chi_p}) \\ \vdots \\ \mathbb{S}\mathbb{S}_N(k_r, k_\chi + lk_{\chi_p}) \end{bmatrix} + \mathbf{n}(k_r, k_\chi) \\
&= \sum_l \mathbf{s}(k_r, k_\chi + lk_{\chi_p}) + \mathbf{n}(k_r, k_\chi)
\end{aligned}
\tag{5.1}
$$

where $\mathbf{n}(k_r, k_\chi)$ is additive white noise. In the above, k_r is a wavenumber that corresponds to the fast-time or range dimension and is related to the frequency of the signal via $k_r = 2\omega/c$, where ω is the radial frequency (in radians per second) of the signal and c is the speed of light. In addition, k_χ is a wavenumber corresponding to the slow-time or along-track dimension and is related to the Doppler frequency, f_ξ, via $k_\chi = 2\pi f_\xi / v_e$. The pulse repetition frequency (PRF), or Doppler sampling frequency, f_{ξ_p}, is related to k_{χ_p} through $k_{\chi_p} = 2\pi f_{\xi_p} / v_e$.

5.2.2.1 Space-time sampling diagram

It is also useful to introduce, at this point, a very useful and simple tool known as the space-time sampling diagram.

The space-time sampling diagram is simply a plot showing the phase-centre positions as a function of time. It aids in determining the properties of a multi-channel HRWS system. For example, Figure 5.3 illustrates a two-antenna system where, when using a specific PRF and identical antenna patterns, increased spatial sampling points compensate for a reduction in temporal sampling. Signal

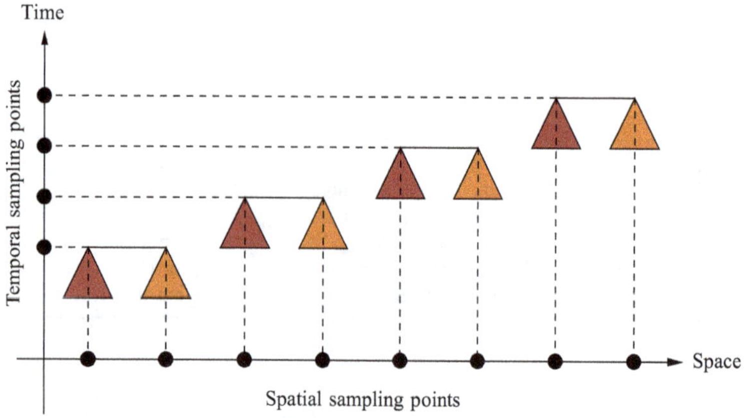

Figure 5.3 Ideal sampling condition where the spatial sampling points are uniform

processing with these 'optimal' design parameters is relatively straightforward, circumventing a need for more advanced techniques that can handle non-ideal PRFs, as described in such papers as [1,3,11,12].

Current incarnations of such an optimal configuration include the commercially available ultrafine and extra-fine modes of RADARSAT-2. Unfortunately, restricting the PRF limits the elevation angle for data acquisition in a space-based SAR, thus reducing the available swaths that can be collected at any given time.

Figure 5.4 provides a second example of a space-time sampling diagram for a real RADARSAT-2 mode. In this case, two antennas measure the scene simultaneously on even numbered pulses, while another two antennas measure the scene simultaneously on odd numbered pulses. This type of toggled measurement is possible with current phased-array systems like RADARSAT-2 and ALOS PALSAR-2, by splitting the receive antenna into two different receiving paths and changing the sub-array weightings from pulse to pulse, see Section 5.5.1. As the figure shows, for the chosen PRF and system phase-centre (PC) positions, spatial sampling is not uniform; therefore, simple interleaving techniques cannot be applied. The added phase-centre positions offer improved GMTI capability [13,14], and, depending on the chosen PRF can also be used for HRWS imaging. Furthermore, as discussed in [15], if the PRF strays from the ideal PRF, then one must determine how best to process the data.

5.2.3 Uniform spatial sampling

To motivate the material in the rest of the chapter, this section re-introduces the simplest case of uniform spatial sampling using two identical antenna patterns, each steered to broadside as illustrated in Figure 5.5. The amplitudes of the antenna patterns are given by $D_0(f_\xi)$ and $D_1(f_\xi)$, while the along-track positions of each antenna are given by α_0 and α_1 in a reference frame that moves with the

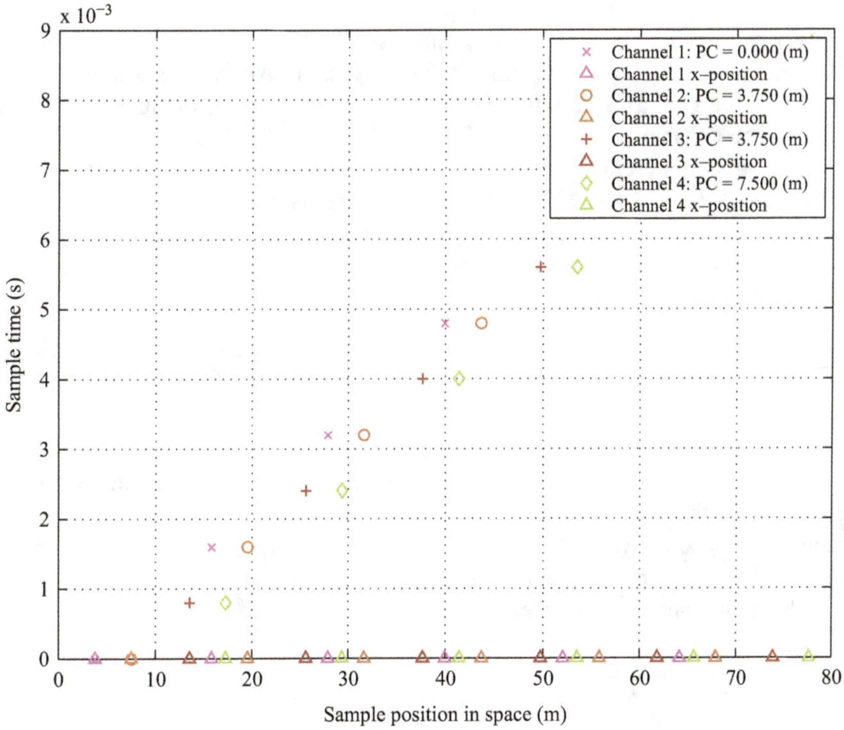

Figure 5.4 Non-ideal sampling where the spatial sampling points are non-uniform

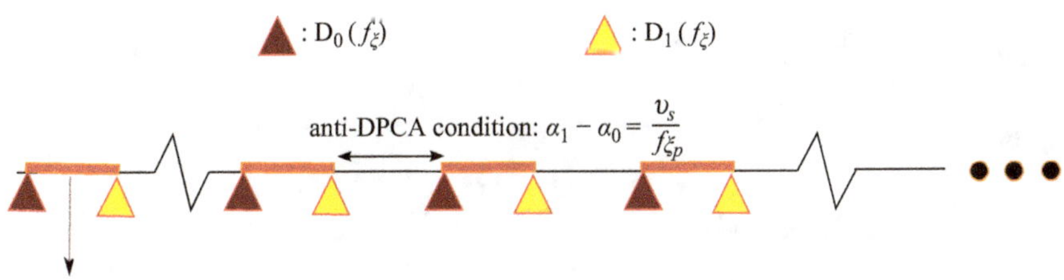

Figure 5.5 Ideal sampling condition for a two-channel system

radar platform. Either the PRF or the along-track positions are selected to ensure that the 'so-called' anti-displaced-phase-centre antenna (anti-DPCA) condition $\alpha_1 - \alpha_0 = v_{\mathfrak{s}}/f_{\xi_p}$ is observed, where, $v_{\mathfrak{s}}$ is the magnitude of the satellite velocity vector in the Earth-centred, Earth-fixed (ECEF) co-ordinate system. This condition yields the uniform spatial sampling. Section 5.2.5 discusses the DPCA configuration as it applies to GMTI.

The anti-DPCA configuration allows relatively straightforward data processing through interleaving of the temporally under-sampled data. Once the measurements have been interleaved, they are suitable for a standard SAR processor to ingest and process with a PRF given by the multiplication of the number of channels and the PRF of each channel.

Several commercially available SAR imagery products from both ALOS PALSAR-2 [16] and RADARSAT-2 [15] are derived from the uniform spatial sampling configuration described in this section. These systems both use a dual-receive phased-array with PRFs chosen to satisfy the anti-DPCA-condition configuration.

5.2.4 Uniform spatial sampling with an across-track baseline for a narrowband system

The situation becomes slightly more complicated when the data are again collected with the ideal PRF, but now with an array-squinted configuration that introduces an across-track baseline, as illustrated in Figure 5.6. In this case, there exists a path length difference between the first and second antennas and before SAR processing, the difference must be compensated.

The path length can be calculated to give an extra phase given by

$$\vartheta = \frac{4\pi(\alpha_1 - \alpha_0)\sin\theta_s}{\lambda} \tag{5.2}$$

The expression can be cast into an equivalent form using the notion of a Doppler centroid (the frequency value corresponding to the peak of the azimuth antenna pattern), see [17]. With the relation,

$$f_{\xi_{dc}} = \frac{2v_s \sin\theta_s}{\lambda} \tag{5.3}$$

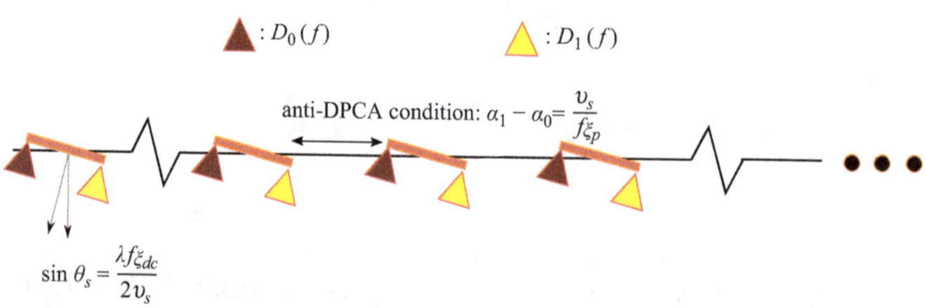

Figure 5.6 Ideal sampling condition with an across-track baseline component for a dual-channel system

the extra phase for the second channel can be written as

$$\vartheta = \frac{2\pi f_{\xi_{dc}}(\alpha_1 - \alpha_0)}{v_{\mathfrak{s}}} \tag{5.4}$$

For an N channel system, a new sequence is created by interleaving the samples from the N antennas; however, each antenna dataset is first multiplied by

$$e^{2i\pi\frac{f_{\xi_{dc}}\alpha_k}{v_{\mathfrak{s}}}}$$

where α_k is the effective, two-way, along-track component of the phase-centre position of antenna k relative to some arbitrary origin.

5.2.5 Uniform spatial sampling with an across-track baseline for a wideband system

The expression in (5.2), yields a physical interpretation of the complication that arises with a non-zero across-track baseline for a wideband system. In this expression, the path length correction depends on the wavelength which for a wideband system cannot be assumed constant. Accurate processing requires computation of a phase correction for each of the range of wavelengths contained in the wideband pulse.

As will be discussed later in the chapter, the presence of an across-track baseline for a wideband system complicates the signal processing. The suite of algorithms designed to operate in the range-Doppler domain, which have already been published in the literature, and will be reviewed in the following section, benefit from only requiring a single-processing filter for each Doppler frequency. That is, the processing filters are invariant with range. This ideal feature no longer applies in the wideband case when there is an across-track component of the baseline. Filters which vary across Doppler frequency as well as across range frequency need to be computed and applied to the measured data.

5.2.6 Non-uniform spatial sampling

Further complication arises when the data are not collected uniformly across space. Practically, it is important to consider this case because the set of possible radar imaging opportunities is much reduced by the requirement of uniform sampling. Figure 5.4 illustrates an example of a non-uniform space-time sampling diagram with the added complexity of toggling.

A pioneering approach to addressing the non-uniform sampling problem for SAR is presented in [1] with other approaches presented in [4,5]. This section introduces the general framework for processing non-uniformly sampled data using linear filters, an approach which incorporates the three approaches just mentioned. As a point of interest, note that [18] reports on compressive sensing as an alternative to linear filtering for HRWS. As the suitability and performance of this approach has yet to match the already established capabilities of linear filtering, it will not be discussed further.

5.2.6.1 Linear filtering to extract signal components

The linear filtering approach applies a multi-dimensional filter in the Doppler domain so that a scalar reconstructed signal, denoted by $\mathcal{ZZ}_R(k_r, k_\chi)$, is created via

$$\mathcal{ZZ}_R(k_r, k_\chi + lk_{\chi_p}) = \mathbf{b}^\dagger{}_l(k_r, k_\chi)\mathbf{z}(k_r, k_\chi) \tag{5.5}$$

The challenge is to find the vectors $\mathbf{b}_l(k_r, k_\chi)$ that yield the desired quality of signal.

There are a few approaches to finding appropriate $\mathbf{b}_l(k_r, k_\chi)$. One method seeks to minimize the square error between the reconstructed signal and a predetermined desired signal. This method is referred to as the minimum mean square error (MMSE) method. The method attempts minimization in the presence of additive noise. When the noise is ignored, one arrives at another popular method called the projection method. The projection method has been popularised by a series of excellent papers in [1–3]. The derivation of these methods is discussed in Section 5.4.

The projection method has been successfully applied to data collected by TerraSAR-X and RADARSAT-2. Reference [6] reports on the first spaceborne demonstration of the projection approach with measured SAR data. Data were collected using the dual-receive aperture mode of TerraSAR-X and down-sampled in azimuth to generate under-sampled data. The paper shows how the under-sampled data could be processed to eliminate the interfering azimuth ambiguities and recover high-resolution imagery.

5.2.6.2 Optimal configurations for a band-limited signal

Suppose that the measured signal is given by

$$\mathbf{z}(k_r, k_\chi) = \sum_{l \in \mathcal{L}} \mathbf{s}(k_r, k_\chi + lk_{\chi_p}) + \mathbf{n}(k_r, k_\chi) \tag{5.6}$$

where the set $\mathcal{L}$ contains only a finite number of elements (i.e. the signal is band-limited). Suppose, further, that this set contains N or fewer elements. An optimal configuration occurs when $\mathbf{s}^\dagger(k_r, k_\chi + lk_{\chi_p})\mathbf{s}(k_r, k_\chi + mk_{\chi_p}) = 0$ when $l \neq m$. Physically, there are two ways to achieve this configuration. In the first configuration, one might have antenna patterns that point in different directions or have support over Doppler frequencies that are mutually exclusive. For instance, one could have a configuration where $\mathcal{SS}_l(k_r, k_\chi) = 0$ if $k_\chi \notin [k_{\chi_p}(l - 1/2), k_{\chi_p}(l + 1/2)]$.

Another configuration has the vectors orthogonal to each other through the design of the complex vector elements. As we shall see in the following material, one instance of this configuration is where the system makes uniform spatial sampling such as in Figure 5.3 although the generalization is to N antennas. In previously published literature this configuration, which depends on the PRF, is referred to as the anti-DPCA condition where the DPCA condition is well known to be ideal for the application of GMTI. One immediate conclusion here, and as illustrated in Figure 5.7, is that optimal GMTI configurations are not ideal for HRWS and vice versa.

The desire is to transform the N elements of the vectors $\mathbf{z}(k_r, k_\chi)$, which are aliased over the main band, into an unambiguous scalar signal over multiple bands.

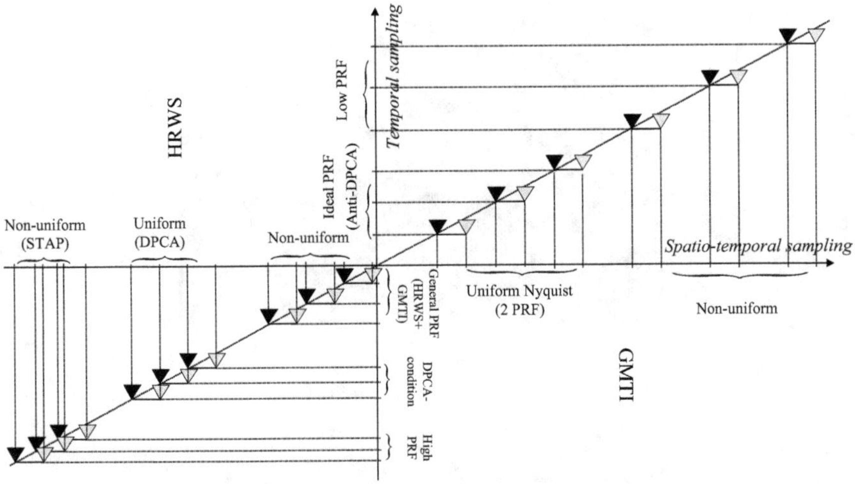

Figure 5.7 Illustration of different sampling regimes for HRWS and GMTI

5.2.6.3 SNR-resolution trade-off

When the set of vectors $\{\mathbf{s}(k_r, k_\chi + lk_{\chi_p})|l \in \mathcal{L}\}$ are not mutually orthogonal, it may still be possible to construct a set of vectors, $\{\mathbf{b}_m(k_r, k_\chi)|m \in \mathcal{L}\}$ such that the inner product $\mathbf{b}^\dagger{}_m(k_r, k_\chi)\mathbf{s}(k_r, k_\chi + lk_{\chi_p})$ is non-zero if $m = l$ and is zero when $l \neq m$. Effectively, this method extracts only a component of the measured signal vectors, while the power of the noise signal remains unchanged leading to a reduced signal-to-noise ratio (SNR). In fact, as any two vectors in $\{\mathbf{s}(k_r, k_\chi + lk_{\chi_p})|l \in \mathcal{L}\}$ become collinear, the SNR goes to zero.

From the viewpoint of the space-time sampling diagram, as the distance between any two spatial sampling points becomes zero, the reconstruction method yields ever degraded SNR. This is only true, of course, if each antenna is identical and points in the same direction. From a signal processing point of view, it is not surprising that multiple samples made at the same sampling point do not change the sampling frequency. Different HRWS processing methods yield different results in this situation. In particular, the projection technique may lead to seriously degraded SNR.

A 2009 campaign using RADARSAT-2 configured in its dual-receive mode yielded a data set that is down-sampled and processed with the projection method in [1–3]. As illustrated in Figure 5.8, these data were collected in a configuration where application of the projection method comes at the cost of much reduced SNR. The figure also presents an image processed with the MMSE filter, presented in Section 5.4. This second method does not suffer from the same loss in SNR.

In summary, when choosing the HRWS signal processing method, it is important to consider the SNR or radiometric resolution as well as the geometric resolution.

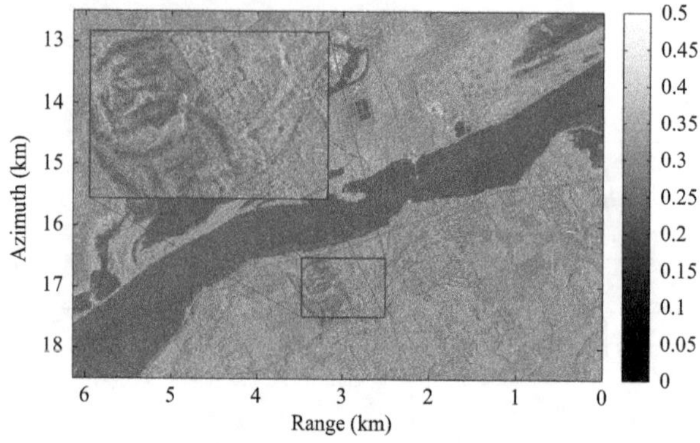

(a) Data processed with MMSE method.

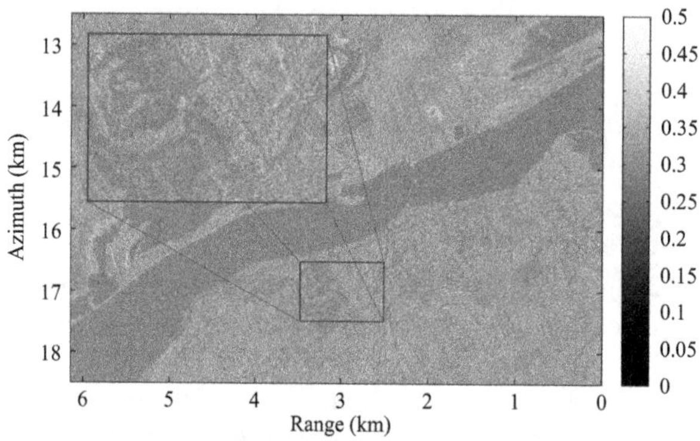

(b) Data processed with the projection method.

Figure 5.8 Multi-channel processed SAR imagery quad-channel mode data. The image is of East Ottawa, with the Ottawa river running diagonally from lower left to upper right. The zoom shows a more detailed image of the Camelot Golf and Country Club

5.2.6.4 Point-spread function distortions

Under the constraint of maintaining the noise level, the reconstructed scalar signal can assume a response quite different from the usual sinc(·) shape. As well, if N, the number of channels, is less than the number of elements in $\mathcal{L}$, then there will be residual aliasing in the measured signal which can manifest in the SAR image as azimuth ambiguities. Different reconstruction algorithms, of which there are several as discussed in, for example [19] will yield different results for a given system.

5.2.7 Section summary

This section introduced the concept of the swath-to-resolution ratio and higher resolution, wider swath. Methods to improve the swath-to-resolution ratio using a multi-channel system were introduced. Through the application of linear filters, the section set the general framework for transforming a vector of under-sampled signals into a scalar, adequately sampled signal. It discussed some trade-offs between the sampling configuration and the resulting image quality and, by considering path length differences, it introduced the complication that arises from an across-track baseline. It was argued that for some special cases of multi-channel space-based SARs, the HRWS processing can be simplified. Among these are narrowband data measured with the anti-DPCA condition and wideband data measured with an along-track-aligned system where the element patterns have gains that are independent of k_r i.e. the antenna gains have flat profiles across the frequencies measured.

The next section derives the signal model for each element of the SAR vector measurement with special attention given to the antenna patterns and phase-centre positions as these are critical in the HRWS signal processing algorithms.

5.3 SAR signal model

This section derives the signal model for a single channel in the wavenumber domain. The signal model applies to wideband systems and to systems with large squints as the squint will be important for the case of an across-track baseline component. Special care is taken to ensure that the antenna pattern gain functions are accurately represented as they play an important role in the HRWS signal processing section to follow. The material in this section reviews some concepts already published in the SAR literature [9,20–23] but with greater care given to the antenna patterns and with, in some cases, simplified derivations.

As introduced in [24], let the parameterised satellite position is given by $\mathfrak{s}(\xi)$, while the point scatterer position is given by $\mathfrak{p}_{\mathfrak{s}}$. In the chosen co-ordinate system, the point scatterer does not move.

A desired baseband waveform, $p(\tau)$, is mixed up to the carrier frequency, ω_0, to yield the transmitted signal which can be represented as the real part of

$$z(\tau) = p(\tau)e^{i\omega_0\tau} \tag{5.7}$$

This signal travels down to the surface, and a delayed version returns to the radar after reflection from the terrain. For any given ξ, the distance that the signal must travel is given by

$$2|\mathfrak{s}(\xi) - \mathfrak{p}_{\mathfrak{s}}| = 2|\mathbf{r}(\xi)| = 2r(\xi) \tag{5.8}$$

If the wave propagates with speed c, then the measured return signal is proportional to

$$z\left(\tau - \frac{2r(\xi)}{c}\right) \tag{5.9}$$

5.3.1 Scene measurement through multiple antenna patterns

The antenna pattern plays a critical role in multi-channel signal processing. Simple antennas, such as the dipole antenna, and simple antenna models, such as a uniformly excited aperture or the uniformly spaced phased-array factor, yield directivity patterns that depend on the wavelength and have beamwidths and gain patterns which depend on the dimensions of the antenna in a given co-ordinate system. For an antenna in the far field, the effective antenna pattern may be represented as

$$\mathrm{A}_n[\hat{\mathbf{r}}(\xi), \mathbf{p}_n(\xi)] = e^{2\frac{2\pi i}{\lambda}\hat{\mathbf{r}}^\dagger(\xi)\mathbf{p}_n(\xi)}\mathrm{D}_n[\lambda, \hat{\mathbf{r}}(\xi)] \tag{5.10}$$

where λ is the narrowband wavelength, $\mathbf{p}_n(\xi)$ represents the co-ordinate of the phase-centre of antenna n in a reference frame centred on the radar platform, $\mathrm{D}_n(\cdot)$ is the two-way antenna pattern amplitude (the product of the transmit antenna pattern and the receive antenna pattern) and $\hat{\mathbf{r}}(\xi)$ is the look vector defined as $\hat{\mathbf{r}}(\xi) = \mathbf{r}(\xi)/r(\xi)$. If the two-way antenna pattern depends on two different phase-centres, located at $\mathbf{p}_{n\mathrm{Tx}}(\xi)$ and $\mathbf{p}_{n\mathrm{Rx}}(\xi)$, then we define

$$\mathbf{p}_n(\xi) = \frac{1}{2}\left[\mathbf{p}_{n\mathrm{Tx}}(\xi) + \mathbf{p}_{n\mathrm{Rx}}(\xi)\right] \tag{5.11}$$

The wideband generalization sees different values for the carrier wavelength in (5.10) so that for a uniformly excited antenna, as λ increases, the antenna can measure over a larger range of angles. Conversely, as λ decreases, the antenna beamwidth narrows.

By defining $2\pi/\lambda = (\omega + \omega_0)/c = k_r/2$, one can write the argument to the antenna pattern as

$$\mathrm{A}_n[k_r, \hat{\mathbf{r}}(\xi)] = e^{ik_r\hat{\mathbf{r}}^\dagger(\xi)\mathbf{p}_n(\xi)}\mathrm{D}_n[k_r, \hat{\mathbf{r}}(\xi)] \tag{5.12}$$

where now, k_r is not constant, but free to wander over a range of values to accommodate a wideband model.

The measured return signal is the superposition of the signal from various locations in the terrain, thus, for some ξ, the overall return signal is given by

$$ss_n(\tau, \xi) = \int \mathrm{A}_n[k_r, \hat{\mathbf{r}}(\xi), \mathbf{p}_n(\xi)]\frac{g(\mathfrak{p}_\mathfrak{s})}{r^2(\xi)}z\left(\tau - \frac{2r(\xi)}{c}\right)d\mathfrak{p}_\mathfrak{s} \tag{5.13}$$

where $g(\mathfrak{p}_\mathfrak{s}) \in C$ denotes a random function which represents the scattering response of the point scatterer at $\mathfrak{p}_\mathfrak{s}$, and the term $r^2(\xi)$ in the denominator accounts for the fact that in the far field, electric fields decay at a rate given by the inverse of the range, and the expression accounts for a two-way propagation. Without loss of generality, other constant factors from the radar equation have been omitted. In any practical calculation that requires computation of the SNR, an account can be made for these factors.

5.3.2 Demodulation

If the return signal is mixed with $e^{-i\omega_0\tau}$, and if filters are applied to remove unwanted images, one obtains the demodulated signal

$$ss_n(\tau,\xi) \rightarrow ss_n(\tau,\xi) = \int \mathrm{A}_n[k_r, \hat{\mathbf{r}}(\xi), \mathbf{p}_n(\xi)]\frac{g(\mathfrak{p}_s)}{r^2(\xi)}p\left(\tau - \frac{2r(\xi)}{c}\right)\cdot e^{-i\omega_0\frac{2r(\xi)}{c}}\mathrm{d}\mathfrak{p}_s \tag{5.14}$$

The fast-time Fourier transform (FT) of $ss_n(\tau,\xi)$ may be computed as

$$Ss_n(\omega,\xi) = P(\omega)\int e^{-i\frac{2[\omega+\omega_0]}{c}r(\xi)}\mathrm{A}_n[k_r, \hat{\mathbf{r}}(\xi), \mathbf{p}_n(\xi)]\frac{g(\mathfrak{p}_s)}{r^2(\xi)}\mathrm{d}\mathfrak{p}_s \tag{5.15}$$

Recall that $k_r = \frac{2(\omega+\omega_0)}{c}$, and define $k_{r'} = \frac{2\omega}{c}$ and $k_{r_0} = \frac{2\omega_0}{c}$, to yield the processed fast-time frequency signal

$$\mathcal{S}s_n(k_r,\xi) = \mathcal{P}(k_{r'})\int e^{-ik_r r(\xi)}\mathrm{A}_n[k_r, \hat{\mathbf{r}}(\xi), \mathbf{p}_n(\xi)]\frac{g(\mathfrak{p}_s)}{r^2(\xi)}\mathrm{d}\mathfrak{p}_s \tag{5.16}$$

where $\mathcal{P}(k_{r'}) = |P(ck_{r'}/2)|$, and the absolute value has been computed by multiplying by the known conjugate phase of $P(\cdot)$. Indeed, range or pulse compression is achieved through multiplication by the conjugate phase. For instance, if the pulse is a chirp, then it is at this point that the chirp phase is removed.

5.3.3 Space-based SAR – orbital motion

This section derives the signal characteristics assuming angular motion. There are a couple of reasons for presenting with this approach: first, the mathematics for computing the stationary phase approximation are simplified (as compared, e.g. to [25]). Second, the approach permits a relation between the physics and the mathematics.

The approach is based upon the idea that the satellite has an instantaneous angular velocity vector $\boldsymbol{\omega}_s(\xi)$ in an ECEF reference frame. Euler's rotation theorem guarantees the existence of an instantaneous angular velocity vector. This section also assumes that, during the period of collection of the synthetic aperture, $\boldsymbol{\omega}_s(\xi)$ remains constant so that we can instead denote the angular velocity vector as $\boldsymbol{\omega}_s$. The direction of $\boldsymbol{\omega}_s$ indicates the direction of the rotation axis, while its amplitude ω_s denotes the rate of rotation. Figure 5.9 illustrates a three-dimensional reference frame with $\hat{y}$ axis in the direction of $\boldsymbol{\omega}_s$ and $\hat{x}$ and $\hat{y}$ in the plane of rotation, with directions relative to some arbitrarily chosen point. The Earth is not shown in the figure because the instantaneous angular velocity vector in an ECEF frame depends on the Earth's angular velocity vector and the satellite angular velocity vector in an inertial reference frame; thus, the relative position of the Earth is not well defined although for most SAR orbits, the orbit plane would slice through the centre of the Earth. Note that the geometry is suitable for an airborne circular SAR. The red curve represents the path along which the satellite travels and neither the direction of travel nor the look direction matter to the analysis. If the satellite is going

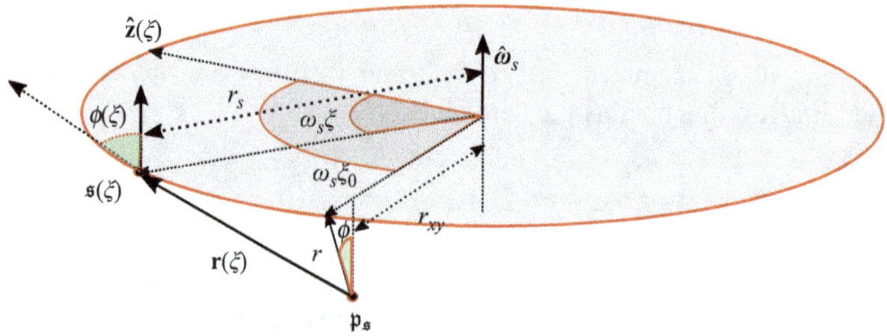

Figure 5.9 Orbital motion. The y-axis points in the direction of the rotation axis, the z-axis and x-axis lie in the plane in which the satellite rotates. The choice of the direction for the z-axis in the rotation plane is arbitrary and not important for the analysis. In a slight abuse of notation, $\phi = \phi(\xi_0)$ and $r = r(\xi_0)$

clockwise, as viewed from above, then the illustrated target is in a left-looking direction, if the satellite is going counter clockwise, then the target is seen from a right-looking perspective. Lastly, $\omega_s\xi_0$, is the particular value of the parameter for which the range vector has minimum magnitude.

The satellite equations of motion can be written via

$$\mathfrak{s}(\xi) = r_s \begin{bmatrix} \cos(\omega_s\xi) \\ \sin(\omega_s\xi) \\ 0 \end{bmatrix} \tag{5.17}$$

A particular scatterer on the ground has co-ordinates given by

$$\mathfrak{p}_\mathfrak{s} = \begin{bmatrix} x \\ y \\ z \end{bmatrix} \tag{5.18}$$

The range vector is given by

$$\begin{aligned} r(\xi) &= |\mathfrak{s}(\xi) - \mathfrak{p}_\mathfrak{s}| \\ &= \sqrt{{r_s}^2 + x^2 + y^2 + z^2 - 2r_s r_{xy}\cos(\omega_s\xi - \omega_s\xi_0)} \end{aligned} \tag{5.19}$$

where, $r_{xy}^2 = x^2 + y^2$ and $\sin\omega_s\xi_0 = y/r_{xy}$. One can now apply the relation $\cos(\alpha) = 1 - 2\sin^2(\alpha/2)$ to see that

$$\begin{aligned} r(\xi) &= \sqrt{(r_s - r_{xy})^2 + z^2 + 4r_s r_{xy}\sin^2(\omega_s\xi/2 - \omega_s\xi_0/2)} \\ &= r\sqrt{1 + \frac{4r_s r_{xy}}{r^2}\sin^2(\omega_s\xi/2 - \omega_s\xi_0/2)} \\ &= r\sqrt{1 + \tan^2\Theta_s(\xi - \xi_0)} \\ &= r\sec\Theta_s(\xi - \xi_0) \end{aligned} \tag{5.20}$$

where

$$\tan\Theta_s(\xi) = -2\sqrt{r_s r_{xy}}/r\sin(\omega_s\xi/2) \tag{5.21}$$

and, with a slight abuse of notation, we have defined $r^2 = r^2(\xi_0) = (r_s - r_{xy})^2 + z^2$. Equation (5.21) can be inverted to yield

$$\xi = -\frac{2}{\omega_s}\sin^{-1}\left[\frac{r\tan\Theta_s(\xi)}{2\sqrt{r_s r_{xy}}}\right] \tag{5.22}$$

The above expression will play a role in the computation of the FT, since one needs an expression for ξ in the FT kernel, $e^{-2\pi i\xi f_\xi}$.

5.3.3.1 Stationary phase to compute the slow-time Fourier transform

The FT over ξ is given by

$$\begin{aligned}\mathcal{S}S_n(k_r, f_\xi) &= \int \mathcal{S}s_n(k_r, \xi)\exp(-2\pi i\xi f_\xi)d\xi \\ &= \mathcal{P}(k_{r'})\iint A_n[k_r, \hat{\mathbf{r}}(\xi), \mathbf{p}_n(\xi)]\frac{g(\mathfrak{p}_{\mathfrak{s}})}{r^2(\xi)}e^{-i\Psi(\xi)}d\mathfrak{p}_{\mathfrak{s}}d\xi\end{aligned} \tag{5.23}$$

where the phase of the integrand is given by

$$\begin{aligned}\Psi(\xi) &= -k_r r(\xi) - 2\pi f_\xi\xi = -k_r r(\xi) - 2\pi f_\xi(\xi - \xi_0) - 2\pi f_\xi\xi_0 \\ &= -k_r r\sec\Theta_s(\xi - \xi_0) + \frac{4\pi f_\xi}{\omega_s}\sin^{-1}\left[\frac{r}{2\sqrt{r_s r_{xy}}}\tan\Theta_s(\xi - \xi_0)\right] - 2\pi f_\xi\xi_0\end{aligned} \tag{5.24}$$

By the rules of calculus,

$$\begin{aligned}\dot{\Psi}(\xi) &= -k_r r\sin\Theta_s(\xi - \xi_0)\sec^2\Theta_s(\xi - \xi_0)\dot{\Theta}_s(\xi - \xi_0) \\ &\quad + \frac{2\pi f_\xi r}{\omega_s\sqrt{r_s r_{xy}}}\frac{\sec^2\Theta_s(\xi - \xi_0)\dot{\Theta}_s(\xi - \xi_0)}{\sqrt{1 - \frac{r^2}{4r_s r_{xy}}\tan^2\Theta_s(\xi - \xi_0)}}\end{aligned} \tag{5.25}$$

where the dot notation signifies differentiation with respect to ξ.

The stationary point is calculated by setting the right side of (5.25) to zero which yields

$$\begin{aligned}k_r\sin\Theta_s(\xi - \xi_0) &= \frac{2\pi f_\xi}{\omega_s\sqrt{r_s r_{xy}}}\frac{1}{\sqrt{1 - \frac{r^2}{4r_s r_{xy}}\tan^2\Theta_s(\xi - \xi_0)}} \\ &= \frac{k_\chi}{\sqrt{1 - \frac{r^2}{4r_s r_{xy}}\tan^2\Theta_s(\xi - \xi_0)}}\end{aligned} \tag{5.26}$$

where k_χ is defined as

$$k_\chi = \frac{2\pi f_\xi}{\omega_s \sqrt{r_s r_{xy}}} = \frac{2\pi f_\xi}{r_s \omega_s \sqrt{r_{xy}/r_s}} = \frac{2\pi f_\xi}{v_\mathfrak{s} \sqrt{r_{xy}/r_s}} = \frac{2\pi f_\xi}{v_e} \tag{5.27}$$

With the above relation, $\chi = v_e \xi$. After some manipulation, and the reader can check this, (5.26) can be inverted to yield

$$\sin^2\Theta_s(\xi_s - \xi_0) = \frac{1}{k_r^2(1-\rho^2)} \left(\frac{k_r^2 + k_\chi^2}{2} - \sqrt{\left(\frac{k_r^2 - k_\chi^2}{2}\right)^2 - \rho^2 k_r^2 k_\chi^2} \right) \tag{5.28}$$

where $\rho = r/\sqrt{r_s r_{xy}}$. Notice that when $\rho = 0$, then $k_r \sin \Theta_s(\xi_s - \xi_0) = k_\chi$.

5.3.3.2 Simplification for large orbital radius

For a large orbital radius, one may make the approximation that $\rho \ll 1$ yielding the following equations

$$\xi \approx \frac{-r}{v_e} \tan \Theta_s(\xi) \tag{5.29}$$

$$\sin \Theta_s(\xi_s - \xi_0) \approx k_\chi / k_r \tag{5.30}$$

$$r(\xi_s) \approx r k_r \Big/ \sqrt{k_r^2 - k_\chi^2} \tag{5.31}$$

$$\Psi(\xi_s) \approx -r\sqrt{k_r^2 - k_\chi^2} - k_\chi \chi_t \tag{5.32}$$

$$\ddot{\Psi}(\xi_s) \approx -\frac{v_e{}^2}{r} \frac{[k_r^2 - k_\chi^2]^{\frac{3}{2}}}{k_r^2} \tag{5.33}$$

where $\chi_t = v_e \omega_s \xi_0 / \omega_s$. The above leads to the stationary phase approximation:

$$\begin{aligned} \mathcal{SS}_n(k_r, k_\chi) = \mathcal{P}(k_{r'}) \frac{(k_r^2 - k_\chi^2)^{\frac{1}{4}}}{v_e k_r} \int & \mathrm{A}_n[k_r, \hat{\mathbf{r}}(\xi_s), \mathbf{p}_n(\xi_s)] \\ & \cdot \frac{g(\mathfrak{p}_\mathfrak{s})}{\sqrt{r_0^3}} e^{-ir\sqrt{k_r^2 - k_\chi^2} - ik_\chi \chi_t} \mathrm{d}\mathfrak{p}_\mathfrak{s} \end{aligned} \tag{5.34}$$

For reference, the stationary phase procedure is an asymptotic approximation given by

$$\int f(t) e^{i\Phi(t)} \mathrm{d}t = e^{i\Phi(t_s)} f(t_s) \sqrt{\frac{2\pi i}{\ddot{\Phi}(t_s)}} \tag{5.35}$$

where t_s is a stationary point, or a point such that $\dot{\Phi}(t_s) = 0$. The approximation becomes ever more accurate for increasing values of the second derivative $\ddot{\Phi}(t_s)$ [26].

5.3.4 Relation between the look vector and the velocity vector

The antenna pattern arguments depend on inner products between the look vector, which changes over ξ, and the position and orientation vectors of each antenna which in the most general setting, may also change over ξ. This section relates these look vectors which change over ξ, to quantities that change over **k**.

By taking the derivative with respect to ξ of

$$r(\xi) = \sqrt{\mathbf{r}^{\dagger}(\xi)\mathbf{r}(\xi)} = r \sec \Theta(\xi) \tag{5.36}$$

one obtains

$$\begin{aligned} \frac{\mathbf{v}^{\dagger}{}_s(\xi)\mathbf{r}(\xi)}{r(\xi)} &= r \sin\, \Theta(\xi) \sec^2 \Theta(\xi)\dot{\Theta}(\xi) \\ \mathbf{v}^{\dagger}{}_s(\xi)\hat{\mathbf{r}}(\xi) &= r \sin\, \Theta(\xi) \sec^2 \Theta(\xi)\dot{\Theta}(\xi) \\ &= -v_e \sin\, \Theta(\xi) \end{aligned} \tag{5.37}$$

In the above, the derivative of $\mathbf{r}(\xi)$ with respect to ξ is given by $\mathbf{v}_s(\xi)$. Thus, at the stationary point,

$$\hat{\mathbf{v}}_s^{\dagger}(\xi_s)\hat{\mathbf{r}}(\xi_s) = \cos\, \theta(\xi_s) = -\kappa \sin\, \Theta(\xi_s) = -\frac{\kappa k_\chi}{k_r} \tag{5.38}$$

where

$$\kappa = \frac{v_e}{v_s} \tag{5.39}$$

and we have assumed that $v_s(\xi) = v_s$ remains constant over ξ. Also, $\theta(\xi)$ is the angle between $\hat{\mathbf{v}}_s(\xi)$ and $\hat{\mathbf{r}}(\xi)$.

5.3.4.1 Fixed antennas relative to track

One can represent the look direction vector and some vector fixed relative to the platform as

$$\hat{\mathbf{r}}(\xi) = \alpha(\xi)\hat{\mathbf{v}}_s(\xi) + \beta(\xi)\hat{\boldsymbol{\omega}}_s + \gamma(\xi)\hat{\mathbf{z}}(\xi) \tag{5.40}$$

$$\mathbf{p}_n(\xi) = \alpha_n\hat{\mathbf{v}}_s(\xi) + \beta_n\hat{\boldsymbol{\omega}}_s + \gamma_n\hat{\mathbf{z}}(\xi) \tag{5.41}$$

where $\hat{\mathbf{z}}(\xi)$ is in the direction $\hat{\mathbf{v}}_s(\xi) \times \hat{\boldsymbol{\omega}}_s$. The lack of an argument in the coefficients of the unit vectors in $\mathbf{p}_n(\xi)$ ensures that the vector is fixed relative to the platform co-ordinate system.

Now, $k_r\hat{\mathbf{r}}^{\dagger}(\xi_s)\hat{\mathbf{v}}_s(\xi_s) = k_r\alpha(\xi_s) = -\kappa k_\chi$. Further, $\mathbf{r}^{\dagger}(\xi)\hat{\boldsymbol{\omega}}_s = \text{Const} = r(\xi)\cos\,\phi(\xi) = r\cos\,\phi$[3]; hence, $k_r\hat{\mathbf{r}}^{\dagger}(\xi_s)\hat{\boldsymbol{\omega}}_s = k_r\beta(\xi_s) = \cos\,\phi\sqrt{k_r^2 - k_\chi^2}$. These relations yield $k_r\gamma(\xi_s) = \sqrt{\sin^2\phi[k_r^2 - k_\chi^2] + k_\chi^2(1 - \kappa^2)}$.

[3]For left looking the sign would change to give $\mathbf{r}^{\dagger}(\xi)\hat{\boldsymbol{\omega}}_s = -r\cos\,\phi$.

Let

$$k_r\hat{\mathbf{r}}(k_r,k_\chi,\phi)=\begin{bmatrix}-\kappa k_\chi\\ \cos\phi\sqrt{k_r^2-k_\chi^2}\\ \sqrt{\sin^2\phi[k_r^2-k_\chi^2]+k_\chi^2(1-\kappa^2)}\end{bmatrix}\tag{5.42}$$

The antenna pattern at the stationary point can be written as

$$\begin{aligned}\mathrm{A}_n[k_r,\hat{\mathbf{r}}(\xi_s),\mathbf{p}_n(\xi_s)]&=\mathrm{A}_n\left[k_r,\hat{\mathbf{r}}(k_r,k_\chi,\phi),\mathbf{p}_n(\xi_s)\right]\\&=e^{ik_r\hat{\mathbf{r}}^\dagger(k_r,k_\chi,\phi)\mathbf{p}_n}\mathrm{D}_n\left[k_r,\hat{\mathbf{r}}(k_r,k_\chi,\phi)\right]\end{aligned}\tag{5.43}$$

where, in the last line, we have suppressed the arguments to the antenna position vector, $\mathbf{p}_n(\xi_s)=\mathbf{p}_n$, as it is fixed relative to the track.

5.3.5 Stationary phase solution

By combining all the computed terms, the stationary phase approximation to the wavenumber domain representation of the signal is given by

$$\begin{aligned}\mathcal{SS}_n(k_r,k_\chi)=&\ C_0\mathcal{P}(k_{r'})\frac{(k_r{}^2-k_\chi^2)^{\frac{1}{4}}}{v_e k_r}\\&\times\int\mathrm{A}_n\left[k_r,\hat{\mathbf{r}}(k_r,k_\chi,\phi),\mathbf{p}_n\right]\frac{g[\mathfrak{p}_\mathfrak{s}]}{\sqrt{r^3}}e^{-i\chi_t k_\chi-ir\sqrt{k_r^2-k_\chi^2}}\,d\mathfrak{p}_\mathfrak{s}\end{aligned}\tag{5.44}$$

where $d\mathfrak{p}_\mathfrak{s}=dxdydz$.

One may make a variable transformation according to

$$x=(r_s-r\sin\phi)\cos\left[\frac{\omega_s}{v_e}\chi_t\right]\tag{5.45}$$

$$y=(r_s-r\sin\phi)\sin\left[\frac{\omega_s}{v_e}\chi_t\right]\tag{5.46}$$

$$z=-r\cos\phi\tag{5.47}$$

Under the conditions that

$$\frac{\partial v_e}{\partial r}\ll 1\tag{5.48}$$

$$\frac{\partial v_e}{\partial\phi}\ll 1\tag{5.49}$$

the transformation introduces the Jacobian $rv_e/v_\mathfrak{s}$. With this transformation, $g[\mathfrak{p}_\mathfrak{s}]\rightarrow g(\chi_t,r,\phi)$

$$\begin{aligned}\mathcal{SS}_n(k_r,k_\chi)=&\ C_0\mathcal{P}(k_{r'})\frac{(k_r^2-k_\chi^2)^{\frac{1}{4}}}{v_\mathfrak{s}k_r}\\&\times\int\mathrm{A}_n\left[k_r,\hat{\mathbf{r}}(k_r,k_\chi,\phi),\mathbf{p}_n\right]\frac{g[r,\chi_t,\phi]}{\sqrt{r}}e^{-i\chi_t k_\chi-ir\sqrt{k_r^2-k_\chi^2}}\,drd\chi_t d\phi\end{aligned}\tag{5.50}$$

5.3.6 SAR processing with adequate sampling

This sub-section shows how the Ω-K algorithm proceeds for SAR processing of an adequately sampled signal in (5.50). This, ultimately could be the approach to processing the HRWS data after the HRWS signal processing algorithms of Section 5.4 have been applied to transform the aliased vector data into a scalar unambiguous signal.

The measured signal is transformed into the 2-D frequency domain to provide a new data set that is represented by (5.50). These data are then interpolated according to a Stolt interpolation

$$\begin{bmatrix} k_s \\ k_\chi \end{bmatrix} = \begin{bmatrix} \sqrt{k_r^2 - k_\chi^2} \\ k_\chi \end{bmatrix} \tag{5.51}$$

The operation yields

$$\mathcal{SS}_n\left[k_s = \sqrt{k_r^2 - k_\chi^2}, k_\chi\right] = C_0 \mathcal{P}\left(\sqrt{k_s^2 + k_\chi^2} - k_{r_0}\right)\frac{1}{v_s}\sqrt{\frac{k_s}{k_s^2 + k_\chi^2}} \times \int \mathrm{A}_n\left[k_s, \hat{\mathbf{r}}(k_s, k_\chi, \phi), \mathbf{p}_n\right] \int \frac{g[r, \chi_t, \phi]}{\sqrt{r}} \exp\left(-i\left[\chi_t k_\chi + r k_s\right]\right) \mathrm{d}r \mathrm{d}\chi_t \mathrm{d}\phi \tag{5.52}$$

The above shows that the interpolated data (on the left) represent a 2-D FT of $g[r, \chi_t, \phi]$, over the first two variables, after it has been windowed in the frequency domain by the antenna pattern, integrated over ϕ and modulated by other slowly varying terms. Recovery of the weighted integration over ϕ follows from an inverse FT with weighting as required to equalise the effects of the antenna pattern and other factors.

When the data are not adequately sampled, it is not possible to identify the frequencies on the left hand side to apply the Stolt interpolation. Therefore, before applying the Stolt interpolation, a multi-channel processing algorithm should first attempt to resolve the aliasing problem of the under-sampled data.

5.3.6.1 Examples of the Stolt-interpolated data

The parameters listed in Table 5.2 are used in the simulation of the point scatterer signal. It is assumed that the data are adequately sampled. Two simulations are

Table 5.2 Satellite simulation parameters

f_0	1.405 GHz
B	500 MHz
L	2 m
f_p	8 KHz
h	690 km
v_s	7,500 m/s
v_e	7,221 m/s

presented. In the first, there is no across-track baseline component, while in the second, there is an across-track baseline component which may be a result of the configuration of the antenna or may be due to a moving point scatterer. Figure 5.10(a) illustrates a measured antenna response for a point scatterer imaged with no across-track baseline component before Stolt interpolation, while Figure 5.10(b) shows the same point scatterer after Stolt interpolation. It can be seen that the response is independent of the range wavenumber.

Figure 5.11(a) shows the antenna response for a point scatterer that has been measured with an across-track baseline component corresponding to a squint of 4.15 degrees before Stolt interpolation, while Figure 5.11(b) shows the point scatterer response after Stolt interpolation. It is clear that in this case, there is a dependence on both the range and azimuth wavenumbers and as we shall see, for wideband systems, this will require the application of a 2-D HRWS reconstruction filter.

5.3.7 Narrowband approximation

As a narrowband SAR offers some simplifications to the HRWS signal processing, this sub-section derives the narrowband approximation from the wideband signal.

The narrowband model assumes a single transmitted wavelength or that $k_r/k_{r_0} \approx 1$.

As computed in Appendix 5.6, this simplification leads to the following range-Doppler signal

$$\begin{aligned} s\mathcal{S}_n(r', k_\chi) &= C_0 \frac{(k_{r_0}^2 - k_\chi^2)^{\frac{1}{4}}}{v_s k_{r_0}} \mathrm{A}_n[k_{r_0}, \hat{\mathbf{r}}(k_{r_0}, k_\chi, \phi), \mathbf{p}_n] \\ &\times \int e^{-i\chi_t k_\chi} e^{-ir\sqrt{k_{r_0}^2 - k_\chi^2}} \frac{g(r, \chi_t, \phi)}{\sqrt{r}} e^{ik_{r_0} r'} \mathrm{p}\left[r' - \frac{r k_{r_0}}{\sqrt{k_{r_0}^2 - k_\chi^2}} \right] d\chi_t dr d\phi \end{aligned} \tag{5.53}$$

where $\mathrm{p}(\cdot)$ represents the inverse FT of $\mathcal{P}(\cdot)$.

Consider a single point scatterer located at (r_0, χ_{t_0}, ϕ). That is $g(r, \chi_t, \phi) = g(r_0, \chi_{t_0}, \phi)\delta(r - r_0)\delta(\chi_t - \chi_{t_0})\delta(\phi - \phi_0)$. In this case, one obtains

$$\begin{aligned} s\mathcal{S}_n(r', k_\chi) &= C_0 \frac{(k_{r_0}^2 - k_\chi^2)^{\frac{1}{4}}}{v_s k_{r_0}} \mathrm{A}_n[k_{r_0}, \hat{\mathbf{r}}(k_{r_0}, k_\chi, \phi_0), \mathbf{p}_n] \\ &\times e^{-i\chi_{t_0} k_\chi} e^{-ir_0\sqrt{k_{r_0}^2 - k_\chi^2}} \frac{g(r_0, \chi_{t_0}, \phi_0)}{\sqrt{r_0}} e^{ik_{r_0} r'} \mathrm{p}\left[r' - \frac{r_0 k_{r_0}}{\sqrt{k_{r_0}^2 - k_\chi^2}} \right] \end{aligned} \tag{5.54}$$

The argument of $\mathrm{p}(\cdot)$ reflects range migration. The only term that varies from channel to channel is the antenna pattern (the only n-dependence), and this depends solely on k_χ when $\mathbf{p}_n$ aligns with $\hat{\mathbf{x}}(\xi)$. In the narrowband case, even if the

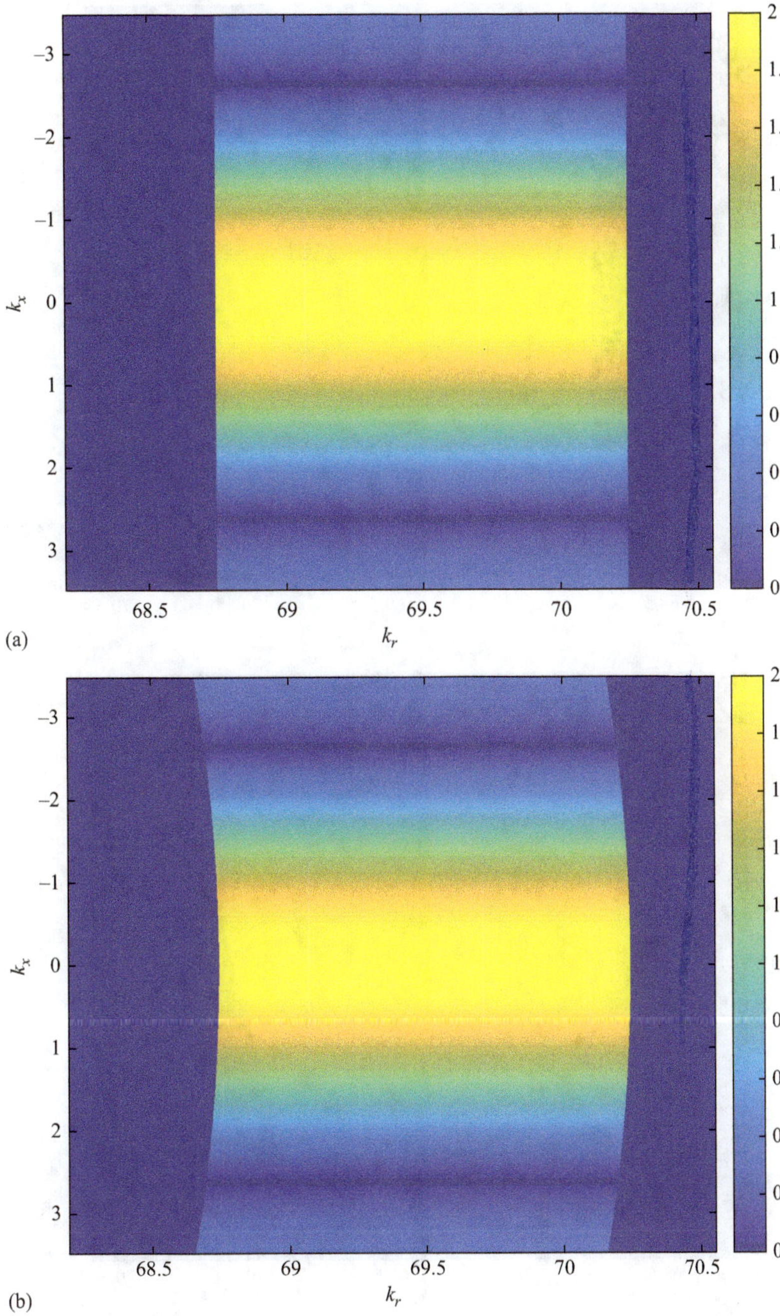

Figure 5.10 (a) Magnitude (stretched by an arbitrary constant for visualization) of the response of a point scatterer with no across-track baseline. (b) Magnitude (stretched by an arbitrary constant for visualization) of the response of a point scatterer with no across-track baseline after Stolt interpolation

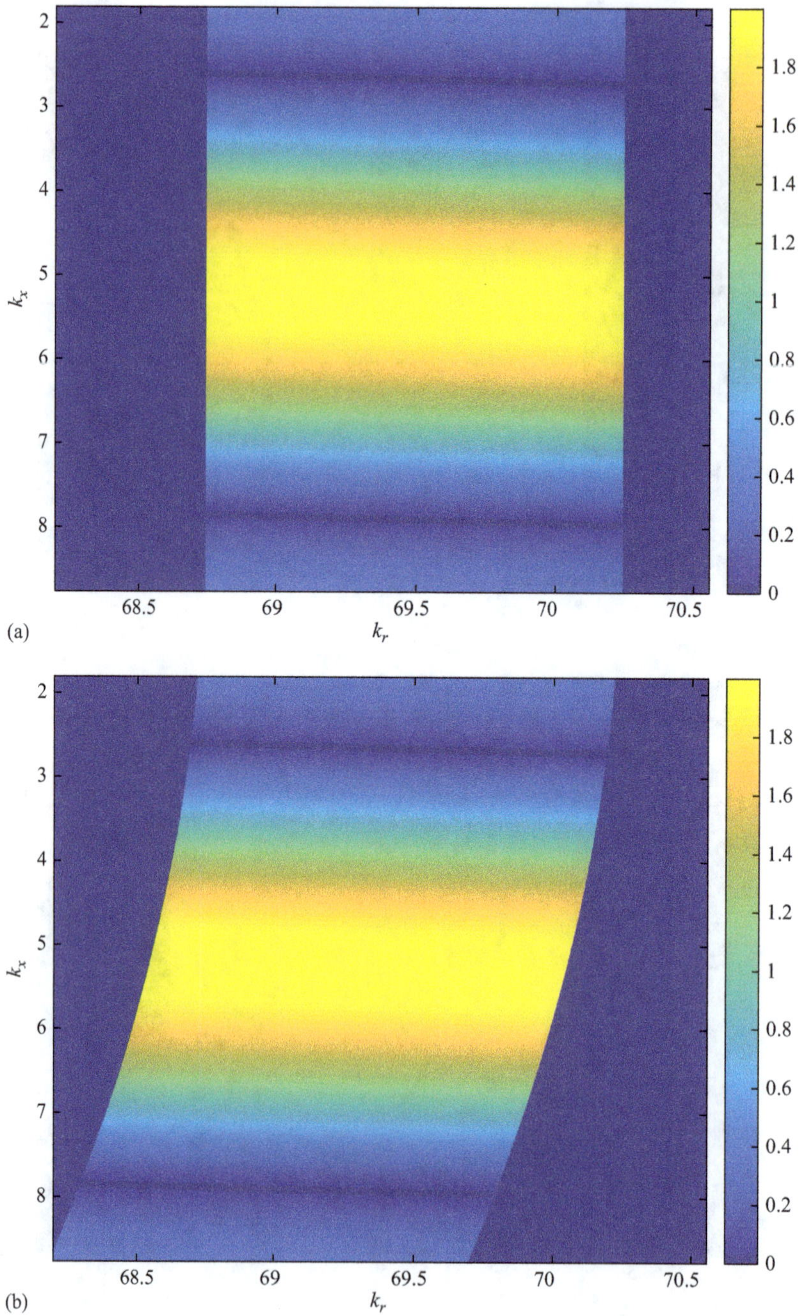

Figure 5.11 *(a) Magnitude (stretched by an arbitrary constant for visualization) of the response of a point scatterer with an across-track baseline component. It is clear that the response for a given k_χ depends on k_r. (b) Magnitude (stretched by an arbitrary constant for visualization) of the response of a point scatterer with an across-track baseline component after Stolt interpolation*

alignment is not perfect, the antenna pattern does not depend on r' so the HRWS filters that we will derive in Section 5.4 are range independent although there is a slight dependence on the depression angle. The computation and application of a single-processing filter reduces the processing demand.

Note that the derivation in Appendix 5.6 does not apply the method of stationary phase. Compare, for instance, with [22] where the stationary phase approximation is applied twice. This is worth mention because the stationary phase approximation is asymptotic in nature and relies, for accuracy, on a large second derivative in the phase function.

5.3.8 Section summary

This section derived the signal model for a single channel in the wavenumber domain. The signal model applies to wideband systems and to systems with large squints as the squint will be important for the case of an across-track baseline component. Special care was taken to ensure that the antenna pattern gain functions are accurately represented as they play an important role in the HRWS signal processing section to follow. The signal model applies equally to a space-based or air-breathing SAR.

5.4 Multi-channel processing for HRWS

With the signal represented in the 2-D frequency domain, we are finally ready to derive the HRWS processing filters.

This section derives signal-processing methods to reconstruct an optimal scalar spectrum for wideband SAR imaging. It is assumed that the antennas are fixed relative to track – i.e. that one can model the signal according to (5.50). Further, it is assumed that the antenna positions relative to track are known. The relative motion parameters are also known, and the objective is to reproduce a high-resolution image from the multi-channel signal which is aliased according to the PRF.

As introduced in Section 5.2.5, the goal of this section is to determine appropriate choices for $\mathbf{b}_l(k_r, k_\chi)$ such that the reconstructed, scalar signal,

$$\mathcal{ZZ}_R(k_r, k_\chi + lk_{\chi_p}) = \mathbf{b}^\dagger{}_l(k_r, k_\chi)\mathbf{z}(k_r, k_\chi) \tag{5.55}$$

is as free from azimuth ambiguities as possible and that it does not suffer catastrophic losses in SNR.

5.4.1 Matrix-vector model for the aliased signal

The derivation of the filters is aided by re-writing the multi-channel signal in matrix-vector notation.

Recall that an N-channel system measurement can be represented as

$$\begin{aligned}\mathbf{z}(k_r, k_\chi) &= \sum_{l\in\mathcal{L}} \mathbf{s}(k_r, k_\chi + lk_{\chi_p}) + \mathbf{n}(k_r, k_\chi) \\ &= \sum_{l\in\mathcal{L}} \int \mathbf{a}(k_r, k_\chi + lk_{\chi_p}, \phi) v(k_r, k_\chi + lk_{\chi_p}, \phi) \mathrm{d}\phi + \mathbf{n}(k_r, k_\chi)\end{aligned} \tag{5.56}$$

where

$$\mathbf{a}(k_r, k_\chi, \phi) = \begin{bmatrix} \mathrm{A}_1\left[k_r, \hat{\mathbf{r}}(k_r, k_\chi, \phi), \mathbf{p}_1\right] \\ \mathrm{A}_2\left[k_r, \hat{\mathbf{r}}(k_r, k_\chi, \phi), \mathbf{p}_2\right] \\ \cdots \\ \mathrm{A}_N\left[k_r, \hat{\mathbf{r}}(k_r, k_\chi, \phi), \mathbf{p}_N\right] \end{bmatrix} \tag{5.57}$$

and

$$v(k_r, k_\chi, \phi) = \frac{C_0 \mathcal{P}(k_{r'})(k_r^2 - k_\chi^2)^{\frac{1}{4}}}{v_s k_r} \int \frac{g[r, \chi_t, \phi]}{\sqrt{r}} e^{-i\chi_t k_\chi - ir\sqrt{k_r^2 - k_\chi^2}} \,\mathrm{d}r\mathrm{d}\chi_t \tag{5.58}$$

The summation can be incorporated into a matrix multiplication to yield

$$\mathbf{z}(k_r, k_\chi) = \int \mathbf{H}(k_r, k_\chi, \phi)\mathbf{v}(k_r, k_\chi, \phi)\mathrm{d}\phi + \mathbf{n}(k_r, k_\chi) \tag{5.59}$$

where the matrix $\mathbf{H}$ is composed of the vectors $\mathbf{a}$ according to

$$\mathbf{H}(k_r, k_\chi, \phi) = \begin{bmatrix} \cdots & \mathbf{a}(k_r, k_\chi - k_{\chi_p}, \phi) & \mathbf{a}(k_r, k_\chi, \phi) & \cdots \end{bmatrix} \tag{5.60}$$

and

$$\mathbf{v}(k_r, k_\chi, \phi) = \begin{bmatrix} \vdots \\ v(k_r, k_\chi - k_{\chi_p}, \phi) \\ v(k_r, k_\chi, \phi) \\ \vdots \end{bmatrix} \tag{5.61}$$

The matrix $\mathbf{H}$ may be fat, skinny or square[4] according to the number of channels and the degree of aliasing of a band-limited signal.

5.4.2 A cost function for HRWS processing

Let us re-write the aliased signal as

$$\mathbf{z}(k_r, k_\chi) = \int \mathbf{H}(k_r, k_\chi, \phi)\mathbf{v}(k_r, k_\chi, \phi) + \boldsymbol{\nu}(k_r, k_\chi, \phi)\mathrm{d}\phi \tag{5.62}$$

[4] A matrix of size $m \times n$ is square, if $m = n$, skinny if $m > n$ and fat if $m < n$.

where

$$\int \boldsymbol{\nu}(k_r, k_\chi, \phi)\mathrm{d}\phi = \mathbf{n}(k_r, k_\chi) \tag{5.63}$$

In fact, in the above, ideally $\boldsymbol{\nu}(k_r, k_\chi, \phi) = \boldsymbol{\nu}(k_r, k_\chi)\mathrm{Rect}_{B_\phi}(\phi)$, i.e. is constant and non-zero only over the support of integration over ϕ.

5.4.2.1 Extraction of the aliased signal components

For the time being, let us assume that we only consider a particular value of ϕ and that we wish to multiply the measured signal at this point by some matrix, $\mathbf{B}(k_r, k_\chi, \phi)$, such that $\mathbf{B}(k_r, k_\chi, \phi)\mathbf{H}(k_r, k_\chi, \phi) = \mathbf{D}(k_r, k_\chi, \phi)$ where $\mathbf{D}(k_r, k_\chi, \phi)$ is some desired diagonal matrix. This desired matrix, resulting from the product, should ideally be one that provides good radiometric resolution. A practical choice is, for instance, the average antenna pattern. Note that if $\mathbf{D}(k_r, k_\chi, \phi)$ is diagonal, then it is straightforward to read the individual aliased components of the signal as they will just correspond to the diagonal element of $\mathbf{D}(k_r, k_\chi, \phi)$ multiplied by the corresponding row of $\mathbf{v}(k_r, k_\chi, \phi)$.

Although a particular value of ϕ has been assumed, the range of angles over which ϕ varies for a typical SAR measurement is quite small, and the processing filters are quite insensitive to changes over this small range. As far as the HRWS processing filters are concerned, the mean value provides a suitable surrogate for the value itself. Of course, one could create filters for a range of different ϕ if desired.

In particular, because, heuristically, we are interested in the signal measured by the *average* antenna pattern, let the diagonal elements of $\mathbf{D}(k_r, k_\chi, \phi)$ be given by

$$D_{ll}(k_r, k_\chi, \phi) = \sqrt{\sum_n |\mathrm{D}_n[\hat{\mathbf{r}}_l(k_r, k_\chi, \phi)]|^2} \tag{5.64}$$

where

$$\hat{\mathbf{r}}_l(k_r, k_\chi, \phi) = \hat{\mathbf{r}}(k_r, k_\chi + lk_{\chi_p}, \phi) \tag{5.65}$$

For different values of l, the range look vector points in different directions. Our choice of $\mathbf{D}(k_r, k_\chi, \phi)$ contains elements corresponding to the signal returned from the different directions corresponding to l (weighted by the average antenna pattern). This provides a mechanism to extract the unambiguous signal.

The minimum square error approach to computing $\mathbf{B}(k_r, k_\chi, \phi)$ seeks to minimize the following cost function:

$$J_0 = |\mathbf{B}(k_r, k_\chi, \phi)\mathbf{H}(k_r, k_\chi, \phi) - \mathbf{D}(k_r, k_\chi, \phi)|^2 \tag{5.66}$$

where $|\cdot|^2$ denotes the Frobenius norm.

5.4.2.2 Amplified additive noise

Before attempting to find minimum values for J_0, one should recognize that if we pre-multiply $\mathbf{z}(k_r, k_\chi)$ by $\mathbf{B}(k_r, k_\chi, \phi)$, then the additive noise term is also pre-multiplied by $\mathbf{B}(k_r, k_\chi, \phi)$ and this might adversely change the SNR. Thus, one should simultaneously try to minimize the following cost function:

$$J_1 = \mathcal{E}\{|\mathbf{B}(k_r, k_\chi, \phi)\boldsymbol{\nu}(k_r, k_\chi, \phi)|^2\} \tag{5.67}$$

5.4.2.3 Blended cost function

In the event that the solution to minimizing each cost function is different, a fair trade would see the construction of a tuneable hybrid cost function given by

$$J_2 = \varrho J_0 + (1 - \varrho)J_1 \tag{5.68}$$

where $\varrho \in (0, 1]$. The solution to this problem is computed in Appendix 5.6 yielding the MMSE filters given by

$$\begin{aligned}\mathbf{B}(k_r, k_\chi, \phi) &= \mathbf{D}(k_r, k_\chi, \phi)\mathbf{H}^\dagger(k_r, k_\chi, \phi)\\ &\quad\times\left[\mathbf{H}(k_r, k_\chi, \phi)\mathbf{H}^\dagger(k_r, k_\chi, \phi) + \frac{1-\varrho}{\varrho}\mathbf{R}_n(k_r, k_\chi, \phi)\right]^{-1}\end{aligned} \tag{5.69}$$

5.4.2.4 Projection filters

HRWS SAR was first made popular through work from the Deutsches zentrum für Luft- und Raumfahrt, Germany in a series of publications [1–3], which proposed, mainly, the use of projection filters. This section shows that these projection filters are a special case of the blended cost function, J_2. To relate the adopted notation to that of [3], we make the following simplifications and substitutions:

$$2\mathbf{p}_n(\xi) \leftrightarrow x_n\hat{\mathbf{x}}(\xi) + 0\hat{\mathbf{v}}_s(\xi) + 0\hat{\boldsymbol{\omega}}_s \tag{5.70}$$

$$\kappa k_\chi \leftrightarrow 2\pi f_\xi / v_s \tag{5.71}$$

Set $\varrho = 1$, and assume that there exists a *square* or *fat* matrix $\mathbf{H}^{-1}(k_r, k_\chi, \phi)$ such that $\mathbf{H}^{-1}(k_r, k_\chi, \phi)\mathbf{H}(k_r, k_\chi, \phi) = \mathbf{I}_{n\times n}$ where $\mathbf{H}(k_r, k_\chi, \phi)$ is an $m \times n$ matrix. The condition of square or fat for the left inverse translates into the condition that the number of antenna channels is greater than or equal to the number of aliased bands in the band-limited signal.

With these assumptions, and with the further simplification that all antenna patterns are identical, (5.69) evaluates to

$$\begin{aligned}\mathbf{B}(k_r, k_\chi, \phi) &= \mathbf{D}(k_r, k_\chi, \phi)\mathbf{H}^{-1}(k_r, k_\chi, \phi)\\ &= \mathbf{D}(k_r, k_\chi, \phi)[\mathbf{P}(k_r, k_\chi, \phi)\mathbf{D}(k_r, k_\chi, \phi)]^{-1}\\ &= \mathbf{P}^{-1}(k_r, k_\chi, \phi)\end{aligned} \tag{5.72}$$

where the element in the mth row and nth column of $\mathbf{P}(k_r, k_\chi, \phi)$ is given by

$$e^{ik_r \hat{\mathbf{r}}_n^\dagger(k_r,k_\chi,\phi)\mathbf{p}_m} = e^{-2\pi i \frac{(f_\xi + n f_{\xi_p})x_m}{2v_s}} \tag{5.73}$$

The above expression can be compared with (31) and the transpose of (39) in [3] to see the equivalence with the projection method. It should be noted that there is an additional phase term in (31) of [3] which is typically very small for baselines that are small compared to the range (a far-field approximation). Further, the model in [3] derives from a parabolic slow-time model while, here, the signal derives from a hyperbolic model. In very high-resolution systems, the requirement for applying the hyperbolic model becomes ever more important for accurate compression. Indeed, the requirement for application to very high-resolution systems lies behind the rather detailed presentation of the hyperbolic signal model in Section 5.3.

Also note that in this special case, the matrix $\mathbf{P}(k_r, k_\chi, \phi) = \mathbf{P}(f)$, i.e. it does not depend on the range, the range wavenumber or the incidence angle. As such, a single 1-D HRWS filter (across Doppler frequency) can be computed and applied across the entire 2-D range-frequency Doppler-frequency domain. This property does not depend on the signal being narrowband.

5.4.2.5 HRWS filters in the narrowband case

Recall from Section 5.3.7 that in the narrowband case,

$$\mathbf{a}(k_r, k_\chi, \phi) \to \mathbf{a}(k_{r0}, k_\chi, \phi) \begin{bmatrix} \mathrm{A}_1\left[k_{r0}, \hat{\mathbf{r}}(k_{r0}, k_\chi, \phi), \mathbf{p}_1\right] \\ \mathrm{A}_2\left[k_{r0}, \hat{\mathbf{r}}(k_{r0}, k_\chi, \phi), \mathbf{p}_2\right] \\ \cdots \\ \mathrm{A}_N\left[k_{r0}, \hat{\mathbf{r}}(k_{r0}, k_\chi, \phi), \mathbf{p}_N\right] \end{bmatrix} \tag{5.74}$$

This leads to

$$\mathbf{H}(k_r, k_\chi, \phi) \to \mathbf{H}(k_{r0}, k_\chi, \phi) = \begin{bmatrix} \cdots & \mathbf{a}(k_{r0}, k_\chi - k_{\chi_p}, \phi) & \mathbf{a}(k_{r0}, k_\chi, \phi) & \cdots \end{bmatrix} \tag{5.75}$$

and HRWS filters given by

$$\begin{aligned} \mathbf{B}(k_{r0}, k_\chi, \phi) &= \mathbf{D}(k_{r0}, k_\chi, \phi)\mathbf{H}^\dagger(k_{r0}, k_\chi, \phi) \\ &\quad \times \left[\mathbf{H}(k_{r0}, k_\chi, \phi)\mathbf{H}^\dagger(k_{r0}, k_\chi, \phi) + \frac{1-\varrho}{\varrho}\mathbf{R}_n(k_{r0}, k_\chi, \phi)\right]^{-1} \end{aligned} \tag{5.76}$$

These filters do not depend on the range wavenumber; thus, only a 1-D processing filter needs to be computed and applied to the data yielding, as promised, a simplification to HRWS signal processing.

5.4.3 Section summary

This section developed the HRWS signal processing methods to construct a scalar signal with reduced or eliminated azimuth ambiguities and acceptable SNR from a vector of aliased signals. The section derived a set of filters that depends on a variable parameter ϱ which controls the level of azimuth ambiguity (or residual aliasing) and the SNR. The choice of $\varrho = 1$ leads to the projection filters of [1–3]. In the general wideband case, the HRWS-processing filters are k_r and k_χ dependant, but for systems with no across-track baseline or narrowband systems, the processing filters depend only on k_χ.

5.5 Multiple channels from sub-arrays

As we have seen from the previous section, the qualities of a multi-channel measurement for high-resolution wide-swath SAR are determined by the types and positions of the multiple antennas. The geometrical structure of the multi-antenna system is captured in the covariance matrix

$$\mathbf{R}(k_r, k_\chi, \phi) = \mathbf{H}(k_r, k_\chi, \phi)\mathbf{H}^\dagger(k_r, k_\chi, \phi) + \frac{1-\varrho}{\varrho}\mathbf{R}_n(k_r, k_\chi, \phi) \tag{5.77}$$

which, in turn, depends on the relations between each pair of antennas.

The content of this section applies to the practical and already realised phased-array radar. With recent advances in T/R module technology, the phased-array SAR provides the simplest and most easily implementable multi-channel SAR.

Let us assume the far-field approximation and a set of coherent antennas (on the same clock) each with a one-way amplitude antenna pattern given by $\mathrm{E}_n\left[k_r, \hat{\mathbf{r}}(k_r, k_\chi, \phi)\right]$. Further, each of these antennas has an azimuth position given by $\mathbf{p}_n$. The nth far-field, two-way sub-array azimuth antenna pattern, which incorporates the displacement of the antenna and the antenna element pattern, is given by

$$\begin{aligned}\mathrm{A}_n(k_r, k_\chi, \phi) = &\sum_{m\in\mathcal{T}_n} \alpha_{nm}\mathrm{E}_m\left[k_r, \hat{\mathbf{r}}(k_r, k_\chi, \phi)\right]\exp(ik_r\hat{\mathbf{r}}^\dagger(k_r, k_\chi, \phi)\mathbf{p}_m/2)\\ &\times \sum_{m'\in\mathcal{R}_n} \beta_{nm'}\mathrm{E}_{m'}\left[k_r, \hat{\mathbf{r}}(k_r, k_\chi, \phi)\right]\exp(ik_r\hat{\mathbf{r}}^\dagger(k_r, k_\chi, \phi)\mathbf{p}_{m'}/2)/\\ &\sqrt{\sum_{m'\in\mathcal{R}_n}|\beta_{nm'}|^2}\end{aligned} \tag{5.78}$$

where the sets of discrete indexes $\mathcal{T}_n, \mathcal{R}_n$ define the sub-arrays used for transmit and receive, respectively, and $\alpha_{nm}, \beta_{nm'}$ record weightings that are applied before forming the sub-arrays. The weighting by the root sum of squares of the receive weighting coefficients serves to ensure that there is no artificial power introduced as per the radar equation, see, for instance [27].

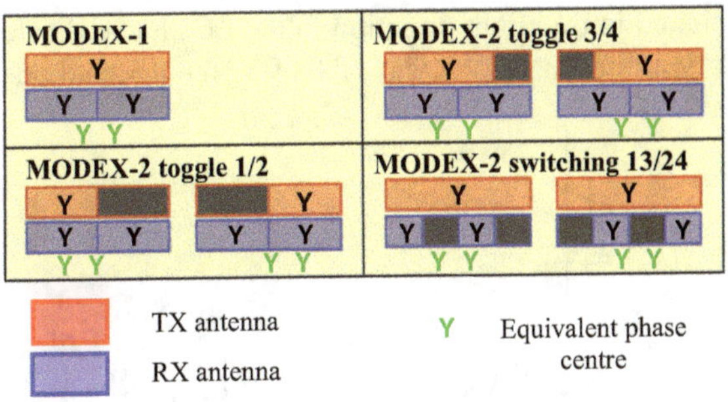

Figure 5.12 Different modes through different sub-array weightings on RADARSAT-2

5.5.1 Uniform phased-array antenna configuration

For a uniform phased-array, with $\mathrm{E}_n[k_r, \hat{\mathbf{r}}(k_r, k_\chi, \phi)] = E[k_r, \hat{\mathbf{r}}(k_r, k_\chi, \phi)]$, and $\mathbf{p}_n = nd\hat{\mathbf{p}}$, where d is the distance between elements, one finds that

$$\begin{aligned}\mathrm{A}_n(k_r, k_\chi, \phi) &= E^2[k_r, \hat{\mathbf{r}}(k_r, k_\chi, \phi)] \sum_{m\in\mathcal{T}_n} \alpha_{nm}\exp\left(ik_r m d\hat{\mathbf{r}}^\dagger(k_r, k_\chi, \phi)\hat{\mathbf{p}}/2\right)\\ &\quad\times \sum_{m'\in\mathcal{R}_n} \beta_{nm'}\exp\left(ik_r m' d\hat{\mathbf{r}}^\dagger(k_r, k_\chi, \phi)\hat{\mathbf{p}}/2\right)\Big/\sqrt{\sum_{m'\in\mathcal{R}_n} |\beta_{nm'}|^2}\\ &= E^2[k_r, \hat{\mathbf{r}}(k_r, k_\chi, \phi)] \frac{\mathbf{T}^\dagger{}_{x_n}\mathbf{e}(k_r, k_\chi)\mathbf{R}^\dagger{}_{x_n}\mathbf{e}(k_r, k_\chi)}{|\mathbf{R}_{x_n}|}\end{aligned} \tag{5.79}$$

where $\mathbf{T}_{x_n}$ and $\mathbf{R}_{x_n}$ are column vectors formed from the $\{\alpha_{nm}\}$ and $\{\beta_{nm}\}$ (augmented with zeros as required), respectively, and where

$$\mathbf{e}(k_r, k_\chi) = \begin{bmatrix} \exp\left(i0dk_r\hat{\mathbf{r}}^\dagger(k_r, k_\chi, \phi)\hat{\mathbf{p}}/2\right) \\ \exp\left(i1dk_r\hat{\mathbf{r}}^\dagger(k_r, k_\chi, \phi)\hat{\mathbf{p}}/2\right) \\ \vdots \\ \exp\left(i[N-1]dk_r\hat{\mathbf{r}}^\dagger(k_r, k_\chi, \phi)\hat{\mathbf{p}}/2\right) \end{bmatrix} \tag{5.80}$$

Figure 5.12 illustrates how different modes of RADARSAT-2 are generated through the use of different sub-array weightings.

5.5.2 Uniform phased-array for yaw-steered systems with no moving targets

This configuration applies to systems such as RADARSAT-2, ALOS PALSAR-2 or TerraSAR-X which all employ antenna steering across the orbit to ensure that the antenna is aligned in the flight direction in the ECEF coordinate system. When imaging stationary clutter, $\mathbf{p}_n(\xi) = nd\hat{\mathbf{x}}(\xi) + 0\hat{\mathbf{v}}_s(\xi) + 0\hat{\omega}_s$ and one finds

$$\mathbf{e}(k_r, k_\chi) = \mathbf{e}(k_\chi) = \begin{bmatrix} \exp\left(-i0d/2\kappa k_\chi\right) \\ \exp\left(-i1d/2\kappa k_\chi\right) \\ \vdots \\ \exp\left(-i[N-1]d/2\kappa k_\chi\right) \end{bmatrix} \tag{5.81}$$

A further simplification is provided when the k_r dependence of the element pattern, $E(k_r, \hat{\mathbf{r}}(k_r, k_\chi, \phi))$, can be ignored, i.e. that the antenna has uniform gain across the fast-time frequencies that it can measure. In this case, the antenna pattern becomes

$$\mathrm{A}_n(k_r, k_\chi, \phi) = E^2[k_\chi, \phi] \frac{\mathbf{T}^{\dagger}{}_{x_n}\mathbf{e}(k_\chi)\mathbf{R}^{\dagger}{}_{x_n}\mathbf{e}(k_\chi)}{|\mathbf{R}_{x_n}|} \tag{5.82}$$

This shows that, for yaw-steered systems with no moving targets, processing requires only a 1-D HRWS filter, even in the wideband case.

5.5.3 Simulation

Given a model of the element pattern for a phased-array, simulations for combinations of sub-arrays for HRWS can be generated. The performance of the HRWS modes in terms of the azimuth ambiguity levels, SNR, azimuth resolution and sidelobe levels can be simulated. By varying the ϱ parameter and the sub-array weightings, the most suitable processing filters for a given requirement can be determined. A particular example of a plot that can be used for such an analysis is presented in Figure 5.13. This plot illustrates the point spread function (PSF) that is expected from a dual-receive RADARSAT-2 mode. The peak azimuth ambiguity levels are below −22 dB, while the azimuth 3 dB resolution is less than 5 m. The plot shows that peak sidelobe levels are also very low at −27 dB.

5.5.4 Example of HRWS data and signal processing

This sub-section shows an example of HRWS imagery derived from RADARSAT-2 dual-receive mode data. In fact, the simulation of the previous sub-section relates to this mode, so we have a prediction performance for the azimuth ambiguity levels and azimuth resolution. The intention is to demonstrate the validity of HRWS processing as well as to illustrate some of the effects of signal measurement and processing.

Data were collected using a low PRF of approximately 860 Hz, well below the Nyquist rate for standard stripmap data. The space-time sampling diagram, illustrated in Figure 5.14, shows that the data are collected in a spatially highly non-uniform pattern. The data are processed using a HRWS filter (MMSE filter with $\varrho = 0.5$, see Section 5.4) to yield 5 m azimuth resolution imagery with a swath of 120 km. Although other studies have demonstrated HRWS processing with down-sampled satellite-derived data [6], the data of Figure 5.15 were actually measured with a reduced PRF.

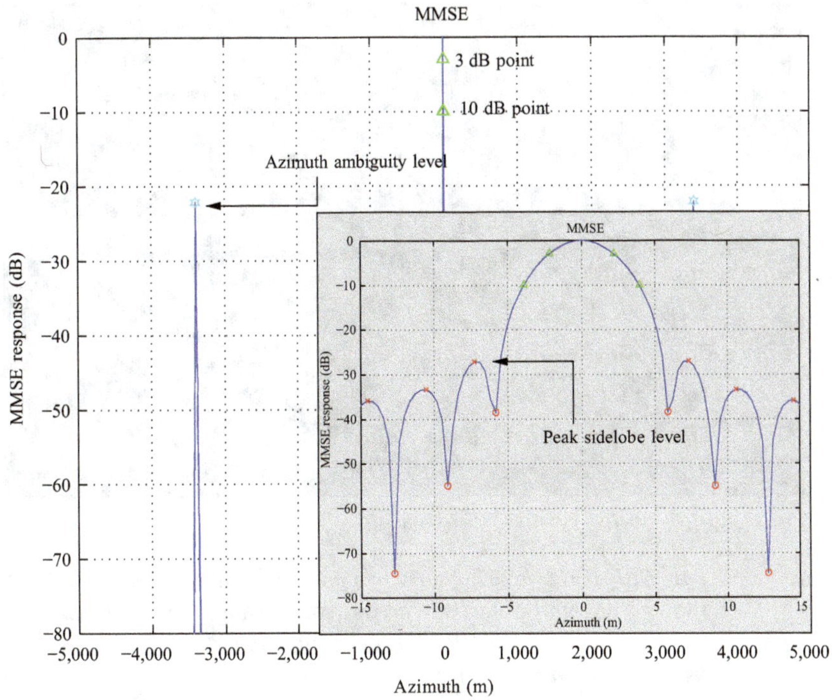

Figure 5.13 Point spread function for an example RADARSAT-2 mode

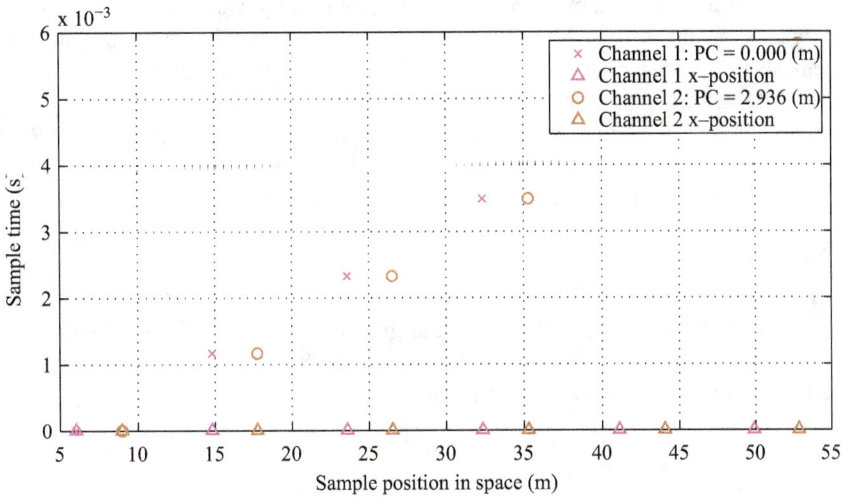

Figure 5.14 Space-time sampling diagram for first space-based low PRF mode on RADARSAT-2. PC stands for the two-way effective phase-centre

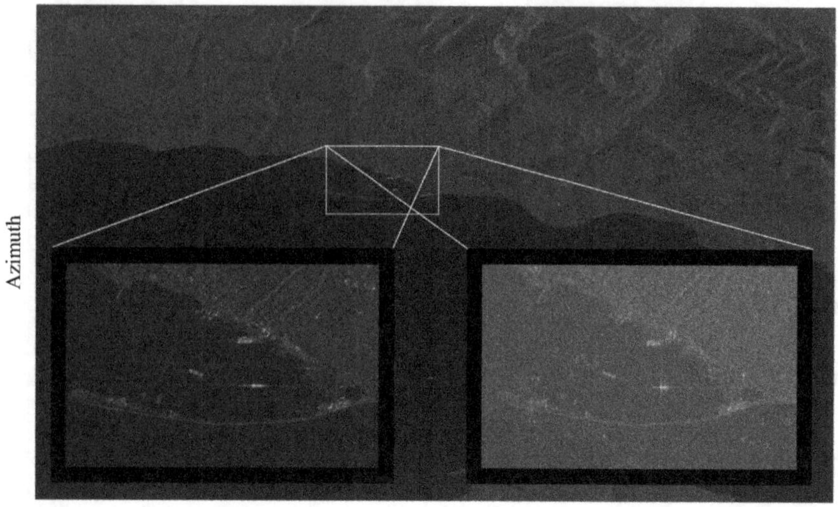

Figure 5.15 *Low PRF RADARSAT-2 data of Vancouver Island. The inset (left) shows a zoom of a harbour area created using a multi-channel measurement and an HRWS processing method, while the inset (right) shows a zoom of the same area created by a single-channel system (of equivalent size) using classical SAR processing*

Figure 5.15 also shows a zoomed-in area around a harbour. There are two insets representing this zoomed-in area. The inset on the left is derived from the HRWS processing algorithm, while the inset on the right is created by simulating the image that would have been collected by a single antenna system of equivalent size to the union of the two sub-antennas. The simulated image is created by simply combining the measurements from each antenna; i.e. the two signals recorded on the right side of Figure 5.1 are added together. It is clear that the HRWS processed image suffers less from the azimuth ambiguities coming from the town adjacent to the water. Further, the HRWS image seems to enjoy better radiometry. With only the addition of the switch, and an increased data recording demand, new imaging modes become feasible for the system.

5.5.5 Section summary

This section discussed HRWS imaging using phased-array approach for multi-channel SAR. It showed that along-track aligned phase-centres allow for simplified HRWS processing. It also demonstrated how HRWS performance could be simulated from various sub-array combinations.

5.6 Chapter summary

This chapter has discussed the principle of HRWS. As has already been demonstrated with orbiting satellite systems, the approach provides a means for imaging wide areas in high resolution at the expense of a multi-antenna system, increased data recording capacity and downlink bandwidth. The chapter argues that this concept should be generalized to higher resolution, wider-swath with application of the added degrees of freedom, offered by the multiple channels, to other modes such as spotlight and high-resolution stripmap.

The approach for reconstruction focused, mainly, on the MMSE method as it is quite general and includes the projection approach as a special case. As argued in [4], it also inherently improves the geometrical and radiometrical resolution simultaneously due to the proper weighting by the antenna pattern and a less aggressive ambiguity prescription as compared to the projection technique. Further, the MMSE method is robust against matrix inversion problems which can render the projection technique intractable.

Further discussion on the comparison between various signal-processing methods can be found in [19].

The derived signal, on which to apply the reconstruction prescriptions, was presented with significant detail, since future systems may operate over very wide bandwidths or beamwidths and may include instances where the data are collected with an across-track baseline. In this case, the analysis shows that reconstruction has to be computed at each two-dimensional frequency point. However, if the data are not collected with a squint, the analysis shows that the processing filters can be considered independent of the range variable.

5.A.1 Simplification of the signal model for narrowband systems

Expand (5.50) around $k_r = k_{r'} + k_{r_0}$ and take the inverse range FT to obtain narrowband result. In the narrowband approximation, we utilize the relation

$$\frac{k_\chi}{k_{r'} + k_{r_0}} = \frac{k_\chi}{k_{r_0}}\left(1 - \frac{k_{r'}}{k_{r_0}} + \ldots\right) \approx \frac{k_\chi}{k_{r_0}} \tag{5.83}$$

which gives

$$\begin{aligned} k_r\hat{\mathbf{r}}(k_r, k_\chi, \phi) &= \begin{bmatrix} \sqrt{\sin^2\phi[k_r^2 - k_\chi^2] + k_\chi^2(1-\kappa^2)} \\ -\kappa k_\chi \\ \cos\phi\sqrt{k_r^2 - k_\chi^2} \end{bmatrix} \\ \rightarrow k_{r_0}\hat{\mathbf{r}}(k_{r_0}, k_\chi, \phi) &= \begin{bmatrix} \sqrt{\sin^2\phi[k_{r_0}^2 - k_\chi^2] + k_\chi^2(1-\kappa^2)} \\ -\kappa k_\chi \\ \cos\phi\sqrt{k_{r_0}^2 - k_\chi^2} \end{bmatrix} \end{aligned} \tag{5.84}$$

In the hyperbolic expression involving the range, one must exercise more care as the range values can be quite large. The hyperbolic term in (5.50) as

$$\begin{aligned} r\sqrt{k_r^2 - k_\chi^2} &= \sqrt{(k_{r_0} + k_{r'})^2 - k_\chi^2} \\ &= rk_{r_0}\sqrt{\left(1 + \frac{k_{r'}}{k_{r_0}}\right)^2 - \frac{k_\chi^2}{k_{r_0}^2}} \\ &\approx rk_{r_0}\sqrt{1 - \frac{k_\chi^2}{k_{r_0}^2}} + \frac{rk_{r'}}{\sqrt{1 - \frac{k_\chi^2}{k_{r_0}^2}}} \\ &= r\sqrt{k_{r_0}^2 - k_\chi^2} + \frac{rk_{r_0}}{\sqrt{k_{r_0}^2 - k_\chi^2}} k_{r'} \end{aligned} \tag{5.85}$$

where we have used

$$f(u) = \sqrt{(1+u)^2 - v^2} \approx \sqrt{1 - v^2} + \frac{1}{\sqrt{1 - v^2}} u + \mathrm{O}(u^2) \tag{5.86}$$

One can now compute the inverse FT with respect to fast time to see

$$\begin{aligned} s\mathcal{S}_n(r', k_\chi) &= C_0 \frac{(k_{r_0}^2 - k_\chi^2)^{\frac{1}{4}}}{v_\mathfrak{s} k_{r_0}} \mathrm{A}_n\left[k_{r_0}, \hat{\mathbf{r}}(k_{r_0}, k_\chi, \phi), \mathbf{p}_n, \mathcal{A}_n\right] \int e^{-i\chi_t k_\chi} e^{-ir\sqrt{k_{r_0}^2 - k_\chi^2}} \\ &\times \frac{g(r, \chi_t, \phi)}{\sqrt{r}} \int \mathcal{P}(k_{r'}) \exp\left(-ir \frac{k_{r_0}}{\sqrt{k_{r_0}^2 - k_\chi^2}} k_{r'}\right) e^{ik_r r'} \mathrm{d}k_r \mathrm{d}\chi_t \mathrm{d}r \mathrm{d}\phi \\ &= C_0 \frac{(k_{r_0}^2 - k_\chi^2)^{\frac{1}{4}}}{v_\mathfrak{s} k_{r_0}} \mathrm{A}_n\left[k_{r_0}, \hat{\mathbf{r}}(k_{r_0}, k_\chi, \phi), \mathbf{p}_n, \mathcal{A}_n\right] \int e^{-i\chi_t k_\chi} e^{-ir\sqrt{k_{r_0}^2 - k_\chi^2}} \\ &\times \frac{g(r, \chi_t, \phi)}{\sqrt{r}} e^{ik_{r_0} r'} \int \mathcal{P}(k_{r'}) \exp\left(-ir \frac{k_{r_0}}{\sqrt{k_{r_0}^2 - k_\chi^2}} k_{r'}\right) e^{ik_{r'} r'} \mathrm{d}k_{r'} \mathrm{d}\chi_t \mathrm{d}r \mathrm{d}\phi \\ &= C_0 \frac{(k_{r_0}^2 - k_\chi^2)^{\frac{1}{4}}}{v_\mathfrak{s} k_{r_0}} \mathrm{A}_n\left[k_{r_0}, \hat{\mathbf{r}}(k_{r_0}, k_\chi, \phi), \mathbf{p}_n, \mathcal{A}_n\right] \int e^{-i\chi_t k_\chi} e^{-ir\sqrt{k_{r_0}^2 - k_\chi^2}} \\ &\times \frac{g(r, \chi_t, \phi)}{\sqrt{r}} e^{ik_{r_0} r'} \mathrm{p}\left[r' - \frac{rk_{r_0}}{\sqrt{k_{r_0}^2 - k_\chi^2}}\right] \mathrm{d}\chi_t \mathrm{d}r \mathrm{d}\phi \end{aligned} \tag{5.87}$$

where $\mathrm{p}(\cdot)$ represents the inverse FT of $\mathcal{P}(\cdot)$.

5.A.2 Minimization of the cost function

This section seeks to find the solution $\mathbf{B}(k_r, k_\chi, \phi)$ which minimizes the following cost function

$$J_2 = \varrho|\mathbf{B}(k_r, k_\chi, \phi)\mathbf{H}(k_r, k_\chi, \phi) - \mathbf{D}(k_r, k_\chi, \phi)|^2 + (1-\varrho)\mathcal{E}\{|\mathbf{B}(k_r, k_\chi, \phi)\nu(k_r, k_\chi, \phi)|^2\} \tag{5.88}$$

It is aesthetically pleasing and no loss in generality to suppress the arguments in the above and seek a solution that minimizes

$$J_2 = \varrho|\mathbf{BH} - \mathbf{D}|^2 + (1-\varrho)\mathcal{E}\{|\mathbf{B}\nu|^2\} \tag{5.89}$$

Upon expanding the square terms and applying the expectation operator, one finds

$$\begin{aligned} J_2 &= \mathrm{Tr}[\varrho\mathbf{BHH}^\dagger\mathbf{B}^\dagger - \varrho\mathbf{BHD}^\dagger - \varrho\mathbf{DH}^\dagger\mathbf{B}^\dagger + \varrho\mathbf{DD}^\dagger + (1-\varrho)\mathbf{BR}_n\mathbf{B}^\dagger] \\ &= \mathrm{Tr}[\mathbf{B}(\varrho\mathbf{HH}^\dagger + (1-\varrho)\mathbf{R}_n)\mathbf{B}^\dagger - \varrho\mathbf{BHD}^\dagger - \varrho\mathbf{DH}^\dagger\mathbf{B}^\dagger + \varrho\mathbf{DD}^\dagger] \\ &= \mathrm{Tr}[\mathbf{BRB}^\dagger - \varrho\mathbf{BHD}^\dagger - \varrho\mathbf{DH}^\dagger\mathbf{B}^\dagger + \varrho\mathbf{DD}^\dagger] \\ &= \mathrm{Tr}[(\mathbf{B} - \varrho\mathbf{DH}^\dagger\mathbf{R}^{-1})\mathbf{R}(\mathbf{B} - \varrho\mathbf{DH}^\dagger\mathbf{R}^{-1})^\dagger - \varrho^2\mathbf{DH}^\dagger\mathbf{R}^{-1}\mathbf{HD}^\dagger + \varrho\mathbf{DD}^\dagger] \end{aligned} \tag{5.90}$$

By the linearity of the trace operation, the solution is seen to be

$$\mathbf{B}(k_r, k_\chi, \phi) = \mathbf{D}(k_r, k_\chi, \phi)\mathbf{H}^\dagger(k_r, k_\chi, \phi) \times \left[\mathbf{H}(k_r, k_\chi, \phi)\mathbf{H}^\dagger(k_r, k_\chi, \phi) + \frac{1-\varrho}{\varrho}\mathbf{R}_n(k_r, k_\chi, \phi)\right]^{-1} \tag{5.91}$$

References

[1] Krieger, G., Gebert, N. & Moreira, A. Unambiguous SAR Signal Reconstruction from Nonuniform Displaced Phase Center Sampling. *IEEE Geoscience and Remote Sensing Letters*, **1**, 260–264. ISSN: 1545-598X (2004).

[2] Gebert, N. and Krieger, G. Azimuth Phase Center Adaptation on Transmit for High-Resolution Wide-Swath SAR Imaging. *IEEE Geoscience and Remote Sensing Letters*, **6**, 782–786. ISSN: 1545–598X (2009).

[3] Gebert, N. *Multi-Channel Azimuth Processing for High-Resolution Wide-Swath SAR Imaging*. PhD thesis (University of Karlsruhe, 2009).

[4] Sikaneta, I., Gierull, C. and Cerutti-Maori, D. Optimum Signal Processing for Multichannel SAR: With Application to High-Resolution Wide-Swath Imaging. *IEEE Transactions on Geoscience and Remote Sensing*, **52**, 6095–6109. ISSN: 0196-2892 (2014).

[5] Li, Z., Wang, H., Su, T. and Bao, Z. Generation of Wide-swath and High-resolution SAR Images from Multichannel Small Spaceborne SAR Systems.

IEEE Geoscience and Remote Sensing Letters, **2**, 82–86. ISSN: 1545-598X (2005).

[6] Kim, J.-H., Younis, M., Prats-Iraola, P., Gabele, M. and Krieger, G. First Spaceborne Demonstration of Digital Beamforming for Azimuth Ambiguity Suppression. *IEEE Transactions on Geoscience and Remote Sensing*, **PP**, 1–12. ISSN: 0196-2892 (2012).

[7] Krieger, G. MIMO-SAR: Opportunities and Pitfalls. *IEEE Transactions on Geoscience and Remote Sensing*, **52**, 2628–2645. ISSN: 0196-2892 (2014).

[8] Fiedler, H., Boerner, E., Mittermayer, J. and Krieger, G. Total Zero Doppler Steering-a New Method for Minimizing the Doppler Centroid. *IEEE Geoscience and Remote Sensing Letters*, **2**, 141–145. ISSN: 1545-598X (2005).

[9] Cumming, I. G. and Wong, F. H. *Digital Processing of Synthetic Aperture Radar Data: Algorithms and Implementation*. ISBN: 1-58053–058-3 (Artech House Remote Sensing Library, Norwood, MA, 2005).

[10] Soumekh, M. *Synthetic Aperture Radar Signal Processing With MATLAB Algorithms* (Wiley and Sons, New York, 1999).

[11] Gebert, N., Krieger, G. and Moreira, A. Digital Beamforming for HRWS-SAR Imaging: System Design, Performance and Optimization Strategies in Geoscience and Remote Sensing Symposium, 2006. *International Geoscience and Remote Sensing Symposium*, Toulouse, France, pp. 1836–1839, July 2006.

[12] Gebert, N., Krieger, G. and Moreira, A. Digital Beamforming on Receive: Techniques and Optimization Strategies for High-Resolution Wide-Swath SAR Imaging. *IEEE Transactions on Aerospace and Electronic Systems*, **45**, 564–592. ISSN: 0018-9251 (2009).

[13] Cerutti-Maori, D., Gierull, C. and Ender, J. Experimental Verification of SAR-GMTI Improvement Through Antenna Switching. *IEEE Transactions on Geoscience and Remote Sensing*, **48**, 2066–2075. ISSN: 0196-2892 (2010).

[14] Cerutti-Maori, D., Sikaneta, I. and Gierull, C. H. Optimum SAR/GMTI Processing and Its Application to the Radar Satellite RADARSAT-2 for Traffic Monitoring. *IEEE Transactions on Geoscience and Remote Sensing*, **PP**, 1–14. ISSN: 0196–2892 (2012).

[15] Fox, P. A., Luscombe, A. P. and Thompson, A. A. RADARSAT-2 SAR Modes Development and Utilization. *Canadian Journal of Remote Sensing*, **30**, 258–264 (2004).

[16] Kankaku, Y., Suzuki, S. and Osawa, Y. ALOS-2 Mission and Development Status in *2013 IEEE International Geoscience and Remote Sensing Symposium – IGARSS* (Melbourne, Australia, July 2013), 2396–2399.

[17] Thompson, A. A. & McLeod, I. H. The RADARSAT-2 SAR Processor. *Canadian Journal of Remote Sensing*, **30**, 336–344 (2004).

[18] Cerutti-Maori, D., Prünte, L., Sikaneta, I. and Ender, J. High-resolution Wide-swath SAR Processing with Compressed Sensing in *2014 IEEE Geo-science and Remote Sensing Symposium* (July 2014), 3830–3833. doi:10.1109/IGARSS.2014.6947319.

[19] Cerutti-Maori, D., Sikaneta, I., Klare, J. and Gierull, C. MIMO SAR Processing for Multichannel High-Resolution Wide-Swath Radars. *IEEE Transactions on Geoscience and Remote Sensing*, **52**, 5034–5055. ISSN: 0196−2892 (2014).

[20] Cumming, I., Neo, Y. and Wong, F. Interpretations of the Omega-K Algorithm and Comparisons with Other Algorithms *in Geoscience and Remote Sensing Symposium, 2003. IGARSS '03. Proceedings. 2003 IEEE International **3*** (Toulouse, France, July 2003), 1455–1458.

[21] Franceschetti, G. and Lanari, R. *Synthetic Aperture Radar Processing* (CRC Press, Washington, 1999).

[22] Raney, R. K. A New and Fundamental Fourier Transform Pair *in Geoscience and Remote Sensing Symposium, 1992. IGARSS '92. International **1*** (Houston, TX, USA, May 1992), 106–107. doi:10.1109/IGARSS.1992.576640.

[23] Bamler, R. A Comparison of Range-Doppler and Wavenumber Domain SAR Focusing Algorithms. *IEEE Transactions on Geoscience and Remote Sensing*, **30**, 706–713 (1992).

[24] Ender, J. H. G. Signal Theoretical Aspects of Bistatic SAR *in Geoscience and Remote Sensing Symposium, 2003. IGARSS '03. Proceedings. 2003 IEEE International **3*** (Toulouse, France, July 2003), 1438–1441. doi:10.1109/IGARSS.2003.1294137.

[25] Franceschetti, G. and Lanari, R. *Synthetic Aperture Radar Processing* (CRC Press, Washington, 1999).

[26] Spiegel, M. R. *Theory and Problems of Complex Variables* (Schaums, New York, 1964).

[27] Gierull, C. H. and Sikaneta, I. C. Potential Marine Moving Target Indication (MMTI) Performance of the RADARSAT Constellation Mission (RCM) *in Proceedings of the 2012 EUSAR Conference* (Nuremberg, Germany, Apr. 2012).

Chapter 6
SAR interferometry

Andrea Monti-Guarnieri, Fabio Rocca**
*and Stefano Tebaldini**

Abstract

SAR interferometry has been developed to estimate the precise elevation of surfaces that reflect radar signals, thus generating digital elevation models, which can be accurate to the metre. Further, it is possible to monitor slow motion of temporally stable targets with millimetre accuracy. Topics as interferometric data processing, performance evaluation, single- and multi-baseline acquisitions, mechanical stability of the targets (coherence) and differential InSAR stacks processing (persistent scatterers, SBAS) are introduced and discussed. The major applications for land, solid Earth and infrastructure monitoring are presented. If the radar faces targets like forests and glaciers, where electromagnetic wave penetration allows backscatter at different depths, it is then possible to separate the layers using SAR tomography, also discussed. Finally, an overview is made of the most challenging future interferometric systems, like passive bistatic companions or geosynchronous SAR.

6.1 Introduction

Interferometric Synthetic Aperture Radar (InSAR) is a SAR data processing technique aimed at estimating any variation of the phase component of two or more SAR images acquired along the same (nominal) orbit. The new generation of high-resolution SAR systems deployed since 2006 (RADARSAT-2, ALOS, TerraSAR-X, CSK and Sentinel-1A) are designed also to sustain InSAR applications (orbit stability, spatial resolution, definite look angles).

Synthetic Aperture Radar Interferometry (InSAR) from satellite platforms allows mapping topography to less than a few metres and surface deformation to the millimetre. It is widely used in seismology and vulcanology, for oil and gas reservoirs and gas storage monitoring and for infrastructure surveillance.

*DEI POLIMI, Italy

Reliable tools and methodologies to infer geophysical parameters from InSAR outcomes have been developed.

In this chapter, the technological tools to develop InSAR will be introduced and discussed, and the main applications will be summarized, starting from the creation of digital elevation models (DEMs). Then, the effects of the temporal stability of the targets will be considered, namely the interferometric coherence will be introduced. The impact will be then studied of the compensation of the flat Earth contribution to the interferometric phase, i.e. the induced spectral shift of the data due to the multiplication by a spatial sinusoid, that has a frequency dependent on the baseline. It is then recalled that the wrap-around of the spectrum due to this frequency shift should be avoided, with a proper prior expansion of the data resolution. The correlation between the two images, namely the coherence is then introduced and its diagnostic impact briefly discussed. Co-registration of the image couples be it based on the complex data or on the moduli is then discussed. Multiple aperture interferometry (MAI) is then introduced, and the Delta k technique, be it in range or azimuth is recalled. The temporal changes of the path length are then analysed, either due to terrain motion or to the tropospheric water vapour. The statistics of this all important phase contribution are discussed. Further phase noise sources due to volume penetration (to be recalled again in the sections devoted to tomography), as well as soil moisture changes are discussed. Finally, the methodologies to use interferometry for terrain motion analyses, i.e. the persistent scatterers (PSs) or small baseline subsets (SBAS) are introduced. The possibility of three-dimensional (3D) measurements is discussed. Finally, applications to studies of subsidence due to fluid injection or extraction are discussed. The effects on infrastructures are finally recalled. Then the effect of penetration in the medium is studied, and tomography is briefly recalled. The formation of the cross track antenna is discussed and the applications to real data (forests and ice) are discussed. A short discussion on the possible future of InSAR concludes the chapter.

6.2 InSAR generalities

The InSAR technique is based on the exploitation of the phase components of two SAR images of the same area [1–3]. Each focused SAR image (SAR focusing is not discussed in this chapter) is composed of a real and an imaginary part, or equivalently, each pixel in a SAR image has an amplitude component, related to the amount of the energy backscattered by each surface target, and a phase component. The latter is related to the travel path length and to the electromagnetic reflectivity: it is composed of a large number of integer periods and the measured fractional phase component. We describe the InSAR technique here for satellite SAR systems, but the same methods are also applied, with small adjustments, to InSAR data acquired by aircraft. The result of the application of InSAR is the so-called 'interferogram', that is the pixel-to-pixel difference of the phase components of two SAR images covering the same targets.

A satellite SAR can observe the same area from slightly different look angles. It can be done simultaneously (two radars, or one transmitter and two receivers, should be mounted on the same or close by platforms) or at different times by exploiting repeated orbits. The distance between the two satellites' antenna phase centres in the plane perpendicular to the orbit is called the 'interferometer baseline' (see Figure 6.1) and its projection perpendicular to the slant range is called the 'normal baseline'.

The normal baseline is one of the key parameters of SAR interferometry. The SAR interferogram is generated by cross-multiplying, pixel-by-pixel, the first SAR image by the complex conjugate of the second. Thus, the interferogram amplitude is the amplitude of the first image multiplied by that of the second one, whereas its phase (called interferometric phase) is the phase difference between the two images.

The speed of electromagnetic waves in a vacuum is constant: once the relative positions of the source and the target are known within a wavelength, then the travel path from the source to the target and back can be determined within a small fraction of the wavelength, using the phases of the return. The resolvable fraction depends on the emitted power and the radio frequency bandwidth. The frequency bandwidth determines the range resolution and the ability to separate neighbouring targets. In the case that only one stable target is present in a SAR resolution cell and its return is identifiable, its two-way travel path can be determined to a fraction of the wavelength, λ, with a dispersion much smaller than the range resolution. The variance of the one-way travel path σ_d^2 is related to the measured phase variance

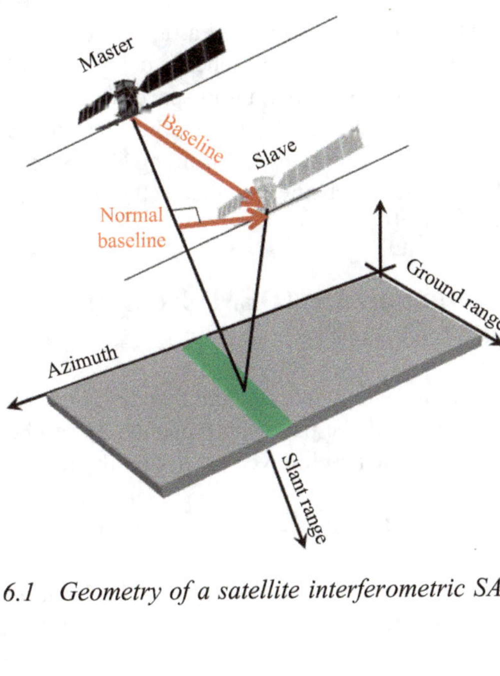

Figure 6.1 Geometry of a satellite interferometric SAR system

through half the wavelength:

$$\sigma_d^2 = \left(\frac{\lambda}{4\pi}\right)^2 \sigma_\varphi^2 \approx \left(\frac{\lambda}{4\pi}\right)^2 \frac{1}{2\text{SNR}_i} \approx \left(\frac{\lambda}{4\pi}\right)^2 \frac{1}{\text{SNR}} \tag{6.1}$$

SNR being the signal-to-noise ratio of the single image (roughly half that of the interferogram SNR_i). Thus, provided that the reflectivity of the target is significant, e.g. SNR = 10, the phase dispersion is less than the dispersion of the travel path (in wavelengths) due to the SNR, and it is a very small (~2.5%) fraction of the wavelength itself. In other words, if a target is visible against the background, its distance in phase cycles can be measured very well [3], better than using complex amplitudes. This is due to the bandpass character of the radar signal. More on this topic will be discussed in Section 6.7.

However, two problems remain: the absolute phase of the target is unknown, and while the medium between source and target is indeed mostly vacuum, it also includes the atmosphere and water vapour at lower elevations [4,5]. Therefore, we need more information to calculate the effective travel path. In order to get rid of these two effects we have to change the measurement from absolute phase to doubly differential, in time and space. If we consider two subsequent passes of the satellite, assume that the reflectivity of the target is stable, and that the target position stays the same in the two passes, then the difference of the two phase returns will not depend on the phase of the target. This interferometric measurement will measure only the changes of the target distance from one pass to the next. This would work, if we knew exactly both the medium delays (atmospheric phase term) and the position of the satellite with a precision smaller than λ; however, this turns out to be difficult, if not impossible [86]. Therefore, we measure the travel path changes with respect to that of a reference point in the image. The effects of the uncertainties of the atmospheric delay and of the position of the satellite, now referenced to a fixed point in space, will be reduced even if they still affect the final accuracy. In conclusion, we have to differentiate in time to remove the unknown phase of the target, assuming it is stable, and in space to mitigate the effects of inhomogeneities of the medium and the uncertainties in the orbit of the satellite [6–10].

6.3 Digital elevation models (DEM) from the interferometric phase

After some calculations, and taking into account that the distance from the targets is much larger than the baselines, the path length change Δr can be approximated as the scalar product between the baseline vector, $\vec{B}$, and the range versor, $\vec{u}_r$:

$$\Delta r = |\vec{R}_s| - |\vec{R}| \approx \vec{B} \cdot \vec{u}_r \approx 2\frac{B_n q_s}{R} \tag{6.2}$$

It depends on the range R and the vector components of the displacement of the source B_n and target q_s that are orthogonal to both the line of sight (LOS) and to the orbit.

The interferometric phase change is approximately:

$$\Delta\phi = \frac{2\pi\Delta r}{\lambda} = \frac{4\pi}{\lambda}\frac{B_n q_s}{R} \qquad (6.3)$$
$$R \gg B_n,\ q_s$$

The component due to q_s can be split into two contributions: one dependent on the height q over a horizontal plane and the second dependent on the slant range abscissa s on that horizontal plane. For small elevations, both depend on the angle of the incident radar beam with respect to the vertical, θ in Figure 6.2

$$q_s = \frac{q}{\sin\theta} - \frac{s}{\tan\theta} \qquad (6.4)$$

This second component has to be removed if we wish that only the topographic height change q is reflected in an interferometric phase change. Notice that this removal entails the multiplication of the range data times a sinusoid as the abscissa s is proportional to the range and thus the phase shift is proportional to the range. A spectral shift ensues that will be discussed later. Then, we name altitude of ambiguity q_a the altitude change correspondent to a 2π phase change: it is inversely proportional to the baseline:

$$q_a = \frac{\lambda R \sin\theta}{2B_n} \qquad (6.5)$$

Equations (6.3) and (6.5) can be combined to get a linear relationship between elevation and phases:

$$\Delta\phi = 2\pi\frac{q}{q_a} \qquad (6.6)$$

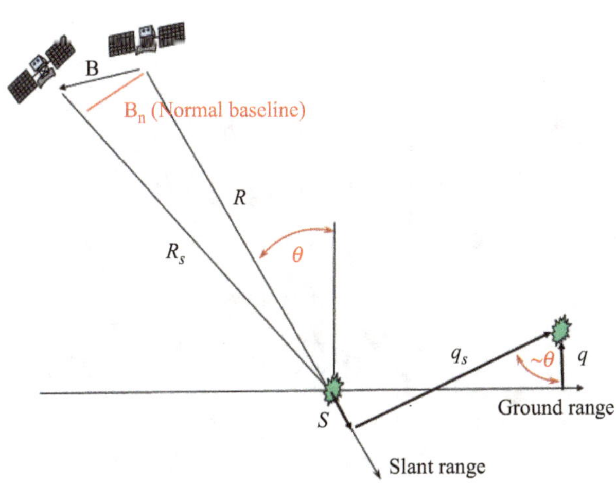

Figure 6.2 Geometric parameters of a satellite interferometric SAR system

So, we are able to interferometrically measure target heights locally (and then globally) over a ground plane (the DEM) provided that the targets are stable and that the atmospheric effects are small, which may not be true for repeat-pass interferometry but is always true for single-pass interferometry.

We notice that (6.6) is an approximation to be used for a rough performance evaluation. However, if two simultaneous SAR observations are given of the same target, the combination of the two target ranges (two spheres), and one target Doppler (namely, one of the two squint angles with respect to broadside, geometrically correspondent to a cone) can be elegantly inverted to provide analytically the exact location of the target in the 3D [11].

6.4 Phase unwrapping and DEM generation

The flattened interferogram provides a measurement of the relative terrain altitude that is ambiguous. Indeed, the phase variation between two points on the flattened interferogram provides a measurement of the actual altitude variation plus an integer number of altitudes of ambiguity (equivalent to an integer number of 2π phase cycles) as shown in Figure 6.3. The process of adding the correct number of altitudes of ambiguity to the interferometric fringes is called phase unwrapping. Notice that the phases can be at times extremely noisy, be it due to small amplitudes of the returns or to the mechanical instability of the scatterers. Thus, to extract topography from a single interferogram is unadvisable. If more than one interferogram is available, and even better if they have different geometries, say, ascending and descending orbits are used, then the combination can lead to the unique determination of the height. In the simplest case of the same geometry and

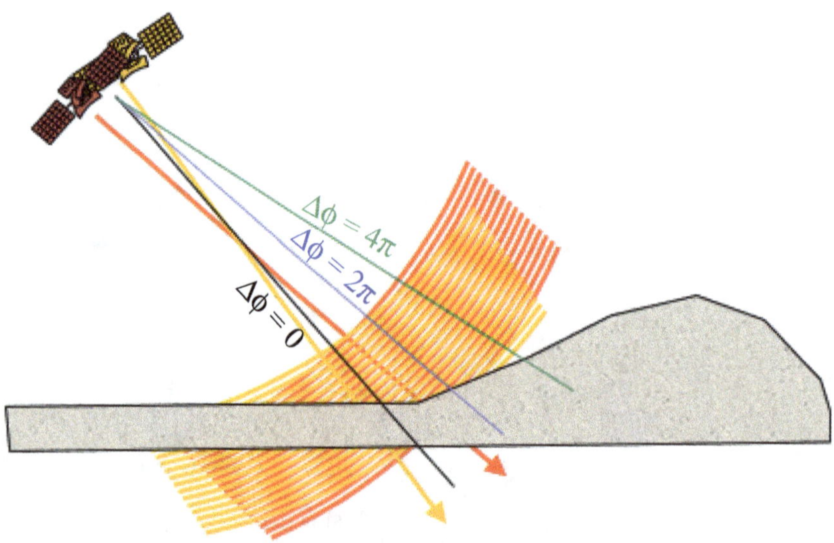

Figure 6.3 Differential phase representation in the zero Doppler plane

different baselines, the Chinese remainder theorem can be used to obtain a unique answer that, however, will only be as good as the used phases. In general, multiple geometries with a few single passes interferograms, or many multipass interferograms have to be combined to yield good topographies. This has been achieved on a global scale first with the NASA Shuttle Radar Topography Mission [12] flown in 2000 (last global data release in late 2015, 30 m ground resolution, absolute height error lower than about 8 m with 90% probability) and then with the Tandem X mission of the DLR [13] (flown from 2010 and completed in 2015) that with a 12 m resolution yields an absolute vertical precision better than 4 m [14].

However, in case of need, there are several techniques and codes for phase unwrapping [2,3,15,16]. Most of these minimize the norm of the error between the gradients of the reconstructed phase field and those measured by the wrapped phases, see Figure 6.4.

Single interferogram phase unwrapping does not have a unique solution and *a priori* information should be exploited to get the right solution. This is obvious as a tower as wide as a pixel and as tall as desired can always be hypothesized to exist. Then, in order to propose a reasonable solution one could use, say, the norm L_2 (least squares). One would then minimize the mean square difference between the rewrapped unwrapped phase (the phase gradients along azimuth and range due to topography, but made smaller than π, and larger than π) and the phase changes measured from the data [15]. The ensuing topography is easy to find but has little

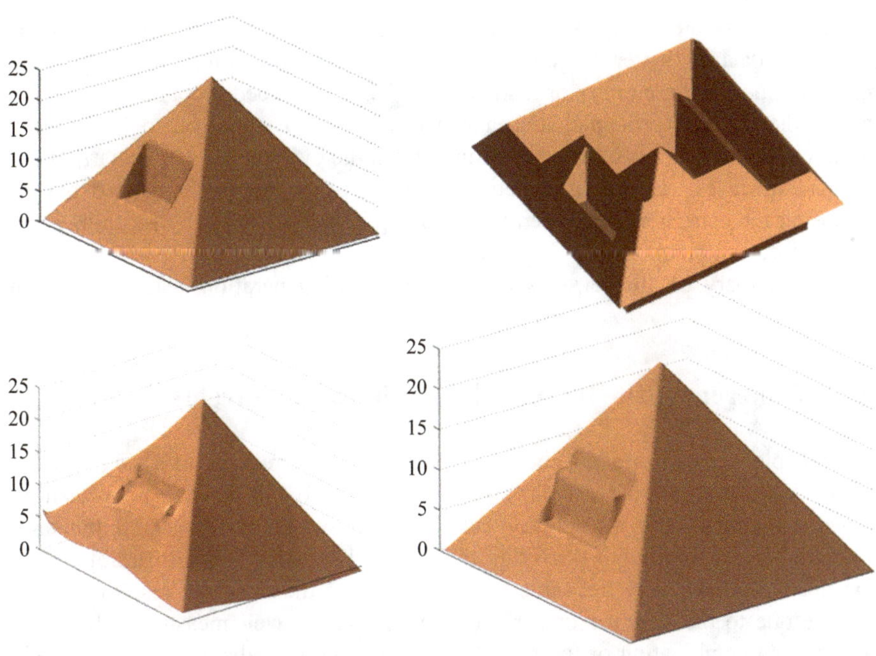

Figure 6.4 Top row: real and wrapped topography. Bottom row: L_2 (left) and L_0 norm (right) unwrapping

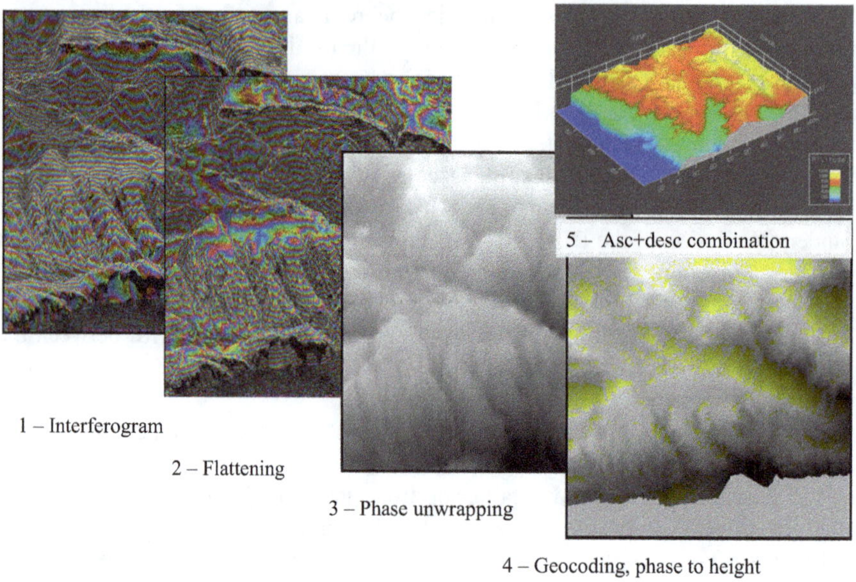

Figure 6.5 Main processing steps for DEM generation from single baseline pairs [16]

significance, as the wrapped topographical phases do not even match exactly the measured data. A better choice, even if more expensive computationally, is to minimize L_0 i.e. the number of instances where the wrapped topography does not match the data. In principle, the easy way to unwrap would be to find a path between any two points of the map along which the cycle count can be carried out without mistakes, at least hypothetically. In [17], unwrapping is carried out by imposing the consistence between the topography reconstructed by ascending and descending combinations.

A summary of all the steps involved in DEM generation is represented in Figure 6.5.

6.5 Coherence: the temporal stability of the targets

In principle, we can measure the interferometric phase relative to a single bright target, such as a man-made radar corner cube reflector. In general, many natural targets are contained in a resolution cell, and therefore the return is a combination of many small contributions, making the received signal a random variable. If all the imaged objects stay the same from one pass to the next, the resulting phase change due to the baseline or surface motions can be well measured. However, the random combination of the returns of the reflectors within the cell may yield a low total signal return, and might therefore be sensitive to even small amounts of additive random noise. Further, some of the targets in the cell may disappear

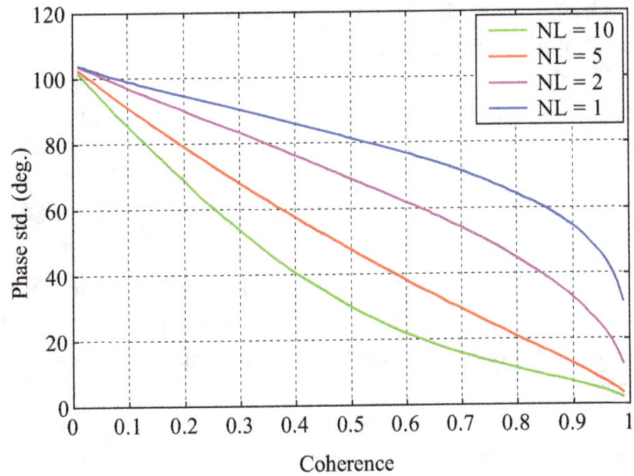

Figure 6.6 Interferometric phase dispersion (degrees) as a function of the modulus of the coherence for different number of looks (NL)

or change amplitude or phase. So, the correlation between the two images could be quite small or even zero in the case of water, or fluids, or tall vegetation. In general, this correlation coefficient, that is indicated as coherence, can be quite useful for diagnostic purposes, just to be able to appreciate the mechanics of the scattering mechanisms. Notice also that in order to estimate a correlation coefficient, we need more than one sample (with one sample the correlation is biased to 1). In substance, the estimate of the coherence is biased if the number of samples is too small. With N samples, the bias and the variance of the estimate of the coherence γ are approximately [2,3]:

$$\hat{\gamma} \approx \sqrt{\frac{\pi}{4N}}; \quad \sigma_{\hat{\gamma}}^2 \approx \frac{(1-\gamma^2)^2}{2N} \qquad N \gg 4 \tag{6.7}$$

In order to reduce the dispersion of the interferometric phase, one of two actions can be taken: either only stable points with high amplitudes are considered, or the complex returns of N (typically $N > 4$) neighbouring resolution cells have to be added together, before calculating their phase change, assuming it to be very similar. In the case that the returns of the neighbouring cells are independent, the number of combined returns is named number of looks. As the look returns are not quite the same in the two passes, the dispersion of the average phase is provided by [18,19] (see Figure 6.6):

$$\sigma_\varphi = \frac{1}{\sqrt{2N}} \frac{\sqrt{1-\gamma^2}}{\gamma} \tag{6.8}$$

where γ is the absolute value of the coherence, namely the amplitude of the correlation coefficient between the returns in the corresponding resolution cells in the two passes. Due to noise, be it of mechanical or electronic causes, the modulus of the coherence is never one and it decreases with time, as more and more changes of the targets take place.

The loss of coherence, or decorrelation can be directly related to SNR:

$$\gamma = \frac{1}{1 + \mathrm{SNR}^{-1}} \rightarrow 1 - \gamma = \frac{\mathrm{SNR}^{-1}}{1 + \mathrm{SNR}^{-1}} \approx \mathrm{SNR}^{-1} \tag{6.9}$$

If we combine (6.8) and (6.9) we get the phase standard deviation in (6.1): that is if the SNR is large and thus γ is close to one, and we have enough looks (homogeneous samples) to average, the same phase accuracy is obtained from point or distributed targets.

The coherence has an important diagnostic power [16,20–28]. Besides random noise, the changes with time of the scattering properties of the target are relevant to determine its coherence. Water bodies have low coherence because of their constantly moving surface and therefore appear black in the usual coherence images. Motion and change of vegetation also affect coherence. Leaf motion will usually cause a total loss of coherence but this does not imply that areas of vegetation will always appear with zero coherence. Radiation will often penetrate the foliage, at least partially, and can be backscattered by the terrain underneath or by the trunk and the branches of the trees, which are mechanically much more stable and will therefore contribute to the coherence. In general, trees, if deciduous, will show high coherence during the winter season when there are no leaves and less coherence in summer due to foliage effects [22]. Similarly, different types of vegetation will show different 1 day coherence values depending on the height of the plant and on the length of the leaf that, if too short, could be practically transparent to the C-band radiation of satellites [23].

Multitemporal interferometric analysis of the coherence and the amplitude of the backscatter can therefore contribute to the detection and classification of forests and of vegetation in general [24], see Figure 6.7. The joint use of coherence and the amplitude of the backscatter allows for a better image segmentation. While the amplitude of the returns depends on the electromagnetic structure of the target, the coherence is mostly related to its mechanical stability. For instance, in open vegetated fields, the level of the coherence is approximately linearly related to the biomass and to the height of the crops [23]. Other causes of coherence loss should be properly taken into account, for example, the wind speed could be considered using meteorological information [25]. Another application of coherence is forested – non-forested area segmentation, for example to find the extent of forest fires. In addition, areas of freeze and thaw in permafrost regions can be detected, and deciduous forests can be separated from the coniferous ones [28]. The penetration of the radiation in dry ice can be evaluated using the volumetric effect and the change of coherence with baseline [29–31]. In general, seasonal effects can be appreciated, using the regular series of satellite images available in selected

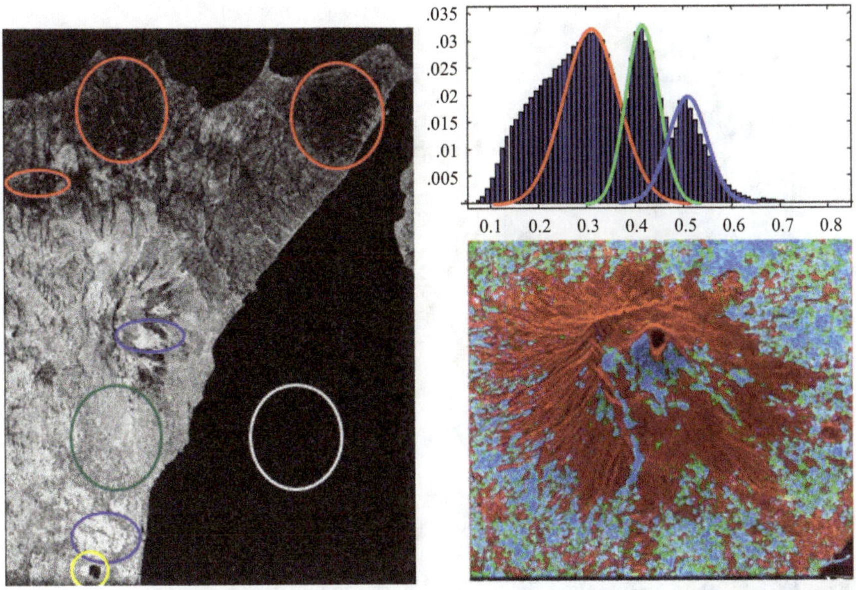

Figure 6.7 *C-band 1-day repeat pass SAR coherence (left), and 35 days coherence (bottom right) and histogram (top right). White: Zero coherence is achieved for water and fast moving targets. Red: very small coherence for densely vegetated areas. Green: medium coherence is achieved for agricultural fields (up to one month). Blue: exposed rocks, dry terrain and urban areas show high coherence (even after several years). Yellow: de-correlation due to man activities*

locations [32]. Thus, multi-temporal techniques make it possible to identify the seasonalities of the coherence connected to plant growth and to the visibility of the terrain in the background; they lead to segmentation techniques with results not so far from those obtainable with optical techniques, in good weather [33]. Examples of multi-temporal coherences are shown in Figure 6.8, C-band, and Figure 6.9, X-band. Notice the quite different time constant, ~40 days in C-band [34], and ~1 day in X-band [35].

It is important to remark that polarimetry adds a new dimension to coherence and interferometry and new satellite missions are being studied [36,37] that will make use of polarimetric interferometry to optimize the investigation of forests, identifying their height above ground. Lower frequencies like P- and L-band that penetrate forests will be used also for tomographic applications, namely to estimate the vertical structure of the penetrated medium.

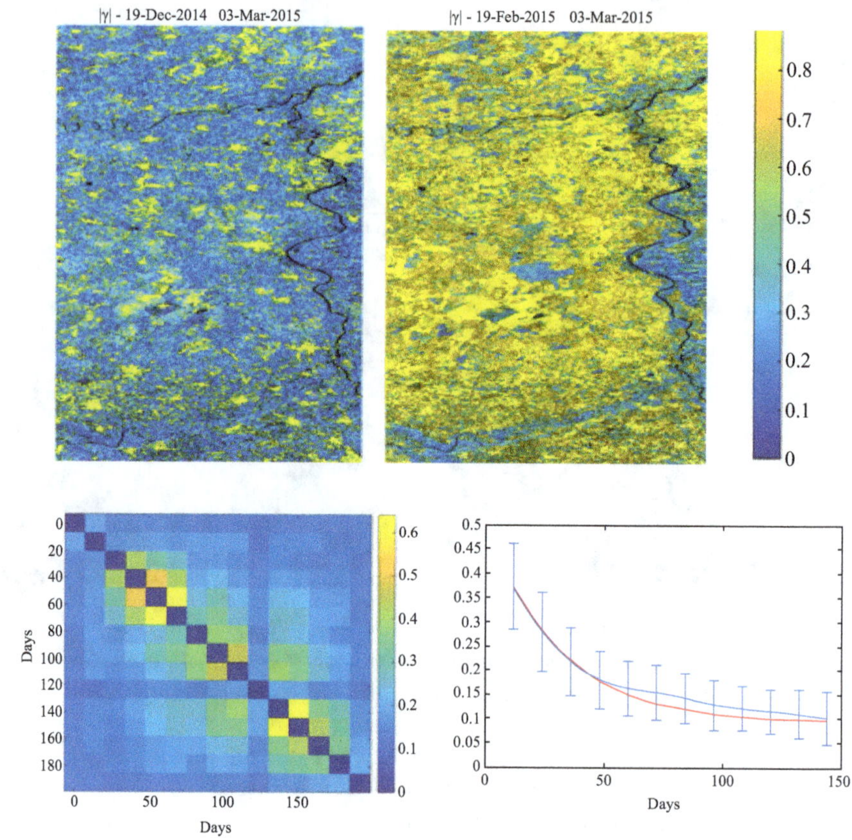

Figure 6.8 C band multi-temporal coherence, Sentinel-1, Val Padana land. Top: absolute values colour coded with coherence, 8-Dec-14; 3-Mar-15 (left) and 19-Feb-15; 3 Mar-15 (right). Bottom left: median of coherences measured in pairs from 10-Oct-2014 to 20-Apr-2015. Right temporal de-correlation (mean and standard deviation as error-bar) and exponential fitting: the decay time is 30 days

6.6 Baselines, coherence and wavenumber shift

The presence of the orbit-separation baseline induces an interferometric fringe gradient on a horizontal plane; this is the flat plane correction discussed in Section 6.3. If the frequency of this sinusoid reaches 2π across each ground resolution cell of length δ, the spectral shift equals the data spectrum and the correlation coefficient and therefore the coherence will be zero. The baseline that corresponds to the annihilation of the amplitude of the interferogram is named critical baseline, B_{cr}. We can compute its value by imposing $\Delta r = \lambda$ in (6.3) with $q = 0$ for

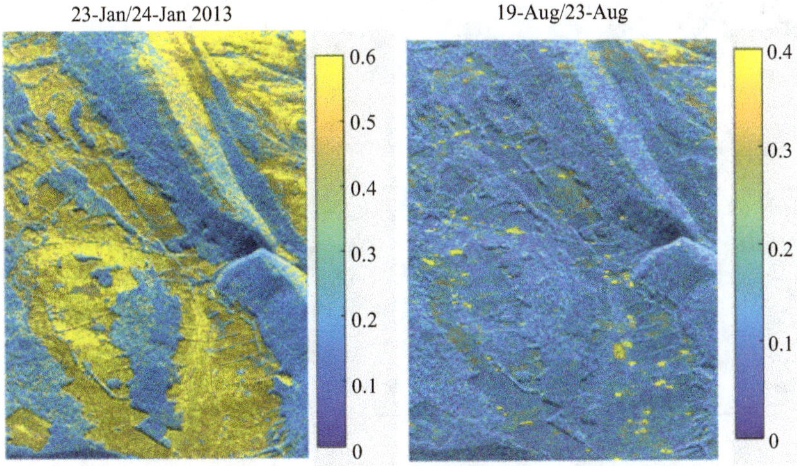

Figure 6.9 X band absolute values colour coded with coherence, in different seasons (CSK data courtesy of ASI) over a vegetated area in Neuchatel (CH). Elevation of ambiguity is 45 m (left) and 100 m right. Notice that vegetation de-correlates in just 1 day

a ground length δ:

$$B_{cr} = \frac{\lambda R}{2\delta \cos \theta} \tag{6.10}$$

With baselines shorter than B_{cr}, in order to carry out the subtraction of the phase due to the horizontal plane fringes, we will have to multiply one of the two images used to produce an interferogram with a complex sinusoid and, as stated, to shift its spectrum with respect to the original one. The complex sinusoid makes one cycle in a resolution cell if the baseline is critical. Hence again, this shift is equal to the radiofrequency bandwidth when the baseline is critical. Once again, we then see that if the spectra of the two images will become frequency disjoint after the shift, the image correlation (the coherence) will drop to zero. For baselines over critical values, only slopes nearly oriented along the LOS could be coherent. Remember that the resolution of the data should be expanded by interpolating on a finer mesh prior to multiply times the flat earth correction. Otherwise, the spectrum would wrap around.

For sub-critical baselines, if we limit the frequency spectra of the two images to the overlapping part, as in Figure 6.10, the coherence will be restored, even if the spatial resolution of the single images is reduced, see Figure 6.11 [38,42]. However, notice that the full resolution interferogram carries no further information: non-overlapping spectral components are indeed independent.

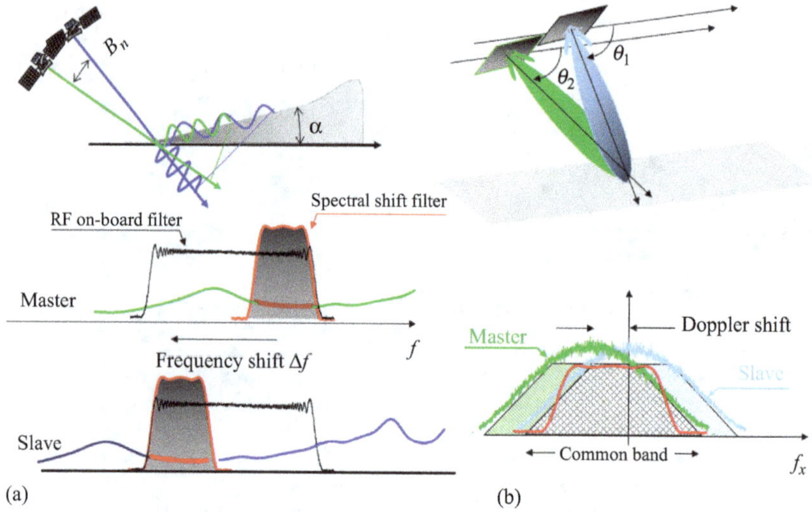

Figure 6.10 Range spectral shift (a) and azimuth common band (b) filtering. In range, the change in the view angle introduces a frequency shift in the reflectivity (a, top). In azimuth, the shift is due to a change in the squint angle (b, top). The filters required to remove the uncorrelated spectral contributions are shown in the middle and bottom plots

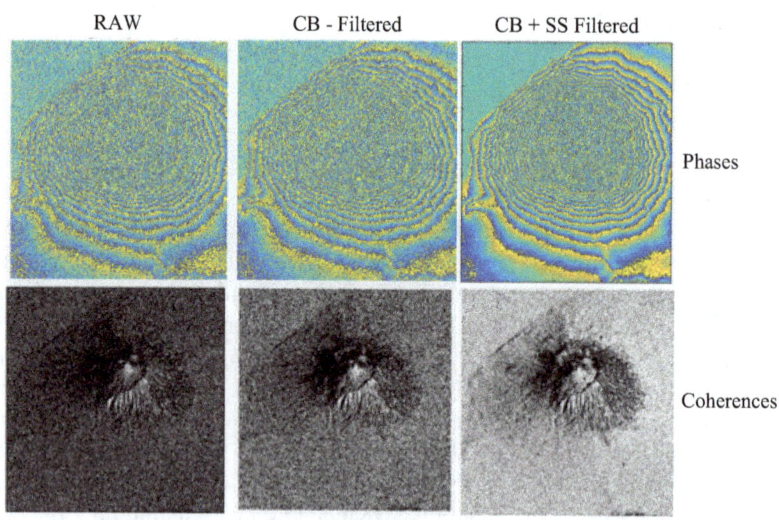

Figure 6.11 Effects of range and azimuth filtering on the final interferogram. A simulated interferogram of Mt. Vesuvius has been generated (top left: phases, bottom left: coherence) by assuming no filtering, with a baseline half of the critical. The decorrelation is entirely due to geometry and to the azimuth spectral shift, (a Doppler centroid difference of PRF/3 was assumed). The other two pictures represent the interferogram phases and coherence maps achieved by performing azimuth Common Band filtering [39], and azimuth CB filtering + range spectral shift filtering

6.7 Co-registration

The co-registration step is fundamental in the interferogram generation, as it ensures that each ground target contributes to the same range, azimuth cell in both the master and the slave image.

In an ideal case of perfect parallel orbits and aligned acquisitions, the co-registration would only need to compensate for the different geometry due to the different view angle. This would be compensated by a proper cross-track stretching of one image. In practice, the co-registration should account also for: orbits crossings/skewing, different sensor attitudes and topographic heights, different sampling rates (maybe due to different PRF, sensor velocities, etc.) and along track shifts. The required transformation is mainly composed of:

- a small rotation of the two images;
- a range and an azimuth 'stretch' (change in sampling rate);
- further second order effects.

By co-registering the **amplitude** images, it is possible to measure motion along the range and azimuth directions using also the pixel offsets. The dispersion of the estimate of the displacement can be obtained from the following formula [40–42]:

$$\sigma_d = \frac{\sqrt{3}}{\pi}\sqrt{\frac{1-\gamma^2}{2M\gamma}} \tag{6.11}$$

where σ_d is the dispersion of the estimate of the displacement, normalized to the sampling interval, γ is the coherence and M is the number of the independent pixels whose **moduli** are used for the estimate. Notice how this dispersion is larger with respect to that related to the phase and thus the wavelength, to be recalled in the next section, which is much smaller than the pixel size.

In order to better calculate the range shifts, the Delta k technique (in range) has been proposed [43–45]. Practically, the range bandwidth is split in three sub-bands. The lowest and highest frequency sub-bands of the two images are made to interfere (L = Low1 * Low2' and H = High1 * High2'). Then the phase difference between the two interferograms (phaseH–phaseL) is a function of the relative phase versus frequency ramp, i.e. the relative delay and thus the co-registration error. The results of this method are practically as good as the theoretically optimal ones that identify the delay from the maximum of the cross correlation between images 1 and 2. The advantage, with respect to the cross correlation maximum, is in the removal of phase shifts common to both sub-bands like topographic or atmospheric effects [44]. This advantage comes at the price of a slightly lower sensitivity and rapid decay if the coherence becomes too low. Then, it is better to enhance the smoothing of the two initial interferograms, but thus one loses part of the common phase independence advantage, as that common phase has to be constant in the entire window. Anyway, range delays are also measurable coherently with the phase shifts, as observed.

Not so in the case of azimuth shifts, where this methodology is useful and is named MAI. However, in order to make this methodology more effective, the equivalent azimuth resolution should be increased. This could be done by squinting (physically or in processing by sub-looks generation) the radar beam forward and aft and thus emulating a shorter transmitting antenna. So, MAI can be done if the same area is imaged twice with two different squints. This happens in the so-called TOPS acquisition mode used in Sentinel 1 [45–47], where the antenna is progressively squinted forward in bursts to accelerate acquisition and thus allow multiple swaths. The overlapping edges of the bursts or the overlapping edges of the swaths are seen twice, but with different squints, and there the equivalent azimuth resolution could reach 2 m with respect to the usual 20 m. Along these overlaps, a high quality of the azimuth co-registration is obtained, and thus a good recovery of the NS motion component, as will be seen in Section 6.12.

In the Figure 6.12 we show the dispersion of the estimate of the azimuth shift ($10\log_{10}\sigma_d$ [m] versus coherence) in the case of Sentinel 1 burst overlap (1 every 20 km, in IW mode) as a function of the total number of pixels in the window. The total number of pixels in a burst overlap is about 200,000 so that 1 mm dispersion should be obtained if the motion is constant there and the coherence is about 0.6. The sub-swath overlap is 2 km, namely 400 pixels, but in this case, the Doppler shift is approximately constant along azimuth and could also be small. On the average, it will be about one-half of that of the inter-burst and therefore the dispersion increases. However, in a length of 20 km the sub-swath overlap contains $4*10^5$ pixels. If this overlap is exploited, it should be possible to reach a dispersion lower than 2.8 mm, more than 50% of the time. So, the peculiarity of the TOPS mode yields an interesting technique to measure the N-S components of the motion, on a tectonical motion scale.

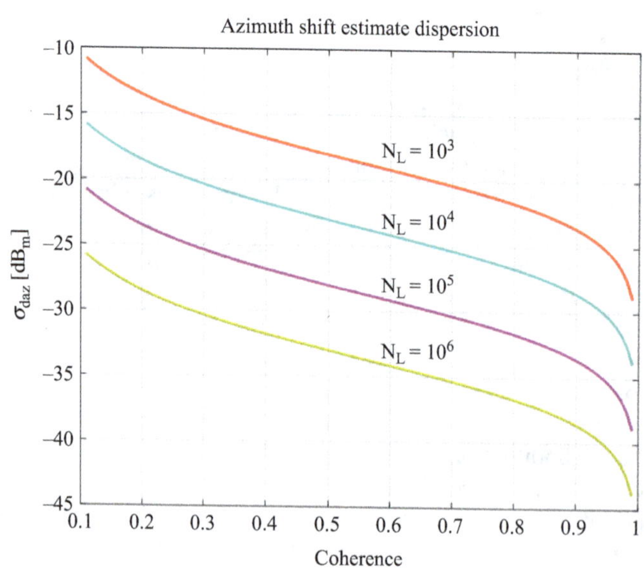

Figure 6.12 Standard deviation (dB m) of the azimuth (along track) shifts measured by MAI in the case of Sentinel-1 IW TOPS mode, as function of the coherence and for different number of looks

6.8 Terrain motion measurement through the interferometric phase

Let us now suppose that some of the point scatterers on the ground slightly change their relative position in the time interval between two SAR observations (as, e.g., in case of subsidence, landslides, earthquakes ...). In such cases the following additive phase term, independent of the baseline, appears in the interferometric phase.

$$\Delta\phi_d = \frac{4\pi}{\lambda} d \tag{6.12}$$

where d is the relative displacement of the target projected on the slant range direction. It is thus evident that after interferogram flattening, i.e. after the removal of the flat terrain phase contribution and the ensuing frequency shift, the interferometric phase contains both topographic and motion contributions:

$$\Delta\phi = \frac{2\pi}{q_a} q + \frac{4\pi}{\lambda} d \tag{6.13}$$

q_a being the elevation of ambiguity, (6.5), that ranges for tens to hundreds of metres in space-borne SARs [3]. The sensitivity of SAR interferometry to terrain motion is then much larger than that to the altitude difference. In C-band, 2.8 cm motion component in the slant range direction would generate a λ interferometric phase variation. As an example, the differential interferogram of the Bam earthquake that struck the Kerman province of south-eastern Iran on 26 Dec 2003, is shown in Figure 6.13.

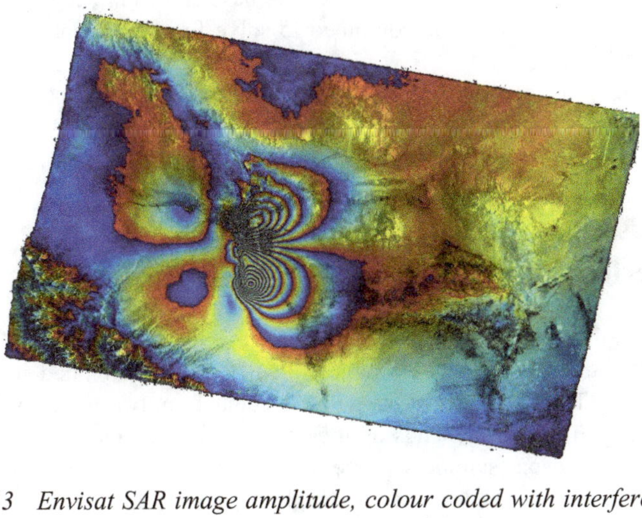

Figure 6.13 Envisat SAR image amplitude, colour coded with interferogram phase, of BAM area, achieved by combining two co-seismic images: 3 December 2003 and 11 February 2004, with a baseline close to zero

In the general case, (6.13) can be rephrased by splitting the differential interferometric phase ϕ_{int} into five terms [6,48]:

$$\Delta\phi = \varphi_f + \varphi_{\text{topo}} + \varphi_{\text{displ}} + \varphi_{\text{atm}} + \varphi_{\text{err}} \tag{6.14}$$

They are, respectively, the 'flat Earth', the topographic phase, the displacement phase, the atmospheric term and the error phase. The first term deals with the SAR acquisition geometry, see (6.4), and can be easily removed thanks to the very precise knowledge of the orbital position and trajectory. The topographic phase is related to the normal baseline, see (6.3–6.6), and can be removed, together with the first, by exploiting an precise DEM – or by planning nearly zero baselines (as in Ground Based Radar, or for the ESA satellite Sentinel 1).

The third term in (6.14) is due to displacement of the ground surface that causes changes in the sensor-to-ground distance, the second contribution in (6.13). This component is the target of the differential interferogram that provides a map of satellite-to-target distance change. Although DInSAR is the actual processing technique used for measuring displacements in most geophysical applications, the more general term InSAR is often used in the recent geophysical literature for interferograms that are corrected for topography. Another term that distinguishes the type of InSAR is 'repeat-pass interferometry' where the data are acquired on repeated passes of the radar, as opposed to 'single-pass interferometry' where two radar antennas are used to acquire SAR data from two locations in one pass for measuring topography (across track interferometry) or target motion (along track interferometry). The DInSAR technique measures the projection of the displacement vectors along the satellite LOS. This means that DInSAR measures the result of the projection of the North, East and Up components of three-dimensional surface displacements into the LOS. However, we have shown in Section 6.7 how two different LOS measures can be combined to solve for the 2D components of the displacements.

The same principle of MAI has been exploited in single pass airborne SAR with two antennas – properly displaced along track and two beams for each antenna to provide 2D estimation of ocean currents [49].

6.9 The atmospheric contribution to the interferometric phase

Signal delays through the water vapour in the atmosphere results in InSAR phase changes [the atmospheric phase screen (APS)] that are correlated in space and time [5,50–52]. The rms value of the delay is about 1 cm two-way, non-dispersive with frequency. The APS depends on the time of day [day more variable (worse) than night], the season (summers worse than winters), the latitude (tropics worse than polar regions) and the topographic height (valleys worse than mountain tops). It can be partially estimated using numerical weather predictions, satellite measurements and GNSS corrections [53–56]. Weather models predict well the vertical profile of water vapour when the stratification is stable. Satellite instruments,

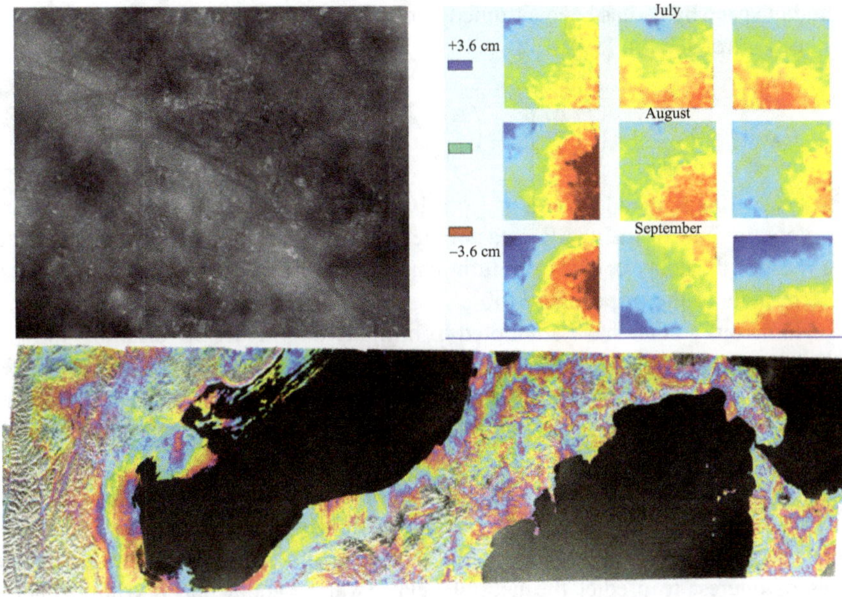

Figure 6.14 Top row: atmospheric phase screens computed from PSInSAR [48], left: superposed to the amplitude image of Paris; right: evaluated in a 30×30 km area near Bologna (It) in different months[1]. Bottom: Sentinel-1 12 days (1/3/2015, 13/3/2015) interferogram over Italy, DEM compensated[2]. The residual phases (superposed to amplitudes) are mostly due to differential water vapour

including MODIS and MERIS, can measure total column water vapour at say tens of metres spatial resolution during the day where there are no clouds, while GNSS is good for total column water vapour measurements with high temporal resolution at continuous GNSS stations, but with an about 7 km spatial resolution [56], as all the satellites in view are used to average the atmospheric contribution.

Examples of APSs derived from InSAR are shown in Figure 6.14.

The correlation function of the APS is approximately exponentially decreasing with space and time but vastly changing with weather conditions. Thus, the following quantities represent rough estimates. The correlation length is about 10 km [57]. Emardson *et al.* [58] used GNSS to show that (on average) the error from the troposphere (1σ standard deviation of path delay in mm) is: $\sigma = m \cdot L^{\alpha}$ [mm], where m, α are typically 2.5 and 0.5 and L is the separation of the observations in km. So 10 mm is a typical error at ~20 km, but at 100 km, a length scale we need for inter-seismic deformation, the error is more like 25 mm. The time-space variogram, that is the mean delay power of two measures displaced in time by

[1]CSK Data are courtesy of ASI.

[2]© [2015] Copernicus. Reprinted, with permission, from http://www.esa.int/spaceinimages/Images/2015/06/Italy_interferogram, id 341143.

Δ_t and in space by Δ_r has been assumed isotropic in the atmospheric turbulence and has been introduced in [59]:

$$\sigma^2_{\mathrm{APS}}(\Delta_t, \Delta_r) = \sigma_0^2 \sqrt{\left(\frac{\Delta_t}{t_0}\right)^2 + \left(\frac{\Delta_r}{r_0}\right)^2} \tag{6.15}$$

where $r_0 \sim 10$ km, $t_0 \sim 10$ h and σ_0 from 100 to 800 mm^2, fitting the statistics from the literature and from InSAR measurements, see Figure 6.15. It should be noted that most of the delay comes from turbulence within the boundary layer, the lowest 1–2 km parts of the troposphere [60].

Numerous studies have attempted to mitigate the impact of variations in neutral tropospheric delay on interferometric observations. As observed before, the wet delay can be estimated independently using data from space-borne radiometers such as MERIS or MODIS [61] and the full neutral delay from continuous GNSS measurements [56,60]. These are generally insufficient since radiometers only work in cloud-free, daylight conditions and the GNSS-network station density is too low in most parts of the world to form a sufficiently dense correction. Attempts have also been made to use high-resolution numerical weather models for particular areas of interest to predict the neutral delay (water vapour and dry atmospheric

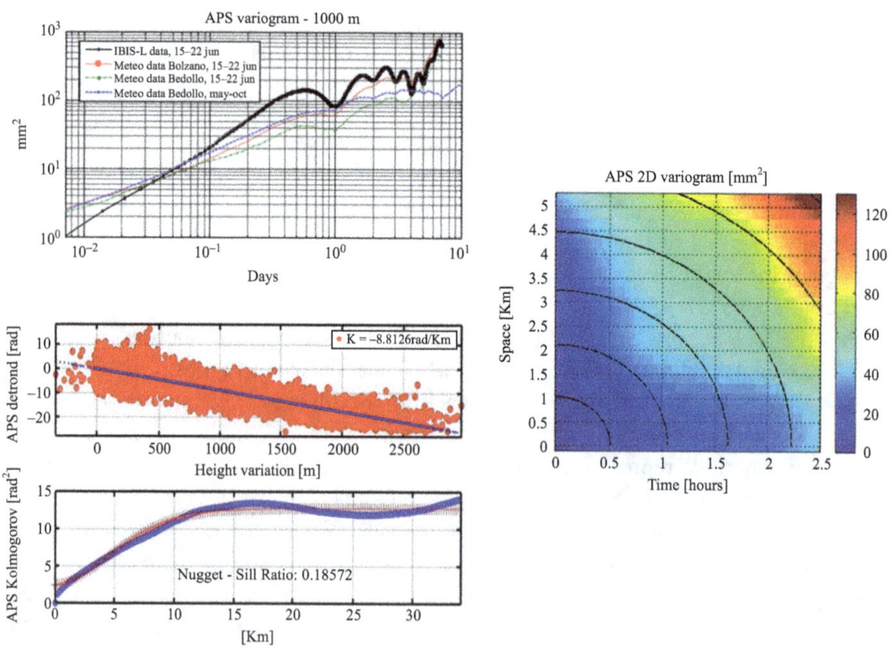

Figure 6.15 Left: time (top, Ku-band and model) and space statistics (bottom, C-band) of atmospheric delay. Right: example of a time-space variogram estimated in a Ku-band Ground Based Radar data campaign, compared with the model (6.15)

pressure) [58]. The results from these are promising for the longer wavelength components of the delay field – Foster *et al.* [62] suggest that up to 60% of the long-wavelength signal can be explained by weather models – but recovering short-wavelength features has proven difficult.

The alternative approach has been to use the InSAR data itself to estimate the APS, either through simple stacking [63], or time series methods such as permanent scatterers [6,48], and SBAS [8]. These methods rely on the assumption that the atmospheric delay is random in time, and then it becomes feasible to remove it from a sequence of InSAR acquisitions spanning many epochs. The error in linear rate is strongly dependent on the revisit time and length of observation. The unwanted random delay can be reduced with statistical averaging or stacking; about 20 acquisitions are needed to get to about 1 mm-level delay. Missions like Sentinel-1, with their short revisit times will reach this number of passes within a few months. Notice that, as current space-borne SAR missions are sampling the troposphere with intervals of days and with interferometric techniques, there is no way to retrieve the absolute multiple of $\lambda/2$ from time to time, then integration with GNSS, MERIS or other measures is needed for meteorological applications [61]. This could be solved in possible future geostationary SARs, that have continuous time observation capabilities [64].

6.10 Other phase noise sources

In PS Interferometry (PSInSAR) [6,48], very high quality performances are achieved by assuming that only one dominant stable scatterer was present in each resolution cell. This is seldom the case in reality, whereby most of the interferometric stacking approaches [8,65,66,72] bases on distributed targets, where many elementary scatterers are present in each resolution cell, which eventually may change in the time interval between two SAR acquisitions. The main effect of the presence of many scatterers per resolution cell and their change in time is the introduction of phase noise.

Three main contributions to the phase noise should be taken into consideration:

1. Phase noise due to temporal change of the scatterers, as discussed in Section 6.5.
2. Phase noise due to spatial changes of the viewpoint, that is, different look angle, as discussed in Section 6.6.
3. Phase noise due to volume scattering. We can roughly make two different cases:
   - 3.1 A finite number of targets is placed randomly in a volume – that is not they are not lying on any planar surface;
   - 3.2 Homogeneous scatterers are distributed in a volume, such as for snow or dry terrain.

In the first case, the rough surfaces, decorrelation can be evaluated, for a uniform distribution of scatterers within a vertical box of size Δq, [38,67] as:

$$1 - \gamma_R = 1 - \operatorname{sinc}\left(\frac{\Delta q}{q_a}\right) \approx \frac{1}{6}\pi^2\left(\frac{\Delta q}{q_a}\right)^2 \tag{6.16}$$

where the approximation holds for Δq much smaller than the elevation of ambiguity, q_a. Similarly, for a Gaussian distribution of the terrain heights over a plane with standard deviation σ_q, the decorrelation is:

$$1 - \gamma_z = 1 - \exp\left(-\frac{\sigma_q^2}{2q_a^2}\right) \approx 2\pi^2 \frac{\sigma_q^2}{q_a^2} \tag{6.17}$$

This can be a major limitation in providing DEM on rough surfaces, like dense and tall forests [68].

The case of a volume filled with *homogeneous scatterers* over a ground layer (Random Volume Over Ground [69]) has been approached in [70]. For infinite volume depth, the complex coherence describes a circle in the complex plane, see Figure 6.16:

$$\gamma_z = \frac{1}{1 + j\pi(d/q_{\text{na}})} \tag{6.18}$$

d being the one-way power-penetration depth and q_{na} the ambiguity elevation in (6.5):

$$q_{\text{na}} = q_a \sqrt{\frac{n^2 - \sin^2\theta}{n^2 \cos\theta}} \tag{6.19}$$

corrected by a term dependent on the refraction index ($q_{\text{na}} \sim q_a$ if $n \sim 1$). Notice that no decorrelation is found when the penetration is much smaller than the

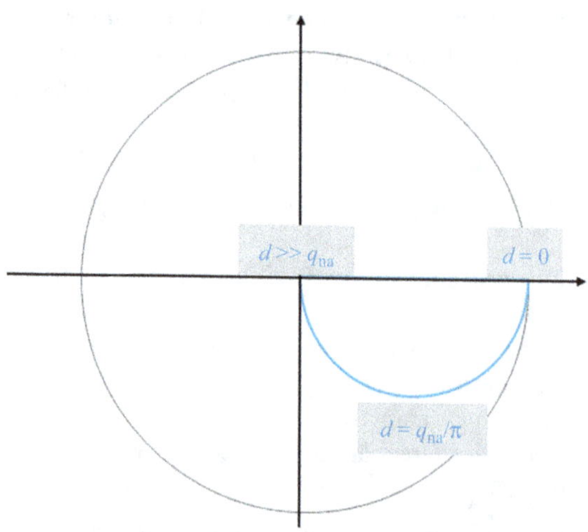

Figure 6.16 The coherence due to homogenous medium describes a blue circle in the complex plane (the unit circle is represented for reference) according to the ratio between the one-way power-penetration depth, d, and the elevation of ambiguity, q_{na} [70]

ambiguity elevation, $d \ll q_{na}$, whereas total decorrelation results when penetration exceeds $q_{na.}$ For small penetration, the decorrelation:

$$1 - \gamma_z = 1 - \frac{1}{\sqrt{1 + \pi^2 \left(d^2 / q_{na}^2\right)}} \approx \pi^2 \frac{d^2}{q_{na}^2} \tag{6.20}$$

compares with the volumetric cases in (6.16) and (6.17).

Phase shifts and amplitudes losses come together. This could be a very good insight for soil moisture extraction [71,74].

Finally, as we have pointed out in Section 6.6, coherence can be totally recovered, by spectral shift filtering [38], for targets on a flat surface and baseline less than critical. When the baseline exceeds the critical, coherence is always zero, and no interferometric information is obtained out of a distributed target. This does not apply for point targets that is PSInSAR.

6.11 Multipass methodologies: persistent scatterers and small baseline

In order to better measure slow ground motion and to be able to remove the atmospheric artefacts, numerous multi pass methodologies have been introduced, and first the permanent (or persistent) scatterer methodology PSInSAR [6,48,67] and the SBAS [8]. We need long time series of observations to be able to evaluate the stability of the targets. Long-term, high-coherence interferograms are rare, see Figures 6.7 and 6.8, as sooner or later many pixels will undergo some changes (due to vegetation growth, soil moisture changes or local ground motion) that lead to random changes in the returned phase. Nonetheless, there will usually be many individual pixels that remain unchanged even for decades. Typically, these correspond to poles, reinforced concrete, outcropping rocks, etc. However, in order to identify such stable points, we first have to remove the effects of the APS that randomizes all phases, and this is done using a non-linear bootstrap process. First, a reference point that is *a priori* deemed stable is determined. Then, its phase is forced to be constant, throughout all the images, just by adding a proper, image dependent, phase shift to compensate for the locally changing water vapour delay. If this assumption holds true, then the neighbouring stable points, due to the spatial correlation of the water vapour, will also appear as stable. Thus, a network of stable points (PSs) is progressively identified. This network allows estimating locally and then interpolating the low spatial frequencies of the water vapour disturbance. Once estimated, the APS can be removed too, and then many more PSs will appear that allow a long-term estimate of the motion, relatively to those additional pixels. This rather simple situation is made more complex by the limited stability of the satellites, in the past decades. The orbital tubes have been rather wide and therefore these significant baselines added another random phase shift due to the imperfectly known height of the target. Hence, not only the motion of the target with time but also its height has to be estimated. As the baselines are known, this adds another parameter to those to be estimated. To have reliable detection, and for a good estimate of motion and height for each pixel, the use of at least 15–20 images is indicated. Keeping the orbital tubes narrow, the impact of topography becomes lower, and the use of lower quality DEMs can be acceptable. On the other hand, using wider tubes, the height of the PS provides useful additional information. Further, the high short-term coherence may induce blunders. Persistence may be traded with short time coherence.

If narrow orbital tubes are maintained, the use of SBAS methods for long-term motion estimation becomes feasible. In this method, the effect of topographically induced bias is lowered, and only the atmospheric bias remains to be controlled. Further, if the interferograms can be chained in a continuous sequence, the water-vapour related phase that adds in one interferogram will subtract in the next, so that the sum of the phases of all the interferograms will only be affected by the atmospheric delay in the first and last acquisition. It will be even better if the weights of the interferograms to be combined are tapered down at the extremes, so that the impact of the first and last APS vanishes. Moreover, as the common bandwidth is always significant due to the small baselines, once the small residual topographic phase contributions have been removed, the interferograms may be smoothed and therefore the number of pixels sharing the same interferometric phase will grow. The dispersion of the phase decreases, and the interferograms, notwithstanding the presence of the spatially smooth atmospheric disturbance, can be unwrapped more easily. Hence, the absolute phase changes can be calculated. Finally, the interferograms can be stacked, the total motion calculated and the atmospheric disturbance averaged out from the easily unwrapped smoothed data.

If the baselines are widely dispersed, phase unwrapping and spatial smoothing becomes impossible as the fringes become very dense. In this case, the data have to be grouped into several sets, each corresponding to a small interval of baselines. The coherence within the set will increase, but the phase histories of the different sets have to be connected, as the epochs of the interferograms will be sparse.

The two methodologies, SBAS and PSInSAR, yield similar results when the baselines have small dispersion: the PSs methodology is computationally more expensive, but has the advantage of a denser set of stable points and a better estimate of the atmosphere. In the case that the PSs are very dense, as in desert or urban areas, the two methodologies yield very similar results [72].

In a new version of the PS algorithm, SqueeSAR [67,73], the temporal smoothing can be obtained by considering the covariance matrix of the pixels in the same spatial window along all the takes. If one strong scatterer is persistent, then the covariance matrix has a strongly dominant eigenvalue, and the phases of the correspondent eigenvector will carry the motion information. In the case that we have some temporal changes, we may approximate the covariance matrix neglecting all its eigenvalues, but the largest. The phases of the correspondent eigenvector can be assumed to be a good estimate of the phase history of the data in that window, as if the target imaged in the window corresponded to a single PS. The more the pixels in the window, the better will be the estimate, as long as the pixels share the same motion. Hence, adaptive windows, based on the commonality

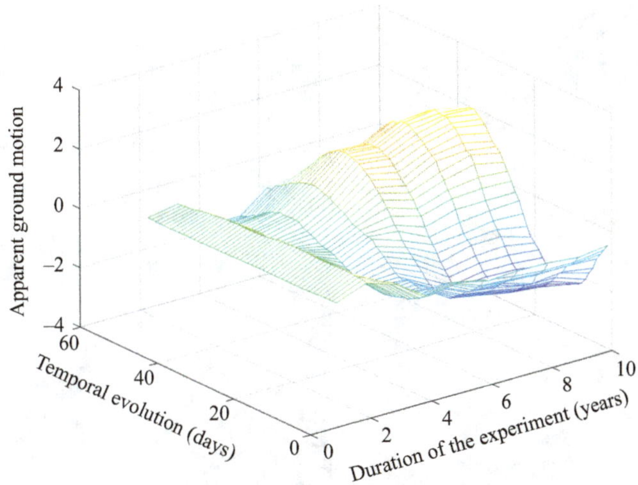

Figure 6.17 Phase behaviour of a cell containing a PS and Distributed Scatterers (DS) (2000 looks). The motion of the PS is sinusoidal, and the DS do not move. The decorrelation time for the DS is 40 days and the total duration of the experiment is on the x-axis. Notice that the PS prevails (even if its amplitude is 23 dB lower than that of the DS) if the total duration of the experiment exceeds 4–5 years

of the nature of the neighbouring pixels have to be used [67]. However, care has to be taken, as PS and Distributed Scatterers with different time behaviour may coexist in the same window. Then, the phases of the eigenvector will follow the PS or the DS depending obviously on their relative amplitude but also on the coherence decay of the distributed scatterers. In Figure 6.17, we display the average phase behaviour with time of the sum of a PS (a sinusoid with time, 50 images) and of distributed scatterers (2000 looks). The simulation has been carried out for C-band where the de-correlation time of DS is 40 days approximately. The total duration of the experiment (in years) is on the x-axis, while the temporal behaviour of the phase (z-axis) is shown along the y-axis. The figure shows the average of 100 runs. Notwithstanding that the DS to PS power ratio is 23 dB; still the PS prevails if the duration of the experiment is longer than 5–6 years.

6.12 3D displacement measurements

Using ascending and descending orbits, as well as different incident angles, it is possible in principle to retrieve the full 3D character of the ground motion, as the matrix yielding the 3D components from those recovered from the different LOS, has full rank. However, the sensitivity to the East West component is about 6 dB lower than the vertical component and the North South by up to -10 dB, at

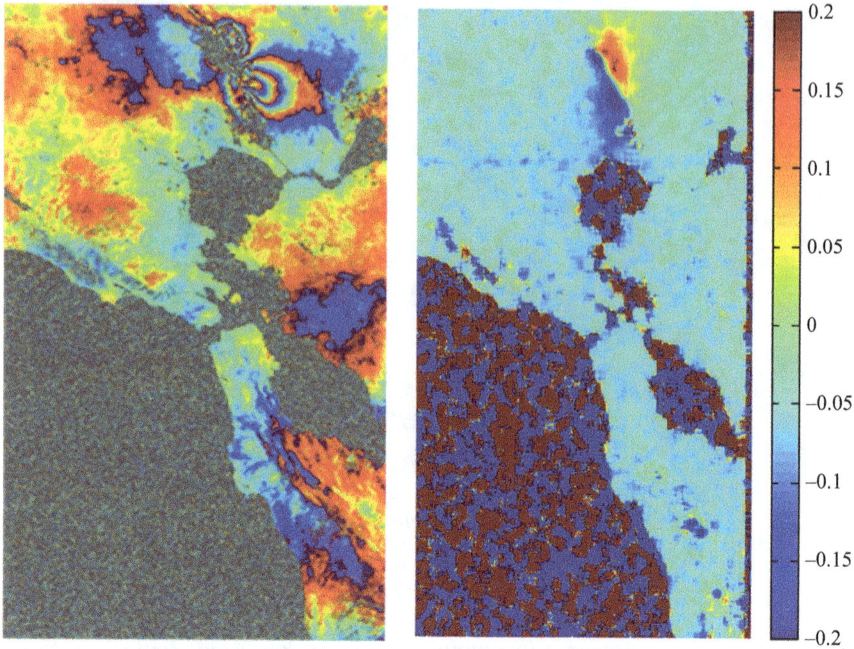

Figure 6.18 Estimation of the 2D (across-track and along-track) components of the Napa quake (24-Aug-2014) deformations by Multi Aperture Interferometry applied to the coseismic Sentinel-1 Stripmap Mode image pair: 07-Aug-2014 and 31-Aug-2014 [47]

mid-latitudes [75]. The N-S motion is retrieved using co-registration methodologies (see (6.11)). Using high resolution satellites like TerraSAR-X or CSK in the SPOT mode, with resolution of less than 1 m, it should be possible to reach centimetre precision with 1400 looks i.e. on areas of the order of 36 × 36 m. With a 200 × 200 m resolution, the dispersion would fall to 2 mm, quite interesting for all applications. The synergy between different types of platforms is therefore to be welcomed for 3D motion applications. Further, MAI can be used [76], and an example is given in Figure 6.18 referred to Napa quake observed by Sentinel-1. A very interesting result has been shown in the case of the Illipel Chilean earthquake, where use is made of the burst overlap double squint of Sentinel-1 [46].

6.13 Applications of differential InSAR: land subsidence and infrastructure monitoring

One of the most relevant application of multi pass techniques is the measurement of slow movements in urbanized areas where the coherent targets are more numerous and easily identified and studied [77–80]. The driving causes of subsidence

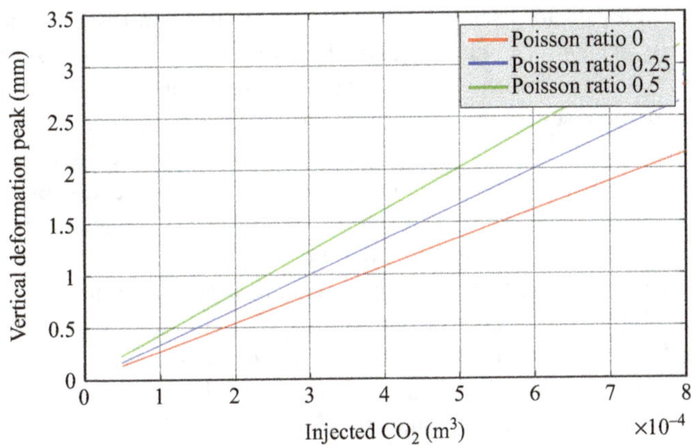

Figure 6.19 Vertical displacement as a function of the injected/extracted fluid volumes

measured in urban areas includes: (1) subsidence due to rise and fall of the water table or extraction/injection of water or other fluids; (2) subsidence from soil compaction in manmade fill; (3) subsidence measured over buildings and infrastructures as the consequence of the deformation of the substrate; and (4) subsidence due to subsurface mining or tunnelling have been compared with the vertical displacement rates obtained by levelling of benchmarks surrounded by coherent pixels. The standard deviation of the difference between SAR and levelling data projected in the radar LOS is about $\sigma = 2.0$ mm/year.

The soil compaction due to the expansion of the city over alluvial valleys is the main factor controlling the observed subsidence. In more detail, these driving factors are the in situ effective stress conditions, the related compressibility and viscous characteristics of the loaded soils, the thickness of the compressible stratum, the time since loading and the entity of loading.

In the case of significant fluid extraction from or injection into a reservoir, the surface deforms measurably over time. Thus, deformation studies prove useful in deducing spatial and temporal changes in the reservoir pressure. As many measurements can be made over time, the progression of the deformation can illustrate the progression of the fluid in the reservoir, and therefore allows inferring data on the permeability of the reservoir itself. The minimum variation in volume, at 2 km reservoir depth, able to generate a displacement of about 1 mm for different values of Poisson's ratio, is shown in Figure 6.19 where it is also apparent that the ground motion is not very sensitive to the rigidity of the medium [81,82]. Hence, diffusive travel times are not very sensitive to the exact geo-mechanical properties of the overburden. InSAR represents the most suitable and cost effective technical choice for this monitoring. By and large, the volume is conserved at least as an order of magnitude and the correspondent subsidence has a radius equal to the depth of the reservoir (Figure 6.19).

It is also relevant to discuss the needs for timeliness in reservoir deformation monitoring. In [81], in the case of extraction of oil, for Steam Assisted Gravity Drainage, it is shown that the fluid extraction causes micro-seisms. They can be monitored and their sources located in the active areas of steam injection in the reservoir. The uplift of the terrain follows the micro-seisms after about 5 to 6 days. A shorter revisit time and fast delivery of the acquired SAR images can be fundamental to ensure a safe monitoring activity dedicated to mitigate the risk of fault reactivation due to fluid injection or extraction.

InSAR data were used as tool for CO_2 sequestration monitoring for the first time in the In Salah Gas Storage Program [82–84]. This represents a CO_2 sequestration effort associated with an active gas reservoir and roughly, 1 million metric tons of CO_2 per year are re-injected in the flanks of the gas reservoir. The field is part of a large structural anticline, intersected by several faults. The reservoir itself is about 20 m thick, and horizontal wells extract the natural gas. Satellite data were gathered to monitor the CO_2 injection. 26 and 19 images were exploited using data from two Envisat tracks. After 842 days, a peak subsidence of about 5 mm was observed, displaced to the east of the three horizontal wells. The high quality of the measurements revealed reactivation of a fault with offsets of a few millimetres. The fault triggering may indicate the cracking of the cap rock, and it could be followed by a possible CO_2 leakage, if the crack gets to the surface. Surface deformation is also useful to predict and model the behaviour of a producing reservoir. In the same location, subsidence related to the gas production was observed, before the injection of CO_2.

6.13.1 Infrastructure monitoring

InSAR can offer a useful contribution to the analysis and the study of the area of interest during the feasibility and preliminary project phase of any infrastructure development [85,92]. Measuring surface movements over time allows the reconstruction of the local surface dynamics to be used for the planning of linear infrastructures like railways or highways.

The Cassia-Monte Mario tunnel, North of Rome, was monitored in its construction phase. The tunnel construction was suspended in 1984. In upgrading works in November 2004, a sudden mudflow created fractures and displacements up to the surface, with a subsidence of tens of centimetres. InSAR was used to assess the extent of the area affected by the ground deformation, approximately 1 km^2. Ascending and descending geometries were considered and a total of 306 satellite radar images were used from ERS-1, ERS-2 and RADARSAT-1.The area was found to be relatively stable between 1992 and 2000, with displacement rates of about 2–3 mm/year, see Figure 6.20. Displacement rates increased between 2003 and 2008, with velocities ranging between 15 and 30 mm/year. A full stabilization was reached in 2006. Displacements occurred mostly along the vertical, and horizontal motions pointed towards the tunnel.

A second example is given in Figure 6.21, where the Alps base tunnel velocity maps in the years 2008–2011 are shown.

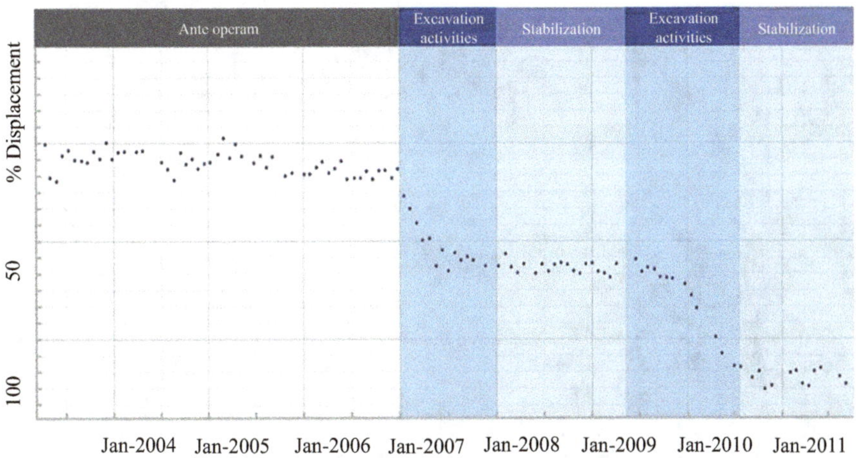

Figure 6.20 Ground motion for the Cassia Monte Mario tunnel (image courtesy of Alessandro Ferretti, TRE)

6.14 SAR tomography of penetrable media

Synthetic aperture radar (SAR) tomography (TomoSAR) is an emerging technology to provide access to the 3D structure of illuminated media by jointly focusing multiple SAR acquisitions. The use of SAR data to derive 3D information was first suggested in the 1990s in [95,96], whereas the first large scale demonstration using airborne data was given in 2000 by Reigber and Moreira [97]. Since then, TomoSAR has received increasing attention by different research groups worldwide. Possible applications include 3D urban reconstruction and monitoring [98,99], forestry [100–102], 3D analysis of snow-pack [103], ice sheets [104,110] and glaciers [106].

In the urban situation, as hard targets do not allow penetration, TomoSAR allows the separation of targets that are superposed within the same radar cell, in a laid over situation. This situation, typical of urban canyons and skyscrapers, is analysed in depth in Chapter 8.

In this section, we discuss the alternate case of medium penetration, achievable say in forests, vegetated areas and glaciers too, if lower frequencies like L-band (1.25 GHz) or P-band (435 MHz) are used. The discussion is supported by results from studies carried out in the frame of airborne and ground-based campaigns by the European Space Agency (ESA).

6.14.1 TomoSAR imaging principles

The basic principle of TomoSAR is relatively simple and well known. By flying the Radar along multiple trajectories nearly parallel to each other, it is possible to create a 2-D synthetic aperture, resulting in the possibility to focus the signal not

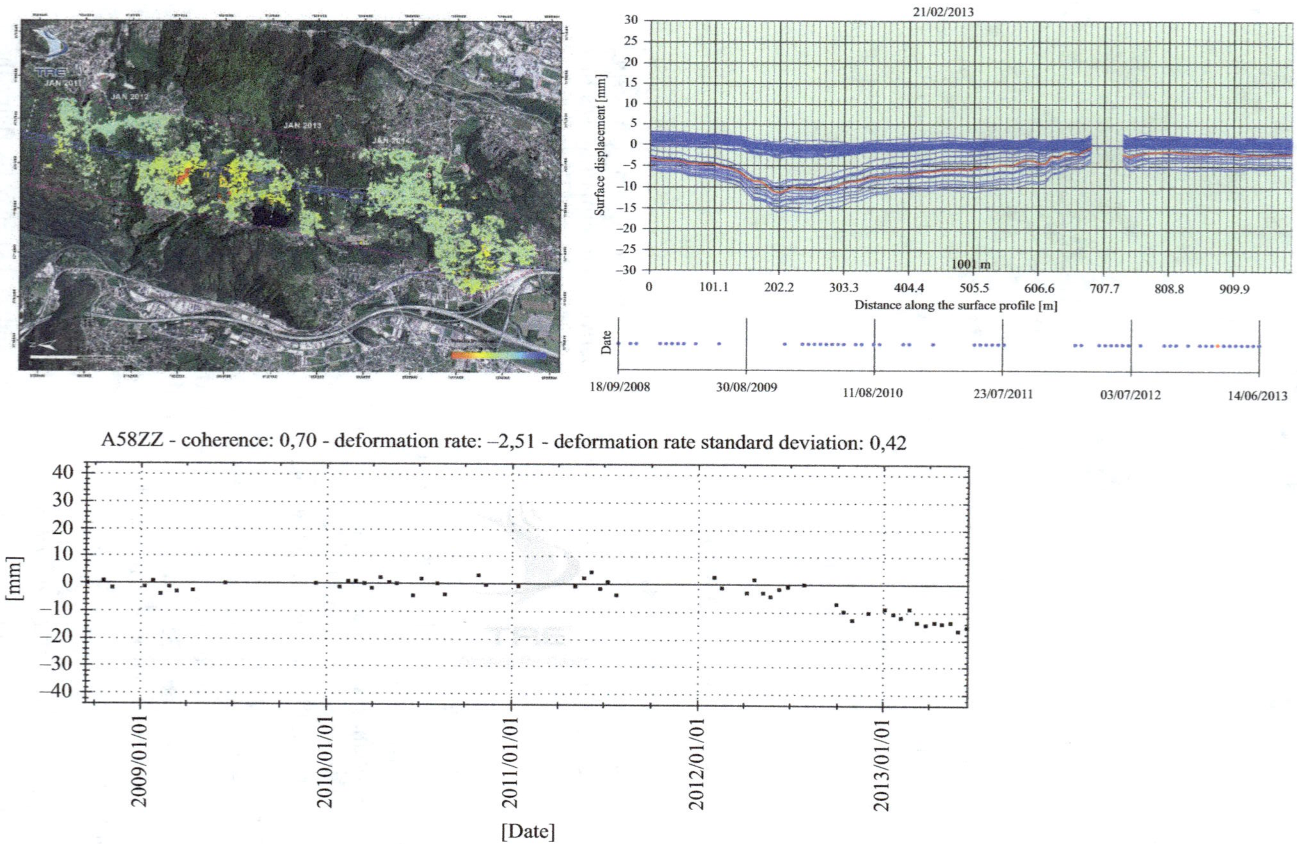

Figure 6.21 Alps base tunnel velocity maps (image courtesy of Alessandro Ferretti, TRE)

only in the range, azimuth plane, but also in the whole 3-D space [97,98]. Assuming that each image within the data stack has been resampled on a common master grid, and that phase terms due to platform motion and terrain topography have been compensated for, the following model holds:

$$d_n(r,x) = \int P(r,x,e)\exp\left(-j\frac{4\pi}{\lambda}b_n e\right)de \tag{6.21}$$

where $d_n(r, x)$ is the pixel value at the range and azimuth coordinates (r, x) in the nth single look complex image of the data stack; b_n is the normal baseline relatively to a common master image; λ is the carrier wavelength; e is the elevation coordinate; and $P(r,x,e)$ is the average scene complex reflectivity within the slant range, azimuth resolution cell [105] as depicted in Figure 6.22. It follows after (6.21) that the SAR multi-baseline data and the scene complex reflectivity within each resolution cell constitute a Fourier pair. Hence, the latter can be retrieved by (inverse-) Fourier transforming the data with respect to the normal baseline:

$$\hat{P}(r,x,e) = \sum d_n(r,x)\exp\left(j\frac{4\pi}{\lambda}b_n e\right) \tag{6.22}$$

A sketch of this concept is shown in Figure 6.22. The SAR resolution cell is split along elevation into several tomographic resolution cells. The geometrical resolution

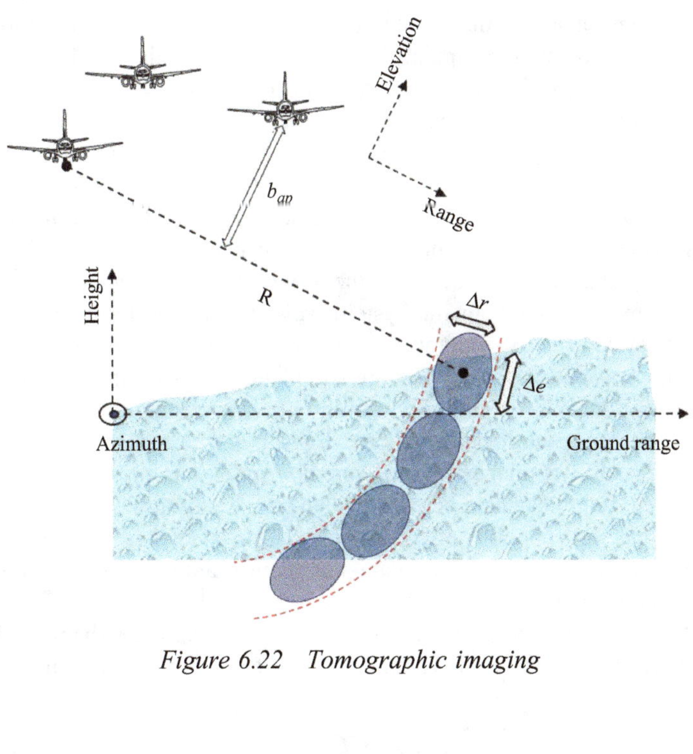

Figure 6.22 Tomographic imaging

in range and azimuth direction is the same as conventional 2D SAR, that is:

$$\Delta r = \frac{c}{2B}; \quad \Delta x = \frac{\lambda R}{2L_s} \tag{6.23}$$

where c is the wave velocity in vacuum, B is the pulse bandwidth, L_s is the synthetic aperture length (in azimuth) and R is the stand-off distance from the imaged target. Resolution in elevation depends on the total length of the synthetic aperture in elevation, usually referred to as baseline aperture b_{ap}, see also (6.5):

$$\Delta e = \frac{\lambda R}{2b_{ap}} \tag{6.24}$$

6.14.2 Real data processing

TomoSAR signal processing involves a number of challenging aspects as compared to conventional 2D SAR focusing.

In the first place, the quality of tomographic imaging is significantly determined by the relative positions among the flights. The achievement of fine vertical resolution and absence of sidelobes is often a difficult task to accomplish when the number of baselines is limited, which pushes for the employment of enhanced imaging techniques. In most cases, tomographic imaging is carried out by decoupling focusing in the range-azimuth plane from focusing in elevation. This approach allows casting tomographic processing in terms of a one dimensional spectral estimation problem, enabling the employment of signal processing techniques that provide enhanced resolution capabilities and sidelobe rejection [111,112].

In the case of high 3D resolution data, however, a relative range and azimuth cell migration is expected to occur from one pass to another. In this case tomographic processing needs to be carried out in all three dimensions at the same time [105,114,115].

Another fundamental requirement to enable 3D data focusing from repeat pass data is that the knowledge of the antenna phase centre position in all flights is accurate enough to predict variations of the distance travelled by the wave to within an accuracy much better than the system wavelength. The location accuracy of navigational systems is usually adequate to correctly predict the distance variation over a synthetic aperture along a single flight line, thus ensuring correct signal focusing in the 2D space. Yet, it might fail to provide the required sub-wavelength accuracy concerning the location of one flight line with respect to another. This results in focused SAR images to be affected by space varying phase disturbances, commonly referred to as phase screens, which may easily turn out to be critical for TomoSAR imaging, possible effects ranging from sidelobes to complete defocusing. For this reason TomoSAR data-sets require quite often a pre-processing phase calibration step [113].

Finally, the issue has to be addressed of the effects of the changes of reflectivity from one pass to the next. Then, the tomography appears blurred, as happens with medical scans when the target moves. However, there is a solution, and that is of acquiring in single pass each baseline, as in single pass interferometry. This is the rationale for future bistatic interferometers such as Tandem-L and SAOCOM-CS [107,108]. By measuring single-pass InSAR coherence at different baselines, one can measure the autocorrelation function of the data with respect to baseline. Then, given the autocorrelation function, one gets the power spectrum by Fourier transformation, hence the distribution of the power of the reflectivity along the vertical direction.

6.14.3 Applications

The application of TomoSAR to forestry has been increasingly considered in the last 10 years by many research teams worldwide, mostly in the frame of future spaceborne missions such as BIOMASS, Tandem-L and SAOCOM-CS. An example of the sensitivity of TomoSAR to forest vertical structure is provided in Figure 6.23, drawn from [109]. In this case, TomoSAR was used to produce intensity associated with different heights above the terrain (top row). Then, each intensity map was analysed by evaluating how it correlates to above ground biomass (bottom row). The results clearly indicated that contributions from the ground level are poorly or negatively correlated to above ground biomass (AGB) and strongly vary with topographic slopes, whereas the 30-m layer is significantly correlated to AGB (for all polarizations) [109]. This finding is easily interpreted by considering that ground scattering is contributed by many other factors than biomass, such as terrain topography, moisture, extinction. Accordingly, TomoSAR helps retrieve the part of the signal that is most informative about forest above ground biomass [109].

Recent experiments at X- and Ku-Band have shown that TomoSAR observations can be used to recover the vertical structure of snow-pack with a resolution of few centimetres [103]. That study revealed the presence of inner layers, which were shown to match direct measurements of hand hardness. TomoSAR was also observed to be sensitive to wave propagation velocity within the snow pack, which was shown to correlate to the apparent tilt of the bottom layers, see Figure 6.24.

TomoSAR was also employed to illuminate alpine glaciers (at L-Band) and ice sheets (at P-Band) [104,105,110]. In the study in [100], in particular, a direct assessment of TomoSAR capabilities was provided by comparing the results against 200 MHz GPR data. Results clearly indicated sensitivity to internal structures such as firn bodies and crevasses, and a penetration down to the bedrock 50 m below the ice surface. An example is shown in Figure 6.24.

6.14.4 Polarimetric SAR tomography

In the frame of a tomographic campaign, the addition of polarimetric information allows producing 3D polarimetric analysis. As an example, the study in [116] demonstrated the occurrence of ground-trunk double bounce interactions in tropical forests, by analysing the complex correlation between HH and VV at the ground level.

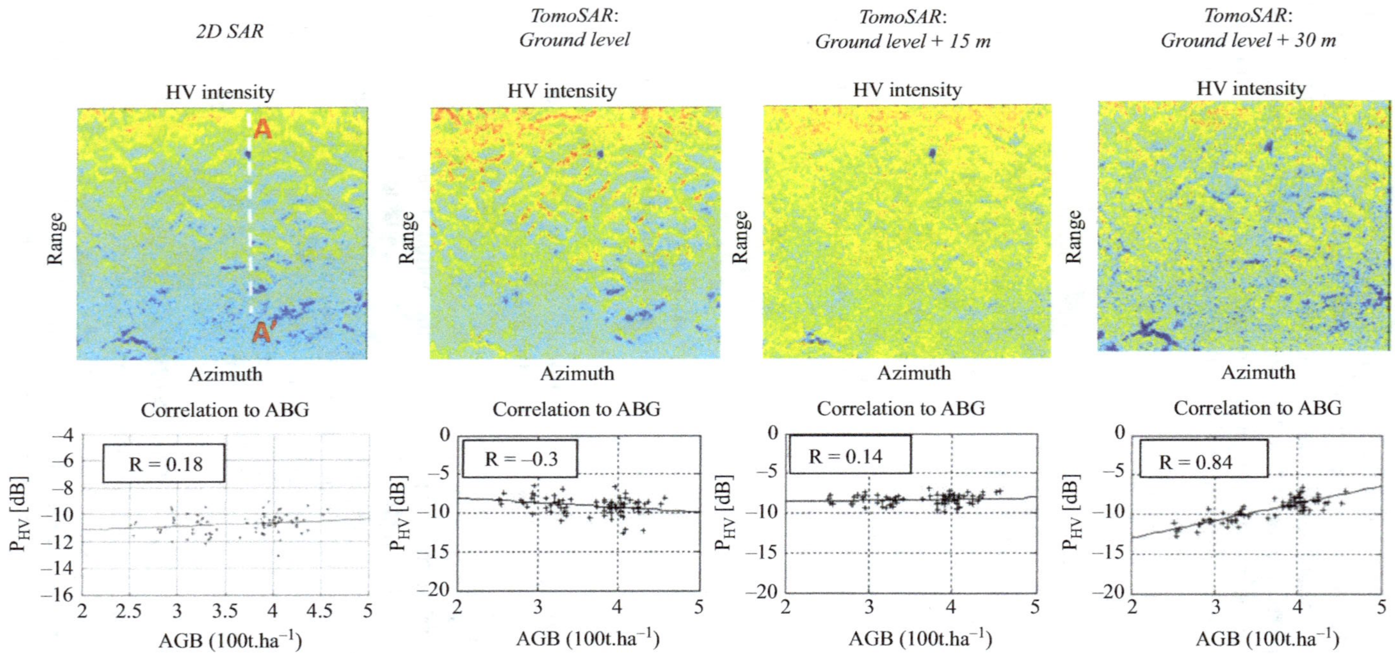

Figure 6.23 *Top row: 2D intensity (left), and TomoSAR intensity corresponding to the ground level, 15 m above the ground and 30 m above the ground. Bottom row: correlation to above ground biomass for each intensity panel. SAR data by ONERA. Campaign: TropiSAR (ESA). Carrier wavelength: P-Band. Picture drawn from [109]*

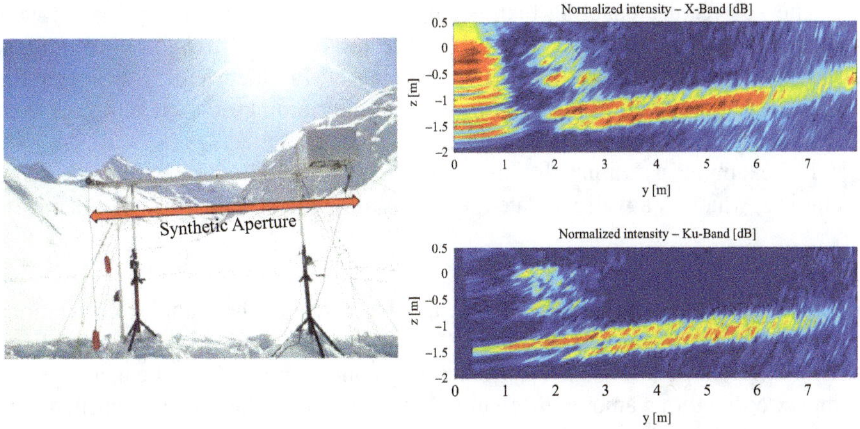

Figure 6.24 Left: set-up of the ground-based TomoSAR experiment in the Rotmoos valley, Austrian Alps. Right: vertical sections of the snowpack at X-Band (top) and Ku-Band (bottom). TomoSAR data were collected by the University of Rennes 1, IETR. Campaign: AlpSAR (ESA). Carrier wavelength: X-Band and Ku-Band. Picture drawn from [103]

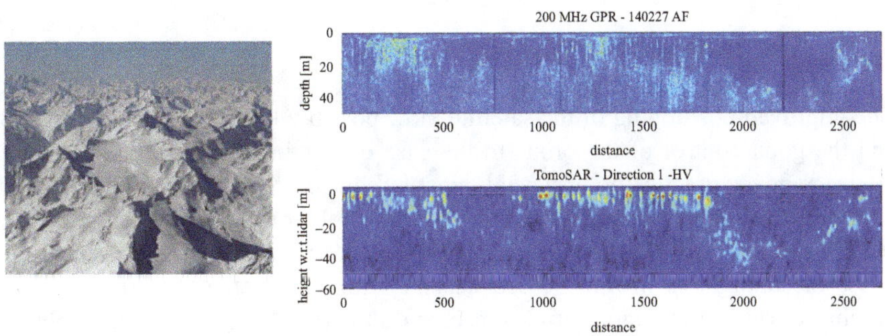

Figure 6.25 Left: the Mittelberferner glacier, Austrian Alps. Top right: GPR transect at 200 MHz. Bottom right: L-Band TomoSAR transect. GPR data were collected by the University of Heidelberg. SAR data by MetaSensing. Campaign: AlpTomoSAR (ESA). Carrier wavelength: L-Band. Picture drawn from [105]

Moreover, the availability of polarimetric tomographic data provides a further, and most powerful, method of analysing forested areas. Indeed, it is shown in [117] that the joint exploitation of baseline *and* polarimetric diversity can be used to decompose the signal in ground-only and volume-only contributions, hence allowing to image them separately. An example is shown in Figure 6.25.

The key feature enabling the decomposition is to notice that the data covariance matrix among all acquisitions (all polarizations and all flights) can be written, under large hypothesis, as a Sum of Kronecker Products (SKP) between the matrices corresponding to the polarimetric and interferometric correlations of each scattering mechanism [117] (Figure 6.26).

For example, assuming that the signal is contributed by ground and volume scattering, which is a well validated hypothesis for forest scenarios, we obtain:

$$\mathbf{W} = \mathbf{C_g} \otimes \mathbf{R_g} + \mathbf{C_v} \otimes \mathbf{R_v} \tag{6.25}$$

where $\mathbf{W}$ is the data covariance matrix, i.e. the matrix of the complex correlations among all polarizations and all flights; $\mathbf{C_g}$ is the matrix of the complex correlations among different polarizations for ground-only contributions; $\mathbf{C_v}$ is the matrix of the complex correlations among different polarizations for volume-only contributions; $\mathbf{R_g}$ is the matrix of the complex correlations among different flight for ground-only contributions; $\mathbf{R_v}$ is the matrix of the complex correlations among different flight for volume-only contributions.

The work in [117] showed that the SKP model in (6.25) provides equations sufficient to extract the matrices C_g, C_v, R_g, R_v, hence decomposing the data into ground-only and volume-only contributions. The SKP model is also shown to be fully consistent with the PolInSAR model proposed in [118] for the case of single baseline data.

6.15 The future of InSAR

The improved positioning of the satellite, now possibly known to centimetres [86] and the good control of the orbit (in the case of Sentinel 1, the orbital radius is 50 m) renders practically irrelevant most of the co-registration, interferogram generation, phase unwrapping and geocoding problems that affected the early InSAR studies. The 5 cm precision of Sentinel-1 orbit knowledge prevented any phase discontinuities in the focused data in TOPS mode [88,89]. Further, short baselines lead to topography irrelevance, and thus most of the previous problems will be only related to the ever-changing atmosphere. Finally, the availability of Numerical Weather Predictions will remove most of the low spatial frequencies of the APS, leaving only its unpredictable turbulent component.

The future developments of InSAR depend on successful deployment of forthcoming missions. At present, the existing constellations are in X-band with 4 CSK satellites staggered in a 16 day orbit to offer 1 to 15 day revisit, the future 3 TerraSAR/TanDEM-X/Paz spacecraft, in C-band the Canadian Radarsat2, the Indian RISAT 1, the L-band missions Alos-2 and the anticipated SAOCOM constellation and SAOCOM bistatic companion.

Indeed, the use of two platforms to achieve one pass interferometry has been promoted by DLR with the TanDEM-X experiment that allowed the retrieval of the world DEM with high quality as observed before [14]. In this case, both platforms were active and basically identical. However, interesting results may be achieved

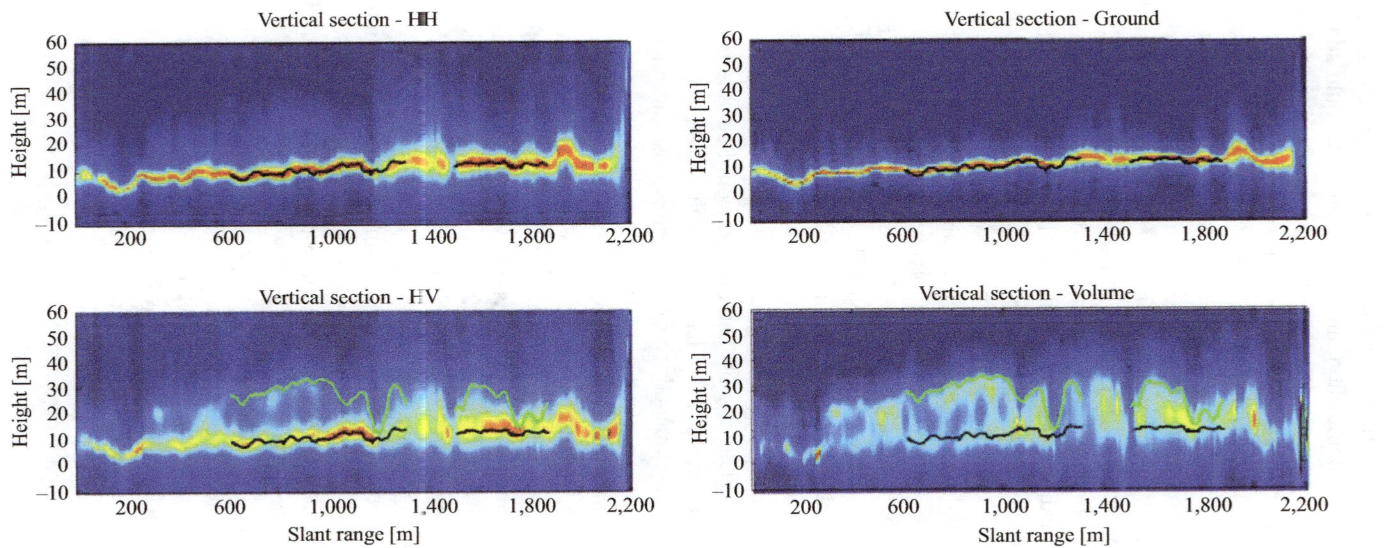

Figure 6.26 TomoSAR vertical sections for HH (top left), HV (bottom left), Ground-only contributions (top left), Volume-only contributions (bottom left). The green and the black lines refer to forest height and terrain elevation as measured by a previous Lidar survey. SAR data by DLR. Campaign: BioSAR (ESA). Carrier wavelength: P-Band. Picture drawn from [113]

adding a receiving satellite that flies in formation with the active one. This would be the case of the SAOCOM companion, an experiment that would add a receiver to the L-band Argentinean system SAOCOM. The advantage of L-band would be in the penetration of the forests, at least the boreal ones, retrieving not only the DEM of the ground underneath, and also that of the canopy, but also the structure of the forest [47,87]. This would entail the use of SAR tomography, namely a multi-pass survey, where the same volumetric structure is seen with different baselines and therefore different altitudes of ambiguity. In the case of a 'frozen forest' that may well be the case for boreal ones in winter, the combination of the layers with different phases due to the progressive baselines, would allow the cancellation of layers and therefore the extraction of the vertical structure [95]. Basically, one has as many equations and thus layers as many passes. Airborne experiments carried out so far have shown the importance of this methodology for the recovery of the above ground biomass [93]. Future missions in L- and P-bands [36,37] will make use of these frequencies to estimate biomass globally.

To date, most InSAR studies of seismic deformation have been focused on single faults or single tracks of InSAR data. With the launch of Sentinel-1, we will have high-quality, coherent InSAR data covering all areas of significant seismic hazard. This offers the potential for InSAR data to be combined with GNSS to produce regional strain maps [88]. These could significantly increase the spatial resolution and quality of models of global strain and would be able to provide an independent mean of assessing seismic hazard, which would not be reliant on incomplete historical or instrumental records of previous earthquake activity [90,91].

Sentinel-1A, in its Wide Swath Interferometric mode will be able to acquire systematically InSAR data, operating with the ESA usual 5.7 cm wavelength. It will acquire data for the whole solid earth surface every 12 days (under current plans) or 6 when the twin platform Sentinel-1B will be made available. The acquisition will be systematic, over swaths 250 km wide, with a resolution of 5 (range) $\times$ 20 (azimuth) m. Both HH and HV polarization data (or VV and VH) will be available, enhancing the sensing capabilities of the system.

The Wide Swath is meant for interferometry and is of particular interest for geophysical applications. The impact of the wide swath is significant for tectonic applications in particular, as long wavelength atmospheric artefacts will be under better control. In fact, in the usual strip map, the low order polynomials that fit the interferometric phases are removed, as they depend both on orbital errors and atmosphere. So, regional tectonic motions are difficult to observe. With wider swaths, wave numbers down to 1 cycle/100 km will be useable, and numerical weather predictions will help their control, to separate them from orbital effects [89]. The continuity and regularity of acquisitions will enforce the role of SAR data for the accurate quantification of deformation rates in multi-temporal InSAR techniques [46]. A lower impact of decorrelation and a better filtering of atmospheric signal will result in a higher spatial density of measurement points. Further, the coherence will be improved thanks also to a tighter orbital control that will maintain the orbital tube within 100 m. For instance, if the phase component due to water vapour with period 100 km has amplitude of 1 cm, and at least 50% of it can be removed using NWP, a 5 mm peak error is left, to be abated in 1 year to less than 1 mm using the 30 available measurements. Therefore, wide area deformation rates of more than 1 mm/year/100 km should be made clearly visible in 1 year.

The availability of very high resolution images will also help for the analysis of local ground motions, for instance, for volcanological applications. SAR platforms to be available further ahead in time may be such that wider swaths and therefore frequent revisit times will be joined by high resolution.

Another reasonable prediction is that of a quasi-geostationary SAR [64], that maybe will be able to reuse the illumination of the terrain due to TV and digital communication signals. Such a system would synthesize the antenna with its relative motion as the geosynchronous orbit is never perfect, but there is always an apparent elliptical wander of the satellite, with diameter say of 100 km. A C-band system could observe with a wide beam an entire continent, yielding say every 20 min water vapour measurements with a resolution of about 100 m, and thus contributing to the weather predictions. The same system, using a 12 h observation time and thus a wider synthetic antenna, would create twice daily images at say 10 m resolution, at a rather low cost, whereas a constellation of a few mini-satellites could achieve day-and-night, continuous imaging capabilities with metric resolution [94].

Acronyms

Acronym	Description
AGB	above ground biomass
ALOS	advanced land observing satellite
APS	atmospheric phase screen
CSK	Cosmo SkyMed
DEM	digital elevation model
DInSAR	differential interferometric SAR
DS	distributed scatterers
ERS	European Remote Sensing
GNSS	global navigation satellite systems
GPR	ground penetrating radar
InSAR	interferometric SAR
MAI	multiple aperture interferometry
MERIS	medium resolution imaging spectrometer
MODIS	moderate resolution imaging spectroradiometer
NWP	numerical weather prediction
PS	persistent scatterers
PSInSAR	persistent scatterer interferometry
S/C, SC	spacecraft

SAOCOM	satellites for observation and communications
SAR	synthetic aperture radar
SBAS	small baseline subsets
SNR	signal-to-noise ratio
TanDEM-X	TerraSAR-X add-on for digital elevation measurement
TomoSAR	tomographic SAR
TRE	Tele Rilevamento Europa
ZTD	zenith total delay
ZWD	zenith wet delay

References

[1] P. A. Rosen, S. Hensley, I. R. Joughin, *et al.*, 'Synthetic aperture radar interferometry,' *Proceedings of the IEEE*, vol. 88, no. 3, pp. 333–382, 2000.
[2] R. Bamler and P. Hartl, 'Synthetic aperture radar interferometry,' *Inverse Problems*, vol. 14, pp. R1–R54, 1998.
[3] A. Ferretti, A. Monti Guarnieri, C. Prati, F. Rocca, and D. Massonnet, *InSAR Principles*, ESA TM-19. ESA, 2007.
[4] L. J. Porcello, 'Turbulence-induced phase errors in radars,' *IEEE Transactions on Aerospace and Electronic Systems*, vol. 6, no. 5, pp. 636–644, 1970.
[5] R. F. Hanssen. *Radar Interferometry: Data Interpretation and Error Analysis*. Kluwer Academic Publishers, Dordrecht, 2001.
[6] A. Ferretti, C. Prati, and F. Rocca, 'Permanent scatterers in SAR interferometry,' *IEEE Transactions on Geoscience and Remote Sensing*, vol. 39, no. 1, pp. 8–20, 2001.
[7] A. Hopper, H. Zebker, and B. Kampes, 'A new method for measuring deformation on volcanoes and other natural terrains using INSAR persistent scatterers,' *Geophysical Research Letters*, vol. 31, p. L23611, 2004.
[8] P. Berardino, G. Fornaro, R. Lanari, and E. Sansosti, 'A new algorithm for surface deformation monitoring based on small baseline differential SAR interferograms,' *IEEE Transactions on Geoscience and Remote Sensing*, vol. 40, no. 11, pp. 2375–2383, 2002.
[9] A. Monti Guarnieri and S. Tebaldini, 'On the exploitation of target statistics for SAR interferometry applications,' *IEEE Transactions on Geoscience and Remote Sensing*, vol. 46, no. 11, pp. 3436–3443, 2008.
[10] S. Tebaldini and A. Monti Guarnieri. *Geoscience and Remote Sensing New Achievements, Chapter Methods and Performances for Multi-Pass SAR Interferometry*. InTech, Rijeka, Croatia, 2010.
[11] E. Sansosti, 'A simple and exact solution for the interferometric and stereo SAR geolocation problem,' *IEEE Transactions on Geoscience and Remote Sensing*, vol. 42, no. 8, pp. 1625–1634, 2004.
[12] http://www2.jpl.nasa.gov/srtm/SRTM_D31639.pdf.

[13] http://elib.dlr.de/98029/1/isrse36_TanDEM-X_Status-final.pd.

[14] Worlddem http://www.geo-airbusds.com/worlddem/.

[15] D. C. Ghiglia and M. D. Pritt. *Two-Dimensional Phase Unwrapping: Theory, Algorithms, and Software*. John Wiley & Sons, Inc, New York, 1998.

[16] C. Prati, F. Rocca, A. Monti Guarnieri, and P. Pasquali, 'Report on ERS-1 SAR interferometric techniques and applications,' ESA Study Contract Report 1994/6.

[17] M. Eineder and N. Adam, 'A maximum-likelihood estimator to simultaneously unwrap, geocode, and fuse SAR interferograms from different viewing geometries into one digital elevation model,' *IEEE Transactions on Geoscience and Remote Sensing*, vol. 43, no. 1, pp. 24–36, 2005.

[18] E. Rodriguez and J. M. Martin, 'Theory and design of interferometric synthetic aperture radars,' *IEE Proceedings-F*, vol. 139, no. 2, pp. 147–159, Apr. 1992.

[19] H. A. Zebker and J. Villasenor, 'Decorrelation in interferometric radar echoes,' *IEEE Transactions on Geoscience and Remote Sensing*, vol. 30, no. 5, pp. 950–959, Sept. 1992.

[20] D. Just and R. Bamler, 'Phase statistics of interferograms with applications to synthetic aperture radar,' *Applied Optics*, vol. 33, no. 20, pp. 4361–4368, 1994.

[21] M. Cattabeni, A. Monti-Guarnieri, and F. Rocca. Estimation and improvement of coherence in SAR interferograms. In *International Geoscience and Remote Sensing Symposium*, Pasadena, CA, USA, 8–12 August 1994, pp. 720–722, 1994.

[22] U. Wegmüller and C. Werner, 'Retrieval of vegetation parameters with SAR interferometry,' *IEEE Transactions on Geoscience and Remote Sensing*, vol. 35, no. 1, pp. 18–24, 1997.

[23] S. Takeuchi, S. Yuzo, and M. Yoshimura, 'A comparative study of coherence information by L-band and C-band SAR for detecting deforestation in tropical rain forest,' *IGARSS* 2001, vol. 5, pp. 2259–2261, Sydney, Australia, Jul. 2001.

[24] M. Engdahl and M. Borgeaud, 'The use of ers-1/2 tandem interferometric coherence in the estimation of agricultural crops heights,' *IEEE Transactions on Geoscience and Remote Sensing*, vol. 9, pp. 1799–1806, 2001.

[25] J. T. Koskinen, J. T. Palliainen, J. M. Hyyppa, M. E. Engdahl, and M. T. Hallikainen, 'The seasonal behavior of interferometric coherence in boreal forest,' *IEEE Transactions on Geoscience and Remote Sensing*, vol. 39, no. 4, pp. 820–829, 2001.

[26] B. Moeremans and S. Dautrebande, 'The Use of ERS SAR Interferometric Coherence and PRI Images to Evaluate Crop Height and Soil Moisture,' Fringe 99, ESA SP-478.

[27] K. J. Ranson and G. Sun, 'Forest Disturbance Characterization Using ERS Tandem Data,' Fringe 99, ESA SP-478.

[28] C. Schmullius, A. Holz, U. Marschalk, A. Roth, J. Vietmeier, and W. Wagner, 'Operational Readiness of ERS SAR Interferometry for Forest Mapping in Siberia,' Fringe 99, ESA SP-478.

[29] H. Rott and A. Siegel, 'Glaciological studies in the Alps and in Antarctica using ERS interferometric SAR,' Proc. Fringe 1996, ESA SP-406, vol. 2, pp. 149–159.

[30] E. Weber Hoen and H. Zebker, 'Penetration depths inferred from interferometric volume decorrelation observed over the Greenland Ice Sheet,' *IEEE Transactions on GARS*, vol. 38, no. 6, pp. 2571–2583, 2000.

[31] D. J. Weydahl, 'Analysis of ERS tandem coherence from glaciers, valleys and fjord ice in Svalbard,' *IEEE Transactions on GARS*, vol. 39, pp. 2029–2039, 2001.

[32] J. I. H. Askne, P. B. G. Dammert, L. M. H. Ulander, and G. Smith, 'C-band repeat-pass interferometric SAR observations of the forest Geoscience and Remote Sensing,' *IEEE Transactions*, vol. 35, pp. 25–35, 1997.

[33] P. B. G. Dammert, J. I. H. Askne, and S. Kuhlmann, 'Unsupervised segmentation of multitemporal interferometric SAR images,' *IEEE Transactions on GARS*, vol. 37, no. 5, pp. 2259–2271, Sept. 1999.

[34] F. Rocca, 'Modeling interferogram stacks,' *IEEE Transactions on Geoscience and Remote Sensing*, vol. 45, no. 10, pp. 3289–3299, 2007.

[35] D. D'Aria, A. Leanza, A. Monti-Guarnieri, and A. Recchia, 'Decorrelating targets: models and measures' In *Proc. IGARSS* 2016, pp. 1–4.

[36] M. Younis, S. Huber, C. T. Herrero, *et al.*, Shimada Tandem-L instrument design and SAR performance overview Geoscience and Remote Sensing Symposium (IGARSS), 2014 IEEE International Year: 2014 pp. 88–91.

[37] S. Quegan, J. Chave, J. Dall, *et al.* The science and measurement concepts underlying the BIOMASS mission Geoscience and Remote Sensing Symposium (IGARSS), 2012 IEEE International Year: 2012 pp. 5542–5545.

[38] F. Gatelli, A. Monti Guarnieri, F. Parizzi, P. Pasquali, C. Prati, and F. Rocca, 'The wavenumber shift in SAR interferometry,' *IEEE Transactions on Geoscience and Remote Sensing*, vol. 32, no. 4, pp. 855–865, Jul. 1994.

[39] G. Fornaro and A. Monti Guarnieri, 'Minimum mean square error space-varying filtering of interferometric SAR data,' *IEEE Transactions on Geoscience and Remote Sensing*, vol. 40, no. 1, pp. 11–21, Jul. 2002.

[40] C. Cafforio and F. Rocca, 'Methods for measuring small displacements of television images,' *IEEE Transactions on Information Theory*, vol. 22, no. 5, pp. 573–579, Sept. 1976.

[41] R. Bamler, 'Interferometric Stereo Radargrammetry: Absolute Height Determination from ERS-ENVISAT Interferograms,' Proc. IGARSS 2000.

[42] R. Bamler and M. Eineder, 'Accuracy of differential shift estimation by correlation and split-bandwidth interferometry for wideband and Delta-k SAR systems,' *IEEE Transactions on Geoscience and Remote Sensing Letter*, vol. 2, no. 2, pp. 151–155, Apr. 2005.

[43] F. De Zan, P. Prats-Iraola, and M. Rodriguez-Cassola, 'On the dependence of delta-k efficiency on multilooking,' *IEEE GRSL*, vol. 12, no. 8, p. 1745, 2015.

[44] F. De Zan, 'Accuracy of incoherent speckle tracking for circular Gaussian signals,' *IEEE Geoscience and Remote Sensing Letters*, vol. 11, no. 1, pp. 264–267, 2014.

[45] P. Prats-Iraola, R. Scheiber, L. Marotti, S. Wollstadt, and A. Reigber, 'TOPS interferometry with TerraSAR-X,' *IEEE Transactions on Geoscience and Remote Sensing*, vol. 50, no. 8, pp. 3179–3188, 2012.

[46] R. Grandin, E. Klein, M. Métois, and C. Vigny, 'Three-dimensional displacement field of the 2015 Mw 8.3 Illapel earthquake (Chile) from across and along-track Sentinel-1 TOPS interferometry,' *Geophysical Research Letters*, 43, 2016, doi:10.1002/2016GL067954.

[47] S. Mancon, S. Tebaldini, A. Monti Guarnieri, and D. Giudici, 'Phase calibration and baseline verification in TOPSAR and Stripmap interferometry,' *IEEE Trans Geoscience and Remote Sensing*, vol. 55, no. 1, pp. 134–147, Jan. 2017.

[48] A. Ferretti, C. Prati, and F. Rocca, 'Nonlinear subsidence rate estimation using Permanent Scatterers in differential SAR,' *IEEE Transactions on GARS*, vol. 38, pp. 2202–2212, 2000.

[49] S. J. Frasier and A. J. Camps, 'Dual-beam interferometry for ocean surface current vector mapping,' *IEEE Transactions on Geoscience and Remote Sensing*, vol. 39, no. 2, pp. 401–414, 2001.

[50] L. J. Porcello, 'Turbulence-induced phase errors in synthetic-aperture radars,' *IEEE Transactions on Aerospace and Electronic Systems*, vol. AES-6, no. 5, pp. 636–644, Sept. 1970.

[51] L. Iannini and A. Monti Guarnieri, 'Atmospheric phase screen in ground-based radar: statistics and compensation,' *IEEE Geoscience and Remote Sensing Letters*, vol. PP, no. 99, pp. 536–540, 2010.

[52] E. Pichelli, R. Ferretti, D. Cimini, *et al.*, 'InSAR water vapour data assimilation into mesoscale model MM5: technique and pilot study,' *IEEE Journal of Selected Topics in Applied Earth Observations and Remote Sensing*, vol. 8, no. 8, 2015.

[53] F. G. Ulmer and N. Adam, 'A synergy method to improve ensemble weather predictions and differential SAR interferograms,' *ISPRS Journal of Photogrammetry and Remote Sensing*, vol. 109, pp. 98–107, 2015.

[54] M. Bevis, S. Businger, T. A. Herring, C. Rocken, R. A. Anthes, and R. H. Ware 'GPS meteorology: remote sensing of atmospheric water vapour using the global positioning system,' *Journal of Geophysical Research*, vol. 97, no. D14, pp. 15787–15801, 1992.

[55] F. Meyer, R. Bamler, R. Leinweber, and J. Fischer, 'A Comparative analysis of tropospheric water vapour measurements from MERIS and SAR,' Geoscience and Remote Sensing Symposium, 2008. IGARSS 2008. IEEE International, Boston, MA, 2008, pp. IV-228–IV-231.

[56] F. Onn and H. A. Zebker, 'Correction for interferometric synthetic aperture radar atmospheric phase artifacts using time series of zenith wet delay observations from a GPS network,' *JGR-Solid Earth*, vol. 111, 2006.

[57] A. Rucci, A. Ferretti, A. Monti Guarnieri, and F. Rocca, 'Sentinel 1 SAR interferometry applications: The outlook for sub millimeter measurements,' *Remote Sensing of Environment*, vol. 120, pp. 156–163, 2012.

[58] T. R. Emardson, M. Simons, and F. H. Webb, 'Neutral atmospheric delay in interferometric synthetic aperture radar applications: Statistical description and mitigation,' *Journal of Geophysical Research*, vol. 108, no. B5, 2231, May 2003.

[59] A. Recchia, A. M. Guarnieri, A. Broquetas, and J. Ruiz-Rodon, 'Assessment of atmospheric phase screen impact on Geosynchronous SAR,' Geoscience and Remote Sensing Symposium (IGARSS), 2014 IEEE International, Quebec City, QC, 2014, pp. 2253–2256.

[60] E. Hammann, A. Behrendt, F. Le Mounier, and V. Wulfmeyer, 'Temperature profiling of the atmospheric boundary layer with rotational Raman lidar during the $HD(CP)^2$ Observational Prototype Experiment,' *Atmospheric Chemistry and Physics*, vol. 15, no. 5, pp. 2867–2881, 2015.

[61] Q. Zeng, Y. Li, and X. Li, 'Correction of tropospheric water vapour effect on ASAR interferogram using synchronous MERIS data,' IGARSS 2007. Barcelona, 2007, pp. 2086–2089.

[62] J. Foster, M. Bevis, and W. Raymond, 'Precipitable water and the lognormal distribution,' *Journal of Geophysical Research*, vol. 111, p. D15102, 2006. doi:10.1029/2005JD006731.

[63] T. J. Wright, B. Parsons, P. C. England, and E. J- Fielding, 'InSAR observations of low slip rates on the major faults of western Tibet,' *Science*, vol. 305, no. 5681, pp. 236–239, 2004.

[64] J. Ruiz Rodon, A. Broquetas, A. Monti Guarnieri, and F. Rocca, 'Geosynchronous SAR focusing with atmospheric phase screen retrieval and compensation,' *IEEE Transactions on GRS*, vol. 51, pp. 4397–4404, 2013.

[65] M. Guarnieri and S. Tebaldini, 'On the exploitation of target statistics for SAR interferometry applications,' *IEEE Transactions on Geoscience and Remote Sensing*, vol. 46, no. 11, pp. 3436–3443, 2008.

[66] A. Hooper, H. Zebker, P. Segall, and B. Kampes, 'A new method for measuring deformation on volcanoes and other non-urban areas using InSAR persistent scatterers,' *Geophysical Research Letters*, vol. 31, p. L23611, 2004. doi:10.1029/2004GL021737.

[67] A. Ferretti, A. Fumagalli, F. Novali, C. Prati, F. Rocca, and A. Rucci, 'A new algorithm for processing interferometric data-stacks: SqueeSAR,' *IEEE Transactions on Geoscience and Remote Sensing*, vol. 49, no. 9, pp. 3460–3470, 2011.

[68] F. De Zan, G. Krieger, and P. Lopez-Dekker, 'On some spectral properties of TanDEM-X interferograms over forested areas,' *IEEE Geoscience and Remote Sensing Letters*, vol. 10, no. 1, pp. 71–75, 2013.

[69] K. Papathanassiou and S.R. Cloude, 'Single-baseline polarimetricSAR interferometry,' *IEEE Transactions on Geoscience and Remote Sensing*, vol. 39, no. 11, pp. 2352–2363, 2001.

[70] J. Dall, 'Insar elevation bias caused by penetration into uniform volumes,' *IEEE Transactions on Geoscience and Remote Sensing*, vol. 45, no. 7, pp. 2319–2324, 2007.

[71] F. De Zan, M. Zonno, and P. López-Dekker, 'Phase inconsistencies and multiple scattering in SAR interferometry,' *IEEE Transactions on Geoscience and Remote Sensing*, 2015, vol. 53, no. 12, pp. 6608–6616.

[72] Y. Yan, M.-P. Doin, P. Lopez-Quiroz, *et al.*, 'Mexico City subsidence measured by InSAR time series: joint analysis using PS and SBAS approaches,' *IEEE Journal of Selected Topics in Applied Earth Observations and Remote Sensing*, vol. 5 no. 4, pp.1312–1326, 2012.

[73] G. Fornaro, S. Verde, D. Reale, and A. Pauciullo, 'CAESAR: an approach based on covariance matrix decomposition to improve multibaseline–multitemporal interferometric SAR processing,' *IEEE Transactions on Geoscience and Remote Sensing*, vol. 53, no. 4, pp. 2050–2065, 2015.

[74] F. De Zan, A. Parizzi, P. Prats-Iraola, and P. López-Dekker, 'A SAR interferometric model for soil moisture,' *IEEE Transactions on Geoscience and Remote Sensing*, vol. 52, no. 1, pp. 418–425, 2014.

[75] F. Rocca, 3D Motion recovery with multi-angle and/or left right Interferometry Proceedings of the third International Workshop on ERS SAR Interferometry (FRINGE 2003), Frascati (Italy), 2–5 December 2003 (2003) ESA SP-550.

[76] N. B. D. Bechor and H. A. Zebker, 'Measuring two-dimensional movements using a single InSAR pair,' *Geophysical Research Letters*, vol. 33, p. L16311, 2006. doi:10.1029/2006GL026883.

[77] A. Hooper, 'A multi-temporal InSAR method incorporating both persistent scatterer and small baseline approaches,' *Geophysical Research Letters*, vol. 35, no. 16, p. 16, 2008.

[78] J. C. L. Normand and E. Heggy, 'InSAR assessment of surface deformations in urban coastal terrains associated with groundwater dynamics,' *IEEE Transactions on Geoscience and Remote Sensing*, vol. 53, no. 12, pp. 6356–6370, 2015.

[79] J. Hu, Z. W. Li, Q. Sun, J. J. Zhu, and X. L. Ding, 'Three-dimensional surface displacements from InSAR and GPS measurements with variance component estimation,' *IEEE Geoscience and Remote Sensing Letters*, vol. 9, no. 4, pp. 754–758, 2012.

[80] B. H. Ketelaar, 'Monitoring surface deformation induced by hydrocarbon production using satellite radar interferometry,' Ph.D. dissertation, Delft University of Technology, Delft, The Netherlands, 2008.

[81] S. C. Maxwell, J. Du, and J. Shemeta, 'Passive seismic and surface monitoring of geomechanical deformation associated with steam injection,' *The Leading Edge*, pp. 1176–1184, 2008.

[82] D. W. Vasco, A. Ferretti, and F. Novali, 'Estimating permeability from quasi static deformation: temporal variations and arrival-time inversion,' *Geophysics*, vol. 73, no. 6, pp. O37–O52, Nov. 2008.

[83] S. Raikes, A. Mathieson, D. Roberts, and P. Ringrose, Integration of 3D seismic with satellite imagery at In Salah CO2 sequestration project, Algeria. Proc. SEG Annual Meeting, Las Vegas, pp. 2856–2858, 2008.

[84] H. Klemm, I. Quseimi, F. Novali, A. Ferretti, and A. Tamburini, Monitoring horizontal and vertical surface deformation over a hydrocarbon reservoir by PSInSAR First break, volume 28. 2010, May.

[85] A. Pigorini, M. Ricci, A. Sciotti, C. Giannico, and A. Tamburini, 'La tecnica PSInSAR™ di telerilevamento satellitare applicata al progetto ed alla realizzazione delle infrastrutture ferroviarie,' Ingegneria Ferroviaria anno LXV Ottobre 2010 (10) pag 729, 2010.

[86] U. Balss, C. Gisinger, X. Cong, R. Brcic, S. Hackel, and M. Eineder, *Precise Measurements on the Absolute Localization Accuracy of TerraSAR-X on the Base of Far-Distributed Test Sites*. EUSAR 2014 - 05 June 2014, ISBN (Print) 978-3-8007-3607-2, 2014.

[87] N. Gebert, B. Carnicero Dominguez, M. W. J. Davidson, M. Diaz Martin, and P. Silvestrin, SAOCOM-CS—A passive companion to SAOCOM for single-pass L-band SAR interferometry EUSAR 2014; 10th European Conference on Synthetic Aperture Radar; Proceedings of Year: 2014.

[88] M. Nannini, P. Prats-Iraola, F. De Zan, and D. Geudtner, 'TOPS time series performance assessment with TerraSAR-X data,' *IEEE Journal of Selected Topics in Applied Earth Observations and Remote Sensing*, vol. no.99, pp. 1–17.

[89] F. Ulmer, N. Adam, and A. Synergy, 'Method to improve ensemble weather predictions and differential SAR interferograms,' *ISPRS Journal of Photogrammetry and Remote Sensing*, vol. 109, pp. 98–107, Nov. 2015, ISSN 0924-2716, http://dx.doi.org/10.1016/j.isprsjprs.2015.09.004.

[90] H. Wang and T.J. Wright, 'Satellite geodetic imaging reveals internal deformation of western Tibet,' *Geophysical Research Letters*, vol. 39, pp. L07303, 2012. doi:10.1029/2012GL05122.

[91] B. Kreemer, W. E. Holt, and A. J. Haines, 'An integrated global model of present-day plate motions and plate boundary deformation,' *Geophysical Journal*, vol. 154, no. 1, pp. 8–34, Jul. 2003.

[92] R. F. Hanssen and F. J. van Leijen, Monitoring water defense structures using radar interferometry Radar Conference, 2008. RADAR'08. Year: 2008 pp. 1–4.

[93] D. Ho Tong Minh, S. Tebaldini, F. Rocca, T. Le Toan, L. Villard, and P. C. Dubois-Fernandez, 'Capabilities of BIOMASS tomography for investigating tropical forests,' *IEEE Transactions on Geoscience and Remote Sensing*, vol. 53, no. 2 pp. 965–975, 2015.

[94] A. Monti Guarnieri, O. Bombaci, T. F. Catalano, *et al.*, '*ARGOS: a fractioned geosynchronous SAR*,' *Acta Astronautica,* Available online 2 December 2015, ISSN 0094-5765, http://dx.doi.org/10.1016/j.actaastro.2015.11.022.

[95] P. Pasquali, C. Prati, F. Rocca, *et al.*, A 3-D SAR experiment with EMSL data, Geoscience and Remote Sensing Symposium, 1995. IGARSS '95. 'Quantitative Remote Sensing for Science and Applications', International Year: 1995, pp. 784–786, vol. 1.

[96] K. Knaell and G. Cardillo, 'Radar tomography for the generation of three-dimensional images,' *IEE Proceedings – Radar, Sonar and Navigation*, vol. 142, no. 2, pp. 54–60, Apr. 1995.

[97] A. Reigber and A. Moreira, 'First demonstration of airborne SAR tomography using multibaseline l-band data,' *IEEE Transactions on Geoscience and Remote Sensing*, pp. 2142–2152, 2000.

[98] X. X. Zhu and R. Bamler, 'Very high resolution spaceborne sar tomography in urban environment,' *IEEE Transactions on Geoscience and Remote Sensing*, no. 99, pp. 1–13, 2010.

[99] G. Fornaro, F. Lombardini, and F. Serafino, 'Three dimensional multipass SAR focusing: experiments with long-term spaceborne data,' *IEEE Transactions on Geoscience and Remote Sensing*, vol. 43, no. 4, pp. 702–714, 2005.

[100] S. Tebaldini and F. Rocca, 'Multibaseline polarimetric sar tomography of a boreal forest at P- and L-bands,' *IEEE Transactions on Geoscience and Remote Sensing*, vol. 50, no. 1, pp. 232–246, 2012.

[101] O. Frey and E. Meier, 'Analyzing tomographic sar data of a forest with respect to frequency, polarization, and focusing technique,' *IEEE Transactions on Geoscience and Remote Sensing*, vol. 49, no. 10, pp. 3648–3659, 2011.

[102] Y. Huang, L. Ferro-Famil, and C. Lardeux, 'Polarimetric sar tomography of tropical forests at P-band,' in *Geoscience and Remote Sensing Symposium (IGARSS), 2011 IEEE International*, Jul. 2011, pp. 1373–1376.

[103] L. Ferro-Famil, S. Tebaldini, M. Davy, and F. Boute, '3d sar imaging of the snowpack in presence of propagation velocity changes: Results from the alpsar campaign,' *Geoscience and Remote Sensing Symposium (IGARSS), 2014 IEEE International*, Jul. 2014, pp. 3370–3373.

[104] F. Banda, J. Dall, and S. Tebaldini, 'Single and multipolarimetric p-band sar tomography of subsurface ice structure,' *IEEE Transactions on Geoscience and Remote Sensing*, vol. no. 99, pp. 1–14, 2015.

[105] S. Tebaldini, T. Nagler, H. Rott, and A. Heilig, 'L-band 3d imaging of an alpine glacier: results from the alptomosar campaign,' in *Geoscience and Remote Sensing Symposium (IGARSS), 2015 IEEE International*, Jul. 2015.

[106] ESA, *Report for Mission Selection: Biomass*. Noordwijk, The Netherlands, 2012.

[107] A. Moreira, G. Krieger, I. Hajnsek, *et al.*, 'Tandem-l: A highly innovative bistatic sar mission for global observation of dynamic processes on the earth's surface,' *IEEE Geoscience and Remote Sensing Magazine*, vol. 3, no. 2, pp. 8–23, Jun. 2015.

[108] ESA, *SAOCOM Companion Satellite Science Report*. Noordwijk, The Netherlands, 2015.

[109] D. Ho Tong Minh, T. Le Toan, F. Rocca, S. Tebaldini, M. Mariotti d'Alessandro, and L. Villard, 'Relating p-band synthetic aperture radar

tomography to tropical forest biomass,' *IEEE Transactions on Geoscience and Remote Sensing*, vol. no. 99, pp. 1–1, 2013.

[110] F. Banda, S. Tebaldini, F. Rocca, and J. Dall, 'Tomographic sar analysis of subsurface ice structure in Greenland: first results,' in *Geoscience and Remote Sensing Symposium (IGARSS), 2013 IEEE International*, 2013.

[111] F. Gini, F. Lombardini, and M. Montanari, 'Layover solution in multibaseline SAR interferometry,' *IEEE Transactions on Aerospace and Electronic Systems*, vol. 38, no. 4, pp. 1344–1356, 2002.

[112] E. Aguilera, M. Nannini, and A. Reigber, 'A data adaptive compressed sensing approach to polarimetric sar tomography of forested areas,' *IEEE Geoscience and Remote Sensing Letters*, vol. 10, no. 3, pp. 543–547, 2013.

[113] S. Tebaldini, F. Rocca, M. Mariotti d'Alessandro, and L. Ferro-Famil, 'Phase calibration of airborne tomographic sar data via phase center double localization,' *IEEE Transactions on Geoscience and Remote Sensing*, vol. no. 99, pp. 1–18, 2015.

[114] O. Frey, C. Magnard, M. Ruegg, and E. Meier, 'Focusing of airborne synthetic aperture radar data from highly nonlinear flight tracks,' *IEEE Transactions on Geoscience and Remote Sensing*, vol. 47, no. 6, pp. 1844–1858, 2009.

[115] M. Nannini and R. Scheiber, 'Height dependent motion compensation and coregistration for airborne sar tomography,' in *Geoscience and Remote Sensing Symposium, 2007. IGARSS 2007. IEEE International*, Jul. 2007, pp. 5041–5044.

[116] M. M. D'Alessandro, S. Tebaldini, F. Rocca, 'Phenomenology of ground scattering in a tropical forest through polarimetric synthetic aperture radar tomography,' *IEEE Transactions on Geoscience and Remote Sensing*, vol. 51, no. 8, pp. 4430–4437, 2013.

[117] S. Tebaldini, 'Algebraic synthesis of forest scenarios from multibaseline PolInSAR data,' *IEEE Transactions on Geoscience and Remote Sensing*, vol. 47, no. 12, pp. 4132–4142, 2009.

[118] K. P. Papathanassiou and S. R. Cloude, 'Single-baseline polarimetric SAR interferometry,' *IEEE Transactions on Geoscience and Remote Sensing*, vol. 39, no. 11, pp. 2352–2363, 2001.

Chapter 7

Space-based SAR ground moving target indication

Christoph H. Gierull, Ishuwa Sikaneta* and Delphine Cerutti-Maori***

Abstract

Today's demand for space-borne Synthetic Aperture Radar (SAR) data has grown to the point where significant commercial funding of advanced space-borne radar system development has been being made available. The current generation of commercial space-based SAR imaging satellites, such as RADARSAT-2, Sentinel-1, TerraSAR-X/TanDEM-X (PAZ), COSMO-SkyMED and ALOS-2, operate at a single frequency (L-, C- and X-band) and are based on active phased array antenna technology that offers beam agility and adds polarization diversity. Consequently, these modern satellites are equipped with more than one receive channel (i.e., AD-converter) that can also be utilized to record measurements from multiple apertures in along-track direction. This is the principal prerequisite for a ground moving target indication (GMTI)[1] capability. While space-based SAR GMTI offers many advantages like global ground coverage and access to strategic regions, it also faces several obstacles such as high satellite velocity, Earth rotation and oftentimes small target reflection energy caused by the enormous distances of more than 1,000 km among others.

This book chapter presents the state-of-the-art of space-based SAR-GMTI science and technology with focus on recent advances and the latest direction of research and development (R&D) activities. Owing to an exponential cost jump, technological advances of space-based radars especially with regard to increased power, increased aperture sizes and additional receiver channels have only been somewhat incremental in the last decades. Spacecraft with more than two parallel receive paths are only expected to materialize two generations down the line. Hence, current R&D put emphasis on innovative new concepts trying to

*Defence R&D Canada – Ottawa Research Centre, Canada

**Fraunhofer Institute for High Frequency Physics and Radar Techniques FHR, Germany

[1]Ground MTI also includes the sea surface for vessel detection but due to its prevalence in the community the terminology GMTI is being used throughout the chapter.

circumvent these technological limitations thereby often pushing the resources on existing SAR payloads to their limits.[2] Virtually all of these concepts are accompanied by cutting-edge but complex and resource-hungry signal-processing algorithms that only recently became feasible based on the fast-paced evolution in computing power over the last decade. Many of the presented proof-of-concept studies are considered building blocks of future operational space-based SAR capabilities, for instance, the synergy between high-resolution-wide-swath (HRWS) imaging and motion indication and estimation. This chapter attempts to provide a comprehensive, in-depth overview of the theory and the radar signal-processing techniques required for space-based SAR-GMTI corroborated by real multichannel data from RADARSAT-2.

7.1 Introduction

7.1.1 Background

Up to the present time, the remote detection and tracking of moving targets by radar has primarily been a military concern and operationally supported by specialized airborne sensors. With the rapid evolution of radar technology, it would now be in general feasible to create space-borne sensors to perform dedicated moving target detection and measurement functions. From a military viewpoint, these space-borne systems have the potential to significantly augment existing operational capabilities. From a civilian viewpoint, space-borne moving target measurements can provide land and sea-traffic-monitoring capabilities that may prove very valuable in designing, monitoring and controlling transportation infrastructure, which are major financial investments at all levels of government in all nations. Although all of the processes needed for functional ground moving target indication (GMTI) radars have been developed and operationally fielded for airborne systems, the impact of differences in: platform velocity, range to target and accessible depression angles between airborne and space-borne radars among others are not fully comprehended and still require a certain optimization in order to be transitioned into space-based systems. At the end, prohibitive capital costs, complexity, available technology and design risks have all combined to preclude the construction and launch of a satellite constellation purely dedicated to GMTI for now.

In comparison, synthetic aperture imaging radars are much less complex than full GMTI radars, as the latter must acquire and process or acquire and downlink considerably more data than a Synthetic Aperture Radar (SAR) system during an identical observation time. An always-growing public demand for available imagery products and lower cost satellite technologies, including active-phased array antennas, has made investment more appealing evidenced by more than ten commercial SAR satellites currently in orbit. However, the air to space path followed by SAR systems is unlikely to occur for GMTI systems. In the foreseeable future, a

[2]Most of the concepts and modes were not envisaged to be implemented on the spacecraft prior to launch.

principle GMTI capability, even if somewhat limited, can be incorporated as a subset of functions into a SAR system and is able to provide operationally relevant information for special applications.

7.1.2 Adding MTI on space-borne SAR

Instead of deploying a fully functional space-borne GMTI, a polarimetric SAR system that has two parallel receiver channels and associated data recorders and downlinks can be easily converted into a two-channel displaced phase centre antenna (DPCA) radar if a full corporate feed is used for each of the polarization channels. The essential conversion is a radio-frequency (RF) switch that allows the two halves of the antenna to be routed to the two receiver channels. If an active antenna [distributed transmit/receive (TR) modules] design is used, additional beamforming controls can be imposed to minimize azimuth sidelobes and to match the two subbeams. Provided that the radar is designed to operate as an interleaved pulse (phase coherent) polarimeter, the SAR pulse repetition frequency (PRF) will be in a range suitable for DPCA or space-time adaptive processing (STAP). Recently, studies on HRWS have attracted much attention as it has been recognized and proven that multiple antenna apertures in the flight direction allow one to lower the PRF significantly without increasing the risk of azimuth ambiguities. The lower PRF permits a much wider swath without abandoning spatial resolution. It is expected that a HRWS capability is one of the distinctive features on the next generation of SAR spacecraft and will thereby inevitably also push the development and utility of along-track interferometric applications.

7.1.3 MODEX on RADARSAT-2

RADARSAT-2's moving object detection experiment (MODEX) has been designed to develop, validate and demonstrate an experimental space-based SAR-GMTI mode to routinely detect, measure and monitor moving vehicles on the Earth's surface [1]. Although MODEX is an experimental research and development (R&D) mode, the results over the years have been so positive and promising that the operational community is starting to pay attention to the potential of this value-added information.

RADARSAT-2 executes baseline GMTI through the MODEX-1 mode. For every pulse, the whole antenna transmits radiation, then the front and back halves simultaneously but separately measure the echoes. When data are measured in the MODEX-1 mode, signal-processing methods create imagery and transform the data to form two measurements from the same point in space, but at slightly different times. These data can then be statistically analysed using the image-based techniques of along-track interferometry (ATI) [2], DPCA [3], the more recently proposed adaptive constant false alarm rate (CFAR) [4] and the two-step detector [5]. Since the launch of RADARSAT-2 in 2008, several important lessons have been learned regarding the different physical characteristics of space-based GMTI data when compared to airborne data. One might suppose that successes in describing the statistical distributions of clutter measurements made from an airborne system directly relate to clutter measurements made from a space-based radar. This is, however, not the case for most commercial systems because of the different measurement environment for a space-based SAR. These are as follows: (1) the signal-to-noise ratio (SNR) is significantly lower than that for an airborne system resulting in the need to exploit the entire coherent processing gain in azimuth and (2) the dwell time on the target in the order of a second and motion compensation is not required resulting in a higher correlation between the noise-free clutter measurements. The astounding 'cleanliness' of the space-based SAR data allows important statistical core descriptors such as the covariance matrix of the clutter measurements to be almost perfectly theoretically predictable. Figure 7.1 exemplifies the excellent agreement of the measured eigenvalue distribution of an estimated MODEX-1 two-by-two covariance matrix and the analytically predicted values.

Over the course of the studies, innovative approaches have been developed, called MODEX-2, to increase the number of digital channels or degrees-of-freedom by using the flexible programmability of the radar payload. For RADARSAT-2, up to four data channels can be recorded by changing the transmit (Tx) or receive (Rx) antenna configuration from pulse to pulse [6]. The additional channels improve target parameter estimation but make signal processing more complex and generally reduce the SNR and the available swath width [7]. Techniques to recover some of the lost coverage while keeping the GMTI capability high are currently a topic of active research and promising results are introduced in Section 7.5. Currently, implemented modes onboard of RADARSAT-2 are the toggled transmit modes 1/2 and 3/4 (alongside a not-much-used receiver switching mode) in which the alternating parts of the transmit aperture are turned off between subsequent pulses [6].

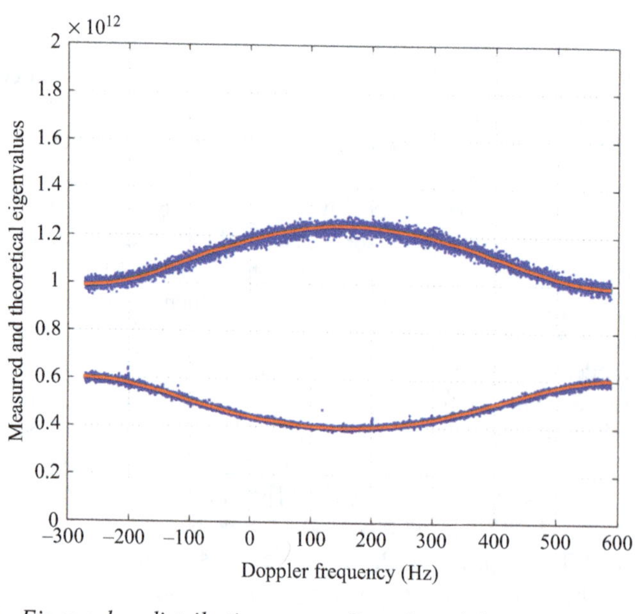

Figure 7.1 Eigenvalue distribution across Doppler of the sample covariance matrix and the theoretically predicated curves

7.2 Classic imaged-based SAR-GMTI

For SAR, when the radar beam dwells for thousands of pulses and when the PRF, range-gate-delay and swath are selected to minimize range and azimuth ambiguities, the same data stream can be processed, through separate processing paths, to produce SAR images, temporal SAR interferometric images and GMTI outputs. The classical two-channel techniques, such as DPCA and ATI, use the entire synthetic aperture time to increase the clutter-to-noise ratio (CNR) thereby improving detectability.[3] Image-based techniques can be regarded as simple and robust methods to quickly produce a kind of 'velocity image' in which motion or moving pixels are colour-coded to allow reliable identification by a human operator/analyst.

In the SAR image domain, n adjacent complex pixels for two receive channels may be modelled as mutually statistically independent random vectors

$$\mathbf{Z}(k) = \begin{bmatrix} Z_1(k) \\ Z_2(k) \end{bmatrix} = \boldsymbol{\mu}(k) + \Delta \cdot \mathbf{C}(k) + \mathbf{N}(k), \quad k \in [1, n] \tag{7.1}$$

The clutter $\mathbf{C}$ and the independent thermal noise $\mathbf{N}$ are commonly considered identically complex Gaussian distributed $\sim \mathcal{N}_2^{\mathbb{C}}(\mathbf{0}, \mathbf{R}_{C,N})$. The vector $\boldsymbol{\mu}$ represents a potential moving target signal, and the scalar $\Delta \in [0, \infty)$ denotes a multiplicative, independent texture random variable (RV) embedding the statistical notion of a randomly varying pixel radar cross section (RCS) (heterogeneity) throughout the SAR image. In other words, since the variance of the SAR data is quite different for rural terrain as compared to urban areas, the texture RV tries to statistically describe these fluctuations, i.e., to suitably elongate the theoretical probability distribution for homogeneous clutter where $\Delta \equiv 1$.

Let us assume that all interference pixels possess the same clutter-plus-noise covariance matrix:

$$\begin{aligned} \mathbf{R} &= \mathcal{E}(\Delta\mathbf{C} + \mathbf{N})(\Delta\mathbf{C} + \mathbf{N})^* = \mathbf{R_C} + \mathbf{R_N} \\ &= \sigma_C^2 \begin{bmatrix} 1 & \rho_C \\ \rho_C & 1 \end{bmatrix} + \sigma_N^2 \begin{bmatrix} 1 & 0 \\ 0 & 1 \end{bmatrix} = (\sigma_C^2 + \sigma_N^2) \begin{bmatrix} 1 & \rho \\ \rho & 1 \end{bmatrix} \end{aligned} \tag{7.2}$$

where $\mathcal{E}$ denotes the expectation operator, the symbol * conjugate complex transpose, and σ_C^2, σ_N^2 the clutter and noise power levels, respectively. The clutter correlation coefficient of $\rho_C < 1$ accounts for residual system aberrations, for instance caused by non-linear RF components, etc. yielding a combined coefficient $\rho \leq 1$. Note that it has been assumed that $\mathcal{E}\Delta^2 = 1$ [5]. It is further assumed that $\mathbf{R}$ is known, for instance accurately estimated with all data in the image, Figure 7.1. The correlation ρ may be assumed real valued when the clutter is stationary and the two channels are already properly coregistered and balanced [8]. Clutter motion, such as ocean currents for instance, can be incorporated as a phase of the coefficient relating to a common radial velocity of the background.

[3]However, the classic methods are not optimum for GMTI in a sense that they do not focus on the moving target but rather on the stationary background.

7.2.1 SAR displaced phase centre antenna

The classical two-phase-centre DPCA methodology was originally developed to improve the detection performance of airborne GMTI radars that are subject to strong background clutter [3]. The test statistic T determining the presence of a moving target is given by the sum of the magnitudes of the pixel differences:

$$T_{\text{DPCA}} = \sum_{k=1}^{n} |Z_1(k) - Z_2(k)|^2 \tag{7.3}$$

The test exploits the fact that the coregistered and balanced stationary background clutter should be considerably reduced (provided a correlation near one), revealing targets that have slightly moved during the time taken for the aft channel to reach the exact location of the fore channel from the previous pulse.

A common approximation in the literature, particularly for airborne radars, is to assume a large CNR thereby neglecting the additive thermal noise altogether. This assumption leads to a simplified classical multiplicative or compound clutter model $\mathbf{Z} = \Delta\mathbf{C}$, which has been extensively used to model different heterogeneous terrain.[4] For ocean clutter dealing with different local wind conditions, often a gamma-probability density function (pdf) has been adopted for the texture RV:

$$f_\Delta(\delta) = \frac{2(\nu)^\nu}{\Gamma(\nu)} \delta^{(2\nu-1)} \exp(-\nu\delta^2) \qquad \delta \geq 0 \tag{7.4}$$

leading to the famous K-distribution. Its test statistic yields the pdf

$$f_{\text{DPCA}}(t;\nu,n) = \frac{2}{\Gamma(n)\Gamma(\nu)} \left(\frac{n\nu}{\sigma_C^2}\right)^{(n+\nu)/2} t^{((n+\nu)/2)-1} K_{n-\nu}\left(2\sqrt{\frac{n\nu}{\sigma_C^2}t}\right) \tag{7.5}$$

where $\Gamma(\cdot)$ denotes the gamma function and $K(\cdot)$ denotes the modified Bessel function [9] giving the pdf its name. The texture parameter ν describes the degree of heterogeneity, i.e., the smaller ν the more heterogeneous the scene. In contrast to this sea-clutter model, the square-root of the inverse chi-square distribution, showing a longer tail, has been found to fit better than the gamma distribution for SAR-GMTI over land [2]. Its pdf is given by

$$f_\Delta(\delta) = \frac{2(\nu-1)^\nu}{\Gamma(\nu)} \delta^{-(2\nu+1)} \exp\left(-\frac{\nu-1}{\delta^2}\right) \tag{7.6}$$

resulting in a 'computationally more favourable' scaled F-distribution for the DPCA test

$$f_{\text{DPCA}}(t;\nu,n) = \frac{1}{B(n,\nu)} \left(\frac{n}{\nu-1}\right)^n \frac{t^{n-1}}{(1+(n/(\nu-1))t)^{n+\nu}} \tag{7.7}$$

[4]Note this assumption is statistically equivalent to texturing the noise, $\mathbf{Z} = \Delta(\mathbf{C}+\mathbf{N})$, provided that clutter and noise are independent Gaussian distributed.

in which $B(a,b)$ describes the beta function. As both texture pdfs tend toward a delta-function for large values of ν, (7.5) and (7.7) converge to the χ^2-density function representing homogeneous clutter. The desired CFAR-threshold η can be determined by solving $\int_0^\eta f_T(t;\nu,n)dt = 1 - P_{\text{fa}}$. This is numerically much less challenging for the F- than the K-distribution because of the transcendental functions. To solve the integral for η, the parameters n and ν must be accurately known, or more practically, be estimated (adapted) from the background clutter data themselves [5].

For the simplified compound clutter model, the probability of detection P_{d} for a deterministic point target, characterized by the parameter ω, can be determined numerically through evaluating the integral

$$P_{\text{d}}(\eta;\omega,\nu) = \frac{n^n(\nu-1)^\nu}{(\omega+\nu-1)^{n+\nu}B(n,\nu)} \cdot \int_\eta^\infty \frac{t^{n-1}}{\left(1+\dfrac{n}{\omega+\nu-1}t\right)^{n+\nu}} \times {}_2\text{F}_1\left(\frac{n+\nu}{2},\frac{n+\nu+1}{2};n;\frac{4n\omega t}{(\omega+\nu-1)^2(1+(n/(\omega+\nu-1))t)^2}\right)dt, \tag{7.8}$$

where ${}_2\text{F}_1(a,b;c;z)$ is the Gauss hypergeometric function.

For space-based SAR, however, the noise cannot be neglected and, consequently, the noise cannot be multiplied by the texture RV. Despite the mathematical intractability to derive the corresponding pdfs analytically, the texture models can still be applied to space-based data. As presented in [10], the false alarm rate $P_{\text{fa}}(\eta)$ for DPCA can be numerically computed via

$$P_{\text{fa}}(\eta) = \frac{1}{\Gamma(n)}\int_0^\infty \Gamma\left(n,\frac{n\sigma^2}{\bar{\sigma}^2(\delta)}\eta\right)f_\Delta(\delta)d\delta \tag{7.9}$$

with $\sigma^2 = \mathbf{u}^*(\mathbf{R}_\text{C}+\mathbf{R}_\text{N})\mathbf{u}$ and $\bar{\sigma}^2(\delta) = \mathbf{u}^*(\delta^2\mathbf{R}_\text{C}+\mathbf{R}_\text{N})\mathbf{u}$ and $\mathbf{u} = [1,-1]'$. Equation (7.9) can be used to determine the CFAR threshold applying any arbitrary texture for which a pdf $f_\Delta(\delta)$ exists.

In cases of an extended target such as large container vessels, where the returns are made up of many independent similarly distributed scattering centres whose combined response is Gaussian distributed according to the central limit theorem, Swerling's models are arguably most widely used to characterize the target's statistics. In this case, the probability of detection, P_{d}, for the general clutter-plus-noise model has been computed in closed form [10]:

$$P_{\text{d}}(\eta,s) = \sum_{\mu=0}^\infty \frac{\Gamma(\mu+s)}{\Gamma(s)\Gamma(\mu+1)\Gamma(n+\mu)}\mathcal{E}_\Delta\left\{\frac{\Gamma(n+\mu,(n\sigma^2/\bar{\sigma}^2(\Delta))\eta)(\xi/\bar{\sigma}^2(\Delta))^\mu}{(1+(\xi/\bar{\sigma}^2(\Delta)))^{\mu+s}}\right\} \tag{7.10}$$

where $s \in \{1,L,2,2L\}$ denotes the chosen Swerling model and $\xi = \frac{L}{s}\mathbf{u}^*\mathbf{R}_\text{S}\mathbf{u}$. The expression $\mathcal{E}_X\{g(X)\} = \int g(x)f(x)\ dx$ implies taking the expectation with respect to the RV X, $\mathbf{R}_\text{S}$ is the target covariance matrix and $L \leq n$ the number of summed

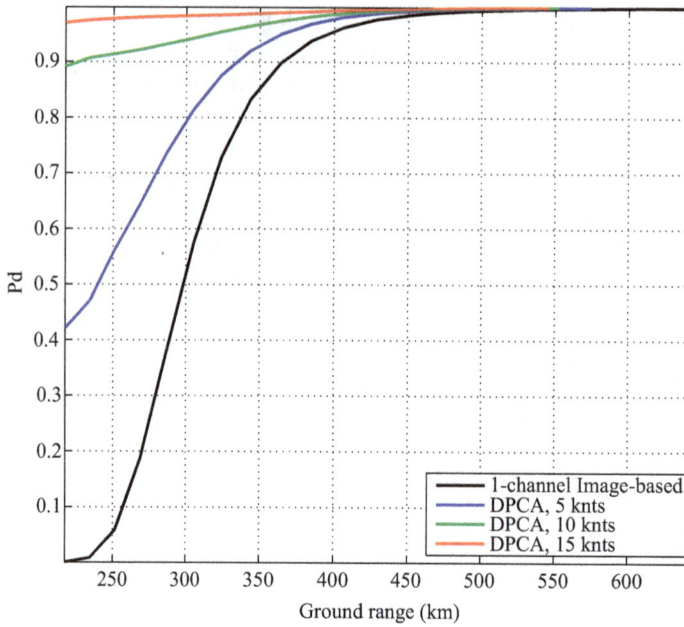

Figure 7.2 P_d of DPCA on RADARSAT-2 for a 25 m Swerling II vessel

pixels containing the target. Note that (7.9) and (7.10) have been successfully used to compute the receiver operating characteristics (ROC) for a wide range of clutter, target and texture models including discrete texture distributions [11]. A deterministic point target signal model is included in (7.10) as the special case when $s \to \infty$ [10].

Exemplified in Figure 7.2 is the achievable detectability of a DPCA mode on RADARSAT-2 for a small vessel in comparison to the simple detection based on the image intensity alone. The ship's incidence angle dependent RCS, and by virtue of the radar equation therewith the SNR as a function of ground range, corresponds to a Swerling II model ($s = L$) with the mean value calculated according to a ship length of 25 m [12]. The clutter RCS (as well as the thermal receiver noise) is complex normal distributed with its mean RCS given by sea reflectivity model CMOD-IFR2 [13] for sea state 5 with 11 m/s along-track wind direction. Accordingly, the texture RV has been chosen as Γ-distributed with parameter $\nu = 5$. For $n = 2$, (7.9) was used to establish the appropriate threshold η that yields a desired $P_{fa} = 2.5 \times 10^{-9}$. It becomes apparent that due to the strong reflection of the high seas reliable detection in near ground range (<300 km) is nearly impossible when solely based on the image intensity (black). In contrast, if the vessel moves with a typical cruising speed of ten knots (ground across-track velocity component), the chance to detect it becomes larger than 90% even close to nadir due to the ability to suppress the strong sea clutter effectively. For higher incidence angles, the sea return fades, and the system becomes noise limited at which point clutter suppression is irrelevant, and DPCA does not offer any advantage anymore. In contrast, application of DPCA is potentially harmful as it might reduce the target power for slow moving targets.

7.2.2 SAR along-track interferometry

In contrast to DPCA, SAR ATI exploits the interferometric phase, i.e., the phase difference between the channels and ignores the magnitude information. The interferometric phase, ϑ, for a multilook cell (i.e., an assembly of n individual statistically independent pixels) is defined as the phase of the coherent sum:

$$\vartheta = \arg\left\{\sum_{k=1}^{n} Z_1(k)Z_2(k)^*\right\} \tag{7.11}$$

For the simplified compound model and in the absence of a target, i.e., $\mathbf{Z} = \varDelta\mathbf{C}$, the argument in (7.11) becomes independent of the real-valued common multiplier,[5] and hence, the multilook marginal phase pdf of clutter plus noise is identical to the homogeneous pdf derived in [14]:

$$f_\vartheta(\vartheta) = \frac{\Gamma(n+1/2)(1-\rho^2)^n\rho\cos(\vartheta-\psi)}{2\sqrt{\pi}\Gamma(n)(1-\rho^2\cos^2(\vartheta-\psi))^{n+1/2}} + \frac{(1-\rho^2)^n}{2\pi} \times {}_2\mathrm{F}_1(n,1;1/2;\rho^2\cos^2(\vartheta-\psi)) \tag{7.12}$$

The angle $\psi \neq 0$ may incorporate potential internal clutter motion such as ocean surface currents. Figure 7.3(a) shows a RADARSAT-2 SAR image of an area around the city of Fargo ND, together with the computed histogram of the ATI phase averaged over four looks. As expected for a stationary scene, the ATI phase is centred around zero. The perfect fit of the pdf (7.12) illustrates the invariance toward multiplicative moderately heterogeneous clutter when the unknown parameters are adaptively estimated [15].

The statistics of the interferometric phase when a moving target signal is superimposed upon the simplified textured clutter (no noise) has first been investigated in [2].[6] Although ATI has been shown to improve robustness against texture, no clutter suppression is done, and a main drawback with regard to target detection is the limited support between $-\pi$ and π, which in many cases precludes a sufficiently small P_{fa}. In other words, if the P_{fa} is too small, the threshold approaches $\pm\pi$ and no meaningful detection is possible. To alleviate this problem, adding an appropriate constant prior to the phase computation has been introduced in [16]. Other variants propose the combination of the ATI phase and the ATI-magnitude to improve performance [2,17].

[5]Since $\varDelta^2$ is common for all n pixels, it can be promoted outside the sum and cancels out when the argument is calculated.

[6]When the noise cannot be neglected, the computation of the statistics of ϑ becomes exceedingly difficult, and no analytical expression has been found to date.

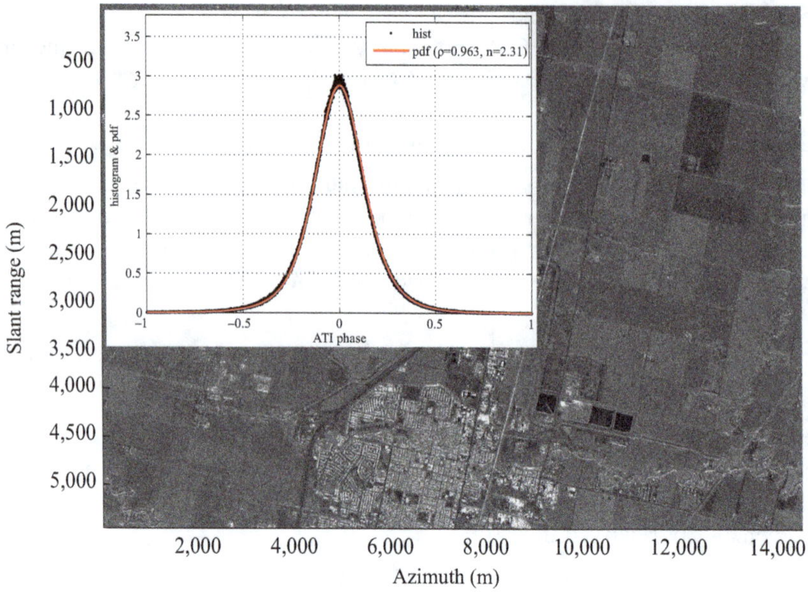

(a) Histogram and pdf for heterogenous clutter.

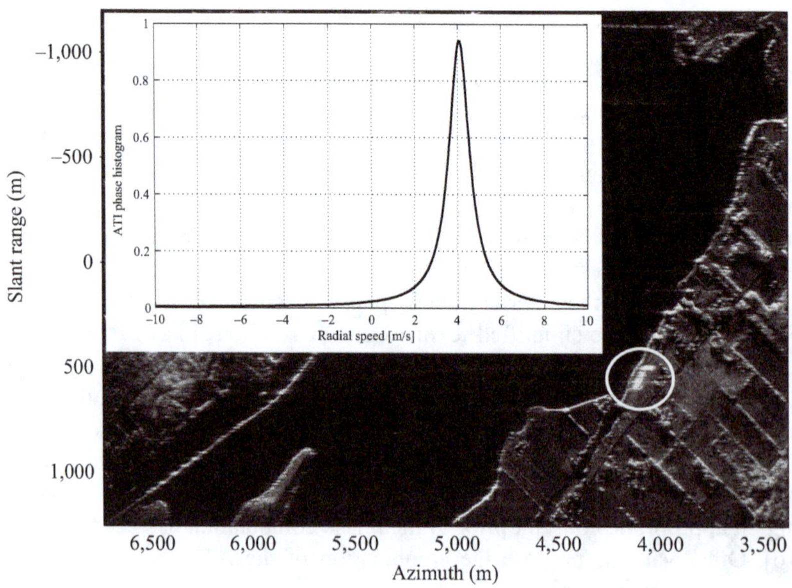

(b) Velocity estimation of an inland vessel.

Figure 7.3 ATI phase estimation

Although often not the preferred choice as detector, due to its simplicity, the ATI phase has established itself as the classic technique to estimate the across-track velocity component for two-channel SAR-GMTI systems. In a first order approximation, the phase difference relates to the target's slant across-track velocity component v_s via

$$\vartheta \cong \frac{2\pi}{\lambda} d \frac{v_s}{v_a} \tag{7.13}$$

where λ denotes the wavelength, v_a the platform or antenna velocity and d the distance between the two receiving antennas. For land targets, the ATI velocity estimate is usually heavily compromised by the strong background clutter contained in the same resolution cell and therefore oftentimes [depending on the signal-to-clutter ratio (SCR)] yields a rather biased and unreliable estimate [18]. In contrast, for ships, the clutter background is oftentimes comparatively weak especially in low sea states such that the ATI phase presents a viable option to derive velocity information of vessels [19,20]. Figure 7.3(b) demonstrates this ability based on a SAR image of a tanker which was travelling in the centre of the St. Lawrence Seaway near Iroquois, Eastern Canada. The circle in the bottom right corner highlights the smeared and displaced image of the ship caused by the motion-induced Doppler-shift. At the time of imaging, the ships automated identification system (AIS) system reported an across-track velocity of 4.028 m/s. The inlet shows the ATI phase histogram of all vessel pixels translated into radial speed via (7.13) confirming an excellent agreement with AIS.

7.2.3 Strong non-homogeneous clutter

For a theoretical ideal measurement system with fully correlated clutter, i.e., $\rho_C = 1$ in (7.2), between the channels, DPCA would always remove the clutter in its entirety regardless of heterogeneity caused by the texture fluctuation.[7] Any real radar system, however, is subject to some hardware errors and channel imbalances resulting in residues of large RCS stationary targets in the difference image. These remaining residues increase the number of false targets drastically rendering application of pure DPCA sometimes less useful for space-based GMTI. Different techniques have been developed to overcome this problem.

7.2.3.1 Two-step detector

It can mathematically be shown that the smallest eigenvalue of the sample covariance matrix and the ATI phase are statistically independent RVs (for $n = 2$) [21] and that the ATI phase and DPCA are to 'a large degree independent' [5]. This

[7]Because $\sigma^2 = \mathbf{u}^* \mathbf{R} \mathbf{u} = \sigma_N^2 \mathbf{I}$ in (9) for $\rho_C = 1$ and $u = [1, -1]'$.

property can be exploited to achieve a large P_d while keeping the false alarm rate acceptably low, because

$$\begin{aligned} P_{fa} &= 2\int_{\eta_1}^{\infty}\int_{\eta_2}^{\pi} f_{T_{DPCA},\vartheta}(t,\vartheta)dtd\vartheta \\ &\cong 2\int_{\eta_1}^{\infty} f_{T_{DPCA}}(t)dt \cdot \int_{\eta_2}^{\pi} f_{\vartheta}(\vartheta)d\vartheta = P_{fa,1}\cdot P_{fa,2} \end{aligned} \tag{7.14}$$

which means that the total P_{fa} can be significantly lowered by applying two subsequent tests where each individually shows a rather moderate false alarm rate. Equivalently, the detection probability is also given as the product $P_d = P_{d,1}\cdot P_{d,2}$ provided statistical independence. Figure 7.4 demonstrates this feature on the RADARSAT-2 image over Fargo, cf. Figure 7.3(a). Due to the fact that DPCA and ATI are not fully independent, a realistically achievable $P_{fa} \cong 10^{-6} < 10^{-4}\cdot 10^{-3}$ is expected leading to on average of less than eight false detections in the image for this particular example. Despite the fact that no ground truth was available, the false detections on the strongly backscattering building complexes in the centre of the image have disappeared. In contrast, reducing the P_{fa} of DPCA alone to 10^{-7} (red circles in Figure 7.4(a)) still leads to too many false alarms, thereby indicating that the simple compound model does not sufficiently map the reality particularly in the tail of the pdf [5]. The remaining detections in Figure 7.4(b) possess a relatively small amplitude in the SAR image, strongly indicating that these are in fact true targets.

7.2.3.2 Fully adaptive CFAR

An entirely different approach attempts to circumvent the use of a multiplicative texture RV all together. The fully adaptive CFAR detector presented in [4] proposes to estimate the ratio between the clutter power and residual clutter power after suppression, e.g., using DPCA, for each individual image pixel. The main idea is based on introducing a measurement distortion random matrix $\mathbf{E}$ that is independent of the clutter and noise. In contrast to (7.1), the new model reads

$$\bar{\mathbf{Z}}(k) = \boldsymbol{\mu}(k) + \mathbf{E}\mathbf{C}(k) + \mathbf{N}(k) \tag{7.15}$$

Let us assume a diagonal structure of the matrix consisting of magnitude and phase distortions, i.e.,

$$\mathbf{E} = \begin{bmatrix} e^{\alpha_1 + j\kappa_1} & 0 \\ 0 & e^{\alpha_2 + j\kappa_2} \end{bmatrix} \tag{7.16}$$

with uncorrelated Gaussian distribution for the variables α_i and κ_i such that $\mathcal{E}\,\mathbf{E} = \mathbf{I}$. In this case, the distribution of the modified DPCA test $\bar{T}_{DPCA} = \sum_{k=1}^{n} |\bar{Z}_1(k) - \bar{Z}_2(k)|^2$ has been analytically derived in [4]. It is important to realize that this model features the desired practical properties, (a) perfect clutter cancelation is only achievable if the channel distortion variances of α_i and κ_i vanish

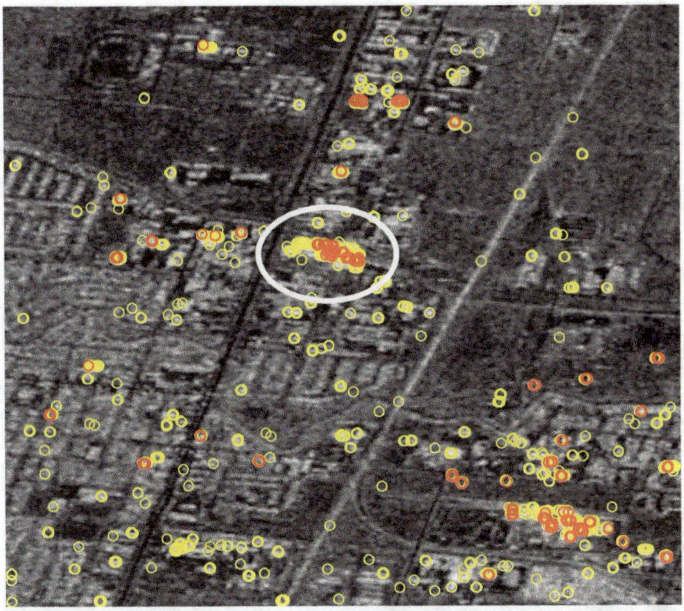

(a)

(b)

Figure 7.4 *SAR image of the fore channel with superimposed moving target detections (red denotes positive and green negative sign of the ATI phase). (a) DPCA for* $P_{fa} = 10^{-4}$ *(yellow) and for* 10^{-7} *(red) and (b) DPCA for* $P_{fa} = 10^{-4}$ *plus ATI for* $P_{fa} = 10^{-3}$

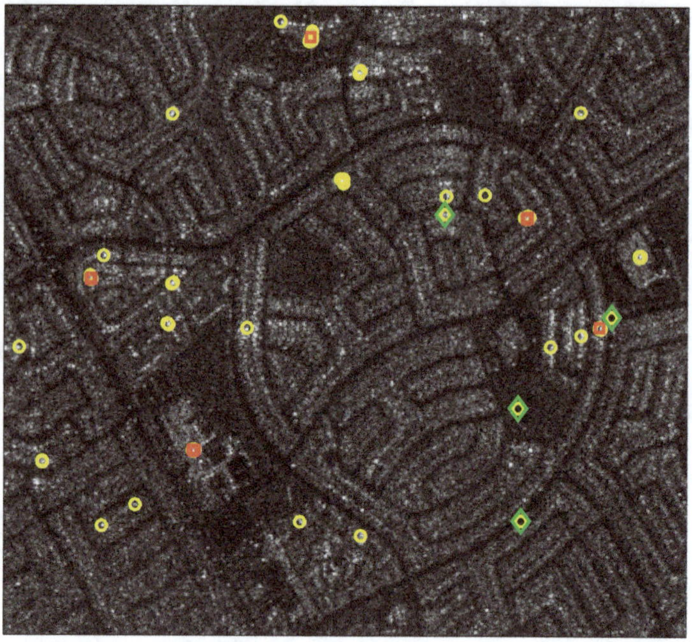

Figure 7.5 CFAR detections for $P_{fa} = 10^{-6}$. (Yellow circles) Homogeneous CFAR. (Red squares) Texture model. (Green diamonds) Fully Adaptive CFAR

and (b) some remnants of stationary scatterers with very large RCS values remain in the processed SAR imagery if they do not vanish.

The CFAR threshold η is given as the numerical solution of the rapidly converging series

$$P_{\text{fa}}(\eta) = e^{-a_0\hat{\sigma}_C^2 + a_1/4\hat{\sigma}_N^2} \sum_{\mu=0}^{\infty} \frac{\Gamma(\mu+1, (\eta/4\hat{\sigma}_N^2))}{\Gamma(\mu+1)^2} \left(\frac{a_0\hat{\sigma}_C^2 + a_1}{4\hat{\sigma}_N^2} \right)^{\mu} \tag{7.17}$$

Alongside the estimates for the clutter and noise power levels $\hat{\sigma}_C^2$ and $\hat{\sigma}_N^2$, respectively, the parameters a_0 and a_1 have to be determined using the original and the DPCA clutter suppressed data; for details see [4]. Figure 7.5 illustrates the effectiveness of the fully adaptive CFAR on RADARSAT-2 data in a heterogeneous urban residential area. Detections using the homogeneous model (no texture $\Delta = 1$) are represented as yellow circles. As predicted, there are excessively many false alarms which are primarily colocated with high RCS clutter. CFAR using DPCA and the texture model reduces the number of false alarms as shown by the red squares. However, other targets that do not colocate with large RCS clutter are also eliminated. In the absence of ground truth data, these targets are assumed to be real moving targets. Fully adaptive CFAR detections are indicated by green diamonds, which show that all of the large RCS false alarms are eliminated, while detection of the remaining likely real targets is preserved.

7.3 Coherent multichannel SAR-GMTI

7.3.1 Spatial diversity via aperture switching and toggling

In the absence of space-based SAR systems with more than two channels, it has been successfully demonstrated on RADARSAT-2 that one can successfully increase the number of degrees-of-freedom to up to four by cyclically varying the antenna aperture between subsequent transmit pulses. The basic underlying idea here is to increase the number of digitally to-be-processed data sets via innovative concepts that can be programed on the spacecraft while in orbit.[8] In its most basic form, this is achieved by turning off different parts of the aperture during transmit or receive provided one is willing to accept some loss in SNR [6,22]. Different partitioning of the whole aperture will shift the phase centre locations and as a consequence creates more independent receive channels in a time-multiplexed fashion. These modes have been dubbed MODEX-2 on RADARSAT-2. As an example, Figure 7.6 depicts six different switching and toggling schemes for two subsequent pulses. Each coloured box depicts either the active transmitting apertures (red) or the active receive apertures (blue), where the hinted sensor symbols describe the individual phase centres. While five modes are examples with full transmit aperture, $\mathbf{T}_{\text{toggle}}$ represents a toggle mode where the energy is transmitted alternately from either half of the antenna. $\mathbf{T}_2$ represents the classical non-switching two-channel ATI mode, and $\mathbf{T}_{\text{synth}}$ indicates a receiver switching modes which synthesizes a third phase centre by summation of both receiver halves during the second pulse. The blue symbols underneath show the resulting transmit-receive phase centres, which illustrate the increased number of digital channels.

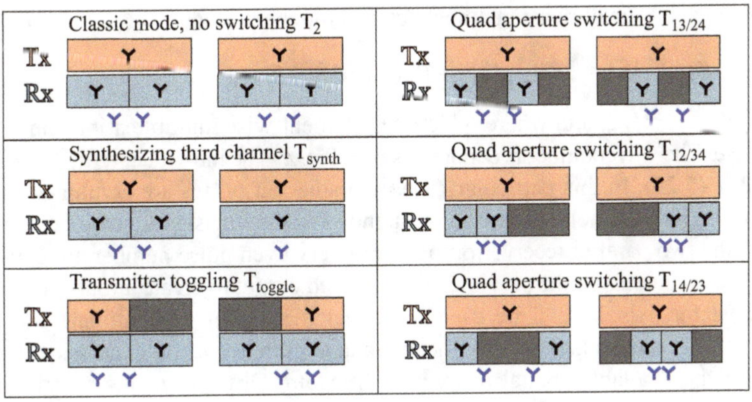

Figure 7.6 Different switching strategies for two-channel radar systems

[8]The costs involved for physically modifying the RF-frontend (combining network, extra ADC-and recording unit etc.) are currently simply prohibitive and preventing a full hardware solution in the near future.

This multichannel capability requires the development of new advanced signal-processing techniques that are founded on the principles of STAP [23,24]. It quickly becomes apparent that the special challenges of space-based SAR require 'long' dwell time on potential targets in order to permit any meaningful detectability and estimation performance of practically relevant moving objects. This requirement is primarily driven by the relative low transmit power and enormous stand-off ranges in comparison to airborne radar systems. In other words, it is essential from space to utilize the entire coherent processing interval (CPI) available (synthetic aperture time) in order to accumulate sufficient signal power in relation to the noise. This necessitates the use of the raw data as basis for the signal processing rather than the already stationary world focused SAR data. In virtually all relevant cases, it is sufficient to consider range-compressed data that are also sometimes referred to as phase history data.

7.3.1.1 Point target signal model

For a SAR system with sufficient range resolution,[9] let a deterministic moving target signal be described by its unknown parameter vector $\boldsymbol{\xi} = [a, v_x, v_s, x_b]'$, in which $a \in \mathbb{C}$ is the complex amplitude containing its RCS, v_x the along-track velocity, v_s the across-track velocity in slant range and x_b its azimuth location at broadside time t_b. Restraining ourselves to a system such as RADARSAT-2 with two parallel receive channels (without loss of generality) as well as constant target velocities (i.e., no acceleration), then the signal for M discrete transmit pulses can be written as the concatenated vector

$$\mathbf{s}(\boldsymbol{\xi}) = \left[\mathbf{s}_1'(\boldsymbol{\xi}), \ldots, \mathbf{s}_{M/2}'(\boldsymbol{\xi})\right]' \in \mathbb{C}^{2M \times 1} \tag{7.18}$$

in which each four-dimensional subvector component reads

$$\mathbf{s}_m(\boldsymbol{\xi}) = \exp\{-j2\beta \mathbf{R}_m(\boldsymbol{\xi})\} \odot \mathfrak{D}_{\mathrm{Tx}}[u_m(\boldsymbol{\xi})] \odot \mathfrak{D}_{\mathrm{Rx}}[u_m(\boldsymbol{\xi})] \quad \in \mathbb{C}^{4\times 1} \tag{7.19}$$

for $m = 1, \ldots, M/2$, and where $\odot$ denotes element-wise multiplication and $'$ vector transpose. The wavenumber is denoted as $\beta = 2\pi/\lambda$. The individual terms are defined in (7.22)–(7.25). In this particular case, two adjacent pulses are combined to form the four digital channels that are subsequently fed into the signal processor. In other words, the two parallel receiver outputs for every even pulse number are combined with the two outputs of the previous odd-sampled output to create a four-dimensional signal vector, which shall be used to cancel the clutter and estimate the target parameters. The idea is to modify the antenna pattern between the pulses in order to increase the available spatial degrees-of-freedom. Obviously, this rearranging of the data increases the effective pulse repetition interval (PRI $= \Delta t$) by a factor of two (or generally by the cycle length L), i.e., the larger dimensionality has been traded off with a loss of effective unambiguous PRF. Or, per contra, the PRF must be doubled which in turn will reduce the available swath width.

[9]For which the target's range location can be assumed known.

The TR-module settings of the N columns in flight direction on Tx, and Rx respectively, can be written as transformation matrices $\mathbf{T}_{\mathrm{Tx}}$ and $\mathbf{T}_{\mathrm{Rx}}$ describing the front-end combiner network. For instance, the conventional MODEX-1 mode with receive on either half of the antenna is given to

$$\mathbf{T}_{\mathrm{Rx}} = \frac{1}{\sqrt{N/2}} \begin{pmatrix} \mathbf{1}_{\frac{N}{2}} & \mathbf{0}_{\frac{N}{2}} & \mathbf{0}_{\frac{N}{2}} & \mathbf{0}_{\frac{N}{2}} \\ \mathbf{0}_{\frac{N}{2}} & \mathbf{1}_{\frac{N}{2}} & \mathbf{0}_{\frac{N}{2}} & \mathbf{0}_{\frac{N}{2}} \\ \mathbf{0}_{\frac{N}{2}} & \mathbf{0}_{\frac{N}{2}} & \mathbf{1}_{\frac{N}{2}} & \mathbf{0}_{\frac{N}{2}} \\ \mathbf{0}_{\frac{N}{2}} & \mathbf{0}_{\frac{N}{2}} & \mathbf{0}_{\frac{N}{2}} & \mathbf{1}_{\frac{N}{2}} \end{pmatrix} \in \mathbb{R}^{2N\times 4} \tag{7.20}$$

in which $\mathbf{1}_\mu$ and $\mathbf{0}_\mu$ are vectors of length μ containing all ones or zeros, respectively. For MODEX-2 $\frac{1}{2}$ for instance,

$$\mathbf{T}_{\mathrm{Tx,tog}\frac{1}{2}} = \frac{1}{N} \begin{pmatrix} \mathbf{1}_{\frac{N}{2}} & \mathbf{1}_{\frac{N}{2}} & \mathbf{0}_{\frac{N}{2}} & \mathbf{0}_{\frac{N}{2}} \\ \mathbf{0}_{\frac{N}{2}} & \mathbf{0}_{\frac{N}{2}} & \mathbf{0}_{\frac{N}{2}} & \mathbf{0}_{\frac{N}{2}} \\ \mathbf{0}_{\frac{N}{2}} & \mathbf{0}_{\frac{N}{2}} & \mathbf{0}_{\frac{N}{2}} & \mathbf{0}_{\frac{N}{2}} \\ \mathbf{0}_{\frac{N}{2}} & \mathbf{0}_{\frac{N}{2}} & \mathbf{1}_{\frac{N}{2}} & \mathbf{1}_{\frac{N}{2}} \end{pmatrix} \in \mathbb{R}^{2N\times 4} \tag{7.21}$$

Other interesting candidates considered for RADARSAT-2 are the three-quarter toggling mode $\mathbf{T}_{\mathrm{Tx,tog}\frac{3}{4}}$ or the quarter-aperture receiver switching modes described in [6]. Consequently, the corresponding antenna pattern for two combined pulses in (7.19) can be written as a function of the time-varying azimuthal cosine $u(k\Delta t)$ with $k \in \mathbb{Z}$

$$\mathfrak{D}_{\mathrm{Tx,Rx}}[u_m(\boldsymbol{\xi})] = \mathbf{T}^*_{\mathrm{Rx,Tx}} \begin{bmatrix} \mathbf{d}[u((2m-1)\Delta t, \boldsymbol{\xi})] \\ \mathbf{d}[u(2m\Delta t, \boldsymbol{\xi})] \end{bmatrix} \tag{7.22}$$

where $\mathbf{d}(\cdot)$ denotes the direction-of-arrival or steering vector of the antenna array

$$\mathbf{d}[u(k\Delta t)] = \begin{bmatrix} D_{-N/2}(u(k\Delta t))e^{j\beta x_{-N/2}u(k\Delta t)} \\ \vdots \\ D_0(u(k\Delta t)) \\ \vdots \\ D_{N/2}(u(k\Delta t))e^{j\beta x_{N/2}u(k\Delta t)} \end{bmatrix} \tag{7.23}$$

$D_\mu(\cdot)$ indicates the one-way diagram and x_μ the position of the μth antenna element. For a space-based SAR, the slow time-dependent range from the antenna origin to the target, and the azimuth cosine at pulse time $k\Delta t$ might be approximated to

$$\begin{aligned} R_0(k\Delta t, \boldsymbol{\xi}) &\cong R_b + v_s(k\Delta t - t_b) + \frac{v_{\mathrm{rel}}^2 - v_{bz}v_a}{2R_b}(k\Delta t - t_b)^2 \\ u(k\Delta t, \boldsymbol{\xi}) &\cong -\frac{v_x - v_a - v_{bz}}{R_b}(k\Delta t - t_b) \end{aligned} \tag{7.24}$$

where $v_{bz}v_a = R_b a_z \cos(\vartheta)$, v_a denotes the satellite velocity in the Earth-centred Earth-fixed (ECEF) coordinate system, R_b the slant range at broadside time, a_z the gravitational acceleration and $v_{\text{rel}}^2 = (v_x - v_a)^2 + v_s^2$ [25]. With (7.24), the range-dependent vector of the chirp-term in (7.19) becomes

$$\mathbf{R}_m(\boldsymbol{\xi}) = \begin{bmatrix} R_0((2m-1)\Delta t, \boldsymbol{\xi}) \\ R_0(2m\Delta t, \boldsymbol{\xi}) \end{bmatrix} \otimes \begin{bmatrix} 1 \\ 1 \end{bmatrix} \quad \in \mathbb{C}^{4\times 1} \tag{7.25}$$

when $\otimes$ stands for the Kronecker-product. It should be noted that this statistical model is not explicitly defined using the notion of phase centre locations but rather generally via arbitrary complex antenna diagrams.

7.3.1.2 Statistical clutter model

The expected signal from one normalized clutter scatterer located at azimuth location x_b can be described via (7.18) using the parameter vector $\boldsymbol{\xi}_c = [1; 0; 0; x_b]$. The total measured clutter signal is then given by integration over all possible clutter contributions for a given range. This integration corresponds to a convolution with the reflectivity function of the underlying scene. Modelling this reflectivity distribution as a stationary stochastic process (i.e., the statistical properties of the reflectivity are invariant against azimuthal shifts), the composite clutter output can be fully described via its matrix-valued covariance function [20]. Assuming a spatially white discrete stochastic process with constant clutter power σ_C^2, i.e., homogeneous clutter, the space-time clutter covariance matrix in the time domain can be practically (numerically) approximated to any arbitrary accuracy via a sum of a large number of distributed point scatterers

$$\mathbf{C} = \sum_{l_b} \sigma_C^2 \mathbf{s}(\boldsymbol{\xi}_c)\mathbf{s}(\boldsymbol{\xi}_c)^* + \sigma_N^2 \mathbf{I}_{2M} \quad \in \mathbb{C}^{2M\times 2M} \tag{7.26}$$

The summing index $l_b \in \mathbb{Z}$ represents the clutter scatterers located at $x_b = l_b \Delta x$, with Δx being an arbitrary small scatterer spacing on the ground (smaller than the azimuth resolution cell). The large matrix contains the cross-correlations between each channel and all $2M$ measurement times (pulses). The second term takes into account the white sensor noise independent between channels and time samples. The matrix $\mathbf{C}$ is of block-Toeplitz structure and generally possesses full rank.

7.3.2 Imaging STAP (iSTAP)

In order to test for the presence of a moving target $\mathbf{s}$ in clutter plus noise $\mathbf{W} = \mathbf{C} + \mathbf{N}$. The following composite hypothesis test problem shall be considered:

$$\begin{aligned} &\text{Hypothesis } \mathcal{H}: \mathbf{Z} = \mathbf{s}(\boldsymbol{\xi}) + \mathbf{W} \\ &\text{Alternative } \mathcal{A}: \mathbf{Z} = \mathbf{W} \end{aligned} \tag{7.27}$$

The maximum likelihood quotient (MLQ) test for this problem reads

$$\Lambda(\mathbf{Z}) = \underset{\boldsymbol{\xi}}{\operatorname{argmax}} \frac{f_{\mathbf{z}}(\mathbf{z}, \boldsymbol{\xi}, \mathcal{H})}{f_{\mathbf{z}}(\mathbf{z}, \mathcal{A})} \gtrless \eta \tag{7.28}$$

where $f_{\mathbf{Z}}(\cdot)$ denotes the pdf of the measured data vector under the hypothesis and alternative, respectively, and η is the detection threshold. Assuming a multivariate complex Gaussian distribution i.e., $Z \sim \mathcal{N}_{2M}^{\mathbb{C}}(\mathbf{s}(\boldsymbol{\xi}), \mathbf{C})$ with pdf $f_{\mathbf{Z}}(\mathbf{z}, \boldsymbol{\xi}) = \pi^{-2M}|\mathbf{C}|^{-2M} \exp\{(\mathbf{z} - \mathbf{s}(\xi))^* \mathbf{C}^{-1}(\mathbf{z} - \mathbf{s}(\boldsymbol{\xi}))\}$, taking the logarithm on both sides, and maximize analytically with respect to the complex amplitude a leads to the test

$$\ln \Lambda(\mathbf{Z}) = \underset{\boldsymbol{\xi}'}{\arg\max} \frac{\left|\mathbf{s}(\boldsymbol{\xi}')^* \mathbf{C}^{-1} \mathbf{Z}\right|^2}{\mathbf{s}(\boldsymbol{\xi}')^* \mathbf{C}^{-1} \mathbf{s}(\boldsymbol{\xi}')} \gtrless \eta \tag{7.29}$$

with the remaining parameter vector $\boldsymbol{\xi}' = [v_x, v_s, t_b]'$. Note that any linear invertible transformation $\mathbf{F}^*\mathbf{Z}$ with $\mathbf{F} \in \mathbb{C}^{2M \times 2M}$ applied to the test (7.27) will not change the optimum test statistic in (7.29), since $\mathbf{F}$ cancels out. This is obviously the case if the columns of $\mathbf{F}$ represent the discrete Fourier-transform at $M/2$ distinct frequency bins between −PRF/2 and PRF/2. In other words, the optimum test is identical regardless of it being performed in the time or Doppler domain.

Although the test is invariant against linear transformations, the statistical properties of the transformed data vector may change considerably. It has been shown in [26] that the spectral density matrix (discrete Fourier-transform of the covariance matrix, $\mathbf{R} = \mathbf{F}^*\mathbf{CF}$) converges to a block diagonal structure if the number of time samples is large enough (always the case in SAR), i.e.,

$$\mathbf{Z}(\mathbf{f}, \boldsymbol{\xi}) := \mathbf{F}^*\mathbf{Z}(\boldsymbol{\xi}) \sim \mathcal{N}_{2M}^{\mathbb{C}}\left(\begin{bmatrix} \mathbf{S}(f_1, \boldsymbol{\xi}) \\ \mathbf{S}(f_2, \boldsymbol{\xi}) \\ \vdots \\ \mathbf{S}(f_{M/2}, \boldsymbol{\xi}) \end{bmatrix}, \begin{bmatrix} \mathbf{R}(f_1) & 0 & \cdots & 0 \\ 0 & \mathbf{R}(f_2) & 0 & \vdots \\ \vdots & 0 & \ddots & 0 \\ 0 & \cdots & 0 & \mathbf{R}(f_{M/2}) \end{bmatrix}\right) \tag{7.30}$$

in which $\mathbf{S}(f_k, \boldsymbol{\xi}) \in \mathbb{C}^{4 \times 1}$ is the Fourier-transformed signal vector in (7.18) at frequency bin f_k for $k = 1, \ldots, M/2$. Inserting (7.30) into (7.29), it is easy to verify that the optimum test statistic in the Doppler domain becomes sum of the smaller dimensional block matrices

$$\boldsymbol{\xi}' = \underset{[v_x, v_s, t_b]}{\arg\max} \frac{\overbrace{\left|\sum_k \overbrace{\mathbf{S}(f_k, v_x, v_s)^*}^{\text{De-chirping/ beamforming}} \; \overbrace{\mathbf{R}(f_k)^{-1}}^{\text{Clutter suppression}} \; \overbrace{\mathbf{Z}(f_k)}^{\text{Data in Doppler}} e^{-j2\pi f_k t_b}\right|^2}^{\text{Discrete Fourier Transform}}}{\sum_k \mathbf{S}(f_k, v_x, v_s)^* \; \mathbf{R}(f_k)^{-1} \; \mathbf{S}(f_k, v_x, v_s)} \gtrless \eta \tag{7.31}$$

revealing the mechanism behind optimum SAR-GMTI processing: For each individual Doppler frequency bin f_k, the clutter in the measured data vector is suppressed by application of the inverse covariance matrix at this frequency, then dechirped with velocity components v_x and v_s and summed over all channels (beamformed). Maximizing with respect to both velocities assures matching to the true underlying speed values. The resulting frequency samples are subsequently Fourier-transformed back into the time domain, the location t_b of the maximum peak sought and finally after a suitable normalization compared to η. Hence, the name integrated or imaging STAP (iSTAP) [7]. The optimum achievable signal-to-noise-plus-clutter ratio (SNCR) for iSTAP given a perfectly matched moving target signal (known velocity components) becomes:

$$\mathrm{SNCR_{opt}} = \sum_k \mathbf{S}(f_k, v_x, v_s)^* \ \mathbf{R}(f_k)^{-1}\mathbf{S}(f_k, v_x, v_s) \tag{7.32}$$

The achievable SCNR after clutter suppression is the most critical parameter determining the P_d. Remembering that the signal vector $\mathbf{S}$ also depends on the toggling and switching matrices according to (7.22), it can be used to analyse and compare the performance of different radar configurations regarding detectability of moving vehicles.

Figure 7.7 illustrates the optimally achievable iSTAP SNCR versus across-track velocity v_s for RADARSAT-2 given an assumed point target with SNR = 20 dB and an identical CNR = 20 dB. As expected, the classic MODEX-1 mode (denoted by T_2) approaches this SNR outside of the clutter band, while the toggling modes are characterized by a nearly constant loss in power of 2.5 dB for toggle 3/4 and about 6 dB for toggle1/2. This loss is caused by the decreased transmit power as well as reduced transmit gain. A slight widening of the clutter band due to the reduced aperture (increased antenna beamwidth) is also observable.

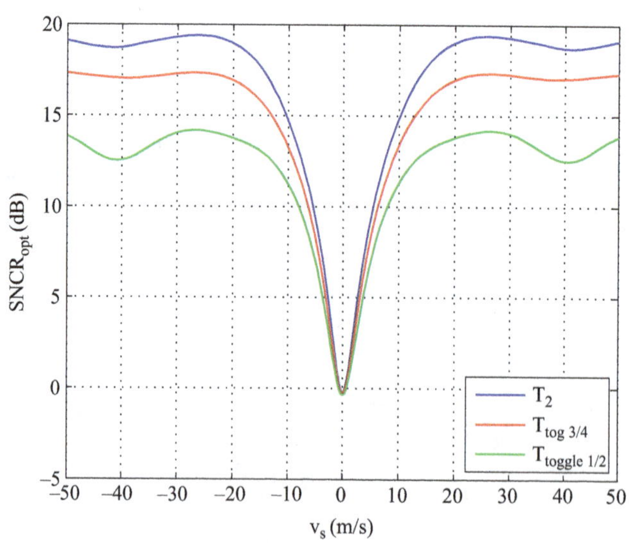

Figure 7.7 Optimum iSTAP SNCR for RADARSAT-2 with a CPI of 2048 pulses

7.3.3 Extended DPCA (EDPCA)

The optimization in (7.31) is very time consuming, as the clutter suppression has to be done for each individual frequency bin and the remaining target energy integrated over the entire Doppler band to achieve a sufficient target SNR.[10] What if we could integrate the energy (at least) approximately before clutter cancelation? Identical to the classic methods DPCA and ATI, one possibility is to apply the azimuth compression with the stationary world matched filter and to coregister the data first only to refine the test afterwards. This procedure can as well be expressed as a matrix transformation of the data vector $\tilde{\mathbf{Z}} = \mathbf{H}^*\mathbf{Z}$, where the columns of $\mathbf{H}$ describe the azimuth reference function for particular scatterer locations and also remove the time delay between the channels [27].

It can be shown that the transformed covariance matrix $\tilde{\mathbf{C}} = \mathbf{H}^*\mathbf{C}\mathbf{H}$ for Nyquist sampling has again a block-diagonal structure but compared to (7.30) with identical blocks $\mathbf{C}_0$, i.e., $\mathbf{Z} \sim \mathcal{N}_{\mathbb{C}}^{2M}\left(\tilde{\mathbf{S}}(\xi), \mathbf{C}_0 \otimes \mathbf{I}_M\right)$, resulting in the test statistic

$$\underset{v_x, v_s}{\operatorname{argmax}} \frac{\left|\sum_k \tilde{\mathbf{S}}_k(v_x, v_s)^* \mathbf{C}_0^{-1} \tilde{\mathbf{Z}}_k\right|^2}{\sum_k \tilde{\mathbf{S}}_k(v_x, v_s)^* \mathbf{C}_0^{-1} \tilde{\mathbf{S}}_k(v_x, v_s)} \gtrless \eta \tag{7.33}$$

in which $\tilde{\mathbf{S}} = \left[\tilde{\mathbf{S}}_1, \ldots, \tilde{\mathbf{S}}_{M/2}\right] = \mathbf{H}^*\mathbf{s}$. In other words, if we coregister and stack up the four SAR images, clutter suppression can be achieved by applying the inverse of one constant 4×4 matrix for each pixel in the image stack. The result has to be summed over the extent of the target. For the special case of only two channels, it can be shown that $\mathbf{C}_0\mathbf{S}_k \cong [1 \; -1]$. Hence, this approach can be regarded as an extension of the classic two-channel DPCA to an arbitrary number of channels, and therefore has been dubbed EDPCA.

Although $\mathbf{C}_0$ is now constant throughout the SAR image, EDPCA is still computationally expensive as the detection, and estimation of the target parameters must be performed simultaneously for all pixels within the image by searching over an extended continuous range of velocities. On way to reduce the computational load is to separate the detection from the estimation by detecting over only a very limited number of discrete velocities first. The actual more accurate velocity estimation can be performed after detection, e.g., using the maximum likelihood estimator (MLE)

$$\hat{v}_x, \hat{v}_s = \underset{v_x, v_s}{\operatorname{argmax}} \frac{\left|\tilde{\mathbf{S}}_{k_c}(v_x, v_s)^* \mathbf{C}_0^{-1} \tilde{\mathbf{Z}}_{k_c}\right|^2}{\tilde{\mathbf{S}}_{k_c}(v_x, v_s)^* \mathbf{C}_0^{-1} \tilde{\mathbf{S}}_{k_c}(v_x, v_s)} \tag{7.34}$$

where k_c denotes the location in which the target was detected.[11] Additionally, although the correct chirp rate of the reference function depends on v_x according to

[10]Optimization in the time domain (7.29) is equally resource-hungry as the space-time covariance matrix $\mathbf{C}$'s dimension is typically in the order of thousands rows and columns.

[11]Note if the azimuth reference functions in $\mathbf{H}$ involves no phase ramp, i.e., $v_s = 0$ (7.24), then the target will appear displaced at k_c and not be located at the true broadside position k_b.

(7.24), for realistic ground moving object velocities and space-based SAR; however, it can oftentimes be neglected and set to zero without compromising the compression gain observably. Consequently, in practice (7.34) may only be maximized over v_s.

Extended experimental benchmark tests with RADARSAT-2 have been carried out over the last several years, whereby global positioning system (GPS) equipped ground truth was applied to confirm the theoretically predicted performance, e.g., [7,22]. For instance, this way the theoretical prediction has been confirmed that RADARSAT-2 is capable of detecting a midsize car with 90% probability when travelling with a 40 km/h radial velocity in a semiurban area ($P_{\text{fa}} = 10^{-6}$). Figure 7.8 shows two SAR images of Ottawa, Ontario overlaid onto a topographic map and a LANDSAT photo and a digital street map. It includes in the centre a stretch of the busy Highway 417 connecting Ottawa and Montréal. Each colour-coded dot represents an automatically repositioned EDPCA detection. It is easily confirmed that the data on the right side are taken on the descending path in the morning, as many more targets are travelling toward downtown Ottawa during a regular business day rush hour. Many lower speed targets < 50 km/h (yellow, light blue) are detected on adjacent feeder streets and country roads. It is also interesting to see that the relocation error on the ascending path (left side) is larger than on the ascending path. The reason is that the data takes for the latter were conducted using the MODEX-2 $\frac{3}{4}$ mode while the former employed MODEX-1. It has been shown in [7] that a lower accuracy for MODEX-1 compared to MODEX-2 is theoretically expected.

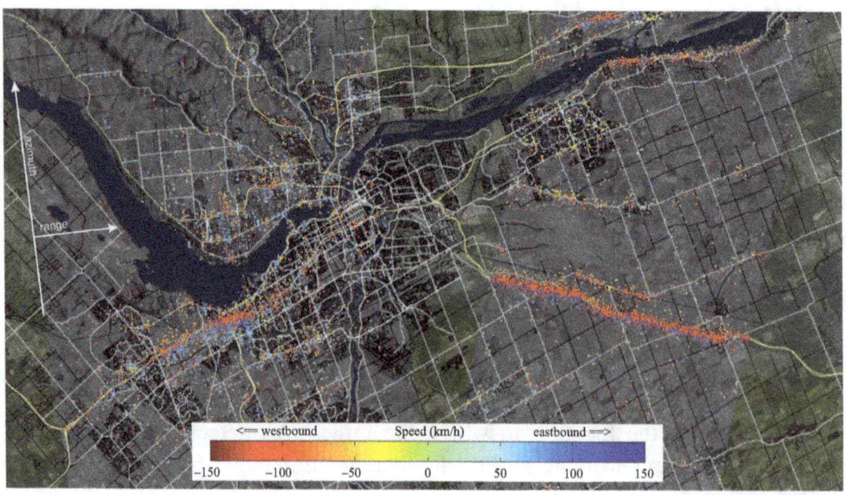

Figure 7.8 EDPCA detected and relocated moving targets; Left MODEX-1, right MODEX-2 $\frac{3}{4}$. Each dot represents a detection and the colour its estimated across-track ground velocity

7.4 Parameter estimation

The second prime objective of SAR-GMTI besides the actual detection is to determine the target's properties such as its velocity components in along-track and across-track direction, its RCS, its true location during imaging or even a potential acceleration. In estimation theory, commonly the Cramér–Rao bound (CRB) is analysed to provide a theoretical prediction of the accuracy with which an unknown parameter is statistically quantifiable. The CRB is a lower bound for the variance of any unbiased estimator and hence suited to judge the minimum possible estimation error. It is important to realize that the CRB describes a lower error bound for a given statistical model and does not depend on the flavour of any particular estimator. Although this bound may not reflect the practically achievable error values, it is, nevertheless, an excellent tool to compare relative strengths and weaknesses of different radar configurations under the same prerequisites, e.g., the various GMTI modes implemented on RADARSAT-2. In the context of this chapter, it is of particular interest to understand how the estimation accuracy of all target parameters is in fact affected by the true underlying values, for instance, how the standard deviation of the reposition error (nicely visible as a scattering cloud around the Queensway in Figure 7.8) is impacted by the true target speed and RCS. By extending the analysis for factored STAP provided in the literature, e.g., [6], this section focuses on the analysis of the performance limits of the iSTAP model, i.e., exploitation of the entire available synthetic aperture in contrast to only a small portion (linear phase approximation).

7.4.1 Cramér–Rao bound (CRB)

Based on the realization $\mathbf{z}$ of the multivariate complex Gaussian distributed random vector $\mathbf{Z} = \mathbf{s}(\boldsymbol{\xi}) + \mathbf{W}$ with pdf $\mathbf{Z} \sim \mathcal{N}_{2M}^{\mathbb{C}}(\mathbf{s}(\boldsymbol{\xi}), \mathbf{C})$ containing several unknown target parameters combined in vector $\boldsymbol{\xi}$ and a known covariance matrix $\mathbf{C}$, the $\mu\nu$th element of the Fisher information matrix J

$$J_{\mu\nu} = \mathcal{E}\left[\frac{\partial}{\partial \xi_\mu} \ln f_{\mathbf{Z}}(\mathbf{z}, \boldsymbol{\xi}) \frac{\partial}{\partial \xi_\nu} \ln f_{\mathbf{Z}}(\mathbf{z}, \boldsymbol{\xi})\right] \tag{7.35}$$

has been calculated in [6] to be given as the quadratic product

$$J_{\mu\nu} = 2\Re\{\mathbf{s}_\mu(\boldsymbol{\xi})^* \mathbf{C}^{-1} \mathbf{s}_\nu(\boldsymbol{\xi})\} \tag{7.36}$$

The indices at the signal vector denote the derivative with respect to parameter ξ_μ and ξ_ν. It is entirely valid to compute the CRB in either the time domain or in the Doppler domain. While deriving the Fourier-transform of the signal vector and the covariance matrix analytically using the method of stationary phase, [7,28], provides deeper insights into underlying physical mechanisms, the calculation of all derivatives is tedious, and the expressions become rather cumbersome. In contrast, it is mathematically less challenging to derive the required derivatives of (7.18) and (7.19) via

$$\begin{aligned}\mathbf{s}_\mu(\boldsymbol{\xi}) &= \frac{\partial \mathbf{s}(\boldsymbol{\xi})}{\partial \xi_\mu} = \left(\frac{\partial \mathbf{s}_m(\boldsymbol{\xi})}{\partial \xi_\mu}\right)_{m=1,\ldots,M/2} \\ &= \Bigg(-j\beta \frac{\partial \mathbf{R}_m(\boldsymbol{\xi})}{\xi_\mu} \exp\{-j\beta \mathbf{R}_m(\boldsymbol{\xi})\} \odot \mathfrak{D}_{\mathrm{Tx}}[u_m(\boldsymbol{\xi})] \odot \mathfrak{D}_{\mathrm{Rx}}[u_m(\boldsymbol{\xi})] \\ &\quad + \exp\{-j\beta \mathbf{R}_m(\boldsymbol{\xi})\} \odot \frac{\partial \mathfrak{D}_{\mathrm{Tx}}[u_m(\boldsymbol{\xi})]}{\partial \xi_\mu} \frac{\partial u_m(\boldsymbol{\xi})}{\partial \xi_\mu} \odot \mathfrak{D}_{\mathrm{Rx}}[u_m(\boldsymbol{\xi})] \\ &\quad + \exp\{-j\beta \mathbf{R}_m(\boldsymbol{\xi})\} \odot \mathfrak{D}_{\mathrm{Tx}}[u_m(\boldsymbol{\xi})] \odot \frac{\partial \mathfrak{D}_{\mathrm{Rx}}[u_m(\boldsymbol{\xi})]}{\partial \xi_\mu} \frac{\partial u_m(\boldsymbol{\xi})}{\partial \xi_\mu}\Bigg)_{m=1,\ldots,M/2}\end{aligned} \tag{7.37}$$

leading to relatively simple terms after taking into consideration (7.22)–(7.25). Splitting the complex constant $a = \alpha \exp(j\varphi)$ up into an unknown amplitude α and an unknown phase φ, the Fisher information matrix reads

$$\mathbf{J} = 2\Re \begin{bmatrix} \alpha^2 \mathbf{s}_{v_s}^* \mathbf{C}^{-1} \mathbf{s}_{v_s} & \alpha \mathbf{s}_{v_s}^* \mathbf{C}^{-1} \mathbf{s} & j\alpha^2 \mathbf{s}_{v_s}^* \mathbf{C}^{-1} \mathbf{s} & \alpha^2 \mathbf{s}_{v_s}^* \mathbf{C}^{-1} \mathbf{s}_{t_b} & \alpha^2 \mathbf{s}_{v_s}^* \mathbf{C}^{-1} \mathbf{s}_{v_x} \\ \alpha \mathbf{s}^* \mathbf{C}^{-1} \mathbf{s}_{v_s} & \mathbf{s}^* \mathbf{C}^{-1} \mathbf{s} & 0 & \alpha \mathbf{s}^* \mathbf{C}^{-1} \mathbf{s}_{t_b} & \alpha \mathbf{s}^* \mathbf{C}^{-1} \mathbf{s}_{v_x} \\ -j\alpha^2 \mathbf{s}^* \mathbf{C}^{-1} \mathbf{s}_{v_s} & 0 & \alpha^2 \mathbf{s}^* \mathbf{C}^{-1} \mathbf{s} & j\alpha^2 \mathbf{s}^* \mathbf{C}^{-1} \mathbf{s}_{t_b} & j\alpha^2 \mathbf{s}^* \mathbf{C}^{-1} \mathbf{s}_{v_x} \\ \alpha^2 \mathbf{s}_{t_b}^* \mathbf{C}^{-1} \mathbf{s}_{v_s} & \alpha \mathbf{s}_{t_b}^* \mathbf{C}^{-1} \mathbf{s} & -j\alpha^2 \mathbf{s}_{t_b}^* \mathbf{C}^{-1} \mathbf{s} & \alpha^2 \mathbf{s}_{t_b}^* \mathbf{C}^{-1} \mathbf{s}_{t_b} & \alpha^2 \mathbf{s}_{t_b}^* \mathbf{C}^{-1} \mathbf{s}_{v_x} \\ \alpha^2 \mathbf{s}_{v_x}^* \mathbf{C}^{-1} \mathbf{s}_{v_s} & \alpha \mathbf{s}_{v_x}^* \mathbf{C}^{-1} \mathbf{s} & -j\alpha^2 \mathbf{s}_{v_x}^* \mathbf{C}^{-1} \mathbf{s} & \alpha^2 \mathbf{s}_{v_x}^* \mathbf{C}^{-1} \mathbf{s}_{t_b} & \alpha^2 \mathbf{s}_{v_x}^* \mathbf{C}^{-1} \mathbf{s}_{v_x} \end{bmatrix} \tag{7.38}$$

in which for notational simplicity the dependence of $\mathbf{s}$ on $\boldsymbol{\xi}$ has been omitted. Knowing $\mathbf{J}$, the covariance of any unbiased estimator $\hat{\boldsymbol{\xi}}$ is bounded by

$$\mathcal{E}\left(\hat{\boldsymbol{\xi}}(\mathbf{Z}) - \boldsymbol{\xi}\right)\left(\hat{\boldsymbol{\xi}}(\mathbf{Z}) - \boldsymbol{\xi}\right)' \geq \mathbf{J}^{-1} \tag{7.39}$$

and in particular the individual variances are bounded by the corresponding diagonal element of the inverse Fisher matrix

$$\sigma_{\xi_\mu}^2 := \mathrm{var}(\xi_\mu) \geq \left(\mathbf{J}^{-1}\right)_{\mu\mu} \tag{7.40}$$

Figure 7.9 shows the CRBs, more specifically the square root of the numerically evaluated expression in (7.40), for the relocation error and the along-track velocity versus a varying true across-track (slant range) velocity v_s for a given[12] $v_x = 0$. The power levels are identical to those used in Figure 7.7, and the transformation matrices represent three different modes on RADARSAT-2. The CRBs in Figure 7.9(a) reveals an unexpected behaviour of iSTAP compared to the factored

[12]It can be shown that the curves do not change much for $v_x \neq 0$.

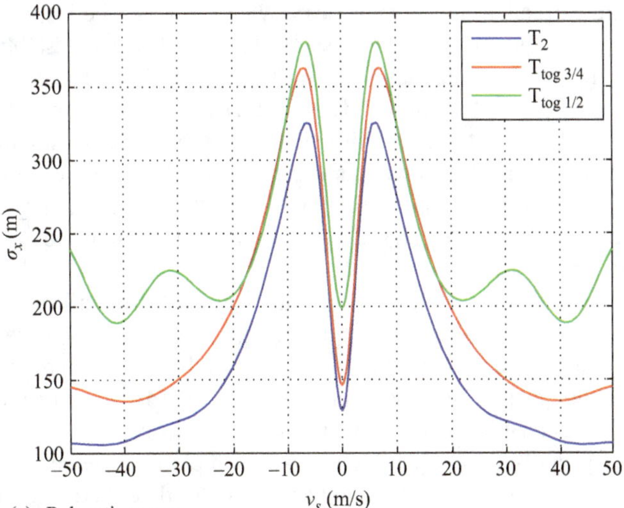

(a) Relocation error.

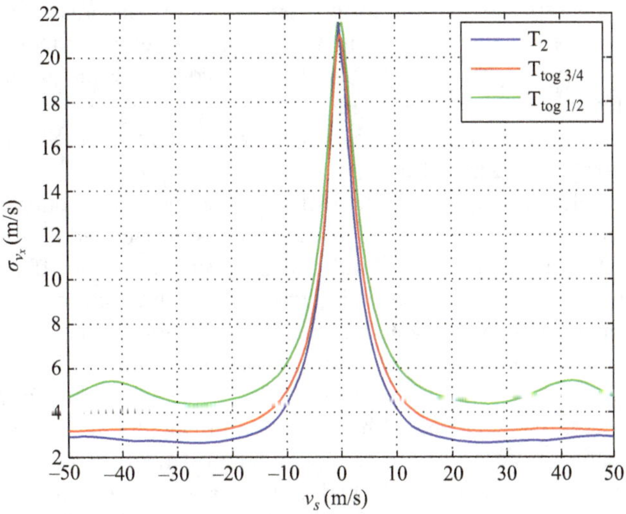

(b) Along-track velocity.

Figure 7.9 CRBs of iSTAP of different MODEX modes using a CPI of 0.5 s (2,048 pulses)

STAP curves published as Figure 13 in [6]. While in general all three modes increase the absolute estimation accuracy due to the prolonged CPI, the classic MODEX-1 mode improves disproportionately for this example and even surpasses that of the MODEX-2 modes. Note, since the CPI and PRF for this analysis have been kept constant for comparison purposes, the MODEX-2 modes sustain an unjust SNR loss caused by the reduction in Tx power and aperture. In practice, the reduced aperture broadens the antenna beam which enlarges the Doppler bandwidth and consequently requires an increased sampling PRF. This increased PRF together with a larger dwell time on the target permits recovery of most of the lost SNR. In other words, in reality, the implemented MODEX-2 modes on RADARSAT-2 typically still slightly outperform MODEX-1 regarding the minimum parameter estimation accuracy, cf. [25] as well as Figure 7.8 of this chapter. Nevertheless, it can be concluded that the published finding that MODEX-2 drastically outperforms MODEX-1 for regular factored STAP, [6], does not hold anymore and diminishes when iSTAP gets applied.

The CRB for the along-track velocity in Figure 7.9(b) confirms this behavior and in addition shows a quasi-independence on v_s over the whole velocity range outside the clutter band. It predominantly depends on the achievable SNCR, Figure 7.7. Although it is intuitively clear that the chirp rate estimation accuracy improves for longer CPIs, CRB values in the order of only a few m/s are surprisingly low,[13] especially for a space-based SAR with platform velocities in the order of 7,000 m/s. It is important to note, however, that to date there is no convincing, evidence-based, experimental verification that the theoretical results for v_x hold in practice. This is especially owing to the complexity of benchmark tests, which in order to guarantee a statistically meaningful sample size would require to deploy many GPS-equipped ground truth targets that cover all ranges of speeds in different directions, reflection strengths, etc. Furthermore, robust and practically implementable estimators for v_x are currently also not available and must be developed, compare Figure 7.13(b).

One possible physical explanation of why the CRBs for both velocity components improve so drastically for iSTAP in contrast to factored STAP is depicted in Figure 7.10. Therein the target and clutter spectra for two different CPIs are plotted. Regardless of the chosen CPI, the clutter contributions come from all directions illuminated by the antenna beam and therefore the spectrum in both cases spans the entire clutter bandwidth (modulated in amplitude by the antenna pattern). If the whole synthetic aperture is used to Fourier-transform the target signal, its spectrum looks identical but is shifted in Doppler due to an across-track velocity component. Using, however, only a small part of the available synthetic aperture, for instance, a portion in which a linear approximation of the hyperbolic range/phase history is justified, results in a delta-function-like spectrum such that the target energy mainly resides in only one Doppler cell. The shaded areas of the respective target spectra outside of the clutter band may be heuristically interpreted as containing the information available for target parameter estimation and as such clearly illustrate the superiority of iSTAP versus factored STAP.

7.4.2 Optimized CRB via time-multiplexed antenna tapering

Instead of simply turning on and off various TR-modules to manipulate the antenna pattern for several pulses, modern phased arrays permit variable attenuator and

[13]Remember, no meaningful estimation of v_x is possible for factored STAP.

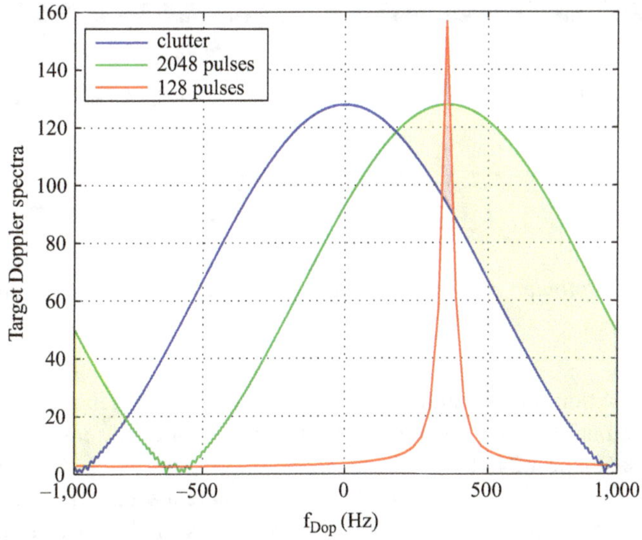

Figure 7.10 Target and clutter spectra for the different CPIs

phase shifter settings. Having this in mind, one question naturally arises: what are the best amplitude and phase settings for all TR-modules along the aperture that optimize the CRB on one hand, and on the other, do not jeopardize detectability (SNR)? Instead of a constant value set for each transmit pulses, one may combine settings for subsequent pulses similar to the existing MODEX-2 modes (two pulses). For example on RADARSAT-2, this optimization problem consists in principle of 128 unknowns; 16 attenuator settings between zero and one and 16 phase shifter settings for Tx and Rx over two pulses. To maximize SNR, RADARSAT-2 always transmits in saturation (no attenuation on Tx), which leaves 96 degrees-of-freedom remaining. Mathematically, one possible cost function is the iSTAP CRB as a function of all complex entries in the transformation matrices $\mathbf{T}_{\mathrm{Tx}}$ and $\mathbf{T}_{\mathrm{Rx}}$ over a range of velocities $v_s \in \{-50, 50\}$ m/s, i.e.,

$$\underset{\mathbf{T}_{\mathrm{Tx}},\,\mathbf{T}_{\mathrm{Rx}}}{\operatorname{argmin}} \sum_{v_s} \sigma_{v_s}^2(\boldsymbol{\xi}, \mathbf{T}_{\mathrm{Tx}}, \mathbf{T}_{\mathrm{Rx}}), \qquad \text{under constraint} \qquad \mathbf{T}_{\mathrm{Rx}}^{*}\mathbf{T}_{\mathrm{Rx}} = \mathbf{I} \tag{7.41}$$

The matrix constraint in (7.41) is intended to avoid an inadvertent amplification of the thermal noise power. A numerical global minimization with a genetic algorithm in MATLAB© after about 1 day of computation yielded the settings schematically depicted in Figure 7.11.

The upper row depicts the TR-module settings for two consecutive transmit pulses and the bottom row the ones on receive, whereby the little gaps mark the border between fore and aft antenna wing. The colour of the boxes indicates the

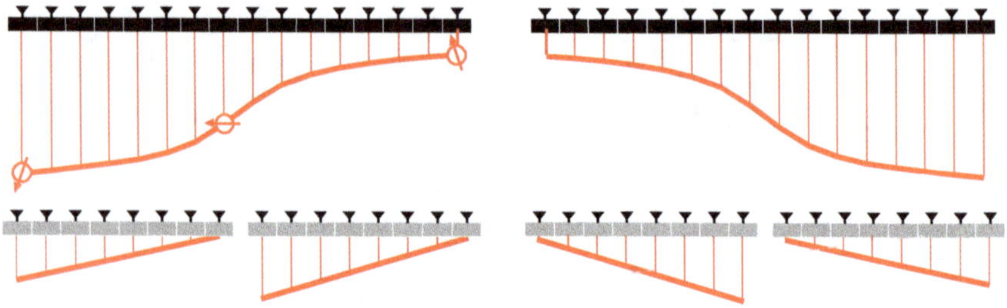

Figure 7.11 Optimum antenna tapering on Tx and Rx for two subsequent pulses

level of attenuation (black stands for 1 and grey for $\frac{1}{\sqrt{8}}$). The length of the red lines describes the phase shifter positions with the longest one on transmit representing 163° and the smallest 24° [29].

Figure 7.12 presents the resulting two-way antenna pattern associated with these TR-settings. Since the settings are mirror-inverted between pulses, four distinct phase responses arise but only two distinct magnitude patterns. It is evident that the optimization demands slight squints in opposite directions. In addition, one magnitude pattern appears to be the derivative of the other, a favourable property for bearing estimation well known from the classic adaptive monopulse estimation of direction-of-arrival (DOA) angles via sigma-delta approaches [20,30]. Interestingly, the optimization does not require any antenna amplitude tapering over the array but exclusively a phase tapering (non-linear on Tx and linear on Rx with varying slopes).

The resulting optimized CRB for both velocity components is compared to that of MODEX-1 in Figure 7.13(a). While the iSTAP CRB for the across-track velocity remains virtually identical,[14] the one for the along-track component v_x has visibly improved owing to the opposite beam squinting which increases the overall dwell time on the target and therewith the chirp rate estimation accuracy [31]. Also included in the figure are the standard deviations of the estimates for 2,500 Monte-Carlo simulations using the iSTAP MLE

$$\hat{v}_x, \hat{v}_s = \underset{v_x, v_s}{\operatorname{argmax}} \frac{\left| \sum_k \mathbf{d}(f_k, v_x, v_s)^* \mathbf{R}(f_k)^{-1} \mathbf{Z}(f_k) e^{-j2\pi f_k \hat{t}_c} \right|^2}{\sum_k \mathbf{d}(f_k, v_x, v_s)^* \mathbf{R}(f_k)^{-1} \mathbf{d}(f_k, v_x, v_s)} \tag{7.42}$$

where $\hat{t}_c$ denotes the location at which the target peak appears in the time domain. Although Figure 7.13(a) might give the impression that the MLE appears always to be a sound estimator for the along-track velocity, this is not generally true. First, the simultaneous optimization with respect to both velocities is prohibitive in

[14]Note, in absolute terms, the performance gain of iSTAP versus factored STAP is still enormous, [6] Figure 7.13.

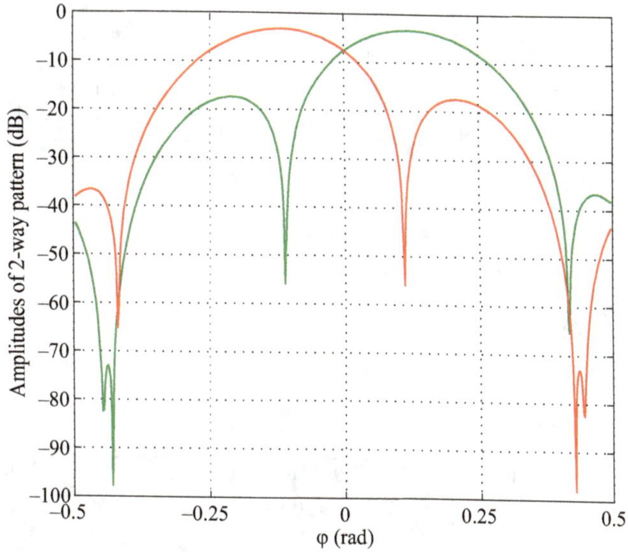

(a) Magnitude.

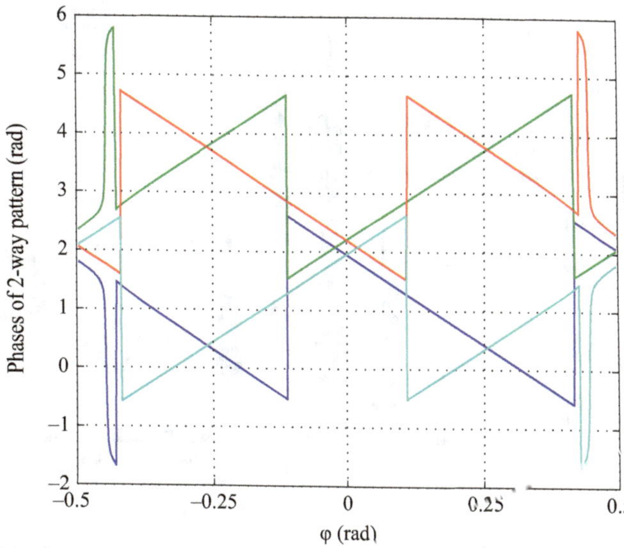

(b) Phase.

Figure 7.12 Magnitude and phase of the two-way antenna pattern for the optimized TR settings

computational time and resources and, second, estimates are sometimes unreliable. This is likely caused by an unfavourable form of the underlying likelihood function that needs to be maximized. This can be recognized in Figure 7.13(b), which shows the undulated nature of the likelihood function that results in a multimodal

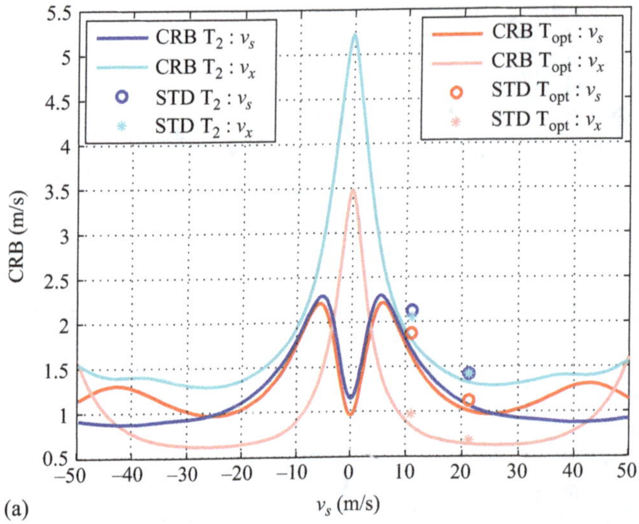

(a)

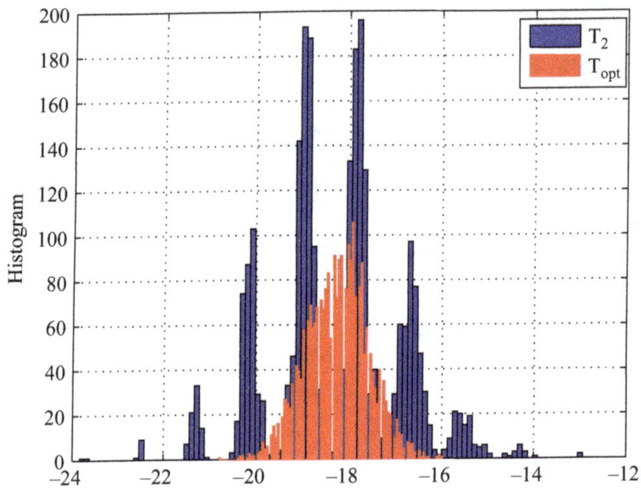

(b)

Figure 7.13 CRBs & MLEs of MODEX-1 and optimum MODEX for a CPI of 1 s (4,096 pulses). (a) RADARSAT-2 CRBs of iSTAP for v_x and v_y and (b) histogram of MLE for v_x based on Monte-Carlo simulations

histogram particularly for MODEX-1. For an individual estimate, this increases the chance of large deviations from the truth with a rather high probability. The exact reason for this is not understood at the moment and a topic of active research alongside the development of suitable estimators.

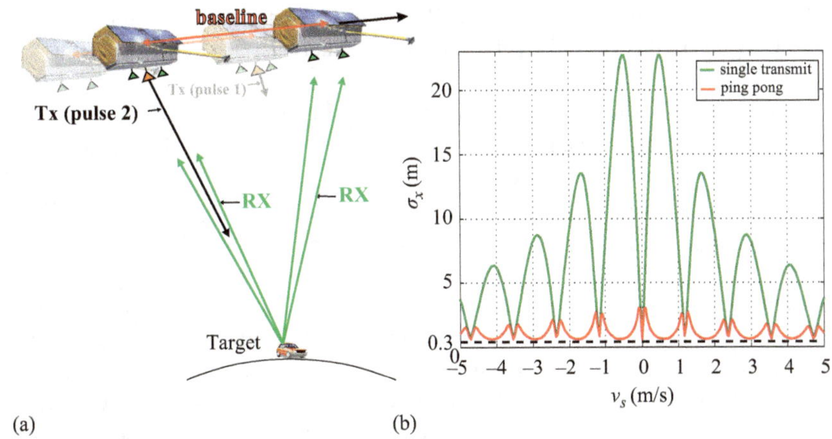

Figure 7.14 Tandem SAR-GMTI. (a) Basic geometry for tandem SAR-GMTI and (b) standard deviation of relocation error for TanDEM-X with two conceptional GMTI schemes

7.4.3 Improved CRB using multistatic configurations

Although creating additional spatial diversity via time-multiplexed antenna aperture partitioning can improve the CRB considerably especially for factored STAP, the absolute estimation accuracy may still not be sufficient for many practical applications. Preliminary analysis has shown that for factored STAP satisfactory results are only achievable with a fully occupied aperture in the order of hundreds of metres. Such large apertures confine the clutter into a narrow Doppler band and also provide a higher location estimation accuracy [32]. Since launching apertures of this size into space is infeasible, a multistatic satellite constellation (potentially consisting of small-sats) flying in close formation might become a viable solution in the future. The simplest coherently working constellation consists of two back-to-back flying satellites similar to the TerraSAR-X/TanDEM-X pair currently in orbit.

Figure 7.14(a) depicts two basic tandem-GMTI concepts in which (a) either only one satellite transmits, while both satellites receive simultaneously and coherently[15] or (b) the transmit pulse is alternated between the two satellites from pulse to pulse (ping-pong mode) [34]. Figure 7.14(b) shows the resulting factored STAP CRB for such a system based on Tandem-X parameters assuming a baseline between the two satellites of 200 m [35]. The CRB in the plot is zoomed into a narrow band between ±5 m/s, since the values drop rapidly outside this range. Figure 7.14(b) compares the standard deviation of the reposition error in metres for

[15]Note, this is different from non-coherent approaches recently proposed for large time gaps when the target displacement between the two separate SAR images can be determined and therefore the velocity inferred [33].

TanDEM-X operating in the one-satellite transmit mode and in the ping-pong mode. Caused by the very sparsely distributed receiver phase centres, many blind velocities appear.

In absence of clutter (noise-only case), the black dashed line represents the absolute minimal error given by

$$\sigma_x = \sqrt{\frac{M}{8\pi^2}\frac{R_b}{\sqrt{\text{SNR}}}\frac{\lambda}{\sqrt{\sum_m x_m^2}}} \tag{7.43}$$

to which both curves converge quickly beyond $v_s \pm 10$ m/s. M is the number of antenna elements ($M = 12$ columns for TerraSAR-X/TanDEM-X) and x_m denotes their locations in metres. For TanDEM-X, this lowest theoretical error bound can be seen to be in the order of one foot for a chosen SNR = 27 dB. On one hand, it is obvious that such high SNR is only realistic for large targets like trucks, trains or ships but was chosen to demonstrate the gain compared to the results in the literature [6]. On the other hand, it also states that iSTAP may not be required if the SNR is sufficient and a large baseline available. For the standard mode, the largest relocation error is about 25 m, while, in contrast, the ping-pong mode reduces the error to less than 3 m for all target velocities owing to the practically doubled maximum phase centre separation. These are indeed only theoretically derived ideal values that are likely not reached in practice due to imperfections such as coherence loss between two free-running oscillators, lack of precise knowledge of relative attitude and timing information, etc. It is nevertheless easily imaginable that systems of this kind may become a ‘game changer’ for space-based SAR-GMTI in the future.

Accuracies of this order will permit even tactical decisions by reliably repositioning of even and slowly moving targets correctly onto their respective roads without any *a priori* street context. First experimental investigations/verification on the achievable performance of coherent multistatic SAR-GMTI systems have been reported in [36], and more are currently being conducted during the science phase of the TerraSAR-X/TanDEM-X mission after completion of the primary mission goal (world-wide high-resolution topographic maps).

A more sophisticated concept, based on a minimum redundant satellite configuration, has been proposed by Ender [37]. In this, a constellation of four X-band satellites each carrying a 12-m aperture flying in a close formation with distances given by 1, 3 and 2 units of the combined total aperture (e.g., 12, 36 and 24 m). This specific partitioning results in a quasioptimum performance, especially does not create the blind velocities (grating lobes) occurring in the tandem-configuration. The analysis of the CRB anticipates a relocation accuracy in the order of metres for very realistic target and clutter conditions.

It is also interesting to note that such a non-equidistant array in flight direction may be utilized, for pure imaging purposes, with a much reduced PRF permitting a considerably enlarged swath width while maintaining the geometric resolution, for details see Section 7.5.2. A first analysis has been reported in [38]. Not investigated to date has been the performance gain when iSTAP is applied to a coherent SAR constellation, be it tandem or fully multistatic. Furthermore, it appears also not inconceivable to be able in the near future to launch a single spacecraft with an extendible boom structure carrying the four receive antennas mounted according to the minimum redundant spacings [38]. This has been successfully accomplished during the Space Shuttle Radar Tomography Mission (SRTM), where the second receive-only antenna was mounted on a 60 m long, extractable, stiff boom structure creating the required across-track interferometric baseline [39]. Although such an aperture length might still not be sufficient for effective GMTI when factored STAP is applied, the performance improvement of iSTAP has the potential to overcome this deficiency thereby offering itself as a viable alternative for near-future operational applications. The capital costs of a monostatic SAR-GMTI system including the more complex structure would likely still be orders of magnitudes less than that for close-formation flying satellite constellations. Not to mention the to-date unsolved practical, extremely challenging command and control problems associated with a small fleet of satellites to assure a minimum collision risk while maintaining precise relative attitudes to keep sufficient coherence between the receivers.

7.5 Wide-area SAR-GMTI

Many countries have vast and sparsely populated coast lines and have both national interest and international obligations to monitor maritime vessel traffic in ocean approaches. MODEX-like SAR-GMTI modes are not suitable to accomplish this task due to relatively small coverage in the order of 50 km. A classic SAR imaging approach to enlarge the swath size while giving up spatial resolution is known as ScanSAR. ScanSAR modes come in several flavours. The most commonly used mode is an imaging mode and uses a small number of radar elevation beams activated in bursts within the synthetic aperture time of the radar beam. In the past, they have been primarily designed to map the sea surface providing information for oceanographic and meteorological applications. Ships that are present in the covered radar swath are embedded in the sea surface returns and typically only larger vessels can be reliably detected based on their bright return compared the background.

Using current technologies, a space-based SAR system can also be optimized for ship detection by deploying a specialized, multibeam ScanSAR mode. These typically divide the radar synthetic aperture time into much larger number of subbeams and tailor the radar operating parameters for each individual subbeam, thereby optimizing ship detection at the cost of reduced or lost sea surface information.[16] RADARSAT-2's Maritime Satellite Surveillance Radar (MSSR) mode has been designed to detect ships longer than 25 m in open-ocean over a 450 km swath in sea surface conditions up to sea state five and is also being developed for its successor RADARSAT Constellation Mission (RCM) [40].

[16]The large SAR images are typically of low quality and discarded after vessel detection.

However, both scanSAR modes stand out by their low-to-medium resolutions. Generally, they work well in the deep ocean but sometimes show limitations under challenging conditions in which strong clutter contamination prohibits reliable detection of smaller vessels, such as low incidence angles, higher sea states, littoral zones or in ice-infested waters. Those deficiencies make the addition of GMTI to scanSAR an attractive solution to overcome an operational capability gap and augmenting specialized vessel detection modes like MSSR. The main objectives of any wide-area SAR-GMTI mode include the detection of smaller vessels in higher sea states, permitting the detection of small vessels at lower resolutions, allowing higher P_d for the same false alarm rate or discriminating between vessels and stationary detections such as icebergs, rocks, etc. In addition, any multichannel GMTI mode also extends originally specified capabilities, in particular, provides estimation of ship speeds and possibly heading or may even potentially improve ocean current estimation.

7.5.1 ScanSAR GMTI

Recently, the combination of scanSAR imaging acquisition with two-channel GMTI processing has been explored with the objective to establish whether the coherence between the beams and channels can be maintained as prerequisite to effective clutter suppression. Of practical interest for GMTI in the maritime, environment is to establish whether or not the highly volatile sea surface decorrelates at a faster rate than the time it takes the aft receiver channel to reach the location of the fore channel at the previous transmit pulse. Were this the case, the strong sea surface clutter could not be suppressed and GMTI would be useless.

A two-beam dual-channel experimental scanSAR mode has been designed and implemented on RADARSAT-2 in 2010 enlarging the swath about three-fold to 150 km. Because of the bursted nature of scanSAR, fewer pulses are available for integration in each subbeam. As a result, the achievable spatial resolution in azimuth is coarser than for stripmap SAR images, which depends on the burst duration and the overlap between subsequent bursts. In general, the bandwidth of the transmit chirp is also limited to compensate for the increased data rate and to balance azimuth and range resolutions. Targets located at different along-track positions are illuminated by a different portion of the non-uniform antenna pattern in azimuth. Hence, the clutter signal is azimuth dependent, i.e., modulated by the antenna pattern (known as scalloping), whereas the receiver noise is still statistically white. First, MODEX-scanSAR data-takes have been acquired on October 15th, 2010, over a region in Northern Germany including the German bight known for heavy merchant ship traffic [41]. After channel-co-registration with an *a priori* assumed phase centre displacement the interchannel coherence was measured to only 0.5 to 0.6 far below values required for any meaningful clutter suppression. This decorrelation is mainly caused by unavoidable channel imbalances in the system transfer function of the electronic components as well as unequal antenna pattern in both directions (different squint angles per wing per subbeam). Since the data were properly Nyquist-sampled, an adaptive two-dimensional channel

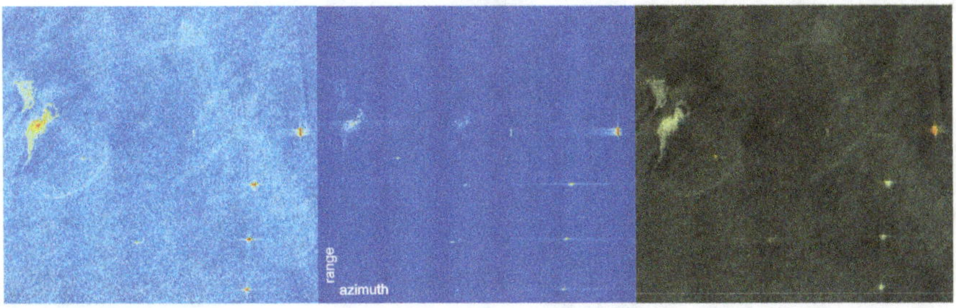

Figure 7.15 Left: scanSAR image of the surroundings of Helgoland island in the German bight, centre: DPCA scanSAR image, right: ATI phase superimposed onto the SAR image

balancing and calibration scheme, called DIGIBAL [8], has been adopted improving the interchannel coherence to above 0.97, values very close or even exceeding those achieved for regular MODEX modes.[17]

Figure 7.15 shows a small portion of a scanSAR image around the German island of Helgoland trying to demonstrate the most prominent added values of scanSAR-GMTI. The figure on the left shows the SAR image including the island, a few high RCS targets (most likely vessels but no ground truth was available) as well as strongly reflecting ocean features. The DPCA image in the centre confirms that the stationary backscatter, and the strong sea surface reflections can be effectively suppressed. In this example by notable 15 to 20 dB. As a result, moving targets remain in the DPCA image and can be more easily detected after clutter suppression especially visible for the one near the island. As well, low RCS ships turn up, indirectly confirmed by the appearance of the ambiguous echoes from the vessels to the right that had been mostly concealed by the sea surface in the right image. These ambiguities possess only a fraction of the SNR and therefore behave like small boats. The composite of the ATI phase difference and the SAR image permits easy determination of their across-track velocity indicated by different colours. For instance, the vessel at the right border and the one near the island move with a considerable velocity upwards while the three ships in the centre right seem to be anchored or travelling in purely along-track direction.

Some unanswered questions, however, remain for future investigations. It is not clear what the implications of the scalloping are for the statistical detectors; the effect is observable in the DPCA image in the centre. Since most operational algorithms are adaptive, however, i.e., are based on dividing the entire image into much smaller subimages in order to cope with any fluctuation in the surface texture, scalloping is not expected to create grave problems. Furthermore, the achieved velocity estimation accuracy needs to be related to the theoretical CRB as well as provided AIS information whenever available. Last but not least, the performance

[17]Note that these channel imbalances vary over subbeams and must to be individually calibrated.

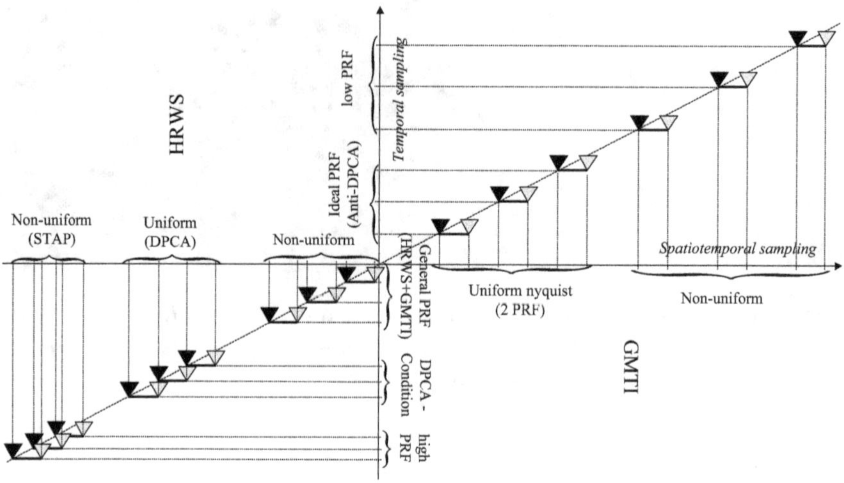

Figure 7.16 Illustration of different sampling regimes for a two-channel system

gain of STAP or iSTAP in conjunction with scanSAR should be determined, in particular when compared to the classic DPCA.

7.5.2 HRWS-GMTI

Although scanSAR in conjunction with GMTI demonstrates the ability to enlarge the coverage considerably, its foremost disadvantages with respect to the latter is the reduced SNR and reduced resolution (lower probability of detection) as well as the background modulation (scalloping) due to the bursting nature of scanSAR. The relatively new TOPSAR mode might eliminate the scalloping affect but still suffers from the loss of detectability and its rather complex antenna beam steering concept. Increasing the swath width without elevation scanning and bursting is only achievable by reducing the PRF in order to avoid the appearance of range ambiguities.[18] A reduced PRF for a given aperture size, however, heightens the risk of emerging azimuth ambiguities. An elegant way to overcome this inherent SAR shortfall and to relax some of the coverage constraint issues is to implement a hybrid real/synthetic aperture configuration known as HRWS mode [42].

7.5.2.1 Ideal PRF

Conceptually, a multiaperture SAR carrying a linear equispaced array of N real receiver channels in the flight direction may be undersampled with a N-times lower PRF in such a way that the missing temporal samples are perfectly replaced by the spatial samples recorded simultaneously in all channels. This scenario is illustrated for a two-channel system in Figure 7.16, second case from the right. In this simplest

[18]Provided that the transmit power is large enough to maintain a sufficient SNR over the entire swath.

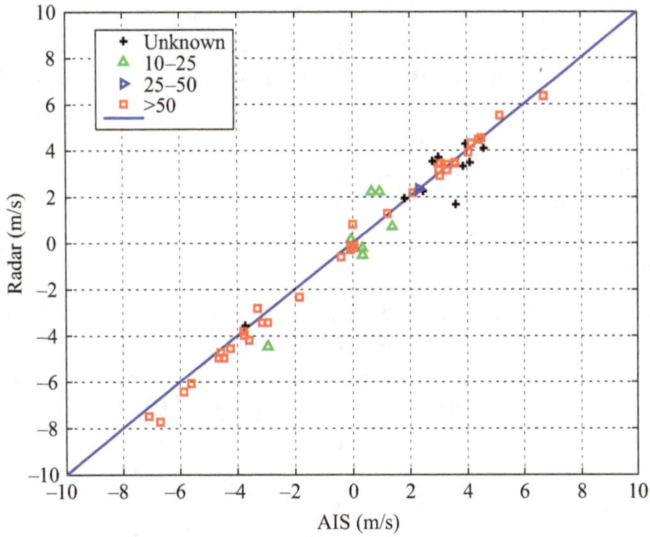

Figure 7.17 SAR radial velocity estimates versus AIS information

scenario, the 'ideal PRF'[19] has to be adjusted to the platform velocity.[20] Assuming identical antenna pattern for all channels, the resulting interleaved one-dimensional data set can be processed by a standard stripmap processor. The three cases on the left in Figure 7.16 describe typical GMTI configurations with varying degrees of channel overlap between subsequent pulses resulting in channel correlations capable of suppressing the clutter before target detection. This highlights the close relationship between GMTI and HRWS.

Examples of commercially available HRWS modes of such kind are the ultrafine and extrafine wide modes on RADARSAT-2. Also the operationally utilized MSSR mode on RADARSAT-2 capitalizes on the two parallel receive channels in combination with the ideal PRF to optimize the vessel detection and ocean surveillance performance [40]. However, since MSSR has been optimized for vessel detection and maximum coverage, it is strictly not an imaging mode. Notwithstanding the lost ability to suppress clutter when the anti-DPCA condition is used (no interchannel correlation), parameter estimation is not greatly impaired; thus, if a moving target is identified by some other means, one can still apply the GMTI estimation algorithms to obtain good estimates of the vessel motion parameters. For instance, it has recently been shown that radial speed information can be extracted for targets that are easily identifiable because of their large response relative to clutter—such as vessels on typically low RCS ocean surface. Figure 7.17 demonstrates practically achieved estimations for the across-track velocity component

[19]Sometimes, also called anti-DPCA condition.

[20]The ideal PRF varies over latitude due to Earth rotation.

from 50 vessels using single look complex (SLC) MSSR mode data over Gibraltar in August 2014 and compares them to land-based AIS information. Marked by different signs and colour are different ship sizes, and the blue diagonal identifies perfect agreement. The targets have been manually selected and their across-track velocities estimated by the EDPCA algorithm (34) for a chosen $v_x = 0$. A very good agreement can be observed even for vessels smaller than 25 m as well as for several anchored ships ($v_s = 0$).[21]

7.5.2.2 Low PRF

To increase the coverage even further, the PRF needs to be reduced accordingly (Figure 7.16 far right). The more the PRF deviates from its ideal value, the higher the ambiguity levels rise caused by the wide gaps of the non-equidistant spatio-temporal sampling rate. As part of a research project at DRDC, an experimental low-PRF HRWS mode was designed and implemented on RADARSAT-2 using a very low PRF below 1 kHz, a setting not envisaged prior to launch. In order to overcome the resulting severe undersampling, i.e., being able to still create useful imagery, advanced signal processing must be applied. The underlying idea is to exploit the information contained in the spatial dimension (different parallel receive channels $Z_1(f), \ldots, Z_N(f)$) about the highly ambiguous signal to reconstruct a one-dimensional unambiguous signal extending outside the sampled Doppler band of length PRF $\left\{-\frac{f_p}{2}, \frac{f_p}{2}\right\}$. The desired reduction in ambiguity level is being traded off against an increased noise floor. Let a set of to-be-determined column vectors be denoted as $\{\mathbf{v}_k(f)\}$ for $k \in \mathbb{Z}$. The prescription for multichannel SAR Doppler-unwrapping processing is to construct a scalar function of Doppler for each band k given by

$$y(f + kf_p) = \mathbf{v}_k^*(f)\mathbf{z}(f) \tag{7.44}$$

such that, when $y(f)$ multiplied by the correct azimuth chirp factor $\exp(-j\Omega(f + kf_p))$ and transformed back into the time domain, yields the desired 'unfolded' azimuth compressed signal. One promising processing method, [43], is based on optimizing a cost function that does not neglect the noise contribution. For instance, one can try to find the set of vectors at each Doppler frequency $\{\mathbf{v}_k(f)\}$ that, applied to the measured ambiguous random vector $\mathbf{Z}(f)$, best approximates the unambiguous contribution $Z_k(f) := Z(f + kf_p)$ in a minimum mean square sense (MMSE). Mathematically formulated it reads

$$\min_{\{\mathbf{v}_k(f)\}_{k\in\mathbb{Z}}} \mathcal{E} \int_{-f_p/2}^{f_p/2} \sum_l |\mathbf{v}_l^*(f)\mathbf{Z}(f) - Z_l(f)|^2 \; df \tag{7.45}$$

[21]The azimuth resolution for this data set is about 24 m.

The measured data vector in the Doppler domain is modelled as the sum of the signal contributions from all Doppler bands plus the thermal noise **N**

$$\mathbf{Z}(f) = \sum_k Z_k(f)\mathbf{g}_k(f) + \mathbf{N}(f) \qquad f \in \left\{ -\frac{f_p}{2}, \frac{f_p}{2} \right\} \tag{7.46}$$

in which $\mathbf{g}_k$ represents a known normalized DOA vector for each Doppler band k. Assuming the individual contributions $Z_k(f)$ to be mutually independent and spectrally white Gaussian RVs (representing the spatially fluctuating nature of the reflecting surface), the covariance matrix computes to

$$\mathbf{R}(f) = \mathcal{E}\mathbf{Z}(f)\mathbf{Z}(f)^* = \sum_k \sigma_C^2 \ |\phi_k(f)|^2 \mathbf{g}_k(f)\mathbf{g}_k(f)^* + \sigma_N^2 \mathbf{I} \tag{7.47}$$

in which $\phi_k(f) = \phi(f + kf_p)$ is a presumably known mean antenna pattern.[22] The optimization in (7.45) has an analytical solution:

$$\mathbf{v}_k^{\text{opt}}(f) = |\phi_k(f)|^2 \ \mathbf{R}(f)^{-1}\mathbf{g}_k(f) \tag{7.48}$$

This prescription (and variants) has been successfully applied to two-channel experimental HRWS RADARSAT-2 data sets [43]. Another impressive example is presented in Figure 7.18. For the image on the left, both channels of a conventional MODEX-1 mode data set over East Ottawa have been downsampled by a factor of two (every other sample has been discarded).

The emerging azimuth ambiguities clearly illustrates why no conventional SAR acquires data below the Nyquist sampling rate. The MMSE processed image on the right is capable of recovering the almost full performance, particularly improving the radiometric resolution alongside the much reduced ambiguity levels of point-like targets. This is impressive considering that the spatiotemporal sampling shows a gap of about 150% in relation to the phase centre spacing. If RADARSAT-2 were to have more transmit power, this extremely low PRF could support a swath width of more than 250 km depending on the incidence angle.

Nevertheless, when attempting to conduct low-PRF HRWS in conjunction with GMTI, it has been concluded that both applications are optimized by specific system configurations each of which consume a limited number of degrees-of-freedom [44]. Multichannel HRWS SAR systems are most appropriate for maritime domain awareness, e.g., high-resolution vessel imaging and identification, a vessel motion estimation (VME) and an improved ship detection performance in littoral zones. For instance, Figure 7.19 compares the HRWS image of a vessel produced with MMSE prescription (7.48) for an extrafine low-PRF RADARSAT-2 data set over Gibraltar to the corresponding classic SAR image of the same vessel. The latter would have been obtained using a system of equivalent aperture but equipped

[22]The scaled sum of all individual two-way pattern in the end determines the shape of the HRWS point spread function.

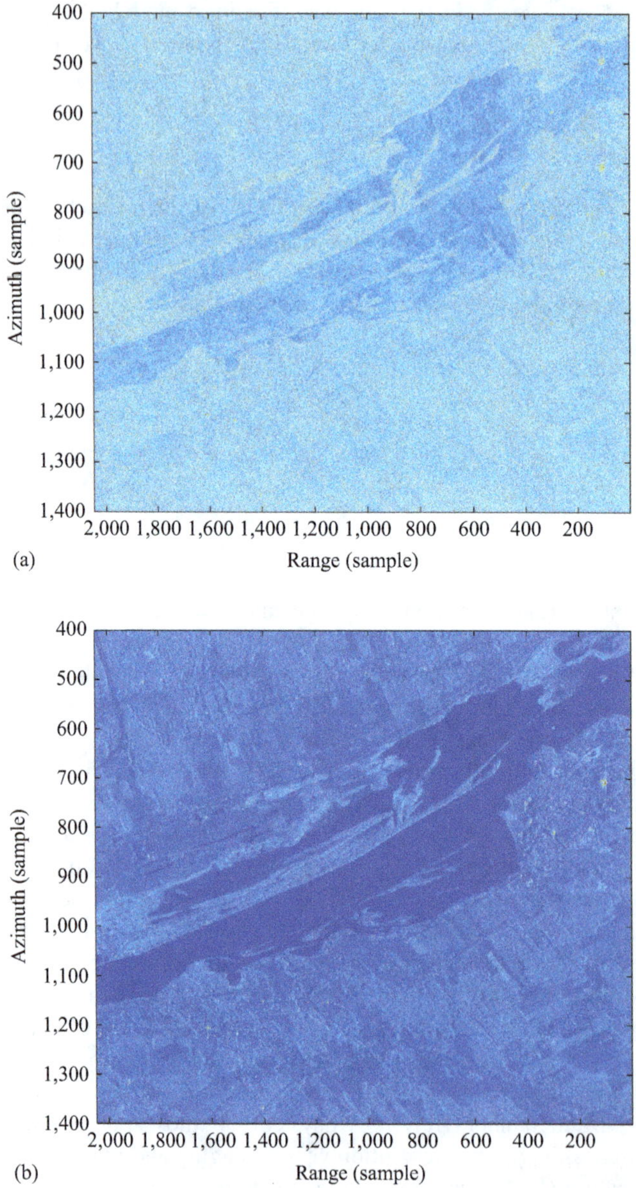

Figure 7.18 Subsampled MODEX data set and MMSE processed version. (a) Classic image and (b) standard deviation of relocation error for TanDEM-X with two conceptional GMTI schemes

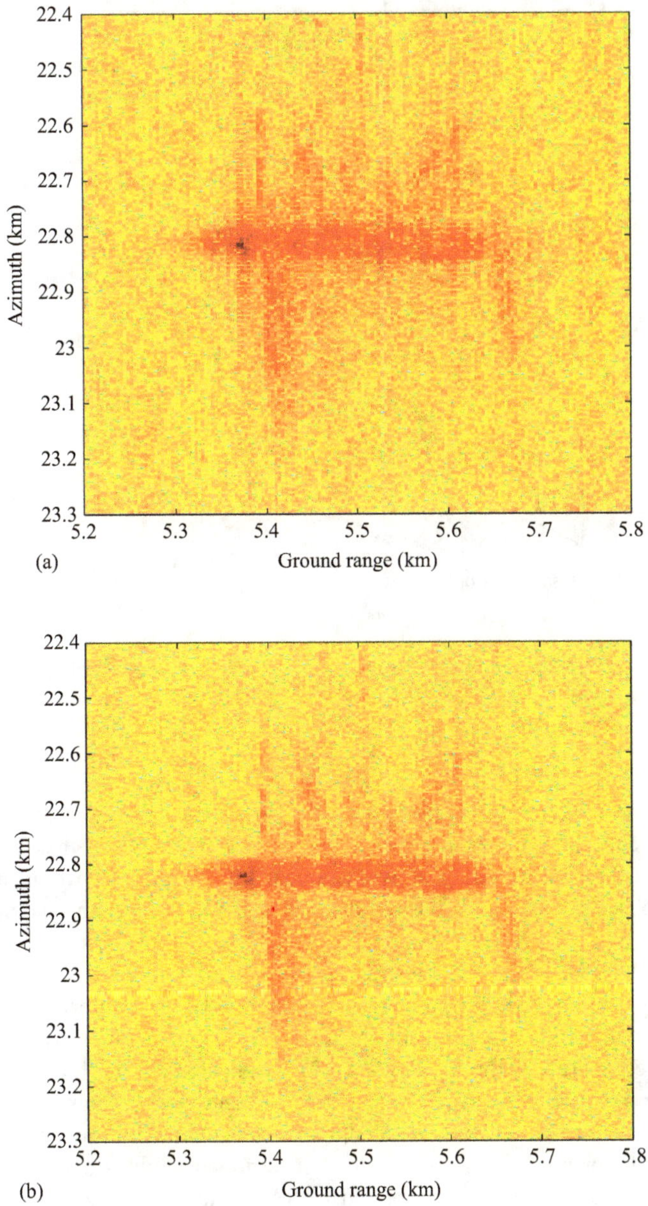

Figure 7.19 Combined channel and HRWS imagery of an extrafine low-PRF data. (a) Single channel image and (b) HRWS image

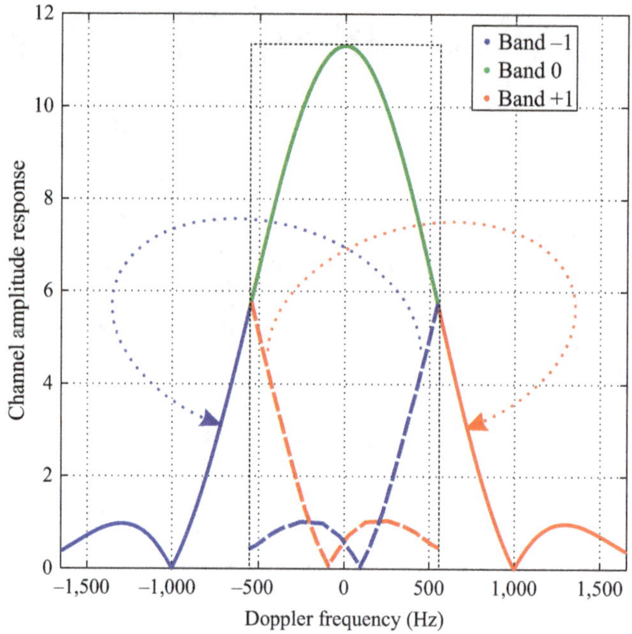

Figure 7.20 Illustration of three backfolded and reconstructed unambiguous Doppler bands

only with one digitizer and recording chain. The single-channel system is synthesized by a coherent sum of the dual-channel measurements in the same way an analogue hardware combining network onboard the spacecraft would have done. It can be observed that both the radiometric and the geometric resolutions have been improved by the MMSE processing. The physical explanation behind this feat is revealed by looking at the illustration of the reconstruction of the ambiguous Doppler spectrum by HRWS processing (Figure 7.20). First, after recuperating the outer Doppler bands through the MMSE, the Doppler bandwidth has been increased (compared to the ambiguous width indicated by the black dashed box, this can be more or less significant). A higher bandwidth relates to an improved geometric resolution. Second, depending on the shape of the antenna pattern (typically sinc or $\cos^2$-like), the unfolded spectrum resembles a smooth Doppler-windowing function commonly applied to SAR data during azimuth processing. A spectral window (e.g., Taylor weighting) is meant to reduce the sidelobes of the point spread function thereby improving the radiometric resolution. Another beneficial feature of HRWS processing, not shown here, is a macroscale 'deghosting' of the entire SAR imagery. Since the energy residing in the ambiguous Doppler band is recollected and used to form the HRWS SAR image of the vessel, the duplicated ghost vessels at the Doppler ambiguous locations vanish or are at least much suppressed.

7.5.2.3 Littoral zones

In contrast to reconstructing the unambiguous signal to alleviate the shortfalls of a low-PRF HRWS SAR mode, one can attempt to cancel only certain ambiguity contributions while keeping other Doppler bands unperturbed. One practically relevant application is the elimination or suppression of ambiguity returns of strong stationary land scatterers at the transition between land and water surfaces [45]. These land ambiguities are typically much stronger than the returns from the sea thereby heavily compromising vessel detection in littoral zones. An example is shown in Figure 7.21, clearly demonstrating that the two vessels on the upper right are much more discernable after land ambiguity suppression HRWS processing. It is important to note that only the Doppler ambiguities on one side of the spectrum corresponding to the water–land transition direction need to be suppressed, for instance, using a projection matrix orthogonal to the contributions related to the positive ambiguous Doppler bands $k = 1, 2, \ldots$. This will lead more degrees-of-freedom remaining to maintain good image quality. It is also worth mentioning that in contrast to GMTI processing, the vessels might be stationary,[23] since only higher Doppler zones are eliminated instead of exploiting the interchannel phase.

7.6 Conclusions and outlook

Most algorithms derived and successfully deployed in airborne SAR-GMTI systems in the past have been proven not be directly transferable to space-based SAR. As a consequence, innovative new concepts have been developed that exploit the flexible programmability of phased array antennas on spacecraft. After full commissioning of MODEX on RADARSAT-2, it has been proven that an add-on GMTI mode on a SAR satellite is feasible and capable of producing meaningful and partly operationally relevant information. For land GMTI, it has been theoretically predicted and experimentally confirmed that RADARSAT-2 can detect and reliably reposition street-bound vehicles moving across-track with at least 40 km/h in a semiurban environment. Comprehensive theoretical analysis of the performance limits for the along-track velocity has revealed the potential of iSTAP to estimate speeds down to walking pace. However, such accuracies have not been experimentally proven yet, and the development of suitable estimators is a hot topic in current research. The biggest potential of short-term impact is in the area of maritime domain awareness and belongs to the outfitting of the operational wide-area ship detection modes with a VME capability.

Although the investigation of multichannel SAR in the past has been motivated by GMTI, recent studies show a wider range of applications. As a matter of fact, due to the linearity of a multichannel measurement, it is always, without any loss, possible to recreate what would have been measured by a single antenna of equivalent length. Thus, there can be no loss in capability or performance. The main trade-offs to consider are the increased cost of the RF-switching hardware, typically negligible compared to the overall mission costs, and the increased data

[23]In fact, the two vessels in Figure 7.21 are anchored according to their AIS information.

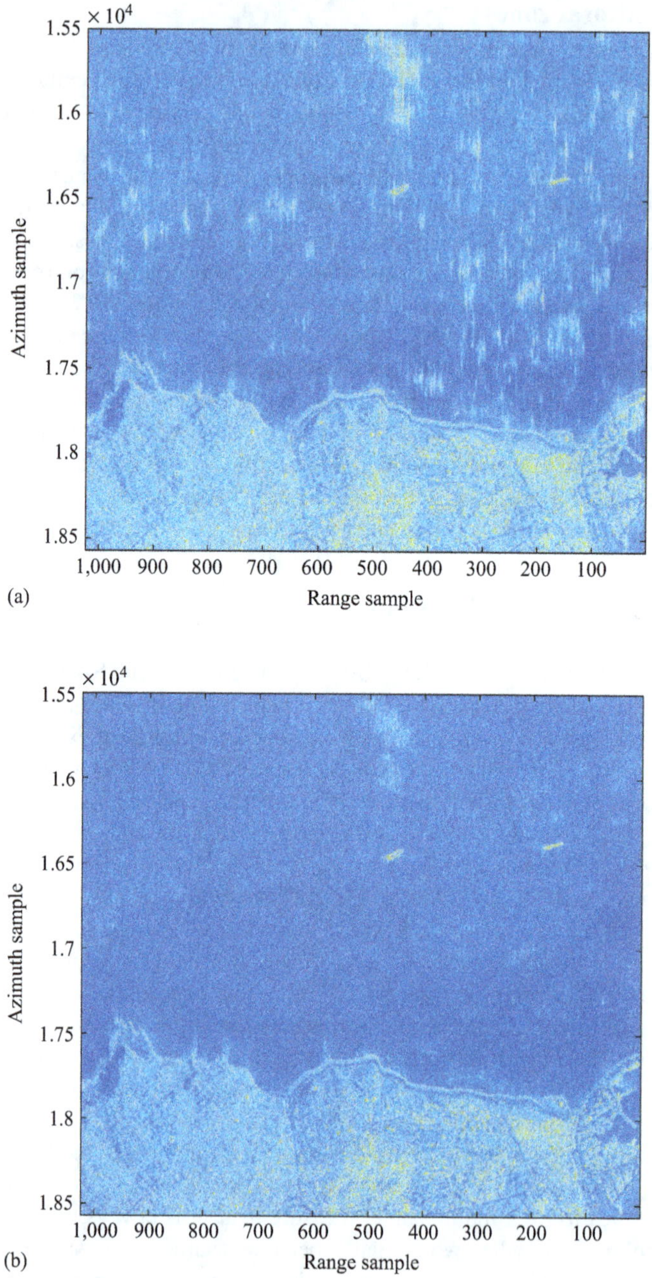

Figure 7.21 Land ambiguity suppression at littoral zones through modified HRWS processing. (a) Conventional image and (b) after ambiguity suppression

volume and subsequent downlink demands. In contrast, this incremental cost offers improved strategic imaging capability through better wide-area surveillance coupled with accurate target velocity estimation, and improved tactical imaging capability through high resolution, high-sensitivity imaging with clutter suppression and GMTI capability.

Due to the demonstrable benefits, there is no doubt that following space-based SAR generations will possess multichannel capability. Consequently, future R&D trends are expected to be directed toward multiaperture HRWS systems combined with high-power active phased arrays and digital elevation beamforming. In contrast to current low-resolution MSSR modes optimized solely for ship detection, these future systems will serve multiple objectives in parallel; high-res wide-coverage for oceanographic purposes like ice charting as well as vessel detection, identification and estimation. These advanced systems will be so effective that a small number of them are sufficient to cover gigantic areas in short revisit times, especially if placed in different orbital planes. High bandwidth together with multiple channels drives up the requirements for data rate and volume exponentially resulting in ever increased demands for downlink capacity forcing exceedingly expensive new ground infrastructure. Therefore, another cross branch of SAR-GMTI research will emphasize on the development of real-time, space-hardened onboard processor hardware based on FPGA technology and robust algorithms trying to minimize the downlink requirements. In terms of space-based SAR-GMTI requirements for more tactical decision-making with the demand for slow, low RCS mover detection and an estimation accuracy in the sub-metre range, R&D will continue to look toward extendible antenna structures to increase aperture and gain as well as coherently cooperating bi- and multistatic constellations flying in close formation. New areas of R&D still in infancy are SAR-GMTI in combination with MIMO using many different transmitted waveforms to create higher diversity as well as polarimetric GMTI which may offer the ability to measure different physical reflection mechanisms of detected moving objects and may allow discrimination between various types of targets.

References

[1] C. H. Gierull, C. Livingstone. 'SAR-GMTI concept for RADARSAT-2', R. Klemm (Ed.), *The Applications of Space-Time Processing*, (IEE Press, Stevenage, UK, 2004).

[2] C. H. Gierull. 'Statistical analysis of multilook SAR interferograms for CFAR detection of ground moving targets', *IEEE Transactions on Geoscience and Remote Sensing*, 2004, **42(4)**, pp. 691–701.

[3] C. E. Muehe, M. Labitt. 'Displaced-phase-center antenna technique', *Lincoln Laboratory Journal*, 2000, **12(2)**, pp. 281–296.

[4] I. C. Sikaneta, C. H. Gierull. 'Adaptive CFAR for space-based multichannel SAR–GMTI', *IEEE Transactions on Geoscience and Remote Sensing*, 2012, **50(12)**, pp. 5004–5013, [Online]. Available: http://ieeexplore.ieee.org/stamp/stamp.jsp?arnumber=6203573.

[5] C. H. Gierull, I. C. Sikaneta, D. Cerutti-Maori. 'Two-step detector for RADARSAT-2's experimental GMTI mode', *IEEE Transactions on Geoscience and Remote Sensing*, 2013, **51(1)**, pp. 436–454, [Online]. Available: http://ieeexplore.ieee.org/stamp/stamp.jsp?arnumber=6232455.

[6] J. H. G. Ender, C. H. Gierull, D. Cerutti-Maori. 'Improved space-based moving target indication via alternate transmission and receiver switching', *IEEE Transactions on Geoscience and Remote Sensing*, 2008, **46(12)**, pp. 3960–3974, [Online]. Available: http://ieeexplore.ieee.org/stamp/stamp.jsp?arnumber=4683338.

[7] D. Cerutti-Maori, I. Sikaneta, C. H. Gierull. 'Optimum SAR/GMTI processing and its application to the radar satellite RADARSAT-2 for traffic monitoring', *IEEE Transactions on Geoscience and Remote Sensing*, 2012, **50(10)**, pp. 3868–3881, [Online]. Available: http://ieeexplore.ieee.org/stamp/stamp.jsp?arnumber=6171842.

[8] C. H. Gierull. 'Digital channel balancing of along-track interferometric SAR data', DRDC Ottawa TM 2003-024, Defence Research and Development Canada, March 2003.

[9] K. K. D. Ward, R. J. A. Tough, S. Watts. *Sea Clutter: Scattering, the K Distribution and Radar Performance,* volume 20, (IET Radar, Sonar and Navigation, Stevenage, UK, 2006).

[10] C. H. Gierull. 'Numerical recipes to determine the performance of multi-channel GMTI radars', Technical Memorandum TM 2011-230, DRDC Ottawa, December 2011, [Online]. Available: http://cradpdf.drdc-rddc.gc.ca/PDFS/unc116/p536465_A1b.pdf.

[11] C. H. Gierull, I. Sikaneta. 'Improved SAR vessel detection based on discrete texture', *Proc. of EUSAR*, pp. 1–4, (Hamburg, Germany2016).

[12] P. Vachon, J. W. M. Campbell, C. J. Bjerkelund, F. W. Dobson, M. T. Rey. 'Ship detection by the radarsat sar: Validation of detection model predictions', *Canadian Journal of Remote Sensing*, 1997, **23(1)**, pp. 48–59.

[13] Y. Quilfen, A. Bentamy. 'Calibration/validation of ERS-1 scatterometer precision products', *Proc. of IGARSS'94*, volume 2, pp. 945–947, (Pasadena, CA1994).

[14] I. R. Joughin, D. P. Winebrenner, D. B. Percival. 'Probability density functions for multilook polarimetric signatures', *IEEE Transactions on Geoscience and Remote Sensing*, 1994, **GRS-32**(May **(3)**), pp. 562–574.

[15] I. C. Sikaneta, C. H. Gierull. 'Parameter estimation for the phase statistics in interferometric SAR', *Geoscience and Remote Sensing Symposium, 2002. IGARSS '02. 2002 IEEE International*, volume 3, pp. 1735–1737, (2002), [Online]. Available: http://ieeexplore.ieee.org/stamp/stamp.jsp?arnumber=1026237.

[16] R. Kohlleppel, C. H. Gierull. 'Enhancement of along-track interferometry for ground moving target indication', *Geoscience and Remote Sensing Symposium, 2008. IGARSS 2008. IEEE International*, volume 2, pp. II–229, (2008).

[17] S. Chiu. 'A constant false alarm rate (CFAR) detector for RADARSAT-2 along-track interferometry', *Canadian Journal of Remote Sensing*, 2005, **31(1)**, pp. 73–84.

[18] C. H. Gierull. 'Moving target detection with along-track SAR interferometry—a theoretical analysis', DRDC Ottawa TR 2002-084, Defence Research and Development Canada, August 2002.

[19] M. V. Dragosevic, C. Livingstone. 'Demonstration of RADARSAT-2 moving object detection experiment (MODEX) capabilities for maritime surveillance', *International Radar Symposium (IRS'09)*, (2009).

[20] C. H. Gierull. 'Analysis of the multi-mode feedhorn concept for multi-channel SAR-GMTI', *IEEE Transactions on Geoscience and Remote Sensing*, 2011, **49(10)**, pp. 3611–3621.

[21] I. C. Sikaneta, C. H. Gierull, J.-Y. Chouinard. 'Metrics for SAR-GMTI based on eigen-decomposition of the sample covariance matrix', *Proc. Int. RADAR 2003*, (Adelaide, South Australia 2003).

[22] D. Cerutti-Maori, C. H. Gierull, J. H. G. Ender. 'Experimental verification of SAR-GMTI improvement through antenna switching', *IEEE Transactions on Geoscience and Remote Sensing*, 2010, **48(4)**, pp. 2066–2075.

[23] R. Klemm. *Space-Time Adaptive Processing*, (IEE Press, Stevenage, UK, 1998).

[24] R. Klemm. *Applications of Space-Time Adaptive Processing*, volume 14, (IET, 2004).

[25] D. Cerutti-Maori, I. Sikaneta, C. H. Gierull. 'Comparison of the GMTI capability of RADARSAT-2 MODEX-1 and MODEX-2 modes—optimum signal processing and theoretical investigation of the GMTI performance', Technical Report DRDC Ottawa TR 2011-270, Defence Research & Development Canada—Ottawa, Canada, March 2011.

[26] I. C. Sikaneta, J. H. G. Ender. 'On the optimality of post-doppler STAP', Technical Report DRDC Ottawa TR 2012-084, Defence Research & Development Canada—Ottawa, Canada, January 2012.

[27] D. Cerutti-Maori, I. Sikaneta. 'A generalization of DPCA processing for multi-channel SAR/GMTI radars', *IEEE Transactions on Geoscience and Remote Sensing*, 2013, **51(1)**, pp. 560–572.

[28] M. Rashid. 'Cramer-Rao lower bound derivation and performance analysis for space-based SAR-SMTI', Master's thesis, Carleton University, Canada, June 2016.

[29] C. H. Gierull, I. C. Sikaneta. 'Potential marine moving target indication (MMTI) performance of the RADARSAT constellation mission (RCM)', *9th European Conference on Synthetic Aperture Radar (EUSAR 2012)*, pp. 404–407, (2012), [Online]. Available: http://ieeexplore.ieee.org/stamp/stamp.jsp?arnumber=6217090.

[30] U. Nickel. 'Overview of generalized monopulse estimation', *IEEE Aerospace and Electronic Systems Magazine*, 2006, **21(6)**, pp. 27–56.

[31] I. Sikaneta, C. H. Gierull. 'Vessel motion estimation (VME) from multi-channel, low-PRF imagery', DRDC-Report of Invention, 7 May 2015, 1416-15/002CA (DSTER 5-5).

[32] W.-D. Wirth. *Radar Techniques Using Array Antennas (IEE Radar, Sonar, Navigation & Avionics Series)*, 10, (IET, 2001).

[33] S. V. Baumgartner, G. Krieger. 'Large along-track baseline SAR-GMTI: first results with the TerraSAR-X/TanDEM-X satellite constellation',

Geoscience and Remote Sensing Symposium (IGARSS), 2011 IEEE International, pp. 1319–1322, (IEEE, 2011).

[34] G. Krieger, M. Zink, M. Bachmann, *et al.* 'TanDEM-X: a radar interferometer with two formation-flying satellites', *Acta Astronautica*, 2013, **89**, pp. 83–98.

[35] C. H. Gierull, D. Cerutti-Maori, J. H. G. Ender. 'Ground moving target indication with tandem satellite constellations', *IEEE Geoscience and Remote Sensing Letters*, 2008, **5(4)**, pp. 710–714, [Online]. Available: http://ieeexplore.ieee.org/stamp/stamp.jsp?arnumber=4656448.

[36] S. Suchandt, H. Runge. 'First results of TanDEM-X along-track interferometry', *Proc. of IGARSS'12*, pp. 1908–1911, (Munich, Germany 2012).

[37] J. H. G. Ender. 'Space-based SAR/MTI using multi-static satellite configurations', *Proc. of EUSAR*, pp. 337–340, (Cologne, Germany 2002).

[38] L.-P. Rousseau, C. H. Gierull, J.-Y. Chouinard. 'Performance analysis of HRWS/GMTI for space-based SAR using sparse arrays', *International Radar Symposium (IRS'16)*, (2016).

[39] D. Geudtner, M. Zink, C. Gierull, S. Shaffer. 'Interferometric alignment of the X-SAR antenna system on the space shuttle radar topography mission', *IEEE Transactions on Geoscience and Remote Sensing*, 2002, **40(5)**, pp. 995–1006.

[40] P. W. Vachon, C. Kabatoff, R. Quinn. 'Operational ship detection in Canada using RADARSAT', *Geoscience and Remote Sensing Symposium (IGARSS), 2014 IEEE International*, pp. 998–1001, (IEEE, 2014).

[41] L.-P. Rousseau, C. H. Gierull, J.-Y. Chouinard. 'First results from an experimental ScanSAR-GMTI mode on RADARSAT-2', *IEEE Journal of Selected Topics in Applied Earth Observations and Remote Sensing*, accepted August 2015.

[42] N. Gebert, G. Krieger, A. Moreira. 'High resolution wide swath SAR imaging with digital beamforming–performance analysis, optimization, system design', *EUSAR 2006*, 2006.

[43] I. C. Sikaneta, C. H. Gierull, D. Cerutti-Maori. 'Optimum signal processing for multichannel SAR: with application to high-resolution wide-swath imaging', *IEEE Transactions on Geoscience and Remote Sensing*, 2014, **52(10)**, pp. 6095–6109, [Online]. Available: http://ieeexplore.ieee.org/stamp/stamp.jsp?arnumber=6733361.

[44] D. Cerutti-Maori, I. C. Sikaneta, C. H. Gierull. 'Detection and imaging of moving objects with high-resolution wide-swath SAR systems', *Proc. of EUSAR*, (Berlin, Germany 2014).

[45] J. Kim, M. Younis, P. Prats-Iraola, M. Gabele, G. Krieger. 'First spaceborne demonstration of digital beamforming for azimuth ambiguity suppression', *IEEE Transactions on Geoscience and Remote Sensing*, 2013, **51(1)**, pp. 579–590.

Chapter 8
Interferometric and tomographic SAR

Gianfranco Fornaro[1] and Antonio Pauciullo[1]

Abstract

Multipass differential synthetic aperture radar (SAR) interferometry (DInSAR), with its capability to accurately monitor ground displacements, has dramatically pushed the applications of imaging radars in many fields, particularly in the area of environmental risk monitoring and security.

Advanced DInSAR techniques operating at reduced resolution allow monitoring very wide areas whereas persistent scatterers interferometry (PSI) and more recently SAR tomography for very high resolution sensors, have shown to be powerful methods for providing 3-D point clouds representing buildings and infrastructures as well as in the monitoring of their possible slow temporal deformations. In this chapter, a detailed description of these techniques is provided, starting on the basic principles of the SAR interferometry and highlighting the relationship between interferometric and tomographic approaches.

8.1 Introduction

The principle of the synthesis of a large aperture is fundamental in Synthetic Aperture Radar (SAR) image formation to enable the capability of producing very high-resolution images of a ground scene with remote sensing systems operating in the microwave region [1–3]. The key point in SAR imaging is the use of data acquired with different along-track phase centres, which are made available by exploiting the movement of the real antenna along a track. This feature allows the possibility of carrying out a sharpening of the beam in the *along-track* direction at the data processing stage, so to achieve an *azimuth* resolution comparable to that of optical sensors.

Either by exploiting repeated passes of a single antenna system, or by having at disposal at least another antenna on a single platform, interferometric SAR (InSAR) exploits the possibility to acquire data from different phase centres

[1]Institute for Electromagnetic Sensing of the Environment (IREA), National Research Council (CNR), Italy

characterized by the presence of offsets in the plane orthogonal to the azimuth direction, determined by the baseline vectors. InSAR thus provides the possibility to image the scene from slightly different incidence angles. The angular diversity enables the measurement of the height distribution of ground scatterers. InSAR therefore provides the 3D localization of the ground targets present in a SAR image [4–7].

The chapter first of all provides a summary of the main characteristics of InSAR mode. It addresses the description of the InSAR system from a geometric point of view, giving emphasis to the introduction of the key parameters, such as the sensitivity to the height and the accuracy in the height estimation, as well as to the description of decorrelation, which is the source of noise in coherent systems, and of the data processing aspects aimed at mitigating their effects [8].

The imaging in any weather condition and by day or night is an important prerequisite for the observation of dynamical processes. Spaceborne SAR sensors allow synoptic imaging as well as systematic data acquisition: The latter is a key aspect for differential interferometric SAR (DInSAR) systems to reveal Earth displacements to a centimetre accuracy level [9–11]. DInSAR, which exploits a specific InSAR configuration characterized by repeated passes with reduced angular diversity, i.e. reduced baselines, is the technique that has opened many and probably among the most important applications of SAR in the area of natural hazards. In fact, DinSAR has been widely used for the analysis of large deformations such as, for example, those generated by major earthquakes or pre-eruptive volcanic activities.

A further, chief, development of InSAR/DInSAR is associated with methods widely known as advanced DInSAR (A-DInSAR) or PSI techniques [12–17], which are able to process large InSAR image stacks acquired over repeated orbits. These techniques are capable to achieve monitoring of ground displacements to sub-centimetre level of accuracy over wide regions, with important applications to the area of environmental risk monitoring and security. DinSAR principles and summaries of the key features of A-DInSAR methods like Small BAseline Subset (SBAS) [12,15], belonging to the class of so-called coherent stacking interferometry (CSI) working at reduced resolution (i.e. with multi-look data), and the persistent scatterers (PSs) technique [16,17] working at full resolution are outlined.

Within the framework of the processing of coherent (interferometric) SAR stacks, SAR tomography (TomoSAR) has provided a further key contribution: it revisits and extends SAR interferometry into a 3D [18,19], or even more (multi-D) [20,21] SAR imaging context. The principle of TomoSAR is rather simple and can be easily explained by using the analogy to antenna synthesis in the azimuth direction explained before: by exploiting acquisitions carried out with multiple baselines, TomoSAR allows the formation of a thin beam also in the cross-azimuth (i.e. cross track) plane or better along the elevation direction. A spaceborne or airborne fine beam 'radar scanner' can be therefore synthesized: this permits an accurate scanning of the scene in 3D, thus extending traditional interferometric approaches.

3D SAR imaging (focusing) was demonstrated for the first time with an airborne system [18] and was subsequently extended to the spaceborne case [19]. It enables a deeper and advanced investigation of the scattering of complex scenes, such as urban areas or forest characterized by the presence of 'structured' scattering, i.e. a scattering mechanism that is distributed or concentrated in multiple points along the vertical direction.

The 3D focusing concept can be extended along the time axis with 4D (space-time) SAR imaging, also known as differential TomoSAR to allow the monitoring of slow temporal deformation of ground scatterers [20,21].

Prospective distortions due to the radar imaging geometry frequently lead in urban areas to interference of the response of multiple scatterers relevant to different ground structures that are imaged in the same image pixel (layover). TomoSAR (3D or 4D) allows, therefore, extending the classical PSI technique by allowing the possibility to detect the presence of the response of multiple scatterers interfering within the same (azimuth-range) pixel due to the layover effect. With respect to PSI, it allows also achieving improvements of the performances in the detection of single dominant scatterers (not affected by layover) as well as in the estimation of the parameters of interest in PSI processing [i.e. the height and deformation mean velocity (DMV)].

The chapter is organized as described in the following. Section 8.2 provides the basic equations related to the interferometric system that allows describing the relation between the scene topography and the interferometric phase. In Section 8.2.1, the relation associating the (ground) point height to the phase of the corresponding image pixel is derived, together with the phase-to-height sensitivity and some characteristic parameters as the ambiguity height. Moreover, the basic (ground) point to (image) pixel interferometric relations are extended to the case of a planar topography thus allowing to relate the terrain slope to the linear variation of the interferometric phase with the range. Section 8.2.2 then summarizes the decorrelation phenomenon, including the baseline decorrelation that is intrinsic to the interferometric system and, for a finite range resolution, a consequence of the presence of the range variation of the phase. The basic concepts of DInSAR for the measurement of displacements are then addressed in Section 8.2.3, thus concluding the description of a basic interferometric system based on two acquisitions. Section 8.3 is dedicated to the extension to an interferometric multi-acquisition system. Key elements of two well-known approaches, i.e. CSI and PSI, tailored to the monitoring of deformation in the presence of distributed and concentrated scattering, respectively, are described. A simple tool based on a two-step, or better a two-scale analysis, for the measurement of small scale (low resolution) and large scale (full resolution) combining CSI and PSI is also described. A feature of PSI is to allow the accurate 3D localization of scatterers at full resolution: PSI is then generalized in the context of TomoSAR in Section 8.4. TomoSAR is then extended with reference to deformation monitoring, in the so-called differential TomoSAR or multi-D SAR focusing. Description of tomographic inversion algorithms as well as investigation of the PS detection in the context of TomoSAR are then addressed.

8.2 SAR interferometry

8.2.1 Basic concepts

We refer to Figure 8.1, which shows the reference geometry of a repeat pass InSAR system in the plane orthogonal to the orbit (flight) direction, i.e. orthogonal to the azimuth (x). Here, we assume two passes of a SAR system mounted on-board a satellite or airborne platform at given times t_1 and t_2 and known positions $\mathbf{S}_1$ and $\mathbf{S}_2$: the first pass is assumed as reference (master) and the two passes are separated by a baseline vector $\mathbf{b}$ determined by b (baseline length) and α_b (baseline tilt angle). The Cartesian reference system is referred to the scene centre: α (the so-called incidence angle) is the angle between the direction corresponding to the wave vector of the impinging radiation and the vertical direction; ϑ is the look angle of the scatterer defined in the tilted Cartesian system, see Figure 8.1. It is worth noting that for a spaceborne system, the look angle differs from the so-called off nadir angle, see again Figure 8.1.

We have assumed the presence of a target P at height z, the unit vector $\hat{\mathbf{r}}$ defines the radar line of sight (los): r is the range of the scatterer at the first antenna, $r + \delta r_T$ is the range of the scatterer to the second (slave) antenna.

A single SAR image is two dimensional (2D) and hence it can provide the measurement, in addition to the azimuth, only of the range (r in Figure 8.1) with

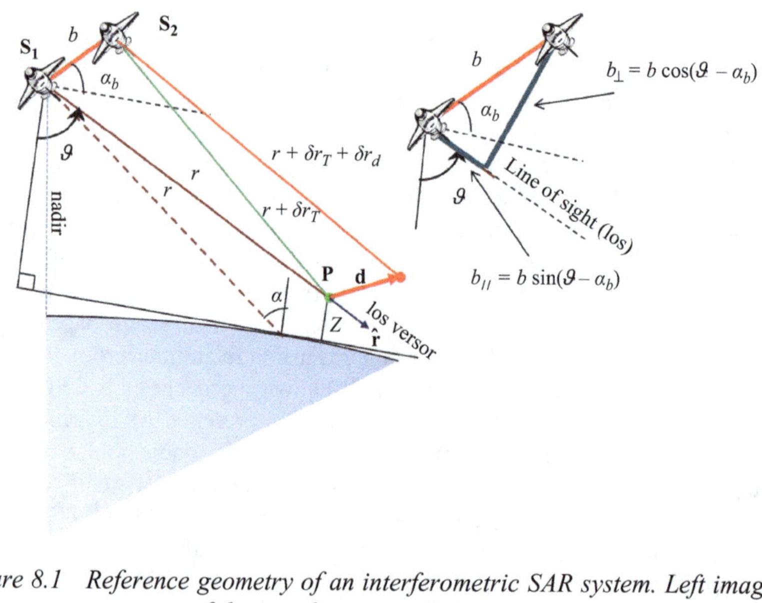

Figure 8.1 Reference geometry of an interferometric SAR system. Left image: geometry of the interferometric SAR acquisition in the plane orthogonal to the flight direction. Right image: relevant to the decomposition of the baseline in to parallel and perpendicular to the target line of sight components

accuracies specified by the sensor resolution. Measuring the range alone cannot provide unambiguous localization of the scatterer: No information is provided about ϑ, which is the third coordinate to be known to achieve knowledge about the position of the scatterer in 3D.

The look angle can be estimated by exploiting at least another image acquired with a different imaging geometry thus providing the measurement of a second range of the target, i.e. $r + \delta r_T$. Similarly to the human vision system, which compares the variation of the projection of the object at the two eyes, the key quantity that allows achieving sensitivity to the target height (z) is δr_T, i.e. the range (travelled path) variation from one image to the other. The intersection of the two circles orthogonal to the azimuth and centred on $\mathbf{S}_1$ and $\mathbf{S}_2$ with radii r and $r + \delta r_T$ provides the localization of the target, and hence the measurement of the height of the target.

In order to translate this concept into mathematical relations, we apply the Carnot theorem to the triangle $\mathbf{S}_1\mathbf{P}\mathbf{S}_2$ in the hypothesis of $r \gg \delta r_T$ and $r \gg b$ thus obtaining

$$\delta r_T \cong -b\sin(\vartheta - \alpha_b) = b_{//} \tag{8.1}$$

which is the fundamental equation relating the path difference to the look angle ϑ knowing the baseline vector. Note that the path difference can be zeroed by zeroing the baseline length.

Letting $H = r\cos\alpha$ to be the height of the master sensor, the measurement of ϑ provides the direct measurement of the height:

$$z = r\cos\alpha - r\cos\vartheta \tag{8.2}$$

Equations (8.1) and (8.2) provide the basic relations between the topographic path difference δr_T and z. Stereoscopic measurement of δr_T by range difference in radar imaging (stereometry), or very accurate interferometric estimation of δr_T by means of phase measurement, allows estimating ϑ from (8.1) knowing the interferometric geometry, i.e. the baseline length b and baseline tilt angle α_b. Given the incidence angle α on the reference ellipsoid and the range r, the measurement of the look angle ϑ can be translated into the target height via (8.2) [4,22].

It is worth noting that the above procedure for the determination of the height of the target is valid in the case of simultaneous acquisitions. In the case of quasi-simultaneous or repeat pass acquisitions, the target can undergo displacements, thus implying the presence of possible range variations that add to the topographic contribution. With the aim to introduce a general model valid in the presence of displacements, and hence also for differential interferometry that is the subject of the next section, we assume in Figure 8.1 that the target can show a displacement $\boldsymbol{d}$ between the two passes: This implies the presence of range variation δr_d along the los (positive when the target is moving towards the sensor) in addition to the topographic path variation δr_T. Depending on the case under analysis, i.e. interferometry or differential interferometry, in $\delta r = \delta r_T + \delta r_d$, we will set in $\delta r_d = 0$ or $\delta r_d \neq 0$, respectively.

The determination of δr_T or, better, of $\delta r = \delta r_T + \delta r_d$, with an accuracy of a fraction of the wavelength λ (i.e. centimetric at microwaves) can be achieved by exploiting fact that electromagnetic waves undergo a phase change proportional to the travelled distance during propagation.

We refer to the expression of the focused signal, for sake of simplicity only along the range, after the data compression. Letting $\gamma_{t_1}(r)$ and $\gamma_{t_2}(r)$ be the scene reflectivity, i.e. the function that models the backscattering properties at the two (possibly different) acquisition times t_1 and t_2, and $f_p(r)$ the post-focusing point spread function (PSF), but for inessential (from the interferometric viewpoint) amplitude factors, we have the following expressions for the focused signal at the first and second antenna (after proper registration, i.e. range alignment of the response) [7]:

$$y_1(r') = \int dr\, \gamma_{t_1}(r) e^{-j\frac{4\pi}{\lambda}r} f_p(r'-r) = e^{-j\frac{4\pi}{\lambda}r'} \int dr\, \gamma_{t_1}(r) e^{-j\frac{4\pi}{\lambda}(r-r')} f_p(r'-r)$$

$$= \gamma_1(r') e^{-j\frac{4\pi}{\lambda}r'} \tag{8.3}$$

$$y_2(r') = \int dr\, \gamma_{t_2}(r) e^{-j\frac{4\pi}{\lambda}(r+\delta r)} f_p(r'-r)$$

$$= e^{-j\frac{4\pi}{\lambda}(r'+\delta r')} \int dr\, \gamma_{t_2}(r) e^{-j\frac{4\pi}{\lambda}(r-r'+\delta r-\delta r')} f_p(r'-r) = \gamma_2(r') e^{-j\frac{4\pi}{\lambda}(r'+\delta r')} \tag{8.4}$$

where $\delta r' = \delta r(r = r')$. Generally, assuming the absence of any tapering of the spectrum of the data leads to a PSF equal to $f_p(r) = R_r^{-1}\text{sinc}(r/R_r)$, where $\text{sinc}(r) = \sin(\pi r)/(\pi r)$ [2] and R_r is the range resolution.

It is at this point worth to comment on the difference between the variables r and r', which represent the range of the scatterer and the radar sampled range, i.e. the range of the image pixel, respectively. The two variables are indistinguishable in the case of an infinite resolution, i.e. when $R_r \to 0$, thus translating the capability of the SAR system to achieve, in this case, a perfect reconstruction of the backscattering.

The path difference can be estimated by extracting the argument of the Hermitian product implemented via the complex beating of the two backscattered signals (interferogram):

$$\phi(r) = \text{Arg}\{y_1(r) y_2^*(r)\} = \frac{4\pi}{\lambda}\delta r(r) + \varphi_n(r) = \frac{4\pi}{\lambda}\delta r_T + \frac{4\pi}{\lambda}\delta r_d + \varphi_n$$

$$= \phi_T + \varphi_d + \varphi_n \tag{8.5}$$

where the dependence on range of the terms in the last equalities has been neglected for the sake of brevity, Arg{} is the function that extracts the (unrestricted) argument of a complex number.

The term φ_n accounts for the unavoidable additive phase noise contribution corrupting the measurements in Equations (8.3) and (8.4) as well as the effects of the phase change between γ_1 and γ_2. The latter is referred to as decorrelation of the radar echoes and accounts for the effects of the difference between the two integrals in (8.3) and (8.4) which coherently sum different complex terms due to possible temporal changes and to the presence of the term $\delta r - \delta r'$, which vanishes only in the case of infinite resolution, that is a Dirac impulse $f_p(r)$.

It is worth to note that, strictly speaking, (8.5) is valid for simultaneous or quasi-simultaneous acquisitions. The latter is the case in which the antennas are mounted on the same platform or on a formation of satellites flying in a closed configuration. In such a case, the configurations are referred to as along-track interferometry and the term δr_d, due to the displacement, can allow detecting fast-moving targets and measuring their radial velocity [23,24] or measuring extended surface displacements [25] such as those associated with currents for maritime applications.

In the presence of a significant time delay between the two acquisitions, as in the case of repeated passes in differential interferometry for monitoring slow, long-term deformation over the time of ground scenes, an additional phase term must be considered to take into account for the variations of the atmospheric conditions [26]. This aspect is better described in Section 8.2.3 addressing the use of differential interferometry for monitoring slow temporal deformation.

By using (8.1), (8.2) and (8.5), it is possible to evaluate the sensitivity of the interferometric system to the height:

$$g_z = \left|\frac{\partial \phi_T}{\partial z}\right| = \frac{4\pi}{\lambda}\left|\frac{\partial \delta r_T}{\partial z}\right| = \frac{4\pi}{\lambda}\left|\frac{\partial \delta r_T}{\partial \vartheta}\frac{\partial \vartheta}{\partial z}\right| = \frac{4\pi}{\lambda}\left|\frac{b_\perp}{r \sin \vartheta}\right| \tag{8.6}$$

where $b_\perp = b \cos(\vartheta - \alpha_b)$ is the orthogonal baseline component, see Figure 8.1.

Equation (8.6) describes an important parameter associated with the accuracy of the interferometric system: it highlights that the larger the orthogonal baseline component, the higher the sensitivity of the system to the height and hence, for a fixed level of phase noise, the higher the accuracy of the system in the estimation of the target height.

The phase is extracted from complex numbers; the measured version of ϕ is, therefore, for each image pixel, a number belonging to the interval $(-\pi, \pi)$. The image corresponding to the wrapped phase has sharp variations (fringes) between pixels located where the phase changes from $-\pi$ to π and vice versa: the whole excursion of the wrapped phase from $-\pi$ to π (or vice versa) is referred to as an *interferogram.*

The wrapping, applied to each pixel, is a non-linear and surjective operator. Hence, the action aimed at inverting this operator, referred to as phase unwrapping (PhU), is, generally, a very difficult task in the whole processing chains leading to the final digital elevation model (DEM). Several PhU algorithms have been proposed in the literature [27–29]: Among them, a widely used approach is framed in the context of linear programming and specifically is based on a minimum cost flow approach [27]. An exhaustive description of such methods is outside the scope of this chapter.

The height variation that generates a phase change of 2π of the topographic component, i.e. ϕ_T, is referred to as ambiguity height or also phase cycle [6,7]:

$$z_{2\pi} = \frac{\lambda r \sin\vartheta}{2b_\perp} = \frac{2\pi}{g_z} \qquad z_{2\pi} = 85\frac{100}{b_\perp}[\mathrm{m}] \quad \text{for ERS/Envisat} \tag{8.7}$$

Accordingly, larger baselines provide lower ambiguity heights, i.e. an increase of the number of fringes for a given topography variation. This choice allows in principle achieving higher accuracy in the DEM reconstruction, see (8.6). Nevertheless, besides making PhU much more challenging, a baseline increase leads as well as to an increase of the decorrelation of radar echoes [8], especially for the class of systems characterized by a limited spatial resolution: This point will be addressed in more detail in the analysis of decorrelation of radar echoes which is the subject of the next section.

So far, we have mainly described the interferometric system on the basis of a single pixel, i.e. the relation between the scatterer height and the corresponding interferometric phase in the image pixel. It is however instructive to evaluate the variation of the topographic path difference with the range to allow relating the phase to canonical topography profiles. This analysis is preliminary to the evaluation of the decorrelation of radar echo described in the next section. Moreover, it allows also demonstrating that even in the absence of any topography, the interferometric phase shows an intrinsic variation with the range due to the change of the look angle over the range. This contribution is usually referred to as flat Earth contribution.

We first of all refer to Figure 8.2 and derive a linearized version of the relations in (8.1) and (8.2) relating the topographic path difference δr_T to the topography z. We there assume that the analysis is carried out locally and therefore we refer to the (slant) Cartesian reference system depicted in green in Figure 8.2 with the los axis

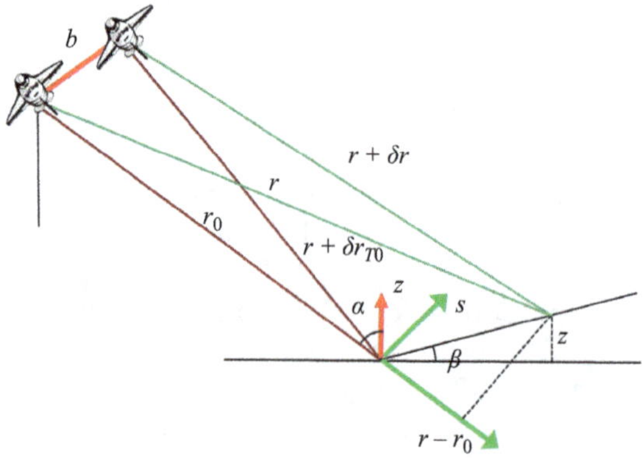

Figure 8.2 Reference geometry for the evaluation of the behaviour of the range path variation with the range for a planar terrain with a generic slope β. r_0 is the reference range and r is the generic range

corresponding to the range and with the axis s (slant height) orthogonal to the range: the latter is usually referred to as elevation.

The relation between the height and the slant height is

$$s = \frac{z}{\sin \alpha} \tag{8.8}$$

By considering the sensitivity equation, i.e. (8.6), specified for $r = r_0$ we have

$$\left.\frac{\partial \delta r_T}{\partial z}\right|_{r=r_0} = -\frac{b_\perp}{r_0 \sin \alpha} \rightarrow \frac{\partial \delta r_T}{\partial s} = -\frac{b_\perp}{r_0} \tag{8.9}$$

that, after integration, provides

$$\delta r_T(r) = -\frac{b_\perp}{r_0} s(r) + \text{cost} \tag{8.10}$$

To this end, we refer to Figure 8.2 where we have a reference point at range r_0 on the ground whose path difference is indicated as δr_{T0} : the aim is to evaluate δr_T at the generic range r as a function of the terrain slope. We assume the scene topography to be planar with a ground slope equal to β (see Figure 8.2).

The topography can be described in the slant reference system (s, r) by $s = s(r)$: for the assumed planar terrain the latter can be specified as

$$s = (r - r_0)\cot(\alpha - \beta) \tag{8.11}$$

At this stage, the substitution in (8.10) of the expression of the planar topography profile in (8.11) provides

$$\delta r_T(r) = \delta r_{T0} - \frac{b_\perp}{r_0}(r - r_0)\cot(\alpha - \beta) \tag{8.12}$$

where δr_{T0} is the range variation at $r = r_0$. (8.12) shows that $\delta r_T(r)$ is a linear function of the range r, with an angular coefficient that depends on the terrain slope: even in the case of a flat Earth ($\beta = 0$), the range path difference changes (approximately) linearly with the range. The angular coefficient of $\delta r_T(r)$ increases for β approaching α. The condition $\beta = \alpha$ (layover limit) provides an infinite angular coefficient: this situation corresponds to a planar terrain whose slope is such that the terrain profile and the los direction are orthogonal. The image of the whole plane would be in this case located at a single range (r_0) and the path difference would be therefore characterized by an infinite derivative in $r = r_0$.

Starting from (8.5) and (8.10), it is possible at this point to derive (up to an inessential constant factor) the general expression of the interferometric phase as a function of the topography and deformation:

$$\phi = \frac{4\pi}{\lambda}\delta r + \varphi_n = -\frac{4\pi}{\lambda}\frac{b_\perp}{r_0}s(r) + \frac{4\pi}{\lambda}\delta r_d + \varphi_n \tag{8.13}$$

Equation (8.13) clearly shows the different sensitivities of the interferometric system due to the topography and to the displacement. In particular, the deformation term δr_d is in a ratio with the wavelength (λ), whereas for the topography, the elevation s is in a ratio with the term $\lambda r_0/b_\perp$. Measurements of the phase with accuracy of a fraction of the wavelength (centimetres at microwaves) provide centimetre-to-millimetre accuracy in the path difference. The same cm/mm accuracy is on one hand directly translated to the deformation measurement; on the other hand, as for the topography, the presence of the term $r_0/b_\perp$ (typically of the order of 10^3) leads to a final accuracy in the estimation of the slant height ranging from decametres to metres.

8.2.2 Decorrelation of radar echoes

As already pointed out, the phase noise in interferometric systems is not only related to the unavoidable additive thermal noise but also to the change of the backscattering characteristics of the scene between the two acquisitions. Changes in the relative phase between γ_1 and γ_2 in (8.3) and (8.4) lead in fact to the presence of a noise contribution in the interferometric phase ϕ in (8.5), which is referred to as decorrelation.

An important term introduced to quantify the degree of decorrelation, or better of correlation between the two interferometric signals, is the cross-correlation index χ, which is a measure of the linear predictability of two random variables [30]:

$$\chi = \frac{E(y_1 y_2^*)}{\sqrt{E(|y_1|^2)E(|y_2|^2)}} = \rho \exp(j\bar{\varphi}) \tag{8.14}$$

where $E(\cdot)$ is the statistical mean operator, $\bar{\phi}$ is the expected interferometric phase; the modulus of χ, i.e. ρ, belongs to the interval $[0, 1]$ and is referred to as coherence. The closer the coherence to 1, the higher the degree of correlation [7,8,31]. The limit value $\rho = 1$ indicates that the two (zero mean) random variables y_1 and y_2 are one linearly predictable from the other by a (complex) scaling with a minimum mean square error equal to zero [30]. Therefore, the higher the value of ρ, the lower the noise contribution [8]. The factor $1 - \rho$ measures on the other hand the decorrelation degree between the two acquisitions.

Under the circular (zero-mean) Gaussian assumptions for both the signal describing the scene backscattering and the additive noise, independent to each other, it can be shown that the limit case $\rho = 0$ leads to a probability density functions (pdf) describing the statistical behaviour of the measured (wrapped) interferometric phase uniformly distributed in $(-\pi, \pi)$ whereas for $\rho = 1$, the pdf is a Dirac impulse thus indicating the absence of phase noise fluctuations. Details about the derivation of the pdf are outside the scope of the chapter and can be found in [8,31]. Meanwhile, it is worth for the subsequent analysis to distinguish and comment upon the different sources of decorrelation.

The coherence is a product of several factors [8] translating the different decorrelation sources, the most important ones being associated with the change of the imaging geometry (ρ_{spatial}) due to the spatial orbit offset (spatial baseline or simply baseline), the (possible) temporal separation between the acquisitions (ρ_{temporal}) and the thermal noise [8]:

$$\rho = \rho_{\text{spatial}}\,\rho_{\text{temporal}}\,\rho_{\text{thermal}} \tag{8.15}$$

Spatial decorrelation is the translation of the speckle noise in the interferometric context. We refer to (8.3) and (8.4): Here, it can be noted that the integrals involve an integration over the resolution cell of the function describing the backscattering. Even in the case of absence of any temporal changes in the scene backscattering function (i.e. $\gamma_{t1} = \gamma_{t2}$), the change of the imaging angle determines a change of the phase factor included in the integral (see the presence of the exponential terms with the presence of an uncommon term $\delta r - \delta r'$). As shown before for a planar topography, this term can be well approximated with a linear function of the range. Accordingly, the exponential term with argument $\delta r - \delta r'$ in (8.4) leads to a spectral shift of the backscattering function and therefore to the presence of uncommon reflectivity spectral components in the spectra of the received signals [32,33]. These components are generally incoherent and hence responsible for the presence of decorrelation.

To quantify the spatial decorrelation for a canonical scattering case, it can be assumed that the distributed scattering, invariant with the time, is described by a white process. Mathematically, the function $\gamma_{t1} = \gamma_{t2}$ in the integrals in (8.3) and (8.4) is a zero mean random process with an impulsive autocorrelation (absence of correlation between scatterers at different range). Under these conditions, it can be shown that the spatial decorrelation (on a planar terrain) increases linearly with an increase of the slope of the linear function $\delta r(r) - \delta r'$ [32]. More specifically, letting $B_r = 2B_t/c$ be the spatial range bandwidth [m^{-1}], where B_t is the transmitted bandwidth and c is the speed of light, it results that [7,8]

$$\rho_{\text{spatial}} = 1 - \frac{|\xi_0|}{B_r} \quad \text{for} \quad |\xi_0| \leq B_r \quad \text{and } 0 \quad \text{elsewhere} \tag{8.16}$$

$$\xi_0 = \frac{2b_\perp}{\lambda r_0}\cot(\vartheta - \beta) \tag{8.17}$$

where r_0 is the reference range and β is the terrain slope in Figure 8.2; ξ_0 [m^{-1}] is the wavenumber shift. It is evident that the larger the baseline, the higher the spatial decorrelation; moreover, the closer β is to ϑ ($\beta = \vartheta$ is the layover limit), the lower is the spatial coherence.

The value b_{cr} of $b_\perp$ leading to $\xi_0 = B_r$, i.e. to a wavenumber shift equal to the spatial bandwidth (m^{-1}), is called the critical baseline:

$$b_{cr} = \frac{\lambda r_0 B_r}{2\cot(\alpha - \beta)} = \frac{\lambda r_0}{2R_r \cot(\alpha - \beta)} \tag{8.18}$$

where $R_r = 1/B_r$ is the range resolution. For high resolution (HR) SAR sensors such as ERS and Envisat, operating at C-band, the critical baseline is slightly above

1 km; for very high resolution (VHR) SAR sensors, such as COSMO-SkyMed and TerraSAR-X operating at X-band, the critical baseline is above 4 km. For VHR systems as well as for Sentinel-1, for which the range resolution was increased with respect to ERS and Envisat, the spatial decorrelation is therefore typically negligible.

Equation (8.16) can be rewritten as

$$\rho_{\text{spatial}} = 1 - \frac{|b_\perp|}{b_{cr}} \quad \text{for} \quad |b_\perp| \leq b_{cr} \;\; \text{and } 0 \quad \text{elsewhere} \tag{8.19}$$

The spatial decorrelation can be contrasted by implementing a filtering, called common band filtering, which is stationary for a planar terrain [32] and non-stationary for a rolling topography [33]. This filtering is effective on distributed scattering (such as associated with rural areas) but is not recommended for an interferometric analysis at full resolution in urban areas where the presence of dominant scatterers tends to correlate the backscattering harmonics [33]. In this case, the hypothesis of 'out of band' decorrelation is no longer valid and the common band filtering can even lead to a degradation of the phase quality.

The temporal decorrelation is due to the change of the scene backscattering properties over time. This decorrelation source is of main importance in repeat pass interferometry, especially with systems operating at higher frequencies (for instance X-band). It is obviously critical not only over rapidly changing scenes, such as sea, but also in vegetated area where the growth, and in general the change of vegetation, leads to strong variation of the backscattering coefficient. Temporal decorrelation is typically modelled with an exponential decay [8], where the decay factor depends mainly on the scene and specifically on the land coverage.

The last key contribution to decorrelation, thermal decorrelation, is due to the presence of thermal noise in the receiving apparatus. It is particularly evident in areas characterized by very low scene backscattering [5]. Other decorrelation sources are associated with variations of the imaging aspect angle (Doppler Centroid decorrelation) and to processing artefacts [5,8].

The correlation index, and therefore the coherence, can be estimated on real data [31] by substituting the statistical averaging $E(\cdot)$ with a spatial averaging operator $\langle\cdot\rangle$, typically referred to as spatial multi-looking:

$$\hat{\chi} = \frac{\langle y_1 y_2^* \rangle}{\sqrt{\langle |y_1|^2 \rangle \langle |y_2|^2 \rangle}} = \hat{\rho} \exp\left(j\hat{\phi}\right) \tag{8.20}$$

where $\hat{\chi}$ and $\hat{\rho}$ are the estimated correlation index and coherence, respectively, whereas, $\hat{\varphi}$ is the averaged (multi-look) phase. Note that $\langle y_1 y_2^* \rangle$ is referred to as multi-look (i.e. spatial averaged) complex interferogram: the complex (coherent) multi-looking implemented in the numerator of (8.20) is therefore used also to filter out the noise from the single look (complex) interferogram $y_1 y_2^*$ at the expense of a spatial resolution loss. Multi-look is generally carried out locally, i.e. by averaging close pixels for instance via a boxcar filter. Coherence estimation is affected by bias [34,35]. Modern algorithms use advanced (adaptive) spatial multi-looking to account for the presence of non-stationary, or better inhomogeneous backscattering [36–39].

8.2.3 Differential interferometry

SAR interferometry is aimed at the estimation of the DEM; in this case, the acquisition is carried out in such a way to image the scene at the same or close by times so that $\delta r_d = 0$. This can be achieved either by using a single system with two SAR antennas or by exploiting two passes repeated at fractions of seconds. In the first case, the antennas are mounted either on an airborne or on special spaceborne platform: the latter was the case for the generation of the worldwide DEM by the Shuttle Radar Topography Mission (SRTM). To the second class belongs the Tandem-X mission for the generation of an upgraded version of the SRTM global DEM [3].

Differently from InSAR for DEM reconstruction where, to increase the coherence, the acquisitions should be simultaneous, differential interferometry is a configuration where the interferometric acquisitions are obtained by two repeated passes over the same orbit [9]. The aim is to measure the interferometric component associated with the displacement, i.e. δr_d. Equation (8.5) shows that a single interferogram contains both the topographic and deformation components: the former term is negligible only in the case of a spatial baseline close to zero, i.e. when the two data takes are acquired with passes repeated 'almost' over the same orbit, which is however typically not the case of real acquisitions.

To eliminate the topographic component, DInSAR relies on the use of an external DEM which is exploited, knowing the orbits, to evaluate an estimate $\delta\hat{r}_T$ of δr_T. Letting $\delta r_{eT} = \delta r_T - \delta\hat{r}_T$ be the contribution to the path difference related to the error in the knowledge of the DEM, we can consider the following new expression for the phase signal:

$$\varphi = \frac{4\pi}{\lambda}\delta r_{eT} + \frac{4\pi}{\lambda}\delta r_d + \varphi_a + \varphi_n, \tag{8.21}$$

where φ_n is, as before, the stochastic term associated with the phase noise including thermal noise and decorrelation. It is worth noting that, since acquisitions are often acquired also with large temporal separations in the DInSAR case (days to months), the effect of temporal decorrelation is typically the main source of noise.

All the contributions in (8.21) are now analysed in details.

As for the displacement component, we refer to Figure 8.3 relevant for the description of a differential interferometric system with two repeated passes of a satellite at different times in the presence of a displacement of the point **P** at range r identified by a vector $\boldsymbol{d}$. Assuming again the far field case ($r \gg \delta r_d$), the following approximation holds

$$\delta r_d = \boldsymbol{d} \cdot \hat{\boldsymbol{r}} \tag{8.22}$$

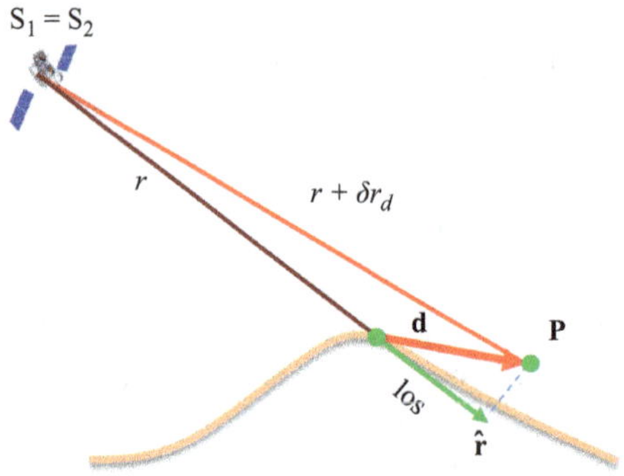

Figure 8.3 Relevant for the description of a differential interferometric SAR (DInSAR) system with two repeated passes of a satellite at different times in the presence of a displacement d is the displacement δr_d is the range variation induced by the displacement

which states that the measurable path difference related to the deformation is the component of the displacement $\boldsymbol{d}$ along the los direction, which is given by the versor $\boldsymbol{d} \cdot \hat{\boldsymbol{r}}$.

With respect to (8.5), (8.21) includes the additional term φ_a associated with the fact that acquisitions are carried out at different times (epochs). When waves propagate through the atmosphere, they are delayed according to the refraction index [26] that typically shows significant variations above distances of the order of a few hundred of metres: the delay is interpreted as an increase of the range which in turn produces a variation of the phase. In case of simultaneous acquisitions, such a phase disturbance affects almost exactly in the same way the two images and therefore it cancels at the stage of signals beating for the extraction of the interferogram. In repeat pass acquisitions with temporal differences typically ranging from several days to months and even years, the refraction index shows significant variations: φ_a is therefore the phase associated with variation of the atmosphere between the two acquisitions. Such term, usually referred to as atmospheric propagation delay (APD) or atmospheric phase screen, can be stochastically characterized [26] and typically shows (at C-band) spatial variations within the scale of a few hundreds of metres. As for the magnitude, the absolute delay of the radiation may reach the order of a metre and more. Whereas, such a delay is important for accurate geo-localization of SAR images, for instance for imaging geodesy [40], for interferometry constant phase factors are inessential: Only spatially variable components of the delay are of concern and these have magnitudes of the order of a few (typically $<3, 4$) multiples of 2π.

The other component, δr_{eT}, is the residual topography. The image corresponding to φ in (8.21) is called differential interferogram as it is obtained by subtracting to the interferogram in (8.5), the contribution associated with the external DEM, usually referred to as synthetic interferogram ϕ_{syn}:

$$\varphi = \phi - \phi_{\text{syn}} = \phi - \frac{4\pi}{\lambda}\delta\hat{r}_T, \tag{8.23}$$

where $\delta\hat{r}_T = \hat{r}_{T2} - \hat{r}_{T1}$, $\hat{r}_{T1}$ and $\hat{r}_{T2}$ are the (topographic dependent) range to the first and second antenna evaluated from the external DEM. The synthetic phase ϕ_{syn} is evaluated by computing, on the grid of pixels of the master image (with respect to which the interferometric signals are evaluated), the difference between the distances of the ground points described by the external DEM to the so-called zero Doppler satellite positions at the first and second antenna [41], the latter being the positions where the target los is orthogonal to the platform velocity vector.

It is clear that, by using a single differential interferogram, a single measurement described (8.21) is available, which alone is not sufficient to distinguish between the components of residual topography (i.e. the DEM error), the wanted deformation component and the atmospheric delay. Accordingly, a first condition to achieve the measurement of the displacement δr_d is that the residual topography δr_{eT} in (8.24), i.e. the DEM error, must be limited, typically a fraction or at most equal to the ambiguity height. Large deformation components, as for instance those associated with large earthquakes, typically involve displacements of several tens of centimetres (i.e. several times $\lambda/2$ and therefore leading to phase variations much larger than 2π): these displacements are therefore normally clearly distinguishable from a residual topography components generated from DEMs with an accuracy higher (i.e. numerically lower) than the ambiguity height (the latter leading to a variation of the phase of 2π). Earthquakes (co-seismic) displacements maps can be therefore easily measured with the DInSAR technology and have provided to geologists information about displacements associated with tectonic movements with a spatial density (i.e. coverage) that cannot be achieved by any other classical technique (levelling, GPS, etc.).

Satellites are able to regularly repeat orbits over the time. As a consequence, stacks of multi-pass acquisitions characterized by angular and temporal diversity are available in remote sensing archives for many areas of the Earth surface and can be processed, as explained in the next session, to increase the accuracy in revealing displacements as well as to allow monitoring deformations over the time; the latter being an option not possible with standard two pass DInSAR.

8.3 Multi-pass differential interferometry

In order to achieve a much higher accuracy in the estimation of the deformation and thereby capture even subtle (mm/yr) movements, modern A-DInSAR processing algorithm jointly process all the images in stacks of multi-temporal acquisitions: this processing allows in fact discriminating between the atmospheric and deformation phase components as well as the residual (w.r.t. the external DEM) topography contribution.

Two different classes of A-DinSAR methods are typically used: CSI [12–15] and PSI [16,17]. In addition to the possibility to estimate and remove the atmosphere, such methods are also able to extract the information about the displacements at the different times (epochs): these are usually referred to as time series.

The first class of algorithms, CSI, essentially analyses stacks of interferograms showing sufficiently high coherence values. To measure the coherence and as well as to mitigate, via the averaging, the phase noise in the interferograms, they make use of spatial multi-looking: spatial multi-looking allows also down-sampling the data in azimuth and range and therefore this class of algorithms is particularly suitable for the analysis of wide areas at lower resolution (small scale). To this class of algorithms belong the well-known small BAseline subsets or simply small BASeline (SBAS) approach in [12] proposed as an extension of the least square interferometric approach in [42]. In this case, the interferograms are selected by bounding up the spatial and the temporal separations (baselines) between the acquisition pairs to limit the decorrelation associated with distributed scattering and with the possible scattering changes over the time, typically occurring in rural areas with sparse vegetation. Conversely, the second class of algorithms analyses interferograms at the highest spatial resolution to determine the deformation of single dominant scatterers, typically associated with man-made structures (dihedral and trihedral corresponding to walls edges, corners, masts, gratings, etc.). In this case, to achieve also a high accuracy in the localization of scatterers, no limitation on the spatial baseline is introduced. To this class of algorithms belong the PSs technique [16,17].

For both approaches, we refer to Figure 8.4 where it is assumed that the satellite is acquiring the N images at ordered time instants (epochs), collected in the vector $\mathbf{t} = [t_1, \ldots, t_N]^T$ ($[\cdot]^T$ is the transpose operator), also referred to as temporal baselines, and characterized by spatial baselines, collected in the vector $\mathbf{b}_\perp = [b_{\perp 1}, \ldots, b_{\perp N}]^T$: temporal and spatial baselines are referred to a reference (master) acquisition, for example acquisition #2 in Figure 8.4. Generally, the master image is selected as the geometric barycentre of the points of the scatter plot of all acquisitions in the baseline domain, i.e. the domain showing the spatial baseline distribution versus the times (temporal baselines). All images are supposed to be registered, i.e. aligned, to the reference (master) image so that the response of a ground target is located in the same pixel in the different images [41].

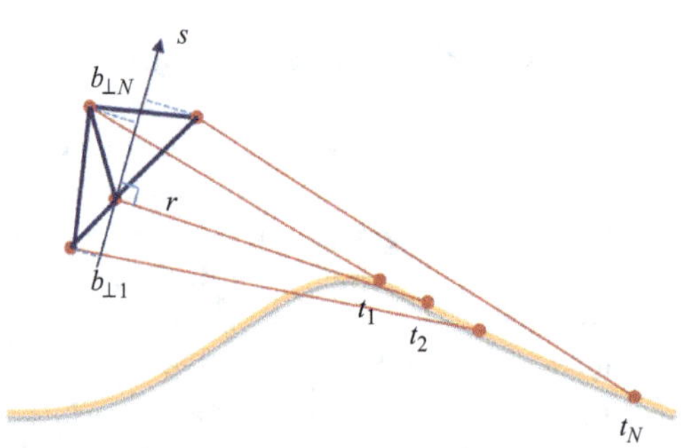

Figure 8.4 Multi-pass DInSAR geometry with a target undergoing a sequence of displacements. The second image is assumed as reference (master) image

8.3.1 Coherent stacking interferometry

Hereafter, we describe the basic concept of CSI methods. Let us suppose that the phase values corresponding to the N acquisitions, for each azimuth-range pixel, are collected in a N-dimensional vector

$$\varphi = [\varphi_1, \ldots, \varphi_N]^{\mathrm{T}} \tag{8.24}$$

each element having the form of the phase of signals as in (8.3) for the master and (8.4) for the slaves.

Due to the presence of the backscattering coefficients $\gamma_1, \ldots, \gamma_N$, see again (8.3) and (8.4), the phase values in (8.24) are affected by the presence of a random phase that, for underlying backscattering mechanism modelled as zero-mean circularly complex Gaussian random process, is uniformly distributed in the $(-\pi, \pi)$ interval. Even for perfectly correlated backscattering coefficients, the measurements of the absolute phases to the different antennas would be in any case affected by a common random variable with a phase uniformly distributed in the $(-\pi, \pi)$ interval.

Anyway, since only relative phase values are of importance for interferometry, the problem of the common unknown (random) phase value can be overcome by referring the phase vector in (8.24) to the phase of one image (reference image): the latter is generally either the barycentre of the acquisitions in the baseline domain or the first image so to achieve, in the case of temporally ordered stacks, deformations are referred to the first available time instant. In the following, we will therefore assume that the phase values are all referred to a reference image, whose index is generic and indicated as n_0 so that $\boldsymbol{\varphi} = [\varphi_1, \ldots, \varphi_N]^{\mathrm{T}} \leftarrow \boldsymbol{\varphi} = [\varphi_1 - \varphi_{n_0}, \varphi_2 - \varphi_{n_0}, \ldots, \varphi_N - \varphi_{n_0}]^{\mathrm{T}}$ in the vector in (8.24). The latter substitution, as well as all the associated ones interesting the related vectors (for instance the substitution that brings to zero the n_0-th element of $\mathbf{b}_\perp$) will be omitted for sake of notation simplicity.

We also assume that the phase signals have been compensated by the ranges $\hat{r}_{T1}, \ldots, \hat{r}_{TN}$ estimated from an available external DEM. Therefore, according to what described in the previous sections, the phase vector in (8.24) can be modelled as

$$\boldsymbol{\varphi} = -\frac{4\pi}{\lambda}\frac{\mathbf{b}_\perp}{r}s + \frac{4\pi}{\lambda}\mathbf{d} + \boldsymbol{\varphi}_a + \boldsymbol{\varphi}_n \tag{8.25}$$

where $\mathbf{d} = [\delta r_{d1}, \delta r_{d2}, \ldots, \delta r_{dN}]^T$ collects the displacements of the target along the los observed at the N antennas, i.e. the time series; $\boldsymbol{\varphi}_a$ and $\boldsymbol{\varphi}_n$ collect the APD and phase noise contribution, respectively; $s = z/\sin\vartheta$, s and z being the slant and topographic heights, respectively, now residual with reference to the external DEM exploited to compensate the known topographic contribution. This latter operation is usually referred to as *Zero* (spatial) *Baseline Steering* (ZBS) because, in the case of perfect knowledge of the DEM, it would allow nullifying the first term at the right hand side in (8.25) as if the spatial baselines were all zero. Obviously $\mathbf{d}$ has, as all other vectors, a zero element in the position of the reference image (n_0).

A remark is now in order. In fact, the phase vector in (8.25) collects the interferograms between the available acquisitions and the same reference image. Although measurable directly from the data, these interferograms are assumed to be unknown in the CSI multi-interferogram analysis because of several drawbacks related to the presence of large baseline interferograms.

First of all, phases are considered to be unwrapped. Even though working with residual topographic phases, the ZBS operation is usually carried out with respect to an inaccurate DEM, thus leaving, on large spatial baseline interferograms, a significant residual contribution that could impair the unwrapping step. Second, interferograms corresponding to large temporal and spatial baselines are typically affected by strong decorrelation.

In order to circumvent these problems, the vector in (8.24), modelled according to the expression in (8.25), is thought as the unknown of the CSI approach, which have to be estimated starting on a set of say M interferograms (the measures) generated according to an acquisition paring involving the presence of small baselines.

To further mitigate the decorrelation phenomena, which typically affects significantly rural areas, the M interferograms are also multi-looked, i.e. generated by using the averaging operator in (8.20), generally via a boxcar filtering, thus leading to a loss of spatial resolution.

The measured interferograms are generated by assuming redundancy, i.e. $M \geq N - 1$: typically $2N \leq M \leq 3N$. In Figure 8.5, it is shown that the acquisition distribution of a real satellite system in the so-called spatial–temporal baseline domain: baselines are referred to the master image. The figure shows a graph in which each node is an acquisition and each edge is an interferogram.

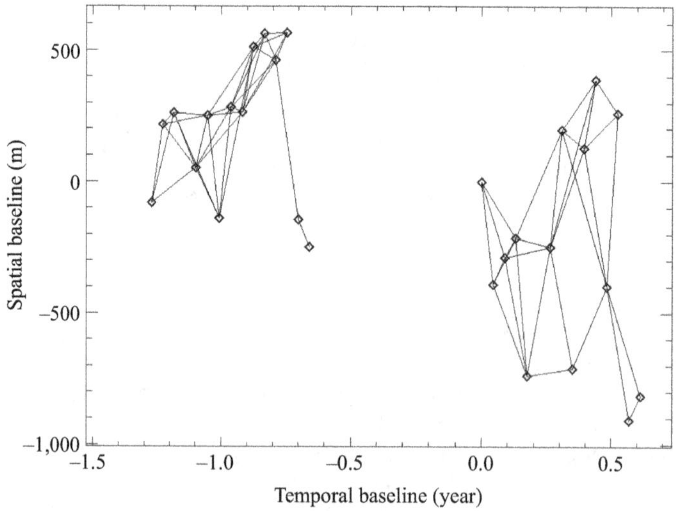

Figure 8.5 The graph corresponding to the pairing of acquisitions in the spatial/ temporal baseline domain. Example from a real passes of a SAR sensor with two separate subsets

The redundancy is introduced also with the aim to control the errors at the PhU stage, which represents one of the most critical steps in the processing chain.

For each azimuth-range pixel, the (unwrapped) values of the measured interferograms are stacked in the M-dimensional vector:

$$\Delta\varphi = [\Delta\varphi_1, \ldots, \Delta\varphi_M]^{\mathrm{T}} \tag{8.26}$$

It is now necessary to relate the M (unwrapped) redundant interferograms values in (8.26) to the N unknown phase values in (8.24), corresponding to the phase difference between the acquisitions and the reference image. To this end, the following incidence matrix describing the pairing between the acquisitions for the interferogram generation, i.e. the graph in the acquisition domain where the arcs (edges) are the interferometric pairs and the end nodes of the acquisitions, can be introduced:

$$\mathbf{B} = \begin{bmatrix} 0 & -1 & 0 & +1 & \ldots \\ 0 & 0 & +1 & 0 & \ldots \\ \ldots & \ldots & \ldots & \ldots & \ldots \\ \ldots & \ldots & \ldots & \ldots & \ldots \end{bmatrix} \tag{8.27}$$

Each row of the $M\times N$ matrix $\mathbf{B}$ describes one single interferogram pairing with $+1$ and -1 located at the positions of the master and slave acquisition of the generic interferogram: For instance in (8.27), it is assumed that the first interferogram is generated by pairing the second (slave) and fourth (master) acquisition; the second interferogram has the third acquisition as master image.

Some considerations about the incidence matrix in (8.27) are now in order. First of all, it is rank deficient by construction [43]: this rank deficiency is a consequence of the fact that the unknown vector in (8.24) collects interferometric values. Since the latter are essentially phase variations from the masters to the slaves acquisition, they result to be insensitive to the presence of a constant phase. Moreover, the incidence matrix does not account for the implicit information carried out by the phase vector in (8.24), i.e. the fact that such a vector is referred to reference acquisitions. Accordingly, the following $(M+1) \times N$ augmented version of the matrix in (8.27) can be considered:

$$\mathbf{B}_{\mathrm{a}} = \begin{bmatrix} \mathbf{B} \\ \mathbf{b}_0 \end{bmatrix} \tag{8.28}$$

where $\mathbf{b}_0$ is introduced to account for the constraint $\varphi_{n_0} = 0$ that sets to zero, the phase on the reference image n_0, thus solving the problem of the intrinsic rank deficiency.

Thanks to the definition of $\mathbf{B}_{\mathrm{a}}$, it is possible to write the equation relating, in each azimuth-range pixel, the interferograms (elements of $\Delta\varphi$) and the acquisitions (elements of φ):

$$\mathbf{B}_{\mathrm{a}}\varphi = \begin{bmatrix} \Delta\varphi \\ 0 \end{bmatrix} \tag{8.29}$$

Equation (8.29) can be therefore inverted to achieve the unknown phases at the different antennas, modelled as in (8.25), from the interferometric measurements in (8.26). Due to the redundancy in the construction of the interferograms, the system in (8.29) is overdetermined i.e. we have more equations than unknowns ($M > N$). The inversion is therefore carried out in the least square sense via the use of the pseudo-inverse [43]:

$$\hat{\boldsymbol{\varphi}} = \mathbf{B}_{\mathrm{a}}^{+} \begin{bmatrix} \Delta\boldsymbol{\varphi} \\ 0 \end{bmatrix} \tag{8.30}$$

$(\cdot)^{+}$denoting the pseudo-inverse operator.

It should be noted that, when a group of acquisitions are disconnected due to possible constraints of the maximum spatial and temporal baseline (see Figure 8.5), thus generating different subsets, the augmented matrix $\mathbf{B}_{\mathrm{a}}$ in (8.29) is still rank deficient. The use of the singular value decomposition (SVD) allows in this case to regularize the pseudo-inverse $\mathbf{B}^{+}$ to achieve the minimum L^2 norm solution [12,43]. An effective way is to change the unknowns in terms of phase variations between ordered times, i.e. $\varphi_2 - \varphi_1, \ldots, \varphi_N - \varphi_{N-1}$ and then apply the SVD procedure so to achieve a minimum L^2 norm solution on the phase variations between consecutive epochs, i.e. a minimum kinetic energy solution like. An integration step is in this case necessary to retrieve the original vector $\boldsymbol{\varphi}$.

Once the estimate $\hat{\boldsymbol{\varphi}}$ of the unknown vector $\boldsymbol{\varphi}$ in (8.24) has been obtained through the inversion in (8.30), the separation of the components present in (8.25), i.e. s, $\mathbf{d}$ and $\boldsymbol{\varphi}_a$, can be then carried out by exploiting their deterministic/statistical properties. This fact is better explained in the following.

The first vector component in (8.25) has a well-known structure (signature) that follows precisely the spatial baseline distribution $\mathbf{b}_\perp$, which in turns is known from the orbital state vectors (satellite positions). In addition, the second vector component in (8.25) can be modelled as $\mathbf{d} = v\mathbf{t} + \mathbf{d}_{\mathrm{NL}}$, where v is the DMV and $\mathbf{d}_{\mathrm{NL}}$ the vector accounting for the non-linear deformation. Accordingly, the residual topography s and the DMV v can be jointly estimated via a least squares inversion of the model in (8.25):

$$\begin{bmatrix} \hat{s} \\ \hat{v} \end{bmatrix} = \frac{1}{1-\rho_{bt}^2} \begin{bmatrix} -r\dfrac{\mathbf{b}_\perp^T}{\|\mathbf{b}_\perp\|^2} + \rho_{bt}\dfrac{r\mathbf{t}^T}{\|\mathbf{t}\|\|\mathbf{b}_\perp\|} \\ -\rho_{bt}\dfrac{\mathbf{b}_\perp^T}{\|\mathbf{b}_\perp\|\|\mathbf{t}\|} + \dfrac{\mathbf{t}^T}{\|\mathbf{t}\|^2} \end{bmatrix} \frac{\lambda}{4\pi}\boldsymbol{\varphi} \qquad \rho_{bt} = \frac{\mathbf{b}_\perp^T\mathbf{t}}{\|\mathbf{t}\|\|\mathbf{b}_\perp\|} \tag{8.31}$$

where ρ_{bt} is the correlation index between the spatial and temporal baselines' distribution. It is interesting to note that in the presence of strong correlation between the baseline vectors, the estimation problems show ambiguities [44].

The subtraction from the estimated phase of both the residual topography and the linear part of the displacement vector leads to

$$\hat{\boldsymbol{\varphi}}_d = \hat{\boldsymbol{\varphi}} + \frac{4\pi}{\lambda}\frac{\mathbf{b}_\perp}{r}\hat{s} - \frac{4\pi}{\lambda}\hat{v}\mathbf{t} = \frac{4\pi}{\lambda}\mathbf{d}_{\mathrm{NL}} + \boldsymbol{\varphi}_a + \boldsymbol{\varphi}_n \tag{8.32}$$

The separation between the displacement and APD vectors is carried out by considering their statistical characterization in terms of spatial and temporal variability: $\mathbf{d}_{NL}$ shows typically a spatial and temporal correlation, whereas φ_a is correlated along the space but uncorrelated with respect to the time. Accordingly, a simple spectral filtering implemented in the wavenumber/frequency domain is typically sufficient to achieve the measurement of the time series in each pixel. Seasonal variations of the APD cannot be, however, handle with this simple filtering.

It is worth pointing out that the most critical part of the processing procedure is represented by the unwrapping step. A correct unwrapping is of key importance to preserve the spectral characteristics of the signal, which is fundamental for the separation of the different components. The unwrapping is therefore carried out on a suitable sparse grid where the phase information is less affected by the noise. To this end, as already specified, the stack of available coherence images is used to distinguish between reliable and noisy areas so to define the sparse grid where the interferometric analysis is performed, i.e. where deformation measurements are provided as output.

Interferograms can be singularly unwrapped on the sparse grid of pixels. However, the availability of a set of interferograms can be profitably exploited to achieve more robust estimation of the unwrapped phase. A first improvement to the unwrapping is given by the introduction of a model for the variation of $\Delta\boldsymbol{\varphi}$ over spatial arcs connecting close pixels (spatial differences). In [14,15], it is explained that an estimation of the variation of $\Delta\boldsymbol{\varphi}$ over spatial arcs connecting nearby pixels can be carried out on the wrapped interferograms by modelling the phase as a sum of just two components, one associated with the residual topography and the other with the average deformation rate, i.e. the mean deformation velocity, that is the linear component of the deformation. As a matter of fact, the operation of spatial differentiation, which, together with the variation among the acquisitions associated with the interferometric pairing, leads to a double differentiation (along the times and along the space), allows mitigating the effect of non-linear phase components associated with the deformation and with the atmosphere. The retrieved (initial) estimation of the residual topography and mean deformation velocity can be subtracted from the wrapped interferograms to mitigate the interferometric phase variations and hence to 'help' the subsequent independent unwrapping of each interferogram [14,15].

A further, more complex improvement can be carried out by exploiting the redundancy of pairing in the interferogram generation, in other words by exploiting the fact that $\Delta\boldsymbol{\varphi}$ should not deviate from the range space $\mathcal{R}(\mathbf{B}_a)$ of the matrix $\mathbf{B}_a$, which is achieved through the projector operator $\mathbf{B}_a\mathbf{B}_a^+$. The description of advanced PhU is as well as outside the scope of the chapter. The reader can refer [45–48] for a description of the so-called 3D PhU: computational cost is in this case a problem that can play an important role due to the need to jointly handle the PhU of the whole interferometric stack.

After the unwrapping, a further check is carried out by analysing the residual vector between the starting (wrapped interferograms) and $e^{j\mathbf{B}_a^+\mathbf{B}_a\Delta\boldsymbol{\varphi}}$ [12].

8.3.2 Persistent scatterers interferometry

Persistent scattering interferometry (PSI) is tailored to the analysis of man-made structures: the PS approach [16,17] was the first method that demonstrated the capability of DInSAR to achieve mm/yr deformation measurements. It carries out an analysis of the data at full resolution (i.e. without any spatial multi-looking) and makes use of a model on complex data to determine (by exploiting the multi-view nature of available stacks) the residual topography, which is necessary to accurately localize the scatterer centre, as well as its temporal evolution (by exploiting the multi-temporal characteristic of available stacks).

The phase model exploited in PSI is essentially the same as CSI, that is, the one in (8.25). However, all the measured interferograms are generated with respect to the same reference (master) image, thus leading the matrix **B** in (8.27) to degenerate in the identity matrix. Differently from CSI, interferograms are not spatially averaged and baselines are not limited: both choices are due to the fact that, being PSs usually associated with strong scatterers (man-made targets such as buildings, masts, etc., as well as natural targets as exposed rocks), typically they are not significantly affected by decorrelation.

Data are considered at the full resolution: coherence cannot be hence estimated via spatial averaging and therefore a proxy for the selection of the grid of PS, which must be used for the PhU and for the estimation of the atmospheric component, is necessary. In the PS approach, the so-called PS Candidates (PSCs) are selected by analysing the amplitude stability. It can be shown [16,17] that, for a scatterer whose amplitude pdf follows a Rice distribution, the standard deviation of the phase is approximately equal to the ratio (referred to as dispersion index, D_A) between the standard deviation and the mean of the amplitude values observed at the different antennas. The dispersion index is therefore evaluated on the stack of amplitude images after the registration step. Pixels with D_A below a fixed threshold (typically set to 0.25) are identified as candidate PS and used to define the sparse grid of pixel where the PhU is carried out.

The unwrapped interferograms φ, modelled as in (8.25), are first calibrated with respect to the estimated and interpolated atmosphere contribution $\hat{\varphi}_a$. Subsequently, the full-resolution calibrated phase $\varphi_c = \varphi - \hat{\varphi}_a$ is modelled as

$$\varphi_c = -\frac{4\pi}{\lambda}\frac{\mathbf{b}_\perp}{r}s + \frac{4\pi}{\lambda}(v\mathbf{t} + \mathbf{d}_{\mathrm{NL}}) + \varphi_n \tag{8.33}$$

where s is the residual (w.r.t. the external DEM) elevation of the scatterer and the displacement vector $\mathbf{d} = v\mathbf{t} + \mathbf{d}_{\mathrm{NL}}$ has been decomposed into the linear part, described by the slope v of the time series w.r.t. the epoch $\mathbf{t}$, and the (los) non-linear deformation components.

PSI techniques assumes $\|\mathbf{d}_{\mathrm{NL}}\| \ll \lambda/(2\sqrt{N})$ and estimates, pixel by pixel on complex data at full resolution, s and v by analysing the following normalized, i.e. in the (0,1) interval, index measuring the un-modelled residues [16,17]:

$$C(s, v) = \frac{1}{N}\left|\sum_{n=1}^{N} \exp\left[j\varphi_{cn} + j\frac{4\pi}{\lambda}\left(\frac{b_{\perp n}}{r}s - t_n v\right)\right]\right| \tag{8.34}$$

φ_{cn} being the nth element of the calibrated (wrapped) phase vector $\boldsymbol{\varphi}_c$.

This quantity is a measure of temporal coherence, i.e. persistency property of the scatterer: the elevation s (residual with respect to the external DEM) and the velocity v of the scatterer are determined via the following maximization:

$$(\hat{s}, \hat{v}) = \arg\max_{s,\, v} \; C(s, v) \tag{8.35}$$

which is typically carried out with an exhaustive search in a predefined interval of elevations and velocity. Only pixels with an index $C(\hat{s}, \hat{v})$ above a fixed threshold are labelled as PSs and therefore the algorithm provides the temporal series as a product.

It is worth noting that, to achieve high accuracy in the topography estimation and to exploit as well as the high angular correlation properties of strong scatterers, the PSI approach does not impose any predefined limit on the spatial baseline. It is as well as easy to understand that all the interferometric beatings are involved to estimate the elevation and DMV in the maximization of $C^2(s, v)$, which is equivalent to (8.35).

The following test, measuring the fit of the model to the data in (8.34), is specifically implemented to discriminate between the presence (hypothesis $\mathcal{H}_1$) and absence (hypothesis $\mathcal{H}_0$) of a PS:

$$C(\hat{s}, \hat{v}) \underset{\mathcal{H}_0}{\overset{\mathcal{H}_1}{\gtrless}} T \tag{8.36}$$

Application of PSI to several contexts has shown that $C(s, v)$ is an extremely powerful (coherent) multi-interferogram testing statistics for the detection of PS directly on complex data. The test in (8.36) provides, at full resolution, the detection of a much higher density of PS compared to the PSC test based only on the amplitude described at the beginning of the section. Moreover, it allows achieving a high accuracy in the localization of PS and in the monitoring of deformation.

PSI exploits only the phase information to estimate the target parameters; moreover, PSI assumes the presence of only one (dominant) scatterer in each resolution cell. These limitations lead to a loss of performances in areas where layover is frequent (for instance urban areas): the latter can be tackled by adopting a TomoSAR processing described starting from Section 8.4.

A final remark on the unwrapping step and atmospheric phase contribution is now in order. The strategy of generating interferograms with respect to a single master acquisition leads to the need to unwrap large baselines interferograms, which may be rather critical. A much easier processing strategy is, as with CSI, to unwrap on the sparse grid of PSCs, a set of full-resolution interferograms where the baselines are in any case limited. The acquisition phases (i.e. the phases that are all referred to the same master image) should be then estimated from the unwrapped interferometric phase evaluated on the small baseline measurements by inverting the matrix **B**, which is no longer an identity matrix. An alternative approach based on a two-step approach is discussed in the next subsection.

8.3.3 The two-step A-DInSAR approach

The two step, or two-scale, A-DInSAR approach [49,50] takes the inspiration from the capability of CSI to perform a classical data and straightforward processing at lower resolution of coherent interferograms and from the effectiveness of the index C used in PSI to identify PSs at full resolution via the test in (8.36).

As already stated, the most critical step in A-DInSAR processing is represented by the PhU. In the PS method, the absence of multi-look and the use of large baselines, together with the fact that PSC as well as PS is generally localized on buildings and ground elements where the topography shows significant variations, are elements that complicate the PhU step necessary to compensate the atmospheric pattern and the non-linear deformation components. The two-step approach allows taking benefit of the advantages of CSI and PSI.

The strategy of the two-step approach is the use of the low-resolution product to calibrate the complex full-resolution data. Since low- and full-resolution quantities are both involved in such approach, in the following, when necessary, we differentiate the low resolution from the residual full-resolution version of the same variable by exploiting the subscripts L and F, respectively.

Letting φ be the phase vector generated at full resolution and letting $\hat{\mathbf{d}}_{\mathrm{L}}$ and $\hat{\varphi}_a$ be the deformation time series and the APD estimated at low resolution on multi-look data via the CSI. The full-resolution phase vector can be thus calibrated with respect to the low resolution time series and APD:

$$
\begin{aligned}
\varphi_c &= \varphi - \frac{4\pi}{\lambda}\hat{\mathbf{d}}_{\mathrm{L}} - \hat{\varphi}_a = -\frac{4\pi}{\lambda}\frac{\mathbf{b}_\perp}{r}s + \frac{4\pi}{\lambda}\mathbf{d}_{\mathrm{F}} + \varphi_n \\
&= \frac{4\pi}{\lambda}\frac{\mathbf{b}_\perp}{r}s + \frac{4\pi}{\lambda}(v_{\mathrm{F}}\ \mathbf{t} + \mathbf{d}_{\mathrm{F,NL}}) + \varphi_n
\end{aligned}
\tag{8.37}
$$

where $\mathbf{d}_{\mathrm{F}}$, v_{F} and $\mathbf{d}_{\mathrm{F,NL}}$ are the full-resolution residual (w.r.t. the low resolution) los deformation vector, DMV (scalar) and non-linear los deformation vector, respectively. Notice that the subtraction of $\hat{\mathbf{d}}_{\mathrm{L}}$ allows implementing a sort of steering to zero of the temporal baselines (temporal ZBS) with reference to the low resolution (small scale) result. The full-resolution (large scale) analysis is therefore carried out only on the residual phase signals, thus limiting residual non-linear component only to possible local non-linear displacements with respect to the nearby pixels. The temporal coherence indicator in (8.34) and (8.35) is then used by considering the elements of φ_c in (8.37) to identify the elevation and full-resolution DMV of PS. By doing so, it has exploited the capability of CSI to achieve estimation and compensation of the atmosphere over wide areas and, at the same time, the powerfulness of PSI full-resolution analysis to detect and measure deformations and residual topography of PS on complex data. As for PS, the latter can be substituted by the TomoSAR method to improve detection and parameter estimation of PS as well as to distinguish multiple scatterers interfering in the same pixel [51] as described in the next section.

8.4 SAR tomography

Similarly to the interferometric description, we start from the 3D case and then extend the analysis to the time (4D) case. The multi-acquisition SAR geometry involved in TomoSAR is depicted in Figure 8.6 in the slant range r – slant height (elevation) s plane. Assuming parallel antennas tracks (typically obtained by multiple passes with current satellite and airborne technologies), as for the interferometric case, $b_{\perp n}$ in Figure 8.6 ($n = 1, \ldots, N$) is the orthogonal component of the baseline connecting the nth antenna to the master one.

In the case of non-simultaneous acquisitions, we assume that an interferometric processing at low resolution has been carried out and that the data have been calibrated by compensating non-linear deformations and atmospheric phase delay as described in Section 8.3.3. Accordingly, supposing for now the absence of residual displacements, the complex return at the nth antenna associated with the response of a point-like scatterer at elevation s (elementary scattering contribution along the elevation) is obtained by exploiting the calibrated phase expression in

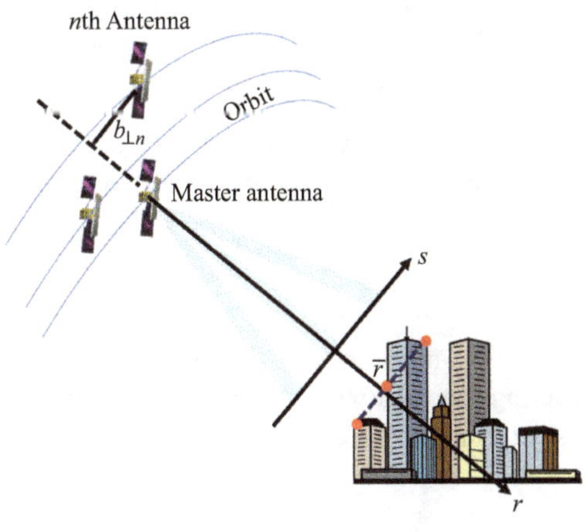

Figure 8.6 Tomographic SAR and reference geometry. The image schematically shows the occurrence of layover with the interference of scatterers, at the same range $\bar{r}$, coming from facades and roofs

(8.37) specialized for $\mathbf{d}_{\mathrm{F}} = 0$: $\gamma(s)\exp\left(-j\frac{4\pi}{\lambda}\frac{b_{\perp n}}{r}s\right)$. Notice that the backscattering coefficient $\gamma(s)$is supposed to be temporally invariant.

We assume linearity (Born approximation) [52] and therefore, in the presence of a distribution of scatterers along the elevation, the complex data, say g_n, can be expressed as the superposition of multiple elementary scattering contributions along s, see Figure 8.6:

$$g_n = \int_{I_S} \gamma(s) e^{-j2\pi\frac{2b_{\perp n}}{\lambda r}s} ds + w_n \tag{8.38}$$

where I_S is the scene extension in the slant height direction, λ the radar wavelength and w_n the noise term in the focused pixel accounting for thermal noise, which have a strong impact in low backscatter areas.

The use in (8.38) of a symbol for the data different from (8.3) and (8.4), i.e. g_n instead of y_n, is associated with the need to highlight the process of data calibration, i.e. the above-described compensation of the non-linear deformation and atmospheric phase delay.

The 3D tomography problem consists of the estimation, in each image pixel, of the backscattering profile $\gamma(s)$ starting from the N samples g_n which are modelled as in (8.38) [18,19,52].

Equation (8.38) shows that the data are noisy samples of the Fourier transformation of the backscattering profile, usually taken on an irregular grid given by the frequencies $\xi_{sn} = 2b_{\perp n}/(\lambda r)$, where the subscript s stands for spatial frequency. The solution of the tomography problem involves therefore a spectral analysis of the received data at the different antennas.

It is instructive at this point to make some considerations on the limitations in the reconstruction of the scene backscattering profile inherent to the imaging system.

The imaging system is able to observe just a spectral window of the backscattering profile equal to $I_B = \max_n \xi_{sn} - \min_n \xi_{sn} = 2B/(\lambda r)$, where B is the total baseline span. Such a spectral window imposes the so-called Rayleigh limit on the resolution that can be achieved on the reconstruction procedure [18,19,52,53]:

$$R_s = \frac{1}{I_B} = \frac{\lambda r}{2B} \tag{8.39}$$

Moreover, by assuming a uniform distribution of baselines with a spacing $\Delta b = B/N$, the sampling rate $2\Delta b/(\lambda r)$ corresponding to the acquired spectral samples of the backscattering profile defines the maximum unambiguous elevation interval

$$I_{S\,\max} = \frac{\lambda r}{2\Delta b} \tag{8.40}$$

This is the maximum scene extension that, according to the Nyquist criterion, can be imaged without introducing aliasing phenomena [52].

In the case of uniform sampling, the Rayleigh resolution in (8.39) and the maximum unambiguous elevation interval in (8.40) are related to each other by

$$R_s = \frac{I_{S\,\max}}{N} \tag{8.41}$$

which summarizes the limits of the information provided by the imaging system.

TomoSAR allows reconstructing unambiguously at most an extension $I_{S\,\max}$ of the scene backscattering profile with a resolution, for basic inversion algorithms, equal to ΔR_s. However, it is worth to note that possible *a priori* information $I_S < I_{S\,\max}$ on the scene extension can be exploited to improve the resolution inherently provided by the imaging system. Indeed, in this case, it is possible to reconstruct via proper tomographic algorithms the backscattering profile with a resolution (super-resolution) up to

$$\Delta s = \frac{I_S}{N_u} \tag{8.42}$$

where $N_u \leq N$ accounts for a decrease of the subspace dimension where the signal is 'reliable', i.e. above the noise contribution in (8.38), see [52] for more details. The trade-off between the increase of *a priori* information (i.e. a reduction of the support I_S) and the decrease of N_u may allow achieving a degree of super-resolution [52].

It is finally worthwhile to underline that, for an irregularly spaced baseline distribution, the above considerations can be still applied by exploiting the average spacing of the distribution in place of Δb.

Turning back to the modelling of the acquired signals, in the more general and frequent case of acquisitions carried out at different time t_n (multiple passes system), the data g_n should take into account also the residual movement at full resolution of the imaged scene along the los. To this aim, as for the 3D case, we write the elementary scattering contribution along the elevation at the nth antenna by exploiting the calibrated phase expression in (8.37) where, this time, the term $\mathbf{d}_\mathrm{F}$ is no longer negligible: i.e. we consider the term $\gamma(s)\exp\left(-j2\pi\xi_n s + j\frac{4\pi}{\lambda}d_\mathrm{F}(t_n, s)\right)$. Notice that $d_\mathrm{F}(t_n, s)$ is the nth element of the vector $\mathbf{d}_\mathrm{F}$, i.e. the residual (full resolution) deformation at the time t_n of the scatterer located at elevation s. By carrying out, at each elevation, the Fourier expansion of the second exponential term with respect to the time t_n and by absorbing its spectrum in $\gamma(s)$, the expression in (8.38) can be written in the more general form [21]:

$$g_n = \iint_{I_S I_V} \gamma(s, v) e^{-j2\pi\left(\frac{2b_{\perp n}}{\lambda r}s - \frac{2t_n}{\lambda}v\right)} ds dv + w_n \tag{8.43}$$

where v is the Fourier pair variable of the velocity frequencies $\xi_{vn} = 2t_n/\lambda$, describing the velocity spectrum of the deformation at full resolution, which we assume to belong to the interval I_V. The function $\gamma(s, v)$, given by the product of $\gamma(s)$ and the spectrum of the exponential term associated with $\mathbf{d}_\mathrm{F}$, describes the backscattering distribution in the (s, v) plane, i.e. along the elevation (s) and along the frequency of the harmonic corresponding to the motion (v). Of course, (8.43) reduces to (8.38) not only when a single pass system is exploited, but also in the absence of any deformation during the observation time interval.

It is worth noting that the variable v in (8.43) is the frequency of the harmonic describing the possible motion and hence it is different from the slope v_F associated with the linear part of the deformation term in (8.37), usually referred to as DMV. However, the two quantities result to be the same in the presence of a pure linear motion with slope (mean velocity) $\bar{v}$, which leads to a spectrum of the deformation term impulsive in $\bar{v}$.

Equation (8.43) represents the 4D (differential) tomography model [21] introduced for the first time in [20], whose inversion with respect to the backscattering distribution provides the solution of the corresponding tomography problem.

The data in (8.43) are samples of the FT of the backscattering distribution, taken on an irregular grid over both the baseline and temporal domain. Accordingly, considerations similar to those carried out in the 3D case lead to define the Rayleigh limit for the velocity resolution:

$$R_v = \frac{\lambda}{2T} \tag{8.44}$$

and the maximum unambiguous velocity interval:

$$I_{V\,\max} = \frac{\lambda}{2\Delta t} \tag{8.45}$$

where T in (8.44) and Δt in (8.45) are the temporal span and the average temporal separation of the acquisitions, respectively.

A discretization of the continuous operators (8.38) and (8.43) is now in order. With reference to the more general model in (8.43), let us consider a 2D grid of $L = L_s \times L_v$ discrete points (s_l, v_l), hereafter called bins, over the elevation/velocity domain $I_s \times I_v$. Moreover, let us define the vector $\boldsymbol{\gamma} = [\gamma_1, \ldots, \gamma_L]^{\mathrm{T}}$ collecting, for each pixel, the samples $\gamma_l = \gamma(s_l, v_l)$ of the backscattering distribution picked up on the considered grid. Finally, for each bin on the elevation/velocity grid, let us define the *steering vector* $\mathbf{a}_l = \left[a_{l,1}, \ldots, a_{l,N}\right]^{\mathrm{T}}$, whose elements are defined as

$$a_{l,n} = \frac{1}{\sqrt{N}} \exp\left[-j2\pi\left(\frac{2b_{\perp n}}{\lambda r} s_l - \frac{2t_n}{\lambda} v_l\right)\right] = \frac{1}{\sqrt{N}} \exp\left(-j2\pi \xi_n^T \mathbf{p}_l\right) \tag{8.46}$$

where $\xi_n = [\xi_{sn}, -\xi_{vn}]^{\mathrm{T}}$ collects the spatial and velocity frequencies corresponding to the nth acquisition and $\mathbf{p}_l = [s_l, v_l]^{\mathrm{T}}$ is the lth bin on the elevation/velocity domain.

Accordingly, the FT operator in (8.43) can be rewritten in the discrete case as

$$\mathbf{g} = \mathbf{A}\boldsymbol{\gamma} + \mathbf{w} \tag{8.47}$$

where $\mathbf{g} = [g_1, \ldots, g_N]^{\mathrm{T}}$ and $\mathbf{w} = [w_1, \ldots, w_N]^{\mathrm{T}}$ are the vectors collecting, for each pixel, the measured complex data and the noise contribution, respectively, and

$\mathbf{A} = [\mathbf{a}_1, \ldots, \mathbf{a}_L]^{\mathrm{T}}$ is the $N \times L$ system matrix collecting the steering vectors associated with each bin on the elevation/velocity grid. The elevation/velocity grid will be referred to as *tomographic domain*. In general, L is chosen at least equal to (typically larger than) N to preserve the details of the reconstructed backscattering function contained in the measurements.

The tomographic model in (8.47) can be linked to the classical PS framework by assuming the backscattering to be concentrated in bins of the tomographic domain. With reference to the application to urban areas, and particularly with high-frequency systems (for instance C-band and X-band sensors), due to frequent occurrence of the layover, the backscattering profile $\boldsymbol{\gamma}$ in (8.47) can be in fact assumed to be a sparse vector composed by a number $K \ll L$ of non-zero elements, each of them representing a PS with a complex amplitude γ_k and located at the elevation/velocity bin (s_k, v_k). The effect of layover is schematically depicted in Figure 8.6. Under this assumption, by indexing from 1 to K the non-zero elements of the vector $\boldsymbol{\gamma}$ and the corresponding steering vectors in the matrix $\mathbf{A}$, the model in (8.47) reduces to

$$\mathbf{g} = \sum_{k=1}^{K} \gamma_k \mathbf{a}_k + \mathbf{w} \tag{8.48}$$

which is referred to as fully coherent (concentrated, i.e. point-like and time invariant) scattering model.

It is worth noting that the model in (8.47) represents also the discrete version of the 3D model in (8.38). The difference is just on the underlying discretization domain, which is, for the 3D model, a 1D grid on the elevation interval I_S. Accordingly, in this case, it results $L = L_S$ and $v_l = 0$ in the expression (8.46) of the steering vectors or, in other words, just the first component of both the vectors ξ_n and $\mathbf{p}_l$ have to be considered.

8.5 Multi-dimensional tomography imaging methods

TomoSAR imaging consists of the estimation of the backscattering vector $\boldsymbol{\gamma}$ starting from the available data $\mathbf{g}$, which are related each other through the linear model in (8.47). Therefore, an inversion of the operator $\mathbf{A}$ in the system in (8.47) is required.

In the following, some of the most common exploited inversion techniques [53,54] are summarized. Each of them is characterized by a different trade-off between simplicity, computational efficiency, side-lobes reductions and super-resolution capability, and may be well suited for specific applications.

8.5.1 Beamforming

A very simple and robust inversion of (8.47) can be carried by using the conjugate operator associated with the system matrix **A:**

$$\hat{\boldsymbol{\gamma}} = \mathbf{A}^{\mathrm{H}} \mathbf{g} \tag{8.49}$$

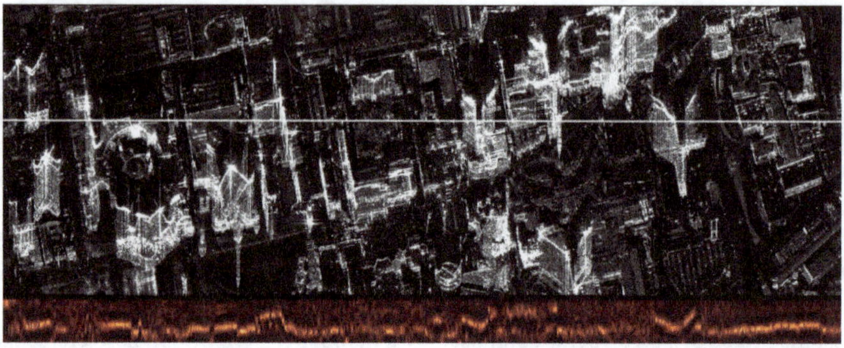

Figure 8.7 Classical azimuth-range focused representation by averaging all the images of a stack of data acquired in the high-resolution spotlight mode (1 m spatial resolution) of the TerraSAR-X sensor over an area in Las Vegas. Image on the top: classical greyscale figure in azimuth (horizontal) and range (vertical). Image on the bottom: azimuth-elevation profile achieved with the tomographic processing with a beamforming approach at the range defined by the horizontal line depicted in the upper image on a vertical excursion of about 150 m. Courtesy of DLR on SLC data

$(\cdot)^H$ being the Hermitian (conjugate and transpose) operator. The reconstruction algorithm is referred to as beamforming (BF) [52,55,56] and has an equivalent interpretation in the framework of matched filtering.

To understand the rationale of the above method of inversion, it is useful to let $L = N$ and assuming a uniform spectral sampling according to the Nyquist criterion. In this case, indeed, the system matrix $\mathbf{A}$ becomes the unitary ($\mathbf{A}^H = \mathbf{A}^{-1}$) DFT matrix and, thus, the BF algorithm in (8.49) provides (but for the noise contribution) the correct reconstruction of $\boldsymbol{\gamma}$ [52].

According to (8.49), BF corresponds to a 're-phase-and-sum' array processing that forms a 'beam' in the elevation/velocity directions.

In Figure 8.7, it is shown an example of tomographic reconstruction achieved via the BF. The data set is composed by HR spotlight mode (1 m spatial resolution) images of the TerraSAR-X sensor over an area in Las Vegas: more details on the data set can be found in [55]. The greyscale image in the top of the Figure 8.7 shows an amplitude reconstruction of the backscattering achieved via the classical SAR focusing: azimuth is horizontal, range is vertical. The image in the bottom shows the reconstruction in the azimuth-elevation domain achieved by a 3D focusing with the BF: a scaling of each vertical line to the mean value has been applied to improve the visualization. In Figure 8.8, it is shown an example of reconstruction of the backscattering in the elevation/velocity plane achieved by a 4D BF. The greyscale image represents a zoom of the image in Figure 8.7 on the Hotel Mirage: the axes have been rotated to fit approximately a North–West geometry, the azimuth is vertical and the range is horizontal increasing from left to

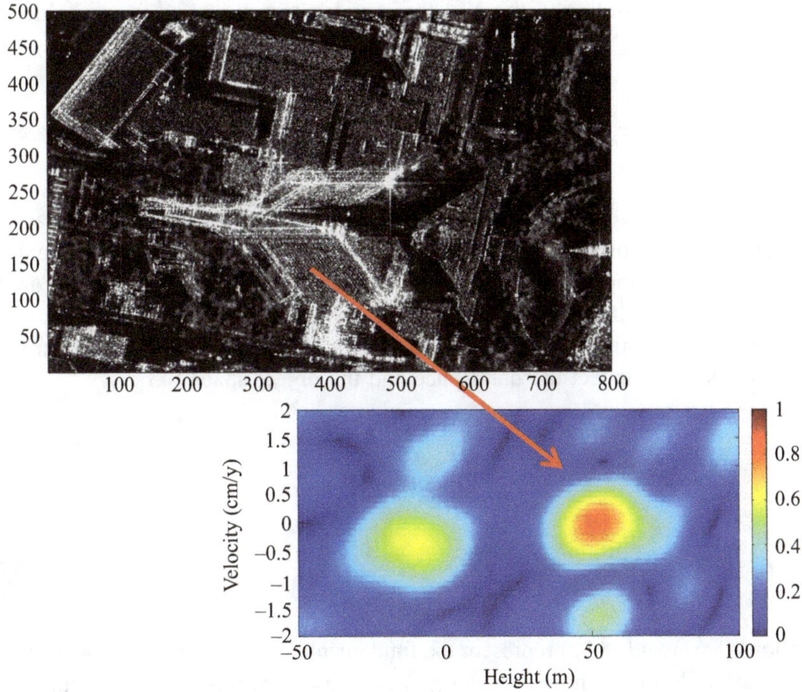

Figure 8.8 Example of elevation velocity focusing for the Las Vegas TerraSAR-X data set of Figure 8.7 over the Mirage Hotel in Las Vegas. Top image: azimuth-range greyscale image. Lower image: distribution of the backscattering in the elevation velocity plane showing the presence of two scatterers. Courtesy of DLR on SLC data

right. In this image, the effect of layover is clearly shown which causes a folding of the façade and roof towards the near range (left). The application of the BF provides the possibility to focus the backscattering in the elevation velocity plane: The image in the bottom shows an example of reconstruction in the area of the façade of the hotel with the presence of two scatterers, located at approximately 60 m distance in the height.

In practical applications, as baselines are far from being uniformly distributed, BF may give poor reconstruction performances in terms of sidelobes and leakage of the PSF. Strategies alternative to plain BF have been adopted to improve the resolution below the inherent Rayleigh limit (super-resolution) and the side-lobes suppression. However, as shown in Section 8.5.5, although simple, BF is a fast, robust and effective inversion scheme for the detection of PSs.

8.5.2 Singular value decomposition

SVD allows one to analyse the properties of a discrete and linear transformation represented by a matrix, implementing an effective regularized and ‘controlled’ inversion [43]. Accordingly, the $N \times L$ system matrix $\mathbf{A}$ in (8.47) can be decomposed as

$$\mathbf{A} = \sum_{n=1}^{N} \sigma_n \mathbf{u}_n \mathbf{v}_n^{\mathrm{H}} \tag{8.50}$$

where the N-length vectors $\mathbf{u}_n$ and the L-length vectors $\mathbf{v}_n$ are orthonormal basis for the N-dimensional column subspace (i.e. the data space) and the L-dimensional row subspaces (i.e. the object or unknown space), respectively, of $\mathbf{A}$, whereas σ_n are the corresponding singular values.

The decomposition in (8.50) allows writing the following two fundamental relationships that connect the data space and the object space [43]:

$$\mathbf{g} = \sum_{n=1}^{N} \sigma_n \mathbf{u}_n \mathbf{v}_n^{\mathrm{H}} \boldsymbol{\gamma} \tag{8.51}$$

$$\boldsymbol{\gamma} = \sum_{n=1}^{N} \frac{1}{\sigma_n} \mathbf{v}_n \mathbf{u}_n^{\mathrm{H}} \mathbf{g} \tag{8.52}$$

Equations (8.51) and (8.52) represent the fundamental result of the SVD analysis: the first equation describes how the data $\mathbf{g}$ is composed starting from the unknown $\boldsymbol{\gamma}$. It states that all the vectors $\mathbf{u}_n$ concur to the composition of the data, although each contribution is weighted by the associated singular value σ_n, thus leading to a weakening of the mapping depending on the magnitude of the singular value.

Singular values are inverted in the inversion formula (8.52). In a real case where data are corrupted by the noise, the presence of low singular values highlights the involvement of critical (weak) directions where the signal could be even overwhelmed by the noise. Accordingly, those directions have to be identified and properly handled with during the inversion process, to avoid high instabilities in the output reconstruction as a resulting from possible noise amplification.

With reference to the 3D case, low singular values are generally observed in the presence of a non-uniform baseline distribution, which causes an uneven distribution of the spectral samples. For a uniform baseline distribution, the behaviour of the singular values of the system matrix results to be dependent on the relation between the maximum unambiguous scene extension $I_{S\max}$ defined in (8.40) and the scene extension I_S. The condition $I_S > I_{S\max}$ produces a spatial aliasing of the reconstruction along the elevation. The spectrum is said to be under-sampled and the singular values (generally ordered according to the magnitude) do not show a clear decay, thus translating the peculiarity of the acquisition system to induce a loss of information. The condition $I_S = I_{S\max}$ corresponds to a sampling at the Nyquist limit: in this case, the singular values are all constant in such a way to provide, accordingly to (8.51) and (8.52), the equivalence $\mathbf{A}^{\mathrm{H}} = \mathbf{A}^{-1}$.

In the presence of a large dynamic range of the singular values, the ratio between the maximum and minimum singular values (conditioning number) can increase considerably thus translating the presence of an ill-conditioning of the matrix **A** that must be accounted for at the inversion stage to avoid noise amplification [43].

Based on the above considerations, the solution space can be restricted by considering only $N_t < N$ singular vectors corresponding to high singular values: in this way, only sufficiently strong directions of the decomposition are kept at the inversion stage. This procedure is usually referred to as Truncated SVD (TSVD) [52,56] and provides the solution

$$\hat{\boldsymbol{\gamma}} = \sum_{n=1}^{N} \mu_n \mathbf{v}_n \mathbf{u}_n^{\mathrm{H}} \mathbf{g} \tag{8.53}$$

where

$$\mu_n = \begin{cases} 1/\sigma_n & n = 1, \ldots, N_t \\ 0 & \text{elsewhere} \end{cases} \tag{8.54}$$

When $I_S < I_{S\max}$, that is when the spectrum of the scene is oversampled (data redundancy), the TSVD in (8.53) and (8.54) can take benefit of the available *a priori* information about the unknown support of the scene, allowing also a slight super-resolution imaging. A more exhaustive discussion about this topic can be found in [52].

In the 4D case, due to the fact that only a single baseline is available at each acquisition, the operator in (8.50) is typically characterized by the absence of low singular values: this translates the fact that such a single antenna repeat pass system typically does not bring significant redundancy in the measurements for the estimation of the distribution of the backscattering in the tomographic domain. In these cases, SVD does not allow improving significantly the BF reconstruction.

8.5.3 Capon filter

An approach that allows significantly improving the BF results by counteracting sidelobe increase and leakage at the expense of a loss of spatial resolution is framed in the context of the adaptive spectral estimation, in particular employing the Capon method [57,58]. The (elevation) spatial leakage associated with the irregular baselines is significantly reduced through the adaptive nulling, which interestingly results, jointly with a sensible reduction of the anomalous sidelobes, in some elevation super-resolution capabilities w.r.t. the Rayleigh limit.

The Capon method provides an estimate of each component γ_l of the backscattering vector $\boldsymbol{\gamma}$ through a minimum output power filter $\mathbf{f}_l$, which is a solution of the following constrained minimization problem

$$\begin{aligned} &\mathbf{f}_l = \arg\min_{\mathbf{f}} \mathrm{E}\left(\left|\mathbf{f}^{\mathrm{H}}\mathbf{g}\right|^2\right) \\ &\text{subject to } \mathbf{f}^{\mathrm{H}}\mathbf{a}_l = 1 \end{aligned} \tag{8.55}$$

where $\mathrm{E}(\cdot)$ is again the statistical expectation operator.

It can be shown that the solution of the problem in (8.55) is [57]

$$\mathbf{f}_l = \frac{\mathbf{C}_\mathrm{g}^{-1}\mathbf{a}_l}{\mathbf{a}_l^\mathrm{H}\mathbf{C}_\mathrm{g}^{-1}\mathbf{a}_l} \tag{8.56}$$

where $\mathbf{C}_\mathrm{g} = E(\mathbf{g}\mathbf{g}^\mathrm{H})$ is the covariance matrix of the data. The backscattering estimation is thus given by

$$\hat{\gamma}_l = \mathbf{f}_l^\mathrm{H}\mathbf{g} \tag{8.57}$$

where $\mathbf{f}_l$ is the Capon filter in (8.56).

It is interesting to note that for uncorrelated data, i.e. $\mathbf{C}_\mathrm{g} = \mathbf{I}$ (flat data spectral power), the Capon filtering leads to $\mathbf{f}_l = \mathbf{a}_l$, i.e. to the classical BF (matched filter).

The advantage of the Capon filter is the achievement of high super-resolution for line spectra (i.e. concentrated scatterers along s) [57,58]. However, a disadvantage of this filter is related to the need to estimate the data covariance matrix. This estimation is carried out via spatial averaging (i.e. multi-look), thus leading to a loss of spatial resolution. Another drawback of Capon is that the square norm of $\mathbf{f}_l$ is not unitary as for the beamforming: this may lead to large gains and distortion, that may impair further processing such as, for instance, the subsequent detection stage described in Section 8.5.5 as well as to self-cancellation in the presence of mis-calibrations [57–59].

8.5.4 Compressed sensing

Compressed sensing (CS) is a technique recently introduced also in the SAR context, used in linear inversion problems for signal recovery: It takes benefit of the hypothesis according to which the signal to be reconstructed have (in some basis) a sparse representation, i.e. a small number of non-zero entries. Under certain assumptions on the system matrix, the signal can be reconstructed from a small number of measurements [60,61].

TomoSAR in urban areas is a favourable application scenario for CS due to the fact that, for typical operating frequencies, the scattering occurs only on some scattering centres associated with ground, façades and roofs of ground structures. From a mathematical point of view, CS looks for the best (in the square norm sense) solution of

$$\mathbf{g} = \mathbf{A}\boldsymbol{\gamma} \tag{8.58}$$

under the hypothesis that of $\boldsymbol{\gamma}$ has only $K \ll L$ non-zero elements ($K = ||\boldsymbol{\gamma}||_0$, where $||\cdot||_0$ is the L^0 norm), starting from a number N of measurements which generally is much lower than L. This latter condition is due to the fact that a fine sampling of the output is required to correctly achieve super-resolution reconstructions. In other words, CS looks for the solution of the following problem

$$\begin{aligned} &\hat{\boldsymbol{\gamma}} = \arg\min_{\boldsymbol{\gamma}} ||\boldsymbol{\gamma}||_0 \\ &\text{subject to } ||\mathbf{g} - \mathbf{A}\boldsymbol{\gamma}||_2 < \varepsilon \end{aligned} \tag{8.59}$$

where $||\cdot||_2$ is the Euclidean (L^2) norm.

It can be shown that, under certain conditions relating *N*, *K* and *L*, the so-called basis pursuit de-noising (BPDN) problem [60,61], which can be formulated as

$$\begin{aligned} &\hat{\boldsymbol{\gamma}} = \arg\min_{\boldsymbol{\gamma}} ||\boldsymbol{\gamma}||_1 \\ &\text{subject to } ||\mathbf{g} - \mathbf{A}\boldsymbol{\gamma}||_2 < \varepsilon \end{aligned} \tag{8.60}$$

$||\cdot||_1$ being the L^1 norm, admits the same solution of (8.59). The advantage is that, from a numerical point of view, the BPDN problem in (8.60) is more tractable than the one in (8.59). As matter of fact, BPDN frames in the context of linear programming, where efficient solvers can be found.

CS has been successfully applied in the context of TomoSAR showing the capability to outperform classical BF in the super-resolution of scatterers in 3D and 4D SAR processing [62,63].

8.5.5 Detection of concentrated scatterers

TomoSAR can be exploited to analyse the vertical structure of the scattering in cases where the radiation penetrates the surface, such as for instance forest and ice frequently in conjunction with polarimetry [64,65]. In such cases, algorithms based on SVD and Capon, if spatial azimuth-range resolution requirements are not stringent, may be used. Nevertheless, as already described, even in the presence of surface scattering, scene including complex areas with vertical structures such as urban areas may involve the presence of scattering coming from a certain number of ground points interfering in the same pixel, see Figure 8.6. In this case, TomoSAR allows solving the interference but a further step of processing is required after whatever previously described inversion algorithm, in order to detect and, hence, locate and monitor single PS and multiple (interfering) PS. In addition, it is also necessary to keep as low as possible the occurrence of signal misinterpretation, i.e. false detection of PS, even in relatively low signal-to-noise ratio conditions.

To this aim, an assumption on the target properties is first of all required. In the following, we assume temporal invariant and concentrated (point-like) elevation scattering centres, thus matching the fully coherent model in (8.48). It is worth to underline that such an assumption is quite realistic with reference to the application to urban areas, and particularly with high frequency systems (for instance C-band and X-band sensors).

Standard PSI techniques carry out the identification of single targets via a threshold comparison of a decision variable, obtained by matching the observed phase values to a multi-baseline/multi-temporal linear model, see (8.34) and (8.35). TomoSAR provides a framework to improve the detection of scatterers showing a sufficient temporal coherency of the response, i.e. a persistency of the scattering, in the context of the detection theory [66,67]. In this way, the degree of misinterpretation is represented by the false-alarm probability, which is controlled at design stage of the detector.

In particular, with reference to the signal model in (8.48), the detection problem for single targets can be formulated in terms of the following binary hypothesis test

$$\begin{aligned} &\mathcal{H}_0 : \mathbf{g} = \mathbf{w} \\ &\mathcal{H}_1 : \mathbf{g} = \gamma\, \mathbf{a}(\mathbf{p}) + \mathbf{w} \end{aligned} \tag{8.61}$$

where γ is the target backscattering coefficient and $\mathbf{a}$ is the steering vector corresponding to the unknown position $\mathbf{p}$ of the target over the tomographic domain.

Under the typical Gaussian statistical modelling, the Generalized Likelihood Ratio Test (GLRT) for the problem in (8.61) is [66]

$$\max_{\mathbf{p}} \frac{|\mathbf{g}^{\mathrm{H}}\mathbf{a}(\mathbf{p})|}{||\mathbf{g}||\,||\mathbf{a}(\mathbf{p})||} \underset{\mathcal{H}_0}{\overset{\mathcal{H}_1}{\gtrless}} T \tag{8.62}$$

where T is the detection threshold, belonging to the $[0, 1]$ interval and sets according to the desired level of false alarm. Interestingly, as far as single targets are concerned, the test statistic in the left-hand side (LHS) of (8.62) picks the highest peak of the normalized BF reconstruction and the argument of the maximization in (8.62) is the maximum likelihood estimation (MLE) of the position $\mathbf{p} = [s, v]^{\mathrm{T}}$ [66]. Moreover, the detector in (8.62) results to ensure the constant false-alarm rate property with respect to the noise power level.

The detection strategy exploited by the standard PSI technique in (8.36) can be recast in a form similar to that in (8.62):

$$\max_{\mathbf{p}} \frac{\left|\mathbf{g}_{\varphi}^{\mathrm{H}}\mathbf{a}(\mathbf{p})\right|}{||\mathbf{g}_{\varphi}||\,||\mathbf{a}(\mathbf{p})||} \underset{\mathcal{H}_0}{\overset{\mathcal{H}_1}{\gtrless}} T \tag{8.63}$$

where $\mathbf{g}_{\varphi}$ is obtained from $\mathbf{g}$ by dropping the amplitude information from each element. Comparison of the probability of detection provided by (8.62) and (8.63) for the same fixed level of false alarm leads to the conclusion that the tomographic approach allows achieving an increase of the number of detected single targets, as well as to a better estimation of the PS parameters, i.e. $\mathbf{p}$ [64].

An extension to the case of double scatterers, based on the sequential use of the detection scheme in (8.62), has been proposed in [67]. In this case, one of the following three hypotheses

$$\begin{aligned} &\mathcal{H}_0 : \mathbf{g} = \mathbf{w} \\ &\mathcal{H}_1 : \mathbf{g} = \gamma_1 \mathbf{a}(\mathbf{p}_1) + \mathbf{w} \\ &\mathcal{H}_2 : \mathbf{g} = \gamma_1 \mathbf{a}(\mathbf{p}_1) + \gamma_2 \mathbf{a}(\mathbf{p}_2) + \mathbf{w} \end{aligned} \tag{8.64}$$

has to be selected ensuring a given level of false alarm, which is properly defined to accommodate the presence of multiple hypotheses. In (8.64), γ_1 and γ_2 are the backscattering coefficients of the two targets, whereas $\mathbf{p}_1$ and $\mathbf{p}_2$ are their positions on the tomographic domain. By assuming without loss of generality $|\gamma_1| > |\gamma_2|$, a sequential procedure [67] can be considered. First, an estimate of the position $\mathbf{p}_1$, thus of the steering $\mathbf{a}(\mathbf{p}_1)$, of the most powerful (first) scatterers according with the maximization on the LHS of (8.62) can be achieved. Subsequently, the first scattering contribution is cancelled by projecting the signal $\mathbf{g}$ onto the orthogonal complement of $\mathbf{a}(\mathbf{p}_1)$ [67]. The new data are then tested with the GLRT rule in (8.62), allowing to decide for the presence of the second scatterer. Indeed, in presence of both scatterers, being the position $\mathbf{p}_1$ correctly estimated and the steering of the second target exactly orthogonal to the one of the first targets, the new (projected) data are perfectly purged from the contribution γ_1 and, thus, the problem of detecting the presence of γ_2 can still be reformulated in terms of a binary test like that in (8.62). Accordingly, the detector can either declare the presence of two targets or continue by exploiting again the GLRT rule in (8.62), this time applied to $\mathbf{g}$. In this latter case, the hypotheses $\mathcal{H}_0$ and $\mathcal{H}_1$ are tested and the procedure ends by declaring the presence of one or zero scatterers.

The scheme in [67] is able to detect targets whose separation in the tomographic domain is above the Rayleigh resolution but does not provide super-resolution (below Rayleigh) capability, due to the projection of the data in the subspace orthogonal to the first target direction. In [68], a more effective sequential exploitation of the GLRT rules has been implemented, which avoids any orthogonal projection and, thus, allows one to reach also some degree of super-resolution. However, when the scatterers separation is above the Rayleigh resolution, the two approaches in [67,68] provide the same performances.

The above detection scheme allows the identification of a cloud of scatterers. Each of them is characterized by two coordinates which provides an estimate of the elevation, which allows the possibility of accurately geolocating the target on the ground and of DMV. A proper filtering of the data in the elevation velocity domain, based on the Fourier representation in (8.43) can also allow the separation of the different time series. Such a topic is, however, out of the scope of this chapter: additional details can be found in [21].

In Figure 8.9 is shown an example of reconstruction of the 3D cloud points of the Mirage Hotel achieved by exploiting the scheme in [67] to process the data set cited also for the real data results discussed in the previous sections. A HR (synthetic aperture) radar scanner from the space is therefore implemented to accurately reconstruct scatterers on the ground: with the expected future increase of the azimuth and range spatial resolution, the TomoSAR approach could provide a cost-effective alternative to classical laser scanning for imaging ground structures. The advantages of spaceborne TomoSAR rely in the possibility to provide 3D maps of whole cities as well as the capability to monitor long-term deformations. An example of monitoring of deformation is shown in Figure 8.10 for the COSMO-SkyMed constellation that was able after the Earthquake in l'Aquila in 2009 to acquire a stack of 33 images in only 6 months [69]. Application of PS and TomoSAR to specific structures like bridges has clearly shown the possibility to monitor even subtle thermal dilations up to an accuracy of 1 mm [70,71].

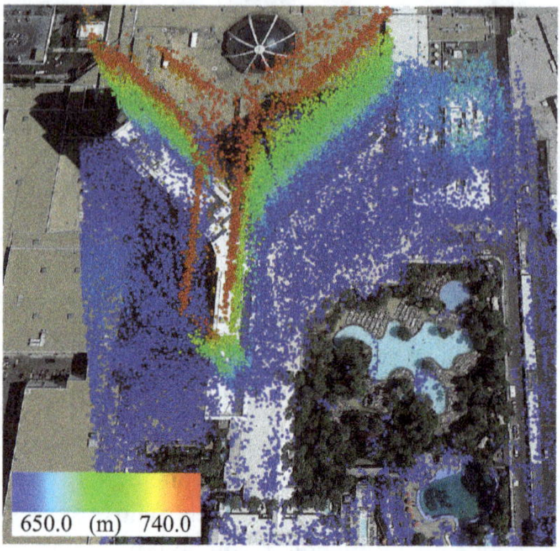

Figure 8.9 Example of localization of detected scatterers with the tomographic processing of for the Mirage Hotel in Las Vegas. Courtesy of DLR on the TerraSAR-X data set

8.5.6 Further aspects on multi-look processing of interferometric SAR data

The main features of TomoSAR imaging are full-resolution processing and exploitation of a model that requires a preliminary processing step, namely data calibration. On one hand, the Capon filter, as well as others algorithms not addressed in this chapter [57,58], requires the estimation of the data covariance matrix and, thus, the implementation of a multi-look; in any case, data are assumed to fit a model defined by the steering vectors thanks to the implementation of the calibration stage at the data pre-processing stage. On the other hand, PSI works at full resolution but may be implemented at the stage of data calibration. In any case, a final fit (only based on the phase information) of the signal to the steering vector is carried out.

TomoSAR and PSI coherent detection are tailored to fully coherent (persistent) scatterers, i.e. concentrated (point like) and stable (in the electromagnetic sense) scattering mechanisms characterized by a high angular and temporal correlation degree. PSI can be extended to implement multi-looking so to better handle partially coherent target; TomoSAR ability to separate targets in layover can be as well as extended to uncompensated data, i.e. prior to the data calibration. We start by referring to the case of a single target and assume that the data have not been calibrated but for the implementation of the ZBS with respect to the available DEM. Moreover, we refer to the expression of the data in (8.3) and (8.4) in a fixed-range coordinate, extended to the case of multiple (i.e. $N > 2$) acquisitions, to express the signals at the different antennas. All the acquired data can be

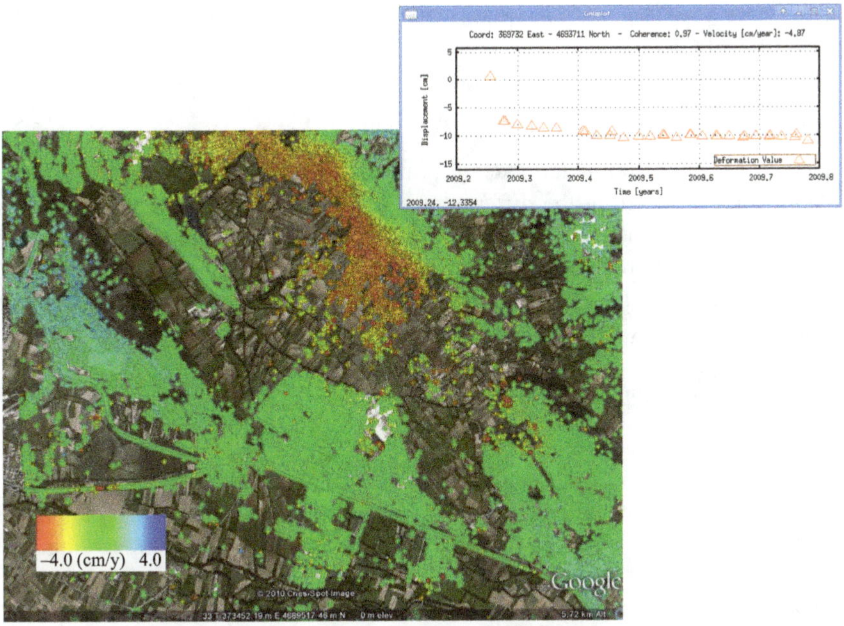

Figure 8.10 Example of monitoring deformation with the 4D (space time) technique. The lower image shows the post-seismic mean deformation velocity achieved with differential SAR tomography technique in the area of Paganica. Deformations have been referred to a reference point coincident with the reference point of a levelling network. The plot in the top image shows the time series extracted for a relevant point: The co-seismic jump is clearly visible as well as the typical exponential decay of post-seismic deformations. The stack of 33 images has been acquired by the COMSO-SKYMED constellation in just 6 months. Courtesy of ASI on raw data

collected in vector $\mathbf{y}$ that can be modelled (for the time being in the absence of noise) as

$$\mathbf{y} = \boldsymbol{\alpha}(\boldsymbol{\varphi}) \circ \boldsymbol{\gamma} \tag{8.65}$$

where $\circ$ denotes the Hadamard (element-wise) product, $\{\boldsymbol{\alpha}(\boldsymbol{\varphi})\}_k = \exp(j\varphi_k)$ is the kth element of the complex vector $\boldsymbol{\alpha}(\boldsymbol{\varphi})$ accounting for the phases $\boldsymbol{\varphi} = [\varphi_1, \ldots, \varphi_N]^{\mathrm{T}}$ at the different antennas and $\boldsymbol{\gamma} = [\gamma_1, \ldots, \gamma_N]^{\mathrm{T}}$ collects the different values of the target backscattering coefficients in (8.3) and (8.4), which can show variation (due to the decorrelation phenomena) at the different antennas. Note that no specific structure of $\boldsymbol{\varphi}$ has been assumed.

The covariance matrix of the data in (8.65) is

$$\mathbf{C}(\boldsymbol{\varphi}) = \boldsymbol{\alpha}(\boldsymbol{\varphi})\boldsymbol{\alpha}^{\mathrm{H}}(\boldsymbol{\varphi}) \circ \mathbf{C}_{\boldsymbol{\gamma}} \tag{8.66}$$

where $\mathbf{C}_\gamma = \mathrm{E}(\boldsymbol{\gamma}\boldsymbol{\gamma}^H)$ is the covariance matrix of the target backscattering coefficients. It can be written as

$$\mathbf{C}_\gamma = \overline{\sigma^2}\mathbf{M} \tag{8.67}$$

where $\overline{\sigma^2}$ is the (arithmetic) average value of the target radar cross sections $\sigma_n^2 = \mathrm{E}(|\gamma_n|^2)$ observed at the different antennas and $\mathbf{M}$ is the so-called *structure matrix*, which accounts for the (normalized) decorrelation effects experienced by the target at the different antennas.

In the case of fully coherent targets, we have $\{\mathbf{C}_\gamma\}_{n,m} = \sigma^2\ \forall n, m$ and, hence, the covariance matrix (8.66) can be written as

$$\mathbf{C}(\boldsymbol{\varphi}) = \sigma^2\boldsymbol{\alpha}(\boldsymbol{\varphi})\boldsymbol{\alpha}^{\mathrm{H}}(\boldsymbol{\varphi}) \tag{8.68}$$

where σ^2 is the constant value of the target radar cross section . It is worth to note again that the elements of $\boldsymbol{\varphi}$ are now not supposed to be structured according to the elevation and DMV model in (8.33). This is to emphasize the fact that calibration of the data is not assumed to be implemented at this stage of the processing: The off-diagonal elements of $\mathbf{C}$ represent in practice the interferograms between pairs of images.

The model in (8.68) is suitable for most man-made targets widely present in urban areas, whereas it is less satisfactory for natural targets, characterized by distributed scattering (hence affected by angular decorrelation). In this case, the most general expression in (8.66) and (8.67) should be considered.

When the vector $\boldsymbol{\gamma}$ in (8.65) accounts, as usually happens, for the speckle over the different acquisitions, it can be demonstrated [6] that the structure matrix $\mathbf{M}$ is real with all the elements along the main diagonal equal to 1: more specifically, $\mathbf{M}$ becomes coincident with the coherence matrix (i.e. the matrix made by elements representing the coherence between the different acquisitions). Mathematically, $\{\mathbf{M}\}_{n,m} = \rho_{n,m}$, where $\rho_{n,m}$ is the coherence between the acquisitions n and m, and $\sigma_n^2 = \sigma^2\ \forall n$. In this case, the structure matrix $\mathbf{M}$ can be obtained from the covariance matrix $\mathbf{C}$ in (8.66) by normalizing the modulus of each element to the average target radar cross section σ^2.

Finally, to account for the presence of additive white noise with variance σ_w^2, the following substitutions should be carried out in (8.66): $\overline{\sigma^2} \leftarrow \overline{\sigma^2} + \sigma_w^2$ and $\mathbf{M} \leftarrow (\mathbf{M}\mathrm{SNR} + \mathbf{I})/(\mathrm{SNR} + 1)$, where $\mathrm{SNR} = \overline{\sigma^2}/\sigma_w^2$.

For reduced resolution analysis, an extension of PSI techniques named SqueeSAR [36] has been recently developed to be able to handle also decorrelating targets. Starting from a number of say P of statistically homogenous pixels, the corresponding data vectors, say $\mathbf{y}_1, \ldots, \mathbf{y}_P$, are modelled as uncorrelated complex zero-mean Gaussian random vectors with covariance matrix given by (8.66) and (8.67). The SqueeSAR algorithm performs the MLE $\hat{\boldsymbol{\varphi}}$ of the vector $\boldsymbol{\varphi}$ as [36]:

$$\hat{\boldsymbol{\varphi}} = \arg\min_{\boldsymbol{\eta}} \mathrm{tr}\left[\mathbf{C}^{-1}(\boldsymbol{\eta})\hat{\mathbf{C}}\right] = \arg\min_{\boldsymbol{\eta}} \mathrm{tr}\left[(\boldsymbol{\alpha}(\boldsymbol{\eta})\boldsymbol{\alpha}^{\mathrm{H}}(\boldsymbol{\eta}) \circ \mathbf{M}^{-1})\hat{\mathbf{C}}\right] \tag{8.69}$$

where

$$\hat{\mathbf{C}} = \frac{1}{P}\sum_{k=1}^{P} \mathbf{y}_k \mathbf{y}_k^{\mathrm{H}} \tag{8.70}$$

is the multi-look covariance matrix estimated from the InSAR data and the operator $\mathrm{tr}[\cdot]$ extracts the trace of a matrix. Interesting it is to note that $\mathrm{tr}[\mathbf{A}^{\mathrm{H}}\mathbf{B}]$ can be interpreted as scalar product, precisely the Frobenius inner product, between the matrices **A** and **B**.

The MLE in (8.69) is performed at data calibration stage and, therefore, does not require the fit of the data to a specific model. It provides the filtered phase data stack $\boldsymbol{\alpha}(\hat{\boldsymbol{\varphi}})$, where $\boldsymbol{\alpha}$ is defined as in (8.66), which can be interpreted as an 'equivalent' PS corresponding to the decorrelating scatterer mechanism in the selected data, thus allowing to extend the PSI analysis to the case of a decorrelated target.

In contrast to other low-resolution interferometric techniques, like SBAS, SqueeSAR is able to match adaptively the data. Accordingly, it does not require any *a priori* hard limitation of the baselines to counteract decorrelation, since the interferometric information is extracted by properly weighting all the available interferograms.

However SqueeSAR, although providing an extension of the PSI to the case of decorrelated targets, is an interferometric technique: It selects only one scattering mechanism associated with distributed targets.

Starting from TomoSAR processing, a technique alternative to SqueeSAR has been proposed with the name of Component extrAction and sElection SAR (CAESAR) [72,73].

Differently from SqueeSAR, it allows the extraction and selection of multiple scattering mechanisms on multi-look InSAR data at level of interferogram generation, i.e. as SqueeSAR, prior to the data calibration. CAESAR is based on the application of the principal component analysis [73] to identify the principal components of the data: mathematically, it decomposes the estimated data covariance matrix in orthogonal dyads corresponding to the orthogonal scattering mechanism:

$$\hat{\mathbf{C}} = \sum_{k=1}^{N} \lambda_k \mathbf{u}_k \mathbf{u}_k^{\mathrm{H}} \tag{8.71}$$

where $\mathbf{u}_k$ and λ_k are the kth eigenvector and the corresponding eigenvalue.

The eigenvectors of such a decomposition represent a filtered version of the interferograms where different scattering mechanisms are separated. As SqueeSAR, CAESAR does not assume any structure for the scattering as in Tomography and as in the PS detector in (8.63), thus having capability to process also not calibrated data. CAESAR frames in this sense in the context of interferometry. The first eigenvector $\mathbf{u}_1$ maximizes (w.r.t. the vector $\mathbf{u}$) the quadratic form $\mathbf{u}^{\mathrm{H}}\hat{\mathbf{C}}\mathbf{u} = \mathrm{tr}[\mathbf{u}\mathbf{u}^{\mathrm{H}}\hat{\mathbf{C}}]$ subject to $\mathbf{u}^{\mathrm{H}}\mathbf{u} = 1$ [53], which can be read as already mentioned as a scalar product. As a consequence, the covariance matrix $\lambda_1 \mathbf{u}_1 \mathbf{u}_1^{\mathrm{H}}$ corresponding to the first scattering mechanism provides the maximum scalar product with the covariance matrix $\hat{\mathbf{C}}$ estimated from the data. The presence of other high eigenvalues translate the presence of additional scattering mechanism with a significant power [72,73]. Accordingly, differently from SqueeSAR, CAESAR allows also extracting multiple components associated to **eigen-interferograms** corresponding to the different scattering mechanisms prior to the calibration of the data at the interferogram generation stage [73].

Acknowledgements

The authors wish to thank the Italian and German Space Agencies, ASI and DLR, for providing the data that have been processed to generate results presented in this chapter. The authors wish as well as to thank Diego Reale the support in the generation of the images included in this chapter.

References

[1] Wiley, C. A.: 'Synthetic aperture radars: a paradigm for technology evolution', *IEEE Trans. Aerosp. Electron. Syst.*, 1985, 21, (3), pp. 440–443.

[2] Franceschetti, G., Lanari, R.: 'Synthetic Aperture Radar Processing' (CRC-PRESS, Boca Raton, March 1999).

[3] Moreira, A., Prats-Iraola, P., Younis, M., Krieger, G., Hajnsek, I., Papathanassiou, K.P.: 'A tutorial on synthetic aperture radar', *IEEE Geosci. Remote Sens. Mag.*, 2013, 1, (1), pp. 6–43.

[4] Rosen, P. A., Hensley, S., Joughin, I. R., *et al.*: 'Synthetic Aperture Radar Interferometry', *Proc. IEEE*, 2000, 88, (3), pp. 333–382.

[5] Bamler, R., Hartl, P.: 'Synthetic Aperture Radar Interferometry', *Inverse Problems*, 1998, 14, pp. R1–R54.

[6] Fornaro, G., Pascazio, P.: *SAR Interferometry and Tomography: Theory and Applications* (Academic Press Library in Signal Processing Volume 2, Elsevier Ltd. 2013).

[7] Fornaro, G., Franceschetti, G.: 'SAR Interferometry', Chapter IV in Franceschetti, G., Lanari, R.: *Synthetic Aperture Radar Processing*, (CRC-PRESS, Boca Raton, Marzo 1999).

[8] Zebker, S. H., Villasenor, J.: 'Decorrelation in interferometric radar echoes', *IEEE Trans. Geosci. Remote Sens.*, 1992, 30, pp. 950–959.

[9] Gabriel, A. K., Goldstein, R. M., Zebker, H. A.: 'Mapping small elevation changes over large areas: Differential radar interferometry', *J. Geophys. Res.*, 1989, 94, (B7), pp. 9183–9191.

[10] Goldstein, R. M., Engelhardt, H., Kamb, B., Froclich, R. M.: 'Satellite RADAR interferometry for monitoring ice-sheet motion—application to an Antarctic ice stream', *Science*, 1993, 262, (5139), pp. 1525–1530.

[11] Massonnet, D., Rossi, M., Carmona, C., Adragna, F., Peltzer, G., Feigl, K., Rabaute, T.: 'The displacement field of the Landers earthquake mapped by radar interferometry', *Nature*, 1993, 364, (6433), pp. 138–142.

[12] Berardino, P., Fornaro, G., Lanari, R., Sansosti, E.: 'A new algorithm for surface deformation monitoring based on small baseline differential SAR interferograms', *IEEE Trans. Geosci. Remote Sens.*, 2002, 40, (11), pp. 2375–2383.

[13] Blanco-Sanchez, P., Mallorqui, J., Duque, S., Monnells, D.: 'The coherent pixels technique (CPT): an advanced DInSAR technique for nonlinear deformation monitoring', *Pure Appl. Geophys.*, 2008, 165, (6), pp. 1167–1193.

[14] Mora, O., Mallorqui, J. J., Broquetas, A.: 'Linear and nonlinear terrain deformation maps from a reduced set of interferometric SAR images', *IEEE Trans. Geosci. Remote Sens.*, 2003, 41, (10), pp. 2243–2253.

[15] Fornaro, G., Pauciullo, A. , Serafino, F.: 'Deformation monitoring over large areas with multipass differential SAR interferometry: a new approach based on the use of spatial differences', *Int. J. Remote Sens.*, 2009, 30, (6), pp. 1455–1478.

[16] Ferretti, A., Prati, C., Rocca, F.: 'Nonlinear subsidence rate estimation using permanent scatterers in differential SAR interferometry', *IEEE Trans. Geosci. Remote Sens.*, 2000, 38, pp. 2202–2212.

[17] Ferretti, A., Prati, C., Rocca, F.: 'Permanent scatterers in SAR interferometry', *IEEE Trans. Geosci. Remote Sens.*, 2001, 39, (1), pp. 8–20.

[18] Reigber, A., Moreira, A.: 'First demonstration of airborne SAR tomography using multibaseline L-band data', *IEEE Trans. Geosci. Remote Sens.*, 2000, 38, (5), pp. 2142–2152.

[19] Fornaro, G., Lombardini, F., Serafino, F.: 'Three-dimensional multipass SAR focusing: experiments with long-term spaceborne data', *IEEE Trans. Geosci. Remote Sens.*, 2005, 43, (4), pp. 702–714.

[20] Lombardini, F.: 'Differential tomography: a new framework for SAR interferometry,' *IEEE Trans. Geosci. Remote Sens.*, 2005, 43, (1), pp. 37–44

[21] Fornaro, G., Reale, D., Serafino, F.: 'Four-dimensional SAR imaging for height estimation and monitoring of single and double scatterers', *IEEE Trans. Geosci. Remote Sens.*, 2009, 47, (1), pp. 224–237.

[22] Sansosti, E.: 'A simple and exact solution for the interferometric and stereo SAR geolocation problem', *IEEE Trans. Geosci. Remote Sens.*, 2004, 42, (8), pp. 1625–1634.

[23] Suchandt, S., Runge, H., Breit, H., Steinbrecher, U., Kotenkov, A., Balss, U.: 'Automatic extraction of traffic flows using TerraSAR-X along-track interferometry', *IEEE Trans. Geosci. Remote Sens.*, 2010, 39, (2), pp. 807–819.

[24] Chapin, E., Chen, C. W.: 'Along-track interferometry for ground moving target indication', *IEEE Aerosp. Electron. Syst. Mag.*, 2008, 23, (6), pp. 19–24.

[25] Romeiser, R., Breit, H., Eineder, M., *et al.*: 'Current measurements by SAR along-track interferometry from a Space Shuttle', *IEEE Trans. Geosci. Remote Sens.*, 2005, 43, (10), pp. 2315–2324.

[26] Hanssen, R. F.: *Radar Interferometry: Data Interpretation and Error Analysis*, Remote Sensing and Digital Image Processing, vol. 2 (Kluwer Academic Publishers, New York, 2001).

[27] Costantini, M.: 'A novel phase unwrapping method based on network programming', *IEEE Trans. Geosci. Remote Sens.*, 1998, 36, (3), pp. 813–821.

[28] Ghiglia, D. C., Pritt, M. D.: *Two-Dimensional Phase Unwrapping, Theory, Algorithms, and Software*, (Wiley-Interscience, New York (USA), 1998), ISBN 0-471-24935-1.

[29] Chen, C. W., Zebker, H. A.: 'Two-dimensional phase unwrapping with use of statistical models for cost functions in nonlinear optimization', *J. Opt. Soc. Am. A*, 2001, 18, 338–351.

[30] Papoulis, A.: *Probability, Random Variables, and Stochastic Processes*, (McGraw-Hill Kogakusha, Tokyo, 1965).

[31] Touzi, R., Lopes, A., Bruniquel, J., Vachon, P. W.: 'Coherence estimation for SAR imagery', *IEEE Trans. Geosci. Remote Sens.*, 1999, 37, pp. 135–149.

[32] Gatelli, F., Monti Guarnieri, A., Palizzi, F., Pasquali, P., Prati, C., Rocca, F.: 'The wavenumber shift in SAR interferometry', *IEEE Trans. Geosci. Remote Sens.*, 1994, 32, pp. 855–865.

[33] Fornaro, G., Monti Guarnieri, A.: 'Minimum mean square error space-varying filtering of interferometric SAR data', *IEEE Trans. Geosci. Remote Sens.*, 2002, 40, pp. 11–21.

[34] Touzi, R., Lopes, A., Bruniquel, J., Vachon, P. W.: 'Coherence estimation for SAR imagery', *IEEE Trans. Geosci. Remote Sens.*, 1999, 37, (1), pp. 135–149.

[35] López-Martínez, C., Pottier, E., 'Coherence estimation in synthetic aperture radar data based on speckle noise modeling', *Appl. Opt.*, 2007, 46, (4), pp. 544–558.

[36] Ferretti, A., Fumagalli, A., Novali, F., Prati, C., Rocca, F., Rucci, A.: 'A new algorithm for processing interferometric data-stacks: SqueeSAR', *IEEE Trans. Geosci. Remote Sens.*, 2011, 49, (9), pp. 3460–3470.

[37] Deledalle, C. A., Denis, L., Tupin, F.: 'NL-InSAR: Nonlocal interferogram estimation', *IEEE Trans. Geosci. Remote Sens.*, 2011, 49, (4), pp. 1441–1452.

[38] Sica, F., Reale, D., Poggi, G., Verdoliva, L., Fornaro, G.: 'Nonlocal adaptive multilooking in SAR multipass differential interferometry', *IEEE J. Sel. Topics Appl. Earth Observ. Remote Sens.*, 2015, 8, (4), pp. 1727–1742.

[39] Deledalle, C., Denis, L., Poggi, G., Tupin, F., Verdoliva, L., 'Exploiting patch similarity for SAR image processing: the nonlocal paradigm', *IEEE Signal Proc. Mag.*, 2014, 31, (4), pp. 69–78.

[40] Eineder, M., Minet, C., Steigenberger, P., Cong, X., Fritz, T.: 'Imaging Geodesy—Toward Centimeter-Level Ranging Accuracy With TerraSAR-X', *IEEE Trans. Geosci. Remote Sens.*, 2011, 49, (2), pp. 661–671.

[41] Sansosti, E., Berardino, P., Manunta, M., Serafino, F., Fornaro, G.: 'Geometrical SAR Image Registration', *IEEE Trans. Geosci. Remote Sens.*, 2006, 44, (10), pp. 2861–2870.

[42] Usai, S.: 'A new approach for long term monitoring of deformations by differential SAR interferometry', Ph.D. thesis, Delft Univ. of Technol., Delft, The Netherlands, 2001.

[43] Golub, G. H., Van Loan, C. F.: *Matrix Computations* (Johns Hopkins Univ. Pr., 1996).

[44] Reale, D., Fornaro, G., Pauciullo, A.: 'Extension of 4-D SAR imaging to the monitoring of thermally dilating scatterers', *IEEE Trans. Geosci. Remote Sens.*, 2013, 51, (12), pp. 5296–5306.

[45] Pepe, A., Lanari, R.: 'On the extension of the minimum cost flow algorithm for phase unwrapping of multitemporal differential SAR interferograms', *IEEE Trans. Geosci. Remote Sens.*, 2006, 44, (9), pp. 2374–2383.

[46] Costantini, M., Malvarosa, F., Minati, F.: 'A general formulation for redundant integration of finite differences and phase unwrapping on a sparse multidimensional domain', *IEEE Trans. Geosci. Remote Sens.*, 2012, 50, (3), pp. 758–768.

[47] Agram, P., Zebker, H.: 'Edgelist phase unwrapping algorithm for time-series InSAR analysis', *J. Opt. Soc. Am. A*, 2010, 27, (3), pp. 605–612.

[48] Fornaro, G., Pauciullo, A., Reale, D.: 'A null-space method for the phase unwrapping of multi-temporal SAR interferometric stacks', *IEEE Trans. Geosci. Remote Sens.*, 2011, 49, (6), pp. 2323–2334.

[49] Lanari, R., Mora, O., Manunta, M., Mallorqui, J. J., Berardino, P., Sansosti, E.: 'A small-baseline approach for investigating deformations on full-resolution differential SAR interferograms', *IEEE Trans. Geosci. Remote Sens.*, 2004, 42, (7), pp. 1377–1386.

[50] Reale, D., Nitti, D. O., Peduto, D., Nutricato, R., Bovenga, F., Fornaro, G.: 'Postseismic deformation monitoring with the COSMO/SKYMED constellation', *IEEE Geosci. Remote Sens. Lett.*, 2011, 8, (4), pp. 696–700.

[51] Ferretti, A., Bianchi, M., Prati C., Rocca, F.: 'Higher-order permanent scatterers analysis', *EURASIP J. Adv. Signal Proc.*, 2005, 20, pp. 1–12.

[52] Fornaro, G., Serafino, F., Soldovieri, F.: 'Three-dimensional focusing with multipass SAR data', *IEEE Trans. Geosci. Remote Sens.*, 2003, 43, (3), pp. 507–517.

[53] Stoica, P., Moses, R. L.: *Introduction to Spectral Analysis* (Prentice-Hall, Englewood Cliffs, USA, 1997).

[54] Fornaro, G., Lombardini, F., Pauciullo, A., Reale, D., Viviani, F.: 'Tomographic processing of interferometric SAR data: developments, applications, and future research perspectives', *IEEE Sig. Proc. Mag.*, 2014, 31, (4), pp. 41–50.

[55] Reale, D., Fornaro, G., Pauciullo, A., Zhu, X., Bamler, R.: 'Tomographic imaging and monitoring of buildings with very high resolution SAR data', *IEEE Geosci. Remote Sens. Lett.*, 2011, 8, (4), pp. 661–665.

[56] Bertero, M.: *Linear Inverse and Ill-posed Problems* (Advances in Electronics and Electron Physics, Academic Press, 1989).

[57] Gini, F., Lombardini, F., Montanari, M.: 'Layover solution in multibaseline SAR interferometry', *IEEE Trans. Aerosp. Electron. Syst.*, 2002, 38, (4), pp. 1344–1356.

[58] Lombardini, F., Montanari, M., Gini, F.: 'Reflectivity estimation for multibaseline interferometric radar imaging of layover extended sources', *IEEE Trans. Signal Process.*, 2003, 51, pp. 1508–1519.

[59] Lombardini, F., Pardini, M., Fornaro, G., Serafino, F., Verrazzani, L., Costantini, M.: 'Linear and adaptive spaceborne three-dimensional SAR tomography: a comparison on real data', *IET Radar Sonar Navig.*, 2009, Special Issue, 3, (4), pp. 424–436.

[60] Candes, E. J., Wakin, M-B.: 'An introduction to compressive sampling', *IEEE Signal Proc. Mag.*, 2008, 25, (2), pp. 21–30.

[61] Donoho, D. L.: 'Compressed sensing', *IEEE Trans. Inf. Theory*, 2006, 52, (4), pp. 1289–1306.

[62] Budillon, A., Schirinzi, G.: 'Three-dimensional SAR focusing from multi-pass signals using compressive sampling', *IEEE Trans. Geosci. Remote Sens.*, 2010, 49, (1), pp. 488–499.

[63] Zhu, X., Bamler, R.: 'Very high resolution spaceborne SAR tomography in urban environment', *IEEE Trans. Geosci. Remote Sens.*, 2010, 48, (12), pp. 4296–4308.

[64] Tebaldini, S.: 'Single and multipolarimetric SAR tomography of forested areas: a parametric approach', *IEEE Trans. Geosci. Remote Sens.*, 2010, 48, (5), pp. 2375–2387.

[65] Nannini, M., Scheiber, R., Moreira, A.: 'First 3D reconstructions of targets hidden beneath foliage by means of polarimetric SAR tomography', *IEEE Geosci. Remote Sens. Lett.*, 2009, 47, (2), pp. 531–543.

[66] De Maio, A., Fornaro, G., Pauciullo, A.: 'Detection of single scatterers in multi-dimensional SAR imaging', *IEEE Trans. Geosci. Remote Sens.*, 2009, 47, (7), pp. 2284–2297.

[67] Pauciullo, A., Reale, D., De Maio, A., Fornaro, G.: 'Detection of double scatterers in SAR tomography', *IEEE Trans. Geosci. Remote Sens.*, 2012, 50, (9), pp. 3567–3586.

[68] Budillon, A., Schirinzi, G.: 'GLRT based on support estimation for multiple scatterers detection in SAR tomography', *IEEE J. Sel. Topics Appl. Earth Observ. Remote Sens.*, 2016, 9, (3), pp. 1086–1094.

[69] Reale, D., Nitti, D. O., Peduto, D., Nutricato, R., Bovenga, F., Fornaro, G.: 'Post-seismic deformation monitoring with the COSMO/SKYMED constellation', *IEEE Geosci. Remote Sens. Lett.*, 2011, 8, (4), pp. 696–700.

[70] Monserrat, O., Crosetto, M., Cuevas, M., Crippa, B.: 'The thermal expansion component of persistent scatterer interferometry observations', *IEEE Geosci. Remote Sens. Lett.*, 2011, 8, (5), pp. 864–868.

[71] Fornaro, G., Reale, D., Verde, S.: 'Bridge thermal dilation monitoring with millimeter sensitivity via multidimensional SAR imaging', *IEEE Geosci. Rem. Sens. Lett.*, 2013, 10, (4), pp. 677–681.

[72] Fornaro, G., Pauciullo, A., Reale, D., Verde, S.: 'Multilook SAR tomography for 3-D reconstruction and monitoring of single structures applied to COSMO-SkyMed data', *IEEE J. Sel. Topics Appl. Earth Observ. Remote Sens.*, 2014, 7, (7), pp. 2776–2785.

[73] Fornaro, G., Verde, S., Reale, D., Pauciullo, A.: 'CAESAR: an approach based on covariance matrix decomposition to improve multibaseline-multitemporal interferometric SAR processing', *IEEE Trans. Geosci. Remote Sens.*, 2014, 53, (4), pp. 2050–2065.

Chapter 9
Bi- and monostatic SAR-GMTI

Ingo Walterscheid, Diego Cristallini* and Robert Kohlleppel**

Abstract

This chapter presents a novel simultaneous monostatic and bistatic ground moving target indication (GMTI) mode for improved target detection and imaging capability. The mode uses an airborne multichannel radar system and a stationary transmitter. Both systems transmit simultaneously on adjacent frequency bands, and the airborne multichannel system receives both its monostatic echoes and the bistatic returns. Geometrical diversity between the monostatic and the bistatic measurements enhances moving target-detection capabilities. Moreover, for movers detected in both datasets, an estimation of the target velocity vector (i.e., velocity and direction of motion) can be performed. By simply extracting a single-channel dataset, this also allows correct focusing of moving targets both in monostatic and in bistatic datasets, if SAR-GMTI capability is required. Consequently, situational awareness over the observed area is greatly improved. The effectiveness of the proposed technique is analyzed both from a theoretical point of view and by means of an ad-hoc experiment conducted by the Fraunhofer Institute for High Frequency Physics and Radar Techniques (FHR) in fall 2013.

9.1 Introduction

Requirements of current radar systems are becoming ever more demanding. For example, in the context of multichannel ground moving target indication (GMTI), the radar should not simply declare the presence of a target, but also accurately estimate its position on the ground, its direction of motion and possibly it should also provide an image of the target so that it can be recognized and classified. All these capabilities have to be available while keeping the radar invulnerable to jamming, and, ideally, at low-cost. To fulfil all these requirements, we consider not only monostatic radars, but also bistatic and multistatic systems where several radars cooperate to increase the overall situational awareness of a given area.

*Fraunhofer FHR, Germany

In fact, by jointly exploiting two or more systems, the overall information that can be extracted is greater than the net sum of the information provided by each system alone. For the end-user, this means more detected movers in the surveyed area, and more information on the movers (where they are, where they are going). Such improved information might open new possibilities toward target classification and identification.

Interest in bistatic radar systems has grown considerably in the last few decades. In particular, the peculiarities of a bistatic configuration for multichannel GMTI have drawn the attention of the space time adaptive processing (STAP) community, thus producing a wide and comprehensive literature on bistatic STAP, see [1–4] as overall references on STAP in general and [5–9] for more detailed analysis in the bistatic case. It is clear that a bistatic system has to face drawbacks not present in its monostatic counterpart such as synchronization and range-dependent clutter spread.

In this chapter, we will focus on the joint simultaneous exploitation of monostatic and bistatic data to improve moving target detection and imaging capabilities in synthetic aperture radar (SAR)-GMTI. We will show how this approach is able to: (i) detect more targets in a given area, regardless of their direction of motion; (ii) relocate them on the ground and estimate their direction of motion; (iii) provide an inverse SAR (ISAR)-like image of the movers. In particular, capabilities (ii) and (iii) will be guaranteed for those targets that can be detected both in the monostatic and in the bistatic data.

It is well known that target-detection capabilities of any airborne multichannel radar system are highly dependent, among other factors, on the target motion. Specifically, the target (with a given radar cross section, RCS) has to move with a minimum **radial**[1] velocity, also referred to as minimum detectable velocity (MDV). This means that no target detection is possible for targets moving tangential to the radar platform. This is simply due to the lack of Doppler-induced modulation in the received target echoes. It would be advantageous, from a user point of view, to overcome this reliance on a particular direction of motion. A single system struggles to meet this requirement, as there will always be directions of target motion blind to the radar (e.g., the directions parallel to the iso-range lines). A potential solution is the so-called scan-MTI approach [10], where the continuous scanning of a given area from different aspect angles during platform motion recovers some Doppler modulation. The problem of moving targets detection perpendicular to the line of sight (LOS) is even more important in the context of SAR-GMTI. In this case, not only the movers are not detected, but they also appear smeared in the resulting SAR image [11]. To solve this problem, several approaches starting from a single-channel SAR system have been proposed [12–18]. In particular, in [15] a bank of focusing filters is developed, with each branch matched to a different possible tangential target velocity value. The output

[1]Radial velocity is defined as the component of the relative motion between radar and target projected along the line of sight between target and radar positions.

image of a given branch will show the moving targets at the corresponding velocity correctly focused upon a background of blurred stationary returns. Detection of targets moving perpendicular to the LOS is then performed by simple constant false alarm rate thresholding. It is clear that the detection capability of this approach resides in the match-to-mismatch azimuth focusing gain, since no stationary background cancelation is performed. Therefore, only targets with high signal-to-clutter ratio (SCR) are expected to be reliably detected. The same approach can be extended to the multichannel SAR case, where a preliminary clutter cancelation step significantly improves the target-detection capabilities [19,20]. An experimental application of this approach, where target imaging is jointly performed together with multichannel clutter rejection, can be found in [21,22]. Also in this latter case, a bank of focusing filters is foreseen to account for different possible target along-track velocities. Also in [23,24], the problem of jointly combining moving target detection and imaging is addressed. In particular, in [24], the target motion parameters are estimated by means of Wigner–Ville distribution. An application of SAR-GMTI to the bistatic case can be found in [25]. In this chapter, we propose a different solution to detection and imaging of moving targets that is based on a simultaneous monostatic and bistatic acquisition. In particular, by properly defining the two geometries, targets moving perpendicular to the monostatic target-radar LOS (indistinguishable from stationary scene) will exhibit a Doppler modulation in the bistatic dataset different from the stationary scene thus being easily detectable, and vice versa (see Figure 9.1).

As a consequence, targets can be detected regardless of their direction of motion, with an evident improvement of the overall situational awareness over the observed area. Moreover, the two target velocity components on the ground can be estimated for those targets detected in both monostatic and bistatic datasets, which is the key information needed for correct moving target focusing.

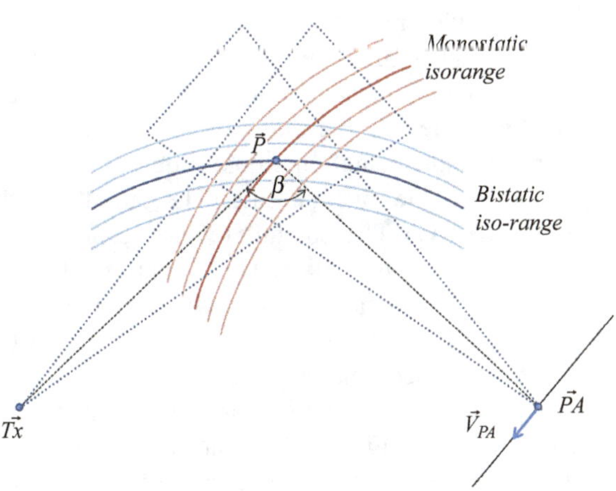

Figure 9.1 Crossing of monostatic and bistatic iso-ranges. © 2015 IEEE. After [26]

This chapter is organized as follows: in Section 9.2, the geometry for simultaneous monostatic and bistatic SAR-GMTI is defined, and in Section 9.3, the detection and localization performances of this approach are analyzed from a theoretical point of view. Section 9.4 is devoted to the problem of association between the monostatic and the bistatic detections. The overall processing scheme of the proposed technique is then described in Section 9.5. Section 9.6 presents some experimental results both concerning the detection and the imaging, and finally in Section 9.7, we draw our conclusions.

9.2 Geometry for joint monostatic and bistatic SAR-GMTI

The considered simultaneous monostatic and bistatic SAR-GMTI uses an active multichannel radar mounted on a flying moving platform, and an additional stationary transmitter. The multichannel radar should be able to receive simultaneously both its own echoes (monostatic), plus the bistatic returns sent by the stationary transmitter and reflected by the area under surveillance. The simplest way to obtain simultaneous orthogonal transmissions is to multiplex them in adjacent frequency bands. Although this solution is simple, it still requires sufficient receiver bandwidth to allow room for the two signals. This configuration allows both monostatic and bistatic multichannel GMTI by using a single active multichannel system and an additional, stationary transmitter. In this work, we only consider the case of a stationary transmitter. It is however evident that the same technique can be also applied by mounting the additional transmitter onto a moving platform. This second solution is expected to complicate the Doppler modulation of clutter returns in the bistatic data, but it clearly offers deployment and reconfiguration advantages in an operational scenario.

Moreover, alternative solutions for simultaneous monostatic and bistatic GMTI are possible by using only one transmitter and two (multichannel) receivers, one of which works in passive-only mode. Obviously, such an alternative solution would be absolutely equivalent in terms of geometry diversity gain, with the advantage of requiring only one transmitted signal. The use of one system in passive-only mode would certainly be interesting for covert operation. On the other hand, such an alternative implementation would require two multichannel receivers, which might increase the overall cost. The simultaneous multichannel monostatic and bistatic SAR-GMTI technique requires a proper definition of the geometry. In fact, the geometry has to guarantee both proper diversity between the monostatic and the bistatic acquisitions and LOS conditions over the area of interest. If a ground-based stationary transmitter is used, it can be conveniently located on top of a hill, but also solutions on a helicopter or on an airship are possible. A sketch of the resulting bistatic geometry is depicted in east-north-up (ENU) Cartesian coordinates in Figure 9.2. Antenna footprints on the ground for the ground-based transmitter and for the receiver are coloured in red and light blue, respectively. Defining the target motion on ground (i.e., (x, y) plane of Figure 9.2) as $\vec{V}_P = [\, V_{P,x} \quad V_{P,y} \quad 0]^T$ and knowing the target location $\vec{P}$, the corresponding

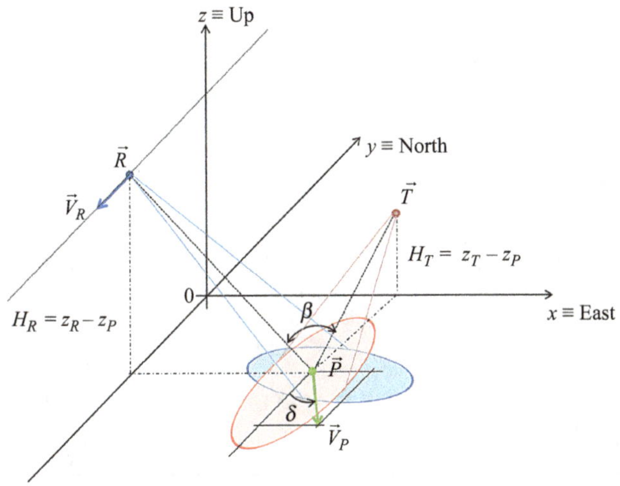

Figure 9.2 Bistatic geometry in a Cartesian coordinate system

monostatic and bistatic target Doppler frequencies are given by the following equation:

$$\begin{cases} f_{D,P}^{(m)} = \dfrac{2}{\lambda_m}\vec{u}_R^{\dagger}(\vec{V}_R - \vec{V}_P) \\ f_{D,P}^{(b)} = \dfrac{1}{\lambda_b}\vec{u}_R^{\dagger}(\vec{V}_R - \vec{V}_P) - \dfrac{1}{\lambda_b}\vec{u}_T^{\dagger}\vec{V}_P \end{cases} \tag{9.1}$$

where λ_m is the carrier wavelength for the monostatic signal, λ_b the bistatic one, $\vec{u}_R = (\vec{P} - \vec{R})/R_{\vec{P}\to\vec{R}} = [u_{R,x}\, u_{R,y}\, u_{R,z}]^{\dagger}$ is the LOS unit vector from the airborne platform position $\vec{R}$ to the target position $\vec{P}$ with $R_{\vec{P}\to\vec{R}}$ the distance from the platform to the target position and $\vec{u}_T = (\vec{P} - \vec{T})/R_{\vec{P}\to\vec{T}} = [u_{T,x}\, u_{T,y}\, u_{T,z}]^{\dagger}$ is the LOS unit vector from the transmitter position $\vec{T}$ to the target position $\vec{P}$. By solving (9.1), both target velocity components $V_{P,x}$ and $V_{P,y}$ can be estimated for those targets detected in the two datasets. The estimated target velocity vector, other than increasing the situational awareness of the moving targets in the scene, can be also exploited to provide correct SAR focusing of the movers, as will be shown in the following.

9.3 Detection and localization performances

In this section, the detection performance of the proposed technique is analyzed from a theoretical point of view. Simulations have been conducted for a geometry similar to the one sketched in Figure 9.2, and with reference to a sample idealized scenario similar to the one used in the real experiment described later in the paper. The main system parameters used for simulations are listed in Table 9.1. Specifically,

Table 9.1 Main system parameters

Parameter	Value
Bistatic angle, β	80 deg
Flight height w.r.t. scene, H_R	835 m
Transmitter height w.r.t. scene, H_T	80 m
Range from $\vec{R}$ to $\vec{P}$ ($R_{\vec{R}\to\vec{P}}$)	1,700 m
Range from $\vec{T}$ to $\vec{P}$ ($R_{\vec{T}\to\vec{P}}$)	1,250 m
PAMIR velocity, $\lvert\vec{V}_R\rvert$	100 m/s
PAMIR antenna beamwidth (azimuth)	6 deg
Stationary antenna beamwidth (azimuth)	27 deg
Parallel receiving channels, N	3
Pulse repetition frequency, PRF	5,000 Hz
Transmitted bandwidth (single sensor)	150 MHz
Number of pulses per CPI, M	512
Number of bins used in ABPD	3

the theoretical detection performances are evaluated in terms of signal-to-interference-plus-noise ratio (SINR) after STAP filtering for the monostatic and for the bistatic cases, respectively. The signal model and derivation to obtain the SINR is here derived for the monostatic case, while the derivation for the bistatic case is analogue and straightforward. We assume to have a receiving multichannel system equipped with N parallel receiving channels each one associated with an antenna sub-array, being the sub-arrays aligned in the direction of motion of the platform and forming a uniform linear array (ULA). The radar coherently combines the echoes from M consecutive pulses transmitted with a pulse repetition frequency $\mathrm{PRF} = 1/\mathrm{PRT}$. The single space-time snapshot of the cell under test (CUT) where the target is present is assumed to be in the form of a $MN \times 1$ vector which can be written as

$$\mathbf{x} = \alpha_P \mathbf{s} + \alpha_c \mathbf{c} + \mathbf{n} \tag{9.2}$$

where α_P is the complex amplitude of the target, $\mathbf{s}$ is the target space-time steering vector, α_c is the complex clutter amplitude, $\mathbf{c}$ is the clutter space-time component assumed Gaussian distributed with zero mean value and covariance matrix $\mathbf{R}_c$ and $\mathbf{n}$ is the thermal noise component also assumed Gaussian distributed with zero mean and covariance matrix $\mathbf{R}_n = \sigma_n^2 \mathbf{I}_{MN}$ (being σ_n^2 the thermal noise power, which is assumed 1 for simplicity in the following). The target space and time steering vectors can be written as follows:

$$\mathbf{a} = \left[\exp\left\{-j\frac{2\pi}{\lambda_m} n d u_R\right\}\right]_{n=1}^{N} \tag{9.3}$$

and

$$\mathbf{b}(\vec{V}_P) = \left[\exp\left\{j2\pi m f_{D,P}^{(m)} \mathrm{PRT}\right\}\right]_{m=1}^{M} \tag{9.4}$$

In the previous expressions, d is the interchannel distance, and u_R refers to the projection of the receiver-target LOS vector onto the ULA direction. In addition, $f_{D,P}^{(m)}$ refers to the monostatic Doppler frequency of the moving target as calculated in (9.1). As one can see, the Doppler frequency $f_{D,P}^{(m)}$ is directly dependent on the target velocity vector, namely on the velocity modulus V_P and on the direction of motion δ. This dependency is highlighted as $(\vec{V}_P)$ in the previous expression of the temporal steering vector **b**. The target space-time steering vector can be written as the Kronecker product of the spatial and temporal components as

$$\mathbf{s}(\vec{V}_P) = \mathbf{a} \otimes \mathbf{b}(\vec{V}_P) \tag{9.5}$$

where the dependency on the target velocity vector $(\vec{V}_P)$ is further made explicit. In the following, the adjacent bin post-Doppler (ABPD) STAP technique [27] is considered. To this end, let us define the following vector $\mathbf{d}(\vec{V}_P)$ as the $1 \times L$ vector containing the values of the L most adjacent Doppler bins to $f_{D,P}^{(m)}$ (modulo PRF). Then, the following $M \times L$ matrix can be defined

$$\mathbf{F}(\vec{V}_P) = \left[\exp\{j2\pi m\mathbf{d}(\vec{V}_P)\mathrm{PRT}\}\right]_{m=1}^{M} \tag{9.6}$$

From (9.6), the ABPD space-time transformation matrix for the Doppler bins of interest can be calculated as

$$\mathbf{T}(\vec{V}_P) = \mathbf{I}_N \otimes \mathbf{F}(\vec{V}_P) \tag{9.7}$$

and consequently the target space-time steering vector can be calculated as follows:

$$\mathbf{s}_T(\vec{V}_P) = \mathbf{T}(\vec{V}_P)^H \mathbf{s}(\vec{V}_P) \tag{9.8}$$

such as the clutter plus noise covariance matrix as

$$\mathbf{R}_T(\vec{V}_P) = \mathbf{T}(\vec{V}_P)^H (\mathbf{R}_c + \mathbf{R}_n)\mathbf{T}(\vec{V}_P) \tag{9.9}$$

The SINR after ABPD STAP filtering can then be expressed as a function of the target velocity vector $\vec{V}_P$ as

$$\mathrm{SINR}(\vec{V}_P) = |a_P|^2 \mathbf{s}_T(\vec{V}_P)^H \mathbf{R}_T(\vec{V}_P)^{-1} \mathbf{s}_T(\vec{V}_P) \tag{9.10}$$

Figure 9.3 reports the monostatic and bistatic SINRs for the considered experimental setup (defined as before, see also (110) in [2]), for an overall target signal-to-noise ratio (SNR) = 15 dB and for a clutter-to-noise ratio (CNR) of about 30 dB both in the monostatic and the bistatic cases. No adaptivity losses in the covariance matrix estimations have been considered.

It is interesting to see the deep notches present in the monostatic and bistatic SINRs for the target directions of motion tangential to the corresponding iso-ranges. These notches occur exactly for the directions tangential to the monostatic and

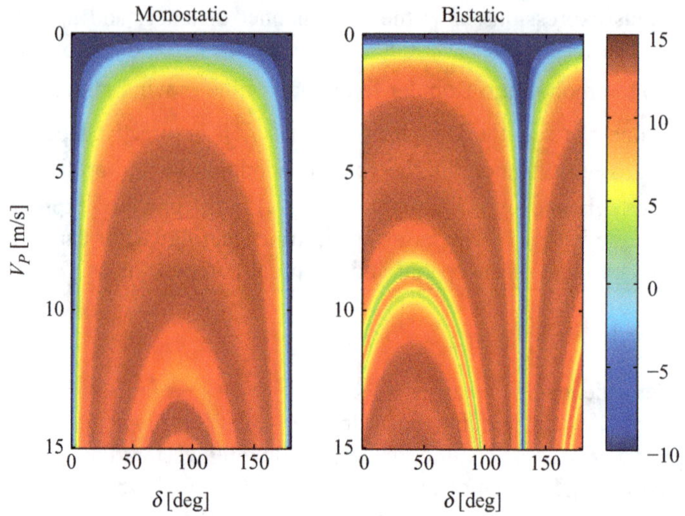

Figure 9.3 SINR after ABPD against target motion parameters [dB]. Colourbar has been limited in the interval [−10 dB, 15 dB] for better visualization

bistatic iso-ranges. By knowing the expected width of these notches, this analysis also helps in the definition of a proper bistatic geometry. As a result, the simultaneous processing of mono- and bistatic data guarantees, in principle, the detection of targets having a minimum velocity $V_P = ||\vec{V}_P||$ independently of their direction of motion δ. Moreover, there are large regions in the (V_P, δ) plane where the target can be detected both in the monostatic and in the bistatic datasets simultaneously.

The achievable target localization accuracy in both monostatic and bistatic configurations has been analyzed, and results are shown in Figure 9.4. Specifically, Figure 9.4 reports $3\sqrt{\sigma_{\text{CRB}}}R_{\vec{R}\to\vec{P}}$, where σ_{CRB} is the Cramer–Rao bound (CRB) of target DOA estimation, and it is obtained as follows (see also (34) in [28]) using the $N = 3$ parallel receiving channels, and a single bin post-Doppler STAP processing.[2]

Let us define as $\mathbf{a}_u$ the derivative of the space target steering vector with respect to the directional cosine u. The 'derivative' of the space-time target steering vector can be then written as

$$\mathbf{s}_u(\vec{V}_P) = \mathbf{a}_u \otimes \mathbf{b}(\vec{V}_P) \tag{9.11}$$

and after sub-optimal STAP reduction transformation as

$$\mathbf{s}_{u,T}(\vec{V}_P) = \mathbf{T}(\vec{V}_P)^H \mathbf{s}_u(\vec{V}_P) \tag{9.12}$$

[2]Please note that CRB are evaluated starting from single bin post-Doppler STAP, since this is the implementation used later on with real data in estimating the target DOA. On the other hand, the single-bin post-Doppler can be simply seen as the previously mentioned ABPD with $L = 1$.

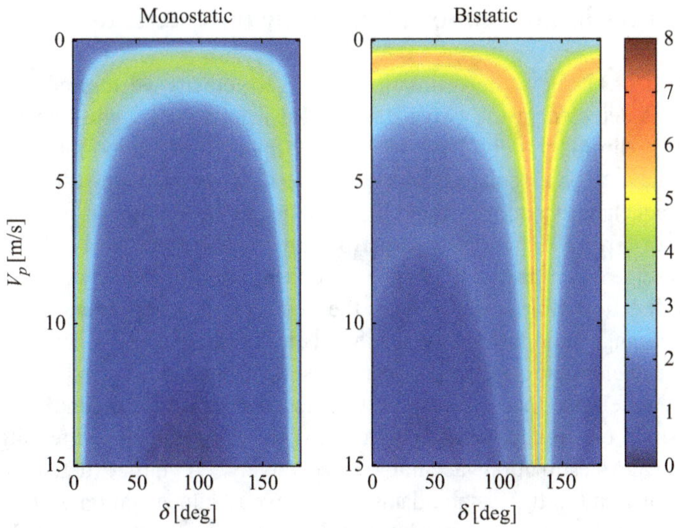

Figure 9.4 Localization accuracy in monostatic (left) and bistatic (right) cases [m]

By defining the following scalar quantities,

$$\begin{aligned} p &= \mathbf{s}_T(\vec{V}_P)^H \mathbf{R}_T(\vec{V}_P)^{-1} \mathbf{s}_T(\vec{V}_P) \\ q &= p\mathbf{s}_{u,T}(\vec{V}_P)^H \mathbf{R}_T(\vec{V}_P)^{-1} \mathbf{s}_{u,T}(\vec{V}_P) + \\ &\quad - \left|\mathbf{s}_{u,T}(\vec{V}_P)^H \mathbf{R}_T(\vec{V}_P)^{-1} \mathbf{s}_T(\vec{V}_P)\right|^2 \end{aligned} \tag{9.13}$$

the CRB can be calculated as

$$\sigma_{\text{CRB}} = \frac{1}{2|u_P|^2} \frac{p}{q} \tag{9.14}$$

For more details on the derivation of the CRB, the reader can refer to the expressions in [28].

In other words, Figure 9.4 reports the crossrange dimension of the localization ellipse where the target is located with a probability of 0.997 (i.e., 3σ), under the assumption that the error is Gaussian distributed. As is apparent, the level of accuracy is very high. This can be explained by several reasons. First, as is well known, the CRB gives an upper bound of the performance of any unbiased estimator. Second, no adaptivity losses in the estimation of the covariance matrices have been considered, which clearly deteriorate the achievable DOA estimation accuracy in the real scenario. Third, the target range is fairly small (see Table 9.1), which inevitably helps in the conversion of DOA estimate to localization on the ground. Despite these issue, it is reasonable to assume for a strong target (remember that we assumed a SNR = 15 dB) a real localization accuracy level comparable (if not better) than the expected size of the moving target (e.g., a car, a truck, or a tank).

9.4 Association of monostatic and bistatic detections

Relations in (9.1) allow the estimation of target velocity vectors components for those targets detected in the two datasets. However, to solve the linear system in (9.1), two preliminary steps have to be performed. First, monostatic and bistatic detections have to be associated. This cannot be done simply using target ranges and Doppler frequencies (monostatic and bistatic), due to the ambiguity between Doppler and angle of arrival. Second, the target position vector $\vec{P}$ has to be determined, to invert and solve (9.1) for $\vec{V}_P$. In [29], the two tasks are fulfilled by simply estimating the target direction of arrival (DOA) in the two datasets independently, and then by using target ranges and DOA estimates to geolocate the detection on the ground. In general, we declare a detection (being monostatic or bistatic) to be located within a corresponding ellipse (named from now on localization ellipse) centred in the estimated position on ground, and having dimensions proportional to the range resolution and DOA estimation accuracy, respectively. A target detected in both the monostatic and in the bistatic dataset will have two localization ellipses that can be associated if the two corresponding localization ellipses intersect, see Figure 9.5. An analysis of the geolocalization accuracy can be found in [29]. It should be noted that no data sharing between platforms is necessary, since both the monostatic and bistatic detections are obtained from a single receiving antenna.

This association principle is generally valid for isolated targets having sufficient strong SINR after GMTI filtering, while the association might be ambiguous if multiple close targets are present (we denote this case as densely populated scenario). Examples of densely populated target scenarios are shown in Figure 9.6 for crossing-targets and for one target passing another, respectively. In both cases, one should note that the small spacing between the targets compared to the ellipse

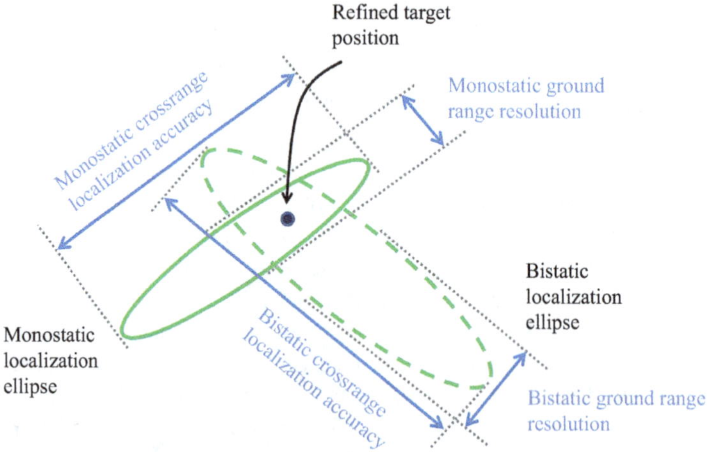

Figure 9.5 Monostatic and bistatic localization ellipses association in a single target case. © 2015 IEEE. After [26]

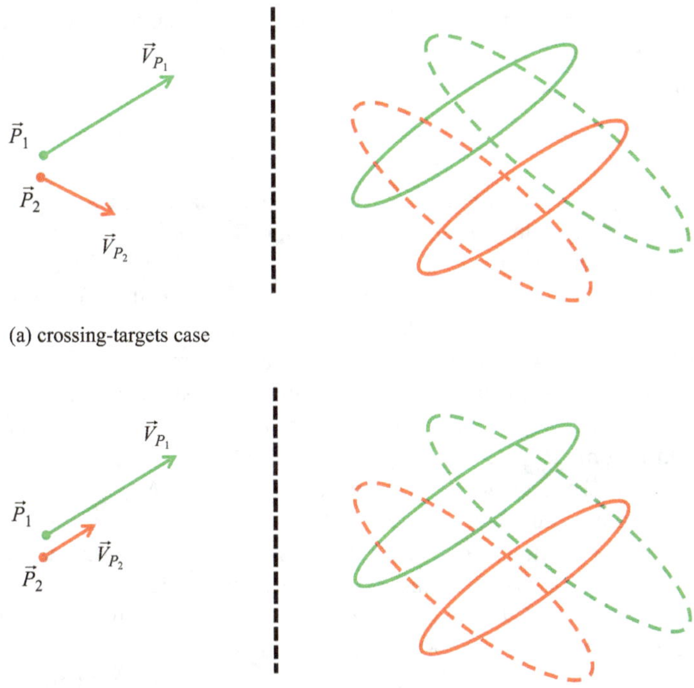

(a) crossing-targets case

(b) over-taking targets case

Figure 9.6 Examples of densely populated target scenarios. Dashed lines refer to bistatic location error ellipses, solid lines to monostatic location error ellipses. © 2015 IEEE. After [26]

dimension produces multiple intersections, thus making the detection associations ambiguous.

9.4.1 Direct association in densely populated target scenarios

As a first step, we extend and formalize the association principle described above to the generic situation of a densely populated target scenario, for the single coherent processing interval (CPI) case. The derived association logic has to take into account not only the potential presence of multiple closely spaced targets, but also the possibility that some missed detections may occur. As before, monostatic and bistatic detections are geolocalized prior to association.

The association can be seen as an assignment problem [30], Section 6.5, p. 342, where pairs of detections (or observations) are created. The creation of each pair has a cost that represents the likelihood that the two observations really originate from the same target. If multiple targets are present, the association problem can be solved by minimizing the overall cost. The costs of all possible pairs are represented in a cost matrix, where each row corresponds to a monostatic detection and each column corresponds to a bistatic detection. Thus, the matrix element (i,j)

gives the cost of assigning bistatic observation j to monostatic observation i. The cost for assigning bistatic detection j to monostatic detection i is taken to be the squared statistical distance

$$d_{i,j}^2 = \Delta\vec{x}_{i,j}^T\left(\mathbf{R}_{x,m} + \mathbf{R}_{x,b}\right)^{-1}\Delta\vec{x}_{i,j} \tag{9.15}$$

where $\Delta\vec{x}_{i,j} = \vec{x}_{m,i} - \vec{x}_{b,j}$ is the difference of the Cartesian positions of the monostatic and bistatic detections after geolocation and $\mathbf{R}_{x,m}$ and $\mathbf{R}_{x,b}$ are the respective geolocation error covariance matrices. This cost is similar to the square of the statistical distance between the predicted measurement and the measurement in a Kalman filter [30], Section 6.2.1, (6.7), p. 329. Given the cost matrix, the assignment problem can be solved with the Munkres algorithm [31]. It is important to notice that the solution of the Munkres algorithm will associate each monostatic detection to one and only one bistatic detection, leading to a minimum of the sum $\sum_i d_{i,j}^2$, even if some $d_{i,j}^2$ are very large and it is unlikely that they originate from the same target. To prevent this erroneous association, N_m dummy detections should be added as columns to the cost matrix, where N_m is the number of monostatic detections. If a monostatic detection is associated with a dummy detection, this shall signify that no matching bistatic detection could be found, i.e., no assignment. The cost of forming a pair between a monostatic detection and a dummy detection is set to c_{na} (cost of no assignment).

To find a numerical value of c_{na}, we consider the case where there are no false alarms, that a single target is present, and that the target is detected both in the monostatic and bistatic dataset. In this case, the detections are correctly associated with each other, if $d_{1,1}^2 < c_{\text{na}}$. Assuming that the geolocation error is Gaussian, the squared statistical distance $d_{1,1}^2$ obeys a chi-square distribution with two degrees of freedom [32], Section 1.4.17, p. 57. Thus, the probability of correct association P_{ca} is given by $P_{\text{ca}} = F(c_{\text{na}}, 2)$, where F is the cumulative probability function of the chi-square distribution. Choosing a probability for the correct association (for this specific case) and solving this equation yields a value of c_{na}.

For illustration, Table 9.2 presents the structure of the cost matrix for a case with two monostatic and three bistatic detections. As in this example, some monostatic and/or bistatic observations may be left unassigned, either because they cannot reasonably originate from the same target or because the number of monostatic and bistatic observations is not equal.

Table 9.2 Structure of a cost matrix for the association of three bistatic detections with two monostatic detections

	Bistatic detections			**Dummy detections**	
Monostatic detections	1	2	3	1	2
1	$d_{1,1}^2$	$d_{1,2}^2$	$d_{1,3}^2$	c_{na}	c_{na}
2	$d_{2,1}^2$	$d_{2,2}^2$	$d_{2,3}^2$	c_{na}	c_{na}

Before proceeding, several comments are in order. First, for each monostatic detection, we tried to find a matching bistatic one. It is clear that a dual association scheme (i.e., for each bistatic detection find a matching monostatic one) would have been equivalent from a conceptual point of view. However, this might have an impact on the dimension of the cost matrix if one dataset has significantly more detections than the other. Second, one should notice that this direct association approach can be readily extended to multiple successive CPIs, each of them having different sets of monostatic and bistatic detections. By doing so, the successive CPIs will be treated independently of each other, performing independent associations from one CPI to the next. Clearly, this does not consider that targets are likely to have similar positions and velocities in successive CPIs. By using this latter temporal correlation, a better exploitation of the information coming from successive CPIs might help in correctly associating target detections. For this reason, in the next section, a tracking-based association scheme is presented, while the direct association will be considered as a benchmark for performance analysis.

9.4.2 Association by target tracking

In this second association scheme, we aim to better exploit the information coming from multiple successive CPIs, in order to better solve the association problem. To do this, we rely on a tracking algorithm which takes as inputs both the monostatic and the bistatic detections (after geolocation) and tries to extract target tracks. In other words, detections are not firstly associated and then eventually passed to a tracker (as in the case of the direct association), but tracks are extracted having as joint inputs monostatic and bistatic detections. The situation is sketched in Figure 9.7. Among different tracking algorithms, we propose the global nearest neighbour (GNN) [30], Section 6.4, because it provides explicit association of detections to tracks. In the considered implementation of GNN, we consider multidimensional target states comprising the two-dimensional target location and the two-dimensional target velocity vector. The state propagation and update of the state by measurements is identical to that of an (extended) Kalman filter. At every time step (every CPI in our case), the target state is firstly predicted (or propagated) using *a priori* information and then updated according to the measurements of the current CPI. The updated target state is then used to make the

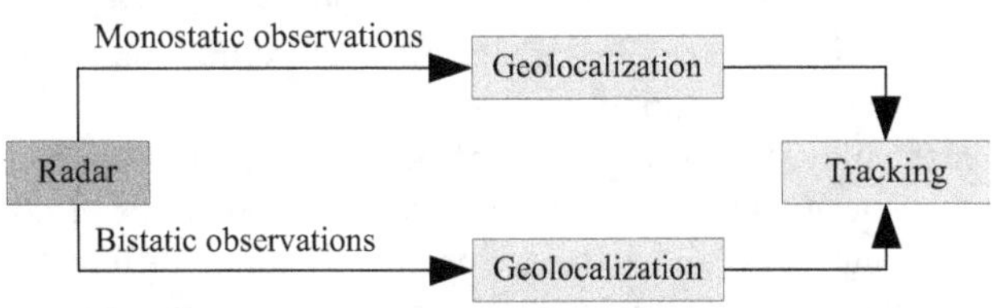

Figure 9.7 Processing flowchart for 'association by target tracking'. © 2015 IEEE. After [26]

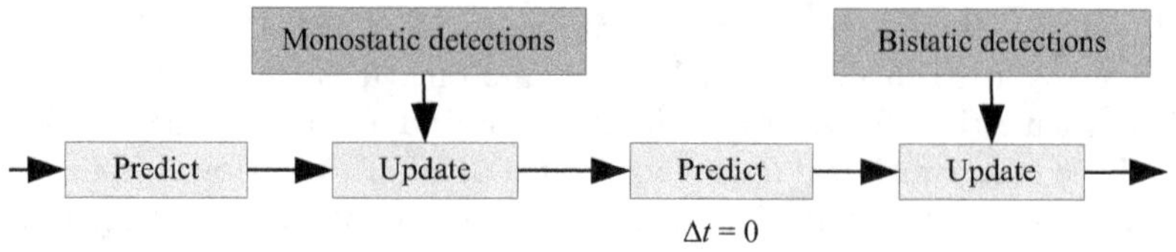

Figure 9.8 Flowchart of the tracking algorithm. One cycle of the recursion. © 2015 IEEE. After [26]

prediction for the next time step and so on. In order to account for simultaneous monostatic and bistatic detection, the single recursion of the GNN is slightly modified, as illustrated in Figure 9.8. In particular, the target state is predicted using *a priori* knowledge, and then a first update is calculated using the information of the monostatic detections. This updated target state is not propagated, but it is updated a second time using now the bistatic detections. The lack of propagation between the first and the second updates accounts for the simultaneity between monostatic and bistatic observations. In the practical implementation, this lack of propagation can be simply realized by propagating the target state for a time interval $\Delta t = 0$ s.

9.4.3 Simulation results

In bistatic target detection and tracking, the performance depends greatly on the acquisition geometry. It is therefore difficult to give universally valid performance measures. Instead, this section uses a case study to provide a comparison of the two association methods.

This case study deals with a realistic geometry which was in fact encountered during an experimental acquisition with PAMIR [33]. The transmitter is stationary on a hill and illuminates a road which is located in a valley below the transmitter. An overview of the scenario is given in Figure 9.9. Two targets (namely P_1 and P_2) moving on adjacent lanes of the road are simulated for 20 s when the overtaking takes place. The velocity of target P_1 is $V_{P_1} = 40$ m/s (144 km/h), while target P_2 is moving at $V_{P_2} = 30$ m/s (108 km/h). The distance between the lane centres is 3.75 m. At the start of the simulation, the overtaking target P_1 is trailing target P_2 by 91 m. The targets localization ellipses after 9 s of simulation time are shown in Figure 9.10 both for the monostatic (solid line) and for the bistatic (dashed line). From Figure 9.10, it is clear that the targets are so close that incorrect assignments of measurements may occur during the overtaking.

For this case study, the performance of both the direct association and of the tracking-based association is evaluated in terms of probability of correct assignment (P_{ca}) via Monte Carlo simulations. Specifically, three different situations are considered: (a) both targets P_1 and P_2 are detected in both monostatic and bistatic datasets; (b) target P_1 is detected only in the monostatic dataset, while target P_2 is detected only in the bistatic dataset; (c) target P_1 is detected in both datasets, while

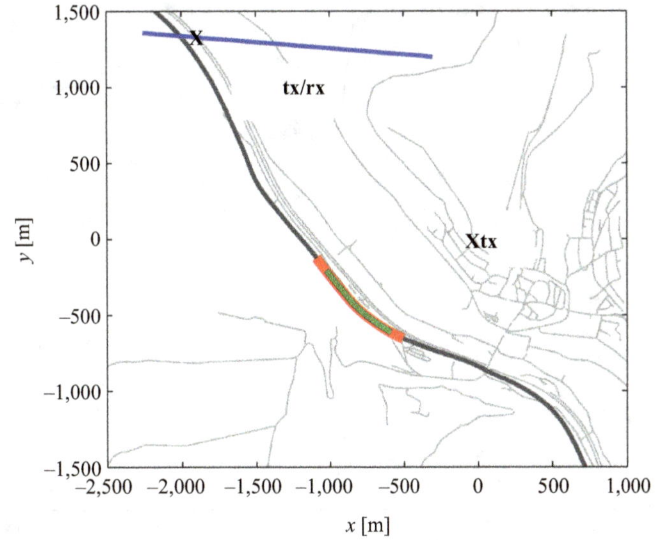

Figure 9.9 Acquisition geometry of the considered case study. © 2015 IEEE. After [26]

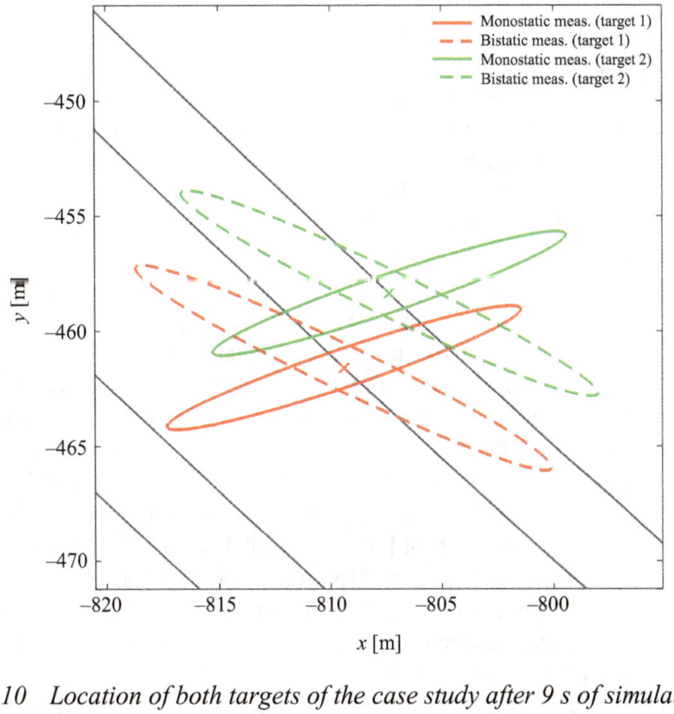

Figure 9.10 Location of both targets of the case study after 9 s of simulation time. The bistatic and monostatic location measurements will lie with 90% probability within the respective error ellipse. © 2015 IEEE. After [26]

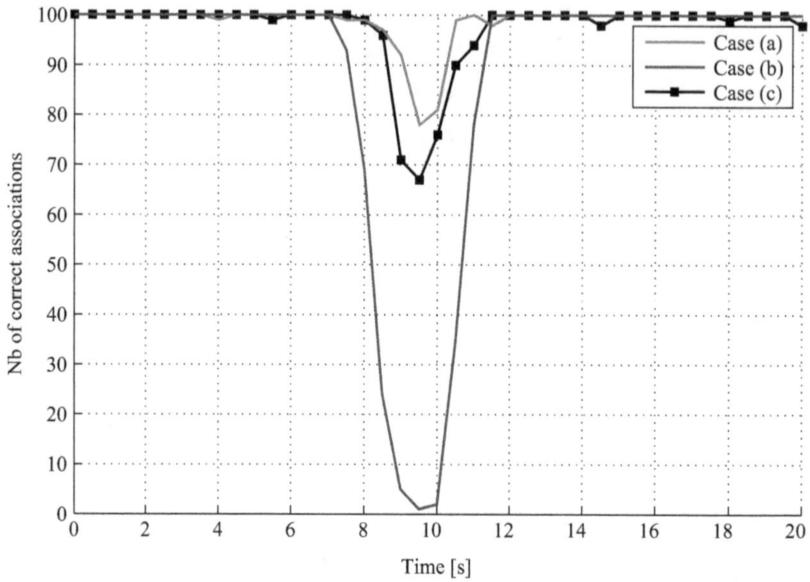

Figure 9.11 Percentage P_{ca} of correct association for the direct assignment method of case (a), (b) and (c).© 2015 IEEE. After [26]

target P_2 is detected only in the monostatic data. In the Monte Carlo simulation, the geolocated monostatic position measurement and the geolocated bistatic position measurement are random variables. As long as the direct association is concerned, the cost of no assignment was set to $c_{\mathrm{na}} = 13.8$ which corresponds to $P_{\mathrm{ca}} = 0.999$.

Figure 9.11 shows the probability of correct association P_{ca} against simulation time for the direct association strategy. As is apparent, P_{ca} is virtually one for all three cases both at the beginning and at the end of the simulation when the targets are far apart. However, during the process of overtaking at around 9 s of simulation time, the targets are close to each other and P_{ca} drops. In case (a) P_{ca} drops only to 0.8, since both measurements of a target are available. In case (b), however, only one measurement for each target is available, and thus the other measurement is incorrectly assumed to originate from the same target. Case (c) is an intermediate case.

The same Monte Carlo simulation has also been repeated for the tracking-based association scheme. For the assignment of observations to tracks, the distance measure as given in (6.29) of [30] was used. The cost of starting a new track for a given observation was set to 50. The track to observation association that is performed by the GNN-tracking algorithm can be converted to an assignment between monostatic and bistatic observations. In an analysis with 100 Monte Carlo runs, it was found that the assignment was always correct.

The superior performance of the tracking-based association scheme is due to the fact that observations are not assigned directly to each other but to tracks.

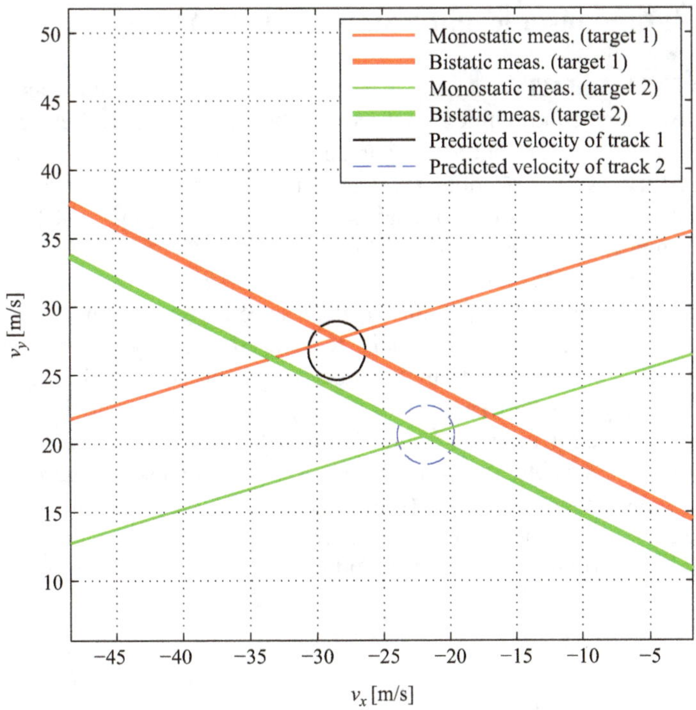

Figure 9.12 Visualization of the velocity measurement and association by target tracking: the velocity vector of target P_1 is confined to a position on the solid, red line by the monostatic radial velocity measurement and to a position on the dashed, red line by the bistatic range rate measurement. Thus, its true velocity vector is given by the intersection of both lines. The same reasoning holds for target P_2. In a realistic scenario, confusion arises due to the presence of two targets, since the association is unclear. In association by tracking, this problem is alleviated by the predicted target state. As this example shows, a target detection can be associated with a track based on the velocity measurement. © 2015 IEEE. After [26]

As tracks come with an estimate of the velocity, in addition to the location of the observation also the velocity measurement can be used for the assignment (see Figure 9.12). In this case study, the measured velocity is different between target P_1 and P_2 and is thus successfully used in the tracking algorithm for the target assignment. Furthermore, when tracking is used, the predicted target location is more accurate than using a single target observation. Thus, in association by target tracking, the task is to assign an inaccurate observation to an accurate prediction while two inaccurate observations have to be assigned to each other in the direct association scheme.

9.5 Joint monostatic and bistatic SAR-GMTI

The overall SAR-GMTI processing is sketched in Figure 9.13. Simultaneous short CPIs are extracted from multichannel monostatic and bistatic datasets. STAP processing is performed for both datasets in parallel, by using sub-optimal algorithms like the ABPD. Target detections are declared by thresholding test statistics of the STAP filtered data (in our case, the adapted matched filter, AMF [34]). After that, DOA is estimated for each detection. This can be conveniently done by resorting to so-called AB-STAP [35–37].

Now, using range and DOA information, detected targets can be relocated on ground and then associated according to one of the association methods of Section 9.4. Based on the data from the associated monostatic and bistatic detections, the corresponding target velocity vector is computed by solving the linear system in (9.1). This estimate can be used to correctly focus the moving target in the monostatic and in the bistatic datasets, as it is shown on the left and right sides Figure 9.13. Here, the entire monostatic and bistatic (single-channel) SAR datasets are processed to image the stationary scene, by using any conventional SAR focusing technique (e.g., chirp scaling algorithm [38], $\omega - k$ algorithm [39], backprojection [40] and so on).

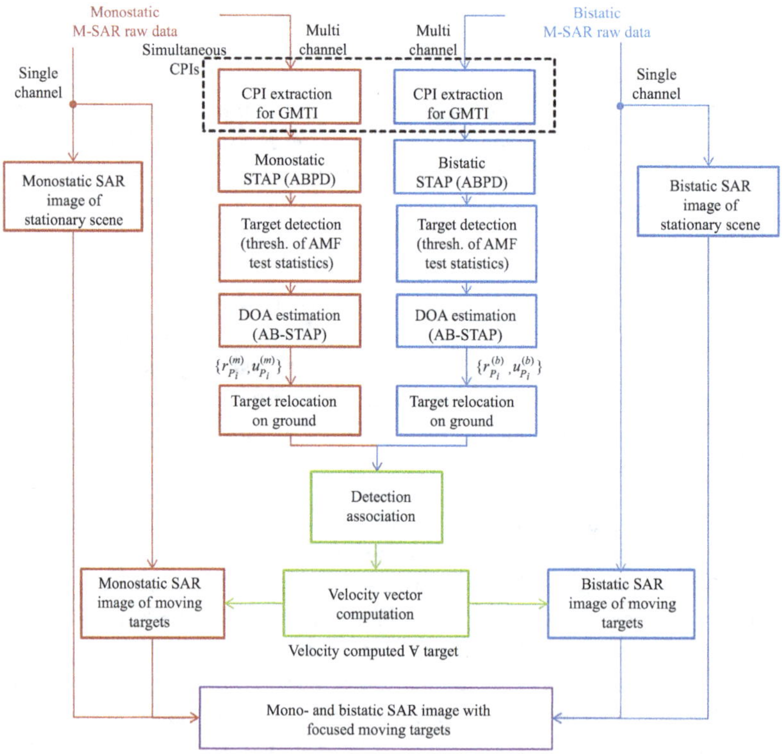

Figure 9.13 Processing block diagram

The same monostatic and bistatic single-channel SAR datasets are then reprocessed for each detected moving target, but now matching the SAR focusing to the estimated target movement. The final result will then be the SAR image of the stationary scene (both monostatic and bistatic) together with relocated and focused images of the moving targets. Here, the correct imaging of the moving targets has a double value. First, it proves the correctness of the moving target velocity vector estimation, since moving target imaging can be done only if target motion is accurately known. Second and more important, simultaneous moving target imaging in both monostatic and bistatic geometries opens the doors to target classification. Even if the resolution might be lower compared to ISAR systems, the size of the moving target can be estimated with good accuracy. This information together with the estimated target velocity vector and with the SAR image of the stationary scene can greatly improve target classification capabilities.

9.6 Experimental results

A simultaneous multichannel monostatic and bistatic SAR-GMTI experiment has been conducted in Enkering, Germany. A stationary ground-based transmitter is located on top of a hill illuminating the valley below with a depression angle of about 4 deg. The transmitter is equipped with a horn antenna of about 27 deg beamwidth which has a fixed steering during the experiment. Figure 9.14 shows a picture taken from the transmitter position, while Figure 9.15 shows the road map of the valley taken from OpenStreetMap transformed in ENU coordinates.

Figure 9.14 View of the scene from the transmitter position. © 2014 VDE VERLAG, Berlin, Offenbach. After [41]

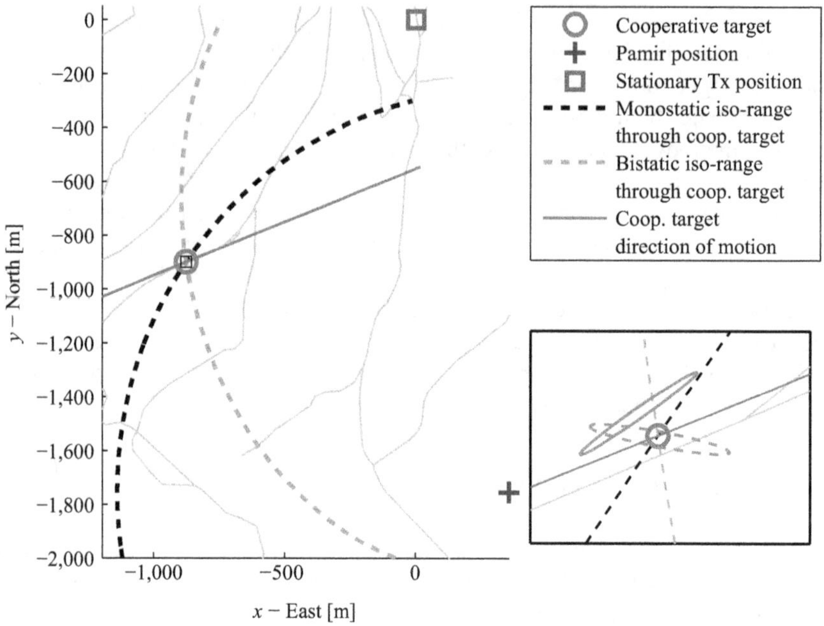

Figure 9.15 Road map of the observed scene in ENU. Zoom on cooperative target with monostatic (solid) and bistatic (dashed) localization ellipses

As a receiver, we used the multichannel system PAMIR mounted on the Transall aircraft in a pure side-looking configuration. Due to wind on the day of the experiment, a crab angle of about 7 deg is experienced. The aircraft flew almost parallel to the orographic orientation of the valley. The main system parameters are the ones already listed in Table 9.1. Both radars transmit simultaneously on adjacent frequency bands slightly above 9 GHz. PAMIR collects with three parallel receiving channels both the monostatic and the bistatic returns. Interchannel calibration has been preliminary conducted separately on both monostatic and bistatic multichannel datasets. The multichannel received spectra at intermediate frequency before calibration and bistatic data synchronization are shown in Figure 9.16. As is apparent, the spectrum is symmetrical, due to the real nature of the signal. Moreover, the monostatic and the bistatic signals are clearly visible on adjacent frequencies. Specifically, the monostatic signal is centred at about 125 MHz, while the bistatic one is centred at 325 MHz. Transmitted signal power in the monostatic case has been preliminarily scaled in order to guarantee a comparable level of the monostatic and bistatic signals in reception, thus avoiding problems with the dynamic range of the analogue-to-digital converters. Despite that, the monostatic signal still appears stronger than the bistatic one. One can also clearly appreciate the different levels of the signals on the three receiving channels, which have to be properly taken into account in the preliminary interchannel calibration stage. Local oscillators at the ground-based stationary transmitter and at the receiver are freely

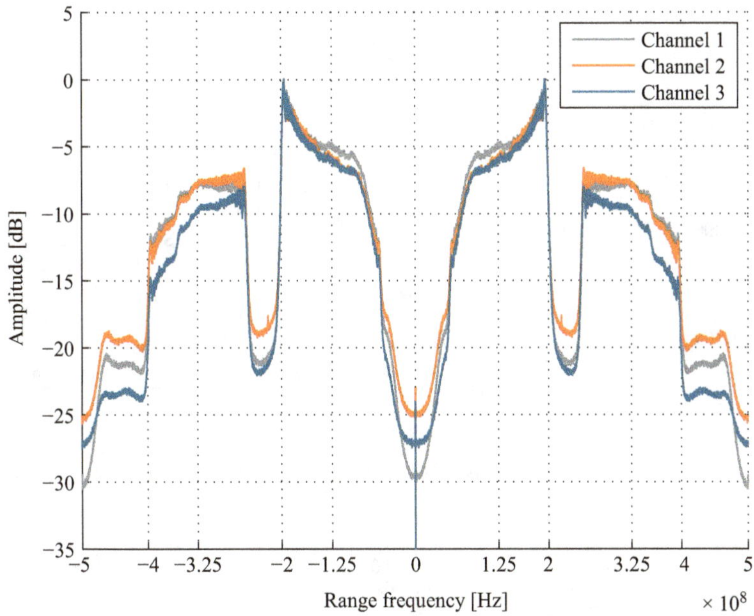

Figure 9.16 Monostatic and bistatic signal spectrum at intermediate frequency

running. For this reason, bistatic data need proper synchronization before processing. A detailed description of the synchronization procedure is reported in the Appendix A.

A cooperative moving target (namely a car, BMW Series 3 Touring) has been used during the experiment and equipped with high accuracy GPS device. This cooperative target moves along a road, see Figure 9.15 (the same road is also clearly visible in Figure 9.14). Figure 9.15 also shows the position of the PAMIR, of the stationary transmitter, as well as the monostatic and the bistatic iso-ranges passing through the cooperative target position. In addition, two synthetic targets (namely S1 and S2) have been inserted in the monostatic and in the bistatic datasets, close to the cooperative target, but moving along the monostatic and the bistatic iso-range lines, respectively (Table 9.3). A sketch of the geometries and of the directions of motion of the synthetic targets is reported in Figures 9.17 and 9.18 superimposed onto the road map of the Enkering valley.

9.6.1 Moving target detection

A CPI of $M = 512$ pulses has been extracted simultaneously for both datasets, according to the block diagram in Figure 9.13. CPI duration ensures an effective STAP filtering via ABPD and also avoids target range migration through several range/Doppler cells. After performing multichannel data calibration and range compression, the range/Doppler maps look as in Figures 9.19 and 9.20, where the position of the cooperative moving target has been highlighted in the circle

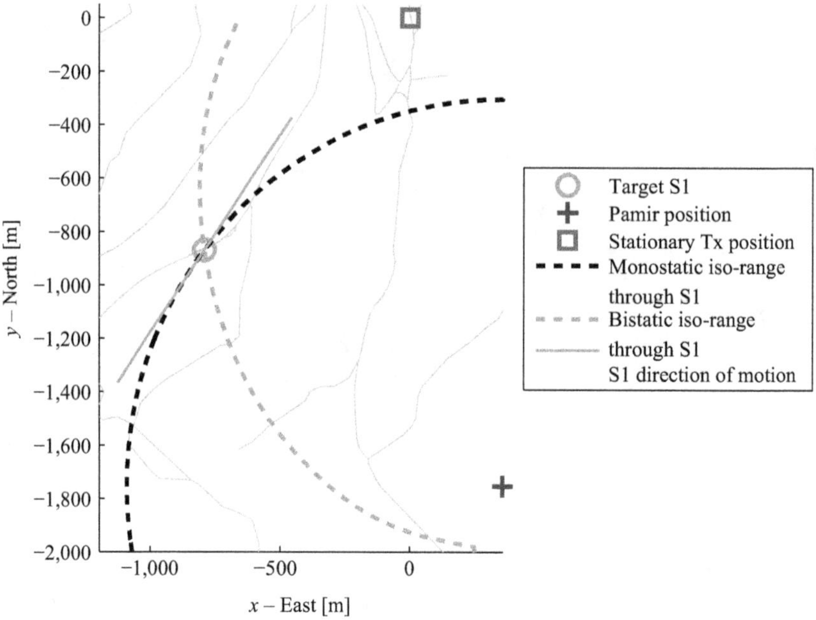

Figure 9.17 Position and direction of motion of synthetic target S1 in ENU

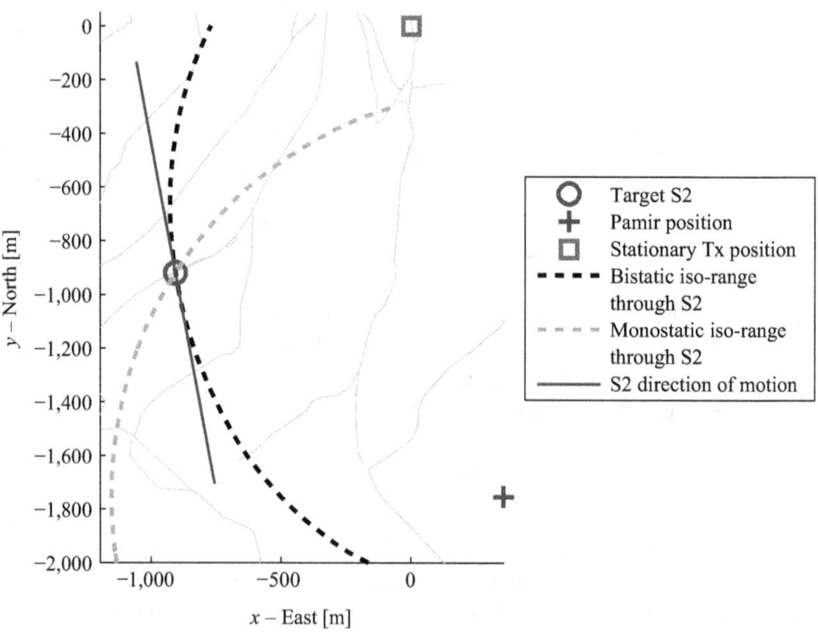

Figure 9.18 Position and direction of motion of synthetic target S2 in ENU

Table 9.3 Parameters for the synthetic targets

Parameter	Target S1	Target S2
$R^{(m)}$ (m)	1,672.9	1,728.7
$R^{(b)}$ (m)	2,852.2	3,022.2
$f_{D,P}^{(m)}$ (Hz)	−252.1	378.3
$f_{D,P}^{(b)}$ (Hz)	−317.1	49.2
Monostatic SNR (dB)	~19	~19
Bistatic SNR (dB)	~19	~19
$\vec{P}$ (m)	$\begin{bmatrix} -792.0 \\ -869.7 \\ -82.0 \end{bmatrix}$	$\begin{bmatrix} -906.0 \\ -919.7 \\ -82.0 \end{bmatrix}$
$\vec{V}_P$ (m/s)	$\begin{bmatrix} -3.37 \\ -4.97 \\ -0.03 \end{bmatrix}$	$\begin{bmatrix} 1.51 \\ -7.86 \\ -0.05 \end{bmatrix}$

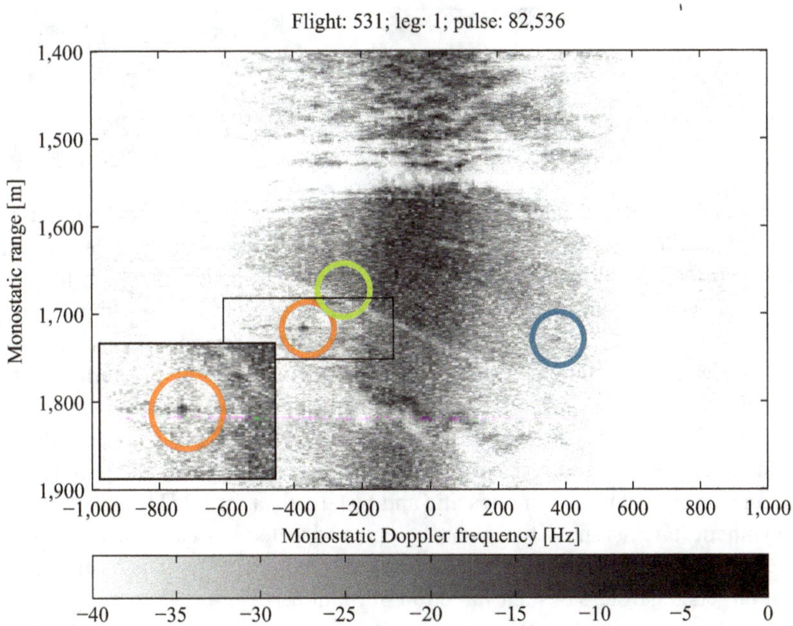

Figure 9.19 Monostatic range/Doppler map before STAP [dB]. Orange circle: cooperative target; green circle: target S1; blue circle: target S2

(see also zoom in the bottom left), for both the monostatic and the bistatic datasets. As is apparent, the target is located on the skirts of both the monostatic and of the bistatic clutter region, thus requiring STAP for proper detection.

ABPD STAP has been applied separately to the two datasets. Interference covariance matrices have been estimated by means of sliding window training by using as

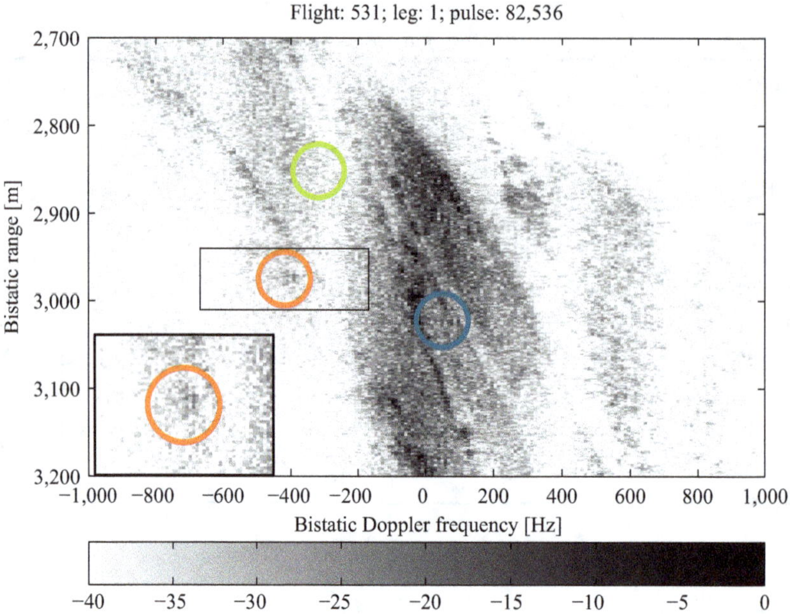

Figure 9.20 Bistatic range/Doppler maps before STAP [dB]. Orange circle: cooperative target; green circle: target S1; blue circle: target S2

secondary data 80 range gates adjacent to the CUT. Moreover, the ten nearest range gates to the CUT have been considered as guard gates, and thus excluded from secondary data. STAP filter outputs consist of AMF test statistics, which have been compared to a fixed threshold of 15 and 13 dB to declare target detections in specific range/Doppler cells for the monostatic and bistatic case, respectively. AMF test statistics after STAP filtering are shown in Figures 9.21 and 9.22 with overlaid red dots indicating the detections. The cooperative target is detected in both datasets, while target S1 is detected only in the bistatic dataset, and target S2 is detected only in the monostatic dataset. This example clearly shows the advantages of the proposed joint simultaneous exploitation of monostatic and bistatic data for GMTI. In particular, the two synthetic targets show the overall improved detection capability that can be reached only jointly exploiting the two dataset simultaneously, thus ensuring detection of targets regardless of their direction of motion.

In addition, targets like the cooperative one which is detected in both datasets can be also imaged. To do this, DOA estimation of these targets is conducted in both datasets using AB-STAP. Target parameter estimation (in particular, target range and DOA) allows target relocation on the ground, leading to the two localization ellipses depicted in Figure 9.15. As is apparent, the two localization ellipses intersect and, from the few detections in Figures 9.21 and 9.22, we are clearly in a *single target case* scenario. This allows monostatic and bistatic detections association according to previous considerations. Table 9.4 shows a comparison of the estimated cooperative target parameters with analogous quantities directly

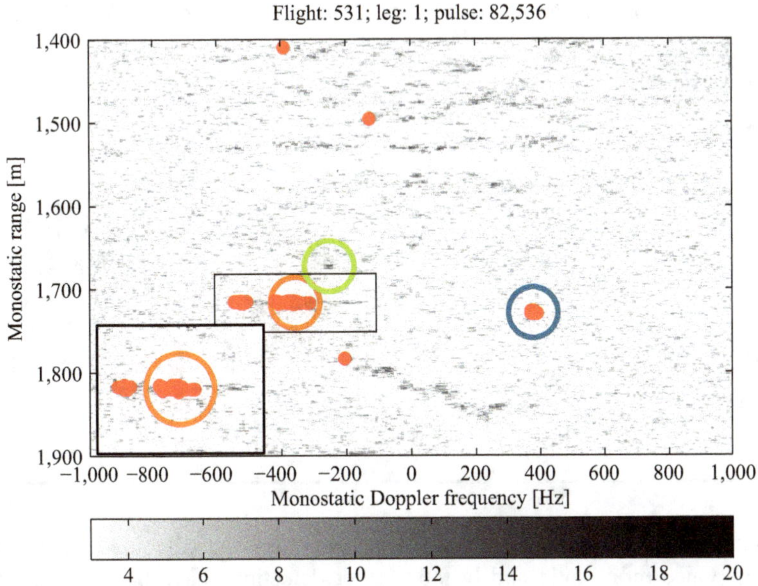

Figure 9.21 Monostatic range/Doppler maps after STAP [dB], with overlaid detections (red dots). Orange circle: cooperative target; green circle: target S1; blue circle: target S2

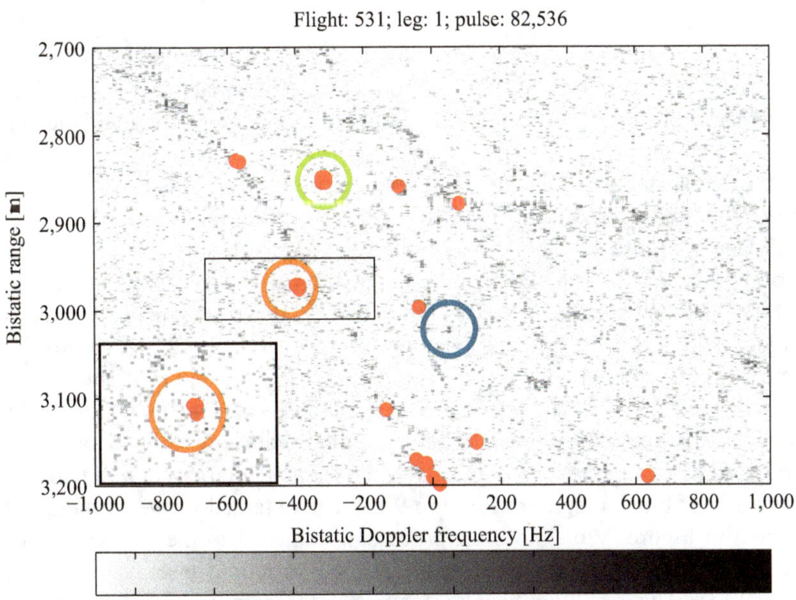

Figure 9.22 Bistatic range/Doppler maps after STAP [dB], with overlaid detections (red dots). Orange circle: cooperative target; green circle: target S1; blue circle: target S2

Table 9.4 Parameters for the cooperative target (n.a. = not applicable)

<table>
<tr><th>Target parameter</th><th>Measure GPS</th><th>Monostatic estimate</th><th>Bistatic estimate</th></tr>
<tr><td>$R^{(m)}$ (m)</td><td>1,716.7</td><td>1,716.5</td><td>n.a.</td></tr>
<tr><td>$R^{(b)}$ (m)</td><td>2,975.1</td><td>n.a.</td><td>2,976.0</td></tr>
<tr><td>$f_{D,P}^{(m)}$ (Hz)</td><td>−320.1</td><td>−330</td><td>n.a.</td></tr>
<tr><td>$f_{D,P}^{(b)}$ (Hz)</td><td>−418.3</td><td>n.a.</td><td>−410.2</td></tr>
<tr><td>u_{DOA}</td><td>0.0759</td><td>0.0791</td><td>0.0802</td></tr>
<tr><td>$\vec{P}$ (m)</td><td>$\begin{bmatrix}-876.0\\-900.0\\-82.0\end{bmatrix}$</td><td>$\begin{bmatrix}-879.7\\-895.5\\-82.0\end{bmatrix}$</td><td>$\begin{bmatrix}-875.6\\-900.4\\-82.0\end{bmatrix}$</td></tr>
<tr><td>$\vec{V}_P$ (m/s)</td><td>$\begin{bmatrix}-8.93\\-3.54\\0.05\end{bmatrix}$</td><td colspan="2">$\begin{bmatrix}-8.65\\-2.22\\0\end{bmatrix}$</td></tr>
</table>

calculated from GPS measurements. Target positions and velocities are expressed in ENU. Once monostatic and bistatic target detections have been relocated and associated, the corresponding Doppler frequencies can be used to solve the linear system in (9.1), and to estimate the two target velocity components $V_{P,x}$ and $V_{P,y}$. The estimated target velocity vector is then exploited to provide correct SAR focusing of the movers, as will be shown in the following.

9.6.2 Imaging

The imaging is performed using monostatic and bistatic single-channel datasets. A backprojection processor [40] is used to focus the stationary scene. The obtained SAR image is presented in Figures 9.23 and 9.25 for the monostatic and the bistatic cases, respectively. A valley (see Figure 9.14) with a road from the north to south-west can be clearly seen. Shadows are evident both in the monostatic and in the bistatic SAR images. Specifically, observing the trees at the centre on the bistatic SAR image of Figure 9.25, both shadows from the transmitter and from the receiver can be clearly recognized. During the acquisition time, the cooperative moving target travelled down the main road in the middle section of the image (red rectangle in Figure 9.23 and corresponding enlarged zoom in Figure 9.24). The position of the mover is highlighted with a white circle in Figure 9.24, and its motion direction is also depicted with a white arrow. The unfocused and shifted moving target response is marked with a circle. The same monostatic raw SAR dataset is then reprocessed matching the SAR focusing to the estimated target motion parameters, see left part of block diagram in Figure 9.13. The obtained image of the mover is presented in Figure 9.26. The SAR processing matched to the estimated moving target motion is also applied to the single-channel bistatic dataset (see right part of Figure 9.13). The resulting bistatic focused SAR response of the cooperative moving target is shown in Figure 9.27. Combining the monostatic and the bistatic moving target imaging capability, information on the size of the target can be

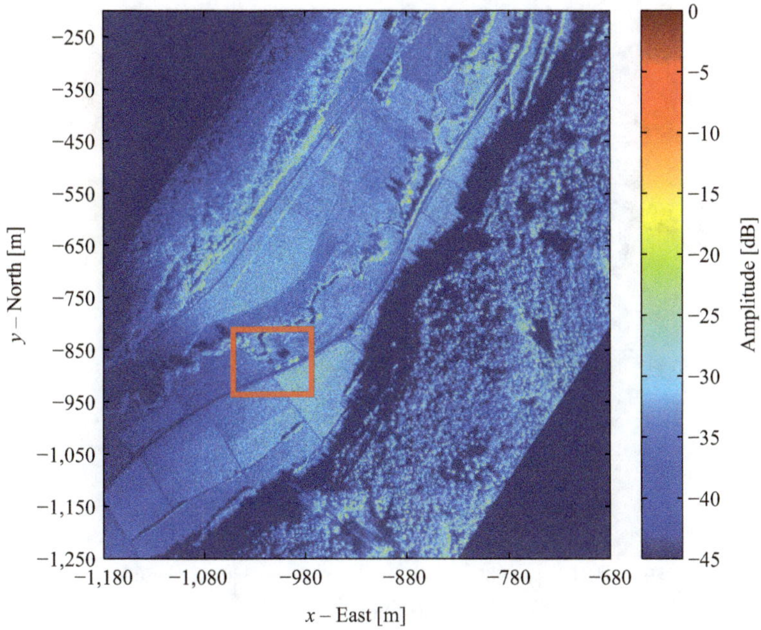

Figure 9.23 Monostatic SAR image in Cartesian ENU coordinates [dB]. © 2014 VDE VERLAG, Berlin, Offenbach. After [41]

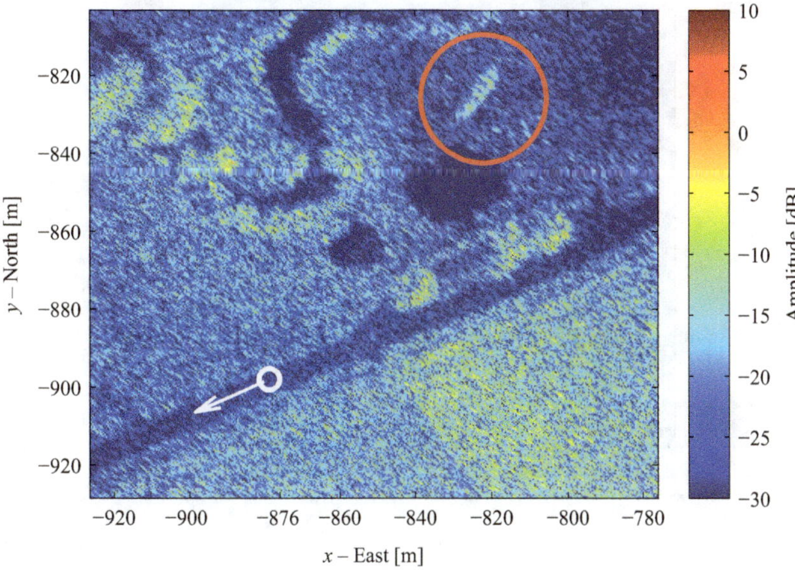

Figure 9.24 Monostatic SAR image—zoom over the unfocused moving target [dB].© 2014 VDE VERLAG, Berlin, Offenbach. After [41]

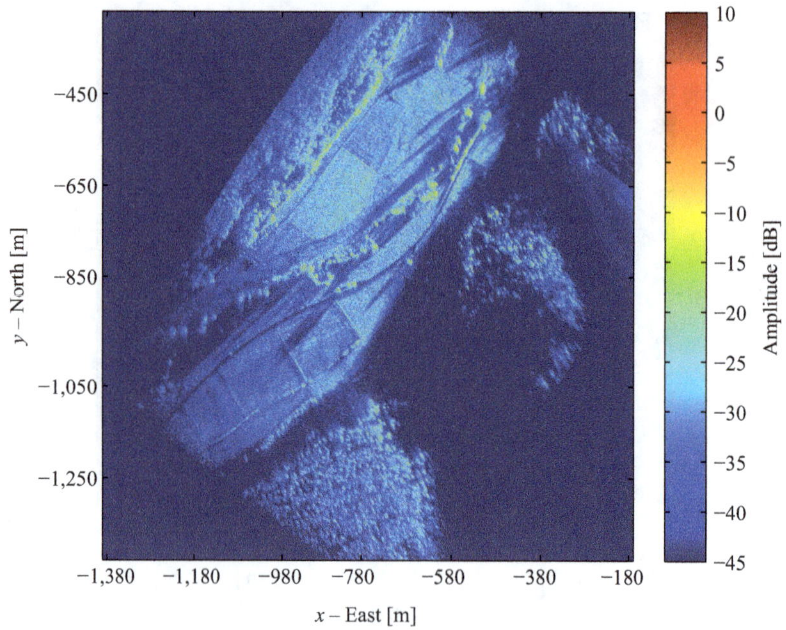

Figure 9.25 Bistatic SAR image in Cartesian ENU coordinates [dB]

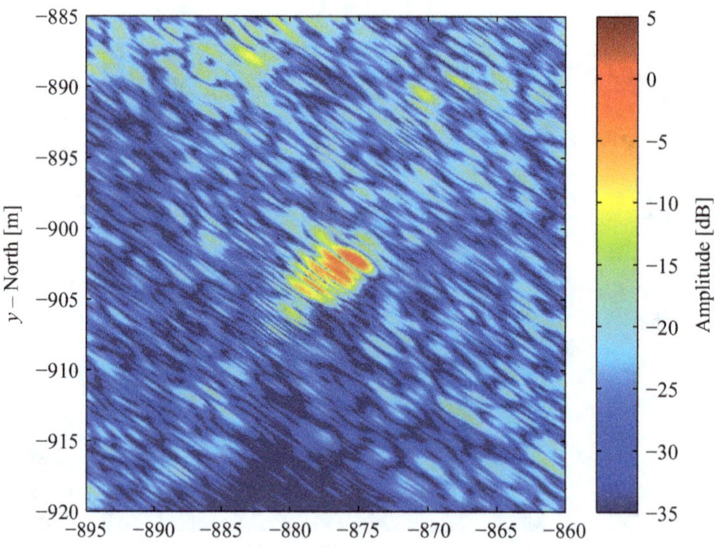

Figure 9.26 Monostatic SAR image of the moving target in ENU coordinates processed matching estimated moving target motion parameters [dB]

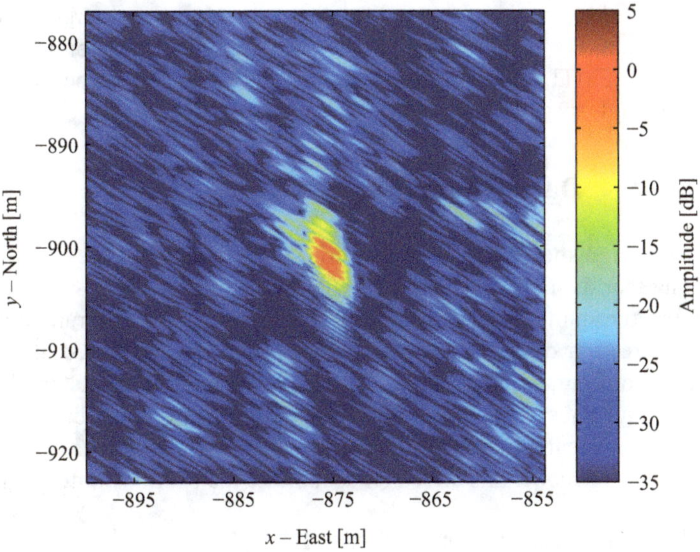

Figure 9.27 Bistatic SAR image of the moving target processed matching estimated moving target motion parameters [dB]

retrieved, which might help with target identification. Please note that the shift between the position of the unfocused moving target (red circle), and its true position (white circle) in Figure 9.24 occurs in the along-track direction. This direction does not appear horizontal in the figure, since the image has been projected in the ENU Cartesian coordinates of Figure 9.2.

9.7 Conclusions

In this chapter, we presented a simultaneous monostatic and bistatic multichannel technique for improved SAR-GMTI. This can be simply obtained adding to a conventional multichannel SAR-GMTI system a stationary transmitter operating on adjacent frequency band. By theoretical simulations and by validation with real data, we showed that such a configuration can greatly improve the situational awareness over the observed area. In particular, targets moving with a minimum velocity can be detected regardless of their direction of motion in at least one dataset. Thanks to the different Doppler modulations observable in the monostatic and in the bistatic datasets and to the availability of multichannel GMTI processing, this approach is valuable especially for detections in strong interference scenarios. Moreover, for those targets which can be detected in both datasets, we showed that an estimation of the target velocity vector on the ground is possible. In this context, we analyzed some issues that may arise in the association of detections in densely populated scenarios. Finally, we used the estimated target velocity vector for (monostatic and bistatic) moving target SAR imaging. This information can help toward target recognition and classification. For these reasons, the proposed simultaneous configuration is a very appealing GMTI technique which dramatically overcomes the intrinsic limitations of a single system, at very low additional costs.

Appendix A: Data synchronization

Synchronization of the bistatic data is a crucial preliminary processing step. The reason being that the radio frequency (RF) local oscillators at the stationary transmitter and at the multichannel receiver (i.e., PAMIR) are freely running. This is expected to create a drift on the received data, as deeply analyzed also in other bistatic radar configurations, as in [42]. That is, due to a slight mismatch of the local oscillators frequencies at the transmitter and at the receiver, the bistatic data are acquired with an apparent delay that is to a good approximation linearly increasing over slow-time. To model this effect, let us indicate the RF transmitted signal as

$$s_{\mathrm{TX}}^{\mathrm{RF}}(t) = s_{\mathrm{TX}}^{\mathrm{BB}}(t)\exp\{j[2\pi f_{\mathrm{TX}}t + \phi_{\mathrm{TX}}(t)]\} \tag{A.1}$$

where $s_{\mathrm{TX}}^{\mathrm{BB}}(t)$ is the transmitted signal in base band (BB), t is the fast-time, f_{TX} is the RF of the local oscillator at the transmitter and $\phi_{\mathrm{TX}}(t)$ is a time varying phase error accounting for all non-ideal behaviours of the transmitting oscillator. Indicating with R_{direct}, the LOS range from the transmitter to the receiver, the direct RF received signal can be written as

$$\begin{aligned} s_{\mathrm{RX}}^{\mathrm{RF}}(t) &= \varepsilon\, s_{\mathrm{TX}}^{\mathrm{RF}}\left(t - \frac{R_{\mathrm{direct}}}{c}\right) \\ &= \varepsilon\, s_{\mathrm{TX}}^{\mathrm{BB}}\left(t - \frac{R_{\mathrm{direct}}}{c}\right) \\ &\quad \exp\left\{j\left[2\pi f_{\mathrm{TX}}\left(t - \frac{R_{\mathrm{direct}}}{c}\right) + \phi_{\mathrm{TX}}\left(t - \frac{R_{\mathrm{direct}}}{c}\right)\right]\right\} \end{aligned} \tag{A.2}$$

where ε accounts for all attenuations. After down conversion to BB, the received signal can be expressed as

$$\begin{aligned} s_{\mathrm{RX}}^{\mathrm{BB}}(t) &= s_{\mathrm{RX}}^{\mathrm{RF}}(t)\exp\{-j[2\pi f_{\mathrm{RX}}t + \phi_{\mathrm{RX}}(t)]\} \\ &= \varepsilon\, s_{\mathrm{TX}}^{\mathrm{BB}}\left(t - \frac{R_{\mathrm{direct}}}{c}\right)\exp\left\{-j2\pi f_{\mathrm{TX}}\frac{R_{\mathrm{direct}}}{c}\right\} \\ &\quad \exp\{j2\pi(f_{\mathrm{TX}} - f_{\mathrm{RX}})t\} \\ &\quad \exp\left\{j\left(\phi_{\mathrm{TX}}\left(t - \frac{R_{\mathrm{direct}}}{c}\right) - \phi_{\mathrm{RX}}(t)\right)\right\} \\ &= \varepsilon\, s_{\mathrm{TX}}^{\mathrm{BB}}\left(t - \frac{R_{\mathrm{direct}}}{c}\right)\exp\left\{-j2\pi f_{\mathrm{TX}}\frac{R_{\mathrm{direct}}}{c}\right\} \\ &\quad \exp\{j2\pi\Delta f t\}\exp\{j\phi_{\mathrm{res}}\} \end{aligned} \tag{A.3}$$

where f_{RX} is the RF of the local oscillator at the receiver, $\phi_{\mathrm{RX}}(t)$ accounts for receiver's oscillator non-idealities, $\Delta f = f_{\mathrm{TX}} - f_{\mathrm{RX}}$ is the frequency offset between the two local oscillators and ϕ_{res} is a random process, which can be modelled as the sum of a stationary phase term and of a random walk component as described in [42] and references therein. In the following, we will only focus on the effects of Δf, since ϕ_{res} plays a role in the long term. The received signal in (A.3) can be extended to the slow-time domain ($m = 1, \ldots, M$ is the index of the corresponding pulse), so that after range compression one gets

$$\begin{aligned}\tilde{s}_{\mathrm{RX}}^{\mathrm{BB}}(t, m) = \varepsilon\, \tilde{s}_{\mathrm{TX}}^{\mathrm{BB}}\left(t - \frac{R_{\mathrm{direct}}[m]}{c} + \Delta\mathrm{PRI} \cdot m\right) \\ \exp\left\{-j2\pi f_{\mathrm{TX}} \frac{R_{\mathrm{direct}}[m]}{c}\right\} \\ \exp\{j2\pi\Delta f[t + \mathrm{PRI} \cdot m]\}\exp\{j\phi_{\mathrm{res}}\}\end{aligned} \tag{A.4}$$

where $\tilde{s}_{\mathrm{TX}}^{\mathrm{BB}}$ is the range compressed BB transmitted signal, PRI is the pulse repetition interval, $\Delta\mathrm{PRI} = (\Delta f / f_{\mathrm{TX}})\mathrm{PRI}$ and where the LOS range from transmitter to receiver is now a function of the slow-time index m. The lack of synchronization manifests itself in two effects. The first is an additional delay $\Delta\mathrm{PRI} \cdot m$ in the envelope of $s_{\mathrm{RX}}^{\mathrm{BB}}$, while the second is the additional phase offset due to the term $\mathrm{PRI} \cdot m$ in the second exponential, which accounts for the accumulating phase offset over multiple PRIs. The first effect is visible only for significant frequency offsets Δf, or for very high range resolutions, while the second term is always visible, as it directly affects the phase of the received signal. In particular, this latter term is multiplied by Δf thus creating a linear phase ramp in the slow-time which results in an apparent Doppler centroid. Please note that both the signal delay and the linear phase ramp sum up with the behaviour of $R_{\mathrm{direct}}[m]$. That is, the total Doppler centroid that one sees on the unsynchronized data will contain a term due to real linear variation of $R_{\mathrm{direct}}[m]$ over slow-time, plus a fictitious term related to unsynchronized oscillators. The same applies for the signal delay. The aim of the data synchronization is to compensate for the fictitious signal delay and Doppler centroid, while preserving the other (true) components. To do that, accurate GPS positions of the transmitter and of the receiver over slow-time are required.

Figure 9.28 shows a zoom of the range compressed bistatic data in slow- and fast-time domain. In particular, it is shown the direct signal sent by the stationary transmitter. Data are also clipped in the slow-time domain, to focus only in the time acquisition interval in which the direct signal is received at its strongest level. A green line is superimposed, which approximates the direct signal in the least square sense. In Figure 9.29, the same data is compared also with the direct signal delay measured from the GPS (red line). From the discussions above, it is clear that data should follow the red line, being the different slope between red and green lines an effect of lack of synchronization. As is apparent, the effect on direct signal delay is very marginal, but still visible due to the high range resolution of the system. The so-called coarse synchronization (see [43]) can be readily performed by compensating for the different slopes between true (red) and estimated (green) behaviour of

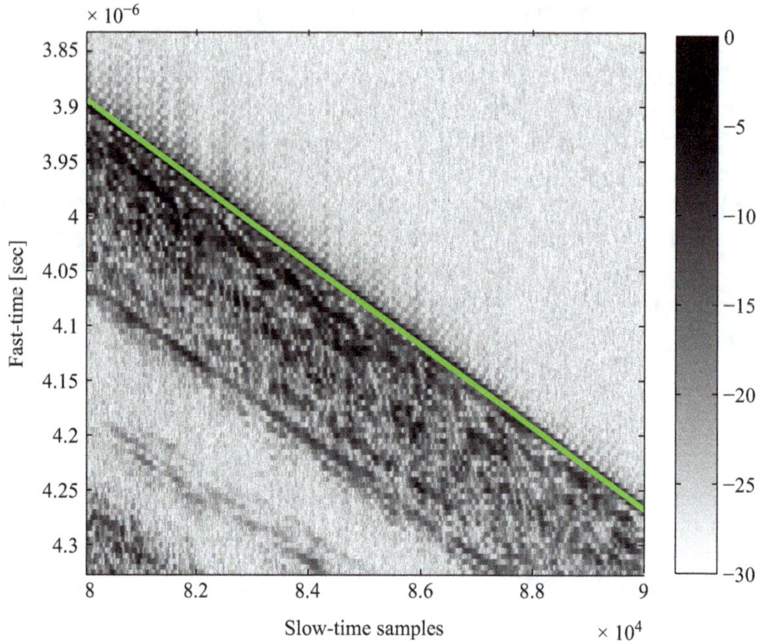

Figure 9.28 Range compressed direct signal before synchronization [dB]. Direct signal least square approximation (green line)

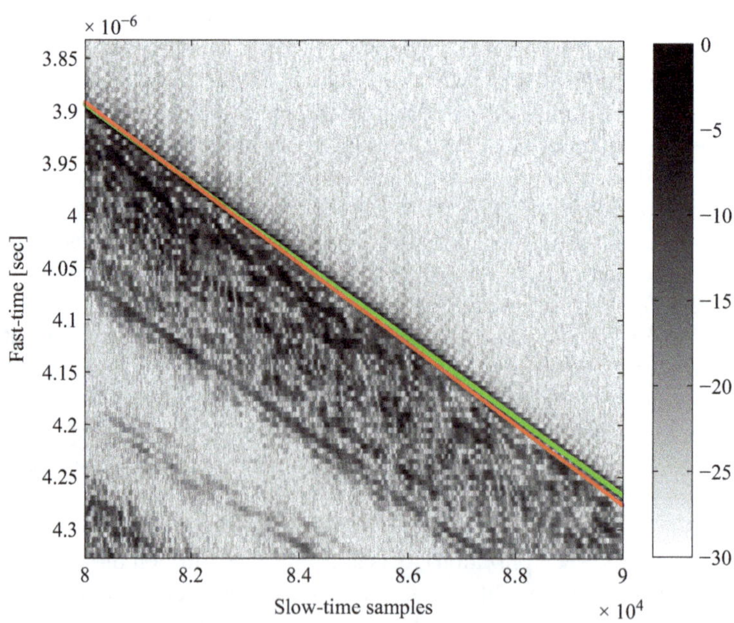

Figure 9.29 Comparison with GPS measured direct signal delay (red line). Range compressed direct signal before synchronization [dB]

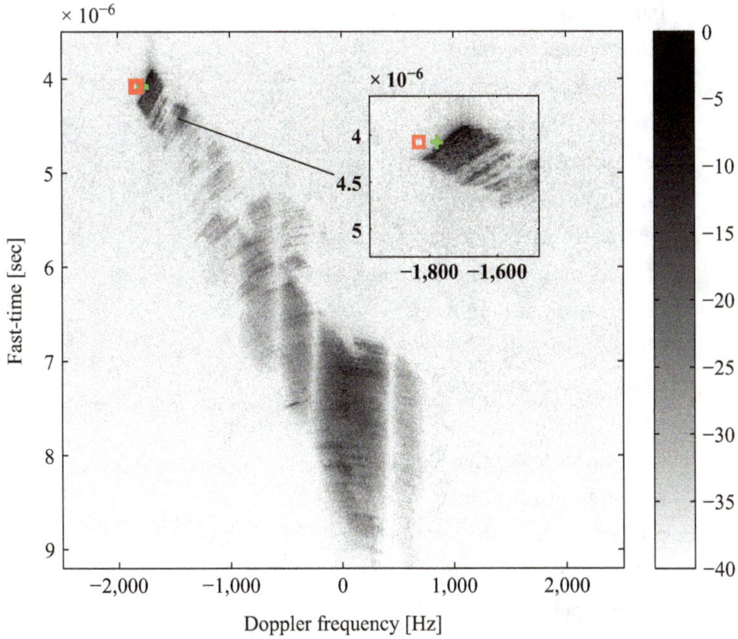

Figure 9.30 Bistatic range/Doppler map after fine synchronization [dB]

direct signal delay. After this step, bistatic data can be transformed in the range/Doppler domain, where the second effect (i.e., the additional fictitious Doppler centroid) is visible. Figure 9.30 shows the bistatic data after Doppler Fourier transform, where now the entire range extent is shown. The zoom highlights the part of the spectrum where the direct signal is located. A green marker is located in the direct signal range/Doppler position estimated from the data (i.e., corresponding to the green line in Figure 9.28), while a red marker is placed in the GPS-expected position (i.e., corresponding to the red line in Figure 9.29). After compensating for the relative shift between the two markers positions, bistatic data are synchronized.

A final comment is in order with reference to Figure 9.30. The direct signal is smeared over several range/Doppler cells. This is due to the variation of $R_{\text{direct}}[m]$ in the long slow-time interval here considered. In other words, the smearing is simply due to the range walk, which is not compensated here. We have chosen such a long slow-time interval to better show the effects of lack of synchronization also in the signal delay (see Figures 9.28 and 9.29).

Glossary

ABPD	adjacent bin post-Doppler
AB-STAP	AB space time adaptive processing
AMF	adaptive matched filter

BB	base band
CNR	clutter-to-noise ratio
CPI	coherent processing interval
CRB	Cramér–Rao bound
CUT	cell under test
DOA	direction of arrival
ENU	east-north-up
GMTI	ground moving target indication
GNN	global nearest neighbour
GPS	global positioning system
ISAR	inverse synthetic aperture radar
LOS	line of sight
MDV	minimum detectable velocity
MTI	moving target indication
PAMIR	phased array multifunctional imaging radar
PRF	pulse repetition frequency
PRI	pulse repetition interval
RCS	radar cross section
RF	radio frequency
SAR	synthetic aperture radar
SCR	signal-to-clutter ratio
SINR	signal-to-interference-plus-noise ratio
SNR	signal-to-noise ratio
STAP	space time adaptive processing
ULA	uniform linear array

References

[1] Klemm R. *Principles of Space-Time Adaptive Processing*. London, UK: IEE, 2002.

[2] Ward J. *Space-time adaptive processing for airborne radar*. MIT Lincoln Laboratory, Lexington, MA, USA, Tech. Rep., 1994.

[3] Guerci J. *Space-time Adaptive Processing for Radar*, Artech House radar library. Norwood, MA, USA: Artech House, Incorporated, 2003.

[4] Melvin W. 'A STAP overview', *IEEE Aerospace and Electronic Systems Magazine*, 2004, **19**, 1, pp. 19–35.

[5] Borsari G. 'Mitigating effects on STAP processing caused by an inclined array'. *Proceedings of the 1998 IEEE Radar Conference*, May 1998, pp. 135–140.

[6] Himed B., Michels J., Zhang Y. 'Bistatic STAP performance analysis in radar applications'. *Proceedings of the 2001 IEEE Radar Conference*, May 2001, pp. 198–203.

[7] Melvin W., Callahan M., Davis M. 'Comparison of bistatic clutter mitigation algorithms for varying geometries'. *2005 IEEE International Radar Conference*, May 2005, pp. 98–103.

[8] Melvin W., Davis M. 'Adaptive cancellation method for geometry-induced nonstationary bistatic clutter environments'. *IEEE Transactions on Aerospace and Electronic Systems*, 2007, **43**, Apr (2), pp. 651–672.

[9] Colone F. 'Spectral slope-based approach for mitigating bistatic space-time adaptive processing clutter dispersion'. *IET Radar, Sonar Navigation*, 2011 **5**, Jun (5), pp. 593–603.

[10] Cerutti-Maori D., Klare J., Brenner A., Ender J. 'Wide-area traffic monitoring with the SAR/GMTI system PAMIR'. *IEEE Transactions on Geoscience and Remote Sensing*, 2008, **46**, 10, pp. 3019–3030.

[11] Raney R. 'Synthetic aperture imaging radar and moving targets'. *IEEE Transactions on Aerospace and Electronic Systems*, 1971, **7**, May (3), pp. 499–505.

[12] Kirscht M. 'Detection and imaging of arbitrarily moving targets with single-channel SAR'. *Proceedings of the 2002 IEEE Radar Conference*, Oct 2002, pp. 280–285.

[13] Dias J., Marques P. 'Multiple moving target detection and trajectory estimation using a single SAR sensor'. *IEEE Transactions on Aerospace and Electronic Systems*, 2003, **39**, Apr (2), pp. 604–624.

[14] Pastina D., Battistello G., Aprile A. 'Change detection based GMTI on single channel SAR images'. *7th European Conference on Synthetic Aperture Radar (EUSAR 2008)*, Jun 2008, pp. 1–4.

[15] Cristallini D., Lombardo P., Pastina D., Mennella A. 'Chirp scaling based detection of moving targets in SAR images'. 2009 *IEEE International Geoscience and Remote Sensing Symposium (IGARSS)*, **5**, July 2009, pp. 340–343.

[16] Vu V. T., Sjogren T., Pettersson M., Gustavsson A., Ulander L. 'Detection of moving targets by focusing in UWB SAR – theory and experimental results'. *IEEE Transactions on Geoscience and Remote Sensing*, 2010, **48**, 10, pp. 3799–3815.

[17] Pastina D., Buratta L., Turin F., Cristallini D. 'Exploiting COSMO-SkyMed spotlight SAR images for GMTI applications'. *2011 IEEE CIE International Conference on Radar (RADAR)*, Oct 2011, **2**, pp. 1918–1921.

[18] Pastina D., Buratta L., Turin F. 'Detection of ground moving targets in COSMO-SkyMed SAR images'. *2012 IEEE International Geoscience and Remote Sensing Symposium (IGARSS)*, Jul 2012, pp. 3827–3830.

[19] Cristallini D., Colone F., Pastina D., Lombardo P. 'Integrated clutter cancellation and high-resolution imaging of moving targets in multi-channel SAR'. *European Radar Conference 2009*, Sept 2009, pp. 57–60.

[20] Cristallini D., Pastina D., Colone F., Lombardo P. 'Efficient detection and imaging of moving targets in SAR images based on chirp scaling'. *IEEE Transactions on Geoscience and Remote Sensing*, 2013, **51**, 4, pp. 2403–2416.

[21] Cerutti-Maori D., Sikaneta I., Gierull C. 'Optimum SAR/GMTI processing and its application to the radar satellite RADARSAT-2 for traffic monitoring'. *IEEE Transactions on Geoscience and Remote Sensing*, 2012, **50**, 10, pp. 3868–3881.

[22] Sjogren T., Vu V. T., Pettersson M., Wang F., Murdin D., Gustavsson A., Ulander L. 'Suppression of clutter in multichannel SAR GMTI'. *IEEE Transactions on Geoscience and Remote Sensing*, 2014, **52**, 7, pp. 4005–4013.

[23] Barbarossa S. 'Detection and imaging of moving objects with synthetic aperture radar. Part 1: Optimal detection and parameter estimation theory'. *IEE Proceedings F (Radar and Signal Processing)*, 1992, **139**, 1, pp. 79–88.

[24] Barbarossa S., Farina A. 'Detection and imaging of moving objects with synthetic aperture radar. Part 2: Joint time-frequency analysis by Wigner-Ville distribution'. *IEE Proceedings F (Radar and Signal Processing)*, 1992, **139**, 1, pp. 89–97.

[25] DiPietro R., Fante R., Perry R. 'Space-based bistatic GMTI using low resolution SAR'. *Proceedings of IEEE Aerospace Conference 1997*, Feb 1997, **2**, pp. 181–193.

[26] Kohlleppel R., Cristallini D. 'Association of detections using simultaneous monostatic and bistatic GMTI'. *IEEE Radar 2015*, May 2015, Arlington VA, USA, pp. 864–869.

[27] DiPietro R. 'Extended factored space-time processing for airborne radar systems' *1992 Conference Record of The Twenty-Sixth Asilomar Conference on Signals, Systems and Computers*, Oct 1992, **1**, pp. 425–430.

[28] Ender J., Gierull C., Cerutti-Maori D. 'Improved space-based moving target indication via alternate transmission and receiver switching'. *IEEE Transactions on Geoscience and Remote Sensing*, 2008, **46**, 12, pp. 3960–3974.

[29] Cristallini D., Walterscheid I. 'Joint monostatic and bistatic STAP for improved SAR-GMTI capabilities'. *IEEE Transactions on Geoscience and Remote Sensing*, 2016, **54**, 3, pp. 1834–1848.

[30] Blackman S., Popoli R. *Design and Analysis of Modern Tracking Systems*. Artech House, 1999.

[31] Bourgeois F., Lassalle J.-C. 'An extension of the Munkres algorithm for the assignment problem to rectangular matrices'. *Communications of the ACM*, 1971, **14**, 12, pp. 802–804.

[32] Bar-Shalom Y., Li X. R., Kirubarajan T. *Estimation with Applications to Tracking and Navigation*. John Wiley & Sons, Inc., 2001

[33] Brenner A., Ender J. 'Demonstration of advanced reconnaissance techniques with the airborne SAR/GMTI sensor PAMIR'. *IEE Proceedings – Radar, Sonar and Navigation*, 2006, **153**, Apr (2), pp. 152–162.

[34] Robey F., Fuhrmann D., Kelly E., Nitzberg R. 'A CFAR adaptive matched filter detector'. *IEEE Transactions on Aerospace and Electronic Systems*, 1992, **28**, 1, pp. 208–216.

[35] Lombardo P., Colone F. 'A dual adaptive channel STAP scheme for target detection and DOA estimation'. *Proceedings of the 2003 International Radar Conference*, Sept 2003, pp. 115–120.

[36] Cristallini D., Colone F., Lombardo P. 'Dual adaptive channel STAP: Theory and experimental results'. *2011 Proceedings International Radar Symposium (IRS)*, Sept 2011, pp. 757–762.

[37] Colone F., Cristallini D., Cerutti-Maori D., Lombardo P. 'Direction of arrival estimation performance comparison of dual cancelled channels space-time adaptive processing techniques'. *IET Radar, Sonar Navigation*, 2014, **8**, 1, pp. 17–26.

[38] Raney R., Runge H., Bamler R., Cumming I., Wong F. 'Precision SAR processing using chirp scaling'. *IEEE Transactions on Geoscience and Remote Sensing*, 1994, **32**, Jul (4), pp. 786–799.

[39] Cafforio C., Prati C., Rocca F. 'SAR data focusing using seismic migration techniques'. *IEEE Transactions on Aerospace and Electronic Systems*, 1991, **27**, Mar (2), pp. 194–207.

[40] Soumekh M. *Synthetic Aperture Radar Signal Processing: with MATLAB Algorithms*. New York, NY, USA: Wiley-Interscience, 1999.

[41] Cristallini D., Walterscheid I. 'SAR-GMTI enhanced with simultaneous monostatic and bistatic detections'. *Proceedings of 10th European Conference on Synthetic Aperture Radar (EUSAR 2014)*, 2014, pp. 989–992.

[42] Lopez-Dekker P., Mallorqui J., Serra-Morales P., Sanz-Marcos J. 'Phase synchronization and doppler centroid estimation in fixed receiver bistatic SAR systems'. *IEEE Transactions on Geoscience and Remote Sensing*, 2008, **46**, 11, pp. 3459–3471.

[43] Cristallini D. 'First results on bistatic MTI activity at Fraunhofer FHR'. *2013 International Conference on Radar*, Sept 2013, pp. 470–475.

Chapter 10
Multistatic and MIMO ISAR techniques

Debora Pastina[1] *and Marta Bucciarelli*[2]

Abstract

The exploitation of ISAR data simultaneously acquired by multiple radar systems is considered in this chapter in order to enhance the quality of ISAR images of moving targets with respect to the conventional single-sensor case, thus making ISAR images more effective when used for target classification and recognition. In particular, multi-sensor data are exploited in order to increase the cross-range resolution of ISAR images of rotating targets and to improve the accuracy in the estimation of the target motion.

The distributed (multi-sensor) ISAR technique is devised for two different cases: (i) MIMO case with each platform carrying an active radar, that transmits and receives RF waveforms, (ii) multistatic case with a single platform carrying an active radar (transmitting and receiving) and the remaining platforms equipped with passive sensors (namely receiving only). For such distributed imaging system: (a) the PSF is derived showing the capability at providing an increase of the cross-range resolution up to the number of platforms in the multistatic case and even higher in the MIMO case; (b) the required focusing technique is also presented and discussed following a decentralized approach; (c) multi-sensor based target motion estimation techniques are considered showing the performance improvement with respect to the conventional single-sensor case. This distributed ISAR system could be of great benefit in applications where the target rotation angle is insufficient to guarantee the desired resolution. A typical case is the imaging of ship targets with rotation induced by the sea swell structure under low sea state conditions.

Results obtained against synthetic ISAR data are presented; moreover, experimental data collected in an anechoic chamber against different targets on a rotating platform are processed by following the presented distributed ISAR technique to validate the approach.

[1]DIET Department, University of Rome 'La Sapienza', Italy
[2]Formerly DIET Department, University of Rome 'La Sapienza', Italy. Now Sympas S.r.l., Italy

10.1 Introduction

As well-known radar images of man-made targets can be obtained by using the ISAR (Inverse Synthetic Aperture Radar) technique: a wide bandwidth waveform is transmitted to achieve fine range resolution; coherently processing the echoes returned from the target at different view angles gives fine cross-range resolution, [1,2]. In ISAR, the angular aperture under which the target is viewed by the radar is due to the motion of the target itself. Specifically, ISAR techniques exploit the rotational motion of a target with respect to its centre as observed by a nearly stationary sensor platform. This is for example the case of airborne or spaceborne systems imaging ship targets interested by yaw, pitch and roll rotation motions induced by the sea.

Considering the specific nature of ISAR imaging, the following drawbacks can be easily pointed out: (a) the focusing and scaling of ISAR images require knowledge of the target motion usually unknown; (b) achievable images provide a 2D projection of man-made targets having a 3D structure; (c) achievable cross-range resolution of ISAR products is target dependent.

Moving from point (a), the knowledge of target translation and rotation motion is typically required to motion compensate, to scale and also to focus (depending on resolution/target size) the ISAR image. Even if some classification/recognition techniques can be designed to cope with unscaled images (as [3,4] for ship targets), the availability of images scaled in the homogeneous range and cross-range domain largely simplifies the procedures allowing the extraction of relevant target features and avoiding database-matching approaches, thus reducing the computational load. Unfortunately, when dealing with a non-cooperative target, the information relevant to its motion is unknown; moreover, even for cooperative targets properly equipped to share the navigation data with the remote control, the information might not be available in real time with the desired accuracy. Therefore, the target motion has to be recovered directly from the received signal. Many autofocus and cross-range scaling techniques have been developed in the past for single-channel ISAR systems. Particularly, for cross-range scaling, these techniques usually assume that the rotational motion of the target is confined to a plane during the coherent processing interval (CPI) and therefore they can experience significant losses when this assumption is not verified and the target develops 3D motion. Moreover, even when such assumption applies, the achievement of good accuracy in the cross-range scaling can impose severe constraints on target size, signal-to-background ratio and needed CPI. Obviously, point (b), classification/recognition procedures would highly benefit also from the knowledge of the 3D structure of the imaged target, while this information is lost in the projection mechanism proper of conventional single-channel ISAR systems. Even if some techniques could be applied in order to select proper image times in order to achieve specific (such as top or side) views of the target, [5], there is no guarantee that target motion conditions would allow such decoupling enabling the extraction of some 3D-related information and features. Obviously, information and features that can be extracted from ISAR images also depend on the quality of these images and particularly on their resolution, [6]. Unfortunately, since the achievable cross-range resolution depends on the intrinsic motion characteristics of the target (and specifically on its overall change of view angle) as highlighted in point (c), there is no way to *a priori* assign the desired value of cross-range resolution. Moreover, depending on the particular conditions, the achievable resolution can be very poor. With reference to the case of ship target imaging, this condition can occur in presence of a low sea state inducing very limited rotations. This is made even worse by the typical use of a short CPI that is required by the constraint of keeping the ship rotation axis fixed. These circumstances can have particularly negative implications as ISAR images are often used to feed NCTR (Non-Cooperative Target Recognition) procedures.

Obviously, the use of multi-channel systems or even of multiple imaging sensors observing the same target with different acquisition geometries and the joint exploitation of the acquired data seem to be a viable solution to overcome the aforementioned issues. Indeed in the last years, the ISAR community has devoted an increasing attention to multi-channel, multistatic and MIMO radar systems, and a number of contributions can be found in the recent literature. Without claiming to be complete, some references are cited in the following.

With specific reference to autofocus/motion compensation exploiting multiple acquisitions, an image entropy based technique has been proposed in [7], and a multilateration based approach has been considered in [8]. With specific regard to rotation motion estimation, few contributions can be found in literature concerning the use of multi-sensor ISAR data: in some cases, the sensors are close so that they give rise to the same image projection plane (IPP) (i.e. co-located antennas), while in other cases the sensors are separated so that the IPP can change with the considered sensor (i.e. separated antennas). In particular, in [9–11], formations of separated sensors with different aspect angles have been proposed for the motion estimation of target undergoing rotation confined to a plane. Moving to the three-dimensional rotation case, interferometric approaches making use of co-located antennas placed onto orthogonal baselines have been considered in [12–14]; in [15], the use of at least three sequences of ISAR images to estimate the aspect rotation angle has been proposed. The case of separated antennas has been instead considered in [16] proving the capability of a formation of sensors with proper angular diversity at providing the estimation of the three rotation-motion components. 3D target re-construction has been achieved mainly by resorting to different interferometric approaches, as for example in [13,14,17–21], even if also some MIMO-related contribution can be found, as for example [22] where MIMO techniques are used jointly with sparse signal recovery or [23] combining MIMO and interferometric approaches. In contrast, MIMO-based approaches have been usefully applied for target imaging purposes: as an example, with the aim at avoiding motion compensation usually required in ISAR, MIMO radar systems have been considered in [24] for the narrowband case exploiting two perpendicular arrays and in [25] for the wideband case using two distributed arrays. Noticeably, multi-sensor, both multistatic and MIMO, systems and related techniques can be usefully applied to counteract the cross-range resolution target motion dependency. In this frame, following the approaches in [26–33] and showing some of the results therein, this chapter describes the possibility of exploiting the data acquired by multiple radar sensors carried by multiple air platforms. If the sensors are appropriately spaced, each scatterer can be globally observed from a much wider observation angle than for a single aperture, and therefore, an ISAR image with a higher resolution can be obtained. Obviously, the same approach could also be used to reduce the CPI required to get a desired cross-range resolution [34].

Two reference scenarios for the sensors belonging to a formation of S platforms will be considered in this chapter: in the first scenario (named multistatic distributed ISAR), only one platform carries an active radar system and the remaining $(S-1)$ carry receiving only sensors, while in the second one (named MIMO distributed ISAR), all the S air platforms carry an active radar that autonomously transmits and receives the radar waveforms. In this latter case, we consider the active sensors additionally able to separately receive also the waveforms transmitted from the other transmitters. A pictorial view of these two reference scenarios is presented in Figure 10.1. Figure 10.1(a) sketches the multistatic-distributed ISAR (DISAR) case where the flying formation is composed of a helicopter carrying the transmit/receive hardware and a few lightweight UAVs used to carry the receive-only devices. Figure 10.1(b) presents the MIMO distributed ISAR case with several hovering helicopters carrying an active radar system each.

The chapter is organized as follows: after describing the distributed ISAR system and geometry (Section 10.2), the imaging capability of the conceived system will be analysed and discussed in Section 10.3 where the distributed ISAR Point Spread Function (DPSF) is derived (Section 10.3.1) to clearly demonstrate the enhanced imaging capability of the multistatic/MIMO ISAR system, the corresponding performance theoretically investigated for cases of special interest (Section 10.3.2) and experimentally validated (Section 10.3.3). Section 10.4 introduces the required focusing technique proposing and discussing the different available approaches for the distributed ISAR image formation (Section 10.4.1) and again theoretically and experimentally analysing the achievable performance (respectively, in Section 10.4.2 and 10.4.3). The exploitation of multi-sensor ISAR

Figure 10.1 Distributed ISAR scenarios in the (a) multistatic case and in the (b) MIMO case. ©[2011] IEEE. Reprinted, with permission, from [30]

data for the sake of target motion estimation is considered in Section 10.5 showing the improvement available with respect to conventional single-sensor techniques. Finally, some conclusions and discussions on advantages/limitations and on future perspectives concerning the considered topic close the chapter.

10.2 Distributed ISAR system

In this section, the distributed ISAR geometry and concept are firstly introduced (Section 10.2.1); then specific discussions concerning the MIMO ISAR configurations are reported (Section 10.2.2).

10.2.1 DISAR geometry and concept

As stated above, the considered distributed ISAR geometry consists essentially of a set of S air platforms, each one equipped with a sensor, characterized by either: (i) transmitting, or (ii) receiving, (iii) or both transmitting and receiving capabilities. All sensors are assumed to carry an antenna appropriately steered towards the moving target to be imaged by exploiting its own rotational motion.

As usual in ISAR literature, we identify an arbitrary reference point in the target, hereafter named target fulcrum, and decompose the motion of the target into a translation of the fulcrum and a rotation of the target body around this point (hypothesis of modelling the target as a rigid body). By referring to the distributed ISAR signal after motion compensation, we can assume any relative translational motion between the platforms and the target fulcrum already compensated (namely platforms and fulcrum considered stationary) and, therefore, we can focus on the target rotation. As a consequence, the technique here discussed can be usefully applied regardless the platforms are moving or stationary.

The target of interest (e.g. a ship) is modelled as a set of K scatterers with position vectors in the target centred (0, XYZ) reference system, Figure 10.2, given

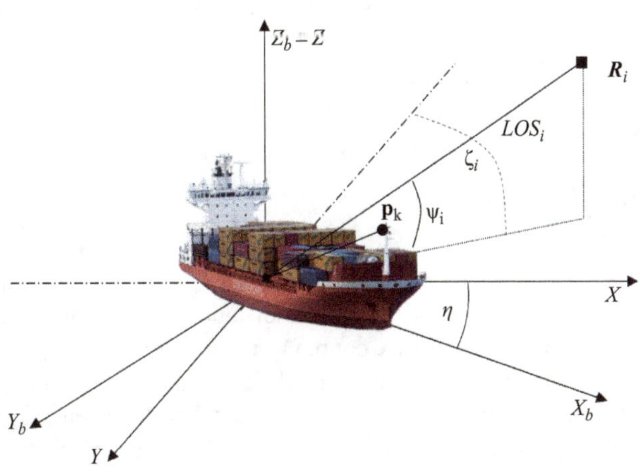

Figure 10.2 DISAR system geometry

by $\mathbf{p_k}(t) = p_k \cdot \hat{\mathbf{p}}_\mathbf{k}(t)$, $k = 1, \ldots, K$, where the vector length $p_k = \|\mathbf{p_k}(t)\|$ is constant, while the unit vector $\hat{\mathbf{p}}_\mathbf{k}(t)$ changes with time due to rotation motions. The same kth scatterer in a reference system integral with the target, (0, $X_bY_bZ_b$) in Figure 10.2, is described by a fully constant position vector $\mathbf{p_k^b}$ with unit vector $\hat{\mathbf{p}}_\mathbf{k}^\mathbf{b}$.

As usual, rotation motion can be taken into account by introducing roll (rotation around X_b axis with rate ω_{roll}), pitch (around Y_b with rate ω_{pitch}) and yaw (rotation around Z_b with rate ω_{yaw}). At the generic time instant, the unit vector $\hat{\mathbf{p}}_\mathbf{k}(t)$ can thus be written as

$$\hat{\mathbf{p}}_\mathbf{k}(t) = \left[\mathbf{M}_\boldsymbol{\eta}(t)\mathbf{M}_\mathbf{yaw}(t)\mathbf{M}_\mathbf{pitch}(t)\mathbf{M}_\mathbf{roll}(t)\right] \hat{\mathbf{p}}_\mathbf{k}^\mathbf{b} \tag{10.1}$$

where $\mathbf{M}_\boldsymbol{\eta}$, $\mathbf{M}_\mathbf{yaw}$, $\mathbf{M}_\mathbf{pitch}$ and $\mathbf{M}_\mathbf{roll}$ are the rotation matrices changing with time and accounting, respectively, for heading (angle η in Figure 10.2), yaw, pitch and roll angles [35].

The sketch in Figure 10.2 also defines the position vector $\mathbf{R}_i$ of the ith platform, $i = 1, \ldots, S$ (more precisely of the phase centre of the corresponding antenna): as for the target, the platform position vector $\mathbf{R}_i = R_i \cdot \hat{\mathbf{r}}_i$ is written by means of its unit vector $\hat{\mathbf{r}}_i = \mathbf{R}_i / R_i$:

$$\hat{\mathbf{r}}_\mathbf{i} = [\cos(\psi_i)\sin(\zeta_i), \quad -\cos(\psi_i)\cos(\zeta_i), \quad \sin(\psi_i)]^T \tag{10.2}$$

where ζ_i and ψ_i are, respectively, the aspect and grazing angle identifying the line of sight (LOS) of sensor i (Figure 10.2) and any dependence on time t has been neglected assuming the translation motion already compensated.

To account for each acquisition with transmission (TX) from system i and reception (RX) from system j, the concept of equivalent sensor n can be introduced. Among all the acquisitions performed by the S systems in the formation, monostatic acquisitions are linked to real sensors, while each bistatic acquisition can be associated to a fictitious equivalent sensor: therefore, it can be stated that a formation of S active and passive radar systems leads to a formation of $N \geq S$ equivalent sensors, where N accounts for the number of possible TX/RX couples. Each equivalent sensor n can be geometrically identified in the (0, XYZ) reference system using the following quantities:

$$\begin{aligned} &\xi_n = \frac{\psi_i + \psi_j}{2}, \ \chi_n = \frac{\psi_i - \psi_j}{2}, \ \alpha_n = \frac{\zeta_i + \zeta_j}{2}, \ \beta_n = \frac{\zeta_i - \zeta_j}{2}, \\ &R_n = \frac{R_i + R_j}{2} \ \hat{\mathbf{r}}_n = \frac{\hat{\mathbf{r}}_i + \hat{\mathbf{r}}_j}{2} \end{aligned} \tag{10.3}$$

where α_n and ξ_n represent, respectively, the aspect and the grazing angles of the equivalent sensor n, while β_n and χ_n are the bistatic angles in aspect and grazing, respectively, arising if in the TX/RX couple i/j the transmitting sensor i is different from the receiving sensor j. R_n is the equivalent monostatic distance associated to the equivalent sensor, whose position is defined by the vector $\hat{\mathbf{r}}_n$. We notice explicitly that for monostatic acquisitions we have $I = j$, so that the sum of the unit vectors $\hat{\mathbf{r}}_n = \hat{\mathbf{r}}_i$. In contrast, in the bistatic case (namely $i \neq j$), $\hat{\mathbf{r}}_n$ is not strictly a unit vector.

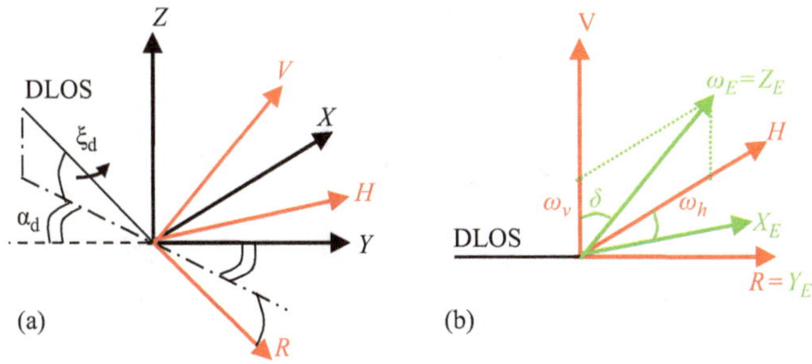

Figure 10.3 Change of reference systems (a) from XYZ to HRV and (b) from HRV to $X_EY_EZ_E$. ©[2012] IET. Reprinted, with permission, from [33]

The joint exploitation of the signals acquired by the N equivalent sensors in the formation also requires the choice of a reference LOS, hereafter called 'Distributed LOS (DLOS)', to identify the point of view of the corresponding DISAR image. The DLOS is here determined according to an aspect-related angle α_d and a grazing-related angle ξ_d which are the average of the α_n and ξ_n angles ($n = 1, \ldots, N$), respectively; correspondingly, the left handed (O,HRV) reference system related to the DLOS is defined (Figure 10.3(a)), [5], where R axis represents the DLOS direction, H axis is normal to R axis and belongs to the (X,Y) plane and, finally, V axis is normal to the (R,H) plane. The motion characteristics of the target can be then described in this last reference system leading to the definition of the effective rotation vector $\boldsymbol{\omega}_\mathbf{E} = [\omega_h, 0, \omega_v]$ as the projection on the H-V plane of the rotation vector $\boldsymbol{\omega}$ describing the motion of the target, being ω_h and ω_v its horizontal and vertical components (Figure 10.3(b)). Rotation around R axis is not considered since, at least to first order, it does not contribute to the phase history of the generic target scatterer.

Since to first order the image formation in the Doppler/cross-range domain is provided by means of $\boldsymbol{\omega}_\mathbf{E}$, a generic 3D motion can be represented by an equivalent 1D motion in a reference system where $\boldsymbol{\omega}_\mathbf{E}$ represents one of the axes; therefore, we can refer to a new reference system, hereafter called (0, $X_EY_EZ_E$), where the vertical axis Z_E is set to be co-incident with the effective rotation axis (i.e. the unit vector associated to $\boldsymbol{\omega}_\mathbf{E}$) so that the target can be thought as interested by an 'equivalent yaw motion' with angular velocity $\omega_{\text{eff}} = \sqrt{\omega_h^2 + \omega_v^2}$. Transformation from HRV to $X_EY_EZ_E$ system (Figure 10.3(b)) can be accomplished through a clockwise rotation of the HRV system by an angle δ around the R axis, where $\delta = \tan^{-1}(\omega_h/\omega_v)$. In this reference, (X_E, Y_E) represents the image plane being X_E the cross-range axis. It has to be underlined that if the sensors in the formation are displaced in the X_EY_E plane, the formation can be considered ideal in terms of distributed processing (as it will be shown in the following section). Here we want to highlight a couple of specific sensors' alignments that will be largely considered in the following. In the case of a dominant vertical motion ($\omega_h = 0$), the X_EY_E

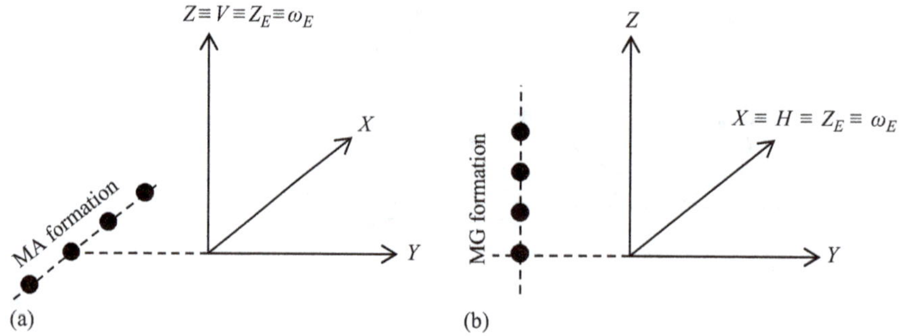

Figure 10.4 *Multi-aspect (a) and multi-grazing (b) formations. ©[2012] IET. Reprinted, with permission, from [33]*

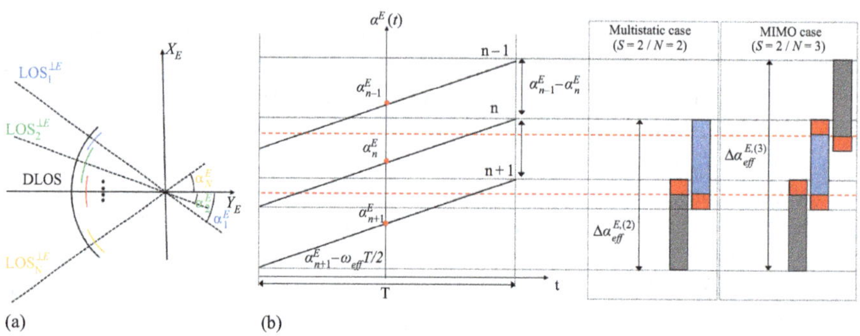

Figure 10.5 *(a) Variation of the view angle α_n^E as observed by equivalent sensors of the formation and (b) detail for the bistatic and MIMO case with $S = 2$. (b) ©[2010] IEEE. Reprinted, with permission, from [26]*

plane corresponds to the *HR* plane and the ideal sensor formation is referred to as Multi-Aspect (MA), Figure 10.4(a), with the sensors in the *XY* plane and observing the target with different aspect angles α_n but with null grazing angles ($\xi_n = 0\ \forall n$). The opposite case is represented by a target mainly undergoing horizontal motion ($\omega_v = 0$): in that case, the $X_E Y_E$ plane corresponds to the *VR* plane and the sensors are ideally displaced in the *XZ* plane having null aspect ($\alpha_n = 0\ \forall n$) but different grazing angles ξ_n, and the formation is referred to as Multi-Grazing (MG), Figure 10.4(b).

The cross-range resolution of the DISAR image is achieved exploiting the global change of the view angle as defined in the $X_E Y_E$ plane (Figure 10.5(a)). For each LOS_n, we can define the orthogonal projection on the $X_E Y_E$ plane $\text{LOS}_n^{\perp E}$ described by the aspect angle α_n^E. The global change of the view angle for the *k*th scatterer, observed by sensors *n* and $n-1$, is thus given by:

$$\Delta\alpha_{\text{eff}}^{E,(n,n-1)} = \omega_{\text{eff}}\, T + \alpha_{n-1}^E - \alpha_n^E \tag{10.4}$$

where T is the time aperture corresponding to the CPI associated to each acquisition.

Therefore, by considering all the N equivalent sensors, the global change of the view angle is written as follows:

$$\Delta\alpha_{\text{eff}}^{E,(N)} = \sum_{n=2}^{N} \Delta\alpha_{\text{eff}}^{E,(n,n-1)} = \omega_{\text{eff}}T + \alpha_1^E - \alpha_N^E \tag{10.5}$$

and the achievable cross-range resolution is

$$\Delta cr^{(N)} = \frac{\lambda}{2\Delta\alpha_{\text{eff}}^{E,(N)}} \tag{10.6}$$

leading to an ideal cross-range resolution improvement with respect to the single-sensor case

$$\gamma = \frac{\Delta\alpha_{\text{eff}}^{E,(N)}}{\omega_{\text{eff}}T} = 1 + \frac{\left(\alpha_1^E - \alpha_N^E\right)}{\omega_{\text{eff}}T} \tag{10.7}$$

Particularly, Figure 10.5(a) shows the variation of the aspect angle α_n^E for the different sensors in the formation, while Figure 10.5(b) details the case of two flying platforms ($S=2$) for both multistatic and MIMO cases. From Figure 10.5(a), it is easy to observe that the synthesis of the global aperture (black line in Figure 10.5(a)) is possible if and only if there is a super-position between subsequent angular coverages (coloured lines in Figure 10.5(a)), namely, it is necessary to guarantee the absence of gaps in the overall observation (even if specific techniques could be considered to cope with the presence of gaps); this implies an upper bound in the allowed angular separation between the adjacent equivalent sensors, that is $0 \le \alpha_{n-1}^E - \alpha_n^E \le \omega_{\text{eff}}T$. It is clear that when the equivalent sensors are maximally separated in angle, that is when $\alpha_{n-1}^E - \alpha_n^E = \omega_{\text{eff}}T\ \forall n$, the maximum improvement $\gamma = N$ is obtained.

Figure 10.5(b) details the case of a formation of two platforms ($S=2$). In the basic multistatic case, a single sensor (SS) transmits and both sensors in the formation receive, giving rise to a monostatic and a bistatic acquisition corresponding to a couple of equivalent sensors ($N=2$) associated to the couple of view angles α_n^E and α_{n+1}^E. In this case, the maximum increase in observation angle is directly of a factor $\gamma = N = 2$, which is easily obtained from (10.7) by setting $\alpha_n^E - \alpha_{n+1}^E = \omega_{\text{eff}}\,T$ for the equivalent sensors. The basic MIMO case, where two transmit–receive platforms are used ($S=2$), gives rise to a total of three equivalent sensors ($N=3$), corresponding to two monostatic acquisitions and to one bistatic acquisition. This case is modelled by the two angles of view of the two monostatic radar sensors, α_{n-1}^E and α_{n+1}^E, and by their central angle $\alpha_n^E = (\alpha_{n-1}^E + \alpha_{n+1}^E)/2$. In this case, the maximum increase in observation angle is directly of a factor $\gamma = N = 3$, which is easily obtained from (10.7) by setting $\alpha_{n-1}^E - \alpha_{n+1}^E = 2\omega_{\text{eff}}\,T$ for the real sensors.

Table 10.1 Optimized MIMO configurations. ©[2010] IEEE. Reprinted, with permission, from [26]

S	N	Sequence of angular spacing	N/S
2	3	2	1.5
3	5	2,2	1.6667
4	9	2,4,2	2.2500
5	13	2,4,4,2	2.6000
6	17	2,4,4,4,2	2.8333
7	21	2,4,4,4,4,2	3.0000
		2,2,6,6,2,2	
8	27	2,2,6,6,6,2,2	3.3750
		2,4,2,10,2,4,2	
9	33	2,2,6,6,6,6,2,2	3.6667
10	41	2,4,2,10,4,10,2,4,2	4.1000
11	45	2,2,2,8,8,8,8,2,2,2	4.0909
		2,2,6,4,8,8,4,6,2,2	
		2,2,6,6,6,6,6,6,2,2	
		2,4,2,10,4,4,10,2,4,2	
12	55	2,4,4,2,14,2,14,2,4,4,2	4.5833
		2,4,2,10,4,10,4,10,2,4,2	
13	65	2,4,2,10,4,10,10,4,10,2,4,2	5.0000
14	73	2,4,2,10,4,10,8,10,4,10,2,4,2	5.2143

10.2.2 MIMO ISAR formations configuration

As is apparent, the MIMO configuration is an especially interesting case, since it seems to actually allow reaching an improvement in cross-range resolution greater than the number of available platforms. This is possible because the number of equivalent sensors N is generally greater than the number of real sensors S and is obtained by using both the monostatic and the bistatic acquisitions, provided that adequate orthogonal waveforms are exploited. The optimization of the positions of a set of MIMO sensors to provide maximum angular coverage without gaps has been considered for obtaining synthetic aperture images in [36]. For small number of real sensors S and referring to the ideal formation, Table 10.1 reports the number of effective sensors N available for a continuous aperture without gaps.

To obtain the largest possible contiguous global view angle $\Delta\alpha_{\text{eff}}^{E,(N)} = N\,\omega_{\text{eff}}T$, using the S sensors for the MIMO ISAR application, it is necessary to appropriately displace the real sensors. In particular, the angular displacement required between each couple of adjacent real sensors is reported in the third column of Table 10.1, expressed in integer numbers of $(\omega_{\text{eff}}T)$. This derives from the direct application of the results of [36] to the MIMO ISAR case. As an example, to obtain the global effective angular spacing of $\Delta\alpha_{\text{eff}}^{E,(9)} = 9\,\omega_{\text{eff}}T$ for the MIMO ISAR with $S = 4$ real sensors, the angular spacing must be assigned as follows: 2 $\omega_{\text{eff}}T$ between sensors #1–#2 and sensors #3–#4, and 4 $\omega_{\text{eff}}T$ between sensors #2 and #3. Figure 10.6 shows the map of real and virtual apertures for MIMO DISAR using the

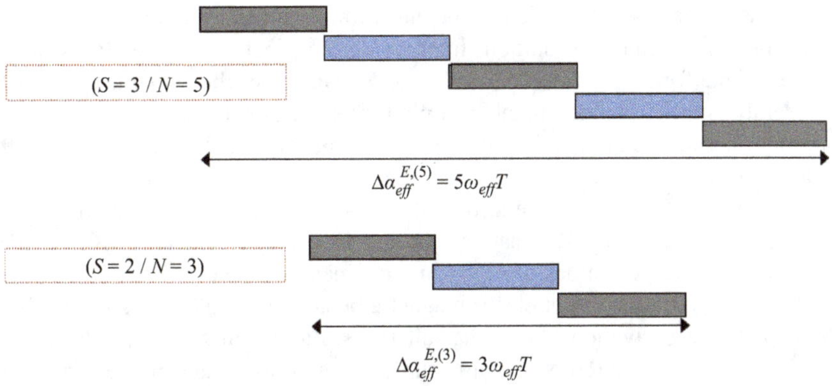

Figure 10.6 Map of real and virtual apertures (angles of view) for MIMO distributed ISAR for ($S = 2$, $N = 3$) and ($S = 3$, $N = 5$)

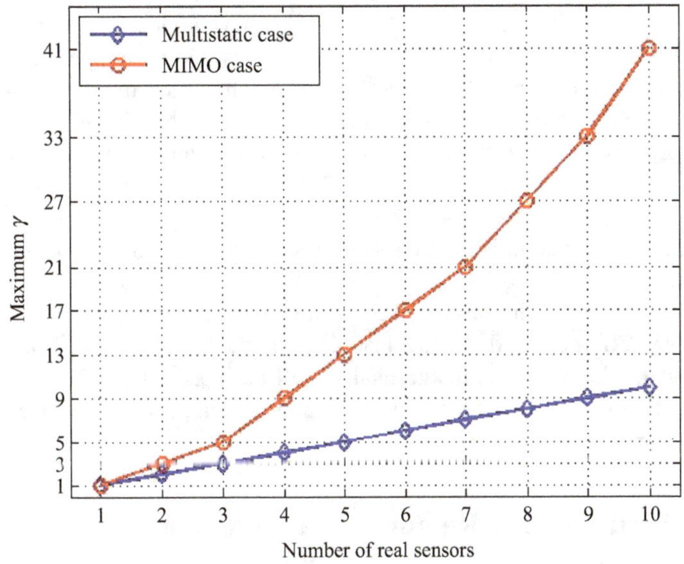

Figure 10.7 Maximum value of cross-range resolution improvement γ in the multistatic and MIMO cases. ©[2010] IEEE. Reprinted, with permission, from [26]

configurations in Table 10.1 for the cases ($S = 2$, $N = 3$) and ($S = 3$, $N = 5$). In particular, the grey and the blue rectangles, respectively, represent the real and the fictitious angular apertures used in the distributed configuration, where the term 'real aperture' indicates the equivalent aperture for the sensors operating in monostatic configuration, whereas 'fictitious aperture' corresponds to bistatic configurations based on different transmit and receive real sensors. Correspondingly, in Figure 10.7, the maximum value of γ is shown for both the multistatic and the

MIMO distributed ISAR. As it is apparent, in the multistatic case, the maximum achievable cross-range resolution improvement γ increases linearly with the number of platforms S, since $N = S$. In the MIMO case, the maximum γ can be considerably higher than the number of platforms S, since the gain N of the MIMO configuration, that is the number of equivalent sensors considered, increases as the number of platforms increases. The quantification of the improvement made available by the MIMO configuration with respect to the multistatic configuration is provided by the ratio N/S that is reported in the last column of Table 10.1. As apparent for larger values of S, the improvement increases.

A comment on the target electromagnetic behaviour is in order, since we have assumed in the derivation above that all the scatterers in the image behave as perfect ideal point scatterers showing the same amplitude and phase during the aperture time T for all monostatic and bistatic observation angles. A number of studies have addressed this point in detail with specific attention at the differences between monostatic and bistatic scattering, among which for example [37,38]. As apparent, this hypothesis is not always verified, especially when operating at high resolution and experiencing a wide change in the view angle. However, in our case, we aim at reaching moderate resolutions (the aim is just to compensate for a low sea state, remaining typically at resolutions much larger than the wavelength, and not to obtain extremely high resolutions), so that the change in the bistatic angle is quite limited over the set of N acquisitions (certainly smaller than $\Delta\alpha_{\text{eff}}^{E,(N)}$). For example, in a typical application in X-band, even looking for a fairly high resolution of 0.5 m, the upper bound of the bistatic angle would be 0.03 radians. In such a case, even for complex targets such as ships, aircrafts or ground vehicles, we can assume to be in the pseudo-monostatic RCS (radar cross-section) region, [39]. Under this hypothesis, the bistatic RCS of the target can be assumed equal to the monostatic RCS measured on the bisector on the bistatic angle. Obviously, a moderate degradation of the image quality will be present if the operative conditions do not comply with the previous assumptions (namely in presence of a wide change in the view angle).

10.3 Distributed ISAR point spread function

In this section, the DPSF is first analytically derived (Section 10.3.1); theoretical achievable performance is then investigated in Section 10.3.2, while experimental validation of theoretical results by means of experimental DISAR data is reported in Section 10.3.

10.3.1 Theoretical DISAR PSF

Using the sensors' position vectors and all the angles defined in the above section, recalling that the apex E states that the corresponding quantities are defined in the $X_E Y_E Z_E$ reference system linked to the effective rotation vector of the target, and assuming that the platform distance is much larger than the target size (namely $R_i \gg p_k$ for $i = 1, \ldots, S$, $k = 1, \ldots, K$ so that $|\mathbf{R_i} - \mathbf{p_k}(t)| \approx R_i - \hat{\mathbf{r}}_i \cdot \mathbf{p_k}(t)$, [2]), the signal received by the *n*th equivalent sensor from the *k*th scatterer can be modelled as

$$s_{k,\,n}(t) = e^{-j\frac{4\pi}{\lambda}\left[R_n - \left(\mathbf{r}_\mathbf{n}^E \cdot \mathbf{p}_\mathbf{k}^E(t)\right)\right]}\mathrm{rect}_T(t). \tag{10.8}$$

where $\mathbf{p}_\mathbf{k}^E(t)$ is the position vector of the *k*th scatterer in the $X_E Y_E Z_E$ reference system written as

$$\mathbf{p}_\mathbf{k}^E(t) = p_k \begin{bmatrix} \cos\left(\theta_k^E - \omega_{\mathrm{eff}} t\right) \cdot \sin\left(\phi_k^E\right) \\ \sin\left(\theta_k^E - \omega_{\mathrm{eff}} t\right) \cdot \sin\left(\phi_k^E\right) \\ \cos\left(\phi_k^E\right) \end{bmatrix} \tag{10.9}$$

where $\theta_\mathbf{k}^E$ and $\phi_\mathbf{k}^E$ are the azimuth and elevation angles measured, respectively, from X_E and Z_E axes and the 'unit vector' $\mathbf{r}_\mathbf{n}^\mathbf{E}$

$$\mathbf{r}_\mathbf{n}^E = \frac{\mathbf{r}_i^E + \mathbf{r}_j^E}{2}$$
$$= \begin{bmatrix} \cos\left(\xi_n^E\right)\cos\left(\chi_n^E\right)\sin\left(\alpha_n^E\right)\cos\left(\beta_n^E\right) - \sin\left(\xi_n^E\right)\sin\left(\chi_n^E\right)\cos\left(\alpha_n^E\right)\sin\left(\beta_n^E\right) \\ -\cos\left(\xi_n^E\right)\cos\left(\chi_n^E\right)\cos\left(\alpha_n^E\right)\cos\left(\beta_n^E\right) - \sin\left(\xi_n^E\right)\sin\left(\chi_n^E\right)\sin\left(\alpha_n^E\right)\sin\left(\beta_n^E\right) \\ \sin\left(\xi_n^E\right)\cos\left(\chi_n^E\right) \end{bmatrix} \tag{10.10}$$

is the above equation obtained from (10.2) and (10.3) expressed in the equivalent reference system.

From (10.8) and (10.9), we notice that the variations of the phase depend on both the horizontal and vertical components of the motion through the parameter ω_{eff}. The expression in (10.8) can be rewritten, using a MacLaurin expansion truncated at first order of the distance $\left(\mathbf{r}_\mathbf{n}^E \cdot \mathbf{p}_\mathbf{k}^E(t)\right)$, as

$$s_{k,n}(t) \approx e^{-j\frac{4\pi}{\lambda}\left[R_n - \left(\mathbf{r}_\mathbf{n}^E \cdot \mathbf{p}_\mathbf{k}^E(0)\right)\right]} \cdot e^{j2\pi f_d^{k,n} t}\mathrm{rect}_T(t), \tag{10.11}$$

where

$$f_d^{k,n} = \frac{1}{2\pi}\frac{d}{dt}\left[\frac{4\pi}{\lambda}\left(\mathbf{r}_\mathbf{n}^E \cdot \mathbf{p}_\mathbf{k}^E(t)\right)\right]\Bigg|_{t=0} \tag{10.12}$$

is defined as the Doppler frequency of the *k*th scatterer of the target as viewed by the *nth* equivalent sensor.

The above echo model for the distributed signal is the basis to analytically derive the DPSF and to develop a proper processing technique able to provide the increased cross-range resolution. A diagram showing the principle highlighting the main slow-time/cross-range processing steps required to coherently combine the signals to obtain the improved cross-range resolution is sketched in Figure 10.8.

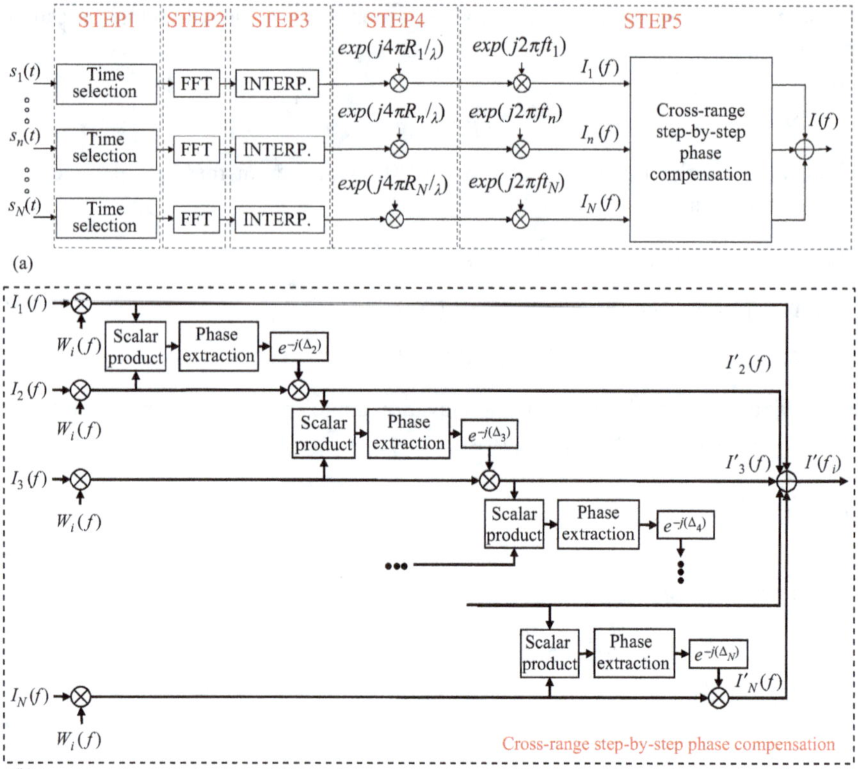

Figure 10.8 Principle of (a) DISAR and (b) cross-range strip by strip phase compensation. In (a) the cross-range strip by strip phase compensation is not shown. ©[2012] IET. Reprinted, with permission, from [33]

Slow-time/cross-range processing comprises five main steps:

1. Time selection which is needed when the observations from the different sensors are partially overlapped. This operation reduces the acquisition time for the individual sensor, discarding a part of the acquired data from time interval T down to T_n. The new instant of centre aperture is defined as ΔT_n [26].
2. Fourier transform which allows to go into the Doppler/cross-range domain and to obtain N low resolution images.
3. Interpolation which sets the same IPP for the N low-resolution focused images, being the same scatterer re-aligned on the same Doppler frequency f_d^k, that can be expressed as in (10.12) for the DLOS in the formation whose position is identified by the aspect angle $\alpha_d^E = 0$ and the grazing angle $\xi_d^E = 0$.

4. Phase alignment which compensates the different phase terms depending on R_n due to the different slant-range distances of the N equivalent sensors to the scene centre.
5. Coherent combination which consists of three operations. The first is performed by multiplying each low-resolution image by a phase term corresponding to an appropriate time shift t_n to align the time selected acquisitions, as the corresponding data were acquired by a single monostatic equivalent sensor. The obtained images ($I_n(f)$, $n = 1,\ldots,N$) are then aligned in phase through a phase compensation procedure (Figure 10.8(b)) and finally are coherently summed to obtain the distributed ISAR image with improved cross-range resolution.

Applying the processing in Figure 10.8(a) with parameters matched to the whole 3D motion of the target, the following expression can be obtained for $I_n(f)$ image $n = 1, \ldots, N$:

$$I_n(f) = K_{k,n} \cdot T_n \cdot \mathrm{sinc}\left[\pi\left(f - f_d^k\right)T_n\right] \cdot e^{-j2\pi\left(f-f_d^k\right)(t_n - \Delta t_n)} \tag{10.13}$$

where $K_{k,n}$ represents a phase term depending on both the considered equivalent sensor (n) and the considered scatterer (k). It could be proven that when all the equivalent sensors are located on the (X_E, Y_E) plane (ideal formation) then $K_{k,n} \equiv K_k$ and, thus, the images can be directly combined to get the final image, i.e. skipping the cross-range strip-by-strip phase compensation. In contrast, in all other cases, the change of $K_{k,n}$ with the considered sensor has to be properly taken into account: directly summing the I_n images without a prior phase alignment would prevent the achievement of the DPSF with the desired characteristics. Moreover, this effect is spatially variant: scatterers belonging to the IPP (*i.e.* $\phi_k^E = \pi/2$) will be almost ideally focused, while scatterers outside the IPP ($\phi_k^E \neq \pi/2$) will suffer a degradation depending on their position.

Phase alignment is obtained via the procedure in Figure 10.8(b) aimed at assuring phase continuity between adjacent acquisitions: this procedure works adaptively on the low resolution images I_n without requiring the knowledge of the 3D shape of the target. For each Doppler frequency f_i (for each Doppler/cross-range bin in the image):

1. the corresponding sub-images are extracted by multiplying images I_n $n = 1, \ldots, N$ by a window function $W_i(f)$ of size w centred around f_i;
2. the phase difference between adjacent observations $\Delta^i_{n,n+1}$ is estimated by evaluating the scalar product between the corresponding sub-images and by extracting the phase [40];
3. the phase compensation term is obtained as $C_{n,i} = \exp\left(-j\sum_{m=1}^{n}\Delta^i_{m-1,m}\right)$ being $\Delta^i_{0,1} = 0$ and applied.

The phase compensation procedure provides the sequence of phase aligned images, $I'_n(f) = I_n(f)\big|_{K_{n,k}=K}$, available for the coherent combination. Therefore, under the assumption of equal acquisition time T and angular spacing $\Delta\alpha^E = \alpha^E_{n-1} - \alpha^E_n$ between the equivalent sensors in $X_E Y_E Z_E$, the DPSF can be computed as

$$I(f) = \sum_{n=1}^{N} I'_n(f) = K\left[\frac{(N-1)\Delta\alpha^E}{\omega_{\text{eff}}} + T\right] \text{sinc}\left[\pi\left(f - f_d^k\right)\left(\frac{(N-1)\Delta\alpha^E}{\omega_{\text{eff}}} + T\right)\right] \tag{10.14}$$

with a cross-range/Doppler resolution improvement equal to

$$\gamma = \frac{(N-1)\Delta\alpha^E}{\omega_{\text{eff}} T} + 1 \tag{10.15}$$

as expected from (10.7) and with also a corresponding peak gain due to the coherent integration of the signals acquired from the sensors in the formation. It is worth noting that Equation (10.14) represents the exact DPSF when an ideal formation is involved, while it represents an approximated solution for the other cases valid under the assumption of a correct phase alignment.

10.3.2 Theoretical performance analysis

To confirm and validate the proposed DPSF analysis, a comparison between the theoretical DPSF characteristics and the output of the elaboration over synthetic data is presented. For the following analysis, we have considered a wavelength of 3 cm, a PRF of 1,000 Hz and an integration time $T = 0.16$ s, corresponding to a single-sensor Doppler resolution of 6.25 Hz. The approach is tested on a simulated point-like scatterer of a ship target (a scatterer located on the mast, [23.39, 0, 30.25] m, or in proximity to the bow, [71.33, 0, 3.213] m) generically interested by a 3D rotational motion.

The cases considered in the simulations differ in terms of target motion characteristics and geometry of acquisition of the formation. The specific motion parameters are listed in Table 10.2: six different study cases labelled as A ... F are considered. The formation comprises two systems: at least one among them is active and eventually able to receive and separate also the echo from the

Table 10.2 Simulated study cases description

Simulation	Description	
A	Vertical motion:	$\omega_h = 0.0000$ deg/s, $\omega_v = 2.4544$ deg/s
B	Main vertical, low horizontal motion:	$\omega_h = 0.0286$ deg/s, $\omega_v = 2.4544$ deg/s
C	Main vertical, high horizontal motion:	$\omega_h = 0.8577$ deg/s, $\omega_v = 2.4544$ deg/s
D	~Horizontal motion:	$\omega_h = 2.1441$ deg/s, $\omega_v = 0.0050$ deg/s
E	Main horizontal, low vertical motion:	$\omega_h = 2.1441$ deg/s, $\omega_v = 0.0281$ deg/s
F	Main horizontal, high vertical motion:	$\omega_h = 2.1441$ deg/s, $\omega_v = 0.6268$ deg/s

transmission of other active systems. In particular, if the second system is a receiving-only device, the bistatic DISAR is investigated, while if it is active the MIMO DISAR is tested. Each one of the motion study cases would require a different ideal alignment of the sensors in the formation: particularly in the A case, the ideal formation is multi-aspect, in the D case, the ideal formation is multi-grazing, finally, in the cases B/C and E/F projections of the MA and MG formations, respectively, on the plane orthogonal to the effective rotation vector should be considered. In all cases, the two platforms are located at a distance of 30 km from the target with an angular separation between sensors such that a maximum resolution improvement with respect to the SS case is expected after the DISAR processing, specifically $\gamma = 2$ in the bistatic case and $\gamma = 3$ in the MIMO case.

With reference to the mast scatterer, Figure 10.9(a) and (b) shows for the motion conditions A, C and D, F respectively the DPSF achieved without the cross-range strip-by-strip phase compensation step if considering the sensors displaced according the correspondent ideal formation. This analysis allows us to prove the ability of the DISAR at achieving the theoretical expected cross-range resolution improvement when ideal formations are involved. As is apparent, they are achieved with respect to the SS case (black lines), both in the bistatic case (red lines) and in the MIMO case (blue lines).

Figure 10.10 shows for the same motion characteristics of Figure 10.9, the DISAR PSFs forcing the formation to be MA (Figure 10.10(a)) and MG (Figure 10.10(b)): as is apparent, there is a very good agreement between the output of the whole procedure comprising the cross-range strip-by-strip phase compensation (red and green lines) and what one would achieve without this last elaboration step but considering the ideal formations (blue and black curves). Indeed, only slight differences in the sidelobe region are experienced, that would be greatly mitigated if narrower angular differences were considered between the sensors in the formation.

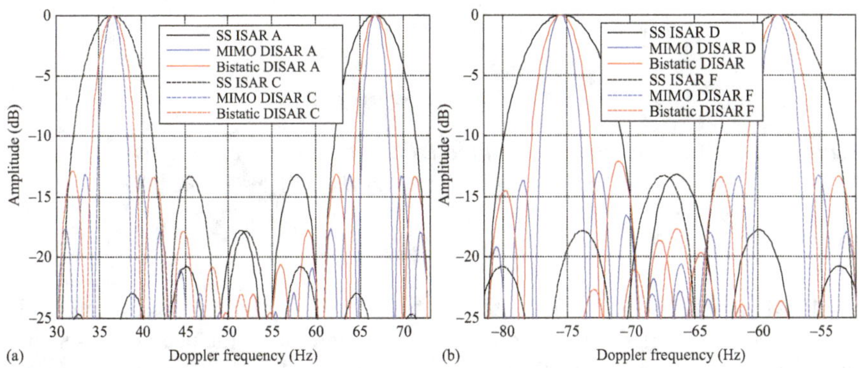

Figure 10.9 DPSFs for the mast scatterer with ideal formations in the dominant (a) vertical motion and (b) horizontal motion cases

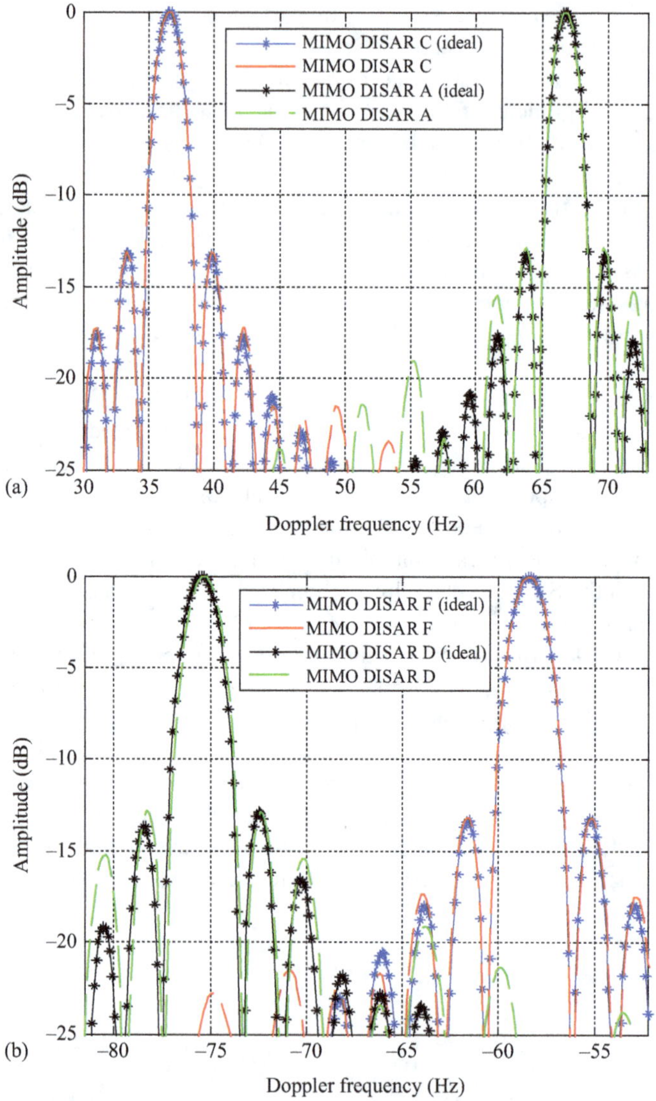

Figure 10.10 Comparison between the DPSFs for the mast scatterer achieved using the ideal and (a) the MA or (b) the MG formations

Finally, Figure 10.11 shows the spatially variant characteristics before the cross-range strip-by-strip phase compensation, highlighting the need of this step to guarantee a homogeneous PSF all over the target. In Figure 10.11(a), the output is shown for the A, B and C cases for the mast scatterer and the same MA formation

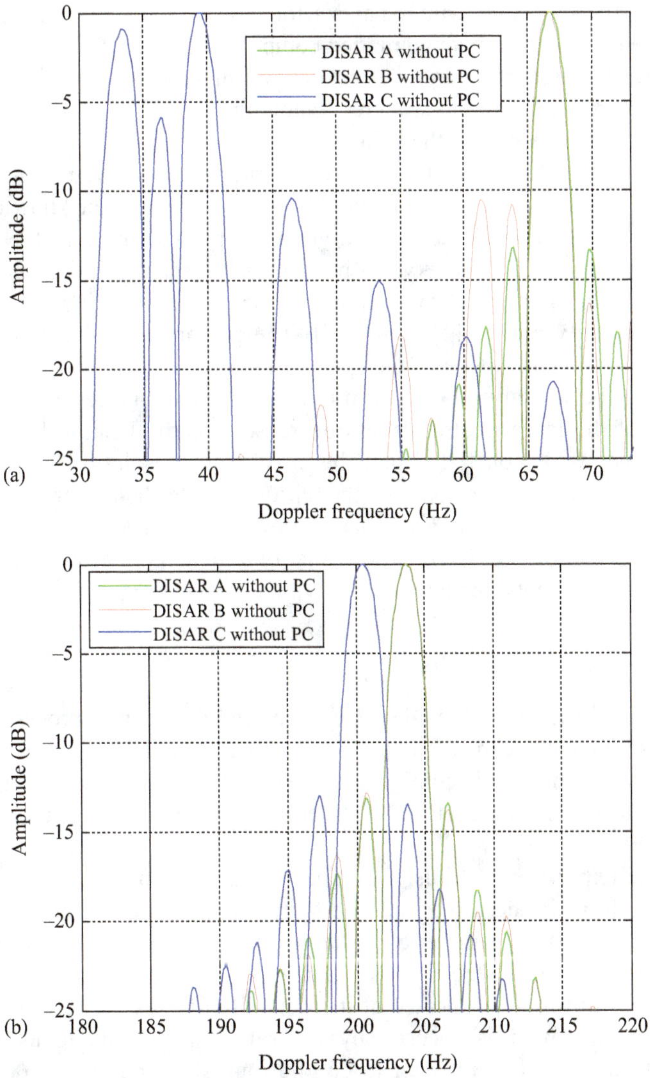

Figure 10.11 Analysis of the space variant effects: DPSFs without phase compensation for (a) mast scatterer and (b) bow scatterer

used for Figure 10.10(a). As is apparent passing from a case where the formation is ideally matched to the motion of the target (A, green curve) to the case were there is a limited contribution of the horizontal motion (B, red curve) only slight effects are experienced in the area of the sidelobes. On the other end, the DPSF completely splits when the horizontal component of the motion increases (C, blue curve).

Figure 10.11(b) shows the same analysis for bow scatterer: as it is apparent the effect of the increasing horizontal motion component on the DPSF characteristics is very limited. This spatially variant effect derives directly from the higher impact of pitch and roll rotations on the scatterers belonging to the mast with respect to the ones belonging to the hull of the ship.

It has to be noticed that the coherent exploitation of the multi-sensor data requires the synchronization among the different sensors and the knowledge of the target rotation rate and of the acquisition geometry. The impact of synchronization errors in multistatic imagery has been already analysed in [41] and, to solve the problem, different synchronization schemes have been proposed and their performance analysed in [42] or in [43]. The performance degradation arising from errors in the knowledge of the target rotation motion and of the acquisition geometry has been analysed in [26] in terms of cross-range resolution loss, misplacement of the target scatterers and decreasing of the Peak-to-Sidelobe ratio. This analysis has demonstrated that the cross-range resolution improvement is basically maintained under non-ideal conditions, while the performance in terms of both the Peak-to-Sidelobe Ratio and the misplacement of the scatterers show a limited degradation provided that the rotation rate and the acquisition geometry are known with acceptable accuracy.

10.3.3 Experimental validation

With the aim of validating the DISAR concept and the devised processing technique, an experimental campaign has been carried on: DISAR data have been acquired in a MA multistatic configuration against a target rotating on a turntable in an anechoic chamber at the SELEX Galileo (now Leonardo SpA) base in Caselle (Turin, Italy).

In these experimental trials, a compact range system is used to generate a planar wave front in the test zone. The system includes a parabolic reflector, a system of feeds and a positioner of the target under test. The reflector is an offset parabolic reflector P/N 5755 made by Scientific Atlanta. A spherical wavefront from the transmitting feed located at its focus illuminates the reflector, then the signal from the target is focused by the reflector into its focus where the receiving feed is located. In order to build a multistatic configuration, a second receiving feed has been added 60 cm from the first one, providing the bistatic channel with bistatic angle β equal to 4.3° in addition to the monostatic receiving channel. An HP 8510C Network Analyzer has been used for the measurements acquisition. The system transmits a stepped frequency waveform in the Ku-band (16.5 GHz) with an equivalent overall bandwidth of 3 GHz and a frequency step $\Delta f = 3.75$ MHz.

The target used for these trials is sketched in Figure 10.12. As we can see, a cross-shaped metallic structure with 4 m-long arms supports a set of metallic cylinders of diameter 2 cm; the heading of this structure with respect to the direction of the incident wave can be set *a priori*. Finally, this target is placed

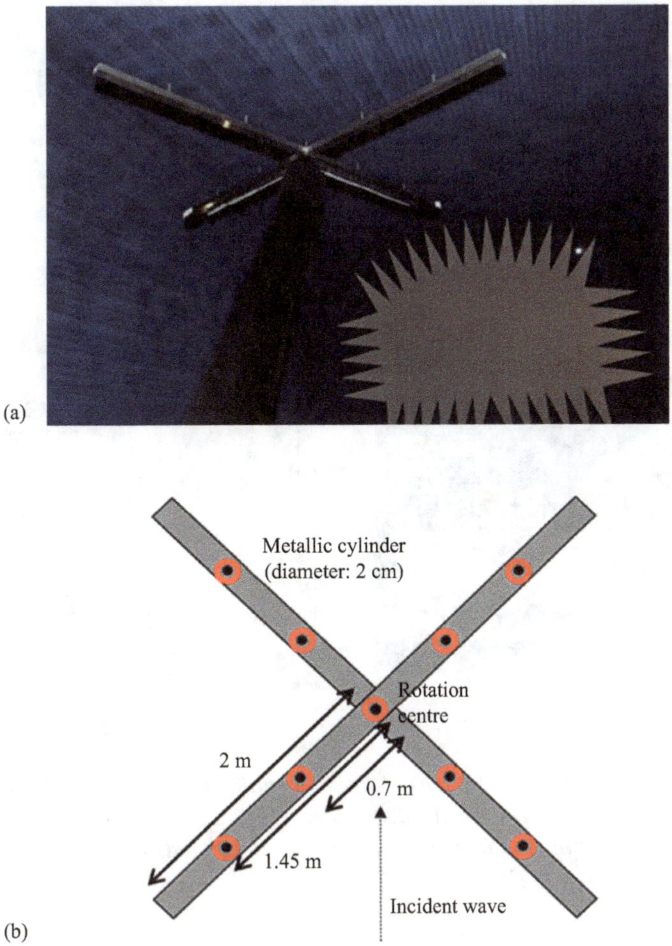

Figure 10.12 Target: (a) real structure and (b) scheme. (a) Courtesy of Selex Galileo (now Leonardo SpA). (b) ©[2011] IEEE. Reprinted, with permission, from [32]

on a turntable rotating with an angular separation burst to burst of about 0.1°. As is apparent, the experimentation allowed us to test a specific case of our more general problem that is an MA multistatic formation whose spatial displacement is ideally matched to the target motion which is a 1D rotation around the vertical axis. Therefore, we expect the distributed technique to provide nearly ideal results in terms of achievable cross-range resolution improvement and PSF even without the implementation of the cross-range strip-by-strip phase compensation.

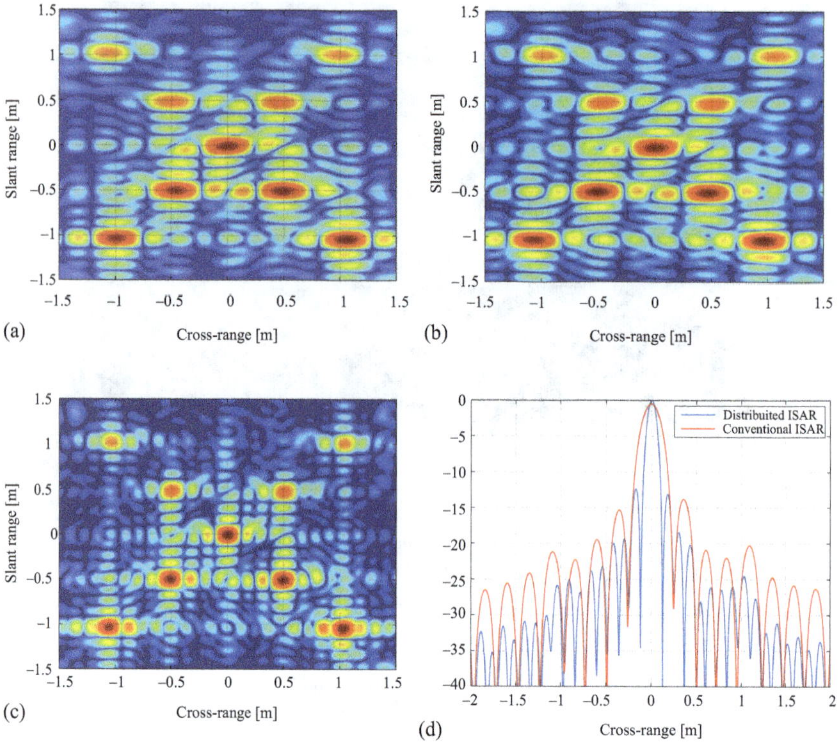

Figure 10.13 Normalized images – (a) monostatic (12 × 24) cm^2; (b) bistatic (12 × 24) cm^2; (c) distributed (12 × 12) cm^2; (d) cross-range cut at sr = 0 m. ©[2011] IEEE. Reprinted, with permission, from [32]

The processing of these acquired data aims at proving the capability of the distributed system to provide an ISAR product with increased cross-range resolution. In order to achieve the maximum improvement, data acquired over an observation angle equal to $\beta/2$ are separately selected at each receiving channel: therefore, for a carrier frequency of 16.5 GHz, in the DISAR image, we expect a cross-range resolution $\Delta r_{cr}^{(2)} = \lambda/(2\beta)$ nearly equal to 12 cm. Moreover, only a portion (1.24 GHz) of the available bandwidth is considered to show a DISAR image with a square resolution cell and the initial heading of the target structure is set to 22.5°. Figure 10.13 shows the obtained images for conventional ISAR (both monostatic and bistatic channels) with $\Delta r_{cr} = 0.24$ m and multistatic ($S = 2$) DISAR case with $\Delta r_{cr}^{(2)} = 0.12$ m; the comparison of the cross-range PSFs is reported in Figure 10.13(d) showing a very good agreement between theoretical and experimental results. Results concerning extended targets will be shown in Section 10.4.3.

10.4 Distributed ISAR image formation

The focusing of a target image with improved cross-range resolution from distributed ISAR data obviously requires an ad hoc bi-dimensional (2D) processing technique able to properly combine the radar signals acquired from the multiple sensors. The 1D (azimuth) processing chain in Figure 10.8 assumes the availability of the azimuth history of the same range bin from the multiple acquisitions and shows the conceptual steps required to coherently combine the multi-sensor data in order to obtain the improved cross-range resolution. However, the correction of the migration through range and Doppler resolution cells is not considered in the presented scheme, which is more a proof of principle introduced to show the validity of the distributed ISAR concept and derive the corresponding cross-range PSF.

The aim of this section is to describe the 2D-distributed ISAR-focusing technique that can be used to obtain the image of the target with the increased cross-range resolution. Therefore, suitable 2D-focusing techniques for distributed ISAR data are discussed in Section 10.4.1, the corresponding theoretical performance (under ideal and non-ideal conditions) analysed in Section 10.4.2 and, finally, the experimental validation is shown in Section 10.4.3.

10.4.1 DISAR focusing techniques

As is typical in ISAR imaging, we assume that chirped radar waveforms are transmitted and a dechirping approach is used on reception. Therefore, the dechirping with the corresponding deskew for each equivalent sensor is in any case the first step of the technique.

The dechirping in range with its deskew is quite standard (see e.g. [2,44]) for the multistatic ISAR case. In contrast, in the MIMO case, the procedure is slightly more complex, since we have to take into account the different transmitted waveforms. Specifically, with S platforms, S almost orthogonal waveforms are needed. For the case studies here considered $S \leq 4$ and the waveforms are selected from the set sketched in Figure 10.14(a) showing an up-chirp, a down-chirp and two triangular frequency modulated waveforms. The data collected by each equivalent sensor should therefore be pre-processed according to the specific waveform used by the corresponding transmitter: assuming that in the multistatic case an up-chirp is used

- the dechirp and deskew step concerning sensors using an up-chirp are unchanged with respect to the multistatic case;
- a time-reversal operation is added after dechirping and deskewing for sensors using a down-chirp;
- the dechirp and deskew step is replaced by the processing chain in Figure 10.14(b) for those sensors using the triangular frequency modulated waveforms: the time scale operation is required to provide an output signal with time duration equal τ (instead of $\tau/2$) as for those sensors using the up- or down-chirp waveforms.

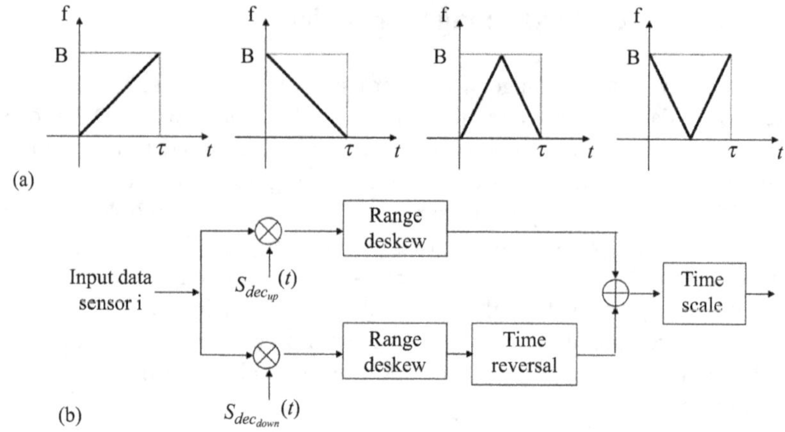

Figure 10.14 (a) Waveforms (in frequency domain) for the MIMO distributed ISAR with N = 4 platforms; (b) range dechirp and deskew processing chain for the triangular frequency modulated waveforms. ©[2010] IEEE. Reprinted, with permission, from [26]

In any case, we can observe that also matched filtering could be used for range compression in place of dechirping. Therefore, the processing schemes proposed in this section can be easily generalized to cope with arbitrary orthogonal waveforms transmitted by a formation of *S* platforms (without any prior assumption on the value of *S*) simply by replacing each dechirp and deskew block with the cascade of the filter matched to the specific waveform used by the corresponding transmitter and a Fourier transform in range domain.

By operating after dechirping and deskew, two different approaches can be followed to obtain a full 2D-distributed ISAR scheme:

- *Centralized technique for distributed ISAR focusing (CT-DISAR)*: the radar data from the *N* equivalent sensors must be properly pre-processed, coherently combined and finally focused using an appropriate 2D-processing scheme.
- *Decentralized technique for distributed ISAR focusing (DT-DISAR)*: *N* low resolution ISAR images, as collected by each single equivalent sensor, are first focused and then coherently combined to achieve the enhanced resolution image.

The approach proposed here is a decentralized technique for multi-angle ISAR focusing, based on a modified version of the polar format algorithm (PFA), [44]. The decentralized approach has the main advantage of providing not only the final improved ISAR image but also the low resolution ISAR images corresponding to the different equivalent sensors. The processing chain is shown in Figure 10.15 for the case of *N* equivalent sensors. It is possible to see how the operation of the traditional PFA-focusing algorithm is integrated with the steps needed by the distributed processing.

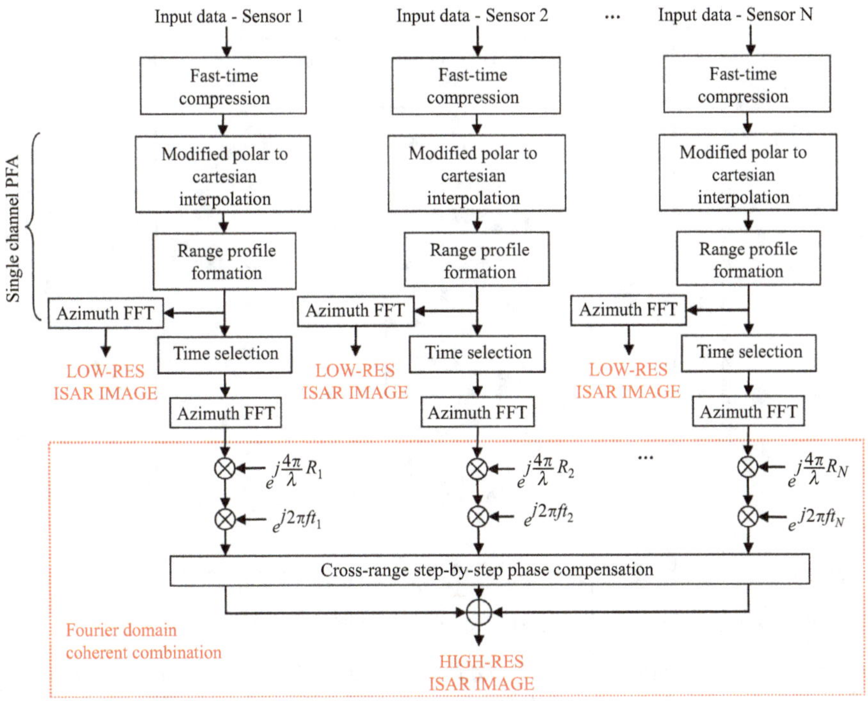

Figure 10.15 Modified decentralized processing chain for D ISAR. ©[2012] IET. Reprinted, with permission, from [33]

For each branch in the scheme (i.e. for each equivalent sensor), the processing is organized in the following steps:

1. Fast-time compression and if needed range Fourier transform (if the fast-time compression takes place via matched filtering instead of dechirping).
2. Modified Polar-to-Cartesian interpolation which removes the range and Doppler migration from SS data (as in conventional PFA) and sets the same IPP for the N low-resolution images (specific for distributed processing).
3. Range profile formation which transforms the data to the range-compressed and azimuth-frequency domain.

   At this point, the low resolution single-sensor ISAR image can be obtained as intermediate output by applying an azimuth Fourier transform to the data coming from step 3. The following steps refer specifically to the processing of the multi-sensor data:
4. Time selection as explained in Section 10.3.1.
5. Azimuth Fourier transform which allows to go in Doppler/cross-range domain.
6. Phase alignment which compensates the different phase terms due to the different slant-range distances of the N equivalent sensors to the scene centre.
7. Coherent combination which consists of three operations, namely time alignment, phase alignment and coherent integration as explained in Section 10.3.1.

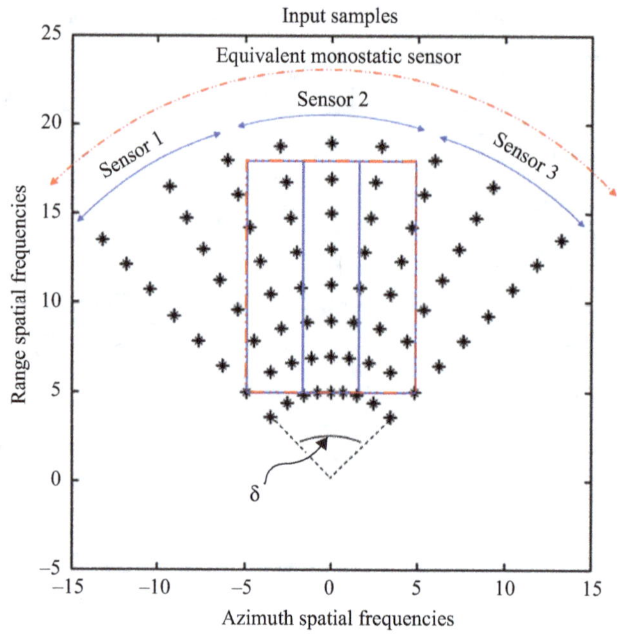

Figure 10.16 Output grid for the modified polar-to-Cartesian interpolation. ©[2011] IEEE. Reprinted, with permission, from [30]

In order to perform the Modified Polar-to-Cartesian interpolation the input polar grid, Figure 10.16, must take into account the angular displacement of the data with respect to the angle of the distributed equivalent acquisition; moreover, the output rectangular grids must not be defined separately for all the images but they need to be defined for all the equivalent images such that the alignment of all the output grids (blue solid line in Figure 10.16) corresponds to the case of a monostatic acquisition whose angular aperture is equal to the distributed equivalent aperture (red dashed line in Figure 10.16).

10.4.2 Theoretical performance analysis

The DT-DISAR focusing technique is here applied to synthetic data from an extended target consisting of a ship with about 500 scatterers, length 120 m and different levels of super-structure; a top view and a side view of the model of the target are shown in Figure 10.17. A formation of $S = 4$ flying platforms is considered: the different analysed study cases correspond to different characteristics of the target motion and, as a consequence, to different ideal alignments of the platforms in the formation; the consideration of ideal alignments allows the application of DT-DISAR without the cross-range strip-by-strip phase compensation. As already pointed out, for the distributed configuration with two or more active platforms to work (MIMO case), a set of ideally orthogonal waveforms is

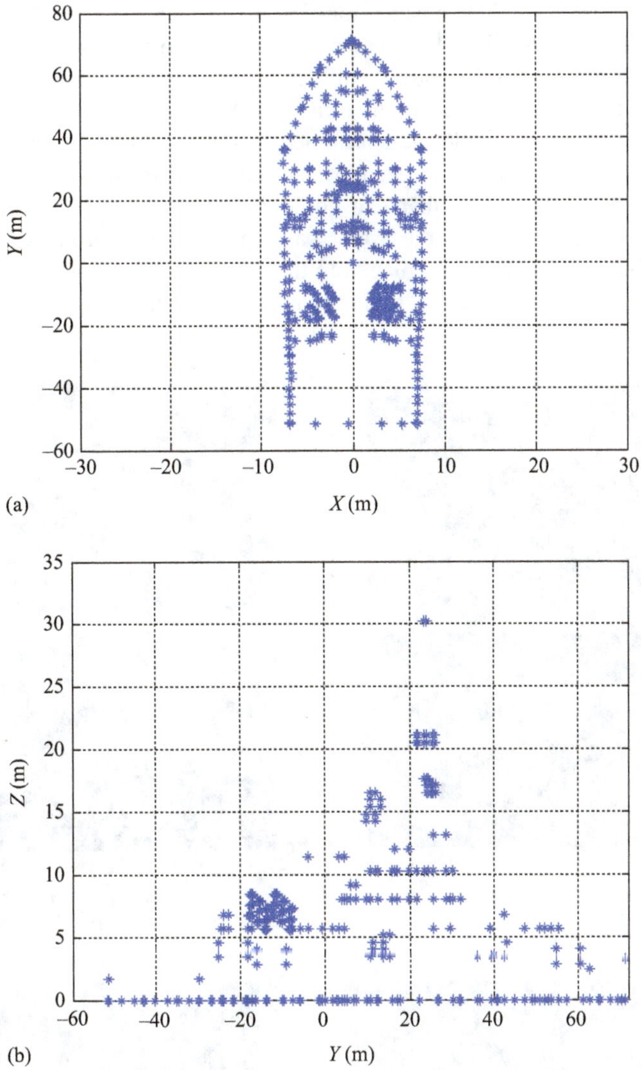

Figure 10.17 (a) Top view and (b) side view of the model of the ship target. (b) ©[2011] IEEE. Reprinted, with permission, from [30]

needed: in practice, the set of linearly modulated waveforms with bandwidth B and pulse length τ sketched in Figure 10.14(a) is exploited; this will introduce non-ideal effects in the final image due to the presence of a floor of not compressed residual signal. However, these disturbance effects will be presumably insufficient to actually affect the quality of the image in terms of PSLR and slant range resolution.

In the first study case, the ship experiences a dominant yaw motion with constant rotation rate $\omega = 3.6°/s$, amplitude of the sinusoidal yaw component $A_y = 1°$ and frequency $f_y = 0.21$ Hz. The observation time (coincident with the CPI) is 0.08 s, PRF = 250 Hz and $B = 300$ Hz. In the conventional monostatic case the cross-range resolution is nearly 2.1838 m, while the slant range resolution is 0.5 m. This corresponds to a typical case, in which the slant range resolution, set as a system parameter once the bandwidth is determined, is quite high, while the cross-range resolution, depending on the target intrinsic motion characteristics, is poor, thus providing a rectangular cell that can compromise the correct extraction of the features of the target or, in the worst case, prevent the correct classification. Figure 10.18(a) shows the ship imaged in the conventional way (SS), while in the zoom we focus in particular on the scatterers around the point (0, 56) m in the

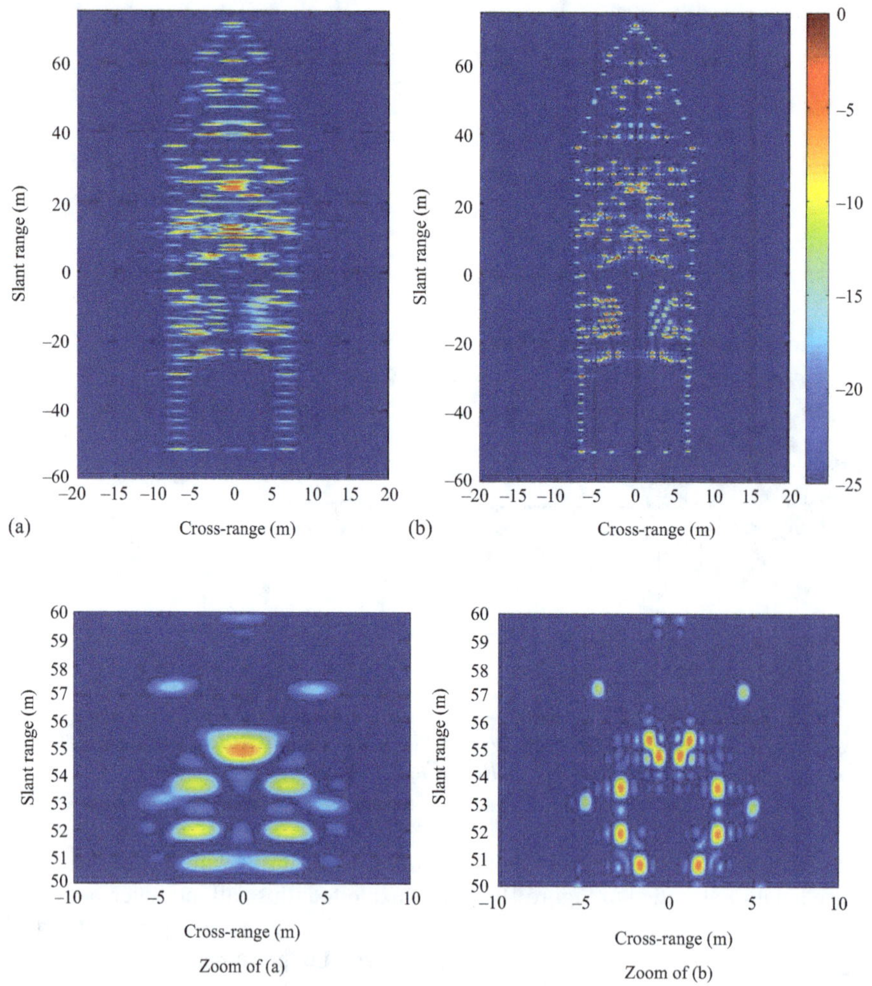

Figure 10.18 First study case ISAR images: (a) Conventional ISAR ($S = 1$) with CPI = 0.08 s and (b) DISAR in multistatic case ($S = 4$)

image. Figure 10.18(b) shows the output of the multistatic DISAR case with $S = 4$ platforms, displaced according to the ideal MA alignment and spaced in order to achieve the maximal cross-range resolution improvement, that in this case is equal to four. It is clearly visible how the cross-range resolution improves, providing a resolution cell that is nearly square (0.55 m × 0.5 m); the inspection of the zooms shows how the resolution improvement allows the separation between the different scatterers of the target.

In the second study case, the ship experiences the same motion but the CPI is set to be half of the previous one, that is 0.04 s, leading to a cross-range resolution of nearly 4.3675 m in the conventional case. Figure 10.19(a) shows the ship imaged in

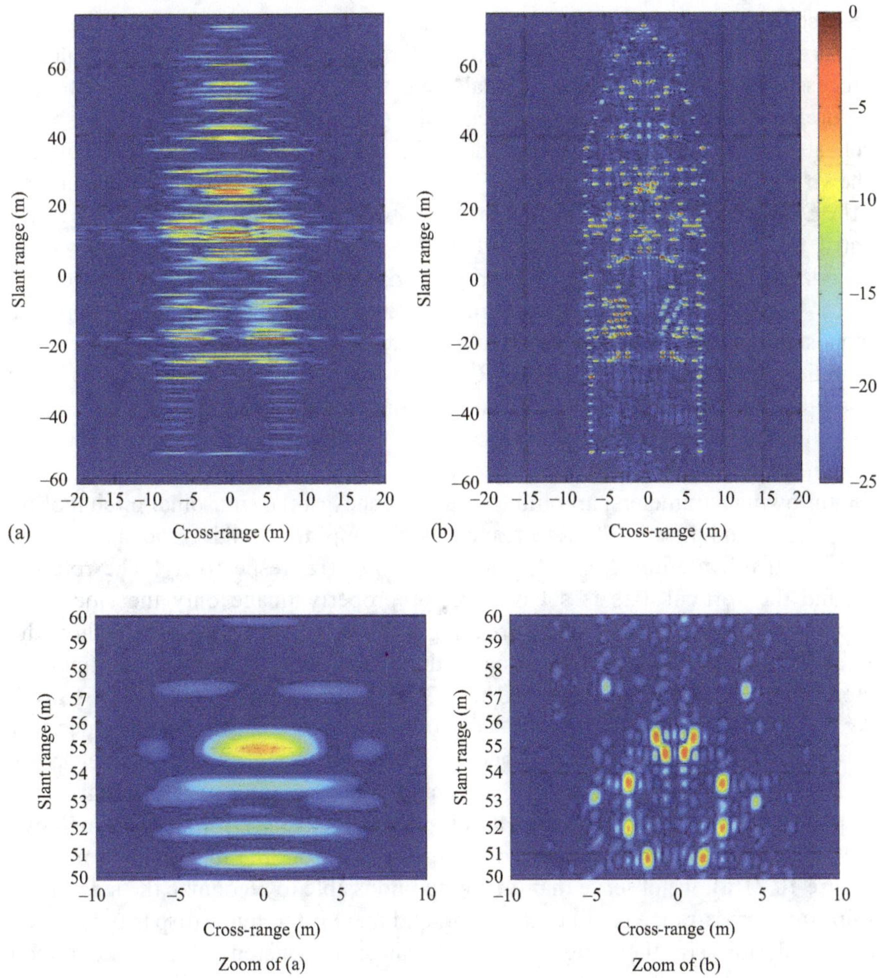

Figure 10.19 Second study case ISAR images: (a) Conventional ISAR ($S = 1$) with CPI = 0.04 s and (b) DISAR in MIMO case ($S = 4$)

the conventional way and again in the zoom we focus on the scatterers around the point (0, 56) m. Figure 10.19(b) shows the output of the MIMO DISAR case with $S=4$ platforms, displaced according to the ideal MA alignment and spaced in order to achieve the maximal cross-range resolution improvement, that in this case is equal to 9. It is clearly visible how the cross-range resolution improves, implying a resolution cell that is nearly square (0.48 m $\times$ 0.5 m); again the inspection of the zooms shows how the resolution improvement allows the separation between the scatterers in the image. In contrast to the multistatic case, where a single waveform was employed, a floor of uncompressed signal is evident the MIMO case due to the use of thc four chirped waveforms in Figure 10.14(a).

From a comparison between Figure 10.18 and Figure 10.19, it is apparent how the MIMO DISAR allows a greater improvement in cross-range resolution with respect to the multistatic case: in particular, it allows achieving almost the same performance of the multistatic DISAR in terms of output cross-range resolution processing a shorter (half) time interval. This is also confirmed by the inspection of Figure 10.20, where the cross-range cuts of a group of scatterers in the slant range cell at ~56 m (Figure 10.20(a)) and of the fore scatterer (Figure 10.20(b)) are shown for the monostatic case with CPI = 0.04 s (black curve), the corresponding MIMO case with four platforms (green curve) with a resolution improvement factor equal to 9, the monostatic case with CPI = 0.08 s (blue curve) and the corresponding multistatic case with four platforms with a resolution improvement factor equal to 4. In particular, Figure 10.20(a) shows how, starting from the monostatic case where the presence of a couple of scatterers cannot be declared for both the simulated CPIs, after DISAR both in the multistatic and the MIMO cases two different scatterers are clearly visible. Moreover, an inspection of Figure 10.20 (b), allows us to confirm the theoretical expected values of cross-range resolution.

In the third study case, the target has only a horizontal motion described by the following parameters: amplitude and frequency of the sinusoidal pitch motion $A_p=0.25°$ and $f_p=0.178$ Hz, respectively, amplitude and frequency of the sinusoidal roll motion $A_r=1.25°$ and $f_r=0.091$ Hz, respectively. The rotation around the vertical axis is set to zero, to properly image only the side view component. The used parameters are typical of low/medium sea state: to show the effectiveness of the proposed approach, the CPI is set to 0.7 s, thus resulting in a cross-range resolution nearly equal to 4.37 m when conventional ISAR is used. Again the slant range resolution is set to 0.5 m. Figure 10.21 shows the side view image of the ship target as obtained by the conventional single-channel ISAR technique and DISAR in the multistatic and MIMO cases with $S=4$ platforms, displaced according to the ideal MG alignments and spaced in order to achieve the maximal cross-range resolution improvement: from the monostatic image in Figure 10.21(a), we observe that it is quite impossible to recognize the side of the ship, since cross-range resolution is poor, and there is too much disparity between the resolutions in the range and cross-range dimensions. Results obtained using a formation of $S=4$ flying platforms are shown in Figure 10.21(b) for the multistatic case with a cross-range resolution improvement equal to four and in Figure 10.21(c) for the MIMO case with a cross-range resolution improvement

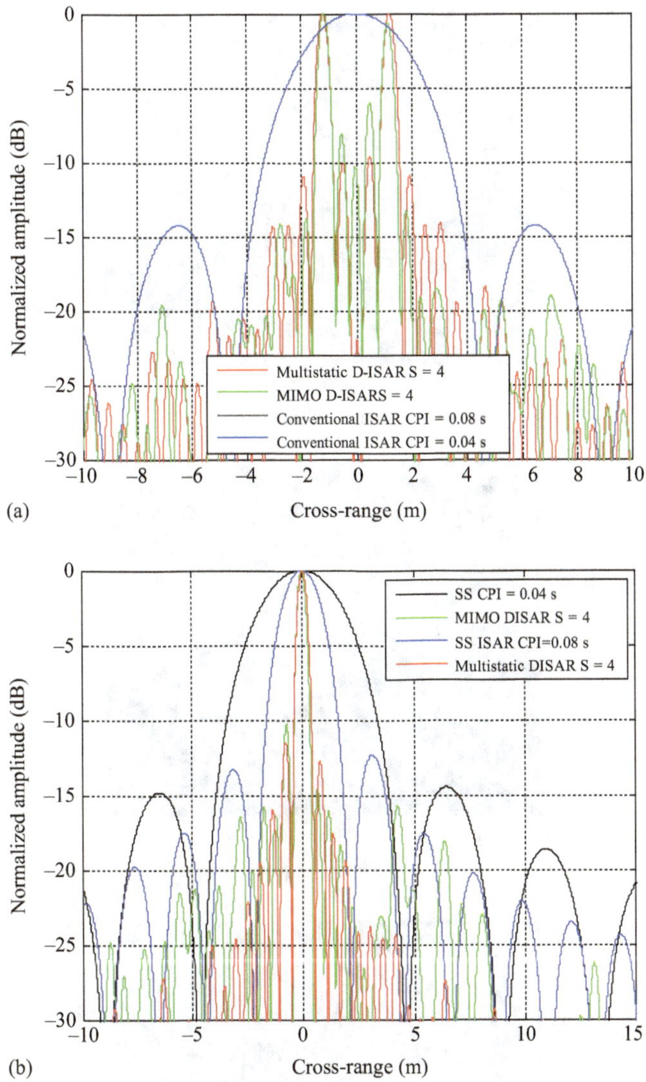

Figure 10.20 Cross-range cut comparisons: (a) slant range cell at ~56 m and (b) fore scatterer

equal to nine. To show the effectiveness of the technique, we observe the zooms of the images around the point in slant-range/cross-range (13.7, 4.3) m, smallest frames in Figure 10.21, and the cross-range cuts at slant-range set nearly equal to 13.7 m, Figure 10.22(a). It is evident how the cross-range resolution improves and therefore, moving from the conventional to the MIMO DISAR image, the scatterers progressively separate until it is possible to recognize four distinct point scatterers. Finally, an inspection of the cross-range cut of the isolated scatterer of the slant

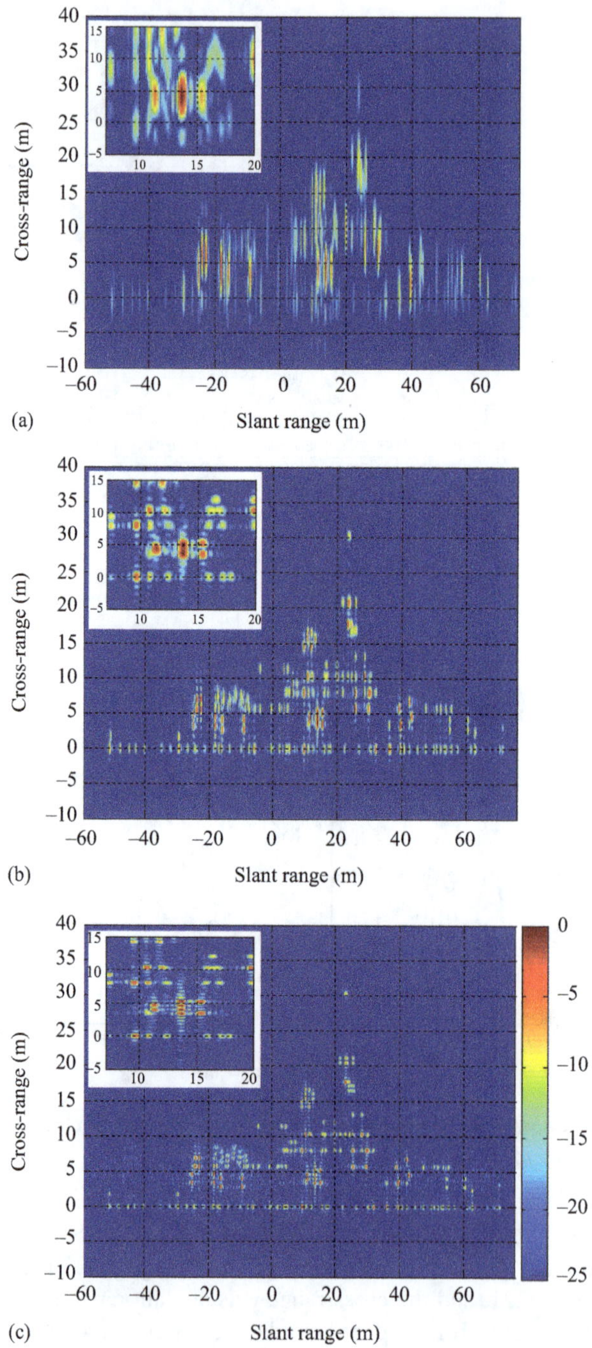

Figure 10.21 *Third study case ISAR images: (a) Conventional ISAR (S = 1), (b) DISAR in multistatic case (S = 4) and (c) DISAR in MIMO case (S = 4). ©[2011] IEEE. Reprinted, with permission, from [30]*

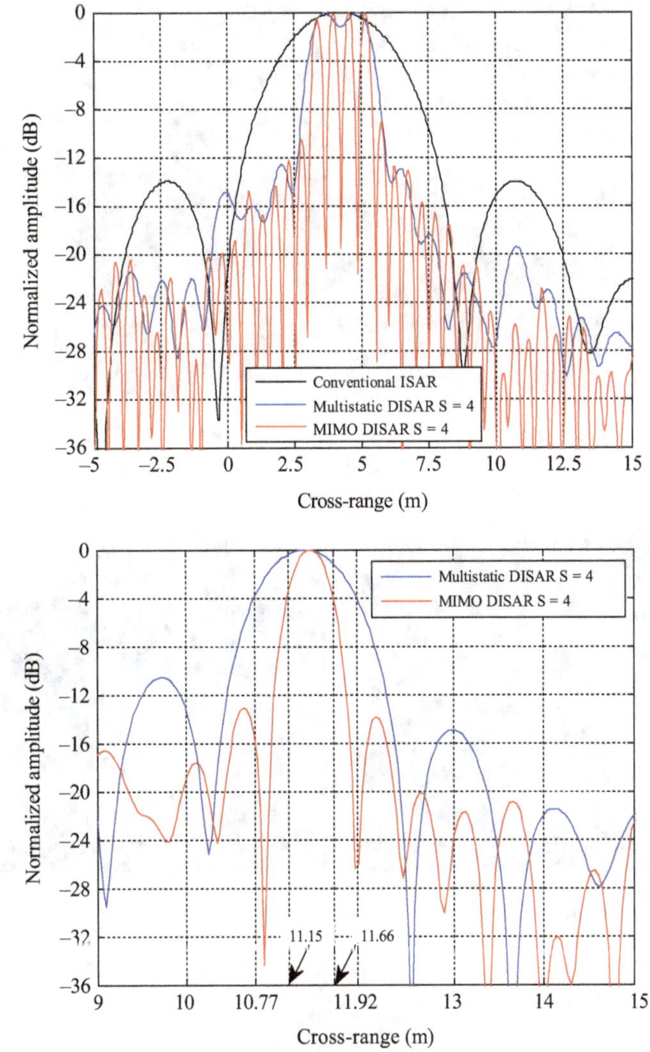

Figure 10.22 Cross-range cut comparison between conventional ISAR and multistatic/MIMO multi-grazing DISAR. ©[2011] IEEE. Reprinted, with permission, from [30]

range cell ∼−4 m in Figure 10.22(b) allows us to confirm the theoretical expected values of cross-range resolution.

10.4.3 Experimental validation

From the same experimental campaign described in Section 10.3.3, some experimental results are here presented obtained by applying the proposed focusing technique to DISAR data acquired against an ATR 42 model (Figure 10.23) at

Figure 10.23 1:20 scale ATR 42 in the anechoic chamber. Courtesy of Selex Galileo (now Leonardo SpA)

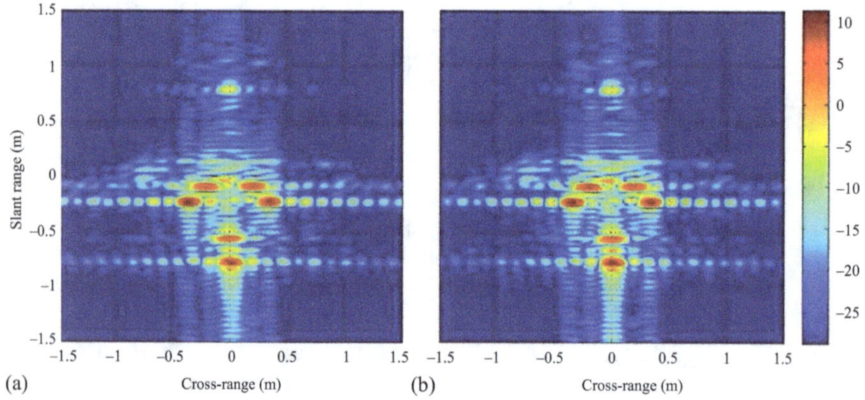

Figure 10.24 Aircraft model images (a) Conventional monostatic ISAR, (b) DISAR

Ku-band. Such multi-sensor data acquired against a complex target can be first exploited in order to experimentally prove the validity of the assumption of the pseudo-monostatic RCS region. To this purpose, Figure 10.24 compares the image resulting when an overall illumination angle equal to $\Theta = 4.3°$ is selected from the monostatic channel (Figure 10.24(a)) to the one obtained by selecting an illumination angle equal to 2.15° from the monostatic and 2.15° from the bistatic channel, so that a distributed overall aspect angle equal to 4.3° is again obtained (Figure 10.24(b)). An excellent correspondence between the two of them can be observed: this confirms the validity of the hypotheses of coherency of the scatterers echoes and of stability of the positions of the scattering centres in the case of bistatic angles of the order of few degrees.

As an example, to show the effectiveness of the DISAR concept and focusing technique, Figure 10.25 shows the conventional monostatic ISAR image

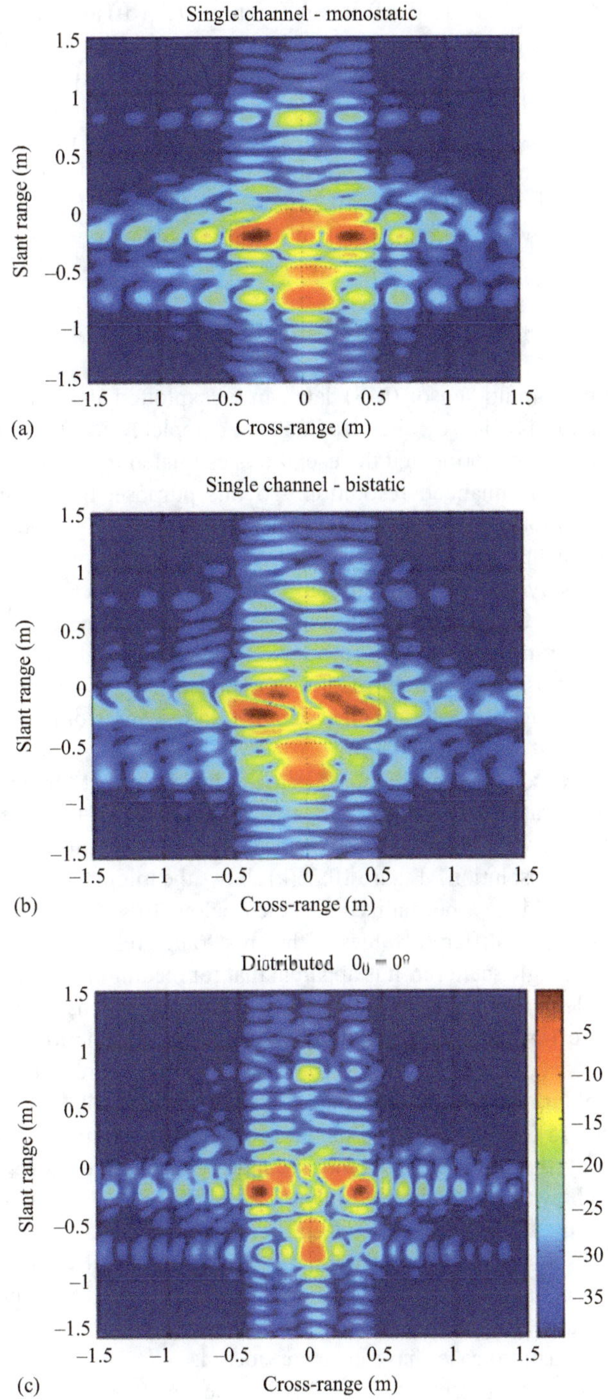

Figure 10.25 Aircraft images: (a) monostatic (12 × 24) cm²; (b) bistatic (12 × 24) cm²; (c) distributed (12 × 12) cm²

(Figure 10.25(a)), the bistatic ISAR image (Figure 10.25(b)) and the DISAR image (Figure 10.25(c)) obtained by jointly exploiting both monostatic and bistatic acquisitions. As evident by comparing these figures, the exploitation of multi-sensor data allows the improvement of the cross-range resolution of a factor two thus allowing the discrimination of features not appreciable in the standard single-sensor case. These results clearly demonstrate that DISAR could be useful not only to improve resolution (as in Figure 10.25) but also to lower the required aperture time to achieve a desired resolution (as in Figure 10.24).

10.5 Motion estimation based on distributed ISAR data

The availability of multi-sensor (MS) data can be exploited also for target rotation motion estimation, [10,11,16], thus leading to a complete processing chain allowing the estimate of the motion and the use of this estimated information to focus the DISAR image with enhanced resolution. To this purpose, both maximum likelihood (ML) and model free (based on contrast in amplitude CO_A and intensity CO_I or entropy He) techniques have been proposed and their performance analysed in [11,16] under many different conditions. Here, in order to clearly highlight the advantages coming from the multiple sensors, the focus will be on the analysis of the performance improvement with respect to the conventional single-sensor case, [45]. To this purpose, an ideal formation is assumed and the performance in the estimation of the effective rotation rate is compared moving from the Cramer–Rao lower bounds reported in [11,45].

The single-sensor case results in [45] clearly highlight that the accuracy of the estimated rotation rate depends on the target size, on the signal-to-noise ratio (SNR, evaluated after coherent integration) and on the coherent processing interval (and thus on the cross-range resolution). Figure 10.26(a) shows the minimum target size versus SNR required to achieve a normalized accuracy below 30% (in blue), 20% (in red) and 10% (in green) for different values of the cross-range resolution when the target consists of just a single scatterer. It is apparent that for medium/low values of SNR to obtain acceptable performance, we need targets with large dimension or a high value of time aperture (cross-range resolution). When these assumptions fail, possible alternative approaches can be constituted by ad hoc techniques properly developed for specific kinds of target incorporating some *a priori* knowledge of the target structure, see for example [5], or by multi-sensor techniques exploiting the data acquired by multiple radar systems. Considering this second option, Figure 10.26(b) compares performance achievable with the SS to that achievable by a MIMO formation with $S=2$ ($N=3$), angular separation between the two platforms 5° and coherent processing interval set so that the conventional single-sensor ISAR images have cross-range resolution equal to 50 cm for the same kind of target used for Figure 10.26(a). For a fair comparison, SS, MS and SS_{av} (separate exploitation and average over the sensors in the formation) cases have been considered.

Different colours represent different SNR values, while different symbols refer to the different estimation approaches. As it is apparent, the normalized accuracy

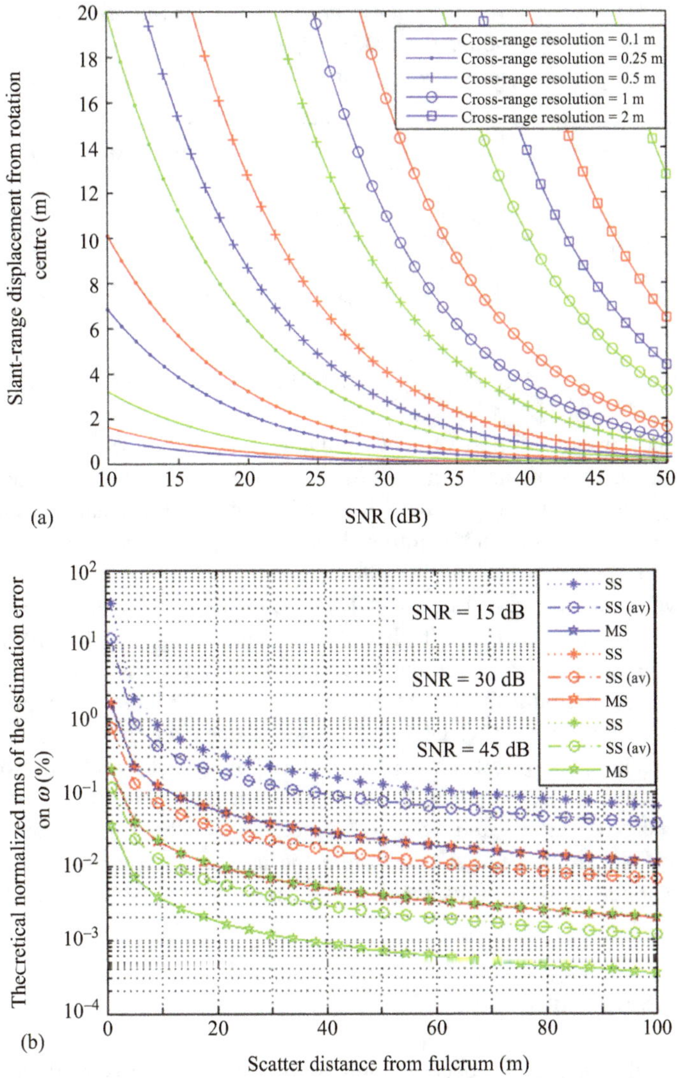

Figure 10.26 (a) Minimum target size required by single-sensor techniques to achieve a given normalized accuracy as a function of SNR for different cross-range resolution values; (b) normalized accuracy for both single- and multi-sensor cases as a function of target size d for several SNR values. (b) ©[2012] IEEE. Reprinted, with permission, from [11]

improves as the target size (i.e. scatterer distance from fulcrum) increases; moreover, fixing the size, the error decreases when SNR increases. It is also visible how the MS approach outperforms SS/SS_{av} techniques, implying that, for a specific SNR, a given normalized accuracy can be achieved with targets of considerably

Table 10.3 Normalized estimation error for the aircraft model [%]. ©[2012] IEEE. Reprinted, with permission, from [11]

	SS	SS_{av}	MS ($\alpha_{AB} = 4.3°$)
ML	116.00	32.14	7.74
CO_A	128.00	82.13	−4.16
CO_I	123.80	53.56	1.19
H_e	127.40	61.30	−6.55

smaller size moving from SS, SS_{av} to the MS case, thus allowing to go beyond SS limits.

Such performance improvement of the MS over the SS case theoretically quantified by means of the Cramer–Rao lower bound can be also experimentally observed again referring to the data acquired in the experimental campaign previously described. As an example, some results concerning the ATR42 (Section 10.4.3) are reported in the following.

An overall angle interval of about 1.74° has been considered from the monostatic and the bistatic channels (to have 30 cm cross-range resolution), being the bistatic angle equal to 4.3°, as in the previous analysis. For the estimation process, the range cell containing the scatter corresponding to the nose of the aircraft has been selected and ML, CO_A, CO_I and He techniques for the SS, SS_{av} and MS estimation techniques have been applied, [11,16]. Results, in terms of normalized estimation error, are reported in Table 10.3. From this table, it can be seen the ineffectiveness of the SS and SS_{av} techniques due to the small overall angular aperture and the limited target size, while the use of just two sensors allows to recover good performance, thus enabling the achievement of focused and properly scaled images.

10.6 Conclusion

The focus of the chapter has been on multi-sensor ISAR systems for target imaging with enhanced quality. Particularly, the distributed ISAR concept has been presented in order to exploit the data acquired by multiple radar systems carried by multiple air platforms to increase the cross-range resolution of ISAR images. The distributed ISAR technique has been devised for the two different multistatic and MIMO cases that correspond to potential application scenarios for which this technique is especially suitable. Having in mind the application of ship-target imaging in the presence of very limited rotation (low sea state), the multistatic case corresponds to the use of a hovering helicopter (which can also act as a stand-off platform) and a set of UAV (with receiving only sensors) flying appropriately in formation, while the MIMO case can correspond for example to the use of a set of helicopters (with transmit and receive devices).

The improvement of the cross-range resolution that can be offered using multiple sensors has been quantified. In particular, it has been shown that this improvement in the multistatic case is upper limited by the number of real sensors in the formation, while in the MIMO case, higher improvements are possible by jointly exploiting monostatic and bistatic acquisitions. The required processing techniques have been presented and the corresponding performance theoretically evaluated and experimentally validated. Finally, the exploitation of the multi-sensor data for motion estimation purposes has been discussed showing that the multiple observations could allow to overcome some of the basic limitations of conventional single-channel techniques. The achievable ISAR products with enhanced quality, in terms of cross-range resolution and scaling, could be fruitfully exploited by automatic classification/recognition procedures for surveillance and monitoring applications.

Acknowledgements

We wish to thank Prof. Pierfrancesco Lombardo of University of Rome 'La Sapienza' for his precious and fundamental contribution. We also acknowledge the collaboration of Dr. Fabrizio Santi of University of Rome 'La Sapienza' on the motion estimation topic and on the live DISAR data processing.

We are also pleased to acknowledge the support of Selex Galileo, now Leonardo SpA, CTO Innovative Projects Analysis (Pomezia, Italy) with special thanks to Mrs. Chiara Spina and Mr. Giovanni Cocca. Finally, we wish to thank Mr. Andrea Delogu of SELEX Galileo, now Leonardo SpA (Turin, Italy) for the valuable co-operation in the experimental campaigns.

Glossary

1D	one dimensional
2D	two dimensional
3D	three dimensional
CO_A	contrast in amplitude
CO_I	contrast in intensity
CPI	coherent processing Interval
CT-DISAR	centralized technique for distributed ISAR
DISAR	distributed inverse synthetic aperture radar
DLOS	distributed line of sight
DPSF	distributed point spread function
DT-DISAR	decentralized technique for distributed ISAR
FFT	fast Fourier transform
H	horizontal
He	entropy

IPP	image projection plane
ISAR	inverse synthetic aperture radar
LOS	line of sight
MA	multi-aspect
MG	multi-grazing
MIMO	multiple input multiple output
ML	maximum likelihood
MS	multi-sensor
NCTR	non-cooperative target recognition
PFA	polar format algorithm
PRF	pulse repetition frequency
PSF	point spread function
PSLR	peak-to-side-lobe ratio
R	radial
RCS	radar cross-section
RF	radio frequency
RX	reception, receiver
SNR	signal-to-noise ratio
SS	single sensor
SS_{av}	single-sensor averaged
TX	transmission, transmitter
UAV	unmanned aerial vehicle
V	vertical

References

[1] Walker J.L. 'Range – Doppler imaging of rotating objects'. *IEEE Trans. Aerosp. Electron. Syst*. 1980, vol. 16(1), pp. 23–52.

[2] Wehner D.R. *High-Resolution Radar* (Artech House, Boston, 1995).

[3] Musman S., Kerr D., Bachmann C. 'Automatic recognition of ISAR ship images'. *IEEE Trans. Aerosp. Electron. Syst*. 1996, vol. 32(4), pp. 1392–1404.

[4] Pastina D., Spina C. 'Multi-feature based automatic recognition of ship targets in ISAR images'. *IET Radar Sonar Nav*. 2009, vol. 3(4), pp. 406–423.

[5] Pastina D., Spina C. 'Slope-based frame selection and scaling technique for ship ISAR imaging'. *IET Signal Proc*. 2008, vol. 3(3), pp. 265–276.

[6] Vespe M., Baker C.J., Griffiths H.D. 'Automatic target recognition using multi-diversity radar'. *IET Radar Sonar Nav*. 2007, vol. 1(6), pp. 470–478.

[7] Brisken S., Martorella M. 'Multistatic ISAR autofocus with an image entropy-based technique'. *IEEE Aerosp. Electron. Syst. Mag*. 2014, vol. 29(7), pp. 30–36.

[8] Brisken S., Martorella M., Mathy T., Wasserzier C., Worms J.G., Ender J.H.G. 'Motion estimation and imaging with a multistatic ISAR system'. *IEEE Trans. Aerosp. Electron. Syst.* 2014, 50(3), pp. 1701–1714.

[9] Yeh C.-M., Xu J., Peng Y.-N., Wang X.-T. 'ISAR image fusion with two separated aspect observation'. *Proc. of IEEE 2009 Radar Conf.*; Pasadena, CA (USA), May 2009, pp. 1–5.

[10] Pastina D., Bucciarelli M., Spina C. 'Multi-sensor rotation motion estimation for distributed ISAR imaging'. *Proc. of 6th European Radar Conf.*; Rome, Italy, October 2009, pp. 282–285.

[11] Santi F., Bucciarelli M., Pastina D. 'Target rotation motion estimation from distributed ISAR data'. *Proc. of IEEE 2012 Radar Conf.*; Atlanta, GA (USA), May 2012, pp. 659–664.

[12] Zhang Q., Yeo T.S., Du G., Zhang, S. 'Estimation of three-dimensional motion parameters in interferometric ISAR imaging'. *IEEE Trans. Geosci. Remote Sens.* 2004, vol. 42(2), pp. 292–300.

[13] Wang G., Xia X.-G., Chen V.C. 'Three-dimensional ISAR imaging of maneuvering targets using three receivers'. *IEEE Trans. Image Process.* 2001, vol. 10(3), pp. 436–447.

[14] Ma C., Yeo T.S., Zhang Q., Tan H.S., Wang, J. 'Three-dimensional ISAR imaging based on antenna array'. *IEEE Trans. Geosci. Remote Sens.* 2008, vol. 46(2), pp. 504–515.

[15] Suwa K., Wakayama T., Iwamoto, M. 'Three-dimensional target geometry and target motion estimation method using multistatic ISAR movies and its performance'. *IEEE Trans. Geosci. Remote Sens.* 2011, vol. 49(6), pp. 2361–2373.

[16] Santi F., Pastina D., Bucciarelli M. 'Multi-sensor ISAR techniques for motion estimation of pitching, rolling and yawing targets'. *Proc. of IEEE 2013 Radar Conf.*; Ottawa, Canada, May 2013, pp. 1–6.

[17] Zhang Q., Yeo, T.S. 'Three-dimensional SAR imaging of a ground moving target using the InISAR technique'. *IEEE Trans. Geosci. Remote Sens.* 2004, vol. 42(9), pp. 1818–1828.

[18] Xu X., Narayanan, R.M. 'Three-dimensional interferometric ISAR imaging for target scattering diagnosis and modelling'. *IEEE Trans. Image Process.* 2001, vol. 10(7), pp. 1094–1102.

[19] Given J.A., Schmidt, W.R. 'Generalized ISAR – part II: interferometric techniques for three-dimensional location of scatterers'. *IEEE Trans. Image Process.* 2005, vol. 14(11), pp. 1792–1797.

[20] Zhao L., Gao M., Martorella M., Staglianò, D. 'Bistatic three-dimensional interferometric ISAR image reconstruction'. *IEEE Trans. Aerosp. Electron. Syst.* 2015, vol. 51(2), pp. 951–961.

[21] Martorella M., Staglianò D., Salvetti F., Battisti, N. '3D interferometric ISAR imaging of noncooperative targets'. *IEEE Trans. Aerosp. Electron. Syst.* 2014, vol. 50 (4), pp. 3102–3114.

[22] Ma C., Yeo T.S., Zhao Y., Feng, J. 'MIMO Radar 3D imaging based on combined amplitude and total variation cost function with sequential

order one negative exponential form'. *IEEE Trans. Image Process*. 2014, vol. 23(5), pp. 2168–2183.

[23] Ma C., Yeo T.S., Tan C.S., Li J.-Y., Shang, Y. 'Three-dimensional imaging using colocated MIMO radar and ISAR technique'. *IEEE Trans. Geosci. Remote Sens*. 2012, vol. 50(8), pp. 3189–3201.

[24] Wang D.-W., Ma X.-Y., Su, Y. 'Two-dimensional imaging via a narrowband MIMO radar system with two perpendicular linear arrays'. *IEEE Trans. Image Process*. 2010, vol. 19(5), pp. 1269–1279.

[25] Wang D.-W., Ma X.-Y., Chen A.-L., Su, Y. 'High-resolution imaging using a wideband MIMO radar system with two distributed arrays'. *IEEE Trans. Image Process*. 2010, vol. 19(5), pp. 1280–1289.

[26] Pastina D., Bucciarelli M., Lombardo, P. 'Multistatic and MIMO distributed ISAR for enhanced cross-range resolution of rotating targets'. *IEEE Trans. Geosci. Remote Sens*. 2010, vol. 48(8), pp. 3300–3317.

[27] Pastina D., Lombardo P., Buratti F. 'Distributed ISAR for enhanced cross-range resolution with formation flying'. *Proc. of 2008 European Radar Conf.*; Amsterdam, Holland, October 2008, pp. 37–40

[28] Pastina D., Bucciarelli M., Lombardo P. 'Multi-platform ISAR for flying formation'. *Proc. of 2009 IEEE Radar Conf.*; Pasadena, CA (USA), May 2009, pp. 1–6.

[29] Pastina D., Bucciarelli M., Lombardo P. 'Multi-platform distributed ISAR for surveillance and recognition'. *Proc. of 2009 Int. Radar Conf.*; Bordeaux, France, October 2009, pp. 1–6.

[30] Bucciarelli M., Pastina D. 'Multi-grazing ISAR for side-view imaging with improved cross-range resolution'. *Proc. of 2011 IEEE Radar Conf.*; Kansas City, KS (USA), May 2011, pp. 939–944.

[31] Pastina, D., Bucciarelli, M., Lombardo, P., Spina, C. 'Multi-grazing ISAR imaging method and system providing ISAR side-view images with improved cross-range resolution', Patent WO 2012020439 A1, PCT/IT2010/000370.

[32] Santi F., Bucciarelli M., Pastina D. 'Multi-angle distributed ISAR with stepped-frequency waveforms for surveillance and recognition'. *Proc. of 2011 CIE Int. Radar Conf.*; ChengDu, China, October 2011, pp. 528–532.

[33] Bucciarelli M., Pastina D. 'Distributed ISAR focusing for targets undergoing 3D motion'. 2012 *IET Int. Radar Conf.*; Glasgow, United Kingdom, October 2012, pp. 1–6.

[34] Zhu Y., Su Y., Yu, W. 'An ISAR imaging method based on MIMO technique'. *IEEE Trans. Geosci. Remote Sens*. 2010, vol. 48(8), pp. 3290–3299.

[35] Pastina D., Montanari A., Aprile A. 'Motion estimation and optimum time selection for ship ISAR imaging'. *Proc. of 2003 IEEE Radar Conf.*; Huntsville, Alabama (USA), May 2003, pp.7–14.

[36] Correll, B. 'Efficient spotlight SAR MIMO linear collection configurations'. *IEEE J. Sel. Topics Signal Process*. 2010, vol. 4(1), pp. 33–39.

[37] Simon M.P., Schuh M.J., Woo, A.C. 'Bistatic ISAR images from a time-domain code'. *IEEE Antennas Propag. Mag.* 1995, vol. 37(5), pp. 25–32.

[38] Eigel R.L. Jr., Collins P.J., Terzuoli A.J. Jr., Nesti G., Fortuny, J. 'Bistatic scattering characterization of complex objects'. *IEEE Trans. Geosci. Remote Sens.* 2000, vol. 38(5), pp. 2078–2092.

[39] Willis, N.J. 'Bistatic radar', in Skolnik, M. (Ed). *Radar Handbook* (McGraw Hill, 2008, 3rd edn.), pp. 23.19–23.20.

[40] Pastina D., Santi F., Bucciarelli, M. 'MIMO distributed imaging of rotating targets for improved 2-D resolution'. *IEEE Geosci. Remote Sens. Lett.* 2015, vol. 12(1), pp. 190–194.

[41] Krieger G., Cassola M.R., Younis M., Metzig, R. 'Impact of oscillator noise in bistatic and multistatic SAR'. *IEEE Geosci. Remote Sens. Lett.* 2006, vol. 3(3), pp. 424–428.

[42] Younis M., Metzig R., Krieger, G. 'Performance prediction of a phase synchronization link for bistatic SAR'. *IEEE Geosci. Remote Sens. Lett.* 2006, vol. 3(3), pp. 429–433.

[43] Wang, W. 'Approach of adaptive synchronization for bistatic SAR real time imaging'. *IEEE Trans. Geosci. Remote Sens.* 2007, vol. 45(9), pp. 2695–2700.

[44] Carrara W. G., Goodman R. S., Majewski R. M. *Spotlight Synthetic Aperture Radar* (Artech House, Boston, 1995).

[45] Pastina D. 'Rotation motion estimation for high resolution ISAR and hybrid SAR/ISAR target imaging'. *Proc. of 2008 IEEE Radar Conf.*; Rome, Italy, May 2008, pp. 1–6.

Chapter 11

Focussing moving objects using the VSAR algorithm

Luke Rosenberg[1], *Mark Sletten*[2] *and Jakov Toporkov*[2]

Abstract

Imaging of moving scatterers in a synthetic aperture radar (SAR) image is a challenging problem as targets will both smear and be displaced in azimuth. One technique which overcomes this limitation is the velocity SAR (VSAR) algorithm which was originally proposed by Friedlander and Porat in 1997. This technique provides the full Doppler spectrum at each and every pixel of a SAR image, which then allows for correction of the distortion caused by the radial motion. This chapter describes the signal processing of the VSAR algorithm and presents the first experimental results using both land and airborne multi-channel radar systems developed by the US Naval Research Laboratory (NRL). These examples demonstrate the VSAR correction for a variety of moving backscatter sources, including automobiles, ships, shoaling ocean waves and tidal currents. Two further applications of VSAR are also briefly covered in the chapter. These being the application to target detection and velocity inverse SAR.

11.1 Introduction

High fidelity imaging of moving targets is a well-known challenge for synthetic aperture radar (SAR), especially in ocean environments. Standard SAR processing methods assume the entire scene is stationary, and interpret the Doppler histories in the data accordingly. Any Doppler shifts introduced by target or scene motion will be misinterpreted and the corresponding backscatter will be displaced in the image. The classic example of this is the 'train-off-the-track', in which the signature of a range-travelling train appears displaced from the signature of the track upon which it is running. In addition, any along-track velocity component causes the target to

[1]Defence Science and Technology Group, Australia
[2]Remote Sensing Division, Naval Research Laboratory, USA

smear out due to a Doppler rate mismatch in the image formation process. More serious distortions can occur when vessels are moving on a dynamic ocean surface. Due to the roll, pitch and yaw motions induced by the waves, different parts of the vessel will have different velocities relative to the radar platform, and consequently the vessel's signature will be smeared as well as displaced. Image formation requires coherent integration over a period of seconds, and the orbital velocity and acceleration associated with the motion of water particles can cause smearing and a loss of azimuth resolution or it can result in wavelike patterns [1,2]. This latter effect is known as 'velocity bunching' and will cause a non-uniform azimuth displacement of the scatterers in the SAR image.

There are a number of techniques for ground based moving target indication (GMTI) which utilise multiple-aperture SAR (MSAR) for detecting moving targets, estimating their velocities and then correcting their positions within a SAR image [3–7]. The apertures in these systems are arranged along the flight axis to provide measurements of the scene at slightly different times but from the same vantage point in space, thereby allowing separation of the scene motion from that of the SAR platform. However, the techniques in [3–7] assume the clutter is stationary, and have limited applicability to ocean environments where a wide range of velocity components are present.

One technique which overcomes this limitation is the VSAR algorithm originally proposed by Friedlander and Porat [8,9] and described in Section 11.2. This technique provides the full Doppler spectrum at each and every pixel of a SAR image by means of a fast Fourier transform (FFT) down a 'time stack' of co-located images. Each image in this time stack is formed through standard SAR processing of data collected by a different phase centre. Given the along-track spacing between the phase centres and the aircraft's along-track motion, these images can be adjusted to have slightly different collection times, but a common collection location in space. Thus, the phase progression through the stack at any given pixel is related to radial motion in the scene at that pixel. The FFT provides a measure of the backscatter velocities present in each pixel by transforming the time stack into a 'velocity stack', in which each image corresponds to a different radial velocity. If the backscatter in a given pixel is generated by multiple scatterers moving at different speeds, the Doppler spectrum of that pixel will exhibit multiple peaks, and thus energy will be found in multiple velocity images. This is in contrast to standard along-track interferometric SAR (ATI-SAR), in which a pair of co-located images provides a single velocity estimate at each pixel, namely the pixel's Doppler centroid [10]. While providing more detailed velocity information than ATI-SAR processing, VSAR's full motion spectrum can also be used to perform a much more accurate correction of distortions induced by the scene motion [11].

Clearly VSAR could be used for GMTI purposes as it is able to distinguish targets moving at different radial velocities. However, the focus of this chapter is the application of VSAR for the imaging of moving targets in a SAR scene and is demonstrated using two experimental radar systems. The first system is the ground based NRL focused phased array imaging radar (FOPAIR) [11]. With this radar, MSAR systems with a wide range of aperture numbers were emulated, including the special case of ATI-SAR. An overview of this radar system is given in Section 11.3 with VSAR results of a small boat moving along the surface of the Chesapeake Bay. The results emphasise that VSAR processing does not require a long surface coherence time to produce fine-resolution imagery, unlike a single-channel SAR. Section 11.4 then demonstrates an application of the VSAR algorithm using the airborne MSAR system, [12]. In September 2014, the first field deployment of this system was undertaken over an ocean inlet on the mid-Atlantic coast of the United States. The imagery contains a variety of moving backscatter sources, including automobiles, ships, shoaling ocean waves and tidal currents. A qualitative analysis of the results clearly illustrates VSAR-based correction of these sources of motion and show the significant increase in detectability provided by the Doppler filtering inherent in VSAR processing. Section 11.5 then outlines two further applications of velocity processing—target detection and velocity inverse SAR (VISAR).

11.2 VSAR processing

This section provides a derivation of the VSAR algorithm showing how the azimuthal shift caused by the radial target motion can be corrected. The signal model for a moving target is derived in a similar fashion to [13,14] to illustrate how VSAR processing is achieved. A discussion on the limitations of the VSAR algorithm is then provided with some ideas for overcoming them.

11.2.1 Focussing with a moving target

The radar collection geometry is shown in Figure 11.1 where (x, y, z) corresponds to the ground range, along track (azimuth) and altitude, respectively. To simplify the derivation, we will use a slant plane representation, (r, y), where $r = \sqrt{x^2 + z^2}$. The radar platform is at altitude H and has an N channel linear array which transmits with the first channel and receives with all N. The inter-element spacing d is aligned in the along track direction with the radar platform travelling at velocity v_p. The radar transmits a frequency modulated (FM) chirp waveform with carrier frequency, f_c and chirp rate, f_r

$$p(t) = \text{rect}\left(\frac{t}{T_p}\right)\exp\left[j2\pi\left(f_c t + \frac{f_r}{2}t^2\right)\right] \tag{11.1}$$

where t is the fast time, T_p is the pulse width and rect is the rectangle function. Now consider a scatterer at position (R_0, y_0). If it is moving with velocity (v_r, v_y) and acceleration (a_r, a_y) and the channels are indexed with $n = 0 \ldots N-1$ so that $d_n = nd$, then the received signal from the nth receive element after demodulation can be written as:

$$p_n(t,s) = \text{rect}\left(\frac{t-\tau_n(s)}{T_p}\right)A_n(u(s))\exp[-j2\pi f_c \tau_n(s)]\exp\left[j\pi f_r(t-\tau_n(s))^2\right] \tag{11.2}$$

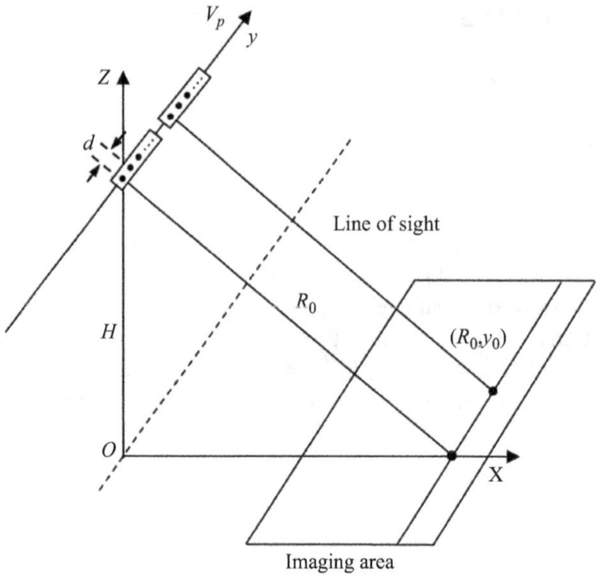

Figure 11.1 VSAR imaging geometry

where $s = s' - y_0/v_p$ is slow-time at the point of closest approach, s' is the slow-time referenced to the origin, c is the speed of light and $A_n(u(s))$ is a complex signal that includes the radar reflectivity. It also comprises the gain from the two-way antenna pattern with aspect angle $u(s) = -sv_p/R_0$ and has a magnitude and phase which is different for each channel due to mismatch between the transmit and receive beampatterns (including squint) as well as other differences in the hardware for each channel. The delay from the moving scatterer/target is as follows:

$$\tau_n(s) = \frac{1}{c}\left(R_0'(s) + R_n'(s)\right) \tag{11.3}$$

where

$$R_n'(s) = \sqrt{(R_0 + q_r(s))^2 + (d_n + q_y(s) - v_p s)^2} \tag{11.4}$$

and the radial and azimuthal motion is represented by $q_r(s) = v_r s + a_r s^2/2$ and $q_y(s) = v_y s + a_y s^2/2$, respectively. Assuming $R_0 \gg d_n$, the bistatic delay can be approximated as follows:

$$\tau_n(s) \approx \frac{2R_n(s)}{c} \tag{11.5}$$

with

$$R_n(s) = \sqrt{(R_0 + q_r(s))^2 + (0.5d_n + q_y(s) - v_p s)^2} \tag{11.6}$$

After pulse compression, the signal becomes

$$g_n(t,s) = A_n(u(s))\exp[-j2\pi f_c \tau_n(s)] W_r(t - \tau_n(s)) \tag{11.7}$$

where W_r is the slant range point spread function. As shown in Appendix A, a Taylor expansion can then be used to approximate the radial distance at the point of closest approach for the nth phase centre, $(R_0, 0.5d_n)$ with instantaneous Doppler components for the radar and target,

$$\begin{aligned} f_{\text{radar}} &= -\frac{2}{\lambda R_0}\left[sv_p^2 - 0.5d_n v_p\right] \\ f_{\text{target}} &= -\frac{2}{\lambda R_0}\left[R_0 v_r + 0.5d_n v_y + s(v_y^2 + R_0 a_r + 0.5d_n a_y - 2v_p v_y)\right] \end{aligned} \tag{11.8}$$

Equation (11.7) can then be written as follows:

$$\begin{aligned} g_n(t,s) \approx A_n(u(s))\exp\Bigg[&-\frac{j4\pi R_0}{\lambda} - \frac{j2\pi}{\lambda R_0}(0.25d_n^2 + 2sv_p(0.5d_n + \delta_{\text{lin}}) \\ &+ s^2(v_p^2 + \delta_{\text{quad}})\Bigg] W_r(t - \tau_n(s)) \end{aligned} \tag{11.9}$$

where the linear term represents a shift in azimuth with the $d_n v_y$ contribution being negligible,

$$\delta_{\text{lin}} = \frac{v_r R_0 + 0.5d_n v_y}{v_p} \approx \frac{v_r R_0}{v_p} \tag{11.10}$$

and the quadratic term

$$\delta_{\text{quad}} = v_y^2 + R_0 a_r + 0.5d_n a_y - 2v_p v_y \tag{11.11}$$

will cause a slight de-focussing effect. With these definitions, the signal model becomes

$$g_n(t,s) \approx A_n'(u(s))b_n(u(s))\exp\left[-\frac{j2\pi}{\lambda R_0}\left(2sv_p\delta_{\text{lin}} + s^2(v_p^2 + \delta_{\text{quad}})\right)\right] \times W_r(t - \tau_n(s)) \tag{11.12}$$

where the complex amplitude is

$$A_n'(u(s)) \approx A_n(u(s))\exp\left[-\frac{j4\pi R_0}{\lambda}\right] \tag{11.13}$$

with the phase term $\pi d_n^2/(2\lambda R_0)$ neglected since $R_0 \gg d_n$ and the spatial signal model for the nth channel can be written as follows:

$$b_n(u(s)) = \exp\left[j\frac{2\pi}{\lambda}d_n u(s)\right] \tag{11.14}$$

The next processing stage is SAR image formation where a number of options are available including chirp scaling, $\omega - k$, backprojection, etc. For simplicity, we will use the range Doppler algorithm, as described in [15].

To transform into the range Doppler domain, the principle of stationary phase can be used with $f_s = -K_a s$ and $K_a = 2v_p^2/(\lambda R_0)$. After matching the phase in azimuth with an appropriate matched filter and then using a modified range cell migration correction to account for the channel offset, the final model can be written as follows:

$$g_n(t,f_s) \approx A'_n\left(u\left(-\frac{f_s}{K_a}\right)\right)b_n\left(u\left(-\frac{f_s}{K_a}\right)\right)W_a(f_s - f_{\delta_{\text{lin}}})W_r\left(\frac{t-2R_0}{c}\right) \tag{11.15}$$

where W_a is the azimuth point spread function and the spatial signal model becomes

$$b_n\left(u\left(-\frac{f_s}{K_a}\right)\right) = \exp[-j2\pi n f_s t_d] \tag{11.16}$$

with $t_d = d/(2v_p)$. Due to the motion of the target, the focus location will be offset in azimuth by δ_{lin} and any blurring due to the quadratic phase term will be present in the azimuth point spread function. At this stage, we recognise that the fast time and Doppler frequency can be mapped to slant range and azimuth respectively with the substitutions, $r = 2t/c$ and $y = v_p f_s/K_a = y' - y_0$, while the spatial term can be related to the radial velocity with $f_s = -2v/\lambda$. This gives the final SAR signal model as follows:

$$g_n(r,y) \approx A'_n\left(u\left(-\frac{y}{v_p}\right)\right)\exp\left[\frac{j2\pi v d_n}{\lambda v_p}\right]W_a(y - y_{\text{lin}})W_r(r - R_0) \tag{11.17}$$

11.2.2 Velocity SAR

The goal of the VSAR processor is to measure the radial motion of the scatterers in the scene and correct their azimuth shift, δ_{lin}. If there is mismatch between the receive channels, either due to hardware differences or the transmit and receive beampattern combination, then channel balancing will be required to remove the channel-specific magnitude and phase differences. However, this must be applied when there is no motion in the scene, otherwise, it will remove the phase due to the target velocity as well. One scheme which has been used successfully is an 'adaptive 2D calibration' [16,17] which is further described in Section 11.4.2. After this step, the channel balanced signal model becomes

$$g_n(r,y) \approx A_0 \exp\left[\frac{j2\pi v d_n}{\lambda v_p}\right]W_a(y - y_{\text{lin}})W_r(r - R_0) \tag{11.18}$$

The next processing step in the VSAR algorithm is to take an FFT across the N channels. This results in N discrete images which represent the frequencies or velocities for each pixel in the stack. They can be represented as follows:

$$v_n = v_{\min} + n\Delta_v, \quad n = 0 \ldots N-1 \tag{11.19}$$

with the velocity resolution given by:

$$\Delta_v = \frac{\lambda v_p}{dN} \tag{11.20}$$

and the maximum negative and positive velocities,

$$\begin{aligned} v_0 \equiv v_{\min} &= -\frac{\pi}{d}\frac{\lambda v_p}{2\pi} = -\frac{\lambda v_p}{2d} \\ v_{N-1} &= \frac{\lambda v_p (N-2)}{2dN} \end{aligned} \tag{11.21}$$

The VSAR algorithm is illustrated in Figure 11.2 with SAR images of a scene containing vehicles (dots) travelling in the range direction along a smooth road (dark stripe). The SAR images will have a common collection location, but different collection times. As shown in Figure 11.2, the basic VSAR step transforms the time stack into a velocity stack by taking an FFT of each pixel over the N channels. In this cartoon, the blue dots represent signatures of vehicles moving towards the radar, while the red dots correspond to vehicles moving in the opposite direction. Variation in the brightness of the ambient area in the velocity stack images indicates that the surrounding clutter is basically stationary and thus most of the associated backscatter is sorted into the central (slow) velocity bins.

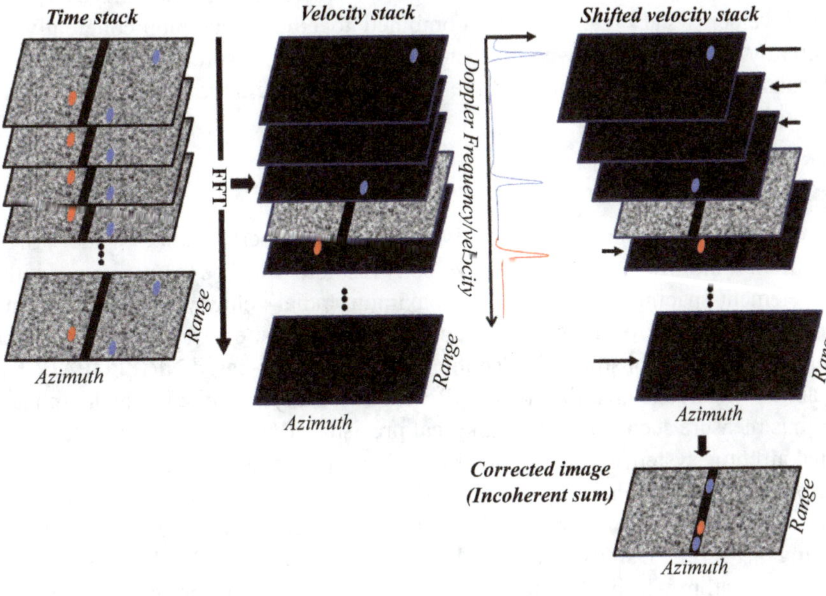

Figure 11.2 Cartoon illustrating transformation of the image time-stack into a velocity stack through an FFT and then azimuthally shifted by an amount proportional to its Doppler velocity. The shifted images can be summed to form a final, corrected image. The coloured dots represent vehicles moving along a road (dark line)

This cartoon also depicts the azimuth displacement mechanism: the signatures of the radially moving vehicles do not lie on the signature of the road, in either the original time-stack or in the derived velocity images. Also due to the relationship in (11.10), the faster vehicles are displaced more than those moving slowly, and the vehicles travelling towards the radar are displaced in one direction, while the shift is reversed for those travelling in the opposite direction.

While the VSAR motion spectrum is obviously useful in itself, it can also be used to correct the azimuth displacement as depicted in Figure 11.2. The resultant velocity image stack is a measure of the radial velocity present in each scatterer. The corrective shift required to align the pixels was given in (11.10). However, we can only measure the velocity at discrete locations, v_n, meaning there will be a small residual shift which in the worst case is half the velocity resolution. This leads to an offset which is a function of the range R_0 and varies with each of the N velocity images,

$$\hat{\delta}_{\text{lin}}(n) = \frac{v_n R_0}{v_p} \tag{11.22}$$

As shown in the cartoon, this corrective shift places the signature of each vehicle in its proper position on the road. For simplicity, each cartoon image is shifted as a whole, whereas in practice, each range gate receives a slightly different shift, as specified by (11.22). To form a single corrected image, the images in the shifted velocity stack can then be combined together either non-coherently or coherently. For this work, we only consider the former since coherent summation will cause a reduction in contrast if the combined scatterers are moving at different velocities.

11.2.3 VSAR limitations

The design of the radar system greatly influences the performance of the VSAR algorithm. If the array is too short, the velocity resolution will be too coarse and if the inter-element spacing is too large, the maximum radial velocity will be too small. Depending on the resolution requirements, the requisite array length may be impractical. As demonstrated in Section 11.3, the ground-based NRL FOPAIR has the ability to emulate an airborne MSAR with a virtually unlimited array length and can thus measure scene velocity with great precision. But the velocity resolution of actual airborne systems is determined by the length of the array and the speed of the platform, both of which have limitations due to practical considerations. For example, consider an X-band system with a platform velocity of 80 m/s and an element spacing of 10.5 cm. Using the relationships in (11.21), with $N = 32$ phase centres, the VSAR algorithm will be able to correct motion which has velocity resolution and maximum absolute (unambiguous) radial velocities of 0.7 and 11 m/s, respectively. However, the length of the array will need to be 3.2 m, which is impractical for many aircraft. An alternative scheme with dual transmitters has therefore been implemented for the airborne MSAR system in Section 11.4 which achieves the same number and spacing of phase centres with half the required array length.

If the radial velocity of a target is greater than the radial velocity which can be measured, then the target position will be aliased. This problem is known as 'azimuth location ambiguity' and potential solutions include a multi-frequency variation to the MSAR [18], a dual-velocity MSAR [19] and the use of non-uniform linear arrays [20]. Another problem identified is the focussing of elevated targets with VSAR. Li and Xia [21] have examined this problem and proposed a VSAR variation which combines multiple frequencies with across track interferometry.

If the velocity resolution is insufficient to resolve two targets of similar velocity, then they could be super-imposed at the same velocity. Similarly, after shifting azimuthally to account for the radial velocity, there is a possibility that more than one target could end up in the same range/azimuth cell in the final combined image. This problem is known as 'velocity layover' and can be minimised through the use of super-resolution spectral estimation techniques to improve the velocity resolution [22,23].

11.3 Ground-based demonstration with the NRL FOPAIR system

This section describes the NRL FOPAIR and how it was used to emulate an MSAR system for the purpose of demonstrating SAR signature correction using the VSAR algorithm. As an MSAR emulator, the NRL FOPAIR provides several advantages. First, as a ground-based radar, compensation for non-ideal platform motion during image formation is unnecessary. Second, as described below, it can emulate MSAR arrays of arbitrary length and can therefore provide a means to quantify the benefits of longer arrays and the finer velocity resolution that they provide.

11.3.1 The NRL FOPAIR system

The NRL FOPAIR system shown in Figure 11.3 is a ground-based, X-band (9.875 GHz) imaging radar that is an updated version of the original FOPAIR designed and built by the University of Massachusetts [24]. On a basic level, the system consists of a single-transmit antenna with a linear array of 64 receive antennas that are sampled individually and sequentially through fast solid-state switches. By sequentially scanning the linear array, a synthetic aperture can be emulated without the need for a moving platform. The system transmits a linear FM chirp waveform with a 220-MHz bandwidth to produce a range resolution of approximately 0.7 m. The receive element spacing is 5.24 cm for a total array length of 3.3 m. A more detailed description of the system can be found in [11].

11.3.2 Emulating an MSAR system

The NRL FOPAIR scans across its array of receive antennas continuously and repeatedly, providing a dense space-time distribution of sampling locations that closely emulates that of an MSAR. Figure 11.4 explains this emulation graphically. Figure 11.4(a) depicts the sampling positions (black dots) for an airborne MSAR

Figure 11.3 NRL FOPAIR system. © 2013 IEEE. Reprinted with permission from [11]

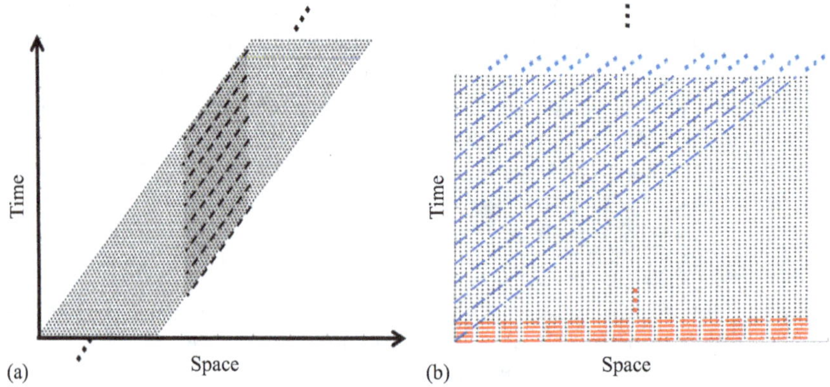

Figure 11.4 Space-time sampling locations (dots) and image formation trajectories (dashed lines) comparison: (a) airborne MSAR and (b) NRL FOPAIR

along with the trajectories (dashed lines) from which data can be processed to form a sequence, or time-stack, of images. The images in this stack all have the same collection location, but have slightly different collection times. It is this time separation between the images that allows measurement of scene motion. The number of images in the MSAR time stack is limited by the number of phase centres in the array (for clarity, only eight trajectories are shown). Figure 11.4(b) is the equivalent graphic for the NRL FOPAIR. As first described in [24], a single

FOPAIR data set can be processed along different space/time trajectories in order to emulate a SAR with different platform velocities. The red lines indicate the actual order in which the receive elements were sampled and indicate 'fast scan' processing trajectories that produce imagery with virtually no motion-induced distortion. The blue lines indicate successive 'slow-scan' trajectories that will produce imagery with a much slower effective platform velocity and consequently more image distortion. The NRL FOPAIR can generate a virtually unlimited number of images, with the limit set only by the duration of the data collection. As the velocity resolution of the VSAR technique is inversely proportional to the time separation between the first and last images in the stack, this means the NRL FOPAIR can measure velocity with extreme precision. In contrast, the airborne MSAR has limited velocity resolution, determined by the number of phase centres and the speed of the aircraft.

11.3.3 VSAR demonstration using FOPAIR

As described in [11], the NRL FOPAIR was used to demonstrate SAR signature correction using the VSAR algorithm. This is illustrated in Figures 11.5 and 11.6, using imagery of a small boat on the calm surface of the Chesapeake bay moving orthogonally to the radar at a speed of 6 m/s. A fast-scan image of the boat is shown in Figure 11.5(a). As mentioned in the previous section, fast-scan FOPAIR processing results in images that are virtually distortion-free, due to the rapid rate at which the array is scanned. This image is therefore a benchmark for evaluating the fidelity of the corrected VSAR images. A slow-scan FOPAIR image is shown in Figure 11.5(b). The distortion caused by motion of the boat and the surrounding water is obvious. The signature is not merely shifted azimuthally relative to the undistorted image, but is also non-uniformly stretched and compressed. Also, the

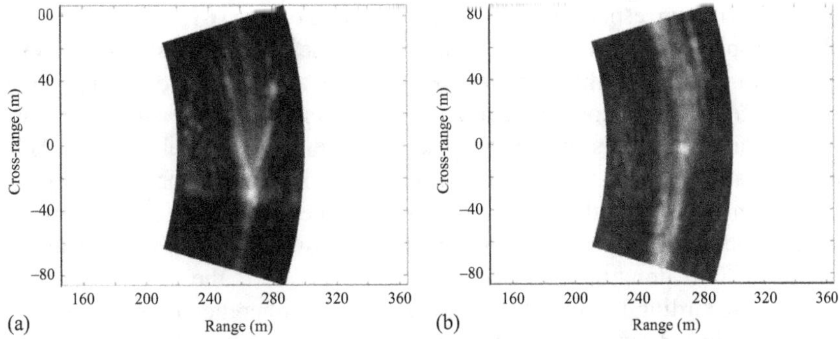

Figure 11.5 FOPAIR images of a small boat (a) fast-scan image (b) emulated SAR image (slow-scan image). © 2013 IEEE. Reprinted with permission from [11]

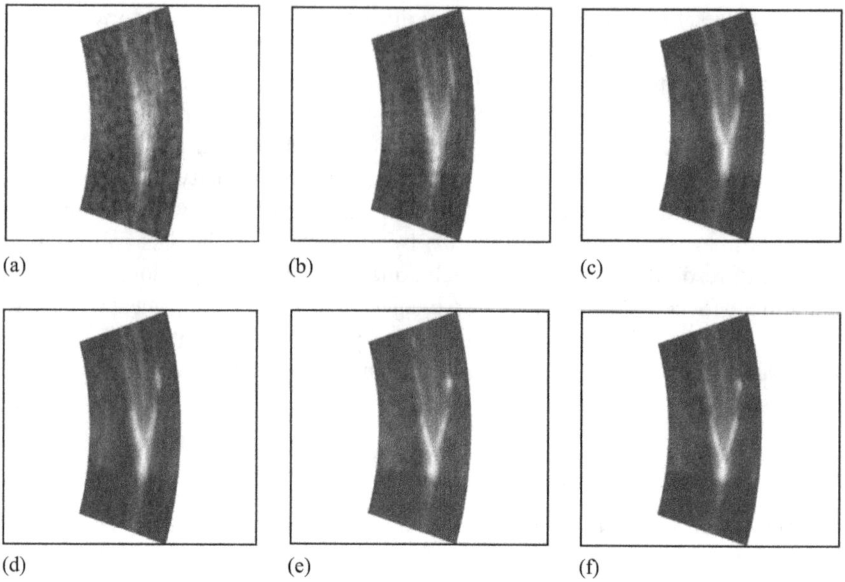

Figure 11.6 VSAR reconstructions of emulated SAR imagery, using progressively larger image stacks: (a) 8-images, (b) 16-images, (c) 32-images, (d) 64-images, (e) 128-images and (f) 256-images. © 2013 IEEE. Reprinted with permission from [11]

SAR image is the result of averaging 256 SAR images and has a lower contrast when compared to the VSAR image formed with 32 velocity images.

The results of applying VSAR corrections to the distorted imagery are shown in Figure 11.6, using 8, 16, 32, 64, 128 and 256 slow-scan images in the time stack with corresponding processing intervals of 91, 101, 122, 163, 245 and 410 ms, respectively. The fidelity of the larger time stack reconstructions is very high. It is notable that these processing intervals are up to two orders of magnitude longer than the coherence time of the backscatter in the signature, which was determined by fast-scan Doppler spectra to be approximately 2–4 ms. While this observation is in contrast to conclusions drawn from the analysis in [9], it is actually consistent with the canonical model for the sea surface which comprises a large number of independent scattering facets moving with different velocities. The VSAR FFT along the image time-stack simply sorts the backscatter within each pixel according to velocity, regardless of the coherence time of the aggregate signal formed by the coherent sum of all the backscatter. A long aggregate coherence time is unnecessary. Instead, the algorithm requires scattering centres with sufficiently long lifetimes and with velocities low enough to keep the scatterers within a given resolution cell long enough to provide an adequate velocity estimate.

11.4 Airborne demonstration with the NRL MSAR system

This section describes the NRL MSAR and presents the first airborne VSAR results which were reported in [12,25,26]. The NRL MSAR is an airborne test bed designed to investigate remote sensing and surveillance applications that exploit multiple along-track phase centres, in particular, applications that require measurement of scene motion. The design of the radar and the sampling technique used to achieve 32 phase centres is described here including details of the channel balancing scheme which was critical to achieve coherency across the array.

The first flights of the NRL MSAR took place in September 2014 over several locations along the mid-Atlantic coast of the USA. An optical image from one of these areas, Oregon Inlet on the Outer Banks of North Carolina is shown in Figure 11.7. The scene includes barrier islands to the North and South of the inlet itself as well as a bridge span. This area was selected for flight testing due to the variety of moving targets in the scene. The VSAR results include the correction of cars on a bridge, an analysis of a breaking wave and the correction of two moving ship signatures.

11.4.1 NRL MSAR system

The NRL MSAR is an airborne system using many components from the NRL FOPAIR radar. It operates at X-band with a centre frequency of 9.875 GHz and uses linear FM chirped waveforms with a bandwidth of 200 MHz to achieve a range resolution of approximately 0.75 m. The peak radiated power is approximately 1.4 kW, while the aggregate pulse repetition frequency (PRF) of 25 kHz and pulse length of 6 μs produce a total average power of 210 W. The system flies on a Saab 340 aircraft using a belly-mounted radome with a nominal depression angle of

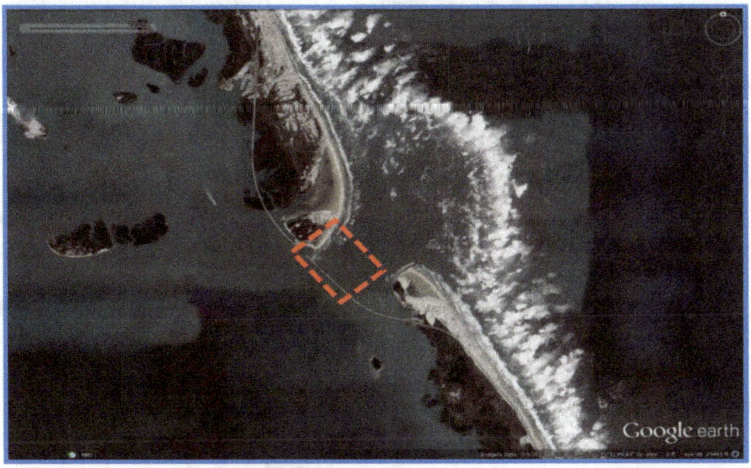

Figure 11.7 Oregon Inlet test site. Red box indicates VSAR test region. Imagery from DigitalGlobe, USDA Farm Service Agency. © 2015 IEEE. Reprinted with permission from [12]

Figure 11.8 Saab 340 aircraft with belly mounted radome. © 2015 IEEE. Reprinted with permission from [12]

Figure 11.9 MSAR in its housing. © 2015 IEEE. Reprinted with permission from [12]

20°, as shown in Figure 11.8. Typical altitude and airspeeds are 914 m (3,000 ft) and 80 m/s (156 knots), respectively.

As operated during its inaugural airborne tests in September 2014, the MSAR system uses 16 receive antennas, equally spaced along the flight axis with a separation of 10.48 cm, and two transmit horns, with a spacing of 125.8 cm. A photograph of the antennas as installed in the Saab radome is shown in Figure 11.9. The long white enclosures below the two transmit horns contain the 16 receive antennas which are all vertically polarised. The aft-mounted horn transmits odd pulses, while the fore-mounted horn transmits even pulses. For the first pair of transmitted pulses, four of the 16 receive antennas are connected to a four-channel receiver and sampled by a high speed data recorder. A bank of microwave switches is then reconfigured to connect the next group of four receive antennas to the receiver and data recorder. After a total of four pairs of pulses (320 μs), this scan sequence is repeated. In this manner, 32 phase centres are generated and sampled at a per-phase-centre PRF of 3.125 kHz, each corresponding to a different combination of transmit and receive antennas. This PRF is sufficient to allow production of 32 SAR images that are free from azimuthal ambiguities. Note that the average power associated with any one of these images is 1/32 of the total average power, or 6.7 W. Since the location of a given phase centre is halfway between the corresponding transmit and receive antennas, the along-track spacing between these phase centres is 5.24 cm, half the physical spacing between the receive elements. With this configuration, there are twice as many phase centres as receive elements, which results in a halving of the velocity resolution and doubling of the maximum unambiguous Doppler velocity which can be measured by the radar. Note that given the spacing between the two transmit horns, four of the phase centres corresponding to the aft-mounted transmit horn are collocated with four produced by the fore-mounted horn. Thus the total number of independent phase centres is 28 with total length of 141.5 cm.

The large number of phase centres and the requirement to revisit each frequently enough to allow alias-free SAR imaging produces a high average data rate. During each pulse repetition interval (PRI) (40 μs), 8192 8-bit samples from each of the four data recorder channels are written to a RAID for a sustained data rate of 819 MB/s. The recorder is based on a 4-channel, 12-bit acquisition card running at a sample rate of 500 MS/s. Selection of the 8192-sample range window within each PRI and reduction of the 12-bit samples down to 8-bits is performed by a field programmable gate array.

Motion of the antennas is measured by a Novatel inertial navigation system mounted in the Saab radome directly behind the antennas. The Novatel ProPak6 OEM638 receiver blends GPS position data with measurements from a Litton LN200 inertial measurement unit to determine the precise positions of all phase centres at a 100 Hz rate. These data are used during image formation in order to compensate for non-ideal aircraft motion.

11.4.2 Pre-processing

Before the VSAR image can be produced, a number of pre-processing steps are required. First, a digital conversion is applied to the real-valued raw data to convert it to in-phase and quadrature baseband samples. This is followed by a low-pass filter designed to remove any signal artefacts outside the frequency band of interest. A chirp scaling algorithm is then implemented to form a SAR image for each phase centre with extensions designed to motion compensate the resulting image [27]. After the images are formed, a final step is required to balance the phase and magnitude between each phase centre. The phase mismatch comprises a linear term due to the along track displacement of the phase centres (see (11.14)) and a non-linear term due to the changing aircraft position along the scene. A third non-linear component of the phase error occurs due to the antenna mismatch and the unbalanced radar hardware. These hardware imbalances also affect the magnitude of the received signal and require calibration using real radar echoes to account for the non-linearity.

The algorithm selected for channel balancing is the 'adaptive 2D calibration' [16,17]. It was proposed for use in the range-frequency/Doppler domain before image formation, but we have found it to be effective when used in the image domain as well. The algorithm must be trained on a scene with no motion present, otherwise the calibration will affect the phase term which is required for subsequent VSAR processing. Typical training regions might include homogenous land, bright returns from corner reflectors or man-made structures. For the examples shown here, a training region was selected around the centre part of the bridge shown in Figure 11.7.

In the azimuth and range spatial frequency domain, the mismatched component of the received signal is denoted $A_n(k_r, k_y)$ for the nth phase centre. The channel model is given by

$$\begin{aligned} A_1(k_r, k_y) &\approx A(k_r, k_y)h_1(k_y)D_1(k_r) \\ A_n(k_r, k_y) &\approx A(k_r, k_y)h_n(k_y)D_n(k_r) \end{aligned} \tag{11.23}$$

where $A(\cdot)$ is a complex signal common to each channel and $h_n(k_y)$ and $D_n(k_r)$ describe channel specific information along each spatial frequency dimension. The information contained along the range frequency, $h_n(k_y)$ is referred to as the 'transfer function', while the azimuth frequency component, $D_n(k_r)$ is primarily due to the antenna beampattern which varies between the different channels. The goal of channel balancing is to minimise the channel specific information so the images have a high coherency and similar magnitude. The ratio between the first and nth channel is given by

$$\frac{A_1(k_r, k_y)}{A_n(k_r, k_y)} \approx \frac{h_1(k_y)}{h_n(k_y)}\frac{D_1(k_r)}{D_n(k_r)} = h_{1,n}(k_y)D_{1,n}(k_r) \tag{11.24}$$

The adaptive 2D calibration algorithm is designed to minimise the square of the difference between $h_{1,n}(k_y)$ and $D_{1,n}(k_r)$ when integrated over the frequency domains. It can be written in terms of the following optimisation [17]:

$$\min_{\{h_{1,n}(k_y), D_{1,n}(k_r)\}} \int_{k_r}\int_{k_y} \left|A_1(k_r, k_y) - A_n(k_r, k_y)h_{1,n}(k_y)D_{1,n}(k_r)\right| dk_r dk_y \tag{11.25}$$

with the solution given as a pair of integral equations

$$\begin{aligned} h_{1,n}(k_y) &= \frac{\displaystyle\int_{k_r} D^*_{1,n}(k_r)A^*_n(k_r, k_y)A_1(k_r, k_y)dk_r}{\displaystyle\int_{k_r} \left|A_n(k_r, k_y)D_{1,n}(k_r)\right|^2 dk_r} \\ D_{1,n}(k_r) &= \frac{\displaystyle\int_{k_y} h^*_{1,n}(k_y)A^*_n(k_r, k_y)A_1(k_r, k_y)dk_y}{\displaystyle\int_{k_y} \left|A_n(k_r, k_y)h_{1,n}(k_y)\right|^2 dk_y} \end{aligned} \tag{11.26}$$

which can be solved iteratively [17]. This algorithm works across the whole image in the range and azimuth spatial frequency domains. As a result, the magnitudes in the image domain do not always balance perfectly. A final step is therefore required to adjust the means in the image domain.

11.4.3 Example dataset 1

The first example dataset demonstrates the correction of cars moving along a bridge and the evolution in velocity of a breaking wave. For reference, an expanded view of the analysis region is shown in Figure 11.10, which was produced by non-coherently summing the 28 original SAR images. There are three cars on the bridge, of which only one is visible and is marked with a red ellipse. After VSAR processing, it is useful to compare the original SAR image with the stationary (DC) velocity image. Figure 11.11 shows the DC image which now reveals a number of stationary scatterers that were masked in the SAR image. The moving scatterers which are now absent include the car to the right of the image, a large breaking wave above the bridge and a reduction in the number of shoaling waves near the shore. The VSAR image is then shown in Figure 11.12 after non-coherently summing the 28 shifted (i.e. corrected) velocity images.

As shown by a comparison of Figures 11.10 and 11.12, VSAR processing corrects the smeared shoaling wave signatures (upper left side of image) by shifting the displaced backscatter contributions back to the actual location of the waves. The resulting signatures in Figure 11.12 have a reduced width that is a more accurate measure of the actual size of the shoaling waves. There are also three faint echoes that have been highlighted and correspond to cars travelling on the bridge, which have been displaced due to their radial velocity. The corrected VSAR image has then shifted these cars back towards the bridge. Clearly the VSAR image also

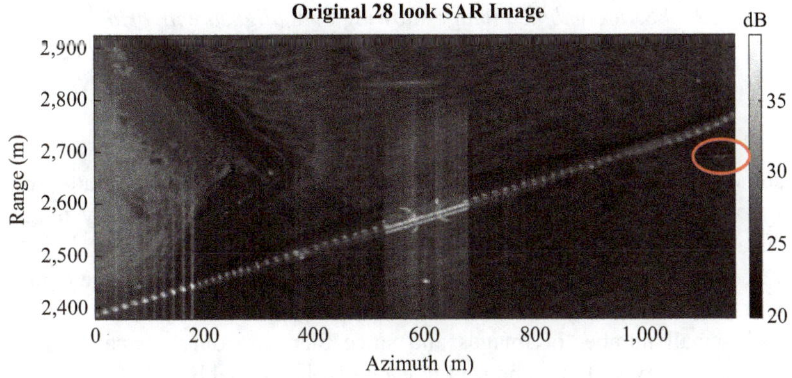

Figure 11.10 28 look SAR image highlighting the difficulty in detecting moving cars. Only the right car is actually visible in this scene

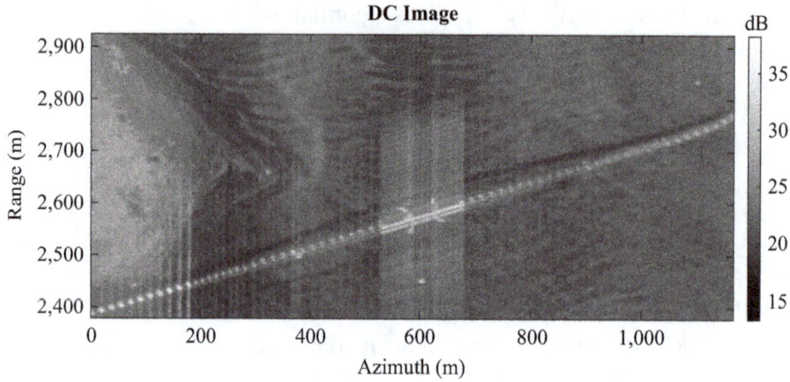

Figure 11.11 VSAR image showing stationary (DC) velocity slice

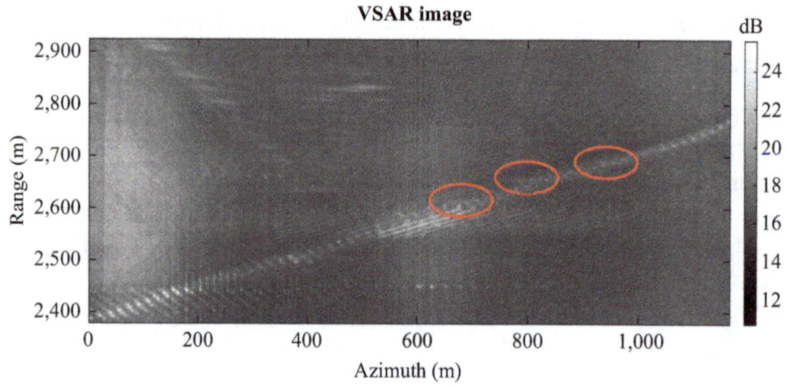

Figure 11.12 VSAR combined image showing cars aligned with bridge. Cars have been relocated on the bridge, but there is a high level of spectral leakage

has some artefacts due to the non-coherent summation. These include spectral leakage where the dominant DC scatterers are present in the velocity images as well as a general loss in contrast across the image.

One way of reducing the spectral leakage and improving the image contrast is to use a different spectral estimate when forming the velocity stack. Due to the relatively small number of images and since there are only a small number of significant velocity scatterers present in each pixel, a model based spectral estimate has been selected. Figure 11.13 shows the result of processing with a fourth order Burg auto-regressive (AR) spectral estimate [28]. When compared to the FFT based method, the contrast has greatly increased and the spectral leakage from the DC image has greatly reduced. Unfortunately, the shifted targets are now very difficult

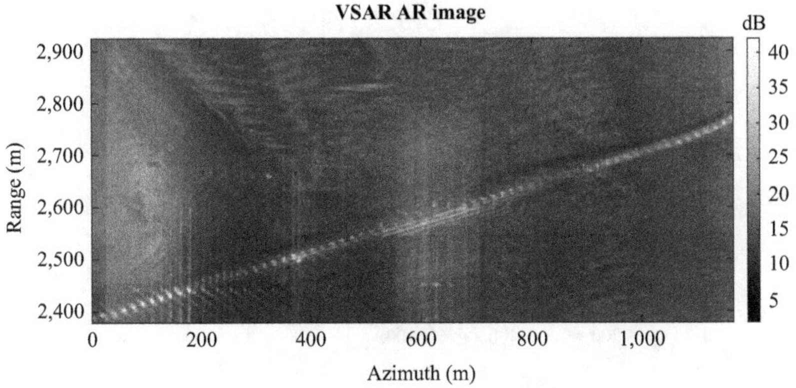

Figure 11.13 VSAR combined image using fourth order Burg AR spectral estimate. Leakage has been reduced, but cars on the bridge are not visible

to see and there are regions in the image where the scatterers on the bridge have been reduced in magnitude.

An alternative method of improving the VSAR image quality is to threshold the magnitude of both the stationary and velocity images. This can be achieved by the following steps:

- Identify strong scatterers in the DC velocity image and mask those pixels in the other velocity images.
- Balance the means of the different velocity images relative to the DC velocity image.
- Examine the non-DC images to identify where the 'velocity noise' is significant. Then create a mask and suppress those pixels which are less than the threshold.

Using these processing steps allows the strongest velocity scatterers to be highlighted through the velocity stack. Through experimentation, the threshold for masking strong scatterers was set as the mean of the DC image, while the significant scatterer threshold can be set between 2 and 6 standard deviations above the mean for each image. By isolating the dominant scatterers in each velocity slice, a composite image, shown in Figure 11.14, can then be formed where the scatterer velocities are overlaid on the DC image. The three targets on the bridge are now clear and it can be seen that not all the cars are travelling in the same direction. This image also highlights the shoaling waves near the shore and the mix of velocities which are present. Note that the two red cars remain smeared over approximately 50 m in azimuth due to the along track component of the vehicles motion over the 2.7 s SAR integration time. The VSAR algorithm can only address distortion caused by radial motion, and thus this additional smearing remains.

The VSAR correction is further demonstrated in Figures 11.15 and 11.16 which show the shifted velocity stack for two sub-sections of the analysis region

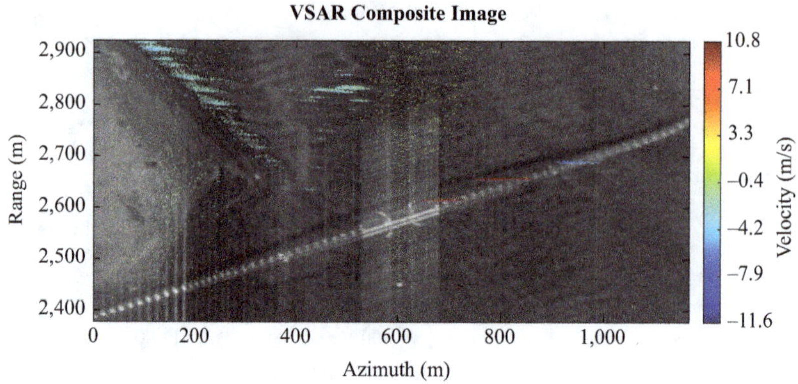

Figure 11.14 VSAR composite image highlighting dominant pixel velocities

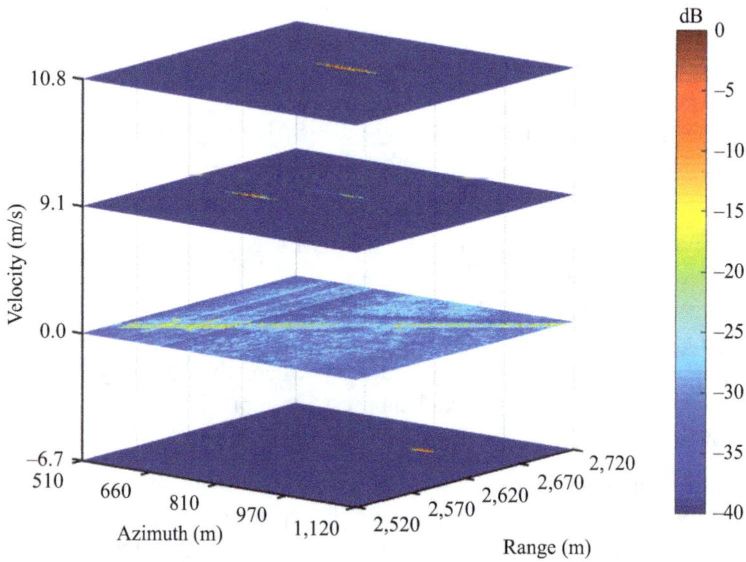

Figure 11.15 Shifted VSAR stack showing the velocities of cars travelling on the bridge. Images formed using fourth order Burg AR spectral estimate

using the Burg AR spectral estimate. The first example shows a zoomed region of the bridge where the cars were corrected. For clarity, only the velocity images corresponding to −6.7, 0, 9.1 and 10.8 m/s are shown. This stack of images shows the (stationary) bridge at 0 m/s with the three cars clearly standing out on their respective slice. After accounting for the look direction of the radar relative to the bridge, the radial velocities correspond to vehicle speeds of 53 ± 7, 73 ± 7 and 86 ± 7 miles per hour (mph), which are within the range of speeds encountered on

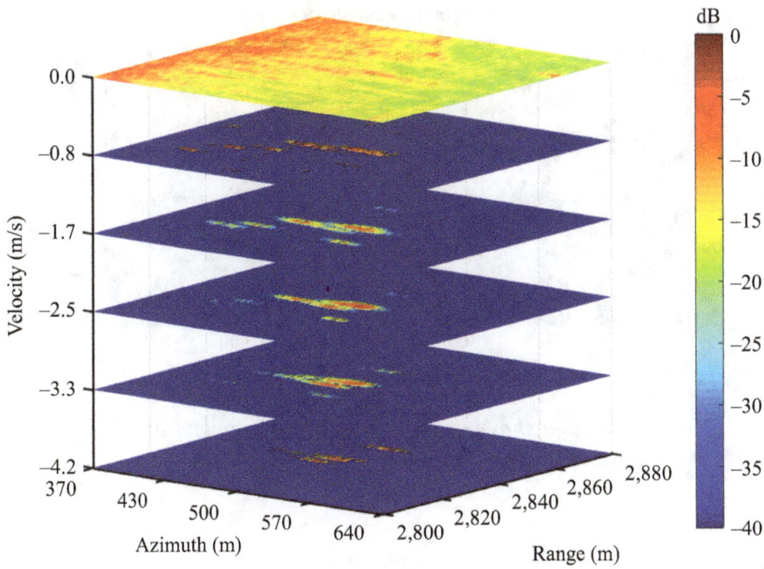

Figure 11.16 Shifted VSAR stack showing the velocity components of a breaking wave. Images formed using fourth order Burg AR spectral estimate

US highways with a posted speed limit of 55 mph. Figure 11.15 also highlights the usefulness of VSAR processing: while the vehicle signatures are weak in the standard SAR imagery and even in the corrected VSAR image, they stand out clearly in their respective velocity images.

The second example illustrates how the velocity stack can be used to investigate the mechanisms behind maritime SAR signatures, which are often complex due to the motion of the sea surface. Figure 11.16 shows a shoaling wave, in which the distribution of backscatter is displayed as a function of velocity. This enables us to study the various stages of breaking (steepening, actively spilling/plunging, turbulent recovery) which tend to move at different speeds and are not observable in a typical SAR image. The VSAR analysis in Figure 11.14 also indicates a larger-scale velocity signature over the inlet channel (see sparse distribution of yellow pixels) that is likely due to a tidal current, either directly or through perturbation of the ambient surface waves over the channel.

11.4.4 Example dataset 2

In the second data set shown in Figure 11.17, there are two targets of opportunity. The top boat has a compact signature in the water and is the better focussed of the two, possibly due to the combination of its azimuth velocity and acceleration over the observation time. In contrast, the bottom boat has a long smeared out signature indicating it is likely travelling along the same direction as the radar. The corrected VSAR image using the Burg AR spectral estimate is then shown in Figure 11.18 where the top boat appears slightly more spread out in azimuth than it was in

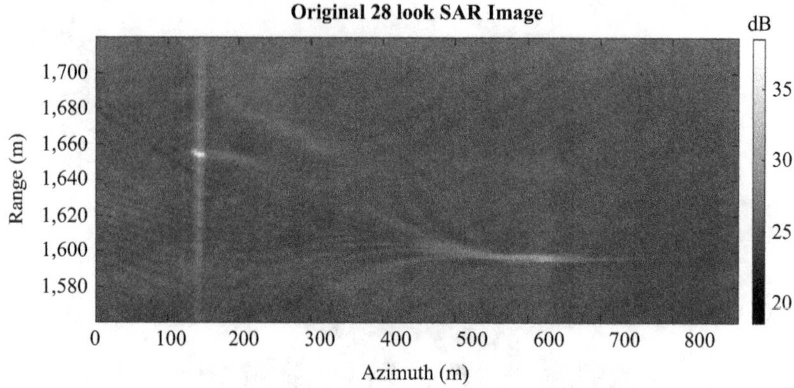

Figure 11.17 SAR image showing the two boats of opportunity. © 2015 IEEE. Reprinted with permission from [26]

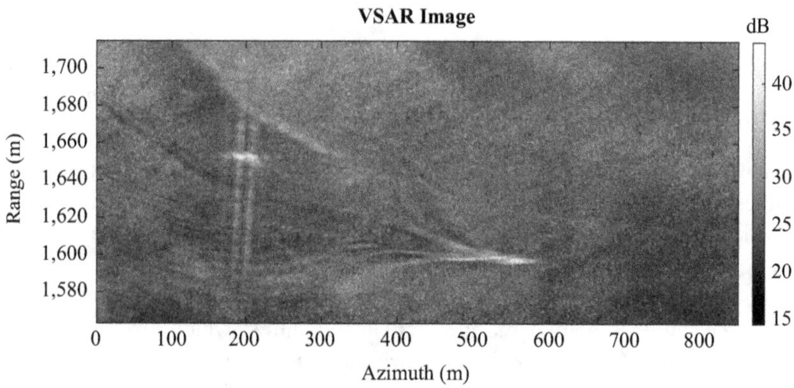

Figure 11.18 VSAR image showing correction of the two boats of opportunity

Figure 11.17 and has shifted by 30 m, indicating it was moving with a significant radial velocity. The smearing in range implies an acceleration and corresponds to different velocity components which have been non-coherently combined when forming the VSAR image. The second boat at the bottom of Figure 11.18 shows a large improvement in its signature with the characteristic 'V' now apparent. This is further shown in Figure 11.19 with a series of VSAR images produced from different numbers of channels and hence different velocity resolutions. The VSAR images correspond to 4, 8, 16 and 28 phase centres which correspond to velocity resolutions of 5.04, 2.52, 1.26 and 0.72 m/s. Clearly as the velocity resolution increases, the fidelity of the corrected signature improves, as also illustrated by the ground-based FOPAIR results presented in Section 11.3.3.

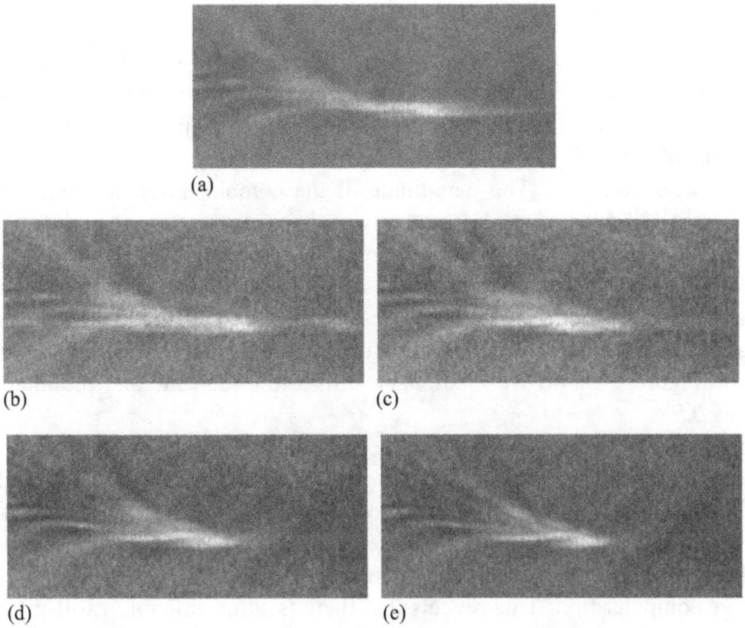

Figure 11.19 *Moving boat example showing improvement with an increasing number of phase centres and finer velocity resolution. The range (vertical) dimension of these images is 150 m while the azimuth (horizontal) span is 425 m: (a) SAR image, (b) $N = 4$, $\Delta_v = 5.04$ m/s, (c) $N = 8$, $\Delta_v = 2.52$ m/s, (d) $N = 16$, $\Delta_v = 1.26$ m/s, (e) $N = 28$, $\Delta_v = 0.72$ m/s. © 2015 IEEE. Reprinted with permission from [26]*

11.5 Applications of velocity processing

This chapter has demonstrated how motion correction can be achieved using the VSAR algorithm. Two other applications are now described which can further exploit this technique. These include target detection and the development of an ISAR counterpart to the VSAR processing.

11.5.1 Target detection

The VSAR algorithm is clearly capable of detecting moving targets and measuring their velocity. However, (11.19) dictates the discrete velocity values which may be in error and result in a mismatch of the estimated target position. A parametric method of velocity estimation has been proposed in [29] which first detects the presence of a target and then improves the estimate of its velocity through the use of super-resolution techniques. A follow on paper then incorporated both range and azimuth multi-looking to improve the estimate accuracy [30].

11.5.2 Velocity ISAR

The VISAR algorithm was proposed in [31] as a means of exploiting the structure associated with the motion dynamics of rigid bodies (including translational and roll-pitch-yaw motions) in maritime conditions. It is implemented by first creating ISAR images for each spatial channel and then applying the velocity correction outlined in Section 11.2. The magnitude of the combined velocity stack is then known as the VISAR image. Alternatively the maximum at each pixel can be calculated to obtain the dominant velocity of the target being imaged.

In Figure 11.20, the imaging results from a small boat are presented with 16 channels used for the velocity processing. For this example, the plane flew circular orbits of radius 2.5 km with an altitude of 1.5 km (5000 ft). The 14 m boat shown in Figure 11.20(a) was stationary with a low sea-state and the integration time was 0.5 s.

Figure 11.20(b) and (c) shows the performance of the VISAR algorithm with and without range compensation. If the integration time is small enough and there is not much rotational motion associated with the target, it may suffice to forego the range alignment step. However, for this example, there are noticeable, though subtle, improvements in the focussing of the scatterers when incorporating range compensation. This reveals that there is some amount of roll-pitch-yaw motion present in the boat apart from the relative translational velocity due to the motion of the plane. Figure 11.20(d) then shows the VSAR performance where

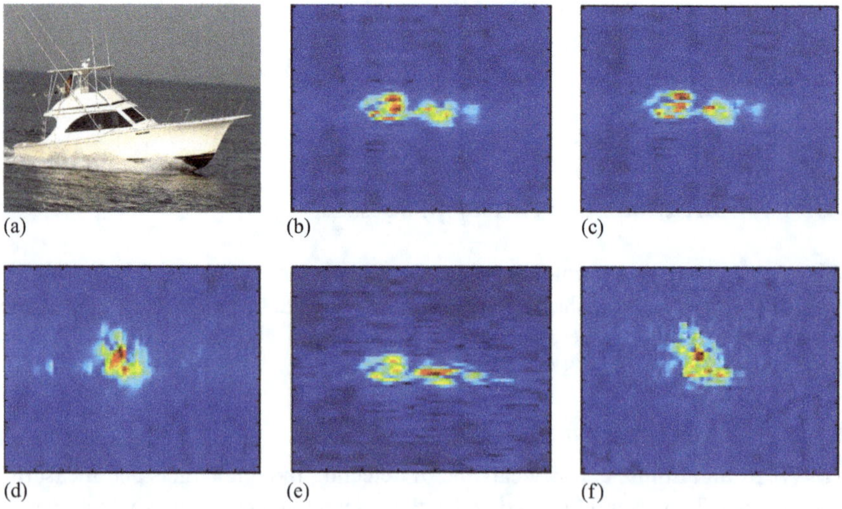

Figure 11.20 Image comparison of small boat including SAR, ISAR, VSAR and VISAR: (a) Small boat, (b) VISAR image without range compensation, (c) VISAR image with range compensation, (d) VSAR image, (e) single-channel ISAR image and (f) single-channel SAR image. © 2015 IEEE. Reprinted with permission from [31]

the images corresponding to each of the sensors are formed via a backprojection algorithm. Since the ISAR algorithm picks up on the roll-pitch-yaw motion of the boat and also dynamically adjusts the image plane based on the relative motion of the boat, it is clear that the VISAR algorithm is able to bring out more features of the boat when compared to VSAR. Furthermore, as expected, the VISAR and VSAR algorithms show a better separation of the dominant scatterers when compared to their single-channel counterparts shown in Figure 11.20(e) and (f), respectively. To further measure the effectiveness of this technique, a larger range of targets will need to be imaged in different sea-states. This data could then feed into an automated ship classification scheme as a final measure of performance.

11.6 Conclusion

This chapter has described how motion can be measured in a SAR image and then corrected through the use of velocity processing. The VSAR algorithm was demonstrated using two of NRL's experimental radar systems—the ground based FOPAIR and the airborne MSAR. Through the use of the FOPAIR testbed, it was shown how the target signature of a moving boat can be improved with an increasing number of channels. With the airborne MSAR system, the VSAR-based correction was demonstrated for cars moving along a bridge, shoaling waves and boats moving both azimuthally and radially.

There are now opportunities to further exploit this technique for other applications such as target detection and identification and classification of both stationary and moving targets. The VISAR algorithm was presented as a first step towards this goal. It was able to clearly reveal extra details of a small boat when compared to the images produced by single channel ISAR and the VSAR algorithm.

Acknowledgements

The authors would like to thank the entire NRL team for contributing to the success of the trials and subsequent data analysis.

Appendix A: Derivation of Doppler components

Assuming that the squint is small, the radial distance to a scatterer at the closest point of approach to the nth phase centre, $(R_0, 0.5d_n)$, moving with velocity, (v_r, v_y) and acceleration, (a_r, a_y) can be written as follows:

$$R_n(s) = \left[(R_0 + q_r(s))^2 + \left(0.5d_n + q_y(s) - v_p s\right)^2\right]^{0.5} \tag{A.1}$$

where $q_r(s) = v_r s + a_r s^2/2$, $q_y(s) = v_y s + a_y s^2/2$ and $d_n = nd$ for $n = 0 \ldots N - 1$ receive channels. To the second order, this can be expanded and rearranged as follows:

$$\begin{aligned} R_n(s) &= \Big[R_0^2 + 0.25d_n^2 + v_p^2 s^2 + 2R_0 q_r(s) + q_r^2(s) + q_y^2(s) \\ &\qquad - d_n v_p s + d_n q_y(s) - 2v_p s q_y(s)\Big]^{0.5} \\ &\approx R_0 + \frac{1}{2R_0}\Big[0.25d_n^2 + v_p^2 s^2 - d_n v_p s + 2R_0 q_r(s) \\ &\qquad + q_y^2(s) + d_n q_y(s) - 2v_p q_y(s) s\Big] \end{aligned} \tag{A.2}$$

The instantaneous Doppler can then be split into a radar and target component

$$f(s) = -\frac{2}{\lambda}\dot{R}(s) = f_{\text{radar}}(s) + f_{\text{target}}(s) \tag{A.3}$$

where

$$\begin{aligned} f_{\text{radar}}(s) &= -\frac{2}{\lambda R_0}\left[v_p^2 s - 0.5 d_n v_p\right] \\ f_{\text{target}}(s) &= -\frac{1}{\lambda R_0}\left[2R_0 \dot{q}_r(s) + \dot{q}_y^2(s) + d_n \dot{q}_y(s) - 2v_p(\dot{q}_y(s)s + q_y(s))\right] \end{aligned} \tag{A.4}$$

and $\dot{q}_r(s) = v_r + a_r s$, $\dot{q}_y(s) = v_y + a_y s$ and $\dot{q}_y^2(s) = 2(v_y + a_y s)(v_y s + a_y s^2/2)$. With these substitutions, the target Doppler can be written as

$$\begin{aligned} f_{\text{target}}(s) &= -\frac{2}{\lambda R_0}\left[R_0(v_r + a_r s) + (v_y + a_y s)(v_y s + a_y s^2/2)\right. \\ &\qquad \left. + 0.5 d_n(v_y + a_y s) - v_p(2v_y s + 3a_y s^2/2)\right] \end{aligned} \tag{A.5}$$

However, all terms which are quadratic or greater will be very small compared to the linear terms and the target Doppler can be further approximated as follows:

$$f_{\text{target}}(s) \approx -\frac{2}{\lambda R_0}\left[R_0 v_r + 0.5 d_n v_y + s(v_y^2 + R_0 a_r + 0.5 d_n a_y - 2v_y v_p)\right] \tag{A.6}$$

References

[1] Alpers, W. R., Ross, D. B., and Rufenach, C. L. 'On the detectability of ocean surface waves by real and synthetic aperture radar', *Journal of Geophysical Research*. 1981; **86**(C7): 6481–6489.

[2] Alpers, W. R. and Rufenach, C. L. 'The effect of orbital motions on synthetic aperture radar imagery of ocean waves', *IEEE Transactions on Antennas and Propagation*. 1979; **AP-27**(5): 685–690.

[3] Ender, J. H. G. 'Space-time processing for multichannel synthetic aperture radar', *Electronics and Communication Engineering Journal*. 1999; **11**(1): 29–38.

[4] Sikaneta, I. C., Gierull, C. H., and Chouinard, J. Y. 'Metrics for SAR-GMTI based on eigen-decomposition of the sample covariance matrix', *International Radar Conference*, Adelaide, Australia. 2003, pp. 442–447.

[5] Cerrutti-Maori, D., Gierull, C. H., and Ender, J. H. G. 'Experimental Verification of SAR-GMTI Improvement through antenna-switching', *IEEE Transactions on Geoscience and Remote Sensing*. 2010; **48**(4): 2066–2075.

[6] Budillon, A., Pascazio, V., and Schirinzi, G. 'Multi-baseline along track SAR interferometric systems for ground moving target indication', *International Geoscience and Remote Sensing Symposium*. Honolulu, HI, 2010, pp. 2924–2927.

[7] Suwa, K., Yamamoto, K., Tsuchida, M., *et al*. 'An experimental study on image based multi-channel SAR-GMTI algorithm', *Asia-Pacific Conference on Synthetic Aperture Radar*. Tsukuba, Japan, 2013, pp. 577–580.

[8] Friedlander, B. and Porat, B. 'VSAR: a high resolution radar system for detection of moving targets', *IEE Proceedings – Radar, Sonar and Navigation*. 1997; **144**(4): 205–218.

[9] Friedlander, B. and Porat, B. 'VSAR: a high resolution radar system for ocean imaging', *IEEE Transactions on Aerospace and Electronic Systems*. 1998; **34**(3): 755–776.

[10] Romeiser, R. 'Current measurement by airborne along-track InSAR: measuring technique and experimental results', *IEEE Journal of Oceanic Engineering*. 2005; **30**(3): 552–569.

[11] Sletten, M. A. 'Demonstration of SAR distortion correction using a ground-based multichannel SAR test bed', *IEEE Transactions on Geoscience and Remote Sensing*. 2013; **51**(5): 3181–3190.

[12] Sletten, M., Menk, S., Toporkov, J., Jansen, R., and Rosenberg, L. 'The NRL multi aperture SAR system', *International Radar Conference*, Alexandria VA, USA. 2015, pp. 192–197.

[13] Frasier, S. J. and Camps, A. J. 'Dual-beam interferometry for ocean surface current vector mapping', *IEEE Transactions on Geoscience and Remote Sensing*. 2001; **39**(2): 401–414.

[14] Sharma, J. J., Gierull, C. H., and Collins, M. J. 'The influence of target acceleration on velocity estimation in dual-channel SAR-GMTI', *IEEE Transactions on Geoscience and Remote Sensing*. 2006; **44**(1): 134–147.

[15] Cumming, I. G. and Wong, F. H. *Digital Processing of Synthetic Aperture Radar Data*. London, UK: Artech House, 2005.

[16] Ender, J. H. G. 'The airborne experimental multi-channel SAR-system AER-II'. *European SAR Conference*. Königswinter, Germany, 1996, pp. 49–52.

[17] Gierull, C. H. Digital channel balancing of along-track interferometric SAR data. *Technical Memorandum 2003-02*. DRDC, Ottawa, 2003.

[18] Wang, G., Xia, X.-G., and Chen, V. C. 'Detection, location, and imaging of fast moving targets using multifrequency antenna array SAR', *IEEE Transactions on Aerospace and Electronic Systems*. 2004; **40**(1): 345–355.

[19] Li, G., Xu, J., Peng, Y.-N., and Xia, X.-G. 'Moving target location and imaging using dual-speed velocity SAR', *IET Radar, Sonar and Navigation*. 2007; **1**(2): 158–163.

[20] Li, G., Xu, J., Peng, Y.-N., and Xia, X.-G. 'Location and imaging of moving targets using nonuniform linear antenna array SAR', *IEEE Transactions of Aerospace and Electronic Systems*. 2007; **43**(3): 1214–1220.

[21] Li, G. and Xia, X.-G. 'Location and imaging of elevated moving target using multi-frequency velocity SAR with cross-track interferometry', *IEEE Transactions on Aerospace and Electronic Systems*. 2011; **47**(2): 1203–1212.

[22] Li, G., Xu, J., Peng, Y.-N., and Xia, X.-G. 'Velocity layover solution in VSAR image', *CIE International Radar Conference*, Shanghai, China. 2006, pp. 1–4.

[23] Shao, J., Tau, R., Zhou, S., and Wang, Y. 'A novel method of signal processing for VSAR system', *International Radar Conference*, Alexandria VA, USA. 2000, pp. 741–744.

[24] McInctosh, R. E., Frasier, S. J., and Mead, J. B. 'FOPAIR: A focused array imaging radar for ocean sensing', *IEEE Transactions of Geoscience and Remote Sensing*. 1995; **33**(1): 115–124.

[25] Sletten, M., Hwang, P., Toporkov, J., Menk, S., Rosenberg, L., and Jansen, R. 'Analysis and correction of maritime signatures with the NRL MSAR', *IEEE International Conference on Geoscience and Remote Sensing*, Milan, Italy. 2015, pp. 2489–2492.

[26] Rosenberg, L. and Sletten, M. 'Analysis of maritime X-band velocity SAR imagery', *International Radar Conference*, Johannesburg, South Africa. 2015, pp. 121–126.

[27] Moreira, A. and Huang, Y. 'Airborne SAR processing of highly squinted data using a chirp scaling approach with integrated motion compensation', *IEEE Transactions on Geoscience and Remote Sensing*. 1994; **32**(5): 1029–1040.

[28] Percival, D. B. and Walden, A. T. *Spectral Analysis for Physical Applications*. Cambridge, UK: Cambridge University Press, 1992.

[29] Xu, J., Li, G., Peng, Y. N., Xia, X.-G., and Wang, Y.-L. 'Parametric velocity synthetic aperture radar: signal modeling and optimal methods', *IEEE Transactions on Geoscience and Remote Sensing*. 2008; **46**(9): 2463–2480.

[30] Xu, J., Li, G., Peng, Y. N., Xia, X.-G., and Wang, Y.-L. 'Parametric velocity synthetic aperture radar: multilook processing and its applications', *IEEE Transactions on Geoscience and Remote Sensing*. 2008; **46**(11): 3488–3502.

[31] Raj, R. G., Jansen, R. W., Lipps, R. D., Sletten, M., and Rosenberg, L. 'Velocity-ISAR: on the application of ISAR techniques to multichannel SAR imaging', *International Radar Conference*, Alexandria VA, USA. 2015, pp. 1050–1055.

Part III

Passive and multistatic radar

Introduction to passive and multistatic radar

Pierfrancesco Lombardo[1]

Due to the recently renewed interest inside the radar community, bistatic and multistatic radar are again a hot topic of research, despite the fact that their history is almost as old as the history of radar itself. This history is also quite well documented and revised, see for example [1–6]. Therefore, revising this whole history is outside of the scope of this introduction. However, a few historic elements are recalled below to show how the contributions presented in this chapter find an especially important place among the novel radar techniques and applications.

In [2] (and references therein), it was noticed that the interest in this topic has had a somewhat cyclic behaviour during these years, with multiple resurgences. The first resurgence was identified to start in the 1950s, while the second one was in the late 1960s. In 2007, the publication of the book 'Advances in bistatic radar', [3], proposed to mark the clear presence of a third resurgence in the bistatic radar topic. Its Part 1 reports on development, testing, and fielding of bi-/multistatic radar systems for various military, scientific and commercial applications: Distant Early Warning-Line gap filling, missile attack warning, planetary exploration, air surveillance, ionospheric measurements and wind measurements. For many of the applications above, the exploitation of illuminators of opportunity is considered to replace the ad-hoc radar transmitter, thus clearly showing the effectiveness of type of sensors usually addressed as Passive Bistatic Radar (PBR) or passive coherent location sensors. Its Part 2 is devoted to bistatic clutter characterization, and to the signal processing techniques for bistatic/multistatic radar applications, there including the long bistatic coherent integration times required for bistatic Synthetic Aperture Radar (SAR) techniques and Space-Time Adaptive Processing (STAP) techniques for moving target indication.

Also back in 2008, [5] was greatly focused on two main areas: bistatic SAR and PBR. Bistatic SAR was largely addressed by considering both spaceborne and airborne applications. Moreover, passive radar was mainly centred on the exploitation of FM-radio and Digital Video Broadcast-Terrestrial (DVB-T) transmitters for air surveillance and on the initial studies for passive SAR with satellite-based illuminators.

Two excellent updates on the state of the art for the bistatic radar have been presented in 2010/2011 in [7,8]. In particular, the discussion in [7] reports on a workshop that was held at University College London in 2009 'to discuss and

[1]DIET University of Rome 'La Sapienza', Italy

identify priorities and initiatives in multistatic and MIMO radar'. Among its ten conclusions, it is interesting to notice: (i) the interest to multiple-input–multiple-output (MIMO) radar as an evolution of multistatic and networked radar, (ii) the need for experimental datasets to assess performance, but also to fully characterize clutter and target echoes, (iii) the potential interest of bistatic and multistatic radar for maritime surface applications, there including (iv) the exploitation of the forward scatter geometry and (v) a high interest for the imaging of building interiors, also exploiting bistatic and multistatic geometries.

Without the pretence to be exhaustive, the novel radar techniques and applications included in this chapter mark some of the advances in the clutter description and signal processing techniques that have taken place in the last decade, namely after publication of [3,5]. In particular, this Part III is focused with specific attention on the novelties in the bistatic clutter description, in the multistatic configurations – including both forward scattering configurations and specific multi-path geometries that are involved in the through-the-wall imaging application – and on the field of passive radar that has been recently object of many research efforts. It is interesting to notice that these topics fall almost entirely among the priorities expressed in [7].

Among these advances, an important role is certainly played by bistatic and multistatic SAR technology and signal processing techniques. In addition to the points made in [3,5], an outstanding review for this specific area was presented in [9]. This reference considers the multiple potentialities offered by the bistatic/multistatic SAR configurations, together with the challenges that were still open in 2006. Since then, major advances have taken place in this area, up to the point to allow conceiving MIMO SAR operations for imaging applications, [10], as well as their use for MIMO SAR moving target detection, [11,12]. All the advances in this area clearly qualify as novel techniques and applications both in Part II – Imaging Radar and in Part III – Passive and Multistatic Radar. As an editorial choice, it was considered more appropriate to present such material in Part II, to keep a unitary view of SAR. Therefore, the reader interested in the bistatic/multistatic SAR topic is encouraged to refer to Part II. As a unique exception to this approach, the results related to passive bistatic SAR are included in this Part III in order to keep here a unitary view of the passive radar advances.

Indeed, the characterization of bistatic clutter echoes has been a major point of interest since the beginning [2,3]. However, its dependence on both transmit and receive incidence angles, together with the transmit and receive polarizations, in addition to the intrinsic dielectric properties and small scale geometrical characteristics, have prevented a complete characterization. The importance of this topic has fostered a significant research activity, especially for sea clutter echoes [13]. This includes both measurement and modelling of the sea clutter echoes, as well as the comparison between simultaneous monostatic and bistatic clutter echoes [14,15]. Among the different, highly qualified research activities, certainly plays a significant role the effort to characterize the bistatic sea clutter, both in terms of average reflectivity and of probability distribution of the echoes amplitude [16–18].

Following directly this introductory chapter, the point of the situation on this effort devoted to bistatic clutter characterization opens Part III. Specifically,

Chapter 12 by Griffiths and Palamà starts from the basic concepts on the clutter characterization in both amplitude and temporal domains and provides an outstanding overview of the progresses in the experimental measurements activity and the state-of-the-art for bistatic clutter modelling.

In the last decade, a significant research activity has also been carried out to investigate the extreme case of the bistatic radar geometry, namely the case where the target is close to the line connecting radar transmitter and radar receiver, namely the so-called baseline. To remark on the specific scattering mechanism involved in this peculiar situation, the term 'forward scattering' is typically used for it, which also emphasizes the opposite extreme geometric situation from the monostatic case of 'back-scattering' [19,20]. While this geometry was exploited to build radar fences since the early days of radar, recently this area has received significant attention and is expected to provide important system potentialities [21].

Significant efforts have been made in the recent temporal frame to characterize this type of radar, typically addressed as forward scatter radar (FSR). In particular, the signal processing techniques to extract the target signature and estimate motion parameters have been addressed in [22], while the phenomenological model of the receive signal has been characterized in [23]. Moreover, specific research activity was devoted to characterize the FSR sea clutter, [24], as well as the target parameter estimation techniques against such type of disturbance, [25].

Chapter 13 by Gashinova, Daniel, Miakinkov and Cherniakov provides a complete presentation of the novel research results obtained in the last decade regarding FSR systems. It authoritatively describes the different aspects of the FSR systems, starting from the phenomenological model, and moving through the characterization of the disturbance environment and the target motion parameter estimation, to eventually consider the FSR target tracking.

It is interesting to notice that the FSR radar can be used as an element in an extended network of FSR sensors, which allows improved motion estimation and localization capabilities, [26]. This opens the way to future research on MIMO FSR systems that are object of current investigation. Among the ongoing research activities, it is also useful to mention that the FSR principle is well suited to the use of sources of opportunity as the radar transmitter, which provides a passive FSR. Preliminary results in this area are reported in [27,28]. However, the topic of passive FSR is expected to be the object of significant advances in the near future.

Although the extreme geometry of FSR applies only to specific configurations, there are many applications where different multistatic geometries play an important role [29,30]. A very specific topic in the last decade is the provision of radar sensors with the capability to detect objects and provide images through the walls of buildings, typically known as through-the-wall radar imaging (TWRI), [31]. In the last few years, the technique to detect changes has been carefully designed and refined [32,33]. It is apparent that a multistatic system can provide increased through-the-wall imaging capabilities. However, even when the sensor is used in a simplified monostatic configuration, the radar signal undergoes multiple propagation mechanisms, including the forward propagation through the wall, the possible back-scattering and multiple bistatic scattering phenomena that take place on internal walls and/or individual scatterers. These multi-path rays need to be appropriately taken into account and exploited to provide an adequate imaging of the building interior [34–37]. This means that TWRI fits nicely in this Part III, dedicated to multistatic radar, since its geometry is intrinsically multistatic.

An outstanding point of the situation on the through-the-wall building interior imaging is presented in **Chapter 14 by Ahmad and Amin**. This chapter nicely completes the presented review of some of the recent advances provided by exploiting bistatic and multistatic geometries. It is very interesting to observe that also this type of application can find implementation based on existing illuminators, so that passive TWRI can be obtained. This is object of very recent and ongoing research activity, [38,39].

The remainder of this Part III is devoted to novel techniques and applications related to the bistatic and multistatic radar systems exploiting sources of opportunity as the radar transmitter, namely the passive radar. Indeed, the principle of passive radar is present from the earliest radar experiments and, through resurgences of bistatic radar, also this specific type of bistatic radar has had many resurgences, the last of which is related to the capability to provide reasonably wide area air surveillance and scientific applications such as ionospheric monitoring, planetary observation, etc. (see for example [3]).

Other than the quoted chapters in [3], it is appropriate to mention that some special issues of technical journals have been dedicated to advances in passive radar, for example [40] and, more recently, [41,42]. Moreover, extensive contributions related to the state of the art on the subject can be found in two chapters of the Principles of Modern Radar book series, respectively devoted to passive radar signal processing techniques, [43], and to passive radar applications, [44].

As is apparent, this field has received significant attention by the radar signal processing community in the last decade and many advances have been made. These include; (i) deriving advanced signal processing techniques to improve the achievable performance for surveillance applications, for example [45,46], as well as (ii) providing specialized processing techniques tailored to expand PBR applications in various directions. Among such techniques certainly appear: short range surveillance, exploitation of space-based illuminators of opportunity, and operation from airborne platforms, which are treated in three chapters of this Part III. Additionally, several very specific PBR signal sources and applications have been investigated; just to quote a few of them: exploitation of over-the-horizon radar transmitters of opportunity, [47,48], providing passive inverse SAR images of moving targets, [49], passive GPS reflectometry, [50–52] and passive radar altimetry for ocean monitoring, [53,54].

An important sign of the interest in novel PBR techniques and applications is clearly given by the number of research and demonstration projects funded by the European Union in the last decade that involve PBR as one of their important elements. Among the others, PBR applications have been funded in air surveillance security, [55], short-range airport security, [56,57], maritime surveillance using ground based illuminators of opportunity, [58,59] and space-based illuminators of opportunity, [60,61]. Moreover, a specific call for maritime border control with explicit mention to PBR surveillance was issued inside the Horizon 2020 EU security call, [62].

As previously introduced, there has been significant activity in the last decade to develop signal processing techniques for short-/mid-range surface target monitoring. While for long-range coverage in some cases even relatively coarse range resolution might be acceptable, for meaningful applications at closer ranges high range resolution is required. Therefore, the exploitation of the narrow-band FM sources must be replaced with wider bandwidth signal sources. After the CELLDAR system, [63], the use of mobile phone basestations has been recently investigated by different research groups, as for example in [64–68], by achieving some interesting results. However, due to power considerations, the coverage is short-range and the range resolution does not make yet its use interesting enough.

In contrast, DVB-T transmitters have both a reasonable power level and a wide bandwidth per single channel, so that they are a promising candidate for medium-range surveillance. Various techniques have been specifically developed for the processing of PBR signals based on DVB-T waveforms, so as to control the ambiguity function sidelobes, [69–72]. Unfortunately, to avoid wasting power, typical DVB-T transmitter antenna setting is tailored to focus the emitted power toward low altitudes, so that air target surveillance has some limitations. In contrast, PBR based on DVB-T signals appear of great interest for surface vehicle surveillance, such as maritime and road traffic. Experimental results in [73,74] confirm its great potentiality for the maritime surveillance application.

For much shorter range applications WiFi signals have been identified as waveforms of interest, since their lower power level goes together with a wide bandwidth. Moreover, they have a widespread presence. The characteristics of the waveform was first analysed in [75], while approaches to control its sidelobes have been presented in [76]. Potentialities and challenges have been summarized in [77], focusing more on the outdoor surveillance of vehicular road targets and moving human targets, while its use for indoor localization of moving persons has been studied in [78].

Advances in the last decade in the short-/medium-range surveillance, including both DVB-T and Wi-Fi-based PBR are presented in **Chapter 15 by Colone**, where both signal processing techniques and application case studies are reported.

A second area of innovation in PBR research is related to the use of space-based illuminators of opportunity. Despite the seminal papers that were published in years 1998, [79], 1999, [80] and 2000, [81], renewed interest was devoted to this type of sensors during the last decade. In particular, the exploitation of the signals emitted by navigation satellites, like GPS, GLONASS and Galileo, has been recently demonstrated to be able to provide imaging capabilities even using stationary receivers, [82]. Despite the low-power levels scattered by the ground, long SAR integration times allow the final power level to be enough to produce reasonable quality images, at least at relatively short range from the receiver. Moreover, the point spread function of this type of sensors has been characterized, [83].

Interesting applications have been recently devised in change detection applications for infrastructural and environmental monitoring, [84,85].

A great attraction of these sensors is in the availability of multiple illuminators that operate from different orbital positions at the same time with almost orthogonal codes, which allows their separation and makes available in principle a multistatic PBR even using a single receiver. The joint exploitation of multistatic PBR images requires fusion techniques to be derived, which is the object of current research efforts, [86,87].

A thorough review of the major results achieved in this field is presented in **Chapter 16 by Antoniou and Cherniakov**, including the challenges and the experimental results obtained, as well as the present and future potentialities.

Remarkable results have been obtained in the last decade also in the airborne exploitation of passive radar, both in terms of algorithms and experiments, [88]. This includes both the capability to form passive SAR images using ground based illuminators of opportunity and the capability to cancel the clutter echoes that occupy a wide frequency bandwidth due to the platform motion.

The problems of cancelling the surface clutter, which is widely spread by the Doppler, is well known in the case of active airborne radar. A major research focus of the past few decades has been devoted to the development of STAP techniques to achieve effective clutter removal by exploiting multiple antennas on board, connected to independent receiving channels, [89–95]. This was later extended to bistatic configurations, [96,97], which is also nicely described in [3], Chapter 11. The application of STAP techniques to passive radar is not straightforward for many reasons, among which is certainly the continuous wave characteristic of the transmitted waveform and its temporal variation, rather than the transmission of sequences of identically shaped radar pulses. An appropriate application of STAP, in its simplified version of the displaced phase centre antenna technique, has been recently demonstrated, [98–100].

The formation of passive SAR images, by exploiting surface-based transmitters of opportunity, has also received a significant attention in this last decade. The specific signal processing techniques required to form such images need to take into account both the bistatic nature of the geometric configuration and the continuously variable transmitted waveform. Various experiments have been recently carried out to verify the capability to properly focus the passive SAR images and to achieve images of good quality, [101–104].

The major recent advances in both airborne SAR and STAP are presented in **Chapter 17 by Kulpa, Gromek and Dawidowicz**, which includes experimental results that allow the validations of the techniques and the demonstration of their effectiveness. Ongoing research shows the possibility to exploit the multistatic geometry, by means of compressive sensing to improve passive radar performance, [105].

A further step forward of the recent years is the capability to exploit the great variety of waveforms of opportunity available, both in terms of geometry and type of sources, [106]. In fact, often the angular distribution of radiation sources makes multistatic passive radar available even with a single receiver position. This allows improved robustness but also localization performance, [107–110] and tracking accuracy, [111,112]. Moreover, it increases the effectiveness against stealth targets, [113].

Similarly, the availability of multiple sources of opportunity available at the same time allows to exploit the advantages of the different waveforms and to avoid most of their drawbacks. As an example, this is the case when jointly exploiting FM-radio, DVB-T and digital audio broadcast illuminators [114,115]. This largely improves the diversity, adding robustness and increasing the overall performance.

A further degree of freedom to be added to the multistatic/hybrid multi-source passive radar is the potential addition of active transmitters, to be used in the absence of the sources of opportunity or in combination with them, [116]. The addition of this further degree, greatly improves the reliability of the overall multi-static system, making it even more appealing for both military and civil applications, where a high standard of safety must be guaranteed.

The advances in this area are presented in **Chapter 18 by Kuschel, Hoffmann and Schroeder**, where experimental results are shown, together with the validation of the multistatic/hybrid approach.

The full exploitation of the multiple transmitters of opportunity is achieved when also using multiple receivers, so that a MIMO passive radar system is obtained. Among the most recent research efforts, there has been an emphasis on the full use of the MIMO passive radar potentialities to achieve still further improvement with respect to the multistatic case, [117], – in target motion parameter estimation, [118] and tracking, [119].

Moreover, it has recently been shown that the MIMO configuration can also provide a potential improvement in terms of detection capability. This improvement has been shown to depend on multiple factors, including the signal-to-noise power ratio (SNR) of the surveillance channel and the quality of the signal reference. To benefit from these potential improvements, both a centralized and a decentralized detection structure have been conceived based on the generalized likelihood ratio test (GLRT), which show different characteristics and trade-offs, [120,121].

A systematic presentation of these recent results on MIMO passive radar detection is reported in **Chapter 19 by Hack, Patton and Himed**. This spans from the GLRT-based algorithms derivation, to a thorough discussion of the achievable performance as a function of reference channel quality and surveillance channel SNR. This chapter sets the theoretical basis for the optimal joint exploitation of all information available in a fully deployed MIMO passive radar system in order to maximize the detection performance. It opens the way to ongoing and future additional research activity aiming at exploiting all types of available information to increase performance. It is expected that future research efforts will start from this basis and proceed to demonstrate the improvements on the field, possibly specialize the approach to the specific waveforms available, and include additional information from the environment.

This also concludes Part III – Passive and Multistatic Radar, which represents an overview of novel radar techniques and applications that have characterized the last decade of research in this area, where the different chapters build a unified frame. This makes it of great interest to engineers, mathematicians, physicists who seek an introduction to the state of the art in multistatic and passive radar and their applications. Although it is not explicitly structured as a basic comprehensive treatment of the subject, it certainly provides enough material to allow a reader with an elementary comprehension of the radar to understand the state of the art of multistatic and passive radar.

As discussed above, this Part III opens with an important update on bistatic clutter characterization, which accounts for the emphasis that this topic has received for its clear impact on the prediction of the bistatic radar performance. After focusing on the recent investigations on the potentialities of the extreme bistatic geometry, namely the forward scatter case, and on the potentialities to exploit the transmitted and scattered signals to achieve interior building imaging, the attention is concentrated on the advances in the techniques and applications for a specific type of bistatic radar, known as passive radar.

Specifically, three main topics are addressed that received a significant attention in these last years: (i) short-/medium-range surveillance, (ii) exploitation of space-based illuminators for passive radar imaging and (iii) airborne passive radar applications, there including passive SAR imaging with surface-based transmitters of opportunity and airborne clutter cancellation using passive radar with multiple antennas.

Finally, the exploitation of multistatic passive radar is considered, there including the recent approaches to jointly exploit waveforms of opportunity of different types and characteristics. This is followed by the presentation of the recent results on the potentialities of a full MIMO passive radar system to optimize the detection probability, which paves the way for a full exploitation of the intrinsic passive radar characteristics.

Each chapter is organized to represent an independent entity so that readers interested in a specific topic can easily select the individual chapter. This makes it appealing also to scientists and engineers that require in-depth knowledge of specific novel techniques or applications. In fact, the presentation of each chapter is based on the recent research work of the authors and their co-workers and includes many references to technical papers published in international journals and conferences.

Clearly, the act of reading this Part III will not provide a comprehensive repertoire of all the knowledge and skills needed to become a scientist able to operate in the multistatic radar field. This would require to read the vast literature on this subject that can be found easily in the internet. However, the individual chapter provides a guide to enter into this literature with an overall vision of the topic. Without the pretence to be comprehensive, due to the high number of important contributions in the Passive and Bistatic Radar area, the selected chapters provide this guidance for some of the main topics in this field that characterized the last decade.

References

[1] Glaser J.I. 'Fifty years of bistatic and multistatic radar'. *IEE Proceedings F: Communications, Radar and Signal Processing*. 1986; **133**(7): 596–603.

[2] Willis N.J. *Bistatic Radar*. (Technology Service Corp., Silver Spring, MD, 2nd ed, 1995 corrected and republished by SciTech Publishing, Inc., 2008).

[3] Willis N.J., Griffiths H.D. (eds) *Advances in Bistatic Radar*. (SciTech Publishing, Raleigh, NC, 2007).

[4] Cherniakov M. (ed) *Bistatic Radar: Principles and Practice*. (John Wiley & Sons, Chichester, UK, 2007).

[5] Cherniakov M. (ed) *Bistatic Radar: Emerging Technology*. (John Wiley & Sons, Chichester, UK, 2008).

[6] Griffiths, H.D. 'Early history of bistatic radar', *EuRAD Conference* 2016, London, 3–7 October 2016.

[7] Griffiths H. 'Workshop on multistatic and MIMO radar'. *IEEE Aerospace and Electronic Systems Magazine*. 2010; **25**(2): 43–45.

[8] Griffiths H. 'Developments in bistatic and networked radar'. *Proceedings of the 2011 IEEE CIE International Conference on Radar*, 2011, pp. 10–13.

[9] Krieger G., Moreira A. 'Spaceborne bi- and multistatic SAR: potential and challenges'. *IEE Proceedings – Radar, Sonar and Navigation*. 2006; **153**(3): 184–198.

[10] Krieger G. 'MIMO-SAR: opportunities and pitfalls'. *IEEE Transactions on Geoscience and Remote Sensing*. 2014; **52**(5): 2628–2645.

[11] Lombardo P., Pastina D., Turin F. 'Ground moving target detection based on MIMO SAR systems'. *IEEE Journal of Selected Topics in Applied Earth Observations and Remote Sensing*. 2015; **8**(11): 5081–5095.

[12] Baumgartner S.V., Krieger G. 'Dual-platform large along-track baseline GMTI'. *IEEE Transactions on Geoscience and Remote Sensing*. 2016; **54**(3): 1554–1574.

[13] Griffiths H.D., Al-Ashwal W.A., Ward K.D., Tough R.J.A., Baker C.J., Woodbridge K. 'Measurement and modelling of bistatic radar sea clutter'. *IET Radar, Sonar & Navigation*. 2010; **4**(2): 280–292.

[14] Palamà R., Greco M.S., Stinco P., Gini F. 'Statistical analysis of bistatic and monostatic sea clutter'. *IEEE Transactions on Aerospace and Electronic Systems*. 2015; **51**(4): 3036–3054.

[15] Ritchie M., Stove A., Woodbridge K., Griffiths H. 'NetRAD: monostatic and bistatic sea clutter texture and Doppler spectra characterization at S-band'. *IEEE Transactions on Geoscience and Remote Sensing*. 2016; **54**(9): 5533–5543.

[16] Arnold-Bos A., Khenchaf A., Martin A. 'Bistatic radar imaging of the marine environment – Part I: Theoretical background'. *IEEE Transactions on Geoscience and Remote Sensing*. 2007; **45**(11): 3372–3383.

[17] Al-Ashwal W.A., Woodbridge K., Griffiths H.D. 'Analysis of bistatic sea clutter – Part I: Average reflectivity'. *IEEE Transactions on Aerospace and Electronic Systems*. 2014; **50**(2): 1283–1292.

[18] Al-Ashwal W.A., Woodbridge K., Griffiths H.D. 'Analysis of bistatic sea clutter – Part II: Amplitude statistics'. *IEEE Transactions on Aerospace and Electronic Systems*. 2014; **50**(2): 1293–1303.

[19] Glaser J.I. 'Bistatic RCS of complex objects near forward scatter'. *IEEE Transactions on Aerospace and Electronic Systems*. 1985; **21**(1): 70–78.

[20] Glaser J.I. 'Some results in the bistatic radar cross section (RCS) of complex objects'. *Proceedings of the IEEE*. 1989; **77**(5): 639–648.

[21] Glaser J.I. 'Forward scatter radar for future systems'. *The Weapon Systems Technology Information Analysis Center (WSTIAC) Quarterly*. 2011; **10**(3): 3–8.

[22] Hu C., Sizov V., Antoniou M., Gashinova M., Cherniakov M. 'Optimal signal processing in ground-based forward scatter micro radars'. *IEEE Transactions on Aerospace and Electronic Systems*. 2012; **48**(4): 3006–3026.

[23] Gashinova M., Daniel L., Sizov V., Hoare E., Cherniakov M. 'Phenomenology of Doppler forward scatter radar for surface targets observation'. *IET Radar, Sonar & Navigation*. 2013; **7**(4): 422–432.

[24] Gashinova M., Kabakchiev K., Daniel L., Hoare E., Sizov V., Cherniakov M. 'Measured forward-scatter sea clutter at near-zero grazing angle: analysis of spectral and statistical properties'. *IET Radar, Sonar & Navigation*. 2014; **8**(2): 132–141.

[25] Kabakchiev H., Behar V., Garvanov I., *et al.* 'Experimental verification of maritime target parameter evaluation in forward scatter maritime radar'. *IET Radar, Sonar & Navigation*. 2015; **9**(4): 355–363.

[26] Pastina D., Contu M., Lombardo P., *et al.* 'Target motion estimation via multi-node forward scatter radar system'. *IET Radar, Sonar and Navigation*. 2016; **10**(1): 3–14.

[27] Martelli T., Colone F., Lombardo P. 'First experimental results for a WiFi-based passive forward scatter radar'. *Proceedings of the 2016 IEEE Radar Conference (RadarConf)*, 2016, pp. 1–6.

[28] Gashinova M., Contu M., De Luca A., *et al.* 'Passive multi-frequency forward-scatter radar measurements of airborne targets using broadcasting signals'. *IEEE Transactions on Aerospace and Electronic Systems*. 2017; **53**(3): 1067–1087.

[29] Erricolo D. 'Experimental validation of second-order diffraction coefficients for computation of path-loss past buildings'. *IEEE Transactions on Electromagnetic Compatibility*. 2002; **44**(1): 272–273.

[30] Erricolo D., Crovella U.G., Uslenghi P.L.E. 'Time-domain analysis of measurements on scaled urban models with comparisons to ray-tracing propagation simulation'. *IEEE Transactions on Antennas and Propagation*. 2002; **50**(5): 736–741.

[31] Debes C., Amin M.G., Zoubir A.M. 'Target detection in single- and multiple-view through-the-wall radar imaging'. *IEEE Transactions on Geoscience and Remote Sensing*. 2009; **47**(5): 1349–1361.

[32] Amin M.G., Ahmad F. 'Change detection analysis of humans moving behind walls'. *IEEE Transactions on Aerospace and Electronic Systems*. 2013; **49**(3): 1410–1425.

[33] Ahmad F., Amin M.G. 'Through-the-wall human motion indication using sparsity-driven change detection'. *IEEE Transactions on Geoscience and Remote Sensing*. 2013; **51**(2): 881–890.

[34] Gennarelli G., Soldovieri F. 'Radar imaging through cinderblock walls: achievable performance by a model-corrected linear inverse scattering approach'. *IEEE Transactions on Geoscience and Remote Sensing*. 2014; **52**(10): 6738–6749.

[35] Leigsnering M., Ahmad F., Amin M., Zoubir A. 'Multipath exploitation in through-the-wall radar imaging using sparse reconstruction'. *IEEE Transactions on Aerospace and Electronic Systems*. 2014; **50**(2): 920–939.

[36] Setlur P., Negishi T., Devroye N., Erricolo D. 'Multipath exploitation in non-LOS urban synthetic aperture radar'. *IEEE Journal of Selected Topics in Signal Processing*. 2014; **8**(1): 137–152.

[37] Leigsnering M., Amin M., Ahmad F., Zoubir A.M. 'Multipath exploitation and suppression for SAR imaging of building interiors: an overview of recent advances'. *IEEE Signal Processing Magazine*. 2014; **31**(4): 110–119.

[38] Chetty K., Smith G.E., Woodbridge K. 'Through-the-wall sensing of personnel using passive bistatic wifi radar at standoff distances'. *IEEE Transactions on Geoscience and Remote Sensing*. 2012; **50**(4): 1218–1226.

[39] Gennarelli G., Amin M.G., Soldovieri F., Solimene R. 'Passive multiarray image fusion for RF tomography by opportunistic sources'. *IEEE Geoscience and Remote Sensing Letters*. 2015; **12**(3): 641–645.

[40] Howland P. 'Special issue on passive radar systems'. *IEE Proceedings – Radar, Sonar and Navigation*. 2005; **152**(3): 106–223.

[41] Farina A., Kuschel H. 'Special issue on passive radar (Part I)'. *IEEE Aerospace and Electronic Systems Magazine*. 2012; **27**(10): 5–59.

[42] Farina A., Kuschel H. 'Special issue on passive radar (Part II). *IEEE Aerospace and Electronic Systems Magazine*. 2012; **27**(11): 4–55.

[43] Lombardo P., Colone F. 'Advanced processing methods for passive bistatic radar systems', in Melvin W.L. and Scheer J.A. (eds) *Principles of Modern Radar: Advanced Radar Techniques* (SciTech Publishing, Inc., 2012), pp. 739–821.

[44] Griffiths, H. 'Passive and bistatic radar', in Melvin W.L. and Scheer J.A. (eds) *Principles of Modern Radar: Radar Applications* (SciTech Publishing, Inc., 2014), pp. 499–540.

[45] Colone F., O'Hagan D.W., Lombardo P., Baker C.J., 'A multistage processing algorithm for disturbance removal and target detection in passive bistatic radar'. *IEEE Transactions on Aerospace and Electronic Systems*. 2009; **45**(2): 698–722.

[46] Smith G.E., Chetty K., Baker C.J., Woodbridge K. 'Extended time processing for passive bistatic radar'. *IET Radar, Sonar & Navigation*. 2013; **7**(9): 1012–1018.

[47] Fabrizio G., Colone F., Lombardo P., Farina A. 'Adaptive beamforming for high frequency over-the-horizon passive radar'. *IET Proceedings – Radar Sonar and Navigation*. 2009; **3**(4): 384–405.

[48] Fabrizio G. *High Frequency Over-the-Horizon Radar: Fundamental Principles, Signal Processing, and Practical Applications* (McGraw Hill Education, New York, 1st ed, 2013).

[49] Olivadese D., Giusti E., Petri D., Martorella M., Capria A., Berizzi F., 'Passive ISAR with DVBT Signals'. *IEEE Transactions on Geoscience and Remote Sensing*. 2013; **51**(8): 4508–4517

[50] Rodriguez-Alvarez N., Akos D.M., Zavorotny V.U., *et al.* 'Airborne GNSS-R wind retrievals using delay–Doppler maps'. *IEEE Transactions on Geoscience and Remote Sensing*. 2013; **51**(1): 626–641.

[51] Nievinski F.G., Larson K.M. 'Inverse modeling of GPS multipath for snow depth estimation – Part I: Formulation and simulations'. *IEEE Transactions on Geoscience and Remote Sensing*. 2014; **52**(10): 6555–6563.

[52] Chew C.C., Small E.E., Larson K.M., Zavorotny V.U. 'Vegetation sensing using GPS-interferometric reflectometry: theoretical effects of canopy parameters on signal-to-noise ratio data'. *IEEE Transactions on Geoscience and Remote Sensing*. 2015; **53**(5): 2755–2764.

[53] Martin-Neira M., Caparrini M., Font-Rossello J., Lannelongue S., Vallmitjana C.S. 'The PARIS concept: an experimental demonstration of sea surface altimetry using GPS reflected signals'. *IEEE Transactions on Geoscience and Remote Sensing*. 2001; **39**(1): 142–150.

[54] Cardellach E., Rius A., Martín-Neira M., *et al*. 'Consolidating the precision of interferometric GNSS-R ocean altimetry using airborne experimental data'. *IEEE Transactions on Geoscience and Remote Sensing*. 2014; **52**(8): 4992–5004.

[55] European FP7 Project ARGUS-3D – 'AiR guidance and surveillance 3D'. http://www.argus3d.eu.

[56] European FP7 Project ATOM – 'Airport detection and tracking of dangerous materials by passive and active sensors arrays'. http://www.atom-project.eu.

[57] European FP7 Project SOS – 'Sensors system for detection and tracking of dangerous materials in order to increase the airport security in the indoor landside area'. http://www.sos-project.eu.

[58] European FP7 Project SEABILLA – 'SEA Border surveILLance'. http://www.seabilla.eu.

[59] European FP7 Project CLOSEYE – 'Collaborative evaluation of border surveillance technologies in maritime environment by pre-operational validation of innovative solutions'. http://www.closeye.eu/.

[60] European FP7 Project DOLPHIN – 'Development of pre-operational services for highly innovative maritime surveillance capabilities'. http://www.gmes-dolphin.eu/.

[61] European H2020 Project SpyGLASS – 'Galileo based passive radar system for maritime surveillance'. http://cordis.europa.eu/project/rcn/193820_en.html.

[62] European H2020 Call for Proposals H2020-BES-02-2015: 'Maritime border security topic 2: affordable and easily deployable technologies for EU coastal border surveillance with reduced impact on the environment'.

[63] BAE Systems' CELLDAR, https://web.archive.org/web/20060308181747/http://www.roke.co.uk/sensors/stealth/celldar.asp.

[64] Tan D.K.P., Sun H., Lu Y., Lesturgie M., Chan H.L. 'Passive radar using global system for mobile communication signal: theory, implementation and measurements'. *IEE Proceedings – Radar, Sonar and Navigation*. 2005; **152**(3): 116–123.

[65] Sun H., Tan D.K.P., Lu Y., Lesturgie M. 'Applications of passive surveillance radar system using cell phone base station illuminators'. *IEEE Aerospace and Electronic Systems Magazine*. 2010; **25**(3): 10–18.

[66] Zemmari R., Broetje M., Battistello G., Nickel U. 'GSM passive coherent location system: performance prediction and measurement evaluation'. *IET Radar, Sonar and Navigation*. 2014; **8**(2): 94–105.

[67] Krysik P., Samczynski P., Malanowski M., Maslikowski L., Kulpa K.S. 'Velocity measurement and traffic monitoring using a GSM passive radar demonstrator'. *IEEE Aerospace and Electronic Systems Magazine*. 2012; **27**(10): 43–51.

[68] Stinco P., Greco M.S., Gini F., Rangaswamy M. 'Ambiguity function and Cramer–Rao bounds for universal mobile telecommunications system-based passive coherent location systems'. *IET Radar, Sonar & Navigation*. 2012; **6**(7): 668–678.

[69] Saini R., Cherniakov M. 'DTV signal ambiguity function analysis for radar application'. *IEE Proceedings – Radar, Sonar and Navigation*. 2005; **152**(3): 133–142.

[70] Palmer J.E., Harms H.A., Searle S.J., Davis L.M. 'DVB-T passive radar signal processing.' *IEEE Transactions on Signal Processing*. 2013; **61**(8): 2116–2126.

[71] Colone F., Langellotti D., Lombardo P. 'DVB-T signal ambiguity function control for passive radars'. *IEEE Transactions on Aerospace and Electronic Systems*. 2014; **50**(1): 329–347.

[72] Zhang X., Liu J., Li H., Himed B. 'Maximum likelihood synchronization for DVB-T2 in unknown fading channels'. *IEEE Transactions on Broadcasting*. 2015; **61**(4): 615–624.

[73] Langellotti D., Colone F., Lombardo P., Sedehi M., Tilli E. 'DVB-T based passive bistatic radar for maritime surveillance'. *Proceedings of the IEEE Radar Conference 2014*, Cincinnati, OH, USA, 19–23 May 2014.

[74] Langellotti D., Colone F., Lombardo P., Tilli E., Sedehi M., Farina A. 'Over the horizon maritime surveillance capability of DVB-T based passive radar'. *Proceedings of the 2014 European Radar Conference (EuRAD 2014)*, Rome, Italy, 8–10 October 2014.

[75] Colone F., Woodbridge K., Guo H., Mason D., Baker C.J. 'Ambiguity function analysis of wireless LAN transmissions for passive radar'. *IEEE Transactions on Aerospace and Electronic Systems*. 2011; **47**(1): 240–264.

[76] Colone F., Falcone P., Bongioanni C., Lombardo P. 'WiFi-based passive bistatic radar: data processing schemes and experimental results'. *IEEE Transactions on Aerospace and Electronic Systems*. 2012; **48**(2): 1061–1079.

[77] Falcone P., Colone F., Lombardo P. 'Potentialities and challenges of WiFi-based passive radar'. *IEEE Aerospace and Electronic Systems Magazine*. 2012; **27**(11): 15–26.

[78] Pastina D., Colone F., Martelli T., Falcone P. 'Parasitic exploitation of Wi-Fi signals for indoor radar surveillance'. *IEEE Transactions on Vehicular Technology*. 2015; **64**(4): 1401–1415.

[79] Prati C., Rocca F., Giancola D., Monti Guarnieri A. 'Passive geosynchronous SAR system reusing backscattered digital audio broadcasting

signals'. *IEEE Transactions on Geoscience and Remote Sensing*. 1998; **36**(6): 1973–1976.

[80] Cherniakov M., Nezlin D., Kubik K. 'Air target detection via bistatic radar based on LEOS communication signals'. *IEE Proceedings – Radar, Sonar and Navigation*. 2002; **149**(1): 33–38.

[81] Cazzani L., Colesanti C., Leva D., *et al.* 'A ground-based parasitic SAR experiment'. *IEEE Transactions on Geoscience and Remote Sensing*. 2000; **38**(5): 2132–2141.

[82] Antoniou M., Zeng Z., Feifeng L., Cherniakov M. 'Experimental demonstration of passive BSAR imaging using navigation satellites and a fixed receiver'. *IEEE Geoscience and Remote Sensing Letters*. 2012; **9**(3): 477–481.

[83] Liu F., Antoniou M., Zeng Z., Cherniakov M. 'Point spread function analysis for BSAR with GNSS transmitters and long dwell times: theory and experimental confirmation'. *IEEE Geoscience and Remote Sensing Letters*. 2013; **10**(4): 781–785.

[84] Liu F., Antoniou M., Zeng Z., Cherniakov M. 'Coherent change detection using passive GNSS-based BSAR: experimental proof of concept'. *IEEE Transactions on Geoscience and Remote Sensing*. 2013; **51**(8): 4544–4555.

[85] Zhang Q., Antoniou M., Chang W., Cherniakov M. 'Spatial decorrelation in GNSS-based SAR coherent change detection'. *IEEE Transactions on Geoscience and Remote Sensing*. 2015; **53**(1): 219–228.

[86] Santi F., Antoniou M., Pastina D. 'Point spread function analysis for GNSS-based multistatic SAR'. *IEEE Geoscience and Remote Sensing Letters*. 2015; **12**(2): 304–308.

[87] Santi F., Bucciarelli M., Pastina D., Antoniou M., Cherniakov M. 'Spatial resolution improvement in GNSS-based SAR using multistatic acquisitions and feature extraction'. *IEEE Transactions on Geoscience and Remote Sensing*. 2016; **54**(10): 6217–6231.

[88] Brown J., Woodbridge K., Griffiths H., Stove A., Watts S. 'Passive bistatic radar experiments from an airborne platform'. *IEEE Aerospace and Electronic Systems Magazine*. 2012; **27**(11): 50–55.

[89] Brennan L.E., Reed I.S. 'Theory of adaptive radar'. *IEEE Transactions on Aerospace and Electronic Systems*. 1973; **9**: 237–252.

[90] Klemm R. 'Adaptive clutter suppression for airborne phased array radars'. *IEE Proceedings F & H Communications, Radar and Signal Processing*. 1983; **130**(1): 125–132.

[91] Ward J. 'Space-time adaptive processing for airborne radar', *Lincoln Laboratory Technical Report 1015*, December 1994.

[92] Klemm R. *Principles of Space-Time Adaptive Processing*, 3rd edition © 2006 IET, London. ISBN 086341 566 0/978-086341-566-1.

[93] Guerci J.R. *Space-Time Adaptive Processing for Radar*. (Artech House Publishers, Norwood, 2003). ISBN 1-58053-377-9.

[94] Klemm R. (ed) *Applications of Space-Time Adaptive Processing*. (IEE Publishing, Stevenage, UK, 2004).

[95] Melvin W.L. 'A STAP overview'. *IEEE AES Systems Magazine – Special Tutorials Issue*. 2004; **19**(1): 19–35.

[96] Klemm R. 'Comparison between monostatic and bistatic antenna configurations for STAP'. *IEEE Transactions on Aerospace and Electronic Systems*. 2000; **36**(2): 596–608.

[97] Melvin W.L., Davis M.E. 'Adaptive cancellation method for geometry-induced nonstationary bistatic clutter environments'. *IEEE Transactions on Aerospace and Electronic Systems*. 2007; **43**(2): 651–672.

[98] Tan D.K.P., Lesturgie M., Sun H., Lu Y. 'Space–time interference analysis and suppression for airborne passive radar using transmissions of opportunity'. *IET Radar, Sonar & Navigation*. 2014; **8**(2): 142–152.

[99] Dawidowicz B., Kulpa K.S., Malanowski M., Misiurewicz J., Samczynski P., Smolarczyk M. 'DPCA detection of moving targets in airborne passive radar'. *IEEE Transactions on Aerospace and Electronic Systems*. 2012; **48**(2): 1347–1357.

[100] Dawidowicz B., Samczynski P., Malanowski M., Misiurewicz J., Kulpa K.S. 'Detection of moving targets with multichannel airborne passive radar'. *IEEE Aerospace and Electronic Systems Magazine*. 2012; **27**(11): 42–49.

[101] Gutierrez Del Arroyo R., Jackson J.A. 'WiMAX OFDM for passive SAR ground imaging'. *IEEE Transactions on Aerospace and Electronic Systems*. 2013; **49**(2): 945–959.

[102] Ulander L.M.H., Frölind P.O., Gustavsson A., Ragnarsson R., Stenström G. 'VHF/UHF bistatic and passive SAR ground imaging'. *Proceedings of the 2015 IEEE Radar Conference (RadarCon)*, 2015, pp. 669–673.

[103] Maslikowski L., Samczynski P., Baczyk M., Krysik P., Kulpa K. 'Passive bistatic SAR imaging – Challenges and limitations'. *IEEE Aerospace and Electronic Systems Magazine*. 2014; **29**(7): 23–29.

[104] Gromek D., Kulpa K., Samczyński P. 'Experimental results of passive SAR imaging using DVB-T illuminators of opportunity'. *IEEE Geoscience and Remote Sensing Letters*. 2016; **13**(8): 1124–1128.

[105] Wu Q., Zhang Y.D., Amin M.G., Himed B. 'High-resolution passive SAR imaging exploiting structured Bayesian compressive sensing'. *IEEE Journal of Selected Topics in Signal Processing*. 2015; **9**(8): 1484–1497.

[106] Farina A., Lesturgie M. 'Special issue on bistatic and MIMO radars and their applications in surveillance and remote sensing'. *IET Radar, Sonar & Navigation*. 2014; **8**(2): 73–166.

[107] Lesturgie M. 'Use of dynamic radar signature for multistatic passive localisation of helicopter'. *IEE Proceedings – Radar, Sonar and Navigation*. 2005; **152**(6): 395–403.

[108] Malanowski M., Kulpa K. 'Two methods for target localization in multistatic passive radar'. *IEEE Transactions on Aerospace and Electronic Systems*. 2012; **48**(1): 572–580.

[109] Greco M.S., Stinco P., Gini F., Farina A. 'Cramer–Rao bounds and selection of bistatic channels for multistatic radar systems'. *IEEE Transactions on Aerospace and Electronic Systems*. 2011; **47**(4): 2934–2948.

[110] Anastasio V., Farina A., Colone F., Lombardo P. 'Cramér–Rao lower bound with Pd < 1 for target localisation accuracy in multistatic passive radar'. *IET Radar, Sonar & Navigation*. 2014; **8**(7): 767–775.

[111] Battistelli G., Chisci L., Morrocchi S., Papi F., Farina A., Graziano A. 'Robust multisensor multitarget tracker with application to passive multistatic radar tracking'. *IEEE Transactions on Aerospace and Electronic Systems*. 2012; **48**(4): 3450–3472.

[112] Choi S., Crouse D., Willett P., Zhou S. 'Multistatic target tracking for passive radar in a DAB/DVB network: initiation'. *IEEE Transactions on Aerospace and Electronic Systems*. 2015; **51**(3): 2460–2469.

[113] Kuschel H., Heckenbach J., Müller St., Appel R. 'Countering stealth with passive, multi-static, low frequency radars'. *IEEE Aerospace and Electronic Systems Magazine*. 2010; **25**(9): 11–17.

[114] O'Hagan D.W., Kuschel H., Ummenhofer M., Heckenbach J., Schell J. 'A multi-frequency hybrid passive radar concept for medium range air surveillance'. *IEEE Aerospace and Electronic Systems Magazine*. 2012; **27**(10): 6–15.

[115] Kuschel H., Ummenhofer M., Lombardo P., Colone F., Bongioanni C. 'Passive radar components of ARGUS 3D'. *IEEE Aerospace and Electronic Systems Magazine*. 2014; **29**(3): 15–25.

[116] Kuschel H., Heckenbach J., Schell J. 'Deployable multiband passive/active radar for air defense (DMPAR)'. *IEEE Aerospace and Electronic Systems Magazine*. 2013; **28**(9): 37–45.

[117] Liu J., Li H., Himed B. 'Two target detection algorithms for passive multistatic radar'. *IEEE Transactions on Signal Processing*. 2014; **62**(22): 5930–5939.

[118] Chiriac V.M., He Q., Haimovich A.M., Blum R.S. 'Ziv–Zakai bound for joint parameter estimation in MIMO radar systems'. *IEEE Transactions on Signal Processing*. 2015; **63**(18): 4956–4968.

[119] Yi J., Wan X., Leung H., Cheng F. 'MIMO passive radar tracking under a single frequency network'. *IEEE Journal of Selected Topics in Signal Processing*. 2015; **9**(8): 1661–1671.

[120] Hack D.E., Patton L.K., Himed B., Saville M.A. 'Centralized passive MIMO radar detection without direct-path reference signals'. *IEEE Transactions on Signal Processing*. 2014; **62**(11): 3013–3023.

[121] Hack D.E., Patton L.K., Himed B., Saville M.A 'Detection in passive MIMO radar networks'. *IEEE Transactions on Signal Processing*. 2014; **62**(11): 2999–3012.

Chapter 12

Bistatic clutter modelling

Hugh Griffiths and Riccardo Palamà**

Abstract

This chapter addresses the study of the properties of bistatic clutter, compared with monostatic with a particular focus on recent results from analysis of real sea clutter data. Clutter is usually defined as the unwanted radar returns from land, sea, rainfall or other phenomena, which may mask the echoes from targets, and its models usually reproduce the normalized radar cross section, spatial and temporal correlation properties, statistical variability and Doppler spectrum. Such models are important in order to develop target detection techniques and evaluate their performance under variable environmental conditions. The properties of monostatic clutter are influenced by radar parameters – i.e. frequency, resolution cell size, incidence angle and polarization – and environmental conditions. In the case of sea clutter, important environmental parameters are: wind and wave direction, level of development of the sea and depth of the sea. On the other hand, bistatic clutter is heavily influenced by the system geometry, i.e. the relative position of the radar receivers with respect to the transmitters, resulting in new degrees of freedom to include in future models.

12.1 Radar clutter

Radar clutter has been studied since the earliest days of radar. Clutter may be defined as the unwanted radar returns from land, sea, rainfall or other phenomena which potentially mask desired echoes from targets. Of course, for a remote-sensing radar, the clutter itself represents the desired echoes.

Clutter is highly variable in its nature. It depends first of all on the properties of the surface, and in particular, on the surface roughness with respect to the radar wavelength. As well as this, it depends on the parameters of the radar: frequency, the resolution cell size, the incidence angle and the polarization. In addition, sea clutter depends on the direction of look of the radar with respect to wind and wave directions, and on the extent to which the sea is developed – that is, the distance and time over which wind will have been interacting with the sea surface. Furthermore, shallow

*University College London, UK

Figure 12.1 Different land and sea surfaces. The nature of the radar clutter from each will be quite different. © 2014 SciTech/IET. Reprinted with permission from [1]

water will influence the wave spectrum. In this respect, there may be a significant difference between sea clutter in littoral regions and that in the open ocean.

Figure 12.1 shows some examples of different land and sea surfaces. It is instructive to imagine the scattering of a microwave radar signal from each of these surface types.

A knowledge of the properties of radar clutter is valuable to the radar designer for several reasons. First, it allows the detection performance as a function of radar, target and clutter properties to be predicted and comparative performance assessments to be made. Second, it allows the detection processing to be optimized. And third, for acceptance purposes, it allows radar performance to be assessed over a range of conditions, since it would generally be impractical to carry out acceptance tests over the full range of conditions.

A great deal of effort has been expended in developing monostatic radar clutter models for these purposes. Such models may be empirical – i.e. based on experimental measurements, or theoretical, based on electromagnetic scattering calculations, or some combination of both – and of course, both approaches should ideally lead to the same results. Models may be developed to reproduce: (i) the normalized radar cross section (NRCS), (ii) spatial and temporal correlation properties, (iii) the statistical variability and (iv) the Doppler spectrum. It is important to realize, though, that such models are an attempt to represent reality, which they may do well, or not so well.[1]

[1]To paraphrase the statistician George Box: 'All models are wrong, but some are useful' [2].

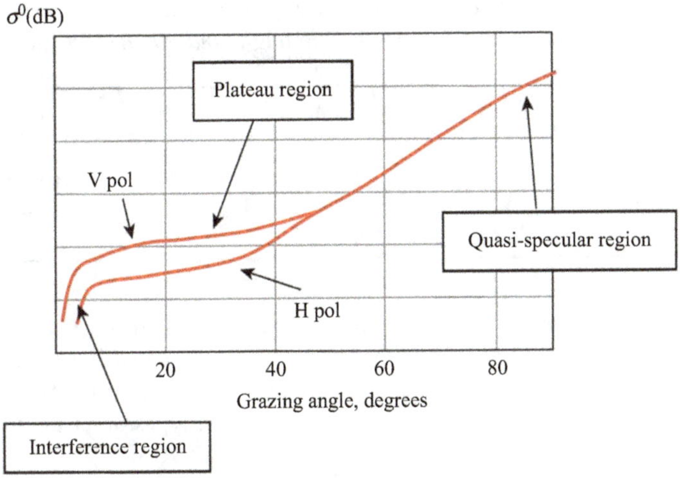

Figure 12.2 Typical variation of clutter reflectivity as a function of grazing angle and polarization. © 2014 SciTech/IET. Reprinted with permission from [1]

Well-regarded books describing these clutter models and measurements for monostatic radar include [3–5].

The rest of the chapter is structured as follows: Section 12.2 provides a brief review of models that have been developed for monostatic radar clutter, then Section 12.3 considers how these may be extended to cover bistatic radar clutter. This is followed by a discussion of the special case of forward scatter. Sections 12.5 and 12.6 cover, respectively, the measurement and modelling of bistatic radar clutter, and the final section provides a summary of the key points and results.

12.2 Clutter models

12.2.1 Mean reflectivity

The mean reflectivity of clutter is characterized in terms of the NRCS $\sigma^0 = \sigma/A$, where σ is the clutter RCS, and A is the clutter cell area. In a monostatic radar, this area is the product of the range resolution $(c/2B)\sec\varphi$ and the azimuth resolution $R\theta_B$, where B is the radar signal bandwidth, φ is the grazing angle and R is the range. In a bistatic radar, the definition of the clutter cell area is more complicated, as discussed in the next section. In a monostatic radar, σ^0 is known as the *backscatter coefficient*. In a bistatic radar, the term backscatter coefficient is not appropriate, so the term *scattering coefficient* is used instead. Other terms that are used are *Normalized Monostatic Radar Cross Section* (NMRCS) and *Normalized Bistatic Radar Cross Section* (NBRCS), respectively.

The variation of σ^0 with grazing angle φ typically follows the form shown in Figure 12.2. Three distinct regions can be identified: at very low grazing angles, the backscatter falls rapidly, and this is known as the *interference region*. As the grazing

Table 12.1 Typical values of σ^0 and γ for different types of surface [1], for 10 degree grazing angle and 10 GHz frequency

	σ^0 dB m²/m²	γdB m²/m²
Smooth water	−53	−45.4
Desert	−20	−12.4
Wooded area	−15	−7.4
Cities	−7	0.6

angle increases, a *plateau region* is reached. Beyond this, there is a rapid rise in backscatter, peaking at normal incidence to the ground (the *quasi-specular region*).

In the plateau region, σ^0 depends approximately on the sine of the grazing angle φ, and a useful guide to values of σ^0 for different types of terrain is given by the constant γ model

$$\gamma = \frac{\sigma^0}{\sin \varphi} \tag{12.1}$$

Some typical values of σ^0 and γ for different types of surface are given in Table 12.1. These correspond to a radar frequency of 10 GHz and a grazing angle of 10°.

12.2.2 Clutter statistics

The overall clutter echo will be the vector sum of contributions from all of the scatterers within the clutter cell. If there is a large number of these, all of comparable amplitude and random in phase, then the central limit theorem indicates that the PDF of the clutter amplitude will be Rayleigh distributed:

$$p(x) = \frac{x}{\sigma^2}\exp\left(-\frac{x^2}{2\sigma^2}\right) \quad x \geq 0 \tag{12.2}$$

However, at low grazing angles and high resolution, it is found that there is a higher probability of a large clutter amplitude, in other words that the clutter is spikier. The presence of large discrete scatterers may give the same effect. Figure 12.3 shows this effect: the two distributions have the same mean value, but the lower one has a higher probability of a high amplitude and hence a longer tail to the distribution.

Several other models have been developed over the years to attempt to represent these distributions, and these are described extensively in the literature [3–5]. The most successful of these are based on a compound approach, in which the overall clutter is represented as the product of an underlying texture and a speckle term.

Of these, the K-distribution and its variants [5] have been particularly successful and widely used. This represents the clutter as the product of an underlying chi-distributed texture and a Rayleigh-distributed speckle. The speckle is represented by

$$p(x|y) = \frac{\pi x}{2y^2}\exp\left(-\frac{\pi x^2}{4y^2}\right) \quad \text{for } 0 < x < \infty \tag{12.3}$$

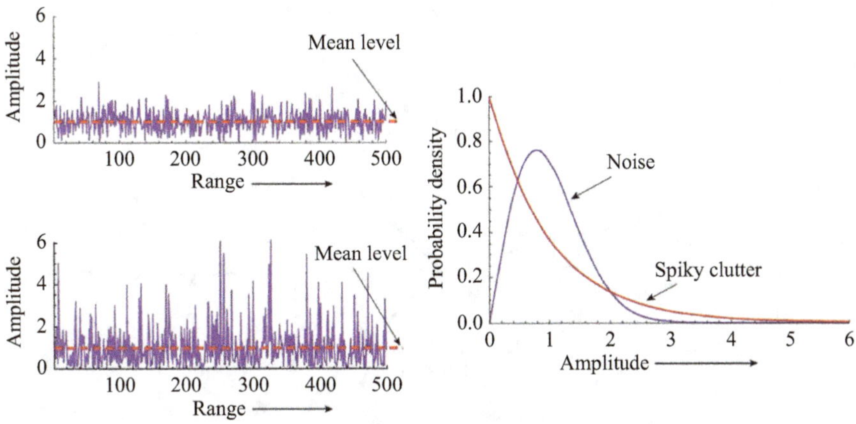

Figure 12.3 The upper left-hand plot shows Rayleigh-distributed noise. The lower left-hand plot shows spiky clutter with the same mean value. The right-hand plot shows the PDFs of the two distributions, and the PDF of the spiky clutter has a longer tail. © 2014 SciTech/IET. Reprinted with permission from [1]

and the texture by

$$p(y) = \frac{2b}{\Gamma(\nu)}(b\nu)^{2\nu-1}\exp(-b^2y^2) \tag{12.4}$$

giving

$$p(x) = \frac{4c}{\Gamma(\nu)}(cx)^{\nu}K_{\nu-1}(2cx) \tag{12.5}$$

where $c = b\sqrt{\pi/4}$ is a scale parameter, ν is a shape parameter, $\Gamma(.)$ is the gamma function and $K_\nu(.)$ is the modified Bessel function of the third kind of order ν. For $\nu = \infty$, the expression reduces to the Rayleigh distribution. Low values of shape parameter ν (<1) indicate spiky clutter.

12.3 Bistatic clutter models

12.3.1 Bistatic geometry

Bistatic clutter depends on all of the same parameters of the surface and of the radar as monostatic clutter, plus the variables associated with the bistatic geometry (Figure 12.4).

Here, θ_i and θ_s are the incidence angles of the transmitter and receiver, respectively. If the transmitter, receiver and clutter patch lie in the same vertical plane, the angle ϕ is 180°, and the geometry is referred to as 'in-plane.' In the more general case, ϕ is not equal to 180°. Also shown is the bistatic angle β at the target or clutter patch formed by the transmitter and receiver, and

$$\beta = \cos^{-1}(\sin\theta_i \sin\theta_s - \cos\theta_i \cos\theta_s \cos\phi) \tag{12.6}$$

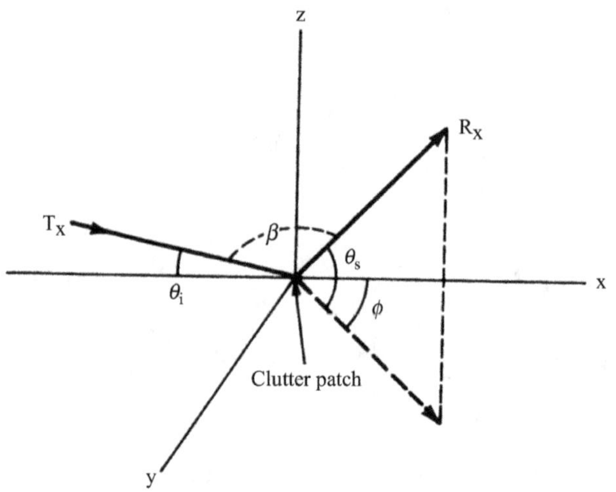

Figure 12.4 Co-ordinate system for bistatic clutter measurements [6]

In the in-plane configuration, $\beta = \theta_s - \theta_i$. At low grazing angles, θ_i and θ_s are both close to 0° and $\beta \approx 180° - \phi$. The monostatic case corresponds to $\theta_i = \theta_s$ and $\phi = 180°$. Note that for the monostatic condition ϕ =180° and not 0° as might be expected. This has been chosen to be consistent with the convention adopted by many of the measurement campaigns.

In addition to these effects, the shape of the clutter cell is more complicated than with monostatic radar. In the monostatic configuration (Figure 12.5(a)) the clutter cell is approximately rectangular, formed by the product of the two-way antenna radiation pattern in the azimuth direction and the compressed pulse in the range direction, projected onto the surface. In the bistatic configuration (Figure 12.5(b)), the clutter cell is the product of the individual beam patterns of the transmit and receive antennas and the pulse, expanding in the form of an elliptical shell, and projected onto the surface. In both cases, it is necessary to know the antenna beam patterns and the compressed pulse shapes quite accurately.

12.3.2 Bistatic sea clutter

Reliable measurements of bistatic clutter are difficult to make. References [7,8] provide lists of bistatic clutter measurement campaigns since the 1960s, which show measurements of a range of different surface types (sea, bare land, land with different types of vegetation), frequencies (from L-band to X-band), polarizations and bistatic geometries. Inevitably, each set of measurements can only hope to cover a small number of parameters, so it is not easy to draw general conclusions from any individual set.

As an example, Figure 12.6 shows the results from some in-plane X-band measurements made by the GEC company in the United Kingdom in the 1960s of the scattering from the sea surface, at horizontal polarization and with a 20 knot wind speed [9]. These measurements used two aircraft – one for the transmitter and one for the receiver – and Continuous Wave (CW) transmissions. There does not appear to have been any attempt to calibrate the measurements.

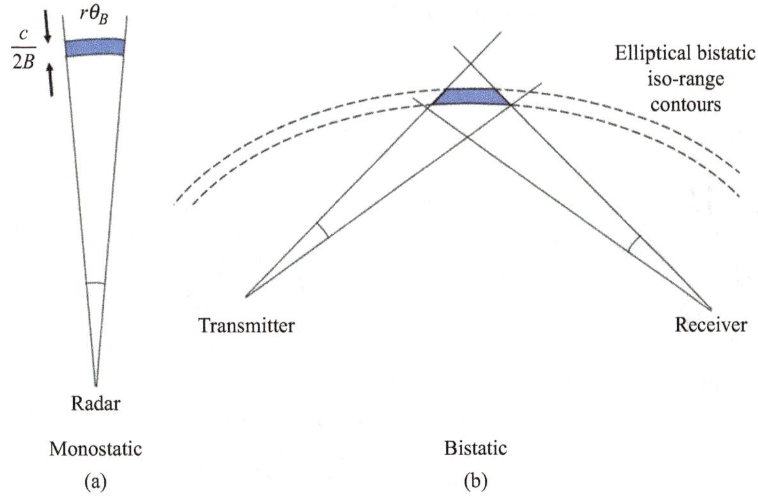

Figure 12.5 The clutter cell area in the bistatic case (b) is given by the intersection of the transmitted pulse expanding in the form of an elliptical shell, and the transmit and receive antenna beams, and is more complicated than in the monostatic case (a)

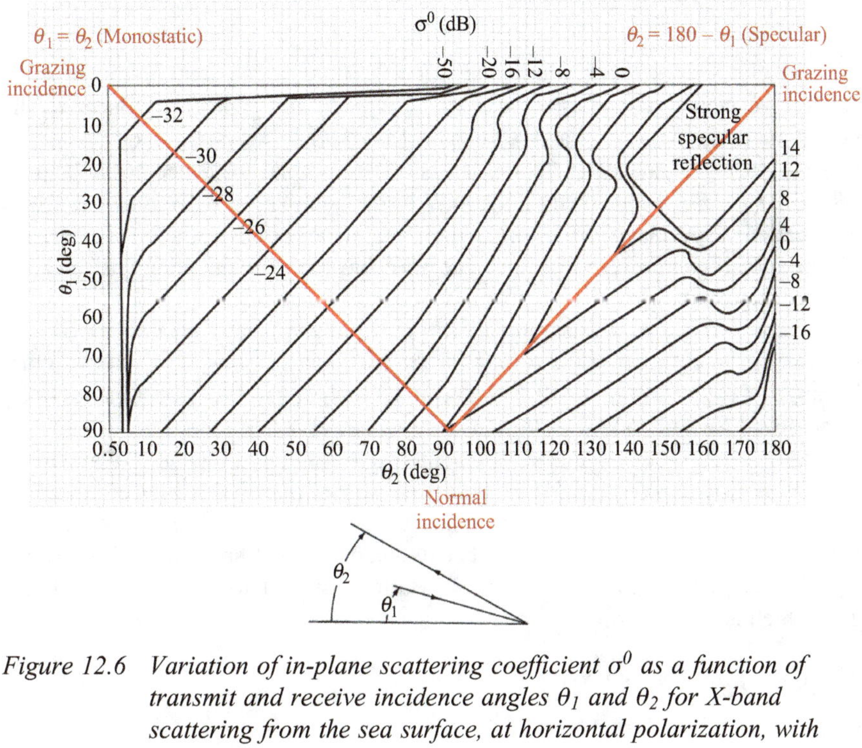

Figure 12.6 Variation of in-plane scattering coefficient σ^0 as a function of transmit and receive incidence angles θ_1 and θ_2 for X-band scattering from the sea surface, at horizontal polarization, with a 20-knot wind [9]

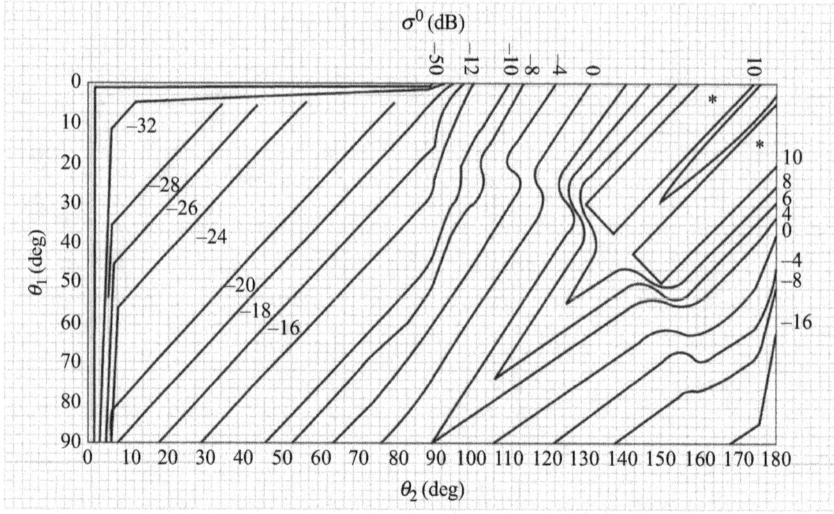

Figure 12.7 Variation of in-plane scattering coefficient σ^0 as a function of transmit and receive incidence angles θ_1 and θ_2 for X-band scattering from the sea surface, at vertical polarization, with a 20-knot wind [9]

The plot shows the scattering coefficient σ^0 as a function of the transmit and receive incidence angles θ_1 and θ_2. If $\theta_1 = \theta_2$ the geometry reduces to the monostatic case, so the left-hand red diagonal line shows the variation of monostatic σ^0 from grazing incidence at the top left corner to normal incidence at the bottom. If $\theta_2 = (180 - \theta_1)$ the geometry is specular, so the right-hand red diagonal line shows the variation of specular σ^0 from normal incidence at the bottom to grazing incidence at the top right. It can be seen that the value of σ^0 has a maximum along this so-called specular ridge, which itself has a maximum at grazing incidence.

Figure 12.7 shows measurements in the same geometry, but now at vertical polarization. In comparing the results for the two polarizations, it can be seen that the range of values in Figure 12.7 is several dB higher, and the largest values of σ^0 are achieved for low values of θ_1 and large values of θ_2.

Other in-plane sea clutter measurements were made by Pidgeon [10] and more recently by Kochanski *et al.* [11]. These have shown that the smaller of the transmitter and receiver grazing angles tends to dominate the resulting NRCS. To develop a model that represents this behaviour, Willis [7] has analysed the Domville data (Figures 12.6 and 12.7) using Barton's model for bistatic land clutter [12], which is of the form:

$$\sigma_B^0 = \gamma\sqrt{\sin(\theta_1)\sin(\theta_2)} \tag{12.7}$$

where γ is as defined in (12.2).

Equation (12.7) is the geometric mean of the NMRCSs (derived using the constant gamma clutter model) that would be obtained by radars positioned at the transmitter and the receiver. This may be improved [13] by using a more accurate description of the mono-static clutter, such as the Georgia Institute of Technology (GIT) model [14]. For the in-plane backscattering case, the NBRCS is given by

$$\sigma_B^0 = \sqrt{\sigma_M^0(\theta_1)\sigma_M^0(\theta_2)} \tag{12.8}$$

where $\sigma_M^0(\theta_1)$ and $\sigma_M^0(\theta_2)$ are the co-polarized GIT NMRCSs from the transmitter and receiver positions, respectively, in m^2/m^2. Figure 12.8 shows contour plots of the scattering predicted by this model at (a) horizontal polarization and (b) vertical polarization.

12.3.2.1 Out-of-plane bistatic sea clutter

In order to extend the model to out-of-plane scattering, we need firstly to account for the effect of different transmitter and receiver azimuth angles in the terms in (12.8). The NMRCS of the sea surface depends on the angle between antenna boresight and the direction of the wind, sea and swell. The effect of this on the equivalent NMRCS from the transmitter and receiver positions in (12.8) can be calculated using the GIT model.

There are then two further effects to be modelled. The first is the reduction of the co-polarized scattered component with increasing azimuth angle difference due to the clutter-scattering polar diagram (e.g. it has been suggested in [15] that sea spikes – large intermittent returns in sea clutter – have a narrow bi-static polar diagram). The second effect of azimuth angle difference is the increase in the cross-polar component. Both of these effects can be attributed to 'skew de-polarization', which can be thought of as projection of the EM fields from the frame of the transmitting antenna to the frame of the receiving antenna. This projection process will decrease the co-polar contribution and will allow the cross-polarized scattered clutter component to enter the co-polar radar channel and therefore to contribute to the NBRCS. For this purpose, we propose the use of multiplying sinusoidal functions to model the above effects. For the co-polar contribution, we propose the use of $|\cos(\varphi)|^m$, where m is adjusted to match the data, the function will have a maximum of 1 for in-plane geometries. As the contribution of the cross-polar component is zero at $\varphi = 0°$ or 180° and maximum at 90° or 270°, a candidate multiplying function $k|\sin(\varphi)|^n$ is chosen, with n and k adjusted to fit the data. For the monostatic cross-polar signal, Long's model [16] is used:

$$\sigma_{Mx}^0(\text{dB}) = 29.8\log(0.5144v_w) + 6\cos(\phi_w) - 84.7 \tag{12.9}$$

where φ_w is the angle between the radar line of sight and the wind, v_w is the wind speed and σ_{Mx}^0 is the cross-polar NMRCS. Thus, assuming that the two contributions add incoherently, the overall model is given by

$$\begin{aligned}\sigma_B^0(\text{dB}) = 10\log_{10}\Big[&\sqrt{\sigma_M^0(\theta_1,0)\sigma_M^0(\theta_2,\varphi)}|\cos(\varphi)|^m \\ &+k\sqrt{\sigma_{Mx}^0(\theta_1,0)\sigma_{Mx}^0(\theta_2,\varphi)}|\sin(\varphi)|^n\Big] \quad k,m,n>0\end{aligned} \tag{12.10}$$

Different combinations of parameters are found to provide best fits to data. In (12.10), m is used to control the slope of σ_B^0 for small azimuth differences.

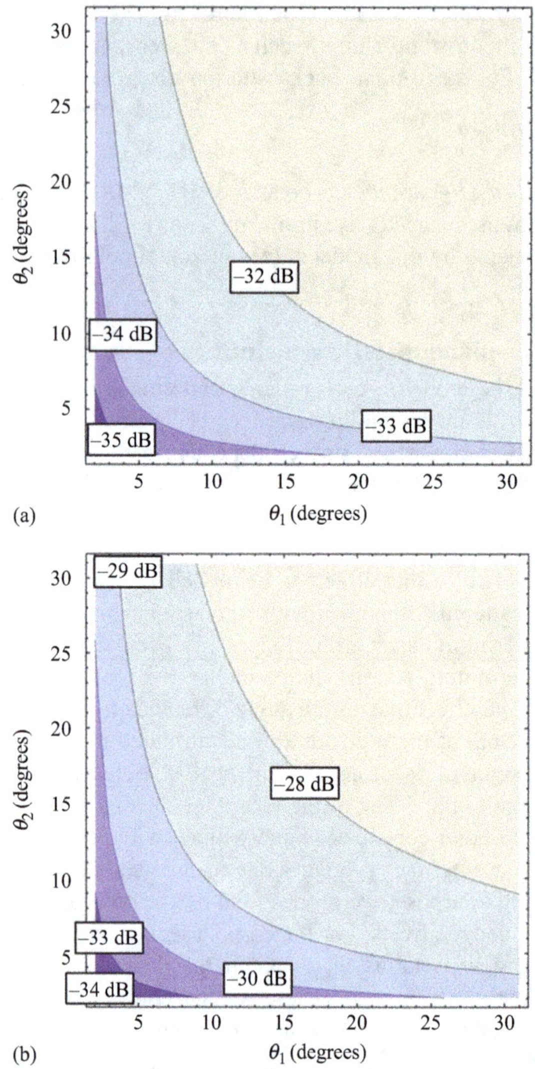

Figure 12.8 In-plane scattering coefficient predicted by the model of (12.8) for a fully developed sea with a 20 kt wind, at (a) horizontal polarization and (b) vertical polarization. © 2010 IET. Reprinted with permission from [13]

n and k are used to control the contribution of the cross-polarized component and, thus, they dominate at larger azimuth differences.

12.3.3 Bistatic land clutter

Figures 12.9–12.11 show equivalent results to those of Figures 12.6 and 12.7 for forested land, urban land and rural land, all at vertical polarization. The same interpretation is appropriate.

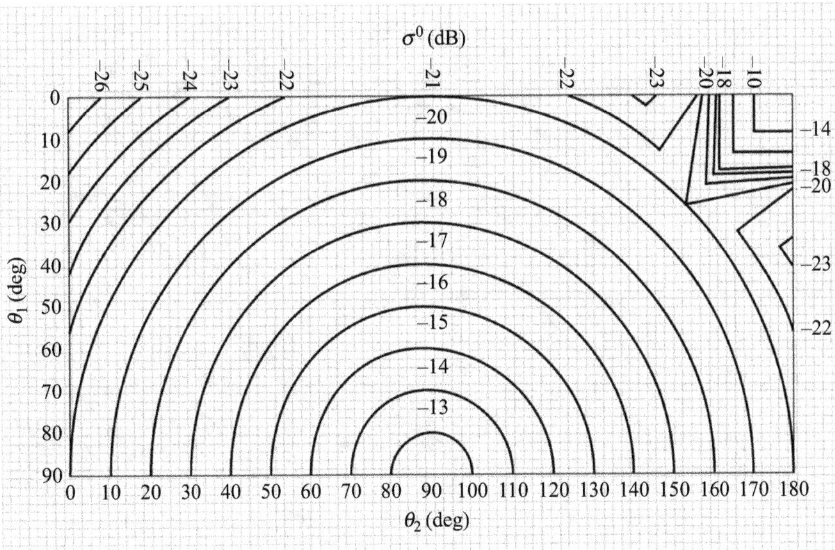

Figure 12.9 Variation of in-plane scattering coefficient σ^0 as a function of transmit and receive incidence angles θ_1 and θ_2 for X-band scattering from forested land, at vertical polarization [9]

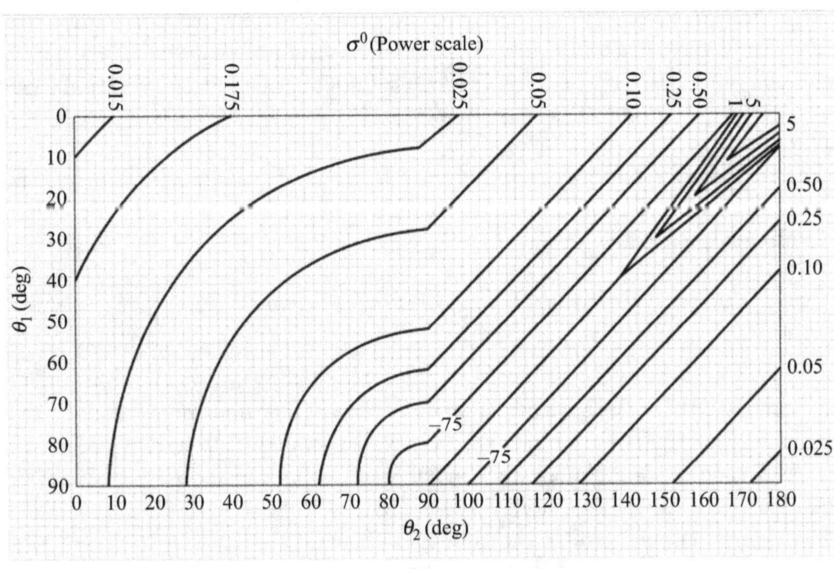

Figure 12.10 Variation of in-plane scattering coefficient σ^0 as a function of transmit and receive incidence angles θ_1 and θ_2 for X-band scattering from urban land, at vertical polarization [9]

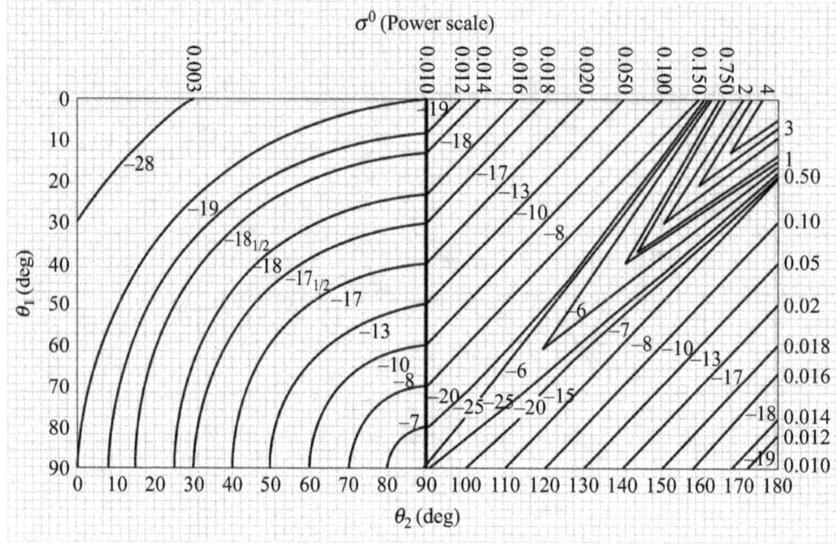

Figure 12.11 Variation of in-plane scattering coefficient σ^0 as a function of transmit and receive incidence angles θ_1 and θ_2 for X-band scattering from rural land, at vertical polarization [9]

In the case of forested land, the values of the in-plane scattering coefficient show a circular symmetry, i.e. they are independent on the bistatic angle, which might be expected, due to the uniform nature of the forested surface. In general, the minimum value is achieved if both the transmit and receive look angles are zero, whereas there are peaks at normal incidence ($\theta_1, \theta_2 = 90$ degrees) and at forward scatter.

Forward scatter contribution is dominant for both urban and rural land scattering. The maximum forward-to-bistatic power ratio is about 25 dB and 31 dB for urban and rural land, respectively.

12.3.4 Statistical properties of bistatic clutter

There have been rather few measurements of the statistical properties of bistatic clutter. Ewell and Zehner [17,18] reported low grazing angle measurements of sea clutter at 9.38 GHz, using a pulsed land-based transmitter and a land-based receiver. Their results indicate that the amplitude distributions of both the monostatic and bistatic clutter were mostly well approximated by lognormal distributions, and also that that there was a tendency for the monostatic clutter echoes to be of larger amplitude than the equivalent bistatic clutter. Yates [19] fitted the compound K-distribution model to measurements of simultaneous monostatic and bistatic SAR imagery of land scenes. Her results showed that the K-distribution gave a good fit, but more importantly, that the shape parameter ν was consistently higher for the bistatic imagery than for the monostatic imagery – in other words,

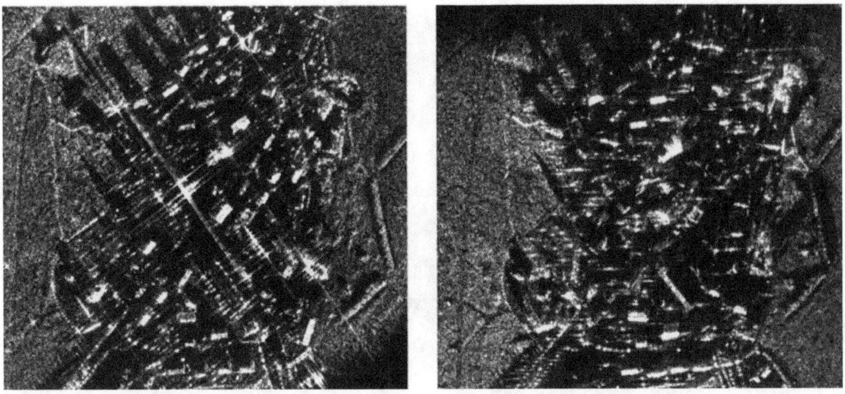

Figure 12.12 Comparison of monostatic (left) and bistatic (right) X-band SAR images of a village target scene [19]

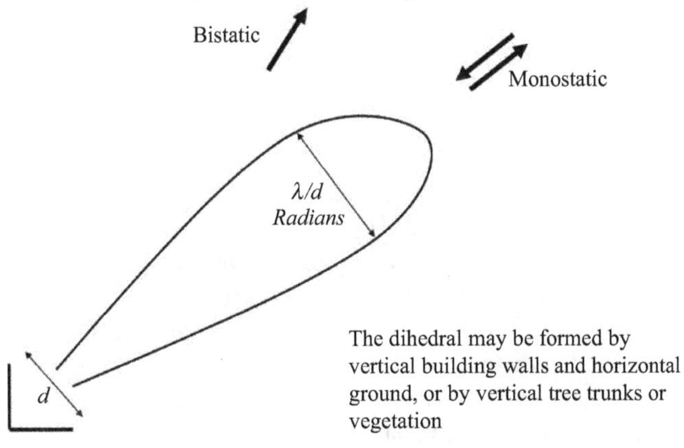

Figure 12.13 Dihedral scattering in bistatic radar

that the bistatic clutter was less 'spiky'. The effect is shown in the SAR images in Figure 12.12. The target scene in this case is a small village in south-west England, and the frequency is in X-band. The monostatic image on the left shows a number of strong scattering features, which are due to the dihedral and trihedrals formed from the vertical walls of buildings and the ground (Figures 12.13 and 12.14). The same features in the bistatic image on the right (in which the bistatic angle is approximately 70°) do not show such strong scattering.

Similar effects have been observed in monostatic and bistatic VHF SAR images of forested regions, in which the dihedrals are formed by vertical tree trunks and horizontal ground [20–22].

If this is more generally true, it would have important implications for the optimum detection of small targets against clutter. These ideas are developed further in Chapter 6 of Part I of Volume 2 on Clutter Diversity.

Figure 12.14 'Urban trihedrals' formed by vertical building walls and horizontal ground

12.3.5 Clutter in passive bistatic radar

There have been rather few measurements of the statistical properties of clutter in passive bistatic radar. In [23], the authors address the analysis of ground clutter data recorded by a passive radar system working at two different bands, i.e. 1,872.2 MHz (with a GSM illuminator of opportunity) and 2,162.6 MHz (with a UMTS transmitter). The measurement site was a rural area near Naples, Italy. Clutter data are fitted to different statistical different distributions, including the K, Weibull and two-state Weibull–Rayleigh model. Results of the goodness-of-fit measurements show that the GSM data are best fitted to the aforementioned two-state model, whereas the UMTS data are closer to the K model.

An FM-based airborne passive radar demonstrator was employed to collect ground clutter data in the South of England [24]. In particular, the presence of a mobile platform allowed several measurements over a range of bistatic angles included between 20 and 160°. Results reveal that the clutter power is minimum for bistatic angle of about 60° and increases by moving from this minimum point.

The work published by Malanowski *et al.* [25] analyses bistatic VHF ground clutter from FM-based passive radar, using three different data sets, associated to the systems developed at the NATO C3 Agency (formerly NC3A, now NATO Communications and Information Agency (NCIA)), the Warsaw University of Technology and the ERA company. The data were pre-processed by filtering the direct signal coming from the transmitter of opportunity, whose sidelobes are otherwise likely to mask the clutter. Range profiles of the normalized clutter reflectivity were computed, showing similar range-dependence for different azimuth pointing angles (circular array antennas were employed to collect data from different beam directions). Statistical analyses also show that in-phase and quadrature clutter data fit the Gaussian distribution only on the 'bell' region, whereas major deviations appear in the tail areas.

Recent work published by researchers from University of Alcalà, Spain [26] addresses the statistical analysis of sea clutter data recorded by a DVB-T (digital television) passive radar demonstrator. The data are split into three different Doppler regions, a first region containing Doppler frequencies between −800 and −100 Hz, a central region associated with the interval (−100, 100 Hz), and a third region, symmetric to the first one. In the first and second regions, square clutter amplitude is best fitted to the exponential distribution, whereas in the third region the log-normal distribution seems to be the closest one.

12.4 Forward scatter

The forward scatter geometry is a special case of bistatic radar, in which the target lies on or close to the bistatic baseline between the transmitter and receiver. This can result in substantial enhancement of the target RCS. However, the clutter level is also very high, both because the scattering at grazing incidence is high and because the clutter cell area is very large. Thus, in a radar which exploits the forward scatter geometry, target detection is likely to be clutter-limited rather than noise-limited.

12.4.1 Target echo signal and clutter

The forward-scattered signal from a target of forward scatter RCS σ_{FS} at range R_T from the transmitter and lying on the baseline between the transmitter and receiver (Figure 12.15) is given by the following equation:

$$P_R = \frac{P_T G_T G_R \lambda^2 \sigma_{FS}}{(4\pi)^3 R_T^2 (d - R_T)^2} \tag{12.11}$$

where P_T is the transmit power, G_T is the transmit antenna gain, G_R is the receive antenna gain, λ is the radar wavelength and d is the baseline range from transmitter to receiver.

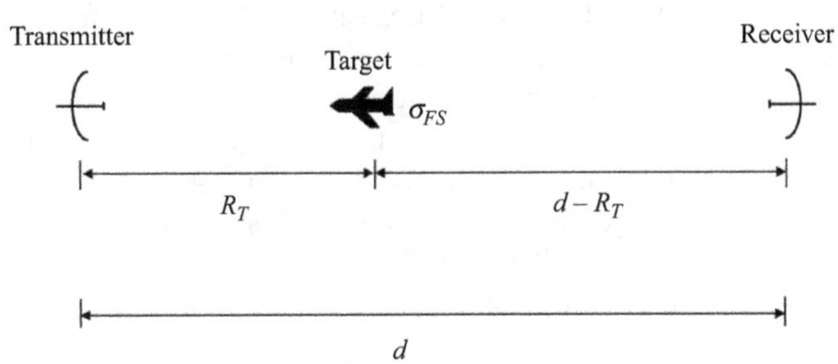

Figure 12.15 Free-space forward scatter geometry

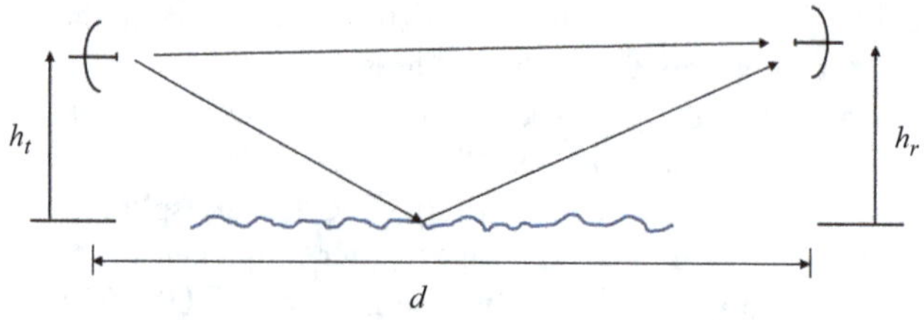

Figure 12.16 Forward scatter over the sea surface. The clutter will consist of a coherent contribution and a non-coherent (diffuse) contribution

Figure 12.17 Glistening and diffuse scattering of sunlight from the sea surface

This has a minimum value when the target lies equidistant between transmitter and receiver, and maximum when the target is either close to the transmitter or close to the receiver.

In the presence of surface clutter, the situation becomes more complicated. We consider the situation where the target is over the sea surface. The clutter signal will in general consist of a coherent contribution and an incoherent contribution (Figure 12.16).

Reference [27] explains that the diffuse scattering originates from an extended region of the sea surface and consists of two main contributions. The first contribution comes from a 'glistening' surface, which consists of small facets that reflect EM energy towards the receiver, depending on their orientation and size. A second contribution to the incoherent component is due to a 'wide-angle' scattering, occurring outside the glistening surface, as a result of capillary waves, spray and breaking waves (similarly to the monostatic sea clutter). Figure 12.17 illustrates this distinction in terms of scattering of sunlight from the sea surface.

In general, the diffuse component depends on the sea surface roughness, thus the computation of this quantity considers mathematical and statistical models of the sea surface.

Figure 12.18 Shadowing

The sea surface height distribution H may be modelled as Gaussian-distributed, i.e. $H \sim N(0, \sigma_H^2)$, where σ_H^2 is the mean square wave height ((246) of [28]), giving

$$\rho_S = \exp\left[-\frac{1}{2}\left(\frac{4\pi\sigma_H \sin\gamma}{\lambda}\right)^2\right] \tag{12.12}$$

where γ is the grazing angle.

From this starting point, the authors develop a model for the sea surface as a two-dimensional Gaussian surface, numerically integrating the values of ρ_S over the glistening surface. Calculation of the wave slope distribution allows the reflectivity to be evaluated as a function of sea state [28]. It is found that unless shadowing is taken into account, this method overestimates the scattering at all but low sea states (Figure 12.18). This can be done by including a 'shadowing factor' which determines the proportion of scatterers that are visible to both transmitter and receiver.

According to [27], the coherent reflection coefficient is given by:

$$\rho_c = \rho_0 \cdot \rho_s \cdot D \tag{12.13}$$

where ρ_0 is the Fresnel reflection coefficient for water which in turn depends on the polarization, incidence angle and the permittivity and permeability of seawater, ρ_S is the additional rough surface reflection coefficient (integrated over a surface much smaller than the glistening area) and D is the divergence factor ([28], p245).

The value of the Fresnel reflection coefficient depends heavily on the polarization and on the grazing angle. In general, the coherent reflection coefficient decreases with increasing grazing angle, and the slope decreases with higher sea states.

12.4.2 Experimental measurements

Researchers from the University of Birmingham, United Kingdom, have made measurements of the forward scattered sea clutter power at low grazing angles at a number of locations, distances and sea states, at 7.5 and 24 GHz [29]. They found that the coherent component of clutter power is given quite accurately by the two-ray propagation model:

$$P_{\text{coherent}} = P_t G_t G_r \frac{h_t^2 h_r^2}{d^4} \tag{12.14}$$

where h_t and h_r are the heights of the transmit and receive antennas, respectively, above the mean sea surface, as shown in Figure 12.16. Figure 12.19 shows the

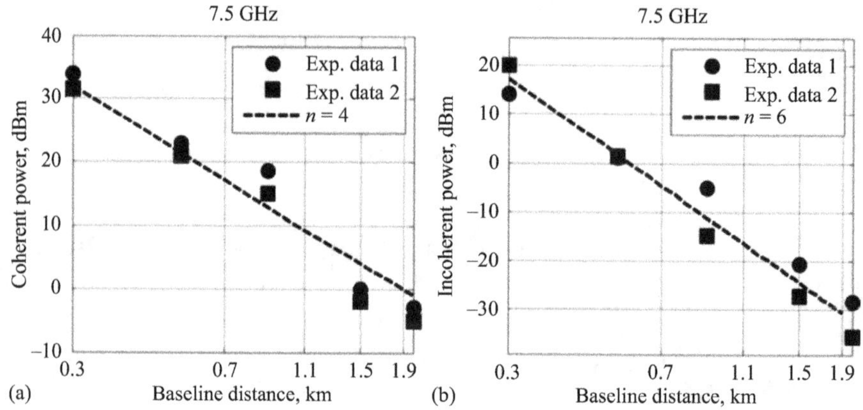

Figure 12.19 Coherent and incoherent received powers as a function of range, measured at 7.5 GHz. © 2014 IET. Reprinted with permission from [29]

behaviour of the coherent and incoherent clutter power as a function of range (d), measured at 7.5 GHz. The values of coherent power decrease with the fourth power of the range, as in (12.14), whereas the incoherent power decreases with higher rate, close to the sixth power of d.

In [29], the authors analyse the spectral and statistical properties of forward-scatter sea clutter data. The coherent clutter component is due to a specular scattering from a stationary micro-surface, which originates a zero-Doppler contribution. On the other hand, the incoherent component is associated to a diffuse scattering from a wide surface, which contains moving particles, thus originating a signal with non-negligible Doppler bandwidth. Hence, the Doppler spectrum of forward-scatter sea clutter shows a strong zero-Doppler component, together with a decreasing behaviour as a function of Doppler frequency – which is well approximated by a slope of 35–40 dB per decade. Furthermore, this behaviour of the clutter Doppler spectrum appears as quite invariant to the radar parameters and sea conditions. From a statistical point of view, clutter samples seem to be Gaussian-distributed, which is probably due to the very large dimension of one clutter cell in forward-scatter geometry, thus containing a large number of strong scatterers. The conclusions about the independence of the Doppler spectrum from radar parameters and environmental conditions and about the Gaussianity of clutter samples are likely to improve target detection performance, but they deserve further investigations in order to be thoroughly validated.

12.5 Bistatic clutter measurements

12.5.1 Practical considerations in bistatic radar trials

It has already been remarked that reliable bistatic clutter measurements are difficult to make. It is necessary to know the exact locations of transmitter and receiver, and to have accurate time and phase synchronization between the nodes – typically achieved by landline connection, or by atomic clocks and/or GPS-disciplined oscillators (GPSDOs).

Other factors include [13]:

- The measurement receiver(s) must be thoroughly characterized, and linear with adequate dynamic range;
- Simultaneous measurements should also be made of monostatic clutter, so that a direct comparison can be made;
- Simultaneous measurements should be made of 'ground truth' and meteorological data (i.e. wind speed and direction, and for sea clutter measurements of sea state, swell direction and wavelength, currents and so on, typically obtained using Waverider buoys, as well as the depth and bathymetry);
- Equally, simultaneous, time-stamped video imagery should be obtained of the sea surface, to provide an indication of breaking wave phenomena;
- The antenna boresights should be accurately calibrated, to be certain that the two antennas are pointing at the correct patch of surface. It is also necessary to know the beam shapes and pulse shapes so as to be able to calculate the NRCS accurately;
- All measurements should be properly calibrated [30]. With monostatic radar measurements, this can be done using a trihedral corner reflector calibration target, which gives a substantial RCS ($\sigma = 4\pi A^2/\lambda^2$, where A is the area), essentially independent of aspect angle. Other approaches include transponders or Van Atta (retro-directive) arrays, but in general, the more complicated the calibration target, the greater the uncertainty in RCS. With bistatic radar, the situation is more difficult, since there are few simple targets that give a large, constant RCS independent of bistatic angle. One good candidate is a large metallized sphere whose radius r is much greater than the wavelength λ, and for which the bistatic RCS is equal to the physical silhouette area, πr^2. If the measurements are all to be made in the horizontal plane, a vertical cylinder will give a larger RCS of $(\pi r l^2 \cos \beta/2)/\lambda$ in the plane perpendicular to the axis of the cylinder, where β is the bistatic angle, and l is the length of the cylinder.
- Polarimetric bistatic calibration is even more complicated and is likely to need a range of different calibration targets.

Some recent measurements of bistatic sea clutter were obtained by researchers at University College London using the NetRAD multistatic radar sensor. This operates at S-band (2.4 GHz) and consists of three independent transmit/receive nodes. In its earlier versions, it was necessary for the nodes to be connected by cable to achieve synchronization, but later, this was achieved by GPSDOs at each node and wireless links for control and data transfer, which allowed greater flexibility and greater separation in their deployment. The GPSDOs were developed by collaborators at the University of Cape Town in South Africa; the wider issues of synchronization in a radar network are discussed in Chapter 2 of Part I of Volume 2.

A very important feature is that this system is able to make simultaneous measurements of bistatic and monostatic scattering from clutter or from a target,

Table 12.2 NetRAD parameters

Parameter	Value
Frequency	2.4 GHz
Transmit power	500 W
Antenna gain	27 dBi
Antenna beamwidth	9° (az) × 11° (el)
Signal bandwidth	50 MHz
PRF	1 kHz

Figure 12.20 Pictures of the NetRAD multistatic radar system in use in trials. Clockwise from top left: transmit equipment in van; operator position; vernier scale for antenna alignment; telescope for antenna alignment; GPSDO

which allows a direct comparison to be made. Table 12.2 lists some of the key parameters of the NetRAD system, and Figure 12.20 shows some aspects of the hardware deployed in trials.

The NetRAD radar was used in 2011 and 2012 to make measurements of bistatic sea clutter and target signatures, in the Western Cape area of South Africa. A full account of the measurements and the results is provided in [31–33], but a short summary is provided below.

In these experiments, two nodes were deployed, one operating monostatically and the other as a bistatic receive-only node (Figure 12.10). The node separation was 2 km. By varying the pointing angle of the antennas, the bistatic angle β could be varied from a relatively small value to close to forward scatter. In this case, results are presented for three values of β: 30, 60 and 90°.

Table 12.3 Summary of results at bistatic angles of 30, 60 and 90° [31]

R_M m	β	σ_M^0 dBm²/m²	σ_B^0 dBm²/m²	CNR$_M$ dB	CNR$_B$ dB	ν_M	ν_B
805	30°	−59.2	−58.7	18.1	21.5	0.93	3.11
417	60°	−47.1	−47.6	36.9	40.2	0.22	0.48
295	90°	−44.1	−55.8	41.1	35.6	0.17	1.04

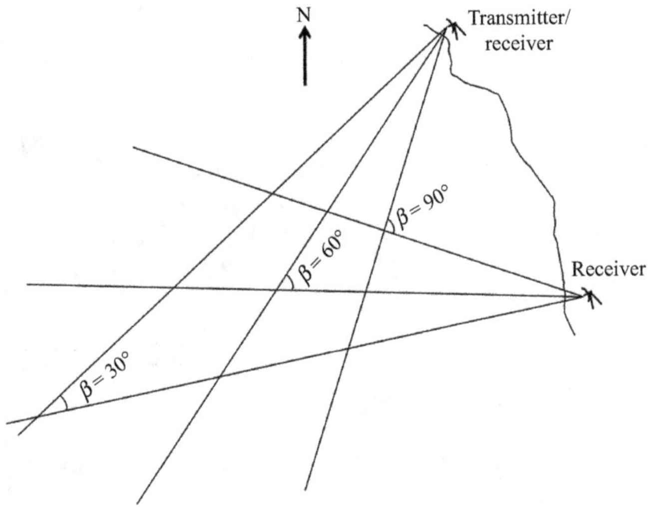

Figure 12.21 Geometry for sea clutter trials. The separation between the two nodes was 2 km. © 2011 IET. Reprinted with permission from [31]

Measurements were made at both horizontal and vertical polarization. Figure 12.22 shows the result of fitting a variant of the K-distribution model ('K+noise') to the measured data at vertical polarization and at a bistatic angle $\beta = 30°$. This variant allows for a finite clutter-to-noise ratio [34]. It can be seen that the fit of the model to these measurements – particularly to the bistatic data on the right – is very good, showing that the K+noise model is a good representation in this case.

Table 12.3 presents results at three values of bistatic angle β, as shown in Figure 12.21 (30, 60 and 90°). The final two columns on the right hand side show the values of monostatic shape parameter ν_M and bistatic shape parameter ν_B for these three angles, and in each case, again, the shape parameter ν is lower for the monostatic clutter than the equivalent bistatic clutter, showing that the bistatic clutter is less spiky (shorter-tailed) than the equivalent monostatic clutter.

In a separate experiment, Figure 12.23 shows fits for monostatic and bistatic clutter, for both HH and VV polarizations, at a bistatic angle of 17°. For both polarizations, it can be seen that the value of the shape parameter ν is lower for the

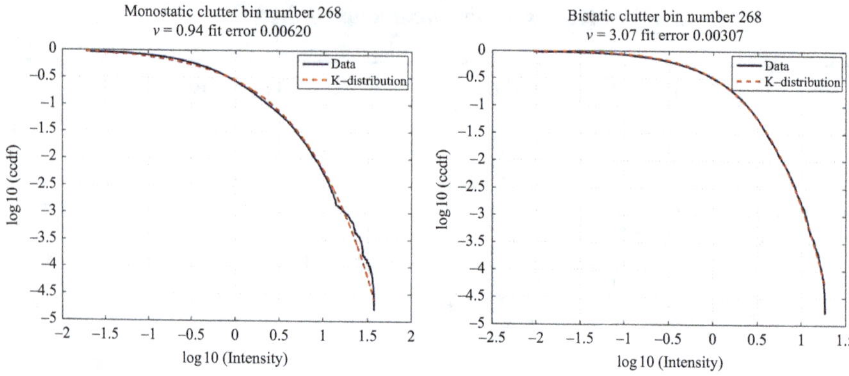

Figure 12.22 Fit of the K+noise statistical clutter model to simultaneously measured monostatic (left) and bistatic (right) sea clutter at a bistatic angle $\beta = 30°$. © 2011 IET. Reprinted with permission from [31]

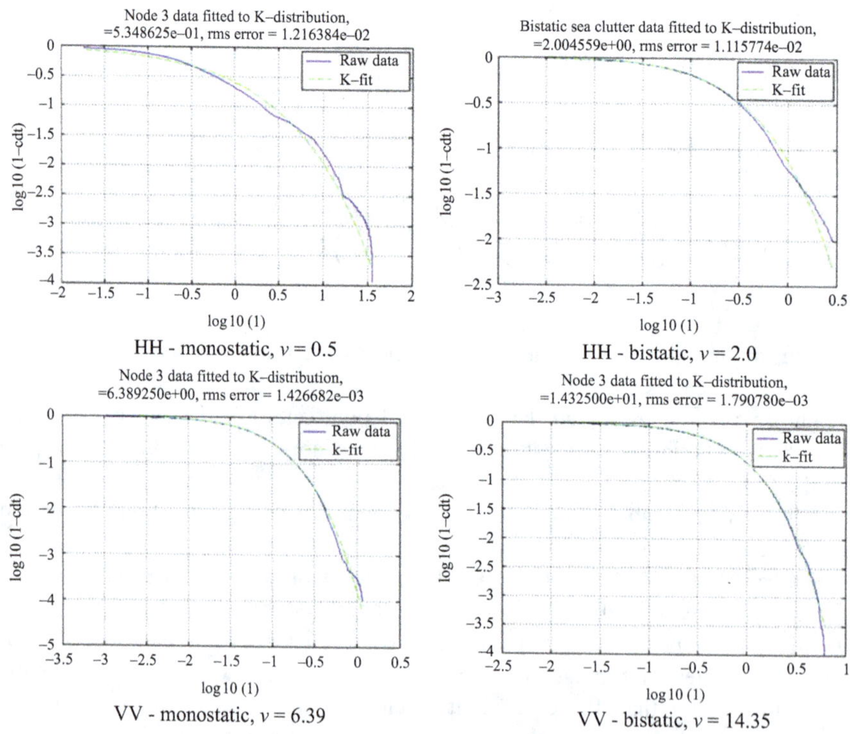

Figure 12.23 Fits of the compound K-distribution model to monostatic and bistatic HH and VV polarization sea clutter data. Bistatic angle $\beta = 17$ degrees. Wind speed $= 13\ ms^{-1}$. © 2011 IEEE. Reprinted with permission from [35]

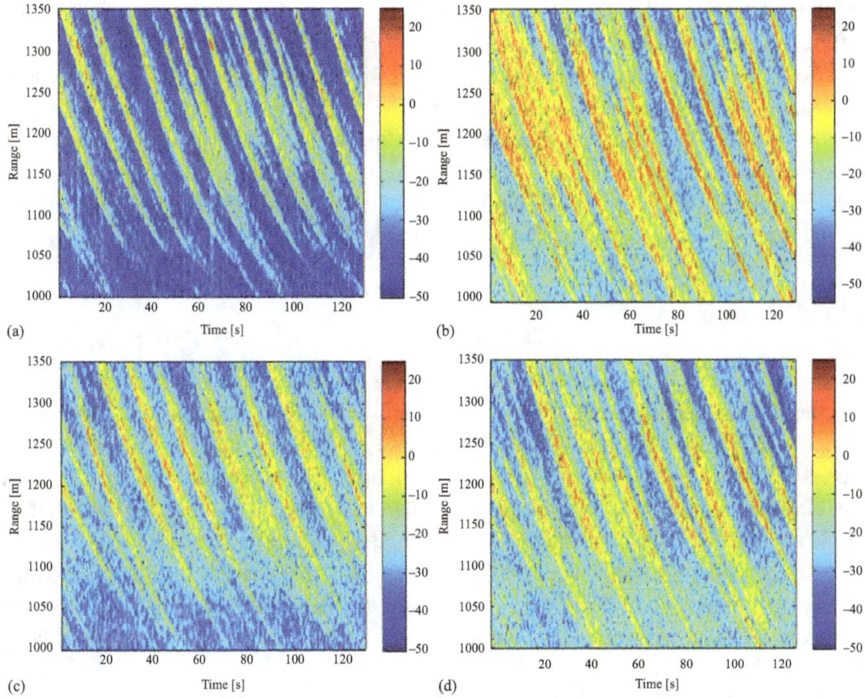

Figure 12.24 Power-range-time maps: (a) bistatic HH, (b) monostatic HH, (c) bistatic VV and (d) monostatic VV. © 2015 IEEE. Reprinted with permission from [36]

monostatic clutter than the equivalent bistatic clutter, showing that the bistatic clutter is less spiky (shorter-tailed) than the equivalent monostatic clutter.

Although this effect has been observed consistently in trials in several locations and sea conditions, it is too early to say that it may be a general result, and more measurements and analysis are certainly needed. However, the significance and potential utility of these results is discussed further in Chapter 6 of Part I of Volume 2 on Clutter Diversity.

12.5.2 Clutter spikes

These data sets have been further analysed by researchers at the University of Pisa [36,37]. In particular, they investigate the distributions of spike width and interval between spikes for simultaneously recorded bistatic and monostatic clutter, at both HH and VV polarization. Figure 12.24 shows the maps of the received power as a function of range and slow time. The raw data are processed through a classification algorithm in order to separate the spiky samples from the background, giving the opportunity of analysing the properties of the sea spikes. They find that both the spike width and, for a lesser extent, the interval between spikes are reasonably well represented by exponential distributions. They observe that the number of spikes is lower in the bistatic data compared to the monostatic, but the difference is quite small for VV polarization.

For the bistatic data, spikes are separated by larger intervals for VV polarization than for HH polarization, even though the width and the number of spikes is smaller for HH data. In other word, the analysis reveals that spikes last longer and are less frequent in the bistatic configuration. A comparison between the two polarizations reveals that the HH-polarized data have shorter and more separated spikes than VV data, both for bistatic and monostatic node.

12.6 Summary

This chapter has reviewed the properties of bistatic radar clutter and drawn together some recent results on the subject. Bistatic clutter is certainly more complicated than monostatic clutter, because it depends not only on the parameters of the radar and of the surface but also on the parameters associated with the bistatic geometry. It is desirable to develop models of the bistatic clutter, to allow detection performance to be evaluated and optimized, and for use in acceptance testing, and models may be developed to represent the NRCS, spatial and temporal correlation properties, the statistical variability, and the Doppler spectrum. Despite this, current understanding of the properties of bistatic clutter is limited at best.

Forward scatter is a special case of the bistatic geometry and can result in substantial enhancement of the target RCS. However, the clutter level is also very high, both because σ^0 at grazing incidence is high, and because the clutter cell area is very large.

An empirical approach to modelling of bistatic sea clutter has shown that the NRCS of both in-plane and out-of-plane sea clutter can be modelled to a fair degree of confidence.

The results of this chapter are developed further in Chapter 6 of Part I of Volume 2 on Clutter Diversity, where it is shown that an understanding of the dependence of the clutter and target signatures on bistatic geometry may be used to determine the optimum geometry for target detection against a clutter background.

Acknowledgements

We are most grateful to the many people who have helped with the work described in this chapter. These include Waddah Al-Ashwal, Chris Baker, Mike Cherniakov, Francesco Fioranelli, Marina Gashinova, Sabrina Greco, Mike Inggs, William Miceli, Matt Ritchie, Stefan Sandenbergh, Andy Stove, Robert Tough, Keith Ward, Simon Watts, Karl Woodbridge and Gill Yates. We are also grateful to the organizations who have funded the work, including THALES UK, the Royal Academy of Engineering, the Engineering and Physical Sciences Research Council, the IET A.F. Harvey Research Prize and the Office of Naval Research (Global).

References

[1] S. Watts, 'Representing clutter', Ch.25 in *Stimson's Introduction to Airborne Radar* (3rd edition, H.D. Griffiths, C.J. Baker and D. Adamy eds.), Scitech Publishing Inc., Raleigh, NC, May 2014.

[2] G.E.P. Box, 'Science and statistics', *Journal of the American Statistical Association*, Vol. 71, No. 356. pp. 791–799, December 1976.

[3] M.W. Long, *Radar Reflectivity of Land and Sea*, Artech House, Dedham, MA, 1983.

[4] J.B. Billingsley, *Low-Angle Radar Land Clutter: Measurements and Empirical Models*, William Andrew Publishing, Norwich, NY, 2002.

[5] K.D. Ward, R.J.A. Tough and S. Watts, *Sea Clutter: Scattering, the K-Distribution and Radar Performance* (2nd edition), IET, Stevenage, 2013.

[6] N.J. Willis, 'Bistatic Radar', chapter 25 in *Radar Handbook*, second edition, (M.I. Skolnik ed.), McGraw-Hill, New York, NY, 1990.

[7] N.J. Willis, 'Clutter', Ch.9 in *Bistatic Radar*, Artech House, Norwood, MA, 1991.

[8] M.W. Weiner, 'Clutter', Ch.9 in *Advances in Bistatic Radar* (N.J. Willis and H.D. Griffiths eds.), Scitech Publishing Inc., Raleigh, NC, 2007.

[9] A.R. Domville, *The Bistatic Reflection from Land and Sea of X-Band Radio Waves*, GEC (Electronics) Ltd., Memo SLM 1802, Stanmore, England, July 1967.

[10] V.W. Pidgeon, 'Bistatic cross section of the sea', *IEEE Transaction Antennas and Propagation*, Vol. AP-14, No. 3, pp. 405–406, May 1966.

[11] T.P. Kochanski, M.J. Vanderhill, J.V. Zolotarevsky and T. Fariss, 'Low illumination angle bistatic sea clutter measurements at X-band', *Proc. IEEE Conference OCEANS'92*, Vol. 1, pp. 518–523, 1992.

[12] D.K. Barton, 'Land clutter models for radar design and analysis', *Proc. IEEE*, Vol. 73, No. 2, pp. 198–204, 1985.

[13] H.D. Griffiths, W.A. Al-Ashwal, K.D. Ward, R.J.A. Tough, C.J. Baker and K. Woodbridge, 'Measurement and modelling of bistatic radar sea clutter', *IET Radar Sonar and Navigation*, Vol. 4, No. 2, pp. 280–292, March 2010.

[14] M.M. Horst, F.B. Dyer and M.T. Tuley, 'Radar sea clutter model', *IEEE AP/S URSI Symp. Digest*, College Park, MD, pp. 6–10, 1978.

[15] K.D. Ward and P. Shepherd, 'Bistatic radar sea clutter experiments and spatial coherence'. *Proc. IEE RADAR-92 Conference*, Brighton, pp. 22–25, 12–13 October 1992.

[16] M.W. Long, 'On the polarization and the wavelength dependence of sea echo', *IEEE Transactions on Antennas and Propagation*, Vol. AP-13, No. 5, pp. 749–754, 1965.

[17] G.W. Ewell and S.P. Zehner, 'Bistatic sea clutter return near grazing incidence', *Proc. RADAR-82 Conference*, London, IEE Conf. Publ. No. 216, pp. 188–192, October 1982.

[18] G.W. Ewell, 'Bistatic radar cross-section measurements', Chapter 7 in *Techniques of Radar Reflectivity Measurements*, 2nd ed. (N.C. Currie ed.), Artech House, Norwood, MA, 1984.

[19] G.A. Yates, 'Bistatic synthetic aperture radar', PhD Thesis, University College London, January 2005.

[20] A. Barmettler, L. Zuberbühler, E. Meier, L. Ulander, A. Gustavsson and P. Wellig, 'Swiss airborne monostatic and bistatic dual-pol SAR experiment at the VHF-band', *Proc. EuSAR Conference 2008*, Friedrichshafen, 2–5 June 2008.

[21] L.M.H. Ulander, B. Flood, P-O. Frölind, *et al.*, 'Bistatic experiment with ultra-wideband VHF-band synthetic aperture radar', *Proc. EuSAR Conference 2008*, Friedrichshafen, 2–5 June 2008.

[22] A. Gustavsson, L.M.H. Ulander, B. Flood, *et al.*, 'Low frequency bistatic SAR measurements', *2012 IEEE International Geoscience and Remote Sensing Symposium*, Munich, pp. 315–318, 22–27 July 2012.

[23] A. De Maio, G. Foglia, N. Pasquino and M. Vadursi, 'Measurement and comparative analysis of clutter for GSM and UMTS passive radars', *IET Radar, Sonar & Navigation*, Vol. 4, No. 3, pp. 412–423, June 2010.

[24] J. Brown, K. Woodbridge, A. Stove and S. Watts, 'VHF airborne passive bistatic radar ground clutter investigation', *IET Int. Radar Conference RADAR 2012*, Glasgow, 22–26 October 2012.

[25] M. Malanowski, R. Haugen, M.S. Greco, D.W. O'Hagan, R. Plsek and A. Bernard, 'Land and sea clutter from FM-based passive bistatic radars', *IET Radar, Sonar & Navigation*, Vol. 8, No. 2, pp. 160–166, February 2014.

[26] N. del-Rey-Maestre, M.P. Jarabo-Amores, D. Mata-Moya, P. Gomez-del-Hoyo and J. L. Bárcena-Humanes, 'Statistical analysis of UHF bistatic radar clutter in coastal scenarios', *2015 European Radar Conference (EuRAD)*, Paris, 2015, pp. 253–256.

[27] R. Mital and V. Gregers-Hansen, 'A computer model for bistatic sea surface microwave reflectivity, NRL report', August 2014.

[28] P. Beckmann and A. Spizzichino, *The Scattering of Electromagnetic Waves from Rough Surfaces*, Pergamon Press, New York, NY, 1963.

[29] M. Gashinova, K. Kabakchiev, L. Daniel, E. Hoare, V. Sizov and M. Cherniakov, 'Measured forward-scatter sea clutter at near-zero grazing angle: analysis of spectral and statistical properties', *IET Radar, Sonar and Navigation*, Vol. 8, No. 2, pp. 132–141, February 2014.

[30] R. Palamà, M.R. Greco, F. Gini, F. Fioranelli. M.R. Ritchie and H.D. Griffiths, 'Copolar calibration of multistatic radar in the presence of multipath', *IEEE Radar Conference* 2016, Philadelphia, PA, pp. 646–651, 2–6 May 2016.

[31] W.A. Al-Ashwal, C.J. Baker, A. Balleri, *et al.*, 'Statistical analysis of simultaneous monostatic and bistatic sea clutter at low grazing angles', *Electronics Letters*, Vol. 47, No. 10, pp. 621–622, 12 May 2011.

[32] W.A. Al-Ashwal, K. Woodbridge and H.D. Griffiths, 'Analysis of bistatic radar sea clutter I: normalized RCS', *IEEE Transactions on Aerospace and Electronic Systems*, Vol. 50, No. 2, pp. 1283–1292, April 2014.

[33] W.A. Al-Ashwal, K. Woodbridge and H.D. Griffiths, 'Analysis of bistatic radar sea clutter II: statistics', *IEEE Transactions on Aerospace and Electronic Systems*, Vol. 50, No. 2, pp. 1293–1303, April 2014.

[34] S. Watts, 'Radar detection prediction in K-distributed sea clutter and thermal noise', *IEEE Transactions on Aerospace and Electronic Systems*, Vol. 23, No. 1, pp. 40–45, January 1987.

[35] W. Al-Ashwal, C.J. Baker, A. Balleri, *et al.*, 'Measurements of bistatic radar sea clutter', *IEEE Radar Conference 2011*, Kansas City, MO, pp. 217–221, 23–27 May 2011.

[36] R. Palamà, M.S. Greco, P. Stinco and F. Gini, 'Statistical analysis of bistatic and monostatic sea clutter', *IEEE Transactions on Aerospace and Electronic Systems*, Vol. 51, No. 4, pp. 3036–3054, October 2015.

[37] R. Palamà, M.S. Greco, P. Stinco and F. Gini, 'Analysis of sea spikes in NetRAD clutter', *EuRAD Conference 2014*, Rome, pp. 109–112, 8–10 October 2014.

Chapter 13
Forward scatter radar

Marina Gashinova[1], *Liam Daniel*[1], *Alexander Myakinkov*[2] *and Mikhail Cherniakov*[1]

Abstract

Forward scatter radar (FSR) is historically thought of as the first type of bistatic radar. Research on FSR has been predominantly focussed on its ability to serve as an electronic fence, and it has already proven the FSR's excellent detection and target motion parameter estimation capabilities. Recently, a wave of interest has emerged in FSR; first, this is a consequence of the introduction of 'stealth' targets. These targets have a significantly reduced radar cross-section (RCS) because of their specific shapes and/or coatings which may greatly suppress backscattering, yet their shadows will still render them perfectly 'visible' to FSR. Second, interest in FSR has appeared because of the establishment of passive coherent location concepts where illuminators of opportunity are used to form a bistatic radar network.

This chapter provides an overview of FSR, theory and phenomenology and further discusses its capabilities and limitations.

13.1 Introduction

Forward scatter radar (FSR) is historically thought of as the first type of bistatic radar. There is however a fundamental difference; while FSR could be viewed as subclass of bistatic radar where the bistatic angle is close to 180°, the physical operational principle is essentially different from that of bistatic radar. In FSR the target signature is formed as a result of interference between the direct path (or leakage) signal and the scattered signal from the target, called shadow radiation. While the target is in motion the scattering mechanism undergoes a fundamental change—bistatic scattering (mainly of the reflective nature) when the receiver is outside the target main shadow lobe which transforms into purely shadow radiation at the time when the target crosses the baseline casting shadow upon the receiver.

[1]Department of Electronic Electrical and Systems Engineering, The University of Birmingham, UK
[2]Nizhny Novgorod State Technical University, Russia

Because of such a mechanism, the research on FSR has predominantly focused on its ability to serve as an electronic fence and FSR's excellent detection and target motion parameter estimation capabilities have already been proven.

Recently a wave of interest has emerged in FSR. Firstly, this is a consequence of the introduction of 'stealth' targets. These targets have a significantly reduced radar cross-section (RCS) because of their specific shapes and/or coatings which may greatly suppress backscattering; yet their shadows will still render them visible to FSR. Secondly, interest in FSR has appeared because of the establishment of passive coherent location (PCL) concept where illuminators of opportunity are used to form a bistatic radar network.

In this chapter, we initially discuss the topology of FSR in relation to mono and bistatic radar, look at the phenomenology of FSR and explain the forward scatter radar cross section and forward scatter effect as a manifestation of shadow radiation. The target signature model is then described in Section 13.3 in fundamental terms of a phase signature modulated by the forward scatter cross section (FSCS) envelope. Power budget analysis for target detection and optimal signal processing for target motion parameter estimation are presented and the performance of experimental FSR systems is demonstrated with experimental results in diverse ground and maritime scenarios. Section 13.4 is dedicated to the analysis and characterisation of FSR clutter and Section 13.5 closes the chain of application domains by giving the principles for airborne target detection and tracking, supported by experimental results.

13.2 Radar topology and electromagnetic wave scattering mechanism

13.2.1 Monostatic, bistatic and forward scatter radar topology

The two main radar types as classified by their configurations are monostatic radar which comprise the majority of modern radars and bistatic radar which are currently the subject of intensive study. In monostatic radar, depicted in Figure 13.1(a), both transmit (Tx) and receive (Rx) antennas are co-located and point towards a region to observe a target (Tg) at a range R_{ms}. In bistatic radar, Figure 13.1(b), the distance between Tx and Rx antennas, d, referred also as the baseline, is comparable with the ranges of the target to the transmitter, R_{t} and to the receiver R_{r}. Three points, Tx, target and Rx, form a bistatic triangle which lie in a bistatic plane.

There are three relevant angles in this plane, θ_{t} and θ_{r}, which are the transmitter and the receiver look angles and $\beta = 180 - (\theta_{\mathrm{t}} + \theta_{\mathrm{r}})$ is the bistatic angle. Within the bistatic radar configurations, there is a special case, where $R_t + R_r \approx d$ and consequently $\beta \approx 180°$, which is known as FSR and is shown in Figure 13.1(c), this subclass of bistatic radar is the focus of this book chapter. Such radars are also known in the literature as line radar, electronic fences and flutter radar [1,2], the historical development and state of the art descriptions can be found in such texts as [1,3]. The forward scatter (FS) topology lends itself towards systems where

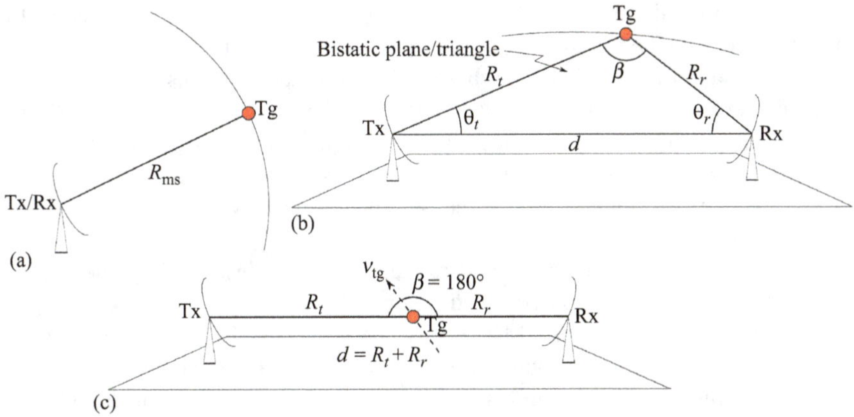

Figure 13.1 Overview of radar topologies. (a) Monostatic radar, (b) bistatic radar and (c) forward scatter radar shown with a target (Tg) crossing the baseline

antennas face each other (when considering directional antennas) and target detection occurs in spatial regions on or very close to the baseline. It is immediately obvious that there is no range resolution in such a radar system – irrelevant of the position of the targets along the baseline, all scattered signals as well as the direct path signal (DPS) from the transmitter reach the receiver at the same time. This observation also follows from the known equation for bistatic radar range resolution, ΔR^{ψ}_{bs}, which is measured along the bi-sector of the bistatic angle and calculated as follows [1 (p.131)]:

$$\Delta R^{\psi}_{bs} \approx \frac{\Delta R_{bs}}{\cos\psi} = \frac{c}{2\Delta F\,\cos(\beta/2)\cos\psi} \tag{13.1}$$

where ψ is the second target's angular deviation from the bistatic bisector of the first target and ΔF is the ranging signal bandwidth. From (13.1), it follows that as $\beta \to 180°$, $\Delta R^{\psi}_{bs} \to \infty$, i.e. the range resolution is dramatically reduced and only wide or ultra-wideband signals introduce some range resolution [4,5]. This obvious drawback of FSR nevertheless gives rise to the reception of non-fluctuating target signals even when scattered from highly manoeuvrable targets. As a result, the maximum coherent analysis time in FSR is equal to the target visibility time T_v. Thus, an absence of range resolution in FSR is partly compensated by enhanced frequency (Doppler) resolution [6,7] which is considered in Section 13.2.2.1.

In FSR, the dominating physical principle of target observation is the forward scattering of electromagnetic (EM) waves (also known in optics as shadowing), as opposed to 'backscattering' in monostatic radar and bistatic radar, where in the absence of a target, the transmitted signal does not arrive at the receiver input unless there is a parasitic leakage signal (LS) present. In contrast, in FSR, the transmitted signal is fundamentally acquired by the receiver, the DPS then serves as a carrier for modulation caused by scattering/shadowing from the target in the forward direction when in the vicinity of the baseline. Target detection and motion parameter estimation is therefore based on DPS perturbations as opposed to received energy increase as in monostatic radar and bistatic radar. Thus in FSR, the DPS presence is a necessary pre-requisite. Such signals can be acquired by different technical ways: direct and scattered signal self-mixing with a square law detector (SLD) [6,8], direct signal strength measurement [9] as well as use of a linear receiver with quadrature channels [10–12].

The effect of EM shadowing and reflection can be described by the modern physical theory of diffraction (PTD) developed by Ufimtsev [13] or by the geometrical theory of diffraction [14]; in this chapter, only the PTD is considered. According to the theory, the total scattered field $\vec{E}_{sc}$ of an object can be described by two constituent fields, the reflected field $\vec{E}_{ref}$ and the shadow field $\vec{E}_{sh}$ [15,16],

$$\vec{E}_{sc} = \vec{E}_{ref} + \vec{E}_{sh} \tag{13.2}$$

and are illustrated in Figure 13.2.

The shadow field is so termed because it is most prominent in the spatial regions which are geometrically shielded (shadowed) by the target, i.e. about the axis directly behind the target in the FS direction. The shadow contour (silhouette) defines the geometric boundary between the illuminated and non-illuminated sides of the object.

A full shadow, i.e. zero field intensity, exists only directly behind the object, so-called umbra region, and is referred to as the geometrical shadow, the phenomena that can be described by means of geometrical optics [13 (p.11)]. In this region, the shadow field is interpreted as a field equal to the incident field but having an opposite phase such that the fields sum is equal to zero $\vec{E}_{sc} = \vec{E}_{inc} + \vec{E}_{sh} = 0$. This kind of shadow is usually observed in optics where the wavelength is many orders less than the object dimension. In FSR, the transmitted signal wavelengths are comparable or of order less than target dimensions and due to diffraction effects the illuminating

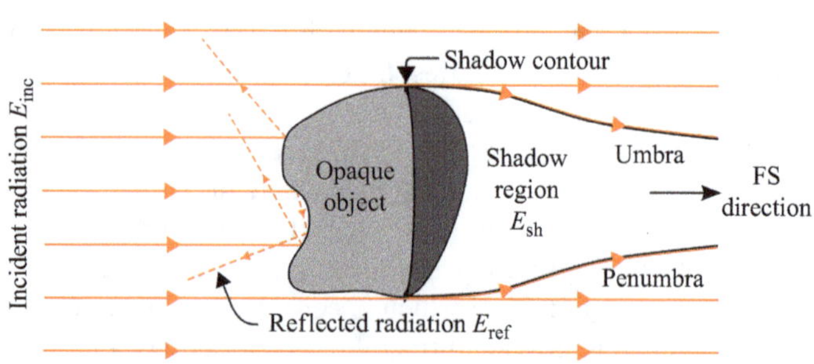

Figure 13.2 Pictorial interpretation of shadow radiation, the shadow field E_{sh} cancels the incident field E_{inc} in regions that are in very close proximity behind the target object

field skirts the obstacles just reducing the strength of the incident field and what is seen behind the object is known in optics as the penumbra. The smaller electrical dimension of the object and the bigger distances between the receiver and the object, the less will be the shadowing effect. In the far field, the shadow field 'on axis' takes the form of a field shifted by $\pi/2$ with respect to the incident field [16]. In [11], the typical ratio between the DPS and the scattered signal levels in the forward direction has been analysed, i.e. the DPS strength modulation due to the target presence in FSR, and this ratio, called signal-to-leakage ratio, was estimated to be -40 to -80 dB.

Summarising from analysis in [13], the shadow field based on the PTD assumptions:

- does not depend on the whole 3D shape of the scattering object and is completely determined by the size and the geometry of the shadow contour (shadow contour theorem).
- is independent of the material properties of the target object assuming of course that this material is not fully transparent for EM waves.

13.2.2 Forward scatter cross-section

When analysing the FS effect, it is worth recalling the mechanisms of EM scattering from objects classified by three regions: optical scattering, where the electrical dimension of the object is large, i.e. geometrical dimension, D, is much greater than the wavelength, λ, $D/\lambda \gg 1$; Mie or resonance region, where $D/\lambda \sim 1$ and, finally, the Rayleigh region where $D/\lambda \ll 1$. In this section, being that the majority of known FSR and related research has been considered in upper Mie and optical region (see Table 13.1), we will consider these cases, but later the Rayleigh region will also be discussed in brief.

In traditional radar, the target echo signal strength is specified by its RCS, σ, for both monostatic radar (σ_m) and bistatic radar (σ_b) cases, which is defined as:

$$\sigma = \lim_{R\to\infty} 4\pi R^2 \frac{|E_{sc}|^2}{|E_{inc}|^2} \left[\mathrm{m}^2\right] \tag{13.3}$$

where E_{inc} and E_{sc} are the electric field strength of the incident wave at the target location and the strength of the scattered electric field in the direction of the observation point correspondingly; R is the target range from the observation point.

In FSR, there is a similar parameter which specifies the signal strength at the receiver – the forward scatter cross-section (FSCS), σ_{fs}, also measured in m^2 or dB m^2.

The FSCS pattern in the Mie and optical region for a target of three-dimensional shape can be calculated by replacing the target with an appropriate plane shape of area A, defined by the shadow contour. This is then replaced by a complementary aperture [19 (p.41)]; the replacement is based on Babinet's principle and is pictorially demonstrated in Figure 13.3.

Table 13.1 EM operational regions of FSR

Reference	Brief description	Frequency (MHz)	Targets	~D/λ	Region
[1 (p.33)]	Radar fences investigated before WWII (Japan)	25–80	Aircraft: 10–20 m of body length	2.0–5.0	Upper Mie/low optical
[2]	Cold war period anti-ballistic missile fences, e.g. AN/FPS 19, AN/FPS 23 (US, Canada)	480–1,350	Long range missiles: 10–20 m	27–75	Optical
[15 (Ch.19)]	Modern anti-stealth Russian radar (RU)	430	Stealth Jet: 10–20 m	20	Optical
[7]	Experimental anti-stealth radar (RU)	900	Fighter Jet: 10–20 m	40	Optical
[17]	Experimental radar for small maritime targets detection	7,000–24,000	Inflatable boat: 3 m	75–225	Optical
[18]	Experimental radar for small ground targets detection	65–430	Typical vehicle, e.g. 4×4: 4 m	1.0–5.0	Mie/upper Mie/lower optical

When the target is directly on the baseline, i.e. $\beta = 180°$, the FSCS tends to its maximum [15 (p.397)],

$$\sigma_{\text{fs}}^{\max} = 4\pi \left(\frac{A}{\lambda} \right)^2 \tag{13.4}$$

coinciding with the monostatic RCS of a flat conductive surface normal to the baseline and with a shape given by the target silhouette with an area A. From (13.4), it is seen that FSCS increases as the fourth power of the target linear size and to a the second power of the carrier frequency. Such rapid increase of forward scattering intensity with frequency is known as the FS effect. At first glance, it highlights advantages of FSR operation at high frequencies; however, the operational frequency choice is actually defined by a number of other parameters considered later.

The monostatic RCS and FSCS of a conductive sphere of radius R_s are shown in Figure 13.4 [20]. It is seen that starting from the Mie region, the mean normalised monostatic RCS does not depend on the wavelength, whereas the FSCS rises by 20 dB/decade. This essential increase in EM scattering in the FS direction has been first derived analytically by Mie [21], and over the years, a large amount of research has been performed in this area both experimentally and theoretically.

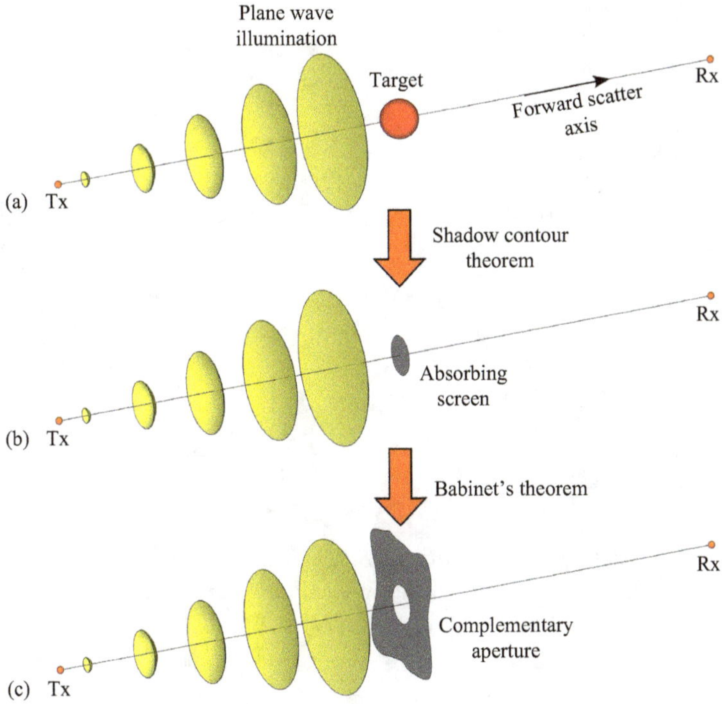

Figure 13.3 Pictorial interpretation of replacement of a complex target with equivalent radiating aperture. (a) Target of a complex 3D shape, (b) equivalent 2D shadow silhouette, (c) silhouette replacement by a complementary aperture in an infinite conductive surface. Wave is planar on incidence with the target

Thus, in the upper Mie and optical region, we may expect $\sigma_{fs} \gg \sigma_{bs}$ and $\sigma_{fs} \gg \sigma_{ms}$, where $\sigma_{bs/ms}$ are the bi/monostatic RCSs. To illustrate this further, Table 13.2 shows the estimated improvement in cross-section at different frequencies for typical targets, where the coefficient $\gamma = \sigma_{fs}/\sigma_m$ is the ratio of the FSCS to its monostatic counterpart.

The physical interpretation of the enhancement of the FSCS in comparison with monostatic RCS follows from the fact that according to the Huygens–Fresnel principle, elementary secondary radiators contribute in phase to a total field at an observation point laying on the baseline if this point is in a far field of the secondary sources. Such a dramatic increase of RCS indicates the main advantage of FSR in improving the radar power budget, meaning that it could be effective for target detection in the case of (i) smaller target size, (ii) larger range to target or (iii) less transmit power. The independence of the FSCS on target material, be it metallic or dielectric, makes this system ideally suited for the detection of stealth targets.

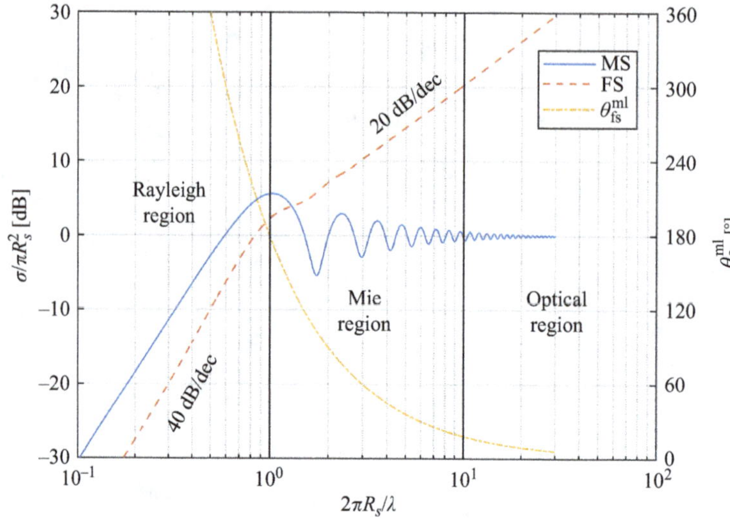

Figure 13.4 Normalised monostatic RCS (labelled MS) and FSCS (labelled FS) of a sphere of radius R_s (left vertical axis) and width of the FSCS main lobe θ_{fs}^{ml} (right vertical axis) [11]

Table 13.2 Comparison of RCS [22 (p.91)] and FSCS for selected targets and frequencies

Carrier frequency (wavelength) [MHz (m)]	**Human 0.5 m^2 RCS A~1 m^2**	**Vehicle 100 m^2 RCS A~8 m^2**	**Tank 100 m^2 RCS A~15 m^2**	**Fighter jet 1 m^2 RCS A~10 m^2**	**Bomber 10 m^2 RCS A~40 m^2**	**Small ship 100 m^2 RCS A~100 m^2**
100 (3)	n/a	n/a	$\gamma=5$	$\gamma=21$	$\gamma=23$	$\gamma=31$
300 (1)	n/a	$\gamma=9$	$\gamma=12$	$\gamma=28$	$\gamma=30$	$\gamma=38$
900 (0.3)	$\gamma=24$	$\gamma=19$	$\gamma=22$	$\gamma=38$	$\gamma=40$	$\gamma=48$
3,000 (0.1)	$\gamma=33$	$\gamma=28$	$\gamma=32$	$\gamma=47$	$\gamma=49$	$\gamma=57$

Equation (13.4) characterises the maximal value of FSCS, another parameter of interest is the FSCS main lobe, which can also be estimated assuming a far-field target/radar configuration. The angular width of the main lobe, θ_{fs}^{ml}, in a certain plane with respect to the target is proportional to the electrical length of the target, i.e. the ratio of the maximal target effective dimension D_{eff} in that plane and the wavelength λ,

$$\theta_{fs}^{ml} = \frac{K\lambda}{D_{eff}} \text{ [rad]} \quad (13.5)$$

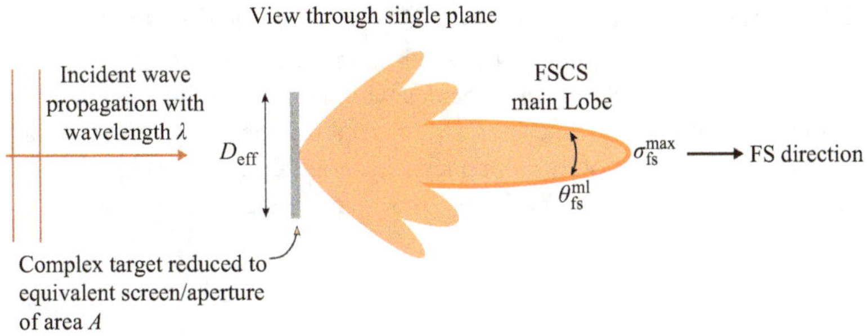

Figure 13.5 Illustration identifying the FSCS main lobe and definitions relating to calculation of the FSCS maximum, σ_{fs}^{max} and main lobe width, θ_{fs}^{ml}, for a complex target which has been reduced to an equivalent screen/aperture

In general, the coefficient K in (13.5) depends on the actual shadow contour shape but for most practical cases [22 (Ch.12)] $K \approx 1$. For clarity, the illustration in Figure 13.5 identifies the main lobe of the FSCS and the parameters of maximum FSCS, σ_{fs}^{max} and main lobe width, θ_{fs}^{ml} from (13.4) and (13.5).

Returning to the sphere of radius R_s, the −3 dB width of the FSCS main lobe is shown in Figure 13.4. We can see that the greater the target electrical dimension, the greater its FSCS, σ_{fs} and the narrower is the region defined by θ_{fs}^{ml}, where the FSCS enhancement can be observed. The width of main lobe of the FSCS where more than 90% of the received signal energy can be concentrated has a vital role in target detection and tracking, where σ_{fs} defines the radar power budget and, hence, the radar detection performance, θ_{fs}^{ml} specifies the FSR coverage area. It is also worth noting that both θ_{fs}^{ml} and σ_{fs} depend on the wavelength and target dimension and for the simplified case of a target of equivalent aperture area A, defined by an effective dimension, D_{eff}, these two parameters can be combined into one equation which highlights the fundamental relationship between the FSCS and operational area of FSR by substituting (13.5) into (13.4), with an example case of a sphere,

$$\theta_{fs}^{ml} \approx \pi\sqrt{\frac{A}{\sigma_{fs}^{max}}} \text{ or } \sigma_{fs}^{max} \approx \frac{A\pi^2}{\left(\theta_{fs}^{ml}\right)^2} \tag{13.6}$$

In spite of the fact that the first sidelobes of the FSCS pattern carry much less energy in comparison with the main lobe they still play significant role at the stage of target trajectory parameter estimation and importantly in automatic target classification. For airborne targets, the FSCS pattern main and sidelobe widths are also important factors when evaluating the FSR coverage area in elevation in order to determine the maximum altitude of a detectable target.

13.2.2.1 Target visibility time and Doppler resolution in FSR

It follows that if the FSCS main lobe width is known, it is possible to evaluate a target visibility time [11,18] which coincides with the target echo coherence interval $\Delta\tau_{fs}$. If a target of effective dimension D_{eff} is crossing the baseline at a distance R_r from the receiver with a velocity component, v_{tg}, normal to the baseline, its visibility time T_v at −3 dB level of the FSCS main lobe can be approximated as,

$$T_v = \Delta\tau_{fs} \approx \frac{\lambda R_r}{2D_{eff} \cdot v_{tg}} \tag{13.7}$$

Hence, the potential Doppler (frequency) resolution is:

$$\Delta f_{fs} = \frac{1}{\Delta\tau_{fs}} = \frac{1}{T_v} = \frac{2D_{eff} \cdot v_{tg}}{\lambda R_r} \tag{13.8}$$

In monostatic radar [15], the target fluctuation spectrum bandwidth Δf_{ms} and corresponding coherency time $\Delta\tau_{ms} \approx 1/\Delta f_{ms}$ also depend on the signal wavelength, target effective dimension D_{eff} and target angular variation rate relevant to the radar boresight $d\varphi/dt$, expressed in °/s, through:

$$\Delta f_{ms} \approx \frac{D_{eff}}{\lambda} \cdot \frac{d\varphi}{dt} \tag{13.9}$$

Examples of the calculated fluctuation spectra bandwidth Δf_{ms} and the maximum coherent integration time $\Delta\tau_{ms}$ for an unmanned aerial vehicle (UAV) (wingspan $D = 12$ m) are presented in Table 13.3 [11] for different wavelengths and aspect angle variation rates alongside corresponding FSR parameters.

It can be seen from the table that the Doppler resolution in FSR is several orders better than its monostatic counterpart. This high Doppler resolution defines FSR capabilities such as target tracking, automatic target classification and even imaging [6,15 (Ch.16 & 17), 23].

Table 13.3 Maximum echo coherent time in monostatic radar and FSR for UAV target [11]

	Monostatic						**FSR**	
$\Delta\varphi/\Delta t$ (°/s) →	**0.2**		**0.4**		**0.8**		**Baseline = 40 km**, v_{tg} **= 50 m/s**	
λ (m) ↓	Δf_{ms}	$\Delta\tau_{ms}$	Δf_{ms}	$\Delta\tau_{ms}$	Δf_{ms}	$\Delta\tau_{ms}$	Δf_{fs}	$\Delta\tau_{fs}$
3.0	0.8	1.25	1.6	0.63	3.2	0.31	0.013	75
1.5	1.6	0.63	3.2	0.31	6.4	0.16	0.026	37.5
0.75	3.2	0.31	6.4	0.16	12.8	0.09	0.053	18.8
0.3	8.0	0.13	16.0	0.06	32.0	0.03	0.13	7.5
0.1	24.0	0.04	48.0	0.02	96.0	0.01	0.4	2.5
0.03	80.0	0.01	160.0	0.006	320.0	0.003	1.3	0.75

13.2.3 Target FSCS pattern

The full FSCS pattern (as opposed to just the main lobe) can be viewed and analysed as the radiation pattern of the secondary planar antenna. The pattern of an arbitrary aperture antenna is shown in Figure 13.6 where the sidelobes are significantly lower than the main lobe. For some targets, these sidelobes may still exceed the peak levels of RCS of bistatic radar or monostatic radar.

In the figure, $\vec{\rho}$ is a radius vector to point P_A on the aperture, $\vec{r}_0$ is the unit vector towards the evaluation point P and $\vec{R}_p$ is the radius vector to that point, A_{sh} is the area of the aperture. Using the notations and co-ordinates depicted in Figure 13.6, the target FSCS σ_{fs} at bistatic angles close to 180° can be calculated as:

$$\sigma_{fs}(\vec{r}_0) = \lim_{R\to\infty} 4\pi R^2 \left(\frac{|E_{sh}|^2}{|E_{inc}|^2} \right) = \frac{4\pi}{\lambda^2} \left| \int_{A_{sh}} \exp[j(2\pi/\lambda)\vec{\rho}\vec{r}_0] dS \right|^2 \tag{13.10}$$

where dS is an elementary area of co-phasal aperture A_{sh}. Equation (13.10) fundamentally describes the full FSCS pattern but for complex targets the integration is complex and typically involves numerical full-wave simulation. Modern software packages for full wave analysis, for example CST Microwave Studio [24] may be used for the 3D FSCS pattern analysis.

Prior to discussion of the FSCS pattern it should be stressed that (13.10) describes directional pattern of a shadow aperture of the target, which can be seen as 'black body' radiation, i.e. radiation of an object which absorbs the incident EM wave. The fundamental analogy to this is the ideal 'stealth' target and, therefore, the only way to detect such a target is to use FSR: any detectable scattering on the

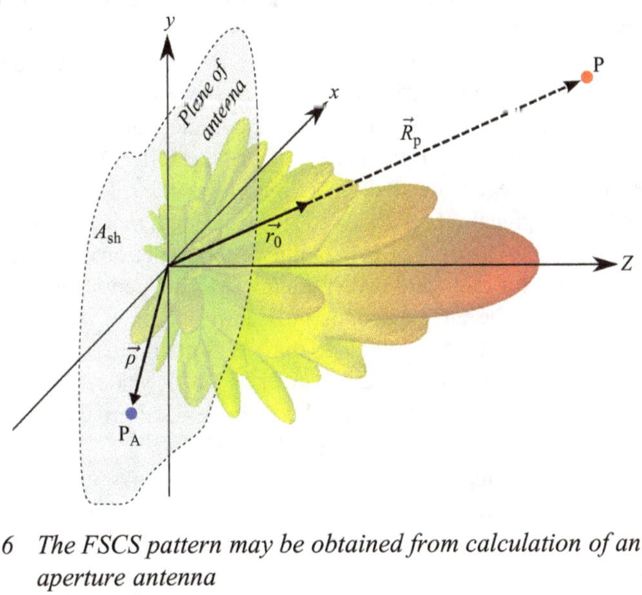

Figure 13.6 The FSCS pattern may be obtained from calculation of an equivalent aperture antenna

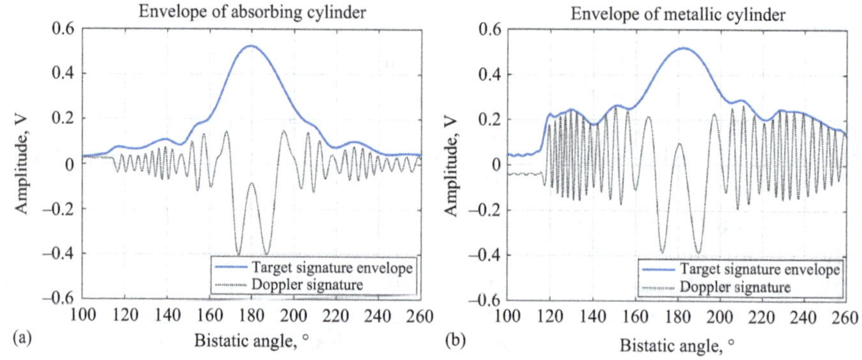

Figure 13.7 Doppler signature and signal envelope (RCS) of (a) a cylinder coated with an absorber and (b) a metal cylinder, as a function of bistatic angle [25]

target shall be attributed to shadowing components even if the target is well outside the baseline. Real targets, even those referred to as 'stealthy' only partly absorb the EM waves and at the receiver both reflected (backscattering and bistatic reflection) and shadow (forward scattering) EM waves components are present, as explained in the previous section. Typically, the FS region is associated with the area where forward scattering components of EM waves dominate above bistatic [1 (p.150–155)]. The boundary between the bistatic and FS region is soft, depends on the material and shape and requires EM analysis for each particular target. The only firm statement which can be made without strict analysis is that the FSCS main lobe shadow component always dominates above the bistatic reflected component. It also may be supposed that the 'stealthier' the target is the wider will be an area of FS domination that will include not only the FSCS main lobe but also its sidelobes. This effect has been demonstrated in [25] where it has been shown experimentally that the magnitude of signals from two identical metallic cylinders one of which was coated with absorbing material differ in the region of bistatic reflection by approximately 8 dB, while in the FS region, related to the mail lobe, the difference is negligible. The envelopes of two signals at the received signal strength indicator (RSSI) receiver output, corresponding to the RCS of both cylinders as a function of bistatic angle, are shown in Figure 13.7 for cylinders which are (a) absorbing and (b) metallic.

The magnitude of the envelope is significantly larger in the region between about 170° and 190° which corresponds to the FSCS main lobe region. Reflections from the coated cylinder within regions of 120°–170° and 190°–240° are significantly smaller than that of the metallic cylinder and in the case of coated cylinder these regions show mainly FSCS sidelobe levels.

In Figure 13.8(a) results of modelling by PTD for shadow radiation pattern of a cylinder (continuous line) and the total physical optics (PO) radiation pattern (dashed line) are shown, highlighting that the shadow radiation is focussed in the FS direction (45°) [13 (p. 239)]. Figure 13.8(b) shows results of full RCS simulation by

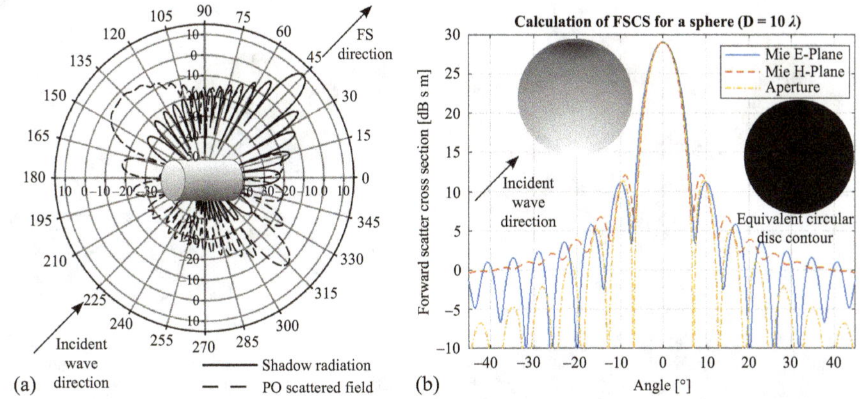

Figure 13.8 The total physical optics scattered far field of (a) cylinder, showing the contribution from shadow radiation which is focussed in the FS direction (45° in this figure) – © [2006] John Wiley and Sons. Reprinted, with permission, from [13] and (b), RCS of sphere simulated according to Mie theory and FSCS from aperture approximation (FS direction is 0°)

Mie theory for a sphere (E and H plane) and FSCS of an aperture approximated as a disc of the same radius (Airy pattern). Clearly, the main and first sidelobes of shadow field/FSCS in these examples dominate above other scattering components.

In [11], the 3D RCS for a side illuminated missile were simulated in CST Microwave Studio at two frequencies of 100 MHz and 1 GHz, corresponding to the upper Rayleigh-low Mie scattering mechanism and optical scattering regions respectively – Figure 13.9(a) and (b). The azimuth plane cross-sections are shown in Figure 13.9(c) and (d). Illumination occurs such that in the figure, the backscattering direction is 0° azimuth and FS corresponds to 180° azimuth angle. For the target length of 3.6 m and the wavelength of 3 m (Rayleigh–Mie region), there are two well-defined maxima of back and forward scattering lobes: 11.7 and 9 dB s m, respectively. This confirms the fact that in the case where the effective size of the object is comparable to the wavelength, the backscattering and forward scattering are of approximately the same intensity. For the same target, illuminated by a 0.3 m wavelength signal, corresponding to the optical scattering region, there are also two pronounced maxima for back and forward scattering, but both lobes are very narrow, of the order of a few degrees, their maximum intensity is greater than for the Rayleigh–Mie case, and the FSCS main lobe is approximately 10 dB greater than that of the backscattering lobe. In the bistatic (side) scattering direction, with azimuth angular ranges away from forward and backscatter and centred around 90° and 270°, the RCS drops by up to 30–35 dB relative to the FS. Assuming that the missile is ‘stealth’ shaped or coated for monostatic radar, the pronounced peak in backscatter direction will be absent, and the target is likely to be detected only in the FS region.

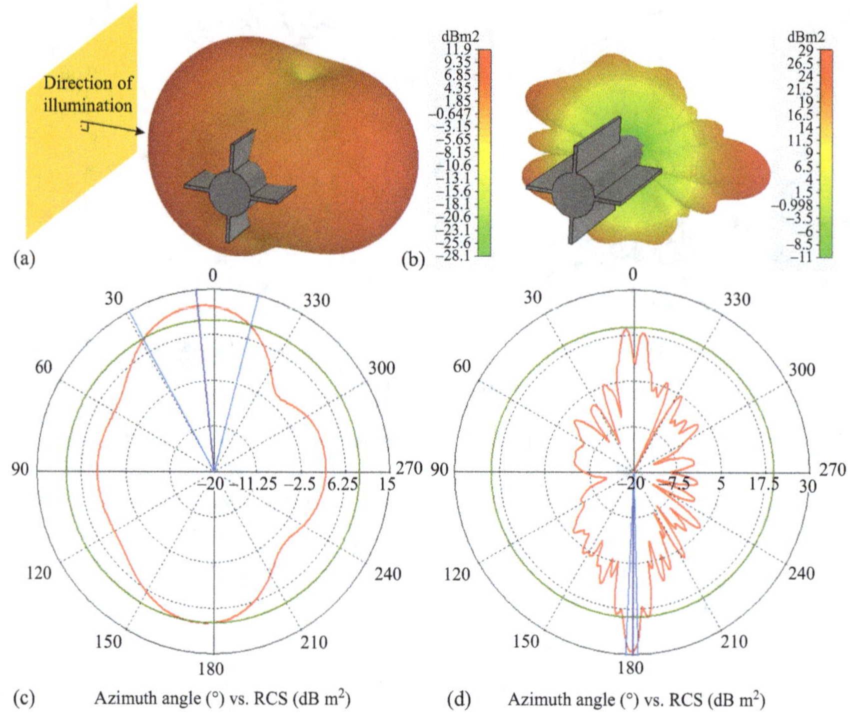

Figure 13.9 Simulated 3D bistatic RCS for missile [11] with (a) 3 m wavelength and (b) 0.3 m wavelength with their corresponding cross-sections (c) and (d). Backscattering angle is 0°, FS angle is 180°

In the same paper, CST Microwave Studio simulation of RCS has been performed for a number of typical targets (shown in Figure 13.10) in order to estimate level of bistatic reflections, where the bistatic angle $\beta = \pi/2$, and forward scattering ($\beta = \pi$) at several frequency bands. Results are summarised in Table 13.4, where frequency bands refer to different scattering mechanisms: Rayleigh (R), Mie (M) and Optical (O). As it can be seen from this table, the significantly larger RCS in FS direction is observed for all cases of Mie and optical scattering. For instance, two targets – a UAV and a 3.6 m long missile, have FSCS of 33 dB m^2 and 23 dB m^2 accordingly at 400 MHz (0.75 m wavelength) – the frequency which is well suited for air target detection. Even without deep analysis, it is seen that these numbers provide an excellent power budget in FSR which are 20–30 dB above the bistatic RCS at 90°.

As was mentioned above for the detailed structure of FSCS pattern is important for the target recognition and imaging in FSR. However, for the majority of radar tasks and to see the general effects of target dimension on radar performance, particularly in terms of radar power budget, we can use rather simple approximation of the target RCS by replacing the target silhouette by a rectangle [26] or composition of rectangles which can be used to approximate any target shadow aperture and has simple and well-known pattern from antenna theory

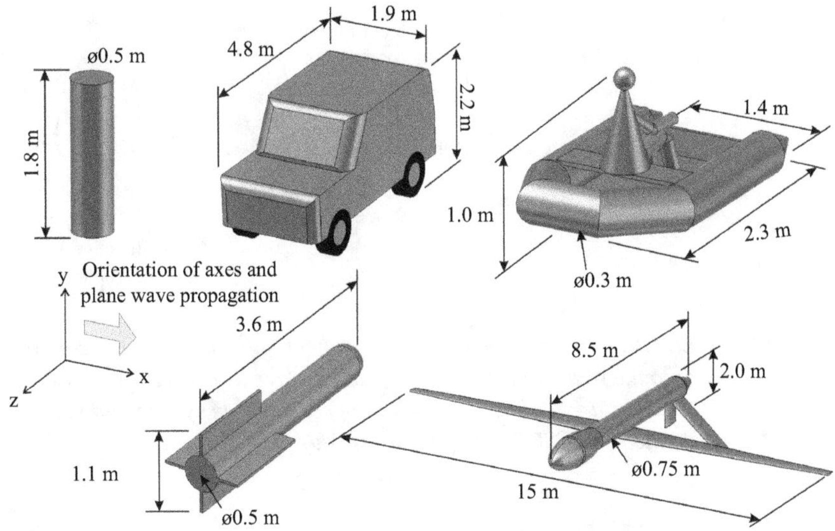

Figure 13.10 Target shapes and dimensions for 3D RCS simulations of which results are shown in Table 13.4 (illumination from left side as shown) [11]

Table 13.4 Bistatic RCS vs FSCS for a selection of targets [11]

	RCS, dB s m				
	Bistatic radar ($\beta = 90°$)/FSR ($\beta = 180°$)				
$\lambda \rightarrow$ $f \rightarrow$	3.0 m 100 MHz	1.5 m 200 MHz	0.75 m 400 MHz	0.3 m 1 GHz	0.1 m 3 GHz
Targets, *L* (m), *H* (m), *V*(m/s)					
Human $L = 0.5$, $H = 1.8$, $V = 1.0$	6.2/7.6 (R)	7.3/11.1 (M)	7.6/15.6 (M)	11.6/22.0 (O)	15.4/33.2 (O)
Vehicle, 4×4 $L = 4.8$, $H = 2.2$, $V = 10$	9.7/22.5 (R/M)	7.6/28.2 (M)	−7.4/34.1 (sub-O)	5.4/41.0 (O)	–
Inflatable boat $L = 2.3$, $H = 1$, $V = 5$	0.6/2.7 (R)	−1.3/6.8 (M)	−9.4/13.0 (Sub-O)	−6.6/21.4 (O)	−7.3/40.3 (O)
Missile $L = 3.6$, $H = 0.5$, $V = 200$	2.9/8.9 (R/M)	1.6/15.3 (M)	−8.0/22.6 (Sub-O)	−6.1/29.0 (O)	−3.2/45.0 (O)
UAV $L = 8.4$, $H = 2$, $V = 50$	4.4/17.6 (M)	2.0/23.5 (O)	−1.6/32.9 (O)	−3.0/60.0 (O)	–

[22 (p.584–602)]. Figure 13.11(a) graphically defines the shadow aperture as an effective aperture projected onto the plane normal to the illuminating wave direction, with effective area A_{eff} and length l_{eff}. With the notation of Figure 13.11(b), the 3D FSCS of the rectangular secondary antenna representing

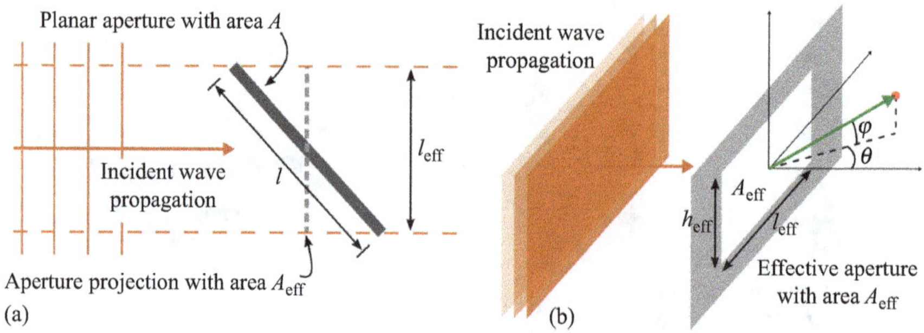

Figure 13.11 Effective shadow aperture as a projection of the shadow contour onto the plane normal to the incident wave direction (a) – for clarity the aperture is shown as solid; the co-ordinate system and variables defined for FSCS calculation are shown in (b)

shadow contour of the target is expressed as a function of the two angles specifying the direction of scattering:

$$\sigma_{fs}(\theta,\varphi) = 4\pi\frac{A_{eff}^2}{\lambda^2}\left(\frac{\sin((\pi l_{eff}/\lambda)\sin\theta)}{(\pi l_{eff}/\lambda)\sin\theta}\right)^2\left(\frac{\sin((\pi h_{eff}/\lambda)\sin\varphi)}{(\pi h_{eff}/\lambda)\sin\varphi}\right)^2 \tag{13.11}$$

This simple method of replacing a complex shape target by an equivalent rectangular plate is very convenient for looking at generalised system behaviour and will be used further in this chapter.

13.3 Power budget, signature of moving target and optimal signal processing in FSR

13.3.1 Power budget analysis

The power budget analysis for FSR [28–30] has a lot in common with both monostatic [30] and bistatic radar [1]. There are two basic models of EM wave propagation: the free space model and two-ray path (TRP) propagation model which are shown in Figure 13.12(a) and (b) accordingly.

The free space propagation model is applicable for the cases: (i) the height of the antennas is comparable to the separation distance between Tx and Rx; (ii) the radar antennas have a rather narrow main beam (shown as shaded areas in Figure 13.12) and do not illuminate the surface below; (iii) the surface roughness is comparable with the wavelength which results in dominant diffused scattering by the potentially reflecting surface so that there is no single ground specular reflected beam contributing into the total field at the receiver input [17]. In contrast with the two-ray model, the reflected signal may have the same order of magnitude as the DPS but the total field will depend on the phase relationship between them.

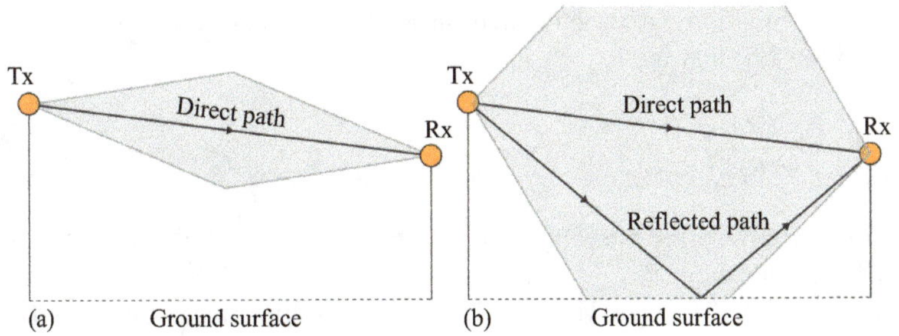

Figure 13.12 Configuration for (a) free space propagation model and (b) two-ray path propagation model when antenna beam intersects the ground

13.3.1.1 Free space propagation model

In FSR, two signals contribute into the total received power and power of each must be individually estimated; the power of leakage (which in the case of the free space model is the DPS), P_{dp}^{fsp}, and power of the signal scattered from the target, P_{tg}^{fsp}, where 'fsp' stands for free space propagation, 'dp' is for the direct path (leakage) signal (Figure 13.12) and 'tg' is for the target signal. The power of the LS can be estimated by the Friis equation [31 (pp.16–18)] which can be presented in one of the following forms:

$$P_{dp}^{fsp} = P_t G_t G_r \left(\frac{\lambda}{4\pi d} \right)^2 \frac{1}{L} = \frac{P_t}{4\pi d^2} \frac{G_t A_r}{L} = \frac{P_t G_t G_r}{L_P L} \quad (13.12)$$

where P_t, G_t, G_r are the transmit power, gains of transmit and receive antennas, correspondingly, A_r is the effective area of the receive antenna, d is the direct path/baseline length, L_P is the propagation loss and L represents the miscellaneous losses which are always present in real systems and environments.

To estimate P_{tg}^{fsp}, the bistatic radar equation can be used, where FSCS, σ_{fs} is used instead of the bistatic RCS:

$$P_{tg}^{fsp} = \frac{P_t G_t G_r \lambda^2 \sigma_{fs}}{(4\pi)^3 R_t^2 R_r^2 L_{tg}} \quad (13.13)$$

and where L_{tg} are the other losses as in (13.12), but in the presence of the target. According to (13.13), P_{tg}^{fsp} reaches a minimum when the transmitter/receiver-to-target ranges are equal, i.e. $R_t = R_r = d/2$ and the radar range equation in this case will take the form:

$$d_{max}^{fsp} = 2 \sqrt[4]{\frac{P_t G_t G_r \sigma_{fs} \lambda^2}{(4\pi)^3 S_{min} L_{tg}}} \quad (13.14)$$

where S_{min} is the minimal required detectable signal level corresponding to a particular signal-to-noise ratio (SNR).

In many situations, it is convenient in FSR to consider the ratio of the target signal and the LS, that is:

$$\frac{P_{\text{tg}}^{\text{fsp}}}{P_{\text{dp}}^{\text{fsp}}} = \frac{d^2 \sigma_{\text{fs}}}{4\pi R_t^2 R_r^2} \tag{13.15}$$

where for a simplicity the miscellaneous loss factors are omitted, but readers should always remember them.

Equation (13.13) looks very similar to its bistatic radar counterpart but in practice has a very important peculiarity which is not only due to essentially bigger RCS. The peculiarity lies where the FSCS is coupled with the system wavelength in the optical region which introduces some restriction in FSR. To demonstrate this, let us first replace σ_{fs} in (13.13) by its maximal value $\sigma_{\text{fs}}^{\max}$ according to (13.4) and assume that the target has a rectangular shadow silhouette of length l_{eff} and height h_{eff} (Figure 13.11), so that:

$$\sigma_{fs}^{\max} = 4\pi \left(\frac{l_{\text{eff}} h_{\text{eff}}}{\lambda} \right)^2 \tag{13.16}$$

Equation (13.13) thus transforms into

$$P_{\text{tg}}^{\text{fsp}} = \frac{P_t G_t G_r \lambda^2}{(4\pi)^3 R_t{}^2 R_r{}^2} \cdot 4\pi \left(\frac{l_{\text{eff}} \cdot h_{\text{eff}}}{\lambda} \right)^2 = \frac{P_t G_t G_r (l_{\text{eff}} \cdot h_{\text{eff}})^2}{(4\pi)^2 R_t{}^2 R_r{}^2} \tag{13.17}$$

which leads to the conclusion that the power of the received signal in FSR does not depend on the carrier frequency, assuming fixed antenna gains.

It is worth recalling here that the required radar performance is ultimately specified by the SNR. The output of an optimal receiver, which is based on a correlator or matched filter, is specified by the energy of the received signal E_{r} normalised to the noise power spectral density (PSD) N_0 as $\text{SNR}_{\text{out}} = 2E_r/N_0$ [33 (p.280)]. Assuming a constant signal power during the observation time $\Delta\tau_{\text{fs}}$, which itself depends on the wavelength as follows from (13.7), the output SNR is:

$$\text{SNR}_{\text{out}}^{\text{fsp}} = \frac{2P_{\text{tg}}^{\text{fsp}} \Delta\tau_{\text{fs}}}{N_0} = \frac{P_{\text{tg}}^{\text{fsp}} \lambda R_r}{l_{\text{eff}} v_{\text{tg}} N_0} \tag{13.18}$$

The notation in this equation corresponds to that of (13.8). Now by combining (13.18) and (13.17):

$$\text{SNR}_{\text{out}}^{\text{fsp}} = \frac{P_t G_t G_r h_{\text{eff}}{}^2 l_{\text{eff}} \lambda}{(4\pi)^2 R_t^2 R_r N_0 v_{\text{tg}}} = \frac{P_t A_t A_r h_{\text{eff}}{}^2 l_{\text{eff}}}{\lambda^3 R_t^2 R_r N_0 v_{\text{tg}}} \tag{13.19}$$

where A_t, A_r are the effective areas of the transmit and receive antennas. Equation (13.19) gives an insight into how the FSR radar parameters affect the system performance for a required SNR, which is inversely proportional to the target-to-receiver range and the square of the transmitter-to-target range. Also it is proportional to the square of the target vertical dimension but linearly related to its length.

While the power in (13.13) reaches its minimum at $R_t = R_r$, (13.19) reaches its minimum at $R_r \approx 0.33d, R_t \approx 0.67d$ and hence the minimum of the SNR will be:

$$\mathrm{SNR}_{\text{out, min}}^{\text{fsp}} = \frac{P_t G_t G_r {h_{\text{eff}}}^2 l_{\text{eff}} \lambda}{0.15(4\pi)^2 d^3 N_0 v_{\text{Tg}}} \tag{13.20}$$

Therefore, the maximum detection range/baseline length $d_{\max}^{\text{fsp}}$ considering the maximum target observation time as defined in Section 13.2.2.1 is:

$$d_{\max}^{\text{fsp}} \approx \sqrt[3]{\frac{P_t G_t G_r {h_{\text{eff}}}^2 l_{\text{eff}} \lambda}{0.15(4\pi)^2 SNR_{\text{out, min}}^{\text{fsp}} N_0 v_{\text{tg}}}} \tag{13.21}$$

The noise PSD can be evaluated as $N_0 = kT \cdot N_f$, where the Boltzmann constant $k = 1.38 \cdot 10^{-23}\ J/K$, T is the receiver temperature and N_f is the noise factor. As an example let us evaluate the maximal baseline for detection of a UAV target (Figure 13.10) flying with a speed of 50 m/s normal to the baseline assuming a noise figure of $N_f = 6$ dB, antenna gains of $G_t = G_r = 6$ dB (corresponding to a single patch antenna), $P_t = 10$ dBW and other losses (not shown in (13.21)) totalling 20 dB. The results of calculation of the maximal baseline for detection over a range of minimum output SNRs at three frequencies are presented in Figure 13.13.

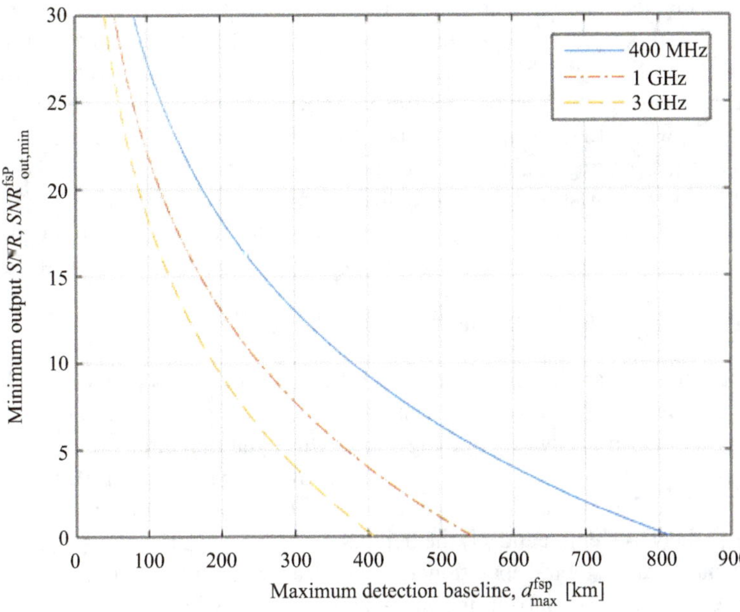

Figure 13.13 Small airborne target maximum detection range (baseline length) in FSR as a function of SNR at three carrier frequencies (free space model)

These numbers, which may seem somewhat excessive, reflect the benefit of a high FSCS and a very long integration time – of the order of seconds for the considered cases. However, it is not likely that such a long integration time can be used in practice. Moreover, it is also important to stress here that at least for a terrestrial system the line-of-sight is rarely present to provide the DPS at the receiver at such a range which is a compulsory condition for FSR operation. In [15 (Ch.19)], an operational FSR with the transmit power of 6 dBW and antenna gains of 18 dB for low-altitude air targets detection and tracking is discussed.

13.3.1.2 Power budget in two-ray path propagation model

In contrast to the free space propagation model, the TRP model takes into account the reflection of the transmitted signal from the surface of land or sea. This model has been developed, comprehensively analysed and well proven experimentally as a part of wireless communication studies, which were reported in numerous papers and books, including the classic book of Kerr [33 (Ch.5)], as well as in radar-related research [31] where the influence of a surface reflection on radar performance has been investigated. For FSR, appropriate material can be found in [28,34].

We will start with a brief description of the TRP model. If the transmitted wave is propagating above the surface, there are two signals (rays) at the reception point, the vector sum of which form a new LS, $E_{\mathrm{lkg}}^{\mathrm{trp}}$. Its magnitude depends on the amplitudes and mutual phase of the two signals; one signal follows the shortest path between the transmitter and the receiver and as in the free space case is referred to as the DPS with electric field strength E_{dp}; the second one is a signal reflected from the surface, E_{ref}– both paths are highlighted in Figure 13.12(b). The latter is a delayed version of the DPS and has a phase shift which depends on the path difference between the two, as well as on the Fresnel reflection coefficient at the air-surface interface, Γ, which in the general case is a complex value. The reflection coefficient depends on the grazing angle, wave polarisation, electrical properties of the surface as well as its roughness and, therefore, the effective reflection coefficient [35] may be used for a rough surface to relax the strict requirement of negligible roughness in the conventional ground reflection model. The amplitude of the combined signal is

$$\vec{E}_{\mathrm{lkg}}^{\mathrm{trp}} = \vec{E}_{\mathrm{dp}} + \vec{E}_{\mathrm{ref}} = \left|\vec{E}_{\mathrm{dp}}\right| \mathrm{e}^{-ikR_{\mathrm{dp}}} + \left|\vec{E}_{\mathrm{ref}}\right| \cdot \Gamma \mathrm{e}^{-ikR_{\mathrm{ref}}} \tag{13.22}$$

where R_{dp} and R_{ref} are the direct path and reflected path distances accordingly. The phase of Γ depends mainly on the surface electrical properties, i.e. electric conductivity and permittivity and grazing angle, while the magnitude depends mainly on the surface roughness. It was shown in a conventional ground reflection model [36 (pp.120–125)], that under the conditions of (i) the ground being smooth with respect to the wavelength and (ii) the heights of the transmit/receive antennas being much smaller than their separation distance, the reflection coefficient does not depend on the wave polarisation and is equal to -1, which is equivalent to the reflection from a flat surface of a perfect conductor where we can expect the maximum signal strength reduction. This extreme approximation nevertheless is very practical as a reference model for the more general case where the surface roughness and electrical properties are unknown *a priori* and depend on presence of vegetation, weather conditions, rain, snow, etc. for land-based radar while for radar operating above the sea surface, it will depend on sea state conditions.

For such a reference model, the LS power at the receiver can be calculated according to:

$$P_{\rm lkg}^{\rm trp} = 4P_tG_tG_r\left(\frac{\lambda}{4\pi d}\right)^2\sin^2\left(\frac{2\pi h_th_r}{\lambda d}\right) \tag{13.23}$$

where $h_{t/r}$ are the antenna heights and other parameters have been previously defined. At very low grazing angles this tends to:

$$P_{\rm lkg}^{\rm trp} = P_tG_tG_r\left(\frac{h_th_r}{d^2}\right)^2 \tag{13.24}$$

Following the same approach, the power scattered from a target can also be evaluated [34]. In this case, the TRP propagation model should be used on both paths: transmitter-to-target and target-to-receiver as in Figure 13.14(a):

$$P_{\rm tg}^{\rm trp} = 16P_tG_tG_r\sigma_{\rm fs}\frac{1}{R_t{}^2R_r{}^2}\frac{\lambda^2}{(4\pi)^3}\sin^2\left(\frac{2\pi h_tz_{\rm tg}}{\lambda R_t}\right)\sin^2\left(\frac{2\pi h_rz_{\rm tg}}{\lambda R_r}\right) \tag{13.25}$$

where $z_{\rm tg}$ is the altitude of a point like target. Then, for near zero grazing (13.25) can be reduced to:

$$\begin{aligned} P_{\rm tg}^{\rm trp} &= 16P_tG_tG_r\sigma_{\rm fs}\frac{1}{R_t{}^2R_r{}^2}\frac{\lambda^2}{(4\pi)^3}\left(\frac{2\pi h_tz_{\rm tg}}{\lambda R_t}\right)^2\left(\frac{2\pi h_rz_{\rm tg}}{\lambda R_r}\right)^2 \\ &= P_tG_tG_r\sigma_{\rm fs}\frac{4\pi}{\lambda^2}\left(\frac{h_tz_{\rm tg}}{R_t{}^2}\right)^2\left(\frac{h_rz_{\rm tg}}{R_r{}^2}\right)^2 \end{aligned} \tag{13.26}$$

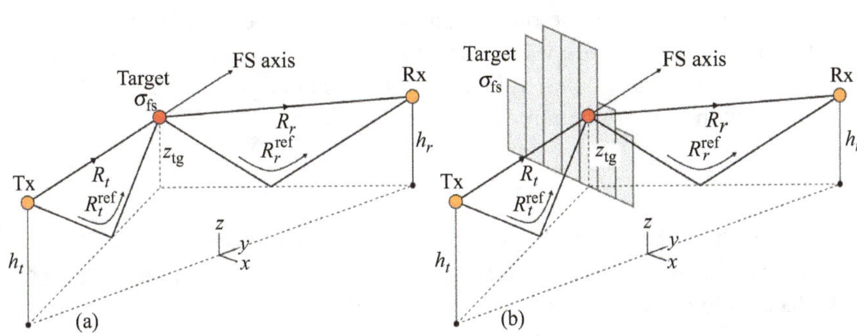

Figure 13.14 Two-ray propagation model for target in FSR considered as (a) a point target and (b) an extended target as composition of strips

In [34], analysis of leakage power and scattered power for a sphere target moving above the ground has been presented, which has demonstrated excellent agreement between that predicted by the TRP model and measured values.

Equation (13.26) can be applied for power budget estimation of a low flying target if its dimensions are smaller than the target altitude z_{tg}, thus acting as a point-like target.

For the case of an extended target, it can be presented as a composition of vertical strips spanning the area of a real target, where z_{tg} will now be defined for the centre point of each strip – Figure 13.14(b). Scattering from each strip of the target could be calculated by (13.26), and the overall received signal power will be a sum of powers scattered from these elementary parts. Further refinement of the model for an extended target should lead to the target aperture being presented by a set of secondary radiators positioned at the centres of elementary rectangular areas of approximately half-wavelength by half-wavelength dimension. This representation is valid when target, transmitter and receiver are in the far field with respect to each other. Thus, the received signal will be the vector sum of fields scattered by each point-like radiator across the shadow aperture of a target [12], power can then be calculated from the received field.

One simplified method involving direct summation of powers has been proposed in [37 (Ch.9)] with application for a maritime monostatic radar. This method will be valid if the target occupies any single Fresnel zone. In this work, the target aperture has been divided into elementary horizontal strips of equal area and the RCS of each part was σ_{fs}/N. The received power from all elementary areas will be the sum of the contributions from each area numbered $1 \leq n \leq N$:

$$P_{tg}^{trp} = \sum_{n=1}^{N} P_t G_t G_r \sigma_{fs} \frac{4\pi}{N\lambda^2} \left(\frac{h_t z_{tg}(n)}{R_t(n)^2} \right)^2 \left(\frac{h_r z_{tg}(n)}{R_r(n)^2} \right)^2 \quad (13.27)$$

where $R_t(n)$ and $R_r(n)$ are ranges to the centre of mass of each elementary strip. When the number of partial areas $N \rightarrow \infty$, the sum in (13.27) becomes an integral. This integral could be calculated analytically under some simplifying assumptions but there is a relatively simple rule of thumb which can be used in practice which leads to a rather accurate result in power budget calculation at least for small surface targets such as cars, boats, animals etc. For such targets, (13.25) and (13.26) can be used with the assumption:

$$z_{tg} \approx 0.5 h_{eff} \quad (13.28)$$

where h_{eff} is the effective height of the planar target equivalent aperture. This rule has been successfully applied and experimentally confirmed in [34,38,39]. Thus, for surface targets, from (13.26), one can use following equation,

$$P_{tg}^{trp} = P_t G_t G_r \sigma_{fs} \frac{4\pi}{\lambda^2} \left(\frac{0.5 h_t h_{eff}}{R_t^2} \right)^2 \left(\frac{0.5 h_r h_{eff}}{R_r^2} \right)^2 \quad (13.29)$$

Following the methodology which has been used to derive (13.18)–(13.21), but now using the TRP model, the equation for the SNR, SNR_{out}^{trp}, at the optimal receiver output will take the form:

$$SNR_{out}^{trp} = \frac{\pi^2 P_t G_t G_r \, h_t^2 h_r^2 l_{eff} {h_{eff}}^6}{\lambda^3 v_{tg} R_t^4 R_r^3 N_0} \tag{13.30}$$

This equation has a minimum when $R_t \approx 0.55d$ and $R_r \approx 0.45d$ and is approximately equal to:

$$SNR_{out,\ min}^{trp} \approx \frac{120\pi^2 P_t G_t G_r \, h_t^2 h_r^2 l_{eff} {h_{eff}}^6}{\lambda^3 v_{tg} d^7 N_0} \tag{13.31}$$

the maximal baseline could then be evaluated as:

$$d_{max}^{trp} \approx \left(\frac{120\pi^2 P_t G_t G_r \, h_t^2 h_r^2 l_{eff} {h_{eff}}^6}{\lambda^3 v_{tg} SNR_{out,\ min}^{trp} N_0} \right)^{1/7} \tag{13.32}$$

As an example, let us estimate the maximal baseline for a target representing a 4×4-type vehicle of $l_{eff} = 5$ m, $h_{eff} = 2$ m, moving with a speed of 10 m/s normal to the baseline, where both antenna heights are $h_t = h_r = 1$ m. Assuming again that $P_t = 10$ dB W, $N_f = 6$ dB, $G_t = G_r = 6$ dB and other unaccounted losses are 20 dB, for a required $SNR_{out,\ min}^{trp} = 13$ dB the calculated max baseline will be ≈ 1.4 km for 400 MHz. The results of calculation of the maximal baseline for detection over a range of SNRs at three frequencies are presented in Figure 13.15, under the assumption of a flat Earth.

Comparison of the results shown in Figures 13.13 and 13.15 demonstrates the essential difference between free space and TRP propagation model scenarios used in FSR in terms of the power budget dependence on frequency. This corresponds well to a case of ground monostatic radar reported in [31].

13.3.2 Target signature in FSR

The basic physical principle of FSR is the shadowing of the DPS by an object when it occurs in a vicinity of the baseline. In Section 13.2.2, it was stated and discussed that the target FSCS is the key parameter which defines the intensity of the shadowing, i.e. amplitude of modulation of the DPS at FSR receiver input due to target presence. Due to the target motion with respect to transmitter and receiver, the received signal in FSR is a specific waveform and the target signature is formed as a super-position of the incident EM field i.e. LS and the shadow radiation from the moving scattering object – the target.

We can assume that the target follows a linear trajectory with constant speed while its FSCS main lobe is directed towards the receiver. This is generally true for the majority of practical applications [11,40]. Shadow radiation is cast upon the receiver according to the width of the shadow lobe while the target moves in

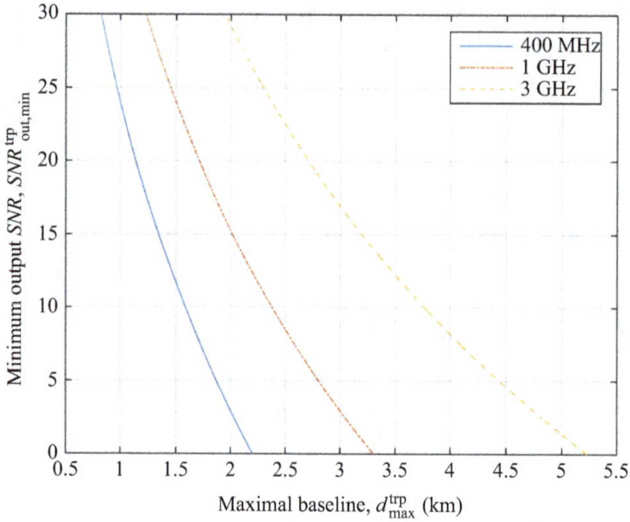

Figure 13.15 4×4 vehicle maximum detection range (baseline length) in FSR as a function of SNR at three carrier frequencies (two-ray path model)

the vicinity of the baseline. The maximum of the shadow radiation corresponds to the case when the target is on the baseline.

13.3.2.1 Phase signature

Let us initially consider a point-like target, which yet casts a shadow on the receiver, and omit any amplitude modulation of the signal caused by propagation loss and by the FSCS pattern, focussing purely on the phase signature of such moving point [11,20,25,41]. Since it is composed of 'shadow radiation', the forward scattered signal is $\pi/2$ phase shifted (imaginary along the FS axis) relative to the LS [43]. Therefore, at the receiver input, a composite signal of the DPS $S_{\text{dp}}(t)$ with an amplitude A_{dp} and delayed scattered signal $S_{\text{tg}}(t)$ from the moving target with amplitude $A_{\text{tg}}(t)$ is acquired,

$$S_r^{\text{in}}(t) = S_{\text{dp}}(t) + S_{\text{tg}}(t) = A_{\text{dp}}\cos(\omega_0 t) + A_{\text{tg}}(t)\sin\left[\omega_0\left(t + \Delta t_{\text{tg}}(t)\right)\right] \qquad (13.33)$$

where ω_0 is the carrier angular frequency and $\Delta t_{tg}(t)$ is the delay time of the signal from the moving target with respect to the direct path, given by:

$$\Delta t_{\text{tg}}(t) = \frac{R_t(t) + R_r(t) - d}{c} \qquad (13.34)$$

where, as defined previously, $R_{t/r}$ are the transmitter/receiver to target ranges, d is the baseline length and c is the speed of light. The initial phases of coherently acquired signals can be omitted without loss of generality.

To extract the Doppler phase signature component, we will consider a principle based on a self-mixing heterodyne receiver as in [11]. The basis of this mixer is a double-sided SLD, followed by a low-pass filter (LPF) of which the bandwidth is defined by the range of Doppler frequencies expected in the signature [15,38]. Thus, the action upon the signal at the receiver input (13.33) will be,

$$\begin{aligned} &S_r^{\text{in}}(t) \overset{\text{SLD}}{\rightarrow} \left[A_{\text{dp}}\cos(\omega_0 t) + A_{\text{tg}}(t)\sin\left[\omega_0\left(t + \Delta t_{\text{tg}}(t)\right)\right]\right]^2 \overset{\text{LPF}}{\rightarrow} \\ &DC - A_{\text{ph}}(t)\sin\left(\omega_0 \Delta t_{\text{tg}}(t)\right) = DC - A_{\text{ph}}(t)\sin\left(\frac{2\pi\left(R_{\text{t}}(t) + R_{\text{r}}(t) - d\right)}{\lambda}\right) \end{aligned} \tag{13.35}$$

where $DC = \left(\left(A_{\text{dp}}^2 + A_{\text{tg}}^2\right)/2\right) \overset{A_{\text{tg}} \ll A_{\text{dp}}}{\approx} \left(A_{\text{dp}}^2/2\right)$ is the power of the DPS, $A_{\text{ph}} = A_{\text{dp}}A_{\text{tg}}$ characterises the amplitude of the phase signature. In terms of Doppler shift,

$$S_r^{\text{in}}(t) \overset{\text{SLD}}{\rightarrow} \left[A_{\text{dp}}\cos(\omega_0 t) + A_{\text{tg}}(t)\sin((\omega_0 + \omega_d)t)\right]^2 \overset{\text{LPF}}{\rightarrow} DC + A_{\text{ph}}(t)\sin\left(\omega_d t\right) \tag{13.36}$$

where ω_d is the angular Doppler frequency shift of the moving target.

Thus,

$$\omega_d t \equiv -\frac{2\pi}{\lambda}\left(R_t(t) + R_r(t) - d\right) + 2\pi n \tag{13.37}$$

where n is an integer, and the last term can be omitted without loss of generality. After removing the leakage (*DC*) component in (13.36), the target Doppler signature extracted at the receiver output is:

$$S_r^{\text{out}}(t) = A_{\text{dp}}A_{\text{tg}}(t)\sin\left[\frac{2\pi}{\lambda}\left(R_t(t) + R_r(t) - d\right)\right] \tag{13.38}$$

For a target crossing the baseline, the phase signature is a two-sided chirp signal, this is well known in radar as a waveform with a good auto-correlation property [43]. An example of this signature and its auto-correlation function are shown in Figure 13.16(a) and (b) correspondingly. This property will be further discussed as a way of target resolution when they follow in a convoy – Section 13.3.3.2.

13.3.2.2 The signature envelope

Following from (13.38), the FS signature could be viewed as a waveform which has frequency (phase) modulated harmonic carrier specified by the Doppler effect and the target motion parameters with an amplitude modulation specified by the FSCS pattern. We will now omit the term A_{dp} to characterise the signature considering it as a scaling coefficient:

$$S_r^{\text{out}}(t) = A_{\text{tg}}(t)\sin\left[\frac{2\pi}{\lambda}\left(R_t(t) + R_r(t) - d\right)\right] \tag{13.39}$$

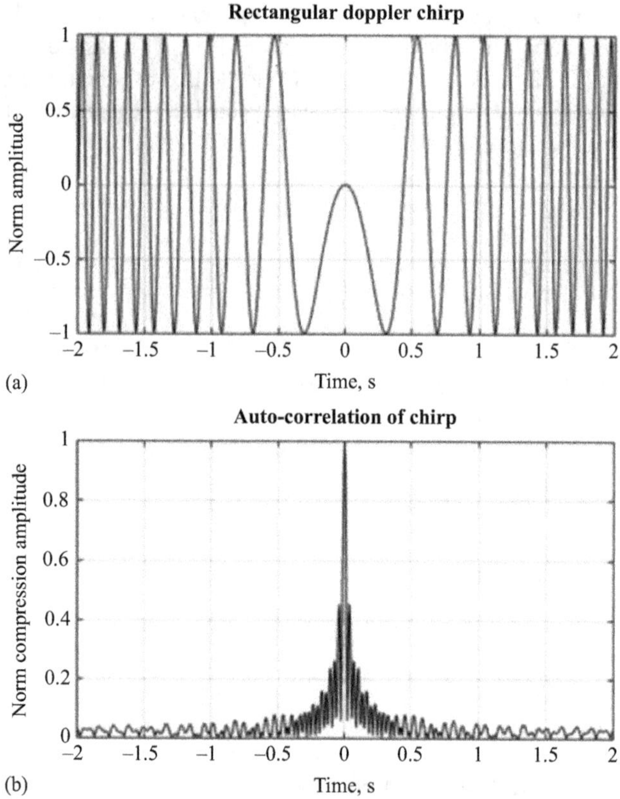

Figure 13.16 Example of a typical two-sided chirp forward scatter Doppler signature (with rectangular envelope) (a) and its auto-correlation function (b)

The actual signature envelope $A_{tg}(t)$ is specified by the target FSCS and is unknown *a priori*. Even if the target and its trajectory are known, when modelling, analytical solutions for the FSCS are only available for a few convex shapes in the optical and sub-optical scattering regions [44–46]. Thus, approximated models from PO or PTD [45–47] or full-wave EM simulation methods [26] must be used for target FSCS estimation. However, the target still can be efficiently detected and its trajectory parameters estimated by means of the signal correlation with a matched phase Doppler signature where any amplitude modulation is omitted [10,28,41]. Then, the envelope of the target signature can be used for target automatic classification and imaging [6,23].

In [25], the nature of FSR target signature has been verified experimentally. A rectangular metal plate and identical plate covered with an absorber to avoid bistatic reflections were used as targets. The experimental results are shown in Figure 13.17(a), (b) and corresponding modelling in (c) using the rectangular aperture as described by (13.11) and extended for moving targets in [41].

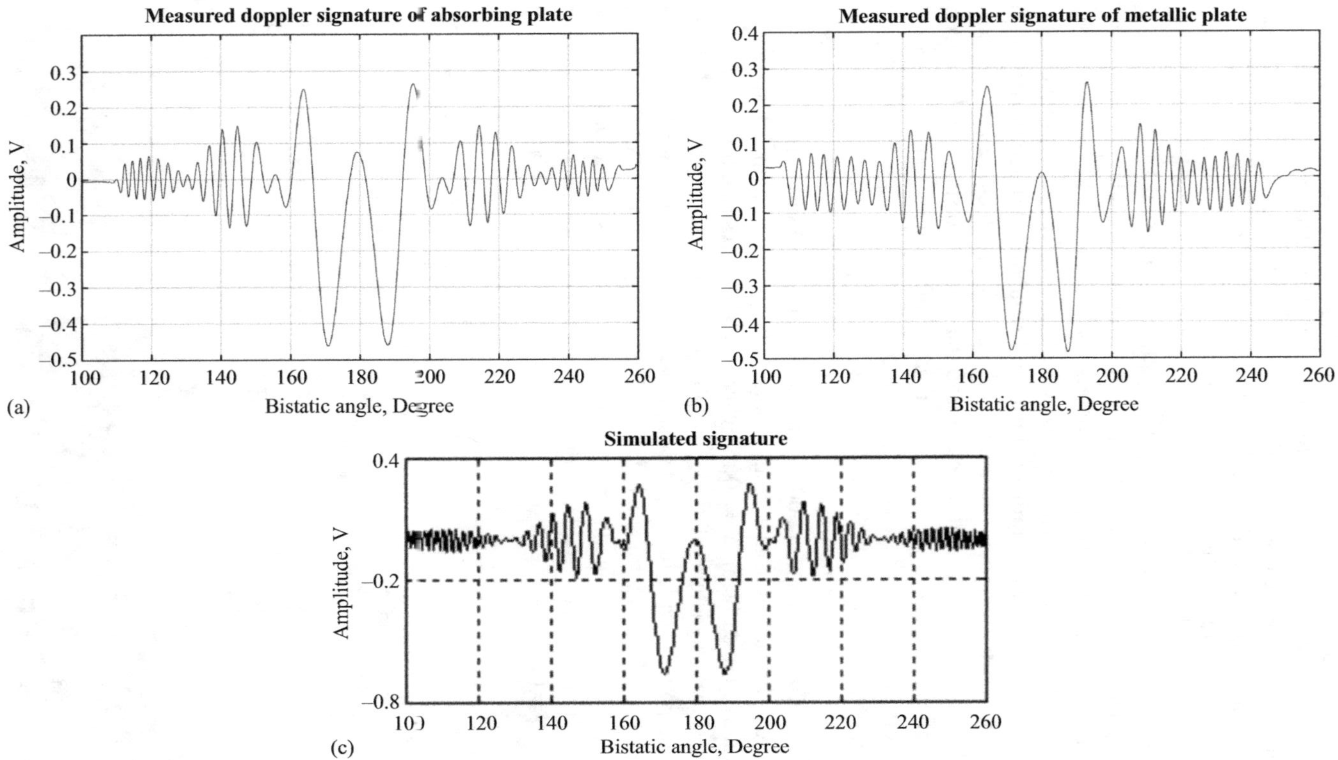

Figure 13.17 Comparison of the measured FS signatures of (a) absorbing plate (b) metal plate of the same size and (c) simulated signature [25]

Evidently, the envelopes and Doppler chirp carriers of measured and simulated signatures agree very well in the range of bistatic angles 140–220° that proves that the FS signature is correctly described by (13.39) as a product of Doppler carrier variation and the envelope specified by the FSCS. The amplitude in the middle part of the signal which lasts while the receiver is within the main shadow scatter lobe of the target does not depend on the material of the target but the size.

13.3.2.3 Examples of measured FSR signatures of real targets

Measured signatures of maritime targets in the upper C-band [25] with sizes defining different diffraction mechanisms are shown in Figure 13.18(a–c). Using the Fresnel parameter $S = D^2/(4\lambda)$, where D is the largest effective size of the target and λ is the wavelength, we will distinguish the Fraunhofer diffraction (far-field) region, $d \gg S$ and the Fresnel diffraction region, $d \sim S$, which define the scattering mechanism from the target in relation to the baseline distance d. However, when considering the time-varying Doppler signature of a moving target, we should not confuse diffraction from the individual target with the Fresnel-like diffraction on the effective inverse aperture defined by the whole path of the moving target which is 'seen' by the radar – as highlighted in Figure 13.19. Signatures were recorded at 7.5 GHz carrier and 300 m baseline. Figure 13.18(a) corresponds to a case of far-field diffraction from a small inflatable boat of size 2.9 m × 1 m (length and height above the surface), $S = 60$ m, where the DPS is essentially bigger than the shadow signal; signature of (b) corresponds to the case where the DPS is somehow comparable with the shadow signal and defines the boundary of the Fresnel to Fraunhofer diffraction zone for medium-size sailing yacht of 5 m length, $S = 160$ m; finally, (c) corresponds to the case of geometrical shadow where a large target essentially fully blocks the DPS, this is Fresnel diffraction from a large motor boat 15 m, $S = 630$ m. The most interesting case from a radar perspective

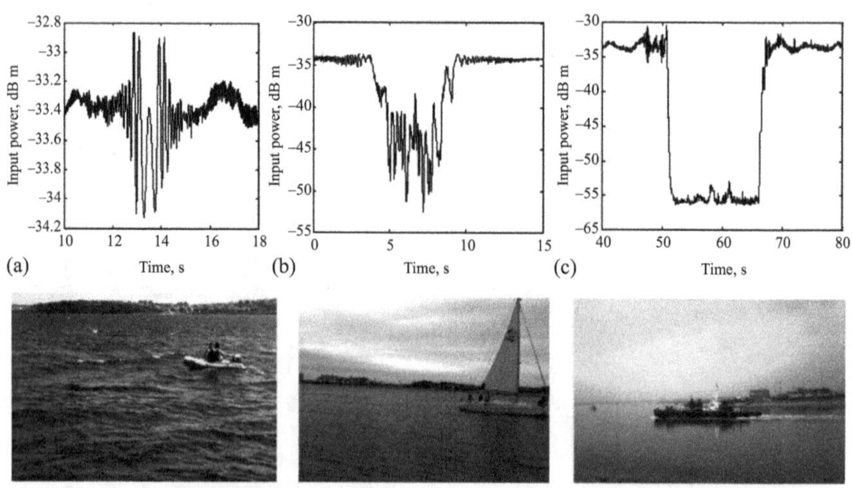

Figure 13.18 Recorded FS signature from targets crossing the middle of an FSR baseline of length d = 300 m [25]. (a) Small inflatable boat; (b) medium size yacht; (c) large motor boat

corresponds to (a) which may be considered as small target detection. In all signatures, the typical Fresnel diffraction behaviour (positive and negative contribution of phases of interfered signals) is visible, for cases (b) and (c) this is at the leading and trailing edges of the signature. Obviously, all three signals are liable to detection and, moreover, (b) and (c) are not difficult targets because their scattered signals are comparable with the LS. However, only the first signal is suitable for the extraction of target motion parameters as its waveform is fully defined by diffraction, describing the specific position and speed of the target passing through the constructive (in phase) and destructive (out of phase) zones over the path.

In Figure 13.20(a), the signature of an A321 aircraft of 44.5 m length crossing the 22 km baseline at 6 km from the receiver is shown for a passive version of FSR where a DVB-T signal of 650 MHz has been used [48]. In this case, S = 1,076 m; Figure 13.20(b) gives the signature of a 4×4 vehicle at 173 MHz and 100 m baseline, $S = 3.2$ m and (c), a human measured at a frequency ~64 MHz at 60 m baseline, $S = 0.2$ m. Thus, all the signatures obtained in the far field have clearly pronounced Doppler double-side chirp carriers with a unique envelope defined by the target silhouette.

Target classification is therefore possible by analysis of the signal envelope. Two approaches have been investigated so far. First is based on parametric analysis of the frequency domain waveform as reported in [49], where principal component analysis was used for automatic target classification. The second approach has been used for target shadow profile re-construction and requires knowledge of the complex envelope which can be extracted if two quadrature channels are used as discussed in [7,23,50].

13.3.3 Optimal signal processing in FSR

One of the objectives of optimal signal processing in FSR (as indeed in any radar) is to maximise SNR which is achieved by matched filtering. For some particular waveforms, signal waveform compression at the output of the matched filter is observed, where the auto-correlation function has a narrow main lobe with intensity

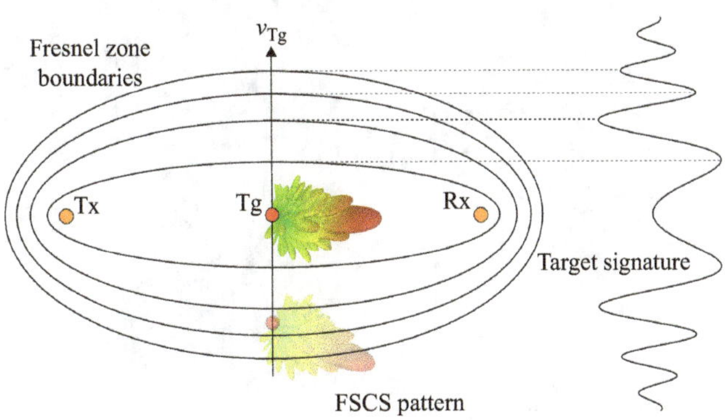

Figure 13.19 Target signature as composition of FSCS of the target forming the signal envelope and Doppler signature formed from passage of target through Fresnel zones

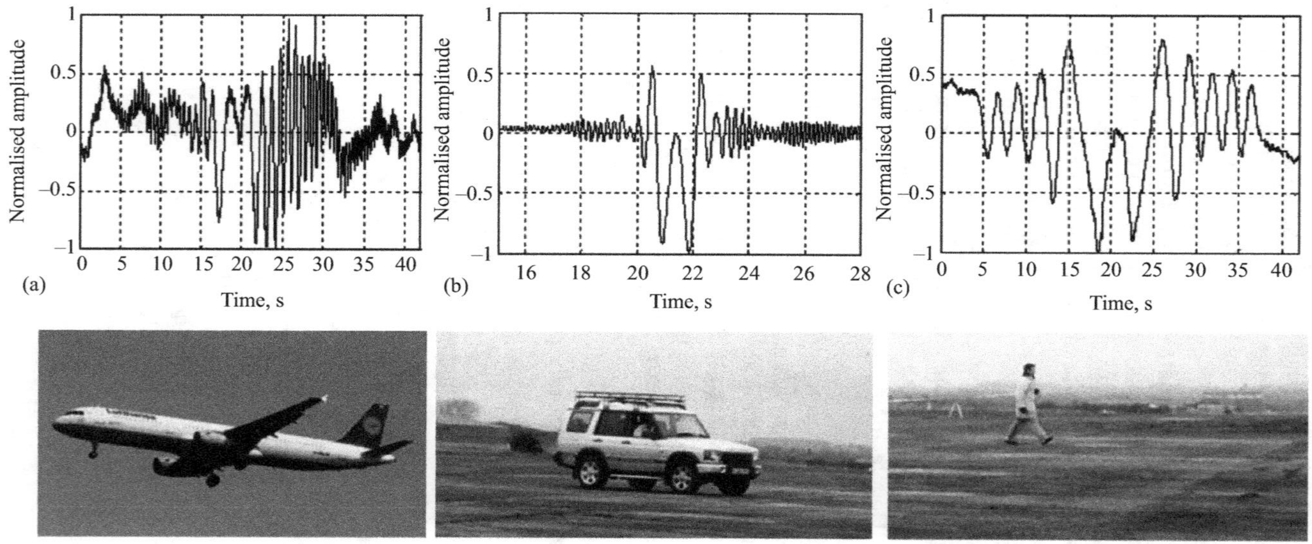

Figure 13.20 *Recorded FSR Doppler signatures. (a) A321 aircraft using 650 MHz DVB-T signal (passive FSR), speed 400 km/h; (b) 4×4 vehicle at 173 MHz, middle crossing of 100 m baseline, speed 32 km/h (c) human being at 64 MHz, middle crossing of 60 m baseline, 6 km/h*

much greater than that of the sidelobes. As shown before (Section 13.3.2.1), in FSR, the received Doppler waveform is a double-sided chirp signal which after matched filtering will demonstrate such a compression [20].

The SNR improvement achieved by matched filtering can be characterised by the signal processing gain G_{SP}. Considering the fundamental case of the target crossing, the baseline at the midpoint, its visibility time will be given by, $T_{max} = (d/v_{tg})\tan\alpha$, where α is the maximal diffraction (excursion) angle of the target from the receiver, and the maximal Doppler deviation during this time will be:

$$f_d^{max} = 2f_c \frac{v_{tg}}{c} \sin\alpha \tag{13.40}$$

Thus, the signal processing gain/time bandwidth product can then be estimated as:

$$G_{sp} = T_{max} \cdot 2f_d^{max} = 4d\frac{f_c}{c}\tan(\alpha)\cdot\sin(\alpha) = 4d\frac{1}{\lambda}\tan(\alpha)\cdot\sin(\alpha) \tag{13.41}$$

where f_c is the carrier frequency, v_{Tg} is the target speed normal to the baseline, c is the speed of light. Using the -3 dB level of the FS main lobe as the reference, this angle can be approximated according to (13.5) as $\alpha \approx (\lambda/2l_{eff})$ giving,

$$G_{sp} = 4d\frac{1}{\lambda}\tan\left(\frac{\lambda}{2l_{eff}}\right)\cdot\sin\left(\frac{\lambda}{2l_{eff}}\right) \tag{13.42}$$

where l_{eff} is the effective target length. It is important to notice that the gain does not depend on v_{tg} due to the inverse dependence of the Doppler variation and observation time with regard to the target speed. It can also be seen that for targets of larger length, for a given frequency, the processing gain is reduced due to the effect of the narrowing FS main lobe width, which works to reduce both the maximal Doppler and the integration time under consideration. Indeed for small angles α, (13.42) can be approximated further by:

$$G_{sp} \approx d\frac{\lambda}{{l_{eff}}^2} \tag{13.43}$$

In summary, the processing gain in FSR offers advantages in the detection of the smaller targets (in length) where it is potentially most required, and less so in the case of the larger length targets which will assumedly have larger maximal FSCS, thus the need for as high a processing gain is in any case reduced. Actual estimations of processing gain for different targets, carrier frequencies and baseline lengths are presented in [41].

13.3.3.1 Target motion parameter estimation

Due to the lack of range resolution in FSR (particularly in the case of continuous wave (CW) signals), target motion/trajectory parameters have to be inferred via other means. This is the other very important output of the matched filter, the ability to extract the target motion parameters as indeed these are the inputs to the matched filtering process and design of matched waveforms. In accordance with the optimal detection theory [51 (Ch.7)], the maximal SNR in the presence of additive white Gaussian noise can be obtained by correlation of the received signal $S_r(t)$ with a reference function $S_{\text{ref}}(t)$, where the reference function is an exact but conjugated copy of the actual target signature for a given target type and trajectory, as given by:

$$S_{\text{opt}}(\tau) = \int_{-T/2}^{T/2} S_r(t) S_{\text{ref}}(t-\tau)dt, \quad S_{\text{ref}}(t) = S_r^*(t) \tag{13.44}$$

where T is the integration time duration equal to the target visibility time as in Section 13.2.2.1. Thus, the optimal filtering process in FSR represents the problem of sequential correlation of the received signal with a set of pre-defined reference waveforms corresponding to the expected target types and motions. The task is to define a set of reference functions for the correlation. As shown in Section 13.3.2.2, the target signature can be described by (13.39), and this therefore is the generalised form of the reference waveform for the matched filtering, $S_{\text{ref}}(t)$. The received signal in FSR depends on the target's speed, trajectory and FSCS which are all unknown *a priori* and in turn define the parameters $R_t(t)$, $R_r(t)$ and $A_{\text{tg}}(t)$ in the equation. Strictly speaking, all these parameters should be estimated when producing the reference waveforms and for a given range of expected velocities and trajectories both $R_t(t)$ and $R_r(t)$, which define the phase signature, can be modelled effectively. However, there is difficulty in describing the amplitude/envelope parameter $A_{\text{tg}}(t)$, which refer to a particular targets FSCS pattern (with small contribution from propagation factors). Thus, reference waveforms would need to be created for not only each set of expected motion parameters, but all possible expected target shapes and sizes as well.

It turns out however that complete knowledge of the signal envelope is not necessary for effective signal compression. The major correlation properties between fast sign alternating functions are mainly specified by their sign functions rather than their amplitude variation; thus, the more slowly varying envelope function $A_{\text{tg}}(t)$ is actually equivalent to a window function which only serves to weight the compression output. This effect has been highlighted by example in [26] and is reproduced in Figure 13.21.

Figure 13.21(a) shows an FSR chirp signal with rectangular, actual FSCS and Kaiser (shape factor 2.5) envelopes. Figure 13.21(b) gives the normalised cross-correlations of the FSCS-windowed chirp with itself (auto-correlation) and with the other two signals. It is noted that even when cross-correlating with the rectangular chirp (no envelope variation), the correlation coefficient only reduces to 0.88. On the other hand, Figure 13.21(c) shows two chirps which differ in instantaneous frequency by 10% and (d) shows the dramatic effect this phase difference has on the correlation coefficient. Thus, the inaccuracy in phase has a much more detrimental effect on the correlation coefficient compared to that of the envelope.

Thus, the reference waveforms for matched filtering, $S_{i,j,k}$ are defined to replicate the phase signature accurately to cover a desired range of velocities and trajectories where i, j are indices of velocity components $(v_{x,i}, v_{y,j})\big|_{i=1..Nx, j=1..Ny}$ and k is the index of baseline crossing points $(y_{c,k})|_{k=1..K}$. Analysis of the selection of velocity increments can be found in [41]. The approximation of the envelope parameter leads to the

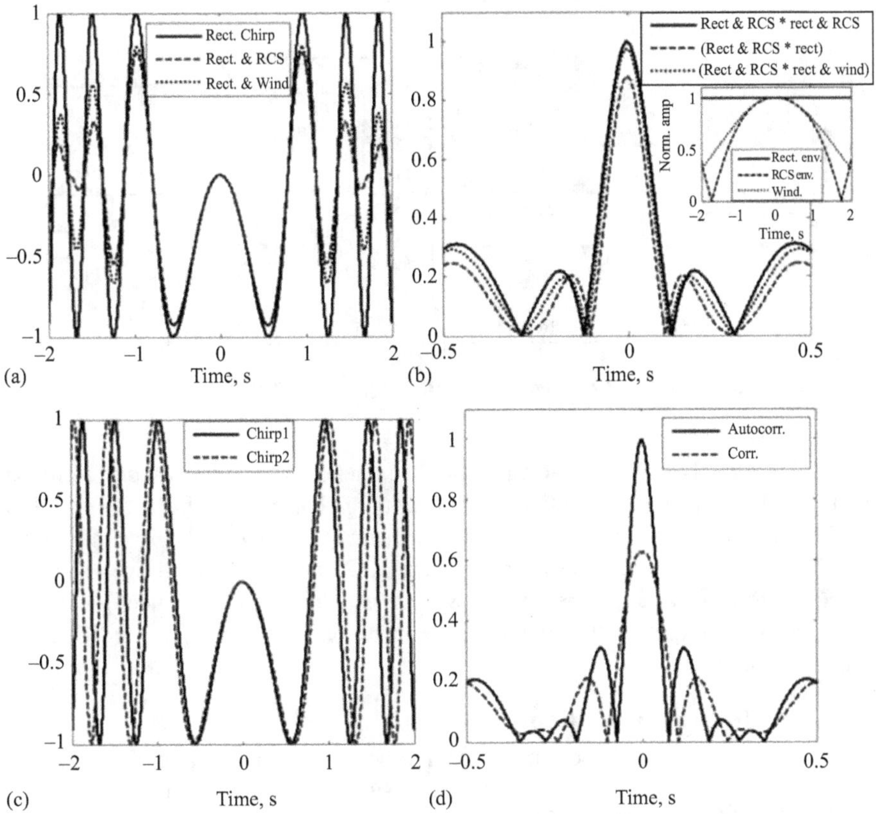

Figure 13.21 Effect of envelope and phase on waveform correlation outputs [25]. (a) shows chirp signals with rectangular, FSCS and Kaiser windowing, (b) gives the normalised correlation outputs of each chirp with the FSCS-windowed chirp. Two chirps differing in frequency by 10% are shown in (c) with the corresponding normalised correlation and cross-correlation outputs in (d)

re-phrasing of this processing to being referred to as quasi-optimal rather than optimal due to the slight processing loss related to the amplitude mis-matches. Figure 13.22 shows the simplified block diagram of such correlation processing, where subscript m indicates the 'matched' signal/parameters and previously undefined HPF and BPF represent high and band pass filters respectively.

The matching waveform delivers a global maximum of all of the correlations, and from this the trajectory and speed are found. Information regarding the shape/size of the target can then be found using profile re-construction algorithms [24]. To provide an example of the output of such processing, Figure 13.23(a–c) show the results of processing of recorded maritime signatures for the same target, a 2.9 m long inflatable boat, but with varying trajectories and baselines [25].

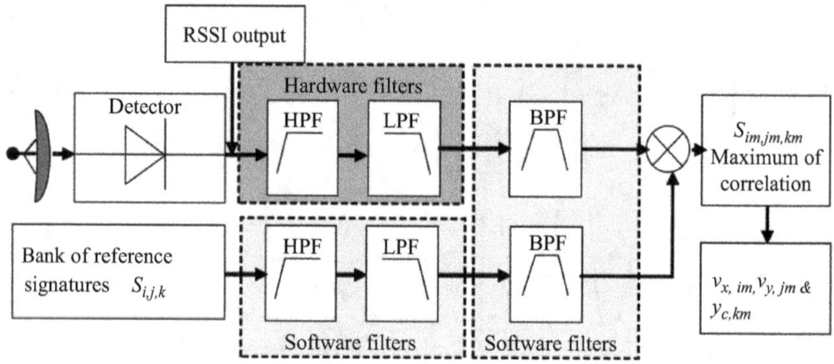

Figure 13.22 Flow chart of matching filtering procedure [25]

Table 13.5 shows the extracted motion parameters (parameters used to form the best matching reference function), along with global positioning system (GPS) tracked values. Crossing points refer to the crossing distance from the baseline centre.

It can be seen, when considering Figure 13.23 and Table 13.5, even in cases when the true signal envelope appears very dissimilar to the reference, good estimations of the target motion parameters can be obtained purely by the Doppler phase matching.

13.3.3.2 Convoy resolution

Another advantage of the optimal signal processing is the improved ability to resolve targets in convoy. The compressed signal has a duration of ΔT that is approximated by the inverse of the bandwidth of the original received signal. So in FSR, relating to the maximal expected Doppler, $f_d^{\max}$, as in (13.40),

$$\Delta T \approx \frac{1}{2f_d^{\max}} \approx \frac{\lambda}{4v_{\text{tg}} \sin\alpha} \tag{13.45}$$

where α is defined in (13.42) and can be approximated for the -3 dB FSCS main lobe width to give:

$$\Delta T \approx \frac{\lambda}{4v_{\text{tg}} \sin(\lambda/2l_{\text{eff}})} \tag{13.46}$$

and could be further simplified in the case of electrically large targets (small main lobe widths). In addition to effect of bandwidth on compressed width, the processing gain itself (and thus integration time) still needs to be large enough to reduce the compression sidelobes to satisfactory levels.

As an example of compression, results are shown in Figure 13.24 for measured signals at 64 and 434 MHz of two humans walking at 6 km/hr separated by 5 m with a baseline of 100 m. The increase in carrier frequency and thus resolution improvement facilitates the ability to separate the two targets; as seen in the 434 MHz results compared to the 64 MHz case.

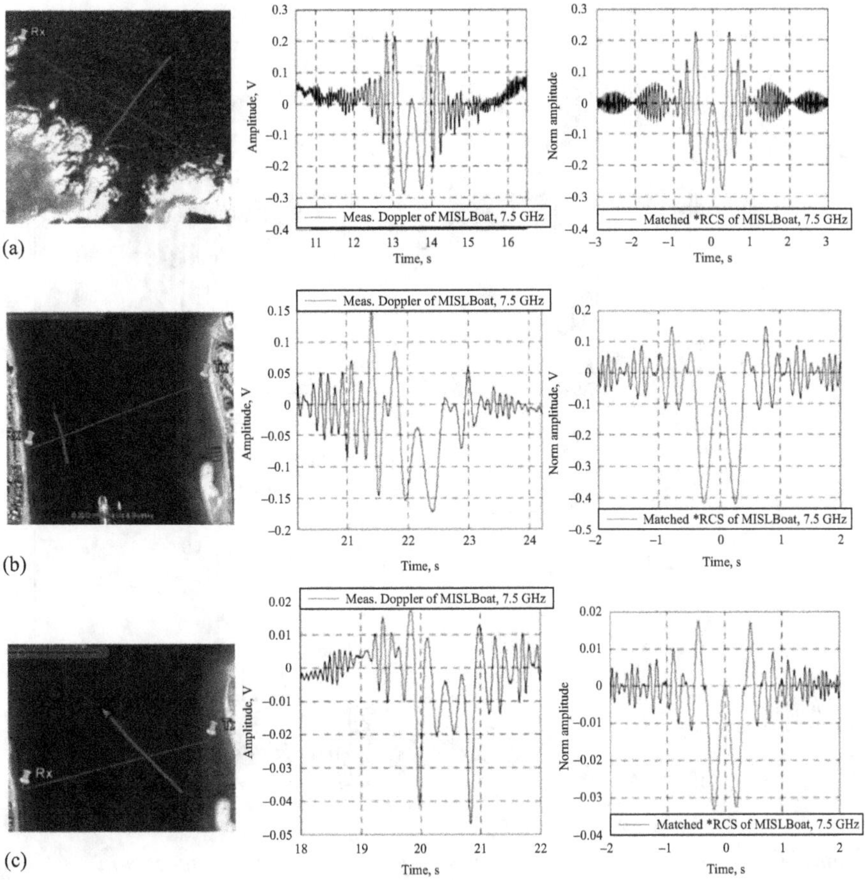

Figure 13.23 Example of measured maritime target signatures and matched waveforms from correlation processing [25]. Left panels shows target trajectory, middle panels are measured signatures, right panels are matched waveforms

Table 13.5 Extracted motion parameters for example target signatures shown in Figure 13.23 [25]

Figure 13.23	Baseline (m)		Speed (km/h)	Crossing angle (°)	Crossing point (m)
(a)	350	GPS	18.5–19.1	~90	12
		Extracted	19.08	90.0	20.0
(b)	285	GPS	12.2–13.0	~90	85.5
		Extracted	13.	90.0	80.0
(c)	262	GPS	16.7–19.1	~60	40
		Extracted	17.64	60.0	60.0

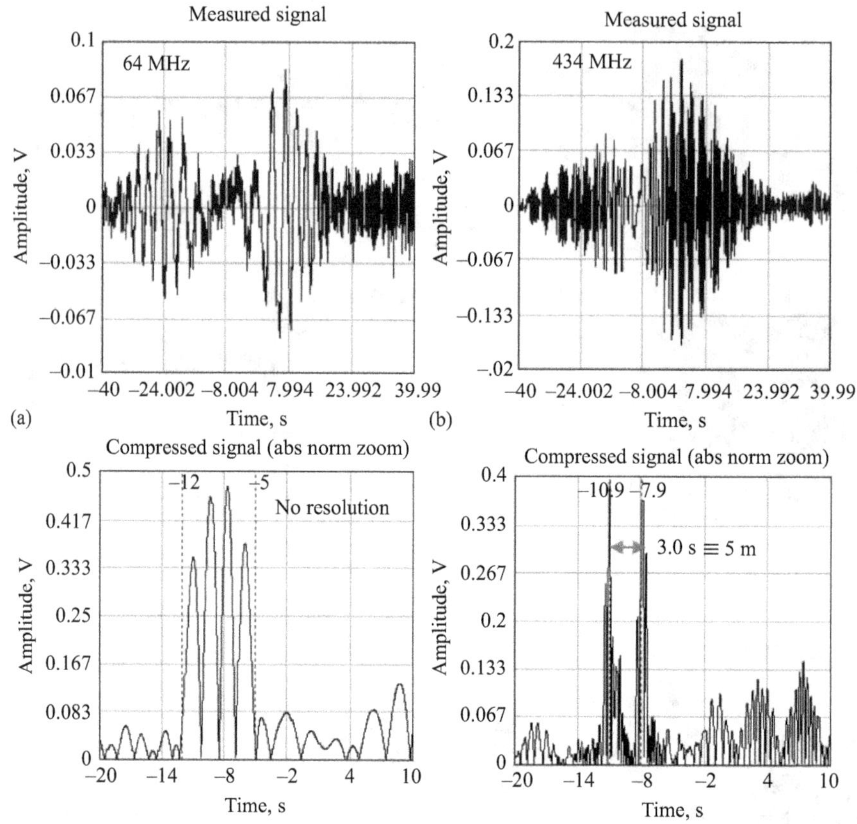

Figure 13.24 Experimentally measured signals of two human targets in convoy and their corresponding compressed signals at two frequencies: (a) 64 MHz and (b) 434 MHz

13.4 Clutter in FSR

Clutter in FSR is related to a dynamic change of the underlying surface and the surrounding environment, which in the case of ground-based FSR is mainly associated with vegetation and for maritime FSR it is the dynamic sea surface. Unlike the monostatic case, where directional antennas and the presence of range resolution lead to clutter being collected from only a limited area, or even in the case of FSR intended for airborne target detection, surface-based target detection in FSR is to be performed in the background of strong Doppler modulated clutter. This clutter is collected from a large volume around and between the transmitter and receiver, where spatially distributed clutter sources will cause backscatter, bistatic and FS signal interference. The main clutter-related problems are associated with: (i) target detection if the spectrum of Doppler modulated clutter overlaps with the target return spectrum; (ii) general non-Gaussian behaviour where long tails of the clutter intensity distribution results in an increased false alarm rate.

Statistical and spectral characteristics of FSR clutter need to be distinguished and studied separately in three main applications of FSR: ground-based target detection, seaborne target detection and airborne target detection, the last being the least affected by clutter. We will therefore give a brief overview of spectral and statistical properties of clutter related to first two cases.

13.4.1 Vegetation clutter

In a ground sensor network, the variations in the generic received signal caused by surrounding wind-blown vegetation may be attributed to one of following mechanisms: multi-path interference, de-polarisation, signal attenuation due to absorption and scattering [52]. As a result, at the reception side, the dominant component of the received signal is accompanied by the time-variant component, which introduces the fading of the received signal. Doppler modulation (magnitude and waveform) caused by this component depends on the wind strength.

In such a context, FSR clutter may be investigated using the statistical approaches used for characterisation of RF channels in communications [53], where wind strength is introduced as an additional parameter. In communication systems theory, fading of the RF channel over forested paths are widely reported and the interested reader can find as such in, for instance, [54–56].

A long-term experimental study has been undertaken in the ultra-high/very high frequency (UHF/VHF) bands, at different sites with terrain profiles varying from a concrete runway to dense woodland during different seasons as reported in [53,57]. The PSD of clutter signals for each operational frequency channel, time variation of average clutter power, dependence of clutter power on baseline range, frequency, wind speed and direction and probability density function (PDF) of clutter amplitudes have been estimated. A few peculiarities in comparison to monostatic clutter have been found in relation to the PSD and to statistical distribution:

Power and power spectrum. The power spectrum width is limited by 1–2 Hz invariantly on carrier frequency; PSD decreases by 40 dB per decade of Doppler frequency, while the total power magnitude increases with the carrier frequency by fourth power as in Figure 13.25. By comparison with wind records, it has been concluded that the wind strength determines the clutter power variations which are non-stationary in the general case.

Statistical properties. Wind strength affects the PDF of the clutter intensity amplitude: for low wind, it is close to Rayleigh, or Weibull distribution with the shape factor of 2, and therefore, the Weibull distribution have been used as a reference. The higher the wind power and the carrier frequency, the smaller is the shape factor. In addition, PDF's of recorded clutter deviate more from the Weibull fit, demonstrating an increase of the 'tail' in the distribution. This indicates potentially worse radar performance in terms of probability of false alarms.

Clutter modelling. A semi-deterministic approach to model clutter has been suggested in [27,57], where clutter has been presented as a sum of scattered signals

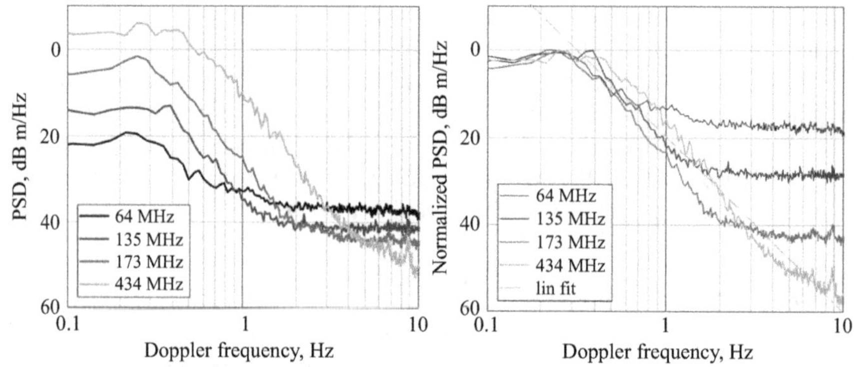

Figure 13.25 Power spectrum density of vegetation clutter in FSR measured at different frequencies (left – absolute values, right – normalised to maximum) – © [2010] IEEE. Reprinted, with permission, from [57]

from elementary oscillators associated with leaves, branches and the trunk of a tree. The statistical behaviour of the oscillators was imitated by randomly distributed initial phases, their location points around a single centre and their amplitudes distributed within a range appropriate for the tree parts. It has been shown that spectral characteristics of the modelled clutter mentioned above are in excellent agreement with results of measurement. Moreover, the statistical distribution has also been in agreement with that of experimentally observed signal and a simulation model of the clutter signal has been developed to be used in a synthetic simulation environment to test FSR radar performance.

13.4.2 Sea clutter

FSR for maritime application, proposed in [5,58] and investigated in [26,39,59,60], is configured as a fence consisting of nodes separated by distances of the order of a kilometre. To form such a fence, the node transceivers are mounted on buoys and therefore low grazing angle with respect to the sea surface is expected. The effect of the sea is 2-fold, (i) it forms an underlying surface as in the TRP propagation model and (ii) complex scattering on the spatially distributed dynamic sea waves is responsible for a sea clutter which will be received from a large illuminated area between buoys.

By its very configuration, the FSR channel is similar to an RF communication channel, and therefore, sea clutter can be described in terms of fading in an RF channel over a sea surface. Forward propagation studies over the sea surface have been dedicated to characterise radio wave propagation [61–68], coastal or ship-to-ship communications [35,63,69–72] as well as radar scattering at low grazing angles in [64–68,70,73,74]. In [17], approaches developed for characterisation of fading of RF channels were applied for analysis of FSR clutter at very low grazing angle. Both spectral and statistic properties were measured and analysed, and results were compared to results of modelling. In particular, the dependence of scattering mechanisms – dominant specular reflections, diffused scattering and partial shadowing – on the sea state has been considered. In summary;

- forward-scatter radar sea clutter measured at very low grazing angles (less than 0.5°) exhibits, to a first approximation, a near constant frequency centred below 1 Hz, with a roll off of approximately 35–40 dB per decade in the Doppler frequency domain – Figure 13.26(a). It has also been shown that Rayleigh is a good fit to the measured clutter intensity distribution – Figure 13.26(b) and (c);
- the clutter spectrum and shape of distribution function are found to be independent of transmit receive/baseline range, sea state and carrier frequency, within the range of limited experimental conditions.

13.5 Air target tracking in CW FSR

This section is dedicated to target tracking in FSR via the maximum likelihood method. The main goal is to obtain the accuracy of maximum likelihood estimations of target co-ordinates. This accuracy is calculated theoretically and mathematical modelling results are also presented.

13.5.1 Target resolution in forward scatter radar

The advantages of using the FS configuration of bistatic radar when detecting low-observable targets, including targets built with stealth technology are described in the previous sections of this chapter. Technical solutions applied when developing FSR for detection and tracking of air targets are based on the principles discussed above, but require some specific clarifications.

One of the main characteristics of any radar is target resolution. For the radars with a pulse probing signal, range resolution is inversely proportional to the signal bandwidth. At the same time, in bistatic radar, to provide range resolution, we should resolve the target scattered signal from the direct transmitter signal. A very particular feature of FS configuration is strong stretching of the coverage along the baseline [15 (p.400), 75]. Thus, the time delay of the scattered signal with respect to direct signal is very small. Simple geometrical calculations show that to provide range resolution of about 100 m with a baseline length of about 40 km, we should use a pulse-probing signal with a band width of more than 100 MHz [15 (p.403)].

Realisation of the transmit and receive positions providing formation and processing of a signal with such bandwidth is both a difficult and expensive solution. FSR with narrowband CW probing signal (quasi-harmonic signal) provides Doppler resolution simultaneously with the possibility of long enough (theoretically, during the whole observation time) coherent signal integration. For an integration time of 1 s, we have Doppler resolution of 1 Hz and the Doppler shifts of the signals scattered by two targets moving with the same velocity but placed at different points inside of the FSR coverage are different [15 (p.403)]. Therefore,

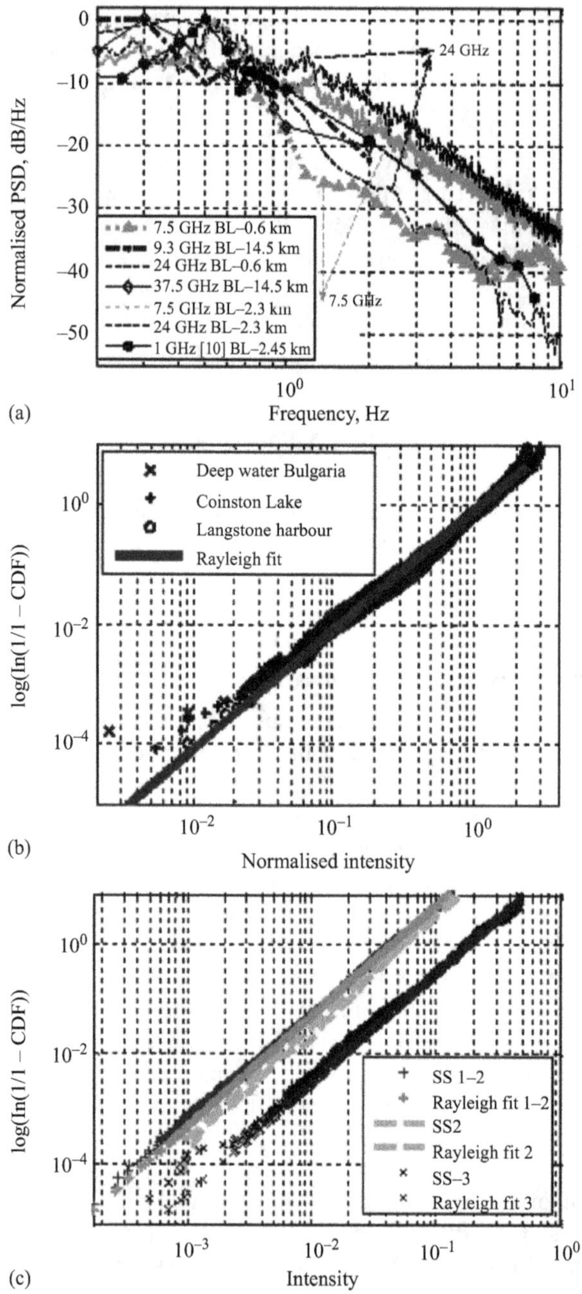

Figure 13.26 (a) Normalised PSDs of FSR sea clutter recorded at varying frequencies, ranges, sea states and test sites. (b) Cumulative distribution function (CDF) of normalised FSR sea clutter from different test sites on Weibull paper and (c) comparison of CDFs for long-term FSR sea clutter measurements in different sea states. SS stands for sea state by Douglas scale (plots collated from [17])

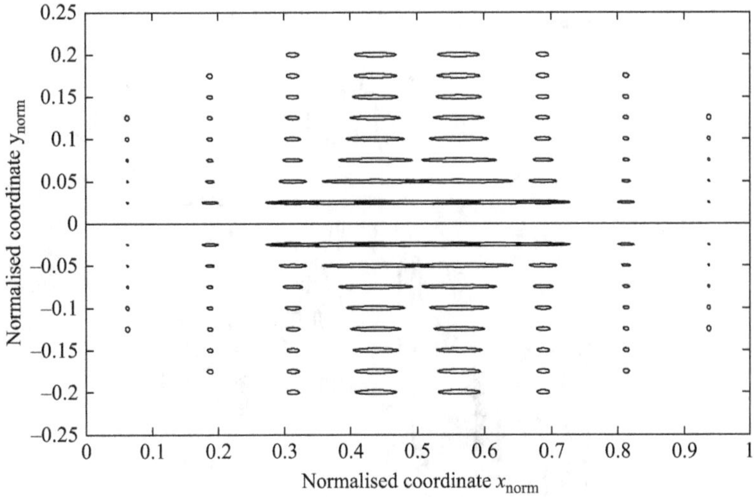

Figure 13.27 Resolution ellipses vs normalised co-ordinates

Doppler resolution in FSR during the tracking of moving targets provides spatial resolution. Simple calculations obtained with the equations concatenating the values of Doppler shift with the target Cartesian co-ordinates [15 (p.412), 76,77] show that in FSR, with the length of baseline equal to 40 km, wavelength of probing signal equal to 0.2 m and integration time of 1 s, we have spatial resolution along the baseline of about 200 m and approximately 20 m across the baseline.

In Figure 13.27, the spatial resolution of CW FSR is presented in the form of ellipses. The axes of these ellipses are equal to the resolution along the x and y co-ordinates. The captures along x and y axes are normalised to the length of the baseline and the ellipses with the smaller axes correspond to the areas where the better resolution is provided. It is seen that the best resolution corresponds to the areas close to transmit and receive positions ($x_{norm} = 0$ or 1, $y_{norm} = 0$), the worst is observed in the middle of the baseline.

Thus, we can make the conclusion that a narrowband CW-probing signal is an optimal solution for ground-based FSR aimed at tracking airborne targets, due to the low cost, simple implementation, low transmitted power and target resolution in Doppler domain equal to the appropriate spatial resolution.

13.5.2 FSCS and coverage when tracking air target

FSR should be able not only to detect targets crossing the baseline but also provide tracking of these targets. The best way to understand the possibilities of air target detection and tracking in FSR is to consider some results of FSCS calculation for a typical air target, together with the experimental results of tracking.

In Figure 13.28, an example of calculation of the azimuth plane FSCS for a helicopter is shown for two wavelengths of 0.3 m and 1 m (curves 1 and 2,

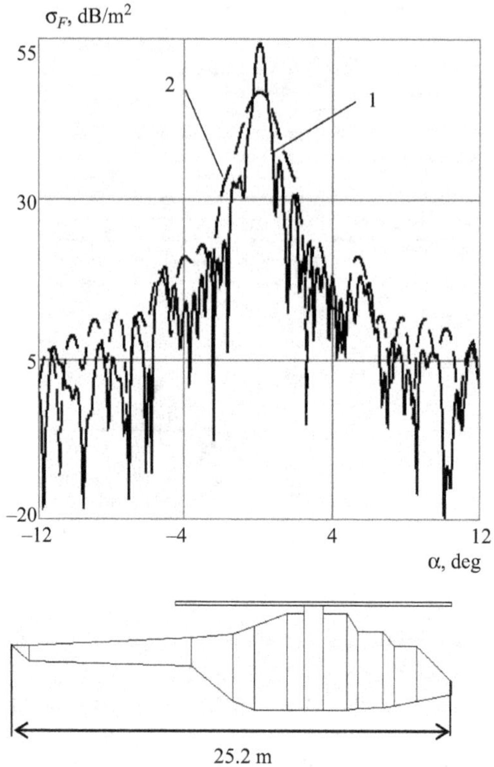

Figure 13.28 An example of helicopter azimuth plane FSCS at two different wavelengths of 0.3 m and 1 m (curves 1 and 2, respectively) – © [2007] John Wiley and Sons. Reprinted, with permission, from [15]

respectively). As it is seen, the width of the main lobe of FSCS is only about 1°. At the same time, the value of the FSCS is large (more than the monostatic RCS which is equal to about 2 m^2) up to values of target azimuth exceeding 10°. Thus, the region, where FSCS is bigger than the monostatic RCS, is much wider than the width of the main lobe.

Experimental results of detection and tracking of the same target are shown in Figure 13.29. The grey region presents FSR coverage for the helicopter calculated via the radar equation for the bistatic system. When calculating this coverage, the helicopter FSCS shown in Figure 13.28 was used. The indices from 1 to 5 are the numbers of the target tracks obtained during the experiment. It is easily seen from Figure 13.29 that target was detected and tracked for values of azimuth of more than 10°. Therefore, detection and tracking of air targets in CW FSR is provided in the region of several sidelobes of the FSCS as well as in the region of the main lobe.

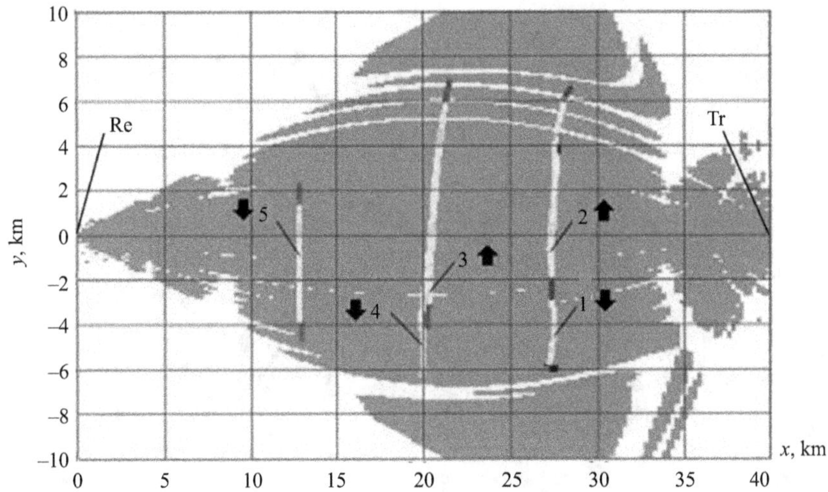

Figure 13.29 Example of target tracking inside of the coverage – © [2007] John Wiley and Sons. Reprinted, with permission, from [15]

13.5.3 Mathematical model of measuring process. Maximum likelihood estimation of trajectory parameters

The three-dimensional bistatic FSR geometry is shown in Figure 13.30. In Figure 13.30, Tr is the transmitting position; Re the receiving position; Tg is the target; AB is the target trajectory; CD is the projection of the target trajectory onto the horizontal plane; x, y, z represent the orthogonal co-ordinate system attached to the receive position; x_0 is the x co-ordinate of the baseline crossing projection; φ is the target trajectory inclination angle towards the baseline in the horizontal plane; ψ is the angle of the target trajectory inclination in the vertical plane; α is the current target azimuth; β is the current target angle of elevation; $\tilde{\beta}_B$ is the bistatic angle.

We will consider three-dimensional bistatic FSR providing estimation of three target Cartesian co-ordinates (x, y, z) and the three components of the target velocity vector (v_x, v_y, v_z). When using a narrowband CW-probing signal, the primary parameters to be measured are the angles of arrival of the target scattered signal (azimuth, α_i and elevation angle, β_i) and Doppler frequency shift $f_{D,i}$ [15 (p.417)], [77–80], where i labels the time instance corresponding to the ith measurement.

When the length of the baseline is equal to several tens of kilometres, the airborne target is usually observed during the time interval of one to several minutes. Each measurement of primary parameters is obtained after target return coherent integration and detection during a time interval which is short enough to neglect the variation of Doppler shift. General techniques to provide measurement

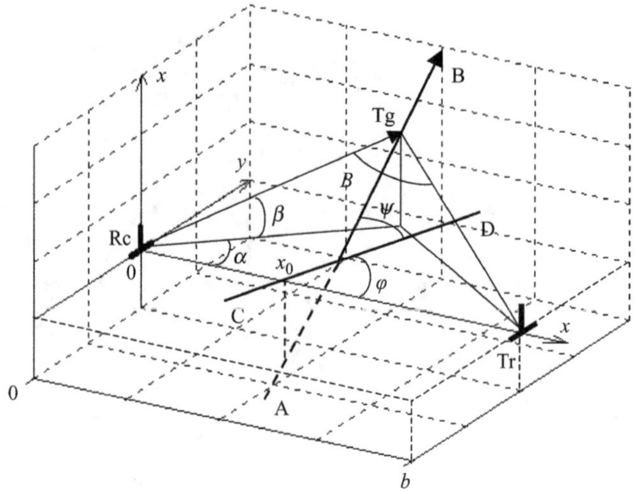

Figure 13.30 Three-dimensional bistatic forward scattering radar geometry – © [2007] John Wiley and Sons. Reprinted, with permission, from [15]

of these parameters are described in [15 (p.417), 81]. When observing typical airborne targets (aircraft, helicopter), the interval of signal integration does not exceed 1 s, the base of scattered signal, B, is close to unity [15 (p.418), 82]. Measurement of Doppler shift could be provided via standard methods of digital spectral analysis (fast Fourier Transform). Measurements of angles could be obtained via monopulse techniques when forming partial beams of the directivity pattern of the antenna in azimuth and elevation planes [15 (p.419)]. To provide angular measurements, an antenna array is used at the receiver.

The sequence of primary measurements forms a vector:

$$\mathbf{z}_n = \left[\hat{f}_{D1}, \hat{\alpha}_1, \hat{\beta}_1, \hat{f}_{D2}, \hat{\alpha}_2, \hat{\beta}_2, \ldots \hat{f}_{Dn}, \hat{\alpha}_n, \hat{\beta}_n\right]^T \tag{13.47}$$

Since the coverage of FSR is narrow enough, the probability of target manoeuvre inside this coverage is small. Therefore, a model of linear motion is the most appropriate. For the case when the trajectory model is assumed to be linear, the tracking vector is written as follows:

$$\mathbf{x}_n = \left[x_n, y_n, z_n, v_x, v_y, v_z\right]^T \tag{13.48}$$

where x_n, y_n, z_n, v_x, v_y, v_z are the values of target Cartesian co-ordinates and their derivatives at observation moment, t_n.

A non-linear system of equations concatenating the values of primary parameters with the values of trajectory parameters appears as follows [15 (p.412)]:

$$f_{Di}(\mathbf{x}_n) = -\frac{1}{\lambda}\left\{ \begin{array}{l} \dfrac{[x_n - v_x(n-i)T]v_x + [y_n - v_y(n-i)T]v_y + [z_n - v_z(n-i)T]v_z}{\sqrt{[x_n - v_x(n-i)T]^2 + [y_n - v_y(n-i)T]^2 + [z_n - v_z(n-i)T]^2}} \\ + \dfrac{[y_n - v_y(n-i)T]v_y - [b - [x_n - v_x(n-i)T]]v_x + [z_n - v_z(n-i)T]v_z}{\sqrt{[b - [x_n - v_x(n-i)T]]^2 + [y_n - v_y(n-i)T]^2 + [z_n - v_z(n-i)T]^2}} \end{array} \right\}$$

$$\alpha_i(\mathbf{x}_n) = \operatorname{arctg}\frac{y_n - v_y(n-i)T}{x_n - v_x(n-i)T}$$

$$\beta_i(\mathbf{x}_n) = \operatorname{arctg}\frac{z_n - v_z(n-i)T}{\sqrt{[x_n - v_x(n-i)T]^2 + [y_n - v_y(n-i)T]^2}} \tag{13.49}$$

A model of the measuring process is based on (13.49). Statistical dependence of the measured values of the primary parameters upon the target trajectory parameters is described by a measurement equation. It can be written in the following compact form [15 (p.421), 77]:

$$\mathbf{z} = \mathbf{h}_n(\mathbf{x}_n) + \Delta\mathbf{z}_n \tag{13.50}$$

$$\begin{aligned} \mathbf{h}_n(\mathbf{x}_n) = [&f_{D1}(\mathbf{x}_n), \alpha_1(\mathbf{x}_n), \beta_1(\mathbf{x}_n), f_{D2}(\mathbf{x}_n), \alpha_2(\mathbf{x}_n), \beta_2(\mathbf{x}_n), \ldots \\ &f_{Di}(\mathbf{x}_n), \alpha_i(\mathbf{x}_n), \beta_i(\mathbf{x}_n), \ldots, f_{Dn}(\mathbf{x}_n), \alpha_n(\mathbf{x}_n), \beta_n(\mathbf{x}_n)]^{\mathrm{T}} \end{aligned} \tag{13.51}$$

where $\mathbf{h}_n(\mathbf{x}_n)$ is a non-linear vector function with elements defined by (13.49); $\Delta\mathbf{z}_n$ is a vector of random primary measurement errors.

For further consideration, we suppose that primary measurement errors are Gaussian with zero mean and variances, $\sigma_f^2, \sigma_\alpha^2, \sigma_\beta^2$, which are considered to be the same for all instances in time. We also suppose that the measurements of the primary parameters are mutually independent at any time shift, as well as the measurements of every parameter being independent at different moments of observation.

Using the accepted Gaussian model of primary measurement errors, the maximum likelihood estimation of the tracking vector, $\hat{\vec{x}}_n$, can be found as:

$$\hat{\mathbf{x}}_n = \arg\min \varphi_n(\mathbf{x}_n) \tag{13.52}$$

where $\phi_n(\mathbf{x}_n) = [\mathbf{z}_n - \mathbf{h}(\mathbf{x}_n)]^T \mathbf{G}_n [\mathbf{z}_n - \mathbf{h}(\mathbf{x}_n)]$; $\mathbf{G}_n = [M\{\Delta\mathbf{z}_n \Delta\mathbf{z}_n^T\}]^{-1}$ is an inverse correlation matrix of the primary measurement error vector, $\Delta\vec{z}_n$.

13.5.4 Potential accuracy of trajectory parameters measurement

When using the maximum likelihood method, the potential accuracy is determined by the Fisher information matrix. For the observation process of the introduced model of the measuring process, the elements of the Fisher matrix of [79,83 (p.325)] are defined as:

$$\mathbf{J}_{lk}(\mathbf{x}_n) = \frac{1}{\sigma_f^2}\sum_{i=1}^{n}\left[\frac{\partial f_{Di}(\mathbf{x}_n)}{\partial x_l}\cdot\frac{\partial f_{Di}(\mathbf{x}_n)}{\partial x_k}\right] + \frac{1}{\sigma_\alpha^2}\sum_{i=1}^{n}\left[\frac{\partial \alpha_i(\mathbf{x}_n)}{\partial x_l}\cdot\frac{\partial \alpha_i(\mathbf{x}_n)}{\partial x_k}\right] + \frac{1}{\sigma_\beta^2}\sum_{i=1}^{n}\left[\frac{\partial \beta_i(\mathbf{x}_n)}{\partial x_l}\cdot\frac{\partial \beta_i(\mathbf{x}_n)}{\partial x_k}\right] \tag{13.53}$$

where x_l, x_k are elements of the vector $\mathbf{x}_n$ ($l, k \in [1,6]$ ($x_1 = x_n, x_2 = y_n, x_3 = z_n, x_4 = v_x, x_5 = v_y, x_6 = v_z$).

Variances $\sigma^2_{x,n}$, $\sigma^2_{y,n}$, $\sigma^2_{z,n}$, $\sigma^2_{vx,n}$, $\sigma^2_{vy,n}$, $\sigma^2_{vz,n}$ of the estimates of trajectory parameters can be calculated as diagonal elements of a matrix inverse to the Fisher matrix. In Figure 13.31, dependences of the normalised root-mean-square deviations of the estimates of co-ordinate x as functions of value of co-ordinate y are shown by curves 1 and 2. These are calculated for $\sigma_f = 1$ Hz; $\sigma_\alpha = 0.5°$ for two different target trajectories with parameters $x_{CR}/b = 0.75$; $\varphi = 45°$ (Figure 13.31(a)) and $x_{CR}/b = 0.25$;,$\varphi = 45°$ (Figure 13.31(b)). Curve 1 corresponds to an altitude of 2 km, while curve 2 is calculated for zero altitude. It was assumed that $b = 40$ km, $V = 200$ m/s, $\lambda = 1$ m, $T = 0.25$ s.

The analysis of the plots shows that practically monotonic reduction of rms deviations of errors of co-ordinate x estimation takes place while the observation time increases (at small values of the angle α they practically coincide with rms deviations of target range estimation errors), but insignificant rise near the baseline is possible. The degradation in accuracy of the estimation of co-ordinate x at the moment of crossing the baseline can be explained by the fact that the Doppler frequency shift on the baseline does not depend on x. The large errors at the initial segment of the trajectory can be explained by the lack of measurements of the

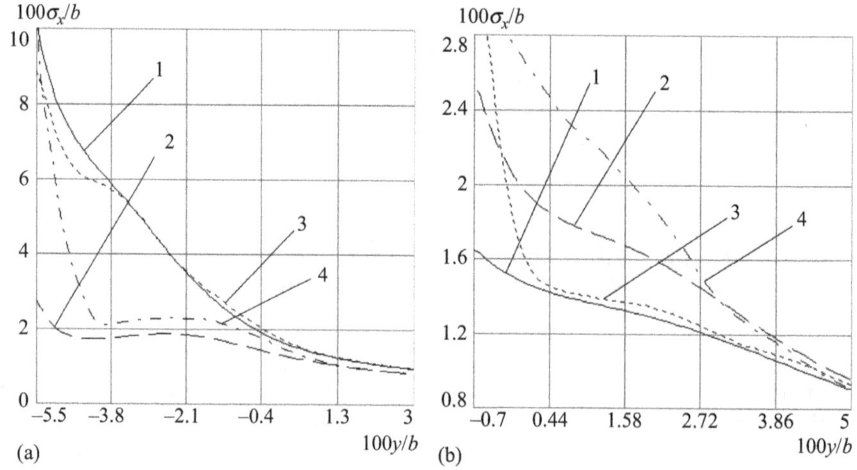

Figure 13.31 Rms errors of target tracking

primary parameters. At the moment the target leaves the coverage, the rms deviation of the error of estimation of co-ordinate x becomes sufficiently small, about 0.5%–1% of the length of the baseline under the conditions described above. From comparison of the results presented in Figure 13.31(a) and (b), we can see that the behaviour of the tracking accuracy in relation to target altitude is different when the target moves closer to the transmit position, Figure 13.31(a), and closer to receive position, Figure 13.31(b). In the first case, the accuracy is worse when the altitude is bigger, while in the second case the inverse situation takes place. There are two reasons for this: first, the rate of change of primary parameters depends on the target location between the transmit and receive sides. This is higher when the target is near to the receiver; second, when target altitude is not equal to zero, the measurements of elevation angle become more informative. This is equivalent to the increasing of the length of the primary measurement vector, $\mathbf{z}_n$, in comparison with the case of zero altitude, which corresponds to the 2-D FSR [15 (p.441)]. So, in the region close to the receiver the second factor intensifies the first one to provide a higher accuracy in 3-D FSR, whilst the inverse situation takes place in proximity to the transmitter region.

The calculation of rms deviation of the error of estimation of co-ordinate y has shown that it is insignificant in comparison with σ_x and does not have a major effect on the accuracy of target position estimation.

The accuracy of altitude estimation in 3-D FSR has also been evaluated. In Figure 13.32, the dependence of normalised rms deviations of altitude estimation

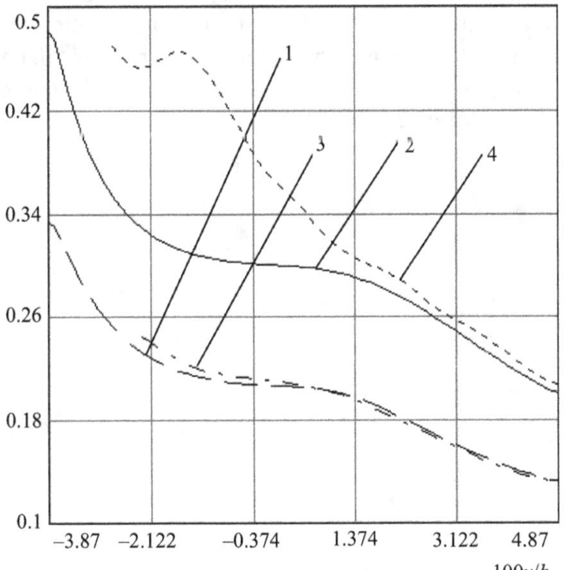

Figure 13.32 Rms deviation of target altitude estimation

errors on the value of co-ordinate y in 3-D FSR is shown. Curve 1 corresponds to the target flight altitude of $h = 1{,}000$ m, curve 2 corresponds to the altitude $h = 2{,}000$ m.

By increasing the altitude, the accuracy of the altitude evaluation falls. This can be easily explained from the viewpoint of geometry – at an equal horizontal range, the linear size of the error region of the altitude estimation increases with the increase of altitude.

13.5.5 Iterative algorithm of co-ordinate estimation

A specific feature of CW FSR is that target co-ordinates cannot be measured directly as well as calculated analytically using measurements of Doppler shift and angles. In practice, iterative algorithms are used in order to obtain the solution of the minimisation problem (13.52), giving the maximal likelihood estimation of the vector of trajectory parameters. The use of the Gauss–Newton algorithm gives [79]:

$$\hat{\mathbf{x}}_n^{i+1} = \hat{\mathbf{x}}_n^i + k_C(\mathbf{H}_n^T\mathbf{G}_n\mathbf{H}_n)^{-1}\mathbf{H}_n^T\mathbf{G}_n\left[\mathbf{z}_n - \mathbf{h}_n(\hat{\mathbf{x}}_n^i)\right] \tag{13.54}$$

Here, $\hat{\mathbf{x}}_n^i$ is the estimate of the vector of trajectory parameters for the ith iteration; $\mathbf{H}_n = \partial\mathbf{h}_n(\mathbf{x})/\partial\mathbf{x}|_{\mathbf{x}=\hat{\mathbf{x}}_n^i}$ is the matrix of derivatives; k_C is the parameter determining the rate of convergence.

In a practical application of the given iterative algorithm, the key problem is to choose an initial approximation, $\hat{\mathbf{x}}_n^0$ [79,80]. Since the minimised function, $\varphi_n(\mathbf{x}_n)$, is essentially non-linear, the application of the Gauss–Newton algorithm guarantees finding the local minimum, which does not always coincide with the absolute minimum.

Suppose there are two triplets of primary measurements, ($f_{D1}, \alpha_1, \beta_1$) and ($f_{Dn}, \alpha_n, \beta_n$), taken at the time interval $\Delta t = (n - 1)T$. Using these triplets of primary parameters, a system of six non-linear equations concatenating the values of the primary parameters and the trajectory parameters can be obtained from (13.49). Taking into account that the bistatic angle, β_B, is close to 180°, the proposed system can be presented in a linearised form:

$$\begin{cases} f_{Dn} = -\dfrac{V_y}{\lambda}\left(\alpha_n + \dfrac{y_n}{b - x_n}\right) \\ \alpha_n = \dfrac{y_n}{x_n} \\ \beta_n = \dfrac{z_n}{\sqrt{x_n^2 + y_n^2}} \\ f_{D1} = -\dfrac{V_y}{\lambda}\left(\alpha_1 + \dfrac{y_1}{b - x_1}\right) \\ \alpha_1 = \dfrac{y_1}{x_1} \\ \beta_1 = \dfrac{z_1}{\sqrt{x_1^2 + y_1^2}} \end{cases} \tag{13.55}$$

The obtained system always has a unique solution determined by the expressions [15 (p.423), 79,80]:

$$\begin{cases} x_n = \left(\lambda f_{Dn} f_{D1}\Delta t + b\alpha_1^2 f_{Dn} - b\alpha_n\alpha_1 f_{D1}\right)/u \\ y_n = \alpha_n x_n \\ z_n = \sqrt{x_n{}^2 + y_n{}^2}\beta_n \\ V_x = b\left(f_{Dn}\alpha_1^2 + f_{D1}\alpha_n^2 - \alpha_n\alpha_1(f_{D1} + f_{Dn})\right)/(u\Delta t) \\ V_y = -\lambda f_{Dn} f_{D1}(\alpha_1 - \alpha_n)/u \end{cases} \tag{13.56}$$

where $u = \lambda f_{Dn} f_{D1}\Delta t/b + f_{Dn}\alpha_1^2 - f_{D1}\alpha_n^2$.

Substitution of the estimates of primary parameters (f_{D1}, α_1, β_1) and (f_{Dn}, α_n, β_n) into (13.56) allows the finding of an initial approximation for the trajectory parameters vector, $\hat{\mathbf{x}}_n^0$. Better results can be achieved if an initial approximation is found using all primary measurements obtained [15 (p.423), 79,80]. Using the least squares method, from the available samples, we can obtain a polynomial approximation of the Doppler shift, f_D, azimuth, α and elevation angle, β. A polynomial of arbitrary degree can be used, but usually it is a linear approximation which yields:

$$\begin{aligned} \tilde{f}_D(t) &= a_f t + b_f \\ \tilde{\alpha}(t) &= a_\alpha t + b_\alpha \\ \tilde{\beta}(t) &= a_\beta t + b_\beta \end{aligned} \tag{13.57}$$

where:

$$\begin{aligned} a_{f(\alpha,\beta)} &= \frac{\sum_{i=1}^{N}(t_i - \hat{m}_t)(\hat{f}_i - \hat{m}_{f(\alpha,\beta)})}{\sum_{i=1}^{N}(t_i - \hat{m}_t)^2} \\ b_{f(\alpha,\beta)} &= \hat{m}_{f(\alpha,\beta)} - a_{f(\alpha,\beta)}\hat{m}_t \end{aligned} \tag{13.58}$$

Substituting f_{D1}, α_1, β_1, f_{Dn}, α_n, β_n in (13.56) for estimates $\tilde{f}_{D1}$, $\tilde{\alpha}_1$, $\tilde{\beta}_1$, $\tilde{f}_{DN}$, $\tilde{\alpha}_N$, $\tilde{\beta}_N$, we obtain the sought after initial approximation.

Using the iterative algorithm (13.54) and initial approximation (13.56), the estimations of the target co-ordinate measurement accuracy were obtained by the mathematical modelling. The statistics of primary measurement errors were accepted as being Gaussian and averaging was made on 1,000 independent realisations. System parameters and trajectories were similar to those that have been described in the calculation of potential accuracy. In Figures 13.31 and 13.32,

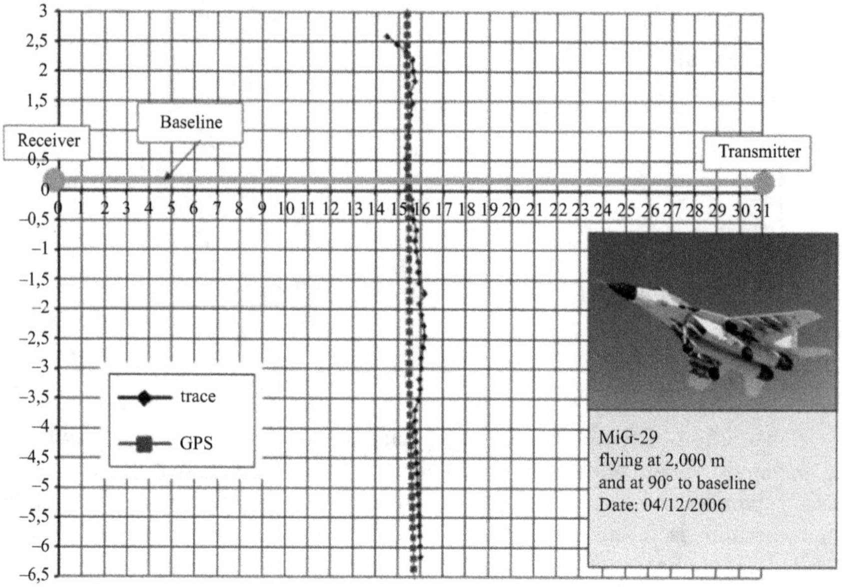

Figure 13.33 Results of tracking of fighter MIG-29

dotted and dash-dotted lines present the results of the modelling. Curves labelled 3 correspond to the target altitude of 1,000 m, curves labelled 4 correspond to an altitude of 2,000 m. It is seen from the comparison of the theoretical curves and curves obtained during modelling, that the described iterative algorithm provides close to maximum potential target tracking accuracy. Parametric ambiguity with respect to the correlation matrix of errors of primary measurements can be eliminated during tracking.

13.5.6 Experimental tracking results

Experiments with the tracking of air targets were carried out on a full-scale basis. In experimental FSR, air target trajectories were estimated using the iterative algorithm considered earlier, on the basis of the maximum likelihood method. A first approximation of the trajectory parameter vector was obtained from the first six primary measurements. The primary measurement interval T was set to be 1 s. The Doppler frequency shift and azimuth measurement variances were estimated during target tracking. The true trajectories were obtained using the data provided by the onboard GPS equipment.

Figures 13.33 and 13.34 show the results of a target-tracking experiment. Thus, the experiment validated the suitability of the maximum likelihood algorithm for establishing trajectory parameters with sufficient accuracy in real time.

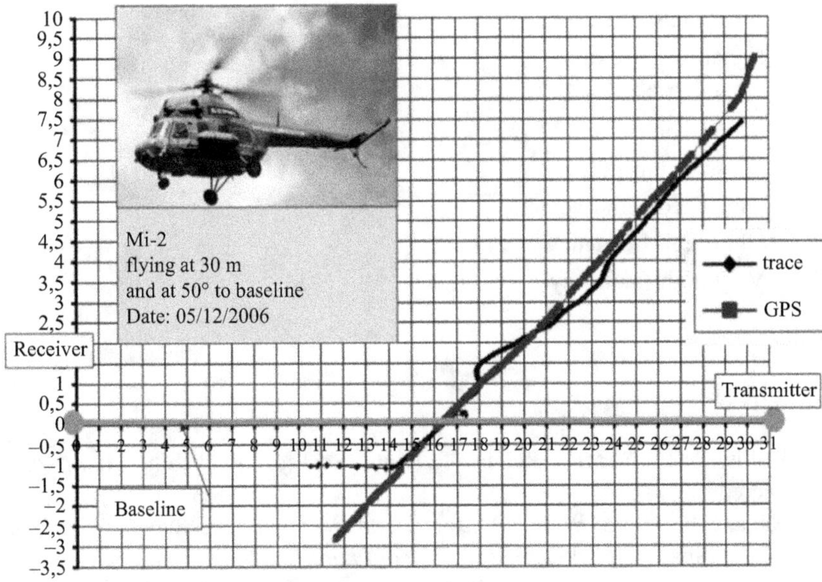

Figure 13.34 Results of tracking of helicopter Mi-2

List of abbreviations

BPF band pass filter
CW continuous wave
DPS direct path signal
EM electromagnetic
FS forward scatter
FSCS forward scatter cross-section
FSR forward scatter radar
HPF high pass filter
LPF low pass filter
LS leakage signal
PDF probability density function
PO physical optics
PSD power spectrum (spectral) density
PTD physical theory of diffraction
RCS radar cross-section
RSSI received signal strength indicator
Rx receive/receiver

SLD square law detector
SNR signal-to-noise ratio
SS sea state
TRP two ray path
Tx transmit/transmitter
UAV unmanned aerial vehicle
UHF Ultra-high frequency
VHF Very high frequency

References

[1] Willis, N.J.: *Bistatic Radar* (Raleigh, SciTech Publishing, Inc., 2005, 2nd edn.).

[2] Naka, F.R., Ward, W.W.: 'Distant Early Warning Line Radars: The Quest for Automatic Signal Detection' *Lincoln Lab. J.*, 2000, 12, (2), pp. 181–204.

[3] Glaser, J.: 'Forward Scatter Radar for Future Systems' *WSTIAC Q.*, 2011, 10, (3), pp. 3–8.

[4] Cherniakov, M., Gashinova, M., Cheng, H., Antoniou, M., Sizov, V., Daniel, L.Y.: 'Ultra Wideband Forward Scattering Radar: Concept and Prospective', 'Radar Syst. 2007 IET Int. Conf.', (2007), pp. 1–5.

[5] Daniel, L.Y., Hoare, E.G., Gashinova, M., Svintsov, A., Cherniakov, M., Sizov, V.: 'Ultra-Wideband Forward Scatter Radar Fence for Maritime Surveillance – Initial Experimental Results', in 'Radar Conf. 2010 IEEE' (2010), pp. 526–531.

[6] Cherniakov, M., Abdullah, R.S.A.R., Jancovic, P., Salous, M., Chapursky, V.: 'Automatic Ground Target Classification using Forward Scattering Radar' *IEE Proc. Radar, Sonar Navig.*, 2006, 153, (5), pp. 427–437.

[7] Chapurskiy, V.V., Sablin, V.N.: 'SISAR: Shadow Inverse Synthetic Aperture Radiolocation', in 'Record of the IEEE 2000 International Radar Conference' (IEEE, 2000), pp. 322–328.

[8] Gashinova, M., Daniel, L., Hoare, E., Kabakchiev, K., Cherniakov, M., Sizov, V.: 'Forward Scatter Radar Mode for Passive Coherent Location Systems', in IEEE Int. Radar Conf. 2013, (1), pp. 235–239.

[9] Sizov, V., Gashinova, M., Zakaria, N., Cherniakov, M.: 'VHF Communication Channel Characterisations Over Complex Wooded Propagation Paths with Applications to Ground Wireless Sensor Networks' *IET Microw., Antennas Propag.*, 2013, 7, (3), pp. 166–174.

[10] Zeng, T., Hu, C., Cherniakov, M., Zhuo, C., Mao, C.: 'Joint Parameter Estimation and Cramer-Rao Bound Analysis in Ground-based Forward Scatter Radar' *EURASIP J. Adv. Signal Process.*, 2012, (1), p. 80.

[11] Gashinova, M., Daniel, L., Hoare, E., Sizov, V., Kabakchiev, K., Cherniakov, M.: 'Signal Characterisation and Processing in the Forward

Scatter Mode of Bistatic Passive Coherent Location Systems' *EURASIP J. Adv. Signal Process.*, 2013, (1), p. 1.

[12] Hristov, S., Daniel, L., Hoare, E., Cherniakov, M., Gashinova, M.: 'Target Shadow Profile Reconstruction in Ground-based Forward Scatter Radar', in 2015 IEEE Radar Conference, pp. 0846–0851.

[13] Ufimtsev, P.Y.: *Fundamentals of the Physical Theory of Diffraction* (Hoboken, John Wiley & Sons, Inc., 2007, 1st edn.).

[14] Pathak, P.H.: 'Techniques for High-Frequency Problems', in Lo, Y.T., Lee, S.W. (Eds.): *Antenna Handbook: Theory, Applications and Design* (New York, Springer Science+Business Media, 1988, 1st edn.), pp. 195–311.

[15] Cherniakov, M.: *Bistatic Radars: Principles and Practice* (Chichester, John Wiley & Sons Ltd, 2007, 1st edn.).

[16] Ufimtsev, P.Y.: 'New Insight into the Classical Macdonald Physical Optics Approximation' *IEEE Antennas Propag. Mag.*, 2008, 50, (3), pp. 11–20.

[17] Gashinova, M., Kabakchiev, K., Daniel, L., Hoare, E., Sizov, V., Cherniakov, M.: 'Measured Forward-scatter Sea Clutter at Near-zero Grazing Angle: Analysis of Spectral and Statistical Properties' *IET Radar, Sonar Navig.*, 2014, 8, (2), pp. 132–141.

[18] Hu, C., Sizov, V., Antoniou, M., Gashinova, M., Cherniakov, M.: 'Optimal Signal Processing in Ground-Based Forward Scatter Micro Radars' *IEEE Trans. Aerosp. Electron. Syst.*, 2012, 48, (4), pp. 3006–3026.

[19] Chernyak, V.S.: *Fundamentals of Multisite Radar Systems* (Amsterdam, Gordon and Breach Science Publishers, 1998, 1st edn.).

[20] Gashinova, M., Daniel, L., Kabakchiev, K., Sizov, V., Hoare, E., Cherniakov, M.: 'Phenomenology of Signals in FSR for Surface Targets Detection' *Radar Syst., IET Int. Conf.*, 2012, pp. 1–6. *(Radar 2012)*.

[21] Mie, G.: 'Beiträge zur Optik trüber Medien, speziell kolloidaler Metallösungen' *Ann. Phys.*, 1908, 330, (3), pp. 377–445.

[22] Balanis, C.A.: *Antenna Theory – Analysis and Design* (New York, John Wiley & Sons, Inc., 1997, 2nd edn.).

[23] Hristov, S., Daniel, L., Hoare, E., Cherniakov, M., Gashinova, M.: 'Target Shadow Profile Reconstruction in Ground-based Forward Scatter Radar', in '2015 IEEE Radar Conference (RadarCon)' (IEEE, 2015), pp. 0846–0851.

[24] 'CST Microwave Studio', https://www.cst.com/Products/CSTMWS. Last accessed on 14 July 2017.

[25] Gashinova, M., Daniel, L., Sizov, V., Hoare, E., Cherniakov, M.: 'Phenomenology of Doppler Forward Scatter Radar for Surface Targets Observation' *IET Radar, Sonar Navig.*, 2013, 7, (4), pp. 422–432.

[26] Daniel, L., Gashinova, M., Cherniakov, M.: 'Maritime Target Cross Section Estimation for an Ultra-wideband Forward Scatter Radar Network', in '2008 Eur. Radar Conf.' (2008), (October), pp. 316–319.

[27] Sizov, V., Cherniakov, M., Antoniou, M.: 'Forward Scattering Radar Power Budget Analysis for Ground Targets' *IET Radar, Sonar Navig.*, 2007, 1, (6), p. 437.

[28] Long, T., Hu, C., Cherniakov, M.: 'Ground Moving Target Signal Model and Power Calculation in Forward Scattering Micro Radar' *Sci. China Ser. F, Inf. Sci.*, 2009, 52, (9), p. 11.
[29] Sizov, V., Gashinova, M., Rashid, N.E., Chen, J., Cherniakov, M.: 'Forward Scattering Micro Radar Efficiency Analysis for Different Landscapes', in *'6th EMRS DTC Tech. Conf.'* (2009), (1), p. 7.
[30] Nathanson, F.E.: *Radar Design Principles* (Mendham, SciTech Publishing, Inc., 1999, 2nd edn.).
[31] Parsons, J.D.: *The Mobile Radio Propagation Channel* (Chichester, John Wiley & Sons Ltd, 2000, 2nd edn.).
[32] Skolnik, M.J.: *Introduction to Radar Systems* (New York, McGraw-Hill, 2001, 3rd edn.).
[33] Kerr, D.E.: *Propagation of Short Radio Waves* (New York, Dover Publications, Inc., 1965, 1st edn.).
[34] Sizov, V., Cherniakov, M., Antoniou, M.: 'Forward Scattering Radar Power Budget Analysis for Ground Targets' *IET Radar, Sonar Navig.*, 2007, 1, (6), p. 437.
[35] Ament, W.S.: 'Toward a Theory of Reflection by a Rough Surface' *Proc. IRE*, 1953, 41, (1), pp. 142–146.
[36] Rappaport, T.S.: *Wireless Communications: Principles and Practice* (Upper Saddle River, Prentice Hall PTR, 2001, 2nd edn.).
[37] Briggs, J.: *Target Detection by Marine Radar* (London, The Institution of Electrical Engineers, 2004, 1st edn.).
[38] Cherniakov, M., Salous, M., Kostylev, V., Abdullah, R.: 'Analysis of Forward Scattering Radar for Ground Target Detection', in 'EURAD 2005 Conf. Proc. – 2nd Eur. Radar Conf., 2005' (2005), pp. 165–168.
[39] Kabakchiev, K., Daniel, L.Y., Sizov, V., Hoare, E.G., Gashinova, M., Cherniakov, M.: 'Received Signal Characterization in Forward Scatter Radar for Maritime Application', in 'Radar Symposium (IRS), 2011 Proceedings International' (2011), pp. 67–72.
[40] Antoniou, M., Sizov, V., Hu, C., *et al.*: 'The Concept of a Forward Scattering Micro-Sensors Radar Network for Situational Awareness', 'Proc. 2008 Int. Conf. Radar, Radar 2008' (2008), pp. 171–176.
[41] Hu, C., Sizov, V., Antoniou, M., Gashinova, M., Cherniakov, M.: 'Optimal Signal Processing in Ground-Based Forward Scatter Micro Radars' *IEEE Trans. Aerosp. Electron. Syst.*, 2012, 48, (4), pp. 3006–3026.
[42] Hiatt, R., Siegel, K., Weil, H.: 'Forward Scattering by Coated Objects Illuminated by Short Wavelength Radar' *Proc. IRE*, 1960, 48, (9), pp. 1630–1635.
[43] Levanon, N., Mozeson, E.: *Radar Signals* (Hoboken, John Wiley & Sons, Inc., 2004, 1st edn.).
[44] Ufimtsev, P.Y.: 'The 50-Year Anniversary of the PTD: Comments on the PTD's Origin and Development' *IEEE Antennas Propag. Mag.*, 2013, 55, (3), pp. 18–28.
[45] Ruck, G.T., Barrick, D.E., Stuart, K.M., Kirchbaum, C.K.: *Radar Cross Section Handbook: Volume 1* (New York, Plenum Press, 1970, 1st edn.).

[46] Ross, R.A.: 'Forward Scattering from A Finite, Circular Cylinder' *Prog. Electromagn. Res. C*, 2008, 2, pp. 207–215.

[47] Glaser, J.: 'Bistatic RCS of Complex Objects near Forward Scatter' *IEEE Trans. Aerosp. Electron. Syst.*, 1985, AES-21, (1), pp. 70–78.

[48] Marra, M., De Luca, A., Hristov, S., Daniel, L., Gashinova, M., Cherniakov, M.: 'New Algorithm for Signal Detection in Passive FSR', in '2015 IEEE Radar Conference', pp. 218–223.

[49] Abdullah, R., Cherniakov, M., Janovic, P., Salous, M.: 'Progress on Using Principle Component Analysis in FSR for Vehicle Classification', in '2nd International Workshop On Intelligent Transportation' (2005), pp. 7–12.

[50] Myakinkov, A., Kuzin, A., Gashinova, M., Sizov, V., Cherniakov, M.: 'Inverse Forward Scatter SAR', in 'Proceedings of International Conference on Synthetic Aperture Sonar and Synthetic Aperture Radar' (2010), pp. 152–156.

[51] Van Trees, H.L., Bell, K.L., Tian, Z.: *Detection Estimation and Modulation Theory – Part 1: Detection, Estimation, and Filtering Theory* (Hoboken, John Wiley & Sons, Inc., 2013, 2nd edn.).

[52] Billingsley, J.B.: *Low-angle Radar Land Clutter: Measurements and Empirical Models* (Norwich, William Andrew Publishing/Noyes Publishing, 2002, 1st edn.).

[53] Sizov, V., Gashinova, M., Zakaria, N., Cherniakov, M.: 'VHF Communication Channel Characterisations Over Complex Wooded Propagation Paths with Applications to Ground Wireless Sensor Networks' *IET Microwaves, Antennas Propag.*, 2013, 7, (3), pp. 166–174.

[54] Joshi, G.G., Dietrich, C.B., Anderson, C.R., *et al.*: 'Near-ground Channel Measurements Over Line-of-Sight and Forested Paths' *IEE Proc. – Microw., Antennas Propag.*, 2005, 152, (6), p. 589.

[55] Cheffena, M., Ekman, T.: 'Dynamic Model of Signal Fading due to Swaying Vegetation' *EURASIP J. Wirel. Commun. Netw.* 2009, 2009, pp. 1–11.

[56] Cheffena, M., Ekman, T.: 'Modeling the Dynamic Effects of Vegetation on Radiowave Propagation', in '2008 IEEE International Conference on Communications' (IEEE, 2008), pp. 4466–4471.

[57] Gashinova, M., Cherniakov, M., Zakaria, N.A., Sizov, V.: 'Empirical Model of Vegetation Clutter in Forward Scatter Radar Micro-sensors', in '2010 IEEE Radar Conference', pp. 899–904.

[58] Daniel, L., Gashinova, M., Cherniakov, M.: 'Maritime UWB Forward Scattering Radar Network: Initial Study', in '2008 International Conference on Radar' (IEEE, 2008), pp. 658–663.

[59] Kabakchiev, K., Daniel, L., Gashinova, M., Hoare, E., Cherniakov, M., Sizov, V.: 'Radar Parameters Influence on the Clutter in Maritime Forward Scatter Radar', in '2014 11th European Radar Conference' (IEEE, 2014), pp. 113–116.

[60] De Luca, A., Daniel, L., Kabakchiev, K., Hoare, E., Gashinova, M., Cherniakov, M.: 'Maritime FSR with Moving Receiver for Small Target Detection', in '2015 16th International Radar Symposium (IRS)' (IEEE, 2015), pp. 834–839.

[61] Berry, R.: 'Radar Propagation at Very Low Altitude over the Sea', in 'Radar-present and Future' (1973), pp. 140–145.
[62] Williams, P.D.L., Cramp, H.D., Curtis, K.: 'Experimental Study of the Radar Cross-section of Maritime Targets' *IEE J. Electron. Circuits Syst.*, 1978, 2, (4), p. 121.
[63] Fabbro, V., Bourlier, C., Combes, P.F.: 'Forward Propagation Modeling Above Gaussian Rough Surfaces by the Parabolic Shadowing Effect' *Prog. Electromagn. Res.*, 2006, 58, pp. 243–269.
[64] Zujkov, V.A., Pedenko, Y.A., Razskazovsky, V.B.: 'Characteristics of Radio Propagation in Sea Surface Layer, Propagation of Radio Waves of Millimeter and Centimeter Bands', in 'Proc. of IRE AN USSR' (1989), pp. 76–82.
[65] Pedenko, Y.A., Razskazovsky, V.B.: 'Multipath Propagation over Sea, Radiophysical Study of the World Ocean', in *'Proc. of IRE AN USSR'* (1992), pp. 32–50.
[66] Beard, C., Katz, I., Spetner, L.: 'Phenomenological Vector Model of Microwave Reflection from the Ocean' *IRE Trans. Antennas Propag.*, 1956, 4, (2), pp. 162–167.
[67] Beard, C., Katz, I.: 'The Dependence of Microwave Radio Signal Spectra on Ocean Roughness and Wave Spectra' *IRE Trans. Antennas Propag.*, 1957, 5, (2), pp. 183–191.
[68] Smith, J.R., Russell, S.J., Brown, B.E., *et al.*: 'Electromagnetic Forward-Scattering Measurements Over a Known, Controlled Sea Surface at Grazing' *IEEE Trans. Geosci. Remote Sens.*, 2004, 42, (6), pp. 1197–1207.
[69] Freund, D.E., Woods, N.E., Ku, H.-C., Awadallah, R.S.: 'Forward Radar Propagation Over a Rough Sea Surface: A Numerical Assessment of the Miller-Brown Approximation Using a Horizontally Polarized 3-GHz Line Source' *IEEE Trans. Antennas Propag.*, 2006, 54, (4), pp. 1292–1304.
[70] Hristov, T.S., Anderson, K.D., Friehe, C.A.: 'Scattering Properties of the Ocean Surface: The Miller–Brown–Vegh Model Revisited' *IEEE Trans. Antennas Propag.*, 2008, 56, (4), pp. 1103–1109.
[71] Brown, R.M., Miller, A.R.: 'Geometric-Optics Theory for Coherent Scattering of Microwaves from the Ocean Surface', 1974, Technical Report, Naval Research Laboratory, Washington, D.C.
[72] Miller, A.R., Brown, R.M., Vegh, E.: 'New Derivation for the Rough-Surface Reflection Coefficient and for the Distribution of Sea-Wave Elevations' *IEE Proc. H Microwaves, Opt. Antennas*, 1984, 131, (2), pp. 114–116.
[73] Beard, C.I.: 'Coherent and Incoherent Scattering of Microwaves from the Ocean' *IRE Trans. Antennas Propag.*, 1961, 9, (5), pp. 470–483.
[74] Anderson, K., Doss-Hammel, S., Tsintikidis, D., *et al.*: 'The RED Experiment: An Assessment of Boundary Layer Effects in a Trade Winds Regime on Microwave and Infrared Propagation over the Sea' *Bull. Am. Meteorol. Soc.*, 2004, 85, (9), pp. 1355–1365.

[75] Blyakhman, A.B., Runova, I.A.: 'Forward Scattering Radiolocation Bistatic RCS and Target Detection', in 'Proceedings of the 1999 IEEE Radar Conference. Radar into the Next Millennium' (IEEE, 1999), pp. 203–208.

[76] Ryndyk, A.G., Kuzin, A.A., Myakinkov, A.V., Blyakhman, A.: 'Target Tracking in Forward Scattering Radar with Multi-beam Transmitting Antenna', in 'International Radar Conference – Surveillance for a Safer World, 2009. RADAR 2009.' pp. 1–4.

[77] Blyakhman, A.B., Ryndyk, A.G., Myakinkov, A.V., Burov, V.N.: 'Algorithm of Trajectory Tracking the Targets, which are Moving along the Curvilinear Trajectories in the Bistatic Forward-scattering Radar System', in 'IET International Conference on Radar Systems (Radar 2012)' (Institution of Engineering and Technology, 2012), pp. 1–4.

[78] Blyakhman, A.B., Myakinkov, A.V., Ryndyk, A.G.: 'Tracking Algorithm for Three-Dimensional Bistatic Forward Scattering Radar with Weighting of Primary Measurements', in 'European Radar Conference, EURAD 2005.' (IEEE, 2005), pp. 153–156.

[79] Blyakhman, A.B., Ryndyk, A.G., Sidorov, S.B.: 'Forward Scattering Radar Moving Object Coordinate Measurement', in 'Record of the IEEE 2000 International Radar Conference' (IEEE, 2000), pp. 678–682.

[80] Blyakhman, A.B., Myakinkov, A.V., Ryndyk, A.G.: 'Algorithm of Target Tracking for Three-Dimensional Bistatic Forward Scattering Radar', in 'Proc. IV International Radar Symposium' (2004), pp. 309–324.

[81] Myakinkov, A.V., Ryndyk, A.G.: 'Space-time Processing in Three-Dimensional Forward Scattering Radar', in '4th International Conference on Antenna Theory and Techniques' (IEEE, 2003), pp. 355–358.

[82] Myakinkov, A.V.: 'Optimal Detection of High-Velocity Targets in Forward Scattering Radar', in '5th International Conference on Antenna Theory and Techniques, 2005.' (IEEE, 2005), pp. 345–347.

[83] Sage, A.P., Melsa, J.L.: *Estimation Theory with Applications to Communications and Control* (New York, McGraw-Hill, 1971, 1st edn.).

Chapter 14

Radar imaging of building interiors

Fauzia Ahmad[1] and Moeness G. Amin[2]

Abstract

Through-the-wall radar imaging (TWRI) provides the capability to see inside buildings using electromagnetic waves for both defence and civilian applications. Unlike traditional free-space radar operation, the presence of walls introduces unique challenges that need to be addressed to render TWRI a viable technology. In this chapter, we present a review of various approaches for building layout determination and imaging of stationary indoor scenes. These approaches effectively address the associated challenges to provide effective and reliable radar operation. Specifically, we consider both ground-based and airborne radar operation and discuss both non-adaptive time-domain and frequency-domain beamforming approaches for through-the-wall image formation of stationary scenes and feature-extraction-based building layout determination methods.

14.1 Introduction

Through-the-wall radar imaging (TWRI) is an emerging technology that addresses the desire to see inside buildings using electromagnetic (EM) waves for various purposes, including determining the building layout, discerning the building intent and nature of activities, locating and tracking the occupants and even identifying and classifying inanimate objects of interest within the building. TWRI is highly desirable for law enforcement, fire and rescue and emergency relief and military operations [1–5]. Applications primarily driving TWRI development can be divided based on whether information on motions within a structure or on imaging the structure and its stationary contents is sought out. Discrimination of movements from background clutter can be achieved through change detection (CD) or exploitation of Doppler [6–14]. One-dimensional (1-D) motion detection and localization systems employ a single transmitter and receiver and can only provide range-to-motion, whereas two-dimensional (2-D) and three-dimensional (3-D) multiantenna systems can provide more accurate

[1]Department of Electrical and Computer Engineering, Temple University, USA
[2]Center for Advanced Communications (CAC), Villanova University, USA

localization of moving targets. The 3-D systems have higher processing requirements compared to 2-D systems. However, the third dimension provides height information, which permits distinguishing people from animals, such as household pets. This is important, since radar cross section alone for behind-the-wall targets can be unreliable.

Imaging of structural features and stationary targets inside buildings requires at least 2-D and preferably 3-D systems [15–37]. Because of the lack of any type of motion, these systems cannot rely on Doppler processing or CD for target detection and separation. Synthetic aperture radar (SAR)-based approaches have been the most commonly used algorithms for this purpose. Most of the conventional SAR techniques usually neglect propagation distortions such as those encountered by signals passing through walls [38]. These distortions degrade radar imaging and can lead to ambiguities in target and wall localizations. Free-space assumptions no longer apply after the EM waves propagate through the first wall. Without factoring in propagation effects, such as attenuation, reflection, refraction, diffraction and dispersion, imaging of contents within buildings will be severely distorted. As such, image-formation methods, array-processing techniques, target detection and image-sharpening paradigms must work in concert and be reexamined in view of the nature and specificities of the underlying sensing problem.

In addition to exterior walls, the presence of multipath and clutter can significantly contaminate the radar data leading to reduced system capabilities for imaging of building interiors and localization and tracking of targets behind walls. The multiple reflections within the wall result in wall residuals along the range dimension. These wall reverberations can be stronger than target reflections, leading to its masking and undetectability, especially for weak targets close to the wall [39]. Multipath stemming from multiple reflections of EM waves off the targets in conjunction with the walls may result in the power being focused at pixels different than those corresponding to the target. This gives rise to ghosts, which can be confused with the real targets inside buildings [40–42]. Further, uncompensated refraction through walls can lead to localization or focusing errors, causing offsets and blurring of imaged targets [25,32,43]. SAR techniques and tomographic algorithms, specifically tailored for TWRI, are capable of making some of the adjustments for wave propagation through solid materials [25–33,43–48]. While such approaches are well suited for shadowing, attenuation and refraction effects, they do not account for multipath as well as strong reflections from the front wall.

The problems caused by the front wall reflections can be successfully tackled through wall clutter mitigation techniques. Several approaches have been devised, which can be categorized into those based on estimating the wall parameters and others incorporating either wall backscattering strength or invariance with antenna location [32,39,49–55]. In [32,49], a method to extract the dielectric constant and thickness of the non-frequency-dependent wall from the time-domain scattered field was presented. The time-domain response of the wall was then analytically modelled and removed from the data. A similar wall parameter estimation procedure was proposed in [50]. Unlike [32,49], the method in [50] exploits averaged measurements across multiple antenna positions, thereby improving the robustness of the estimation procedure against disturbances due to the scatterers behind the wall and the noise. In [39], a spatial filtering method was applied to remove the dc component corresponding to the constant-type radar return, typically associated with the front wall. The third method, presented in [51–54], was based not only on the wall scattering invariance along the array but also on the fact that wall reflections are relatively stronger than target reflections. As a result, the wall subspace is usually captured in the most dominant singular values when applying singular value decomposition (SVD) to the measured data matrix. The wall contribution can then be removed by orthogonal subspace projection. Both SVD-based and spatial filtering-based wall mitigations were considered for sub-Nyquist sampled data in [56]. In [55], discrete prolate spheroidal sequences were used to represent wall returns, which were then captured by a sparsity-based approach. Subtraction of the captured returns from the reduced set of measurements at each antenna resulted in wall clutter-free data.

Several methods have also been devised for dealing with multipath ghosts in order to provide proper representation of the ground truth. Earlier work attempted to mitigate the adverse effects stemming from multipath propagation [29,57]. Subsequently, research has been conducted to utilize the additional information carried by the multipath returns [42,58–66]. Multipath-exploitation-based TWRI was first attempted in [58], wherein areas in the shadow region of highly attenuating targets, which were not directly illuminated by the radar, were imaged by utilizing multipath returns. Following a similar idea, the work in [42] made use of the energy in the ghost targets resulting from secondary reflections at known interior walls to generate a ghost-free image with improved signal-to-clutter ratio (SCR). Multipath modeling and exploitation has further been considered as an inverse scattering problem in [59,60]. Multipath exploitation within the compressive sensing framework was addressed in [62,63] for sparsely populated stationary indoor scenes. By using proper modeling under known wall locations, sparse reconstruction in this case yielded an image where ghosts were eliminated and their energy was added to the real targets. Compressive-sensing-based multipath exploitation was extended to a general sparse indoor scene of stationary and moving targets in [64]. Offerings of multipath exploitation for target classification in TWRI applications were examined in [61]. It is noted that the aforementioned approaches require a radar system comprising a real or synthesized array aperture. Multipath exploitation for sensing indoor stationary scenes can be also performed using a single-antenna ultrawideband (UWB) radar system [65,66]. The concept of exploitation is embedded in using the resolvable multipath to create virtual radar units at different locations dictated by the positions of both targets and walls and utilizing the embedded directivity in UWB antennas.

For determination of the building interior layout, feature extraction methods allow detection, classification and localization of building elemental structures such as walls, ceilings and corners. Based on the locations of elemental structures, the building layout can be synthesized. Different feature extraction methods can be used for extracting building elements, exploiting either radar images or raw radar data [17,21,22,67,68]. In addition to the feature extraction methods, other approaches have been presented in the literature, which use a model of the complete building structure to predict the radar measurements [15,16,20,69]. In such model-based approaches, a detailed EM simulation is performed for forward prediction of the measurements based on the chosen building layout. The layout is updated based on the difference between the predicted measurement vector and the actual measurement vector. These algorithms use finite elements methods together with, for instance simulated annealing or jumped diffusion algorithm, for updating the building layout.

In this chapter, we focus on determining the building interior layout and imaging of stationary indoor scenes. We present a review of various techniques that address the unique challenges associated with attaining these objectives in urban operations. Section 14.2 deals with imaging of stationary indoor scenes, whereas feature-extraction-based building layout determination is discussed in Sections 14.3–14.5. More specifically, Section 14.2 discusses time-domain and frequency-domain data independent beamforming approaches for through-the-wall image formation of stationary scenes. The presented approaches account for refraction effects on signals propagating through walls. However, due to space limitations, wall clutter mitigation and multipath suppression/exploitation are not discussed. Interested readers are referred to [32,39,42,49,51–55,58–66] for further details on these aspects of imaging of stationary indoor scenes. Section 14.3 presents a framework for 3-D feature estimation from multielevation SAR apertures for determining the building interior layout. A pattern-matching approach that uses an image descriptor known as the correlogram is discussed in Section 14.4. Use of overcomplete dictionaries (OCD) in conjunction with sparse reconstruction for building interior layout estimation is presented in Section 14.5. Model-based building layout estimation is not discussed in this chapter. For an in-depth treatment of these techniques, see [15,16,20,69]. Concluding remarks are provided in Section 14.6.

14.2 Beamforming for imaging stationary indoor scenes

TWRI systems generally employ beamforming-on-receive approach, which utilizes several wide-beam transmit elements to sequentially illuminate the entire region of interest with the return signal sampled by a receive array for each instance of transmission. Beamforming-on-receive is the preferred method for through-the-wall image formation, since it permits various types of signal compensation for wall propagation effects to be easily accommodated in the beamforming process [25,43]. A number of algorithms, both data-independent and data-adaptive schemes, can be used to realize the beamforming operation [24,25,29–31,33,43,70,71]. Data-independent approaches include backprojection and direct frequency-domain beamforming, whereas data-adaptive approaches include the Capon method. For coherent radar imaging, the latter is employed through an approach called high-definition vector imaging, which was originally proposed for general SAR applications in [72] and investigated specifically for TWRI applications in [44,70,71]. Data-independent approaches are discussed in detail in this section. See [43,44,70,71] for detailed discussion of data-adaptive beamforming methods for TWRI.

We assume the use of wideband waveforms and monostatic or near-monostatic arrays, be it a physical aperture, a synthesized aperture, or a combination thereof. Moreover, far-field operating condition is assumed, which implies that the electric field of the propagated radar signal is represented by transverse plane waves. This approximation is applicable for the most part to TWRI systems, because typical operational wavelengths range between a few centimetres to 50 cm, while stand-off distances from exterior building walls and interior features of interest are on the order of 10 to 100 m. The far-field condition further implies that plane wave refraction at wall-air interfaces is described by Snell's law only. All scatterers are considered to be ideal point targets, for which the amplitude and phase responses are independent of angle and frequency.

14.2.1 Data-independent beamforming

Let N_t be the number of transmitters and N_r be the number of receivers. Let the mth transmitter illuminate the scene with a modulated wideband signal $s(t)e^{-j2\pi f_c t}$, where f_c is the carrier frequency and $s(t)$ is the baseband signal. Assuming P point targets in the region of interest, with the pth target located at position $\mathbf{r}_p$, the received baseband signal at the nth receiver can be expressed as

$$\tilde{y}_{mn}(t) = \sum_{p=1}^{P} A_p s(t - \tau_{mn}(\mathbf{r}_p))e^{-j2\pi f_c \tau_{mn}(\mathbf{r}_p)}, \quad n = 1,\ldots,N_r, \quad m = 1,\ldots,N_t \tag{14.1}$$

where A_p is the complex reflectivity of the pth target and τ_{mn} is the pulse propagation time between the mth transmitter, pth target and the nth receiver.

The region of interest is divided into a regular grid with a finite number of grid points, say Q, with the qth grid point located at position $\mathbf{r}_q$. Then, the complex image amplitude corresponding to the qth grid point is given by [25,30,43]

$$I(\mathbf{r}_q) = \sum_{m=1}^{N_t}\sum_{n=1}^{N_r} \left[w_{mn} y_{mn}(t + \hat{\tau}_{mn}(\mathbf{r}_q)) * h(t)\right]_{t=0} \tag{14.2}$$

where the operator '$*$' denotes convolution, w_{mn} is a weighting applied to the nth received signal with the mth transmitter active, $\hat{\tau}_{mn}(\mathbf{r}_q)$ is the estimated total propagation delay of the signal from the mth transmitter to the grid point at $\mathbf{r}_q$ and then back to the nth receiver, $h(t) = s^*(-t)$ is the impulse response of the filter matched to the transmitted waveform and the superscript '$*$' denotes complex conjugate. This process is repeated for all grid points to form an image of the region of interest. The weights w_{mn} are primarily used to control the sidelobe response of the beamformer, but they can also include factors to compensate for power variations in the signal returns due to normal geometric attenuation.

Equation (14.2) constitutes the backprojection algorithm, which is a time-domain approach. The use of a time-domain backprojection algorithm offers the following key advantages for through-the-wall imaging applications [43]:

- Since the algorithm makes no geometrical approximations, it can inherently accommodate near-field imaging scenarios.
- The use of the estimated signal propagation delay, $\hat{\tau}_{mn}(\mathbf{r}_q)$, in the image formation process provides a convenient mechanism to compensate signals for additional propagation delays caused by the presence of walls, either analytically as described in Section 14.2.2, with model-based corrections, and/or through autofocusing techniques.
- Direct incorporation of $\hat{\tau}_{mn}(\mathbf{r}_q)$ also inherently accommodates the use of non-linear or non-planar arrays, which may result from design choices or, for the case of SAR processing, from the path of the moving radar platform.

Alternatively, image formation can be carried out in the frequency-domain through evaluation of the following expression [24,43],

$$I(\mathbf{r}_q) = \sum_{m=1}^{N_t}\sum_{n=1}^{N_r}\sum_{k=1}^{K} w_{mn} Y_{mn}(f_k) S^*(f_k) \exp(j2\pi f_k \hat{\tau}_{mn}(\mathbf{r}_q)) \tag{14.3}$$

where $Y_{mn}(\cdot)$ is the Fourier transform of the nth received signal with the mth transmitter active, $S(\cdot)$ is the Fourier transform of the wideband baseband transmit signal, K is the number of frequency samples measured by the radar and f_k is the kth measured frequency. This frequency-domain approach offers the same advantages as backprojection for TWRI applications.

14.2.2 Compensation of wall propagation effects

First, consider the case of homogeneous walls. At frequencies typically employed for beamforming in TWRI (0.5–3 GHz), solid walls can be idealized as a slab of a dielectric material with sufficient accuracy. Assuming the wall parameters, such as its thickness and material dielectric constant, to be known, the propagation delay estimates, $\hat{\tau}_{mn}(\mathbf{r}_q)$, in the presence of a homogeneous wall can be obtained as follows.

For simplicity of analysis, consider the N_r-element receive array with inter-element spacing Δ_0, placed against a homogeneous wall of thickness d and dielectric constant ε_r, as shown in Figure 14.1. Since it can be readily shown that the path lengths are invariant to the wall location between the array and the grid point, the results herein are also applicable to the general case of a non-zero stand-off distance. It is noted, however, that the radiation properties of the antennas change when in close proximity to the wall, which should be taken into account for accurate imaging. Using the notation defined in Figure 14.1, Snell's law dictates that $\sin\beta = \sqrt{\varepsilon_r}\sin\beta'$. Then, with this relation and using Taylor series expansion for ρ_n, the one-way path length $R_n(\rho,\theta)$ between the nth receiver and the grid point at $\mathbf{r}_q = (\rho,\theta)$ can be expressed as [43]

$$\begin{aligned} R_n(\rho,\theta) \approx {} & \rho\left\{1 + \frac{d}{\rho}[\gamma + (1-\cos\theta)\mu]\right\} - \rho\left[\sin\theta + \frac{\sin(2\theta)}{2}\frac{d}{\rho}\mu\right]\left(\frac{n\Delta_0}{\rho}\right) \\ & + \frac{\rho}{2}\left[\cos^2\theta - \cos\theta(3\sin^2\theta - 1)\frac{d}{\rho}\mu\right]\left(\frac{n\Delta_0}{\rho}\right)^2 \end{aligned} \tag{14.4}$$

where

$$\gamma = \sqrt{\varepsilon_r} - 1, \mu = \sqrt{\gamma(\gamma+2)} - \gamma$$

As the ratio d/ρ tends to zero (i.e., for a thin wall or at long ranges), (14.4) takes the form

$$R_n(\rho,\theta) \approx \rho\left\{1 - \left(\frac{n\Delta_0}{\rho}\right)\sin\theta + \left(\frac{n\Delta_0}{\rho}\right)^2\frac{\cos^2\theta}{2}\right\} \tag{14.5}$$

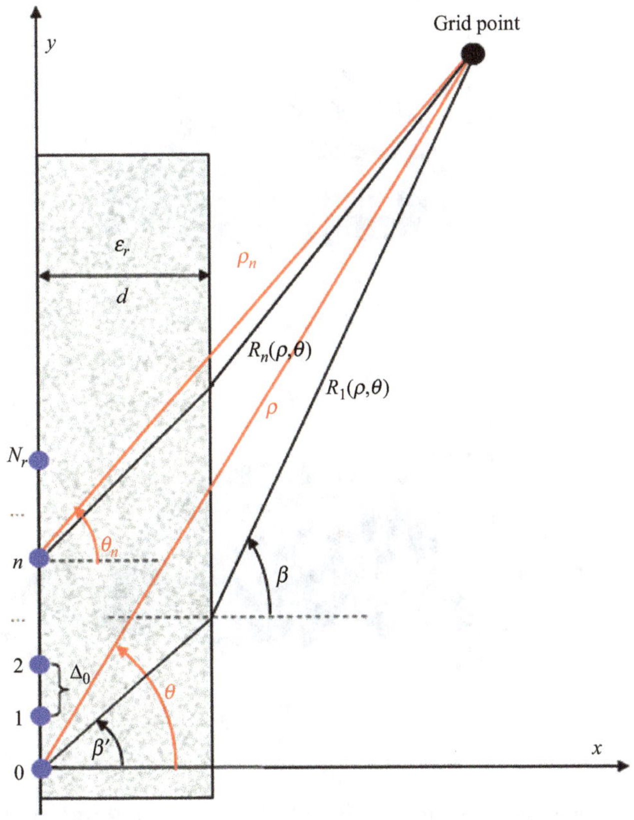

Figure 14.1 Through-the-wall imaging geometry for zero stand-off distance using a linear receive array. ©CRC Press. Adapted, with permission, from [43]

which is the free-space expression for $R_n(\rho, \theta)$. The one-way delay $R_m(\rho, \theta)$ between the mth transmitter and the grid point can be computed in a similar manner. Then, the propagation delay estimate $\hat{\tau}_{mn}(\mathbf{r}_q)$ is given by

$$\hat{\tau}_{mn}(\mathbf{r}_q) = \frac{R_m(\rho, \theta) + R_n(\rho, \theta)}{c} \tag{14.6}$$

where c is the speed of light.

In practice, the wall characteristics, namely, its thickness and dielectric constant, may not be available *a priori*. Various methods have been devised to estimate the wall parameters from radar data [32,49]. Alternative techniques, ranging from autofocus of the image [47] to comparing imaging from sub-arrays [45,73], have also been devised for reliable imaging in the presence of walls with unknown characteristics.

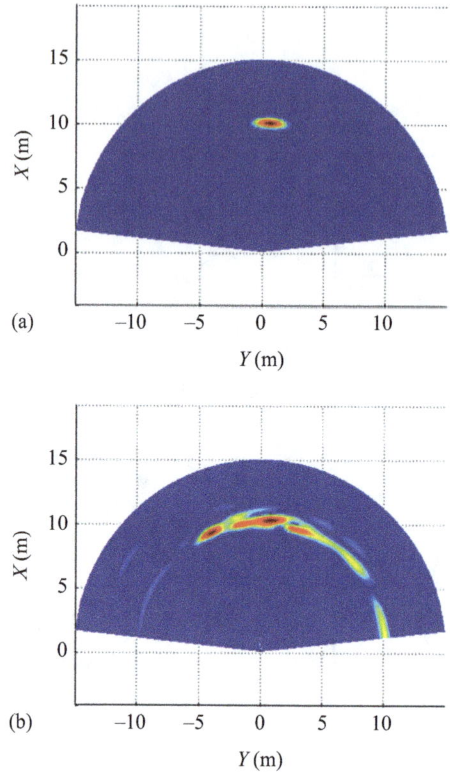

Figure 14.2 Imaging results (a) free-space propagation (no wall is present), (b) hollow brick wall. ©CRC Press. Reprinted, with permission, from [43]

For non-homogeneous walls with an internal structure of air gaps, such as cinder block or hollow bricks, the signal propagation delay can vary rapidly from one element to the other across the array, especially, when the array is used in contact with the wall. This is because some elements may be close to the void, while for others, the propagation may occur through one of the filled volumes of the structure. From a phenomenological point of view, this problem is mainly encountered for building materials that are highly variant with respect to translation of the array parallel to the wall. Figure 14.2 shows the imaging results corresponding to a person at a range of a few metres through air and through a hollow brick wall [43]. Clearly, the hollow brick wall destroys the focusing almost completely, compared to the case of free-space propagation. This is attributed to the mismatch between the assumed propagation delay in the image formation algorithm and the actual propagation delays, as depicted in Figure 14.3, resulting in poor image focusing. The observed propagation delay is much more complex, since each channel experiences a rather different propagation path. By time shifting the

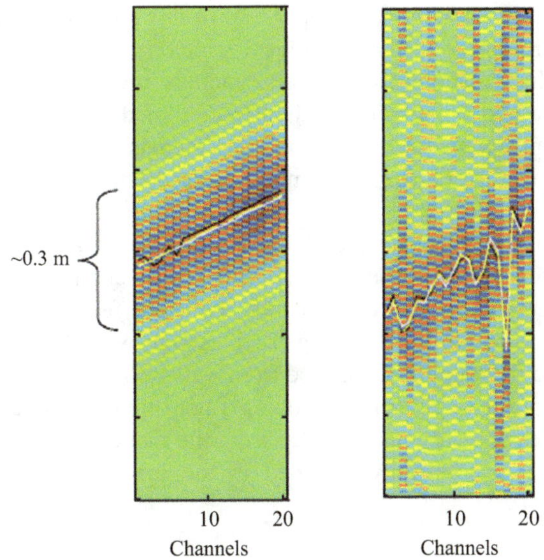

Figure 14.3 Range profiles of raw data from a 20-channel monostatic array for a target at 45° *bearing from boresight, under free-space propagation (left figure) and through a hollow brick wall (right figure). ©CRC Press. Reprinted, with permission, from [43]*

data samples in each channel based on the measured relative propagation delays shown in Figure 14.3, the data can be realigned. Subsequent image formation using realigned data produces a focused image (not shown). It should be noted that in real-time application, it may be difficult to apply this propagation delay measurement due to its high computational requirement and the fact that automatic tracking of propagation delay may not be robust enough in real environment. If propagation tracking is incorrect, the result may be even worse than the original unadjusted image. For walls with periodic structures, such as the hollow brick wall, a model-based approach to compensating the receiver channel signals may be somewhat effective if wall material is known [43].

14.3 SAR imaging using attributed scattering centre features for characterization of building interior structure

Radar images of building interiors may be obtained using beamforming techniques based on a point target model, as detailed in Section 14.2. However, the radar responses to corners, walls and ceilings are not well represented by a point target model, resulting in classification confusion. Further, point-target-model-based beamforming causes the energy in the multipath reflections to be focused at imaging locations where no targets are present, leading to false alarms that clutter the image. As such, building features can be more reliably extracted using an imaging approach exploiting scatterer models [17,67]. In this approach, the building structural elements are represented by canonical scatterers, such as corners and planar surfaces. For these canonical scatterers, models are available describing the amplitude, phase and polarimetric characteristics of the radar responses [74,75]. Exploitation of such *a priori* defined scattering models provides robustness against clutter and multipath reflections, assuming these do not match the defined scattering models.

In this section, we consider building structure estimation from an airborne SAR system. Dominant scattering features, such as trihedrals, dihedrals and plates, are used to infer building interior structure. Trihedral features describe dominant scattering from corners in the building (such as wall-wall-floor corners), whereas dihedrals capture the dominant response from wall-floor structures. Direct specular scattering from plates (e.g., walls or floors) is not often seen in airborne data collection geometries and are thus excluded from the model in this section. Note that plate scattering can be readily included in the model for ground-based radar measurements.

14.3.1 Canonical scattering models

We consider a SAR system which collects coherent backscatter measurements $Y(f, \phi, \theta_i)$ on circular apertures over the full $[0, 2\pi]$ range of azimuth angles ϕ and at two closely spaced elevation angles $\{\theta_1, \theta_2\}$, as shown in Figure 14.4. Note that the model and algorithm presented below can be readily extended to the case of a small number (>2) of closely spaced elevation angles [67]. The backscatter measurements are collected over a band of frequencies $f \in [f_c - B/2, f_c + B/2]$, where B is the bandwidth. Assuming P canonical scatterers comprising the structure of the building being interrogated, the SAR measurement corresponding to frequency f, azimuth angle ϕ and the elevation angle θ_i can be expressed as [67]

$$Y(f, \phi, \theta_i) = \sum_{p=1}^{P} T_p(f, \phi) G_p(f, \phi) A_p e^{-j(4\pi f/c) R(\mathbf{r}_p, \phi, \theta_i)} \qquad (14.7)$$

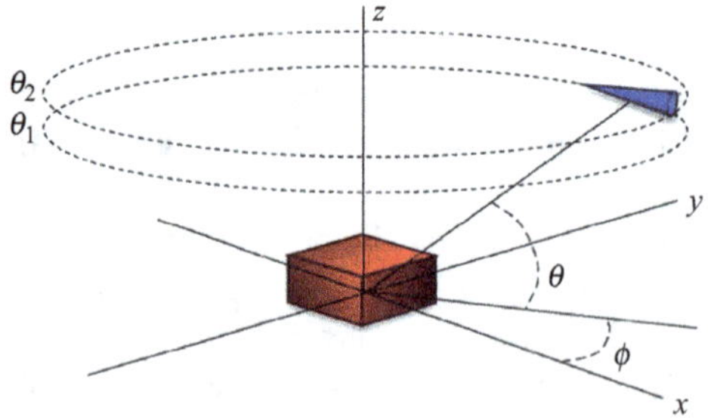

Figure 14.4 Circular SAR data collection geometry. ©IEEE. Reprinted, with permission, from [17]

where A_p is the amplitude of the pth scattering centre, $T_p(f,\phi)$ is the frequency and aspect dependent propagation transfer function through the dielectric layers, $G_p(f,\phi)$ models the frequency and azimuth dependence of the pth scattering centre and $R(\mathbf{r}_p,\phi,\theta_i)$ is the range to the pth scattering centre at building-centric coordinate position $\mathbf{r}_p = (x_p, y_p, z_p)$ relative to the scene centre

$$R(\mathbf{r}_p,\phi,\theta_i) = x_p\cos(\phi)\cos(\theta_i) + y_p\sin(\phi)\cos(\theta_i) + z_p\sin(\theta_i) \tag{14.8}$$

In general, A_p, T_p and G_p are all functions of elevation angle θ as well. However, since we assume that the elevation difference is small, so that $\theta_1 \approx \theta_2$, we drop the θ dependence for notational brevity. In general, each scattering centre is subject to a different propagation transfer function $T_p(f,\phi)$ to account for different number of walls that are in the propagation path.

The scattering response for a dihedral of length L, positioned flat on the xy-plane at an azimuth pose angle of ϕ_r (indicating the direction of the surface normal) is given by [67]

$$G_{di}(f,\phi,\theta) = \sqrt{j\frac{f}{f_c}}\ \text{sinc}\left(\frac{\pi f}{c}L\sin(\phi-\phi_r)\right) \tag{14.9}$$

This model is derived under the assumption of a single dihedral made up of perfectly conducting faces. In a practical building, however, the dihedral is not only formed by multilayer dielectric materials, but the two faces could be made of different materials. The dispersion and delay caused by these materials is captured by the term $T_p(f,\phi)$ in (14.7).

The trihedral scattering model comprises returns from three orthogonal plates aligned with the coordinate axes. The peak scattering occurs at $\phi = \pi/4$ and $\theta = \tan^{-1}(1/\sqrt{2}) \approx 35.26°$. The frequency and azimuth response for the trihedral is given for $\theta \in [0,\pi/2]$ and $\phi \in [0,\pi/4]$ as [67]

$$G_{tri}(f,\phi,\theta) = \sqrt{j\frac{f}{f_c}}H(\phi,\theta) \tag{14.10}$$

where $H(\phi,\theta)$ is a function defined on $(\phi,\theta) \in [0,\pi/2] \times [0,\pi/2]$ with a maximum at $(\phi,\theta) = (\pi/4, \tan^{-1}(1/\sqrt{2}))$ and that decreases smoothly to zero as either ϕ or θ approaches 0 and $\pi/2$ [76]. The specific form of H depends on the particular shapes of the trihedral faces. However, the beamwidth of this response exceeds 45° in both azimuth and elevation for common face shapes, such as rectangles or triangles. The primary difference between the dihedral and trihedral scattering models is the significantly larger azimuthal persistence of trihedral scattering. The dihedral azimuthal mainlobe width is approximately $2\lambda/L$, which for a 6-m wall at L-band amounts to approximately 5°. On the other hand, trihedral scattering persists for more than 45° in azimuth. This difference in azimuthal persistence between the two scattering centres is exploited for classification.

14.3.2 Feature extraction

We now address the problem of estimating the relevant parameters to describe the scattering primitives in (14.7) from a set of measurements $Y(f, \phi, \theta_i), i = 1, 2$. We are interested in finding the parameter vector Θ which solves the following optimization problem,

$$\arg\min_{\Theta} \sum_{i=1}^{2} \int_f \int_\phi \left| Y(f, \phi, \theta_i) - \sum_{p=1}^{P} T_p(f, \phi) G_p(f, \phi) A_p e^{-j(4\pi f/c)R(\mathbf{r}_p, \phi, \theta_i)} \right|^2 df\, d\phi \tag{14.11}$$

where $\Theta = \{\Theta_p\}_{p=1}^{P}$ and each Θ_p contains the parameters (location, type, reflection function G_p and transmitting path function T_p) that characterize the pth scattering primitive. However, this optimization problem is highly non-linear and direct minimization of (14.11) is computationally intractable in general. Therefore, the high-dimensional problem in (14.11) is decomposed into a set of smaller dimensional minimization problems that can be solved independently. The decomposition is accomplished as follows [67]. First, for each elevation pass i, we form 2-D subaperture images using a subset of the measured circular SAR data. Scattering features appear as localized responses in these images, although for TWRI, there is significant spatial spread and also significant spatial overlap in these responses. Second, non-parametric preprocessing is used to form an initial deconvolution of the scattering primitive responses to localize and isolate energy from individual primitives in the image planes. Once the energy is localized, we decompose the image into regions in which the response of only a small number of primitives (usually only one) is isolated. We then perform the non-linear minimization independently for each scattering primitive in each subaperture. In this way, we decompose the high-dimensional minimization of (14.11) into many small-order, computationally tractable minimization problems. The outline of the feature extraction algorithm is provided in Figure 14.5 and each step is detailed below.

1. *Subaperture imaging*: A subaperture image $I_{i,m}(\tilde{x}, \tilde{y})$ is formed by first interpolating the measurements from a subaperture of width Δ centred at azimuth angle ϕ_m and at elevation θ_i, followed by application of a windowing function and performing a 2-D inverse Fourier transform. The parameters $\tilde{x}$ and $\tilde{y}$ denote the corresponding downrange and crossrange dimensions of that ground-plane image. Subaperture imaging requires accurate radar phase coherence only over azimuth extents corresponding to the subapertures, and not over the entire 360° aperture. Further, the azimuth angle(s) for which a scattering centre is present or persistent provides valuable information about the orientation and scattering type and is helpful for hypothesizing building layout in subsequent processing stages.
2. *Height processing*: Scattering primitives that appear in subaperture images are projections into the 2-D image plane from 3-D scattering primitive locations. As a result, the next step is to use interferometric processing to estimate 3-D locations and then to separate scattering primitives by height. Imaging with

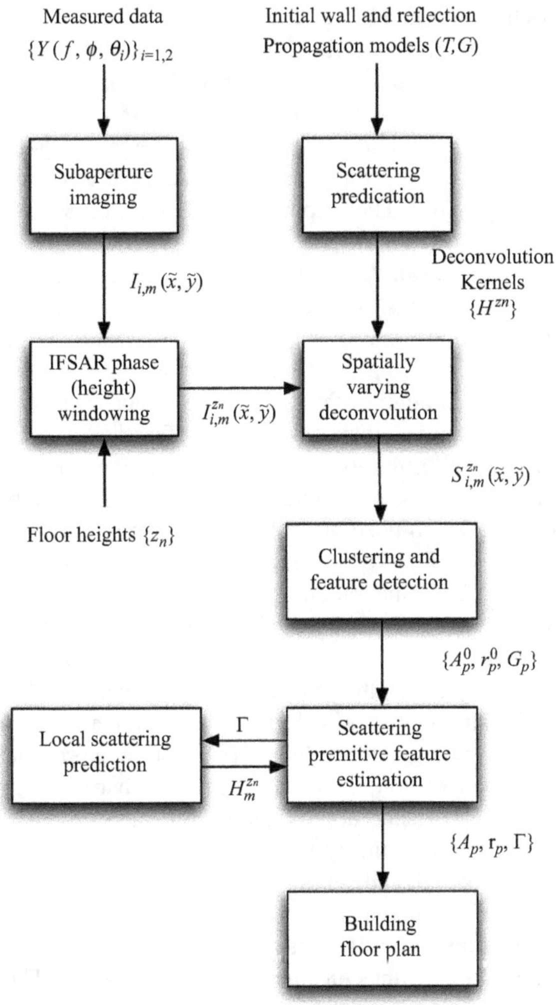

Figure 14.5 Processing steps for extracting building features. ©2009 IEEE. Reprinted, with permission, from [17]

SAR data produces complex-valued reflectivity maps, whose phase contains modulation terms related to scene height and elevation angle of the radar. Therefore, the relative phase of the images $(I_{1,m}, I_{2,m})$ can be used to separate returns from the different floors of the interrogated building. For the considered airborne collection geometry, the dominant reflectors in the scene are located at floor-wall junctions [77,78]. Scattering primitives from different heights in the building exhibit significantly different response profiles; this is primarily because T_p differs significantly for scattering primitives at different floors in the building, since primitives at lower floors include transmissions through multiple walls or floors above them. Assuming that the nominal prior information about the number of floors N_f and their associated heights z_n is available, each set of subaperture images can be filtered using a window function on the relative phase, centred at the hypothesized height of interest, z_n, to isolate returns from scattering centres at different floors. If prior information about the number of floors and their heights is not available, these could be estimated from histogram peaks of the relative complex phase of the subaperture images [67]. The output of the height processing step is the set of height-filtered subaperture images $I_{i,m}^{z_n}(\tilde{x},\tilde{y})$.

3. *Non-parametric deconvolution*: Owing to the reflections and transmissions through walls and ceilings, the backscattered responses from dielectric primitives formed from floors and walls are dispersed in both range and crossrange. In addition, the windowing in the subaperture imaging step causes further spatial spreading. To address this spreading, a radar system function can be developed in the image domain that captures these effects. This function can then be utilized in a non-parametric sparse-reconstruction algorithm for deconvolving the radar system function, thereby forming initial estimates for scatterer amplitudes and positions. The radar system function is based on two approximations [67]. First, we assume that images can be segmented to separately isolate trihedral, dihedral and plate primitives. One simple way to do so is to separately treat cardinal-angle subaperture images and off-cardinal-angle subaperture images, with the assumption that returns from the buildings on cardinal angles are dominated by returns from plates and dihedrals, while returns from buildings off-cardinal angles are dominated by trihedrals. More sophisticated processing can use hypothesis testing approaches (e.g., [79]) to achieve this decomposition. Second, we assume that scattering primitives at the same height experience approximately the same through-the-wall transmission distortions. Thus, the system function $H_m^{z_n}$ can be determined from an initial hypothesis of wall and ceiling dielectric compositions and knowledge of collection and imaging geometry. We note that the system function depends on heights z_n. This is because reflectors at lower floors of a building are subjected to larger number of ceilings in their propagation paths. The deconvolution process aims to reconstruct the images $S_{i,m}^{z_n}$ from the measured images $I_{i,m}^{z_n}(\tilde{x},\tilde{y})$, given the knowledge of the deconvolution kernel $H_m^{z_n}$. Since $S_{i,m}^{z_n}$ consists of a few non-zero elements, the deconvolution problem is formulated as a sparse reconstruction problem as [80,81]

$$\arg\min_{S}\left\{||\boldsymbol{I}-\mathbf{H}\boldsymbol{S}||_2^2+\mu||\boldsymbol{S}||_1\right\} \tag{14.12}$$

where $\boldsymbol{I}$ is the vector of grid points of the height filtered subaperture image $I_{i,m}^{z_n}$, $\boldsymbol{S}$ is the vector of deconvolved image grid points, $\mathbf{H}$ is the Toeplitz matrix that represents convolution with the given system function $H_m^{z_n}$, $||\cdot||_2$ is the 2-norm and $||\cdot||_1$ denotes the 1-norm. Fractional norms can also be used instead of the 1-norm [80]. The parameter μ is either determined based on the expected SCR, or crossvalidation-based methods can be used for selecting μ [82,92]. The deconvolved sparse images $S_{i,m}^{z_n}$ isolate individual primitive responses by height, subaperture and image location and permit a decomposition of the optimization problem in (14.11) into several smaller dimension problems.

4. *Primitive detection and parameter estimation*: In the final stage of the feature-extraction algorithm, the primitive features are estimated using the initial estimates of primitive type, location and orientation obtained from deconvolved subaperture images, along with initial estimates of transmission and reflection response functions in a non-linear Quasi-Newton method [67]. More specifically, we isolate subregions of high energy in the subaperture images and assign estimated primitive locations. Dihedrals and plates are assumed to persist for a single subaperture image set, and their initial parameters are obtained from a single azimuth subaperture. On the other hand, trihedrals persist across multiple subapertures, and combined processing across subapertures is required. To this end, multiple subaperture images corresponding to each quadrant of off-cardinal azimuth angles and each floor are processed separately to isolate responses from trihedral corners oriented in one of four primary directions in that floor. Wide azimuthal persistence of trihedrals results in multiple detections at close spatial location from the different subaperture images belonging to the same quadrant. A clustering algorithm is then employed to assign the detections $\{A_p, \mathbf{r}_p\}$ into M clusters $C(\mathbf{r}_m)$, which are used to initialize the following non-linear minimization problem that is solved for the trihedral features for each of the four quadrants,

$$\min_{A_p,\mathbf{r}_p,\Gamma} \sum_{i=1}^{2} \sum_{m \in M_j} \left| I_{i,m}^{z_n} - H_m^{z_n}(\Gamma) * \sum_p A_p \delta(\tilde{x} - \tilde{x}^l(\mathbf{r}_m), \tilde{y} - \tilde{y}^l(\mathbf{r}_m)) \right|^2 \tag{14.13}$$

Here M_j is the set of subaperture indices m corresponding to the jth quadrant, $j = 1, 2, 3, 4$ and $(\tilde{x}^l, \tilde{y}^l)$ is the projected layover image grid point corresponding to the 3-D location $\mathbf{r}_m$. Note that the iteration of the wall parameters Γ has been incorporated in this minimization to account for any ambiguities. Dihedral and plate parameter estimation is a special case of the trihedral processing and does not require clustering.

14.3.3 Illustrative example

We present results of interior corner estimation of a two-story building with dimensions 23.6 × 13.2 × 6.0 m [17]. Two monostatic, single polarization, circular SAR measurement apertures are synthesized using the National Electromagnetics Consortium (NEC) Basic Scattering Code (BSC) [83], at elevation angles 45° and 45.2°. The outside walls and the lower and upper floor ceilings in the building are modelled as rebar-reinforced concrete, and interior walls are modelled as cinder-block walls. Measurements are obtained for the 0.5 to 1.5 GHz frequency band and horizontal polarization. Noise is added to simulate a background clutter that is 40 dB below the peak amplitude of the largest scattering return. Feature extraction and hypothesis refinement is implemented using 20 subaperture images formed at 10° increments. We present trihedral feature extraction and localization results; for trihedrals, we do not use the images centred at the four cardinal angles of the building. We used hypothesized floor heights centred at 0 and 3 m for the height-filtering procedure. For the sparse reconstruction step, the required system function can be obtained from *a prior* estimate of the outside wall/ceiling structure of the building, or from an isolated scattering response from the measured SAR image. We have used NEC-BSC and single (double) rebar ceiling models for the upper (lower) floor reflectors to compose $T^{z_n}(f,\phi)$ and a trihedral with layered dielectric walls to approximate $G_{tri}(f,\phi)$. To test robustness to model mismatch, we used nominal values for the rebar parameters and dielectric coefficients that differed by approximately 20% from the actual values used in the simulation.

Figure 14.6 depicts the scattering centre detections from all subapertures after the height filtering and sparse reconstruction step (output of Step 3). The estimates

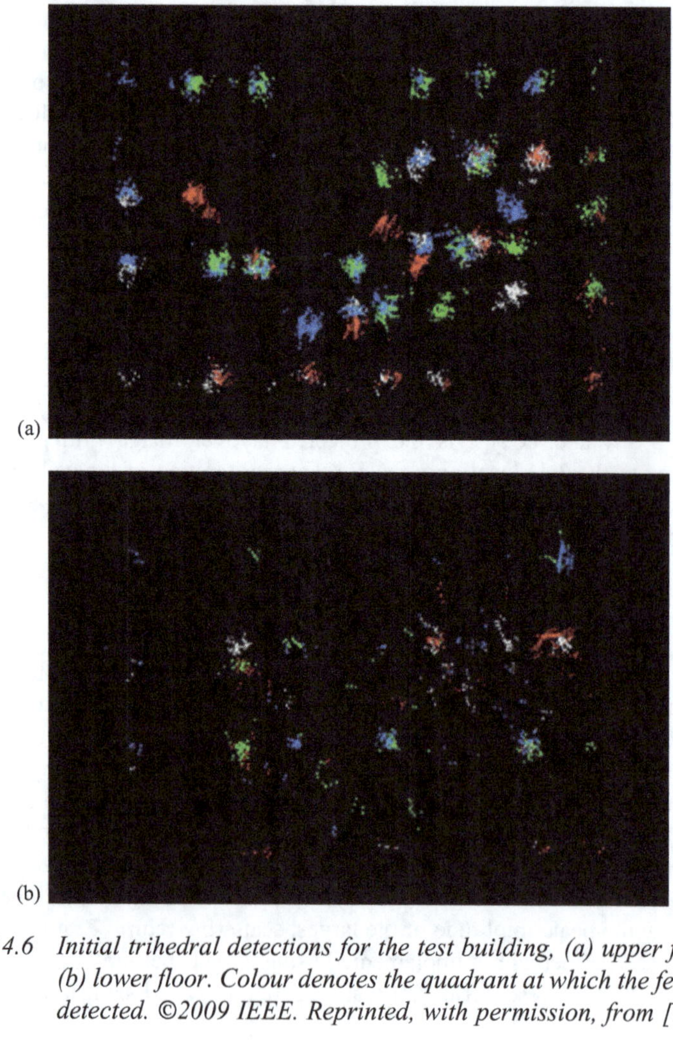

Figure 14.6 Initial trihedral detections for the test building, (a) upper floor, (b) lower floor. Colour denotes the quadrant at which the feature was detected. ©2009 IEEE. Reprinted, with permission, from [17]

are provided for lower floor and upper floor trihedral corner features. After the estimation step, one obtains a much sparser set of trihedral scattering primitive features, as shown in Figure 14.7. The actual building layout estimate is overlaid on the estimated trihedral features. The trihedrals are denoted as one of the four primitive icons (denoting trihedral orientation), and each icon corner is positioned at the final location estimate of the primitive. The *z* estimates (not visible in the top view depiction in Figure 14.7) are all within 25 cm of the true heights for these primitives. Features are well aligned with building corners, and no false-alarm trihedral detections are seen at locations that do not have floor-wall-wall corners. In addition, the location estimates show strong consistency for the differently coloured icons, which use independent data sets in forming the estimates; primitive estimates from, say, northeast-pointing trihedrals are obtained using disjoint data and independent estimation processing than northwest or southeast corners, yet align well with each other in location.

(a)

(b)

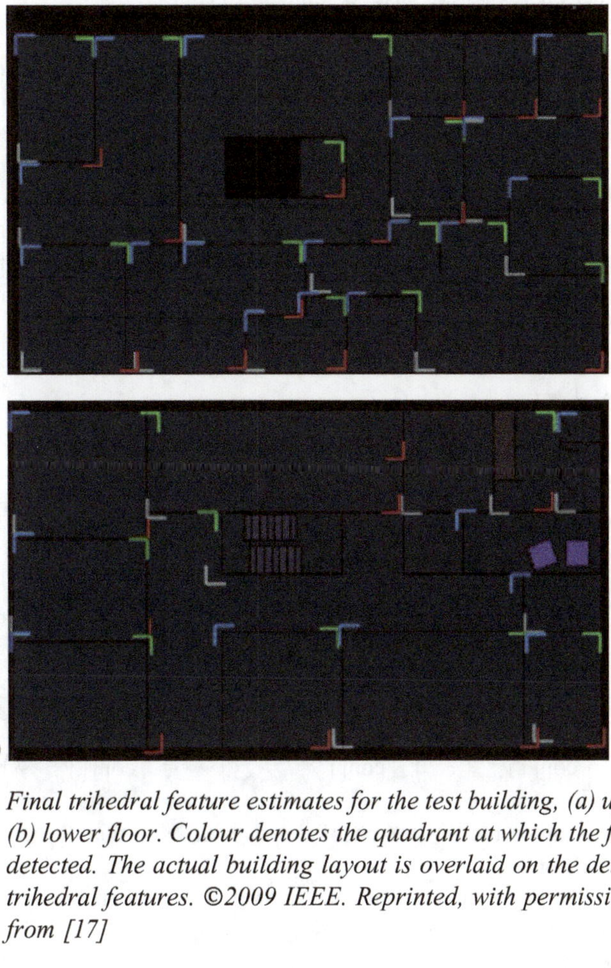

Figure 14.7 Final trihedral feature estimates for the test building, (a) upper floor, (b) lower floor. Colour denotes the quadrant at which the feature was detected. The actual building layout is overlaid on the detected trihedral features. ©2009 IEEE. Reprinted, with permission, from [17]

14.4 Correlogram-based pattern matching for building feature extraction

An alternate approach for building feature extraction is based on image pattern matching strategy [22], which employs a novel image descriptor, namely the correlogram [84]. The correlogram encodes information about the spatial correlation of the image complex amplitudes. The pattern matching strategy assumes flexibility in ground-based radar operation, which allows proper angular illuminations, thereby avoiding the front wall returns and preserving the corner features created by the wall junctions. This can be achieved using squint beamforming or broadside beams with tilted aperture [85,86]. Estimating locations of corners, by comparing the known correlogram of the scattering response of an isolated canonical corner reflector with the correlogram of the received radar returns within a correlation matching framework, allows the inference of building interior structure. The feature-based nature of the image pattern matching strategy enables corner separation from other indoor scatterers, such as humans [22].

For convenience, we consider a monostatic linear array for imaging and assume that the region of interest is divided into a uniform grid of Q grid points. The corresponding 2-D image is assumed to be obtained using beamforming. We use the infinity norm to measure the distance between image grid points, i.e., we define the distance between the q_1th grid point at $\mathbf{r}_{q_1}$ and q_2th grid point at $\mathbf{r}_{q_2}$ as $\max\{|x_{q_1} - y_{q_1}|, |x_{q_2} - y_{q_2}|\}$. For a reference grid point, each distance defines a set of grid points equidistant from that grid point. Let the complex image amplitudes corresponding to the set of N_q grid points located at distance $\bar{d}$ from the qth grid point be denoted as $I^{(\bar{d})}(\mathbf{r}_{q_i})$, $i = 1, \ldots, N_q$. Then, the correlogram of the qth grid point is defined as,

$$\gamma_q^{(\bar{d})} \triangleq \frac{1}{N_q} \sum_{i=1}^{N_q} I(\mathbf{r}_q) \operatorname{conj}\left(I^{(\bar{d})}(\mathbf{r}_{q_i})\right) \tag{14.14}$$

where $\operatorname{conj}(\cdot)$ is the complex conjugate function. From (14.14), we can build the correlogram matrix as [87],

$$\hat{\mathbf{R}}_q \triangleq \begin{bmatrix} \operatorname{conj}\left(\gamma_q^{(\bar{d}=0)}\right) & \gamma_q^{(\bar{d}=1)} & \cdots & \gamma_q^{(\bar{d}=D-1)} \\ \operatorname{conj}\left(\gamma_q^{(\bar{d}=1)}\right) & \operatorname{conj}\left(\gamma_q^{(\bar{d}=0)}\right) & \cdots & \gamma_q^{(\bar{d}=D-2)} \\ \vdots & \vdots & \ddots & \vdots \\ \operatorname{conj}\left(\gamma_q^{(\bar{d}=D-1)}\right) & \operatorname{conj}\left(\gamma_q^{(\bar{d}=D-2)}\right) & \cdots & \operatorname{conj}\left(\gamma_q^{(\bar{d}=0)}\right) \end{bmatrix} \tag{14.15}$$

where D determines the dimension of the matrix. Note that $\hat{\mathbf{R}}_q$ is positive semi-definite by definition.

The isolated presence of a canonical corner at the qth grid point will generate a particular reference correlogram matrix, which is denoted as $\mathbf{R}_q^{\text{ref}}$. The pattern matching procedure is based on a scan which reacts only when the reference scatterer is present. Based on this assumption, an error function is required to measure the reference corner contribution contained in the given sample correlogram matrix of each grid point. A detector that best suits the space generated by correlogram matrices is derived by forcing a positive semidefinite residual correlogram matrix $\hat{\mathbf{R}}_q - \kappa(q)\mathbf{R}_q^{\text{ref}}$. The problem can be formulated as [22]

$$\max_{\kappa(q)\geq 0} \quad \kappa(q)\text{s.t.} \quad \hat{\mathbf{R}}_q - \kappa(q)\mathbf{R}_q^{\text{ref}} \succeq 0 \tag{14.16}$$

If $\hat{\mathbf{R}}_q - \kappa(q)\mathbf{R}_{(q)}^{\text{ref}}$ must be positive semidefinite, so must be $\mathbf{I}_D - \kappa(q)\hat{\mathbf{R}}_q^{-1}\mathbf{R}_q^{\text{ref}}$, where $\mathbf{I}_D$ denotes an identity matrix of size $D \times D$. Thus, using the eigen-decomposition of $\hat{\mathbf{R}}_q^{-1}\mathbf{R}_q^{\text{ref}}$ defined by $\mathbf{U}\Lambda\mathbf{U}^H$,

$$\mathbf{I}_D - \kappa(q)\mathbf{U}\Lambda\mathbf{U}^H \succeq 0 \Rightarrow \mathbf{I}_D - \kappa(q)\Lambda \succeq 0 \tag{14.17}$$

where Λ is a diagonal matrix of the eigenvalues of $\hat{\mathbf{R}}_q^{-1}\mathbf{R}_q^{\text{ref}}$, $\mathbf{U}$ is the unitary matrix containing the eigenvectors of $\hat{\mathbf{R}}_q^{-1}\mathbf{R}_q^{\text{ref}}$ and the superscript 'H' denotes Hermitian. Note that if (14.17) is satisfied for the maximum eigenvalue, then it is satisfied for all eigenvalues. Therefore, the condition that always ensures positive semidefinite residual correlogram is given by [22],

$$\lambda_{\max}^{-1}\left(\hat{\mathbf{R}}_q^{-1}\mathbf{R}_q^{\text{ref}}\right) - \kappa(q) = 0 \tag{14.18}$$

Thus, the solution to (14.16) is given by the inverse of the maximum eigenvalue of $\hat{\mathbf{R}}_q^{-1}\mathbf{R}_q^{\text{ref}}$. That is,

$$\kappa(q) = \lambda_{\max}^{-1}\left(\hat{\mathbf{R}}_q^{-1}\mathbf{R}_q^{\text{ref}}\right) \tag{14.19}$$

The values of $\kappa(q), q = 0, 1, \ldots, Q-1$, directly indicate the presence of a potential corner at a given grid point.

14.4.1 Illustrative example

For illustration, we simulate the three-room building geometry depicted in Figure 14.8(a) in FEKO Geometric Optics solver in conjunction with the OPTFEKO options [88]. The walls are 20 cm thick and made of solid concrete with $\varepsilon_r = 6$. A perfectly conducting sphere of 15 cm diameter is located at (0.02, 4.24) m. A stepped-frequency signal consisting of 201 frequencies covering the 1 to 2 GHz frequency band was used for interrogating the scene. A ground-based eight-element monostatic line array with an interelement spacing of 53 cm was used. An oblique illumination of the scene is used to avoid wall returns, while preserving the important corner features. The angular tilt of the array baseline is 25°. In this case, only the upper-left corners of the three rooms are expected to produce strong scattering responses for most, if not all, of the antenna elements. The concave sides

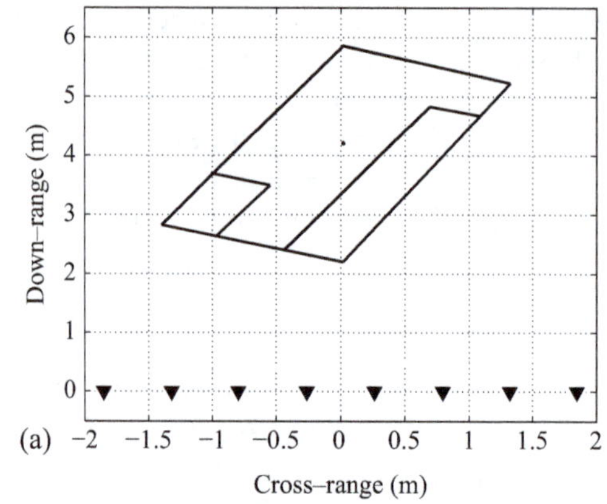

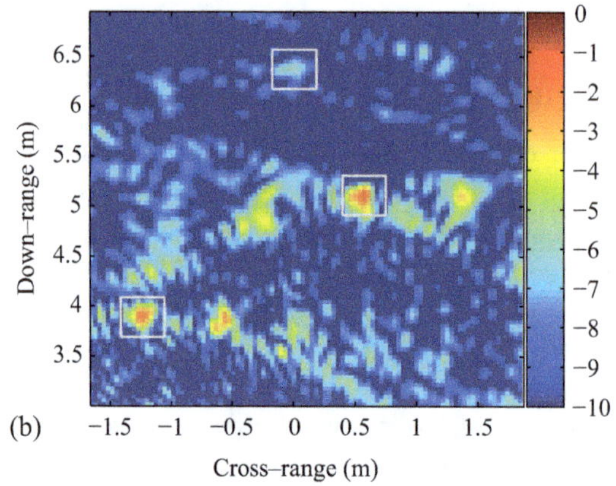

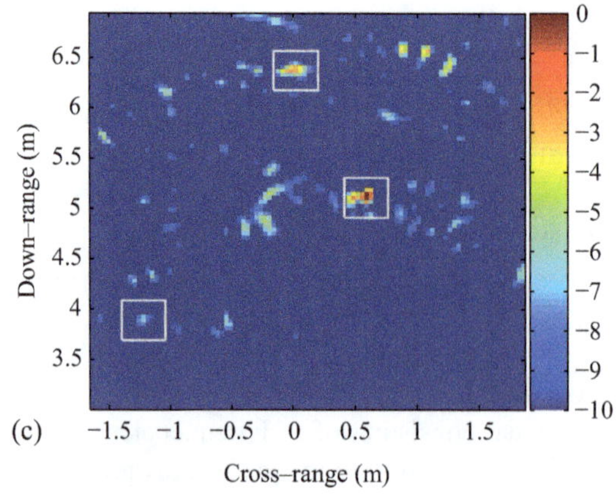

Figure 14.8 (a) Geometry of the simulated scene; (b) beamformed image of the scene; (c) correlogram matching result. ©2014 IEEE. Reprinted, with permission, from [22]

of the remaining corners are either facing away from the array or only visible to a small number of antenna elements. We, therefore, focus on the detection of these three corners.

The region to be imaged is chosen to be 3.5 (crossrange) × 3.9 (downrange) m^2, centred at (0.11, 4.97) m and is divided into 100 × 100 pixels. Figure 14.8(b) shows the beamformed image corresponding to the measured scene, using all 201 frequencies. In this figure and all subsequent images in this section, we plot the image intensity with the maximum intensity value in each image normalized to 0 dB. Although the corners of interest are visible in the image (indicated by white rectangles), it is difficult to discriminate them from other scatterers and clutter. Note that the imaged scatterers appear at biased locations compared to the ground truth. This is because the change in propagation speed within the walls has not been accounted for in beamforming. Figure 14.8(c) shows the image obtained with the correlogram matching approach using only 101 uniformly selected frequencies. More specifically, Figure 14.8(c) depicts the values of $\kappa(q)$ for $D = 5$. Clearly, the image shown in Figure 14.8(c) has detected the three corners and is less cluttered than the beamformed image in Figure 14.8(b).

The superior performance of the correlogram matching technique in terms of scatterer discrimination is confirmed by a comparison of the target-to-clutter ratio (TCR), provided in Table 14.1. The TCR of an image $I(\mathbf{r}_q)$ is defined as the ratio between the highest pixel intensity value of the true corner location area to the maximum pixel intensity value of the clutter area. The corner reflector area comprises the highest pixel values of the different detected corner areas, which are manually selected in close vicinity to the true upper-left corner positions (11 × 11 pixel rectangular box centred at the real corner position). The clutter area consists of the remaining pixels that are not included in the detected corner areas. Smaller values of TCR would increase the chance of misclassifications in subsequent thresholding-based detection schemes [89,90]. From Table 14.1, we observe that the correlogram matching method significantly surpasses the TCR values of the beamformed image for $D = 5$ and 15, leading to an enhanced detection performance. The performance of the proposed method is clearly linked to the dimensionality D of the correlogram matrix, which should be sufficiently large to capture the image-domain response of a corner reflector and small enough not to infringe over neighbouring corners. Increasing D adds more information to the correlogram matrix, resulting in higher TCR values as seen in Table 14.1 for $D = 5$ and $D = 15$. However, a higher D not only increases the computational complexity but also invades neighbouring scatterers, resulting in lower TCR values as seen in Table 14.1 for $D = 30$.

Table 14.1 Target-to-clutter ratio

Method	$D = 5$	$D = 15$	$D = 30$
Backprojection	1.48 dB		
$\kappa_M(k, l)$	6.11 dB	8.28 dB	0.76 dB

14.5 Building feature extraction using overcomplete dictionaries

Another approach for feature-based building layout determination is based on the use of OCD in a linear measurement model relating the image grid point characteristics to the radar measurements [68,91]. In an OCD, the response to a specific image grid point is assumed to be a weighted superposition of the responses to different canonical scatterer types, such as walls, trihedral corners and dihedral corners. As a result, the radar measurement model has a coefficient vector (scene reflectivity map) of higher dimension than the measurement vector. This yields an underdetermined system and extraction of the building features needs to be performed based on sparse representation algorithms. As such, for OCD processing even when the data are Nyquist sampled, sparse representation algorithms are required due to the simultaneous processing of multiple scattering models. This approach first performs some kind of inversion for obtaining the representation vector. After detection, plot extraction and classification, the obtained building features are fed into a mapping algorithm for obtaining the building layout.

Assuming P different canonical scatterer types, P different bases can be constructed with the pth scatterer model resulting in a dictionary $\mathbf{A}_p$. These different models are used to form a single, large dictionary matrix as,

$$\mathbf{A} = [\mathbf{A}_1 \quad \mathbf{A}_2 \quad \ldots \quad \mathbf{A}_P] \tag{14.20}$$

Using the overcomplete dictionary matrix $\mathbf{A}$ that contains $\overline{K}$ atoms for columns, the radar measurements can be represented as a linear combination of these atoms. The representation of the measurement vector $\mathbf{y}$ may either be exact

$$\mathbf{y} = \mathbf{A}\mathbf{x} \tag{14.21}$$

or approximate

$$||\mathbf{y} - \mathbf{A}\mathbf{x}||_l \leq \varepsilon \tag{14.22}$$

Here, the vector $\mathbf{x}$ contains the atom coefficients of the measurements $\mathbf{y}$ and is referred to as the representation of $\mathbf{y}$ and ε is a parameter reflecting the anticipated noise level in the measurements. In approximation methods, typical norms used for measuring the deviation correspond to l-norms for $l = 1, 2$, and ∞, with $l = 2$ being used most widely. In case of OCD, the linear representation system is underdetermined, requiring constraints to be set on the solution. The sparsest solution (solution with the fewest number of non-zero coefficients) for the representation is the solution of either

$$\min_{\mathbf{x}} ||\mathbf{x}||_0 \quad \text{subject to} \quad \mathbf{y} = \mathbf{A}\mathbf{x} \tag{14.23}$$

or

$$\min_{\mathbf{x}} ||\mathbf{x}||_0 \quad \text{subject to} \quad ||\mathbf{y} - \mathbf{A}\mathbf{x}||_2 \leq \varepsilon \tag{14.24}$$

where $|| \cdot ||_0$ is the zero norm, counting the non-zero entries of a vector. Extraction of the sparsest representation is a NP-hard problem. Algorithms for finding approximating solutions, for instance based on the 1-norm, have been extensively investigated (see [92] and references therein). The proper choice of ε is very important for accurate reconstruction. When the measurement noise statistics are known, ε can be chosen to be slightly larger than the standard deviation of the noise [93]. If the noise statistics are not known *a priori*, crossvalidation can be used for selecting ε [82].

14.5.1 OCD design

For the building structure determination problem, there are two approaches for building the OCD, namely, using known atoms or using adaptive dictionary techniques [68].

1. *Knowledge-based dictionaries*: This approach includes the use of physical models. Each atom represents a signal based on a physical assumption, e.g., the radar response of a wall or a trihedral corner. This allows the result to be easily interpreted. Each entry of the representation vector is directly related to a known model. The output yields the desired information. The suitability of the models to describe the signals defines how well the signal is represented. The models can be based on free-space propagation or purely on geometry. These models consider mainly the phase of the signal. The polarimetric information (both amplitude and phase) can also be included in the model.
2. *Adaptive knowledge-based dictionaries*: The suitability of the models might be compromised by measurement errors or errors of the environmental model. For instance, intervening walls might not have been included in the scattering model or the motion of the radar may not have been measured with sufficient accuracy. As a result, the atoms will change when intervening walls are taken into account or autofocusing techniques will change the motion model and hence the atoms. These new atoms will become specific for the measurements considered and are no longer as general as the pure knowledge-based atoms. The entries in the representation vector will still have the same physical interpretation as the non-adaptive representation. That is, the entries still represent the presence of planar surfaces, dihedral and trihedral corners. In order to account for and combat such modelling errors, simultaneous optimization can performed for obtaining the representation vector and the phase errors resulting from model mismatch [94,95].

14.5.2 Atom definition

In this section, we discuss the practical implementation of the atoms for the OCD. The approach taken here is the use of known atoms based on physical assumptions; each atom is tuned to the backscatter characteristics of a specific building elemental structure. Walls, i.e., planar surfaces and corners are canonical scatterers for which models exist describing the amplitude, phase and polarization characteristics of the backscatter. These specific characteristics can be exploited to extract the canonical scatterers from the radar data. Here, the phase change induced by the different types of scatterers is used for detection, classification and localization. In 3-D radar data, canonical scatterers induce different phase changes over the aperture in elevation and azimuth. Considering only specular reflections, the phase change characteristics of planar surfaces, dihedral corners and trihedral corners are as follows [91]:

- Planar surfaces are formed by walls that are roughly parallel to the aperture. Large planar walls induce a linear phase change over the aperture in azimuth and in elevation.
- Horizontal dihedral corners induce a linear phase change in azimuth and a quadratic phase change in elevation. Horizontal corners are formed by the junction of a wall and the ceiling or floor of a room.
- Vertical dihedral corners induce a quadratic phase change in azimuth and a linear phase change in elevation. Vertical corners are formed by the junction of two walls of a room.
- Trihedral corners induce a quadratic phase change in azimuth as well as in elevation. Trihedral corners are formed by the junction of two walls and the floor or ceiling of a room.

Walls perpendicular to the aperture are difficult to detect, since they only give rise to diffuse scattering in the direction of the radar. The presence of perpendicular walls needs to be deduced from the detected room corners.

The ideal phase change for a certain type of scatterer is affected by transmission through walls, but the polynomial degree of the phase change is preserved [68,91]. For reasons of clarity, the discussion here is limited to atom definition in the 2-D plane. This is sufficient to obtain the building floor plan, since it can be deduced from the location of walls and corners in a B-scan. Consequently, only phase changes related to planar walls, dihedral corners and point-like scatterers are treated. In the 2-D plane, trihedral and dihedral corners cannot be distinguished because they give rise to the same quadratic phase change. Trihedral corners are present at the intersection of two walls and the floor or ceiling of a room. Thus, extraction of trihedral corners provides primarily information on the height of a room. To extract trihedral corners, the representation process needs to be performed sequentially on a B-scan and an E-scan. When the representation indicates the presence of a dihedral corner for corresponding imaging points in both scans, it is likely that a trihedral corner is present. Another option is to define full 3-D atoms, but this will significantly increase the dimensions of the representation problem and, in turn, the required processing load. The atoms are, therefore, based on the free-space phase changes; the effect of propagation through one or more walls is neglected. Moreover, the defined atoms are based on specular reflections from walls and corners. This approach mitigates multipath reflections assuming these do not match the defined phase changes.

The application of an atom tuned to a quadratic phase change enables detection and classification of dihedral and trihedral corners. The part of the phase change parabola that is observable in the radar data depends on the orientation of the corner

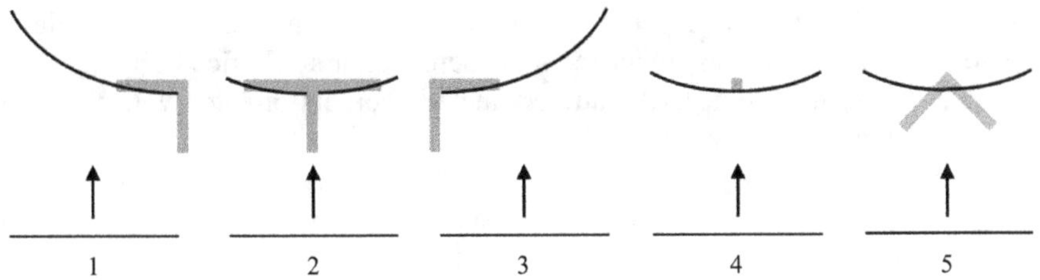

Figure 14.9 Schematic of five quadratic phase atoms. For different scatterer types (grey), the associated quadratic phase change (black curve) over the aperture (horizontal lines) is indicated. The arrows denote the radar line of sight. ©2016 IEEE. Reprinted, with permission, from [91]

relative to the aperture, as is schematically shown in Figure 14.9. For dihedral corners that are oriented such that one side is perpendicular to the aperture, only half of the phase parabola is observable. Thus, to discriminate left (no. 1) and right (no. 3) dihedral corners, distinction must be made between forward-looking and backward-looking atoms, each defining one half of the phase parabola. The forward-looking phase change (no. 1) is defined as [68,91]

$$\varphi_i(x) = \frac{4\pi}{\lambda}\sqrt{R_0^2 + x^2}, \quad \text{with } -L_a \leq x \leq 0 \tag{14.25}$$

where x is azimuth position, L_a is the aperture length and R_0 is the shortest distance between the aperture and the scatterer. The backward-looking phase change (no. 3) is defined as [68]

$$\varphi_i(x) = \frac{4\pi}{\lambda}\sqrt{R_0^2 + x^2}, \quad \text{with } 0 \leq x \leq L_a \tag{14.26}$$

For a double T-shaped corner (no. 2), the full phase parabola is observable in the radar data. This type of scatterer can be extracted by applying a third atom defining the full phase parabola. However, to prevent classification confusion, the full-parabola atom is not applied in the sparse representation with an OCD; it is assumed that a T-shaped corner is present when the representation yields a certain minimum value for both the forward-looking and backward-looking atoms at a single point in the representation grid. The ratio between the two values is an indication of the angle of the T-shaped corner with respect to the aperture [96].

As indicated in Figure 14.9, point-like scatterers (no. 4) also induce a full parabolic phase change. By applying the defined forward-looking and backward-looking atoms, point-like scatterers and T-shaped corners are indistinguishable. Objects inside the building, such as furniture, may act as point scatterer and lead to false alarms as far as the building map is concerned. This problem can be overcome by first detecting the building interior walls and subsequently by searching for corners only in the range bins where walls have been found [21,97]. This approach also reduces the processing load, since the search for corners is done on a limited number of grid points. A potential disadvantage of this approach is that building structural elements, such as free-standing concrete pillars in office buildings, may be missed. Note that point-like scatterers include corners oriented with the open side toward the aperture (no. 5).

Planar walls that are roughly parallel to the radar aperture induce a linear phase change over the aperture. In practice, the exact orientation of walls relative to the aperture is unknown and several atoms tuned to different wall orientations need to be defined, see Figure 14.10. To obtain orthogonal atoms, the difference in angle $\Delta\theta$ between the wall orientations defined in successive atoms is determined by the frequency resolution. The minimum angular step size $\Delta\theta$, given the frequency resolution, can be readily derived with the aid of Figure 14.10 as [68,91]

$$\tan(\Delta\theta) \geq \frac{\lambda}{2L_a}. \tag{14.27}$$

The linear phase change can be defined as

$$\varphi_i(x) = \frac{4\pi}{\lambda} R_0 + x\tan(i\Delta\theta), \quad \text{with } -\frac{L_a}{2} \leq x \leq \frac{L_a}{2} \tag{14.28}$$

The defined linear and parabolic atoms are of the following format

$$\mathbf{A}_i = R_0^\alpha \exp(j\varphi_i), \quad \text{with } \alpha > 0 \tag{14.29}$$

where α is an attenuation parameter and φ_i is the phase change, defined by either (14.25), (14.26), or (14.28). The attenuation parameter is used to compensate signal attenuation. It may be tuned to compensate free-space losses ($\alpha = 2$) or to compensate attenuation due to propagation through walls ($\alpha > 2$). For sparse representation, the attenuation parameter α plays an important role because sparse representation is an optimization procedure considering all grid points simultaneously. By setting the α to a higher value, points of the representation grid close to the aperture are penalized, enabling detection and classification of canonical

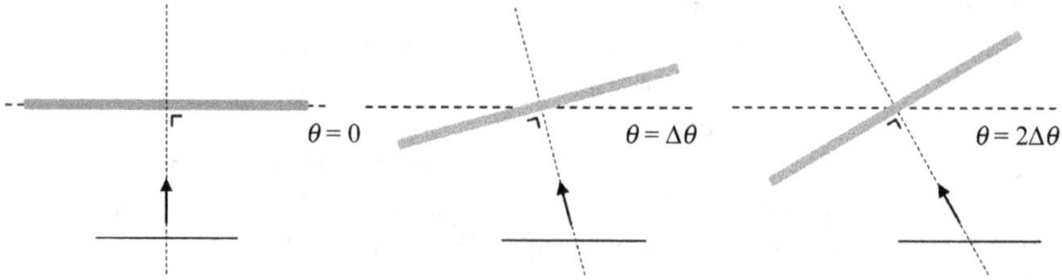

Figure 14.10 Schematic of three linear phase atoms. Different atoms need to be defined for different angles between the wall (grey) and the aperture (black line). The arrows denote the radar line of sight. Note that specular reflection of the wall is considered. ©2016 IEEE. Reprinted, with permission, from [91]

scatterers deeper inside the building. In order to avoid the classification confusion, the atoms need to be orthogonal. However, e.g., due to oversampling, the atoms within a single model are not orthogonal. Also, the quadratic and linear phase change atoms will be correlated, since a parabola can be piece-wise approximated by line segments. The level of correlation depends on the measurement geometry and the location of the grid point under test.

14.5.3 Illustrative example

The OCD approach is assessed using real data measurements obtained with the SAPPHIRE through-the-wall radar [68]. SAPPHIRE is a vehicle borne stand-off radar system that acquires 3-D data measurements using a vertical array by moving it parallel to the building being interrogated. The building under test is a three-story building, whose layout is schematically shown in Figure 14.11. The rooms on both sides of the hallway contain furniture items, except for room no. 3 which only contains a trihedral radar reflector on a tripod at 1.5 m height in the middle of the

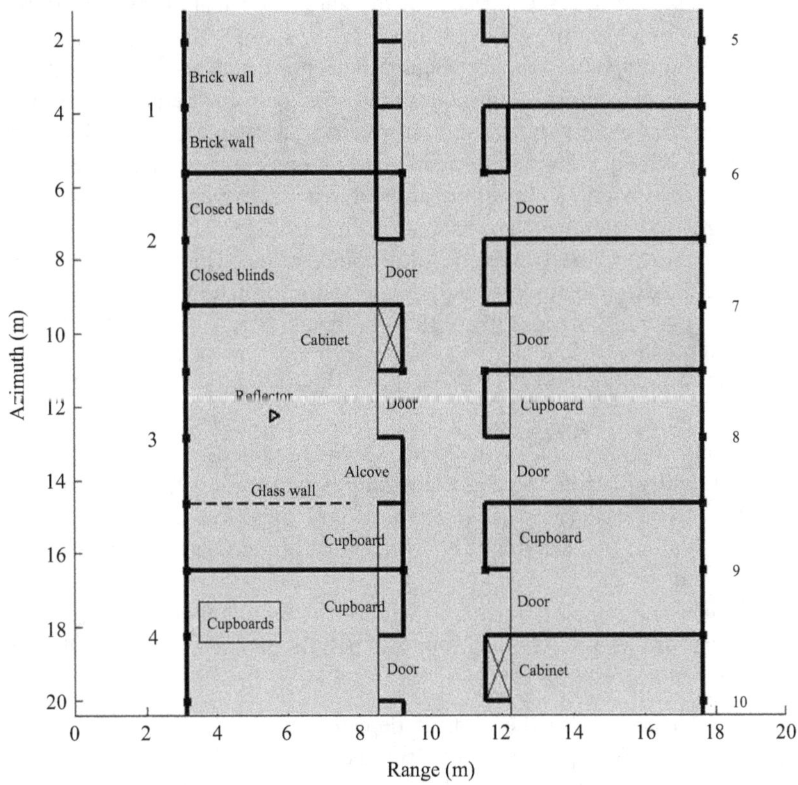

Figure 14.11 Schematic of the test building. ©CRC Press. Reprinted, with permission, from [68]

room. The sides of the hallway consist of concrete pillars and brick walls forming open alcoves, cupboards, or closed spaces. The radar track is along the azimuth axis. A uniform image grid with a step size equal to one-half the system resolution is assumed. The sparse reconstruction algorithm used in this example is complex approximate message passing (CAMP) (see [98] and references therein), which provides a noisy representation for each scattering type. The noisy representations are then used as input for detection, plot extraction and classification algorithms.

1. *Reflectivity maps*: The data used as input to CAMP is resampled down to the Nyquist rate in order to have an underdetermined problem when applying CAMP to obtain the representation vector. Only data up to 10 m in range were used in order to keep the computational burden sufficiently low. The output of CAMP is a noisy representation vector for all three different models included in the OCD. These representations can be used as equivalent to reflectivity maps (See Figure 14.12).
2. *Classified scatterers*: A fixed-threshold detector is applied to the noisy representations. Due to the choice of a grid size smaller than the resolution (correlated adjacent grid points) and detection of sidelobes, the detections appear in groups. As such, the second step of the automated feature extraction is plot extraction or clustering of detections. To this end, the detections are first clustered in range to remove sidelobe detections and correlated grid points. Considering detections in adjacent range bins, only the detection corresponding to the grid point with the highest reflection is retained, while all others are removed. Subsequently, the remaining detections are clustered in azimuth and the detections within a cluster are aligned to a single range bin. The plots after clustering and alignment are shown in Figure 14.13. The extracted plots are input to the third step, namely, plot combination. This step is required to translate the extracted plots to building layout information suitable to input to a building mapping process. The plot combination is based on the following rules [68]:
   - An uninterrupted row of at least four similar corners is interpreted as a wall.
   - Coinciding left and right corners are interpreted as a double T-shaped corner. Corners are seen as coinciding if they are less than four grid points (i.e., two resolution cells) apart in any direction. If several plots are coinciding, the left and right corners that are closest are combined into a double corner.
   - The remaining plots related to left corners are combined into a single left corner if they are less than four grid points apart. This rule applies to right corners as well.

The result after combining the plots is depicted in Figure 14.14. The scatterers classified as walls correspond to planar surfaces. The classified double corners are related to point-like scatterers, such as the concrete pillars and the reference reflector. The other classified single and double corners seem either due to objects giving high reflections or possible classification confusion in populated areas.

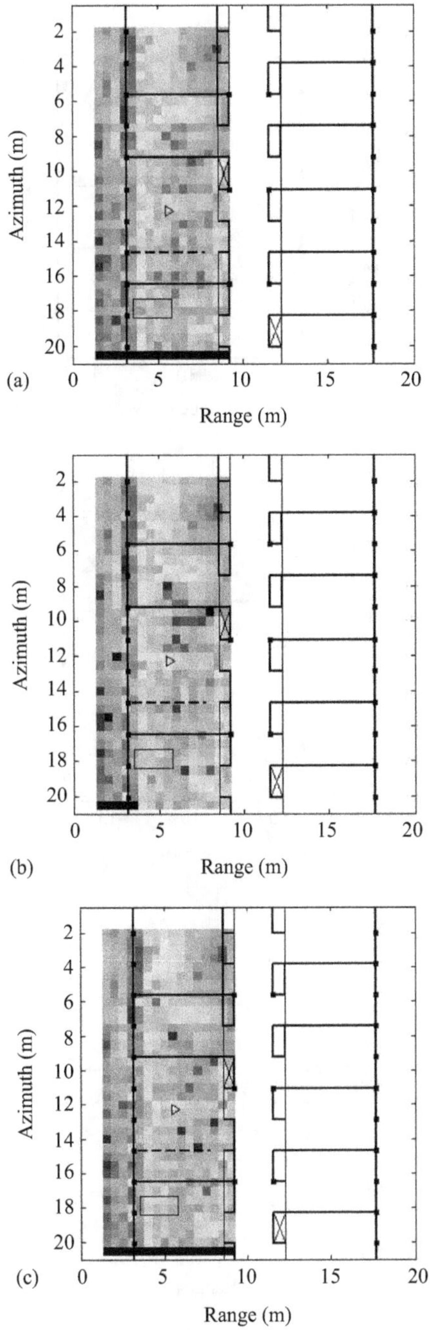

Figure 14.12 Reflectivity maps using OCD. (a) The forward-looking model, (b) the backward-looking model and (c) the linear model. The actual building floor plan is overlaid on the representation results. ©2016 IEEE. Reprinted, with permission, from [91]

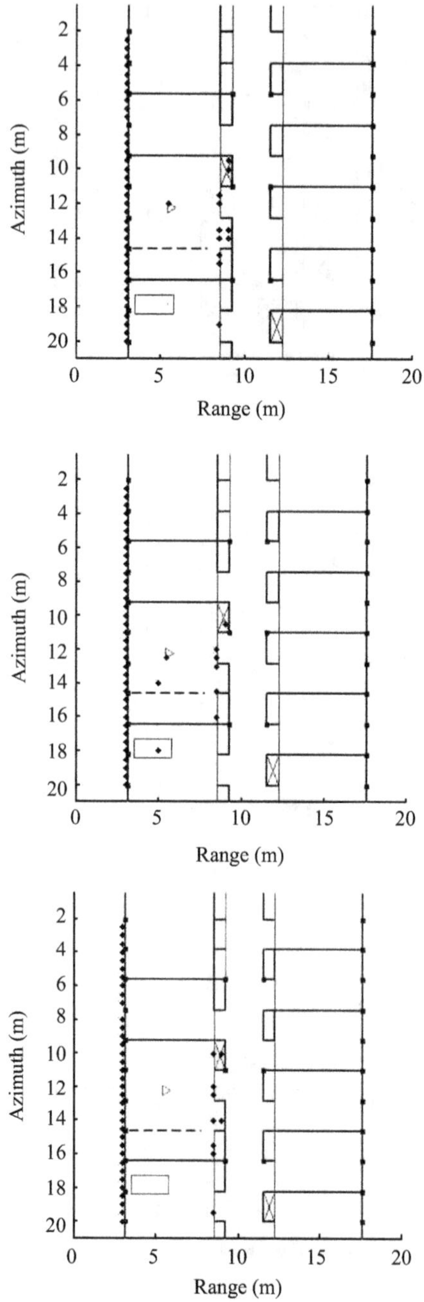

Figure 14.13 Plots after clustering and alignment: the forward-looking model (top), the backward-looking mode (centre) and the linear model (bottom). The actual building floor plan is overlaid on the detection results. ©2016 IEEE. Reprinted, with permission, from [91]

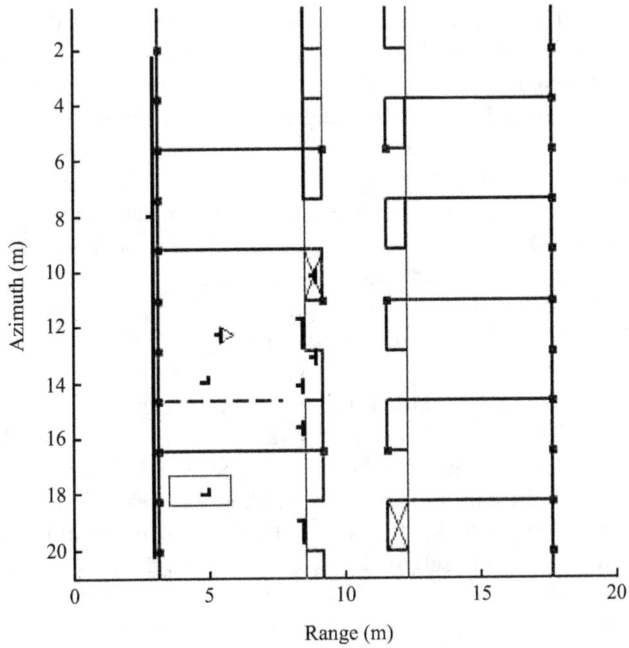

Figure 14.14 Combined plots providing building feature information suitable for input to a building mapping algorithm. The actual building floor plan is overlaid on the detection results. ©2016 IEEE. Reprinted, with permission, from [91]

14.6 Conclusion

In this chapter, we presented a review of important approaches for building layout determination and imaging of stationary indoor scenes. These approaches address the unique challenges associated with imaging through walls. First, non-adaptive time-domain and frequency-domain beamforming algorithms were presented which accounted for propagation effects through walls. Beamforming incorporating wall effects compensation was shown to successfully detect and accurately localize the targets. Second, a feature-based approach was presented for estimating scattering primitives that can be used to infer information about the interior structure of a building. The approach was based on decomposing the radar backscatter response from a building into returns from discrete trihedral and dihedral scatterers formed by wall-floor interactions, as well as flat plate returns. The problem was posed as one of estimating the parameters of a scattering primitive model. This high-order non-linear maximization problem was decomposed into a set of lower-order problems by a combination of subaperture azimuth imaging and height filtering from SAR interferometry. In addition, a non-parametric sparse reconstruction procedure was used to initialize the iterative non-linear maximization procedures.

Using numerical data from a two-story building, extracted trihedrals were shown to be in good agreement with building structure, with all upper-floor and most lower-floor corners detected with high accuracy. Third, a pattern matching strategy was presented for imaging of interior building structure. Under assumption of oblique illumination, the technique made use of a correlogram and correlation matching framework to detect corners. The scheme is applicable under both full and reduced data volume and was shown to provide reliable determination of building layouts using numerical EM data. Finally, the use of OCD for building interior feature extraction was detailed, wherein simultaneous use of the outputs of three different models were used to locate and classify the walls, dihedral corners and point-like scatterers. The performance of the OCD approach was assessed using real data measurements.

Although the progress reported in this chapter is significant and noteworthy, there are many challenges still facing the radar technology for building interior imaging. The presence of false positives and ghosts, which are the products of multipath, render it difficult to determine building features. Further, the relatively limited bandwidth does not always lead to sufficient resolution of key scatterers, a situation that does not lend itself to effective feature extraction. Complex buildings, comprising multiple floors, each with populated scenes present the ultimate challenge in building interior imaging, specifically with long stand-off distances, irrespective of whether ground-based or airborne platforms are employed. All these factors necessitate more research and development efforts to be conducted for effective and reliable building interior imaging using radars.

List of acronyms

TWRI	through-the-wall radar imaging
EM	electromagnetic
CD	change detection
1-D	one-dimensional
2-D	two-dimensional
3-D	three-dimensional
SAR	synthetic aperture radar
SVD	singular value decomposition
SCR	signal-to-clutter ratio
UWB	ultrawideband
NEC	National Electromagnetics Consortium
BSC	basic scattering code
TCR	target-to-clutter ratio
OCD	overcomplete dictionaries
CAMP	complex approximate message passing

References

[1] Amin, M.G.: *Through-the-Wall Radar Imaging* (CRC Press, Boca Raton, FL, 2011).

[2] Amin, M.G.: *Compressive Sensing for Urban Radar* (CRC Press, Boca Raton, FL, 2015).

[3] Amin, M.G., Sarabandi, K. (Eds.): 'Special issue on remote sensing of building interior', *IEEE Trans. Geosci. Remote Sens.*, **47**, (5), pp. 1270–1420.

[4] Borek, S.E.: 'An overview of through the wall surveillance for homeland security'. *Proc. 34th Applied Imagery and Pattern Recognition Workshop*, Washington, DC, USA, October 2005, **6**, pp. 19–21.

[5] Burchett, H.: 'Advances in through wall radar for search, rescue and security applications'. *Proc. Inst. of Eng. and Tech. Conf. Crime and Security*, London, UK, June 2006, pp. 511–525.

[6] Martone, A.F., Ranney, K., Le, C.: 'Noncoherent approach for through-the-wall moving target indication', *IEEE Trans. Aerosp. Electron. Syst.*, 2014, **50**, (1), pp. 193–206.

[7] Amin, M.G., Ahmad, F.: 'Change detection analysis of humans moving behind', *IEEE Trans. Aerosp. Electron. Syst.*, 2013, **49**, (3), pp. 1410–1425.

[8] Ram, S.S., Ling, H.: 'Through-wall tracking of human movers using joint Doppler and array processing', *IEEE Geosci. Remote Sens. Lett.*, 2008, **5**, (3), pp. 537–541.

[9] Lai, C.P., Narayanan, R.M.: 'Ultrawideband random noise radar design for through-wall surveillance', *IEEE Trans. Aerosp. Electron. Syst.*, 2010, **46**, (4), pp. 1716–1730.

[10] Greneker, E.F.: 'RADAR flashlight for through-the-wall detection of humans'. *Proc. SPIE Targets Backgrounds: Charact. Representation IV Conf.*, Orlando, FL, USA, 1998, **3375**, pp. 280–285.

[11] Hunt, A.R.: 'Use of a frequency-hopping radar for imaging and motion detection through walls', *IEEE Trans. Geosci. Remote Sens.*, 2009, **47**, (5), pp. 1402–1408.

[12] Ahmad, F., Amin, M.G., Zemany, P.D.: 'Dual-frequency radars for target localization in urban sensing', *IEEE Trans. Aerosp. Electronic Syst.*, 2009, **45**, (4), pp. 1598–1609.

[13] Maaref, N., Millot, P., Pichot, C., Picon, O.: 'A study of UWB FM-CW radar for the detection of human beings in motion inside a building', *IEEE Trans. Geosci. Remote Sens.*, 2009, **47**, (5), pp. 1297–1300.

[14] Soldovieri, F., Solimene, R., Pierri, R.: 'A simple strategy to detect changes in through the wall imaging', *Progress in Electromagnetics Research M*, 2009, **7**, pp. 1–13.

[15] Lavely, E.M., Zhang, Y., Hill III, E.H., *et al.*: 'Theoretical and experimental study of through-wall microwave tomography inverse problems', *J. Franklin Institute*, 2008, **345**, (6), pp. 592–617.

[16] Subotic, N., Keydel, E., Burns, J., *et al.*: 'Parametric reconstruction of internal building structures via canonical scattering mechanisms'. *Proc. IEEE Int. Conf. Acoust. Speech Signal Process.*, Las Vegas, NV, USA, March-April 2008, pp. 5189–5192.

[17] Ertin, E., Moses, R.L.: 'Through-the-wall SAR attributed scattering center feature estimation', *IEEE Trans. Geosci. Remote Sens.*, 2009, **47**, (5), pp. 1338–1348.

[18] Le, C., Dogaru, T., Nguyen, L., Ressler, M.A.: 'Ultrawideband (UWB) radar imaging of building interior: Measurements and predictions', *IEEE Trans. Geosci. Remote Sens.*, 2009, **47**, (5), pp. 1409–1420.

[19] Aftanas, M., Drutarovsky, M.: 'Imaging of the building contours with through the wall UWB radar system', *J. Radioeng.*, 2009, **18**, (3), pp. 258–264.

[20] Nikolic, M.M., Ortner, M., Nehorai, A., Djordjevic, A.R.: 'An approach to estimating building layouts using radar and jump-diffusion algorithm', *IEEE Trans. Antennas Propag.*, 2009, **57**, (3), pp. 768–776.

[21] Lagunas, E., Amin, M.G., Ahmad, F., Najar, M.: 'Determining building interior structures using compressive sensing', *J. Electron. Imaging*, 2013, **22**, (2), pp. 021003.

[22] Lagunas, E., Amin, M.G., Ahmad, F., Najar, M.: 'Pattern matching for building feature extraction', *IEEE Geosci. Remote Sens. Lett.*, 2014, **11**, (12), pp. 2193–2197.

[23] Ahmad, F., Amin, M.G., Dogaru, T.: 'Partially sparse image reconstruction of stationary indoor scenes', *EURASIP J. Adv. Signal Proc.*, 2014, 2014:100, pp. 1–15.

[24] Ahmad, F., Frazer, G.J., Kassam, S.A., Amin, M.G.: 'Design and implementation of near-field, wideband synthetic aperture beamformers', *IEEE Trans. Aerosp. Electron. Syst.*, 2004, **40**, (1), pp. 206–220.

[25] Ahmad, F., Amin, M.G., Kassam, S.A.: 'Synthetic aperture beamformer for imaging through a dielectric wall', *IEEE Trans. Aerosp. Electronic Syst.*, 2005, **41**, (1), pp. 271–283.

[26] Song, L.P., Yu, C., Liu, Q.H.: 'Through-wall imaging (TWI) by radar: 2-D tomographic results and analyses', *IEEE Trans. Geosci. Remote Sens.*, 2005, **43**, (12), pp. 2793–2798.

[27] Ahmad, F., Amin, M.G.: 'Noncoherent approach to through-the-wall radar localization', *IEEE Trans. Aerosp. Electron. Syst.*, 2006, **42**, (4), pp. 1405–1419.

[28] Soldovieri, F., Solimene, R.: 'Through-wall imaging via a linear inverse scattering algorithm', *IEEE Geosci. Remote Sens. Lett.*, 2007, **4**, (4), pp. 513–517.

[29] Ahmad, F., Amin, M.: 'Multi-location wideband synthetic aperture imaging for urban sensing applications', *J. Franklin Institute*, 2008, **345**, (6), pp. 618–639.

[30] Ahmad, F., Zhang, Y., Amin, M.G.: 'Three-dimensional wideband beamforming for imaging through a single wall', *IEEE Geosci. Remote Sens. Lett.*, 2008, **5**, (2), pp. 176–179.

[31] Amin, M.G., Ahmad, F.: 'Wideband synthetic aperture beamforming for through-the-wall imaging', *IEEE Signal Process. Mag.*, 2008, **25**, (4), pp. 110–113.

[32] Dehmollaian, M., Sarabandi, K.: 'Refocusing through building walls using synthetic aperture radar', *IEEE Trans. Geosci. Remote Sens.*, 2008, **46**, (6), pp. 1589–1599.

[33] Yoon, Y-S., Amin, M.G.: 'High-resolution through-the-wall radar imaging using beamspace MUSIC', *IEEE Trans. Antennas Propag.*, 2008, **56**, (6), pp. 1763–1774.

[34] Dehmollaian, M., Thiel, M., Sarabandi, K.: 'Through-the-wall imaging using differential SAR', *IEEE Trans. Geosci. Remote Sens.*, 2009, **47**, (5), pp. 1289–1296.

[35] Yang, Y., Fathy, A.: 'Development and implementation of a real-time see-through-wall radar system based on FPGA', *IEEE Trans. Geosci. Remote Sens.*, 2009, **47**, (5), pp. 1270–1280.

[36] Zhang, W., Hoorfar, A., Thajudeen, C., Ahmad, F.: 'Full polarimetric beamforming algorithm for through-the-wall radar imaging', *Radio Sci.*, 2011, **46**, (5), RS0E16.

[37] Soldovieri, F., Ahmad, F., Solimene, R.: 'Validation of microwave tomographic inverse scattering approach via through-the-wall experiments in semicontrolled conditions', *IEEE Geosci. Remote Sens. Lett.*, 2011, **8**, (1), pp. 123–127.

[38] Soumekh, M.: 'Synthetic Aperture Radar Signal Processing with Matlab Algorithms' (John Wiley and Sons, New York, NY, 1999).

[39] Yoon, Y-S., Amin, M.G.: 'Spatial filtering for wall-clutter mitigation in through-the-wall radar imaging', *IEEE Trans. Geosci. Remote Sens.*, 2009, **47**, (9), pp. 3192–3208.

[40] Burkholder, R.: 'Electromagnetic models for exploiting multi-path propagation in through-wall radar imaging'. *Proc. IEEE Int. Conf. Electromagnetics in Advanced Applications*, Torino, Italy, September 2009, pp. 572–575.

[41] Dogaru, T., Le, C.: 'SAR images of rooms and buildings based on FDTD computer models', *IEEE Trans. Geosci. Remote Sens.*, 2009, **47**, (5), pp. 1388–1401.

[42] Setlur, P., Amin, M., Ahmad, F.: 'Multipath model and exploitation in through-the-wall and urban radar sensing', *IEEE Trans. Geosci. Remote Sens.*, 2011, **49**, (10), pp. 4021–4034.

[43] Alli, G., DiFilippo, D.: 'Beamforming for through-the-wall radar imaging', in Amin, M.G. (Ed.): *Through-the-Wall Radar Imaging* (CRC Press, Boca Raton, FL, 2011), pp. 81–120.

[44] Ahmad, F., Amin, M.G., Kassam, S.A.: 'A beamforming approach to stepped-frequency synthetic aperture through-the-wall radar imaging'. *Proc. 1st IEEE Int. Workshop Computational Advances in Multi-Sensor Adaptive Process.*, Puerto Vallarta, Mexico, December 2005, **1**, pp. 24–27.

[45] Wang, G., Amin, M.G., Zhang, Y.: 'A new approach for target locations in the presence of wall ambiguity', *IEEE Trans. Aerosp. Electron. Syst.*, 2006, **42**, (1), pp. 301–315.

[46] Ahmad, F., Amin, M.G.: 'Performance of autofocusing schemes for single target and populated scenes behind unknown walls'. *Proc. SPIE Radar Sensor Technology XI Conf.*, Orlando, FL, April 2007, **6547**, pp. 654709.

[47] Ahmad, F., Amin, M.G., Mandapati, G.: 'Autofocusing of through-the-wall radar imagery under unknown wall characteristics', *IEEE Trans. Image Process.*, 2007, **16**, (7), pp. 1785–1795.

[48] Soldovieri, F., Prisco, G., Solimene, R.: 'A multi-array tomographic approach for through-wall imaging', *IEEE Trans. Geosci. Remote Sens.*, 2008, **46**, (4), pp. 1192–1199.

[49] Thajudeen, C., Zhang, W., Hoorfar, A.: 'Time-domain wall parameter estimation and mitigation for through-the-wall radar image enhancement'. *Proc. Progress in Electromagnetics Research Symp.* (EMW publishing, Cambridge, MA, 2010).

[50] Solimene, R., Soldovieri, F., Prisco, G., Pierri, R.: 'Three-dimensional through-wall imaging under ambiguous wall parameters', *IEEE Trans. Geosci. Remote Sens.*, 2009, **47**, (5), pp. 1310–1317.

[51] Chandra, R., Gaikwad, A.N., Singh, D., Nigam, M.J.: 'An approach to remove the clutter and detect the target for ultra-wideband through-wall imaging', *J. Geophys. Eng.*, 2008, **5**, (4), pp. 412–419.

[52] Riaz, M., Ghafoor, A.: 'Through-wall image enhancement based on singular value decomposition', *Int. J. Antennas Propag.*, 2012, **2012**, Article ID 961829, pp. 1–20.

[53] Ahmad, F., Amin, M.G.: 'Wall clutter mitigation for MIMO radar configurations in urban sensing'. *Proc. 11th Int. Conf. Inf. Sci. Signal Process., and their Applications*, Montreal, Canada, July 2012, pp. 1165–1170.

[54] Tivive, F.H.C., Bouzerdoum, A., Amin, M.G.: 'A subspace projection approach for wall clutter mitigation in through-the-wall radar imaging', *IEEE Trans. Geosci. Remote Sens.*, 2015, **53**, (4), pp. 2108–2122.

[55] Ahmad, F., Qian, J., Amin, M.G.: 'Wall clutter mitigation using discrete prolate spheroidal sequences for sparse reconstruction of indoor stationary scenes', *IEEE Trans. Geosci. Remote Sens.*, 2015, **53**, (3), pp. 1549–1557.

[56] Lagunas, E., Amin, M.G., Ahmad, F., Najar, M.: 'Joint wall mitigation and compressive sensing for indoor image reconstruction', *IEEE Trans. Geosci. Remote Sens.*, 2013, **51**, (2), pp. 891–906.

[57] Mansour, H., Liu, D.: 'Blind multi-path elimination by sparse inversion in through-the-wall-imaging'. *Proc. 5th IEEE Int. Workshop Computational Advances Multi-Sensor Adaptive Process.*, Saint Martin, December 2013, pp. 256–259.

[58] Kidera, S., Sakamoto, T., Sato, T.: 'Extended imaging algorithm based on aperture synthesis with double-scattered waves for UWB radars', *IEEE Trans. Geosci. Remote Sens.*, 2011, **49**, (12), pp. 5128–5139.

[59] Gennarelli, G., Soldovieri, F.: 'A linear inverse scattering algorithm for radar imaging in multipath environments', *IEEE Geosci. Remote Sens. Lett.*, 2013, **10**, (5), pp. 1085–1089.

[60] Gennarelli, G., Riccio, G., Solimene, R., Soldovieri, F.: 'Radar imaging through a building corner', *IEEE Trans. Geosci. Remote Sens.*, 2014, **52**, (10), pp. 6750–6761.

[61] Smith, G., Mobasseri, B.: 'Analysis and exploitation of multipath ghosts in radar target image classification', *IEEE Trans. Image Process.*, 2014, **23**, (4), pp. 1581–1592.

[62] Leigsnering, M., Ahmad, F., Amin, M.G., Zoubir, A.M.: 'Multipath exploitation in through-the-wall radar imaging using sparse reconstruction', *IEEE Trans. Aerosp. Electron. Syst.*, 2014, **50**, (2), pp. 920–939.

[63] Leigsnering, M., Amin, M.G., Ahmad, F., Zoubir, A.M.: 'Multipath exploitation and suppression in SAR imaging of building interiors', *IEEE Signal Process. Mag.*, 2014, **31**, (4), pp. 110–119.

[64] Leigsnering, M., Ahmad, F., Amin, M.G., Zoubir, A.M.: 'Compressive sensing based multipath exploitation for stationary and moving indoor target localization', *IEEE J. Sel. Topics Signal Process.*, 2015, **9**, (8), pp. 1469–1483.

[65] Setlur, P., Smith, G.E., Ahmad, F., Amin, M.G.: 'Target localization with a single sensor via multipath exploitation', *IEEE Trans. Aerosp. Electron. Syst.*, 2012, **48**, (3), pp. 1996–2014.

[66] Muqaibel, A.H., Amin, M.G., Ahmad, F.: 'Target localization with a single antenna via directional multipath', *Int. J. Antennas Propag.*, 2015, **2015**, Article ID 510720, pp. 1–10.

[67] Ertin, E., Moses, R.L.: 'Through-the-wall SAR for characterization of building interior structure using attributed scattering center features', in Amin, M.G. (Ed.): *Through-the-Wall Radar Imaging* (CRC Press, Boca Raton, FL, 2011), pp. 379–412.

[68] van Rossum, W., de Wit, J.: 'Overcomplete dictionary design for building feature extraction', in Amin, M.G. (Ed.): *Compressive Sensing for Urban Radar* (CRC Press, Boca Raton, FL, 2015), pp. 49–86.

[69] Weichman, P.B., Lavely, E.M., Hill III, E.H., Zemany, P.D.: 'Through-the-wall microwave building tomography', in Amin, M.G. (Ed.): *Through-the-Wall Radar Imaging* (CRC Press, Boca Raton, FL, 2011), pp. 219–268.

[70] Ahmad, F., Amin, M.G.: 'High resolution imaging using Capon beamformers for urban sensing applications'. *Proc. IEEE Int. Conf. Acoust. Speech Signal Process.*, Honolulu, HI, USA, April 2007, **2**, pp. 985–988.

[71] Yoon, Y.-S., Amin, M.G., Ahmad, F.: 'MVDR beamforming for through-the-wall radar imaging', *IEEE Trans. Aerosp. Electron. Syst.*, 2011, **47**, (1), pp. 347–366.

[72] Benitz, G.R.: 'High-definition vector imaging', *Lincoln Lab. J.*, 1997, **10**, (2), pp. 147–169.

[73] Wang, G., Amin, M.G.: 'Imaging through unknown walls using different standoff distances', *IEEE Trans. Signal Process.*, 2006, **54**, (10), pp. 4015–4025.

[74] Potter, L.E., Moses, R.L.: 'Attributed scattering centers for SAR ATR', *IEEE Trans. Image Process.*, 1997, **6**, (1), pp. 79–91.

[75] Jackson, J., Rigling, B., Moses, R.: 'Canonical scattering feature models for 3D and bistatic SAR', *IEEE Trans. Aerosp. Electron. Syst.*, 2010, **46**, (2), pp. 525–541

[76] Corona, P., Ferrara, G., Agostino, F.D., *et al.*: 'An improved physical optics model for the evaluation of the field backscattered by triangular trihedral corner reflectors'. *Proc. 8th Mediterranean Electrotechnical Conf.*, Bari, Italy, May 1996, **1**, pp. 534–537.

[77] Franceschetti, G., Iodice, A., Riccio, D.: 'A canonical problem in electromagnetic backscattering from buildings', *IEEE Trans. Geosci. Remote Sens.*, 2002, **40**, (8), pp. 1787–1801.

[78] Stowell, M., Fasenfest, B., White, D.: 'Investigation of radar propagation in buildings: a 10-billion element Cartesian-mesh FETD simulation', *IEEE Trans. Antennas Propag.*, 2008, **56**, (8) Part 1, pp. 2241–2250.

[79] Akyildiz, Y., Moses, R.L.: 'Structure selection in synthetic aperture radar scattering models'. *Proc. IEEE Int. Conf. Acoust. Speech Signal Process.*, Istanbul, Turkey, June 2000, **5**, pp. 3021–3024.

[80] Moses, R., Potter, L., Cetin, M.: 'Wide angle SAR imaging'. *Proc. SPIE Algorithms for Synthetic Aperture Radar Imagery XI Conf.*, Orlando, FL, USA, April 2004, **5427**, pp. 164–175.

[81] Cetin, M., Karl, W.C.: 'Feature enhanced synthetic aperture radar image formation based on nonquadratic regularization', *IEEE Trans. Image Process.*, 2001, **10**, (4), pp. 623–631.

[82] Boufounos, P., Duarte, M., Baraniuk, R.: 'Sparse signal reconstruction from noisy compressive measurements using cross-validation'. *Proc. IEEE Workshop Statistical Signal Process.*, Madison, WI, USA, August 2007, pp. 299–303.

[83] Marhefka, R.J.: *Numerical electromagnetic code-basic scattering code (NEC-BSC version 4.2), Users manual* (The Ohio State University, Columbus, OH, 2000).

[84] Huang, J., Kumar, S., Mitra, *et al.*: 'Image indexing using color correlograms'. *Proc. IEEE Conf. Computer Vision and Pattern Recognition*, Puerto Rico, June 1997, pp. 762–768.

[85] Sevigny, P., DiFilippo, D.: 'A multi-look fusion approach to through wall radar imaging'. *Proc. IEEE Radar Conf.*, Ottawa, Canada, April–May 2013, pp. 1–6.

[86] Amin, M.G., Ahmad, F., Hoorfar, A.: 'Squinted array beamforming for wall back plate detection'. *Proc. Int. Radar Conf.*, Lille, France, October 2014, pp. 1–3.

[87] Shaghaghi, M., Vorobyov, S.: 'Correlogram for undersampled data: Bias and variance analysis', *Proc. IEEE Intl. Conf. Acoust. Speech, Signal Process.*, Kyoto, Japan, March 2012, pp. 3513–3516.

[88] 'FEKO EM Simulation Software Web page', http://www.feko.info, accessed August 2016.

[89] Debes, C., Amin, M.G., Zoubir, A.M.: 'Target detection in single and multiple-view through-the-wall radar imaging', *IEEE Trans. Geosci. Remote Sens.*, 2009, **47**, (5), pp. 1349–1361.

[90] Seng, C., Amin, M., Ahmad, F., Bouzerdoum, A.: 'Image segmentations for through-the-wall radar target detection', *IEEE Trans. Aerosp. Electron. Syst.*, 2013, **49**, (3), pp. 1869–1896.

[91] van Rossum, W., de Wit, J.: 'Extraction of building features from stand-off measured through-wall radar data', *IEEE J. Sel. Topics Appl. Earth Observ. Remote Sens.*, 2016, **9**, (1), pp. 149–158.

[92] Wakin, M.B.: 'Compressive sensing fundamentals', in Amin, M.G. (Ed.): *Compressive Sensing for Urban Radar* (CRC Press, Boca Raton, FL, 2015), pp. 1–48.

[93] Gurbuz, A.C., McClellan, J.H., Scott, W.R.: 'A compressive sensing data acquisition and imaging method for stepped-frequency GPRs', *IEEE Trans. Signal Process.*, 2009, **57**, (7), pp. 2640–2650.

[94] Onhon, N.O., Cetin, M.: 'A nonquadratic regularization-based technique for joint SAR imaging and model error correction'. *Proc. SPIE Algorithms for Synthetic Aperture Radar Imagery XVI Conf.*, Orlando, FL, USA, April 2009, **7337**, pp. 73370C-1–73370C-10.

[95] Leigsnering, M., Ahmad, F., Amin, M.G., Zoubir, A.M.: 'CS based specular multipath exploitation in TWRI under wall position uncertainties'. *Proc. IEEE 8th Sensor Array and Multi-Channel Signal Process.* Workshop, A Coruna, Spain, June 2014, pp. 481–484.

[96] Chen, B., Jin, T., Zhou, Z., Lu, B.: 'Estimation of pose angle for trihedral in ultrawideband virtual aperture radar', *Progress in Electromagnetic Research*, 2013, **138**, pp. 307–325.

[97] Lagunas, E., Amin, M.G., Ahmad, F., Najar, M.: 'Improved interior wall detection using designated dictionaries in compressive urban sensing problems'. *Proc. SPIE Compressive Sensing II Conf.*, Baltimore, MD, USA, May 2013, **8717**, pp. 87170K-1–87170K-7.

[98] Anitori, L., Maleki, A., Otten, M.P.G., *et al.*: 'Design and analysis of compressed sensing radar detectors', *IEEE Trans. Signal Process.*, 2013, **61**, (4), pp. 813–827.

Chapter 15
Short-range passive radar potentialities

Fabiola Colone[1]

Abstract

The potentialities of passive radar (PR) will be illustrated in this chapter with reference to the surveillance of areas of limited extent. Despite the exploitation of the PR principle in short-range applications could appear as a simple task, it is shown that it brings a number of challenging issues that must be solved to benefit from its potential advantages. In addition, the requirements on such sensors could be much more demanding to enable advanced capabilities.

The chapter attempts to give the reader an insight into the real-world applications of short-range PR. The considered applications span from maritime traffic control, to vehicular traffic monitoring, and indoor surveillance. For each case, experimental results are reported, obtained in different scenarios when exploiting different waveforms of opportunity. Walking through these results also gives the chance to describe some technical aspects related to system design issues and signal processing techniques.

15.1 Introduction

Passive radar (PR) has rapidly reached a point of maturity in many long-range surveillance applications, with much of the research rooted in air traffic control (ATC) systems [1–7]. However, in the last years, many advanced solutions and emerging applications have been also proposed; both are aiming to increase the reliability of systems, to improve their potentialities, and hence widen the range of uses.

In particular, a number of studies have looked at the use of PR systems for the detection and tracking of targets at close distance. The list of references provides several examples originated by various research groups (e.g., see [8–45]).

This chapter takes this perspective and illustrates some recent developments in PR sensing for short-range surveillance applications. However, before proceeding further, it is worth clarifying what the words 'short-range' encompass.

[1]Department of Information Engineering, Electronics and Telecommunications (DIET), Sapienza University of Rome, Italy

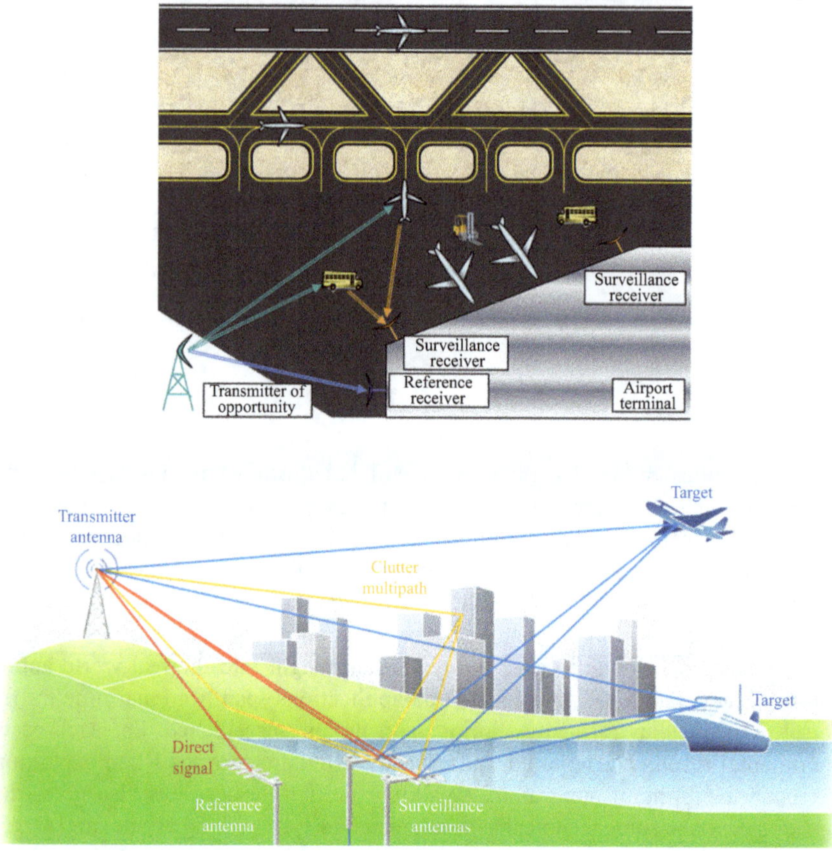

Figure 15.1 Examples of ATC and maritime short-range PR applications

They might refer to bistatic ranges that are short if compared to the wide coverage usually required in typical PR applications such as ATC and maritime surveillance (see Figure 15.1). With particular reference to maritime surveillance applications, we might consider the problem of detecting, localizing and tracking vessels within territorial waters and harbour areas [8–18]. Similarly, in the ATC context, short-range PR sensors might be exploited for the surveillance of specific aerial regions as well as the monitoring of the movement area of the airport [19–22].

In such applications, the transmitters (Txs) for mobile personal communication and network connection have been successfully exploited as illuminators of opportunity; these include the base stations of Global System for Mobile Communications (GSM), Universal Mobile Telecommunications System (UMTS), Worldwide Interoperability for Microwave Access (WiMAX) and Long-Term Evolution (LTE) [8–13,23–25]. However, the transmissions of digital video broadcasting-terrestrial (DVB-T) and digital audio broadcasting (DAB) represent one of the most attractive choices since, by properly tailoring the required signal processing techniques, they can provide both the wide coverage and advanced short-range capabilities [14–22].

Figure 15.2 Examples of vehicular traffic monitoring applications

The word 'short' might also indicate the distance between the targets and the PR receiver (Rx) and this leads us to surveillance applications against terrestrial targets such as the vehicular traffic monitoring both on roads and on railways (see Figure 15.2).

In this case, the opportunistic exploitation of transmissions for mobile communication and networking is especially attractive since they have been proliferating at a very rapid rate for both commercial and private use and nowadays represent a widely accessible source of opportunity in populated areas [8,12,25–32]. In addition, it is worth noticing that the use of such illuminators potentially enables a hybrid active and passive localization of the targets based both on self-reported positions and radar measurements.

It should be noticed that, in the definition above, there is not a constraint on the Tx distance from the observed scene so that also satellite based Txs can be exploited for the considered short-range applications provided that the Rx is located in the proximity of the observed scene. Potential illuminators of opportunity include global navigation satellite systems along with telecommunication satellites for audio and video broadcast, data exchange and personal communications [33–38]. Some of them appear to be a viable solution to guarantee ubiquitous coverage even in sparsely populated regions (there including deserts, open sea, glaciers, etc.) and this would be an invaluable characteristic to be exploited for either surveillance or search and rescue applications.

Finally, the range of interest can be short up to indoor applications such as, for example, people location and navigation along buildings, vehicles navigation or asset tracking (see Figure 15.3). In these applications, the PR sensors are expected to nicely complement the many existing technologies especially when operating against non-cooperative targets. The IEEE 802.11 Standard based (WiFi) Txs are well suited for this very local parasitic exploitation since they provide a limited coverage but potentially wide bandwidth and well-controlled signals useful when aiming at indoor surveillance or at monitoring small external areas [39–45].

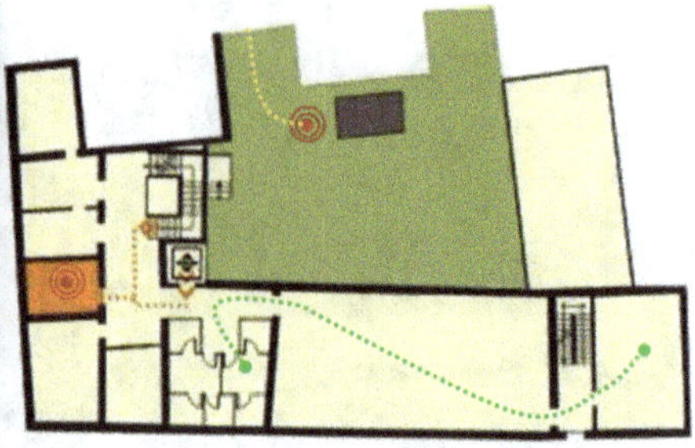

Figure 15.3 Examples of indoor surveillance applications

At first glance, the exploitation of the PR principle in short-range applications could appear as a simple task. However, it brings a number of challenging issues that must be solved to benefit from its potential advantages. In addition, the requirements on the sensors could be much more demanding. Some key-points are briefly illustrated below.

- We might admit that a higher resolution is required to be effective on small areas possibly characterized by dense traffic. This requirement usually determines the exploitation of signals with wider bandwidths and possibly transmitted at higher carrier frequencies, and in turn results in increased system complexity and computational load.
- Similarly, increased accuracy is required for the measurements provided by the PR for the detected target. A viable solution, particularly when monitoring areas of small dimensions, is indeed the use of a network of PR sensors since this allows to widen the set of measurements simultaneously collected. However, the network geometry should be carefully designed to limit the system complexity while guaranteeing the availability of measurements with a sufficient degree of spatial diversity so that a significant improvement can be gained in terms of target localization capability.
- Special attention should be devoted to the availability of the selected transmission of opportunity (this particularly applies when operating indoor or in rural areas) as well as on its control aiming to avoid severe masking effects.
- Nevertheless, when the Rx and the Tx are in close proximity, the considered sensors might experience high power level direct signal and severe multipath so that a more efficient management is required of the Rx dynamic range and effective signal processing techniques should be employed to remove the interference contributions.
- Among the positive aspects, the capability of detecting very small targets in specific areas gives to the considered sensors the role of gap-filler for already deployed (long- or short-range) technologies.
- Moreover these sensors can be provided with advanced capabilities such as accurate target tracking, target imaging or cross-range profiling and automatic classification that represent invaluable tools in the considered scenarios.

In the subsequent sections, these points are addressed with reference to specific examples of real-world applications. Effective processing techniques are illustrated that have been designed to enable the practical operation of the considered systems and to rise to the specific challenges issued by advanced applications. Moreover several experimental results are reported, obtained by the author within test campaigns performed in cooperation with colleagues of industry and academia. Such results are representative of different scenarios and are based on different waveforms of opportunity. The aim is to give the reader an insight into the potentialities and current limitations of short-range PR sensing.

15.2 Maritime surveillance applications

The potential exploitation of PR sensors in maritime surveillance applications has been recently addressed in a number of papers that have appeared in the technical literature [8–18,37,38,46] and is the subject of several studies, R&D projects, or innovation actions conducted by governmental, industrial and academic entities (see e.g., [47–51]).

The integration of such sensors within the sea border surveillance solutions would be desirable especially with the role of gap fillers. In fact, the wide need for maritime surveillance extends from the coastline up to the limit of national territory and even outside the sea border. At long distance from the coast, the use of conventional active sensors with medium/low resolution might be acceptable; moreover, a few sensors transmitting a reasonable power level are enough to cover wide areas. Nevertheless, in such cases, the availability of additional measurements from passive sensors could improve the reliability of the system, increase its localization capability and possibly extend its nominal coverage. In contrast, at a short distance from the coast, the required surveillance must be characterized by a high detection capability also for small vessels, a high spatial resolution and precise tracking. The PR characteristics of low-cost, small size and rapid update might guarantee the implementation of a dense network of sensors to provide complete and continuous coverage especially in coastal areas characterized by cliffs and reliefs close to the sea. This network would have the advantage of limiting the environmental impact and avoiding potential increase of electromagnetic (EM) pollution that are generally not very welcome from the local resident population.

The achievable performance highly depends on the exploited waveform of opportunity. As mentioned in the introduction, in the last years a number of existing Txs have been considered as sources of opportunity for maritime PR surveillance at short ranges. Among the most studied illuminators of opportunity, the base stations for GSM and WiMAX [8–13] and the Txs for DVB-T [14–18].

In this section, we focus on the use of DVB-T broadcast transmissions and discuss specific aspects relevant to the adopted signal processing techniques and the practical implementation of the system, along with the results of experimental tests performed against specific targets of interest.

15.2.1 Signal processing scheme and its peculiarities

As is well known, transmissions for DVB-T guarantee a quite stable-range resolution of about 20 m (equivalent monostatic value) thanks to the Orthogonal Frequency-Division Multiplexing (OFDM) modulation based on persistent sub-carriers spanning a total bandwidth of up to 8 MHz [52]. In addition, the exploitation of a DVB-T based PR for maritime surveillance presents a number of positive implications. For instance, compared to the air traffic surveillance application, the expected radar coverage benefits from the longer integration times allowed by vessels and from the vertical-plane radiation patterns of typical DVB-T Txs that are tailored to avoid wasting too much power above the horizontal. In contrast, the wide horizontal beam and high power level of the broadcast Txs often allows the availability of such signals even at large distance from the coast and this enables their exploitation for short-range surveillance by PR Rxs on moving maritime platforms [18]. Moreover, the low carrier frequencies of the considered waveforms, especially when exploiting channels in the very high frequency (VHF) band and in the lower portion of the ultra high frequency (UHF) band, might allow the detection of targets beyond the normal line-of-sight horizon thanks to anomalous propagation phenomena typical of sea paths where super-refractive conditions are prevalent [46].

However, the considered scenario brings in a number of specific requirements that should be properly fulfilled for PR to be effective in maritime security applications. With particular reference to short-range surveillance, such sensors should have precise localization and tracking characteristics as well as the capability to detect very small targets or in general targets with low radar cross-section (RCS) values. Moreover, these characteristics should be guaranteed against severe clutter returns from both ground and sea. In addition, the system should be able to operate in the presence of a high diversity of targets in terms of both RCS and expected motion.

To fulfil the requirements above, effective signal processing techniques should be adopted at each of the blocks that constitute the basic PR processing chain. For the sake of our example, Figure 15.4 sketches the DVB-T based PR data processing scheme employed to obtain the results reported in Section 15.2.2. Here the main blocks relevant to target detection and localization are briefly described giving particular emphasis to the processing stages that have been properly modified and optimized to be effective in the considered maritime surveillance application.

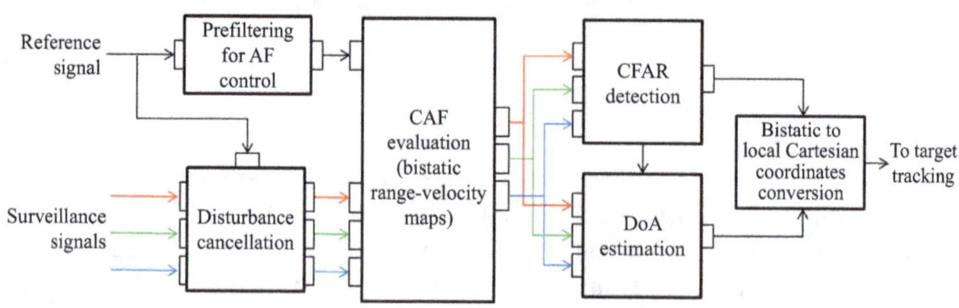

Figure 15.4 Block diagram of the DVB-T based PR processing scheme

As it is well known, in a standard PR processing scheme, the signal collected at the reference channel is first used to remove undesired contributions in the signals collected at the surveillance channels. To this purpose we resort to a modified version of the adaptive cancellation approach presented in [53], the extensive cancellation algorithm (ECA), which operates by subtracting from the surveillance signal $s_{\text{surv}}(t)$ proper scaled and delayed replicas of the reference signal $s_{\text{ref}}(t)$. Specifically, by sampling the received signals at f_s and assuming that the multipath echoes are backscattered from the first K range bins, the output of the ECA is evaluated as:

$$s_o[n] = s_{\text{surv}}[n] - \sum_{k=0}^{K-1} \alpha_k s_{\text{ref}}[n-k] \tag{15.1}$$

The filter coefficients $\boldsymbol{\alpha} = [\alpha_0\ \alpha_1\ \cdots\ \alpha_{K-1}]^T$ are evaluated by resorting to a Least Square (LS) approach that minimizes the power of the signal at the output of the filter:

$$\boldsymbol{\alpha} = \left(\mathbf{S}_{\text{ref}}^{\text{H}} \mathbf{S}_{\text{ref}}\right)^{-1} \mathbf{S}_{\text{ref}}^{\text{H}} \mathbf{s}_{\text{surv}} \tag{15.2}$$

where $\mathbf{s}_{\text{surv}}$ is a $N_B \times 1$ vector containing N_B consecutive samples of the surveillance signal and $\mathbf{S}_{\text{ref}}$ is a $N_B \times K$ matrix whose columns are the delayed versions of the corresponding reference signal fragment. In order to preserve the detection of slowly moving vessels while guaranteeing the removal of the highest interference contributions (mainly the direct signal from the Tx), the filter coefficients are estimated over long batches (hundreds of ms); as a result narrow filter cancellation notches are synthesized thus preserving the low Doppler target echoes. However, when operating against a highly time-varying interference scenario, the setting above contrasts with the need to rapidly adapt the filter coefficients. In such cases, a sliding version of the ECA (S-ECA) can be employed which operates on partially overlapped signals batches and takes advantage of a smooth estimate of the filter coefficients to appropriately counteract the limitations above [54,55].

After the cancellation stage, the output signal is filtered with properly mismatched and Doppler shifted versions of the reference signal to evaluate the bistatic cross-ambiguity function (CAF):

$$C[l,p] = \left| \sum_{n=0}^{N-1} s_o[n] \cdot s_{\text{ref}}^*[n-l] \cdot e^{-j2\pi(pn/N)} \right|^2 \tag{15.3}$$

where

- l is the time bin representing the time delay $\tau[l] = l/f_s$; the time delay can be converted in a bistatic range difference by defining the corresponding range bin as $\Delta R[l] = c \cdot \tau[l]$ c being the speed of light;
- p is the Doppler bin representing the Doppler frequency $f_D[p] = pf_s/N$; the Doppler frequency can be converted in a bistatic velocity by defining the corresponding velocity bin as $\Delta v[l] = \lambda f_D[p]$, being λ the centre wavelength of the exploited transmission.

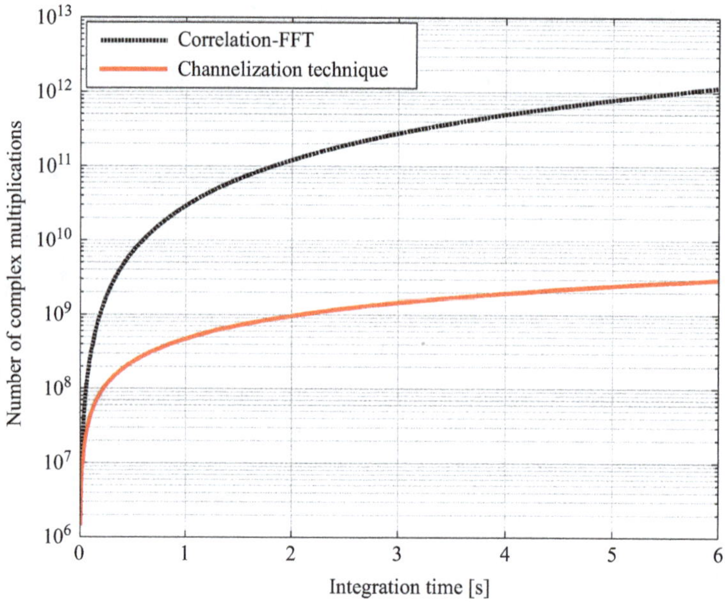

Figure 15.5 Comparison of computational load required by different algorithms for CAF map evaluation

Long coherent processing intervals (CPI), $T_{\text{int}} = N/f_s$, should be exploited at this stage to obtain an acceptable signal-to-noise ratio (SNR) gain; this could effectively improve the detection of small (low RCS) targets or extend the radar coverage. In the considered maritime application, especially at short ranges, the low velocity of many potential targets allows to increase the CPI duration up to a few seconds without experiencing severe migration in both range and Doppler; in turn, increasing the CPI allows to improve the Doppler resolution as well as the capability to discriminate between slowly moving vessels and docked boats.

Obviously, evaluating the CAF over long CPI might represent a very costly operation. As an example, Figure 15.5 reports the number of complex multiplications required to evaluate the range-velocity map as a function of the CPI length using an optimum algorithm. Specifically, the Correlation-FFT technique is employed as it represents the most efficient solution for the considered application, [5,6]. In the reported example, the desired bistatic range-velocity map is assumed to span a surveillance area of 30 km × 60 m/s, the sampling frequency is set to $f_s = 64/7$ MHz and the carrier frequency is $f_c = 650$ MHz. With these parameters, using a CPI equal to 4 s the evaluation of the CAF would require about 5×10^{11} complex operations.

To reduce the computational load and guarantee a reasonable update rate of the target tracks, sub-optimum techniques can be exploited to evaluate the CAF with

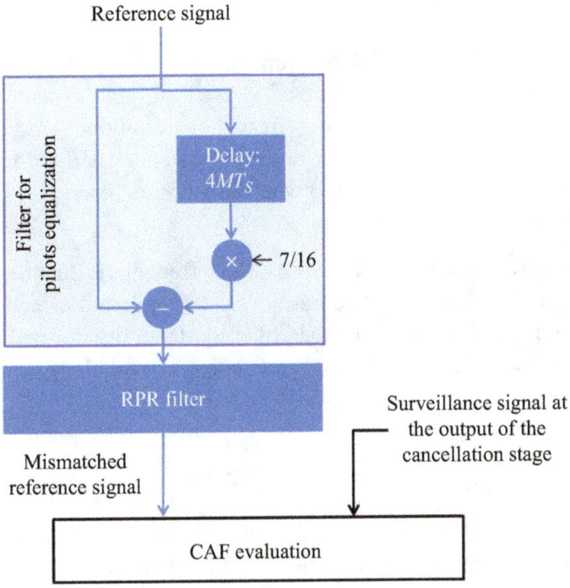

Figure 15.6 Sketch of proposed linear approach for removal of unwanted peaks

only limited SNR loss. For instance, operating with the Channelization technique described in [5,6], a computational load saving of more than two order of magnitude might be obtained in the considered case study (see Figure 15.5) by accepting a 0.42 dB loss. Notice that, operating with the optimum approach, this reduced computational load could be achieved only by reducing the CPI duration down to 0.26 s, which corresponds to an 11.9 dB loss.

As previously mentioned, the reference signal is properly filtered before it is employed for the CAF evaluation (see Figure 15.4). Such mismatching stage is intended to remove the high sidelobes and spurious peaks appearing in the DVB-T signal ambiguity function (AF) [21,56,57] . The AF sidelobe control is an essential processing step for the considered maritime application aiming at preventing severe masking effect by the strong target returns on the weak echoes of targets with low RCS.

To this purpose, we resort to the linear approach presented in [21] which is based on the cascade of the pilots signals equalization and a residual peaks removal (RPR) filter to remove the remaining zero-Doppler peaks (see sketch in Figure 15.6).

To briefly summarize the proposed approach, the sequence of samples of the DVB-T signal $s[n]$ is written as the sum of two contributions:

$$s[n] = s_D[n] + \frac{4}{3} s_P[n] \tag{15.4}$$

where

- $s_D[n]$ is the contribution resulting from the data and TPS subcarriers that are transmitted at power level normalized to 1;
- $s_P[n]$ is the contribution resulting from both continual and scattered pilot subcarriers that are transmitted at boosted power level $E_P = 16/9$; notice that this is explicitly indicated in (15.4) where we assume that $s_P[n]$ has power level normalized to 1.

Since the overall continual and scattered pilots frequency positions are periodic over four OFDM symbols, each of duration T_S, a simple and effective solution for pilots equalization can be obtained by subtracting, from the original signal, its own replica, delayed by an integer multiple of $4T_S$ and scaled by $(1 - 1/E_P) = 7/16$ obtaining:

$$s_{\text{Seq}}[n] = s_D[n] - \frac{7}{16} s_D[n - \text{MN}_{\text{SS}}] + \frac{3}{4} s_P[n] \tag{15.5}$$

where $N_{\text{SS}} = 4T_S f_s$. The amplitude of the pilots contribution in (15.5) is equal to 3/4 and allows equalization of the pilots in the surveillance signal after cross-correlation, while the second term at second member of (15.5) yields a time shifted contribution to the final CAF which does not affect the surveillance area and can be neglected, especially in the considered short-range application.

After pilot equalization stage, all the residual peaks in the desired delay-Doppler area appear at zero Doppler cut and can be removed by the RPR filter whose coefficients $w[n]$ can be obtained by solving the following system of linear equations.

$$R_{\text{out}}[l] = \sum_{n=-L}^{L} w[n] R_{eq}[l - n] = \begin{cases} 1 & l = 0 \\ 0 & |l| \leq L, l \neq 0 \end{cases} \tag{15.6}$$

where L is the maximum delay (in range bins units) to be considered in this filtering stage and $R_{eq}[n]$ is the signal mismatched auto-correlation after pilots equalization. Different approaches are presented in [21] to solve this system and to limit the computational load of the proposed technique so that it can be regarded as a cost effective solution.

It is worth noticing that, when applied against real data, the proposed AF control technique allows to limit the sidelobe level down to 45 dB below the main peak in a wide range-velocity region. Specifically, the proposed approach is effective inside a typical surveillance area defined according to the limits on bistatic delay-Doppler plane (maximum relative bistatic range $R_{\max} = 200$ km, maximum target bistatic velocity less than 600 m/s). However, as is apparent, the allowed peak-free area is much greater than the surveillance region foreseen for a short-range application. Furthermore, being interested in slowly moving target detection, the pilots equalization stage can be neglected as it allows to remove undesired secondary peaks appearing in the AF at quite high Doppler values. Nevertheless, this stage is required anyway when (fast moving) aerial targets are also present in the considered scenario since, if not removed, their secondary peaks might fall within the surveillance region and cause false alarms or masking effects on the maritime targets of interest.

Once the CAF maps have been evaluated at all the available surveillance channels, a conventional cell average constant false alarm rate (CA-CFAR) threshold is separately applied to each map to detect targets with a probability of false alarm (P_{fa}) higher than 10^{-4} to preserve the small target detections. In order to not overburden the tracking stage, an M-out-of-N criterion is adopted to integrate the detection results obtained at the surveillance channels.

Assuming that the surveillance receiving channels are connected to a few (horizontally) aligned antenna elements, an interferometric approach can be used to estimate the direction of arrival (DoA) of the detected target echo. Notice that, in the considered application, assuming that the PR sensor is located on the coastline, it is typically unnecessary to design array configurations providing 360° coverage. In contrast, a linear array of few directive antennas can be employed to improve the Rx gain on target echoes while limiting the size of the antenna system. In addition, when more than two antenna elements are employed, a baseline diversity can be exploited to increase the unambiguous angular sector. Finally, the bistatic range and the DoA information are converted into local Cartesian coordinates.

The above processing scheme has been exploited against the data collected for several experimental tests showing its effectiveness against typical scenarios. Some examples are reported in the following section for the considered short-range maritime application.

15.2.2 Experimental results against small RCS targets

We focus our attention on the experimental results achieved in the framework of the SeaBilla FP7 project [47], a R&D project carried out by Leonardo S.p.A. (formerly Selex ES) in 2011–14, in cooperation with Sapienza University of Rome.

Figure 15.7 describes the DVB-T based PR Rx installation employed for the test campaigns considered in the following [17]. Yagi-Uda antennas have been employed with a gain of 15 dB and main beam width equal to 36° (see Figure 15.7(a)). In particular, for different tests, two or three surveillance antenna elements have been exploited, aligned in the horizontal plane in order to estimate the DoA of the detected target echo so that its position can be evaluated in local Cartesian coordinates.

After a preliminary signal conditioning on the different receiving channels (i.e., pass-band filtering, amplifications and attenuations in order to optimize the input dynamic of the signal), the receiving chain includes a commercial device which allows to acquire and digitalize the received signals on the different physical channels. A software interface has been developed to properly set acquisition parameters such as carrier frequency, sampling frequency, single acquisition duration, desired number of scans, temporal interval between two consecutive scans, etc. Moreover, an external frequency reference of 10 MHz has been used, derived from a commercial global positioning system (GPS) Rx, in order to lock

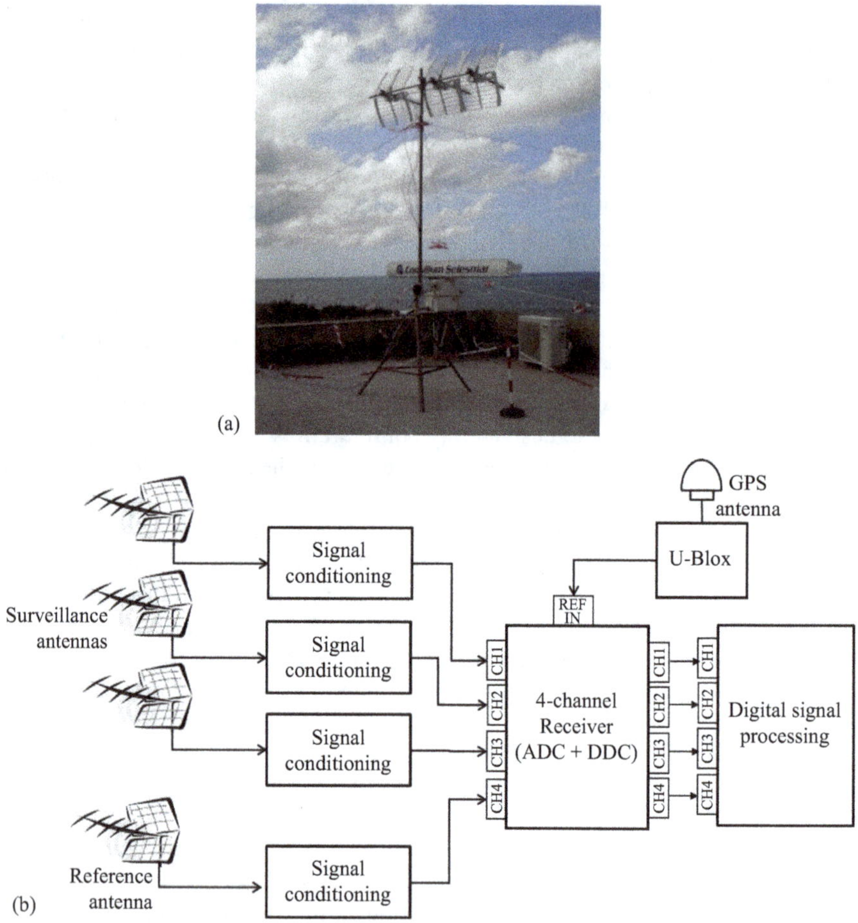

Figure 15.7 DVB-T based PR receiver: (a) example of antennas installation for the test campaign in Livorno and (b) block diagram of the receiving chain. © [2014] IEEE. Reprinted, with permission, from [17]

the PR Rx with a common reference signal. Notice that the carrier and sampling frequencies synchronization also guarantees the effective operation of the AF control technique [21,58]; in fact we recall that most DVB-T Txs use GPS disciplined oscillators in order to guarantee the single frequency network (SFN) feasibility.

The results are reported for two different acquisition campaigns that have been carried on in coastal areas along the western coast of Italy.

In particular, the campaign held in Civitavecchia (Lazio) aimed at verifying the potentialities of DVB-T based PR in detecting very small boats with low RCS in short-range maritime applications. The acquisition geometry is sketched in

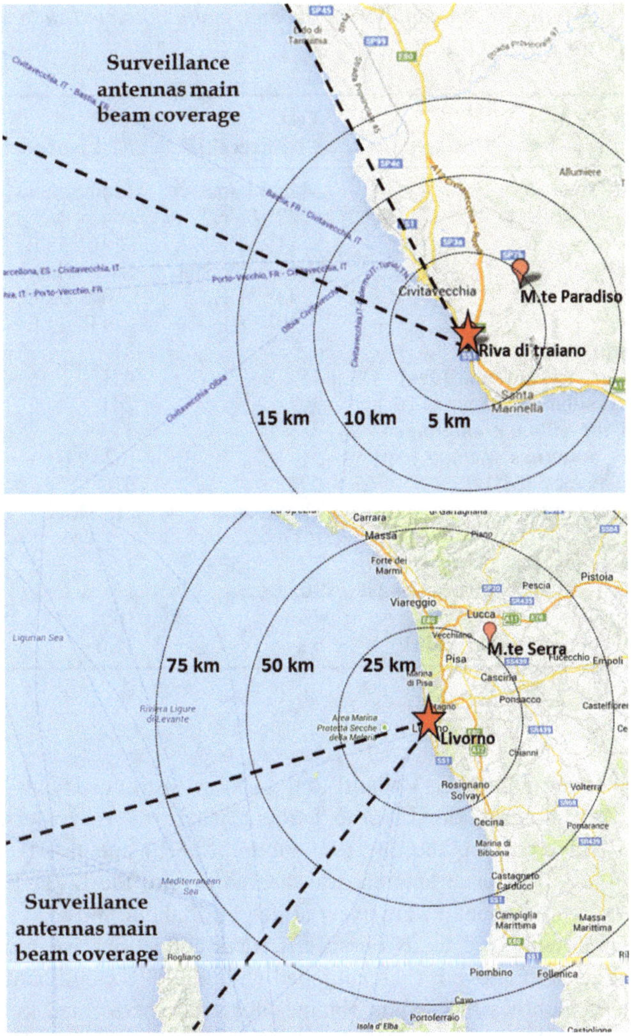

(a) (b)

Figure 15.8 Acquisition geometry for the experimental campaigns of (a) Civitavecchia and (b) Livorno. © [2014] IEEE. Reprinted, with permission, from [17]

Figure 15.8(a) and some details are provided in Table 15.1: the PR Rx was located in the proximity of the tourist port of Riva di Traiano; the DVB-T Tx of Monte Paradiso was exploited as illuminator of opportunity, with a baseline Tx-Rx of about 4.5 km.

The campaign of Livorno (Tuscany) aimed at demonstrating the effectiveness of the conceived sensor in monitoring typical maritime traffic close to the harbour area. In this case the PR Rx has been installed at about 23 m height, on the roof of a

Table 15.1 Details of the acquisition geometry and collected data sets for different tests

	Test Civitavecchia	Test #1 Livorno	Test #2 Livorno
Tx position	Monte Paradiso 42° 05′N, 11° 51′E	Monte Serra 43° 45′N, 10° 33′E	Monte Serra 43° 45′N, 10° 33′E
Rx position	Riva di Traiano 43° 31′N, 10° 18′E	Ist. Vallauri 42° 03′N, 11° 49′E	Ist. Vallauri 42° 03′N, 11° 49′E
Baseline [km]	4,5	31	31
Steering reference antenna [deg]	45	45	45
Steering surveillance antennas [deg]	307	241	241
Number of surveillance antennas	2	3	3
Surveillance antennas spacing [cm]	51	[62; 93]	[62; 93]
Carrier Frequency [MHz]	690	630	626
DVB-T signal mode	8K-Mode, GI ¼	8K-Mode, GI ¼	8K-Mode, GI ¼
Single scan acquisition duration [s]	1	1	1
Interval between successive scans [s]	10	10	6
Number of scans	40	100	62
Total acquisition duration [min]	6.5	16.5	6.0

building belonging to the 'G. Vallauri' Italian Navy Institute. This is located on the coastline, very close to the Livorno harbour where many vessels, of different dimensions, travel during the day (see Figure 15.8(b) and last two columns of Table 15.1). The reference antenna was steered toward the DVB-T Tx of Monte Serra that is located about 31 km away at height of about 860 m.

In the following, we focus on specific data collections wherein cooperative targets have been employed, each equipped with a GPS Rx on board.

Figure 15.9 shows the results obtained for a test performed in Civitavecchia against a small rubber boat, of 4 m length and 2 m in width (see Figure 15.9(a)). Specifically, Figure 15.9(b) reports, in Cartesian coordinates, the detection results obtained by using the processing scheme described in Section 15.2.1 against the collected data set whose details are listed in Table 15.1. A burst of 40 consecutive acquisitions is considered, each of 1 s, displaced by 10 s. During this period, the target goes out from the marina of Riva di Traiano (and thus gets away from the PR Rx site), by navigating very close to the coast (see red track in Figure 15.9(b)). In order to preserve the detection of the small target, we exploited the maximum integration time allowed by the performed acquisitions (1 s); correspondingly, a P_{fa} values equal to 10^{-4} and a '1-out-of-2' integration criterion across the two available surveillance channels have been adopted. The detection rate estimated by counting the raw detections associated to the cooperative target over the sequential scans is 29/40. However, as the detection outputs of multiple consecutive data files

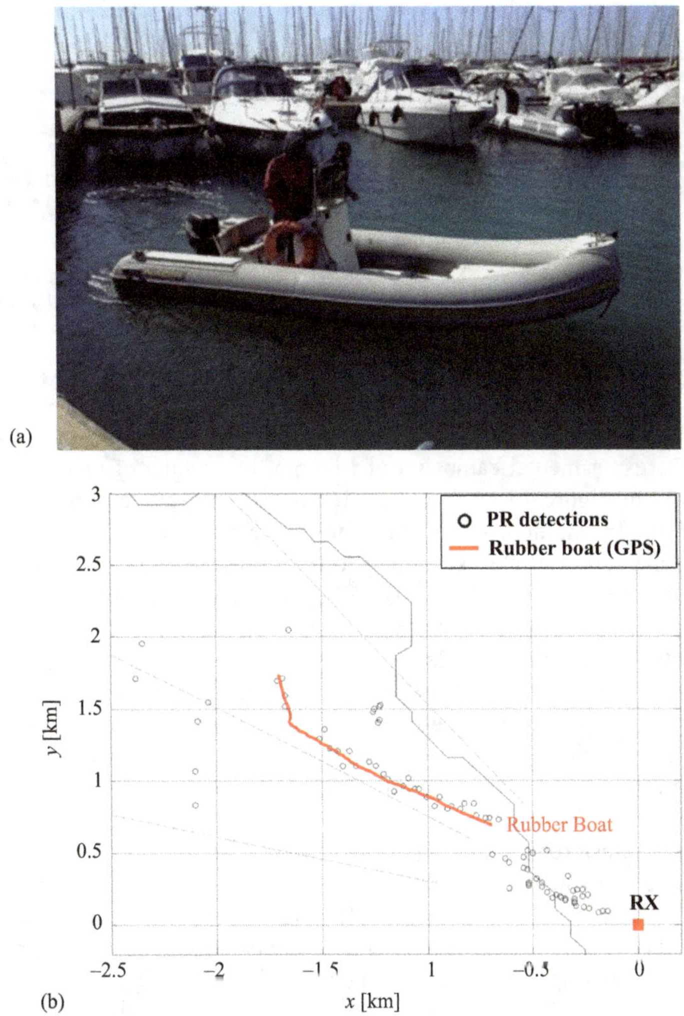

Figure 15.9 Experimental results for the test in Civitavecchia against a rubber boat: (a) picture of the target; (b) comparison of PR detections and the on-board GPS registrations. © [2014] IEEE. Reprinted, with permission, from [17]

are contemporaneously displayed (black plots in Figure 15.9(b)), to obtain a much clearer figure a standard scan-to-scan correlator has been applied against the raw detections obtained at consecutive scans; specifically a 2-out-of-2 criterion is adopted to reduce the number of false alarms and to better identify potential target tracks. Despite this expedient, the results in Figure 15.9(b) clearly show the remarkable continuity of the sensor in detecting and localizing the small vessel up to a few kilometres from the Rx location.

(a) (b) (c)

Figure 15.10 Cooperative targets employed in the experimental tests: Costal Patrol boat (a), Motorboat 1 (b) and Motorboat 2 (c). © [2014] IEEE. Reprinted, with permission, from [17]

For the experimental campaign of Livorno, the results of two different tests are reported in Figure 15.11(a) and (b) using the cooperative targets depicted in Figure 15.10. In addition, during these tests, many targets of opportunity were present in the surveyed area according to the available AIS (Automatic Identification System) registrations. In Figure 15.11, black plots identify the PR detections after the proposed data processing scheme. In this case, a '3-out-of-3' criterion has been adopted to integrate the detection results separately achieved on the three available surveillance channels. The AIS live registrations and the GPS based positions of cooperative targets are reported in blue and red colours, respectively. In addition, the expected coverage of the surveillance antenna main beam is identified with discontinuous lines. Additional details about the acquisition geometry and the exploited signals are listed in Table 15.1.

In the first test, reported in Figure 15.11(a), the Italian Coast Guard patrol boat 'Classe 2000' (labelled as Coastal Patrol—CP) has been used with approximate dimensions 12.5×3.6 m (see Figure 15.10(a)). During the data collection of approximately 17 min (i.e., 100 consecutive acquisitions, each of 1 s, displaced by 10 s), the CP boat approaches the Rx position, by navigating from the open sea toward the coastline. The PR sensor provided 71 correct detections of the CP boat out of the 100 successive scans performed.

The second test, reported in Figure 15.11(b), employed two cooperative vessels belonging to motorboats category (MotorBoat Naphtha—MBN) with length between 6 and 9 m (see Figure 15.10(b)). The registered burst is composed by 62 consecutive acquisitions, each of 1 s, displaced by 6 s. Initially, the MBNs move along parallel directions, very close to each other and at the same velocity, going toward the coast line. After a few nautical miles, the MBN 1 (red track) changes the direction, veering to the south of about 30° and keeping constant the velocity. In this case, we obtained detection rates equal to 52/62 and 48/62 for MBN1 and MBN2, respectively. Also most of the opportunistic targets navigating in the surveyed area are correctly detected by the employed DVB-T based PR. However the dimensions of these targets are always larger than those of the cooperative targets employed. In addition, there are many detections which could be likely to

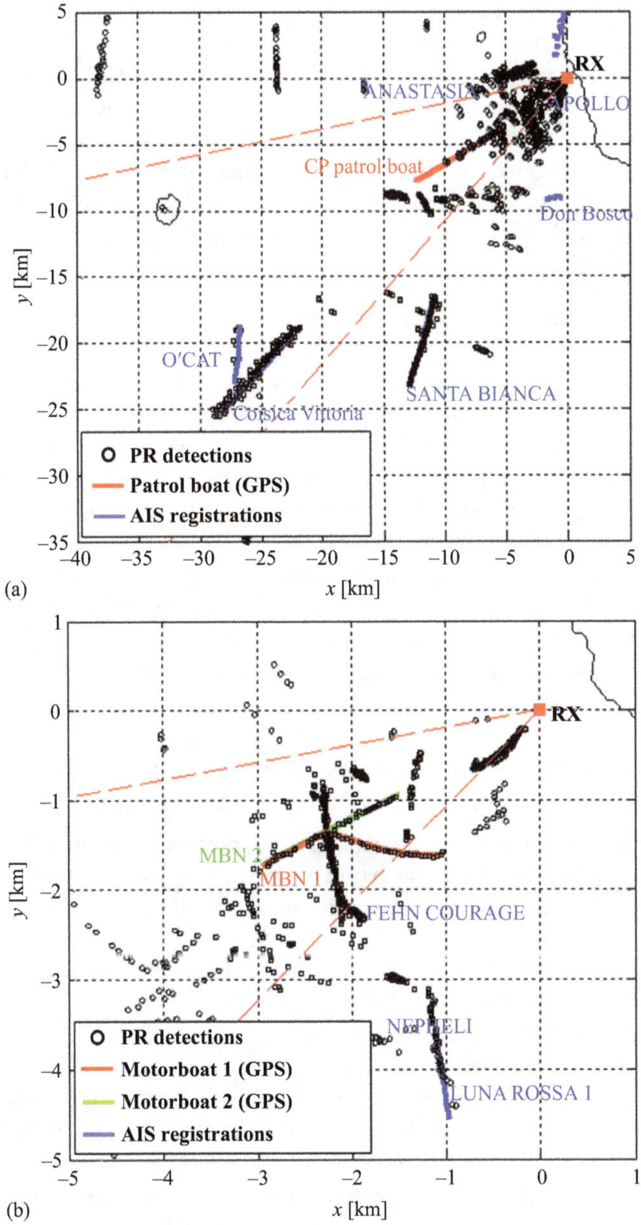

Figure 15.11 Results for the tests in Livorno against small cooperative vessels: (a) Test #1; (b) Test #2. © [2014] IEEE. Reprinted, with permission, from [17]

correspond either to non-cooperative small targets, for which AIS registrations are not available, or to the ghosts due to SFN operation [59].

The obtained results clearly show that the conceived DVB-T based PR allows remarkable detection and localization capabilities even against small targets moving in coastal areas. While the reported analysis does not provide a definitive answer to the considered question, it suggests that the expectations on PR potentialities might be increased to include short- and medium-range maritime surveillance for homeland security or search and rescue applications.

15.3 Vehicular traffic monitoring

The effectiveness of PR has been also successfully demonstrated for the detection, localization and tracking of vehicles or man-made objects moving on the ground in a nearby area [8,12,25–32]. In this section some enabling solutions and experimental results will be reported for vehicular traffic monitoring via PR sensors. Despite the reported examples are mostly based on tests against cars, it is worth mentioning that a wider class of targets of interest could be considered in such application. These might include any motorized road vehicle transporting people or goods, from two wheeled motor vehicles up to trucks. In addition, we might envisage PR exploitation against transport vehicles running on rails and traffic on the manoeuvering area and the aprons of an airport, there including taxiing of aircrafts.

15.3.1 Vehicles detection using different illuminators of opportunity

This sub-section reports and comments on the experimental results that have been obtained in vehicles detection tests based on waveforms of opportunity transmitted by alternative illuminators at different distances from the surveyed scene.

15.3.1.1 Exploitation of ground-based and space-based broadcast transmissions

Broadcast Txs of digital video can be successfully exploited also for vehicular traffic monitoring as they allow a good compromise between range resolution and accessibility. With specific reference to the ground-based DVB-T Txs, the signal processing scheme illustrated in Section 15.2.1 can be exploited, provided that the relevant parameters are properly adjusted for the considered application.

Actually, the result shown in the following is a side result of the experimental tests carried on in cooperation with the colleagues of Leonardo S.p.A. (formerly Selex ES) to demonstrate the effectiveness of DVB-T based PR for ATC [21]. The PR system was fielded on the roof of the Leonardo S.p.A. premises in Rome (see Figure 15.12(a)). It employed a two-channel Rx to collect the signal from a reference antenna steered toward the DVB-T Tx of Monte Cavo (about 22.5 km away) and a single surveillance antenna pointed toward south-west so to include, in its 3 dB pattern, many of the standard departure and arrival routes of the Rome city airports.

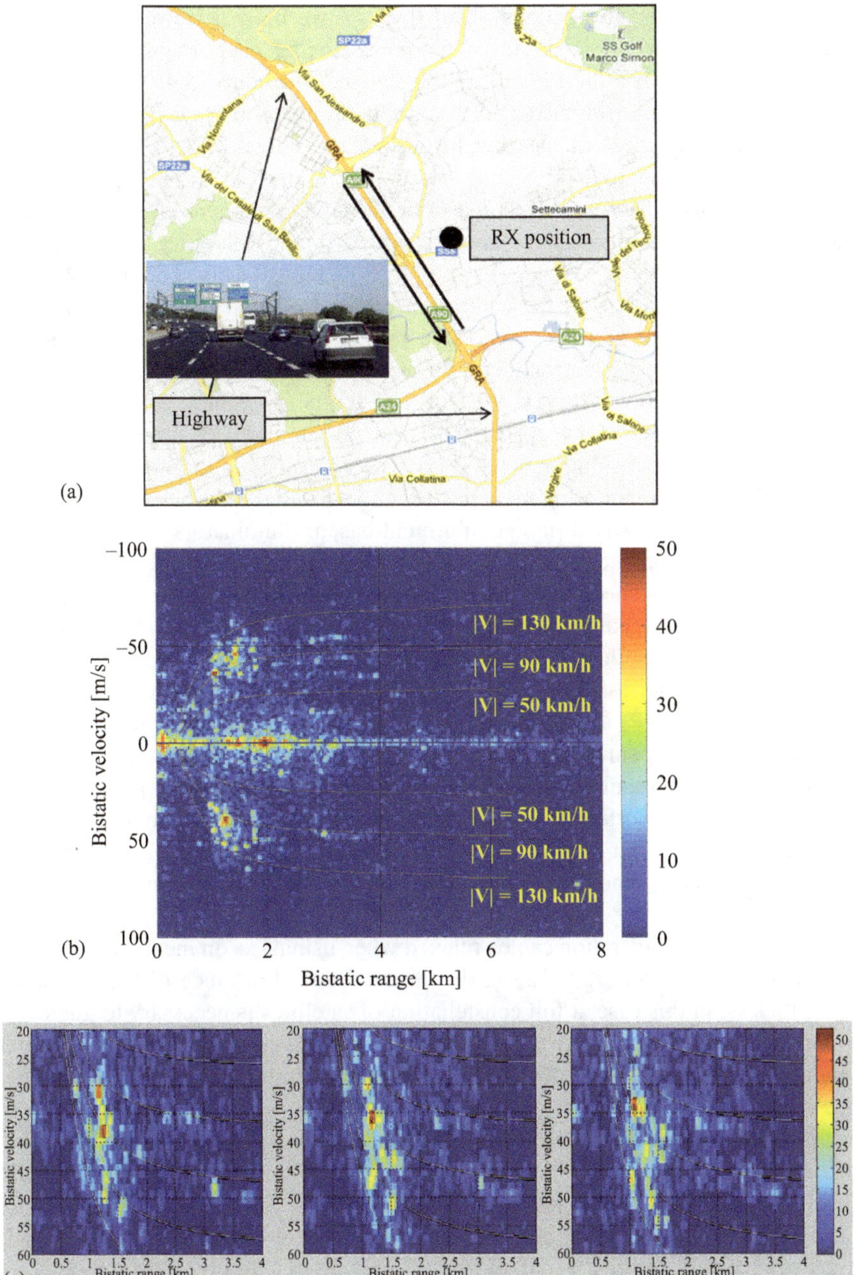

Figure 15.12 Examples of vehicles detection potentialities for a DVB-T based PR: (a) acquisition geometry; (b) range–velocity map at a single frame; (c) enlarged view of a specific region of the map over a sequence of three successive frames. © [2014] IEEE. Reprinted, with permission, from [21]

Continuous data acquisitions of about 5 s have been performed; these have been then split into 14 consecutive fragments, being each fragment separately processed according to the PR processing scheme of Figure 15.4. Along with the good detection results obtained against aircrafts at medium ranges, we noticed the presence of strong peaks appearing in the bistatic range–velocity maps at short ranges (see e.g. the high intensity values in the first 2–3 km of the map in Figure 15.12(b)). These were verified to be in large agreement with the expected range–velocity trajectories of cars moving with different velocities on the highway nearby the Rx location (yellow lines in Figure 15.12(b)). This is quite apparent in Figure 15.12(c) that reports the enlarged view of a portion of the range–velocity map for a sequence of three successive frames of the acquisition. Apparently, a strong peak is present surrounded by other peaks moving on trajectories corresponding to velocities between 90 and 110 km/h for vehicles moving on the highway (maybe there was a good traffic flow on the highway at that time!).

As an alternative solution, satellite-based broadcasting Txs could be in principle exploited for monitoring an area surrounding the ground-based Rx location.

The use of space-based (in lieu of ground-based) illuminators of opportunity in high orbits potentially enables a 'ubiquitous' short-range surveillance capability thanks to the simultaneous coverage of wide areas. In addition, with particular reference to Txs in geostationary earth orbits (GEO), the PR operation could be simplified since it would not require tracking the satellite position.

However, broadcast Txs in GEO have typical level of equivalent isotropic radiated power (EIRP) in the range of 40–50 dBW, yielding a low level EM field close to the Earth's surface. In fact, very long integration times (order of many minutes/hours) are required to yield an acceptable SNR. This allows images to be made of the Earth's surface (also by exploiting a limited satellite motion relative to the Earth) [60, 61], but not detection of moving objects since their motion might limit the integration time to the order of seconds to avoid severe range and Doppler migration.

The above consideration can be relaxed when using Txs on medium Earth orbit (MEO) and low Earth orbit (LEO), due to the reduced distance of these sources. Nevertheless, in this case, a full constellation of satellites is necessary to guarantee a continuous coverage of a specific region and a hand-over capability from a satellite to another has to be developed. Moreover a continuous satellites tracking is necessary.

According to the considerations above, a promising solution has been investigated in [35] based on the use of Txs for DVB-Satellite services to Handhelds (DVB-SH) carried by geostationary satellites. These have been introduced for Satellite Mobile Digital TV broadcast purpose and are designed to allow mobile users to receive satellite TV without large antennas. Therefore, they are expected to operate with sensibly higher EIRP and adequate bandwidth making them suitable as illuminators of opportunity in PR. In addition, adequate waveform characteristics are guaranteed since these transmissions exploit a similar OFDM modulation with respect to their terrestrial counterparts with a slightly narrower bandwidth. In [35], a preliminary evaluation of the achievable coverage has been

Figure 15.13 Parabolic antenna exploited at the reference channel for the experimental DVB-SH based PR Rx

conducted, showing that mid-range detection of aerial targets could be in principle achieved with reliable performance.

Some preliminary experimental tests have been conducted in the framework of a research project carried on by Leonardo S.p.A. (formerly Selex Galileo) in cooperation with Sapienza University of Rome. To this purpose, an ad hoc Rx has been developed based on a super-heterodyne architecture. This also includes the parabolic antenna required to provide the narrow high gain beam at the reference channel (see Figure 15.13).

Unfortunately, only a few successful tests have been performed since the EIRP of the selected transmitter was verified to be much smaller than its nominal specification and this was later verified to be due to a damage occurred to the Tx antenna during the launch.

However, some promising results have been obtained against vehicles of opportunity moving in a nearby street that could be detected against the background as shown in the example reported in Figure 15.14. Specifically, Figure 15.14(a) depicts the acquisition geometry with the Rx deployed on the roof of the building of the Engineering Faculty. Figure 15.14(b) reports two Doppler cuts of the CAF obtained with a CPI of 0.5 s: the two cuts refer to the first two range bins, being the bin spacing equal to 52.5 m. No cancellation has been performed so that both the cuts show a strong peak at zero bistatic velocity. Nevertheless, the Doppler cut extracted at the second range bin clearly reveals the presence of an echo from an object moving away from the Rx location with moderate velocity.

15.3.1.2 Exploitation of base stations for network connection

When very local area surveillance is considered, transmissions for wireless communications and networking could be also exploited. Among them, the Txs of WiFi,

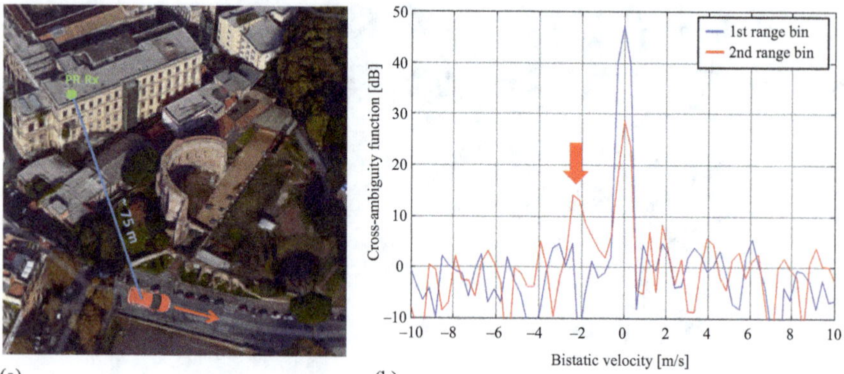

(a) (b)

Figure 15.14 Example of experimental result for a DVB-SH PR against vehicular targets: (a) acquisition geometry; (b) Doppler cuts of the CAF at the first two range bins

WiMAX and LTE are especially attractive for their wider bandwidth and vast accessibility both in private and in public areas. In particular this section focuses on the results obtained with a WiFi-based PR which allows radar coverage up to a few hundreds of metres with range resolutions of 14 or 8 m depending on the adopted modulation that can be either direct sequence spread spectrum (DSSS) or OFDM.

To this purpose, proper processing techniques have been devised to enable the effective exploitation of WiFi signals using the PR principle for the considered vehicular traffic monitoring application [26,27,29,63–65]. Basically, despite the block diagram of Figure 15.4 is suitable for various waveforms of opportunity, the specific technique to be employed at each block and its main parameters should be carefully designed as they are largely related to the Standards that regulate the transmission of the exploited signals.

A complete processing scheme for a WiFi-based PR is shown in Figure 15.15. In this section, we briefly summarize the blocks relevant to target detection, paying special attention to the adjustments and modifications made to guarantee their effectiveness in the case under examination. The remaining blocks of the scheme will be discussed in the following sections of this chapter.

First of all, we observe that in the considered scenarios, the PR sensor is typically installed very close to the WiFi access point (AP) exploited as illuminator of opportunity; therefore, assuming the Tx to be partially cooperative, its signal can be spilled from the Tx antenna path using a directional coupler thus obtaining a quite pure copy of the emitted signal. If this is not the case, conventional approaches can be still exploited to recover the transmitted signal, i.e., using a dedicated antenna or demodulating/re-modulating the signal received at the surveillance channel [26].

Whilst the Tx could be partially cooperative in the sense described above, the transmitted waveform is not within the control of the radar designer, so that high sidelobes or undesired peaks appear in the corresponding AF varying with the adopted WiFi signal modulation [62]. Therefore, the reference signal must undergo

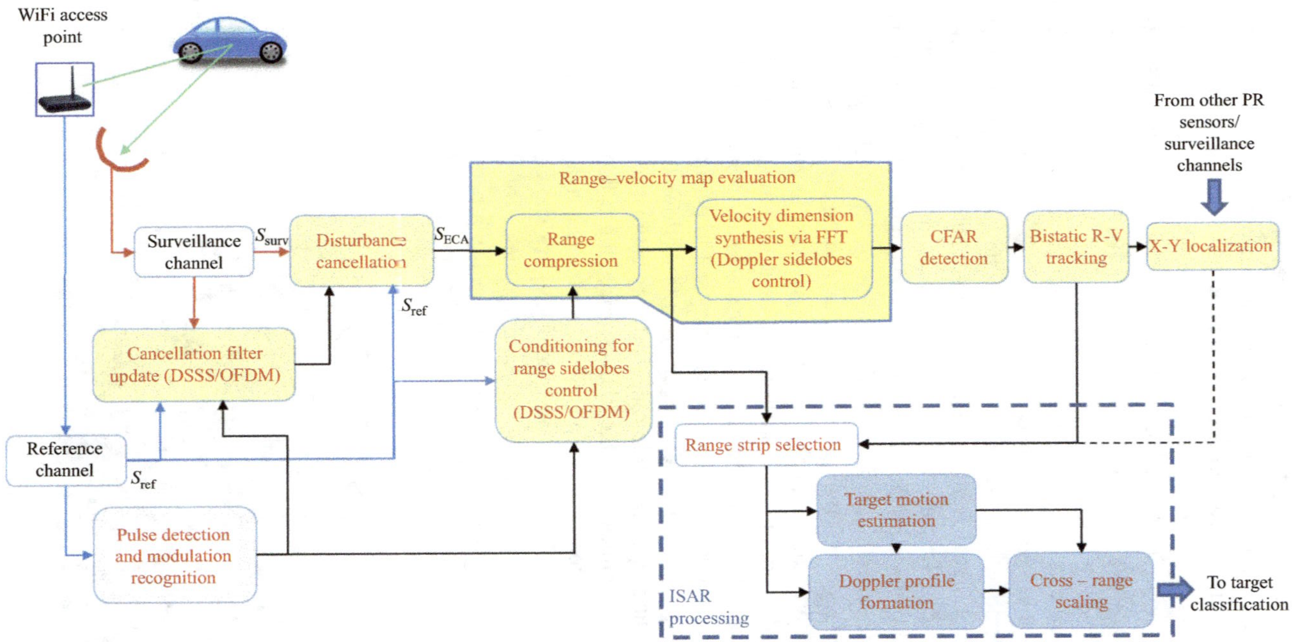

Figure 15.15 WiFi-based PR processing scheme for target detection, localization and ISAR processing

a modulation-dependent conditioning stage aimed at improving the resulting mismatched AF in the range dimension [26,63,64].

It is worth mentioning that the WiFi signal is of a pulsed type so that the evaluation of the range–velocity map can be performed via two separate steps. Specifically, the range compression is obtained by cross-correlating the surveillance and the reference signals on a pulse basis. Then, the velocity dimension is synthesized by applying a fast Fourier transform (FFT) over the pulses available in the CPI; ad hoc designed taper functions can be employed at this stage to control the sidelobes level in the Doppler dimension [65].

Interference cancellation is usually a required stage also in this application, especially when targets with lower RCS values are considered. The ECA can be exploited to remove undesired contributions in the surveillance channel typically due to the direct signal from the Tx and its multipath rays (i.e., bounces on stationary obstacles). In particular, since the ECA filter weights depend on the exploited signal auto-correlation function and this in turn is related to the adopted modulation, the cancellation filter weights should be separately estimated for DSSS and OFDM pulses by averaging over pulses exploiting the same modulation [64].

A CFAR threshold is applied on the CAF map to automatically detect the potential targets according to a specific CFAR detection scheme. This provides a first target localization over the bistatic range-velocity plane. Then, the measures collected at consecutive observations can be used to perform a line tracking over this plane. Using a conventional Kalman algorithm allows to reduce the false alarm rate while yielding more accurate range/velocity estimations.

The results obtained with the above processing scheme have been reported in many contributions showing its effectiveness against typical scenarios [26,27,29,45,63–65]. Some examples from [29] are shown in the following. The corresponding data set has been collected by means of the experimental setup sketched in Figure 15.16; the test was performed in a parking area where a vehicular target was employed moving on a given trajectory (see the left-hand side

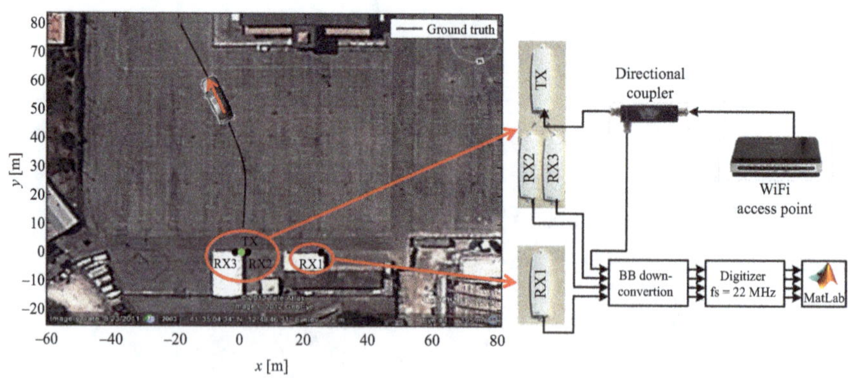

Figure 15.16 Sketch of the experimental setup and the performed test. After [29] © [2014] [IET]

of Figure 15.16). A GPS Rx was mounted on the moving car to collect the ground truth.

A commercial WiFi AP was used as Tx of opportunity. Its antenna output was connected to the transmitting antenna (TX) that was located at the point represented with the coordinates $(x_{TX}, y_{TX}) \equiv (0, 0)$ m; a directional coupler was used to send a -20 dB copy of the transmitted signal to the first receiving channel of a quad-channel receiving system with the aim of collecting the reference signal.

Three surveillance antennas were directly connected to the three remaining receiving channels. Using the three available surveillance channels, we could synthesize two independent PR sensors one of which is equipped with a pair of surveillance antennas providing the target DoA estimation capability. To this purpose, the first standalone surveillance antenna (RX1) was located in $(x_1, y_1) \equiv (25, 0)$ m in a bistatic configuration with respect to the TX; on the other hand antennas RX2 and RX3 were placed a few tens of centimetres below the TX (in a quasi-monostatic configuration), displaced in the horizontal direction by 12 cm which gives a 60° non-ambiguous angular sector for the DoA estimation.

After a fully coherent base-band down-conversion stage, the signals collected at the different receiving channels are sampled at 22 MHz and stored for off-line processing.

The WiFi-based PR processing scheme depicted in Figure 15.15 is applied against the collected surveillance signals separately at each receiving channel. In particular, the ECA is applied with a batch duration equal to 100 ms over a range of 300 m; a CPI of 0.5 s is used to evaluate the CAF map over consecutive portions of the acquired signals (frames) with a fixed displacement of 0.1 s (10 frames per second are thus obtained); and target detection is performed by resorting to a standard cell-average CFAR threshold with a probability of false alarm equal to 10^{-4}.

Figure 15.17 reports the detection results obtained for the two PR systems for a 15 s acquisition (150 frames) compared to the ground truth. Specifically, the output is shown of a standard association stage, based on a nearest neighbour association strategy, aimed only at discarding false alarms; moreover, the detections associated to stationary tracks have been discarded since they are likely to correspond to clutter residues. The raw target plots are reported since the obtained measurements have not been filtered at this stage.

In particular, Figure 15.17(a) reports the results for the PR system based on the standalone antenna RX1, namely PR1. The results for PR2 are shown in Figure 15.17(b) where a 2-out-of-2 criterion has been adopted on the detection results separately obtained against the signals collected by antennas RX2 and RX3.

As is apparent, many plots are obtained that clearly correspond to the target echoes. This clearly shows that the target detected with remarkable continuity along its path by both the PR sensors. In addition, a long plot sequence is detected that is likely to be generated by the double-bounce reflection of the target echo over the building in the upper zone of Figure 15.16 (see the sequence of detections at positive Doppler frequency and decreasing bistatic range between 250 and 150 m). In practical applications, the false targets generated by multipath effect on the

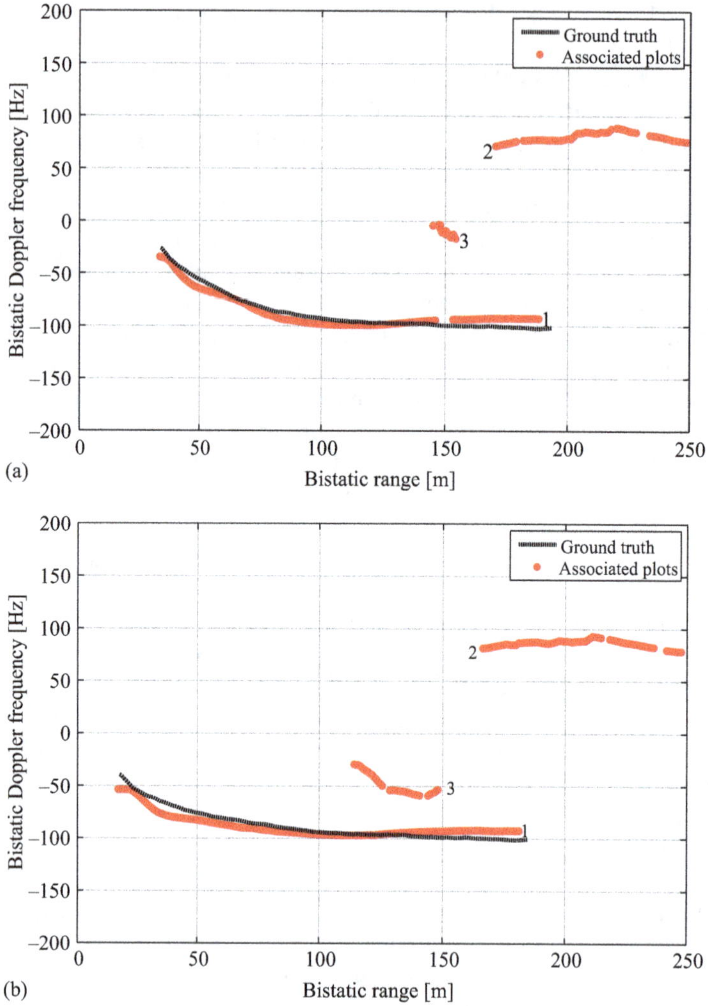

Figure 15.17 Detection results after association stage over the bistatic range-Doppler plane for: (a) PR1; (b) PR2. After [29] © [2014] [IET]

target echo might be reasonably recognized if the *a priori* knowledge of the stationary scene is available, namely the shape and position of the main stationary obstacles forming the observed scene. In fact, based on geometrical considerations and assuming a simple propagation model, it is possible in principle to identify and discard the target tracks that are likely to correspond to multiple-bounce echoes. Combining, the available measurements for the true target echo the target 2D localization can be obtained as described in the following sub-section.

15.3.2 Target localization based on a network of passive sensors

Once the target has been detected on the bistatic range-velocity plane, its localization in local Cartesian coordinates can be obtained by using different sets of measurements. To this purpose, multiple PR sensors can be exploited which allow to collect a set of range/Doppler measurements for the same target which is observed at different bistatic geometries. Moreover, assuming that each sensor uses a couple of surveillance antennas, a simple interferometric approach can be exploited to estimate the DoA of the target echo. This obviously increases the set of available measurements thus potentially improving the localization capability of the conceived system. However, different measurements (range/Doppler/DoA) might be characterized by different accuracies, also affected by the exploited waveform of opportunity, and their impact on the target 2D localization might be highly dependent on the observation geometry [66–68].

In addition, the availability of multiple measurements is typically paid in terms of system complexity and cost. However, we might observe that the implementation of a network of PR sensors for the surveillance of an area of small dimensions is a more feasible task than in the case of wide area coverage (e.g., with reference to synchronization issues, maintenance costs, etc.). Such approach could significantly improve the localization capability of the resulting system provided that the employed sensors are properly dislocated so that the target is observed at different geometries conveying a great diversity of information. In this regard, different strategies have been investigated to identify suitable Rx locations for a multistatic PR aiming at increasing its coverage and the target localization accuracy [68–74].

Correspondingly, alternative target localization approaches have been proposed based on different sets of homogeneous or non-homogeneous measurements [29,75–79]. Depending on the considered measurements, the complexity of the problem to be solved might increase. For example, since the Doppler measurements depend both on the instantaneous position and on the target motion components, including them in the system of equations to be solved would require an extended set of measurements and, therefore, an increased number of sensors.

In this sub-section, we consider the problem of obtaining the target *x*–*y* coordinates from a minimum set of measurements [29]. As is well known, a minimum number of two position dependent measurements is required to perform the 2D target localization. However, different sets of two measurements could be exploited and this choice has implications both on the resulting localization accuracy and on the PR system to be designed there including the number and displacement of the Rxs and the number of receiving antennas for each passive sensor.

For the purpose of our analysis, we refer to the experimental test described in Section 15.3.1.2 (see Figures 15.16 and 15.17) for a WiFi-based PR exploiting two separate sensors [29]. The nearest sequence of plots in Figure 15.17 is selected which clearly corresponds to the employed target. This procedure has been partially automatized by applying a conventional tracking stage on the range-Doppler plane and then manually extracting the track of interest; specifically the plots associated

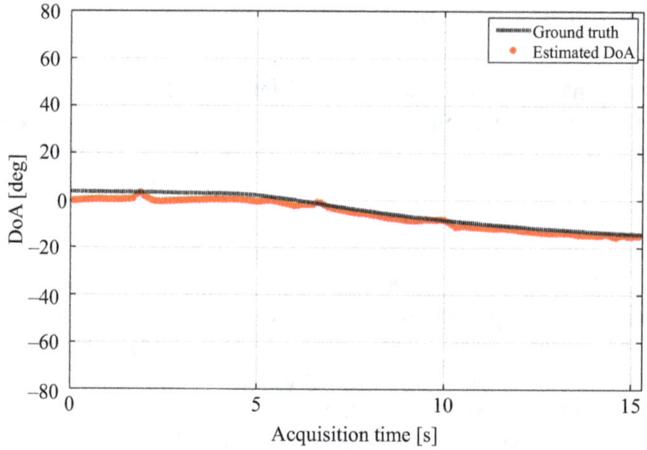

Figure 15.18 Target DoA measures obtained by PR2 in test experiment of Figure 15.16. After [29] © [2014] [IET]

to the selected track are collected which represent a set of bistatic range and Doppler frequency measurements (namely, R_{B1} and f_{D1}), for the considered target, performed by PR1 along the acquisition. Similarly, the set of range/Doppler measures is collected for PR2 (R_{B2} and f_{D2}); in this case, also the phase difference between RX2 and RX3 measured at the target detection point is exploited to obtain the target DoA estimation. The estimated target DoA along the performed acquisition is show in Figure 15.18, compared to the ground truth showing that it is well in line with the test geometry.

Therefore, with this experimental setup, the target 2D localization at a given CPI can be performed by using the following subsets of available measurements:

(a) Two (possibly filtered) bistatic range measurements provided by two PR sensors. For a given PR with known Tx and Rx locations, a bistatic range measurement univocally identifies the ellipse where the target lies. The intersection points of two ellipses provide the estimation of up to four possible target positions over the x–y plane. The ambiguous solutions might be discarded by forcing the target to be within the antennas' main lobe; this presumes some degree of *a priori* knowledge, namely information about the area covered by transmitting and receiving antennas beams should be available. The achievable target localization accuracy depends on both the range measures accuracy and on the additional multiplicative effect caused by the multistatic system geometry on the achievable precision; the latter effect is known as the dilution of precision (DOP) factor in GPS and geomatics engineering [80]. Better performance could be obtained, if the target range measurements are used at the output of a dynamics model-based filtering stage that takes into account the multiple sequential measurements

[29]. To this purpose, a conventional Kalman tracking algorithm can be applied over the range-Doppler plane separately for each PR, by assuming a parabolic motion model for the target which results in a 3×1 system state vector (range, Doppler, Doppler rate). Notice that a range/Doppler tracking stage is usually adopted in PR applications because it allows both to discard false alarms and to improve the range measure accuracy prior to merging the results from multiple PR Rxs. In fact, after the line tracking, also the target bistatic Doppler measurements are indirectly exploited to obtain more accurate range measurements with respect to the raw case. As a consequence, the localization procedure based on the intersection of the corresponding ellipses yields a much more stable estimation of the target position in the x–y plane.

(b) A range measurement and a DoA measurement provided by a single PR sensor. The target position is obtained by intersecting the bistatic ellipse with the line identifying the DoA of the target echo. Notice that ambiguous solutions might be found using this approach, as well as when exploiting two range measurements, because the DoA measurement could be ambiguous itself. With particular reference to our case study, the horizontal displacement of the receiving antennas (12 cm) yields a 60° non-ambiguous sector for the DoA estimation. Nevertheless, since directive antennas are employed, the ambiguous solutions can be discarded by forcing the target to be within the area covered by the antennas' main lobe. Again, the achievable localization accuracy depends both on the accuracy of the exploited measures and on the PR/target geometry. For example, the performance is expected to rapidly get worse as the target moves away from the Rx; basically this is due to the decrease in the target echo power level and to the widening of the uncertainty x–y area caused by a given DoA error.

The results obtained with the above strategies against the experimental test in Figure 15.16 are reported in Figure 15.19. In addition, the first three rows of Table 15.2 summarize the localization performance in terms of positioning error defined as $\varepsilon_P = \sqrt{\varepsilon_x^2 + \varepsilon_y^2}$ where ε_x and ε_y are the errors along the x and y dimensions, respectively.

As is apparent, when using the raw range measurements (white plots in Figure 15.19(a)), localization errors in the order of metres are experienced along the x dimension while the track follows the actual motion of the target along the y dimension (which is oriented along the range axis). In particular, the worst localization performance is experienced when the target is far from the two employed sensors. In fact, in such locations the bistatic ellipses become quite similar and almost tangent, thus increasing the DOP factor along the x dimension.

Better performance is obtained by exploiting the filtered range measurements (blue plots in Figure 15.19(a)). The resulting track is almost identical to the actual path of the car; the slight differences are due to clutter residuals which affect the estimation of the target range and Doppler frequency, especially when it

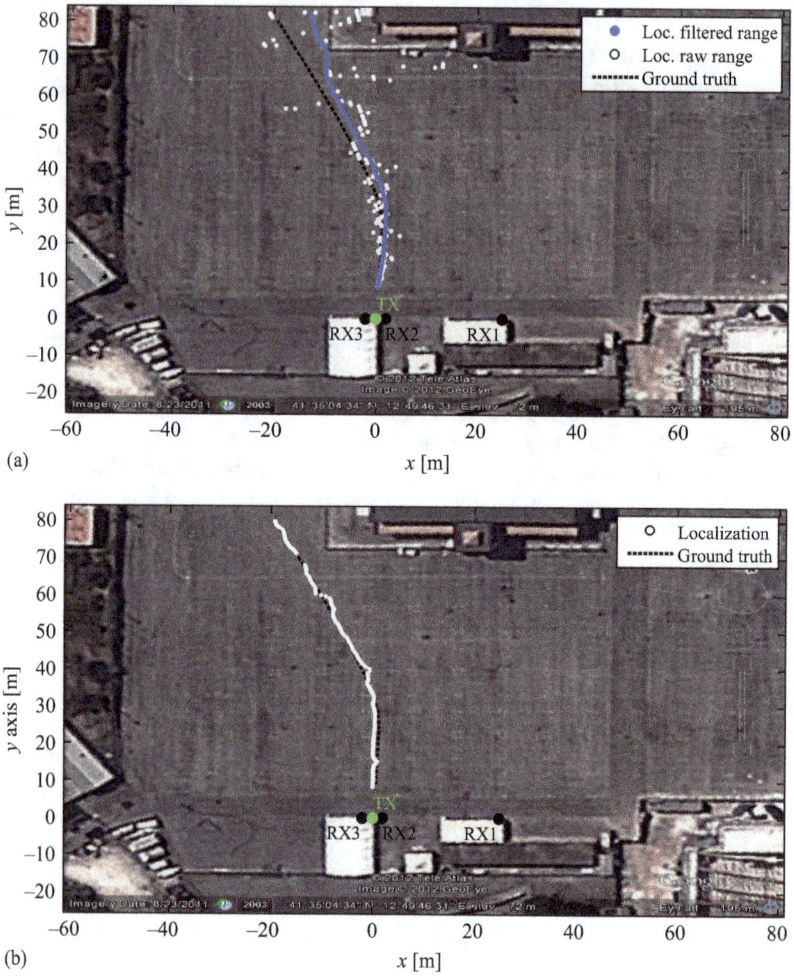

Figure 15.19 Results of target localization experiment using: (a) two bistatic range measures from PR1 and PR2 and (b) a single bistatic range and a DoA measurement from PR2. After [29] © [2014] [IET]

Table 15.2 Target localization performance using different strategies

Exploited measurements and method	Maximum positioning error (m)	Mean positioning error (m)	Positioning error standard deviation (m)
2 raw range measurements	32.65	4.81	5.78
2 filtered range measurements	6.64	2.47	1.53
1 filtered range & 1 DoA measurements	2.56	1.79	0.54
2 filtered ranges & 1 DoA measurements (ML approach)	2.37	1.6	0.49

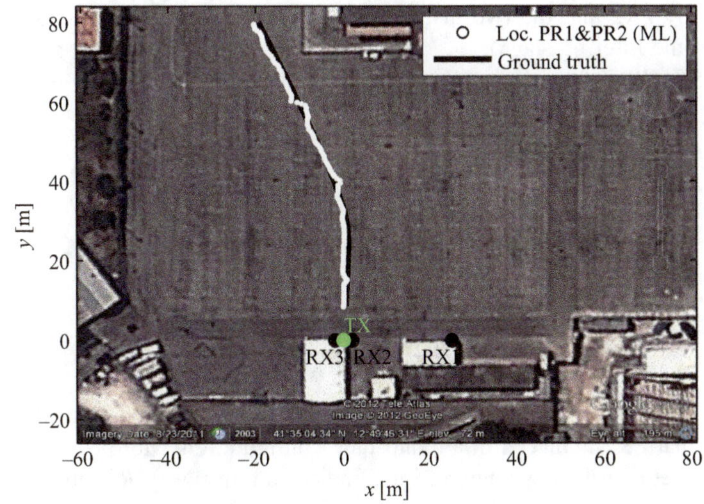

Figure 15.20 Target localization using two range measurements and a single DoA jointly provided by PR1 and PR2. After [29] © [2014] [IET]

approaches the building in the parking area. Correspondingly, the localization accuracy is significantly improved since the standard deviation of the positioning error is reduced by 4.2 m with respect to the previous case (see second row of Table 15.2). Obviously, better results could be obtained by applying a second tracking stage over the $x-y$ plane; however, we are interested in the direct impact of the available measures on the 2D localization capability of the system so that the comparison is performed without a target tracking in the $x-y$ plane.

Finally, the exploitation of the quite accurate phase difference measurement allows to further improve the target localization accuracy (white plots in Figure 15.19(b)). Basically, the maximum positioning error along the target trajectory is now comparable with the target size.

Based on the reported analysis, we might gather that, despite the simple approach exploited for DoA estimation (i.e., phase difference measurement between a couple of surveillance antennas), the target localization based on (at least) one angular measurement yields more reliable performance in the considered case study. This is a quite interesting result because it contrasts with the typical case of other PR systems operating at lower carrier frequencies for which the DoA estimation tends to be unstable and inaccurate. Moreover, this directs the design of a multistatic WiFi-based PR system to the inclusion of few sensors equipped with multiple receiving antennas instead of many lower cost sensors using a single surveillance antenna.

To complete the picture, Figure 15.20 shows the result obtained with the maximum likelihood (ML) approach presented in [29] for jointly using all the measurements provided by PR1 and PR2 (i.e., the target DoA measurement provided by PR2 and the two bistatic range measurements provided by PR1 and PR2 after the filtering stage).

Specifically, we look for the solution of the following system of three non-linear equations in two unknowns:

$$\begin{cases} R_{B1} = f_{R1}(x_t, y_t) + \varepsilon_{R1} = \sqrt{(x_{TX} - x_t)^2 + (y_{TX} - y_t)^2} \\ \qquad + \sqrt{(x_1 - x_t)^2 + (y_1 - y_t)^2} + \varepsilon_{R1} \\ R_{B2} = f_{R2}(x_t, y_t) + \varepsilon_{R2} = 2\sqrt{(x_{TX} - x_t)^2 + (y_{TX} - y_t)^2} + \varepsilon_{R2} \\ \vartheta = f_\vartheta(x_t, y_t) + \varepsilon_\vartheta = \arctan\left(\frac{x_t}{y_t}\right) + \varepsilon_\vartheta \end{cases} \tag{15.7}$$

where (x_{TX}, y_{TX}) are the Tx coordinates, (x_1, y_1) are the coordinates of PR1 and we assumed that PR2 is placed in a monostatic configuration. Moreover, $\mathbf{p} \equiv (x_t, y_t)^T$ are the target coordinates (superscript T indicates transpose), R_{B1} and R_{B2} are the bistatic ranges given by PR1 and PR2 respectively, ϑ is the target angle of arrival provided by PR2, and ε_{R1}, ε_{R2} and ε_θ are the corresponding measurement errors.

Under the hypothesis of small errors affecting the performed measurements, the system in (15.7) is linearized by using a first-order Taylor series approximation about the target tentative position $\mathbf{p}_0 \equiv (x_{t0}, y_{t0})^T$:

$$\mathbf{m} \cong \mathbf{m}_0 + \mathbf{H} \cdot (\mathbf{p} - \mathbf{p}_0) + \boldsymbol{\varepsilon}_M \tag{15.8}$$

where a matrix notation has been adopted based on the following definitions:

$$\mathbf{m} = [R_{B1} \quad R_{B2} \quad \vartheta]^T \tag{15.9}$$

$$\mathbf{m}_0 = [f_{R1}(x_{t0}, y_{t0}) \quad f_{R2}(x_{t0}, y_{t0}) \quad f_\vartheta(x_{t0}, y_{t0})]^T \tag{15.10}$$

$$\mathbf{H} = \begin{bmatrix} \left.\frac{\partial f_{R1}}{\partial x}\right|_{(x_{t0}, y_{t0})} & \left.\frac{\partial f_{R1}}{\partial x}\right|_{(x_{t0}, y_{t0})} \\ \left.\frac{\partial f_{R2}}{\partial x}\right|_{(x_{t0}, y_{t0})} & \left.\frac{\partial f_{R2}}{\partial y}\right|_{(x_{t0}, y_{t0})} \\ \left.\frac{\partial f_\vartheta}{\partial x}\right|_{(x_{t0}, y_{t0})} & \left.\frac{\partial f_\vartheta}{\partial y}\right|_{(x_{t0}, y_{t0})} \end{bmatrix} \tag{15.11}$$

Moreover, let us assume that the joint probability density function of the errors $\boldsymbol{\varepsilon}_M$ is Gaussian with zero mean value and covariance matrix $\boldsymbol{\Sigma}_M$ given by the following equation:

$$\boldsymbol{\Sigma}_M = \begin{bmatrix} \sigma_{R1}^2 & 0 & 0 \\ 0 & \sigma_{R2}^2 & 0 \\ 0 & 0 & \sigma_\vartheta^2 \end{bmatrix} \tag{15.12}$$

where σ_{R1}^2, σ_{R2}^2 and σ_{ϑ}^2 are the variances of the errors ε_{R1}, ε_{R2} and ε_{θ}, respectively. In practice, σ_{R1}^2 and σ_{R2}^2 can be derived from the *a posteriori* error covariance matrix of the Kalman algorithm, separately applied on PR1 and PR2; similarly σ_{ϑ}^2 can be estimated as $\sigma_{\vartheta}^2 = (\lambda/2\pi d \cos\vartheta)^2 \text{SNR}^{-1}$ where ϑ is the estimated target DoA and the SNR is estimated at the target detection point on the range-Doppler map. Notice that it has been reasonably assumed that the cross-covariance between the collected measurements is equal to zero.

Under these hypotheses, the ML estimator for the target position might be derived by solving the following minimization problem:

$$\min_{\mathbf{p}} = \left\{ [(\mathbf{m} - \mathbf{m}_0) - \mathbf{H} \cdot (\mathbf{p} - \mathbf{p}_0)]^T \Sigma_{\mathrm{M}}^{-1} [(\mathbf{m} - \mathbf{m}_0) - \mathbf{H} \cdot (\mathbf{p} - \mathbf{p}_0)] \right\} \tag{15.13}$$

that has the following solution:

$$\mathbf{p} = \left(\mathbf{H}^T \Sigma_{\mathrm{M}}^{-1} \mathbf{H}\right)^{-1} \mathbf{H}^T \Sigma_{\mathrm{M}}^{-1} (\mathbf{m} - \mathbf{m}_0) + \mathbf{p}_0 \tag{15.14}$$

This allows to update the target coordinates with respect to the tentative position and to reiterate the procedure until the displacement $\|\mathbf{p} - \mathbf{p}_0\|$ is within the requirements on the positioning accuracy or the maximum admitted number of iterations is reached. As is apparent, due to the non-homogeneous nature of the exploited measures, the target position updating obtained by resorting to the ML estimator depends not only on the collected measurements and the target-sensors geometry, but also it depends on the accuracy of the available measurements. In fact, different measurements are weighted according to their own accuracies and, in this case study, this suggests to rely mainly on the angular information.

The target instantaneous positions estimated by the resulting PR network in Figure 15.20 are in large agreement with the ground-truth. The quantitative analysis of the performance of the ML estimator jointly exploiting the measurements from PR1 and PR2 is reported in the last row of Table 15.2. As is apparent, the maximum deviation between the ground truth and the estimated target position reduces to approximately 2.4 m. The improvements in the mean error and the error standard deviation are 0.18 and 0.05 m, respectively, with respect to the use of a single range and DoA measurements and much higher with respect to the use of two range measurements. This reveals that the system is able to identify the most reliable measurements and exploit all information in the best possible way. Interestingly, enough, the use of the ML intrinsically avoids that the use of noisy information degrades the global 2D localization accuracy, so that it is generally a viable practical solution.

The ML approach described above can even be nicely extended when the set of exploited measurements is widened by including the Doppler frequencies aiming at estimating also the target's velocity components along the x and y directions, namely v_x and v_y. Since the Doppler frequency of the target echo depends on its velocity components v_x–v_y as well as on the system bistatic geometry, the target motion evaluation requires the joint estimation of both the target velocity and the target position. Specifically, the following two equations should be added to the system in (15.7):

$$\begin{cases} f_{Dl} = \dfrac{1}{\lambda}\left[\begin{array}{l} \dfrac{(x_{\mathrm{TX}}-x_t)v_x}{\sqrt{(x_{\mathrm{TX}}-x_t)^2+(y_{\mathrm{TX}}-x_t)^2}} + \dfrac{(y_{\mathrm{TX}}-y_t)v_y}{\sqrt{(x_{\mathrm{TX}}-x_t)^2+(y_{\mathrm{TX}}-x_t)^2}} + \\ + \dfrac{(x_l-x_t)v_x}{\sqrt{(x_l-x_t)^2+(y_l-x_t)^2}} + \dfrac{(y_l-y_t)v_y}{\sqrt{(x_l-x_t)^2+(y_l-x_t)^2}} \end{array}\right] + \varepsilon_{fl} \\ l=1,2 \end{cases} \tag{15.15}$$

Therefore, a system of five equations in four unknowns should be solved. Proceeding as in (15.8)–(15.14), the solution of (15.15), based on the ML approach, can be written as in (15.14) where the matrix structures are replaced with their augmented versions and the covariance matrix $\mathbf{\Sigma}_{Ma}$ (5×5) includes the variances of the Doppler frequency measurements.

The results are reported in Figure 15.21 along the considered acquisition for both the velocity components compared to the ground truth. For the purpose, of our analysis, we have also reported the velocity components obtained as time derivatives of the estimated target x–y coordinates obtained in Figure 15.20. As is apparent, the joint estimation of the target position and velocity based on the ML approach provides much more accurate results with respect to the case of time

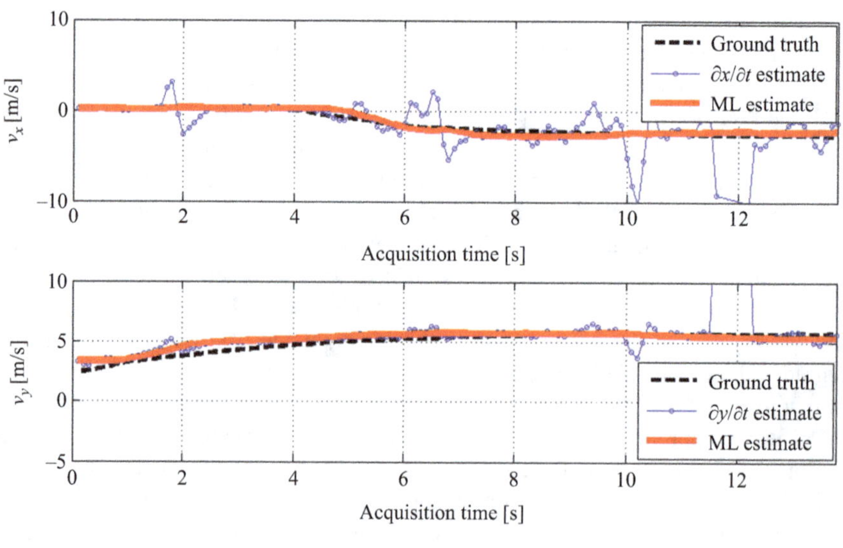

Figure 15.21 Target velocity components estimation. After [29] © [2014] [IET]

derivatives of the position components. This is due to the exploitation of the quite accurate Doppler frequency measurements that directly depend on the target motion parameters. In contrast, the estimated target position yields information about its motion only when observed at consecutive time instants; as a consequence, the localization errors might be doubled when evaluating the target velocity components.

Clearly, different approaches might be exploited for the estimation of the target velocity components. Nevertheless, the reported results show that the target motion parameters can be reasonably estimated based on the measurements provided by a couple of PR sensors dislocated on the surveillance area. This estimate might fruitfully feed the target tracking algorithm over the $x-y$ plane or it can effectively initialize the target motion parameters estimation stage needed for the target cross-range profile formation based on ISAR techniques (see Section 15.5).

While further improvements are expected if a wider network of properly displaced Rxs is used to collect a wider set of range/Doppler/DoA measurements, the reported results clearly demonstrate that there are good potentialities for the WiFi-based PR for short-range surveillance to provide accurate 2D location of vehicular targets, despite the wide antenna beams typically used by such sensors.

15.4 Indoor surveillance applications

In recent years, there has been a growing interest in indoor localization and tracking systems due to the security and public safety issues as well as service matters. Real-world applications include, for example, people location and navigation along buildings, automotive safety, vehicle navigation, or asset tracking.

Along with video surveillance systems, various wireless technologies have been used for indoor localization among which infrared, IEEE 802.11 wireless LAN and ultrasonic. More recently Radio Frequency Identification (RFID) has become a very attractive solution thanks to a number of desirable features, such as contactless communications, high data rate and security, nonline-of-sight readability, compactness and low cost [81,82].

In this section, we examine the potentiality of PR for indoor area monitoring. Although alternative illuminators of opportunity could be in principle considered, we refer to the exploitation of WiFi transmissions as they represent the most obvious choice for indoor applications [39–45].

As a first step, we attempt to give an answer to the following question: which would be the benefits of employing PR for indoor surveillance applications? To answer this question we build on the assumption that PR sensors are not intended to replace existing technologies but rather to complement them at a reasonable cost. Therefore our aim is to understand which could be the potential advantages conveyed by PR systems and how they could compensate for possible limitations of current solutions.

To this purpose, we first observe that a PR sensor is able to operate against non-cooperative targets. This is clearly an advantage with respect to technologies such as RFID, IEEE 802.11 wireless LAN and ultrasonic positioning that require the target objects to be equipped with a cooperative device (e.g., a wireless LAN card, a RFID tag, an ultrasonic transceiver, etc.). For the sake of completeness, it has to be mentioned that active ultrasonic sensors can be used also non-cooperatively; however they are quite sensitive to the presence of external noise sources and obstacles. Thus, compared to the technologies above, PR systems are better suited for specific surveillance applications such as intruder detection, location and tracking of unauthorized vehicles in forbidden areas, etc. Furthermore, since the EM scattering is characteristic of a target, PR in principle enables non-cooperative classification of the observed objects.

In addition, PR sensors based on WiFi transmissions do not require direct line-of-sight as, for example, for the infrared or video cameras; this makes the PR solution effective for covert detection of people moving behind walls [40].

Moreover, PR is not subject to the blind spots and potentially intrusive equipment required by video surveillance, so that it could be used in public areas or private commercial premises.

Finally, based on the PR concept, no extra signal is transmitted and this limits the energy consumption, prevents possible interference with preexisting systems, and makes the sensor free from any issue related to human health.

Among the drawbacks of PR for indoor surveillance applications, the severity of the interference contributions (i.e., direct signal and multipath) is a key point. Such undesired signals, typically mask the target echoes even in the presence of a large range-Doppler separation because of the sidelobes level of the signal AF that is not within the control of the radar designer. Therefore, the signal processing techniques devised for outdoor applications should be carefully tailored to counteract these effects and enable effective target detection and localization. In addition, the surveyed scene might be characterized by a high concentration of targets. As a consequence, proper strategies should be identified to discriminate closely spaced targets moving in the observed area.

In the following sub-sections, the feasibility of PR sensors for indoor surveillance is verified with reference to experimental results obtained for a WiFi-based PR prototype. Specifically, the issues above are addressed and possible solutions are investigated and tested.

15.4.1 Experimental results for indoor target detection and localization

Many experimental tests have been performed by the research group at the DIET Department of Sapienza University of Rome in order to demonstrate the feasibility of WiFi-based PR for indoor surveillance. In particular, such experimental campaigns have been carried on mainly within the framework of two projects funded by the EU under the 7th Framework Programme [83,84]. These projects presented an innovative approach to the challenging tasks required by the airport security system where the WiFi-based PR sensor is jointly employed with other active and passive sensors to enhance the security level in the airport terminal area.

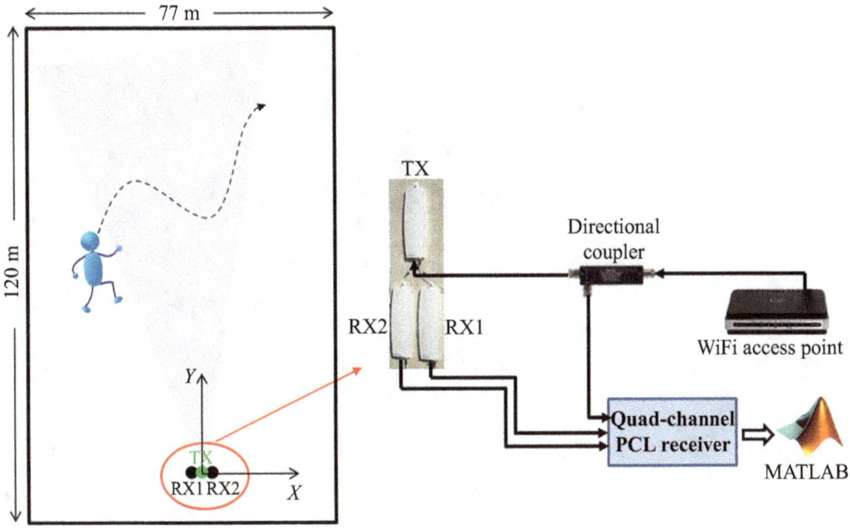

Figure 15.22 Sketch of the experimental setup employed for WiFi-based PR tests in indoor scenarios. © [2015] IEEE. Reprinted, with permission, from [45]

Therefore, in this sub-section we report the results obtained against human targets using the experimental setup depicted in Figure 15.22 [45]. The PR Rx has been fielded in a wide exhibition hall of the 'Nuova Fiera di Roma' whose size is comparable with an airport's terminal (see picture on the left hand side of Figure 15.3). An (X,Y) coordinate system has been defined with the X axis aligned with the short side of the hall and the Y axis oriented toward its centre, being the origin of the system located on the Tx antenna.

The system employed was the same described in Section 15.3.1.2; however, in the reported experiments, a quasi monostatic configuration was adopted, being the PR Rx equipped with two surveillance antennas to provide angular localization capability. About 60 different tests have been performed using one or two human targets walking simultaneously in the area to be surveyed. In the following, for illustrative purposes, we will report the results of the three tests described below:

- Test T1 [Figure 15.23(a)]: a man moves toward the Tx of opportunity with roughly constant velocity along the Y axis (i.e., DoA equal to zero degrees);
- Test T2 [Figure 15.23(b)]: two men walk along crossing paths. In particular, they initially move closer together; once they have come across, they contemporaneously change they walking directions departing from the Y axis with different angles;
- Test T3 [Figure 15.23(c)]: two men move along partially overlapping paths. In particular, they initially walk on the same direction with a constant separation of about 1 m; then they abruptly change their heading, going toward opposite sides of the hall.

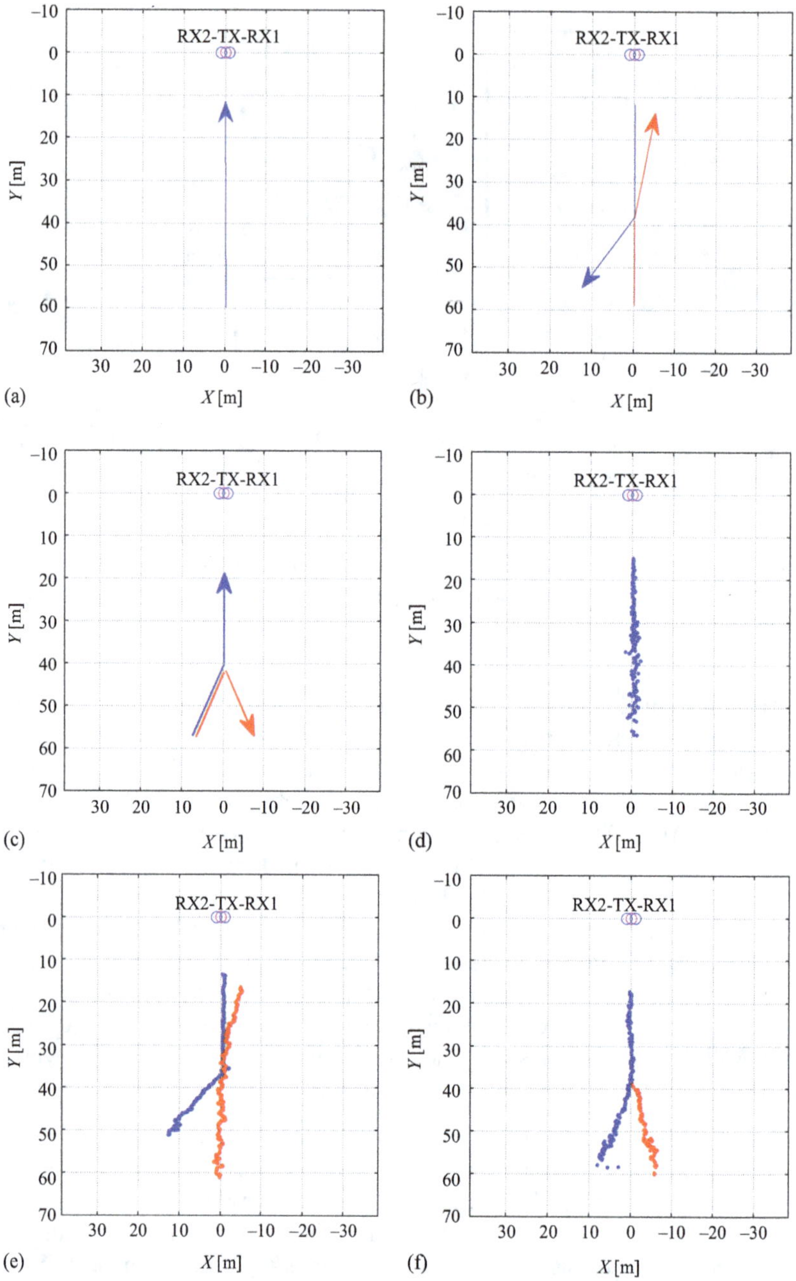

Figure 15.23 Real target trajectories for (a) Test T1; (b) Test T2; and (c) Test T3. PR localization results for (d) Test T1; (e) Test T2; and (f) Test T3. © [2015] IEEE. Reprinted, with permission, from [45]

The data collected for each test covers a length of time equal to 25 s and has been processed according to the WiFi-based PR processing scheme of Figure 15.15. In particular, the ECA is applied with a batch duration equal to 100 ms over a range of 300 m; a coherent integration time of 0.5 s is used to evaluate the bistatic range–velocity map over consecutive portions of the acquired signals (frames) with a fixed displacement of 0.1 s (10 frames per second are thus obtained); and target detection is performed by resorting to a standard cell-average CFAR threshold with a probability of false alarm equal to 10^{-4}. The above settings were experimentally verified to yield remarkable cancellation performance against the very slowly varying characteristics of the environment while enabling the detection of people moving in a building.

The results obtained for the considered tests T1, T2 and T3 are reported in Figure 15.23(d)–(f), respectively. As is apparent, in all cases, only small deviations are observed with respect to the real targets trajectory; these are mainly due to the target range/DoA estimation accuracies and their projection on the $X-Y$ plane. However, we verified that these fluctuations are partly due to the micro-Doppler effects typical of the human body. Incidentally, we might observe that these effects could be fruitfully exploited by a target classification stage at least to distinguish between human targets and man-made objects [43]. Anyway, better results are expected by exploiting multiple PR sensors properly dislocated around the hall; moreover, the achievable positioning accuracy can be increased by applying a second tracking stage on the Cartesian plane.

With reference, to the experimental tests involving two targets (i.e., T2 and T3), we observe that the conceived system is able to correctly identify and track the two men moving in the hall as far as they can be distinguished in the range-velocity plane; in particular, when the targets move toward opposite directions they yield echoes with opposite Doppler frequencies.

In contrast, in test T3, only a single target is recognized and tracked in the first part of the acquisition as the two men walk close each other. Obviously, this is a consequence of the limited resolution provided by the WiFi-based PR, especially in the range dimension. In fact, we recall that the achievable range resolution of the system is limited to several metres, due to the limited frequency bandwidth occupied by the available signals (e.g., 11–18 m, depending on the adopted modulation). Therefore, although the above results clearly show the promising characteristics of the conceived sensor for indoor target localization and tracking in surveillance applications, it is of great interest the study of alternative processing techniques able to provide an improved resolution capability on the observed targets, along the line investigated in [45,85,86].

15.4.2 Resolution improvement via ISAR techniques

Additional resolution in cross-range direction could be achieved by resorting to Inverse Synthetic Aperture Radar (ISAR) techniques, namely by exploiting the motion of the target itself [87]. Therefore in [45,85,86], we have shown that a cross-range resolution considerably higher than the range resolution can be achieved by applying ISAR techniques to targets with a motion component in the cross-range direction.

The main processing steps are sketched in Figure 15.15 in the dashed frame. Once a moving target has been detected, the corresponding range strip is selected from the compressed data and fed in input to the ISAR processing block constituted by the cascade of target motion estimation, cross-range profile focusing and scaling. The target phase history in the ISAR CPI is approximated by an M degree polynomial law. Consequently the target motion is estimated by searching for the set of coefficients that better compensates the migration through both range and Doppler resolution cells and thus provides the best quality of the focused profile according to a specific cost function. Once the target motion parameters are available, the cross-range profile is formed and properly scaled by mapping Doppler frequencies into cross-range distances. Further details are reported in [85] on the search strategy, on the objective function adopted for the autofocus and on the focusing/scaling stage and will be briefly discussed in Section 15.5.

Potential of ISAR techniques in resolving closely spaced human targets moving in a cluttered indoor scenario is here preliminary demonstrated by using live data acquired by means of the PR prototype in Figure 15.22. The PR Rx has been fielded in the canteen of the School of Engineering whose size is approximately equal 11 m wide and 28 m long (see Figure 15.24). In the reported experiments, simplifying conditions have been adopted for the targets motion; consequently the basic processing scheme presented in [85] can be effective without requiring advanced modifications aimed at isolating the contributions of multiple targets present in a given range cell [87]. Specifically, many tests have been performed using two human targets about two m apart both moving in cross-range direction

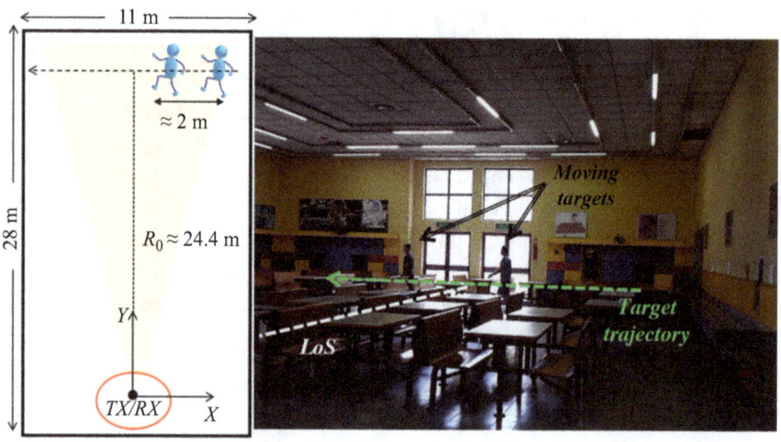

Figure 15.24 Sketch of experimental test. © [2015] IEEE. Reprinted, with permission, from [45]

(i.e., trajectory parallel to X axis in Figure 15.24) with approximately the same speed of 1.3 m/s, distance of minimum approach $R_0 = 24.4$ m and overall test duration equal 10 s [45].

The acquired data has been processed via the processing scheme in Figure 15.15 with ECA applied with a batch duration set to 400 ms over a range of 300 m and the bistatic range–velocity map evaluated over a CPI of 0.5 s. A cut of the bistatic range-Doppler map at the range bin interested by the two human targets is reported in Figure 15.25 (blue curve).

We observe that the two targets give rise to the presence of a single strong peak meaning that the two targets can be detected but not resolved. By increasing the CPI (3 s) and feeding the ISAR processing block (with $M = 3$) the red curve in Figure 15.25 is obtained. We observe that the increased CPI provides an improvement in terms of both Doppler resolution and target peak power. Particularly, the two targets give rise to the presence of two clearly resolved peaks whose Doppler separation (1.635 Hz) scaled accordingly to the estimated cross-range speed (1.313 m/s) provides 1.865 m which is largely in agreement with the actual separation between the targets. The reported results clearly show that the target motion can be fruitfully exploited to improve cross-range resolution so that closely spaced targets can be effectively discriminated. Moreover the characteristics of the considered indoor scenario demonstrate the potential of the proposed approach in cases of practical interest.

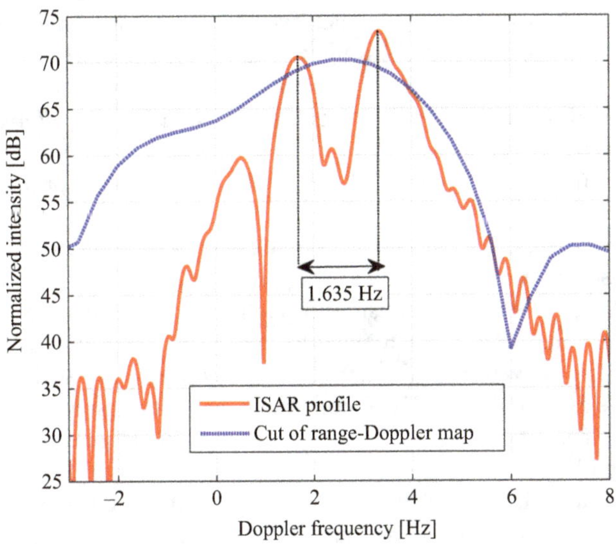

Figure 15.25 Cut of the range-Doppler map (CPI = 0.5 s) and ISAR output (CPI = 3 s) comparison. © [2015] IEEE. Reprinted, with permission, from [45]

15.5 Steps toward target classification: cross-range profiling of targets

In the previous sub-sections, the discussion has been focused on the target detection and localization capability of a PR sensor in different applications. In addition, especially in the considered short-range case, it is of great interest the possibility to obtain high resolution radar images of the observed targets. In fact, the availability of reliable and meaningful target's images may open the doors to non-cooperative target recognition (NCTR) capabilities for the considered sensors.

Several studies have been reported in the technical literature that investigate the possibility to form 2D images of the observed scene when exploiting various signals of opportunity in different applications [60,61,88–100]. In these studies, SAR or ISAR techniques are applied based on the exploitation of the motion of the Tx and/or of the target itself. In some cases, further improved resolution is synthesized by adjoining more frequency channels transmitted by the same emitter [91,94]. In addition, a special attention has been recently devoted to the exploitation of multistatic imaging approaches that are expected to yield enhanced imaging capability thanks to the diversity provided by sparse sensors distribution [95–100].

Unfortunately, for a great variety of signals of opportunity, the range resolution is typically insufficient to obtain meaningful 2D images due to the limited frequency bandwidth. In contrast, for targets with a motion component in the cross-range direction, it is still possible to achieve a higher (cross-range) resolution by applying appropriate ISAR processing schemes [45,85,86,101]. Such approach provides the PR with the advanced capability to obtain high resolution 1D cross-range profiles of moving objects that could anyway enable the automatic classification of designated targets.

To this purpose, an effective ISAR scheme for cross-range profiling has been introduced in [85], tailored for the typical short-range surveillance scenarios of a WiFi-based PR. The main processing steps are sketched in Figure 15.15 and have been briefly summarized in Section 15.4.2.

To explain, the ISAR processing in slightly more details we refer to Figure 15.26 and consider a target moving at a given velocity (V_x, V_y) and passing through the point ($x=0$, $y=y_0=R_0$) at time aperture centre t_0. For the sake of simplicity, we assume that the Tx of opportunity and the passive Rx are located in ($x=0, y=0$) in a quasi-monostatic configuration (the more general bistatic case is considered in [85]).

As is typical in ISAR, the two-way radar-target distance can be approximated around time t_0 as a polynomial law, namely as a Taylor series of order M. In the considered geometry, we have

$$R_t(t) = 2\sqrt{x_t(t)^2 + y_t(t)^2} \approx R_t(0) + 2\sum_{m=1}^{M} \alpha_m \frac{(t-t_0)^m}{m!} \tag{15.16}$$

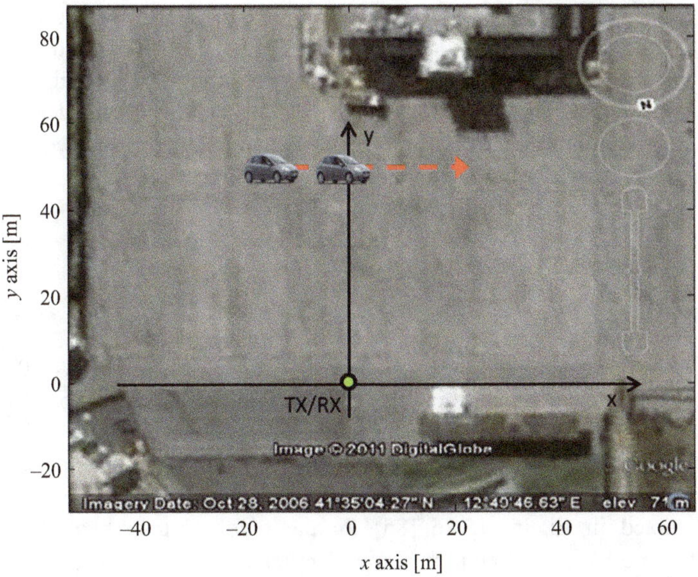

Figure 15.26 Sketch of the acquisition geometry for cross-range profiling tests

where $x_t(t) = V_x(t - t_0)$, $y_t(t) = R_0 + V_y(t - t_0)$ and the coefficients up to the second order can be easily expressed as functions of the target motion parameters:

$$R_t(0) = 2R_0; \quad \alpha_1 = V_y; \quad \alpha_2 = \frac{V_x^2}{R_0} \tag{15.17}$$

As far as the M value is concerned this has to be chosen according to the depth of focus criterion: in our case, for the considered short-range acquisition geometry and motion conditions, it is typically sufficient to set $M = 3$. In addition, we observe that both range and Doppler cell migration can occur. Specifically, in range direction the migration is related mainly to range walk due to the target radial velocity (namely α_1). As far as cross-range (Doppler) direction is concerned, the target phase history, $\varphi_t(t) = 2\pi R_t(t)/\lambda$, in our application is characterized by quadratic and cubic phase terms, that need to be compensated to perform coherent integration over the long CPI T required to achieve the high cross-range resolution.

Unfortunately, the target motion parameters are usually unknown and must be estimated from the received echoes via auto-focus. To this purpose a range strip centred on the range of the detected target is selected from the range compressed data and given as input to the target motion estimation procedure. An effective approach to obtain the parameter estimation is to search for the set of values $(\alpha_1, \alpha_2, \alpha_3)$ that provides the best quality for the focused profile. In [85], the estimated $(\widehat{\alpha}_1, \widehat{\alpha}_2, \widehat{\alpha}_3)$ values are chosen as those values which minimize the entropy of the final profile [102] and an efficient approach is proposed to implement the corresponding search in the 3D parameter space with a limited computational burden. In this regard, we also observe that this auto-focusing stage can be fruitfully initialized by the rough motion estimation provided by the PR tracking stage.

Once the $R_t(0)$ value is known from the selected range bin, and the estimate of $\widehat{\alpha}_2$ is obtained, also the target cross-range velocity can be obtained $\widehat{V}_x$, which is directly used to scale the Doppler axis, thus mapping Doppler frequencies into cross-range distances. Moreover, this allows also to evaluate the global Doppler bandwidth B_D and the azimuth resolution r_a, given by:

$$B_D = \frac{2}{\lambda}\alpha_2 T \qquad r_a = \frac{V_x}{B_D} = \frac{\lambda R_0}{2V_x T} \tag{15.18}$$

For example, under the hypothesis of $R_0 = 50$ m, ($V_x = 4.5$ m/s, $V_y = 0$ m/s) and $T = 6$ s, the obtainable resolution in the azimuth dimension is of the order of tens of centimetres.

Figure 15.27 shows an example of the results obtained for an experimental test performed against vehicular targets in a parking area, using the acquisition geometry depicted in Figure 15.26 [85]. In the reported test, two identical cars (i.e., Fiat Punto Evo) were employed moving along the axis $y = y_0 = 50$ m (dashed line in Figure 15.26) at about $V_x = 4.5$ m/s with a separation of about 3–4 m so that, at the middle of the acquisition, they were almost symmetrically displaced about the point (0 m; 50 m).

Specifically, Figure 15.27 shows the sequence of five profiles focused from subsequent (partially overlapped) frames extracted from the collected signal. An ISAR CPI of 6 s is used, sliding along the 10 s of the globally available acquisition time. The different frames refer to image times spanning from 3 to 7 s with spacing 1 s. Initially, only a single car is present inside the antenna beam, which is clearly apparent from the profile. We observe that as the second car enters in the antenna beam, two almost identical cross-range profiles are obtained: these profiles persist in all the considered frames occupying progressively increasing cross-range positions. If we evaluate the cross-range distance travelled by scatter A between the first frame (image time 3 s) to the last frame (image time 7 s), we get 17.81 m which is well in line with cross-range velocity of about 4.5 m/s and temporal separation between the first and the last frame of 4 s. In addition, we observe that each of the two identical patterns in the reported cross-range profiles is characterized by three main peaks which correspond to the main scattering centres of the two identical cars used for the considered experiment. Thus, the proposed technique allowed us both to separate the two vehicles in the cross-range direction and to identify their main scattering points making it possible to measure their displacement. Basically, by evaluating the distances between homologous points of the cars (profile peaks) we can estimate the cross-range displacement of the two targets (about 6.8 m on average) which is well in line with the test geometry (we recall that the length of each car is 4 m and their separation along the path is about 2–3 m). Incidentally, we notice that the identification of the main scattering centres represents the first step for a target classification procedure since it gives an indication of the target electromagnetic length and signature.

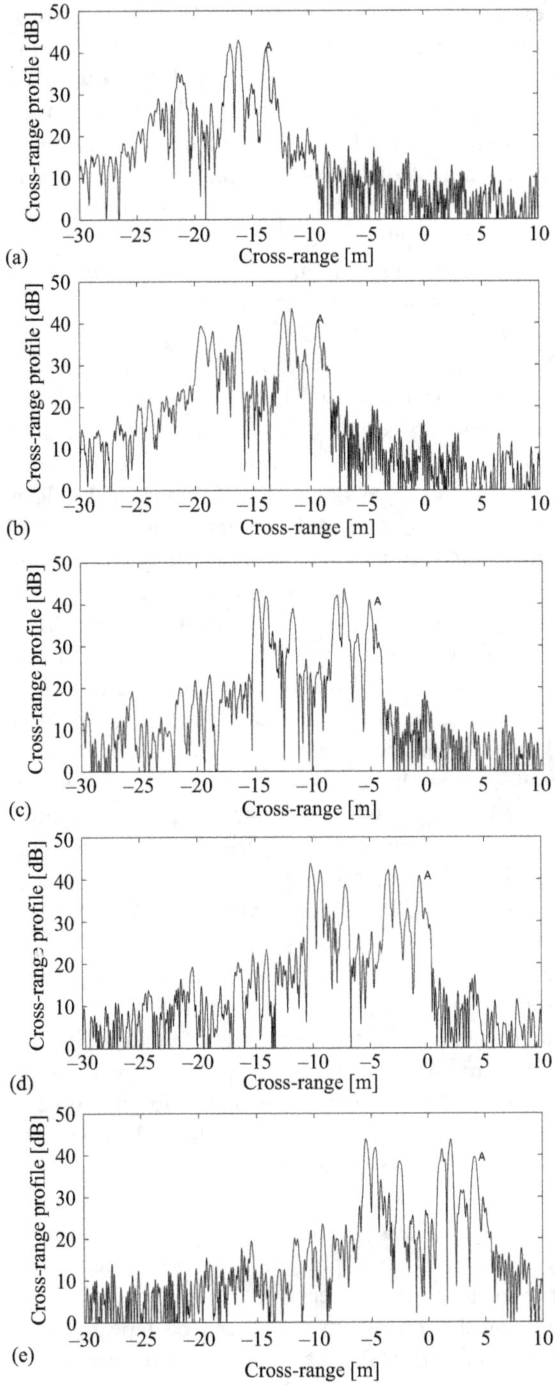

Figure 15.27 Sequence of frames from cross-rage profiling test employing two identical cars. © [2014] IEEE. Reprinted, with permission, from [85]

It is worth explicitly noticing that in order to get an ISAR profile, the target does not have to move strictly in cross-range direction but it can move with both radial and cross-range velocities not null. The only case which prevents the formation of the ISAR profile is when the target is moving strictly in the radial direction: in such case the target motion does not provide the angular aperture needed for the achievement of cross-range resolution. Obviously, the limitation pertaining this very special case could be easily overcome if multiple Rxs (even just two) properly located are used.

Notice that the first blocks in Figure 15.15 are in common with the PR processing scheme for target detection so that the additional required processing steps can be nicely integrated in the basic scheme to provide the final system with advanced ISAR capability. Specifically, in [85] we have shown that the range compressed data after interference cancellation have to be exploited to avoid strong contributions from stationary scatters, i.e., interference cancellation is a required step also in the passive ISAR mode. However, the interference cancellation has a non-negligible effect on the target signal fed in input to the ISAR processing so that it might reduce the quality of the ISAR products. Basically, the adopted cancellation algorithm (e.g., ECA) strongly attenuates the target signal components at low Doppler frequencies since they are recognized as stationary contributions and included in the adaptive estimation of the cancellation weights. As a consequence, a considerable part of the target Doppler spectrum can be lost and this potentially causes a degradation of the ISAR profiling. This degradation can occur in the case of slow-moving targets as well as in the case of targets moving along the cross-range direction, that are the most interesting ones to apply the ISAR processing schemes. As a consequence, the processing techniques described above might yield limited performance especially when accurate profiling is required.

Obviously, different applications impose different constraints and requirements on the quality of the ISAR products so that this potential degradation might be appreciable or not. Particularly when considering human targets, as in Section 15.4.2, the strong requirement is on the resolution capability enabling the separation of the different targets. In contrast, when dealing with man-made targets (such as vehicles), the requirement is on the accurate separation and extraction of the different scattering centres of the same target so that ATR procedures could be enabled. While standard quality ISAR products can allow the separation of different targets, enhanced quality products are essential for classification purposes.

Therefore, in [45] an advanced processing scheme has been presented to obtain an effective removal of interference contributions from the stationary scene while preserving the target echo and, consequently, the quality of the ISAR processing. The proposed approach exploits the output of the ISAR profiling stage to retrieve from it the signal contribution concerning the considered target and uses this recovered component to clean the received surveillance signal. In this way, the estimation of the cancellation filter weights can be repeated using a signal with 'target reduced' characteristics. By iteratively repeating the above steps, progressively a 'target free' signal is made available for filter weights estimation and the low-frequency components of the target returns are better reconstructed. The block

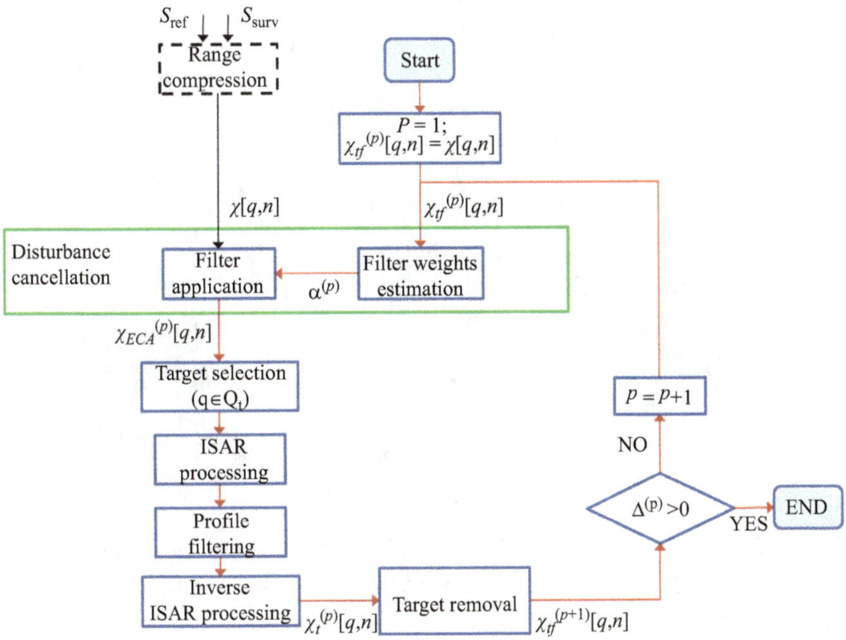

Figure 15.28 Block diagram of the iterative target preserving ISAR profiling technique. © [2015] IEEE. Reprinted, with permission, from [45]

diagram of the proposed iterative target preserving (ITP) technique is sketched in Figure 15.28 while a discussion about the relevant parameters setting and the stop criterion can be found in [45].

As an example, the cross-range profiles obtained with the ITP technique are reported in Figure 15.29 for a single frame of the same experimental test shown in Figure 15.27. In this case, the ITP algorithm stops after 16 iterations. As is apparent, compared to the simple cascade of the ECA and the ISAR processing, ITP technique yields an improvement of few dBs in terms of profile amplitude. In fact, the capability to preserve a greater portion of the target spectrum results in better target motion estimation and target focusing that in turn lead to the formation of more stable profiles to be exploited for the extraction of reliable information on the observed targets. In this regard, Table 15.3 reports the distances between homologous points of the cars (i.e., profile peaks labelled with capital letters) measured from the cross-range profiles obtained with the ECA and the ITP approaches, respectively. Even, if the actual value of this distance is not available, it is expected to be constant independently of the considered pair of peaks. Apparently, for all the three considered pairs, the use of the iterative technique provides less scattered values, when compared to ECA, thus proving a more accurate extraction of the target scattering centres by means of ITP technique. Obviously, the enhanced quality of the achieved ISAR profiles is paid in terms of

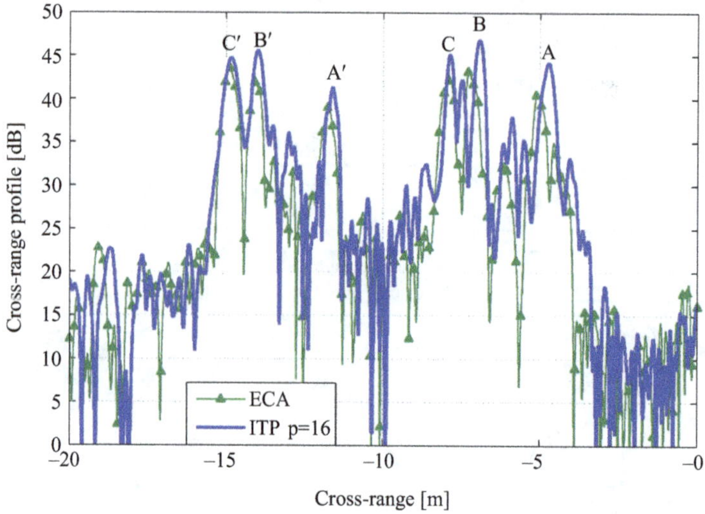

Figure 15.29 Cross-range profiles comparison after a single cancellation stage and the ITP technique. © [2015] IEEE. Reprinted, with permission, from [45]

Table 15.3 Cross-range displacements of homologous points of the profiles

	ECA (m)	ITP (iter $p = 16$) (m)
$D_A = \lvert A' - A \rvert$	6.67	6.88
$D_B = \lvert B' - B \rvert$	6.84	7.08
$D_C = \lvert C' - C \rvert$	7.03	6.97

an increased computational cost of the iterative technique with respect to the basic approach (almost coinciding with the first iteration of the ITP).

15.6 Conclusions

As is hopefully evident from this chapter, the nice characteristics of PR sensors could be successfully exploited in a number of short-range surveillance applications. The focus of the chapter has been primarily on the capability of reliably detecting and accurately tracking targets of potential interest in such applications. These included small vessels and boats in coastal areas, ground vehicles on roads and parking areas and even people moving indoors. Several experimental results were reported to verify the effectiveness of PR sensors when employed in such challenging scenarios. In addition, the possibility to obtain high resolution cross-range profiles of the observed targets has been illustrated; this could be a valuable characteristic for many short-range surveillance applications as it may open the doors to non-cooperative target recognition capabilities for the considered sensors.

Leveraging the reported results and related discussions, we might draw some considerations regarding possible strategies to be implemented in order to provide the conceived sensors with advanced capabilities and improved performance.

The development of a large network of PR sensors is highly recommended and appears to be a quite viable solution for the considered short-range applications. By properly dislocating the Rxs within the area to be surveyed, the resulting system gains the advantage of a high spatial diversity that can be fruitfully exploited to improve the situational awareness. In fact, both the basic functions of target detection, localization and tracking, and the advanced capability of forming images of designated targets significantly benefit from the joint exploitation of different observation geometries; moreover, these might include favourable bistatic geometries with the potential of an increased and much more stable target RCS, such as in the forward scattering mode.

Further enhanced capabilities can be expected if the conceived systems operate in conjunction with other active/passive sensors. In such a network of heterogeneous components, the PR sensors could successfully play their strategic role of gap-fillers.

Moreover, especially in short-range applications based on transmissions for mobile communication and networking, a hybrid active and passive localization of the targets can be foreseen by exploiting both self-reported positions and radar measurements. Specifically, with such approach, PR can be viewed as an inseparable part of a more complex communication system thus embracing the current trend toward shared spectrum access for radar and communications.

Acknowledgements

The author would like to gratefully acknowledge the collaboration of many colleagues at Sapienza University of Rome. I am extremely thankful to Prof. P. Lombardo for sincere and valuable guidance and encouragement extended to me. I am indebted to Dr. D. Pastina for sharing expertise on ISAR and signal processing topics. Still, experimental tests and extensive data analysis would not have been possible if I did not have the support of many young colleagues and students. Among them I would like to gratefully thank Dr. C. Bongioanni, Dr. P. Falcone, Dr. D. Langellotti, Dr. A. Macera, Dr. T. Martelli, Dr. C. Palmarini, Dr. D. Cristallini and Dr. M. Caruso.

I am also grateful to the colleagues of Leonardo S.p.A. (formerly Finmeccanica) for provision of expertise. Particular thanks to Dr. A. Farina for his precious suggestions, L. Timmoneri, S. Immediata, G. Cocca and S. Scafè for their technical support, E. Tilli for leading the reported DVB-T experiments and Dr. M. Sedehi for setting the experimental setup.

Finally, I am glad to acknowledge that parts of this chapter were investigated while working on PR subjects under support from

- Leonardo S.p.A. (formerly Finmeccanica);
- the FP7 European Project ATOM – 'Airport detection and Tracking Of dangerous Materials by passive and active sensors arrays' [83].
- the FP7 European Project SOS – 'Sensors system for detection and tracking of dangerous materials in order to increase the airport security in the indoor landside area' [84].

List of acronyms

ITP	iterative target preserving
NCTR	non-cooperative target recognition
ISAR	inverse synthetic aperture radar
RFID	radio frequency identification
ML	maximum likelihood
DOP	dilution of precision
CFAR	constant false alarm rate
FFT	fast Fourier transform
DSSS	direct sequence spread spectrum
AP	access point
DVB-SH	digital video broadcasting-satellite services to handhelds
MEO	medium earth orbit
LEO	low earth orbit
EIRP	equivalent isotropic radiated power
GEO	geostationary earth orbits
GPS	global positioning system
AIS	automatic identification system
Tx	transmitter
Rx	receiver
SFN	single frequency network
AF	ambiguity function
ATC	air traffic control
CA-CFAR	cell-averaging constant false alarm rate
CAF	cross-ambiguity function
CPI	coherent processing interval
DAB	digital audio broadcasting
DoA	direction of arrival
DVB-T	digital video broadcasting-terrestrial

ECA	extensive cancellation algorithm
EM	electromagnetic
GSM	system for mobile communications
LS	least square
LTE	long-term evolution
OFDM	orthogonal frequency-division multiplexing
PR	passive radar
RCS	radar cross section
RPR	residual peaks removal
S-ECA	sliding extensive cancellation algorithm
SNR	signal-to-noise ratio
UMTS	universal mobile telecommunications system
WiFi	wireless fidelity
WiMAX	worldwide interoperability for microwave access

References

[1] Special Issue on Passive Radar Systems (Guest Editor: P. Howland), *IEE Proceedings on Radar, Sonar and Navigation*. 2005; **152**(3): 105–223.

[2] Special Issue on Passive Radar (Part I) (Guest Editors: A. Farina, H. Kuschel), *IEEE Aerospace and Electronic Systems Magazine*. 2012; **27**(10): 5–59.

[3] Special Issue on Passive Radar (Part II) (Guest Editors: A. Farina, H. Kuschel), *IEEE Aerospace and Electronic Systems Magazine*. 2012; **27**(11): 4–55.

[4] Special issue on bistatic and MIMO radars and their applications in surveillance and remote sensing (Guest Editors: A. Farina, M. Lesturgie), *IET Radar, Sonar & Navigation*. 2014; **8**(2): 73–166.

[5] Lombardo P., Colone F., Farina A. 'Passive radar: Harvesting e.m. radiations for surveillance'. *Tutorial at the IEEE Radar Conference 2010*; Washington DC, USA, May 2010.

[6] Lombardo P., Colone F. 'Advanced processing methods for passive bistatic radar systems', in Melvin W.L., Scheer J.A. (eds.) *Principles of Modern Radar: Advanced Radar Techniques* (SciTech Publishing, Inc., 2012), pp. 739–821.

[7] Griffiths H. 'Passive and bistatic radar', in Melvin W.L., Scheer J.A. (eds.) *Principles of Modern Radar: Radar Applications* (SciTech Publishing, Inc., 2014), pp. 499–540.

[8] Sun H., Tan D.K.P., Yilong L., Lesturgie M. 'Applications of passive surveillance radar system using cell phone base station illuminators'. *IEEE Aerospace and Electronic Systems Magazine*. 2010; **25**(3): 10–8.

[9] Krysik P., Kulpa K. 'The use of a GSM-based passive radar for sea target detection'. *Proc. of the 9th European Radar Conference (EuRAD 2012)*; Amsterdam, Netherlands, Oct. 2012, pp. 142–45.

[10] Zemmari R., Daun M., Feldmann M., Nickel U. 'Maritime surveillance with GSM passive radar: Detection and tracking of small agile targets'. *Proc. of the International Radar Symposium 2013 (IRS 2013)*; Dresden, Germany, Jun 2013, pp. 245–51.

[11] Zemmari R., Broetje M., Battistello G., Nickel U. 'GSM passive coherent location system: Performance prediction and measurement evaluation'. *IET Radar, Sonar and Navigation*. 2014; **8**(2): 94–105.

[12] Wang Q., Hou C., Lu Y. 'An experimental study of WiMAX-based passive radar'. *IEEE Transactions on Microwave Theory and Techniques*. 2010; **58**(12—Part 1): 3502–10.

[13] Chetty K., Woodbridge K., Guo H., Smith G.E. 'Passive bistatic WiMAX radar for marine surveillance'. *Proc. of the IEEE National Radar Conference 2010*; Washington (DC), USA, May 2010, pp. 188–93.

[14] Raout J. 'Sea target detection using passive DVB-T based radar'. *Proc. of the 2008 International Conference on Radar*; Adelaide, Australia, Sept. 2008, pp. 695–700.

[15] O'Hagan D.W., Capria A., Petri D., *et al.* 'Passive bistatic radar (PBR) for harbour protection applications'. *Proc. of the IEEE National Radar Conference 2012;* Atlanta (GA), USA, May 2012, pp. 446–50.

[16] Wan X., Yia J., Zhao Z., Ke H. 'Experimental research for CMMB-based passive radar under a multipath environment'. *IEEE Transactions on Aerospace and Electronic Systems*. 2014; **50**(1): 70–85.

[17] Langellotti D., Colone F., Lombardo P., Sedehi M., Tilli E. 'DVB-T based passive bistatic radar for maritime surveillance'. *Proc. of the IEEE National Radar Conference 2014*; Cincinnati (OH), USA, May 2014, pp. 1197–202.

[18] Ummenhofer M., Schell J., Heckenbach J., Kuschel H., 'O Hagan D.W. 'Doppler estimation for DVB-T based passive radar systems on moving maritime platforms'. *Proc. on the IEEE National Radar Conference 2015*; Arlington (VA), USA, May 2015, pp. 1687–91.

[19] Kuschel H., Ummenhofer M., Lombardo P., Colone F., Bongioanni C. 'Passive radar components of ARGUS 3D'. *IEEE Aerospace and Electronic Systems Magazine*. 2014; **29**(3): 15–25.

[20] Cardinali R., Anniballi E., Bongioanni C., Macera A., Colone F., Lombardo P. 'ARGUS 3D: Security enhancements through innovative radar technologies'. *Proc. of the 2013 Int. Conference on Availability, Reliability and Security (ARES 2013)*; Regensburg, Germany, Sept. 2013, pp. 759–65.

[21] Colone F., Langellotti D., Lombardo P. 'DVB-T signal ambiguity function control for passive radars'. *IEEE Transactions on Aerospace and Electronic Systems*. 2014; **50**(1): 329–47.

[22] Barott W.C., Johnson M.B., Scott K.M. 'Passive radar for terminal area surveillance: Performance feasibility study', *Proc. of the AIAA/IEEE Digital Avionics Systems Conference 2014*; Colorado Springs (CO), USA, Oct. 2014, pp. 3D1-1–13.

[23] Sun H., Tan D.K.P., Lu Y. 'Aircraft target measurements using a GSM-based passive radar'. *Proc. of the 2008 IEEE Radar Conference*; Rome, Italy, May 2008, pp. 1–6.

[24] Krysik P., Samczynski P., Malanowski M., Maslikowski L., Kulpa K. 'Detection of fast maneuvering air targets using GSM based passive radar'. *Proc. of the International Radar Symposium 2012*; Warsaw, Poland, May 2012, pp.69–72.

[25] Tan D.K.P., Sun H., Lu Y., Lesturgie M., Chan H.L. 'Passive radar using global system for mobile communication signal: Theory, implementation and measurements'. *IEE Proceedings: Radar, Sonar and Navigation*. 2005; **152**(3): 116–23.

[26] Colone F., Falcone P., Bongioanni C., Lombardo P. 'WiFi-based passive bistatic radar: Data processing schemes and experimental results'. *IEEE Transactions on Aerospace and Electronic Systems*. 2012; **48**(2): 1061–79.

[27] Falcone P., Colone F., Lombardo P. 'Potentialities and challenges of WiFi-based passive radar'. *IEEE Aerospace and Electronic Systems Magazine*. 2012; **27**(11): 15–26.

[28] Krysik P., Samczynski P., Malanowski M., Maslikowski L., Kulpa K.S. 'Velocity measurement and traffic monitoring using a GSM passive radar demonstrator'. *IEEE Aerospace and Electronic Systems Magazine*. 2012; **27**(10): 43–51.

[29] Falcone P., Colone F., Macera A., Lombardo P. 'Two-dimensional location of moving targets within local areas using WiFi-based multistatic passive radar'. *IET Radar, Sonar and Navigation*. 2014; **8**(2): 123–31.

[30] Zemmari R., Nickel U., Wirth W. 'GSM passive radar for medium range surveillance'. *Proc. of the 6th European Radar Conference (EuRAD 2009)*; Rome, Italy, Sept. 2009, pp. 49–52.

[31] Salah A.A., Abdullah R.S.A.R., Ismail A., Hashim F., Abdul Aziz N.H. 'Experimental study of LTE signals as illuminators of opportunity for passive bistatic radar applications'. *Electronics Letters*. 2014; **50**(7): 545–47.

[32] Evers A., Jackson J.A. 'Analysis of an LTE waveform for radar applications'. *Proc. of the IEEE National Radar Conference 2014*; Cincinnati (OH), USA, May 2014, pp. 200–5.

[33] Cherniakov M., Nezlin D., Kubik K., 'Air target detection via bistatic radar based on LEOS communication signals'. *IEE Proceedings—Radar, Sonar and Navigation*. 2002; **149**(1): 33–8.

[34] He X., Cherniakov M., Zeng, T. 'Signal detectability in SS-BSAR with GNSS non-cooperative transmitter'. *IEE Proceedings—Radar, Sonar and Navigation*. 2005; **152**(3): 124–32.

[35] Cristallini D., Caruso M., Falcone P., *et al.* 'Space-based passive radar enabled by the new generation of geostationary broadcast satellites'. *Proc. of the 2010 IEEE Aerospace Conference*; Big Sky (MT), USA, Mar. 2010.

[36] Gill L.P., Grenier D., Chouinard J.Y. 'Use of XMTM radio satellite signal as a source of opportunity for passive coherent location'. *IET Radar Sonar Navigation*. 2011; **5**(5): 536–44.

[37] Corucci L., Petri D., Berizzi F., Martorella M., Capria A. 'An integrated space-based passive system for maritime operational traffic surveillance'. *Proc. of the 2012 IEEE First AESS European Conference on Satellite Telecommunications (ESTEL)*; Rome, Italy, Oct. 2012, pp. 1–4.

[38] Golabi M., Sheikhi A., Biguesh M. 'A new approach for sea target detection in satellite based passive radar'. *Proc. of the 21st Iranian Conference on Electrical Engineering (ICEE)*; Mashhad, Iran, May 2013, pp. 1–5.

[39] Chetty K., Smith G., Guo H., Woodbridge K. 'Target detection in high clutter using passive bistatic WiFi radar'. *Proc. of the IEEE National Radar Conference*; Pasadena (CA), USA, May 2009.

[40] Chetty K., Smith G., Woodbridge K. 'Through-the-wall sensing of personnel using passive bistatic WiFi radar at standoff distances'. *IEEE Transactions on Geoscience and Remote Sensing*. 2012; **50**(4): 1218–26.

[41] Falcone P., Bongioanni C., Macera A., *et al.* 'Active and passive radar sensors for airport security'. *Proc. of the 2012 Tyrrhenian Workshop on Advances in Radar and Remote Sensing*; Naples, Italy, Sep. 2012, pp. 314–21.

[42] Smith G.E., Chetty K., Baker C.J., Woodbridge K. 'Extended time processing for passive bistatic radar'. *IET Radar, Sonar & Navigation*. 2013; **7**(9): 1012–18.

[43] Tan B., Woodbridge K., Chetty K. 'A real-time high resolution passive WiFi Doppler-radar and its applications'. *Proc. of the 2014 International Radar Conference*; Lille, France, Oct. 2014, pp.1–6.

[44] Qingchao Chen, Tan B., Woodbridge K., Chetty K. 'Indoor target tracking using high Doppler resolution passive Wi-Fi radar'. *Proc. of the 2015 IEEE International Conference on Acoustics, Speech and Signal Processing (ICASSP)*; South Brisbane (QLD), Australia, Apr. 2015, pp. 5565–69.

[45] Pastina D., Colone F., Martelli T., Falcone P. 'Parasitic exploitation of Wi-Fi signals for indoor radar surveillance'. *IEEE Transactions on Vehicular Technology*. 2015; **64**(4): 1401–15.

[46] Langellotti D., Colone F., Lombardo P., Tilli E., Sedehi M., Farina A. 'Over the horizon maritime surveillance capability of DVB-T based passive radar'. *Proc. of the European Radar Conference (EuRAD) 2014*; Rome, Italy, Oct. 2014.

[47] European FP7 Project SEABILLA – 'SEA Border surveILLance', http://www.seabilla.eu.

[48] European FP7 Project CLOSEYE – 'Collaborative evaLuation Of border Surveillance technologies in maritime Environment bY pre-operational validation of innovativE solutions', http://www.closeye.eu/.

[49] European FP7 Project DOLPHIN – 'Development of Pre-operational Services for Highly Innovative Maritime Surveillance Capabilities', http://www.gmes-dolphin.eu/.

[50] European H2020 Project SpyGLASS – 'Galileo based passive radar system for maritime surveillance', http://cordis.europa.eu/project/rcn/193820_en.html.

[51] European H2020 Call for Proposals H2020-BES-02-2015: 'Maritime border security topic 2: affordable and easily deployable technologies for EU coastal border surveillance with reduced impact on the environment'.

[52] Baker C.J., Griffiths H.D., Papoutsis I. 'Passive coherent location radar systems. Part 2: Waveform properties'. *IEE Proceedings—Radar, Sonar and Navigation*. 2005; **152**(3): 160–68.

[53] Colone F., O'Hagan D.W., Lombardo P., Baker C.J. 'A multistage processing algorithm for disturbance removal and target detection in passive bistatic radar'. *IEEE Transactions on Aerospace and Electronic Systems*. 2009; **45**(2): 698–722.

[54] Palmarini C., Martelli T., Colone F., Lombardo P. 'Disturbance removal in passive radar via sliding extensive cancellation algorithm (ECA-S)'. *Proc. of the IEEE Radar Conference 2015*; Johannesburg, South Africa, Oct. 2015.

[55] Colone F., Palmarini C., Martelli T., Tilli E. 'Sliding extensive cancellation algorithm (ECA-S) for disturbance removal in passive radar', *IEEE Transactions on Aerospace and Electronic Systems*. 2016; **52**(3): 1309–26.

[56] Saini R., Cherniakov M. 'DTV signal ambiguity function analysis for radar application'. *IEE Proceedings—Radar, Sonar and Navigation*. 2005; **152**(3): 133–42.

[57] Palmer J.E., Harms H.A., Searle S.J., Davis L.M. 'DVB-T passive radar signal processing', *IEEE Transactions on Signal Processing*. 2013; **61**(8): 2116–26.

[58] Langellotti D., Bongioanni C., Colone F., Lombardo P. 'Impact of synchronization on the ambiguity function shape for PBR based on DVB-T signals'. *Proc. of the 11th International Radar Symposium (IRS2010)*; Vilnius, Lithuania, Jun. 2010.

[59] Daun M., Berger C.R. 'Track initialization in a multistatic DAB/DVB-T network'. *Proc. of the 11th International Conference on Information Fusion*; Cologne, Germany, Jun. 2008.

[60] Prati C., Rocca F., Giancola D., Monti Guarnieri A. 'Passive geosynchronous SAR system reusing backscattered digital audio broadcasting signals'. *IEEE Transactions on Geoscience and Remote Sensing*. 1998; **36**(6): 1973–76.

[61] Cazzani L., Colesanti C., Leva D., *et al.* 'A ground-based parasitic SAR experiment'. *IEEE Transactions on Geoscience and Remote Sensing*. 2000; **38**(5): 2132–41.

[62] Colone F., Woodbridge K., Guo H., Mason D., Baker C.J. 'Ambiguity function analysis of wireless LAN transmissions for passive radar'. *IEEE Transactions on Aerospace and Electronic Systems*. 2011; **47**(1): 240–64.

[63] Falcone P., Colone F., Bongioanni C., Lombardo P. 'Experimental results for OFDM WiFi-based passive bistatic radar'. *Proc. of the 2010 IEEE Radar Conference*; Washington (DC), USA, May 2010, pp. 516–21.

[64] Colone F., Falcone P., Lombardo P. 'Passive Bistatic Radar based on mixed DSSS and OFDM WiFi transmissions'. *Proc. of the 2011 European Radar Conference (EuRAD 2011)*; Manchester, UK, Oct. 2011, pp. 154–57.

[65] Falcone P., Colone F., Lombardo P. 'Doppler frequency sidelobes level control for WiFi-based passive bistatic radar'. *Proc. of the 2011 IEEE Radar Conference*; Kansas City (MO), USA, May 2011, pp. 435–40.

[66] Stinco P., Greco M.S., Gini F., Rangaswamy M. 'Ambiguity function and Cramer-Rao bounds for universal mobile telecommunications system-based passive coherent location systems'. *IET Radar, Sonar & Navigation*. 2012; **6**(7): 668–78.

[67] Gogineni S., Rangaswamy M., Rigling B.D., Nehorai A. 'Cramér-Rao bounds for UMTS-based passive multistatic radar'. *IEEE Transactions on Signal Processing*. 2014; **62**(1): 95–106.

[68] Anastasio V., Farina A., Colone F., Lombardo P. 'Cramér-Rao lower bound with Pd<1 for target localisation accuracy in multistatic passive radar'. *IET Radar, Sonar & Navigation*. 2014; **8**(7): 767–75.

[69] Anastasio V., Colone F., Lombardo P. 'A procedure for effective receiver positioning in multistatic passive radar'. *Proc. of the European Radar Conference (EuRAD 2009)*; Rome, Italy, Sep. 2009, pp. 493–96.

[70] Gumiero F., Nucciarone C., Anastasio V., Lombardo P., Colone F. 'Multistatic passive radar geometry optimization for target 3D positioning accuracy'. *Proc. of the European Radar Conference (EuRAD 2010)*; Paris, France, Sep. 2010, pp. 467–70.

[71] Zelnio A.M., Rigling B.D. 'Detection-based localization of passive radar receivers'. *Proc. of the 2012 IEEE Radar Conference*; Atlanta (GA), USA, May 2012, pp.173–77.

[72] Stinco P., Greco M., Gini F., Farina A. 'Sensor selection in PCL radar systems based on bistatic PCRLB'. *Proc. of the Tyrrhenian Workshop on Advances in Radar and Remote Sensing (TyWRRS 2012)*; Naples, Italy, Sep. 2012, pp.41–45.

[73] Chitgarha M.M., Majd M.N., Radmard M., Nayebi M.M. 'Choosing the position of the receiver in a MISO passive radar system'. *Proc. of the European Radar Conference (EuRAD 2012)*; Amsterdam, Netherlands, Oct. 2012, pp.318–21.

[74] Ivashko I.M., Krasnov O.A., Yarovoy A.G. 'Receivers topology optimization of the combined active and WiFi-based passive radar network'. *Proc. of the European Radar Conference (EuRAD 2014)*; Rome, Italy, Oct. 2014, pp.517–20.

[75] Lesturgie M. 'Use of dynamic radar signature for multistatic passive localisation of helicopter'. *IEE Proceedings Radar, Sonar and Navigation*. 2005; **152**(6): 395–403.

[76] Malanowski M., Kulpa K. 'Two methods for target localization in multistatic passive radar'. *IEEE Transactions on Aerospace and Electronic Systems*. 2012; **48**(1): 572–80.

[77] Norouzi Y., Derakhshani M. 'Joint time difference of arrival/angle of arrival position finding in passive radar'. *IET Radar, Sonar & Navigation*. 2009; **3**(2): 167–76.

[78] Wielgo M., Krysik P., Klincewicz K., Maslikowski L., Rzewuski S., Kulpa K. 'Doppler only localization in GSM-based passive radar'. *Proc. of the 2014 International Radar Conference*; Lille, France, Oct. 2014, pp.1–6.

[79] Subedi S., Zhang Y.D., Amin M.G., Himed B. 'Motion parameter estimation of multiple ground moving targets in multi-static passive radar systems'. *Eurasip Journal on Advances in Signal Processing*. 2014; **2014**(1): 1–14.

[80] Kaplan E., Hegarty C. (eds.). *Understanding GPS Principles and Applications*. 2nd edn. Artech House; 1996.

[81] Liu H., Darabi H., Banerjee P., Jing L. 'Survey of wireless indoor positioning techniques and systems'. *IEEE Transactions on Systems, Man, and Cybernetics, Part C: Applications and Reviews*. 2007; **37**(6): 1067–80.

[82] Sanpechuda T., Kovavisaruch L. 'A review of RFID localization: Applications and techniques'. *Proc. of the IEEE Int. Conf. Electr.Eng./Electron. Comput. Telecommun. Inf. Technol.*; May 2008, pp. 769–72.

[83] European FP7 Project ATOM – 'Airport detection and Tracking Of dangerous Materials by passive and active sensors arrays', http://www.atom-project.eu.

[84] European FP7 Project SOS – 'Sensors system for detection and tracking Of dangerous materials in order to increase the airport Security in the indoor landside area', http://www.sos-project.eu.

[85] Colone F., Pastina D., Falcone P., Lombardo P. 'WiFi-based passive ISAR for high-resolution cross-range profiling of moving targets'. *IEEE Transactions on Geoscience and Remote Sensing*. 2014; **52**(6): 3486–501.

[86] Martelli T., Pastina D., Colone F., Lombardo P. 'Enhanced WiFi-based passive ISAR for indoor and outdoor surveillance'. *Proc. of the 2015 IEEE Radar Conference*; Arlington (VA), USA, May 2015, pp. 974–79.

[87] Wehner D.R. *High-Resoultion Radar*. 2nd edn. Boston, MA: Artech House, 1995, Ch. 7, pp. 341–64.

[88] Çetin M., Lanterman A.D. 'Region-enhanced passive radar imaging'. *IEE Proceedings Radar, Sonar and Navigation*. 2005; 152(3): 185–94.

[89] Cherniakov M., Saini R., Zuo R., Antoniou M. 'Space-surface bistatic synthetic aperture radar with global navigation satellite system transmitter of opportunity: Experimental results'. *IET Radar, Sonar and Navigation*. 2007; **1**(6): 447–58.

[90] Gutierrez del Arroyo J.R., Jackson J.A. 'WiMAX OFDM for passive SAR ground imaging'. *IEEE Transactions on Aerospace and Electronic Systems*. 2013; **49**(2): 945–59.

[91] Olivadese D., Giusti E., Petri D., Martorella M., Capria A., Berizzi F. 'Passive ISAR with DVB-T signals'. *IEEE Transactions on Geoscience and Remote Sensing*. 2013; **51**(8): 4508–17.

[92] Wacks S., Yazici B. 'Passive synthetic aperture hitchhiker imaging of ground moving targets—Part 1: Image formation and velocity estimation'. *IEEE Transactions on Image Processing*. 2014; **23**(6): 2487–500.

[93] Maslikowski L., Samczynski P., Baczyk M., Krysik P., Kulpa K. 'Passive bistatic SAR imaging: Challenges and limitations'. *IEEE Aerospace and Electronic Systems Magazine*. 2014; **29**(7): 23–29.

[94] Hui Ma, Antoniou M., Cherniakov M. 'Passive GNSS-based SAR resolution improvement using joint Galileo E5 signals'. *IEEE Geoscience and Remote Sensing Letters*. 2015; **12**(8): 1640–44.

[95] Wu Y., Munson D.C. 'Multistatic synthetic aperture imaging of aircraft using reflected television signals'. *Proc. SPIE*. 2001; **4382**, pp. 1–12.

[96] Stevens S.R., Jackson J.A. 'Emitter selection criteria for passive multistatic synthetic aperture radar imaging'. *IET Radar, Sonar & Navigation*. 2014; **8**(9): 1267–79.

[97] Santi F., Antoniou M., Pastina D. 'Point spread function analysis for GNSS-based multistatic SAR'. *IEEE Geoscience and Remote Sensing Letters*. 2015; **12**(2): 304–8.

[98] Gennarelli G., Amin M.G., Soldovieri F., Solimene R. 'Passive multiarray image fusion for RF tomography by opportunistic sources'. *IEEE Geoscience and Remote Sensing Letters*. 2015; **12**(3): 641–45.

[99] Turin F., Pastina D. 'Multistatic passive ISAR based on geostationary satellites for coastal surveillance'. *Proc. of the 2013 IEEE Radar Conference*; Ottawa (ON), Canada, Apr. 2013.

[100] Garry J.L., Baker C.J., Smith G.E., Ewing R.L. 'Investigations toward multistatic passive radar imaging'. *Proc. of the 2014 IEEE Radar Conference*; Cincinnati (OH), USA, May 2014, pp. 607–12.

[101] Colone F., Pastina D., Marongiu V. 'VHF cross-range profiling of aerial targets via passive ISAR: Signal processing schemes and experimental results'. *IEEE Transactions on Aerospace and Electronic Systems*. 2017; **53**(1): 218–35.

[102] Li Xi, Liu Guosui, Jinlin Ni 'Autofocusing of ISAR images based on entropy minimization'. *IEEE Transactions on Aerospace and Electronic Systems*. 1999; **35**(4): 1240–52.

Chapter 16

GNSS-based passive radar

Michail Antoniou[1] and Mikhail Cherniakov[1]

Abstract

This chapter is an introduction to passive synthetic aperture radar (SAR) using global navigation satellite systems (GNSS), as illuminators of opportunity. Such systems include the global positioning system (GPS), the Russian GLONASS system, or the forthcoming Galileo constellation. Apart from the traditional benefits of a passive radar system, which include cost efficiency, license-free and covert operation, GNSS-based systems have a number of relative merits compared to passive systems based on terrestrial illuminating sources. One of its most prominent features is the potential for persistent monitoring anywhere in the world due to the global GNSS coverage. In addition, as GNSS guarantee a number of satellites illuminating the same point on Earth from multiple aspect angles simultaneously, this provides an opportunity to enhance radar information space. This can be done either by comparing individual passive SAR images obtained from multiple satellite perspectives, or combining them using multistatic SAR techniques. On the other hand, the relatively low power flux density near the Earth's surface restricts the field of view of applications for such a system to monitoring local areas.

At the same time, and partially due to its rather broad scope, this technology has not yet reached the maturity of terrestrial-based systems. As such, the purpose of this chapter is to present methods and results on the fundamental science and technology behind GNSS-based SAR, as a stepping stone to realizing its full potential. This includes an introduction to the system concept and its fundamental parameters (power budget/resolution), signal processing algorithms for signal synchronization and image formation, as well as proof-of-concept results for advanced techniques such as change detection, multi-perspective and multistatic imaging.

[1]Department of Electronic, Electrical and Systems Engineering, University of Birmingham, UK

16.1 Introduction

For more than two decades, Global Navigation Satellite Systems (GNSS) such as the Global Positioning System (GPS) have found an exciting new application other than navigation: radar and remote sensing. In essence, GNSS are used as illuminators of opportunity in passive radar systems. The receivers can record satellite signal reflections from an observed area and process them in a number of ways to gain different types of information regarding a scene.

One of the most mature GNSS-based remote-sensing technologies is the so-called GNSS reflectometry [1]. This is a passive bistatic radar which has found a plethora of applications, from ocean altimetry [2] to wind speed estimation [3], soil moisture content retrieval [4] and sea ice characterization [5], among many others. The technology has reached a substantial maturity, with a GNSS reflectometer being included as an experimental instrument onboard the UK-DMC satellite [6]. This topic is covered in detail in a number of textbooks [7,8], so it will not be considered further here.

What will be considered instead is another emerging GNSS-based radar system. This system is a passive synthetic aperture radar (SAR), using GNSS signals of opportunity to provide radar images of terrain. The primary objective of such a system is to detect and provide images of objects (man-made, such as buildings or cars, or natural, such as trees, for example) at the background of land or sea, rather than attempting to determine land or sea properties (even though this may also be possible to some degree) as in the case of GNSS reflectometry or indeed active SAR. It can therefore be seen as a complementary system to a GNSS reflectometer in terms of its outputs; however, the system structure, signal processing and range of applications are fundamentally different.

The motivation in using such a technology is that while a GNSS system cannot outclass a dedicated SAR constellation in terms of image quality and sensitivity by default, it possesses unique properties which make it attractive for radar and remote sensing purposes. First of all, since GNSS have been designed for a persistent and global coverage, they possess potential for permanent and continuous monitoring anywhere in the world, including the poles. In addition, a fully operational GNSS constellation guarantees that any point on Earth is illuminated by several satellites (typically 6–8 for a single GNSS constellation) simultaneously from different angles, each one of which emits a number of signals with different frequencies. This property has a number of implications. First, satellites at the most appropriate positions could be selected for forming the optimal bistatic topology to achieve better spatial resolutions while minimizing shadowing effects. Second, multistatic/multi-perspective and multi-frequency SAR techniques are possible, using all satellites in the field of view of the system but with a single receiver, which may substantially enhance image information space. At the same time, the system has the same benefits of a passive system. That means that the receiver itself has an architecture similar, if not identical, to that of a GNSS receiver used for navigation, which enables a silent, low-cost operation with no additional contribution to electromagnetic pollution.

This chapter provides an overview of the theory and practice of GNSS-based SAR. It begins with a brief introduction to the generic monostatic, bistatic and GNSS-based SAR concept in Section 16.2. Subsequently, Section 16.3 gives an overview of GNSS systems as the transmit segment and their signal characteristics. Sections 16.4 and 16.5 investigate the primary parameters of the GNSS-based SAR system that is the power budget and spatial resolution, respectively. Section 16.6 is an overview of the signal-processing algorithms required for GNSS-based SAR image formation, followed by example experimental results and their discussion in Section 16.7. Finally, Section 16.8 shows early results obtained so far in realizing the broader vision of GNSS-based SAR, namely the detection of scene changes, multi-perspective SAR and multistatic SAR techniques for spatial resolution enhancement. As it can be understood research in these areas is far from final; however, the results contained therein can be used as the stepping stone in the ultimate realization of the full system potential.

16.2 Monostatic, bistatic and GNSS-based SAR

16.2.1 Monostatic SAR

Monostatic SAR is one of the major remote-sensing tools today. Only its major principles are covered here to provide a basis for comparison with the GNSS-based SAR concepts later on. For an in-depth analysis of monostatic SAR, there are a number of excellent textbooks such as [9–11].

Systems employing synthetic apertures seek to overcome the constraints of physical antenna arrays by moving a single antenna to different positions in an array, radiating, collecting and storing echo signals in each position (Figure 16.1). In principle, synthetic arrays may be one or two dimensional; one dimensional arrays are commonly used, and these are usually (but not necessarily) approximately linear.

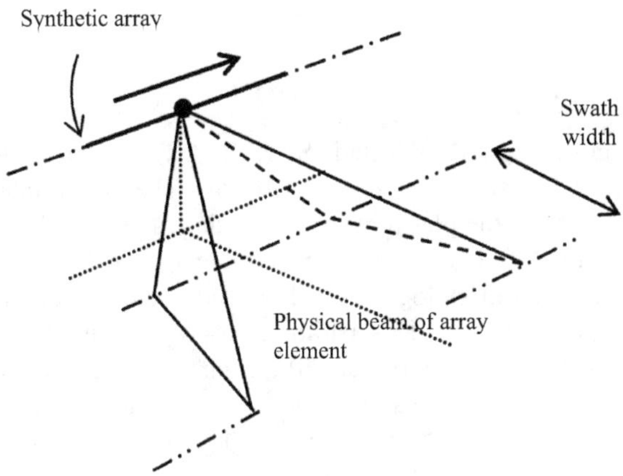

Figure 16.1 Monostatic SAR concept

Typically, an aircraft or spacecraft equipped with a terrain-mapping SAR forms a synthetic array by flying along a linear trajectory while illuminating a swath of the ground beneath it using a small antenna. In monostatic SAR, the same antenna is normally used for signal transmission and echo reception. Very long linear arrays can be formed and, if the array is sufficiently long, its effective length when observing a target will depend on the beamwidth of the physical antenna (array element) and the range of the target, the effective array length being the lateral width of the physical beam at the range of interest.

As an imaging system, it is crucial to know the spatial resolution in the range and azimuth dimensions for a SAR system, as they define the size of an image pixel. 'Resolution' is defined here as the minimum distance two targets should be separated by in order to appear as two returns in the radar image. For a monostatic SAR, image resolution in both range and azimuth is straightforward to calculate. In the slant range direction, the resolution is defined by the transmitted signal bandwidth ΔF as:

$$\delta_{\mathrm{rm}} = \frac{c}{2\Delta F} \tag{16.1}$$

where c is the speed of light. For the azimuth resolution, it can be shown [11] that it is equal to half the lateral dimension D of the physical antenna used for SAR image formation:

$$\delta_{\mathrm{am}} = \frac{D}{2} \tag{16.2}$$

Equation (16.12) states that the shorter the physical antenna, the finer the azimuth resolution. This is in contrast to conventional radar, where long (i.e. narrow-beam) physical antennas are required to provide a sufficiently high angular resolution. This is because a shorter antenna implies a wider beamwidth, therefore, a longer target exposure within it and a longer synthetic, rather than physical, antenna aperture length. For the same reason, the SAR azimuth resolution does not depend on the distance between the sensor and a target.

16.2.2 Bistatic SAR

In contrast to monostatic SAR, in bistatic SAR (BSAR) the transmitter and receiver are onboard different, spatially separated platforms. The only fundamental restriction is for at least one of the platforms to be moving, so that at least one synthetic aperture may be formed. This separation introduces an extra degree of freedom with regard to the system topology, as well as additional complexity. Despite the complexity, however, this spatial separation has opened new horizons to SAR research. For example, in terms of scientific progress, a bistatic acquisition may substantially increase the information space of monostatic SAR acquisitions by considering different target scattering angles, among other things.

A number of different BSAR topologies have been proposed on the theoretical level and/or demonstrated at the experimental level (Figure 16.2), such as: bistatic airborne SAR [12], bistatic spaceborne SAR [13,14], or hybrid spaceborne/airborne

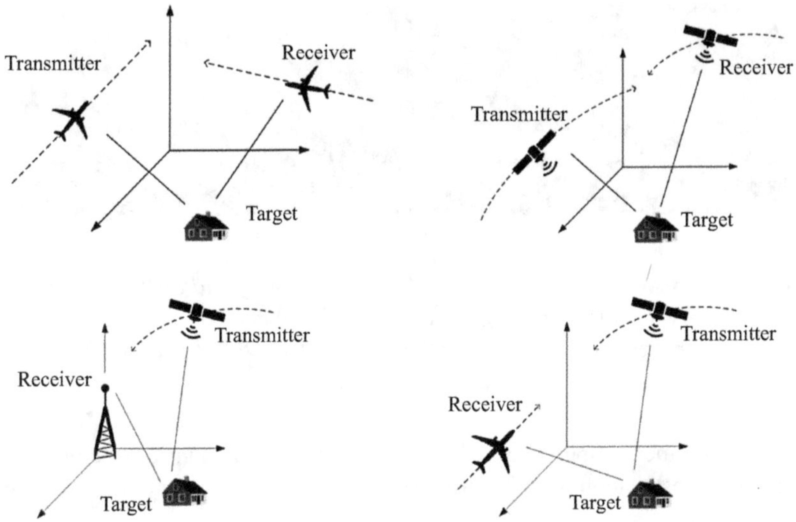

Figure 16.2 Example BSAR topologies

BSAR [15–17]. In any of these cases, either the transmitter or the receiver could be fixed on the ground.

From this plethora of possible topologies, we will focus on hybrid BSAR (bottom of Figure 16.2). Even this special topology offers a number of system configurations, since either the transmitter or the receiver can be onboard an aircraft, a ground-moving vehicle or even fixed on the ground, with the only assumption that one of the two platforms is spaceborne. The spaceborne segment can be a radar satellite or even a transmitter of opportunity. The most typical situation uses radar satellites as the illumination sources.

Perhaps the first record of a hybrid BSAR experiment with a dedicated radar satellite goes back to 1998 [18]. Since then, substantial research has been conducted on this topic by a number of organizations on the theoretical and experimental levels, with [19–22] as some of the notable examples. Figure 16.3 [23] shows an example image obtained by the spaceborne/stationary bistatic topology using the TerraSAR-X radar satellite as the transmitter. It is a detail of the image showing a factory 700 m away from the position of the stationary receiver (Figure 16.3(a)), with the monostatic image in slant range geometry shown in Figure 16.3(b) and an optical view of the scene in Figure 16.3(c) for comparison. Results such as these confirm the fundamental feasibility of hybrid BSAR as well as the motivation for conducting BSAR research in general, since by simple observation, there is a substantial difference between the monostatic and BSAR images of the same scene.

However, despite using dedicated radar satellites in a hybrid SAR configuration is feasible and of great scientific interest, it does suffer from a number of limitations. Perhaps unsurprisingly, these restrictions are similar to those of spaceborne SAR in general. The major limitation is that current commercial SAR cannot provide

Figure 16.3 Bistatic vs monostatic SAR images with hybrid SAR: (a) bistatic result (ground range geometry), (b) monostatic image (slant range geometry), (c) orthophoto. (b) © DLR TerraSAR-X ground segment, (c) © LVermA NRW (GEObasis.nrw) [23]

persistent area monitoring. This is limited by the small amount of satellites in any given SAR constellation and their re-visit cycles, which can vary from 10 days to more than a month.

16.2.3 GNSS-based SAR

One hybrid BSAR variant uses GNSS, rather than radar satellites, as transmitters of opportunity (Figure 16.4). Any GNSS can be used, including the GPS (US), the Global Navigation Satellite System (GLONASS, RU), BeiDou (China) and the Galileo (EU) constellations. The receiver can be stationary or mounted on a surface vehicle or an aircraft and comprises two separate channels. The first channel records the direct satellite signal(s) for signal synchronization purposes, a necessary process in any coherent bistatic radar (more on this in Section 16.6.1). Usually, this channel is equipped with a low-gain antenna to avoid antenna pointing errors and to maximize the number of satellites that can be used. This channel will be referred to as the heterodyne channel (HC) hereafter. The second channel has a high-gain antenna pointed towards the observation area to record satellite signal reflections for image formation. This channel will hereafter be called the radar channel (RC). In the case of an airborne receiver, the synthetic aperture is essentially formed by the aircraft motion only, since the Doppler contribution due to the spacecraft motion over the dwell time on target is insignificant. In the case of a stationary receiver, the synthetic aperture is formed solely from the GNSS motion and, therefore, longer data acquisitions are required to bring azimuth resolution to acceptable levels (Section 16.4).

Apart from its unique advantages, mentioned at the Introduction, this system also has its unique challenges. Since GNSS were not originally intended for remote sensing, their power density near the Earth's surface is substantially low. For this reason, long dwell times on target are needed to increase signal-to-noise ratio (SNR) via coherent integration (Section 16.4); however, even in that case, the system is more suited to local area monitoring. In terms of signal processing, signal synchronization differs to that of traditional passive radar. Furthermore, the asymmetric structure of the system restricts the types of image formation algorithms that can be used.

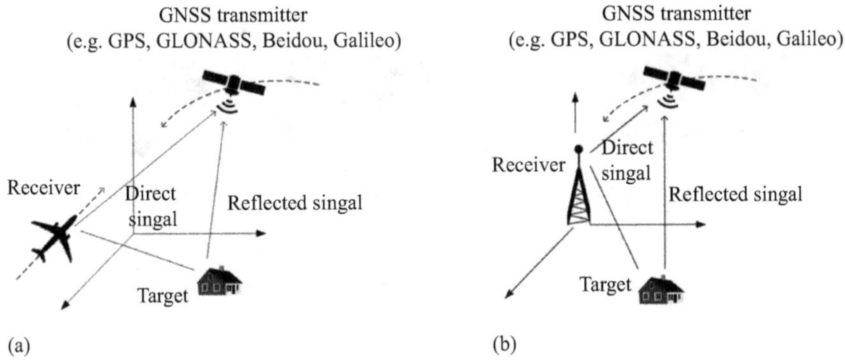

Figure 16.4 GNSS-based SAR with (a) airborne, (b) fixed receiver

16.3 GNSS overview

Out of the four possible GNSS options, two are currently in full operation; these are GPS and GLONASS. The other two, Galileo and Beidou, are still on the deployment stage and scheduled to reach their full operational capacity in 2019 and 2020, respectively. For all GNSS, the major system characteristics are published in detail under their respective interface control documents (ICDs), available online [24–27]. In this chapter, the fundamental parameters of GNSS systems for radar operation will be described and compared for the sake of the following analysis.

16.3.1 GNSS signals

Navigation satellites continuously transmit navigation signals in two or more frequency bands within L-band. From the radar user's perspective, these signals contain ranging codes which can be used for remote sensing, as well as navigation codes which can additionally be used for positioning. Figure 16.5 shows the available GNSS frequency bands. An overview of some of the most popular GNSS signals and their generic structure is provided below.

GPS has two dedicated frequency bands, L1 (1,575.42 MHz) and L2 (1,227.60 MHz), with a new frequency band, L5 (1,176.45 MHz). Within these two bands, GPS satellites transmit two ranging codes: the coarse acquisition (C/A) and/or the precision (P) code. Both the C/A- and the P-codes are gold codes. Satellite codes are uniquely identified by a code division multiple access (CDMA) protocol, meaning that different satellites transmit different codes, whose generation methods are listed in the GPS ICD [26]. The C/A-code on L1 is also referred to as the 'legacy' signal and is broadcast by all satellites. This code is transmitted using binary phase-shift keying (BPSK), as shown in Figure 16.6.

Similarly, some of the more popular GLONASS bands are shown as G1 and G2 in Figure 16.5. Their nominal frequencies are 1,602 MHz and 1,246 MHz, respectively. Like GPS, both C/A- and/or P-codes are transmitted in these bands,

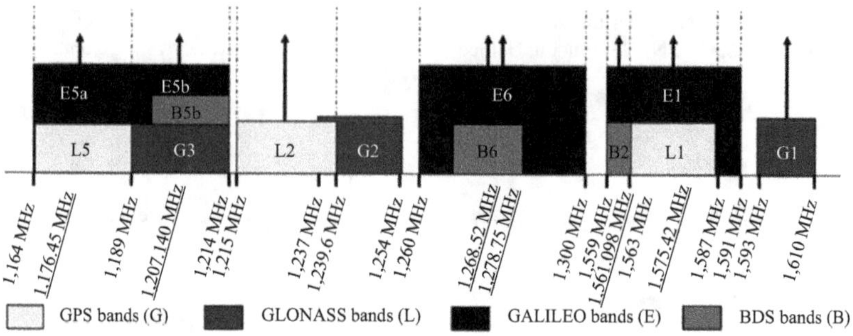

Figure 16.5 GNSS navigation frequency bands

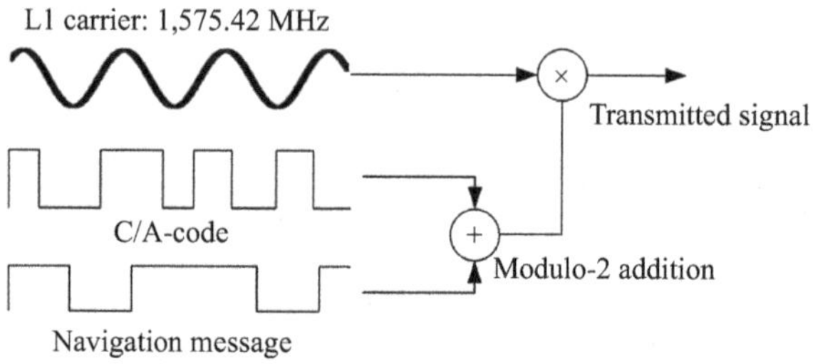

Figure 16.6 GPS legacy signal structure

with a signal structure almost identical to that shown in Figure 16.6. Also similar to GPS, GLONASS-ranging codes are pseudo-random sequences. Unlike GPS, different satellites are uniquely identified by transmitting at a carrier frequency that is slightly offset from the nominal G1 and G2 values, i.e. GLONASS operates on a frequency division multiple access protocol. The frequency for each satellite at the G1 and G2 bands is given by [24]

$$\begin{aligned} f_{K1} &= f_{01} + K\Delta f_1 \\ f_{K2} &= f_{02} + K\Delta f_2 \end{aligned} \tag{16.3}$$

where K is a frequency index, f_{01} and f_{02} are the G1 and G2 nominal frequencies, and Δf_1, Δf_2 are the frequency spacings for the G1 and G2 bands, equal to 562.5 kHz and 437.5 kHz, respectively. A new frequency band, G3, has been proposed and is planned as part of the new GLONASS satellites from 2018.

Even though Galileo is not yet fully developed, perhaps one of its most interesting frequency bands is the E5 band. That is due to its alternative binary offset carrier type of modulation [27], that substantially differs from the general GPS or GLONASS signal structure. The E5 band consists of two sub-bands, namely E5a and E5b, centred 15 MHz to the sides of the E5 carrier of 1,191.795 MHz (Figures 16.7 and 16.8). However, the two sub-carriers for E5a and E5b are actually

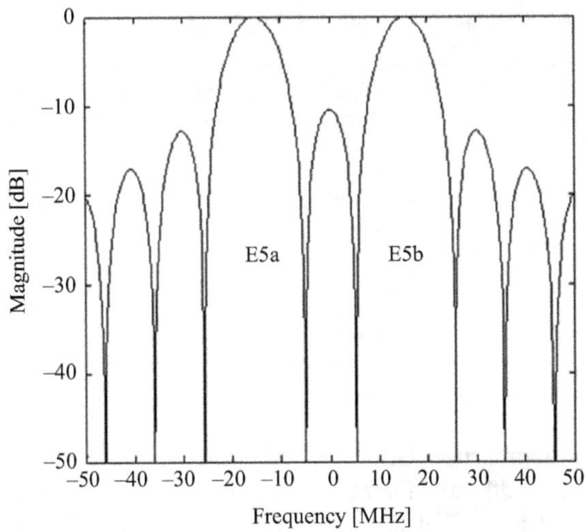

Figure 16.7 Power spectral density of E5 signal

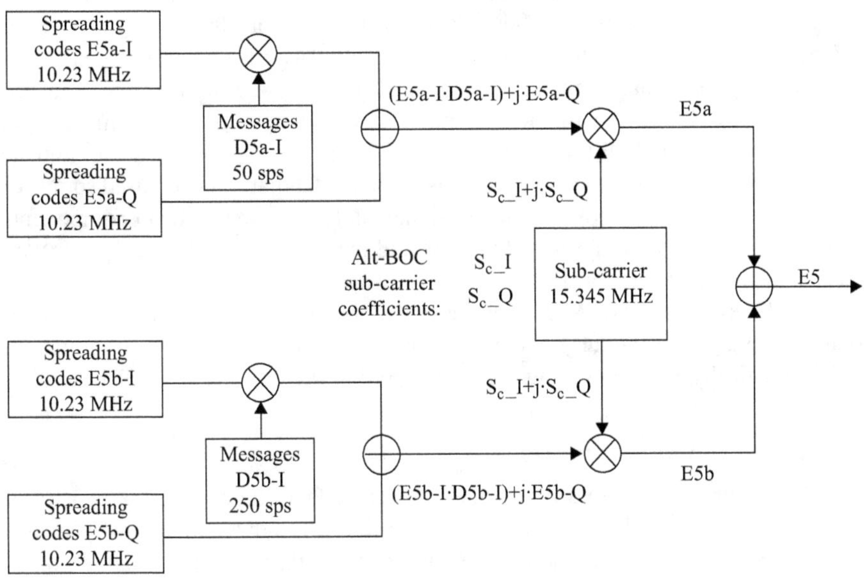

Figure 16.8 E5 signal structure

generated as digital bit streams digitally. Furthermore, in each sub-band, there are two primary codes based on a CDMA protocol which are transmitted in phase quadrature (E5a-I/Q, E5b-I/Q), and with each primary code comes a secondary code. The E5 signal structure is shown in Figure 16.8.

Table 16.1 GNSS major signal characteristics

Signal	Time duration	Bandwidth (MHz)	Navigation message/data rate
GPS C/A-code (L1/L2)	1 ms	1.023	Yes/50 bits/s
GPS P-code (L1/L2)	7 days	10.23	Yes/50 bits/s
GLONASS C/A-code (G1/G2)	1 ms	0.511	Yes/50 bits/s
GLONASS P-code (G1/G2)	1 s	5.11	Yes/50 bits/s
Galileo E5a-I	1 ms	10.23	Yes/50 symbols/s
Galileo E5b-I	1 ms	10.23	Yes/250 symbols/s
Galileo E5a-Q/E5b-Q	1 ms	10.23	No

In choosing appropriate GNSS signals for passive radar applications, the ranging signal bandwidths and their time durations are of significant importance, as well as their complexity. That is because signal time delay is with respect to the start of the code, and the ranging signal bandwidth is directly related to range resolution as shown in (16.1). Table 16.1 shows the major characteristics for the signals described above.

From the table, the GPS P-code is already inappropriate due to its duration of 7 days. Its C/A-code is of a sufficiently short duration, but from (16.1), its bandwidth yields a range resolution of approximately 147 m, that is too coarse for imaging purposes. That rules out the use of GPS for passive SAR for the moment; however, the introduction of the L5 band, fully overlapping with Galileo's E5a band, shows potential. With regard to GLONASS, the P-code is usable, and its details having been published in [28]. This is the signal that has been used so far for GNSS-based SAR. The most promising signals in this group are the Q-components of the Galileo E5. Not only their durations and bandwidths (14.7 m resolution) are acceptable but also they lack navigation message data which makes their signal processing less complex. What is more, it has been shown in a number of publications [29,30] that the full E5 signal bandwidth can be used for image formation, which drastically improves image resolution but at the expense of SNR loss.

16.3.2 GNSS signal power

In order to derive the GNSS-based SAR power budget, the power flux density of the satellites near the Earth's surface is required. The power flux density over a 1 m^2 surface area, and over the entire signal bandwidth, may be calculated via

$$\rho = \frac{\text{EIRP}}{4\pi R_T^2} \tag{16.4}$$

where EIRP is the effective isotropic radiated power of the satellite, and R_T is the range from the satellite to the surface. Note that GNSS are designed to have an almost constant power flux density over their overall beam coverage. What is also often quoted in the GNSS community is the guaranteed minimum power levels near

Table 16.2 GNSS power flux densities and minimum received power levels [31]

Transmitter	Power output (W)	EIRP (dBW)	Orbit altitude (km)	Power density (dBW/m^2)	Minimum power level (dBW)
Galileo E5a/b	50	32	23,222	−126	−157
GPS L1	50	30	20,180	−127	−158
GLONASS L1	50	28	19,130	−128	−161

the Earth. They are given as signal power levels at the output of a linear or circularly polarized low-gain antenna, including some atmospheric attenuation and assuming the satellite is at least 5° over the horizon. The power flux density and the minimum power levels (obtained from [24,26,27]) are listed in Table 16.2. Power flux density is calculated via (16.4), taking R_T as the satellite orbital altitude for simplicity (R_T in general varies with the relative position between the satellite and the surface). GPS, GLONASS and Galileo systems generate more or less same power flux density, with Galileo having 4 dB higher.

16.4 GNSS-based SAR power budget

The receiver records two signals via two separate channels, heterodyne and radar (HC, RC). The HC records the direct satellite–receiver signal for signal synchronization, while the RC records satellite signal reflections from an observation area for imaging.

The power received at the output of the HC antenna is given by [31]:

$$P_{\mathrm{HC}} = \rho A_{eH} \tag{16.5}$$

where A_{eH} is the effective area of the HC antenna. For a low-gain antenna of 6 dB, that is typical for GNSS antennas used for navigation, A_{eH} is approximately 0.013 m^2 at the GNSS frequency bands, and using values for ρ from Table 16.2, the received power for Galileo signals is approximately 3.265×10^{-15} W.

The signal is received at the background of receiver noise, whose power is

$$P_N = kT_s\Delta F \tag{16.6}$$

where k is Boltzmann's constant, and T_s is the operating temperature, in Kelvins. The SNR prior to any signal processing is the ratio of the received power to the noise power, hence

$$\mathrm{SNR} = \frac{P_{\mathrm{HC}}}{P_N} \tag{16.7}$$

Assuming the receiver operates at room temperature (290 K), and the system bandwidth is equal to that of the Galileo E5a band, the SNR at the receiver output is approximately equal to −11 dB, i.e. the direct signal is buried under the noise and therefore cannot be used for signal synchronization purposes. However, using a matched filter, the SNR can be substantially improved. For a phase-coded waveform such as a GNSS signal, it can be shown that the SNR at the matched filter output is equal to:

$$\mathrm{SNR}_{\mathrm{MF}} = \mathrm{SNR} \times T_p \times \Delta F \tag{16.8}$$

where T_p is the ranging code duration. For Galileo, E5a/E5b $T_p = 1$ ms and $\Delta F = 10.23$ MHz, giving a SNR improvement of about 40 dB. This is why signal synchronization is based on matched filtering techniques.

For the RC, calculations are more complex. Following the analysis in [32] and assuming free space propagation losses, the SNR at the output of the SAR image formation processor may be written as:

$$\mathrm{SNR}_{\mathrm{RC}} = \rho \times \frac{\sigma A_{eR}}{4\pi R_R^2} \times T_p \Delta F \times T_D \mathrm{PRF} \times \frac{1}{kT_s \Delta F} \tag{16.9}$$

where σ is the bistatic radar cross-section (RCS), A_{eR} is the effective area of the RC antenna, R_R is the receiver–target range, T_D is the dwell time on target, and PRF is the pulse repetition frequency along the synthetic aperture.

The first factor in (16.9) is the power flux density near the Earth, which can be taken as a constant for a given observation area. The third factor is the signal processing gain due to matched filtering in range, which is identical to the matched filter operation described in (16.8). The fourth factor is the signal processing gain obtained after azimuth signal processing. Note the product of the total dwell time on target with the PRF equals the amount of pulses, N, transmitted over the full aperture. This is assuming the target RCS remains constant over the dwell time and of course that the target scatters coherently over the dwell. Hence, azimuth processing is the equivalent of coherent summation of all transmitted pulses (see Section 16.6.2).

Equation (16.9) can be somewhat simplified by assuming the receiver bandwidth (used for receiver noise calculations) is equal to the ranging signal bandwidth. Moreover, the total pulse duration is usually taken as the pulse repetition interval (PRI) in this case, so $T_p \times \mathrm{PRF} = 1$. Therefore, (16.9) takes its final form as:

$$\mathrm{SNR}_{\mathrm{RC}} = \rho \times \frac{\sigma A_{eR}}{4\pi R_R^2} \times \frac{1}{kT_s} \times T_D \tag{16.10}$$

Equation (16.10) indicates that the SNR for GNSS-based SAR may be substantially increased by increasing the dwell time on target. The same expression is valid for both the airborne and fixed receiver configurations. The only difference lies in the practicalities of these two options. For a fixed receiver setup, the dwell time on target may be from the order of minutes to hours, since GNSS are visible above an area for such intervals, while the receiver is constantly pointed at the area of interest. Typically, integration of several minutes (five or more) is sufficient to obtain a sufficient SNR and azimuth resolution. This integration time is more similar to that of a Geosynchronous SAR [33] where the integration time can be up

Table 16.3 Example SNRs for fixed receiver

RCS (m^2)	10	10	50	250	250
Distance receiver–target (km)	3	5	5	10	15
Dwell time on target (s)	300	1,000	300	1,000	1,000
Signal-to-noise (power ratio-dB)	17	18	20	26	22.5

Table 16.4 Example SNRs for airborne receiver

RCS (m^2)	10	50	50	100	100	250
Distance receiver–target (km)	1	1	2	2	4	5
Dwell time on target (s)	9.6	9.6	19	19	38	47.6
Signal-to-noise (power ratio-dB)	12	19	16	19	16	19

to hours, rather than a spaceborne SAR in low earth orbit (LEO), where dwell times on target may be fractions of a second. Table 16.3 shows the SNR obtained for a fixed receiver, for different target RCSs and at different distances. Table entries have been calculated using (16.10), using a receive antenna gain of 15 dB and additionally assuming 3 dB system losses and a receiver noise figure of 1.5 dB. The transmitter is assumed to be Galileo; however, similar values can be found for GPS and GLONASS simply by subtracting their difference in power flux density, listed in Table 16.2.

Setting a SNR of 12 dB as the target detection threshold shows that GNSS-based SAR operational ranges are in the order of a few kilometres. That is the reason why GNSS-based SAR is more attractive for the persistent monitoring of local areas.

For a fixed receiver, the power budget is more favourable since the dwell time on target can be substantial. For an airborne receiver, dwell time on target and hence power budget are restricted by the RC antenna beamwidths and the aircraft speed. To illustrate this, (16.10) may be slightly re-arranged by re-writing T_D as:

$$T_D = \frac{2R_R \tan(\theta/2)}{V_a} \tag{16.11}$$

where V_a is the aircraft speed, and θ is the RC antenna beamwidth, equal to λ/D where D is the along-track physical dimension of the RC antenna. Combining (16.10) and (16.11), with $D = 1$ m and $V_a = 100$ m/s, that may be towards the speed upper limit for a practical detection range, Table 16.4 shows example SNRs for the airborne receiver case.

The analysis introduced above does not include ways of further power budget improvement. SNR can be improved essentially by using non-coherent integration of signals from more than one transmitting channel and/or more than one satellite.

16.5 Spatial resolution

The definition of bistatic target resolution is identical to that of monostatic target case: the degree to which two or more targets may be separated in one or more dimensions, such as angle, range, velocity (or Doppler), etc. However, unlike monostatic SAR (1) and (2), resolution in BSAR depends on the bistatic data collection geometry. This issue should be taken into special consideration for GNSS-based SAR, due to its asymmetric topology (a spaceborne transmitter with a receiver near the Earth's surface) and additionally due to the fact that special data acquisition geometries that often simplify matters are difficult, if possible at all, to implement in practice.

A general method for BSAR resolution analysis was proposed in [34], by means of the generalized ambiguity function (GAF) approach [35], that is directly applicable to GNSS-based SAR.

Figure 16.9 shows the general BSAR geometry used in the following analysis. L denotes the baseline between the transmitter and the receiver. $\overrightarrow{V_T}$, $\overrightarrow{\omega_T}$, A_T are the transmitter's velocity, angular speed and coverage with respect to the target, and $\overrightarrow{V_R}$, $\overrightarrow{\omega_R}$, A_R are the corresponding receiver parameters. $\overrightarrow{U_T}$, $\overrightarrow{U_R}$ are unit vectors in the transmitter-target and receiver-target directions, respectively.

To obtain the resolution performance, we can refer to the point spread function (PSF) or GAF [35]. The PSF describes the 2-D auto-correlation function of one point target and can be decomposed into the range and azimuth correlation functions, while the GAF is the 2-D correlation function in the delay-Doppler

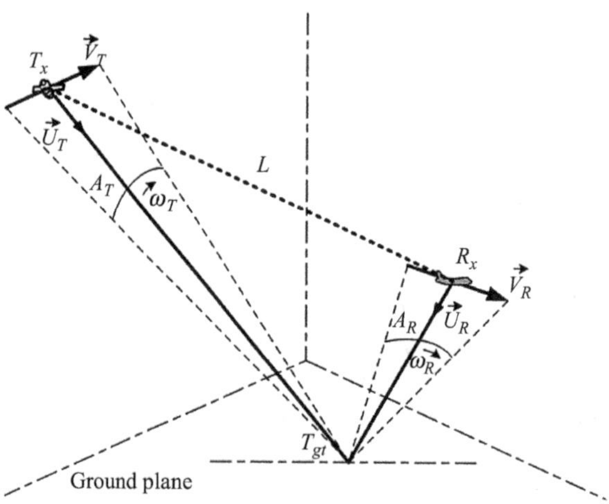

Figure 16.9 Bistatic SAR geometry

domain. Therefore, the PSF and the GAF are representations of the same parameter but in different domains, and one can be transformed to the other via the appropriate domain transformation. The 3 dB widths of GAF in range and azimuth dimensions are respectively the range resolution and azimuth resolution. Considering a ground-based stationary receiver collecting the signals emitted from a GNSS transmitter and reflected by one stationary point target, the two-dimensional bistatic resolution cell of such a system can be derived from the GAF. In the hypothesis of narrowband signal and narrow synthetic aperture, the GAF, $X(A,B)$, is given by the product of two normalized functions, $p(\cdot)$ and $m_A(\cdot)$. The former is the matched filter output of the ranging signal, and the latter is the inverse transform of the normalized received signal magnitude pattern. The derivation is lengthy and will not be shown here, but may be found in detail in [35]. The final result can be found as:

$$|X(A,B)| \approx p\left(\frac{2\cos(\beta/2)\Theta^T(r)}{c}\right) \cdot m_A\left(\frac{2\omega_E \Xi^T(r)}{\lambda}\right) \tag{16.12}$$

where A is the vector position of the desired point reflector to be evaluated, vector B is an arbitrary position of another reflector in the vicinity of A and $r = B - A$; β is the bistatic angle and Θ is a unit vector in the direction of its bisector; $\omega_E = \frac{\left|\overrightarrow{\omega_T} + \overrightarrow{\omega_R}\right|}{2}$ and Ξ are called the equivalent angular speed and motion direction, since a monostatic SAR moving in the direction Ξ with angular speed ω_E would exhibit similar Doppler-based resolution characteristics, c is the speed of light, and λ is the radar wavelength. The superscript 'T' denotes matrix transpose.

The theoretical form of the PSF in (16.12) has been confirmed using experimental data [36]. Figure 16.10 shows the comparison of an experimental PSF, obtained using a point-like target with GLONASS emissions, and the expected PSF from (16.12) for the same bistatic geometry. The colourscale is in dB, with 0 dB representing the highest intensity point in the PSF. The two results are nearly identical, so (16.12) can now be analysed on the theoretical level for brevity but without loss of generality. It should also be stated that (16.12) holds for any BSAR system and is not restricted to GNSS-based SAR.

In GNSS-based SAR, since the transmitted signals are bit sequences, $p(\cdot)$ is a triangular function. On the other hand, due to the azimuth compression of the received signal, $m_A(\cdot)$ is a sinc $(\cdot)$ function, hence the shape of the PSF in Figure 16.10. For a better visualization, Figure 16.11 shows an example PSF computed using (16.12) and assuming Galileo transmissions, along with its range and azimuth cross-sections. The first thing to note is that the sidelobe directions are not orthogonal, which is unlike the monostatic SAR case. This is because the resolution directions Θ and Ξ are not orthogonal but depend on the bistatic acquisition geometry. This means that depending on the data collection geometry, the orientation, as well as the size, of the resolution cell will change.

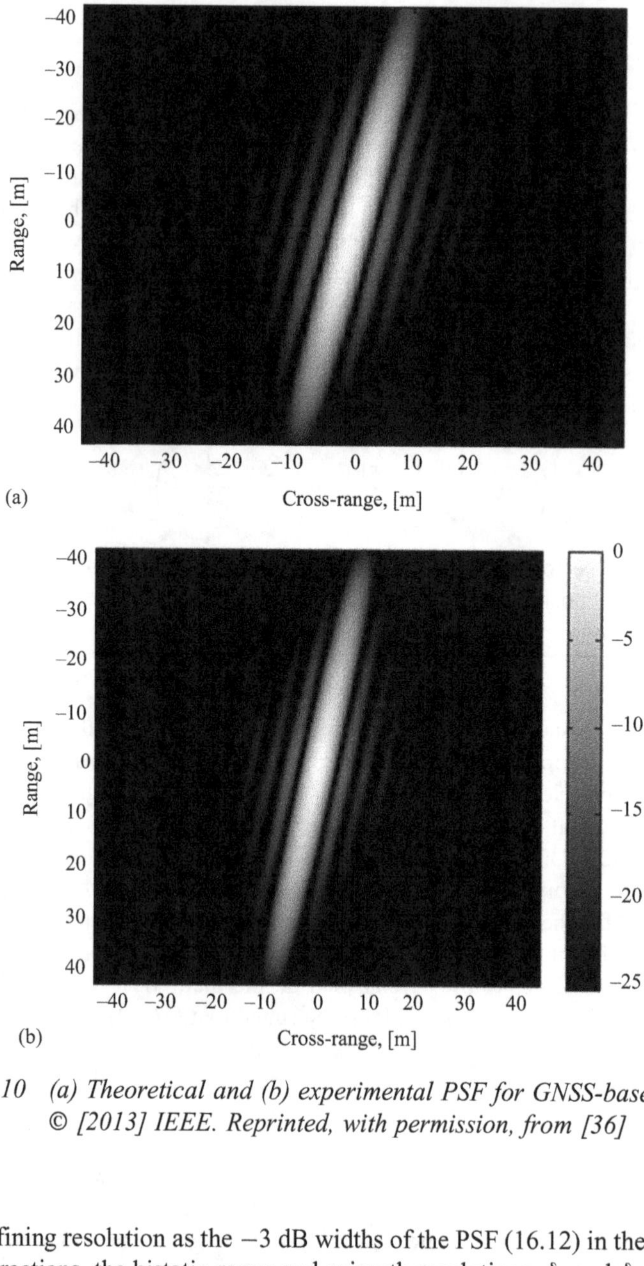

Figure 16.10 (a) Theoretical and (b) experimental PSF for GNSS-based SAR. © [2013] IEEE. Reprinted, with permission, from [36]

By defining resolution as the −3 dB widths of the PSF (16.12) in the range and azimuth directions, the bistatic range and azimuth resolutions, δ_r and δ_a, are along the directions of Θ and Ξ and can be found as [35]:

$$\delta_r = \frac{c}{2\Delta F \cos(\beta/2)}$$
$$\delta_a = \frac{\lambda}{2T_D\omega_E} \qquad (16.13)$$

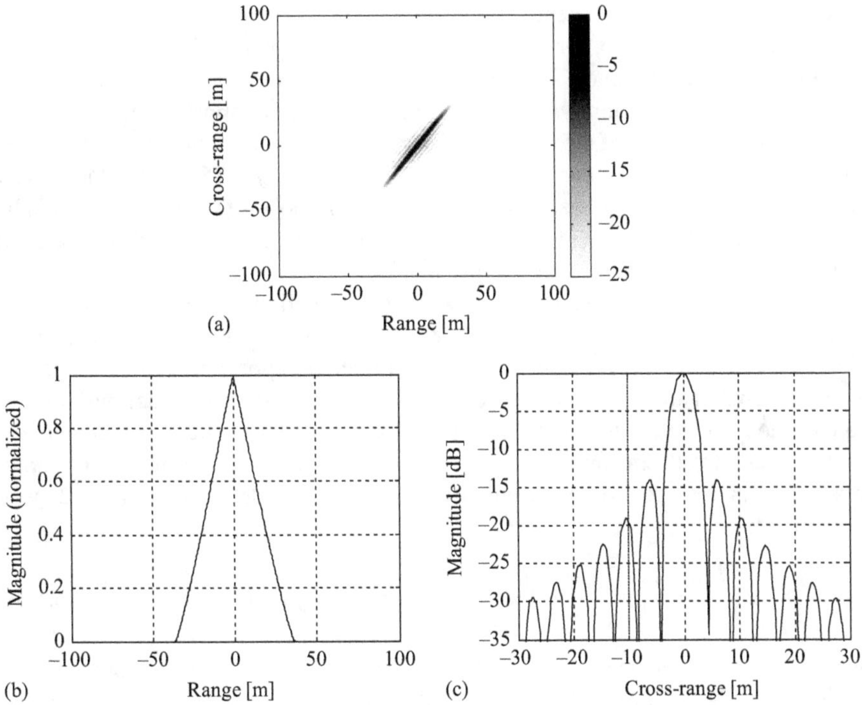

Figure 16.11 (a) Example PSF and its (b) range, (c) cross-range cross-sections

Comparing δ_r with its monostatic equivalent (16.1), it can be seen that the bistatic range resolution is degraded by a factor of $\cos(\beta/2)$. The worst case scenario is when $\beta = 180°$, which corresponds to a forward-scatter geometry where there is no range resolution ($\delta_r = \infty$). The best case scenario is the quasi-monostatic configuration, where the transmitter and receiver are directly behind each other so $\beta = 0°$. In this case, the bistatic resolution is equal to the monostatic one, similar to the general case of bistatic radar [37].

For the azimuth resolution, $T_D\omega_E$ is similar to the length of the aperture along the direction of the BSAR equivalent motion. If an airborne receiver is used, the angular speed of the transmitter relevant to a target is negligible (since the satellite is at a distance of tens of thousands of kilometres from it) compared to that of the receiver. In that case, the azimuth resolution is mainly defined by the receiver motion and can be written as:

$$\delta_a = \frac{\lambda R_R}{V_R T_D} \tag{16.14}$$

On the other hand, if the receiver is fixed, azimuth resolution is only defined by the motion of the satellite and can be calculated as:

$$\delta_a = \frac{\lambda R_T}{V_T T_D} \tag{16.15a}$$

Therefore, to obtain a sufficiently high azimuth resolution in this mode, a long T_D is desirable. As an example, for GNSS, R_T is around 20,000 km (Table 16.2) and V_T, the orbital speed, is around 7 km/s. For a GNSS wavelength of the order of 20 cm, a dwell time of 300 s (5 min) yields an azimuth resolution of approximately 2 m. For a dwell time on target of 0.5 s, which is typical for a monostatic spaceborne SAR in low Earth orbit, the azimuth resolution would be nearly 1.15 km. So, long dwell times on target for GNSS-based SAR serve a dual purpose – to maximize both SNR and azimuth resolution – and for acceptable resolutions in azimuth, dwell times in the order of several minutes are required.

However, it can be shown that very long dwell times on target can also somewhat increase range resolution [38]. The main idea is that over the dwell time the trajectory of the satellite can no longer be approximated as a straight line. In terms of imaging and the PSF, this means that range and azimuth are no longer independent entities but rather they are coupled, and therefore an increase in the total dwell time/aperture length results in both an azimuth and range resolution increase. In order to quantify this, the extended GAF to (16.12) was derived as:

$$\left|\mathrm{X}_g(A,B)\right| = \left|\int p\left(\frac{f_{\mathrm{dc}}}{f_c}\cdot\bar{u}\right) m_A(\bar{u})\exp\left(j2\pi f_{\mathrm{dc}}\bar{u} + j\pi f_{\mathrm{dr}}\bar{u}^2\right)d\bar{u}\right| \tag{16.15b}$$

where $\bar{u} = u - u_c$ is the slow-time vector u minus the time u_c corresponding to the midpoint of the synthetic aperture, f_c is the carrier frequency and f_{dc}, f_{dr} are the Doppler centroid and Doppler rate, respectively.

Based on the extended GAF, the range and azimuth resolution improvements vs the dwell (integration) time on target could be computed, and also cross-checked with experimental results from a GLONASS resolution measurement. The same results could be compared by those expected from (16.12), as shown in Figure 16.12.

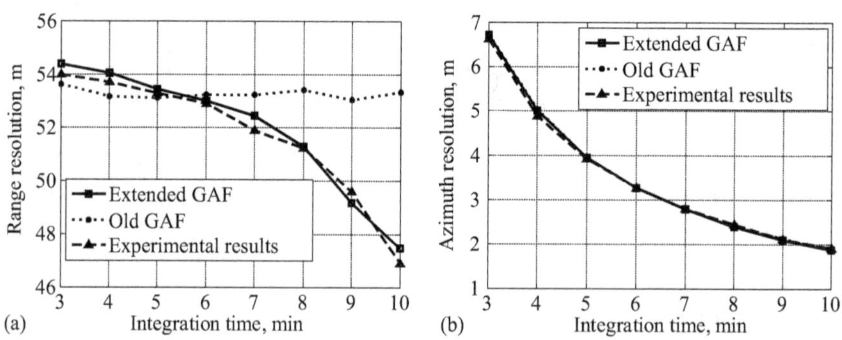

Figure 16.12 (a) Range and (b) azimuth resolution improvements for long dwell times on target.

The figure shows that for shorter integration times, the extended GAF (16.15) and the standard GAF (16.12) nearly coincide, as expected. However, as the integration time increases, the range resolution becomes finer.

Of course, this is but one of the ways to improve range resolution, but there are other techniques employing different principles, and they will be briefly considered in Section 16.7.

16.6 GNSS-based SAR signal processing

16.6.1 Signal synchronization

GNSS-based SAR requires two fundamental types of signal processing – signal synchronization and SAR image formation [39]. Signal synchronization is necessary for any bistatic radar, since the transmitter and receiver are spatially separated, and hence they have different clocks and local oscillators. The purpose of signal synchronization is then to maintain the signal coherence needed for image formation [40,41].

Traditionally, bistatic radar systems have two receiving channels, the HC and the RC, mentioned in Section 16.2.3, with the HC pointed towards the transmitter and the RC towards the observation area, and both channels have common clocks and local oscillators so any receiver errors such as clock slippage and local oscillator drift are common to them. If the SNR in the HC is sufficiently high, the direct signal in the HC may be directly used as the reference signal for matched filtering with the signal in the RC. At the output of this operation, common errors in the two channels are differentially compensated for, and the output of the matched filter is equivalent to the correlation function of the received signal in the range direction.

The above technique is fundamentally possible for high-power illuminators of opportunity such as DVB-T (although even in that case some signal conditioning to suppress signal artefacts such as pilot carriers is needed) or DAB [16]. However, this is not the case for GNSS where the satellites' power density near the Earth's surface is very low, and the SNR in the HC is approximately -11 dB in the ideal case, as calculated in Section 16.4.

For this purpose, signal synchronization in GNSS-based SAR is essentially a tracking algorithm based on matched filtering. The first step is to track the direct signal parameters in the HC, i.e. the delay, Doppler, initial phase and navigation message (if any). Subsequently, a replica of the direct signal is reconstructed locally with the tracked parameters, so it is noise-free. This replica is then used for matched filtering with the RC signal to obtain the range-compressed output signal.

It is important to note at this stage that the tracking process mentioned above is no different to the signal processing performed in GNSS for navigation purposes. Therefore, for signal synchronization in GNSS-based SAR, any GNSS signal tracking option can be adopted, whether it is implemented in hardware (e.g. with delay-/phased-locked loops) or software. For the software-based approach, there are a number of algorithms that can be implemented and described in detail in a number of sources [42–44]. The algorithm selected here is one of the more

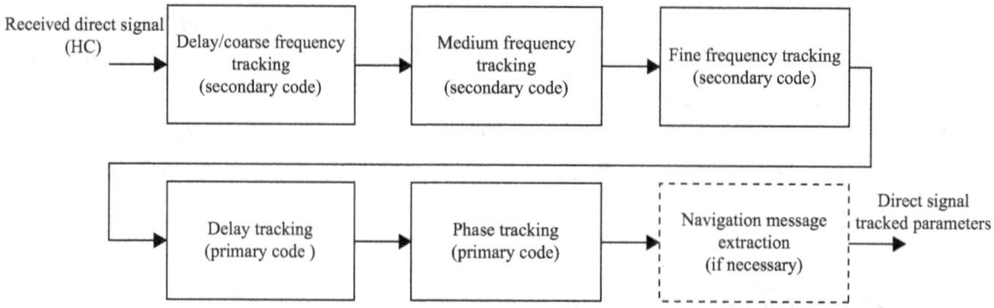

Figure 16.13 BASS algorithm for GNSS-based SAR synchronization [39]

well-known algorithms called the Block Adjustment of Synchronizing Signal (BASS). Its high-level block diagram is shown in Figure 16.13.

As it was shown in Section 16.3.1, different GNSS transmit different signals. However, the generic structure of a GNSS signal is the same and, therefore, the BASS algorithm can be realized for any GNSS. A GNSS signal typically contains two ranging codes modulating a navigation message (if any). Of the two ranging codes, one is a wide bandwidth code desirable for SAR imaging, while the bandwidth of the other is substantially narrower. For simplicity, the former code will be called the 'primary' code, while the other will be called the 'secondary code'. For example, for GLONASS, the primary code would be the P-code, and the secondary code would be the C/A-code. The secondary code and any navigation message could then be viewed as interfering with the primary code and need to be removed to identify the primary code parameters. That is why the first step in the algorithm (top row in Figure 16.13) is to track the secondary, rather than the primary code.

In addition, a single receiving HC antenna will receive direct signals from all the satellites in its field of view, when for BSAR only a single satellite is of interest. However, interference from neighbouring satellites is not problematic, since they normally transmit signals which are orthogonal in terms of their codes or frequencies, and can be neglected.

The signal $Y(t)$ transmitted by a GNSS satellite may be generally written as:

$$Y(t) = P(t)M_P(t)\cos(\omega_c t + \varphi) + D(t)M_D(t)\sin(\omega_c t + \varphi), \tag{16.16}$$

where t is time, $P(t)$ and $D(t)$ are the primary and secondary GNSS ranging codes, respectively, $M_P(t)$ and $M_D(t)$ are the associated navigation messages, ω_c is the signal carrier frequency and φ is the initial signal phase. After quadrature demodulation, the received signal can be written as:

$$\begin{aligned} s(t_n, u) = \; & P[t_n - \tau_{dP}(u)]M_P[t_n - \tau_{dP}(u)]\exp[j(\omega_d(u)t_n + \varphi_{dP}(u))] \\ & + j \times D[t_n - \tau_{dD}(u)]M_D[t_n - \tau_{dD}(u)]\exp[j(\omega_d(u)t_n + \varphi_{dD}(u))] \end{aligned} \tag{16.17}$$

where $t_n \epsilon [0, \text{PRI}]$ denotes fast-time and PRI is the pulse repetition interval, $u \epsilon [-T_D/2, T_D/2]$ is slow-time, and $\tau_{dP,D}(u)$, $\omega_d(u)$ and $\varphi_{dP,D}(u)$ are the instantaneous direct signal time delay, Doppler and initial phase associated with each code,

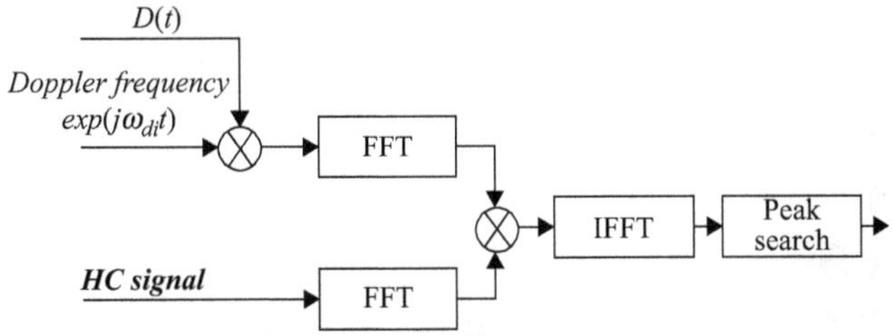

Figure 16.14 Delay/coarse frequency tracking block diagram [39]

respectively, all of which are varying with slow-time. Note that the primary and secondary codes are different in structure and length (i.e. also nearly orthogonal) and, therefore, their time delays and phases are different by a constant value. However, their Doppler frequencies, defined as the derivative of their phases, are equal.

The first stage in the algorithm combines the delay $\tau_{dD}(u)$ and coarse Doppler frequency tracking of the secondary code. These parameters are provided at every PRI, which is equal to the GNSS code duration and is usually 1 ms (Table 16.1). This step is a 2-D search algorithm in delay and Doppler, and comprises a bank of matched filters. The envelope of each filter is a locally generated replica of $D(t)$ for the required satellite, modulated with a different, test Doppler frequency ω_{di} (Figure 16.14).

This signal is used as the reference for matched filtering with the HC signal. For efficiency, matched filtering can be performed in the frequency domain via fast Fourier transforms (FFTs). A peak search on the outputs of the 2-D bank of filters indicates the signal time delay/and Doppler (Figure 16.15). The Doppler search window is from −20 to 20 kHz, which corresponds to the maximum Doppler expected from a GNSS satellite [42]. The separation between the Doppler filters is equal to the PRF (1 kHz); hence, the signal Doppler frequency detected at this stage is called the 'coarse' frequency.

The next step is to detect the Doppler frequency with a higher accuracy in what is called a 'medium' frequency tracking. To do this, 5 ms of data (five PRIs) are taken together, assuming the delay and Doppler variation within them is negligible. The secondary code is removed from them using the delay found in the previous step, so that at the output of this step all that remains is a continuous wave signal (Figure 16.16). A discrete Fourier transform then calculates the medium frequency, which is the Doppler frequency with a resolution of 200 Hz (1/5 ms).

Finally, the fine frequency of the signal may be obtained by compensating the time delay, secondary code and medium frequency from the 5 ms data block and computing the difference in phase between consecutive 1 ms blocks.

Based on all the tracked parameters of the secondary code, its corresponding component can be removed from the received signal (16.17). In addition, at the output of the fine frequency tracking, the direct signal Doppler has been estimated. Note that the navigation message $M_D(t)$ has not been tracked; however, it can be viewed as a random signal with low cross-correlation values with the primary code

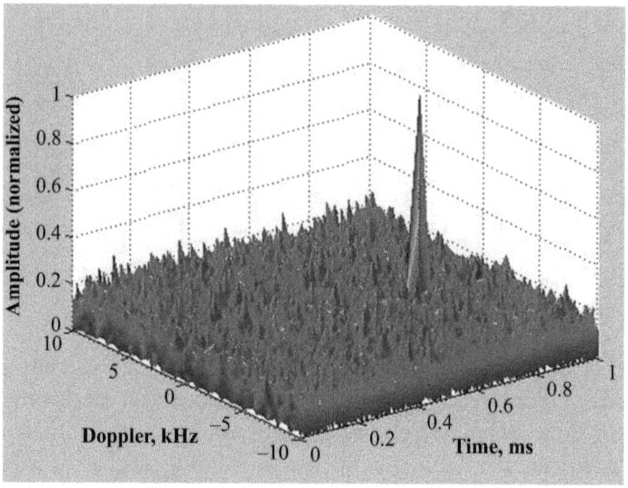

Figure 16.15 Output of delay/coarse frequency tracker for 1 ms (one PRI) experimental data [39]

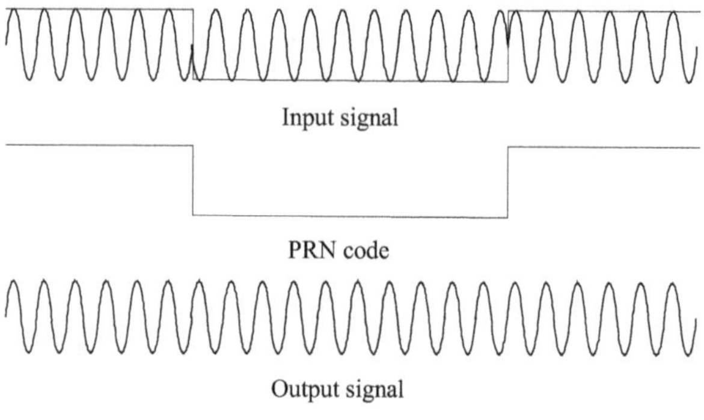

Figure 16.16 Stripping the secondary code for medium frequency tracking [39]

and therefore can be neglected. With this observation, the second term in (16.17) is compensated, and the remaining received signal may be written as:

$$s(t_n, u) = P[t_n - \tau_{dP}(u)]M_P[t_n - \tau_{dP}(u)]\exp[j(\omega_d(u)t_n + \varphi_{dP}(u))] \quad (16.18)$$

In order to track $\tau_{dP}(u)$, matched filtering is used. The reference signal is the envelope of $P(t)$, shifted in Doppler by $\omega_d(u)$ which was estimated in the previous step. Finally, the phase and the navigation message (if one exists) can be extracted after the time-delayed and Doppler-shifted primary code have been stripped from (16.18), as shown in Figure 16.17. The navigation message is a BPSK signal and can be regarded as a phase transition of $\pm\pi$ on $\varphi_{dP}(u)$. Therefore, using a phase transition detector, both the navigation message and the phase can be extracted.

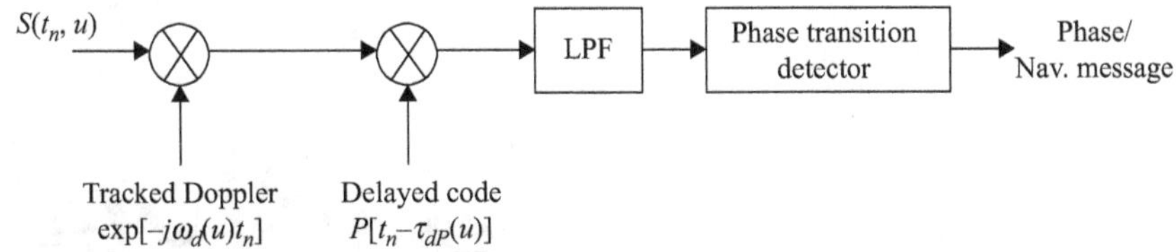

Figure 16.17 Primary code phase/navigation message extraction [39]

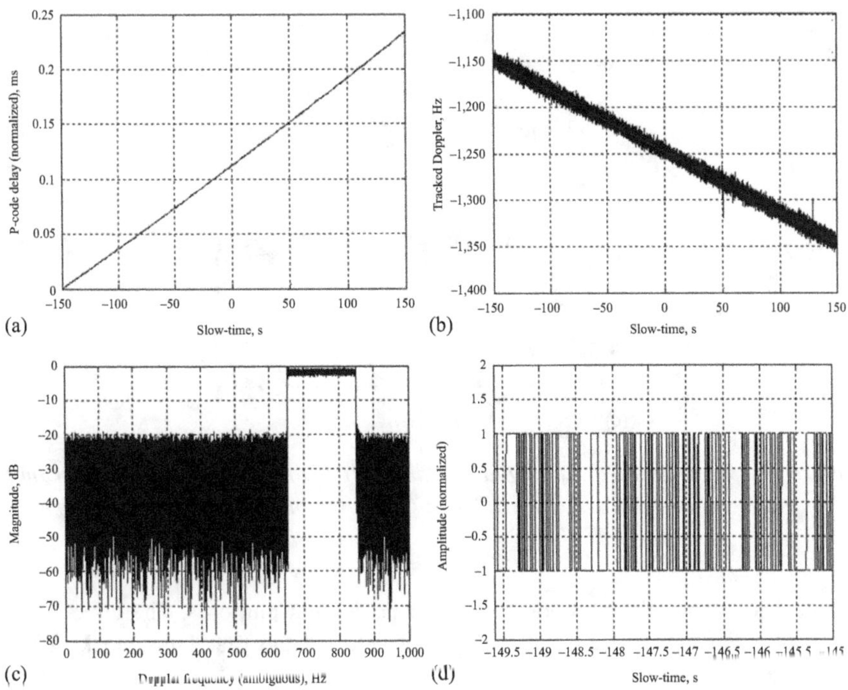

Figure 16.18 Example tracked direct signal (a) delay, (b) Doppler, (c) phase spectrum, (d) decoded navigation message for GNSS-based SAR with a fixed receiver [39]

At the output of the signal synchronization algorithm, the direct signal time delay $\tau_{dP}(u)$, Doppler $\omega_d(u)$, phase $\varphi_{dP}(u)$ and navigation message $M_P(t)$ have been estimated. In the following section, the derived algorithm is confirmed using various experimental data.

Figures 16.18 and 16.19 show some example synchronization outputs from GNSS-based SAR data with fixed and airborne receivers, respectively. The fixed receiver case used GLONASS emissions while in the airborne receiver Galileo was used. Each figure shows the direct signal delay, Doppler, phase spectrum and navigation message. The navigation message sequence at this step may even be

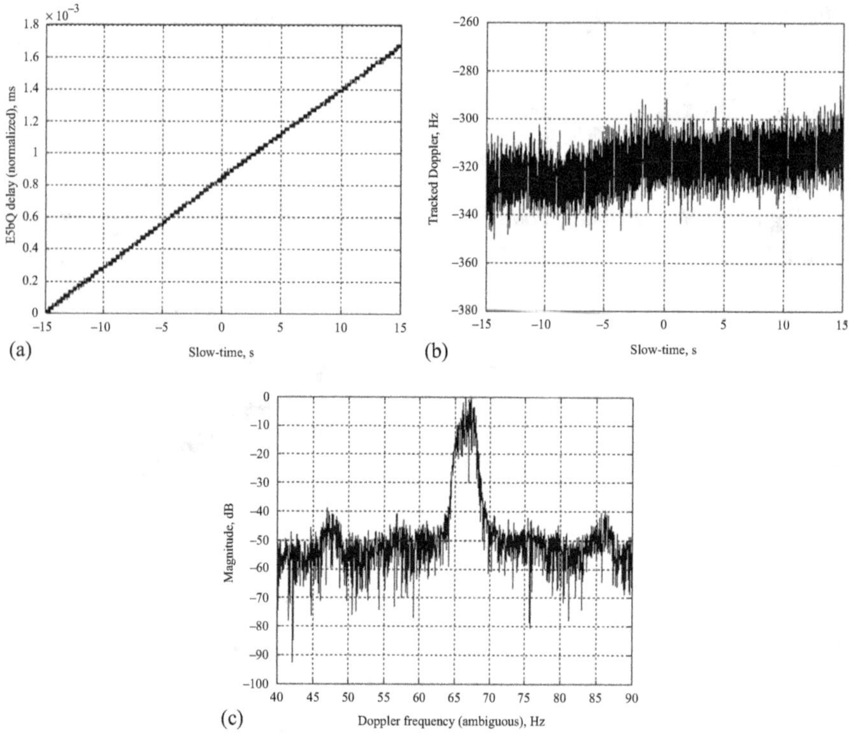

Figure 16.19 Tracked direct signal (a) delay, (b) Doppler, (c) phase spectrum for GNSS-based SAR with Galileo transmitters and an airborne receiver [39]

decoded to enable platform positioning for navigation purposes. The direct signal delay plots are normalized, with the signal delay at the first 1 ms block of data serving as the reference.

The tracked phase spectrum was generated by taking the complex exponential of the tracked phase, followed by an FFT. Effectively, this is the azimuth spectrum of the direct signal. The obtained results shows a near-perfect chirp signal spectrum, which is as expected from the instantaneous phase history of the satellite. Note that the tracked outputs contain the true time delay and Doppler variation, as well as receiver artefacts such as clock slippage, local oscillator drift as well as atmospheric effects on the direct signal.

Figure 16.19 shows the tracked direct signal parameters obtained using a Galileo transmitter (E5b-Q signal) and an airborne receiver during flight. As shown in Table 16.1, the E5bQ signal does not contain a navigation message; hence, the corresponding step in the synchronization algorithm is not applicable. The irregularity in the direct signal Doppler is due to aircraft trajectory deviations during flight.

A number of conclusions can be derived from the experimental results. First, the proposed algorithm can operate irrespective of the selected GNSS transmitter, and irrespective of the topology, even in a dynamic environment where the airplane trajectory deviations affect the direct signal parameters. In terms of the performance, all tracked outputs are obtained with sufficiently high SNR. In the fixed receiver case, the tracked Doppler curve is linear, implying a stable signal Doppler history that resembles a chirp signal. In the moving receiver case, effects of trajectory deviations are visible, an issue which should be dealt with at the image formation stage.

These tracked parameters can be used to create the reference signal required for range compression in SAR image formation, as will be shown in the next section.

16.6.2 Image formation

Following signal synchronization, an image formation algorithm is required to generate imagery of an interrogated scene. In general, BSAR image formation algorithms differ from monostatic ones, which is largely due to the more complex data collection geometries. For that reason, substantial research on this topic has been carried out. Extensions of frequency-based monostatic SAR algorithms to the bistatic case, such as Chirp Scaling [45], Range-Doppler [46], Omega-K/Range-migration and their variants [47–49], and even polar format [50], have been suggested in the literature, and some of them have been experimentally confirmed.

In order to apply a frequency-based algorithm in BSAR, certain assumptions have to be made, so the majority of such algorithms operate on specific BSAR geometries only, where there is some level of control on the platform trajectories relevant to a target scene. Such approaches are difficult, if possible at all, to apply in GNSS-based SAR for a number of reasons. The first reason is that GNSS are satellite constellations, with a number of satellites in various positions, so it is impractical to design special data collection geometries for all of them in real conditions and if persistent area monitoring is required. Another reason is that over the dwell time on target, satellite trajectories can no longer be approximated as straight lines, which makes the application of frequency-based algorithms a very difficult task. That is especially true for fixed-receiver acquisitions. For airborne acquisitions, it is possible to use a Range-Doppler algorithm [51], but the algorithm only works in this scenario.

For these reasons, perhaps the most universal solution for GNSS-based SAR is algorithms operating in the time domain. While they are more computationally inefficient, they can handle any BSAR acquisition geometry without adding any additional degradation to image quality, and hence are more convenient and flexible to use. In this chapter, one of the simplest algorithms in this group, the back-projection algorithm (BPA) is described to illustrate the imaging concept. In the literature, fast BPAs have also been implemented [52,53] and can be directly applied to GNSS-based SAR.

The concept of the BPA is the same as in the monostatic SAR case [9], and its block diagram is shown in Figure 16.20. In the case of an airborne receiver, motion compensation (MoComp) is integrated to the algorithm to account for aircraft trajectory deviations during flight.

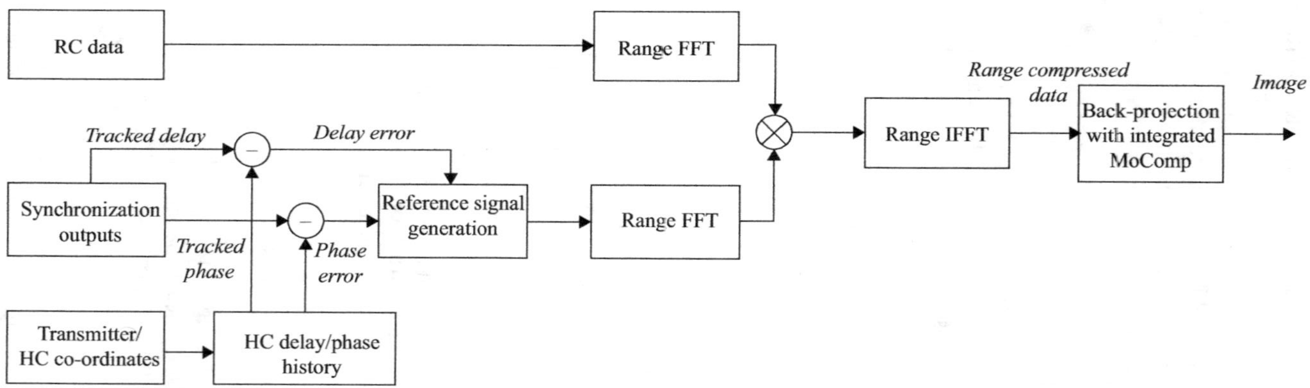

Figure 16.20 BPA block diagram for GNSS-based SAR. MoComp is required only for the case of an airborne receiver [39]

Two major steps are required to achieve image formation in BPA: the range compression and the back-projection integral calculation. For the GNSS-based SAR case, the reference signal for range compression is generated based on the parameters obtained by synchronization (Section 16.6.1), during which the direct signal delay and phase are tracked, both including receiver and errors. The receiver errors are common to both HC and RC channels, and the operational range is in the order of few kilometres, where atmospheric errors can be generally assumed to be the same. Because the satellite/receiver trajectories are known, the expected delay and phase history for the HC is also known. Therefore, phase and delay errors can be extracted from the tracked outputs (e.g. those shown in Figure 16.19) and incorporated to the reference signal for range compression. If the delay and phase errors are t_e and φ_e, respectively, the reference signal for range compression can be written as:

$$s_0(t_n, u) = p[t_n - t_e(u)]\exp[-j\varphi_e(u)] \tag{16.19}$$

In (16.19), $p(t_n)$ includes the GNSS primary code and any navigation message, i.e. $p(t_n) = P(t_n)M_P(t_n)$.

The signal in the RC reflected from a fixed point target in the scene may be written as (after quadrature demodulation and ignoring constant amplitude and phase terms):

$$s_{RC}(t_n, u) = p\left\{t_n - \left[\frac{R_T(u) + R_R(u)}{c} + t_e\right]\right\} \times \exp\left\{-j\left[\frac{2\pi}{\lambda}(R_T(u) + R_R(u)) + \varphi_e\right]\right\} \tag{16.20}$$

where $R_T(u)$ and $R_R(u)$ are the instantaneous transmitter-target and receiver-target ranges, which vary with slow-time as the transmitter/receiver move.

Range compression can be done in the fast-time frequency domain via matched filtering. At the output of the matched filter the signal takes the form:

$$r(t_n, u) = R_x\left[t_n - \frac{R_T(u) + R_R(u)}{c}\right]\exp\left\{-j\frac{2\pi}{\lambda}[R_T(u) + R_R(u)]\right\} \tag{16.21}$$

where $R_x(t_n)$ is the cross-correlation function of the primary code $P(t_n)$. The matched filtering operation removes the delay/phase errors and the navigation message. Also, the time delay and phase histories of each target depend on the radio wave propagation path, as dictated from the transmitter and receiver positions at the time of measurement. In practice, transmitter positions can be extracted for Two-Line Element data, which have been found of sufficient precision to accurately focus the data without the need for specialized post-processing techniques such as autofocus. Following that operation, the computation of the back-projection integral may be performed in a similar manner to the monostatic case, but taking into account the bistatic target time delay rather than the monostatic round trip delay. For the moving receiver case, it is straightforward to integrate a MoComp process to the back-projection integral computation for each image pixel, and the accuracy of this scheme relies on the accurate knowledge of the receiver positions [54].

The implementation of the BPA involves the generation of a rectangular grid of points with Cartesian co-ordinates (x_i, y_j) counted from the receiver position. For each of these points, the BPA back-tracks signal returns at the time delays $t_{ij}(u) = \left[R_{T_{i,j}}(u) + R_{R_{i,j}}(u)\right]/c$, and integrates the data over slow-time, i.e.:

$$f(x_i, y_j) = \int_u r\left[t_{ij}(u), u\right] \exp\left\{j\frac{2\pi}{\lambda}\left[R_{T_{i,j}}(u) + R_{R_{i,j}}(u)\right]\right\} du \qquad (16.22)$$

At the output of this step, the reflectivity of the scene $f()$ at the location (x_i, y_j) is estimated. To obtain the full image, the integral (16.22) is computed over all (x_i, y_j) grid points.

16.7 Experimental results

16.7.1 Fixed receiver

A series of experimental data sets were collected using GLONASS transmitters and a fixed receiver. The receiver was placed at the roof of the Department of Electronic, Electrical and Systems Engineering at the University of Birmingham. The RC antenna was overlooking the area to the west of the building. A satellite photograph (taken from Google Earth) of the observation area is shown in Figure 16.21. The receiving setup is shown in Figure 16.22. The experimental parameters are listed in Table 16.5.

There are two sections of interest within the area. First are four isolated towers approximately 1.2 km away from the receiver, which could serve as reference targets. Second, there are tree lines facing towards the receiver at a range of approximately 850 m from the receiver, which could provide high strength echoes. The experimental radar images, obtained after signal synchronization and image formation (Section 16.6), are shown in Figure 16.23, super-imposed on the photograph of Figure 16.22. The colourscale is in dB, where 0 dB corresponds to the highest intensity echo in the image. The dynamic range has been clipped to 20 dB. All four towers have been detected at the correct location, and the outline of the trees can be directly seen. At the same time, areas where the expected reflectivity is expected to be low (such as grassy areas) appear as dark patches in the image.

16.7.2 Airborne receiver

Moving receiver measurements were taken using Galileo as the transmitter, while the receiver was mounted on an AS355 helicopter (Figure 16.24(a)). The same receiving hardware as in fixed receiver case was used. Trials were conducted around the East Fortune airfield in Scotland (Figure 16.24(b)). The other experimental parameters are in Table 16.6.

This set of data required MoComp processing due to the irregular motion of the helicopter, which was made worse by weather conditions at the time of

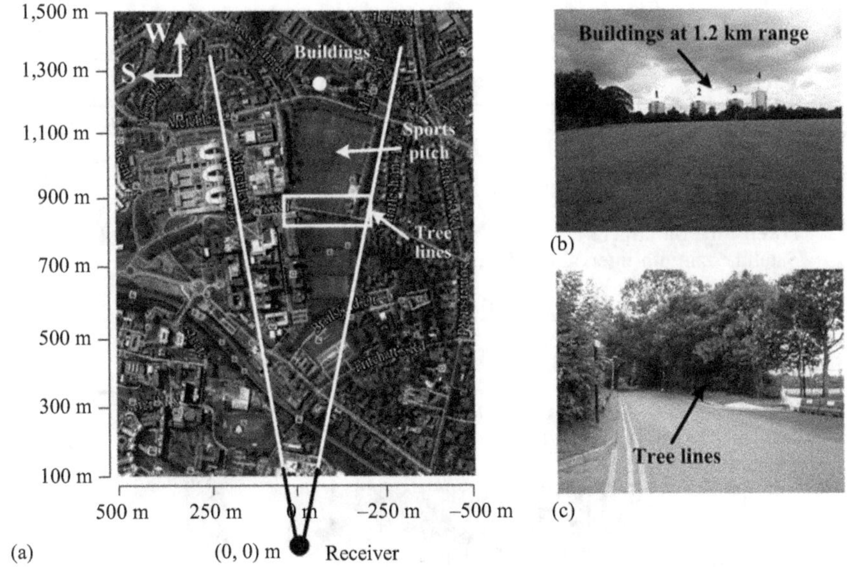

Figure 16.21 (a) Google Earth photograph of target site, (b) towers at far range, (c) tree lines in mid range

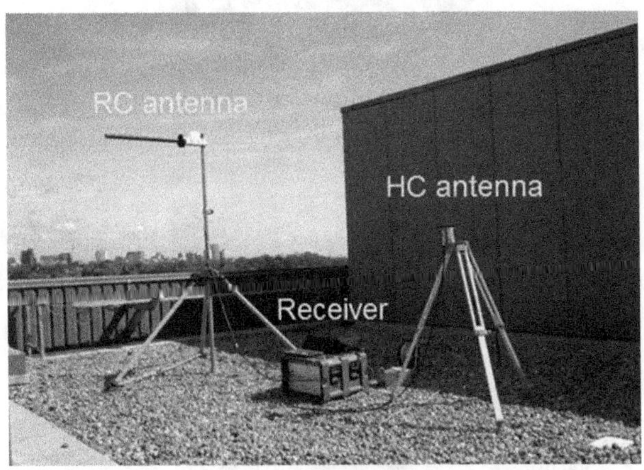

Figure 16.22 Fixed receiver experimental setup

measurement. This can be readily seen from the synchronization results for this data set, shown in Figure 16.19. The helicopter location was recorded with a standard GPS positioner with a 1 Hz update rate, which was not sufficient to sample trajectory deviations. In addition, the helicopter was not equipped with any inertial navigation system and used its own GPS receiver to navigate. For these reasons, it was expected that the obtained imagery would not be as accurate as in the fixed

Table 16.5 Fixed receiver experimental parameters

Parameter	Value
Satellite	GLONASS COSMOS 736
Frequency band	G1
Signal bandwidth	5.11 MHz
PRF	1 kHz
Dwell time on target	200 s
Satellite azimuth interval over dwell time (relevant to north)	187°–189°
Satellite elevation interval over dwell time (relevant to local horizon)	54°–52°
Bistatic angle	63.5°

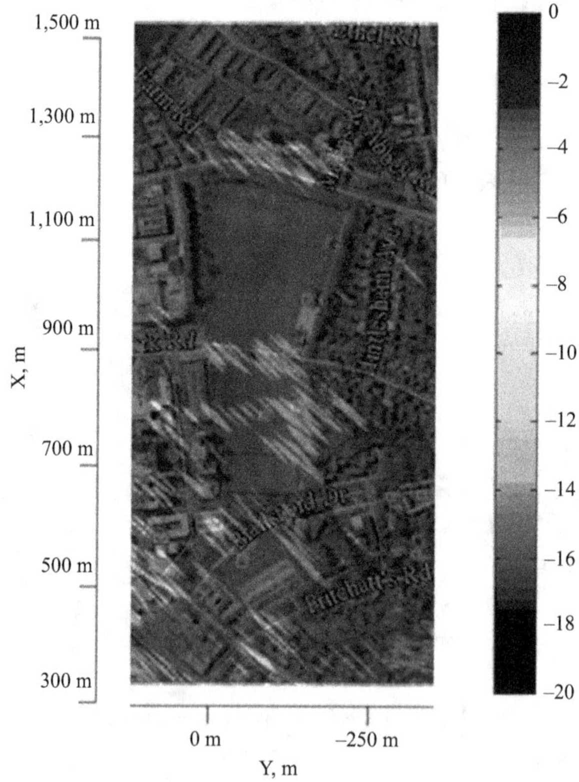

Figure 16.23 GNSS-based SAR image super-imposed on imaging scene

receiver case. The obtained image is shown in, super-imposed on a satellite photograph of the observed area (Figure 16.25).

It is clear from the observed imagery that the image is de-focused. For example, the signal return of an aircraft above the leftmost hangar (Target 1)

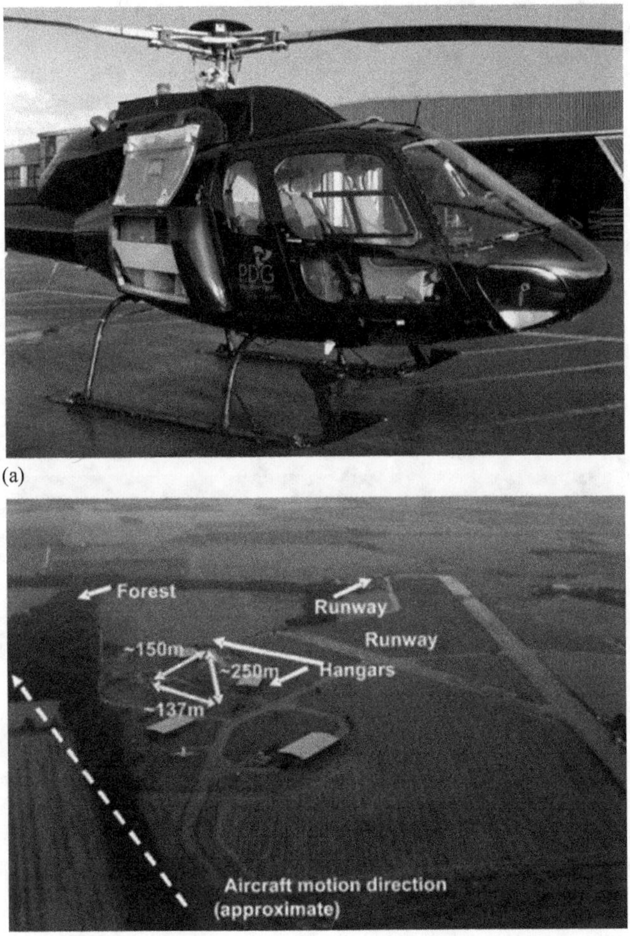

(a)

(b)

Figure 16.24 (a) Receiving system onboard AS355, (b) aerial photograph of imaging scene [39]

Table 16.6 Airborne receiver experimental parameters

Parameter	Value
Satellite	Galileo GIOVE-A
Frequency band	E5b-Q
Signal bandwidth	10.23 MHz
Receiver speed (nominal)	72 km/h
Receiver aperture length	~800 m
Receiver altitude (nominal)	250 m
Satellite elevation interval	70°–80°
Satellite azimuth interval	100°–130°

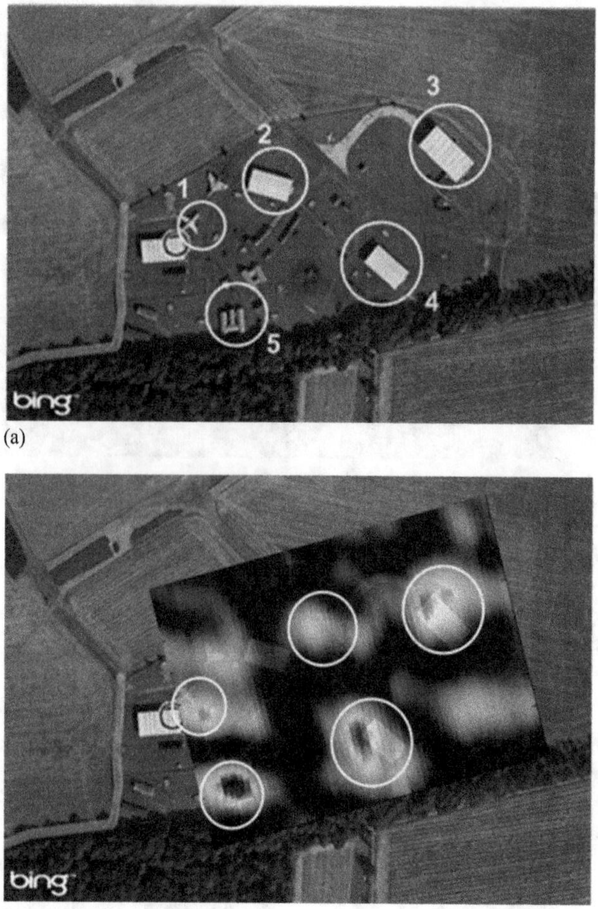

(a)

(b)

Figure 16.25 (a) Satellite photograph of imaging area, including markers for strong reflectors, (b) radar image super-imposed on imaging scene [39]. [Bing^{TM} screen shots reprinted with permission from Microsoft Corporation]

appears to be completely smeared. In addition, signal returns are not registered in their appropriate locations, such as Targets 2 and 4 (aircraft hangars) which appear shifted. Furthermore, Target 3 appears as multiple peaks in the image, implying asymmetric sidelobe levels in the PSF. These artefacts are due to the accuracy and update rate of the GPS receiver onboard the helicopter, as well as the absence of any inertial navigation equipment on it, hence the inadequacy to sample trajectory deviations sufficiently.

Despite de-focus, five main targets have been detected. All of them correspond to buildings (such as hangars) or aircraft which could yield significantly high reflections, such as Targets 4 and 5. Higher intensity parts in the lower right part of the image are due to an occupied car park.

16.8 GNSS-based SAR potential for advanced techniques

The validation of the GNSS-based SAR capability has paved the way for the consideration of advanced SAR techniques. All of them are now at the proof of concept level; however, the obtained results have so far shown clear potential, and they are included in the following sections as the stepping stone for further research in the future.

16.8.1 Coherent change detection

One of the prospective applications for GNSS-based SAR is persistent surface change detection in local landscapes. As mentioned before, GNSS are obviously outmatched by imaging radar satellites in terms of radar performance, such as resolution and power budget. However, there are also reasons to support such a topology for this particular application. A single GLONASS or Galileo satellite has re-visit cycles on the order of 8 to 9 days, which is already faster than most radar satellites for repeat-pass imagery. However, by considering multiple GNSS satellites, an 'effective' re-visit cycle can be formed that is not bound by a single satellite's orbital properties. For example, if eight to nine different satellites are used in daily sequence, the effective re-visit cycle for repeat-pass imagery is 1 day and so on. Also, GNSS satellites transmit in L-band, which is within current trends for Earth observation, and have reasonable resolution cell sizes for these applications. Due to the relatively low power budget, monitoring of local areas is envisaged, which is more similar to the applications encountered in ground-based SAR, where potential targets could be man-made targets (e.g. bridges, dams, etc.) or reference targets installed for earthworks monitoring.

At the first stage, the potential for monitoring changes in a surface can be investigated by applying coherent change detection (CCD) techniques. The literature for CCD in monostatic SAR is vast, and the reader is prompted to [55–57] for some of the seminal papers on this topic. However, CCD has been considered little, if at all, for BSAR in general, let alone a passive BSAR with transmitters which are not radar satellites. For these reasons, a systematic study on the feasibility of such a system is required. As in any SAR-related activity, the starting point in such a study is to consider point targets. The aim of the study is to investigate first of all whether or not phase differences can fundamentally be detected in GNSS-based SAR, and if this is the case, how precisely can these differences be estimated. The latter provides an upper limit in the change detection performance, which may or may not be reached in practice since the real world comprises mostly of distributed targets. However, it does give an understanding on whether or not the technology is fundamentally possible and a springboard on which to extend theories from point targets to extended scenes with distributed targets.

The CCD scheme proposed as a first stage is the coherence estimation of repeat-pass image pairs, as in the monostatic case. That is, consecutive images obtained at the re-visit cycle of a single satellite will be compared at the phase level, and image de-correlation will be translated into actual surface change.

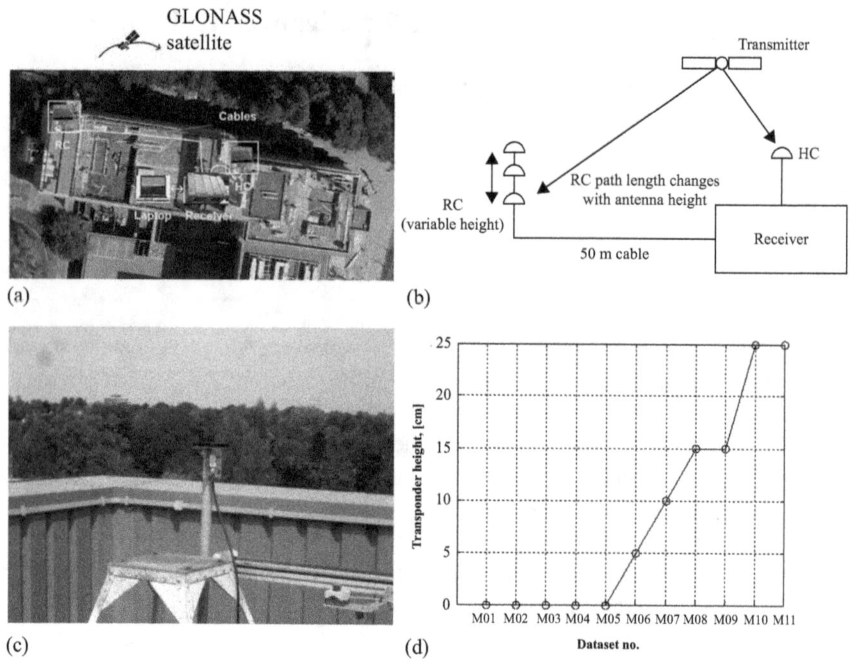

Figure 16.26 Experimental test-bed (a) topology and (b) concept, (c) adjustable height transponder (low-gain antenna and front end), (d) transponder height variation. © [2013] IEEE. Reprinted, with permission, from [36]

Nevertheless, due to the complexity of the system, it is required that some of its major aspects should be examined prior to a further system investigation, to decide whether it may be fundamentally feasible. An initial experimental programme of the system's capacity as a change detector was designed to verify its feasibility on the fundamental 'proof of concept' level. To prove this concept, an experimental programme was built, using a low-gain antenna for the RC acting a semi-active transponder whose height was varied between consecutive satellite passes to emulate a surface change on the point target level (Figure 16.26). A total of 11 data acquisitions were made at the re-visit cycle of a single GLONASS satellite, and the resulting PSFs from the transponder were compared at the phase level to extract the corresponding height displacement.

A comparison between the theoretical and experimental results is shown in Figure 16.27. From the experimental results, the minimum and maximum displacement errors are 0.22 and 1.47 cm, respectively. The average error is calculated to be −0.4 cm, with a standard deviation of 1.15 cm. This is expected as the averaging process partially compensates phase errors. Nevertheless, even in the presence of cable phase noise, the CCD accuracy is sufficiently high to justify further research on the behaviour of more realistic targets in real scenes.

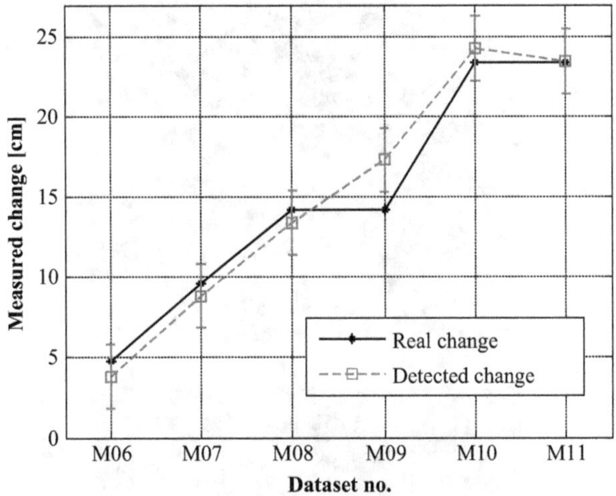

Figure 16.27 Measured displacements via CCD vs real transponder height change. © [2013] IEEE. Reprinted, with permission, from [36]

Another issue to consider is spatial de-correlation between repeat-pass GNSS-based SAR images. This study is needed in the development of this system to monitor temporal changes in a scene. The main challenge is that, unlike monostatic SAR, in a BSAR configuration spatial coherence depends on the bistatic geometry. The theoretical framework to describe spatial coherence for this case was developed by extending well-established monostatic models [58] with real GNSS-based SAR imagery from a test site (Figure 16.28).

To validate that theoretical model, an experimental image was used. A master image was generated using data from the start time of acquisition and for a dwell time T_{sys} (5 min) that was less than the dwell time T (7 min) on target. Then, a set of slave images was generated, with the same T_{sys} but starting N seconds later than the master image, with N from 1 to 30 s. In other words, N was the temporal separation between images, resulting in a spatial de-correlation that could be analytically estimated. This methodology was chosen for three main reasons. First of all, since all images were obtained from the same data set, temporal de-correlation should be minimal, allowing the measurement of only spatial de-correlation. Second, varying the temporal separation between images could allow a visualization of how spatial de-correlation varies with satellite viewing angles. Finally, this scheme simulates a real repeat-pass acquisition, where the repeat-pass image is acquired with some time offset due to practical reasons. The coherence maps obtained between the master image and slave images with offsets of $N = 1$, 10 and 20 s are shown in Figure 16.29. The colourscale shows the coherence between images, ranging from 1 (full coherence) to 0 (no coherence). As expected, when the temporal image separation increases, so does the spatial de-correlation and, therefore, the total image coherence drops.

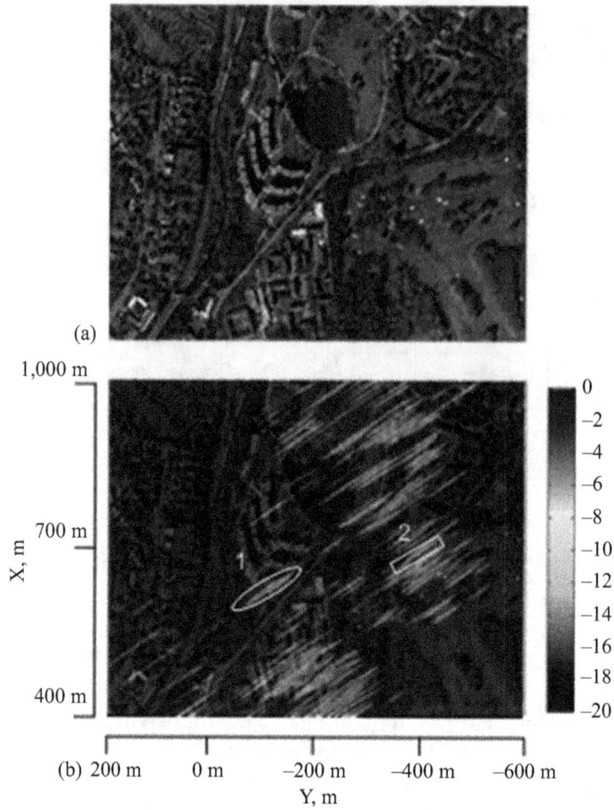

Figure 16.28 (a) Satellite photograph of the observation area, (b) GNSS-based SAR image super-imposed on (a) for spatial de-correlation verification. © [2015] IEEE. Reprinted, with permission, from [58]

To compare experimental spatial de-correlation results with theoretical ones, two representative areas within the image were selected. The first one was located around (630, −116) m (Figure 16.28). In this area, there is a strong compressed echo from a single building, which is a student accommodation hall (Figure 16.30). The second one is located around (694, −398) m and contains a patch of trees, which were also detected with a sufficiently high SNR. The coherence measured at each of these locations is plotted vs N in Figure 16.31 and compared with the derived theoretical model to show a good co-incidence. The analysis on the spatial de-correlation is not restricted to GNSS-based SAR but can be applied to all BSAR CCD systems.

16.8.2 Multi-perspective imaging

In this thematical area, the spatial, rather than temporal, diversity of GNSS-based SAR is being exploited. By selecting different bistatic topologies, diverse

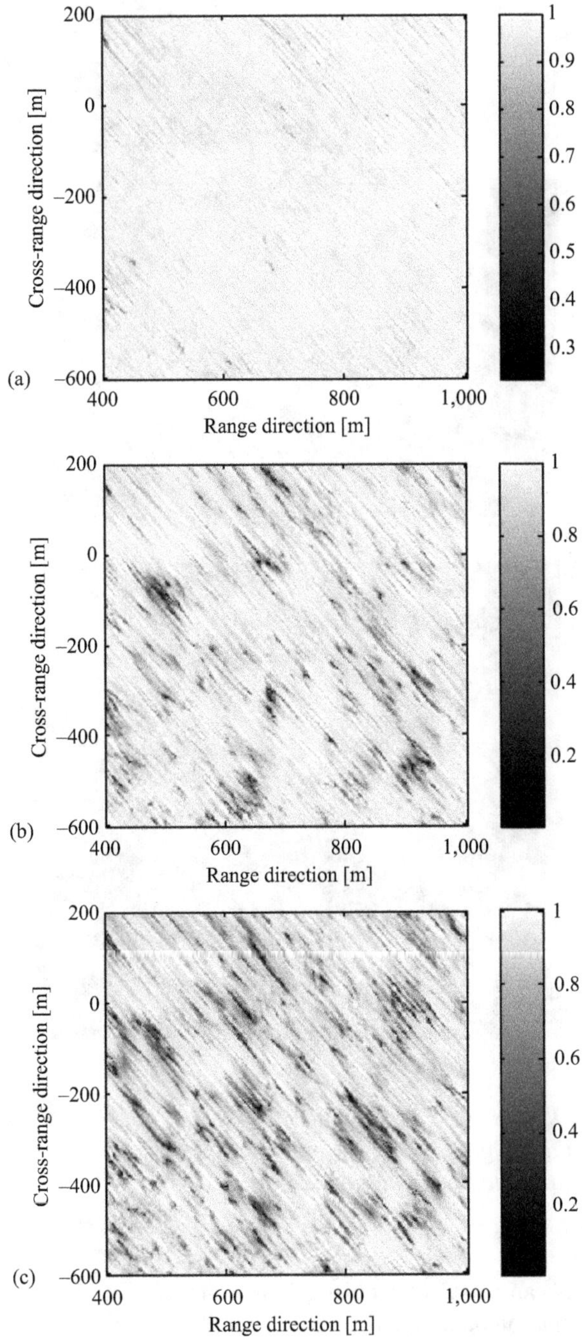

Figure 16.29 Coherence maps for (a) 1-, (b) 10-, and (c) 20-s temporal separation between the master and slave images.

(a)

(b)

Figure 16.30 Objects selected for spatial de-correlation evaluation within the imaging scene: (a) building, (b) trees. © [2015] IEEE. Reprinted, with permission, from [58]

scattering effects can be seen for objects in the same scene. This may increase the amount of information on a given scene and could be used to aid automatic object classification based on their bistatic scattering properties.

Four GNSS-based SAR experiments were conducted with a GLONASS transmitter and a fixed receiver [59], with the experimental setup and target area

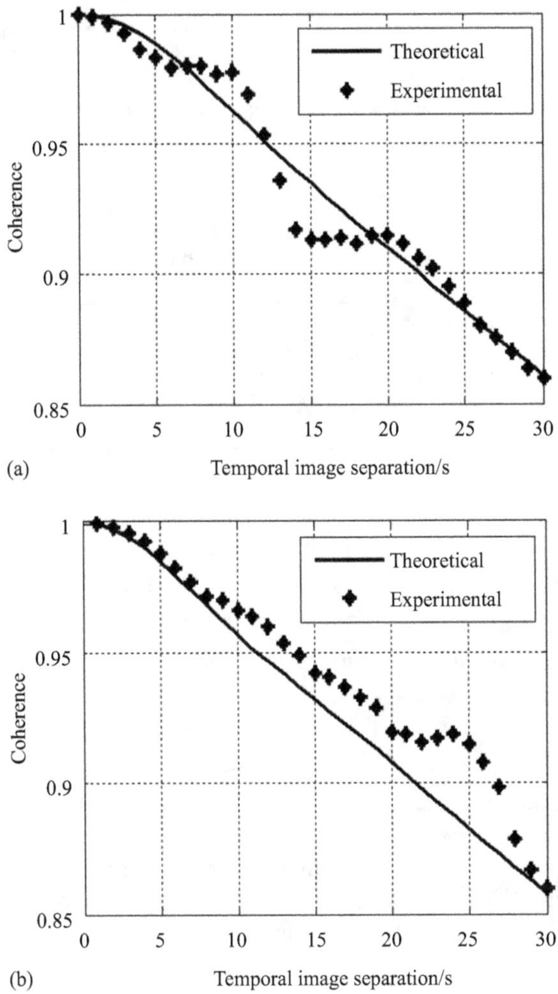

Figure 16.31 Experimental vs theoretically predicted spatial coherence for (a) building area and (b) tree area. © [2015] IEEE. Reprinted, with permission, from [58]

having been shown in Figure 16.21 and Figure 16.22. Each satellite had a different orientation with regard to the imaging scene, shown in Figure 16.32. The four experimental radar images, after signal synchronization and image formation, are super-imposed on the imaging scene and presented in Figure 16.33.

Each image was obtained with a different bistatic configuration and, therefore, resolution cell sizes and orientations varied between images.

The figure shows that bistatic radar images from the same scene are almost completely different and, therefore, they have the potential to increase the amount

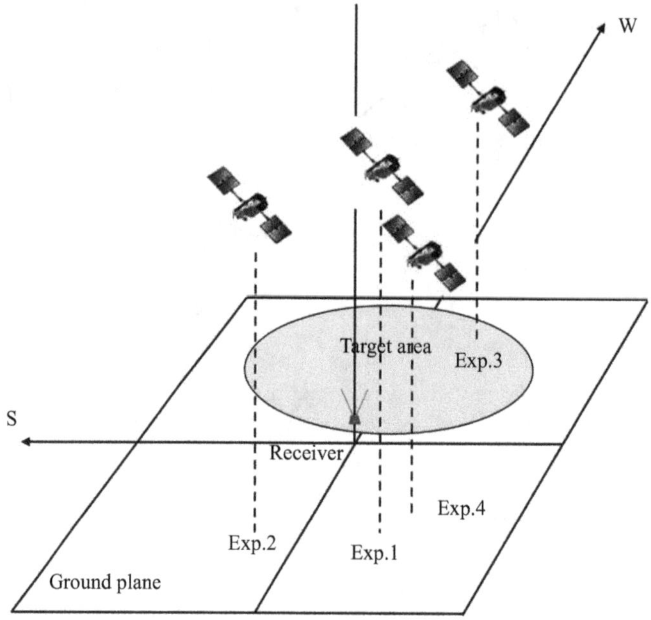

Figure 16.32 Satellite positions for multi-perspective imaging experiment

of area information. This can be done either by treating each image separately, or combining images together.

With regard to the latter, experiments were performed with Beidou-2 [60], where 16 different images taken from different satellite positions were fused. A comparison between fused images and single BSAR images (Figure 16.34) shows that the information detail in the multi-angle image has substantially increased.

16.8.3 Multistatic imaging for spatial resolution improvement

In the previous section, the spatial diversity inherent in GNSS-based SAR was used to increase the amount of detail regarding a scene. In this section, spatial diversity is exploited for another purpose: the increase in image spatial resolution [61]. In Section 16.5, it was shown that the image resolution is restricted to the GNSS ranging signal bandwidth and is further degraded by the bistatic geometry. However, in the same section, it was shown that the PSF orientation also depends on the geometry. The idea is then to combine images obtained from different geometries, so that the resulting multistatic image responses are the common areas between the individual bistatic images.

This idea has first been explored on the PSF level, before moving on to real imaging scenes. As a first image combination approach, a non-coherent addition of

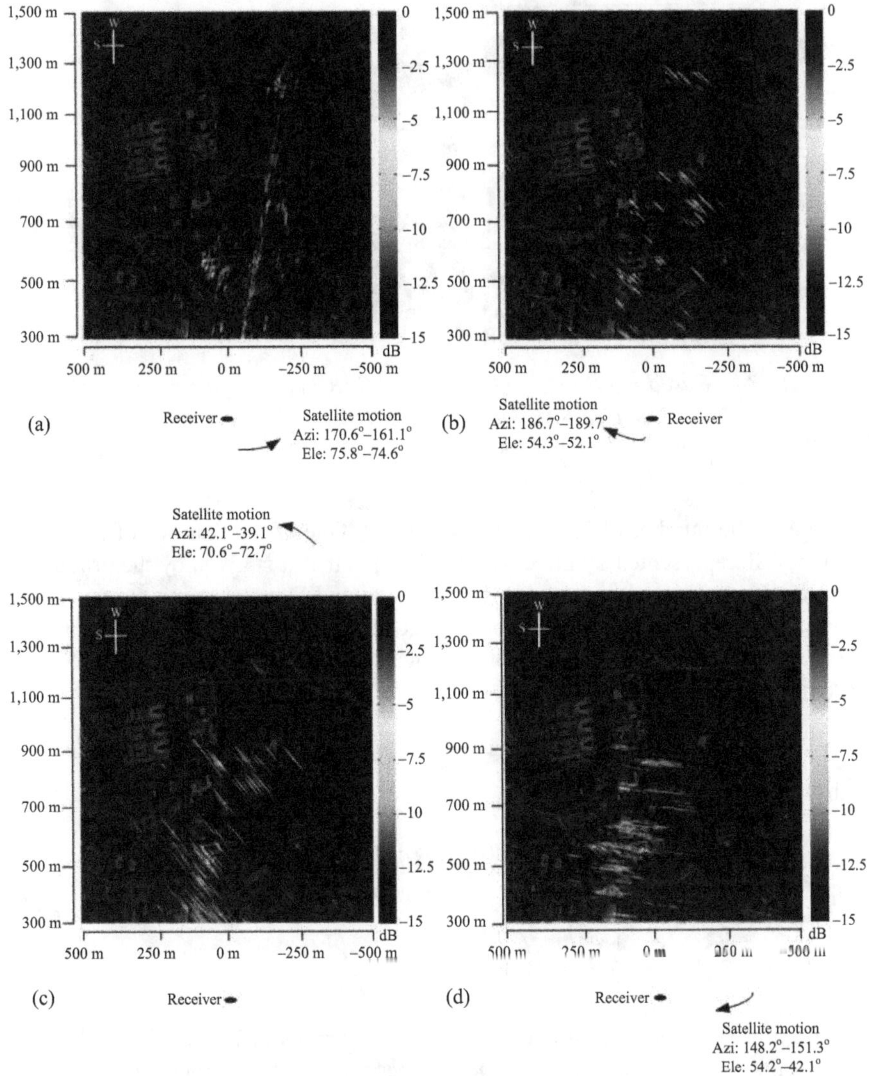

Figure 16.33 GNSS-based SAR images of the same scene from experiments 1–4 in (a)–(d), respectively. © [2013] IEEE. Reprinted, with permission, from [59]

bistatic PSFs is assumed. Using (16.12), the resulting multistatic PSF (MPSF) may be written as:

$$\text{MPSF} : \frac{1}{N}\sum_{n=1}^{N} p\left(\frac{2\cos(\beta_n/2)\Theta_n^T(r)}{c}\right) \cdot m_A\left(\frac{2\omega_{En}\Xi_n^T(r)}{\lambda}\right) \tag{16.23}$$

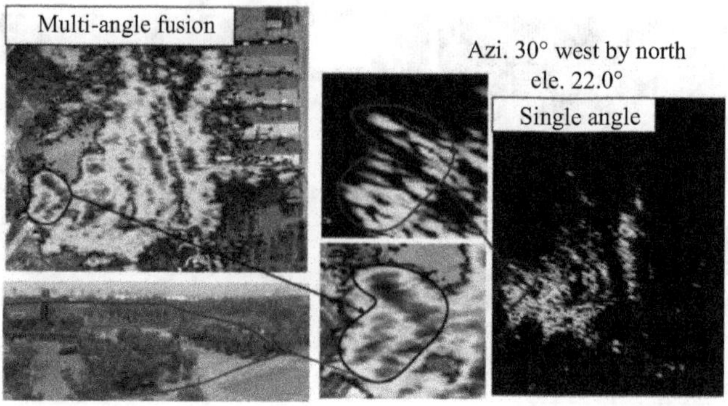

Figure 16.34 Fused multi-angle GNSS-based SAR image vs a single GNSS-based SAR image. © [2014] IEEE. Reprinted, with permission, from [60]

where N is the number of PSFs being combined. One of the features of the single channel PSF represented by the GAF in (16.12) is that it is given by the product of two functions separately pertaining to the range and Doppler domain: even if the range and Doppler directions are not orthogonal, their domains are still separable. For the MPSF in (16.23), this cannot be done in the strict sense, but the MPSF can be approximated as:

$$\text{MPSF} \approx \overline{p}\overline{m} \tag{16.24}$$

with $\overline{p}$ and $\overline{m}$ being:

$$\begin{aligned} \overline{p} &= \frac{1}{N}\sum_{n=1}^{N} p_n = \frac{1}{N}\sum_{n=1}^{N} p\left(\frac{2\cos(\beta_n/2)\Theta_n^T(r)}{c}\right) \\ \overline{m} &= \frac{1}{N}\sum_{n=1}^{N} m_{An} = \frac{1}{N}\sum_{n=1}^{N} m_A\left(\frac{2\omega_{En}\Xi_n^T(r)}{\lambda}\right) \end{aligned} \tag{16.25}$$

In this expression, the range and Doppler domains are again separated. To illustrate the benefits in this combination method, experiments were conducted with point-like targets, in a fixed receiver configuration, and with two different GLONASS satellites, denoted as 'A' and 'B' hereafter, at different aspect angles. The bistatic PSFs were obtained and then combined together with the non-coherent addition described above to get the MPSF. The MPSF was then analytically calculated for the same GLONASS trajectories and compared to the experimental results (Figure 16.35).

Looking at the experimental MPSF in (Figure 16.35), a drastic improvement in image resolution can be observed. This is because, as mentioned before, the MPSF resolution is now defined as the overlapping segment between the individual bistatic PSFs. A comparison with the theoretical expectations also confirms the validity of the MPSF model in (16.24) and (16.25).

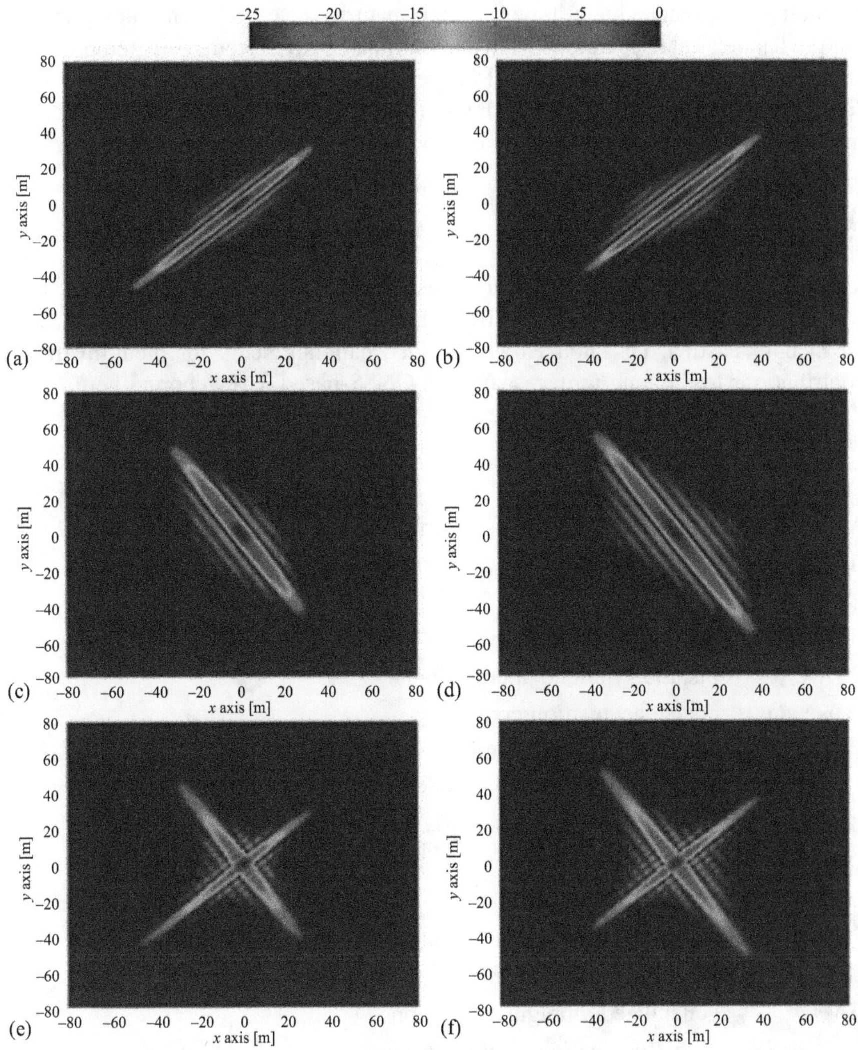

Figure 16.35 *Bi/multistatic PSFs: (a) experimental PSF A, (b) simulated PSF A, (c) experimental PSF B, (d) simulated PSF B, (e) experimental MPSF, (f) simulated MPSF. © [2015] IEEE. Reprinted, with permission, from [61]*

16.9 Summary

This chapter has provided a brief overview of GNSS-based SAR. Its performance as a passive imaging system, i.e. its resolution and power budget, was analysed, and signal-processing algorithms required for image formation were described. This technology has reached a sufficiently high maturity level to begin tapping into its unique features that separate it from conventional SAR systems – its temporal and spatial diversity. The former is ideal for persistent area monitoring, while the latter can be used to increase scene information as well as spatial resolution, which are the first steps in realizing this system as a powerful remote-sensing tool.

Acknowledgements

The authors would like to thank Dr Rajesh Saini, Dr Rui Zuo, Dr Hui Ma, Prof. Cheng Hu, Prof. Zeng Tao, Dr Feifeng Liu, Dr Qilei Zhang, Dr Fabrizio Santi, Dr Debora Pastina, Dr Zhou Hong and Dr Zhangfan Zeng, for their invaluable contributions to the long-term research on GNSS-based SAR reported here.

List of Acronyms

AltBOC	alternative binary offset carrier
BASS	block adjustment of synchronizing signal
BPA	back-projection algorithm
BPSK	binary phase-shift keying
BSAR	bistatic synthetic aperture radar
C/A-code	coarse acquisition code
CCD	coherent change detection
CDMA	code division multiple access
DAB	digital audio broadcasting
DFT	discrete Fourier transform
DVB-T	digital video broadcasting – terrestrial
FDMA	frequency division multiple access
FFT	fast Fourier transform
GAF	generalized ambiguity function
GLONASS	global navigation satellite system
GNSS	global navigation satellite systems
GPS	global positioning system
HC	heterodyne channel
ICD	interface control document
INS	inertial navigation system
MoComp	motion compensation
MPSF	multistatic point spread function
P-code	precision code
PSF	point spread function
RC	radar channel
RCS	radar cross-section

SNR	signal-to-noise ratio
SAR	synthetic aperture radar
UK-DMC	satellite that formed part of the disaster-monitoring constellation

References

[1] Martin-Neira, M.: 'A passive reflectometry and interferometry system (PARIS): Application to ocean altimetry', *ESA J.*, 1993, 17, pp. 331–355.

[2] Martin-Neira, M., Caparrini, M., Font-Rosello, J., Lannelongue, S., Vallmitjana, C.S.: 'The PARIS concept: An experimental demonstration of sea surface altimetry using GPS reflected signals', *IEEE Trans. Geosci. Remote Sens.*, 2001, 39, (1), pp. 142–150.

[3] Zavorotny, V.U. and Voronovich, A.G.: 'Scattering of GPS signals from the ocean with wind remote sensing application', *IEEE Trans. Geosci. Remote Sens.*, 2000, 38, pp. 951–964.

[4] Rodriguez-Alvarez, N., Bosch-Lluis, X., Camps, A. *et al.*: 'Soil moisture retrieval using GNSS-R techniques: Experimental results over a bare soil field', *IEEE Trans. Geosci. Remote Sens.*, 2009, 47, (1), pp. 3616–3624.

[5] Komjathy, A., Maslanik, J., Zavorotny, V.U., Axelrad, P. and Katzberg, S.J.: 'Sea ice remote sensing using surface reflected GPS signals', *IEEE IGARSS*, 2000, 7, pp. 2855–2857.

[6] Clarizia, M.P., Gommenginger, C., Gleason, S., Galdi, C. and Unwin, M.: 'GNSS-R from the UK-DMC satellite for remote sensing of the ocean surface', *IEEE IGARSS*, 2008, 1, pp. I-276–I-279.

[7] Gleason, S. and Gebre-Egziabher, D. (Eds): *GNSS-Applications and Methods* (Artech House, Norwood, MA, 2009).

[8] Jin, S., Cardellach, E. and Xie, F.: *GNSS Remote Sensing: Theory, Methods and Applications* (Springer, Netherlands, 2013).

[9] Soumekh, M.: *Synthetic Aperture Radar Signal Processing with MATLAB Algorithms* (John Wiley & Sons, Inc., New York, 1999).

[10] Carrara, W.G., Goodman, R.S. and Majewski, R.M.: *Spotlight Synthetic Aperture Radar* (Artech House, Norwood, MA, 1995).

[11] Cumming, I.G. and Wong, F.H.: *Digital Processing of Synthetic Aperture Radar Data* (Artech House, Norwood, MA, 2005).

[12] Dubois-Fernandez, P., Cantalloube, H., Vaizan, B. *et al.*: 'ONERA-DLR bistatic SAR campaign: Planning, data acquisition, and first analysis of bistatic scattering behaviour of natural and urban targets', *IET Proc. Radar, Sonar Navig.*, 2006, 153, (3), pp. 214–223.

[13] Rodriguez-Cassola, M., Prats, P., Schulze, D. *et al.*: 'First bistatic spaceborne SAR experiments with Tandem-X', *IEEE Geosci. Remote Sens. Lett.*, 2012, 9, (1), pp. 33–37.

[14] Moreira, M., Krieger, G., Hajnsek, I. *et al.*: 'Tandem-L: A highly innovative bistatic SAR mission for global observation of dynamic processes on the Earth's surface', *IEEE Geosci. Remote Sens. Mag.*, 2015, 3, (2), pp. 8–23.

[15] Wang, R., Loffeld, O., Nies, H. and Ender, J.H.G.: 'Focusing spaceborne/airborne hybrid bistatic SAR data using wavenumber-domain algorithm', *IEEE Trans. Geosci. Remote Sens.*, 2009, 47, (7), pp. 2275–2283.
[16] Prati, C., Rocca, F., Giancola, D. and Monti-Guarnieri, A.: 'Passive geosynchronous SAR system reusing backscattered digital audio broadcasting signals', *IEEE Trans. Geosci. Remote Sens.*, 1998, 36, (6), pp. 1973–1976.
[17] Cazzani, L., Colesanti, C., Leva, D. *et al.*: 'A ground-based parasitic SAR experiment', *IEEE Trans. Geosci. Remote Sens.*, 2000, 38, (5), pp. 2132–2141.
[18] Martinsek, D. and Goldstein, R.: 'Bistatic radar experiment', *EUSAR*, 1998, pp. 31–34.
[19] Rodriguez-Cassola, M., Baumgartner, S.V., Krieger, G., Moreira, A.: 'Bistatic TerraSAR-X/F-SAR spaceborne–airborne SAR experiment: Description, data processing, and results', *IEEE Trans. Geosci. Remote Sens.*, 2010, 48, (2), pp. 781–794.
[20] Walterscheid, I., Espeter, T., Brenner, A.R. *et al.*: 'Bistatic SAR experiments with PAMIR and TerraSAR-X: Setup, processing, and image results', *IEEE Trans. Geosci. Remote Sens.*, 2010, 48, (8), pp. 3268–3279.
[21] Wang, R., Deng, Y., Zhang, Z. *et al.*: 'Double-channel bistatic SAR system with spaceborne illuminator for 2-D and 3-D SAR remote sensing', *IEEE Trans. Geosci. Remote Sens.*, 2013, 51, (8), pp. 4496–4507.
[22] Zeng, T., Wang, R., Li, F. and Long, T.: 'A modified nonlinear chirp scaling algorithm for spaceborne/stationary bistatic SAR based on series reversion', *IEEE Trans. Geosci. Remote Sens.*, 2013, 51, (5), pp. 3108–3118.
[23] Behner, F. and Reuter, S.: 'HITCHHIKER-hybrid bistatic high resolution SAR experiment using a stationary receiver and TerraSAR-X transmitter', *EUSAR*, 2010, pp. 1030–1033.
[24] Russian Institute of Space Device Engineering: 'Global Navigation Satellite System (GLONASS) Interface Control Document', Edition 5.1, 2008.
[25] China Satellite Navigation Office: 'BeiDou Navigation Satellite System Signal In Space Interface Control Document, Open Service Signal', 2013.
[26] Global Positioning Systems Wing (Systems Engineering and Integration): 'Navstar GPS Space Segment/Navigation User Interfaces', IS-GPS-200, Rev. E, 2010.
[27] European GNSS Open Service Signal-in-Space Interface Control Document (OS SIS ICD), Issue 1.1, September 2010.
[28] Lennen, G.R.: 'The USSR's GLONASS P-code Determination and Initial Results', ION GPS, 1989, pp. 77–83.
[29] Ma, H., Antoniou, M. and Cherniakov, M.: 'Passive GNSS-based SAR resolution improvement using joint Galileo E5 signals', *IEEE Geosci. Remote Sens. Lett.*, 2015, 12, (8), pp. 1640–1644.
[30] Ma, H., Antoniou, M. and Cherniakov, M.: 'Passive GNSS-based SAR imaging and opportunities using Galileo E5 signals', *Sci. China Inf. Sci.*, 2015, 58, (1), pp. 1–11.
[31] Zuo, R.: 'Bistatic synthetic aperture radar using GNSS as transmitters of opportunity'. Ph.D. thesis, University of Birmingham, UK, 2012.

[32] He, X., Zeng, T. and Cherniakov, M.: 'Signal detectability in SS-BSAR with GNSS non-cooperative transmitter', *IET Proc. Radar, Sonar Navig.*, 2005, 152, pp. 124–132.

[33] Rodon, J.R., Broquetas, A., Monti-Guarnieri, A., and Rocca, F.: 'Geosynchronous SAR focusing with atmospheric phase screen retrieval and compensation', *IEEE Trans. Geosci. Remote Sens.*, 2013, 51, (8), pp. 4397–4404.

[34] Zeng, T., Cherniakov, M. and Long, T.: 'Generalized approach to resolution analysis in BSAR', *IEEE Trans. Aerosp. Electron. Syst.*, 2005, 41, (2), pp. 461–474.

[35] Cherniakov, M. (Ed.): *Bistatic Radar-Emerging Technology* (Wiley, Chichester, UK, 2008, pp. 228–243).

[36] Liu, F., Antoniou, M., Zeng, Z. and Cherniakov, M.: 'Coherent change detection using passive GNSS-based BSAR: Experimental proof of concept', *IEEE Trans. Geosci. Remote Sens.*, 2013, 51, (8), pp. 4544–4555.

[37] Willis, N.J.: *Bistatic Radar* (Artech House, Boston, MA, 1991, p. 131).

[38] Liu, F., Antoniou, M., Zeng, Z. and Cherniakov, M.: 'Point spread function analysis for BSAR with GNSS transmitters and long dwell times: Theory and experimental confirmation', *IEEE Geosci. Remote Sens. Lett.*, 2013, 10, (4), pp. 781–785.

[39] Antoniou, M. and Cherniakov, M.: 'GNSS-based bistatic SAR: A signal processing view', *EURASIP J. Adv. Sig. Proc.*, 2013, 98, available at http://asp.eurasipjournals.com/content/2013/1/98.

[40] Younis, M., Metzig, R. and Krieger, G.: 'Performance prediction of a phase synchronisation link for bistatic SAR', *IEEE Geosci. Remote Sens. Lett.*, 2006, 3, (3), pp. 429–433.

[41] Lopez-Dekker, P., Mallorqui, J.J., Serra-Morales, P. and Sanz-Marcos, J.: 'Phase synchronisation and Doppler centroid estimation in fixed receiver bistatic SAR systems', *IEEE Trans. Geosci. Remote Sens.*, 2008, 46, (11), pp. 3459–3471.

[42] Tsui, J. B-Y.: *Fundamentals of Global Positioning System Receivers – A Software Approach* (John Wiley & Sons, Inc., 2005, pp. 129–185).

[43] Borre, K., Akos, D.M., Bertelsen, N., Rinder, P., and Jensen, S.H.: *A Software-Defined GPS and Galileo Receiver – A Single Frequency Approach* (Birkhauser, Boston, 2007).

[44] Pany, T.: *Navigation Signal Processing for GNSS Software Receivers* (Artech House, Norwood, MA, 2010).

[45] Wong, F.H., Cumming, I.G. and Neo, Y.L.: 'Focusing bistatic SAR data using the nonlinear chirp scaling algorithm', *IEEE Trans. Geosci. Remote Sens.*, 2008, 46, (9), pp. 2493–2505.

[46] Neo, Y.L., Wong, F.H. and Cumming, I.G.: 'Processing of azimuth-invariant bistatic SAR data using the range-doppler algorithm', *IEEE Trans. Geosci. Remote Sens.*, 2008, 46, (1), pp. 14–21.

[47] Wu, J., Li, Z., Huang, Y., Yang, J. and Liu, Q.H.: 'An omega-k algorithm for translational invariant bistatic SAR based on generalised Loffeld's bistatic formula', *IEEE Trans. Geosci. Remote Sens.*, 2014, 52, (10), pp. 6699–6714.

[48] Walterscheid, I., Ender, J.H.G., Brenner, A.R. and Loffeld, O.: 'Bistatic SAR processing and experiments', *IEEE Trans. Geosci. Remote Sens.*, 2006, 44, (10), pp. 2710–2717.

[49] Wang, R., Loffeld, O., Nies, H. *et al.*: 'Focusing bistatic SAR data in airborne/stationary configuration', *IEEE Trans. Geosci. Remote Sens.*, 2010, 48, (1), pp. 452–464.

[50] Rigling, B.D. and Moses, R.L.: 'Polar format algorithm for bistatic SAR', *IEEE Trans. Aerosp. Electron. Syst.*, 2004, 40, (4), pp. 1147–1159.

[51] Antoniou, M., Cherniakov, M. and Hu, C.: 'Space-surface bistatic SAR image formation algorithms', *IEEE Trans. Geosci. Remote Sens.*, 47, (6), pp. 1827–1843.

[52] Rodriguez-Cassola, M., Prats, P., Krieger, G. and Moreira, A.: 'Efficient time-domain image formation with precise topography accomodation for general bistatic SAR configurations', *IEEE Trans. Aerosp. Electron. Syst.*, 2011, 47, (4), pp. 2949–2966.

[53] Shao, Y.F., Wang, R., Deng, Y. *et al.*: 'Fast backprojection algorithm for bistatic SAR imaging', *IEEE Trans. Geosci. Remote Sens.*, 2013, 10, (5), pp. 1080–1084.

[54] Antoniou, M., Zhou, H., Zeng, Z., Zuo, R., Zhang, Q. and Cherniakov, M.: 'Passive bistatic synthetic aperture radar imaging with Galileo transmitters and a moving receiver: Experimental demonstration', *IET Proc. Radar, Sonar Navig.*, 2013, 7, (9), pp. 985–993.

[55] Preiss, M., Gray, D. and Stacy, N.J.: 'Detecting scene changes using synthetic aperture radar interferometry', *IEEE Trans. Geosci. Remote Sens.*, 2006, 44, (8), pp. 2041–2054.

[56] Preiss, M. and Stacy, N.J.: 'Coherent change detection: Theoretical description and experimental results', DSTO Technical Report, report no. DSTO-TR-1851, August 2006.

[57] Rignot, E.J. and van Zyl, J.J.: 'Change detection techniques for ERS-1 SAR data', *IEEE Trans. Geosci. Remote Sens.*, 1993, 31, (4), pp. 896–906.

[58] Zhang, Q., Antoniou, M., Chang, W. and Cherniakov, M.: 'Spatial decorrelation in GNSS-based SAR coherent change detection', *IEEE Trans. Geosci. Remote Sens.*, 2015, 53, (1), pp. 219–228.

[59] Zeng, Z., Antoniou, M., Zhang, Q. and Cherniakov, M.: 'Multi-perspective GNSS-based passive BSAR: Preliminary experimental results', *International Radar Symposium (IRS)*, 2013, pp. 467–472.

[60] Zhang, T., Tian, W., Zeng, T. and Hu, C.: 'Multi-angle fusion of SS-Bi SAR images using Compass-2/Beidou-2 satellites as opportunity illuminators', *IEEE International Conference on Radar*, 2014.

[61] Santi, F., Antoniou, M. and Pastina, D.: 'Point spread function analysis for GNSS-based multistatic SAR', *IEEE Geosci. Remote Sens. Lett.*, 2015, 12, (2), pp. 304–308.

Chapter 17
Airborne passive radar

Krzysztof Kulpa[1], *Damian Gromek*[1] *and Bartek Dawidowicz*[1]

Abstract

Passive radar is one of the most rapidly developing fields in the radar technology in recent years. The ground-based passive radar technology is now entering a stage of maturity. In the past, in a case of active radars, the technology developed for ground-based sensing was adapted for airborne platforms. The same trend has been observed in the passive radar technology as it is being adapted for moving platforms, mostly the airborne ones. Major aims of this adaptation include, among others, protection of the platforms, detection of airborne and surface moving targets and remote sensing, including SAR and ISAR imaging. The use of mobile PCL systems provides an extended functionality in comparison to the one of the stationary ground-based radars.

The chapter consists of two key sections. The first section presents the passive imaging technique based on a bistatic SAR concept, the second section discusses the main challenges for the airborne passive radars, in particular a cancelation of Doppler spread clutter, and also presents solutions based on the CLEAN techniques, DPCA and space–time adaptive processing. All theoretical considerations are illustrated with the simulation and experimental validation examples.

Passive radar has been one of the most rapidly developing fields in radar technology in recent years. The low-cost nature of passive radar, known also as passive coherent location (PCL), has resulted in strong interest from a significant number of companies and research institutions. As a result of this research, many passive radar demonstrators and commercial products have been developed [1–7].

Existing passive radar systems utilize various different types of signals for target illumination [8]. The most popular signals used by modern passive demonstrators are as follows: FM radio [2,4,6], Digital Audio Broadcasting (DAB) radio [3,9–12], analogue and digital television (DVB-T) [13–16], Global System for

[1]Warsaw University of Technology, Institute of Electronic Systems (ISE), Poland

Mobile Communications, originally Groupe Spéciale Mobile (GSM) networks [17–19] and WiFi signals [20–25].

Most of the known variants of passive radar are stationary, ground-based systems dedicated to the detection and tracking of airborne targets. The last decade of PCL systems development has led this technology towards a state of relative maturity for ground-based operation. These developments have prompted the research conducted for this study, which will focus specifically on passive radar on a moving platform. In the literature available on this subject, initial analyses devoted to PCL mounted on ground-moving vehicles and airborne platforms can be found [26–33] and as for ground-based radar, proper waveform selection is essential [34]. Theoretical analysis presented in open publications suggests that PCL on a moving platform is a feasible concept. The use of mobile PCL systems provides extended functionality in comparison to the stationary ground-based radar. A moving passive radar sensor also provides modes such as ground moving target indication [35,36] and imaging [37–41], synthetic aperture radar (SAR) [42–44] or inverse synthetic aperture radar (ISAR) [45,46].

In classical ground-based passive radar, the scene is illuminated by a stationary transmitter (TX), and the ground returns (ground clutter) have zero Doppler frequency. The moving targets of interest have non-zero Doppler frequency and thus can easily be distinguished from the ground clutter. In the case when the passive radar receiver (RX) is placed on an airborne platform, the echo coming from no-moving ground objects (ground clutter) possess no-zero Doppler frequency. This is a result of the relative velocity between the mobile radar platform and stationary objects on the ground. That Doppler frequency and its change as a function of time can be exploited in passive synthetic aperture imaging of the ground. But in the case when our goal is to detect moving targets, it is more difficult to distinguish between targets of interest and ground clutter. In the case when a target of velocity greater than the radar platform velocity is of interest, the Doppler frequency of the target is out of the range of Doppler frequencies of the ground clutter, and fast targets can be detected after Doppler spread clutter cancellation. But if we are interested in detecting slower targets, the Doppler frequency of the reflections from moving objects is within the clutter bandwidth. Moreover, they are weaker than the ground returns. The objective of the signal processing is to suppress ground clutter, while preserving the echoes from moving objects. This problem is well known in the classic airborne pulse Doppler radars. Typical solutions are based on space-time adaptive processing (STAP)- or displaced phase centre antenna (DPCA)-type algorithms. However, direct implementation of these algorithms is not possible in passive radars, while they were developed for monostatic configuration of pulse-Doppler airborne radars.

In this chapter, the passive SAR image formation methods will be shown in Section 17.1, and moving target detection and ground clutter cancellation will be shown in Section 17.2.

17.1 Airborne passive synthetic aperture radar

The most popular radar technique for Earth surface imaging is the SAR. Since 1951 when Carl Wiley from the Goodyear Aircraft Corporation proposed this technique [47,48], SAR has been intensively developed, and many new concepts of image formation have been proposed and demonstrated [49,50]. Nowadays, SAR is at the stage of technological maturity. SAR is widely used as a sensor on satellites, aircraft and unmanned aerial vehicles (UAVs). The active SAR works in the monostatic configuration where the TX and RX compound one radar device. The bistatic and multistatic configuration has been also intensively studied in last two decades, and numerous papers have been published on bistatic SAR [51,52]. At the present time, active bistatic SAR is fully operational – Tandem X (TerraSAR-X add-on for Digital Elevation Measurement), twin satellite mission, has been launched in 2010 and now high-quality 3D images of the Earth are readily available [53,54].

The bistatic configuration corresponds to significant separation between the TX and the RX. The bistatic configuration can provide different information from the monostatic, since the target is observed from different bistatic angles. Even more interesting is the multistatic approach, when the target is observed from a number of different angles. The number of theoretical consideration and experiments has been conducted on bistatic SAR, but most of them have been dedicated to cooperative or non-cooperative pulse radar illumination [55,56] or satellite illumination [57–67].

An interesting alternative to the active SAR system is a passive one, when the target or the scene is illuminated by illuminators of opportunity. In a similar way to active radar, passive radars were first used in a ground-based configuration for air target detection. The first passive radar experiment was conducted over 70 years ago. In 1935, Arnold Wilkins and Robert Watson Watt demonstrated detection of a bomber aircraft in Daventry, United Kingdom, based only on the illumination from a BBC short-wave broadcast station. And this was not the first time that passive sensing had been used. In 1924, Appleton and Barnett conducted an experiment to measure the height of the ionosphere using a BBC medium wave broadcast TX (770 kHz), located at Bournemouth on the south coast of England. The RX was placed in at Oxford at a distance of some 120 km [68].

At that time only, analogue signal processing was feasible, and full implementation of passive radar using continuous wave signals was impossible. Passive technology was then forgotten for several decades. Nowadays, passive technology is being rediscovered [69–74] due in part to the rapid development of digital technology and computational power. The modern technology allows the use of existing emissions (e.g. FM radio, mobile communication GSM, digital terestial televioion DVB-T and satelite television DVB-S, among others) for object detection. The fundamental advantage of passive technology is the lack of any dedicated emission, so it is impossible to detect a passive radar based on radar emission, and there is no need for any frequency allocation or transmission licence. Passive SAR technology is thus a 'green' technology which does not pollute the electromagnetic environment.

The combination of the three technologies, namely SAR, bistatic (or multistatic) radar and passive radar, leads to the concept of passive airborne SAR (PASAR) – an airborne radar that uses ground-based TXs of opportunity for illumination. A good PASAR illuminator should have high power density at the ground, high bandwidth and favourable ambiguity function performance to ensure

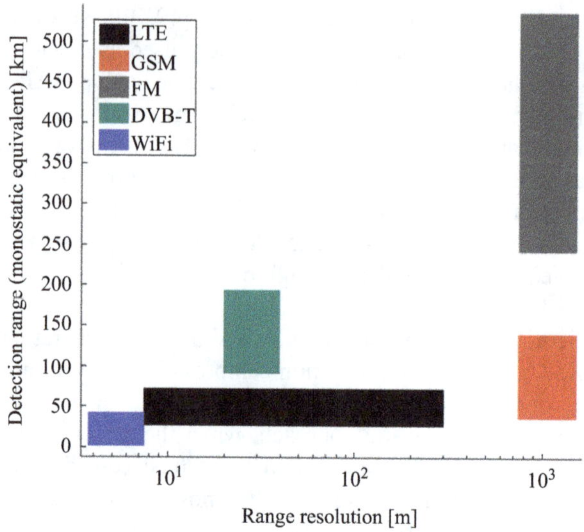

Figure 17.1 The range-resolution map for commercial signals used in passive radar

good quality of SAR images. There are plenty of commercial illuminators (FM, GSM, long-term evolution cellular system (LTE), DVB-T, see Figure 17.1) which can be used for illumination of the ground for SAR purposes [75]. The best candidate for passive SAR imaging purposes seems to be the DVB-T TX, which has relatively high power (10–1,000 kW) and reasonable bandwidth (~7.6 MHz for a single DVB-T channel), giving a good power budget and fine resolution of up to 20 m [76,77]. In many countries, it is possible to find illumination of a wider bandwidth, while multiple channels are often transmitted from the same location. This can improve the range resolution of the passive SAR. But also other illuminators, such as WiMax [78–80], have been taken into consideration.

In the literature, one can find descriptions of such concepts [76,81,82] and theoretical analyses [81–89], but up to now only few preliminary results of experimental trials have been published in the open literature [74,76,90,91]. In [76], the authors present a range-Doppler map of the observed ground area obtained using a DVB-T-based passive radar. Such a method is similar to unfocused SAR processing. In [74,91] are the results of single channel passive SAR processing obtained from the CARABAS SAR system working in a passive mode. The signal processing was based on the autocorrelation function of the received signal. In single antenna airborne SAR, it is also possible to reconstruct the transmitted signal [92] and use the presented passive SAR image formation method based on cross-correlation of reference and surveillance signals.

The considered PASAR radiolocation scenario is depicted in Figure 17.2. The TX of opportunity illuminates the scene, and also an aircraft. The passive radar, placed on board the aircraft consists of two or more antennas and two or more receiving channels. One antenna is directed towards the illuminator (TX of

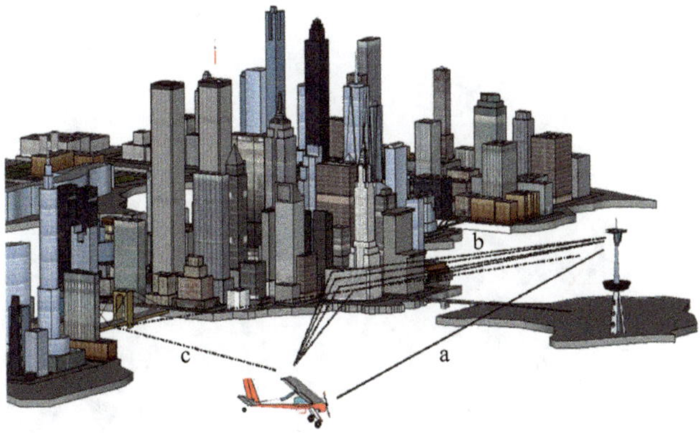

Figure 17.2 Airborne passive SAR concept and its bistatic geometry

opportunity) and receives the reference channel signal. The second antenna is directed towards the observed scene and receives the echo signal (indirect path signal) reflected from the buildings, ground surface and all objects illuminated by the TX and are visible to this antenna. The signals from both channels – reference and imaging – are converted into digital form using ADC converters and passed to the processing unit. It is assumed that both channels are fully synchronized in frequency and phase. The synchronization is not required between the passive radar and the DVB-T TX, since the SAR image is obtained by correlating the signal from the reference and imaging channels.

The received signal consist of four major components: the direct path signal, reflected echoes from stationary targets (multipath signal), echoes from moving objects and noise. The main task of the PASAR radar is ground image formation, so the echoes from moving objects can be treated as a clutter. Due to velocity mismatch moving targets in the final passive SAR, image will be defocused and appear blurred in the image. It is also possible to focus moving targets, but moving target imaging is outside of the scope of this chapter.

The signal received by the antennas can be expressed by the following formula:

$$s_{\mathrm{ref/sur}}(t) = s_a(t) + \sum_k s_{bk}(t) + \xi(t) \tag{17.1}$$

where $s_a(t)$ is the direct path signal, $s_{bk}(t)$ is the echo reflected from kth stationary object and $\xi(t)$ is the Gaussian additive noise.

In the ideal case, the reference antenna receives only the direct signal, and the imaging (surveillance) antenna only echoes from the imaged area. Thermal and ambient noise is present in both channels. In practice, both antennas receive all the above signals, but the amplitudes of the components are significantly different, and the received amplitudes depend on directional gain of antenna in the directions of arrival of the signals.

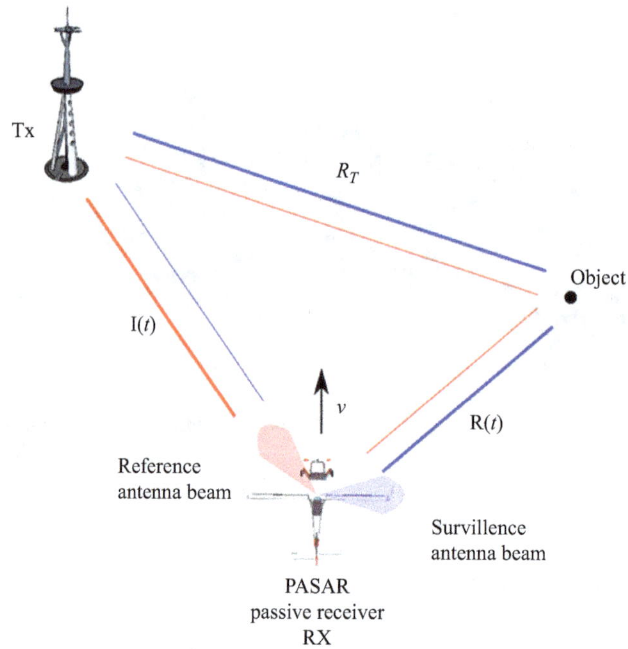

Figure 17.3 The two channel (antenna) configuration of PASAR radar. The blue line represents signal paths which are received by the surveillance channel. The red line represents signal paths which are received by the reference antenna. The thickness of the line represents the amplitude of the signal component

Due to the two receiving channels configuration of the PASAR and the difference in antenna gain of the surveillance and reference antennas in the directions of arrival of each signal component, the amplitudes of the components $(s_{bk}(t)s_a(t))$ will differ in each receiving channel (reference and surveillance). The two antenna configuration of PASAR radar is shown in Figure 17.3.

In most practical cases, the illuminating signal bandwidth B is much smaller than the carrier frequency f_c. In such a case, it is possible to apply the narrowband model of the illuminating signal represented by the following formula:

$$s_T(t) = s(t)\exp\{j2\pi f_c t + j\varphi\} \tag{17.2}$$

where $s(t)$ is the baseband illuminating signal, f_c is the carrier frequency (e.g. DVB-T channel carrier frequency) and φ is the initial phase of the signal.

Considering only the direct path component in the reference channel, the radio-frequency (RF) reference signal can be represented by the following formula:

$$s_{\text{ref}}(t) = As_T(t - \tau_l(t)) = As(t - \tau_l(t))\exp\{j2\pi f_c(t - \tau_l(t)) + j\varphi\} \tag{17.3}$$

where $\tau_l(t)$ is the time delay related to distance between the moving PASAR radar RX and DVB-T illuminator $= l(t)/c$.

The baseband signal in the reference channel (after quadrature demodulation) is given by:

$$s_{\text{ref\_IF}}(t) = As(t - \tau_l(t))\exp\{-j2\pi\tau_l(t) + j\varphi\} \tag{17.4}$$

Similar expressions can be provided for the surveillance signal. The surveillance signal after downconversion to baseband is given by:

$$s_{\text{sur\_IF}}(t) = A_k s(t - \tau_k(t))\exp\{-j2\pi\tau_k(t) + j\varphi\} \tag{17.5}$$

where $\tau_k(t)$ is the time delay related to sum of distances of the kth stationary object to the PASAR radar RX and to the DVB-T illuminator $\tau_k(t) = (R_{T,k} + r(t)_k)/c$.

The passive bistatic SAR image can be formed using the back-propagation algorithm. This algorithm is based on correlation of the received signal with that modelled for each potential scatterer in the imaging plane. Let us consider that the objective is to make a 2-D PASAR image of the Earth's surface, and all scattering points are placed on the XY plane. For each pixel of the final image, it is possible to calculate the complex intensity of the image by computing the scalar product of the received signal and the modelled echo signal. In theory, the modelled echo signal can be created using the illuminating signal and applying the time delay related to the selected scatterer. In practice, the passive SAR radar does not have direct access to the illuminating signal but can use the reference channel signal to model the selected echo signal. Taking into account that the reference signal is delayed by $\tau_l(t)$ and the echo signal from the kth scatterer placed at position (x, y) is delayed by $\tau_k(t)$, the modelled echo signal can be expressed as $s_{\text{ref\_IF}}(t - \tau(t))\exp\{j2\pi\tau(t)\}$ where

$$\tau(t) = \tau_k(t) - \tau_l(t) = \frac{(R_{T,k} + r(t)_k - l(t))}{c} \tag{17.6}$$

The final SAR image can thus be calculated as the scalar product in the form

$$\text{SAR}_{(x,y)} = \frac{1}{T}\int_{-T/2}^{T/2} s_{\text{sur\_IF}}(t)s^*_{\text{ref\_IF}}(t - \tau(t))\exp\{j2\pi\tau(t)\}dt \tag{17.7}$$

where T is the observation time. In most cases, the signal is in digital form, so we can use summation instead of the integral.

Substituting (17.4)–(17.6) into (17.7), finally the SAR pixel value will be given by:

$$\text{SAR}_{(x,y)} = \frac{AA_k}{T}\int_{-T/2}^{T/2} s^2(t - \tau_k(t))dt \tag{17.8}$$

The direct use of (17.7) is computationally expensive. Assuming a DVT signal with a sampling frequency of the order of 10 MHz and an integration time of 10–100 s, it is necessary to integrate 100–1,000 million samples for each PASAR pixel. To obtain a 1-M pixel image, it is necessary to perform 10^{14}–10^{15} complex operations (not counting delay computation). Assuming an available computational power of the order of 10 GFlops, the time required to compute the whole image will be 10^4–10^5 s (3–30 h). As real time applications are of great interest to the user, significant simplification and computational power reductions are therefore required.

To simplify the algorithm, single integration over several seconds can be divided into sum of short intervals (blocks), for which the phase change introduced by the signal delay is almost constant (smaller than $\pi/4$). Under such assumptions, (17.6) can be re-written as follows:

$$\mathrm{SAR}_{(x,y)} = \frac{1}{\mathrm{MT_{PRF}}} \sum_{u=-M/2}^{M/2-1} \exp\{j2\pi\tau(uT_{\mathrm{PRF}})\} \int_{uT_{\mathrm{PRF}}}^{(u+1)T_{\mathrm{PRF}}} s_{\mathrm{sur\_IF}}(t) s^*_{\mathrm{ref\_IF}}(t - \tau(uT_{\mathrm{PRF}}))dt \tag{17.9}$$

where M is the number of blocks (even number) and $\mathrm{MT_{PRF}} = T$.

In the case of an illuminating signal of bandwidth B at carrier frequency f_c and an airborne passive radar platform which moves with constant velocity v, the maximum integration time is given by:

$$T_{\mathrm{PRF(MAX)}} = \frac{c}{8vf_c} \tag{17.10}$$

For example, for an aircraft velocity of 50 m/s and a carrier frequency of 750 MHz, the maximum integration time $T_{\mathrm{PRF(MAX)}} = 1$ ms. The other requirement on T_{PRF} is that the range migration within the integration time is smaller than a single resolution cell. This leads to the condition that $T_{\mathrm{PRF}} \ll (c/8vB)$. Since the narrowband signal model is used ($B \ll f_c$), the condition in (17.10) is more stringent ($T_{\mathrm{PRF}} = (c/8vf_c) \ll (c/8vB)$).

Based on (17.9), the signal-processing chain can be divided into two main steps: the first step is calculation of the cross-correlation function between the surveillance and reference signals. In this step, so-called range profiles are computed. This processing step is performed by computing the following equation:

$$s_{\mathrm{corr}}(u, \tau) = \int_{uT_{\mathrm{PRF}}}^{(u+1)T_{\mathrm{PRF}}} s_{\mathrm{sur\_IF}}(t) s^*_{\mathrm{ref\_IF}}(t - \tau)dt \tag{17.11}$$

The block size of signals used in (17.11) is relatively small – typically 8,000 samples – and only a limited number of time delays, related to the final PASAR image range size, have to be computed. But still direct use of (17.11) is computationally expensive. To reduce the computational costs, a fast convolution algorithm based on the fast Fourier transform (FFT) may be used.

After computing the range data, one can use the simplified back-projection algorithm or apply classical bistatic SAR processing based on range migration compensation and cross-range compression [14,93,94].

Movement of the passive radar RX produces different Doppler frequencies for each ground object separated in the azimuthal direction. The plot of the range

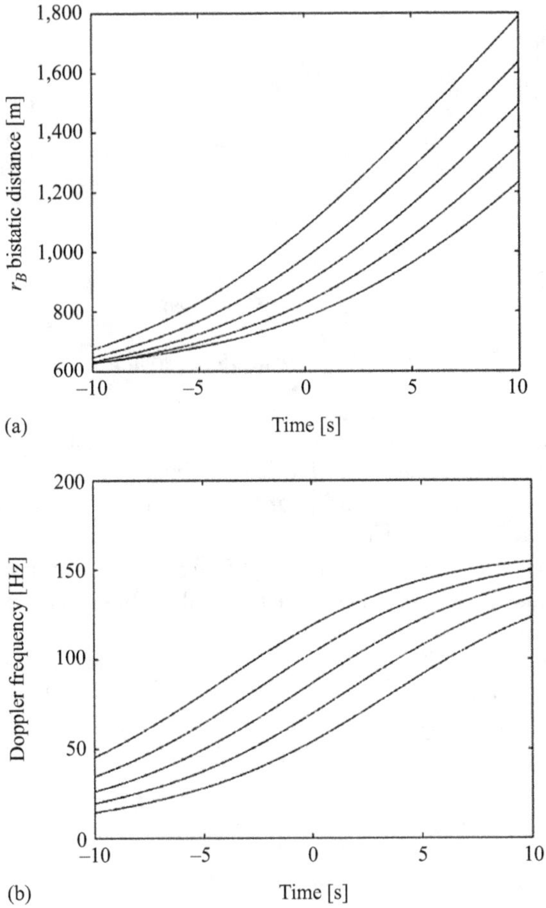

Figure 17.4 (a) Bistatic distance between the transmitter-object-airborne receiver and (b) Doppler frequency for the object

(also phase) and Doppler history for a few single ground point scatterers is presented in Figure 17.4(a) and (b). The geometry under consideration is shown in Figure 17.5. The centre of the DVB-T channel frequency was 600 MHz, simulation time 20 s, the speed of the platform was constant and set to 50 m/s. Point scatterers were separated in the azimuthal direction by 100 m.

As can be seen from Figure 17.4(b), in many cases, the Doppler frequency history for the target has a linear form. The signal in the azimuthal direction for a single scatterer can be treated as a Linear Frequency Modulated signal. The other issue which has direct impact on signal-processing scheme is strong range migration, which always occurs in PASAR. In comparison to active SAR radars, the range migration algorithms always have to be applied in passive SAR radar. The signal processing has to take into account both effects. As a result, matched

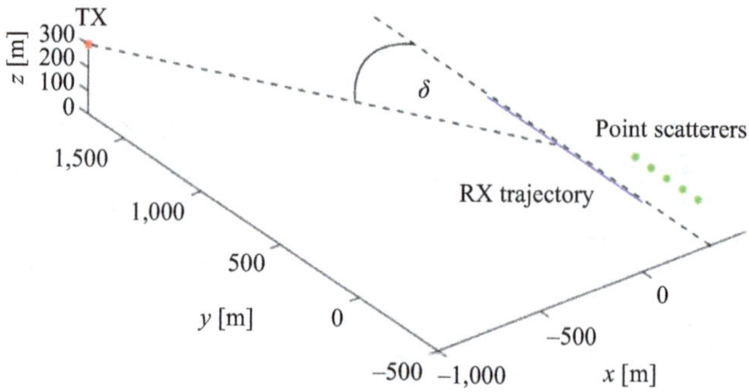

Figure 17.5 Simulation geometry

filtering in the azimuthal direction produces a focused passive SAR image. From the knowledge of the position of the TX and the trajectory of the passive radar, the position of a ground-based stationary target can also be determined. One of the methods for final SAR image creation can be the time domain back-projection algorithm [94] on the artificially created matrix of raw data. In the back-projection algorithm, the final SAR image is obtained by calculating each pixel according to the equation:

$$\mathrm{SAR}(x,y) = \int s_{\mathrm{corr}}\left(u, \tau_{(x,y)}(u)\right) * e^{-j\varphi_{(x,y)}(u)} du \tag{17.12}$$

where

$$\tau_{(x,y)}(i) = \frac{r_{B(x,y)}(u)}{c} \tag{17.13}$$

$$\varphi_{(x,y)}(u) = \frac{2\pi}{\lambda} r_{B(x,y)}(u) \tag{17.14}$$

$$r_{B(x,y)}(u) = R_{T(x,y)} + r_{R(x,y)}(u) - l(u) \tag{17.15}$$

$t_{(x,y)}(u)$ is the time delay of the echo in the ith artificially created pulse $s_{\mathrm{corr}}(u)$ from the point scatterer placed on the flat ground scene at coordinates (x,y) – the flat Earth assumption, and $\varphi_{(x,y)}(u)$ represents the signal phase for that single scatterer. The time delay $t_{(x,y)}(u)$ and signal phase $\varphi_{(x,y)}(u)$ are determined by the bistatic distance $r_{B(x,y)}(u)$.

17.1.1 Signal dynamic range considerations

One of the typical problems of passive radar is the high signal dynamic range. In active pulsed radar, gain control as a function of range is applied to overcome this problem. In passive radar, the strong direct signal and the weak target echoes are received simultaneously, so the RX has to have a very high dynamic range.

The reference and surveillance signal can be received by a single channel. The single channel passive SAR image were shown by Swedish engineers [74,91],

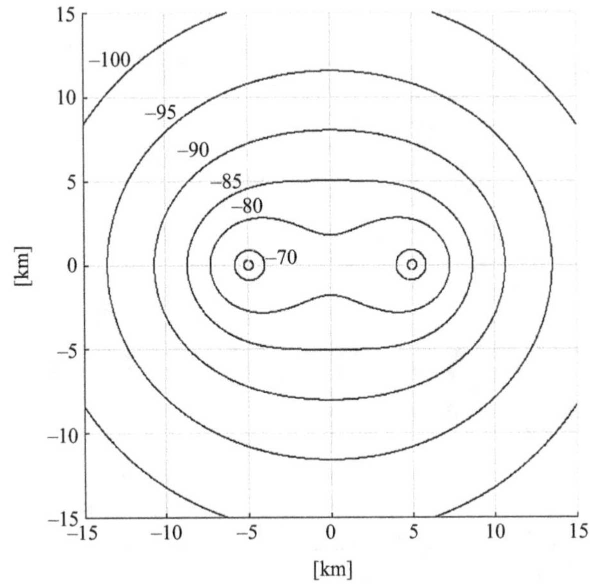

Figure 17.6 Constant signal-to-interference ratio (SIR) map

where signals were collected using the single channel RX of the CARABAS system.

But the requirement for the dynamic range of a single channel passive radar RX is very high – at least 100 dB. A typical active radar dynamic range may be around 50–70 dB. One of the parameters which describes the bistatic passive radar is SIR – signal-to-interference ratio. The SIR is the ratio of the echo signal power and the power of all interfering signals. In our case, the strongest interfering signal is the reference (direct path signal). The SIR parameter determines the required dynamic range of the passive radar for detecting a target characterized by a given radar cross-section at a given distance. In Figure 17.6, the contours of constant SIR parameter for a single channel-passive radar are presented. The contours of constant SIR form so-called Cassini Ovals [95]. The SIR parameter was calculated for a baseline distance between the TX and RX of $L = 10$ km and a target bistatic radar cross-section of $\sigma = 1\ \text{m}^2$.

As can be seen, the SIRC levels are from -70 to -100 dB. That means that the target echo is much smaller than the direct signal and the masking effect is significant. In a two-antenna system, the situation is little better, since the surveillance antenna is usually not directed towards the TX. However, due to the small size of the antenna, the sidelobes are rather high – at a level of perhaps -10 to -20 dB. As a result, in a two-channel system the SIRC is of the order of -50 to -90 dB and the masking effect is still too high to be neglected. To obtain a good quality PASAR image, it is necessary to remove the direct path signal from the surveillance channel. One possible solution is to apply an analogue direct path canceller,

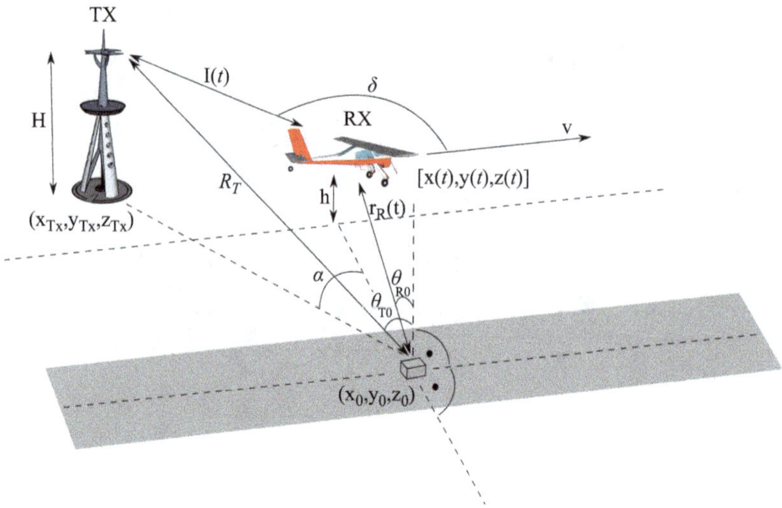

Figure 17.7 StripMap mode of operation in PASAR radar

but such a solution is difficult, expensive and can be not very stable. In case of reception of multiple radio channels, hardware have to be multiplied as well. Better results can be obtained by applying digital direct signal cancellation, such as the CLEAN algorithm [42,96]. The application of this procedure reduces the direct signal by 20–60 dB, which significantly decreases the masking effect and helps to improve the final passive SAR image.

17.1.2 Range and cross-range resolution in airborne passive SAR

Up to now, we have assumed that the platform can have arbitrary motion and that motion can be fully compensated using for example the back-projection algorithm. In most SAR cases straight line, constant velocity platform motion is taken into account, leading to significant simplifications and closed-form expressions for signal processing and resolution prediction.

Let us assume the StripMap mode of operation of a PASAR radar in the geometry presented in Figure 17.7.

The range resolution of the active SAR radar depends on the signal bandwidth – the wider the bandwidth, the better (finer) the range resolution. The range resolution can be expressed by the formula:

$$\Delta R = \frac{c}{2B} \tag{17.16}$$

On flat ground, the ground range resolution depends on the incidence angle of observation of the target and can be expressed by the formula:

$$\Delta R = \frac{c}{2B\sin(\theta_R)} \tag{17.17}$$

In the PASAR radar, the resolution also depends on the bandwidth of the signal. Due to the physical separation between the TX and the airborne RX, and the fact that only the RX is in motion (the TX is stationary), the ground range resolution in PASAR can be determined from the geometrical relation presented in Figure 17.6. The formula is as follows:

$$\Delta R \sim \frac{c}{B(\sin(\theta_R) + \sin(\theta_T)\cos(\alpha))} \tag{17.18}$$

The cross-range resolution in the azimuth direction is determined by the integration time in that direction. The cross-range resolution is depended on the bistatic geometry – angle δ (see Figure 17.6) and the parameter $k = L/R_R$, where L and R_R are the distances between the TX and the airborne RX and the target and the airborne RX at time $t = 0$, $L = l(0)$, $R_R = r_R(0)$.

When k tends to infinity, the cross-range resolution reaches its minimum value (best resolution):

$$\Delta R_{az} = D \tag{17.19}$$

where D is the dimension of the antenna aperture in the azimuth direction. It should be remembered that for the active SAR radar the cross-range resolution is a factor of two better and equal to $D/2$, and it is not dependent on geometry. This factor ½ is related to the motion of both TX and RX. In the passive SAR radar, considered here, the TX of opportunity is stationary and only the airborne passive RX is moving so the factor half is not present in (17.19).

Figures 17.7 and 17.8 present the plots of calculated ground range and cross-range resolution. It can be concluded that both resolutions strongly depend on the bistatic geometry. The other fact which comes from the range resolution plot (Figure 17.7) is that the imaging in passive SAR is not possible in the forward scatter geometry when the TX and imaging target/area are on the same side of the airborne RX. In the forward scatter geometry, the cross-range resolution still exists.

17.1.3 Airborne passive SAR experiments

The theoretical considerations presented in the previous chapters have been verified experimentally. The measurement campaign took place in December 2014. Measurements were made in the vicinity of the city of Sierpc, Poland. The DVB-T TX used as an illuminator of opportunity was located in Rachocin. The GPS coordinates of the TX are 52.89° N, 19.65° E. The parameters of the transmitted signal by the illuminator are as follows:

Carrier Frequency	618 MHz
EIRP (equivalent **i**sotropically **r**adiated **p**ower)	100 kW
Mast height above ground level	250 m
Signal bandwidth	7.8 MHz
Signal modulation	OFDM, 64QAM

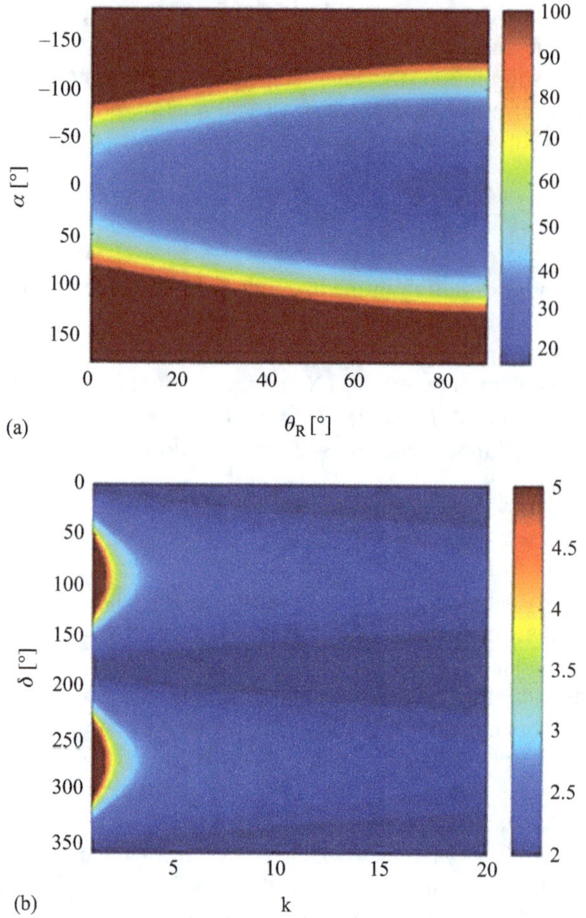

Figure 17.8 (a) ***Passive SAR resolutions****: ground range resolution, colour-bar values from 20 to 100 m, incidence angle* $\boldsymbol{\theta_T}$ *is constant and equal to 85°, (b) Azimuth resolution, colour-bar values from 2 to 5 m, physical length of the aperture = 2 m*

The measurement scenario of the experiment is presented in Figure 17.9. A Polish aircraft PZL-104 called 'Wilga' was used as an airborne passive radar carrier. Figure 17.10 presents the aircraft with installed equipment. The PaRaDe was constructed using Commercial-Off-The-Shelf components such as: DVB-T antennas, a USRP N210 as a data acquisition unit and a portable PC for data recording and processing. As the TX was on the opposite side as the imaging area, two DVB-T antennas were mounted on the opposite aircraft windows.

In the measurement scenario presented in Figure 17.9, the red line represents the flight path of the aircraft given by the GPS/INS device. According to the GPS

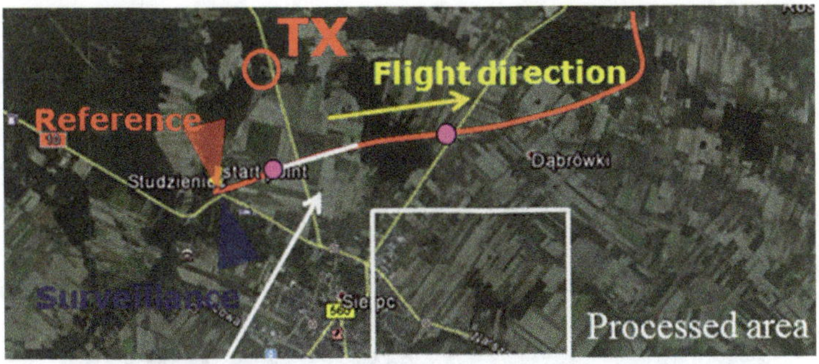

Figure 17.9 Measurement scenario

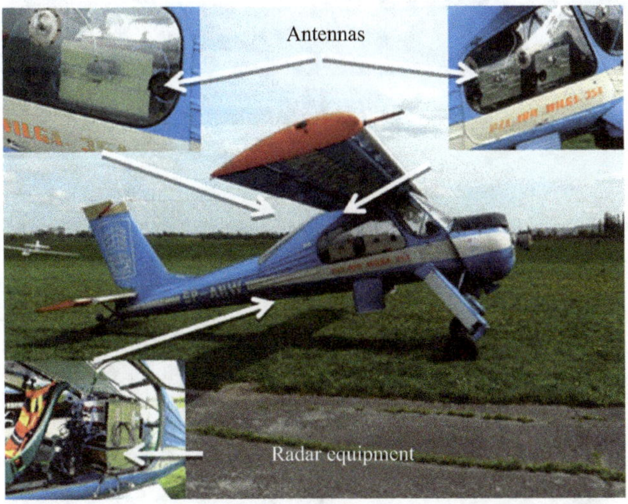

Figure 17.10 Aircraft with passive SAR radar demonstrator equipment

device, the speed of the platform during the flight was around 40 m/s, and the altitude was around 300 m above the ground. The total travelled distance (length of the red line), shown in Figure 17.9, was around 8 km (200 s × 40 m/s). The shortest distance between the flight path and the TX was around 2 km. The approximate localization of the illuminator is marked with a red circle in Figure 17.9. The area which was processed using a back-projection time domain algorithm in order to obtain the SAR images is marked with white rectangle. The pixel size of the image in the back-projection algorithm was set to be 3 m × 3 m. The integration time in azimuth in this case was 40 s (which is around 1.6 km of synthetic aperture), and the integration path is marked with a white line in Figure 17.9. The authors relied on the INS/GPS data provided by the equipment, and no additional motion compensation technique was applied in the signal processing. The grazing angle of the

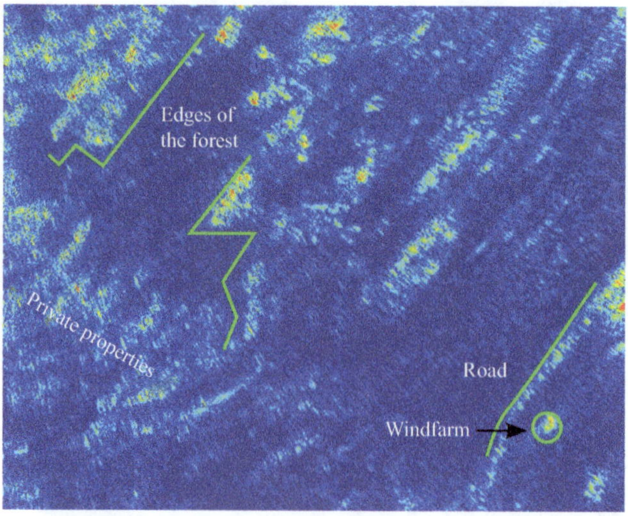

Figure 17.11 DVB-T-based passive SAR image

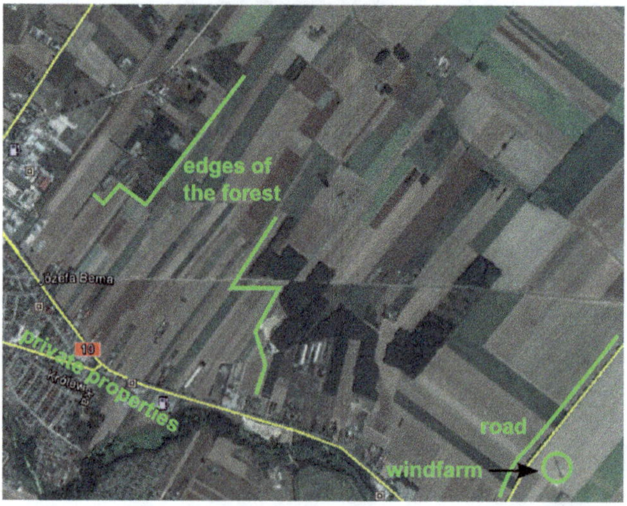

Figure 17.12 Corresponding Google optical image

illumination from the DVB-T TX was around 3.5° for the processed area presented in Figure 17.9.

The resulting passive SAR image is presented in Figure 17.11 with its corresponding Google Earth map image in Figure 17.12. In both figures, the corresponding structures are marked with white lines. Most of the targets which are visible in the Google optical image can be easily identified in the passive SAR

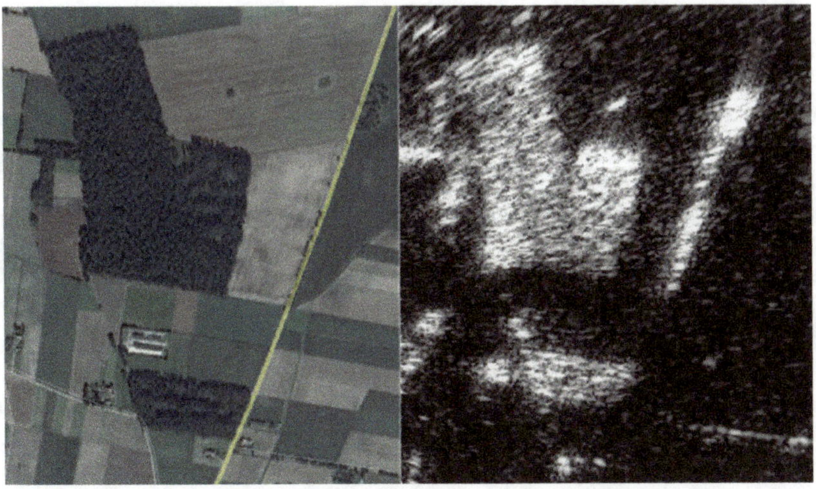

Figure 17.13 PASAR image (right) and corresponding optical image (left)

image. Structures such as the edges of the forest as well as roads, private properties and even a wind farm are visible. Some of the targets which are visible in the optical image have no representation in their SAR counterparts and vice versa. This may be due to various reasons such as the differing time of acquisition of both images. An additional reason might be the strong influence of shadowing between objects. It has to be taken into account that the grazing angle of the illumination is 3.5°. For such low grazing angles of the DVB-T illumination, the terrain profile also has a strong influence on SAR imaging capability.

Another example of passive SAR image is presented in Figure 17.13.

17.1.4 PASAR conclusions

PASAR imaging is a novel technique. The opportunity of using existing DVB-T TXs seems to be promising due to their wide bandwidth and high power. Passive SAR radars provide certain advantages over the active SAR radar, such as a lack of emission of electromagnetic energy, which results in silent operation. Moreover, the relatively low frequency signals can penetrate foliage and other obstacles and can be used for visualization of concealed targets.

In the experiment described above a single DVB-T channel of 7.8 MHz bandwidth was used, giving maximum range resolution of 20 m. In many places, it is possible to exploit several channels (5–25) with total bandwidth of 40–200 MHz giving maximum range resolution 0.7–3 m, comparable with modern active SAR systems. In some cases, TV channels are not consecutive, and frequency gaps exist between channels, so more advanced sparse technique have to be used to obtain a high quality SAR image [89,97,98]. In many places, it is also possible to produce multistatic images [99], when terrain is illuminated from several TXs. A challenge for signal processing is when all TXs work in a single-frequency network. In such cases, it is possible to create several images – one from each TX and to fuse the images (average in the simplest case).

As in active SAR case, the precise knowledge of passive SAR platform flight-path is essential to obtain focused images. For DVB-T case, the required accuracy (~10 cm) is much lower than for microwave active SAR (better than 1 cm for X-band SAR and 1 mm for W-band SAR), but still higher than provided by temporary navigation system. So to obtain the focused PASAR image, precise platform trajectory estimation has to be provided [100], finally accomplished with auto-focusing of the final image [101].

Passive SAR radar can bc used not only for ground surface imaging but also for moving target detection [40,102] and imaging [91,103].

However, the passive SAR radar has some disadvantages as well. Since the illuminator of opportunity is placed at a relatively low altitude (100–300 m) from the ground, some areas can be shadowed and the illumination angle is low or very low – in many cases, below 1°. Another disadvantage is the complex signal processing, which requires high computational power. The dynamic range is not sufficient in some cases, while strong direct echo leakage and strong echoes may mask weak stationary targets. This problem can be countered by applying appropriate signal processing methods, i.e. the adaptive removal of the direct echo or strong scatterers [42,104,105].

An interesting idea is to combine active SAR images from typical radar bands (C, X, Ku, Ka) that do not provide foliage penetration with a passive SAR image at a frequency that does provide foliage penetration. This technology might be widely used in the near future.

17.2 Target detection in airborne passive radar

One of the main tasks of the passive airborne radar is to detect aerial targets in the vicinity of the aircraft. The passive radar detections can further be used for aircraft collision avoidance and for self-protection. An application of this kind is depicted in Figure 17.14. The passive radar placed on the aircraft is equipped with several receiving antennas and forms several beams. One of the beams is directed to the TX and receives the reference signal, while others are used for detecting targets of interests – other aircraft, drones, birds.

There are three classes of potential objects on the scene: ground-based stationary objects, land-based moving objects and airborne moving objects. From the user point of view, one of the most interesting functions of the radar is detection of low velocity objects, both ground-based and airborne. This is usually a challenge in radars on mobile platforms due to the wide Doppler bandwidth of the ground clutter which often overlaps Doppler frequencies from slow moving objects. The same problem exists in mobile passive radar technology. Usually, in the pulse Doppler radars, this is solved by space-time processing techniques. The same approach should be used in mobile passive radar technology. However, utilization of space-time techniques for passive radar technology is not straightforward and requires

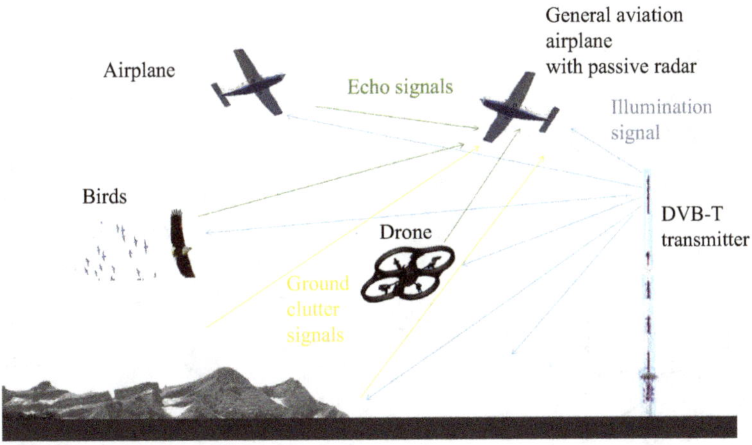

Figure 17.14 Airborne passive radar for self-protection and collision avoidance

additional careful modifications. Apart from this, the other challenge is high computational complexity of the space-time algorithms.

The general case where the both the TX and radar RX are mobile is very complex and applicable only when classical cooperative bistatic airborne radar is considered.

In most practical applications, we are looking for existing illuminations. Two types of illuminators currently provide continuous and wide area illuminations – ground illuminators (DVB-T, DAB, GSM and others) and satellite illuminators (DVB-S, GPS, sat-radio). For the purpose of passive sensing, a relatively high power density at the target is required. As satellite illuminators are of low power, the detection range of passive radars is limited to hundreds of metres. Such low detection range is of low practical value, so only ground-based illuminators can be considered in practical applications.

In this chapter, we have assumed that the TX is stationary and that the RX is in motion. As the main challenge is to remove the Doppler spread ground clutter prior to the target detection, the appropriate approximations of the space-time family of algorithms [9,28] will be presented. The work is focused on two topics: approximation of the classic STAP algorithms [102,106,107–113] and adaptation of the DPCA concept. To apply one of these concepts, it is necessary to build multichannel RX equipped with antenna array.

The basic geometry of passive airborne radar is shown in Figure 17.15.

The TX and the RX are located in different positions. It is assumed that the TX is stationary and that the RX is moving with velocity vector $\overrightarrow{V_{RX}}$. Both TX and RX antennas are looking at the target P, which can be stationary or moving with velocity vector $\overrightarrow{V_P}$. The passive radar is equipped with the multichannel RX

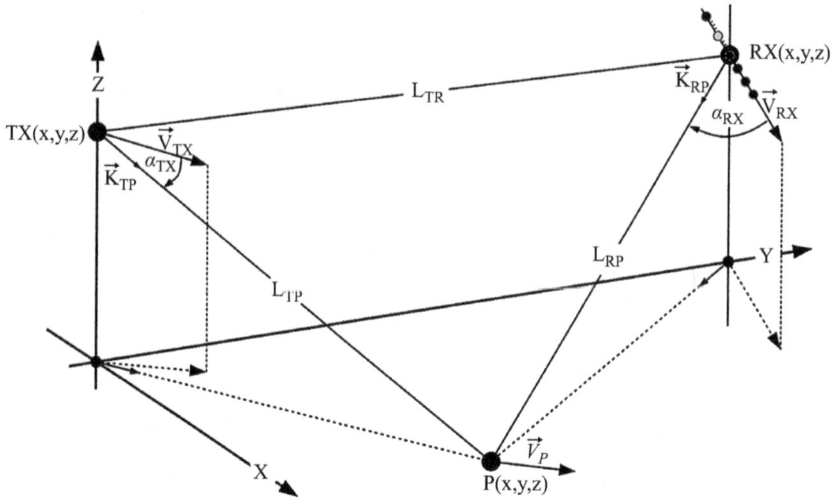

Figure 17.15 Multistatic geometry of moving passive radar

utilizing N antennas R_n where n is the index number of the RX's channels. The distance between the receiving antennas is d, and its mean that equal spaced received antenna array is used. It is also assumed that the TX, nth RX and the target (P) are moving along straight lines with constant velocities $\overrightarrow{V_T}$, $\overrightarrow{V_{Rn}}$ and $\overrightarrow{V_P}$, respectively. The position of a TX and a RX is denoted as $T(x_T, y_T, z_T)$ and $R(x_R, y_R, z_R)$. The position of the mth object P viewed from the nth selected antenna phase centre can be described as $P_m(x_{Pm}, y_{Pm}, z_{Pm})$.

Let us introduce the key directional vectors which will simplify the analysis of the geometry and the core passive radar parameters. A unit vector pointing in the direction of the passive radar to the TX is given by:

$$\overrightarrow{k_{\mathrm{TR}}} = \hat{x}\cos\theta_{\mathrm{RT}}\sin\varphi_{\mathrm{RT}} + \hat{y}\cos\theta_{\mathrm{RT}}\cos\varphi_{\mathrm{RT}} + \hat{z}\sin\theta_{\mathrm{RT}} \tag{17.20}$$

where $\hat{x}$, $\hat{y}$ and $\hat{z}$ are the unit vectors of the Cartesian coordinate system A unit vector pointing in the direction of the TX to the ith object is given by:

$$\overrightarrow{k_{\mathrm{TP}i}} = \hat{x}\cos\theta_{\mathrm{TP}i}\sin\varphi_{\mathrm{TP}i} + \hat{y}\cos\theta_{\mathrm{TP}i}\cos\varphi_{\mathrm{TP}i} + \hat{z}\sin\theta_{\mathrm{TP}i} \tag{17.21}$$

A unit vector pointing from the passive radar to the ith object on the scene is given by:

$$\overrightarrow{k_{\mathrm{TRP}i}} = \hat{x}\cos\theta_{\mathrm{RP}i}\sin\varphi_{\mathrm{RP}i} + \hat{y}\cos\theta_{\mathrm{RP}i}\cos\varphi_{\mathrm{RP}i} + \hat{z}\sin\theta_{\mathrm{RP}i} \tag{17.22}$$

Passive radar measures the bistatic range L, which is the difference between the TX-to-object-to-RX distance $L_{\mathrm{sum}} = L_{\mathrm{TP}} + L_{\mathrm{RP}}$ and TX-to-RX distance (L_{TR}). The bistatic range sum L_{sum} is the total distance travelled by the passive signal from the TX to the scatterer and to the passive radar's antenna reference point and is related to the round-trip time $\tau_{\mathrm{RT}} = (L_{\mathrm{TP}} + L_{\mathrm{RP}})/c$. Objects located on an iso-range surface have constant range sum (round-trip time). This surface is the locus of points for which the sum of the distances to two fixed points is a constant. Hence, this is an ellipsoid of revolution with the TX and RX as focal points. The instantaneous bistatic range for the signal reflected by the ith object and received by the nth antenna can be defined as:

$$L_{RnPi}(t) = L_{TPm}(t) + L_{RnPi}(t) - L_{\mathrm{TR}}(t) \tag{17.23}$$

Isorange surfaces are an important concept in radar as the signal corresponding to one particular range is the resultant contribution of all objects located on the isorange surface associated with the range of interest. When the scatterers are located on a given plane, the intersection of this ellipsoid with this plane defines an isorange contour (or simply isorange) which is an ellipse.

Let us discuss the derivation of the Doppler frequency in the bistatic radar case. The Doppler frequency corresponding to the stationary point P is given by

$$f_d = \frac{\overrightarrow{k_T}\left(\overrightarrow{V_T} + \overrightarrow{V_P}\right)}{\lambda} + \frac{\overrightarrow{k_{Rn}}\left(\overrightarrow{V_{Rn}} + \overrightarrow{V_P}\right)}{\lambda} \tag{17.24}$$

where $\overrightarrow{k_T}$ and $\overrightarrow{k_{Rn}}$ are unit vectors pointing at the stationary object from the TX and the nth RX, respectively. For the sake of simplicity, it may be assumed initially that the signal reception is performed using a single channel RX. Therefore, $\overrightarrow{k_{Rn}} = \overrightarrow{k_R}$ and $\overrightarrow{V_{Rn}} = \overrightarrow{V_R}$. This leads to the equation:

$$f_d = \frac{\overrightarrow{k_T}\left(\overrightarrow{V_T} + \overrightarrow{V_P}\right)}{\lambda} + \frac{\overrightarrow{k_R}\left(\overrightarrow{V_R} + \overrightarrow{V_P}\right)}{\lambda} \tag{17.25}$$

The Doppler characteristics of the ground clutter ($V_P = 0$) can then be described as follows:

$$f_d = \frac{\left|\overrightarrow{V_T}\right| \cos \alpha_T}{\lambda} + \frac{\left|\overrightarrow{V_R}\right| \cos \alpha_R}{\lambda} \tag{17.26}$$

Consider the derivation of angle-of-arrival (AOA) of the received signal in the passive radar. The following analysis is performed under the assumption that the passive signals are narrowband. This means that the modulation bandwidth satisfies the condition that $c/B \gg Nd$ where d is the spacing between antenna sensors. The physical interpretation is that the propagation delay across the N-element uniform linear array (ULA) is manifested as a simple phase shift. This is true for the typical signals used in the passive radar. The spatial steering vector describes the varying phase among the elements resulting from a propagating plane wave emanating from the direction of the scatterer normal to $\widehat{k_{\mathrm{RP}}}(\theta_{\mathrm{RP}}, \varphi_{\mathrm{RP}})$. For the relative time delay $\Delta\tau_n$, representing the time it takes the passive signal to arrive at the nth element

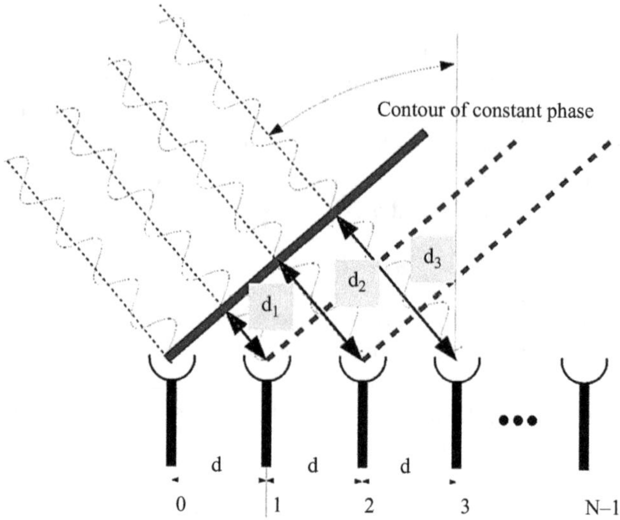

Figure 17.16 Phase shift in N-element ULA

with respect to the reference element of the ULA, the corresponding phase shift is $\gamma_n = -2\pi f_c \Delta\tau_n$. Then $\Delta\tau_n = d_n/c$ where d_n is the distance the passive signal must travel after impinging on the reference element.

For a side-looking ULA (Figure 17.16), $d_n = nd\cos\theta_{\mathrm{RP}}\sin\varphi_{\mathrm{RP}}$. The phase shift at the nth element follows the equation:

$$\psi = \frac{2\pi nd}{\lambda}\cos\theta_{\mathrm{RP}}\sin\varphi_{\mathrm{RP}} \tag{17.27}$$

where λ is the wavelength of the passive signal. The spatial frequency is defined to be

$$\psi = \frac{d}{\lambda}\cos\theta_{\mathrm{RP}}\sin\varphi_{\mathrm{RP}} \tag{17.28}$$

The generalized spatial steering vector is then

$$a(\varphi,\theta) = [1\,\exp(j2\pi\psi)\cdots\exp(j2\pi(N-2)\psi)\,\exp(j2\pi(N-1)\psi)]^T \tag{17.29}$$

Therefore, the phase variation across the array for an arbitrary AOA is linear for the ULA. The spatial steering vector assumes a Vandermonde form because of the ULA geometry and of identical element patterns.

17.2.1 Monostatic case

The monostatic scenario can be perceived as a very special case of a bistatic scenario where the position, velocity and direction of motion of TX and RX coincide.

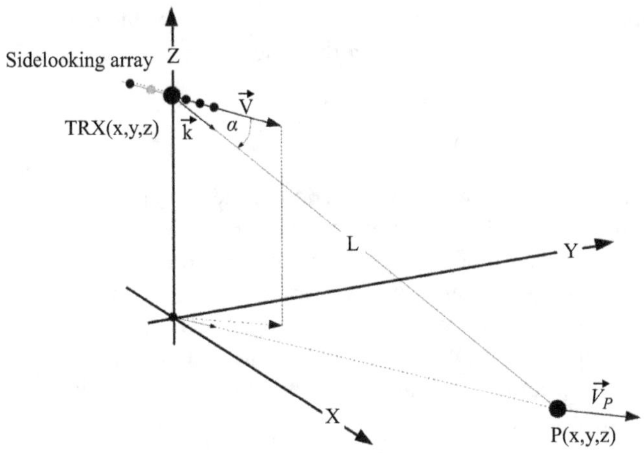

Figure 17.17 Monostatic geometry

It is presented here for completeness as well as for the fact that most of the state-of-the-art ground clutter suppression algorithms were developed for this case [114]. The radar geometry is presented in Figure 17.17, showing a situation where a radar is equipped with a single-channel TX and a multichannel RX utilizing N antennas R_{Xn}.

The distance between the receiving antennas is denoted as d. It is assumed that radar and object on the scene are moving along straight lines with constant velocity $\overrightarrow{V}$ and $\overrightarrow{V_P}$, respectively. The radar antennas are in the side-looking configuration. The positions of the TX/RX (radar) and an object on the radar scene are defined by TRX(x, y, z) and $P(x_P, y_P, z_P)$, respectively. Consider the derivation of the Doppler frequency in the monostatic radar case. The Doppler frequency corresponding to the stationary point P is given by

$$f_d = \frac{\overrightarrow{k_T}\left(\overrightarrow{V} + \overrightarrow{V_P}\right)}{\lambda} + \frac{\overrightarrow{k_{Rn}}\left(\overrightarrow{V} + \overrightarrow{V_P}\right)}{\lambda} \tag{17.30}$$

where $\overrightarrow{k_T}$ and $\overrightarrow{k_R}$ are unit vectors pointing to the stationary object from TX and RX, respectively. For the sake of simplicity, it might be assumed temporarily that the RX is equipped with single channel RX. Therefore, this leads to the equation:

$$\overrightarrow{k_T} = \overrightarrow{k_R} = k \tag{17.31}$$

$$f_d = \frac{2\overrightarrow{k}\left(\overrightarrow{V} + \overrightarrow{V_P}\right)}{\lambda} \tag{17.32}$$

Assuming that $\overrightarrow{V_P} = 0$ and incorporating the formula for the dot product between the two vectors into (17.31), we obtain the model of ground clutter:

$$f_d = \frac{2|V|\cos(\alpha)}{\lambda} \tag{17.33}$$

where the α is the angle between the direction of the velocity vector $\overrightarrow{V}$ and the unit vector $\overrightarrow{k}$ pointing towards the stationary object P.

The airborne passive radar scenario was presented in Figure 17.15. Moreover, the real measurements presented in this chapter are also in line with this geometry. This scenario is valid for most passive radars, especially, for those utilizing transmitters of opportunity like DVB-T and radio (FM and DAB).

In a typical airborne case, the TX is stationary, and thus $\overrightarrow{V_{\text{TX}}} = 0$. Equation (17.30) now has a form:

$$f_d = \frac{\overrightarrow{k_T}\,\overrightarrow{V_P}}{\lambda} + \frac{\overrightarrow{k_{Rn}}\left(\overrightarrow{V_{\text{RX}}} + \overrightarrow{V_P}\right)}{\lambda} \tag{17.34}$$

Assuming that target velocity $\overrightarrow{V_P} = 0$ and incorporating the formula for the dot product between two vectors into (17.33) the model of ground clutter can be obtained:

$$f_d = \frac{2|V|\cos(\alpha)}{\lambda} \tag{17.35}$$

where the α is the angle between the direction of the of the velocity vector $\overrightarrow{V}$ and the unit vector $\overrightarrow{k}$ pointing towards the stationary object P.

Comparing (17.35) with the monostatic case (17.33), it can be seen that the ground clutter characteristics should have a similar character. The only difference between those two formulas is the constant factor 2 and the reason is that in the monostatic case both RX and TX are in motion, while in airborne passive coherent locator (APCL) case only the RX is in motion.

17.2.2 Received signal model

The signal-processing techniques in this description are derived under the assumption that a target in the radar scene may be treated as a point scatterer. This is true when the target size is much smaller than the range resolution cell. The size of the resolution cell is approximately equal to 3 km for FM radio, 100 m for DAB and 20 m for DVB-T signal, respectively. For the sake of simplicity, it is assumed that the reference signal (signal transmitted by the TX of opportunity) is a narrowband signal. Such an assumption is satisfied for almost all analogue and digital transmission standards used in passive radars (FM radio, analogue television, DAB digital radio, DVBT DVB-T, etc.). The transmitted RF signal can be modelled as:

$$x_T(t) = R\{X_T(t)\exp(j2\pi F_T t)\} \tag{17.36}$$

where F_T is the carrier frequency of the transmitted signal and $X_T(t)$ is the complex baseband information signal. Typically in passive radar, a delayed and attenuated (via path (a) in Figure 17.2) copy of this signal is received by the antenna directed towards the TX and used as the reference signal. For the transmitted signal defined by (17.36), the echo of a single reflector (target) appearing at the nth channel of the PCL RX can be described in the baseband by the following expression:

$$X_R(t) = \begin{pmatrix} X_{R1}(t) \\ X_{R2}(t) \\ \vdots \\ X_{Rn}(t) \end{pmatrix} \tag{17.37}$$

where

$$X_{Rn}(t) = AX_{Tn}\left(t - \frac{r_n(t)}{c}\right)\exp\left(j2\pi F_T \frac{r_n(t)}{c}\right), \tag{17.38}$$

and $r_n(t)$ is the instantaneous bistatic range of the target to the nth antenna (i.e. the length of the path (b) which changes as a function of time due to motion of the RX, or path (c) or (d), which changes as a function of time due to the motion of both the RX and the target), A is the complex amplitude of the signal reflected from the target. At typical integration times of the RX the delay $r(t)/c$ may be assumed almost constant for the envelope X_T. The phase of received signal $F_{Tn}(r(t)/c)$ is, however, changing fast enough to cause significant frequency shift (Doppler frequency). This shift is proportional to the time derivative of $r_n(t)$, which is called bistatic velocity.

For simplicity, we will assume that during the observation time the bistatic velocity is approximately constant, so we may substitute $r(t) = r_0 + v_0 t$. Then, (17.38) becomes

$$X_R(t) = A\exp\left(j2\pi F_T \frac{r_0}{c}\right)X_T\left(t - \frac{r_0}{c}\right)\exp\left(j2\pi F_T \frac{v_0}{c}t\right), \tag{17.39}$$

The received signal consists of many such contributions from all the reflecting points in the antenna field of view, and additional noise which we assume to be white Gaussian.

17.2.3 Estimation of the multichannel passive radar parameters

The passive radar is in general able to determine the following parameters of the received signal: bistatic range, bistatic velocity and AOA. Bistatic range is defined by (17.23) as the difference between the range from TX-to-object-to-RX and TX-to-RX.

Bistatic velocity is proportional to the Doppler frequency (17.25). AOA relates to the spatial variation of phase delay across the antenna array and thus defines the spatial frequency.

The signal processing in passive airborne radar is three dimensional, and results can be plotted as two different cross-sections. The first one is called range-velocity plot and enables to estimate the bistatic range and bistatic velocity of the objects. However, the information regarding the angular position is not visible. The second one is called angle-velocity plot. It enables estimation of the objects velocity and angular position in the selected bistatic range cell. The angle-velocity imaging is possible only using in multichannel RXs, where spatial processing is possible.

The passive radar range-velocity processing is based on the correlation of the received signal with a template constructed from the reference signal [21,104,115]. To model the echo of a target expected at a certain bistatic range r and bistatic velocity v, the reference signal must be time-shifted by $\tau = r/c$ and frequency-shifted by $f_d = F_T v/c$, where F_T is the carrier frequency of the signal and c is the speed of light. Since the parameters of the target are unknown, range-velocity surface is created for the expected span of bistatic range and velocity values according to the following formula:

$$Y(r, v) = \int_0^{\Delta T} X_R(t) X_T^*\left(t - \frac{r}{c}\right) \exp\left(j2\pi F_T \frac{v}{c} t\right) dt, \tag{17.40}$$

where X_T is the transmitted signal (reference) and X_R is the received signal, reflected from the radar scene and ΔT is the integration time. The product of $X_R(t)$ by $X_T^*(t - (r/c))$ is called a mixing product $M(t, r)$:

$$M(t, r) = X_R(t) X_T^*\left(t - \frac{r}{c}\right), \tag{17.41}$$

Incorporating (17.40) into (17.41) leads to the equation:

$$Y(r, v) = \int_0^{\Delta T} M(t, r) \exp\left(j2\pi F_T \frac{v}{c} t\right) dt, \tag{17.42}$$

Equation (17.42) can be interpreted as a Fourier transform of the mixing product $M(t, r)$ for the fixed bi-static range r. This leads to the equation:

$$Y(v, r = r_k) = \mathrm{FFT}\{M(t, r_k)\}, \tag{17.43}$$

Equation (17.43) enables computationally efficient implementation of the passive radar processing utilizing FFT algorithms.

The angle-velocity processing is a generalization of the range-velocity processing for a selected range. It enables to present the target expected at a certain bistatic range r, bistatic velocity v and angle α_{RX}. It exploits the advantages of multichannel RX and its ability to estimate the angle-of-arrival of received signal.

The algorithm works as follows: first the reference signal in the RX must be time-shifted by r/c and frequency-shifted by $F_T v/c$. The generalized mixing product $M_n(t, r_k)$ is calculated for the signal received by each nth antenna sensor X_{Rn} and selected bistatic range r.

$$M_n(t, r) = X_{Rn}(t) X_T^*\left(t - \frac{r}{c}\right), \tag{17.44}$$

where n is the index of the antenna sensor. In the second step, the calculated mixing product is space frequency shifted by αn where n is the index of the antenna sensor and α is the expected AOA. Since the parameters of the target are unknown, angle-velocity surface is created for the expected span of bistatic angles α and velocity values v. Therefore, the angle-velocity plot can be computed according to the following formula:

$$\gamma(v,\alpha) = \int_{-\pi/2}^{\pi/2}\int_{0}^{\Delta T} M_n(t,r_0)\exp\left(j2\pi F_T\frac{v}{c}t\right)\exp(j\alpha n)dtdn, \tag{17.45}$$

where N is the total number of elements in the antenna array. Equation (17.45) can be presented as a two-dimensional Fourier transform of the mixing product:

$$\gamma(v,\alpha) = \mathrm{FFT}\{\{M_n(t,r)\}_t\}_{\alpha(n)}, \tag{17.46}$$

The range resolution [115] of the passive radar is inversely related to the bandwidth B of the exploited signal and decline with a larger target bistatic angle β given as

$$\Delta R = \frac{c}{2B\cos(\beta/2)} \tag{17.47}$$

which represents a minimum requirements for target separation in range. The bistatic angle β is the angle between $\overrightarrow{k_{\mathrm{TP}}}$ and $\overrightarrow{k_{\mathrm{RP}}}$ unit vectors. The Doppler resolution is determined by the coherent integration time (CIT) of the collected datacube and can be expressed as

$$\Delta f = \frac{1}{\mathrm{CIT}} \tag{17.48}$$

The bistatic velocity resolution depends on the wavelength λ [95] and can be derived as:

$$\Delta V = \frac{\lambda}{2\mathrm{CIT}\cos(\beta/2)} \tag{17.49}$$

17.2.4 Time datacube

The typical passive radar exploits the continuous wave illumination. Although pulse illumination from other radars can be also used, in this chapter, considerations are limited to continue wave case. As mentioned before, the received antenna array consists of several elementary antennas which can be placed in at any arbitrary position according to the centre of the plane. For the sake of simplicity, only basic N element linear antenna array is considered. All antenna elements are considered to be identical. The signal from each antenna is amplified, filtered, down-converted and sampled resulting in obtaining complex baseband samples. The time interval over which the received signal is collected is referred to as the CIT.

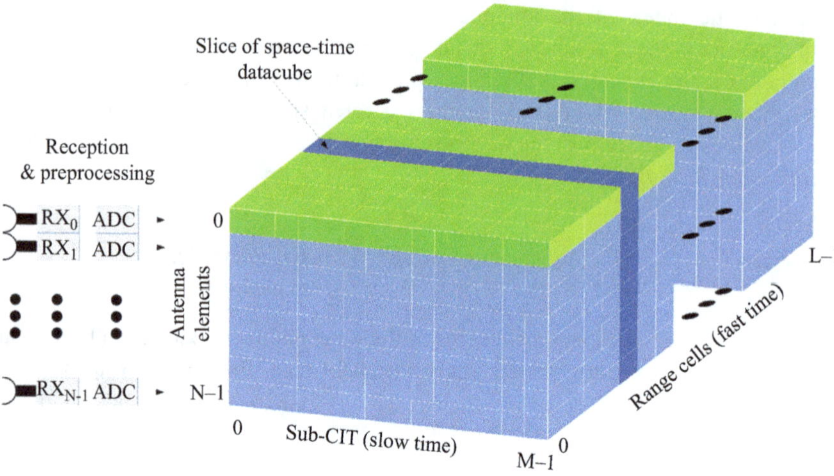

Figure 17.18 Datacube in passive airborne radar

For each antenna element/channel, the CIT temporal dimension is segmented into M sub-CITs. Each sub-CIT has duration T_{sub} and a sub-CIT repetition frequency is equal to $f_{\text{sub}} = 1/T_{\text{sub}}$. For each sub-CIT, there are $T = T_{\text{sub}} * f_{\text{BW}}$ range cells covering the range sum interval, where L is the total number of range cells and f_{BW} being the complex sampling rate. Therefore, this multidimensional data set for MTI signal processing for the airborne passive radar is visualized as the $N*M*L$ cube of complex baseband samples (Figure 17.18), where $n = 0, \ldots, N-1$, $m = 0, \ldots, M-1$ and $l = 0, \ldots, L-1$ are the antenna index, sub-CIT and range cell index, respectively. It is common to refer to the range dimension l as fast-time and the sub-CIT dimension m as slow time. Along the range dimension, the range profile of the received passive signals is obtained by correlation processing, identical as described in previous subchapter – see (17.40). Thus, the range correlation (direct path and received signal) is carried out separately on a sub-CIT basis where the result is a 1_L correlation function r for each sub-CIT.

By lining up different r for each different sub-CIT, a $M \times L$ correlation function matrix R for each element can be formed, i.e. $R = [r_0; r_1; r_2; \ldots; r_{M-1}]$. Figure 17.18 illustrates a pictorial view of the passive CIT datacube. Each row of the datacube corresponds to a spatial sample and each column to a slow time sample, while the L range samples extend in the third dimension. The $N \times M$ matrix which is a slice of the datacube corresponding to the lth range cell

$$X = \begin{bmatrix} X_{l(0,0)} & X_{l(0,1)} & \cdots & X_{l(0,M-1)} \\ X_{l(1,0)} & X_{l(1,1)} & \cdots & X_{l(1,M-1)} \\ \vdots & \vdots & \ddots & \vdots \\ X_{l(N-1,0)} & X_{l(N-1,1)} & & X_{l(N-1,M-1)} \end{bmatrix} \tag{17.50}$$

Vectorizing by stacking each succeeding column one after the other yields the $MN \times 1$ vector X_l termed a space-time snapshot, for the lth range cell.

17.2.5 Target detection in passive airborne radar

Detection of targets are in airborne passive radar is almost the same as detection in classical passive radar. In the same way, as for a stationary radar case, it is necessary to remove strong direct signal and strong ground clutter to be able to detect weak targets echoes. The direct signal can be easily removed using CLEAN-type algorithms, as discussed in [104,116]. More challenging is removal of Doppler spread ground clutter.

The CIT is an important parameter which sets the signal processing gain due to coherent integration, leading to the desired effect of increasing the SNR. The two important constraints affecting the maximum CIT are that of target range cell migration and Doppler cell migration [117–119]. Range cell migration occurs when the range resolution is smaller than the distance travelled by the target during integration, leading to an energy dispersal in the range correlation as it moves through multiple range cells. For no range cell migration (and the assumption of no range migration compensation), the maximum distance the target travels over the CIT is limited by the range cell resolution. Thus, the condition for the maximum CIT without inducing target range cell migration is transformed into the inequality

$$\mathrm{CIT} < \frac{\Delta R}{V_{t(\max)}} \tag{17.51}$$

where $V_{t(\max)}$ is the maximum relative velocity between the target and the passive radar. The inequality $<$ in (1.32) states that the CIT should be considerably smaller than the right-sided term. Doppler migration occurs then the target accelerates through several Doppler cells during integration, leading to an energy dispersal in Doppler correlation. For no Doppler cell migration, the requirement of $\Delta f < a_{t(\max)}/\lambda$ must be satisfied where $a_{t(\max)}$ is the maximum bistatic target acceleration. This leads to the inequality for maximum CIT without inducing target Doppler frequency cell migration as

$$\mathrm{CIT} < \sqrt{\frac{\lambda}{a_{t(\max)}}} \tag{17.52}$$

In summary, the maximum CIT must satisfy

$$\mathrm{CIT} < \min\left(\frac{\Delta R}{V_{t(\max)}}, \sqrt{\frac{\lambda}{a_{t(\max)}}}\right) \tag{17.53}$$

Airborne passive radar target detection can utilize range-velocity processing [2,104,115,120] or angle-velocity processing, but much better results can be achieved using joint range-velocity-angle processing. The detection is carried out by comparison of the absolute value from correlation processing with a selected threshold. The threshold is usually selected according to the Neyman–Pearson (N–P) criterion.

The N–P criterion is the decision rule that chooses the probability of false alarm (P_{FA}) to be fixed at or below some value and then maximizes the probability of detection (P_D). More detailed information can be found [121,122]. The minimum detectable power P_d in passive radar is:

$$P_d = \frac{kTND_0}{t_i} \tag{17.54}$$

where k is Bolztmann's constant, T is the equivalent RX noise temperature, N stands for total losses and D_0 is the detection threshold (usually 13 dB). For DVB-T radar with 16 ms integration time, the minimum echo power is -140 dBm.

One of the fundamental relationships in passive radar is range equation. The Power of the signal reflected from the object on the radar scene can be described as (17.54)

$$P_{Rx} = \frac{P_{Tx}G_{Tx}\sigma_b G_{Rx}\lambda^2}{(2\pi)^3 R_T^2 R_R^2}\varsigma \tag{17.55}$$

where P_{Tx} is the transmitted power, G_{Tx} is the directional gain of the TX's antenna, σ_b is effective bistatic radar target cross-section, G_{Rx} is the directional gain of the RX's antenna, λ is the wavelength, ς is the attenuation coefficient, R_T is a distance between TX and an object, R_R is a range between radar RX and an object.

The maximum achievable signal-to-noise ratio can be described as:

$$\mathrm{SNR}_{in} = \frac{P_R}{P_n} = \frac{P_{Tx}G_{Tx}\sigma_b G_{Rx}\lambda^2}{(2\pi)^3 R_T^2 R_R^2}\frac{1}{kT_0B_R}\varsigma \tag{17.56}$$

where P_n is the power of the noise, k is Boltzmann's constant, T_0 is the effective RX temperature and B_R is the RX's bandwidth. The use of the cross-ambiguity function provides the processing gain equal to the time-bandwidth product:

$$G_P = BT \tag{17.57}$$

In theory, the processing gain could be as high as the designer requires, but in practice, it is limited by several factors. First of all, the integration time is limited by the time in which the target remains in the range and Doppler resolution gate. For the passive radar using the DVB-T illuminator of opportunity, the range gate size is 20 m, and for airborne targets with a maximum relative speed of 1,200 m/s, the maximum integration time is 16 ms.

The maximum integration gain is thus 130,000 (51 dB).

17.2.6 Clutter cancellation in airborne passive radar

Clutter cancellation in airborne radars is usually based on STAP [114]. The idea of STAP is based on the weighted summation of signals from all elementary antennas used in the system and their delayed copies in slow time domain. In passive radar signal processing, there are no pulses and range compression is done via correlation between surveillance and reference signals or computation of cross-ambiguity function using those signals. Slow time in pulse radar is based on time delay between consecutive pulses. In passive radar, slow time processing is based on the computation of correlation or cross-ambiguity functions with arbitrarily chosen integration time and time delay similar to pulse repetition time in pulse radar. So instead of working on range-pulse data set, what is natural for pulse radar, in passive radar, we will work on range-time delayed data sets – delayed range-velocity diagrams (results of cross-ambiguity function computations). In pulse radar, the minimum delay step is equal to the pulse repetition interval, while in passive radar such delay may be selected precisely and adjusted with much bigger precision equal to the sampling interval, what is very important for DPCA processing. Assuming that for nth antenna the summation can be described by a finite response filter with pulse response $h_n(\tau)$ the STAP filter can be described as convolution in the form:

$$Z(r, v) = \sum_{n} Y_n(r, v, \tau) * h_n(\tau), \tag{17.58}$$

In order to design the full STAP filter, it is necessary to find a set of impulse responses $h_n(\tau)$ that minimises $|Z(r, v)|$ when only ground clutter is visible by the radar.

However, the full implementation of the algorithm for STAP processing leads to high complexity solutions. To counter this, the proposal of a more cost-effective algorithm, utilizing the DPCA method is presented. DPCA processing is a technique for countering the clutter spectrum spreading induced by platform motion. This technology was invented originally for the pulse radar.

The basic concept of DPCA technology is to make the antenna appear stationary even though the platform is moving forward. This is performed by electronically shifting the receive aperture backwards during the operation.

In a pulse radar, the return from shifted antenna is subtracted from the unshifted return (delayed by time equal to antenna phase centre displacement divided by platform velocity). The DPCA concept, after some modification, can be utilized in passive radar.

To attenuate the ground clutter for an N-channel system, the most straightforward approach is to find such delay $\Delta\tau = \tau_2 - \tau_1$, so that:

$$Y_n(r, v, \tau_1) = Y_{n-1}(r, v, \tau_2) \quad \text{for } n = 2, 3, \ldots, N. \tag{17.59}$$

This means that the phase centre of antenna element $n-1$ will move to the phase centre of antenna element n after time $\Delta\tau$, as shown in Figure 17.19. As a result, the clutter signal received by antenna $n-1$ for selected r and v should be statistically equal to the clutter signal received by antenna n for the same r and v for $n = 2, 3, \ldots, N$. The clutter cancellation can be achieved if we subtract them from each other. At this point, it is worth noting that a passive radar signal can be approximated with a random process. The complex amplitude of the clutter

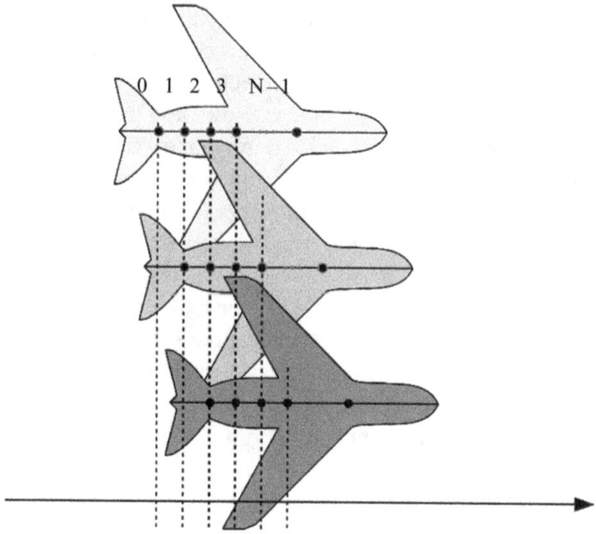

Figure 17.19 The consecutive positions of the platform corresponding to the displacement between the antenna elements

$Y_n(r, v, \tau)$ is generally a random variable and obeys its own pdf. Equation (17.59) therefore should be understood as $E\{Y_n(r, v, \tau_1)\} = E\{Y_{n-1}(r, v, \tau_2)\}$ or $Y_n(r, v, \tau_1) - Y_{n-1}(r, v, \tau_2) = \delta$, where δ is a zero mean random variable.

The condition, equivalent to (17.59) imposes serious limitations on any implementation of the DPCA technique for the pulse radar. Once the element spacing and the PRF are fixed, the platform velocity is also constrained. In practice, even if the condition of appropriate space and time shift is satisfied, the clutter cancellation is still limited due to other factors, such as irregular motion and crabbing of the aircraft, spatial de-correlation among antenna elements, as well as the intrinsic motion of clutter. Moreover, the elementary antenna's spatial characteristic should be exactly this some, while even small deviation of beam width, looking angle or shadowing and multipath from wings and fuselage can decrease the effectiveness of clutter cancellation.

The major difference between pulsed and PCL radar is the flexibility in changing the delay $\Delta\tau$ between signals coming from each antenna. In a pulsed radar, the delay may be adjusted with step equal to the inverse of pulse repetition frequency which is usually only up to a few kilohertz (2–4 kHz). In PCL, the adjustment can be made with accuracy limited by inverse of the sampling frequency, e.g. in FM-radio-based passive radar, it is equal to several hundreds of kilohertz (about 200 kHz) and in DVB-T radar around 10 MHz. This gives a potential advantage to the PCL DPCA technique because processing can be easily adapted to the variable radar velocity and overcome the limitations from (17.59).

The other difference lies in the type of the signal used. PCL radars utilize continuous random-like waveforms from TXs of opportunity instead of deterministic short time duration pulses. Typical carrier frequencies of transmitted signals are usually below 1 GHz. This has an impact on the antenna system design. The typical distance between the antennas can be from 50 cm for DVB-T passive radars up to 1.5–3 m for FM based.

The simplest passive airborne radar can consist of reference antenna and two measurement antennas. In such case, a simple two-point canceller can be applied:

$$\begin{bmatrix} h_1[\tau] \\ h_2[\tau] \end{bmatrix} = \begin{bmatrix} 1 & 0 & \cdots & 0 & 0 \\ 0 & 0 & \cdots & 0 & -1 \end{bmatrix} \tag{17.60}$$

This corresponds to a simple subtraction of two range-velocity surfaces $Y_1(t, v, \tau)$ and $Y_2(t, v, \tau)$ created for first and second antenna with appropriate delay.

The PCL DPCA clutter cancellation filter responses described by (17.60) versus target bistatic velocity and mobile platform velocity are presented in Figure 17.19 (for elementary antenna phase centres displacement equal to $2/3\lambda$). Figure 17.20 presents a cross-section of Figure 17.19 for an assumed platform velocity of 100 m/s.

The clutter attenuation of the simple filter (17.60) is invariant with respect to the bistatic range. For a PCL system with a fixed antenna configuration, the DPCA filter suppression rate depends only on the velocity of the radar platform and the bistatic velocity of the target on the scene. Due to the statistical character of the sounding signal and due to the difference in integration time instant corresponding to the centre of integration time, the attenuation of the filter is limited. This is depicted in Figures 17.19 and 17.20. For longer distances between the antennas, the well-known 'blind speed' problem begins to occur (visible in Figure 17.20 as notches in the filter response). The calculated results are based on the assumption that the receiving antennas have exactly the same radiation patterns. This may not necessarily be true in a real-life situation so the attenuation of ground clutter can be limited.

By increasing the number of elementary antennas in the receiving antenna array, it is possible to use more sophisticated filters. The characteristic of three-point canceller with coefficients [−1, −1, 2] is depicted in Figure 17.21 and characteristic of four point canceller with coefficients [−1, −1, 1, 1] in Figure 17.22.

17.2.7 Measurement campaigns

The concept of the mobile PCL radar has been practically demonstrated by several researchers [26,123]. In this chapter, the result of three measurement campaigns carried out by Warsaw University of Technology (WUT) is shown. During both tests, the multichannel FM-based PaRaDe system, developed at WUT, was used. The first experiment was made using a car as a moving platform [124], the second

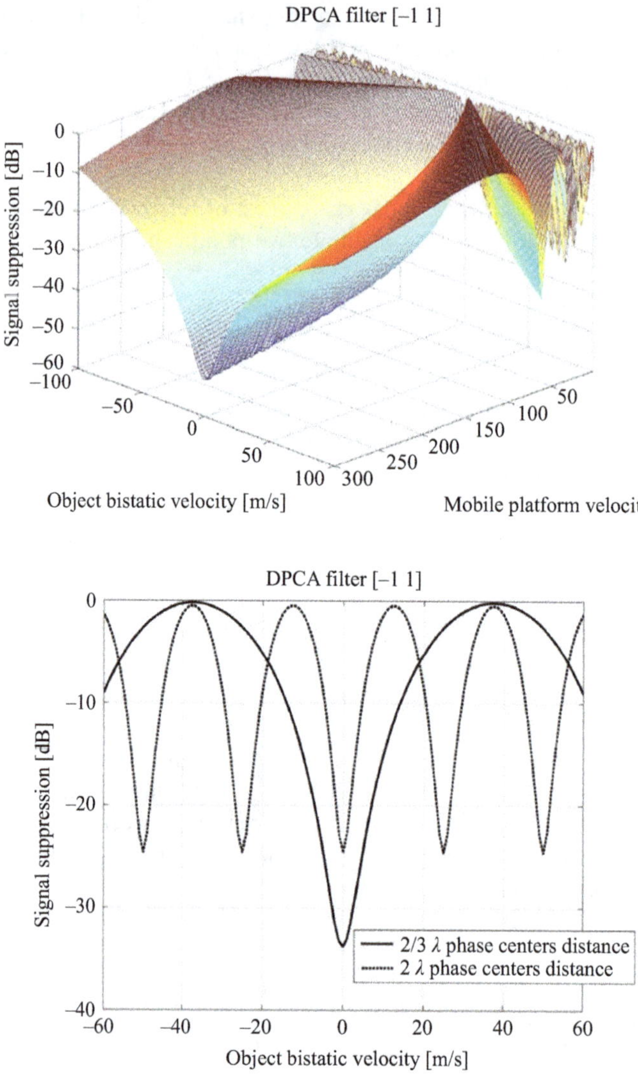

Figure 17.20 PCL DPCA [−1, 1] filter response. Cross-section of the PCL DPCA filter response for radar platform velocity equal to 100 m/s

and third used airborne platform – a Skytruck aircraft [82,125–127]. The measurement campaigns with the PCL system mounted to the car platform were carried out near the landing path of the Okecie Airport (Warsaw) in December 2008. The measurement campaigns with the PCL system mounted to the airborne platform were carried out close to the Polish coastline of the Baltic Sea in May 2008 and April 2009.

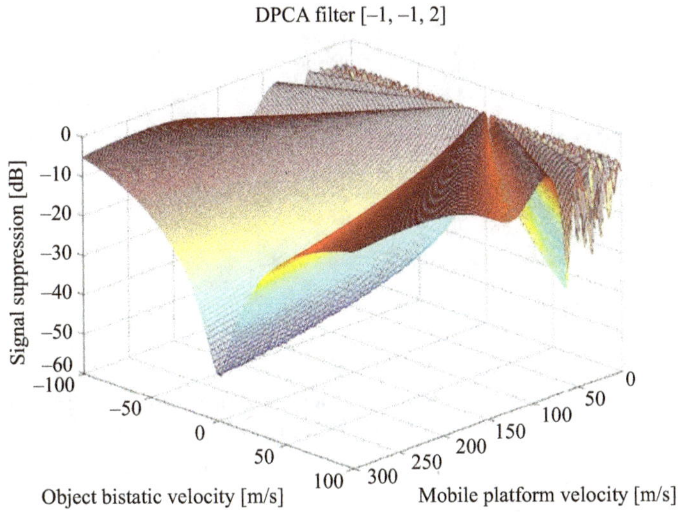

Figure 17.21 PCL DPCA [–1, –1, 2] filter response

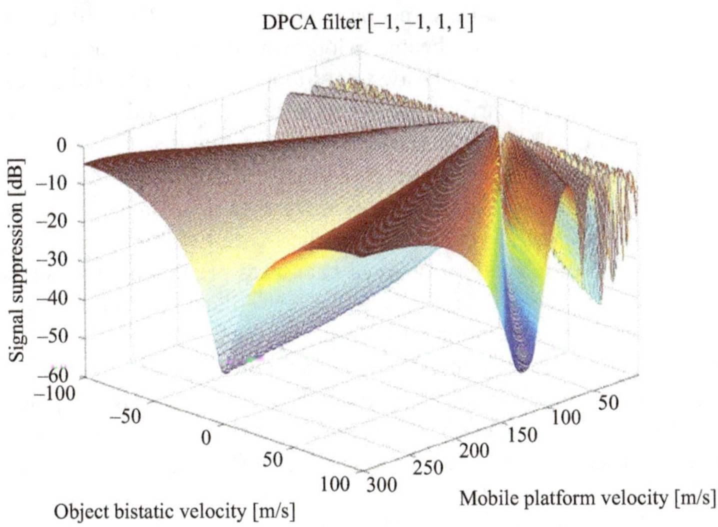

Figure 17.22 PCL DPCA [–1, –1, 1, 1] filter response

The radar platform was equipped with a six element antenna array – three on each side of the moving platform. The signals from the antenna elements were amplified and filtered by commercial FM amplifiers and directly sampling by fast analogue–digital converter. The digital HF signal was digitally downconverted and recorded on the hard drive. To obtain the final results off-line processing, was applied, consisting of beamforming, a CLEAN algorithm removing the direct

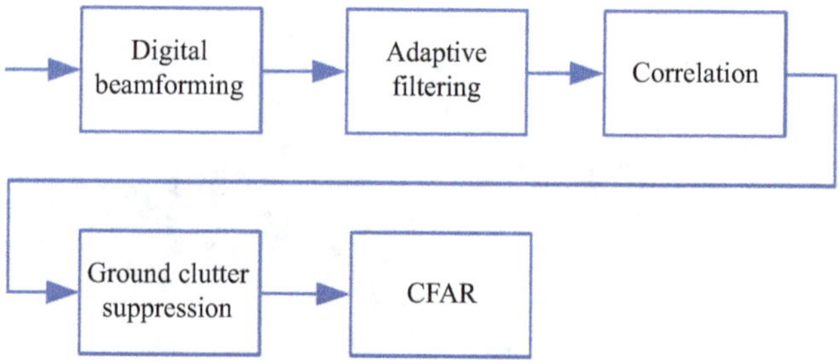

Figure 17.23 Signal-processing block diagram

signal from the measurement beam and space-time processing. The block diagram of the signal processing performed on the PC in shown in Figure 17.23. The processing starts with digital beamforming [128–130]. The signals from all antennas are weighted summed to obtain one reference and several surveillance beams. The coupling between the antenna elements is compensated by a suitable calibration procedure [128].

The beams obtained as a result of the beamforming have relatively low level of sidelobes. In spite of this fact, the direct path interference (DPI) is still the strongest contribution in the surveillance beams. Moreover, the surveillance channel also contains strong clutter echoes. The two contributions: DPI and clutter can completely mask the moving target echoes; therefore, a method for removing the unwanted signal components from the echo signal is required.

The direct signal canceller implemented in the system is based on an adaptive lattice filter [131–136]. The filter consists of two parts: a lattice predictor and a tapped delay line. The lattice predictor is excited with the reference signal, which leads to its decomposition into orthogonal backward prediction error signals. The prediction errors are then removed from the surveillance channel in the tapped delay line by means of correlation and subtraction.

After the DPI, the cross-ambiguity of the reference and surveillance channel is calculated. This procedure involves correlation of a delayed and Doppler-shifted version of the reference signal with the surveillance signal. When the chosen delay and Doppler shift correspond to the parameters of a target echo, a correlation peak appears. The process is repeated for a certain span of delays and frequency shifts, resulting in a two-dimensional range-Doppler surface.

The calculated cross-ambiguity is passed to the clutter suppression processor. Then, the processing results are passed to the CFAR in order to perform target detection. The CFAR is based on a two-dimensional cell averaging algorithm. The position of the detected peak on the range-Doppler surface provides coarse estimates of bistatic range and velocity.

The antenna array of the system consists of a number of half-wave symmetrical dipoles placed vertically. Depending on the configuration, two to eight dipoles are used in different geometrical configurations, and multiple beams are formed digitally without the need for physical relocation of antenna elements.

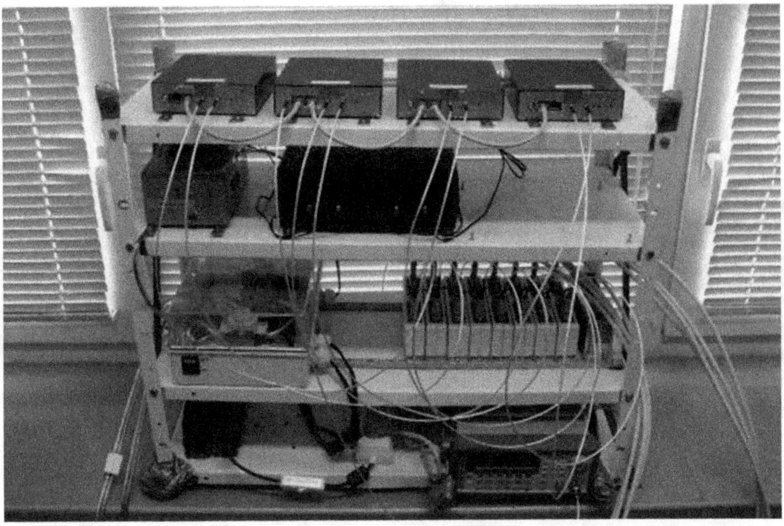

Figure 17.24 PaRaDe acquisition system

The signal from the antennas is amplified and filtered by an 88–108 MHz bandpass filter. The signal digitization is performed directly in RF without analogue down conversion process. Nyquist sampling requires at least 220 MHz sampling frequency. But the useful bandwidth of the whole FM band is 20 MHz, so the under-sampling technique can be exploited with a sampling frequency of at least 40 MHz. In the presented case, the sampling frequency was equal to 55 MHz which means that the signal is under-sampled. The effects of the aliasing are minimized by using bandpass filters. After sampling, the selected radio channel was digitally downconverted to baseband by a quadrature digital demodulator.

The data acquisition module (DAM) can perform demodulation of two radio channels simultaneously. Since the sampling frequency is much higher than the bandwidth of the signal (approximately 100 kHz), low-pass filtering and decimation is performed. The final sampling rate is equal to 200 kHz. The complex IQ samples are sent to a general purpose PC using USB 2.0 interface. A single DAM has two RF inputs and can therefore record signals from two antennas. Multichannel coherent registration can be performed with several DAMs connected in a daisy chain, where one module provides synchronization and triggering for the others. Figure 17.24 showing the PaRaDe system, the eight channel acquisition hardware used in all above-mentioned experiments is presented.

The measurement campaign with the passive radar mounted on a vehicle platform (depicted in Figure 17.25(a) and (b)) was carried out in the vicinity of Warsaw Airport in December 2007 [73]. The main objective of the first phase of the experiment was to investigate the effects of the platform movement on the signals received by passive radar.

(a) (b)

Figure 17.25 PaRaDe car platform: (a) front view, (b) side view

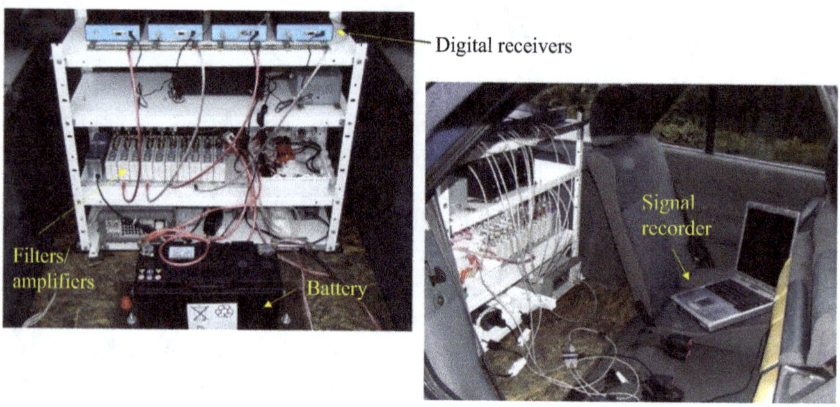

Figure 17.26 PaRaDe system on the car platform

A six-channel RX of the passive radar was mounted inside the car, as shown in Figure 17.26(a) and (b). The antenna array consisting of six half-wave dipoles was installed onto the roof of the car. The reference and echo beams were created digitally by appropriate summation of the signals from antenna array elements. The mobile platform was moving with velocity of 16–25 m/s at the distance of 5–15 km from the taking off/landing routes of aircraft.

According to the analysis presented in previous sections, the clutter signals should be shifted in frequency due to the platform motion, and due to the width of the antenna beam significant spread of the clutter in the frequency is expected. The experiments conducted using the ground vehicle confirmed those predictions [124].

Figure 17.27 shows the cross-ambiguity function of the reference and echo signals. As expected, the clutter is spread in Doppler due to the motion of the

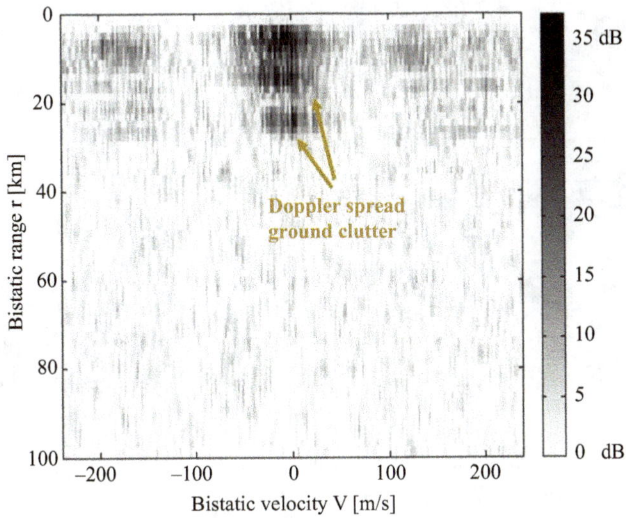

Figure 17.27 Cross-ambiguity function of a real-life signal recorded in the car campaign. Doppler-spread clutter visible

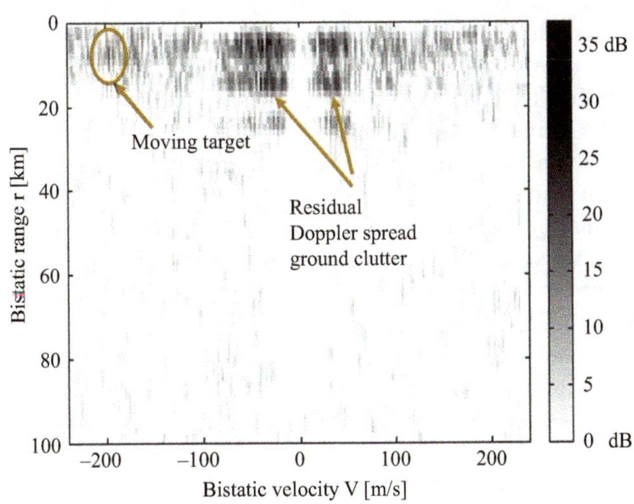

Figure 17.28 Cross-ambiguity function of a real-life signal recorded during the car campaign. Doppler-spread clutter removed

platform. One of the approaches to the problem of the spread clutter is to simply remove it with an adaptive filter. The result of such a procedure is shown in Figure 17.28, where the adaptive filter was applied for bistatic ranges from 0 to 100 km, and bistatic velocities from −240 to +240 m/s. This removed the clutter

(a) (b)

Figure 17.29 PaRaDe system on the airborne platform. (a) BRYZA plane on the runway; (b) FM antenna

effectively, so that a moving target, which previously was covered by the correlation sidelobes of the clutter, is now visible.

This approach is, however, disadvantageous from the point of view of detecting targets with small bistatic velocity, since they can overlap the clutter region and can be removed by the adaptive filter together with clutter. This is not a crucial problem in the case of a car, since the speed is relatively low (approximately 70 km/h); therefore, the clutter spread is not substantial. However, as will be shown later, the clutter spread is much more serious in the case of an airborne radar platform. An alternative solution to using an adaptive filter is to apply a method known from active airborne pulsed radars, such as DPCA or STAP.

An additional problem encountered during the trials was a strong multipath effect. Because the antennas were mounted at a low height (approximately 2 m above the ground), and the radar was moving, a strong variation of the signal amplitude could be observed.

This is a serious problem from the point of view of the reference signal recovery and digital beamforming. In the experiment, an analogue FM radio signal was used, therefore, channel equalization and signal reconstruction are not possible, or at least not very effective. In the case of using DVB-T or other digital modulation schemes, the multipath effects can be eliminated.

In the airborne trial campaign, carried out on the Polish coast of the Baltic Sea, the PCL sensor was mounted to a Skytruck aircraft as shown in Figure 17.29(a) [126]. The radar was equipped with a six-element antenna array. The antenna elements were placed inside the aircraft in window openings, as shown in Figure 17.29(b). There were three elements of the array on each side of the aircraft. In the experiment, two elementary antennas on one side of the aircraft were used as the surveillance antennas, and DPCA algorithm was used for clutter cancellation. The bistatic range-velocity surface calculated for each of the two antennas is presented in Figure 17.30 for recorded data and in Figure 17.31 for simulated data.

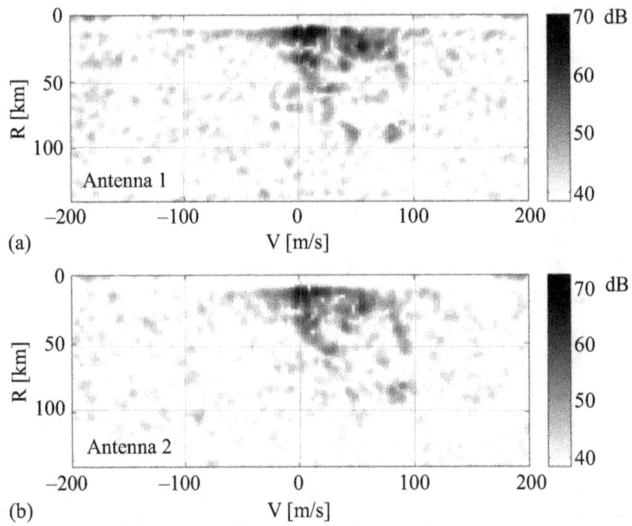

Figure 17.30 Range-velocity surface of a real-life signal obtained for: (a) antenna 1, (b) antenna 2

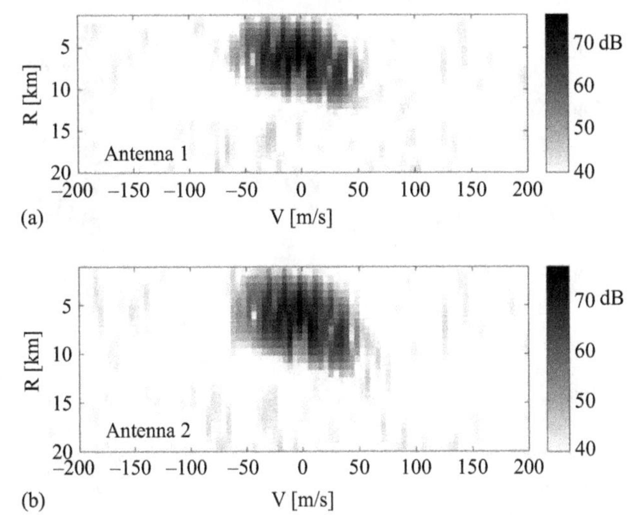

Figure 17.31 Range-velocity surface of a simulated signal obtained for: (a) antenna 1, (b) antenna 2

In the processing, the direct signal was removed using adaptive filter [22,137]. It can be seen that the clutter present in the signal has substantial spread in the bistatic velocity domain, which reached 100 m/s.

In order to remove the Doppler-spread clutter, a modified adaptive filtering could be used [133], but together with clutter slow target echoes will also be

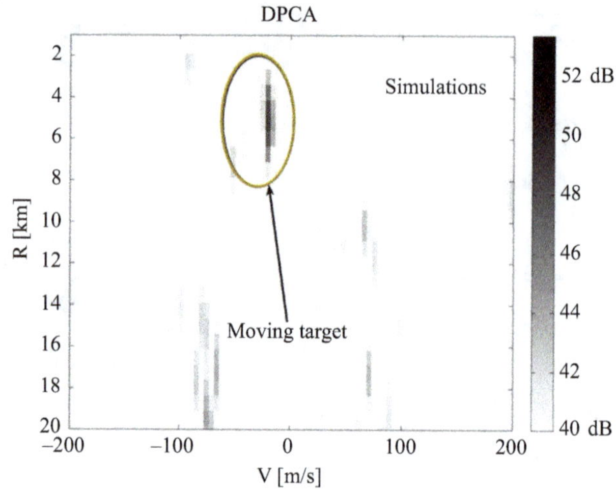

Figure 17.32 Range-velocity surface for a simulated signal obtained after DPCA processing

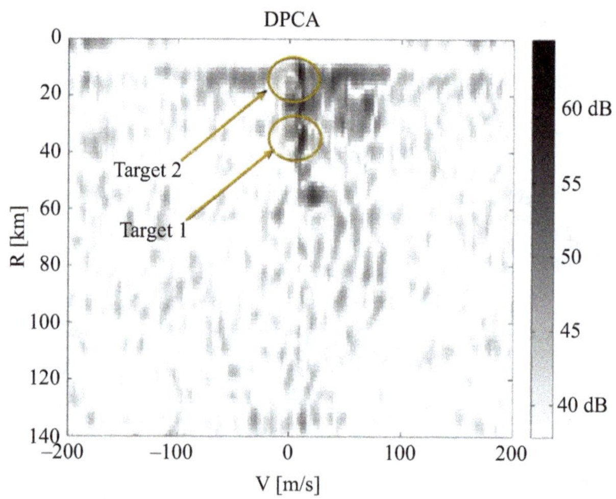

Figure 17.33 Range-velocity surface for a real-life signal obtained after DPCA processing

removed. For this reason, the DPCA algorithm is used instead. The results of the DPCA algorithm on the simulated data are shown in Figure 17.32 and for recorded data in Figure 17.33.

After the application of the DPCA, the ground clutter was suppressed by approximately 30 dB. The suppression ratio is limited by the time and spatial de-correlation of the clutter, as well as different radiation patterns of the two elements used in the DPCA algorithm. However, both in simulations and with real signal, suppression of clutter was sufficient to reveal one moving target for simulation (Figure 17.32) and two real targets (Figure 17.33) which were previously buried within the clutter.

17.2.8 Airborne passive radar conclusions

PCL systems dedicated for use on mobile platforms are very promising solution for future civilian and military application. The main advantage of passive radar mounted to moving platforms is the extension of its functionality in comparison to a classical stationary PCL system. Passive radar on an airborne platform has the potential to detect targets at long ranges (in the order of tens through to hundreds of kilometres), including low RCS targets, with a fast update rate (0.1–5 s) and high accuracy. Furthermore, the multistatic operation of airborne PCL increases the probability of the detection of the targets observed by radar. Low power consumption and the light weight of PCL systems make it possible to be installed in a small aircraft or UAV platform. Moreover, long target illumination allows us to use algorithms for target imaging, such as SAR and ISAR [77].

Mobile passive radar technology is, however, still in its infancy and requires thorough investigation. Any future work on mobile PCL should concentrate on multi-element antenna systems for different bands, DPCA and STAP, ground clutter cancellation techniques, multistatic operation of mobile passive system and sensors networking to exchange the data. Many researchers are now searching for the best possible processing schema for clutter cancellation [113,138] and reference signal acquisition and restoration [139], but it is still space for future algorithm development, test and optimization.

The ground clutter cancellation is not the only challenge for airborne passive radar. For ground-based passive radars, the horizon limits the number of visible illuminators. In most cases, TXs at distances longer than 50–100 km are not visible and do not interfere with echo signals. For the airborne radar placed at the height of 10 km all TXs till the distance 500 km are visible, and some of them can work at this some frequency as closes one, and then interfere with echo signals. To overcome this problem, an extended STAP algorithms designed for interference cancellation are required [140].

References

[1] D. Gould, R. Pollard, C. Sarno, and P. Tittensor. ‘A multiband passive radar demonstrator’, Proc. International Radar Symposium 2006, pages 657–660, Cracow, Poland, 24–26 May 2006.

[2] P. E. Howland, D. Maksimiuk, and G. Reitsma. FM radio based bistatic radar, *IEE Proceedings Radar, Sonar and Navigation*, 152(3):107–115, June 2005.

[3] H. Kuschel, M. Glende, J. Heckenbach, S. Müller, J. Schell, and C. Schumacher, ‘Experimental passive radar systems using digital illuminators (DAB/DVB)’,

Proc. International Radar Symposium 2007, pages 411–417, Cologne, Germany, 5–7 September 2007.

[4] A. Di Lallo, A. Farina, R. Fulcoli, P. Genovesi, R. Lalli, and R. Mancinelli, 'Design, development and test on real data of an FM based prototypical passive radar', Proc. IEEE Radar Conference 2008, Rome, Italy, pages 26–30 May 2008.

[5] M. Malanowski, K. Kulpa, and J. Misiurewicz, 'PaRaDe – Passive radar demonstrator family development at Warsaw University of Technology', Proc. Microwaves, Radar and Remote Sensing, Kiev, Ukraine, pages 22–24 September 2008.

[6] B. D. Nordwall, 'Silent Sentry' a new type of radar. *Aviation Week and Space Technology*, pages 70–71, Nov. 30, 1998.

[7] D. Poullin and M. Flecheux, 'Recent progress in Passive Coherent Location (PCL) concepts and techniques in France', Proc. IEEE Radar Conference 2008, Rome, Italy, pages 26–30 May 2008.

[8] H.D. Griffiths and C.J. Baker, 'The signal and interference environment in passive bistatic radar', Information, Decision and Control, 2007. IDC '07, pages 1–10, 2007.

[9] C. R. Berger, B. Demissie, J. Heckenbach, P. Willett, and S. Zhou, Signal processing for passive radar using OFDM waveforms. *IEEE Journal of Selected Topics in Signal Processing*, 4(1):226–238, 2010.

[10] C. Coleman and H. Yardley, 'DAB based passive radar: Performance calculations and trials', 2008 International Radar Conference, pages 691–694, 2008.

[11] C. Coleman and H. Yardley, Passive bistatic radar based on target illuminations by digital audio broadcasting. *IEE Proceedings Radar, Sonar and Navigation*, 2(October (5)):366–375, 2008.

[12] C. J. Coleman, R. A. Watson, and H. Yardley, 'A practical bistatic passive radar system for use with DAB and DRM illuminators', Radar 2008 International Radar Conference, pages 1–6, May 2008.

[13] P. E. Howland, Target tracking using television-based bistatic radar. *IEE Proceedings Radar, Sonar and Navigation*, 146(3):166–174, June 1999.

[14] D. Feng, H. Xie, D. An, and X. Huang, 'A fast back-projection algorithm for bistatic forward-looking low frequency UWB SAR imaging', 2015 16th International Radar Symposium (IRS), Dresden, pages 1135–1140, 2015.

[15] M. Cai, F. He, and L. Wu, 'Application of UKF algorithm for target tracking in DTV-based passive radar', Image and Signal Processing, 2009. CISP '09. 2nd International Congress, 2009.

[16] R. Saini and M. Cherniakov, DTV signal ambiguity function analysis for radar application. *IEE Proceedings Radar, Sonar and Navigation*, 152(June (3)):133–142, 2005.

[17] H. Guo, S. Coetzee, D. Mason, K. Woodbridge, and C. Baker, 'Passive radar detection using wireless networks', IET International Conference RADAR 2007, pages 1–4, 2007.

[18] H. Sun, D. K. P. Tan, Y. Lu, and M. Lesturgie, Applications of passive surveillance radar system using cell phone base station illuminators. *IEEE AES Magazine*, 25(3):10–18, 2010.

[19] D. K. P. Tan, H. Sun, Y. Lu, M. Lesturgie, and H. L. Chan, Passive radar using Global System for Mobile communication signal: theory, implementation and measurements. *IEE Proceedings Radar, Sonar and Navigation*, 152(3):116–123, June 2005.

[20] K. Chetty, G. Smith, Hui Guo, and K. Woodbridge, 'Target detection in high clutter using passive bistatic wifi radar', Radar Conference, 2009 IEEE, pages 1–5, May 2009.

[21] C. J. Baker, H. D. Griffiths, and I. Papoutsis, Passive Coherent Location radar systems. Part 2: Waveform properties. *IEE Proceedings Radar, Sonar and Navigation*, 152(June (3)):160–168, 2005.

[22] S. Rzewuski, M. Wielgo, K. Kulpa, M. Malanowski, and J. Kulpa, 'Multistatic passive radar based on WIFI – Results of the experiment', 2013 International Conference on Radar, Adelaide, SA, pp. 230–234, 2013.

[23] P. Maechler, N. Felber, and H. Kaeslin, 'Compressive sensing for WiFi-based passive bistatic radar', 2012 Proceedings of the 20th European Signal Processing Conference (EUSIPCO), Bucharest, pages 1444–1448, 2012.

[24] P. Falcone, F. Colone, and P. Lombardo, 'Localization of moving targets with a passive radar system based on WiFi transmissions', IET International Conference on Radar Systems (Radar 2012), Glasgow, UK, pp. 1–6, 2012.

[25] P. Falcone, F. Colone, and P. Lombardo. Potentialities and challenges of WiFi-based passive radar. *IEEE Aerospace and Electronic Systems Magazine*, 27(11):15–26, 2012.

[26] J. Brown, K. Woodbridge, A. Stove, and S. Watts, Air target detection using airborne passive bistatic radar. *Electronics Letters*, Vol. 46, 20 September 2010.

[27] B. Dawidowicz and K. Kulpa, 'Experimental results from PCL radar on moving platform', Proc. International Radar Symposium, Wroclaw, Poland, 2008, pages 305–308, 21–23 May 2008.

[28] K. Kulpa, M. Malanowski, J. Misiurewicz, M. Mordzonek, P. Samczyñski, and M. Smolarczyk, 'Airborne PCL radar: the concept and primary results', Proc. Military Radar, Amsterdam, The Netherlands, 2008, page CD, 27–29 October 2008.

[29] D. K. P. Tan, M. Lesturgie, H. Sun, and Y. Lu, 'Target detection performance analysis for airborne passive bistatic radar', 2010 IEEE International Geoscience and Remote Sensing Symposium, pages 3553–3556, 2010.

[30] H. Wuming and J. Wang, 'Airborne SAR passive radar imaging algorithm based on external illuminator', Synthetic Aperture Radar, 2007. APSAR 2007. 1st Asian and Pacific Conference on, pages 642–645, 2007.

[31] J. Gutierrez del Arroyo and J. A. Jackson, 'Passive SAR imaging using commercial OFDM networks', 2011 MSS Tri-Service Radar Symposium. 2011.

[32] M. Cetin and A. D. Lanterman, Region-enhanced passive radar imaging. *IEE Proceedings Radar, Sonar and Navigation*, 152(3):185–194, 2005.

[33] Q. Wu, Y. D. Zhang, M. G. Amin, and B. Himed, High-resolution passive SAR imaging exploiting structured bayesian compressive sensing. *IEEE Journal of Selected Topics in Signal Processing*, 9(8):1484–1497, 2015, DOI: 10.1109/JSTSP.2015.2479190.

[34] D. K. P. Tan, M. Lesturgie, H. Sun, and Y. Lu, 'Signal analysis of airborne passive radar using transmissions of opportunity', Proceedings of 2011 IEEE CIE International Conference on Radar, 1:169–172, 2011.

[35] D. Cerutti-Maori and J. H. G. Ender, Performance analysis of multistatic configurations for spaceborne GMTI based on the auxiliary beam approach. *IEE Proceedings Radar, Sonar and Navigation*, 153(2):96–103, 2006.

[36] K. Y. Li, U. Pillai, and B. Himed, 'Moving target geolocation in bistatic/passive SAR images using ATI', 2015 IEEE Radar Conference (RadarCon), pages 0045–0050, 2015.

[37] S. Wacks and B. Yazici, 'Bistatic Doppler-SAR DPCA imaging of ground moving targets', 2014 IEEE Radar Conference, 2014, pages 1071–1074, 2014, DOI: 10.1109/RADAR.2014.6875753.

[38] S. Wacks and B. Yazici, 'Passive synthetic aperture radar imaging of ground moving targets', *SPIE Defense, Security and Sensing Conference*, Baltimore, pp. 80510C, April 2012.

[39] S. Wacks and B. Yazici, Passive synthetic aperture hitchhiker imaging of ground moving targets—Part 1: image formation and velocity estimation. *IEEE Transactions on Image Processing*, 23(6):2487–2500, 2014, DOI: 10.1109/TIP.2014.2302682.

[40] G. Li, J. Xu, Y.-N. Peng, and X.-G. Xia, Bistatic linear antenna array SAR for moving target detection, location, and imaging with two passive airborne radars. *IEEE Transactions on Geoscience and Remote Sensing*, 45(3):554–565, 2007.

[41] K. Y. Li, U. Pillai, and B. Himed, 'Exploiting temporal proximity for moving target identification using bistatic/passive SAR', 2016 IEEE Radar Conference (RadarConf), pages 1–6, 2016.

[42] K. Kulpa, P. Samczynski, M. Malanowski, L. Maslikowski, and V. Kubica, The use of CLEAN processing for passive SAR image creation. Radar Conference (RADAR), 2013 IEEE, pages 1,6, April 29–May 3 2013.

[43] M. Cetin and A. D. Lanterman. Region-enhanced passive radar imaging. *IEE Proceedings Radar, Sonar and Navigation*, 152(3):185–194, 2005.

[44] W. Jun, Z. Xinwen, and B. Zheng, 'Passive radar imaging algorithm based on subapertures synthesis of multiple television stations', CIE International Radar Conference, pages 1–4, 2006.

[45] Y. Wu and D. C. Munson Jr., 'Wide-angle ISAR passive imaging using smoothed pseudo Wigner-Ville distribution', IEEE Radar Conference 2001, pages 363–368, 2001.

[46] T. Ziyue, Z. Zhenbo, Z. Lixiao, and S. Shaoying, 'Research on imaging of ship target based on bistatic ISAR', 2009 2nd Asian-Pacific Conference on Synthetic Aperture Radar, Xian, Shanxi, 2009, pp. 997–1000.

[47] C. A. Wiley. Synthetic aperture radars. *IEEE Transactions on Aerospace and Electronic Systems*, AES-21(May (3)):440–443, 1985.

[48] W. G. Carrara, R. S. Goodman, and R. M. Majewski, 'Spotlight Synthetic Aperture Radar', Norwood, MA: Artech House, 1995.

[49] K. Papathanassiou, F. Kugler, and I. Hajnsek, 'Exploring the potential of POL-InSAR techniques at X-band first results and experiments from Tandem-X', Proceedings of the 3rd International Asia-Pacific Conference on Synthetic Aperture Radar, Seoul, South Korea, pages 1–2, 26–30 Sept. 2011.

[50] H. D. Griffiths, C. J. Baker, J. Baubert, N. Kitchen and M. Treagust, 'Bistatic radar using satellite-borne illuminators', RADAR 2002, Edinburgh, UK, pages 1–5, 2002.

[51] L. Maslikowski, P. Samczynski, M. Baczyk, P. Krysik, and K. Kulpa, Passive bistatic SAR imaging—Challenges and limitations. *IEEE Aerospace and Electronic Systems Magazine*, 29(7):23–29, 2014.

[52] M. Rodriguez-Cassola, S. V. Baumgartner, G. Krieger, and A. Moreira, Bistatic TerraSAR-X/F-SAR Spaceborne–Airborne SAR experiment: Description, data processing, and results. *IEEE Transactions on Geoscience and Remote Sensing*, 48(2):781–794, 2010.

[53] M. Martone, P. Rizzoli, B. Brüutigam, and G. Krieger, First 2 years of TanDEM-X mission: Interferometric performance overview. *Radio Science*, 48(Sept.(5)):617–627, 2013.

[54] J. L. Bueso-Bello, M. Martone, P. Prats-Iraola, and B. Bräutigam, First characterization and performance evaluation of bistatic TanDEM-X experimental products. *IEEE Journal of Selected Topics in Applied Earth Observations and Remote Sensing*, 9(3):1058–1071, 2016.

[55] S. Reuter, F. Behner, H. Nies, O. Loffeld, D. Matthes, and J. Schiller, 'Development and experiments of a passive SAR receiver system in a bistatic spaceborne/stationary configuration', 2010 IEEE International Geoscience and Remote Sensing Symposium, pages 118–121, 2010.

[56] V. Kubica and X. Neyt, 'Passive SAR imaging using the ASAR instrument of ENVISAT as transmitter of opportunity', EUSAR 2012, 9th European Conference on Synthetic Aperture Radar, pages 275–278, 2012.

[57] V. Kubica, E. Cristofani, R. Hock, and X. Neyt, 'Strategies for the calibration of an array of patch antennas in passive bistatic SAR imaging', 2011 IEEE Radar Conference, pages 922–927, 2011, DOI: 10.1109/RADAR.2011.5960671.

[58] T. Zeng, T. Zhang, W. Tian, and C. Hu, Space-surface bistatic SAR image enhancement based on repeat-pass coherent fusion with Beidou-2/Compass-2 as illuminators. *IEEE Geoscience and Remote Sensing Letters*, 13(12):1832–1836, 2016.

[59] M. Cherniakov, E. Plakidis, M. Antoniou, and R. Zuo, 'Passive space-surface bistatic SAR for local area monitoring: Primary feasibility study', 2009 European Radar Conference (EuRAD), pages 89–92, 2009.

[60] M. Antoniou, Z. Zeng, L. Feifeng, and M. Cherniakov, Experimental demonstration of passive BSAR imaging using navigation satellites and a fixed receiver. *IEEE Geoscience and Remote Sensing Letters*, 9(3):477–481, 2012.

[61] Z. Zeng, M. Antoniou, Q. Zhang, M. Hui, and M. Cherniakov, 'Multi-perspective GNSS-based passive BSAR: Preliminary experimental results',

2013 14th International Radar Symposium (IRS), volume 1, pages 467–472, 2013.

[62] M. Antoniou, F. Liu, Z. Zeng, V. Sizov, and M. Cherniakov, Coherent change detection using GNSS-based passive SAR: First experimental results. IET International Conference on Radar Systems (Radar 2012), pages 1–5, 2012.

[63] M. Antoniou and M. Cherniakov, 'Experimental demonstration of passive GNSS-based SAR imaging modes', IET International Radar Conference 2013, pages: 1–5, 2013, DOI: 10.1049/cp.2013.0164.

[64] F. Santi, M. Antoniou, D. Pastina, D. Tzagkas, M. Bucciarelli, and M. Cherniakov, 'Passive multi-static SAR with GNSS transmitters: first theoretical and experimental results with point targets', EUSAR 2014; 10th European Conference on Synthetic Aperture Radar, pages 1–4, 2014.

[65] F. Santi, M. Bucciarelli, D. Pastina, M. Antoniou, and M. Cherniakov, 'Passive multistatic SAR with GNSS transmitters and using joint bi/multi-static CLEAN technique', 2016 IEEE Radar Conference (RadarConf), pages 1–6, 2016.

[66] F. Santi, D. Pastina, M. Bucciarelli, M. Antoniou, D. Tzagkas, and M. Cherniakov, 'Passive multistatic SAR with GNSS transmitters: Preliminary experimental study', 2014 11th European Radar Conference, pages 129–132, 2014.

[67] F. Santi, M. Bucciarelli, D. Pastina, M. Antoniou, and M. Cherniakov, Spatial resolution improvement in GNSS-based SAR using multistatic acquisitions and feature extraction. *IEEE Transactions on Geoscience and Remote Sensing*, 54(10):6217–6231, 2016.

[68] E. V. Appleton and M. A. F. Barnett, On some direct evidence for downward atmospheric reflection of electric rays. *Proceedings of the Royal Society*, 109(December):261–641, 1925.

[69] G. Krieger, H. Fiedler, D. Hounam, and A. Moreira, 'Analysis of system concepts for bi- and multi-static SAR missions', Proceedings of the IEEE International Geoscience and Remote Sensing Symposium, Toulouse, France, pages 770–772, 21–25 July 2003.

[70] C. Prati, F. Rocca, D. Giancola, and A. Monti Guarnieri, Passive geosynchronous SAR system reusing backscattered digital audio broadcasting signals. *IEEE Transactions on Geoscience and Remote Sensing*, 36(6):1973–1976, 1998.

[71] H. Ma, M. Antoniou, and M. Cherniakov, Passive GNSS-based SAR resolution improvement using joint Galileo E5 signals. *IEEE Geoscience and Remote Sensing Letters*,12(8):1640–1644, 2015, art. no. 7088585.

[72] A. Evers and J. A. Jackson, 'Experimental passive SAR imaging exploiting LTE, DVB, and DAB signals', 2014 IEEE Radar Conference, Cincinnati, OH, pages 0680–0685, 2014.

[73] J.R. Gutiérrez Del Arroyo and J.A. Jackson, WiMAX OFDM for passive SAR ground imaging. *IEEE Transactions on Aerospace and Electronic Systems*, 49(2), 945–959, 2013, art. no. 6494391.

[74] L. M. H. Ulander, P.-O. Frolind, A. Gustavsson, R. Ragnarsson, and G. Stenstrom, 'VHF/UHF bistatic and passive SAR ground imaging', IEEE International Radar Conference 2015 IEEE, pages 0669–0673, 10–15 May 2015.

[75] S. R. Stevens and J. A. Jackson, Emitter selection criteria for passive multistatic synthetic aperture radar imaging. *IET Radar, Sonar & Navigation*, 8(9):1267–1279, 2014, DOI: 10.1049/iet-rsn.2014.0014.

[76] D. Gromek, P. Samczynski, K. Kulpa, P. Krysik, and M. Malanowski, 'Initial results of passive SAR imaging using a DVB-T based airborne radar receiver', European Radar Conference (EuRAD), 2014 11th, pages 137,140, 8–10 Oct. 2014.

[77] M. K. Bączyk, P. Samczyński, and K. Kulpa, 'Passive ISAR imaging of air targets using DVB-T signals', Proc. of 2014 IEEE Radar Conference, Cincinnati, OH, USA, 19–23 May 2014.

[78] J. R. G. Del Arroyo and J. A. Jackson, WiMAX OFDM for passive SAR ground imaging. *IEEE Transactions on Aerospace and Electronic Systems*, 49(2):945–959, 2013, DOI: 10.1109/TAES.2013.6494391.

[79] J. Gutierrez del Arroyo and J. A. Jackson, 'Collecting and processing WiMAX ground returns for SAR imaging', 2013 IEEE Radar Conference (RadarCon13), pages 1–6, 2013, DOI: 10.1109/RADAR.2013.6586157.

[80] J. Gutierrez del Arroyo and J. Jackson, 'SAR imaging using WiMAX OFDM PHY', 2011 IEEE Radar Conference, pages 129–134. 2011.

[81] A. Dimitrov, L. Todor, and P. Kostadinov. Bistatic SAR/ISAR/FSR: Theory algorithms and program implementation, November 2013, Wiley-ISTE, ISBN: 978-1-84821-574-0.

[82] K. Kulpa, M. Malanowski, P. Samczynski, and B. Dawidowicz, 'The concept of airborne passive radar', 2011 Microwaves, Radar and Remote Sensing Symposium, Kiev, 2011, pages 267–270, 2011.

[83] Edmund G. Zelnio, Frederick D. Garber, 'Algorithms for Synthetic Aperture Radar Imagery XIX', edited by Edmund G. Zelnio, Frederick D. Garber, Proc. of SPIE Vol. 8394, May 1, 2012.

[84] B. Mulgrew, 'Algorithms for passive SAR imaging', IEEE/IET Waveform Diversity and Design, Jan. 2006.

[85] A. D. Lanterman and D. C. Munson, 'Deconvolution techniques for passive radar imaging. Algorithms for Synthetic Aperture Radar IX (Proceedings), volume 4727, 166–177. August 2002. Function for PCL Systems'. NAECON Conference in Dayton, OH, 2009 IEEE, 2009.

[86] L. Wang, I. Y. Son, and B. Yazici, 'Passive imaging using distributed apertures in multiple scattering environments,' *Inverse Problems*, 26: 065002(37 pages), May 2010.

[87] C. E. Yarman and B. Yazici, Synthetic aperture hitchhiker imaging. IEEE Transactions on Image Processing, 17(11):2156–2173, 2008.

[88] C.E. Yarman, L. Wang, and B. Yazici. 'Passive synthetic aperture radar imaging with single frequency sources of opportunity', IEEE International Radar Conference, 2010, pages 949–954. May 2010.

[89] E. Mason, I.-Y. Son, and B. Yazici, 'Passive synthetic aperture radar imaging based on low-rank matrix recovery', 2015 IEEE Radar Conference, 2015.

[90] A. Evers and J. A. Jackson, 'Experimental passive SAR imaging exploiting LTE, DVB, and DAB signals', 2014 IEEE Radar Conference, pages 0680–0685, 2014.

[91] P. O. Frölind, 'Results of airborne passive SAR ground and sea target imaging using DVB-T signals', 2016 IEEE Radar Conference (RadarConf), Philadelphia, PA, pages 1–4, 2016.

[92] C. Berthillot, A. Santori, O. Rabaste, D. Poullin, and M. Lesturgie, 'Improving BEM channel estimation for airborne passive radar reference signal reconstruction', 2015 16th International Radar Symposium (IRS), pages 77–82, 2015.

[93] L. M. H. Ulander, P. O. Froelind, A. Gustavsson, D. Murdin, and G. Stenstroem, 'Fast factorized back-projection for bistatic SAR processing', 8th European Conference on Synthetic Aperture Radar, Aachen, Germany, pages 1–4, 2010.

[94] O. Frey, C. Magnard, M. Rueegg, and E. Meier, 'Non-linear SAR data processing by time-domain back-projection', Synthetic Aperture Radar (EUSAR), 2008 7th European Conference on, pages 1,4, 2–5 June 2008.

[95] N. J. Willis, 'Bistatic Radar', Raleigh NC, USA: SciTech Publishing Inc, 2nd edition, 2005.

[96] K. Kulpa, 'The CLEAN type algorithms for radar signal processing', Microwaves, Radar and Remote Sensing Symposium, 2008. MRRS 2008, pages 152, 157, 22–24 Sept. 2008.

[97] H. Leung and X. Liu, 'Resolution enhancement in passive SAR imaging using distributed multiband fusion', 2013 IEEE Antennas and Propagation Society International Symposium (APSURSI), pages 1026–1027, 2013.

[98] D. Cristallini, P. Wojaczek, and M. Ummenhofer, 'Range resolution improvement for multiple PCL radar systems on moving platforms', 2015 IEEE Radar Conference, pages 394–399, 2015, DOI: 10.1109/RadarConf.2015.7411915.

[99] X. Mao, D. Yimin Zhang, and M. G. Amin, 'Two-stage multi-static passive SAR imaging with reduced complexity', 2014 IEEE Radar Conference, pages 0340–0343, 2014.

[100] J. Palmer, A. Summers, M. Ummenhofer, *et al.* 'Receiver platform motion estimation using terrestrial broadcast transmitters for passive radar', 2015 IEEE Radar Conference, pages 151–155, 2015.

[101] Kuang-Hung Liu and D. C. Munson, 'Autofocus in multistatic passive SAR imaging', *2008 IEEE International Conference on Acoustics, Speech and Signal Processing*, Las Vegas, NV, 2008, pp. 1277–1280.

[102] A. P. Whitewood, B. R. Muller, H. D. Griffiths, and C. J. Baker, 'Bistatic synthetic aperture radar with application to moving target detection', 2003 Proceedings of the International Conference on Radar (IEEE Cat. No.03EX695), pages 529–534, 2003, DOI: 10.1109/RADAR.2003.1278797.

[103] R. F. Ogrodnik, 'Bistatic laptop radar: an affordable, silent radar alternative,' Proceedings of the 1996 IEEE National Radar Conference, Ann Arbor, MI, pages 369–373, 1996.

[104] K. S. Kulpa and Z. Czekała, Masking effect and its removal in PCL radar. *IEE Proceedings-Radar, Sonar and Navigation*, 152(3):174–178, 2005.

[105] K. S. Kulpa, J. Misiurewicz, P. Samczyński, M. Smolarczyk, and M. Mordzonek, 'SAR image enhancement by dominant scatterer removal', Proceedings of the IET RADAR 2007, Edinburgh, Scotland, pages 15–18 October 2007.

[106] M. Alam, K. Jamil, and S. M. Alhumaidi, Target detection using space-time adaptive processing (STAP) and a multi-band, multi-channel software defined passive radar', 2015 European Radar Conference (EuRAD), pages 552–555, 2015, DOI: 10.1109/EuRAD.

[107] Z. Ul Mahmood, M. Alam, K. Jamil, and M. Elnamaky, 'On modeling and hardware implementation of Space-Time Adaptive Processing (STAP) for target detection in passive bi-static radar', 2012 11th International Conference on Information Science, Signal Processing and their Applications (ISSPA), pages 1013–1017, 2012, DOI: 10.1109/ISSPA.

[108] X. Neyt, J. Raout, M. Kubica, *et al.* 'Feasibility of STAP for passive GSM-based radar', 2006 IEEE Conference on Radar, page 6, 2006, DOI: 10.1109/RADAR.2006.1631853.

[109] Q. Wu, Y. D. Zhang, M. G. Amin, and B. Himed, 'Space-time adaptive processing in bistatic passive radar exploiting group sparsity', 2015 IEEE Radar Conference, pages 0886–0890, 2015, DOI: 10.1109/RADAR.

[110] Q. Wu, Y. D. Zhang, M. G. Amin, and B. Himed, Space-time adaptive processing and motion parameter estimation in multistatic passive radar using sparse bayesian learning. 54(2):944–957, 2016.

[111] J. Raout, X. Neyt, and P. Rischette, 'Bistatic STAP using DVB-T illuminators of opportunity', 2007 IET International Conference on Radar Systems, pages 1–5, 2007, DOI: 10.1049/cp:20070670.

[112] X. Neyt, M. Acheroy, and J. G. Verly. 'Maximum-likelihood based range-dependence compensation for coherent multistatic STAP radar', 2007 IEEE Radar Conference, pages 772–777, 2007, DOI: 10.1109/RADAR.2007.374317.

[113] P.-C. Yang, X.-D. Lyu, Z.-H. Chai, D. Zhang, Q. Yue, J.-M. Yang, Clutter cancellation along the clutter ridge for airborne passive radar. *IEEE Geoscience and Remote Sensing Letters*, PP(99):1–5, 2017.

[114] R. Klemm, *Principles of Space-Time Adaptive Processing*, London, UK: IET, 3rd edition, 2006.

[115] H. D. Griffiths and C. J. Baker, *An Introduction to Passive Radar*, Boston, London: Artech House, 2017.

[116] K. Kulpa, 'The clean type algorithms for radar signal processing', Microwaves, Radar And Remote Sensing Symposium, 2008. MRRS 2008, pages 152–157, 2008.

[117] K. Kulpa, Z. Czekała, and M. Smolarczyk, 'Long-time-integration surveillance noise radar', First International Workshop On The Noise Radar Technology (NRTW 2002), Yalta, Crimea, Ukraine, pages 238–243, 18–20 September 2002.

[118] K. Kulpa and J. Misiurewicz, 'Stretch processing for long integration time passive covert radar', Proceedings of 2006 CIE International Conference on Radar, Shanghai, China, 2006.

[119] M. Malanowski and K. Kulpa, 'Analysis of integration gain in passive radar', Proc. Radar 2008, Adelaide, Australia, pages 323–327, 2–5 September 2008.

[120] H. D. Griffiths and C. J. Baker, Passive coherent location radar systems. Part 1: Performance prediction. *IEE Proceedings Radar, Sonar and Navigation*, 152(June (3)):153–159, 2005.

[121] M. I. Skolnik, *Radar Handbook, Third Edition*, New York, McGraw-Hill, 2nd edition, 2008.

[122] R. N. McDonough and A. D. Whalen, *Detection of Signals in Noise*, New York, London: Academic Press; 2nd edition, 1995.

[123] J. Brown, K. Woodbridge, H. Griffiths, A. Stove, and S. Watts, Passive bistatic radar experiments from an airborne platform. *IEEE Aerospace and Electronic Systems Magazine*, 27(11):50–55, 2012.

[124] B. Dawidowicz and K. S. Kulpa, 'Experimental results from PCL radar on moving platform', 2008 International Radar Symposium, Wroclaw, pages 1–4, 2008.

[125] B. Dawidowicz, K. S. Kulpa, M. Malanowski, J. Misiurewicz, P. Samczynski, and M. Smolarczyk, DPCA detection of moving targets in airborne passive radar. *IEEE Transactions on Aerospace and Electronic Systems*, 48(April (2)):1347–1357, 2012.

[126] B. Dawidowicz, P. Samczynski, M. Malanowski, J. Misiurewicz, and K. S. Kulpa, Detection of moving targets with multichannel airborne passive radar. *IEEE Aerospace and Electronic Systems Magazine*, 27(11):42–49, 2012.

[127] B. Dawidowicz, K. S. Kulpa, and M. Malanowski, 'Suppression of the ground clutter in airborne PCL radar using DPCA technique', 2009 European Radar Conference (EuRAD), Rome, pages 306–309, 2009.

[128] M. Malanowski and K. Kulpa, 'Digital beamforming for passive coherent location radar', Proc. IEEE Radar Conference, Rome, Italy, 26–30 May 2008.

[129] M. Malanowski, G. Mazurek, K. Kulpa, and J. Misiurewicz, 'FM based PCL radar demonstrator', Proc. International Radar Symposium 2007, Cologne, Germany, pages 431–435, 5–7 September 2007.

[130] G. Mazurek, M. Malanowski, and K. Kulpa, 'Flexible hardware platform for software radio experiments', Proc. SPIE – Photonics Applications in Astronomy, Communications, Industry, and High-Energy Physics Experiments IV, Jachranka, Poland, volume 6937, pages 69373X-1–69373X-8, 24–26 May 2007.

[131] K. Kulpa, 'Adaptive clutter rejection in bi-static CW radar', Proc. International Radar Symposium 2004, Warsaw, Poland, pages 61–66, 19–21 May 2004.

[132] K. Kulpa, 'Ground clutter suppression in noise radar', Proc. International Conference on Radar, Toulouse, France, 18–22 October 2004.

[133] K. Kulpa, 'Simple sea clutter canceller for noise radar', Proc. International Radar Symposium 2006, Cracow, Poland, pages 299–302, 24–26 May 2006.

[134] M. Malanowski, 'Comparison of adaptive methods for clutter removal in PCL radar', Radar Symposium, 2006. IRS 2006. International, pages 1–4, May 2006.

[135] W. L. Melvin, M. J. Callahan, and M. C. Wicks, 'Adaptive clutter cancellation in bistatic radar', Signals, Systems and Computers, 2000. Conference Record of the Thirty-Fourth Asilomar Conference on, volume 2, pages 1125–1130, 2000.

[136] R. Saini, M. Cherniakov, and V. Lenive, 'Direct path interference suppression in bistatic system: DTV based radar', Proc. International Conference on Radar, Adelaide, Australia, pages 309–314, 3–5 September 2003.

[137] R. Cardinali, F. Colone, C. Ferretti, and P. Lombardo, 'Comparison of clutter and multipath cancellation techniques for passive radar', Radar Conference, 2007 IEEE, pages 469–474, 2007.

[138] D. K. P. Tan, H. Sun, Y. Lu, and M. Lesturgie, 'Signal processing for airborne passive radar', *2014 11th European Radar Conference*, Rome, 2014, pp. 141–144.

[139] H. Gao, W. Zhang, F. Jia, and Z. He, 'The extraction method for direct path wave in the airborne passive radar system', 2016 IEEE 13th International Conference on Signal Processing (ICSP), pages 1589–1592, 2016, DOI: 10.1109/ICSP.2016.7878094.

[140] J. R. Lievsay and N. A. Goodman, 'Multi-transmitter clutter modeling for passive STAP', 2016 IEEE Radar Conference (RadarConf), pages 1–6, 2016.

Chapter 18

Multi-illuminator and multistatic passive radar

Heiner Kuschel[1], *Fabienne Hoffmann*[2] *and Alexander Schroeder*[2]

Abstract

Passive radar (PR) is always dependent on illuminators of opportunity and the waveform they provide. Using multiple different illumination sources as well as the distribution of multiple sensors in multistatic networks can provide some diversity. As examples some multistatic configurations and multi-band passive coherent location (PCL)-systems are sketched indicating the wide range of applications. Furthermore, the properties of illuminators, which can be used for medium range PR processing, like frequency modulated radio, digital audio broadcast and digital video broadcast terrestrial are described. Special signal processing approaches are highlighted and a hybrid PR system and processing concept is explained. The chapter is completed with some measurement results from multistatic experiments and a civilian PCL-network application of a multistatic PCL system.

18.1 Introduction

Since one of the most critical aspects of passive radar (PR) is its dependence on illuminators of opportunity with properties, which cannot be influenced, diversity is a powerful means to alleviate this challenge. The use of multiple different illumination sources as well as the distribution of multiple sensors in multistatic networks can provide this diversity. Multistatic configurations and multi-band PCL-Systems (Passive Coherent Location) exploiting different broadcast networks with different properties and wave forms provide the flexibility to make use of advantageous properties of the illuminating services, while avoiding to be stuck

[1]Fraunhofer Institute for High Frequency Physics and Radar Techniques FHR, Germany
[2]Airbus Defence & Space, Germany

with potential shortfalls. Furthermore, a multistatic geometry providing multiple illumination angles as well as multiple bistatic angles increases the probability of observing an advantageous radar cross section of a target. This chapter will cover a fairly wide range of applications from military low level area coverage to a collision warning sensor network for wind turbines. A hybrid PCL system concept will be described as the future perspective for a versatile PR system. Demonstrator systems reflecting the experiences of the authors will be addressed and measurement results shall illustrate their performance capabilities.

18.1.1 Multistatic PCL configurations

Multistatic PCL configuration can consist of various concepts, which may also require different signal processing approaches [1]. The most general system concept is based on multiple netted receivers exploiting the illumination of multiple spatially distributed transmitters. Dependent on the complexity the receiver provider can master, each sensor can measure the time difference of arrival, direction and bistatic Doppler of echoes excited on one target by multiple transmitters. Thus an over determined target localization procedure can provide a high probability of detection by multiple simultaneous detections of the same target in a network. In reality, however, this is hardly affordable and thus more realistic multistatic system concepts either exploit multiple transmitters with one receiver, preferably featuring direction of arrival and time difference of arrival measurements, or consist of a network of omnidirectional passive receivers exploiting one dominant transmitter. In the latter case, the localization procedure is based on the calculation of intersecting ellipsoids, where the intersections indicate the location of targets. The focal points of the ellipsoids are the locations of the transmitter and the receiver. Here, the target resolution depends on the signal bandwidth and multiple ellipsoid intersections lead to improved accuracy.

In the former case with the knowledge of the transmitter locations and the bearing measurement of the target echoes, the location of targets can be determined by the intersection of the bearing vector with the time difference of arrival ellipsoid. The resolution here depends on the signal bandwidth and the beamwidth of the antenna providing the bearing information.

18.1.2 Multi-band PCL systems with spectrally orthogonal illuminators

In the case of multi-illuminator configurations, it makes a huge difference if the transmitter signals are orthogonal in frequency, like different frequency modulated (FM)-broadcast channels, so that target echoes can unambiguously be assigned to a transmitter. Thus, also in such PCL systems, which exploit the emission of multiple, spectrally orthogonal transmitters, an ellipsoid intersection procedure can be applied for target localization. This is the case when using FM-radio stations of different carrier frequencies as illuminators. It is more complex in the case of digital TV or digital radio where the broadcast signals are transmitted in single frequency networks, where all transmitters transmit coherently the same signal.

After highlighting the illuminator properties and the specific signal processing steps for FM broadcast transmitters as example for multi-frequency networks and digital audio broadcast (DAB)/DVB-T (Digital Audio Broadcast, Digital Video Broadcast-Terrestrial) transmitters representing single frequency networks (see also [2]), a hybrid PR system will be presented, followed by two demonstrator examples.

18.2 Passive radar processing for sensors using FM broadcast transmitters

18.2.1 Illuminator properties

FM radio is the world's most widespread broadcast standard. Its most important characteristics including some consequences for PR operation are shown in Table 18.1.

FM is well suited for PR long-range coverage due to its large transmit power. Compared to other broadcasting types FM transmitters typically show a relatively wide antenna elevation characteristic and depending on their location often do not use a down-tilt of the beam [3]. Figure 18.1 depicts the elevation diagram of a typical FM transmit antenna [4].

Thus, FM provides a good airspace illumination even for high altitude targets. However, as target range accuracy and range resolution depend on the signal bandwidth, both are limited due to the low and varying bandwidth. The ambiguity functions resulting from different kinds of signal content differ clearly. Speech exemplarily leads to low resolution and high sidelobes due to its low bandwidth, whereas reggae results in a narrow peak and low sidelobes. Therefore, as the performance of a FM-based PR depends on the signal content, it is important to choose the operating frequencies carefully.

Nevertheless, there are several ways to enhance the localization accuracy resulting from the range resolution. It can be improved e.g. by multistatic illumination in combination with adequate data fusion and tracking algorithms. Also, several FM-signals at different frequencies but originating from the same transmitter station can be combined to obtain increased bandwidth and thereby to further improve the range resolution [5].

Table 18.1 Typical characteristics of FM transmitters and consequences for passive radar operation

Frequency band	88–108 MHz
Bandwidth	0–150 kHz variable with program content
Range resolution	>2 km dependent on bandwidth
EIRP	10–100 kW
3 dB Beamwidth in elevation	Approx. 9°
Detection capabilities	Detections at far range and high altitudes possible Low resolution

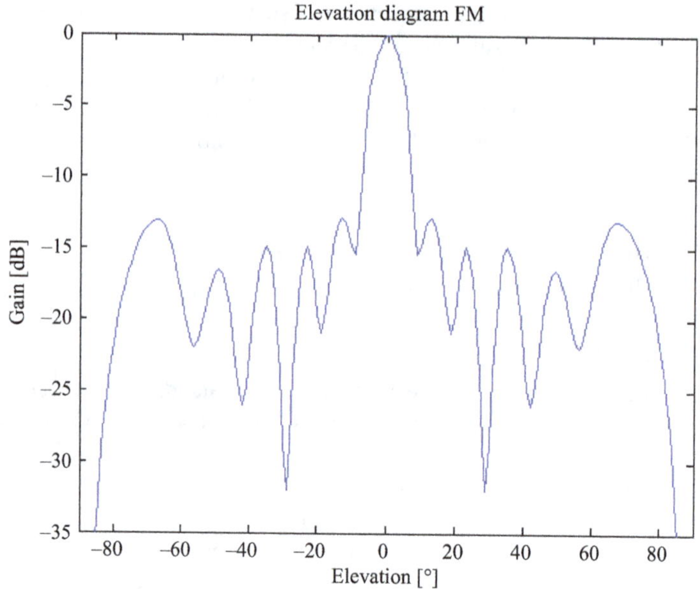

Figure 18.1 Elevation diagram of typical FM transmit antenna

18.2.2 Direct signal suppression

Figure 18.2 shows a rough overview of the single PCL signal processing steps. The single steps are described in the following. Input for the signal processing is IQ-data, consisting among others of the direct signal of the transmitter and target echoes.

In order to reveal targets, which are masked by the much stronger direct signal of the transmitter inside the range Doppler map, two signal processing methods are combined.

With the help of adaptive beamforming the direct signal is suppressed by forming a zero in the direction of the transmitter inside the beamforming pattern. As the direct signal is quite important for the correlation process, in a second beamforming step the direct signal is extracted as well for reference.

In addition, the so-called CLEAN algorithm [6] is used to subtract remaining parts of direct signal and clutter in the final range Doppler map after correlation.

The result can exemplarily be seen in Figure 18.3 which shows a range Doppler map before (left) and after (right) applying the CLEAN algorithm.

Finally, Constant False Alarm Rate (CFAR) detection is used to distinguish between peaks induced by noise and target echoes.

The output of the signal processing are plot data which are still in the bistatic space and consist mainly of bistatic range and bistatic Doppler. By combining the bistatic detections of different transmitters inside tracking with a multi-lateration approach, the position in 3D-space can be calculated. Depending on the number of

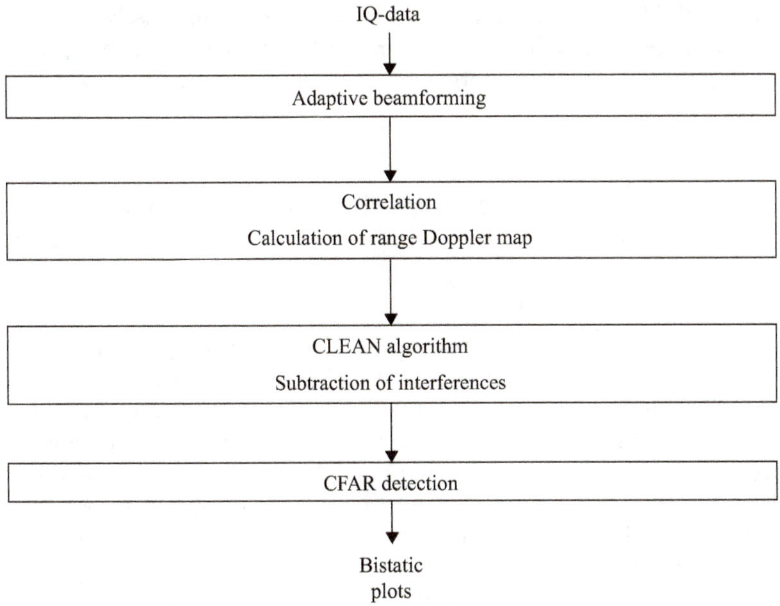

Figure 18.2 Overview of PCL signal processing

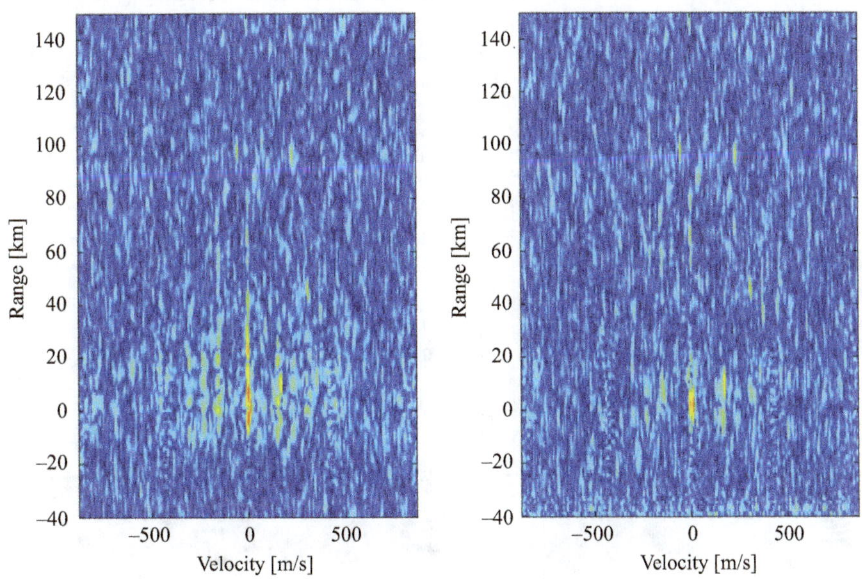

Figure 18.3 Range Doppler map before (left) and after (right) CLEAN

transmitters the localization accuracy is then increased immensely compared to the bistatic range resolution. Therefore, tracking is a rather important part of the PR functionality. For further information on tracking see Part II of Volume 2 of this book.

18.3 Passive radar processing using digital broadcast transmissions

18.3.1 Illuminator properties

When broadcast illuminators of opportunity with digital wave forms are used for PR purposes, namely DAB (DAB+) and DVB-T(2), the field strength produced by a station depends, inter-alia, on effective radiated power, antenna heights, local terrain and tropospheric scattering conditions. The antenna systems usually consist of several individual radiating bays fed as a phased-array. Their radiation characteristics concentrate the energy in the horizontal plane towards the population to be served, minimizing the radiation out into space, which requires the vertical plane radiation pattern to be tilted slightly below the horizontal. This is a common procedure in broadcast engineering and is referred to as beam-tilt. DAB- and DVB-T antennas generally have different elevation beamwidths, which make DAB-illumination more suitable for higher elevation coverage. Examples of elevation diagrams for typical DAB- and DVB-T antennas with beams of 11 and 3°, respectively, are given in Figure 18.4.

Both DAB and DVB-T standards utilize the 'coded orthogonal frequency division multiplex' (COFDM) modulation. The DAB signal (according to EN 300 401, mode 1) is an orthogonal frequency division multiplex (OFDM) [7] coded CW signal, which is subdivided into frames of 96 ms duration. Each frame consists of

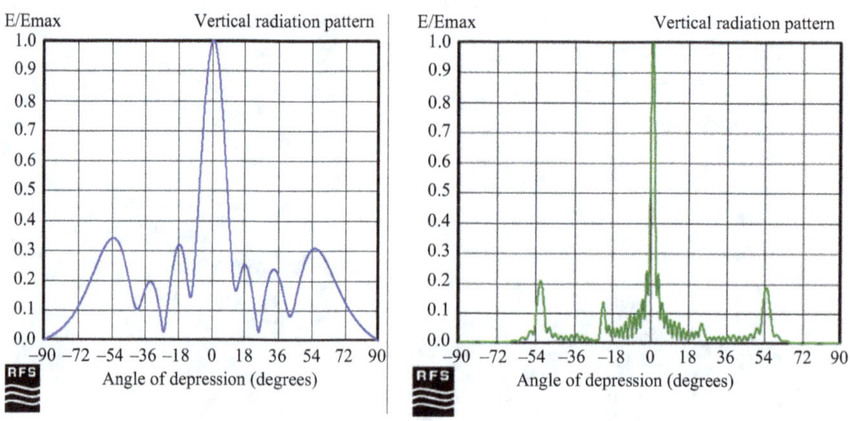

Figure 18.4 DAB (11°, left) and DVB-T (3°, right) antenna elevation pattern, Courtesy of Kevin Thompson, RFS World, Australia

76 sections, called symbols with a length of 1.246 ms, each, plus a so-called null-symbol of 1.296 ms length, where no signal is emitted. It leads the data symbol flow of a frame and is used for a coarse time synchronization of the receiver. The DVB-T standard [8] uses OFDM modulation schemes, where a number of N closely-spaced orthogonal carrier frequencies are used to carry the Moving Picture Experts Group-2 coded video signal. All carriers in one OFDM symbol are transmitted simultaneously. An OFDM-frame is constituted by a set of 68 OFDM-symbols. All data carriers in one symbol are modulated using quadrature phase shift keying, 16-quadrature amplitude modulation (QAM) or 64-QAM modulations.

18.3.2 Single versus multi-frequency networks

Digital broadcast services in some countries are operating in so-called single frequency networks, where all transmitters in the network transmit coherently the same signal at the same time. This requires a highly synchronous network. Synchronization is generally achieved by locking the transmitter oscillators to a global positioning system (GPS) reference, which then can also be used to synchronize the receiver. PR operation in such a single frequency network is only unambiguous for target echoes, which have time delays with respect to the closest transmitter (first) of the network, which are shorter than the delay of the signal of the second closest transmitter (second). Target echoes with longer time delays may have been exited by either the first or the second transmitter and the target location is thus ambiguous. Furthermore, some broadcast networks deliberately introduce delays into the transmissions in order to equalize the coverage. This makes it more complicated to identify the receivers from their time delay pattern at the receiver location. Identification of the transmitter in order to determine its location, however, is vital for target localization using the method of intersecting multiple ellipsoids.

Other broadcast networks operate in so-called multi-frequency networks of transmitters, where all transmitters use their own frequency. This makes it of course easy to identify the source of an echo; however, the synchronization of the receiver to the transmitter signal cannot be accomplished by GPS since there is no need for the transmitter to operate in a GPS disciplined mode. Hence, there can always be a phase jitter on the transmitter signal, which is difficult to compensate and can lead to an increased noise floor.

18.3.3 Signal reconstruction

In order to exploit the DVB-T signal for radar purposes, the decoding of the received signal is required. This ensures that an ideal replica can be synthesized, which resembles the original coded signal as purely as possible. This signal is then continuously updated and used in a correlation processing with the measured signal. In a first encoding step, the received signal is synchronized using a reference sequence. This sequence is a synthesized segment of four DVB-T OFDM symbols containing only the known modulation of the scattered and continual pilots. Correlating both signals results in a detection statistic that has a dominant peak at the synchronization position. The correct location k of each carrier within the received OFDM symbol can be determined.

The received signal in the frequency domain can thus be represented as

$$Y(k) = \sum_{n=0}^{N-1} y(n)e^{-(j2\pi kn/N)}, n = 0, 1, 1, \ldots, N-1 \tag{18.1}$$

$$= X(k) \cdot H(k) + W(k) \tag{18.2}$$

where $H(k)$ denotes the channel response function of the channel and $W(k)$ AWG noise contribution.

Extracting the pilot carrier contribution $Y(k)$ from the signal, the transfer function of the pilots is determined by

$$\hat{H}(l, k_p) = \frac{Y(l, k_p)}{X(l, k_p)} \quad k_p = 1, 2, \ldots, N_p \tag{18.3}$$

where $X(l,\ k_p)$ denotes the well-known pilot carrier modulation in symbol l.

The channel characteristics for the remaining carriers can now be obtained with linear extrapolation.

This algorithm uses the transfer functions from two successive pilot carriers k_p and k_{p+1} to determine the channel response for the data carriers k_d in between.

Let L_d be the number of carriers between two pilots then the transfer function is given by

$$H(k) = \hat{H}(k_p) + \frac{d}{L_d}\left(\hat{H}(k_p + 1) - \hat{H}(k_p)\right) \tag{18.4}$$

where $0 \le d \le L_d$.

After applying the transfer function to each carrier, the resulting reconstructed signal is de-mapped into the digital data stream. Figure 18.5 shows the phase states of a reconstructed OFDM-symbol in a so-called constellation diagram of a 16 QAM DVB-T signal.

In contrast to the DVB-T standards, DAB and DAB+ use a null symbol and a reference symbol for synchronization.

In DAB, each data-frame starts with a null-symbol (no signal) for synchronization, followed by a reference symbol, which can be exploited for signal reconstruction. Comparing the phases of the specified reference symbol with the phases of a measured DAB-signal after all delayed signal contributions have been collected allows the estimation of the impulse response $h(t)$ of the environment.

$$r(t) = s(t) * h(t) \tag{18.5}$$

The knowledge $h(t)$ allows to reconstruct the originally coded digital broadcast signal as it has been transmitted by each transmitter in the net. As the environment may be subject to changes, the estimation of the transfer function has to be updated continuously.

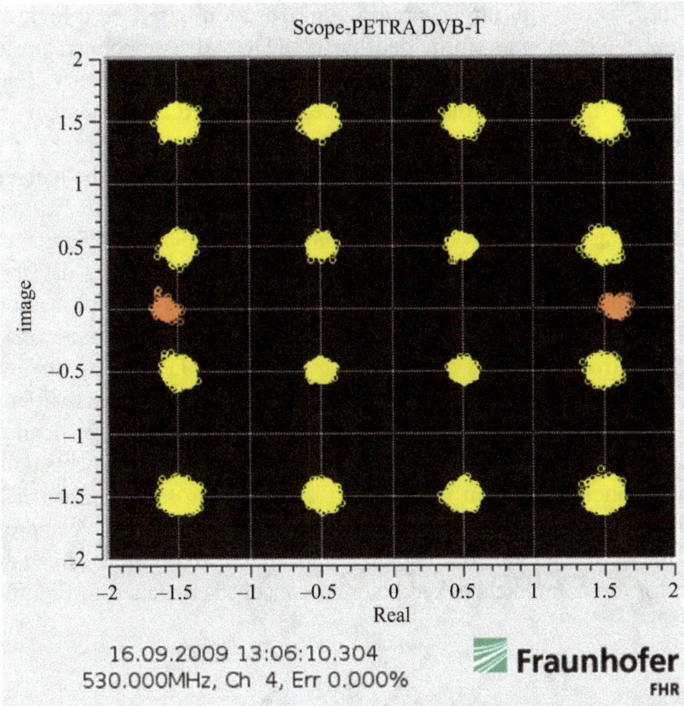

Figure 18.5 Constellation diagram of a 16 QAM symbol after channel estimation

18.4 A hybrid passive radar processing concept

The use of different types of illuminators, ranging from analogue FM-radio over digital broadcasters like DAB and DVB-T to cell-phone base stations, WiFi and satellite borne transmitters has been considered for short and medium range low level air target detection [7–9]. They are characterized by individual strengths and weaknesses with respect to radar requirements. In addition to the signal modulation characteristics, their elevation illumination coverage has been analysed. As technology has now reached a sufficient readiness level, a hybrid PR concept for medium-range air surveillance has been proposed [10] combining the best properties of selected donor networks for extended range and improved target resolution.

The system concept comprises a low frequency target detection component (TDC) using FM-radio (88–108 MHz) emissions, a 'high' frequency target resolution component (TRC) exploiting DVB-T emissions (450–900 MHz) and an intermediate frequency component covering close-in high elevation airspace by utilizing DAB or digital media broadcast signals.

The TRC and the high elevation component (HEC) are being cued by the TDC, which provides longer range at the cost of a coarse resolution and reduced accuracy. Both the TRC and the HEC apply track-before-detect strategies and extended coherent integration with target motion compensation based on a high resolution target Doppler measurement provided by the TDC. All components are equipped with omni-directional array antennas for permanent volume surveillance with multiple beams. In addition, the antenna feeding the TRC, has elements spaced in two vertical planes to allow target elevation estimation. The signal processing front-end applies the software-defined radar principle, sampling the received signal according to the used bandwidth after band filtering in the radio frequency (RF) domain using fast A/D converters with high dynamic range. Target detection, tracking, cued tracking and height estimation are performed in a data processing component comprising clustered high performance off-the-shelf processors.

A circular array of vertical dipoles covering the frequency band of 88 to 240 MHz feeds the front-end units of the TDC and the HEC. Depending on the required degree of mobility/transportability, the chosen number of elements varies between 8 and 16. The TRC, using DVB-T emissions, is fed by two vertically stacked circular arrays of vertical broad band dipoles. The number of elements doubles that of the low-frequency array in both planes, thus leading to a number of elements ranging from 32 to 64. All three circular arrays are designed to have approximately the same diameter and can be mounted stacked on a common extendable mast. A scheme for the complete stacked antenna concept is depicted in Figure 18.6.

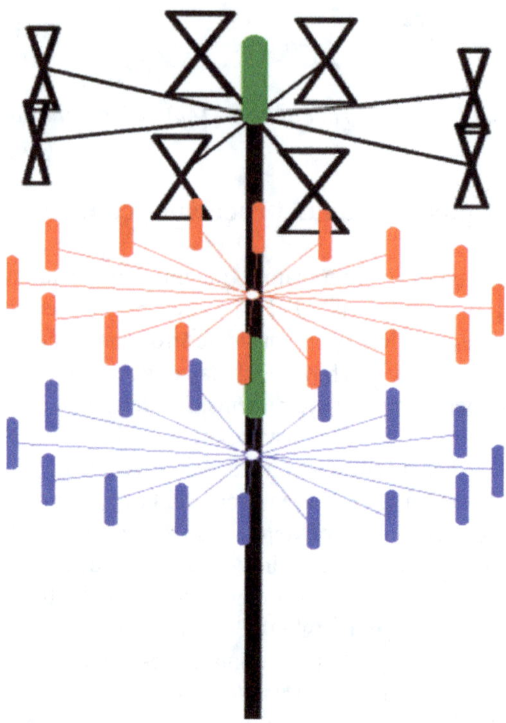

Figure 18.6 Layered multi-frequency antenna concept (black = 88–249 MHz, red and blue = 450–900 MHz, green = Calibration elements)

The principal signal processing concept of PR is based on the cross correlation of a reference which is a replica of the transmitted broadcast signal and a delayed echo signal, which is reflected by a target illuminated by the transmitter.

Both, the reference signal and the echo signal are obtained from the circular array by digital beamforming after direct RF sampling. After proper bandpass filtering and low-noise amplification sampling rates of up to 100 Msamp/s allow the digitization – by under sampling – of the whole FM-spectrum or several DAB or DVB-T channels, respectively, at a time. Parallel processing provides simultaneously a reference channel and target detection channels pointing in pre-determined directions.

The general difference between the TDC using FM-radio signals and the HEC and TRC using DAB and DVB-T signals, respectively, is the generation of the reference signal. In the TDC the reference signal is generated through a separate reference channel, which is directed towards the transmitter for optimum signal quality.

The digital waveforms, COFDM, used in the HEC and TRC contain – in contrast to FM-radio – synchronization information, which is required for proper demodulation. It can be exploited to reconstruct the transmitted signal from the received mix of direct signals and multi-path signals and generate a clean reference signal without multi-path interferences.

Cross-correlation of the reference channel and the target echo channels and a Fourier transform of time-spaced cross correlations provide a detection matrix spread in bistatic range and bistatic range rate (Doppler). The range resolution is determined by the signal bandwidth and the Doppler resolution by the integration time. The detection threshold is determined by a range CFAR, which moves an averaging window of range cells along with the cell under examination to adapt the detection threshold to the local noise floor. A detection is declared, if the threshold is exceeded in three or more neighbouring resolution cells. Such detection clusters are used as seeds for range/Doppler tracks.

As stated before, the range resolution is dependent on the signal bandwidth and determines the time, a target with a given radial velocity, where the term radial applies only to a quasi-monostatic geometry, stays within the range cell and thus the possible coherent integration time. Hence, the rather low bandwidth of FM-radio signals providing range resolutions in the order of several kilometres allows coherent integration times of up to 1 s. In contrast, the comparably high bandwidths of DAB or DVB-T signals of 1.5 and 7.6 MHz, respectively, provide high range resolution but drastically reduced coherent integration times.

If the target motion is known, e.g. from detections in the FM-radio component of a system, procedures can be applied to compensate the motion dependent phase and range walk in the other (DAB, DVB-T) components and extend the coherent integration time beyond the time the target stays within the range cell. Such procedures have been developed e.g. for high range resolution radar applications like ISAR (Inverse Synthetic Aperture Radar) and sound navigation and ranging applications. They, however, require the knowledge of the target's motion. In ISAR applications, where target super-resolution is attempted, a recommended solution is the tracking of a single scattering centre to determine the target motion. In PR, a small resolution cell simply prevents coherent integration of sufficient echo energy to exceed the detection threshold.

With the comparably coarse range resolution FM-radio based PR provides, long integration time in the order of a second and more can be spent to increase the detection range. When cueing e.g. a DVB-T based PR (TRC) with detections from the FM-based TDC, the integration time, consistent with the resolution cell may be extended considerably, provided that motion compensation procedures and multi-hypothesis tracking can be applied for longer detection ranges and larger coverage.

A multi-hypothesis track before detect procedure has to be developed on the basis of the Doppler measurements performed in the TDC.

Let the carrier frequency of the transmitter providing the Doppler measurement be f_1 and the measured Doppler frequency f_{D1}, the corresponding estimated target Doppler shift in the TRC will be

$$f_{D2} = f_{D1} \times \frac{f_2}{f_1} \tag{18.6}$$

with f_2 being the carrier frequency of the DVB-T signal exploited in the TRC. Using (18.6), a Doppler matched cross-correlation can be processed and a track can be started in each TRC range cell falling into the larger TDC range cell. With the target motion the bistatic angle β and the target velocity angle ϑ vary and thus varies with the bistatic Doppler.

$$f_D = 2 \times \frac{v}{\lambda} \times \cos(\vartheta) \times \cos\left(\frac{\beta}{2}\right) \tag{18.7}$$

Thus, knowing the position of the transmitter as well as that of the receiver the bistatic geometry is used to estimate the target Doppler in adjacent range cells under the assumption of a linear flight trajectory and thus allows range cell exceeding coherent integration. Thus, the number of track hypotheses equals the number of TRC range cells filling one TDC range cell. If the other flight trajectories – with varying Doppler e.g. – shall be allowed, the number of hypotheses will increase according to the allowed Doppler gradient. This has to be respected in the range/Doppler tracking.

Once a range/Doppler track has been established and confirmed, the track head is transformed into Cartesian coordinates and a Cartesian tracking algorithm [11] is applied. A detailed description of a multi-hypothesis-tracker and the development of tracking algorithms for PR can be found in Chapter 11 ('Multistatic tracking for passive radar applications') of Part II of Volume 2 of this book.

18.5 A multi-illuminator passive radar system

A multi-illuminator PCL demonstrator/prototype system, which bears the potential to apply the above described hybrid PR has been proposed by Airbus DS, Germany.

Figure 18.7 Passive radar van and operator consoles

All design considerations for the multi-illuminator system of Airbus DS were directed towards a mature and fully mobile assembly. It therefore covers analogue FM broadcast as well as DAB and DVB-T waveforms with a single mast multi-band antenna. The system is able to process multiple FM transmitters simultaneously as well as DAB single frequency networks (SFNs) and DVB-T SFNs in real-time. Electronic equipment and operator's workstations are integrated in a specific PR van which comprises the integrated, extendible antenna mast system. The innovative system design enables full mobility and flexible deployment in all kinds of terrain. Figure 18.7 shows the van in transport mode (left side) and operational mode (middle). For transportation the lifting arm is retracted and the antenna elements are removed by quick fasteners. On-site system installation can be realized within 30 min. An additional picture gives insight into the PR van and shows the operator's work environment with three independent work stations (right).

The system architecture of the multi-band PR system is shown in Figure 18.8. The integrated multi-band antenna system uses 21 elements distributed on three elevation levels. This antenna structure provides 360° azimuth coverage and enables 3D bearing detection. The antenna comprises an integrated calibration system for improved bearing accuracy. The FM and DAB sub-systems use seven antenna elements and seven channel direct sampling and down-conversion

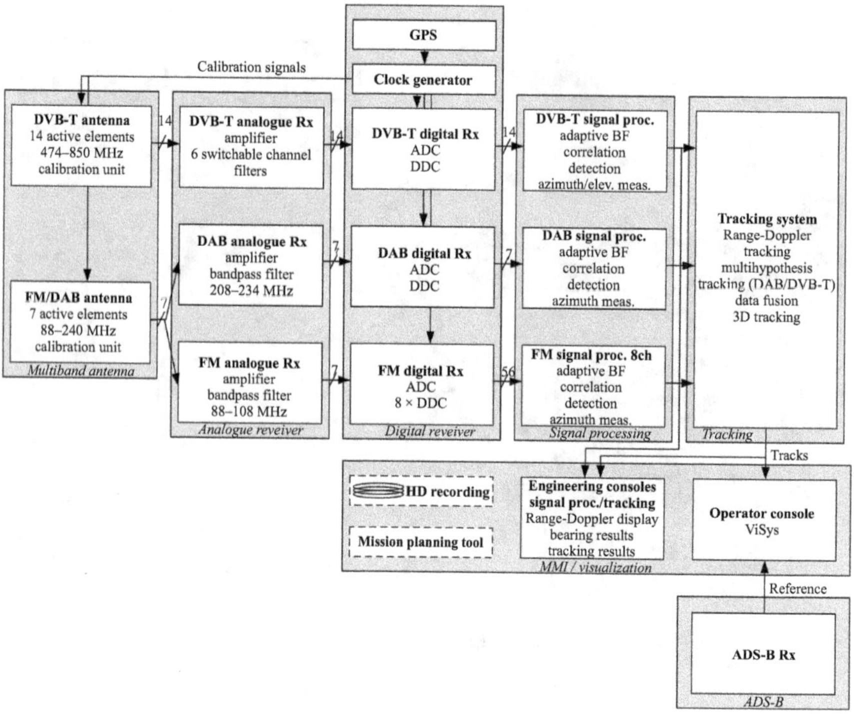

Figure 18.8 PR system architecture

receivers. DVB-T is received and processed by direct down-conversion in 14 channels with antennas on two elevation levels for additional elevation angle bearing detection. Real-time processing is implemented on commercial PC hardware based on standard quad core processors. An enhanced multi-hypothesis tracking system resolves ambiguities resulting from the multi-illuminator situation for digital SFNs and fuses FM, DAB and DVB-T detections and range-range rate-tracks to optimize spatial coverage and localization performance. The MMI (man–machine interface) consists of three separated displays. The signal processing engineering console allows the display of e.g. antenna amplitude levels, the frequency spectrum of the received signals as well as the processed range Doppler maps.

In addition, a tracking engineering console is realized for the display of intermediate tracking results and further internal parameters. The fused output tracks are visualized on the professional and end-user oriented operator console ViSys (visualization system). ViSys is a modern air surveillance MMI which is based on a standard product for the German-Improved Air Defence System (GIADS). For reference purposes, a commercial automatic dependent surveillance – broadcast (ADS-B) receiver is part of the system. Its tracks can also be displayed on ViSys for real-time comparison between actual air picture and the PR performance.

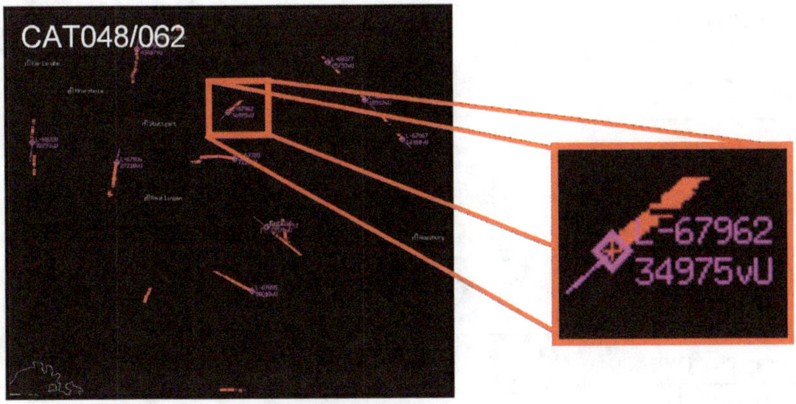

Figure 18.9 Display of passive radar data in CAT048/062 in parallel

To use existing infrastructures like the already mentioned GIADS the PR output data is mapped into standard ASTERIX protocol formats (so-called categories 'CAT' [12]). Active radars typically transfer their data in CAT048 (incl. CAT034) for plots respectively CAT062 for tracks.

Unlike active radar plots the individual bistatic PR plots consisting of bistatic range and Doppler depending on the transmitter do not allow a location of the target in 2D/3D. Thus the transmission of real PR plot data is meaningless to standard multi-sensor tracking systems and operators. Hence, the same tracker output data is used for coding track positions as 'pseudo plots' in CAT048 as well as track positions including velocity and heading as tracks in CAT062.

Since common visualization tools do not provide track history information for CAT062, the additional parallel display of 'pseudo plots' in CAT048 visualizes both, the high update rate as well as the track behaviour including track attributes as heading and speed. Figure 18.9 shows the combined information on ViSys.

The PR is able to provide the ASTERIX data over a Transmission Control Protocol/Internet Protocol data link for the immediate vicinity of the system. For long-distance data transfer a Universal Mobile Telecommunications System (UMTS) interface has been implemented to support mobile operation [3].

Figure 18.10 gives an overview of the target types which have already been successfully used for measurements with the above described sensor. Green-marked fields show the individual waveforms which contributed to the detection process. White fields do not necessarily mean that the target was not detected with the corresponding waveform but show that the test has not been carried out.

The results match clearly with the before analysed characteristics of the different waveforms due to the low-range resolution obtained with FM-signals targets in short distance up to a few kilometres cannot be detected with FM because they are masked by the direct signal. As very low flying or ground and sea targets may be shadowed in higher distances, the FM-waveform is not suitable for those kinds of objects. Besides slowly moving targets like some ground and sea targets, stay in

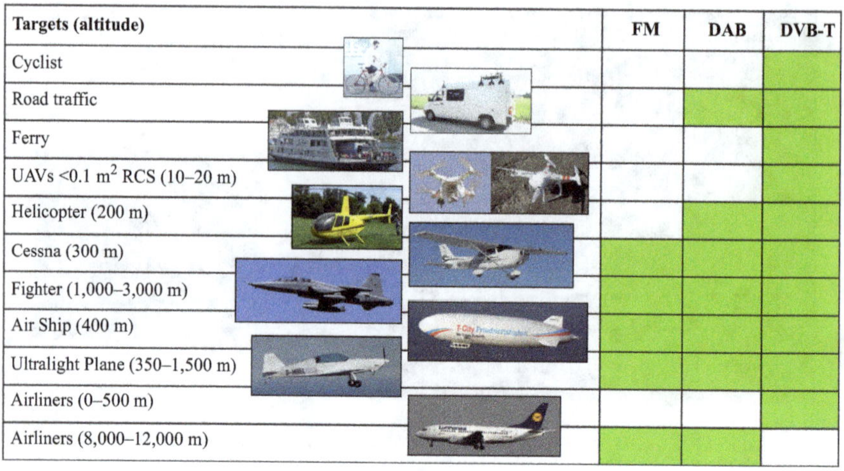

Figure 18.10 Detected target types

one range cell for several processing intervals. This fact adds to the difficulty of detecting and tracking this kind of target with FM. In contrast, air targets at higher altitudes like airliners, fighters, light and ultra-light aircrafts etc. could all be perfectly detected and tracked with FM.

In contrast, the DVB-T waveform with its large bandwidth and therefore high resolution is able to resolve very small targets even in distances down to 10–20 m. The narrow and downwards-tilted beam of the transmit antenna makes it impossible to detect cruising airliners, but is on the other hand advantageous for ground, sea and low flying targets.

DAB does not have its own coverage area where neither FM nor DVB-T contributes. However, it supports and improves the results of the remaining waveforms in different aspects. As its range resolution lies somewhere between FM and DVB-T, it increases the high resolution area in vertical direction and enhances the localization accuracy of mainly FM based tracks in areas where DVB-T cannot contribute.

Altogether, the different waveforms complement each other perfectly and produce a complete air picture of good quality.

The performance of the described sensor has been tested extensively during the last years. Figure 18.11 shows an overview of the recorded tracking data collected all over Germany in several measurement campaigns. The maximum achieved range of the sensor locations differs due to the surrounding transmitter landscape and interference sources. However, detailed analysis of the sensor site before the measurement helps to optimize the exact position and increases the range in many cases.

In addition, trials and demonstrations in Switzerland and Poland took place which also showed the potential of PR even in hilly terrain.

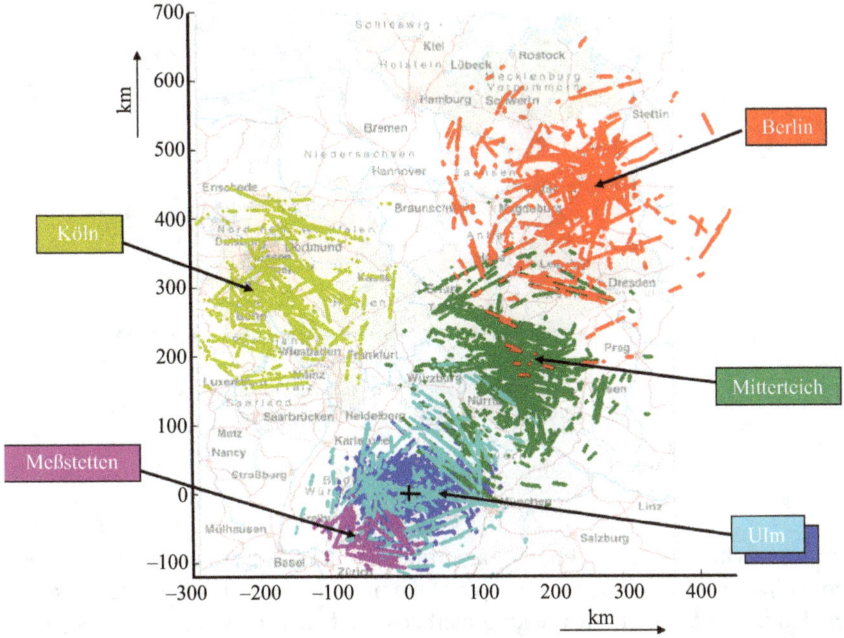

Figure 18.11 Overview of trials in Germany

18.6 Multistatic, multi-illuminator passive radar applications

18.6.1 A multistatic PCL cluster for low-level target gap coverage

One possibility to increase track range or to improve the localization accuracy is given by operating multiple sensors in a sensor cluster network. The data of the single sensors is processed in a central tracker, encrypted and fed into an external network, e.g. MilRADNET (German military radar data network). The sensor cluster network here appears as a unity for the outside world. The number of sensors, their configurations and arrangement depend on the specific task in the area of interest to meet the particular requirements. The principle is shown in Figure 18.12.

There are two different approaches to fuse the data of the single sensors. The data of both systems can be fused in the central tracker on bistatic plot level. This leads to intersections of ellipsoids with different focal points representing the two different sensor positions. This approach is very efficient, especially, for low density transmitter infrastructures. In this case, a single sensor probably could not provide a 3D-air picture due to the limited transmitter-sensor geometry.

The second method is the sensor data fusion based on track-level using a track-to-track correlation algorithm. This method is rather applied in dense-target

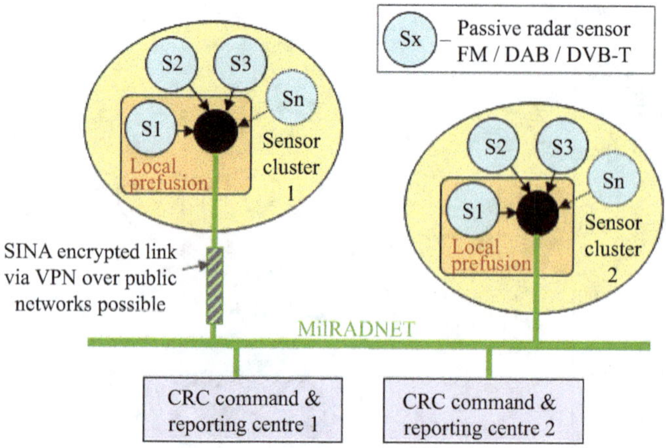

Figure 18.12 Sensor cluster principle

scenarios with a huge amount of contributing transmitters, in order to remove computing load from the global solver of the multi-hypothesis tracking system. For detailed information on PR tracking methods see Part II of Volume 2 of this book.

In order to benefit from sensor cluster operation, a second sensor, the portable PR, was realized as a nearly one-to-one copy of the FM/DAB part of the mobile system. Up to now, the system can be operated either in FM or DAB mode. In sensor cluster operation, the portable system acts as slave and sends its plot or track data to the mobile system which acts as master. Figure 18.13 shows a picture of the mobile and the portable system (right) [13].

Below, a specific sensor cluster measurement scenario is described. The master sensor is located in Bavaria close to the Czech boarder. The slave sensor is operated in Ulm. The distance between both sensors is approximately 240 km. In this trial, the master sensor (blue) processes FM, DAB and DVB-T. The slave sensor (red) operates in the FM band. Distances between processed transmitters and sensors are up to 110 km for FM. Figure 18.14 shows the individual tracks of both sensors operated as stand-alone without any interaction.

Now the data of both systems are fused on plot level. Figure 18.15 highlights the additional benefit resulting from the plot-based central tracking.

The coverage enhancement of operating two sensors together is evident. Track updates generated by the master sensor are marked in turquoise. Updates resulting from the slave sensor are marked red. Mixed tracks are generated in an overlapping area between both sensor positions. Within this area tracks from a single sensor would not remain stable for a long-time period. Plot-based data fusion especially enhances the overall performance in low transmitter density infrastructures, e.g. in areas close to the maximum range of an individual single sensor [13].

The second sensor cluster campaign in 2013 in an alpine valley in Switzerland was motivated by the difficulty to achieve good radar coverage in mountainous areas. During preparation, preliminary measurements were conducted, coverage

Figure 18.13 Picture of the portable 2nd sensor of Airbus DS (right)

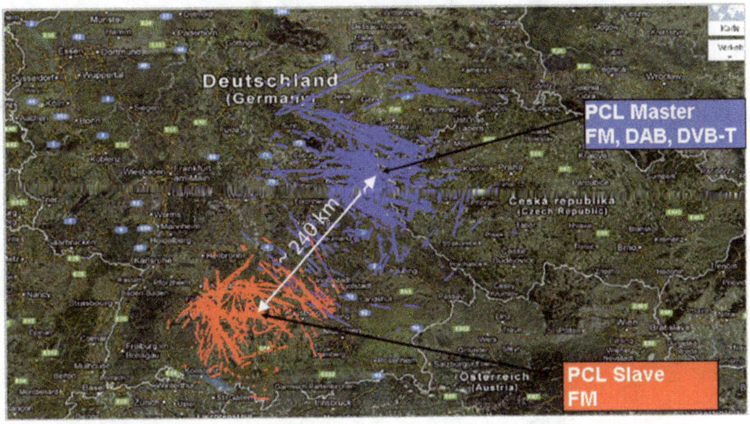

Figure 18.14 Individual sensor tracking results

calculations were performed etc. The preparation showed that the transmitter infrastructure within the valley was challenging in terms of transmitter variety and transmit power. Due to the topography, the transmitters were located at both ends of the valley and caused therefore suboptimal geometries for the multi-lateration task. Those pre-conditions had to be compensated by using two

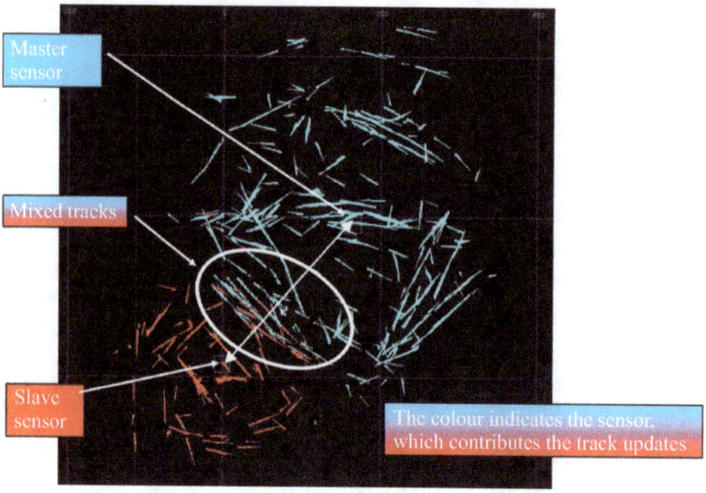

Figure 18.15 Mixed track updates

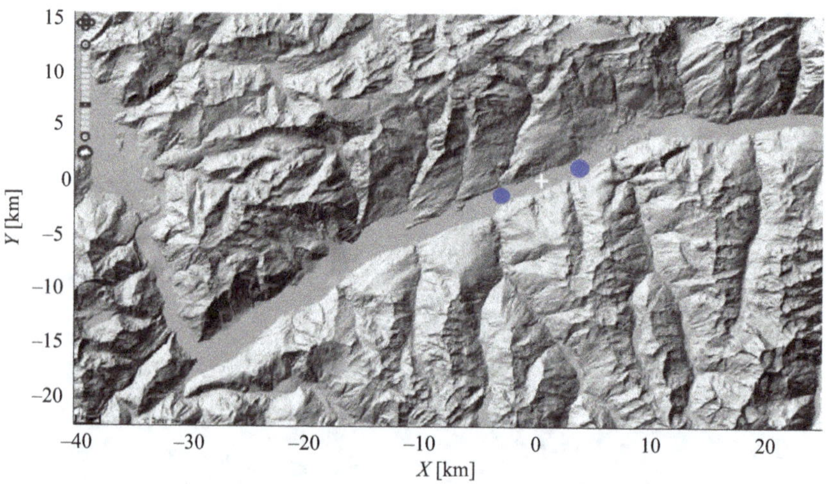

Figure 18.16 Alpine environment and passive radar sensors

dislocated PR sensors in order to enable proper 3D-tracking. Both systems (master and slave, description above) were located at the valley bottom, about 10 km away from each other and connected via a dedicated UMTS link. This enabled the combination of bistatic plot data of the two sensors to a 3D-air-picture in real-time by a global-tracker. The valley, the location of the two sensors (blue) and the virtual sensor cluster position (white cross) are depicted in Figure 18.16.

Figure 18.17 Cooperative target 'PC12'

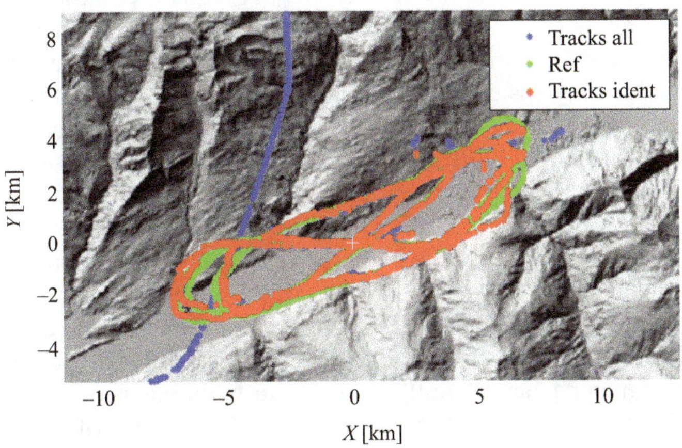

Figure 18.18 Tracks along trajectory of PC12 (red)

For performance evaluation, a PC12 aircraft (equipped with GPS) was flying along dedicated patterns. It is shown in Figure 18.17. In parallel, some targets of opportunity (airliners and VFR traffic) were also detected and evaluated on the basis of their ADS-B emissions.

Some of the results are described in the following. The trajectories were located in a distance of up to 8 km around the virtual sensor cluster position in different flight altitudes.

Figure 18.18 shows the resulting tracks of the co-operative target for a flight altitude of 1,300 m above sea level (ASL) in red, the GPS reference in green and further tracks of targets of opportunity in blue. In total, 84% of the trajectory was covered with an accuracy of approximately 200 m. Due to the narrowness of the valley and the tight curves this is a convincing result.

As most transmitters are installed with a significant beam-tilt towards the ground due to the topographic conditions, low altitude targets were expected to show better performance than higher ones. This is verified by comparing the result described above with the result for the same trajectories in an altitude of 1,900 m ASL where 75% could be covered.

The measurement shows that a length of 16 km and thus half of the valley can be covered by a cluster of two sensors. Therefore, at most four sensors are enough for the whole valley. As an alternative, directive antennas may allow to further reduce the number of sensors [14].

18.6.2 A passive radar based warning sensor network for aviation obstacles

On the other end of the spectrum of applications, a short range PR network in support of air traffic safety in the vicinity of wind farms is proposed.

Wind energy is an important pillar of the renewable energy mix and has been growing continuously during the recent years. Its acceptance among citizens, however, is not only fostered by the environmentally friendly attitude of wind energy, but also restricted by its visual and acoustic appearance, modifying the picture of nature. In particular, at night, when the red collision-avoidance-illuminations print blinking lines on the sky, people feel bothered by this safety feature. Moreover, birds are attracted by the red blinking lights and may suffer fatal collisions with the blades of the wind turbines. Thus, switching on the collision-avoidance-illumination on demand, only, when an aircraft appears within a protection sector of a wind farm is a requirement of the wind energy community. Here, the PARASOL (which is a German acronym 'Passiv Radar basierte Schaltung der Objektbefeuerung für die Luftfahrt') system is proposed, which is based on low cost PR technology. PARASOL stands for PR-based switching of object-illumination for air traffic.

The PARASOL system [15], which is now installed in the wind farm Reussenkoege in northern Germany consists of three sensors, each of which can operate on the frequency range from 450 to 850 MHz to use the DVB-T transmissions of the digital TV networks operated in Germany. Each sensor measures the direct signal coming from the DVB-T transmitter and echo signals coming from the environment. The direct signal is used to re-construct the transmitted signal to obtain a clean replica for cross-correlation with the echo signals. The cross-correlation process is performed for each symbol disregarding the guard interval in order to avoid artefacts induced by signal periodicities. Multiple cross-correlation products are then integrated via fast Fourier processing (FFT) to obtain Doppler information and collect energy from weak target returns. The length of the FFT corresponding to the integration time depends on the expected target-velocity and should not exceed the time the target maintains within the resolution cell. The result of this processing chain is a bistatic range versus bistatic Doppler matrix with the

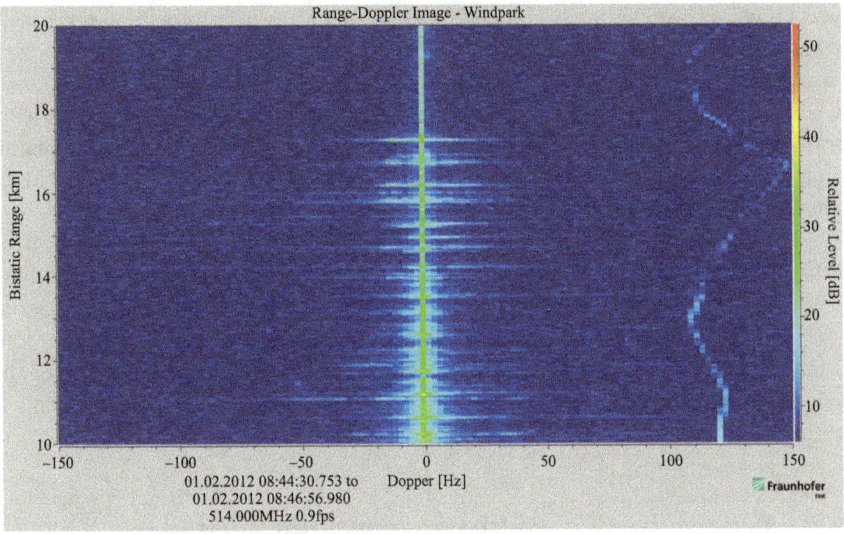

Figure 18.19 Range/Doppler matrix with wind turbine echoes and air target

ground returns distributed along the zero Doppler line and moving target echoes distributed in the positive and negative Doppler planes. Figure 18.19 shows an example of such a range/Doppler matrix [16,17].

A Cartesian target location is determined through the intersection of three time-difference-of-arrival ellipsoids from the three sensors. A PARASOL sensor operates with two parallel channels to process the signals received by two vertically stacked antennas for interferometric height measurement. Each channel consists of an RF-module and an analogue digital converter/field programmable gate array module. Both channels feed into a high-performance PC module.

One of the main challenges is the optimum distribution of the sensors in the wind farm. For practical reasons, the sensors are mounted on the pylons of wind turbines, taking advantage of their internal power supply and communication network. All sensors should have approximately the same coverage to ensure that a target echo is received by all three sensors. Only when an echo of the same target is received by all sensors of the system, ellipsoid intersections can be calculated and the Cartesian location of a target can be determined. The constraints taken into consideration range from the transmitter characteristics over target reflectivity assumptions to signal-to-interference ratio (SIR) induced by the direct transmitted signal. The transmitter-characteristic is primarily described by the transmitted power and the transmitter antenna characteristics in azimuth and elevation. While the azimuth directivity of a DAB or DVB-T broadcast antenna is mostly omnidirectional, the elevation characteristic shows a very narrow beam in the order of 3 to 5°, mostly with a tilt towards the horizon. As a further constraint the receiver-antenna characteristics including the shielding influences of the wind power pylon have to be respected. Such constraints lead to an optimization tool for sensor locations with the task to achieve the coverage of a given area around and above a wind power farm. Figure 18.20 shows an example of two sensor distributions, one of which can cover the southern part (a) and one the northern part (b) of a wind farm located in northern Germany using a close DVB-T transmitter as illuminator. White dots indicate the locations of wind power plants, white dots with circles indicate the location of the PARASOL sensors.

After a location optimization procedure using the above mentioned tool, a set of three sensors has been installed in a wind farm in Schleswig-Holstein in northern Germany. The antenna units have been mounted on the mast of wind turbines in such a way that the direct signal is received strong enough to allow signal reconstruction but is still reduced in power to limit the required dynamic range of the system allowing small-target detection. A photo of the installed antenna unit is depicted in Figure 18.21.

18.7 Perspectives for multistatic multi-illuminator PCL systems

The first section of the previous chapter showed the clear advantages of sensor cluster operation. The coverage for a network of two sensors is not only double but more than that, since ellipsoid intersections of transmitter receiver pairs of both sensors are possible. In addition, the localization accuracy gets improved with the number of sensors, depending of course on the size of the overlapping area. In order to predict the amount of improvement, exact knowledge of the used waveforms and the geometries between the involved sensors and illuminators is required.

Due to those advantages, the future PRs should consist of two or more sensors, each operating in the FM, DAB and DVB-T band, typically exchanging bistatic plot data in order to optimize coverage and accuracy. They should apply the processing scheme described in Chapter 4. With this configuration, it will be possible to cover a medium range with high precision (DAB and DVB-T) and a larger range around the focal point of the sensors with improved FM accuracy. The resulting detection performance in low and medium-altitude scan be comparable with medium-range active air surveillance radars (high accuracy up to 30 km, standard accuracy up to 100 km) while acquisition and maintenance cost are considerably lower and flexibility is higher.

Taking into account its physical characteristic of silent operation without any radio emissions, PR will provide radar performance where active radars are less suitable. This applies in a military environment primarily where operation of a radar shall not be made visible to others. A second major advantage is the use of PR where active radars are simply forbidden to be used, for example in urban environment or harbours. Finally, the absence of radio emissions guarantees operation independent from authority approvals and without disturbance of any other radio based equipment, which will help both fixed installations as well as rapid deployment.

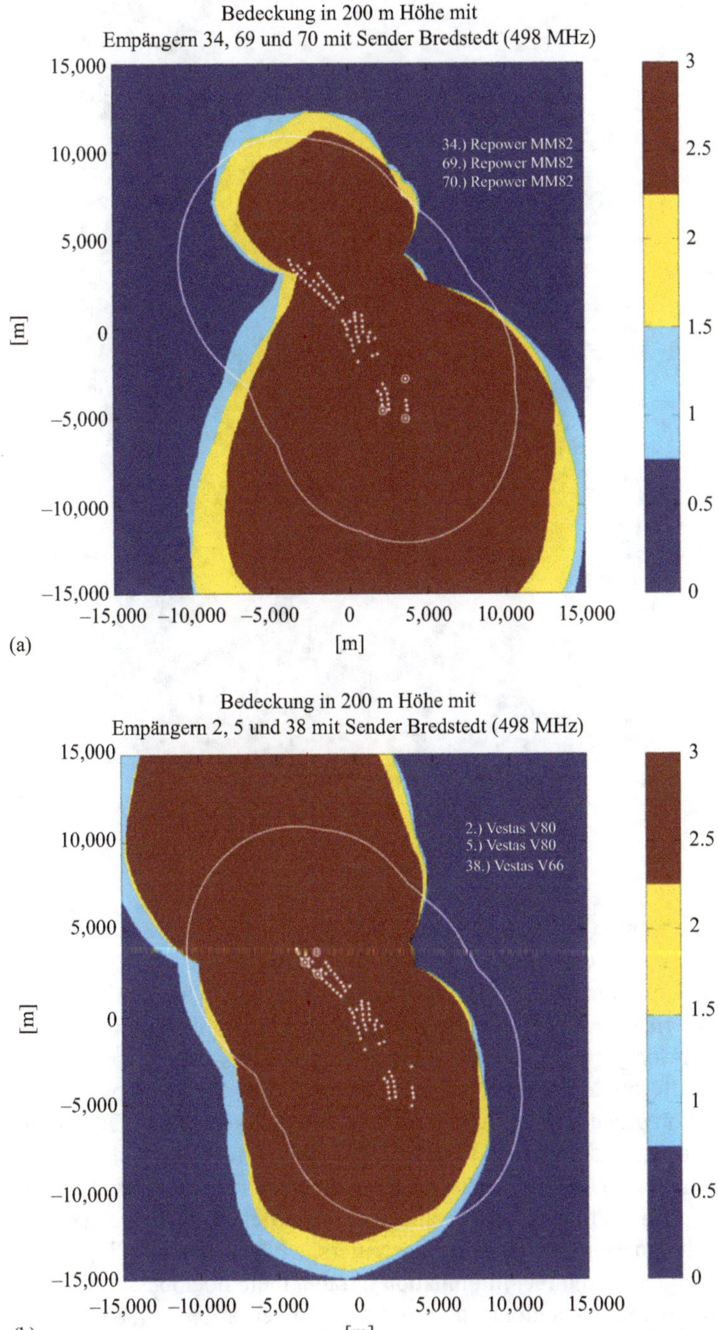

Figure 18.20 (a) Set of 3 sensors covering and (b) set of 3 sensors covering northern southern part of wind farm

Figure 18.21 Antenna unit (2 discone antennas) on mast

For short-range applications, like in the shown example of a wind farm collision-warning system PR systems of the PARASOL type can be cost efficient alternatives to active radar networks, which always require a frequency allocation.

Future fields of application and challenges in multi-illuminator multistatic PR will no longer be restricted to stationary radar networks but rather include moving platforms and clusters of moving platforms. First approaches to motion-compensation procedures, based on on-board inertial systems (reference) have been discussed and trials on naval vessels have been performed. Airborne PR sensors and sensor networks will be considered in future as they may provide advantageous properties in detecting low-level and ground targets, which are well within the range of illumination of broadcast transmitters, while airborne PCL platforms could benefit from reduced direct illumination through their sidelobes. A major challenge will be the special synchronization and the compensation to the platform motion influencing the generation of a motion free replica of the transmitted signal as a reference for cross-correlation.

List of acronyms

DAB	digital audio broadcast
DVB	Tdigital video broadcast terrestrial
PCL	passive coherent location
FM	frequency modulated (broadcast)
EIRP	effective isotropic radiated power
PR	passive radar
CFAR	constant false alarm rate
ERP	effective radiated power
COFDM	coded orthogonal frequency division multiplex
OFDM	orthogonal frequency division multiplex
QPSK	quadrature phase shift keying
QAM	quadrature amplitude modulation
GPS	global positioning system
TDC	target detection component
TRC	target resolution component
HEC	high elevation component
DMB	digital media broadcast
ISAR	inverse synthetic aperture radar
SONAR	sound navigation and ranging
SFN	single frequency network
MFN	multiple frequency network
MMI	man–machine interface
ViSys	visualization system
GIADS	German-Improved Air Defence System
MST	multi-sensor tracker
TCP/IP	Transmission Control Protocol/Internet Protocol
UMTS	Universal Mobile Telecommunications System
MilRADNET	German military radar data network
ADS-B	automatic dependent surveillance – broadcast
ASL	above sea level
PARASOL	Passiv Radar basierte Schaltung der Objektbefeuerung für die Luftfahrt
FFT	fast Fourier transform
RF	radio frequency
ADC	analogue digital converter
FPGA	field programmable gate array

References

[1] D. Poullin, M. Lesturgie 'Multistatic radar using non cooperative transmitters'. *Colloque Radar*, Paris, 1994.

[2] D. Poullin, 'Comparison of the main civilian broadcasters usable for PCL systems FM and COFDM ones (DAB, DVB-T)'. *Proc. of the IRS 2009*, Hamburg, 2009.

[3] M. Edrich, A. Schröder, F. Meyer (Airbus DS), 'Design and performance evaluation of a mature FM/DAB/DVB-T multi-illuminator passive radar system', *IET Radar, Sonar and Navigation*, vol. 8, no. 2, 2014, pp. 114–122.

[4] D. O'Hagan (Fraunhofer FHR), 'Elevation coverage analysis of common PCR illuminators of opportunity', *IQPC Military Sensors*, London, UK, 2009.

[5] P. E. Howland, D. Maksimiuk, and G. Reitsma, 'FM radio based bistatic radar', *IEE Proceedings on Radar Sonar Navigation*, vol. 152, no. 3, 2005, pp. 107–115.

[6] R. D. Fry, D. A. Gray, 'CLEAN deconvolution for sidelobe suppression in random noise radar', *Proc. of the Int. Conf. Radar*, 2–5 September 2008, Adelaide (Australia), pp. 209–212.

[7] K. Fazel und S. Kaiser, *Multi-Carrier and Spread Spectrum Systems*, John Wiley & Sons, Brussels, 2003.

[8] H. D. Griffiths, 'Bistatic radar – principles and practice', *SBMO International Microwave Conference*, vol. 2, pp. 519–526, Sao Paulo, Brazil, August 1993.

[9] H. D. Griffiths, N. R. W. Long, 'Television-based bistatic radar', *IEE Proceedings on Communications, Radar and Signal Processing*, vol. 133 – Part F, no. 7, pp. 649–657, December 1986.

[10] H. Kuschel, J. Heckenbach, D. O'Hagan, M. Ummenhofer, 'A hybrid multifrequency passive radar concept for medium range air surveillance', *MRRS 2011*, Kiew, 08.2011.

[11] A Baruzzi, G. Bournaka, J. Heckenbach, H. Kuschel, 'Experimental validation of Kalman filter based tracking in passive radar systems', *2015 IEEE International Radar Conference*, May 2015, Arlington VA, USA.

[12] Asterix Specifications, http://www.eurocontrol.int/services/specifications-documents (*last visited*: January 26, 2016).

[13] A. Schröder, M. Edrich (Airbus DS), 'Passive radar sensor cluster operation', *IRS*, June 2014.

[14] S. Stroth, A. Schröder (Airbus DS), Luca Quiriconi *et al.* (Swiss Air Force), 'Test and evaluation of a PCL demonstrator in an alpine environment', *PCL Focus Days*, April 2015.

[15] J. Heckenbach, H. Kuschel, J. Schell, M. Ummenhofer, 'Passive radar based control of wind turbine collision warning for air traffic', *PARASOL, IRS 2015*, June 2015, Dresden, Germany.

[16] H. Kuschel, J. Heckenbach, S. Muller, and R. Appel, 'On the potentials of passive, multistatic, low frequency radars to counter stealth and detect low flying targets', *Proc. of IEEE International Radar Conference*, May 2008, pp. 1443–1448.

[17] D. W. O'Hagan, H. Kuschel, and J. Schiller, 'Passive bistatic radar analysis', *Proc. of SPIE*, vol. 7502, August 2009.

Chapter 19
Passive MIMO radar networks

Daniel E. Hack, Lee K. Patton* and Braham Himed***

Abstract

In distributed sensing applications, the nodes of a sensor network cooperatively detect and localize targets of interest. In particular, *active multiple-input multiple-output (MIMO) radar* (AMR) uses multiple transceivers to transmit separable signals and receive the scattered returns, while *passive MIMO radar* (PMR) uses multiple receivers to receive the direct-path (transmitter-to-receiver) and target-path (transmitter-to-target-to-receiver) signals originated by multiple non-cooperative transmitters to detect and localize targets. This chapter surveys recent results in the theory of centralized detection in PMR networks. Generalized likelihood ratio test (GLRT)-based detectors for PMR detection have been developed and their performances are compared to related detectors for AMR and passive source localization (PSL) sensor networks. PMR detection sensitivity and ambiguity are then analyzed as a function of both the target-path and direct-path signals-to-noise ratios (SNRs). The results demonstrate that PMR detection sensitivity and ambiguity approach that of active MIMO radar sensor networks when the direct-path SNRs are sufficiently high. Conversely, PMR detection sensitivity and ambiguity approximate that of passive source localization sensor networks when the direct path SNRs are sufficiently low. In this way, PMR networks unify these related active and passive sensor network architectures in a common theoretical framework.

19.1 Introduction

Passive radar is a type of radar that exploits the emissions from one or multiple non-cooperative transmitters to detect, localize, track and/or image targets. The transmitters are non-cooperative in the sense that their emissions are not controlled by the passive radar system. Typically, the passive radar will know the frequency channels (centre frequency and bandwidth) and modulation types of the exploited

*Matrix Research Inc., USA
**RF Technology Branch, Air Force Research Laboratory, USA

signals, but not their instantaneous modulations. Commonly exploited signal types include terrestrial radio and television broadcasts, cellular communications and emissions from communications and navigation satellites [1,2]. Even though a passive radar system consists of only non-transmitting receiver sites (hence the term *passive*), the combination of a passive radar system and the non-cooperative transmitters exploited by that system may be described as a passive radar *network*.

The transmitter–receiver topology of a passive radar network may be classified as *bistatic*, *multistatic*, or *multiple-input multiple-output* (MIMO). A passive bistatic radar, shown in Figure 19.1(a), consists of one receiver that exploits the signal emitted by one non-cooperative transmitter. This type of passive radar is also commonly termed passive covert radar or passive coherent location. A passive multistatic radar network consists of either one receiver that exploits the signals emitted by multiple distributed non-cooperative transmitters (Figure 19.1(b)), or multiple distributed receivers that exploit the signal emitted by one non-cooperative transmitter (Figure 19.1(c)). Due to the availability of non-cooperative transmitters, the single-receiver multistatic configuration is more common than the single-transmitter configuration. Finally, a passive MIMO radar (PMR) network, shown in Figure 19.1(d), consists of multiple distributed receivers that exploit the signals emitted by multiple distributed non-cooperative transmitters. PMR networks are the most general because they reduce to passive bistatic and passive multistatic radar networks as special cases.

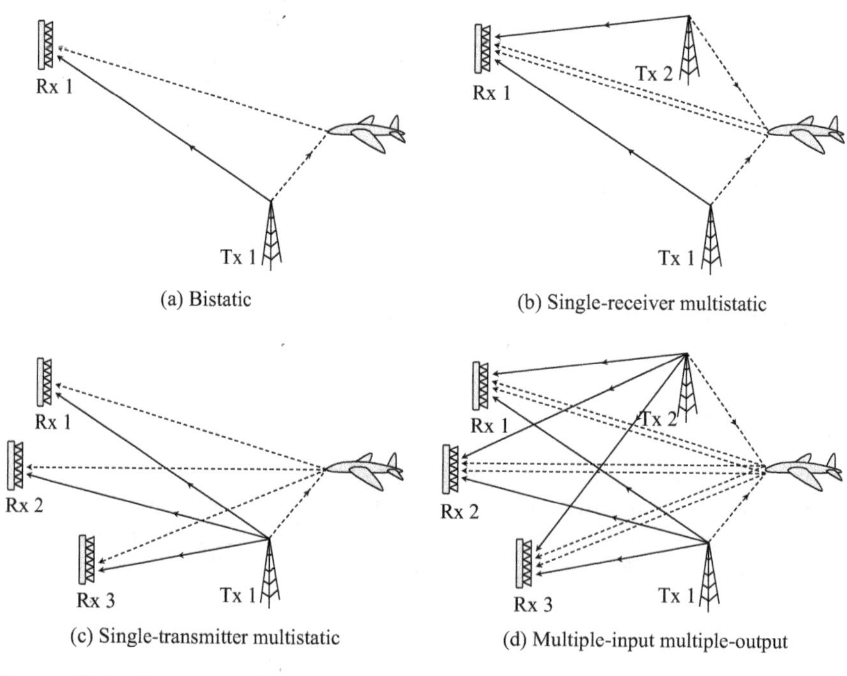

Figure 19.1 Passive radar network topologies. (d) Copyright © 2014 IEEE. Reprinted, with permission, from [6]

The transmitted signals in a PMR network can occupy overlapping or distinct frequency channels. For example, broadcast digital television according to the Digital Video Broadcast-Terrestrial (DVB-T) standard uses a single-frequency network configuration in some European countries, in which the same signal is simultaneously transmitted on multiple distributed transmitters in the same frequency channel. In contrast, frequency modulated (FM) radio uses a multi-frequency network configuration, in which signals are transmitted within a given geographic region on non-overlapping frequency channels. Exploitation of single-frequency networks complicates the data association problem by introducing measurement origin ambiguity in the sense that the transmitter that originated a given measurement is uncertain. This challenge has been addressed in recent passive radar publications [3–5] and is not addressed in this chapter, which considers multi-frequency networks only.

Within each frequency channel, the signal environment observed by the receivers in a PMR network consists of direct-path, clutter-path and target-path signals. This is shown in Figure 19.2. The direct-path signal is a copy of the transmitted signal received via the transmitter-to-receiver propagation channel. The clutter-path signal is the superposition of many transmitted signal copies received via a diversity of bistatic transmitter-to-ground-to-receiver propagation channels. The target-path signal is a copy of the transmitted signal received via the bistatic transmitter-to-target-to-receiver propagation channel. In typical operating conditions, the direct-path signal exceeds the strength of the clutter-path signal, and both the direct-path and clutter-path signals far exceed the strength of the target-path signal. For this reason, the direct-path and clutter-path signals are interference sources that must be mitigated to make target detection possible [7–9]. On the other hand, the direct-path signal also serves a crucial role by providing the passive radar system with an estimate of the otherwise-unknown instantaneous modulation of the transmitted signal. This estimate, though imperfect, enables a passive radar system to approximate the performance of active radar systems, in which the transmit signals are known.

The conventional approach to detection and localization in passive radar networks is *decentralized* in the sense that detection is performed independently by each bistatic (transmitter–receiver) pair, and the resulting detections are fused across pairs to localize targets in Cartesian space [10–13]. First, the direct-path and target-path signals are isolated into *reference* and *surveillance* channels, respectively. This is accomplished by pointing directional antennas at the transmitter and surveillance

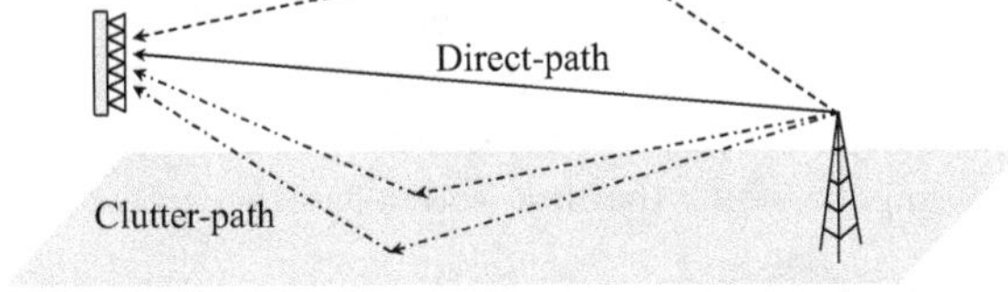

Figure 19.2 Passive radar signal environment

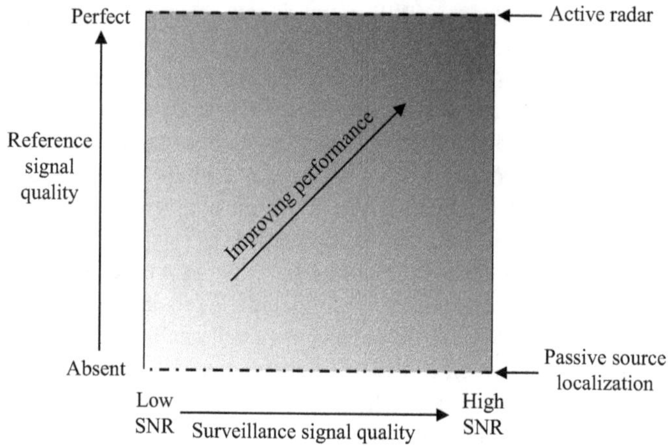

Figure 19.3 PMR performance versus reference and surveillance signal quality

volume, respectively [14,15], or by digital beamforming in multi-channel systems [16–19]. Additional measures are often taken to suppress residual direct-path and clutter-path leakage [7,9,20,21].[1] The cross-ambiguity function (CAF) between the reference and surveillance channels is then computed, and target detection is declared in reference–surveillance CAF (RS-CAF) range-Doppler cells that exceed a threshold value. Finally, the resulting detections are associated across bistatic pairs to localize and track targets in Cartesian space [5,10–12].

The performance of the conventional processing chain depends on many system and scenario parameters including the location and number of transmitters and receivers, the effective radiated power of the transmitters, the target radar cross section (RCS), the noise figure and dynamic range of the receivers and many others. Two fundamental parameters that capture many of these dependencies are the reference and surveillance channel signal-to-noise ratios (SNRs); more precisely, the SNRs of the direct-path and target-path signals within the reference and surveillance channels, respectively. The relationship between this two-dimensional SNR parameter space and performance is illustrated in Figure 19.3. Surveillance signal quality may range from low to high SNR. Similarly, reference signal quality may range from absent (SNR $\rightarrow -\infty$) to perfect (SNR $\rightarrow +\infty$). Intuitively, performance should increase as the reference and/or surveillance signal SNRs increase. The specific way in which this holds true has only recently begun to be studied and understood [6,22,23].

The objective of this chapter is to characterize the detection performance of PMR networks over the entire reference–surveillance SNR space. The approach is to present a centralized generalized likelihood ratio test (GLRT) detector for PMR networks, and to study its detection sensitivity and ambiguity over this space. Centralized processing, in which the time series data from all receivers are jointly

[1]Some of these approaches entail demodulating and remodulating the direct-path signal to provide a higher quality reference signal. This type of processing is not carried out in this chapter.

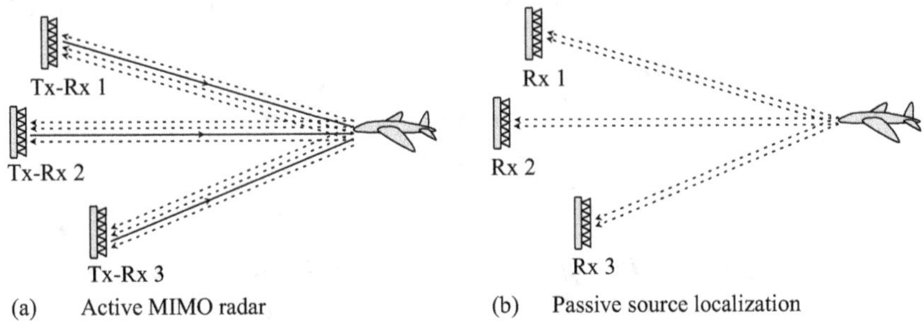

Figure 19.4 Related sensor networks. Copyright © 2014 IEEE. Reprinted, with permission, from [6]

processed to detect targets, generally achieves better performance than decentralized processing, and thereby represents an upper bound on performance. Clutter-path interference is not addressed in this chapter but is assumed to have been previously mitigated using techniques such as those described in [9,20]. GLRTs are presented in parallel for distributed *active MIMO radar* (AMR) and *passive source localization* (PSL) sensor networks.[2] AMR networks (Figure 19.4(a)) use multiple distributed transceivers to transmit known separable signals and receive the scattered returns [24,26]. PSL networks (Figure 19.4(b)) use multiple receivers to intercept signals transmitted by the target itself [27,28]. Conceptually, these networks bound PMR performance by occupying two one-dimensional cuts of the SNR space in Figure 19.3. AMR networks utilize known transmit signals (i.e., perfect references), while PSL networks have only target-path signals (i.e., no references). Examining the performance of AMR and PSL sensor networks allows us to better understand PMR performance by placing it in context of these more familiar sensor networks.

The results of this chapter show that PMR is intermediate to AMR and PSL networks. When the reference signals are high quality, PMR sensitivity and ambiguity approach that of AMR networks. When the reference signals are low quality, PMR sensitivity and ambiguity are similar to that of PSL networks. When the reference signals are of intermediate quality, PMR sensitivity and ambiguity are intermediate to that of AMR and PSL networks and exhibit characteristics of both. In this way, PMR networks unify AMR and PSL sensor networks within the framework shown in Figure 19.5.

The remainder of this chapter is organized as follows: first, signal models are presented for PMR, AMR and PSL sensor networks in Section 19.2. Centralized GLRTs are then presented for each sensor network in Section 19.3. Detection sensitivity is investigated by numerical simulation in Section 19.4, followed by ambiguity in Section 19.5. Finally, the results of this chapter are summarized in Section 19.6.

[2]In this chapter, AMR refers to what is commonly termed distributed, non-coherent, or statistical MIMO radar [24]. This is in contrast to coherent MIMO radar, in which multiple co-located phase centers transmit distinct signals and jointly process the return [25].

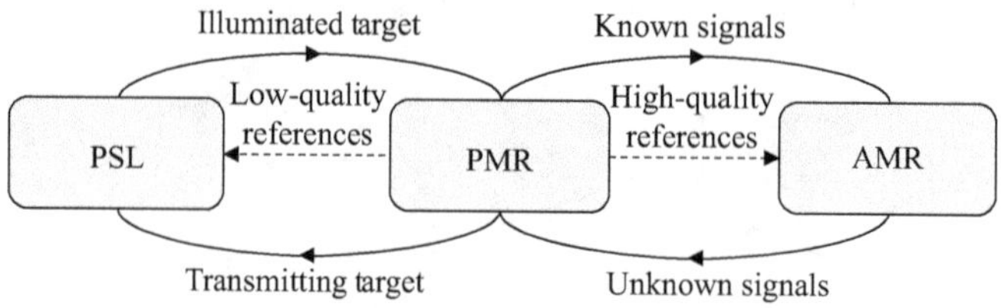

Figure 19.5 Relationships between PSL, PMR and AMR networks

19.2 Signal models

19.2.1 Passive MIMO radar

Consider a PMR network with N_t transmitters and N_r receivers. Figure 19.6 depicts the geometry and signal environment of the ijth bistatic pair, which consists of the ith transmitter and jth receiver. The positions of the ith transmitter, jth receiver and target are denoted by $\mathbf{d}^i$, $\mathbf{r}^j$ and $\mathbf{t}$, respectively, and their velocities by $\dot{\mathbf{d}}^i$, $\dot{\mathbf{r}}^j$ and $\dot{\mathbf{t}}$. All positions are implicitly time-dependent, and the transmitter and receiver positions are assumed known. Let τ_d^{ij} denote the direct-path propagation delay,

$$\tau_d^{ij} = \frac{1}{c} R_0^{ij} = \frac{1}{c}\|\mathbf{r}^j - \mathbf{d}^i\|, \tag{19.1}$$

where c is the speed of light, R_0^{ij} is the transmitter-to-receiver range and $\|\cdot\|$ denotes the vector ℓ^2-norm. Similarly, let ν_d^{ij} denote the direct-path Doppler shift,

$$\nu_d^{ij} = -\frac{1}{\lambda^i} \dot{R}_0^{ij} = -\frac{1}{\lambda^i} \frac{(\mathbf{r}^j - \mathbf{d}^i)^T (\dot{\mathbf{r}}^j - \dot{\mathbf{d}}^i)}{\|\mathbf{r}^j - \mathbf{d}^i\|}, \tag{19.2}$$

where $\lambda^i = c/f_c^i$ is the ith transmitter's wavelength and f_c^i is the ith transmitter's carrier frequency in Hertz. Finally, let τ_t^{ij} and ν_t^{ij}, the target-path propagation delay and Doppler shift, be defined similarly as,

$$\tau_t^{ij} = \tau_{t,1}^i + \tau_{t,2}^j = \frac{1}{c}\left(R_1^i + R_2^j\right) = \frac{1}{c}\left(\|\mathbf{t} - \mathbf{d}^i\| + \|\mathbf{r}^j - \mathbf{t}\|\right) \tag{19.3}$$

$$\begin{aligned} \nu_t^{ij} &= \nu_{t,1}^i + \nu_{t,2}^{ij} = -\frac{1}{\lambda^i}\left(\dot{R}_1^i + \dot{R}_2^j\right) \\ &= -\frac{1}{\lambda^i}\left(\frac{(\mathbf{t} - \mathbf{d}^i)^T(\dot{\mathbf{t}} - \dot{\mathbf{d}}^i)}{\|\mathbf{t} - \mathbf{d}^i\|} + \frac{(\mathbf{r}^j - \mathbf{t})^T(\dot{\mathbf{r}}^j - \dot{\mathbf{t}})}{\|\mathbf{r}^j - \mathbf{t}\|}\right) \end{aligned} \tag{19.4}$$

where R_1^i is the transmitter-to-target range and R_2^j is the target-to-receiver range.

As discussed in Section 19.1, the incident signal at the jth receiver's antenna is the sum of all direct-path, target-path and clutter-path signals across all transmit frequency channels, which are assumed to be non-overlapping in frequency. The incident signals are channelized in frequency, demodulated to baseband and

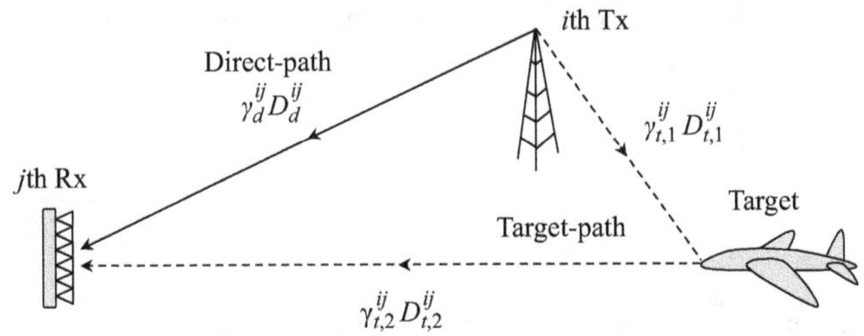

Figure 19.6 Geometry of the ijth bistatic pair in a PMR network

sampled in time. Within each frequency channel, the direct-path signal is isolated by beamforming in the known transmitter direction, and the target-path signal by beamforming in the anticipated target direction while nulling the transmitter direction, thereby forming reference and surveillance channels, respectively [16–19]. Residual clutter-path returns in the reference channel can be mitigated, at least in part, by equalization [9,20]. Similarly, any residual direct-path or clutter-path signals within the surveillance channel can be mitigated via adaptive cancellation [21,29]. It is assumed the target-path and direct-path signals are perfectly isolated by these means, which represents a best case scenario. Let $\mathbf{s}_s^{ij} \in \mathbb{C}^{L\times 1}$ and $\mathbf{s}_r^{ij} \in \mathbb{C}^{L\times 1}$ denote the resulting length-L complex baseband surveillance and reference signals, respectively, on the ith frequency channel of the jth receiver, given by

$$\begin{aligned} \mathbf{s}_s^{ij} &= \gamma_t^{ij} \mathcal{D}_t^{ij} \mathbf{u}^i + \mathbf{n}_s^{ij} \\ \mathbf{s}_r^{ij} &= \gamma_d^{ij} \mathcal{D}_d^{ij} \mathbf{u}^i + \mathbf{n}_r^{ij}, \end{aligned} \tag{19.5}$$

where $\mathbf{u}^i \in \mathbb{C}^{L\times 1}$ is the unknown complex baseband signal emitted by the ith transmitter; $\mathcal{D}_d^{ij} = \mathcal{D}\left(\tau_d^{ij}, \nu_d^{ij}\right) \in \mathbb{C}^{L\times L}$ and $\mathcal{D}_t^{ij} = \mathcal{D}\left(\tau_t^{ij}, \nu_t^{ij}\right) \in \mathbb{C}^{L\times L}$ are delay-Doppler operators that account for the delays and Doppler shifts imparted to the ith transmit signal as it propagates to the jth receiver along the direct and target paths, respectively; γ_d^{ij} and γ_t^{ij} are channel coefficients that account for the complex scaling of $\mathbf{u}^i$ associated with the ijth direct-path and target-path channels, respectively; $\mathbf{n}_s^{ij}$ and $\mathbf{n}_r^{ij}$ are circular Gaussian noise vectors distributed as $\mathcal{CN}(\mathbf{0}_L, \sigma^2 \mathbf{I}_{L\times L})$ with known variance σ^2 that are independent across transmit channels and receivers, i.e.,

$$\mathrm{E}\left\{\mathbf{n}_s^{ij}\left(\mathbf{n}_s^{kl}\right)^H\right\} = \sigma^2 \delta_{i-k}\delta_{j-l}\mathbf{I}_{L\times L}, \tag{19.6}$$

$$\mathrm{E}\left\{\mathbf{n}_r^{ij}\left(\mathbf{n}_r^{kl}\right)^H\right\} = \sigma^2 \delta_{i-k}\delta_{j-l}\mathbf{I}_{L\times L}, \tag{19.7}$$

$$\mathrm{E}\left\{\mathbf{n}_s^{ij}\left(\mathbf{n}_r^{kl}\right)^H\right\} = \mathbf{0}_{L\times L}, \tag{19.8}$$

where $\mathbf{0}_L$ is the length-L zero vector, $\mathbf{0}_{L\times L}$ is the $L \times L$ zero matrix, $\mathbf{I}_{L\times L}$ is the $L \times L$ identity matrix, $(\cdot)^H$ is the Hermitian transpose and δ_x is the Kronecker delta.

The operator $\mathcal{D}(\tau, \nu) \in \mathbb{C}^{L \times L}$ is the unitary linear delay-by-τ and Doppler shift-by-ν operator defined by

$$\mathcal{D}(\tau, \nu) = \mathbf{D}_L\left(\frac{\nu}{f_s}\right)\mathbf{W}^H\mathbf{D}_L\left(-\frac{\tau f_s}{L}\right)\mathbf{W}, \tag{19.9}$$

where

$$\mathbf{D}_L(x) = \text{diag}\left(\left[e^{j2\pi(0)x}, \ldots, e^{j2\pi(L-1)x}\right]\right) \in \mathbb{C}^{L \times L} \tag{19.10}$$

and diag($\mathbf{x}$) for $\mathbf{x} \in \mathbb{C}^L$ is the $L \times L$ square matrix with diagonal elements $\mathbf{x}$ and $\mathbf{W} \in \mathbb{C}^{L \times L}$ is the unitary discrete Fourier transform matrix. Note that the target-path delay-Doppler operator $\mathcal{D}_t^{ij}$ can be factored (ignoring a unimodular complex scale factor) into the product of two operators (see Figure 19.6),

$$\mathcal{D}_t^{ij} = \mathcal{D}_{t,2}^{ij}\mathcal{D}_{t,1}^{i} \tag{19.11}$$

where $\mathcal{D}_{t,1}^{i} = \mathcal{D}\left(\tau_{t,1}^{i}, \nu_{t,1}^{i}\right)$ and $\mathcal{D}_{t,2}^{ij} = \mathcal{D}\left(\tau_{t,2}^{j}, \nu_{t,2}^{ij}\right)$ characterize the delays and Doppler shifts of the first and second legs of the bistatic target-path channel, respectively.

Observe that the coefficients γ_d^{ij} and γ_t^{ij} are multiplicative with $\mathbf{u}^i$ in (19.5). Consequently, the transmit signal $\mathbf{u}^i$ is defined such that $\|\mathbf{u}^i\|^2 = L$ and the composite scaling of $\mathbf{u}^i$ is accounted for by γ_d^{ij} and γ_t^{ij}. The direct-path coefficient γ_d^{ij} accounts for the transmitter's effective radiated power in the receiver direction, spreading and absorptive losses of the transmitter-to-receiver propagation path, and the receiver antenna's effective aperture in the transmitter direction. The target-path coefficient γ_t^{ij} can be factored into the product of two coefficients (see Figure 19.6),

$$\gamma_t^{ij} = \gamma_{t,2}^{ij}\gamma_{t,1}^{ij}, \tag{19.12}$$

where $\gamma_{t,1}^{ij}$ accounts for the transmitter's effective radiated power in the target direction and the spreading and absorptive losses of the transmitter-to-target propagation path and $\gamma_{t,2}^{ij}$ accounts for the target reflectivity, spreading and absorptive losses of the target-to-receiver propagation path and the receiver antenna's effective aperture in the target direction. Both γ_d^{ij} and γ_t^{ij} are assumed to include random phase terms that result from carrier propagation, random target reflectivity, and the use of distributed receivers that are not calibrated to a common phase reference.

Finally, observe that the signal $\mathbf{u}^i$ can represent either a continuous-wave (high-duty cycle) signal or a pulse train, and it is assumed the target reflectivity is constant per transmit–receive pair over the processing interval (i.e., slowly fluctuating). If this assumption is thought to be invalid, the model in (19.5) can be trivially extended to the fast-fluctuating reflectivity case by dividing the surveillance signal into multiple sub-intervals indexed by $k = 0 \ldots K-1$ and giving each its own channel coefficient, $\gamma_{t,k}^{ij}$, and waveform, $\mathbf{u}_k^i$. This does not alter the basic form of the resulting detector.

19.2.2 Active MIMO radar

Figure 19.7 depicts the *ij*th bistatic pair in an AMR network. AMR networks are distinguished from PMR networks by the use of cooperative transmitters that transmit known signals. The transmitters are often co-located with the receivers, although this is not necessarily true. The transmit signals are designed to be orthogonal (or nearly orthogonal) in the time, frequency, Doppler, or code domains, so that they can be isolated on receive [30]. Here, it is assumed the transmit signals are perfectly isolated on receive, regardless of the approach used to implement signal orthogonality.

Similar to the procedure described previously for PMR networks, each bistatic target-path signal is isolated into a surveillance channel by beamforming in the anticipated target direction and applying operations appropriate to the method used to achieve signal orthogonality. The resulting complex baseband surveillance signal associated with the *i*th transmitter at the *j*th receiver is given by

$$\mathbf{s}_s^{ij} = \gamma_t^{ij} \mathcal{D}_t^{ij} \mathbf{u}^i + \mathbf{n}_s^{ij}, \tag{19.13}$$

where the terms in (19.13) are the same as in the PMR case. Note that in the context of pulsed operation, $\mathbf{u}^i$ represents an entire pulse train. Note also that the direct-path signal is excluded in (19.13) in comparison to the PMR signal model in (19.5) [24,26]. This is because the direct-path signal is unnecessary as a signal-processing reference due to the use of known transmit signals. Furthermore, any incident direct-path signal can be excluded from the surveillance channel by spatial nulling during surveillance channel beamforming, or simply by receiver blanking during direct-path signal incidence in the context of pulsed operation.

19.2.3 Passive source localization

A PSL sensor network consists of N_r distributed receivers that intercept N_t target-emitted signals to detect and localize targets. Figure 19.8 depicts the geometry of the *ij*th bistatic pair in a PSL sensor network, which consists of the *i*th transmitted signal and the *j*th receiver. Note that typically only a single emission is intercepted ($N_t = 1$), but multiple frequency-channelized signals are possible. Unlike PMR and AMR networks, PSL networks detect target-emitted rather than target-scattered signals. Like PMR networks but unlike AMR networks, the exact forms of the detected signals are *a priori* unknown to the PSL system. The *j*th receiver beamforms

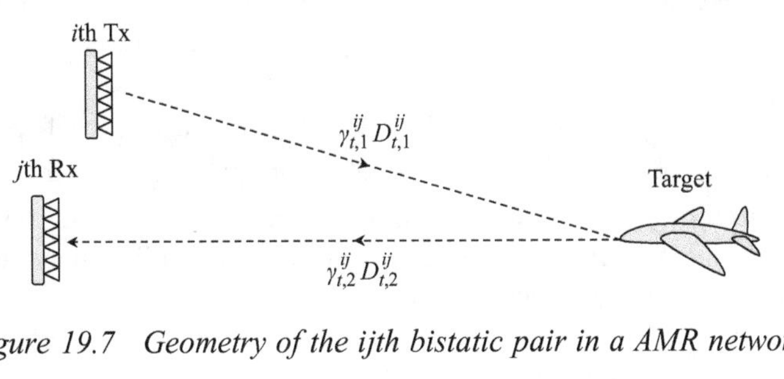

Figure 19.7 Geometry of the ijth bistatic pair in a AMR network

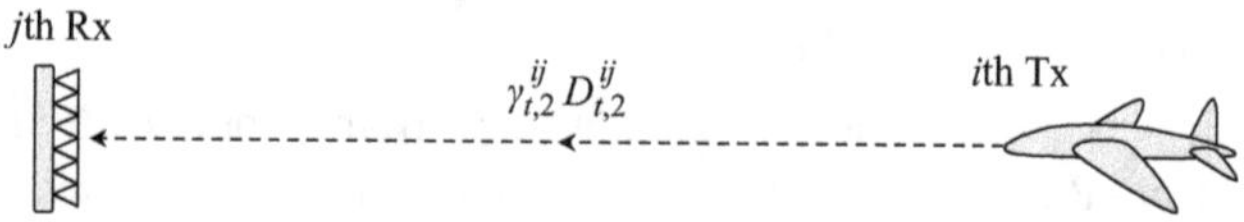

Figure 19.8 Geometry of the ijth transmitter–receiver pair in a PSL network

in the anticipated target direction and channelizes the intercepted signal in frequency. The resulting ith complex baseband signal at the jth receiver is given by

$$\mathbf{s}_s^{ij} = \gamma_{t,2}^{ij}\mathcal{D}_{t,2}^{ij}\mathbf{u}^i + \mathbf{n}_s^{ij}, \tag{19.14}$$

where $\mathbf{u}^i$ is the unknown transmit signal, $\mathcal{D}_{t,2}^{ij}$ accounts for the delay and Doppler shift of the target-to-receiver propagation channel and the channel coefficient $\gamma_{t,2}^{ij}$ accounts for the transmitter's effective radiated power in the receiver direction rather than the target's complex reflectivity as previously for PMR and AMR networks.

19.3 Centralized GLRT detection

19.3.1 Passive MIMO radar

The PMR detection problem involves discriminating between the presence or absence of a target within a hypothesized Cartesian position–velocity cell under test, $(\mathbf{p}, \dot{\mathbf{p}})$, where $\mathbf{p}$ and $\dot{\mathbf{p}}$ are the hypothesized position and velocity, respectively. This problem may be formulated in terms of the received signals in (19.5) as a binary hypothesis test between alternative $(\mathcal{H}_1)$ and null $(\mathcal{H}_0)$ hypotheses,

$$\begin{aligned}
\mathcal{H}_1 :\ & \mathbf{s}_s^{ij} = \gamma_t^{ij}\mathcal{D}_p^{ij}\mathbf{u}^i + \mathbf{n}_s^{ij} \\
& \mathbf{s}_r^{ij} = \gamma_d^{ij}\mathcal{D}_d^{ij}\mathbf{u}^i + \mathbf{n}_r^{ij} \\
\mathcal{H}_0 :\ & \mathbf{s}_s^{ij} = \mathbf{n}_s^{ij} \\
& \mathbf{s}_r^{ij} = \gamma_d^{ij}\mathcal{D}_d^{ij}\mathbf{u}^i + \mathbf{n}_r^{ij},
\end{aligned} \tag{19.15}$$

for $i = 1, \ldots, N_t$ and $j = 1, \ldots, N_r$. Note that $\mathcal{D}_p^{ij} = \mathcal{D}(\tau_p^{ij}, \nu_p^{ij})$, where τ_p^{ij} and ν_p^{ij} are the bistatic delay and Doppler of $(\mathbf{p}, \dot{\mathbf{p}})$ with respect to the ith transmitter and jth receiver. Equation (19.15) is a composite hypothesis test between two composite hypotheses. Under $\mathcal{H}_1$, the joint probability density function of the reference and surveillance channels is parameterized by the unknown channel coefficients γ_t^{ij} and γ_d^{ij} as well as the unknown transmit signal $\mathbf{u}^i$. Similarly, under $\mathcal{H}_0$, this distribution is parameterized by γ_d^{ij} and $\mathbf{u}^i$. It is shown in [6] that a GLRT for this test, in which these unknown parameters are replaced by their maximum likelihood estimates in the likelihood ratio test, is given by

$$\xi_{\text{pmr}} = \frac{1}{\sigma^2}\sum_{i=1}^{N_t}\left(\lambda_1\left(\mathbf{G}_1^i\right) - \lambda_1\left(\mathbf{G}_{rr}^i\right)\right) \underset{\mathcal{H}_0}{\overset{\mathcal{H}_1}{\gtrless}} \kappa_{\text{pmr}}, \tag{19.16}$$

where $\lambda_1(\cdot)$ denotes the largest eigenvalue of its matrix argument, $\mathbf{G}_1^i \in \mathbb{C}^{2N_r \times 2N_r}$ is a Gram matrix with block structure

$$\mathbf{G}_1^i = \begin{bmatrix} \mathbf{G}_{ss}^i & (\mathbf{G}_{rs}^i)^H \\ \mathbf{G}_{rs}^i & \mathbf{G}_{rr}^i \end{bmatrix}, \tag{19.17}$$

$\mathbf{G}_{ss}^i$, $\mathbf{G}_{rr}^i$ and $\mathbf{G}_{rs}^i \in \mathbb{C}^{N_r \times N_r}$ are defined

$$\mathbf{G}_{ss}^i = \begin{bmatrix} \|\tilde{\mathbf{s}}_s^{i1}\|^2 & \langle \tilde{\mathbf{s}}_s^{i1}, \tilde{\mathbf{s}}_s^{i2} \rangle & \cdots & \langle \tilde{\mathbf{s}}_s^{i1}, \tilde{\mathbf{s}}_s^{iN_r} \rangle \\ \langle \tilde{\mathbf{s}}_s^{i2}, \tilde{\mathbf{s}}_s^{i1} \rangle & \|\tilde{\mathbf{s}}_s^{i2}\|^2 & & \vdots \\ \vdots & & \ddots & \vdots \\ \langle \tilde{\mathbf{s}}_s^{iN_r}, \tilde{\mathbf{s}}_s^{i1} \rangle & \cdots & \cdots & \|\tilde{\mathbf{s}}_s^{iN_r}\|^2 \end{bmatrix}, \tag{19.18}$$

$$\mathbf{G}_{rr}^i = \begin{bmatrix} \|\tilde{\mathbf{s}}_r^{i1}\|^2 & \langle \tilde{\mathbf{s}}_r^{i1}, \tilde{\mathbf{s}}_r^{i2} \rangle & \cdots & \langle \tilde{\mathbf{s}}_r^{i1}, \tilde{\mathbf{s}}_r^{iN_r} \rangle \\ \langle \tilde{\mathbf{s}}_r^{i2}, \tilde{\mathbf{s}}_r^{i1} \rangle & \|\tilde{\mathbf{s}}_r^{i2}\|^2 & & \vdots \\ \vdots & & \ddots & \vdots \\ \langle \tilde{\mathbf{s}}_r^{iN_r}, \tilde{\mathbf{s}}_r^{i1} \rangle & \cdots & \cdots & \|\tilde{\mathbf{s}}_r^{iN_r}\|^2 \end{bmatrix}, \tag{19.19}$$

$$\mathbf{G}_{rs}^i = \begin{bmatrix} \langle \tilde{\mathbf{s}}_r^{i1}, \tilde{\mathbf{s}}_s^{i1} \rangle & \langle \tilde{\mathbf{s}}_r^{i1}, \tilde{\mathbf{s}}_s^{i2} \rangle & \cdots & \langle \tilde{\mathbf{s}}_r^{i1}, \tilde{\mathbf{s}}_s^{iN_r} \rangle \\ \langle \tilde{\mathbf{s}}_r^{i2}, \tilde{\mathbf{s}}_s^{i1} \rangle & \langle \tilde{\mathbf{s}}_r^{i2}, \tilde{\mathbf{s}}_s^{i2} \rangle & & \vdots \\ \vdots & & \ddots & \vdots \\ \langle \tilde{\mathbf{s}}_r^{iN_r}, \tilde{\mathbf{s}}_s^{i1} \rangle & \cdots & \cdots & \langle \tilde{\mathbf{s}}_r^{iN_r}, \tilde{\mathbf{s}}_s^{iN_r} \rangle \end{bmatrix}, \tag{19.20}$$

where $\tilde{\mathbf{s}}_s^{ij}$ is the ijth surveillance signal $\mathbf{s}_s^{ij}$ after delay-Doppler compensation to remove the bistatic delay and Doppler shift associated with the hypothesized target position–velocity state,

$$\tilde{\mathbf{s}}_s^{ij} = \left(\mathcal{D}_p^{ij}\right)^H \mathbf{s}_s^{ij}. \tag{19.21}$$

$\tilde{\mathbf{s}}_s^{ij}$ is the ijth reference signal $\mathbf{s}_r^{ij}$ after delay-Doppler compensation to remove the known direct-path delay and Doppler shift,

$$\tilde{\mathbf{s}}_r^{ij} = \left(\mathcal{D}_d^{ij}\right)^H \mathbf{s}_r^{ij} \tag{19.22}$$

and $\langle \cdot, \cdot \rangle$ is the inner product $\langle \mathbf{x}, \mathbf{y} \rangle = \mathbf{x}^H \mathbf{y}$.

19.3.1.1 Relationship to cross-ambiguity function processing

Before moving on to the GLRTs for the AMR and PSL sensor networks, let us first take a closer look at the PMR test statistic ξ_{pmr} in (19.16). Insight into the types of signal processing represented by this statistic is obtained by examining the elements of $\mathbf{G}_1^i$ in (19.17). Consider, first, the off-diagonal block matrix $\mathbf{G}_{rs}^i$, for which the jkth element is given by

$$\left[\mathbf{G}_{rs}^i\right]_{jk} = \langle \tilde{\mathbf{s}}_r^{ij}, \tilde{\mathbf{s}}_s^{ik} \rangle. \tag{19.23}$$

This inner product represents the correlation of the reference signal measured at the jth receiver with the surveillance signal measured at the kth receiver, both with respect to the ith transmit channel, and both after appropriate delay-Doppler compensation. This scalar value can be expressed in terms of the CAF between the aforementioned reference and surveillance signals by expanding (19.23) as follows:

$$\left[\mathbf{G}_{rs}^i\right]_{jk} = \langle \tilde{\mathbf{s}}_r^{ij}, \tilde{\mathbf{s}}_s^{ik} \rangle \tag{19.24}$$

$$= \left(\left(\mathcal{D}_d^{ij} \right)^H \mathbf{s}_r^{ij} \right)^H \left(\left(\mathcal{D}_p^{ik} \right)^H \mathbf{s}_s^{ik} \right) \tag{19.25}$$

$$= \left(\mathcal{D}_p^{ik} \left(\mathcal{D}_d^{ij} \right)^H \mathbf{s}_r^{ij} \right)^H \left(\mathbf{s}_s^{ik} \right) \tag{19.26}$$

$$= \left(e^{j2\pi\nu_d^{ij}\left(\tau_p^{ik}-\tau_d^{ij}\right)} \mathcal{D}\Big(\underbrace{\tau_p^{ik} - \tau_d^{ij}}_{=\Delta\tau_{pd}^{i,jk}}, \underbrace{\nu_p^{ik} - \nu_d^{ij}}_{=\Delta\nu_{pd}^{i,jk}} \Big) \mathbf{s}_r^{ij} \right)^H \mathbf{s}_s^{ik} \tag{19.27}$$

$$= e^{-j2\pi\nu_d^{ij}\Delta\tau_{pd}^{i,jk}} \left(\mathcal{D}\left(\Delta\tau_{pd}^{i,jk}, \Delta\nu_{pd}^{i,jk} \right) \mathbf{s}_r^{ij} \right)^H \mathbf{s}_s^{ik} \tag{19.28}$$

$$= e^{-j2\pi\nu_d^{ij}\Delta\tau_{pd}^{i,jk}} \chi_{rs}^{i,jk} \left(\Delta\tau_{pd}^{i,jk}, \Delta\nu_{pd}^{i,jk} \right), \tag{19.29}$$

where $\Delta\tau_{pd}^{i,jk} = \tau_p^{ik} - \tau_d^{ij}$ and $\Delta\nu_{pd}^{i,jk} = \nu_p^{ik} - \nu_d^{ij}$ denote the relative delay and relative Doppler shift between the ikth hypothesized target-path signal and the ijth direct-path signal, respectively, and $\chi_{rs}^{i,jk}(\Delta\tau, \Delta\nu)$ is the RS-CAF between the ijth reference channel and the jkth surveillance channel at delay $\Delta\tau$ and Doppler $\Delta\nu$,

$$\chi_{rs}^{i,jk}(\Delta\tau, \Delta\nu) = \sum_{l=0}^{L-1} \left[\mathbf{s}_s^{ik}\right]_l \left[\mathbf{s}_r^{ij}\right]_{l-\Delta\tau f_s}^* e^{-j2\pi(\Delta\nu/f_s)\, l}. \tag{19.30}$$

In other words, the inner product $\langle \tilde{\mathbf{s}}_r^{ij}, \tilde{\mathbf{s}}_s^{ik} \rangle$ can be obtained by sampling the RS-CAF between the ith reference signal of the jth receiver and the ith surveillance channel of the kth receiver at the relative delay and Doppler (with respect to the direct-path delay and Doppler) associated with the hypothesized Cartesian position–velocity cell under test, and multiplying by a known unimodular complex scale factor. Observe that once $\chi_{rs}^{i,jk}$ is calculated, the jkth entry of $\mathbf{G}_{rs}^i$ can be populated by appropriately sampling this RS-CAF as the cell under test is varied over a space of hypothesized position–velocity target states. The reference–surveillance correlations represented by the entries of $\mathbf{G}_{rs}^i$ are either *co-receiver* or *cross-receiver* depending on whether the reference and surveillance signals are measured at the same ($j = k$) or different ($j \neq k$) receivers. This connects the GLRT test statistic in (19.16) with conventional passive radar detection processing, which is based upon calculation of co-receiver RS-CAFs for each bistatic pair.

Next, consider the diagonal block matrix $\mathbf{G}^i_{ss}$. The jkth element of this matrix is the inner product between the surveillance signals associated with the ith transmit channel measured at the jth and kth receivers,

$$\left[\mathbf{G}^i_{ss}\right]_{jk} = \langle \tilde{\mathbf{s}}^{ij}_s, \tilde{\mathbf{s}}^{ik}_s \rangle. \tag{19.31}$$

Observe that the diagonal entries ($j = k$) reduce to the surveillance signal energy,

$$\left[\mathbf{G}^i_{ss}\right]_{jj} = \|\tilde{\mathbf{s}}^{ij}_s\|^2 = \|\mathbf{s}^{ij}_s\|^2. \tag{19.32}$$

It can be shown that the off-diagonal terms ($j \neq k$) can be expressed in terms of a CAF between surveillance signals,

$$\left[\mathbf{G}^i_{ss}\right]_{jk} = e^{-j2\pi\nu^{ij}_p \Delta\tau^{i,jk}_{pp}} \chi^{i,jk}_{ss}\left(\Delta\tau^{i,jk}_{pp}, \Delta\nu^{i,jk}_{pp}\right), \tag{19.33}$$

where $\Delta\tau^{i,jk}_{pp} = \tau^{ik}_p - \tau^{ij}_p$ is the time-difference of arrival (TDOA) between the hypothesized target-path signals at the jth and kth receivers, $\Delta\nu^{i,jk}_{pp} = \nu^{ik}_p - \nu^{ij}_p$ is the frequency-difference of arrival (FDOA) between the hypothesized target-path signals at the jth and kth receivers and $\chi^{i,jk}_{ss}(\Delta\tau, \Delta\nu)$ is the surveillance–surveillance CAF (SS-CAF) between the ith surveillance channels of the jth and kth receivers evaluated at TDOA $\Delta\tau$ and FDOA $\Delta\nu$,

$$\chi^{i,jk}_{ss}(\Delta\tau, \Delta\nu) = \sum_{l=0}^{L-1} \left[\mathbf{s}^{ik}_s\right]_l \left[\mathbf{s}^{ij}_s\right]^*_{l-\Delta\tau f_s} e^{-j2\pi(\Delta\nu/f_s)l}. \tag{19.34}$$

Note that the off-diagonal entries represent cross-receiver surveillance–surveillance correlations. Observe that once $\chi^{i,jk}_{ss}$ is calculated, the jkth entry of $\mathbf{G}^i_{ss}$ can be populated by appropriately sampling this SS-CAF as the cell under test is varied over a space of hypothesized target position–velocity states.

Finally, consider the diagonal block matrix $\mathbf{G}^i_{rr}$. The jkth element of this matrix is the inner product between the delay-Doppler compensated reference signals associated with the ith transmit channel measured at the jth and kth receivers,

$$\left[\mathbf{G}^i_{rr}\right]_{jk} = \langle \tilde{\mathbf{s}}^{ij}_r, \tilde{\mathbf{s}}^{ik}_r \rangle. \tag{19.35}$$

As seen previously for the entries of $\mathbf{G}^i_{ss}$, the diagonal entries ($j = k$) of $\mathbf{G}^i_{rr}$ also reduce to the signal energy,

$$\left[\mathbf{G}^i_{rr}\right]_{jj} = \|\mathbf{s}^{ij}_r\|^2 \tag{19.36}$$

and the off-diagonal entries ($j \neq k$) can be expressed in terms of reference–reference CAFs (RR-CAFs). However, the entries of $\mathbf{G}^i_{rr}$ do not depend on the hypothesized cell under test because the reference signals that do not vary between hypotheses.

Therefore, it is more efficient to directly calculate the inner products in (19.35) to populate $\mathbf{G}^i_{rr}$ than to calculate and sample the relevant RR-CAFs.

19.3.2 Active MIMO radar

The AMR detection problem may be formulated as a binary hypothesis test between a simple null hypothesis and a composite alternative hypothesis. Specifically, using the signal model in (19.13),

$$\begin{aligned}\mathcal{H}_1 &: \mathbf{s}^{ij}_s = \gamma^{ij}_t \mathcal{D}^{ij}_p \mathbf{u}^i + \mathbf{n}^{ij}_s \\ \mathcal{H}_0 &: \mathbf{s}^{ij}_s = \mathbf{n}^{ij}_s,\end{aligned} \tag{19.37}$$

for $i = 1, \ldots, N_t$ and $j = 1, \ldots, N_r$. In comparison to the PMR detection hypothesis test in (19.15), each transmit signal $\mathbf{u}^i$ is known rather than unknown, and there is no longer a reference signal under either hypothesis. Regarding the channel coefficients $\{\gamma^{ij}_t\}$ as deterministic unknowns, the resulting GLRT is given by

$$\xi_{\text{amr}} = \frac{1}{\sigma^2} \sum_{i=1}^{N_t} \sum_{j=1}^{N_r} \left| \left(\mathbf{u}^i\right)^H \tilde{\mathbf{s}}^{ij}_s \right|^2 \underset{\mathcal{H}_0}{\overset{\mathcal{H}_1}{\gtrless}} \kappa_{\text{amr}}, \tag{19.38}$$

The statistic ξ_{amr} is formed by: (a) delay-Doppler compensating the surveillance signal to remove the delay and Doppler shift due to bistatic propagation with respect to the hypothesized target position–velocity cell under test; (b) taking the inner product with the known transmit signal; and (c) non-coherently integrating across bistatic transmitter–receiver pairs. These signal-processing operations are commonly found in the structure of AMR detectors [24,26].

19.3.3 Passive source localization

The PSL detection problem may be formulated by a composite hypothesis test that is similar to (19.37) except that the channel is direct rather than bistatic, and the transmit signals $\{\mathbf{u}^i\}$ are unknown rather than known,

$$\begin{aligned}\mathcal{H}_1 &: \mathbf{s}^{ij}_s = \gamma^{ij}_{t,2} \mathcal{D}^{ij}_{p,2} \mathbf{u}^i + \mathbf{n}^{ij}_s \\ \mathcal{H}_0 &: \mathbf{s}^{ij}_s = \mathbf{n}^{ij}_s,\end{aligned} \tag{19.39}$$

for $i = 1, \ldots, N_t$ and $j = 1, \ldots, N_r$. Regarding the channel coefficients $\{\gamma^{ij}_{t,2}\}$ and transmit signals $\{\mathbf{u}^i\}$ as deterministic unknowns, the resulting GLRT is given by [28]

$$\xi_{\text{psl}} = \frac{1}{\sigma^2} \sum_{i=1}^{N_t} \lambda_1\left(\mathbf{G}^i_{ss}\right) \underset{\mathcal{H}_0}{\overset{\mathcal{H}_1}{\gtrless}} \kappa_{\text{psl}}, \tag{19.40}$$

where the Gram matrix $\mathbf{G}^i_{ss}$ is defined in (19.18).

19.3.4 Detector comparisons

Table 19.1 compares the PMR, AMR, and PSL GLRT detectors in terms of the types of correlation exploited by each. As discussed in Section 19.3.1.1, the PMR

Table 19.1 GLRT detector comparison

Network	GLRT test statistic	Ref–surv	Surv–surv	Ref–ref
PMR	$\sigma^{1/2}\sum_{i=1}^{N_t}\left(\lambda_1(\mathbf{G}_1^i)-\lambda_1(\mathbf{G}_{rr}^i)\right)$	Yes	Yes	Yes
AMR	$\sigma^{1/2}\sum_{i=1}^{N_t}\sum_{j=1}^{N_r}\lvert(\mathbf{u}^i)^H\tilde{\mathbf{s}}_s^{ij}\rvert^2$	Yes	No	No
PSL	$\sigma^{1/2}\sum_{i=1}^{N_t}\lambda_1(\mathbf{G}_{ss}^i)$	No	Yes	No

detector exploits both co- and cross-receiver reference–surveillance correlations, cross-receiver surveillance–surveillance correlations and cross-receiver reference–reference correlations. In comparison, the AMR detector exploits only reference–surveillance correlations via matched filtering.[3] In contrast, the PSL detector exploits only cross-receiver surveillance–surveillance correlations through calculation of the elements of $\mathbf{G}_{ss}^i$, which is a sub-block of $\mathbf{G}_1^i$. Since the PMR detector exploits both reference–surveillance and surveillance–surveillance correlations, it is reasonable to expect that it might exhibit characteristics of the AMR and PSL detectors under some circumstances. The simulation results presented in Sections 19.4 and 19.5 explore this relationship with regard to detection sensitivity and ambiguity.

19.3.5 Probability distributions

This section briefly discusses the probability distributions of the GLRT test statistics ξ_{pmr}, ξ_{amr} and ξ_{psl}. The distributions of ξ_{amr} and ξ_{psl} are known exactly, while the exact distribution of ξ_{pmr} is unknown but can be approximated under conditions that are common in practice. First, consider ξ_{amr}. It is shown in [6] that ξ_{amr} follows the non-central and central chi-squared distributions under $\mathcal{H}_1$ and $\mathcal{H}_0$, respectively. Next, consider ξ_{psl}. The matrix $\mathbf{G}_{ss}^i$ is a complex Wishart matrix that is non-central with rank-1 non-centrality matrix under $\mathcal{H}_1$ and central under $\mathcal{H}_0$ [31]. Consequently, the exact distributions of ξ_{psl} can be found under $\mathcal{H}_0$ and $\mathcal{H}_1$ using results provided in [32].

Finally, consider the PMR statistic ξ_{pmr}. The matrices $\mathbf{G}_1^i$ and $\mathbf{G}_{rr}^i$ are complex Wishart distributed with rank-1 non-centralities under both hypotheses, yet the exact distribution of ξ_{pmr} does not appear to be known. However, ξ_{pmr} approaches ξ_{amr} under two conditions that are common in practice [6]. The first condition is that the average direct-path SNR across receivers is high (approximately $\geq$10 dB) with

[3]Note that there is no distinction between co-receiver and cross-receiver reference–surveillance correlations for the AMR detector due to its use of known replicas of the transmitted signal (i.e., perfect references) in the matched filtering operation.

respect to each transmit channel. Mathematically, $\mathrm{SNR}_r^i \geq 10$ dB for all i, where SNR_r^i is the average input direct-path SNR across all reference channels associated with the ith transmitter,

$$\mathrm{SNR}_r^i = \frac{1}{N_r}\sum_{j=1}^{N_r}\frac{|\gamma_d^{ij}|^2}{\sigma^2}. \tag{19.41}$$

Note that $|\gamma_d^{ij}|^2/\sigma^2$ is the input direct-path SNR in the ijth reference channel. The second condition is that the average direct-path-to-target-path power ratio is high (approximately $\geq$10 dB) with respect to each transmit channel. Mathematically, $\bar{\rho}^i \geq 10$ dB for all i, where $\bar{\rho}^i$ denotes the ratio of the average direct-path power to the average target-path power across receivers with respect to the ith transmit channel,

$$\bar{\rho}^i = \frac{\sum_{j=1}^{N_r}|\gamma_d^{ij}|^2}{\sum_{j=1}^{N_r}|\gamma_t^{ij}|^2} = \frac{\mathrm{SNR}_r^i}{\mathrm{SNR}_s^i}, \tag{19.42}$$

where SNR_s^i is the average input target-path SNR across all surveillance channels associated with the ith transmitter,

$$\mathrm{SNR}_s^i = \frac{1}{N_r}\sum_{j=1}^{N_r}\frac{|\gamma_t^{ij}|^2}{\sigma^2}. \tag{19.43}$$

Note that $|\gamma_t^{ij}|^2/\sigma^2$ is the input target-path SNR in the ijth surveillance channel. These two conditions are often satisfied in practice because incident direct-path signals are often quite strong and many orders of magnitude stronger than their respective target-path signals. Therefore, ξ_{pmr} is approximately non-central and central chi-squared distributed under $\mathcal{H}_1$ and $\mathcal{H}_0$, respectively, under these conditions.

It is interesting to note that the distributions of ξ_{amr}, ξ_{psl} and ξ_{pmr} can be shown to depend on only SNR_s^i, SNR_r^i (ξ_{pmr} only), N_t, N_r and signal energy $\|\mathbf{u}^i\|^2 = L$ [6]. Two significant conclusions follow from this result. First, under the mathematical signal model assumed in the preceding development (i.e., single-target environment, perfect isolation of target-path and direct-path signals and perfect interference cancelation), detection sensitivity does not depend on the specific waveform $\mathbf{u}^i$, but only the energy of the received signals. This is a well-known result in active radar, in which the peak output of the matched filter is a function of the signal's energy and not its shape [33]. Second, detection sensitivity does not depend on the specific target-path and direct-path SNRs at individual receivers but only the average target-path and direct-path SNRs across all receivers.

19.4 Detection sensitivity

This section compares the detection sensitivity of the PMR, AMR and PSL GLRTs via numerical simulation. Recall from the discussion in Section 19.3.4 that SNR_s^i, SNR_r^i and signal length L are primary drivers of PMR GLRT detection sensitivity. The following simulations illustrate how the performance of each detector varies with these parameters. After introducing the simulation scenario in Section 19.4.1, SNR_s^i and SNR_r^i are investigated in Section 19.4.2 and L in Section 19.4.3.

19.4.1 Simulation scenario

The simulation scenario consists of $N_t = 2$ transmitters and $N_r = 3$ receivers. The transmit signals are defined by $\mathbf{u}^i = \exp\{j\boldsymbol{\theta}^i\}$, where $\boldsymbol{\theta}^i \in \mathbb{R}^{L\times 1}$ is a random phase vector with independent and identically distributed (i.i.d.) elements uniformly distributed on $[0, 2\pi]$. Note that $\|\mathbf{u}^i\|^2 = L$, as discussed in Section 19.2.1. Each receiver is assumed to isolate the direct-path and target-path signals into reference and surveillance signals, respectively, according to the signal model in (19.5). Each direct-path channel coefficient γ_d^{ij} is randomly drawn from a $\mathcal{CN}(0, 1)$ distribution on each trial under $\mathcal{H}_0$ and $\mathcal{H}_1$, and then the set $\{\gamma_d^{ij} : j = 1, \ldots, N_r\}$ is scaled by a common factor to achieve a desired SNR_r^i according to (19.41). The target-path channel coefficients $\{\gamma_t^{ij}\}$ are similarly drawn and scaled to achieve a desired SNR_s^i according to (19.43) on each $\mathcal{H}_1$ trial. For convenience, it is assumed that $\text{SNR}_s^i = \text{SNR}_s$ for all i, i.e., the average surveillance channel target-path SNR across receivers is the same for each transmit channel. It is similarly assumed that $\text{SNR}_r^i = \text{SNR}_r$ for all i. Note that SNR_r and SNR_s represent the average *input* target-path and direct-path SNRs within the surveillance and reference channels (i.e., before processing gain). Consequently, they are labelled 'Input SNR_r' and 'Input SNR_s' on all figures. Detection thresholds are empirically set to achieve a probability of false alarm (P_{fa}) of 10^{-3} using 10^5 trials under $\mathcal{H}_0$. The probability of detection is calculated using 10^4 trials under $\mathcal{H}_1$ for each test point.

As discussed in Section 19.3.5, the detection sensitivities of the PMR, AMR and PSL GLRT detectors do not depend on the specific transmit waveform $\mathbf{u}^i$, and they depend on the channel coefficients $\{\gamma_d^{ij}\}$ and $\{\gamma_t^{ij}\}$ only through SNR_r and SNR_s. Consequently, the detection curves for these detectors apply for any waveform $\mathbf{u}^i$ such that $\|\mathbf{u}^i\|^2 = L$, and for any channel coefficients that produce a given SNR_r and SNR_s. Note that this implies that the detectors are formulated irrespective of distributions imposed on these variables. Nonetheless, in this simulation, the channel coefficients are randomly varied across trials in order to simulate the effects of non-isotropic target reflectivity and non-phase synchronized receivers.

19.4.2 Dependence on reference and surveillance SNR

Figure 19.9 depicts probability of detection (P_d) curves as a function of SNR_s for $\text{SNR}_r = \{-30, -10, +10\}$ dB and received signal length $L = 100$. The detection

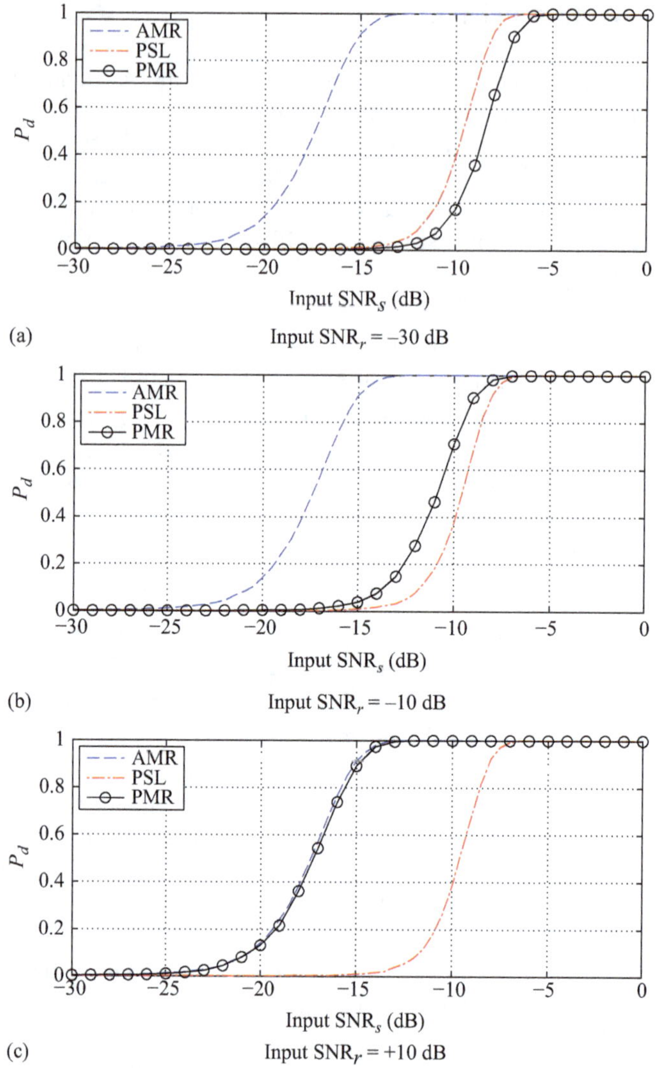

Figure 19.9 P_d curves as a function of SNR_s for a scenario with $N_t = 2$ transmitters, $N_r = 3$ receivers and $L = 100$ samples per received signal

sensitivity of the PMR GLRT improves with increasing SNR_r, while the sensitivities of the AMR and PSL do not depend on SNR_r. This dependence on SNR_r is shown more clearly in Figure 19.10, which depicts the PMR GLRT P_d as a function of both SNR_r and SNR_s. It also depicts $P_d = 0.9$ iso-contours for all considered detectors. This figure shows that the detection sensitivity of the PMR detector appears to asymptote with SNR_r when $SNR_r > 10$ dB and $SNR_r < -30$ dB and varies

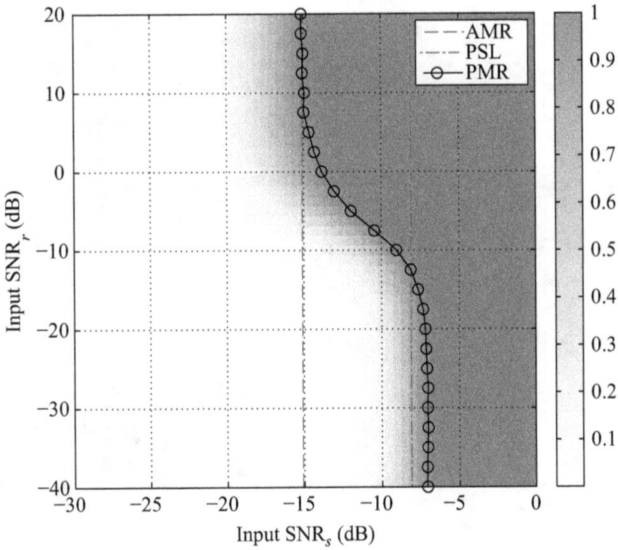

Figure 19.10 *PMR GLRT P_d as a function of SNR_s and SNR_r for $N_t = 2$, $N_r = 3$ and $L = 100$. $P_d = 0.90$ iso-contours are also shown for all considered detectors*

smoothly between these limits at intermediate SNR_r values. In the high-SNR_r asymptotic region ($SNR_r > 10$ dB), the sensitivity of the PMR GLRT is benefitted by exploiting high-quality direct-path references, and its sensitivity asymptotes with the sensitivity of the AMR GLRT. In the low-SNR_r asymptotic region ($SNR_r < -30$ dB), the PMR GLRT no longer benefits from direct-path reference signals and its sensitivity asymptotes near the sensitivity of the PSL GLRT.

19.4.3 Dependence on signal length

Detection sensitivity also depends on the received signal length L. Figure 19.11 depicts a figure identical to Figure 19.10 except that $L = 1{,}000$ rather than $L = 100$, which reflects a processing interval that is ten times longer than previously. In comparison, the PMR and AMR detectors are 10 dB more sensitive in the high-SNR_r asymptotic region than previously. This integration gain is consistent with a coherent integration process. In the low-SNR_r asymptotic region, the PMR and PSL detectors are both approximately 5.4 dB more sensitive than previously, which is consistent with a non-coherent integration process. This distinction between coherent and non-coherent integration may be understood by recognizing that reference–surveillance correlations, which are significant in the high-SNR_r region, involve one noisy surveillance signal and one nearly noiseless reference signal (under high direct-path SNR conditions), while surveillance–surveillance correlations, which are significant in the low-SNR_r region, always involve two noisy surveillance signals.

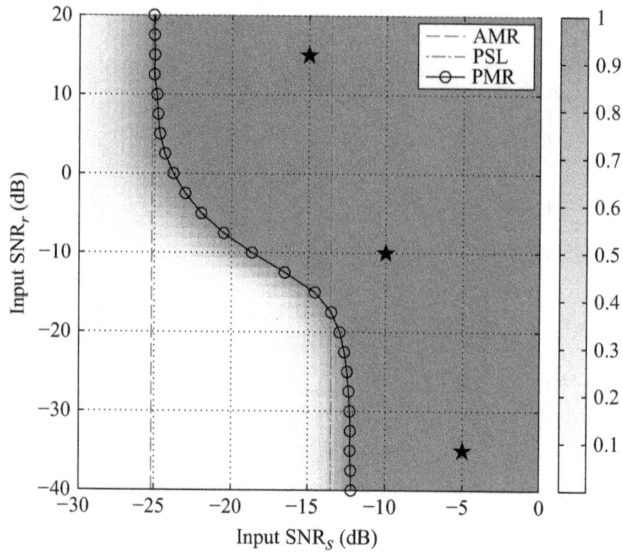

Figure 19.11 PMR GLRT P_d as a function of SNR_s and SNR_r for $N_t = 2$, $N_r = 3$ and $L = 1{,}000$. $P_d = 0.90$ iso-contours are also shown for all considered detectors

Comparison of Figures 19.10 and 19.11 reveals a consistent relationship between the sensitivities of the PMR, AMR and PSL detectors. Figure 19.12 shows that this relationship holds as L is increased in roughly 5 dB increments from $L = 1$ to $L = 10{,}000$. Lines of constant average power ratio between reference and surveillance signals, $\bar{\rho} = \mathrm{SNR}_r/\mathrm{SNR}_s$, are also depicted. This figure shows that PMR GLRT performance may be divided into three regions: a high-SNR$_r$ region ($\mathrm{SNR}_r \geq 0$ dB), where PMR sensitivity asymptotes with AMR sensitivity; a low-$\bar{\rho}$ region ($\bar{\rho} \leq -5$ dB), where PMR sensitivity asymptotes slightly below PSL sensitivity; and a transition region between them.

19.4.4 Discussion

The preceding results have the following interpretations. First, PMR detection sensitivity approaches AMR detection sensitivity when SNR$_r$ is positive, and the PMR detector exploits both co-receiver and cross-receiver reference–surveillance correlations. This defines a high direct-path SNR region in which reference–surveillance correlations are the primary contributors to detection sensitivity, and the sensitivity improves coherently with increasing signal length. This is consistent with the result in Section 19.3.5, which states that the distribution of ξ_{pmr} approaches that of ξ_{amr} under high direct-path SNR and high-$\bar{\rho}$ conditions. Second, PMR detection sensitivity approximates PSL detection sensitivity when the average power ratio, $\bar{\rho}$, falls below approximately −5 dB, i.e., when the

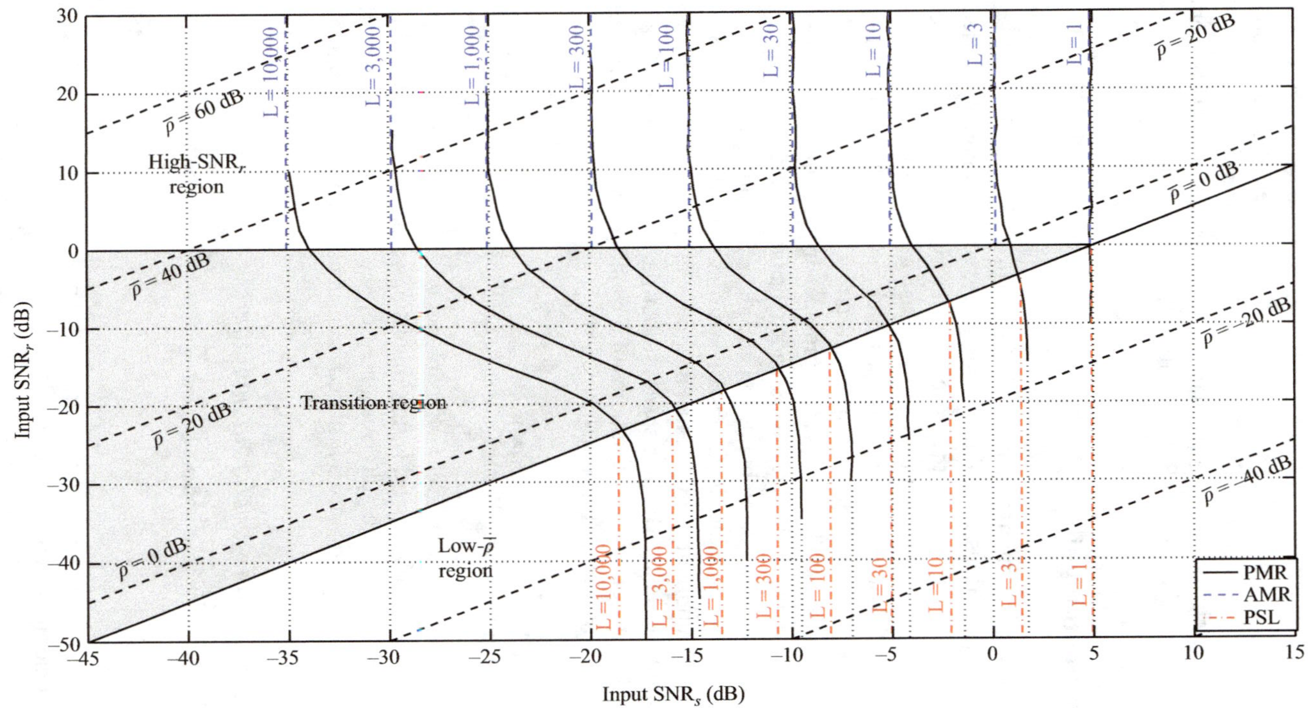

Figure 19.12 *Depiction of PMR GLRT $P_d = 0.9$ iso-contours for $N_t = 2$, $N_r = 3$ and $L =[1, 3, 10, 30, 100, 300, 1{,}000, 3{,}000, 10{,}000]$. The SNR_s values at which the PSL and AMR GLRTs achieve $P_d = 0.9$ are depicted by dash-dotted red lines and dashed blue lines, respectively, for each L. Copyright © 2014 IEEE. Reprinted, with permission, from [6]*

average target-path SNR exceeds the average direct-path SNR by approximately 5 dB. In this region, surveillance–surveillance correlations are the primary contributors to detection sensitivity, and sensitivity improves non-coherently with increasing signal length. Finally, PMR sensitivity varies between PSL and AMR sensitivity when $\text{SNR}_r \leq 0$ dB and $\bar{\rho} \geq -5$ dB. This defines a transition region, in which reference–surveillance and surveillance–surveillance correlations both contribute in varying degrees to PMR sensitivity.

19.5 Detection ambiguity

This section compares the ambiguity of the PMR, AMR and PSL GLRT detectors. In contrast to usual discussions of radar ambiguity, which examine the effect of delay and Doppler mismatch on the matched filter output assuming a given transmit waveform (i.e., waveform range-Doppler ambiguity), this section examines the effect of position and velocity mismatch between the hypothesized target position–velocity state $(\mathbf{p}, \dot{\mathbf{p}})$ and the actual target state $(\mathbf{t}, \dot{\mathbf{t}})$ on the individual GLRT test statistics. This perspective provides insight into how system properties, such as the number of receivers and transmitters and their relative geometries, affect system-level ambiguity in the detection domain, which in centralized detection processing is Cartesian position–velocity space. Nonetheless, system-level ambiguity in Cartesian position–velocity space depends on waveform ambiguity in delay-Doppler space in that the elements of the Gram matrices $\mathbf{G}_{rs}^i$ and $\mathbf{G}_{ss}^i$ can be expressed in terms of the ambiguity function (AF) of the ith transmit waveform $\mathbf{u}^i$, defined by

$$\chi^i(\Delta\tau, \Delta\nu) = \sum_{l=0}^{L-1} \left[\mathbf{u}^i\right]_l \left[\mathbf{u}^i\right]_{l+\Delta\tau f_s}^* e^{j2\pi(\Delta\nu/f_s)\,l}. \tag{19.44}$$

The relationship between system-level ambiguity and waveform ambiguity is discussed in Section 19.5.1. Following this, the simulation scenario is introduced in Section 19.5.2. Finally, the ambiguity responses of the AMR, PSL and PMR detectors in the given scenario are illustrated and compared in the remaining sections.

19.5.1 Dependence on waveform ambiguity

First, consider the jkth element of $\mathbf{G}_{rs}^i$,

$$\left[\mathbf{G}_{rs}^i\right]_{jk} = \langle \tilde{\mathbf{s}}_r^{ij}, \tilde{\mathbf{s}}_s^{ik} \rangle. \tag{19.45}$$

The delay-Doppler compensated reference signal $\tilde{\mathbf{s}}_r^{ij}$ can be expressed using (19.5) and (19.22), and ignoring receiver noise, as

$$\tilde{\mathbf{s}}_r^{ij} = \left(\mathcal{D}_d^{ij}\right)^H \mathbf{s}_r^{ij} = \gamma_d^{ij}\left(\mathcal{D}_d^{ij}\right)^H \mathcal{D}_d^{ij}\,\mathbf{u}^i = \gamma_d^{ij}\,\mathbf{u}^i. \tag{19.46}$$

Similarly, the delay-Doppler compensated surveillance signal $\tilde{\mathbf{s}}_s^{ik}$ can be expressed using (19.5) and (19.21), and ignoring receiver noise, as

$$\tilde{\mathbf{s}}_s^{ik} = \left(\mathcal{D}_p^{ik}\right)^H \mathbf{s}_s^{ik} = \gamma_t^{ik}\left(\mathcal{D}_p^{ik}\right)^H \mathcal{D}_t^{ik}\,\mathbf{u}^i. \tag{19.47}$$

Substituting (19.46) and (19.47) into (19.45), recalling that $\mathcal{D}_p^{ik} = \mathcal{D}(\tau_p^{ik}, \nu_p^{ik})$ and $\mathcal{D}_t^{ik} = \mathcal{D}(\tau_t^{ik}, \nu_t^{ik})$, and simplifying gives

$$\left[\mathbf{G}_{rs}^i\right]_{jk} = \left\langle \gamma_d^{ij}\mathbf{u}^i, \gamma_t^{ik}\mathcal{D}^H\left(\tau_p^{ik}, \nu_p^{ik}\right)\mathcal{D}\left(\tau_t^{ik}, \nu_t^{ik}\right)\mathbf{u}^i\right\rangle \tag{19.48}$$

$$= c_1\left(\gamma_d^{ij}\right)^*\gamma_t^{ik}\left(\mathcal{D}\left(\tau_p^{ik} - \tau_t^{ik}, \nu_p^{ik} - \nu_t^{ik}\right)\,\mathbf{u}^i\right)^H\mathbf{u}^i \tag{19.49}$$

$$= c_1\left(\gamma_d^{ij}\right)^*\gamma_t^{ik}\,\chi^i\left(\tau_t^{ik} - \tau_p^{ik}, \nu_t^{ik} - \nu_p^{ik}\right), \tag{19.50}$$

where c_1 is a unit-modulus phase term. Noting that $\chi^i(\Delta\tau, \Delta\nu)$ peaks when its arguments are zero leads to the following iso-range and iso-Doppler conditions:

$$\tau_p^{ik} = \tau_t^{ik} \qquad \text{and} \qquad \nu_p^{ik} = \nu_t^{ik}. \tag{19.51}$$

In words, $\left[\mathbf{G}_{rs}^i\right]_{jk}$ peaks at hypothesized target position–velocity states for which (a) the hypothesized bistatic delay equals the actual target bistatic delay, and (b) the hypothesized bistatic Doppler equals the actual target bistatic Doppler. This is representative of the processing associated with each bistatic pair in active multistatic and MIMO radar, which localize targets along contours of constant bistatic range and bistatic Doppler [34,35].

Next, consider the jkth element of $\mathbf{G}_{ss}^i$. It can be shown using a procedure similar to that used previously for $\mathbf{G}_{rs}^i$ that

$$\left[\mathbf{G}_{ss}^i\right]_{jk} = c_2\left(\gamma_t^{ij}\right)^*\gamma_t^{ik}\,\chi^i\left(\Delta\tau_p^{i,jk} - \Delta\tau_t^{i,jk}, \Delta\nu_p^{i,jk} - \Delta\nu_t^{i,jk}\right), \tag{19.52}$$

where $\Delta\tau_p^{jk}$ and $\Delta\tau_t^{jk}$ are the TDOAs of the hypothesized and actual target signals with respect to the jth and kth receivers, respectively, given by

$$\Delta\tau_p^{i,jk} = \tau_p^{ij} - \tau_p^{ik} \tag{19.53}$$

$$\Delta\tau_t^{i,jk} = \tau_t^{ij} - \tau_t^{ik} \tag{19.54}$$

$\Delta\nu_p^{i,jk}$ and $\Delta\nu_t^{i,jk}$ are the FDOAs of the hypothesized and actual target signals with respect to the ith transmit channel and jth and receivers, respectively, given by

$$\Delta\nu_p^{i,jk} = \nu_p^{ij} - \nu_p^{ik} \tag{19.55}$$

$$\Delta\nu_t^{i,jk} = \nu_t^{ij} - \nu_t^{ik} \tag{19.56}$$

and c_2 is a unit-modulus phase term. Again, noting that $\chi^i(\Delta\tau, \Delta\nu)$ peaks when its arguments are zero leads to the following iso-TDOA and iso-FDOA conditions:

$$\Delta\tau_p^{jk} = \Delta\tau_t^{jk} \qquad \text{and} \qquad \Delta\nu_p^{i,jk} = \Delta\nu_t^{i,jk}. \tag{19.57}$$

In words, the magnitude of $\left[\mathbf{G}_{ss}^i\right]_{jk}$ peaks when the hypothesized TDOA equals the actual target TDOA and the hypothesized FDOA equals the actual target FDOA. This is representative of PSL networks, which localize targets along contours of constant TDOA and FDOA [27,36].

The previous analysis considered only the elements of $\mathbf{G}_{rs}^i$ and $\mathbf{G}_{ss}^i$, and it is not immediately clear how these entries affect the statistic ξ_{pmr}, which is a function of the largest eigenvalue of $\mathbf{G}_1^i$, of which $\mathbf{G}_{rs}^i$ and $\mathbf{G}_{ss}^i$ are sub-blocks. However, the simulation results presented in this section confirm that the salient ambiguity properties of ξ_{pmr} (as well as those of ξ_{amr} and ξ_{psl}) can be explained in terms of the preceding iso-range/iso-Doppler and iso-TDOA/iso-FDOA conditions.

19.5.2 Simulation scenario

The ambiguity simulation scenario has two stationary transmitters at $[0.5, 4]$ km and $[-0.5, -4]$ km, three stationary receivers at $[-4, 2]$ km, $[-4, 0.5]$ km and $[-4, -2.5]$ km and a target at $\mathbf{t} = [4, 0]$ km. For clarity and brevity, the target is stationary, $\dot{\mathbf{t}} = \mathbf{0}$, rather than moving. The transmitters have carrier frequencies of 8.0 and 8.1 GHz. Complex baseband signals $\mathbf{s}_s^{ij}$ and $\mathbf{s}_r^{ij}$ are simulated according to (19.5) with sampling rate $f_s = 500$ kHz, length $T = 2$ ms and $\mathbf{u}^i = \exp\{j\boldsymbol{\theta}^i\}$, where $\boldsymbol{\theta}^i \in \mathbb{R}^{L\times 1}$ is a random phase vector with i.i.d. elements uniformly distributed on $[0,\ 2\pi]$ and length $L = f_s T = 1{,}000$. Note that $\mathbf{u}^i$ is critically sampled, i.e., $f_s = B$, where B is the bandwidth of $\mathbf{u}^i$ in Hertz; consequently, L equals the time-bandwidth product TB. Figure 19.13 depicts the normalized AF for one of the signals in the present scenario as a function of delay mismatch $\Delta\tau$ and Doppler mismatch $\Delta\nu$. As shown, delay resolution is $1/f_s = 2\ \mu\text{s}$, Doppler resolution is $1/T = 500$ Hz, and the average AF floor is $10\log_{10}(L) \approx 30$ dB below the main-lobe peak. For clarity, the target-path channel coefficients are made equal across all bistatic pairs, $\gamma_t^{ij} = \gamma_t$ for all i and j, so that each bistatic pair contributes equally to the detection statistic and its ambiguity response. Similarly, $\gamma_d^{ij} = \gamma_d$ for all i and j. The specific values of γ_t and γ_d are chosen to achieve desired values of SNR_s and SNR_r, respectively.

19.5.3 AMR ambiguity

First, consider the AMR detector. Figure 19.14 illustrates the ambiguity response of the AMR GLRT as a function of the hypothesized position $\mathbf{p} = [p_x, p_y]$ when the hypothesized velocity is matched to the target velocity, $\dot{\mathbf{p}} = \dot{\mathbf{t}} = \mathbf{0}$. Both isotropic (Figure 19.14(a)) and directional (Figure 19.14(b)) antennas are used to form surveillance channels from the incident target-path signals. In the directional antenna case, the surveillance channels are formed by steering a six-element uniform linear array with approximately $\lambda/2$ element spacing at each hypothesized target position.

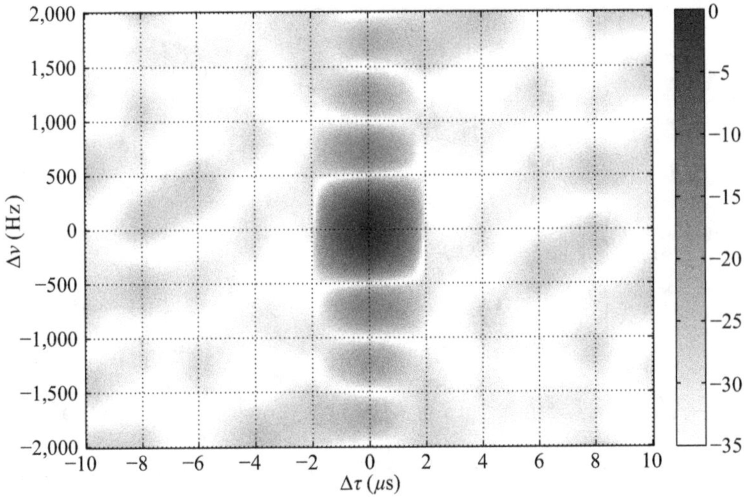

Figure 19.13 Example signal ambiguity function (dB)

The target-path channel coefficients are scaled so that each input SNR is -15 dB after surveillance channel formation, i.e., $|\gamma_t^{ij}|^2/\sigma^2 = -15$ dB. According to Figure 19.11, this SNR is 10 dB higher than the SNR required to achieve 90% probability of detection, which accounts for the well-resolved ambiguity responses in Figure 19.14.

In both the isotropic and directional cases, the response ξ_{amr} peaks at the true target position, and it exhibits ridges along the iso-range ellipses associated with each of the six bistatic pairs in the system (denoted by dashed blue lines). For a given bistatic pair, an ellipse is generated at the hypothesized positions that satisfy the iso-range and iso-Doppler conditions in (19.51). This ellipse has foci at the transmitter and receiver positions and major axis equal to the bistatic range of the target. In the case of isotropic antennas, each ellipse is seen in its entirety; in the case of directional antennas, each ellipse is spatially filtered due to the roll-off of the directional antenna patterns as they are steered away from the true target position. Note that the thickness of the iso-range contours is a function of the waveform bandwidth. Wider bandwidths would result in narrower iso-contours.

19.5.4 PSL ambiguity

Next, consider the PSL detector. Figure 19.15 illustrates the ambiguity response of the PSL GLRT detector as a function of the hypothesized position $\mathbf{p}$ when the hypothesized velocity is matched to the target velocity, $\dot{\mathbf{p}} = \dot{\mathbf{t}} = \mathbf{0}$. In this scenario, the target emits two signals that are intercepted and processed by each receiver. Both isotropic (Figure 19.15(a)) and directional (Figure 19.15(b)) antennas are used to form surveillance channels from the target-path signals as described

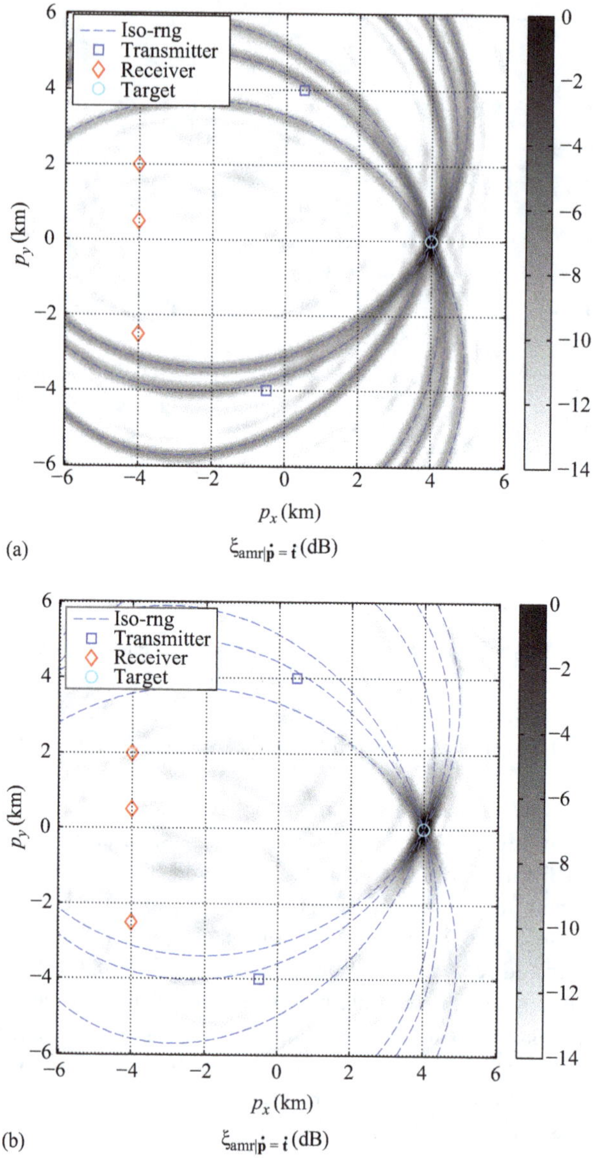

Figure 19.14 AMR ambiguity response using (a) isotropic and (b) directional antennas. Copyright © 2014 IEEE. Reprinted, with permission, from [37]

in Section 19.5.3. The target-path channel coefficients are scaled so that $|\gamma_{t,2}^{ij}|^2/\sigma^2 = -5$ dB, which, according to Figure 19.11, is about 8 dB higher than the SNR required to achieve 90% probability of detection. Consequently, the ambiguity responses are well resolved.

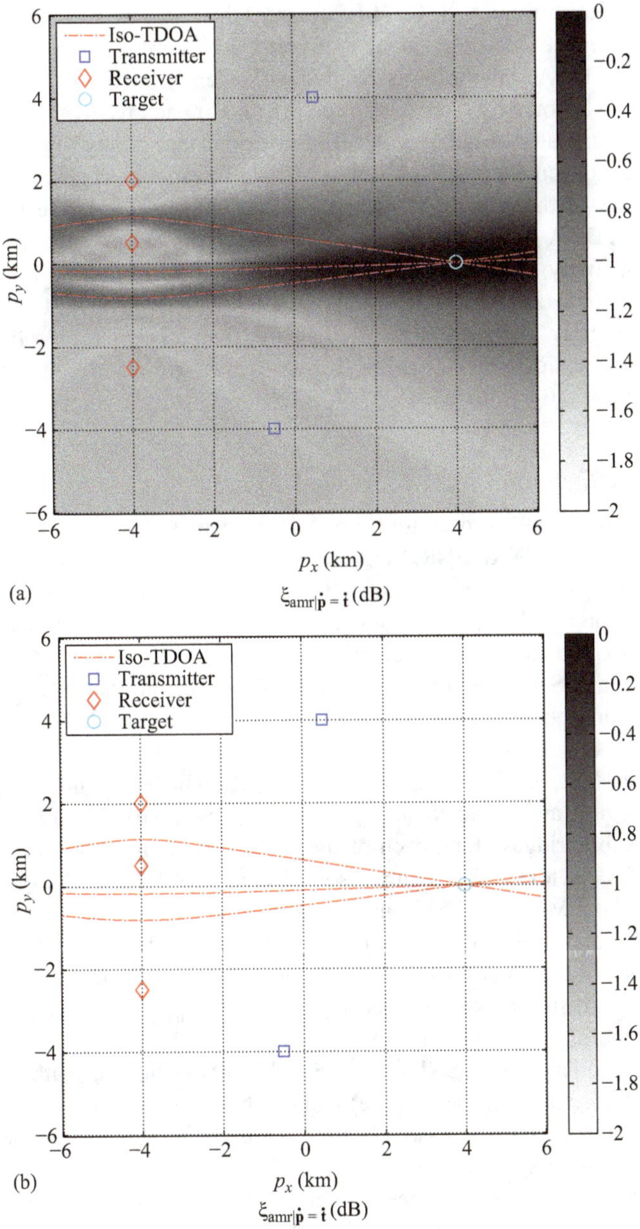

Figure 19.15 PSL ambiguity response using (a) isotropic and (b) directional antennas. Copyright © 2014 IEEE. Reprinted, with permission, from [37]

In both the isotropic and directional cases, the response ξ_{psl} peaks at the true target position. The isotropic antenna response in Figure 19.15(a) exhibits ridges along the iso-TDOA hyperbolas associated with each of the three distinct receiver–receiver pairs (denoted by red dash-dotted lines). For each receiver–receiver pair, a hyperbola is generated at the hypothesized positions that satisfy the iso-TDOA and iso-FDOA conditions in (19.57). This hyperbola has foci at the receiver locations, and the points of the hyperbola have a TDOA with respect to the two receivers equal to the TDOA of the target-path signal. In the directional antenna case, these ridges are spatially filtered around the true target location due to the steering of each antenna array at each hypothesized position. In comparison, the mainlobe of the PSL ambiguity response is larger than the mainlobe of the AMR ambiguity response, and the sidelobes of the PSL response are significantly higher (notice the difference in the dB colour scale).

19.5.5 PMR ambiguity

Finally, consider the PMR detector. Figure 19.16 depicts the ambiguity response of the PMR GLRT at $(\text{SNR}_s, \text{SNR}_r) = (-15, 15)$ dB, $(-10, -10)$ dB and $(-5, -35)$ dB. These points, depicted in Figure 19.11 by black stars, are within the high-SNR_r, transition and low-$\bar{\rho}$ regions, respectively. A directional antenna is used to form reference and surveillance channels according to (19.5). Each point is at least 8 dB above the SNR_s required by the PMR GLRT at that SNR_r to achieve $P_d = 0.90$. Consequently, well-defined peaks in ξ_{pmr} are expected at the true target state $(\mathbf{t}, \dot{\mathbf{t}})$.

Consider, first, $(\text{SNR}_s, \text{SNR}_r) = (-15, 15)$ dB, which is within the high direct-path SNR region and depicted in Figure 19.16(a) and (b). The PMR ambiguity response exhibits ridges along each of the six bistatic ellipses, as seen previously for the AMR detector. In fact, comparison of Figure 19.16(a) with Figure 19.14(b) shows that the PMR and AMR ambiguity responses are nearly identical. This demonstrates that reference–surveillance correlations determine the ambiguity characteristics of the PMR GLRT in this region. As seen previously with detection sensitivity, this result is expected because ξ_{amr} and ξ_{pmr} become equivalent under high-SNR_r and high-$\bar{\rho}$ conditions.

Next, consider $(\text{SNR}_s, \text{SNR}_r) = (-5, -35)$ dB, which is within the low-$\bar{\rho}$ region and depicted in Figure 19.16(e) and (f). In this region, the PMR ambiguity response is aligned with the iso-TDOA contours associated with the receiver–receiver pairs in the system, as seen previously with the PSL ambiguity response. This demonstrates that surveillance–surveillance correlations determine the ambiguity characteristics of the PMR GLRT in this region.

Finally, consider $(\text{SNR}_s, \text{SNR}_r) = (-10, -10)$ dB, which is within the transition region and shown in Figure 19.16(c) and (d). In this case, the ambiguity response shows ridges along both the iso-range ellipses and the iso-TDOA hyperbolas. Thus, both reference–surveillance and surveillance–surveillance correlations contribute to the ambiguity characteristics of the PMR GLRT in this region.

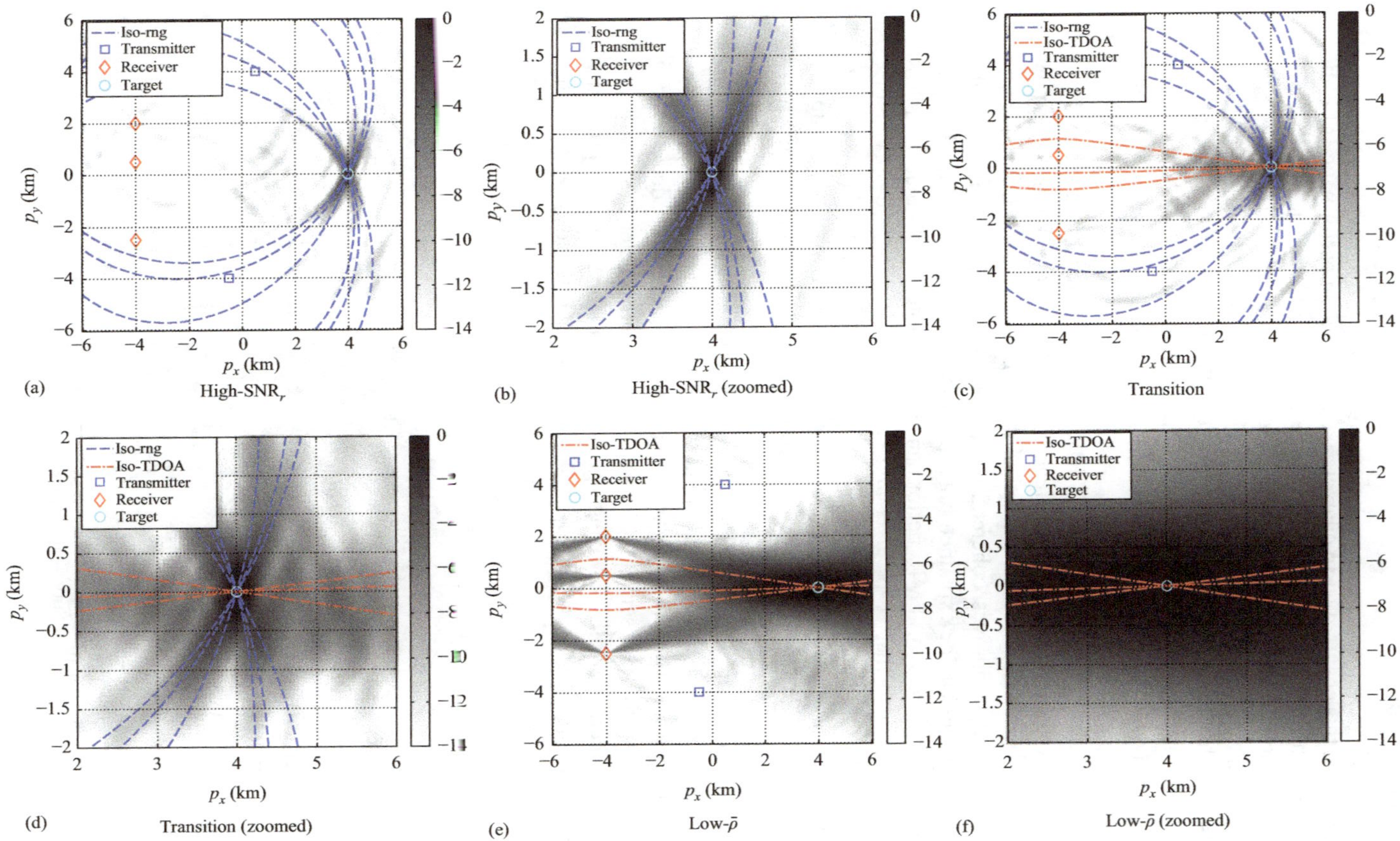

Figure 19.16 PMR ambiguity responses $\xi_{pmr}|_{\dot{\mathbf{p}}=\dot{\mathbf{t}}}$ (dB) in the (a) and (b) high-SNR$_r$, (c) and (d) transition and (e) and (f) low-$\bar{\rho}$ regions. Copyright © 2014 IEEE. Reprinted, with permission, from [37]

19.6 Conclusion

The objective of this chapter was to characterize the detection performance of PMR networks over the entire reference–surveillance SNR space. The manner in which this was accomplished is summarized in Figure 19.17. Specifically, it has been shown through examination of a centralized GLRT for PMR detection, and comparison with centralized AMR and PSL GLRTs, that PMR sensitivity and ambiguity performance can be divided into three regions: a high-SNR_r region; a low-$\bar{\rho}$ region; and a transition region between them. These regions are illustrated in Figure 19.17.

In the high-SNR_r region ($\mathrm{SNR}_r \geq 0$ dB), the PMR system has the benefit of high-quality reference signals, and reference–surveillance correlations determine the sensitivity and ambiguity properties of the GLRT detector. Consequently, detection sensitivity improves coherently with increasing signal length, and the ambiguity properties can be explained in terms of target iso-range and iso-Doppler conditions. In this region, PMR performance asymptotes with that of AMR performance, provided that co-receiver *and* cross-receiver reference–surveillance correlations are exploited. Note that cross-receiver reference–surveillance correlations, which are only possible in centralized processing, account for the sensitivity improvement of centralized over decentralized detection architectures in the high-SNR_r region [6].

In the low-$\bar{\rho}$ region ($\bar{\rho} \leq -5$ dB), the PMR system has access to only low-quality reference signals (or none at all), and surveillance–surveillance correlations determine the sensitivity and ambiguity properties of the GLRT detector in a manner analogous to PSL sensor networks. Consequently, detection sensitivity improves non-coherently with increasing signal length, and the ambiguity properties can be explained in terms of target iso-TDOA and iso-FDOA conditions. Note that

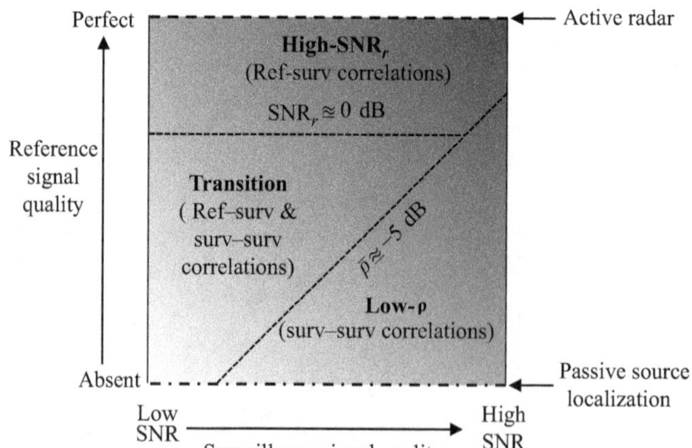

Figure 19.17 PMR performance versus reference and surveillance signal quality

surveillance–surveillance correlation processing requires centralized processing and multiple receiver sites; consequently, it is not available in bistatic or single-site multistatic passive radar topologies. The concept of reference-less PMR networks that exploit only surveillance–surveillance correlations has recently been explored [31,38,39].

Finally, within the transition region, the PMR system has access to reference signals of intermediate quality, and both reference–surveillance and surveillance–surveillance correlations contribute in varying degrees to the sensitivity and ambiguity properties of the GLRT detector. In this region, the sensitivity and ambiguity performance of PMR detection improves smoothly between that of PSL and AMR networks with increasing reference signal quality. In this way, PMR networks can be thought of as intermediate to PSL and AMR networks, unifying them within a common theoretical framework.

References

[1] C. J. Baker, H. D. Griffiths, I. Papoutsis. 'Passive coherent location radar systems. Part 2: waveform properties', *IEE Proc-Radar Sonar Navigation*, **152**(**3**), pp. 160–168, (2005).

[2] H. D. Griffiths, C. J. Baker. 'Passive bistatic radar waveforms', F. Gini, A. De Maio, L. K. Patton (Eds.), *Waveform Diversity & Design for Advanced Radar Systems*, Radar Sonar and Navigation Series 22, chapter 5, pp. 121–147, (Institution of Engineering and Technology, 2012).

[3] K. S. Bialkowski, I. V. L. Clarkson. 'Passive radar signal processing in single frequency networks', *Conference Record of the Forty Sixth Conference on Signals, Systems and Computers (ASILOMAR)*, pp. 199–202, (2012).

[4] D. Poullin, M. Flecheux. 'Passive 3D tracking of low altitude targets using DVB (SFN broadcasters)', *IEEE AES Mag*, **27**(**11**), pp. 36–41, (2012).

[5] M. Daun, U. Nickel, W. Koch. 'Tracking in multistatic passive radar systems using DAB/DVB-T illumination', *Signal Proc*, **92**, pp. 1365–1386, (2012).

[6] D. E. Hack, L. K. Patton, B. Himed, M. A. Saville. 'Detection in passive MIMO radar networks', *IEEE Trans Signal Process*, **62**(**11**), pp. 2999–3012, (2014).

[7] H. D. Griffiths, C. J. Baker. 'Passive coherent location radar systems. Part 1: Performance prediction', *IEE Proc-Radar Sonar Navigation*, **152**(**3**), pp. 153–159, (2005).

[8] H. D. Griffiths, C. J. Baker. 'The signal and interference environment in passive bistatic radar', *Proceedings Information, Decision and Control IDC'07*, pp. 1–10, (2007).

[9] F. Colone, D. W. O'Hagan, P. Lombardo, C. J. Baker. 'A multistage processing algorithm for disturbance removal and target detection in passive bistatic radar', *IEEE Trans Aerosp Electron Syst*, **45**(**2**), pp. 698–722, (2009).

[10] M. Malanowski, K. Kulpa, R. Suchozebrski. 'Two-stage tracking algorithm for passive radar', *Proceedings 12th International Conference on Information Fusion FUSION'09*, pp. 1800–1806, (2009).

[11] M. Malanowski, K. Kulpa. 'Two methods for target localization in multistatic passive radar', *IEEE Trans Aerosp Electron Syst*, **48(1)**, pp. 572–580, (2012).

[12] A. Schroeder, M. Edrich, V. Winkler. 'Multi-illuminator passive radar performance evaluation', *Proceedings on International Radar Symposium (IRS)*, pp. 61–64, (2012).

[13] M. Klein, N. Millet. 'Multireceiver passive radar tracking', *IEEE AES Magazine*, **27(10)**, pp. 26–36, (2012).

[14] C. Bongioanni, F. Colone, S. Bernardini, L. Lelli, A. Stavolo, P. Lombardo. 'Passive radar prototypes for multifrequency target detection', *Proc. SPIE, 6937*, (2007).

[15] D. W. O'Hagan, F. Colone, C. J. Baker, H. D. Griffiths. 'Passive bistatic radar (PBR) demonstrator', *Proc. IET Int. Radar Systems Conf*, pp. 1–5, (2007).

[16] D. Gould, R. Pollard, C. Sarno, P. Tittensor. 'Developments to a multiband passive radar demonstrator system', *Proc. IET Int. Radar Systems Conf*, pp. 1–5, (2007).

[17] R. Zemmari, U. Nickel, W.-D. Wirth. 'GSM passive radar for medium range surveillance', *Proc. European Radar Conf. EuRAD 2009*, pp. 49–52, (2009).

[18] A. Schroeder, M. Edrich. 'CASSIDIAN multiband mobile passive radar system', *Proc. Int. Radar Symp. (IRS)*, pp. 286–291, (2011).

[19] R. Tao, H. Z. Wu, T. Shan. 'Direct-path suppression by spatial filtering in digital television terrestrial broadcasting-based passive radar', *IET Radar, Sonar & Navigation*, **4(6)**, pp. 791–805, (2010).

[20] J. Palmer, S. Searle. 'Evaluation of adaptive filter algorithms for clutter cancellation in passive bistatic radar', *Proc. IEEE Radar Conf. (RADAR)*, pp. 0493–0498, (2012).

[21] R. Zemmari. 'Reference signal extraction for GSM passive coherent location', *Proc. 11th Int. Radar Symp. (IRS)*, pp. 1–4, (2010).

[22] G. Cui, J. Jiu, H. Li, B. Himed. 'Signal detection with noisy reference for passive sensing', *Signal Processing*, **108**, pp. 389–399, (March 2015).

[23] J. Liu, H. Li, B. Himed. 'Analysis of cross-correlation detector for passive radar applications', *IEEE Int. Radar Conference 2015*, pp. 0772–0776, (2015).

[24] E. Fishler, A. Haimovich, R. S. Blum, J. Cimini, L. J. D. Chizhik, R. A. Valenzuela. 'Spatial diversity in radars-models and detection performance', *IEEE Trans. Signal Process.*, **54(3)**, pp. 823–838, (2006).

[25] J. Li, P. Stoica. 'MIMO radar with colocated antennas', *IEEE Signal Process. Mag.*, **24(5)**, pp. 106–114, (2007).

[26] Q. He, N. H. Lehmann, R. S. Blum, A. M. Haimovich. 'MIMO radar moving target detection in homogeneous clutter', *IEEE Trans. Aerosp. Electron. Syst.*, **46(3)**, pp. 1290–1301, (2010).

[27] A. J. Weiss. 'Direct geolocation of wideband emitters based on delay and Doppler', *IEEE Trans. Signal Process.*, **59(6)**, pp. 2513–2521, (2011).

[28] N. Vankayalapati, S. Kay. 'Asymptotically optimal detection of low probability of intercept signals using distributed sensors', *IEEE Trans. Aerosp. Electron. Syst.*, **48(1)**, pp. 737–748, (2012).

[29] F. Colone, R. Cardinali, P. Lombardo, O. Crognale, A. Cosmi, A. Lauri, T. Bucciarelli. 'Space-time constant modulus algorithm for multipath removal on the reference signal exploited by passive bistatic radar', *IET Radar, Sonar & Navigation*, **3(3)**, pp. 253–264, (2009).

[30] M. Xue, J. Li, P. Stoica. 'MIMO radar waveform design', F. Gini, A. D. Maio, L. Patton (Eds.), *Waveform Diversity & Design for Advanced Radar Systems*, pp. 20–26, (The Institution of Engineering and Technology, 2011).

[31] D. E. Hack, L. K. Patton, B. Himed, M. A. Saville. 'Centralized passive MIMO radar detection without direct-path reference signals', *IEEE Trans. Signal Process.*, **62(11)**, pp. 3013–3023, (2014).

[32] M. Kang, M.-S. Alouini. 'Largest eigenvalue of complex Wishart matrices and performance analysis of MIMO MRC systems', *IEEE J. Sel. Areas Commun.*, **21(3)**, pp. 418–426, (2003).

[33] N. Levanon, E. Mozeson. *Radar Signals*, (John Wiley & Sons, Inc., Hoboken, NJ, 2004).

[34] A. Mrstik. 'Multistatic-radar binomial detection', *IEEE Trans. Aerosp. Electron. Syst.,* **AES-14(1)**, pp. 103–108, (1978).

[35] E. Hanle. 'Survey of bistatic and multistatic radar', *IEE Proceedings F Communications, Radar and Signal Processing*, **133(7)**, pp. 587–595, (1986).

[36] P. Chestnut. 'Emitter location accuracy using TDOA and differential Doppler', *IEEE Trans. Aerosp. Electron. Syst.*, **18**, pp. 214–218, (1982).

[37] D. Hack, L. Patton, B. Himed. 'Direct Cartesian localization in passive MIMO radar networks', *Radar Conference (Radar), 2014 International*, pp. 1–6, (2014).

[38] J. Liu, H. Li, B. Himed. 'Two target detection algorithms for passive multistatic radar', *IEEE Trans. Signal Process.*, **62(22)**, pp. 5930–5939, (2014).

[39] K. S. Bialkowski, I. V. L. Clarkson, S. D. Howard. 'Generalized canonical correlation for passive multistatic radar detection', *Proc. IEEE Statistical Signal Processing Workshop (SSP)*, pp. 417–420, (2011).

Index

Note: Page numbers preceded by "*V1*:" and "*V2*:" refer to Volume 1 and Volume 2, respectively.